The Geometry and Growth of Normal Faults

Geological Society books refereeing procedures

The Society makes every effort to ensure that the scientific and production quality of its books matches that of its journals. Since 1997, all book proposals have been refereed by specialist reviewers as well as by the Society's Books Editorial Committee. If the referees identify weaknesses in the proposal, these must be addressed before the proposal is accepted.

Once the book is accepted, the Society Book Editors ensure that the volume editors follow strict guidelines on refereeing and quality control. We insist that individual papers can only be accepted after satisfactory review by two independent referees. The questions on the review forms are similar to those for *Journal of the Geological Society*. The referees' forms and comments must be available to the Society's Book Editors on request.

Although many of the books result from meetings, the editors are expected to commission papers that were not presented at the meeting to ensure that the book provides a balanced coverage of the subject. Being accepted for presentation at the meeting does not guarantee inclusion in the book.

More information about submitting a proposal and producing a book for the Society can be found on its website: www.geolsoc.org.uk.

It is recommended that reference to all or part of this book should be made in one of the following ways:

Childs, C., Holdsworth, R. E., Jackson, C. A.-L., Manzocchi, T., Walsh, J. J. & Yielding, G. (eds) 2017. *The Geometry and Growth of Normal Faults*. Geological Society, London, Special Publications, **439**.

Reilly, C., Nicol, A. & Walsh, J. 2017. Importance of pre-existing fault size for the evolution of an inverted fault system. *In*: Childs, C., Holdsworth, R. E., Jackson, C. A.-L., Manzocchi, T., Walsh, J. J. & Yielding, G. (eds) *The Geometry and Growth of Normal Faults*. Geological Society, London, Special Publications, **439**, 447–463. First published online February 5, 2016, https://doi.org/10.1144/SP439.2

GEOLOGICAL SOCIETY SPECIAL PUBLICATION NO. 439

The Geometry and Growth of Normal Faults

EDITED BY

C. CHILDS
University College Dublin, Ireland

R. E. HOLDSWORTH
University of Durham, UK

C. A.-L. JACKSON
Imperial College, UK

T. MANZOCCHI
University College Dublin, Ireland

J. J. WALSH
University College Dublin, Ireland

and

G. YIELDING
Badley Geoscience Ltd, UK

2017
Published by
The Geological Society
London

THE GEOLOGICAL SOCIETY

The Geological Society of London (GSL) was founded in 1807. It is the oldest national geological society in the world and the largest in Europe. It was incorporated under Royal Charter in 1825 and is Registered Charity 210161.

The Society is the UK national learned and professional society for geology with a worldwide Fellowship (FGS) of over 10 000. The Society has the power to confer Chartered status on suitably qualified Fellows, and about 2000 of the Fellowship carry the title (CGeol). Chartered Geologists may also obtain the equivalent European title, European Geologist (EurGeol). One fifth of the Society's fellowship resides outside the UK. To find out more about the Society, log on to www.geolsoc.org.uk.

The Geological Society Publishing House (Bath, UK) produces the Society's international journals and books, and acts as European distributor for selected publications of the American Association of Petroleum Geologists (AAPG), the Indonesian Petroleum Association (IPA), the Geological Society of America (GSA), the Society for Sedimentary Geology (SEPM) and the Geologists' Association (GA). Joint marketing agreements ensure that GSL Fellows may purchase these societies' publications at a discount. The Society's online bookshop (accessible from www.geolsoc.org.uk) offers secure book purchasing with your credit or debit card.

To find out about joining the Society and benefiting from substantial discounts on publications of GSL and other societies worldwide, consult www.geolsoc.org.uk, or contact the Fellowship Department at: The Geological Society, Burlington House, Piccadilly, London W1J 0BG: Tel. +44 (0)20 7434 9944; Fax +44 (0)20 7439 8975; E-mail: enquiries@geolsoc.org.uk.

For information about the Society's meetings, consult *Events* on www.geolsoc.org.uk. To find out more about the Society's Corporate Affiliates Scheme, write to enquiries@geolsoc.org.uk.

Published by The Geological Society from:
The Geological Society Publishing House, Unit 7, Brassmill Enterprise Centre, Brassmill Lane, Bath BA1 3JN, UK

The Lyell Collection: www.lyellcollection.org
Online bookshop: www.geolsoc.org.uk/bookshop
Orders: Tel. +44 (0)1225 445046, Fax +44 (0)1225 442836

British Library Cataloguing in Publication Data

A catalogue record for this book is available from the British Library.
ISBN 978-1-86239-967-9
ISSN 0305-8719

Distributors
For details of international agents and distributors see:
www.geolsoc.org.uk/agentsdistributors

Typeset by Nova Techset Private Limited, Bengaluru & Chennai, India
Printed and bound by CPI Group (UK) Ltd, Croydon CR0 4YY

Contents

This volume is dedicated to Juan Watterson

Juan on board 'Kimberlit' in Greenland, 1975.

Acknowledgements

The volume editors would like thank all those friends and colleagues who agreed to take the time to review papers for this volume; they include:

Ian Alsop	Mike Hall	Vincent Roche
Christian Beck	David Healy	Atle Rotevatn
Rebecca Bell	Jonathan Imber	Roy Schlische
Wayne Bailey	Michael Kettermann	Martin Schöpfer
Susanne Buiter	Young-Seog Kim	Richard Schulz
Richard Davies	Bob Krantz	Xu Shunshan
Oliver Duffy	Jonathan Long	Kevin Smart
Peter Eichhbul	Randall Marrett	Aisling Soden
David Ferrill	Steve Martel	Roger Soliva
Haakon Fossen	Shankar Mitra	Simon Stewart
Paul Gillespie	Andy Nicol	Bruce Trudgill
Neil Goulty	Casey Nixon	Rich Walker
Bernhard Grasemann	Douglas Paton	Chris Wibberley
Richard Groshong	Alan Roberts	Scott Wilkins

We also thank Angharad Hills at the Geological Society for her invaluable help in navigating us through the editorial process and Jo Armstrong at the Geological Society Publishing House for bringing the volume into production.

Introduction to the geometry and growth of normal faults

CONRAD CHILDS[1]*, ROBERT E. HOLDSWORTH[2], CHRISTOPHER A.-L. JACKSON[3], TOM MANZOCCHI[1], JOHN J. WALSH[1] & GRAHAM YIELDING[4]

[1]*Fault Analysis Group, School of Earth Sciences, University College Dublin, Belfield, Dublin 4, Ireland*

[2]*Department of Earth Sciences, University of Durham, South Road, Durham DH1 3LE, UK*

[3]*Basins Research Group, Imperial College, Prince Consort Road, London SW7 2BP, UK*

[4]*Badley Geoscience Ltd, Hundleby, Spilsby, Lincolnshire PE23 5NB, UK*

**Correspondence: conrad.childs@ucd.ie*

Normal faults are the dominant structures found in extensional sedimentary basins developed in continental rifts and passive margins. The geometry and growth of faults are intimately linked, and much of our understanding of how faults grow is derived directly from observations of fault geometry. The key geometric relationship that has underpinned the study of fault growth since the 1980s is the relationship between fault maximum displacement (D) and fault length (L) as defined by Elliott (1976) and Watterson (1986). This relationship is expressed as $D \alpha L^n$. The value of the exponent n in this relationship has been a topic for discussion for the last 30 years and values ranging between 0.5 and 2.0 have been advocated (e.g. Walsh & Watterson 1988; Cowie & Scholz 1992; Schultz *et al.* 2008). The range of values reflects the natural variation between different areas and uncertainties in data quality and sampling (Gillespie *et al.* 1992; Kim & Sanderson 2005). Irrespective of the value of the exponent, the recognition of a positive correlation between displacement and length suggests that faults grow progressively as their displacement increases (Watterson 1986; Walsh & Watterson 1988).

Since the late 1990s, several studies have been performed in areas where the rate of sedimentation exceeds the fault displacement rate, so that across-fault changes in the thickness of growth strata provide a record of the surface trace length and displacement distribution of faults through time (Morley 1999; Walsh *et al.* 2002; Childs *et al.* 2003; Paton 2006). These studies have shown that evidence for the propagation of normal faults in the geological record is often difficult to find and, in many cases, it appears that the lengths of faults are formed instantaneously within the time resolution of the data. If this is true, then the positive correlation between displacement and length does not define a trend along which individual faults grow. Instead, faults would establish their lengths very early in their development, with initially very low ratios of displacement to length that increase rapidly as displacement accumulates. Which of these models of fault growth is correct remains a matter for debate, as reflected in the content of this volume.

This introduction does not provide a comprehensive review of the literature on this topic, but aims to outline the current models for the growth of faults as a backdrop to the collection of papers in this volume. In the following section, we provide an in-depth description of current models of fault growth. We then discuss the methods by which these can be investigated, highlighting how geometric observations provide constraints on fault growth with reference to the papers published in this volume.

Fault growth models

We consider first the growth of individual faults and later discuss models for the growth of fault systems. Faults have many fractal characteristics (Kakimi 1980; Walsh *et al.* 1991; Marrett & Allmendinger 1992; Wojtal 1994), so that the distinction between faults and fault systems can be dependent on the scale of observation. The processes considered in both – namely, fault initiation, propagation, interaction and linkage – are the same. Despite these considerations, we distinguish between faults and fault systems because the concepts of fault growth were first described in terms of the growth of a single fault and also because the key concerns and the terminology adopted in

From: Childs, C., Holdsworth, R. E., Jackson, C. A.-L., Manzocchi, T., Walsh, J. J. & Yielding, G. (eds) 2017. *The Geometry and Growth of Normal Faults*. Geological Society, London, Special Publications, **439**, 1–9.
First published online September 5, 2017, https://doi.org/10.1144/SP439.24

the literature are different when applied to faults and fault systems.

Growth of an individual fault

The first plots of displacement v. the length of faults were interpreted in terms of the growth of a single fault plane (e.g. Watterson 1986; Walsh & Watterson 1988; Dawers *et al.* 1993). Walsh & Watterson (1989) showed that individual faults may consist of an array of segments that are kinematically the same as a single fault, therefore recognizing a distinction between faults and fault segments. The role of linkage between faults or fault segments in the formation of larger faults became the focus of later studies (e.g. Peacock & Sanderson 1991; Cartwright *et al.* 1995).

Conceptual models of the growth of a single fault from an initial array of smaller faults or fault segments are illustrated in Figure 1 (similar figures can be found in **Nicol *et al.* 2016**; **Morley 2016**; **Finch & Gawthorpe 2017**; **Jackson *et al.* 2017**). Three different models are distinguished based on two criteria. These criteria are: first, whether the initial faults formed as elements within a coherent structure or whether they formed in mechanical and kinematic isolation from one another (coherent v. non-coherent in Fig. 1); and second, whether the faults established their lengths at low displacements or whether there was a protracted period of fault propagation during displacement accumulation (constant-length v. propagating in Fig. 1).

Figure 1a illustrates the non-coherent case with three initially unconnected faults that are broadly aligned along-strike, but are kinematically unrelated to one another. As their displacement increases, the faults become longer until they interact with one another preventing, or at least retarding, further propagation, as reflected by increased displacement gradients at the boundaries between the faults (i.e. at the relay zones). As the displacement increases further, the faults become connected or linked to form a single large fault with associated splays. The displacement distribution on the resultant through-going fault is irregular, with local displacement lows at the sites of former segment boundaries. However, the characteristic feature of this model is the irregularity of the profile of the total, or aggregate, fault displacement, with local displacement highs and lows recording the earlier presence of separate, previously isolated, faults.

Figure 1b, c illustrate the cases in which geometric coherence is maintained throughout the growth of a fault. Geometric coherence is illustrated by the simple triangular profile of aggregate displacement, which demonstrates that the individual fault segments are part of a single larger structure (Walsh & Watterson 1991) even though they are not geometrically connected at the scale of observation (i.e. not hard-linked). Instead, soft linkage allows coherence of the large-scale fault displacement through distributed strain accommodated within the adjacent wall rock. Although the displacement distribution on the through-going fault

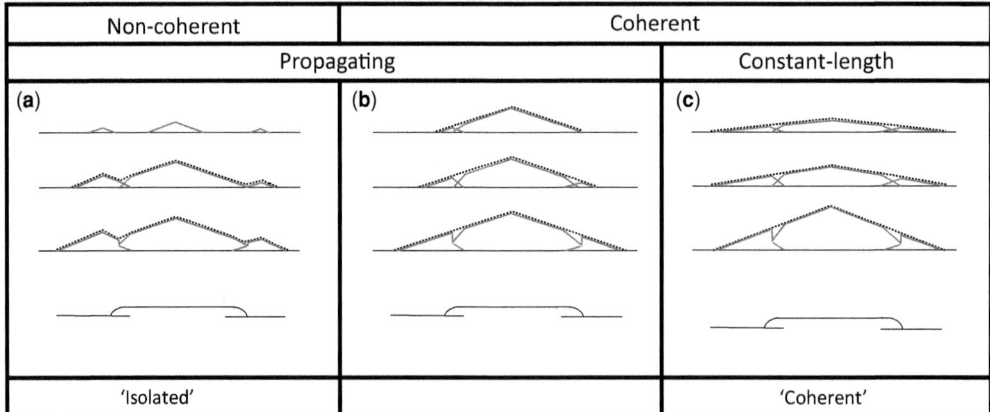

Fig. 1. Displacement-length profiles illustrating three potential growth histories leading to the same fault map pattern. The solid lines are the displacement profiles for individual faults and the dashed lines are profiles of aggregate displacement. The three cases are distinguished on the basis of whether the faults are coherent and whether the faults propagate continuously as displacement increases. The non-coherent propagating case is commonly referred to as the 'isolated' model and the constant-length coherent case is commonly referred to as the 'coherent' model. The propagating coherent case is also possible, but is less commonly invoked. The evolution of the fault traces for parts (a) and (c) are illustrated in Figure 2. The profiles shown are for normal faults in map view, but similar considerations are equally applicable to faults in cross-section.

is again irregular, with lows at the locations of the initial segment boundaries, in this case the profile of aggregate displacement has a simple triangular shape, in which lows on the through-going fault are fully compensated by displacements on splays and bed rotations. In Figure 1b the geometrically coherent fault array propagates as displacement accumulates and new segments formed at the tip of the propagating fault are part of the larger structure from the outset. In Figure 1c the fault trace lengths are established in the first growth increment and subsequent fault propagation occurs only during linkage between segments.

A variety of terms have been applied to these models. The model illustrated in Figure 1a is widely referred to as the 'isolated' fault model, reflecting the fact that individual faults within an array of faults can initiate as kinematically isolated structures at random locations, eventually coalescing to form a large fault by incidental interaction and linkage. This contrasts with 'coherent' fault models, where fault segments initiate as elements within a larger structure. The existence of a larger structure is most clearly illustrated for arrays of fault segments seen in map or cross-section that connect to a single fault in the third dimension (Walsh *et al.* 2003).

Although it is possible to distinguish between faults and fault segments that developed according to the isolated and coherent models based on finite displacement distributions, distinguishing between the constant-length and propagating variants of the coherent model requires the back-stripping of displacements to determine whether there was a protracted period of fault propagation. In principle, arrays of fault segments could progressively propagate laterally (the coherent propagating model of Fig. 1b), but this type of behaviour is much less commonly advocated than the isolated (Fig. 1a) and the constant-length coherent (Fig. 1c) models, which are therefore generally considered to be end-member fault growth cases.

Growth of fault systems

The models and concepts described for the growth of individual faults are broadly equivalent to those used to describe the growth of fault systems. Although discussions on the development of individual faults have largely centred on the difference between the isolated and coherent models (e.g. Jackson & Rotevatn 2013), recent discussions on the growth of fault systems have focused more specifically on the distinction between the isolated and the constant-length coherent model. These two models are illustrated on the schematic maps in Figure 2 and fault displacement distributions from each are shown in Figure 1a and c, respectively. In the isolated model, faults are initially randomly located and kinematically independent of one another. Faults grow by progressive propagation and by the interaction and coalescence of fortuitously aligned smaller faults; one such fault and its displacement evolution is shown in Figure 1a. According to this model, the established positive correlation between fault displacement and length defines the trend that individual faults follow as they accrue displacement and lengthen (Fig. 2f). This model and variants of it have been referred to as the isolated fault model (e.g. **Ghalayini *et al.* 2016**; **Morley 2016**), the standard fault model (Walsh *et al.* 2002) and the increasing length fault model (**Nicol *et al.* 2016**). The term 'isolated fault model' is most widely used in the current literature and describes the defining feature that faults initiate as spatially and kinematically isolated structures.

According to the constant-length, coherent model for fault system evolution, fault trace lengths and interactions between faults are established at low strains and subsequent fault propagation is limited to the linkage between adjacent faults. The associated displacement distribution through time is illustrated in Figure 1c. The D–L growth curve followed by an individual fault in this case is in two portions (Fig. 2f) separated by a bend, where a growing fault changes from rapid propagation at low strain to the accumulation of displacement without significant propagation. This model has been referred to as the alternative model (Walsh *et al.* 2002), the coherent model (**Morley 2016**) and the constant-length model (Curry *et al.* 2016; **Nicol *et al.* 2016**; **Jackson *et al.* 2017**), with the last two becoming the most commonly used in the recent literature. Here we refer to it as the constant-length coherent fault model.

The description of these models is simplified here and, for example, we do not represent the step-wise increase in fault length associated with fault linkage in the isolated fault model in Figure 2e. Similarly, we illustrate these models on two-dimensional map view sections partly for simplicity, but also because the best available constraints on fault growth currently come from studies of syn-faulting sedimentation, which provide information on the surface trace lengths of faults; both models can be considered in three dimensions with faults that propagate out of the plane of observation. For the purpose of clarity, the two models for the growth of fault systems are starkly contrasted here, but conceptually they represent end-members in a range of possible behaviours. For example, as extension increases within fault systems originally subscribing to the constant-length model, there will be a progressive increase in D/L ratios on individual faults, eventually leading to the linkage of adjacent faults. Although this is reminiscent of the isolated fault model, it differs in the crucial respect that the fault

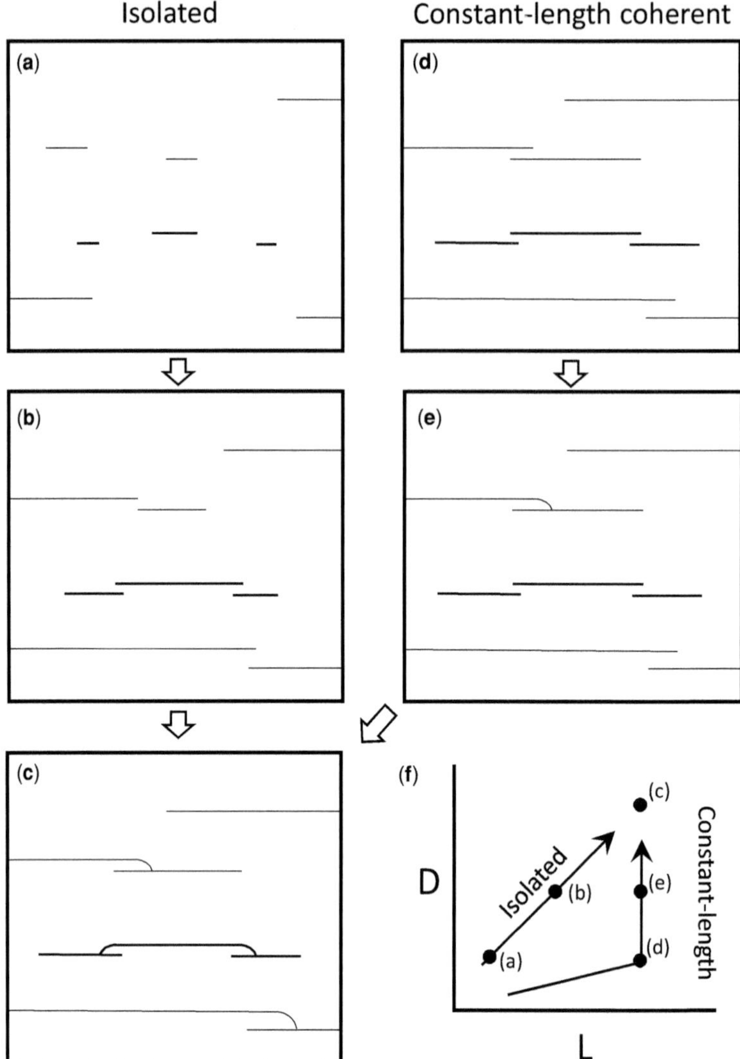

Fig. 2. Schematic maps illustrating how the isolated (**a**, **b**) and constant-length coherent (**d**, **e**) models of fault growth can give rise to the same fault map pattern (**c**). The displacement profiles through time for the faults highlighted in bold are shown in Figure 1.

map is effectively established very early apart from the linkage of adjacent faults at higher strains. Many of the papers in this volume do not specifically address these models of fault growth, but most do make directly relevant geometrical observations on displacement distributions on faults. Several papers directly address these growth models, either from a study of faults that intersect growth strata (**Morley 2016**; **Worthington & Walsh 2016**; **Jackson *et al.* 2017**), from finite displacement profile data (**Ghalayini *et al.* 2016**; **Homberg *et al.* 2017**) or from modelling (**Whipp *et al.* 2016**; **Finch & Gawthorpe 2017**), and invoke one or other of the isolated or constant-length coherent models.

Temporal resolution

The two fault growth models can be viewed as end-members of a spectrum of behaviours. For example, the simple map view cartoon in Figure 2d shows that all faults achieve their length in the first increment of fault growth. There is clearly a finite period in which these faults propagate through the rock volume to establish their lengths, but the duration of this period is difficult to establish. Sediment thickness data may show that faults are established in the first resolvable time increment, but in some cases this increment may be millions of years. Similarly, for the isolated fault model there is a time when all faults have propagated to interact with adjacent faults and the subsequent formation of new fault trace lengths is predominantly by fault linkage. The key distinction between the models is therefore how quickly, or at what strain, the transition from behaviour dominated by fault propagation to behaviour dominated by fault interaction occurs. Taylor *et al.* (2004), for example, suggested that this transition occurred *c.* 700 ka after fault initiation. Analysis of an Australian NW shelf dataset described by Walsh *et al.* (2002) shows that this transition occurred over <500 ka and at very low strains (<3%). In the clay models of **Whipp *et al.* (2016)**, the early stages of fault growth are dominated by isolated faults that propagate and grow towards one another; the development of the model fault systems is considered to adhere to the isolated fault model. However, in these models, *c.* 95% of the fault length is established in a strain interval from 10 to 25% extension, with subsequent growth accommodated by an increase in displacement without further propagation.

Fault-related folding

For most fault systems, geological conditions and/or the available data preclude direct study of fault growth histories. In these situations, it may be possible to distinguish between the isolated and coherent models based on whether mapped displacement profiles resemble that in Figure 1a (or Fig. 1b or c). The key difference between the isolated and coherent models is the shape of the total or aggregate throw profile. Accurate definition of this profile can be difficult, primarily due to the ductile component of displacement that accompanies faulting. The interchange between faulting and folding, or discontinuous and continuous deformation, within geological structures is well established and several contributions to this volume address this topic (e.g. **Childs *et al.* 2016**; **Delogkos *et al.* 2016**; **Fazlikhani *et al.* 2016**; **Ferrill *et al.* 2016**; **Lăpădat *et al.* 2016**; **Homberg *et al.* 2017**). Within normal fault systems, continuous deformation occurs primarily as monoclines related to fault tip propagation and as bed rotations within relay zones separating adjacent fault segments (**Childs *et al.* 2016**; **Lăpădat *et al.* 2016**; **Khalil & McClay 2016**). Using detailed outcrop and seismic mapping, these papers show that arrays of fault segments are often accompanied by associated folding, particularly adjacent to relay zones, highlighting the interchangeability of folding and faulting through time within a single kinematic structure. Bed rotations within relay zones can occur at relay ramps seen in map view or at sites of vertical segmentation (**Roche *et al.* 2016**; **Homberg *et al.* 2017**). Irrespective of the origin of fault-related continuous deformation, it represents an integral part of the total displacement across a fault and failure to include this component can result in false lows on profiles of aggregate displacement and hence to the erroneous attribution to the isolated model.

Boundary conditions and pre-existing structures

Idealized models for the growth of fault systems are described in the context of simplified considerations of a volume of rock subjected to uniform extension (e.g. Fig. 2, **Finch & Gawthorpe 2017**; **Whipp *et al.* 2016**). In reality, such simple boundary conditions rarely apply and, for example, the magnitude of extension may vary along strike, or the extended crust may be strongly heterogeneous (Jackson & Rotevatn 2013; **Ferrer *et al.* 2016**; **Ferrill *et al.* 2016**). The departure from idealized models of fault system evolution is addressed in several contributions to this volume. **Ford *et al.* (2016)** provide detailed evidence for two phases of propagation of faults bounding the Gulf of Corinth. This propagation is attributed to a change in the tectonic control on local extension rate. Many earlier studies (e.g. Morley 2002; Noll & Hall 2006) have demonstrated lateral changes in the focus for extension, which again is generally attributed to changes in boundary conditions. Changes in displacement rates along the lengths of faults may also be driven by internal changes in the geometry of the active fault system as it evolves – for example, as a fault system becomes progressively polarized into one dominant dip direction (Nixon *et al.* 2016; **Finch & Gawthorpe 2017**). **Morley (2016)** shows how displacement profiles for a large boundary fault record progressive migration in activity along strike, pointing to the fact that the resultant displacement distribution is not consistent with either of the simple current models of fault growth. Sandbox modelling of listric fault geometries provides a clear rationale for the migration of the locations of active faulting (**Ferrer *et al.* 2016**).

The constant-length coherent fault model was first formally described for normal faults that reactivate earlier underlying fault systems, where the extension directions of the early and late rifting phases were subparallel (Walsh *et al.* 2002). In this case, the rapid development of very long fault traces with low displacements in the cover sequence occurred because these fault lengths were already established in the pre-rift section during the previous extensional phase and, during subsequent reactivation, propagated upwards to intersect the free surface. Walsh *et al.* (2002) also pointed out, however, that the rapid development of fault lengths could even apply to faults that are not reactivated because there was, and still is, little direct published kinematic evidence for slow progressive growth in fault length. In an earlier study, Morley (1999) described basin-bounding faults in the East African Rift that showed rapid and early propagation, without invoking a role for pre-existing structures. Nevertheless, similar, more recent, studies demonstrating a constant-length coherent model have recognized a significant control of underlying structure (e.g. Paton 2006; Jackson & Rotevatn 2013). There is therefore little doubt that an optimally oriented pre-existing structure favours rapid fault propagation, but this could often be the case given the presence of older structure in many newly formed basins. Conversely, the extent to which the isolated model could apply in 'virgin' crust is difficult to ascertain because many basins are developed in crust that has been deformed during earlier phases of deformation (whether in extension or compression).

The significance of pre-existing structure on fault system evolution, both in extension and compression, is the subject of several contributions to this volume. **Worthington & Walsh (2016)** present an example of a reactivated basement fault in which upwards propagation provides a geometrically coherent array of fault segments characterized by very rapid and, effectively constant, fault trace lengths in the cover. In this case, coherence is achieved by soft linkage in map view and hard linkage into a single fault at depth. **Reilly *et al.* (2016)** provide an example of the reverse reactivation of earlier normal faults and show that only the largest faults in the extensional system become reactivated and that this reactivation is generally along the full fault length. **Morley (2016)** describes how multiple superposed extension events result in both over- and under-displaced faults with displacement variations that are not readily assigned to any one model of fault growth. **Fossen *et al.* (2016)** document how the presence of Caledonian structure and Devonian extensional faults mapped onshore Norway can be extrapolated offshore, where they control the geometries of later Triassic and Jurassic rifting. They discuss how pre-existing structure determines the development of different structural domains within an evolving fault system. By contrast, analogue (**Whipp *et al.* 2016**) and numerical (**Finch & Gawthorpe 2017**) modelling show that different structural domains characterized by opposed fault polarity and separated by distinct boundaries can also emerge from extension of a homogeneous medium.

The selection of papers in this volume show that the boundary conditions and pre-existing crustal structure can have a significant impact on the geometry and growth histories of faults, an issue which explains why it can sometimes be challenging to determine the generic characteristics of fault growth from studies of particular areas with different geological and tectonic settings.

Fault zone structure and evolution

Patterns of minor faults and damage zones associated with normal faults, including correlations between the widths of damage zones and fault displacement, have previously been used to underpin models of fault zone growth (e.g. Shipton & Cowie 2003; Fossen *et al.* 2007). **Skar *et al.* (2016)** find no correlation between damage zone width and displacement for faults with throws in the range 0.6–30 m, but find that the variation in the measured damage zone width on a single fault is greater than between faults of different displacement. **Nicol *et al.* (2016)** make the outcrop observation that fault tips are often characterized by the splaying of faults, but the splays that might be expected along the lengths of faults at former tip locations are absent. This observation suggests that the studied faults established their lengths rapidly and form geometrically coherent arrays. **Roche *et al.* (2016)** recognize a similar increase in fault zone complexity at fault tips and develop a model in which the locally increased thickness of fault zones can be attributed to the arrest of bed-normal propagation at less competent units within a layered sequence. Numerical modelling of the growth of normal faults by **Schöpfer *et al.* (2016)** focuses on the localization of strain within a fracture array established at very low strains. The width of the initial zone of fracture is largely independent of the mechanical properties of the model, but the localization of strain onto a single fault within the zone as displacement accrues is strongly dependent on the mechanical properties of the faulted sequence.

Fault growth studies generally concentrate on how fault surfaces achieve their lengths and displacements. A closely related topic is the internal structure of fault zones and there is considerable overlap between these two topics. This is particularly clear where, for example, relay zones between segments of an array of normal faults are breached and the relay zone is preserved as a splay or a lens within a fault zone (e.g. **Whipp *et al.* 2016**). Studies of the internal structure of fault zones can therefore shed light on fault growth. **Yielding (2016)** details the relationship between the orientations of branch lines between faults, and between slip-surfaces within fault zones, and relates them to style of faulting and extension direction. **Childs *et al.* (2016)** present a model relating the geometry of segment boundaries and, in particular, the associated folding (often reflected as normal drag) to the fault zone internal structure that results as displacement increases. **Gabrielsen *et al.* (2016)** examine the internal structure of faults at higher strains, where former segment boundaries may be preserved as lenses within fault zones.

Geometry and growth of normal faults

This volume of papers was compiled following a conference of the same title held at the Geological Society of London in 2014 in honour of Juan Watterson. The topic of the conference was chosen to reflect Juan's interests and his significant contribution to our understanding of the growth of faults. Juan was a firm believer in the analysis of strain as a means of understanding the evolution of geological structures and he pioneered the development of various quantitative methods related to displacement and strain that explored the nature of fault growth. His view was that stresses and forces can only be speculated for ancient events and, because we deal with large observable strains, much more progress can be made if structural geologists have a cultural disposition towards geometry, displacement and strain. In that sense, he subscribed to the view of Burland (1965) (quoted in Roscoe 1970), that 'stress is a philosophical concept, deformation is a physical reality'. The availability of improved constraints from earthquake studies and advances in numerical modelling techniques are, however, helping us to address stress-related issues and their implications for fault growth on geological timescales. Although Juan would have been supportive, and perhaps at the vanguard, of the adoption of new approaches, he still would have been a strong advocate of the acquisition and analysis of the best three-dimensional geological datasets, ideally from fieldwork, to resolve outstanding scientific questions. In that sense, Juan would have been delighted to read this volume because it reveals interesting and new insights on fault growth by the application of established and new methods. It would not have been in his nature to solicit credit for his work, but the involvement of many of the world's experts in this volume reflects the importance of his contributions and the respect with which he is held.

References

Burland, J.B. 1965. *Deformation of soft clay*. PhD thesis, University of Cambridge.

Cartwright, J.A., Trudgill, B.D. & Mansfield, C.S. 1995. Fault growth by segment linkage: an explanation for scatter in maximum displacement and trace length data from the Canyonlands Grabens of SE Utah. *Journal of Structural Geology*, **17**, 1319–1326.

Childs, C., Nicol, A., Walsh, J.J. & Watterson, J. 2003. The growth and propagation of synsedimentary faults. *Journal of Structural Geology*, **25**, 633–648.

Childs, C., Manzocchi, T., Nicol, A., Walsh, J.J., Soden, A.M., Conneally, J.C. & Delogkos, E. 2016. The relationship between normal drag, relay ramp aspect ratio and fault zone structure. *In*: Childs, C., Holdsworth, R.E., Jackson, C.A.-L., Manzocchi, T., Walsh, J.J. & Yielding, G. (eds) *The Geometry and Growth of Normal Faults*. Geological Society, London, Special Publications, **439**. First published online August 17, 2016, https://doi.org/10.1144/SP439.16

Cowie, P.A. & Scholz, C.H. 1992. Displacement-length scaling relationship for faults: data synthesis and discussion. *Journal of Structural Geology*, **14**, 1149–1156.

Curry, M.A., Barnes, J.B. & Colgan, J.P. 2016. Testing fault growth models with low-temperature thermochronology in the northwest Basin and Range, USA. *Tectonics*, **35**, 2467–2492.

Dawers, N.H., Anders, M.H. & Scholz, C.H. 1993. Fault length and displacement: scaling laws. *Geology*, **21**, 1107–1110.

Delogkos, E., Manzocchi, T. et al. 2016. Throw partitioning across normal fault zones in the Ptolemais Basin, Greece. *In*: Childs, C., Holdsworth, R.E., Jackson, C.A.-L., Manzocchi, T., Walsh, J.J. & Yielding, G. (eds) *The Geometry and Growth of Normal Faults*. Geological Society, London, Special Publications, **439**. First published online November 21, 2016, https://doi.org/10.1144/SP439.19

Elliott, D. 1976. The energy balance and deformation mechanisms of thrust sheets. *Philosophical Transactions of the Royal Society of London A: Mathematical, Physical and Engineering Sciences*, **283**, 289–312.

Fazlikhani, H., Back, S., Kukla, P.A. & Fossen, H. 2016. Interaction between gravity-driven listric normal fault linkage and their hanging-wall rollover development: a case study from the western Niger Delta, Nigeria. *In*: Childs, C., Holdsworth, R.E., Jackson, C.A.-L., Manzocchi, T., Walsh, J.J. & Yielding, G. (eds) *The Geometry and Growth of Normal Faults*. Geological Society, London, Special Publications, **439**. First published online December 13, 2016, https://doi.org/10.1144/SP439.20

Ferrer, O., McClay, K. & Sellier, N.C. 2016. Influence of fault geometries and mechanical anisotropies on the growth and inversion of hanging-wall synclinal basins: insights from sandbox models and natural examples. *In*: Childs, C., Holdsworth, R.E., Jackson, C.A.-L., Manzocchi, T., Walsh, J.J. & Yielding, G. (eds) *The Geometry and Growth of Normal Faults*. Geological Society, London, Special Publications, **439**. First published online March 15, 2016, https://doi.org/10.1144/SP439.8

Ferrill, D.A., Morris, A.P., McGinnis, R.N. & Smart, K.J. 2016. Myths about normal faulting. *In*: Childs, C., Holdsworth, R.E., Jackson, C.A.-L., Manzocchi, T., Walsh, J.J. & Yielding, G. (eds) *The Geometry and Growth of Normal Faults*. Geological Society, London, Special Publications, **439**. First published online March 30, 2016, https://doi.org/10.1144/SP439.12

Finch, E. & Gawthorpe, R. 2017. Growth and interaction of normal faults and fault network evolution in rifts: insights from three-dimensional discrete element modelling. *In*: Childs, C., Holdsworth, R. E., Jackson, C. A.-L., Manzocchi, T., Walsh, J. J. & Yielding, G. (eds) *The Geometry and Growth of Normal Faults*. Geological Society, London, Special Publications, **439**. First published online August 30, 2017, https://doi.org/10.1144/SP439.23

Ford, M., Hemelsdaël, R., Mancini, M. & Palyvos, N. 2016. Rift migration and lateral propagation: evolution of normal faults and sediment-routing systems of the western Corinth rift (Greece). *In*: Childs, C., Holdsworth, R.E., Jackson, C.A.-L., Manzocchi, T., Walsh, J.J. & Yielding, G. (eds) *The Geometry and Growth of Normal Faults*. Geological Society, London, Special Publications, **439**. First published online August 15, 2016, https://doi.org/10.1144/SP439.15

Fossen, H., Schultz, R.A., Shipton, Z.K. & Mair, K. 2007. Deformation bands in sandstone: a review. *Journal of the Geological Society*, **164**, 755–769.

Fossen, H., Fazli Khani, H., Faleide, J.I., Ksienzyk, A.K. & Dunlap, W.J. 2016. Post-Caledonian extension in the West Norway–northern North Sea region: the role of structural inheritance. *In*: Childs, C., Holdsworth, R.E., Jackson, C.A.-L., Manzocchi, T., Walsh, J.J. & Yielding, G. (eds) *The Geometry and Growth of Normal Faults*. Geological Society, London, Special Publications, **439**. First published online February 5, 2016, https://doi.org/10.1144/SP439.6

Gabrielsen, R.H., Braathen, A., Kjemperud, M. & Valdresbråten, M.L.R. 2016. The geometry and dimensions of fault-core lenses. *In*: Childs, C., Holdsworth, R.E., Jackson, C.A.-L., Manzocchi, T., Walsh, J.J. & Yielding, G. (eds) *The Geometry and Growth of Normal Faults*. Geological Society, London, Special Publications, **439**. First published online February 5, 2016, https://doi.org/10.1144/SP439.4

Ghalayini, R., Homberg, C., Daniel, J.M. & Nader, F.H. 2016. Growth of layer-bound normal faults under a regional anisotropic stress field. *In*: Childs, C., Holdsworth, R.E., Jackson, C.A.-L., Manzocchi, T., Walsh, J.J. & Yielding, G. (eds) *The Geometry and Growth of Normal Faults*. Geological Society, London, Special Publications, **439**. First published online April 6, 2016, https://doi.org/10.1144/SP439.13

Gillespie, P.A., Walsh, J.J. & Watterson, J. 1992. Limitations of dimension and displacement data from single faults and the consequences for data analysis and interpretation. *Journal of Structural Geology*, **14**, 1157–1172.

Homberg, C., Schnyder, J., Roche, V., Leonardi, V. & Benzaggagh, M. 2017. The brittle and ductile components of displacement along fault zones. *In*: Childs, C., Holdsworth, R.E., Jackson, C.A.-L., Manzocchi, T., Walsh, J.J. & Yielding, G. (eds) *The Geometry and Growth of Normal Faults*. Geological Society, London, Special Publications, **439**. First published online March 1, 2017, https://doi.org/10.1144/SP439.21

Jackson, C.A.L. & Rotevatn, A. 2013. 3D seismic analysis of the structure and evolution of a salt-influenced normal fault zone: a test of competing fault growth models. *Journal of Structural Geology*, **54**, 215–234.

Jackson, C.A.-L., Bell, R.E., Rotevatn, A. & Tvedt, A.B.M. 2017. Techniques to determine the kinematics of synsedimentary normal faults and implications for fault growth models. *In*: Childs, C., Holdsworth, R.E., Jackson, C.A.-L., Manzocchi, T., Walsh, J.J. & Yielding, G. (eds) *The Geometry and Growth of Normal Faults*. Geological Society, London, Special Publications, **439**. First published online February 7, 2017, https://doi.org/10.1144/SP439.22

Kakimi, T. 1980. Magnitude-frequency relation for displacement of minor faults and its significance in crustal deformation. *Bulletin of the Geological Survey of Japan*, **31**, 467–487.

Khalil, S.M. & McClay, K.R. 2016. 3D geometry and kinematic evolution of extensional fault-related folds, NW Red Sea, Egypt. *In*: Childs, C., Holdsworth, R.E., Jackson, C.A.-L., Manzocchi, T., Walsh, J.J. & Yielding, G. (eds) *The Geometry and Growth of Normal Faults*. Geological Society, London, Special Publications, **439**. First published online March 30, 2016, https://doi.org/10.1144/SP439.11

Kim, Y.-S. & Sanderson, D.J. 2005. The relationship between displacement and length of faults: a review. *Earth-Science Reviews*, **68**, 317–334.

Lăpădat, A., Imber, J., Yielding, G., Iacopini, D., McCaffrey, K.J.W., Long, J.J. & Jones, R.R. 2016. Occurrence and development of folding related to normal faulting within a mechanically heterogeneous sedimentary sequence: a case study from Inner Moray Firth, UK. *In*: Childs, C., Holdsworth, R.E., Jackson, C.A.-L., Manzocchi, T., Walsh, J.J. & Yielding, G. (eds) *The Geometry and Growth of Normal Faults*. Geological Society, London, Special Publications, **439**. First published online September 26, 2016, https://doi.org/10.1144/SP439.18

Marrett, R. & Allmendinger, R.W. 1992. Amount of extension on 'small' faults: an example from the Viking graben. *Geology*, **20**, 47–50.

Morley, C.K. 1999. Patterns of displacement along large normal faults: implications for basin evolution and fault propagation, based on examples from East Africa. *AAPG Bulletin*, **83**, 613–634.

Morley, C.K. 2002. Evolution of large normal faults: evidence from seismic reflection data. *AAPG Bulletin*, **86**, 961–978.

Morley, C.K. 2016. The impact of multiple extension events, stress rotation and inherited fabrics on normal fault geometries and evolution in the Cenozoic rift basins of Thailand. *In*: Childs, C., Holdsworth, R.E., Jackson, C.A.-L., Manzocchi, T., Walsh, J.J. & Yielding, G. (eds) *The Geometry and Growth of Normal Faults*. Geological Society, London, Special Publications, **439**. First published online April 13, 2016, https://doi.org/10.1144/SP439.3

Nicol, A., Childs, C., Walsh, J.J., Manzocchi, T. & Schöpfer, M.P.J. 2016. Interactions and growth of faults in an outcrop-scale system. *In*: Childs, C., Holdsworth, R.E., Jackson, C.A.-L., Manzocchi, T., Walsh, J.J. & Yielding, G. (eds) *The Geometry and Growth of Normal Faults*. Geological Society, London, Special Publications, **439**. First published online March 10, 2016, https://doi.org/10.1144/SP439.9

NIXON, C.W., MCNEILL, L.C. *ET AL*. 2016. Rapid spatio-temporal variations in rift structure during development of the Corinth Rift, central Greece. *Tectonics*, **35**, 1225–1248.

NOLL, C.A. & HALL, M. 2006. Normal fault growth and its function on the control of sedimentation during basin formation: a case study from field exposures of the Upper Cambrian Owen Conglomerate, West Coast Range, western Tasmania, Australia. *AAPG Bulletin*, **90**, 1609–1630.

PATON, D.A. 2006. Influence of crustal heterogeneity on normal fault dimensions and evolution: southern South Africa extensional system. *Journal of Structural Geology*, **28**, 868–886.

PEACOCK, D.C.P. & SANDERSON, D.J. 1991. Displacements, segment linkage and relay ramps in normal fault zones. *Journal of Structural Geology*, **13**, 721–733.

REILLY, C., NICOL, A. & WALSH, J. 2016. Importance of pre-existing fault size for the evolution of an inverted fault system. *In*: CHILDS, C., HOLDSWORTH, R.E., JACKSON, C.A.-L., MANZOCCHI, T., WALSH, J.J. & YIELDING, G. (eds) *The Geometry and Growth of Normal Faults*. Geological Society, London, Special Publications, **439**. First published online February 5, 2016, https://doi.org/10.1144/SP439.2

ROCHE, V., HOMBERG, C., VAN DER BAAN, M. & ROCHER, M. 2016. Widening of normal fault zones due to the inhibition of vertical propagation. *In*: CHILDS, C., HOLDSWORTH, R.E., JACKSON, C.A.-L., MANZOCCHI, T., WALSH, J.J. & YIELDING, G. (eds) *The Geometry and Growth of Normal Faults*. Geological Society, London, Special Publications, **439**. First published online February 5, 2016, https://doi.org/10.1144/SP439.5

ROSCOE, K.H. 1970. The influence of strains on soil mechanics. *Geotechnique*, **20**, 129–170.

SCHÖPFER, M.P.J., CHILDS, C., MANZOCCHI, T. & WALSH, J.J. 2016. Three-dimensional distinct element method modelling of the growth of normal faults in layered sequences. *In*: CHILDS, C., HOLDSWORTH, R.E., JACKSON, C.A.-L., MANZOCCHI, T., WALSH, J.J. & YIELDING, G. (eds) *The Geometry and Growth of Normal Faults*. Geological Society, London, Special Publications, **439**. First published online September 15, 2016, https://doi.org/10.1144/SP439.17

SCHULTZ, R.A., SOLIVA, R., FOSSEN, H., OKUBO, C.H. & REEVES, D.M. 2008. Dependence of displacement–length scaling relations for fractures and deformation bands on the volumetric changes across them. *Journal of Structural Geology*, **30**, 1405–1411.

SHIPTON, Z.K. & COWIE, P.A. 2003. A conceptual model for the origin of fault damage zone structures in high-porosity sandstone. *Journal of Structural Geology*, **25**, 333–344.

SKAR, T., BERG, S.S., GABRIELSEN, R.H. & BRAATHEN, A. 2016. Fracture networks of normal faults in fine-grained sedimentary rocks: examples from Kilve Beach, SW England. *In*: CHILDS, C., HOLDSWORTH, R.E., JACKSON, C.A.-L., MANZOCCHI, T., WALSH, J.J. & YIELDING, G. (eds) *The Geometry and Growth of Normal Faults*. Geological Society, London, Special Publications, **439**. First published online September 26, 2016, https://doi.org/10.1144/SP439.10

TAYLOR, S.K., BULL, J.M., LAMARCHE, G. & BARNES, P.M. 2004. Normal fault growth and linkage in the Whakatane Graben, New Zealand, during the last 1.3 Myr. *Journal of Geophysical Research: Solid Earth*, **109**, B02408.

WALSH, J.J. & WATTERSON, J. 1988. Analysis of the relationship between displacements and dimensions of faults. *Journal of Structural Geology*, **10**, 239–247.

WALSH, J.J. & WATTERSON, J. 1989. Displacement gradients on fault surfaces. *Journal of Structural Geology*, **11**, 307–316.

WALSH, J.J. & WATTERSON, J. 1991. Geometric and kinematic coherence and scale effects in normal fault systems. *In*: ROBERTS, A.M., YIELDING, G. & FREEMAN, B. (eds) *The Geometry of Normal Faults*. Geological Society, London, Special Publications, **56**, 193–203, https://doi.org/10.1144/GSL.SP.1991.056.01.13

WALSH, J.J., WATTERSON, J. & YIELDING, G. 1991. The importance of small-scale faulting in regional extension. *Nature*, **351**, 391–393.

WALSH, J.J., NICOL, A. & CHILDS, C. 2002. An alternative model for the growth of faults. *Journal of Structural Geology*, **24**, 1669–1675.

WALSH, J.J., BAILEY, W.R., CHILDS, C., NICOL, A. & BONSON, C.G. 2003. Formation of segmented normal faults: a 3-D perspective. *Journal of Structural Geology*, **25**, 1251–1262.

WATTERSON, J. 1986. Fault dimensions, displacements and growth. *Pure and Applied Geophysics*, **124**, 365–373.

WHIPP, P.S., JACKSON, C.A.-L., SCHLISCHE, R.W., WITHJACK, M.O. & GAWTHORPE, R.L. 2016. Spatial distribution and evolution of fault-segment boundary types in rift systems: observations from experimental clay models. *In*: CHILDS, C., HOLDSWORTH, R.E., JACKSON, C.A.-L., MANZOCCHI, T., WALSH, J.J. & YIELDING, G. (eds) *The Geometry and Growth of Normal Faults*. Geological Society, London, Special Publications, **439**. First published online March 31, 2016, https://doi.org/10.1144/SP439.7

WOJTAL, S.F. 1994. Fault scaling laws and the temporal evolution of fault systems. *Journal of Structural Geology*, **16**, 603–612.

WORTHINGTON, R.P. & WALSH, J.J. 2016. Timing, growth and structure of a reactivated basin-bounding fault. *In*: CHILDS, C., HOLDSWORTH, R.E., JACKSON, C.A.-L., MANZOCCHI, T., WALSH, J.J. & YIELDING, G. (eds) *The Geometry and Growth of Normal Faults*. Geological Society, London, Special Publications, **439**. First published online July 1, 2016, https://doi.org/10.1144/SP439.14

YIELDING, G. 2016. The geometry of branch lines. *In*: CHILDS, C., HOLDSWORTH, R.E., JACKSON, C.A.-L., MANZOCCHI, T., WALSH, J.J. & YIELDING, G. (eds) *The Geometry and Growth of Normal Faults*. Geological Society, London, Special Publications, **439**. First published online February 22, 2016, https://doi.org/10.1144/SP439.1

The geometry of branch lines

GRAHAM YIELDING

Badley Geoscience Ltd, North Beck House, North Beck Lane, Hundleby, Spilsby, Lincolnshire PE23 5NB, UK
graham@badleys.co.uk

Abstract: A branch line is the line of intersection between two hard-linked fault planes, or between two parts of a single fault plane of more complex geometry. Of interest is whether they provide any information about the kinematic development of the fault system to which they belong. Analysis of branch lines from a variety of normal fault networks, interpreted on seismic reflection datasets, shows that the branch lines are generally aligned parallel to the extension direction. This relationship is shown to be a feature of polymodal (orthorhombic) fault systems produced by three-dimensional strain. Branch lines between bimodal faults (conjugate, with opposing dip) tend to be perpendicular to the slip direction.

The motivation for this short contribution comes from a conversation about fault populations between the author and Juan Watterson, in which it was asserted that 'the orientation of branch lines must tell us something about the direction of slip on the faults' (Watterson pers. comm. 1989). However, after some 10 years of further research into fault geometries and displacements with the Fault Analysis Group at Liverpool University, the opposite assertion was claimed: 'Branch-line orientations are … not simply related to the fault slip direction' (Walsh *et al.* 1999). In this contribution, it is shown that both of these assertions can in fact be true, depending on the nature of the three-dimensional (3D) strain and the symmetry of the fault system.

A branch line is the line of join between two faults (e.g. Boyer & Elliott 1982) and, hence, the branch lines are a fundamental part of the overall fault network. However, very little has been documented in terms of the observed geometry of branch lines. Branch lines are not explicit in conventional cross-section or map representations of structure. Branch *points* are seen. It is, therefore, appropriate to first consider how we can determine the orientation of branch lines in different kinds of datasets. A number of different examples of branch lines are then presented. Finally, the relationship of branch-line orientations to the extension direction is summarized in the context of the geometrical properties of the fault network.

How can we 'map' branch lines?

The inverted commas in this section heading are used deliberately to point out that, even in geology, much of our language is based on a 2D view rather than a 3D one (Krantz *et al.* 2014). The verb 'to map' describes the process of recording the geometry of a structure, by producing a result (a map) that is 2D and horizontal. There is no 3D equivalent, except, perhaps, the cumbersome 'map out in 3D' or 'build a 3D model'. Determining the geometry of branch lines requires 3D information. This might either be assembled from (i) a succession of 2D slices or (ii) from multiple intersections with a high-relief topography, or, preferably, by (iii) true 3D imaging.

An example of 2D slicing to determine 3D fault-network geometry is given by Kristensen *et al.* (2008), who excavated serial cross-sections across minor faults within soft sediments. Fault displacements were 10 cm or less. To capture details such as relay zones and fault-bound lenses, sections were spaced at 0.5–2.5 cm intervals. This allowed branch points to be correlated from one section to the next, to build up the geometry of individual branch lines.

On a larger scale, 2D slices may be assembled in the subhorizontal direction – for example, a succession of structure maps of different stratigraphic boundaries – within the same area. An example of this style of data is given later, from the Gullfaks oil field (Fossen & Rørnes 1996; Fossen & Hesthammer 1998).

In a 3D seismic reflection dataset, a more continuous data volume is available. However, the conventional approach to interpreting subsurface geometry in such a dataset is still basically 2D. To build a 3D model of the fault network, the following workflow is common:

(1) Interpret fault traces (segments or 'sticks') on seismic sections, preferably in two or more directions.

From: Childs, C., Holdsworth, R. E., Jackson, C. A.-L., Manzocchi, T., Walsh, J. J. & Yielding, G. (eds) 2017. *The Geometry and Growth of Normal Faults*. Geological Society, London, Special Publications, **439**, 11–22.
First published online February 22, 2016, https://doi.org/10.1144/SP439.1

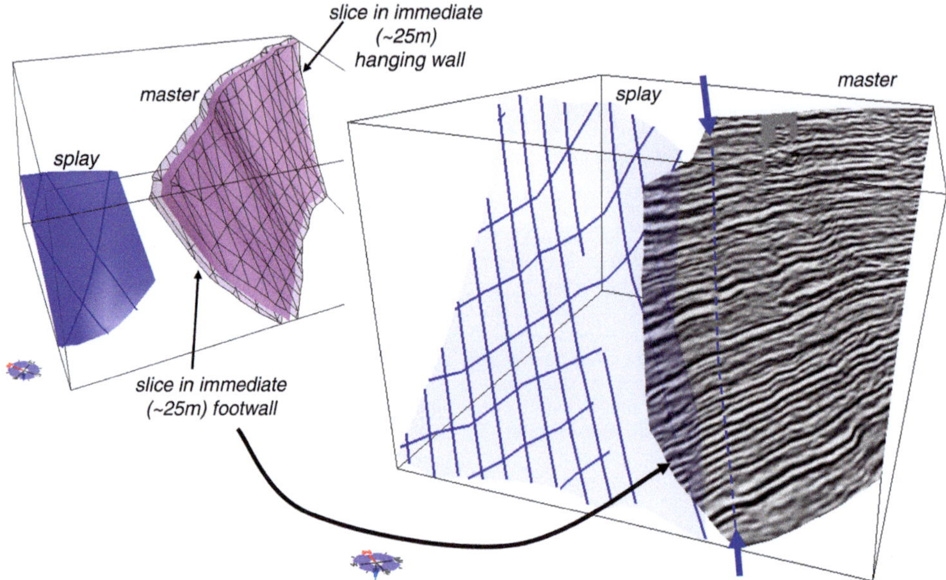

Fig. 1. Illustration of the technique of seismic fault slicing in a 3D seismic reflection dataset (West Cameron). On the left, splay and master fault planes are shown in blue and pink, respectively. Slice planes are shown on each side of the master fault, shifted 25 m into the footwall and hanging wall. On the right, the master fault is shown rendered with the seismic data from the footwall slice. The discontinuity visible in the seismic reflections (arrowed) corresponds to the branch line where the splay joins the master fault.

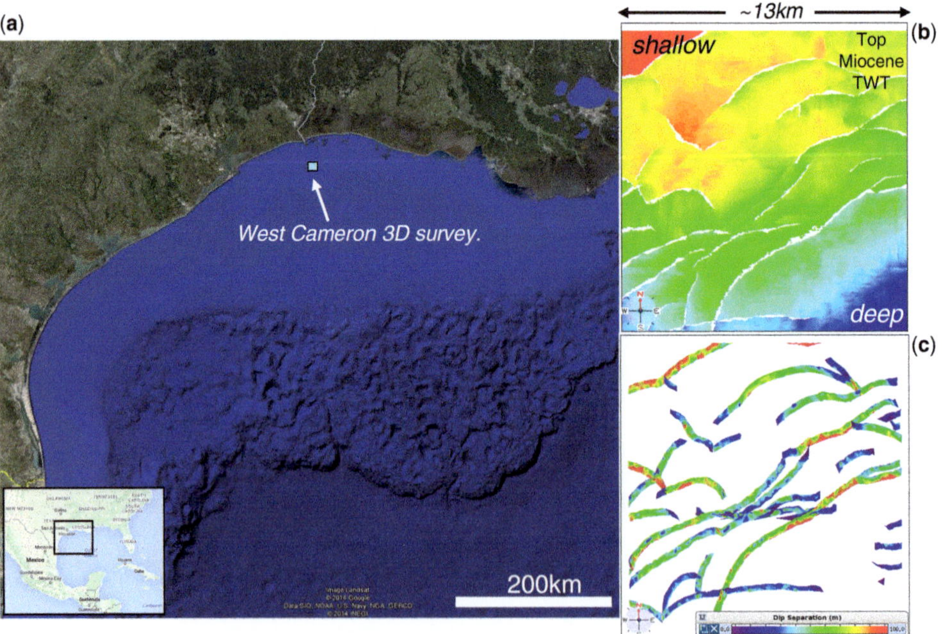

Fig. 2. (**a**) Google Earth view showing location of the West Cameron seismic survey in the northern Gulf of Mexico. (**b**) TWT map of the Top Miocene surface in the West Cameron dataset (red denotes shallow; blue denotes deep). (**c**) Dip separations (red = 100 m, purple = 0 m) on fault surfaces near the Top Miocene horizon.

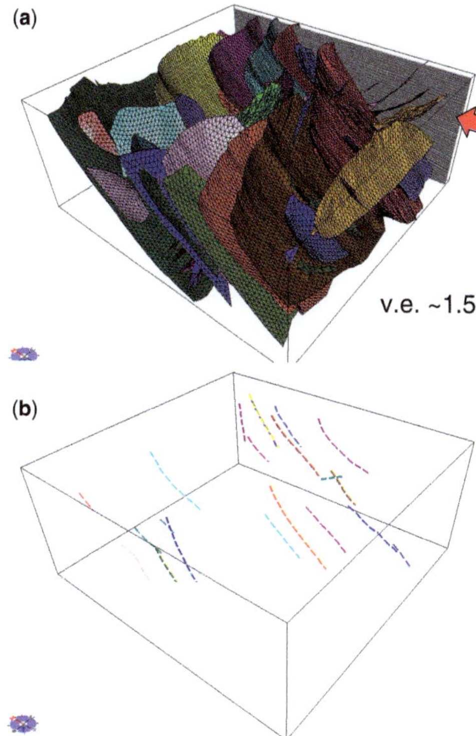

Fig. 3. (**a**) Interpreted fault network in the West Cameron 3D seismic dataset, viewed from the SW. One seismic row is shown at the eastern edge of the survey. The red arrow indicates the approximate position of the Top Miocene reflection (shown in Fig. 2b). (**b**) Same view as (a) but with fault surfaces removed to show the branch lines. v.e., vertical exaggeration.

(2) Model the fault raw data (segments) into fault planes (triangulated or gridded surfaces).
(3) Establish linkages (branch lines) between individual fault surfaces.

The last step, closing the gap between one interpreted fault surface and another, is typically handled in 3D-modelling software in a purely geometrical way. The ‘splay’ fault surface is extrapolated onto the ‘master’ fault, based on its 3D trend as it approaches the master fault. Unfortunately, faults often curve as they approach and interact (e.g. Segall & Pollard 1980), and so, potentially, the resulting intersection may be poorly located by using planar extrapolation.

A more data-driven way to obtain the branch line in 3D seismic reflection datasets is a technique termed ‘seismic fault slicing’ (Brown *et al.* 1987). The 3D geometry of the master fault is used to sample the seismic data immediately adjacent to the fault plane, on the side of the fault where the splay surface is located (Fig. 1). In the figured example, the seismic slice shows the data in the footwall of the master fault, and in doing so provides a view of the splay fault as it approaches the master; the splay is visible as an aligned discontinuity in the reflections in the slice. In the limit of the fault slice being very close to the master, this discontinuity is equivalent to a view of the branch line. In most 3D seismic datasets, even of mediocre quality, this technique provides a much more accurate definition of the branch-line location and geometry.

Example 1: Gulf of Mexico

The Gulf of Mexico continental shelf is characterized by listric normal faults driven by the gravitational collapse of the margin (e.g. Diegel *et al.* 1995). The West Cameron sector is located offshore Louisiana, near the Texas border (Fig. 2). A 3D seismic reflection dataset was acquired by GECO in 1987 about 40 km offshore, and is widely used for research and teaching (e.g. Reymond 1994). Miocene–Pleistocene shelf sequences are cut by an array of mainly SSE-dipping listric normal faults. Observations of channel displacements confirm that the faults have normal-sense offset with no strike-slip components (Reymond 1994). The down-to-south sense of the faulting is indicated by the vertical offsets in an example stratigraphic surface (Fig. 2b), and the dip separations at a comparable level in the fault network are shown in the map of Figure 2c.

A 3D view of the fault network is shown in Figure 3a. The seismic interpretation of the faults was mainly performed on a subset of the survey rows and columns, at a line spacing of 125 m throughout the survey volume (Kneen 2007). These interpreted fault segments were correlated into surfaces by triangulation and gridding. Links between fault surfaces (i.e. the branch lines) were constructed using seismic fault slicing – in fact, the example shown in Figure 1 is from the West Cameron dataset. Figure 3b shows the branch lines with the fault surfaces stripped away. In concert with the dominant southerly dip of the fault surfaces, the branch lines also mainly dip south.

In Figure 4a, the branch lines are displayed in map view, and show a strong north–south alignment, with a mean trend of 176° and mean angular deviation of 32.9° (Fig. 4b). In qualitative terms, this is perpendicular to the approximately east–west trend of the margin (Fig. 2), and is therefore parallel to the expected transport direction, driven by gravitational collapse into the deep-water Gulf of Mexico.

Although the original seismic volume is in the two-way-time (TWT) domain, interpreted structures

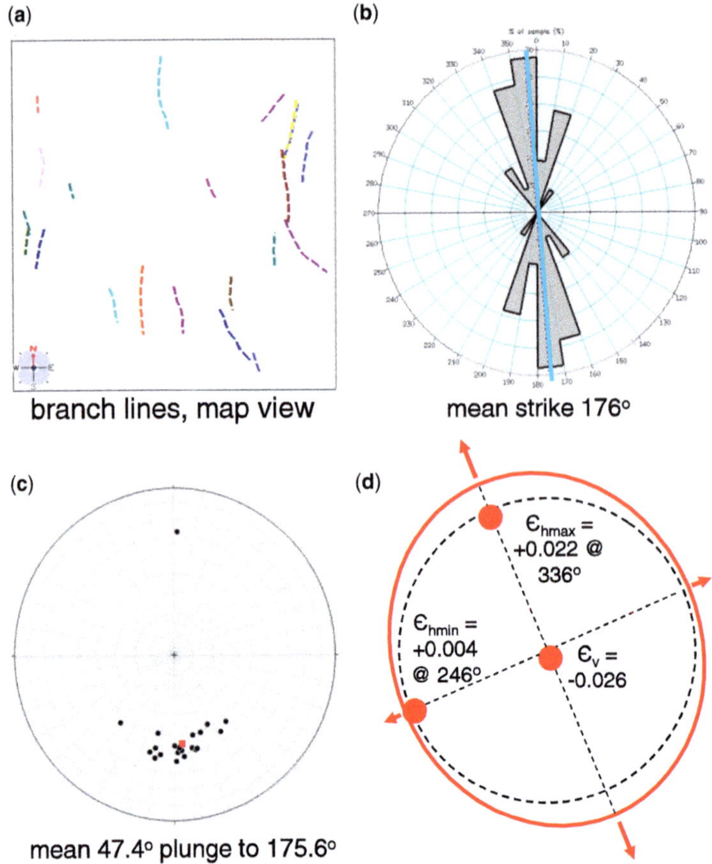

Fig. 4. (**a**) Map view of West Cameron branch lines of Figure 3b. (**b**) Rose diagram of the branch-line strikes. The mean strike is 176° (mean angular deviation is 32.9°). (**c**) Lower-hemisphere equal-area projection of branch-line trends (average trend per branch line). The mean orientation is a plunge of 47.4° towards 175.6°. (**d**) Principal strain axes derived from a Kostrov moment summation of the fault network. $\epsilon_{h_{max}}$ plunges 3° towards 336° and $\epsilon_{h_{min}}$ plunges 0.1° towards 246°.

can be converted into the depth domain using the appropriate interval velocity, which is 3 km s^{-1} for the interpretation shown in Figure 3. The true dip and plunge of the branch lines can then be computed, and they are shown in a lower-hemisphere equal-area plot in Figure 4c. Mean plunge is 47.4° towards 175.6°.

An independent estimate of the extension direction can also be obtained directly from the fault geometry and displacements by Kostrov's moment summation (Kostrov 1974; England & Molnar 1997). For a sample area *A*, and taking fault dip and displacement measurements at an along-strike sampling interval of *s*, then:

$$\epsilon_{ij} = \left(\frac{s}{2A}\right) \sum [(\mathbf{u}/(\sin\theta))(\mathbf{u_i n_j} + \mathbf{n_i u_j})]$$

where θ is the fault dip, **u** is the fault displacement vector and **n** is the fault normal vector. The eigenvectors of ϵ_{ij} are the principal axes of the strain ellipsoid, and the eigenvalues are the principal extensions. Applying this method at the level of the Top Miocene (Fig. 2) gives the strain ellipsoid shown in Figure 4d. The maximum extension direction $\epsilon_{h_{max}}$ is somewhat east of south, with an azimuth of 156°/336°. This is 20° away from the mean of the branch-line azimuths (Fig. 4b), although it lies within the mean angular deviation (33°) from the mean.

Further examples: northern North Sea

The Viking Graben in the northern North Sea is a north–south-trending rift system that was active in the Permo-Trias and Late Jurassic (e.g. Badley *et al.* 1988; Yielding *et al.* 1992; Fraser *et al.* 2002). Extension is generally considered to be

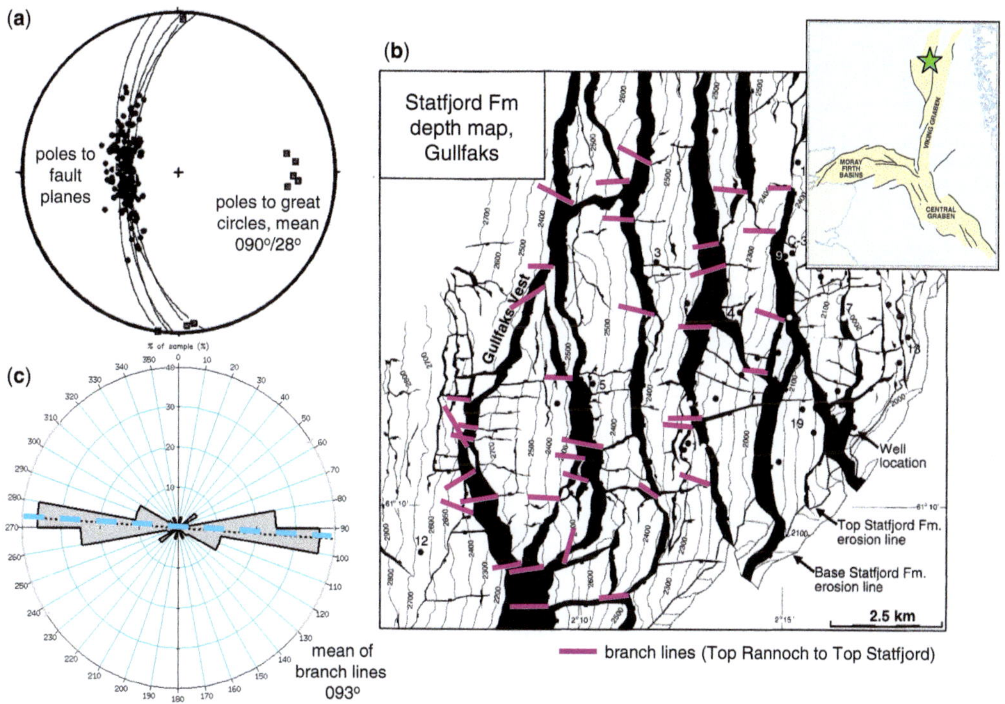

Fig. 5. (**a**) Lower-hemisphere equal-area projection of poles to five main faults in the Gullfaks field, from Fossen & Hesthammer (1998). For each fault, the best-fit circle and its pole are shown; the mean of the five poles plunges 28° towards 090°. Note the cylindrical nature of the fault planes. (**b**) Depth map of the Statfjord Formation, Gullfaks field, from Fossen & Hesthammer (1998). The purple lines are branch lines that have been constructed between the Top Rannoch Formation and the Top Statfjord Formation. Inset shows the location in the North Sea. (**c**) Rose diagram of the branch-line strikes in (b).

broadly east–west (e.g. Roberts *et al.* 1990; Erratt *et al.* 1999), although temporal variations on this average have also been suggested (Færseth *et al.* 1997; Davies *et al.* 2001). Here, we look at three example datasets across the basin, on a transect from west to east. All data are taken from the Lower–Middle Jurassic sedimentary section and, therefore, record the Late Jurassic fault movements.

Fossen & Hesthammer (1998) produced a detailed structural study of the Gullfaks field on the western margin of the Viking Graben, including structure contour maps, stereoplots and graphs. One of their key observations was that the main faults have a cylindical (corrugated) geometry, defined by local poles to the fault planes plotting along great circles (Fig. 5a). The pole of the great circle for each fault defines the fault's axis of curvature. Following similar observations by Koestler *et al.* (1992), Fossen & Hesthammer (1998) argued that these poles must be close to the mean slip direction on these faults. The poles plunge gently eastwards, with a mean of 28° dip towards 090° (Fig. 5a). This direction is also down the dip of the faults, implying that the faults are strongly dip-slip without an oblique component.

The fault patterns at different horizons published by Fossen & Rørnes (1996) and Fossen & Hesthammer (1998) allow the construction of branch-line orientations by identifying the same fault intersection on different maps. In Figure 5b, branch lines determined by this technique are shown superimposed on the structure contour map of the Statfjord Formation. Depth values at the branch lines are poorly constrained from the contour maps, so only azimuths are considered rather than dips, and these are plotted in Figure 5c. The mean orientation is 093°, which is very similar to the slip direction inferred from fault-plane curvature (090°). Hence, the branch lines point down the slip direction.

A more complicated dataset is shown in Figure 6, from the central part of the Viking Graben. Interpretation and depth conversion of a 3D seismic dataset was used to produce a fault framework. In Figure 6a, the fault pattern at a pre-rift horizon is shown with thin black lines, with coloured lines showing the branch lines extracted from the

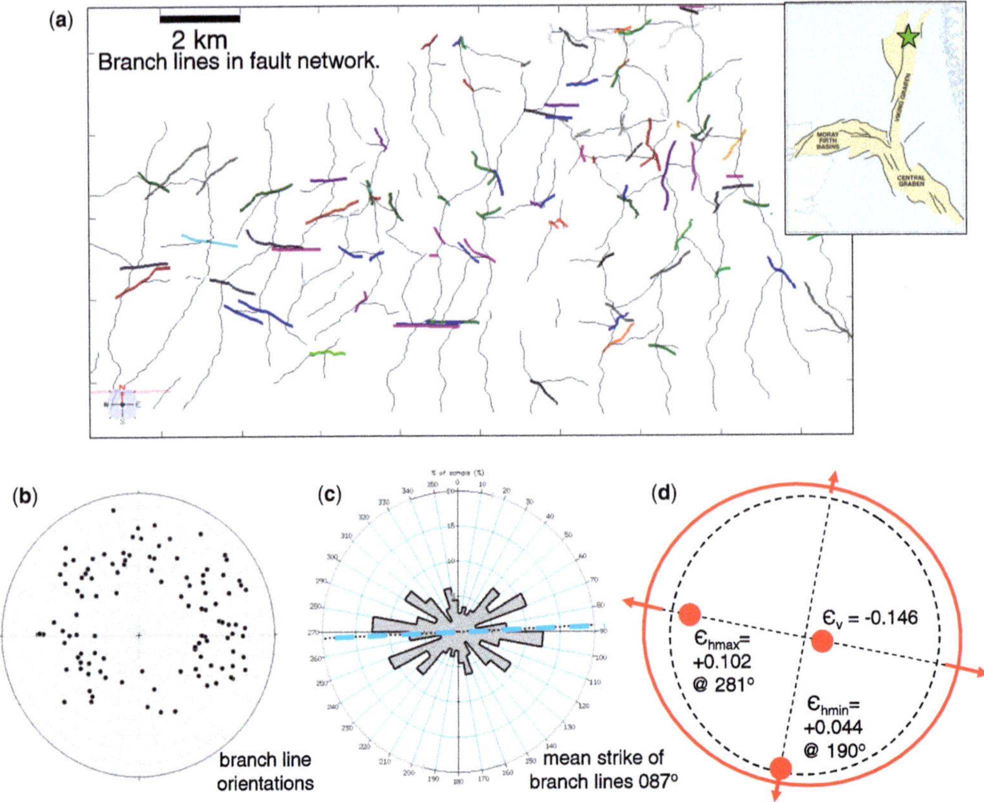

Fig. 6. (**a**) Fault pattern and extracted branch lines from a 3D seismic survey in the Viking Graben, data courtesy of StatoilHydro. (**b**) Lower-hemisphere equal-area projection of branch-line trends (average trend per branch line). (**c**) Rose diagram of the branch-line strikes. The mean strike is 087° (mean angular deviation is 72.2°). (**d**) Principal strain axes derived from a Kostrov moment summation of the fault network. $\epsilon_{h_{max}}$ plunges 11° towards 281° and $\epsilon_{h_{min}}$ plunges 4° towards 190°.

framework. Although the dominant fault orientation is broadly north–south, there are many subsidiary orientations, leading to the occurrence of fault-bound compartments. The branch lines show a similar complexity: their orientations are plotted on a lower-hemisphere projection in Figure 6b, and show plunges to the east and west as well as to the north, precluding the calculation of a simple mean. Plotting only the strikes of the branch lines (Fig. 6c) allows the west- and east-dipping lines to be classed together, and provides a mean strike of 087°.

Using the fault framework of the dataset in Figure 6, it is also possible to extract a strain tensor from a Kostrov moment summation, as described earlier, for the West Cameron dataset. This process assumes dip-slip movement, as was demonstrated above for the Gullfaks dataset. The result is shown in Figure 6d, extracted at a mid-Jurassic pre-rift horizon. The maximum horizontal strain in this case is computed to be slightly more than 10% extension along a trend of 101–281°. Despite there being a much greater spread in branch-line orientations than in the West Cameron dataset, the mean trend of the branch lines, again, has a similar alignment to $\epsilon_{h_{max}}$, in this case approximately east–west. There is a 14° difference between the branch-line mean and the maximum horizontal strain direction.

The third North Sea example comes from the Troll Field on the eastern margin of the Viking Graben. Fault interpretation by Bretan *et al.* (2011) was performed on a 3D seismic reflection dataset and depth-converted using well interval velocities. Branch-line orientations from the fault framework are shown in map view in Figure 7a, in a lower-hemisphere projection in Figure 7b and in a rose diagram in Figure 7c. The fault network is intermediate in complexity between the previous two examples. The mean strike of the branch lines is 081°.

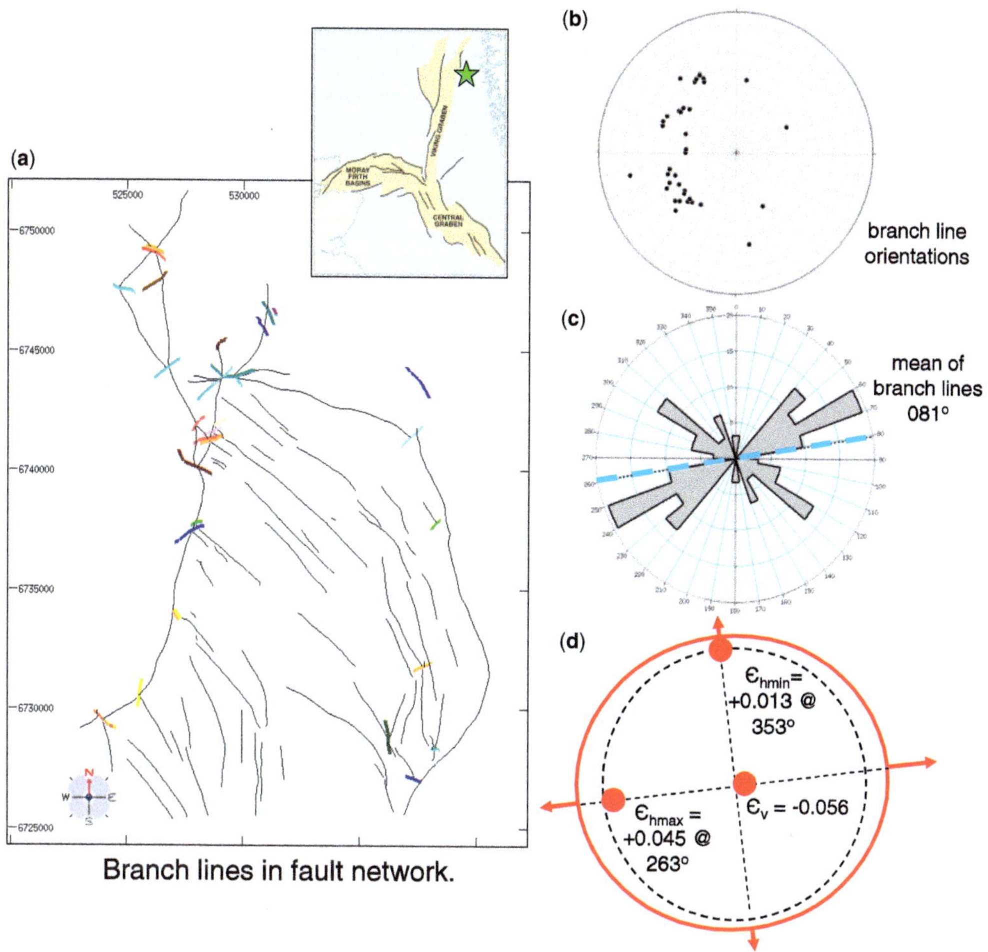

Fig. 7. (**a**) Fault pattern and extracted branch lines from a 3D seismic survey of the Troll field (Bretan *et al.* 2011). (**b**) Lower-hemisphere equal-area projection of branch-line trends (average trend per branch line). (**c**) Rose diagram of the branch-line strikes. The mean strike is 081° (mean angular deviation is 68.3°). (**d**) Principal strain axes derived from a Kostrov moment summation of the fault network. $\epsilon_{h_{max}}$ plunges 6° towards 263° and $\epsilon_{h_{min}}$ plunges 1° towards 353°.

Calculation of regional strain from a Kostrov moment summation, at a Jurassic pre-rift horizon, indicates maximum horizontal extension at an azimuth of 083–263° (Fig. 7d). Hence, branch lines and $\epsilon_{h_{max}}$ are again both aligned approximately east–west (2° difference), perpendicular to the graben axis.

Branch lines from the three North Sea datasets are shown together in Figure 8, along with the corresponding slip direction or strain direction indicators. Both the branch-line observations and the slip/strain indicators suggest approximately east–west extension across the basin, with a spread of 12° for the branch lines and 18° for the slip/strain indicators.

Branch-line orientation and extension direction

Although they can provide an excellent picture of subsurface 3D geometry, seismic reflection data do not, generally, give precise kinematic information, in terms of slip vectors. Channel offsets in the West Cameron dataset, described above, imply approximate dip-slip movement. In the Gullfaks dataset, slip is inferred to be down the fault-surface corrugations.

The existence of branch lines requires faults of different orientations (i.e. non-parallel linked faults). Such fault networks are variously referred to as orthorhombic, polymodal or multimodal

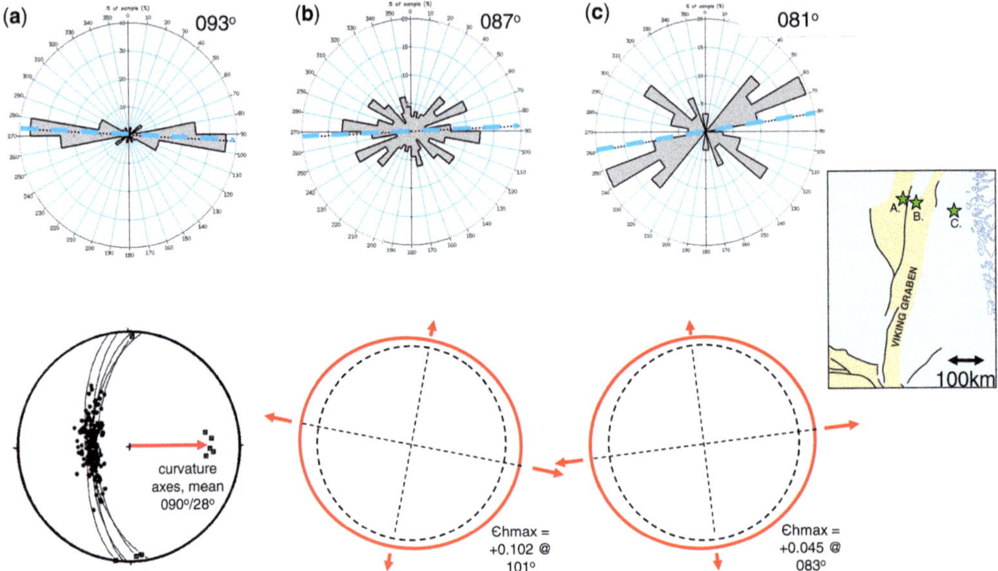

Fig. 8. Synthesis of the three North Sea datasets shown in Figures 5, 6 & 7 (**a**, **b** and **c**, respectively). The upper row shows rose diagrams of the branch-line strikes. The lower row shows the sense of slip (a) or extension direction (b & c), expressed in the range 0°–180°.

(Oertel 1965; Reches 1978; Aydin & Reches 1982; Groshong 1988; Krantz 1988; Healy *et al.* 2006) (see Fig. 9a). They result from 3D (poly-axial) strain. Truly orthorhombic systems have four orientations of coeval faults, arranged as two sets, each of which includes opposing dips. The angle between the two sets is dependent upon the ratio of the principal strains (Krantz 1988). For orthorhombic normal faults, displacement on each set is approximately dip-slip (Oertel 1965). The faults are arranged symmetrically about the strain axes and, therefore, there are four possible fault-intersection directions (Fig. 9a). Depending on the lateral extent of the fault surfaces, the intersection lines can be

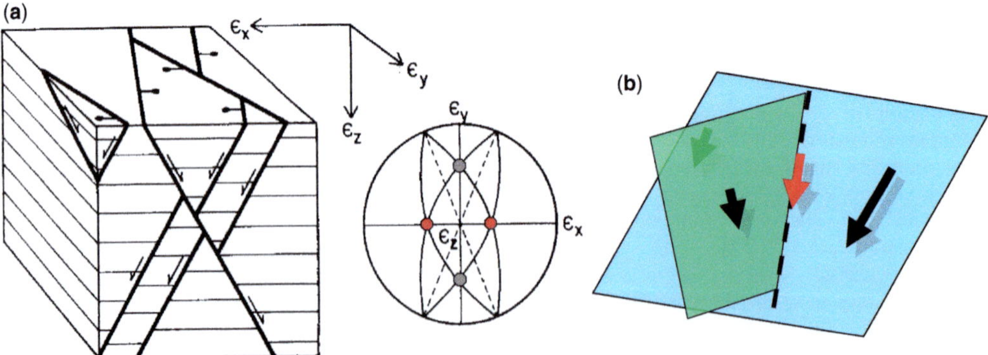

Fig. 9. (**a**) Block diagram and stereonet of orthorhombic normal faults, from Krantz (1988). On the stereonet, the red symbols mark intersections between faults dipping in the same direction; the grey symbols mark intersections between faults dipping in opposite directions. (**b**) Perspective view of two faults (same dip direction) intersecting at a branch line, each with their own slip vector (black arrow). The red arrow represents the local realignment of slip parallel to the branch line.

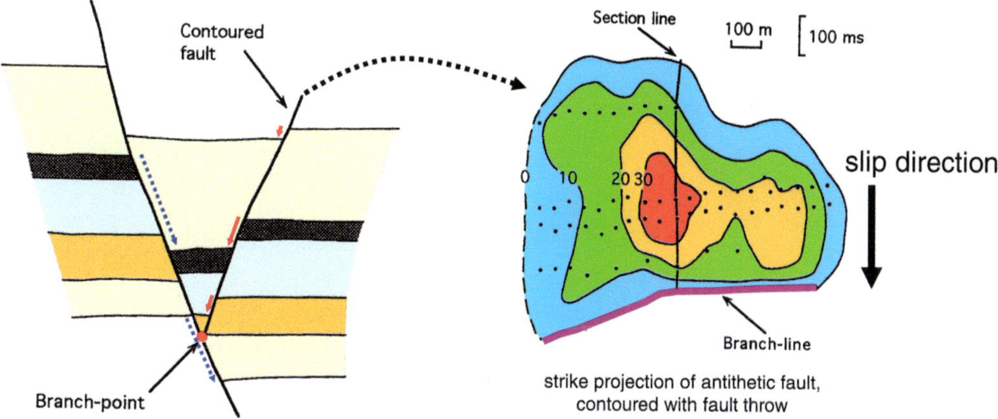

Fig. 10. Cross-section (left) and strike projection view (right) of an antithetic fault terminating downwards at another (conjugate) fault, taken from Nicol *et al.* (1996).

either branch lines (one fault terminates at the other) or cross-cutting locations (both fault surfaces exist on both sides of the line).

Intersections between faults of similar dip direction are aligned in the direction of maximum horizontal strain ϵ_x (red symbols in Fig. 9a). These intersection lines are close to the slip vector on both fault surfaces and, therefore, this geometry can accommodate simultaneous slip on both surfaces with relatively minimal strain in the rock volume around the branch line. Geomechanical modelling (e.g. Maerten 2000) would suggest that, locally, the slip vectors rotate to be parallel to the branch line (Fig. 9b). This kind of branch line was termed a conservative fault junction by King & Yielding (1984).

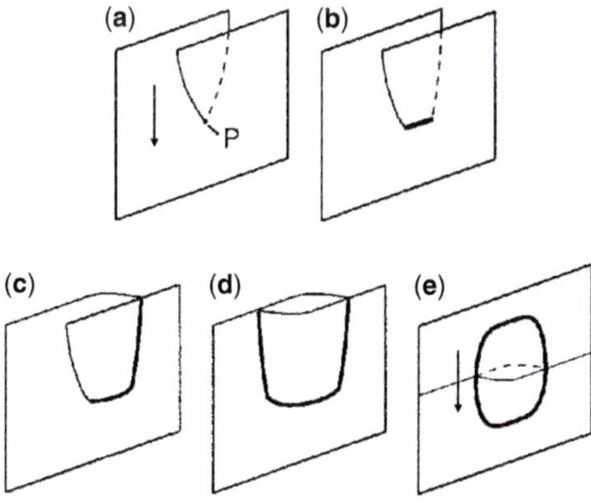

Fig. 11. Stages of development of fault-bounded lens in a fault zone, taken from Walsh *et al.* (1999). P is an initial branch point; branch lines are shown bold. Arrows indicate the direction of fault slip.

Intersections between faults of different dip direction are aligned in the direction of minimum horizontal strain ϵ_y (grey symbols on the stereonet in Fig. 9a). Since the faults are dipping in opposite directions, there must be significant differences in their slip vectors and these intersections are not kinematically compatible.

The Andersonian model of conjugate (bimodal) faults can be regarded as a special case of the orthorhombic (polymodal) symmetry, in which intermediate strain is 0 (i.e. plane strain). The ϵ_y fault intersections are then horizontal. A common observation in conjugate fault systems is that one of the two faults terminates downwards at the other, to give a synthetic–antithetic pair where the antithetic fault dies out downwards at the larger fault. Figure 10 shows a classic example of this from Nicol *et al.* (1996), from a seismic reflection dataset in the Timor Sea. The branch point in cross-section belongs to the subhorizontal branch line in three dimensions. Throw on the antithetic fault decreases downwards to the branch line, which in this case is effectively a tip-line. The throw on the antithetic fault does *not* add downwards to the throw on the main fault because the angle between the vectors is too large. In contrast to the orthorhombic (polymodal) examples discussed above, the branch line is *perpendicular* to the down-dip slip vector.

Another special case that can be recognized is that of internal segmentation within a single fault surface. This description is scale-dependent, in that multiple fault surfaces may exist in detail but, at the broad scale, the fault zone has a common orientation. The geometry of branch lines between normal faults and their sub-parallel splays has been summarized by Walsh *et al.* (1999) using 3D seismic reflection data, and by Kristensen *et al.*

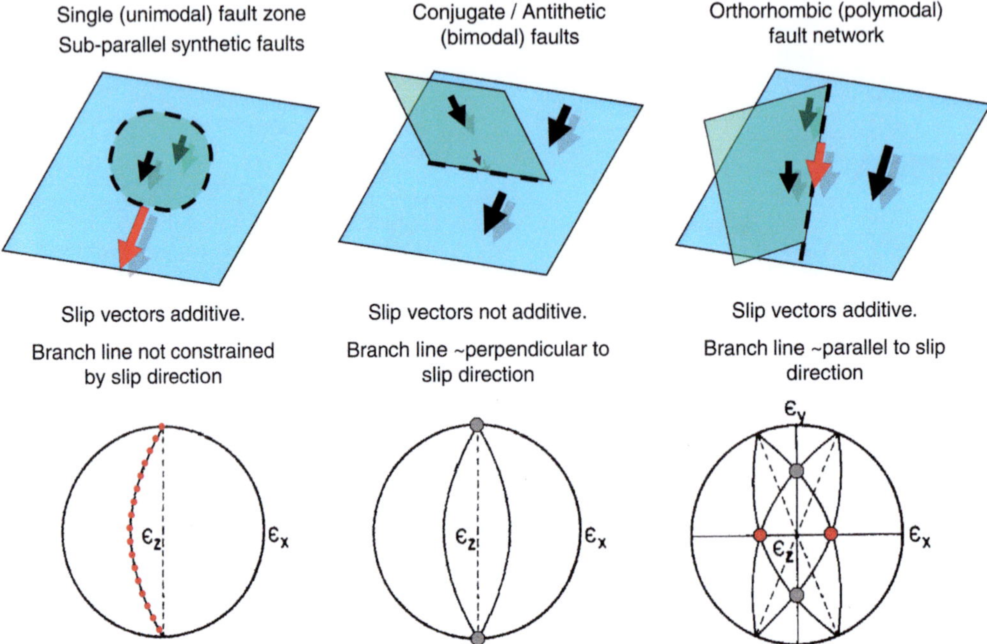

Fig. 12. Summary of the three generalized relationships between the branch-line orientation and slip/extension direction. Left to right: single fault zone (unimodal), conjugate (bimodal) faults, orthorhombic/polymodal fault network. The upper row shows the perspective views of principal (blue) and subsidiary (green) faults; arrows indicate slip vectors (red vector applies at branch line). The lower row shows corresponding stereonets, with red symbols indicating kinematically compatible branch-line orientations and grey symbols indicating kinematically incompatible orientations.

(2008) using serial sections in soft sediment. Walsh *et al.* (1999) argued that these structures grow by failure of relay zones in a propagating multi-strand fault surface (Fig. 11). As seen at Figure 11e, the ultimate branch-line geometry may be a closed loop, with parts of the line both parallel and perpendicular to the down-dip slip vector – the branch-line orientations are not simply related to the slip direction. If angular differences between the different fault strands are small, slip can occur relatively easily regardless of the orientation of the branch line. However, as slip incompatibility increases, there is a greater chance that removal of asperities will act to smooth out the geometry and return the system to compatibility (Childs *et al.* 2009).

Conclusions

The different relationships between branch-line orientation and slip direction (or extension direction) are summarized in Figure 12. They are controlled by whether the normal fault or fault network has one, two or four fault-plane orientations, corresponding to a single propagating fault (unimodal), a conjugate (bimodal) fault pair or an orthorhombic/polymodal fault network. The branch-line loops (single fault zone) can also occur within the bimodal and polymodal networks.

Thus, it is possible for a branch line to have any angular relationship to the slip direction, from parallel to perpendicular or anywhere in-between. On a single propagating fault surface, local irregularities leading to tip bifurcation can ultimately generate closed-loop branch lines, which can remain stable if the enclosed lens has acute-angled edges. The parts of the branch line that are perpendicular and parallel to the slip direction are edge and screw dislocations, respectively.

Where the normal faults have opposed dip (e.g. in conjugate pairs), the subhorizontal branch line between them is perpendicular to the fault slip direction, and acts to terminate the slip field on one of the faults. By contrast, in a polymodal fault pattern produced by 3D strain, branch lines between different fault sets of similar dip are compatible with the slip vector. Analysis of branch lines in a polymodal fault network therefore provides a simple way of estimating the extension direction.

The examples and analysis presented here have deliberately been chosen from relatively simple settings where the observed offsets on the faults have been generated in one phase of deformation involving predominantly dip-slip movement. In more complicated tectonic settings, it should be expected that additional factors could affect the relationship between bulk extension and branch-line geometry. For example, if two separate phases of deformation occurred with two different extension directions, then the early fault surfaces are later reactivated with oblique movement (e.g. Henza *et al.* 2011). In this case, a moment summation would need to be performed carefully using the true slip vectors for each phase, which is unlikely to be possible in a seismic reflection dataset. Analysis of branch-line geometries in an analogue experiment with two known extension directions (like that of Henza *et al.* 2011) might provide an insight into whether the branch line orientations have any relationship to those two directions, and whether they evolve during deformation. A more subtle effect might occur where the orientations of new faults might be influenced by pre-existing basement fault patterns. Arguably, this could be the case in the North Sea examples here, and, perhaps, is the reason why their branch lines show a greater spread than in the West Cameron example.

This contribution was inspired by, and is dedicated to, Juan Watterson. Discussions with Dave Sanderson, Pete Bretan, James Jackson and Dave Healy helped to focus the direction of the first draft of the manuscript. I am grateful to Atle Rotevatn, Richard Groshong and Bob Holdsworth for their constructive comments on the manuscript, which led to an improved revised version. Stereograms were produced in Stereonet software (Cardozo & Allmendinger 2013).

References

Aydin, A. & Reches, Z. 1982. Number and orientation of fault sets in the field and in experiments. *Geology*, **10**, 107–112.

Badley, M.E., Price, J.D., Rambech Dahl, C. & Agdestein, T. 1988. The structural evolution of the northern Viking Graben and its bearing upon extensional modes of basin formation. *Journal of the Geological Society, London*, **145**, 455–472, https://doi.org/10.1144/gsjgs.145.3.0455

Boyer, S.E. & Elliott, D. 1982. Thrust systems. *American Association of Petroleum Geologists Bulletin*, **66**, 1196–1230.

Bretan, P., Yielding, G., Mathiassen, O.M. & Thorsnes, T. 2011. Fault-seal analysis for CO_2 storage: an example from the Troll area, Norwegian Continental Shelf. *Petroleum Geoscience*, **17**, 181–192, https://doi.org/10.1144/1354-079310-025

Brown, A.R., Edwards, G.S. & Howard, R.E. 1987. Fault slicing – A new approach to the interpretation of fault detail. *Geophysics*, **52**, 1319–1327.

Cardozo, N. & Allmendinger, R.W. 2013. Spherical projections with OSXStereonet. *Computers & Geosciences*, **51**, 193–205, https://doi.org/10.1016/j.cageo.2012.07.021

Childs, C., Manzocchi, T., Walsh, J.J., Bonson, C.G., Nicol, A. & Schöpfer, M.P.J. 2009. A geometric model of fault zone and fault rock thickness variations. *Journal of Structural Geology*, **31**, 117–127.

Davies, R.J., Turner, J.D. & Underhill, J.R. 2001. Sequential dip-slip fault movement during rifting: a new model for the evolution of the Jurassic trilete North Sea rift system. *Petroleum Geoscience*, **7**, 371–388, https://doi.org/10.1144/petgeo.7.4.371

Diegel, F.A., Karlo, J.F., Schuster, D.C., Shoup, R.C. & Tauvers, P.R. 1995. Cenozoic structural evolution and tectono-stratigraphic framework of the northern Gulf coast continental margin. *In*: Jackson, M.P.A., Roberts, D.G. & Snelson, S. (eds) *Salt Tectonics: A Global Perspective*. American Association of Petroleum Geologists, Memoirs, **65**, 109–151.

England, P. & Molnar, P. 1997. The field of crustal velocity in Asia calculated from Quaternary rates of slip on faults. *Geophysics Journal International*, **130**, 551–582.

Erratt, D., Thomas, G.M. & Wall, G.R.T. 1999. The evolution of the Central North Sea Rift. *In*: Fleet, A.J. & Boldy, S.A.R. (eds) *Petroleum Geology of Northwest Europe, Proceedings of the 5th Conference*. Geological Society, London, 63–82, https://doi.org/10.1144/0050063

Færseth, R.B., Knudsen, B.E., Liljedahl, T. & Midbue, P.S. & Suderstrum, B. 1997. Oblique rifting and sequential faulting in the Jurassic development of the northern North Sea. *Journal of Structural Geology*, **19**, 1285–1302.

Fossen, H. & Hesthammer, J. 1998. Structural geology of the Gullfaks field, northern North Sea. *In*: Coward, M.P., Daltaban, T.S. & Johnson, H. (eds) *Structural Geology in Reservoir Characterization*. Geological Society, London, Special Publications, **127**, 231–261, https://doi.org/10.1144/GSL.SP.1998.127.01.16

Fossen, H. & Rørnes, A. 1996. Properties of fault populations in the Gullfaks Field, northern North Sea. *Journal of Structural Geology*, **18**, 179–190.

Fraser, S.I., Robinson, A.M. *et al.* 2002. Upper Jurassic. *In*: Evans, D., Graham, C., Armour, A. & Bathurst, P. (eds) *The Millennium Atlas: Petroleum Geology of the Central and Northern North Sea*. Geological Society, London, 157–189.

Groshong, R.H., Jr. 1988. Low-temperature deformation mechanisms and their interpretation. *Geological Society of America Bulletin*, **100**, 1329–1360.

Healy, D., Jones, R.R. & Holdsworth, R.E. 2006. Three-dimensional brittle shear fracturing by tensile crack interaction. *Nature*, **439/5**, 64–67.

Henza, A.A., Withjack, M.O. & Schlische, R.W. 2011. How do the properties of a pre-existing normal-fault population influence fault development during a subsequent phase of extension? *Journal of Structural Geology*, **33**, 1312–1324, https://doi.org/10.1016/j.jsg.2011.06.010

King, G. & Yielding, G. 1984. The evolution of a thrust fault system: processes of rupture initiation, propagation and termination in the 1980 El Asnam (Algeria) earthquake. *Geophysical Journal of the Royal Astronomical Society*, **77**, 915–933.

Kneen, R.D. 2007. *Interpret the fault network imaged in the West Cameron 3D seismic survey*. MSc thesis, Imperial College, London.

Koestler, A.G., Milnes, A.G. & Storli, A. 1992. Complex hanging-wall deformation above an extensional detachment – example: Gullfaks Field, northern North Sea. *In*: Larsen, R.M., Brekke, H., Larsen, B.T. & Talleraas, E. (eds) *Structural and Tectonic Modelling and its Application to Petroleum Geology*. Norwegian Petroleum Society (NPF) Special Publication, **1**. Elsevier, Amsterdam, 243–251.

Kostrov, V. 1974. Seismic moment and energy of earthquakes, and seismic flow of rock. *Physics of the Solid Earth*, **1**, 13–21.

Krantz, B., Ormand, C. & Freeman, B. 2014. Geologists get a closer 'Look' at complexity. *AAPG Explorer*, February 2014, http://www.aapg.org/Publications/News/Explorer/Details/ArticleID/3788/Geologists-Get-a-Closer-'Look'-at-Complexity

Krantz, R.W. 1988. Multiple fault sets and three-dimensional strain: theory and application. *Journal of Structural Geology*, **10**, 225–237.

Kristensen, M.B., Childs, C.J. & Korstgård, J.A. 2008. The 3D geometry of small-scale relay zones between normal faults in soft sediments. *Journal of Structural Geology*, **30**, 257–272.

Maerten, L. 2000. Variation in slip on intersecting normal faults: implications for paleostress inversion. *Journal of Geophysical Research*, **105**, 25 553–25 565.

Nicol, A., Watterson, J., Walsh, J.J. & Childs, C. 1996. The shapes, major axis orientations and displacement patterns of fault surfaces. *Journal of Structural Geology*, **18**, 235–248.

Oertel, G. 1965. The mechanism of faulting in clay experiments. *Tectonophysics*, **2**, 343–393.

Reches, Z. 1978. Analysis of faulting in a three-dimensional strain field. *Tectonophysics*, **47**, 109–129.

Reymond, B. 1994. *Three-dimensional sequence stratigraphy offshore Louisiana, Gulf of Mexico (West Cameron 3D seismic data)*. Thèse de doctorat, Université de Lausanne.

Roberts, A.M., Yielding, G. & Badley, M. 1990. A kinematic model for the orthogonal opening of the late Jurassic North Sea rift system, Denmark–Mid Norway. *In*: Blundell, D.J. & Gibbs, A.D. (eds) *Tectonic Evolution of the North Sea Rifts*. Clarendon Press, Oxford, 180–199.

Segall, P. & Pollard, D.D. 1980. Mechanics of discontinuous faults. *Journal of Geophysical Research*, **85**, 4337–4350.

Walsh, J.J., Watterson, J., Bailey, W.R. & Childs, C. 1999. Fault relays, bends and branch-lines. *Journal of Structural Geology*, **21**, 1019–1026.

Yielding, G., Badley, M.E. & Roberts, A.M. 1992. The structural evolution of the Brent Province. *In*: Morton, A.C., Haszeldine, R.S., Giles, M.R. & Brown, S. (eds) *Geology of the Brent Group*. Geological Society, London, Special Publications, **61**, 27–43, https://doi.org/10.1144/GSL.SP.1992.061.01.04

Interactions and growth of faults in an outcrop-scale system

A. NICOL[1,2]*, C. CHILDS[3], J. J. WALSH[3], T. MANZOCCHI[3] & M. P. J. SCHÖPFER[4]

[1]*GNS Science, PO Box 30368, Lower Hutt, New Zealand*

[2]*Present address: Department of Geological Sciences, University of Canterbury, Private Bag 4800, Christchurch, New Zealand*

[3]*Fault Analysis Group, School of Earth Sciences, UCD, Dublin, Ireland*

[4]*Department for Geodynamics and Sedimentology, University of Vienna, Althanstrasse 14, A-1090 Vienna, Austria*

**Correspondence: andy.nicol@canterbury.ac.nz*

Abstract: Fault growth could be achieved by (1) synchronous increases in displacement and length or (2) rapid fault propagation succeeded by displacement-dominated growth. The second of these growth models (here referred to as the constant length model) is rarely applied to small outcrop-scale faults, yet it can account for many of the geometric and kinematic attributes of these faults. The constant length growth model is supported here using displacement profiles, displacement–length relationships and tip geometries for a system of small strike-slip faults (lengths of 1–200 m and maximum displacements of 0.001–3 m) exposed in a coastal platform in New Zealand. Displacement profiles have variable shapes that mainly reflect varying degrees of fault interaction. Increasing average displacement gradients with increasing fault size (maximum displacement and length) may indicate that the degree of interaction increases with fault size. Horsetail and synthetic splays confined to fault-tip regions are compatible with little fault propagation during much of the growth history. Fault displacements and tip geometries are consistent with a two-stage growth process initially dominated by propagation followed by displacement accumulation on faults with near-constant lengths. Retardation of propagation may arise due to fault interactions and associated reduction of tip stresses, with the early transition from propagation- to displacement-dominated growth stages produced by fault-system saturation (i.e. the onset of interactions between all faults). The constant length growth model accounts for different fault types over a range of scales and may have wide application.

Analysis of outcrop-scale faults over the last 30 years has been central to improved understanding of the growth of fault-zone widths, fault dimensions and fault displacements (e.g. Muraoka & Kamata 1983; Walsh & Watterson 1987; Peacock 1991; Peacock & Sanderson 1991; Dawers *et al.* 1993; Childs *et al.* 1996, 2009; Vermilye & Scholz 1999; Kim *et al.* 2000, 2004; Schultz *et al.* 2008; Nixon *et al.* 2011, 2012; Rotevatn & Fossen 2012; Nicol *et al.* 2013; Roche *et al.* 2016). A growth model has been widely advocated mainly using data for normal faults in which length and displacement accrue synchronously throughout faulting (Watterson 1986; Cowie & Scholz 1992*a*, *b*; Gillespie *et al.* 1992; Cartwright *et al.* 1995; Peacock & Sanderson 1996) (Fig. 1 left). For such a model, increase in fault length is achieved by propagation of individual faults and/or linkage of initially isolated faults (Cartwright *et al.* 1995; Cowie *et al.* 2000; Kim & Sanderson 2005; Bergen & Shaw 2010). Alternatively, fault growth could be achieved by an initial phase of rapid propagation followed by a more prolonged period of displacement accumulation on faults with near-constant lengths (Filbrandt *et al.* 1994; Morley 1999; Poulimenos 2000; Meyer *et al.* 2002; Walsh *et al.* 2002; Nicol *et al.* 2005, 2010; Manzocchi *et al.* 2006; Schlagenhauf *et al.* 2008; Jackson & Rotevatn 2013) (Fig. 1 right). The constant fault length model has been proposed to account for the growth of faults in analogue models and imaged by seismic reflection datasets. It is consistent with fault segmentation in circumstances where segments are kinematically coherent from their initiation and linkage reflects a strain localization process within individual fault zones (Walsh *et al.* 2003; Childs *et al.* 2009).

The two alternative fault-evolution models can be tested using growth strata (e.g. Meyer *et al.* 2002; Childs *et al.* 2003; Jackson & Rotevatn 2013; Tvedt *et al.* 2013) and analogue or numerical

From: Childs, C., Holdsworth, R. E., Jackson, C. A.-L., Manzocchi, T., Walsh, J. J. & Yielding, G. (eds) 2017. *The Geometry and Growth of Normal Faults*. Geological Society, London, Special Publications, **439**, 23–39.
First published online March 10, 2016, https://doi.org/10.1144/SP439.9

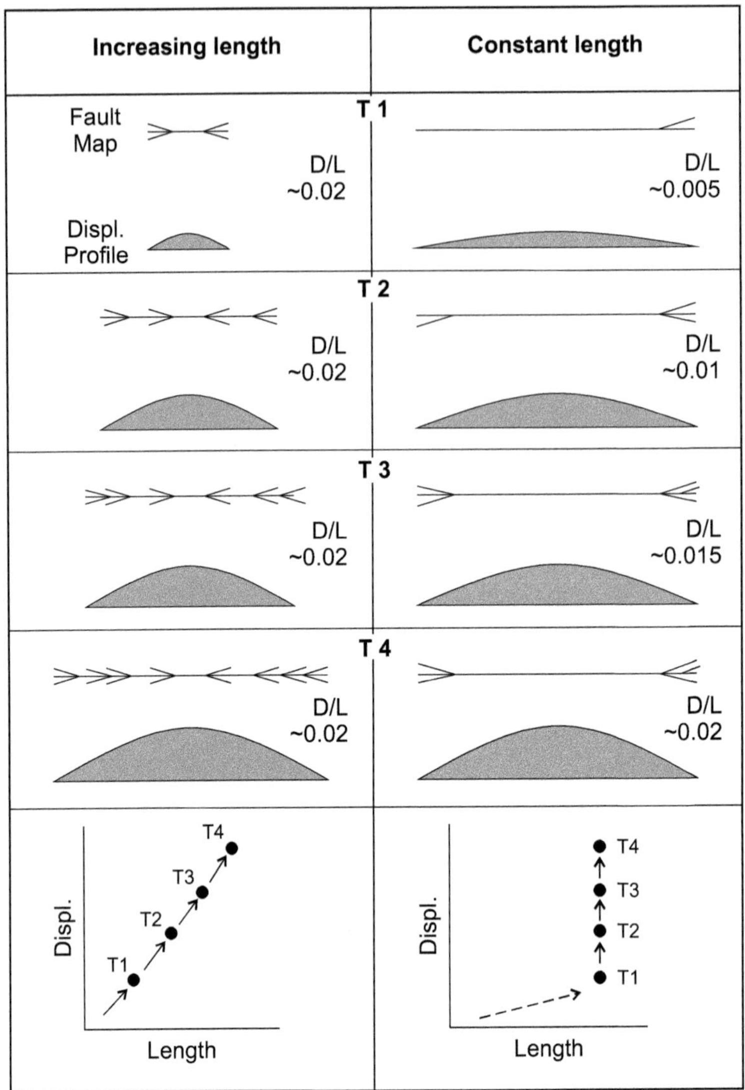

Fig. 1. Schematic diagram illustrating two potential fault-growth models. Diagram shows fault length, tip-zone secondary faulting (fault-trace maps) and along-strike displacement profiles at four times (T1–T4) for increasing fault length (left) and constant length (right) growth models. Displacement–length (*D–L*) ratios are indicated; they are approximately constant for the increasing length model and increase with displacement for the constant length model. Evolution of displacement and length for each model is shown in the lower schematic graphs. The horizontal and vertical scales are dissimilar for both displacement profiles and displacement–length graphs.

modelling (e.g. Filbrandt *et al.* 1994; Cowie *et al.* 2000; Schlagenhauf *et al.* 2008); however, these types of data rarely furnish information on the development of small-scale faults in outcrop. In the absence of growth strata, analysis of finite fault geometries and displacements provides a basis for discriminating between increasing and constant length growth models illustrated in Figure 1. These fault outcrop data are generally interpreted to support the increasing length fault-growth model (e.g. Peacock & Sanderson 1996); however, it is our contention that they can be equally well explained by the constant fault length model. The present paper focuses on presenting the data and interpretations for an outcrop-scale fault system that support the constant fault length model.

Fault geometries and displacements constrain the evolution of an outcrop-scale system of faults

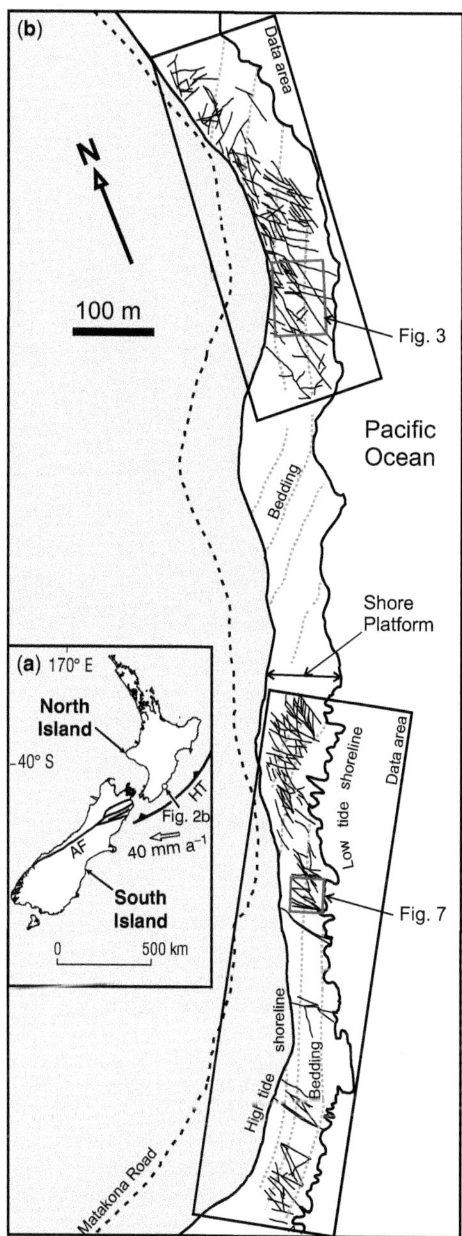

Fig. 2. Study area location map. (**a**) New Zealand location of the outcrop examined and plate boundary setting. The white open arrow shows the present relative plate motion vector of Beavan *et al.* (2002). AF, Alpine Fault; HT, Hikurangi Trough (front of the subduction accretionary prism). (**b**) Map of the shore platform showing faults (thin black lines), bedding (dotted lines) and locations of Figures 3 and 7 (boxes). The shore platform is bounded by the high- and low-tide shorelines. Areas of data collection are shown by the black rectangles; faults outside of these areas are not shown.

with apparent strike-slip displacements from the east coast of New Zealand's North Island (Fig. 2). These faults displace a moderately dipping (*c.* 50°) sequence of well-bedded turbidites exposed in a near-horizontal shore platform. The outcrop geometry is equivalent to small (lengths of 1–200 m and maximum displacements of 0.001–3 m) normal faults cropping out in cliff sections, but with the dimensions of the platform permitting easy access to a large stratigraphic interval and many faults (compared to typical outcrop cross sections). Local variations in the fault displacement profiles support suggestions that all faults interact to varying degrees, with individual faults being both hard- and soft-linked (cf. Walsh & Watterson 1991). The increase in displacement gradients with increasing fault size also supports the view that fault interactions play a critical role in establishing fault dimensions and displacement distributions. The outcrop data, including the concentration of secondary splays and minor faults at the tips of some faults, are consistent with a model in which faults propagate rapidly until they reach saturation and subsequently grow mainly by displacement accumulation with near-constant lengths.

Fault system geometry and kinematics

Outcrop-scale strike-slip to oblique-slip faults displace moderately dipping (*c.* 45–55°) turbidites (Whakataki Formation) in coastal exposures on the east coast of New Zealand's North Island (Figs 2–4a). The Whakataki Formation is an Early Miocene (*c.* 20–24 Ma) deep-water turbidite succession (Johnston 1980) that, at low tide, forms a spectacular wave-cut platform up to 120 m wide and several kilometres long (e.g. Figs 2 & 3). This formation comprises interbedded sandstone and siltstone beds (net to gross locally varies from 20 to 80%) ≤40 cm thick that mainly strike NNE approximately parallel to the coast and dip to the west (Johnston 1980; Edbrooke & Browne 1996; Field 2005) (Figs 2 & 4a). Bedding on the wave-cut platform is displaced by hundreds of small post-depositional faults, with apparent strike slip displacement of <3 m. In addition, some early bedding-parallel slip and thrusts are present, which may have partly formed during folding and associated bedding-parallel shortening (see early thrust in Fig. 3). Late-stage strike-slip faults are the focus of this study. These faults formed post-lithification and displace bedding-parallel (and sub-parallel) faults (e.g. Fig. 3), from which we suggest that they mainly postdate folding and bed rotation. The timing of formation of strike-slip faults is poorly constrained; however, given that the Whakataki Formation was probably buried by up to 3 km (Edbrooke & Browne

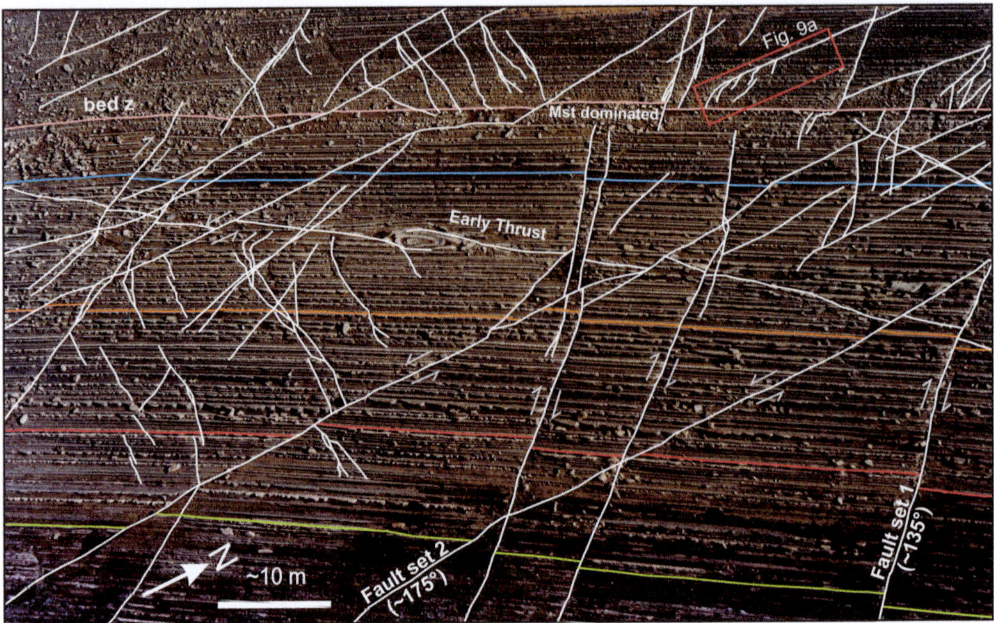

Fig. 3. Interpreted oblique aerial photo of the Whakataki Formation shore platform illustrating the fault system and its relationship to bedding (photograph by Lloyd Homer). White lines indicate faults and coloured lines indicate individual beds mapped across the platform. Arrows indicate the sense of slip on the faults. Location of photo is shown in Figure 2. Location of fault 18 (Fig. 9a) is indicated.

1996; Field 2005) and then tilted prior to strike-slip faulting, a mid-Miocene or younger age is inferred here.

The coastal platform trends approximately 030° and provides map data for the apparent strike-slip faults (Fig. 2). The horizontal outcrop, moderate bed dips and well-defined bed boundaries every *c.* 20–40 cm enable fault lengths, intersection relationships and displacements to be analysed. The fault locations, lengths and apparent strike-slip displacements of bedding have been delineated using high-resolution aerial photographs (both vertical and oblique, *c.* 1:50 to 1:1000 scales) together with outcrop tape and compass measurements. Detailed data for a total of about 150 faults include: fault surface and slip orientations, fault maximum displacements, fault lengths and along-strike displacement profiles. The fault system comprises two dominant fault sets that strike *c.* 135° (fault-set 1) and *c.* 175° (fault-set 2) and cross-cut each other to form conjugates (Figs 3 & 4b). Fault dips are predominantly steep (>60°) and, in many cases, exceed 80°, with fault-set 1 mainly dipping north and fault-set 2 dipping both east and west (Fig. 4b). The faults observed in the platform range in length from *c.* 0.4 to 200 m and in apparent strike slip from 0.001 to 3 m. Spacings of faults mapped in aerial photos generally range from 1 to 40 m with fault intersections and fault tips being common. Cross-cutting and abutting relationships between individual faults indicate that the two main fault sets formed synchronously on geological timescales.

Apparent strike-slip displacements were measured on bedding boundaries (Fig. 3). Slip indicators (slickenside striations) mainly plunge at shallow to moderate angles (mostly ≤50°), at a high angle to bedding (mainly ≥60°) (Fig. 4b, c). Slickenside striations indicate a significant component of strike-slip displacement with the *c.* 135° fault set (1) being right-lateral and the *c.* 175° fault set (2) being left-lateral (the faults are hereafter referred to as strike slip). These faults formed in association with principal horizontal shortening oriented approximately SSE–NNW sub-parallel to the regional direction of shortening along the east coast of the North Island, which is associated with subduction of the Pacific Plate during the late Cenozoic (e.g. Nicol *et al.* 2007) (Fig. 2a).

Fault displacement profiles

Displacement profiles and fault-trace maps have been widely employed in the literature to constrain the mechanics, kinematics and growth of faults over a range of scales (e.g. Muraoka & Kamata

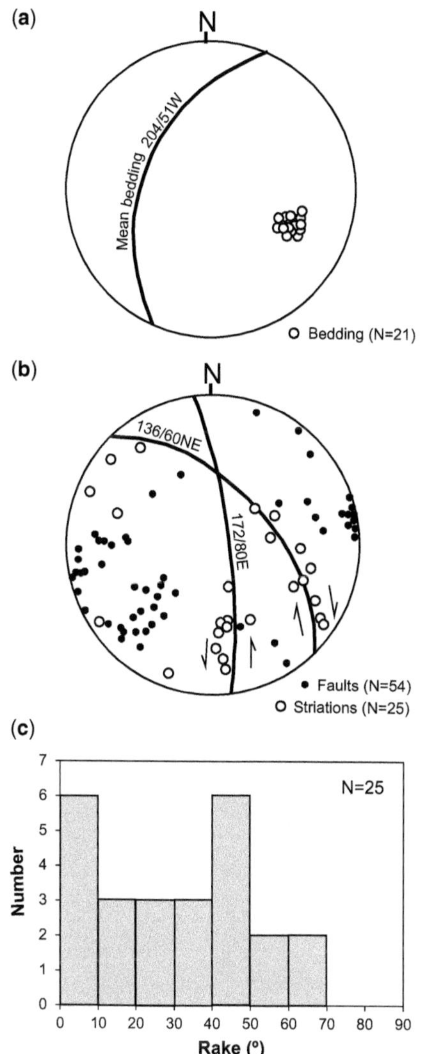

Fig. 4. Equal-area stereonets showing bedding, fault and slickenside striation orientations (a & b) together with frequency histogram of slip rake (c) from the areas enclosed by the dataset rectangles in Figure 2. (**a**) Stereonet of poles to bedding and mean bedding plane orientation. (**b**) Poles to faults, mean orientations for two main fault sets (*c.* 135° set 1 and *c.* 175° set 2) and fault-slip striation orientations. (**c**) Histogram of rake for measured slip data. Number of observations (*N*) is indicated for each plot.

1983; Barnett *et al.* 1987; Walsh & Watterson 1987, 1988; Ellis & Dunlap 1988; Peacock 1991, 2002; Peacock & Sanderson 1991, 1992, 1996; Cowie & Scholz 1992*a*, *b*; Dawers *et al.* 1993; Dawers & Anders 1995; Nicol *et al.* 1996, 2005, 2010; Cowie & Shipton 1998; Manighetti *et al.* 2001, 2004; Wilkins & Gross 2002; Manzocchi *et al.* 2006; Schlagenhauf *et al.* 2008; Bergen & Shaw 2010; Rotevatn & Fossen 2012; Jackson & Rotevatn 2013; Tvedt *et al.* 2013; Roche *et al.* 2016). In this study strike-parallel displacement profiles display variations in strike slip along the length of fault traces confined to the shore platform (Figs 5 & 6). The maximum displacements for the faults profiled range from 7 mm to 37 cm (Fig. 5) with uncertainties on displacement measurements typically in the range of 2–10 mm. The profiles presented in Figure 5 have a number of characteristics that are consistent with the results of many previous studies (see papers referred to previously in this section). These are: (i) the displacement profiles show a range of geometries including near-triangular with a centrally located maximum (Fig. 5; faults 10, 11 and 13), asymmetrical with relatively steep gradients towards one tip (Fig. 5; fault 8 and faults 2, 6 and 7) and flat-topped profiles with high gradients at both tips (e.g. Fig. 5; faults 3 and 5). (ii) Approximately bell-shaped profiles were rarely sampled in the Whakataki Formation outcrop and many of the faults do not display a decrease in displacement gradient approaching fault tips (Fig. 5). (iii) Intersecting or closely spaced faults (e.g. <10 m spacing) often display complementary displacement profiles, with equal and opposite displacement gradients (Fig. 7). (iv) Average displacement gradients from the point of maximum displacement to the fault tip typically increase with fault size (Fig. 8).

Variations in the shapes of displacement profiles and associated displacement gradients can be influenced by a number of factors, including fault and host rock mechanical properties, the magnitudes and orientations of stresses imposed on the fault, the displacement history of the fault and its individual slip increments, and interactions with other faults in the array (e.g. Walsh & Watterson 1987; Peacock & Sanderson 1996; Schultz 1999; Peacock 2002; Manzocchi *et al.* 2006). In multilayer sequences comprising beds with variable rheologies the geometry of the fault surface (strike or dip), the displacement on the primary slip surface and the location of fault tips may be strongly influenced by changes in rock properties (e.g. Muraoka & Kamata 1983; Peacock & Sanderson 1992; Nicol *et al.* 1996; Manighetti *et al.* 2001; Wilkins & Gross 2002; Schöpfer *et al.* 2006; Roche *et al.* 2016). In particular, it is common for displacements to decrease across, and fault tips to be located within, mechanically weak beds dominated by mudstone (e.g. Nicol *et al.* 1996; Wilkins & Gross 2002). In the Whakataki Formation fault tips are largely confined to mudstone beds or to the boundaries of mudstone beds. In addition, some mudstone beds appear to contain a greater number of fault tips than would be predicted if the tips were randomly distributed in the sequence. Inspection of the faults

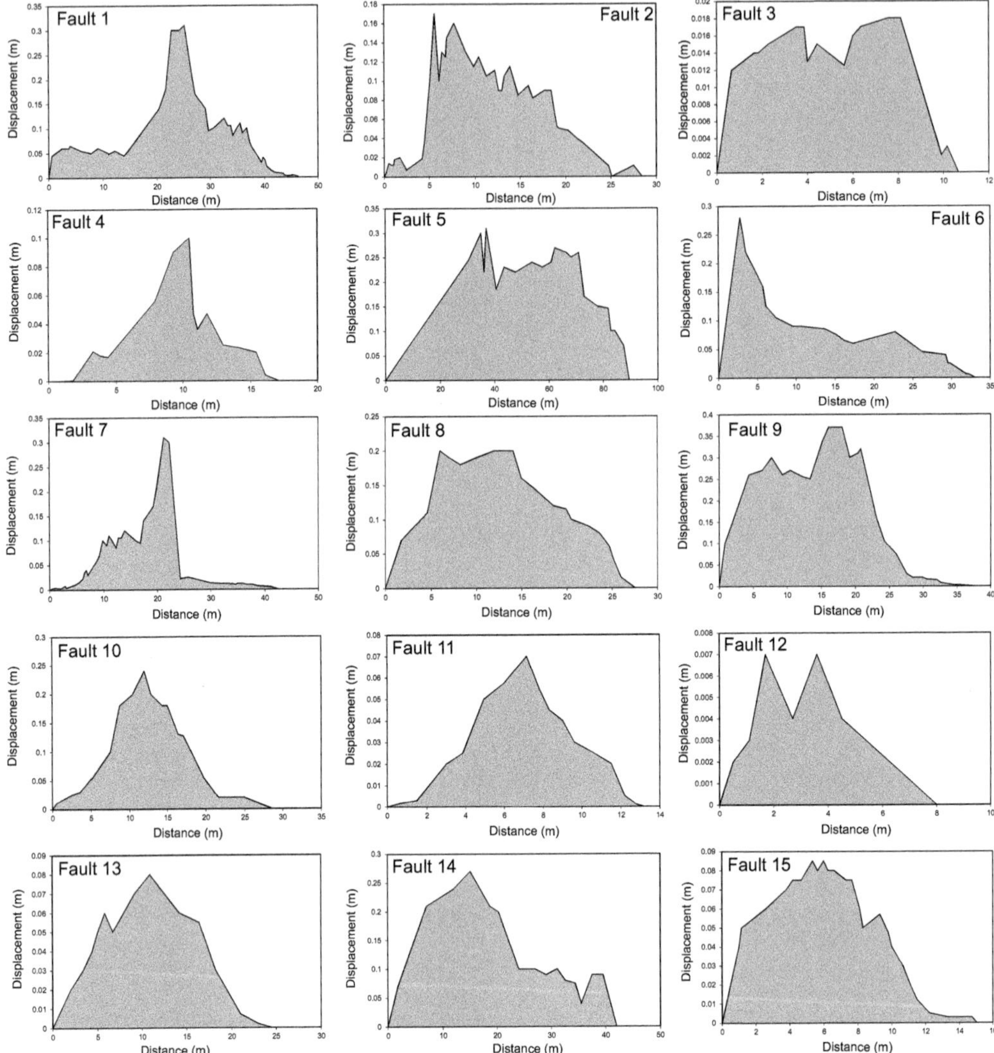

Fig. 5. Along-strike displacement profiles for strike-slip faults entirely contained within the platform. Note that horizontal and vertical scales change between graphs with displacements and lengths varying from 0.007 to 0.37 m and 8 to 90 m, respectively. Locations of faults 8–15 are shown in Figure 7.

in Figure 3, for example, confirms that a similar geometry and density of faults occurs on either side of a mudstone-rich zone that contains bed z; however, in detail 30–40% of individual faults or fault segments terminate in this zone, suggesting that its mechanical properties may have impacted on the continuity (and propagation) of some faults. These observations support the view that bed rheology and architecture can influence fault growth (Peacock & Sanderson 1992; Peacock 2002; Schultz & Fossen 2002; Wilkins & Gross 2002).

The ideal isolated fault is a valuable concept for generating initial models that contribute to improved understanding of faulting processes and evolution (Barnett *et al.* 1987; Walsh & Watterson 1987); however, it is widely recognized that few faults are likely to form in complete kinematic or dynamic isolation from other structures. On both geological (millions of years) and earthquake (<1000 years) timescales faults generally represent interacting elements within kinematically coherent systems (e.g. Peacock 1991, 2002; Walsh & Watterson 1991; Bürgmann *et al.* 1994; Nicol *et al.* 1996, 2010; Stein 1999; Cowie *et al.* 2000; Nixon *et al.* 2014). Fault interactions provide an important (in many cases the primary) control on the shapes of

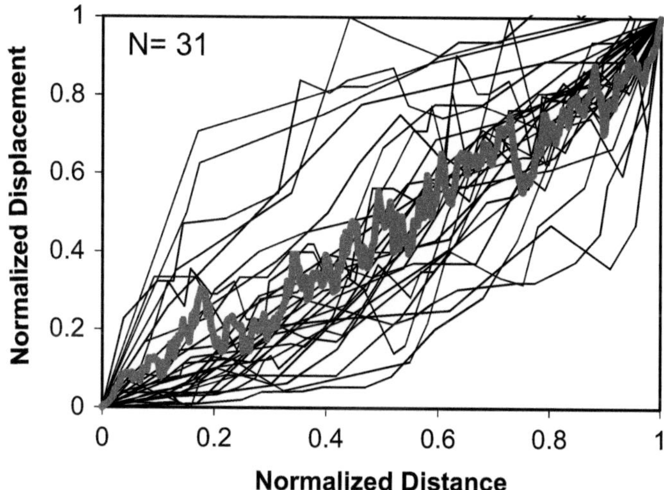

Fig. 6. Normalized displacement profiles ($n = 31$) from maximum displacement to tip for 16 faults (black lines). The plot primarily uses data from Figure 5. The thick line is the mean curve for the entire dataset ($n = 435$ displacement measurements) constructed using a running average with a ten-observation bin size and bin centres at 0.01 increments of normalized length. The near-linear general slope of the mean curve is comparable to many of the underpinning profiles and was not significantly modified using a range of bin sizes (5–50 observations) and bin location increments (0.005–0.1). Profiles above the mean curve are considered to be more influenced by interactions than profiles below the mean.

displacement profiles with the degree of interaction and the magnitude of displacement gradients inferred to be positively related (e.g. Muraoka & Kamata 1983; Peacock 1991, 2002; Nicol *et al.* 1996, 2010; Manighetti *et al.* 2001). This relationship may arise because fault interaction retards fault propagation without significantly impacting the rate of increase in maximum displacement (Peacock & Sanderson 1996). The role of fault interactions in determining the shape of displacement profiles is supported by this study. Here, as is common in other fault systems, flat-topped or highly asymmetrical profiles with high tip displacement gradients are associated with those faults that intersect, or relay displacement to, nearby faults (e.g. faults 2, 3, 5, 6, 7, 12, 14 and 15, Fig. 5). By contrast, the faults with symmetrical profiles that generally have the lowest displacement gradients tend to be those that are farthest from other faults in the system of similar size. These conclusions are supported by the fault map and displacement profiles in Figure 7. Faults 13–15 and 16–17 intersect and produce complementary displacement profiles; displacement is transferred southwards from fault 17 to 16, while the displacements on faults 13 and 15 largely fill the displacement low down on fault 14. Faults 9 and 10 do not intersect in the plane of the platform and appear to be separated by a zone across which displacement is transferred between faults. Such fault interactions are difficult to unequivocally demonstrate on the shore platform for faults with spacings of tens of metres or greater, in part because the faults are small, with changes in displacement and displacement gradients typically being low (typically <1 cm per metre for faults that are not hard-linked). However, fault interactions over length scales of kilometres have been inferred for normal faults with maximum displacements of tens of metres to kilometres and lengths of kilometres or more (e.g. Walsh & Watterson 1991; Nicol *et al.* 1996, 2010). These observations are consistent with the notion that all faults interact and the degree of interaction is positively related to the lateral displacement gradients (i.e. low displacement gradient faults are inferred to indicate less interaction and displacement transfer with nearby faults than high-gradient faults).

In addition to reflecting fault interactions, the shapes of displacement profiles have the potential to provide information about fault propagation (Peacock & Sanderson 1996; Manighetti *et al.* 2001; Childs *et al.* 2003; Manzocchi *et al.* 2006). Finite displacement profiles are often approximately triangular with varying degrees of asymmetry and near-linear displacement gradients from maximum displacement to fault tip (e.g. Walsh & Watterson 1987; Nicol *et al.* 1996; Cowie & Shipton 1998; Manighetti *et al.* 2001, 2004; Manzocchi *et al.* 2006). The triangular-shaped profiles observed for different fault types (mainly strike slip and normal)

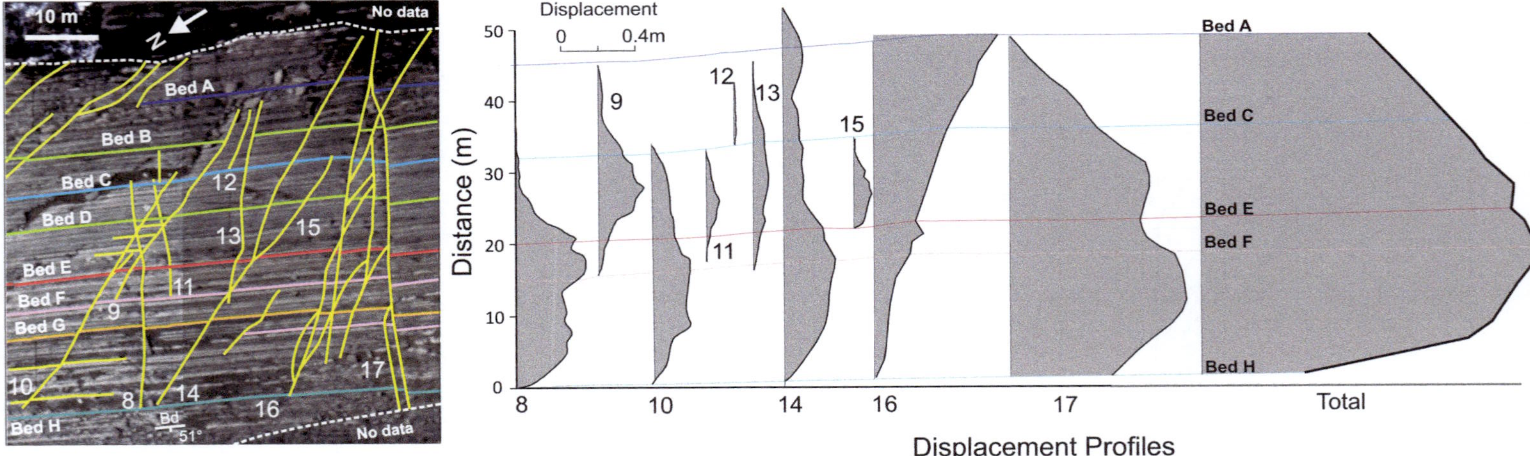

Fig. 7. Fault map and strike-slip displacement profiles projected onto planes normal to bedding. Fault numbers indicated on the map apply to the profiles in this figure and those presented in Figure 5. The displacement scale is the same for all profiles. Beds A, C, E, F and H are shown on both map and profiles.

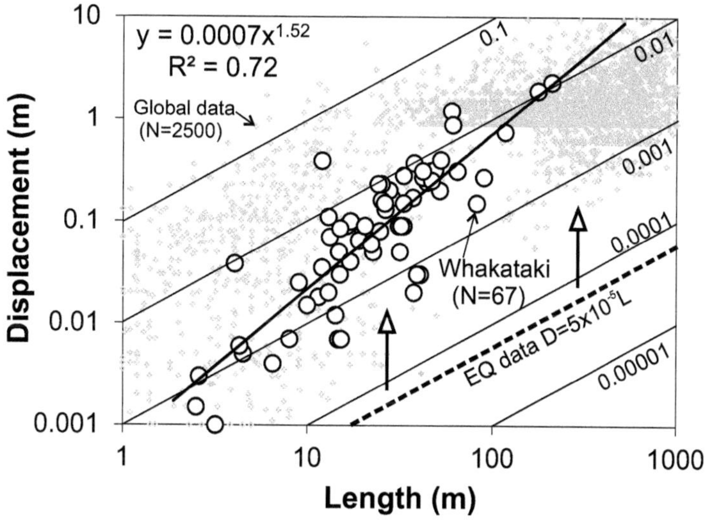

Fig. 8. Displacement–length plot for Whakataki faults sampled from the shore platform (open circles) and a global compilation of faults (grey symbols). The global compilation is primarily from Bailey *et al.* (2005) with additional data from subsequent publications (Lamarche *et al.* 2005; Fossen *et al.* 2007; Bergen & Shaw 2010; Nicol *et al.* 2010; Reilly *et al.* 2016) and unpublished measurements. Uncertainties on the fault length and displacement measured at Whakataki are typically small (<1 m and <1 cm, respectively), and do not impact on the general distribution of the data points on the log–log graph or the first-order conclusions. The line of best fit ($D = 0.0007L^{1.52}$) is a least-squares regression on the *y*-axis variable. Contours labelled 0.00001 to 0.1 are D–L ratios (i.e. 0.5 of displacement gradients). The average line for earthquake data (i.e. $D = 0.00005L$) is dashed and from Wells & Coppersmith (1994). Vertical arrows show growth trends for the constant length model (Walsh *et al.* 2002).

in many fault systems are consistent with the observations of this study, where many of the faults and the average profile (see thick red line on Fig. 6) have approximately linear gradients. Near-triangular displacement profiles do not display the low displacement tails predicted for faults by post-yield fracture mechanics (Cowie & Scholz 1992*b*) or for rapidly propagating faults (Peacock & Sanderson 1996). Low displacement concave-upwards tails are observed for up to *c.* 27% of the faults in Figure 5 (i.e. 8 of 30 tips in Fig. 5; fault 1 right tip, fault 7 left and right tip, fault 9 right tip, fault 10 right tip, fault 11 left tip, fault 13 right tip and fault 15 right tip). In such cases the displacement profiles can be interpreted to reflect fault propagation; however, it is also possible that incremental and finite displacement profiles had comparable shapes (when normalized to their respective maximum displacements), with little propagation during the majority of displacement accumulation. Similarly, the fact that the remaining *c.* 75% of faults have approximately uniform displacement gradients approaching their tips and these fault displacement profiles provide no basis for favouring a fault propagation model over one in which little or no propagation occurs for much of the duration of faulting (Fig. 5).

Mutually cross-cutting relationships at the intersections of the two main fault sets indicate that the faults in this study accrued displacement synchronously (on geological timescales) during multiple slip events. Therefore, the history of fault growth and the shape of incremental displacement profiles is not uniquely constrained by the finite displacement profiles in the shore platform. The semi-elliptical incremental profiles modelled for linear elastic materials (e.g. Pollard & Segall 1987) are rarely replicated in the finite profiles of the Whakataki Formation (e.g. Fig. 5). Field mapping provides little evidence to support the model that departure of finite displacement profiles from semi-elliptical arises due to off-fault deformation (see Manighetti *et al.* 2004); however, we cannot discount the possibility that slip during these discrete events was semi-elliptical. In circumstances where the shape of incremental slip profiles is not known, stochastic earthquake modelling shows that for a range of incremental slip profile shapes (including elliptical), triangular finite profiles and a propagating fault are not compatible with the displacement gradients typically observed for outcrop-scale faults (Manzocchi *et al.* 2006; see next section for further discussion). Stochastic modelling of Manzocchi *et al.* (2006) indicates that the near-triangular profiles widely observed (including in this study; Figs 5 & 6) are consistent with a fault-growth model in which fault lengths are established rapidly and subsequent faulting is dominated by the accumulation of displacement and tip restriction, resulting in little or no propagation.

Displacement–length relationship

The ratio of displacement–length and how it varies with fault size constrains fault-growth models (Walsh & Watterson 1988; Cowie & Scholz 1992*c*; Gillespie *et al.* 1992; Dawers & Anders 1995; Schultz & Fossen 2002; Walsh *et al.* 2002; Wilkins & Gross 2002; Bailey *et al.* 2005; Kim & Sanderson 2005; Manzocchi *et al.* 2006; Schultz *et al.* 2008; Nicol *et al.* 2010; Rotevatn & Fossen 2012). Displacement–length relations are described by the expression $D = cL^n$, where D is displacement, L is fault length, c is a constant defined by the y-intercept and n is a constant defined by the slope of the distribution (e.g. Fig. 8). The most widely accepted value of n is 1; however, displacement–length (D–L) plots for individual fault systems typically reveal slopes (n) of 0.5–1.5 over 2–3 orders of magnitude of displacement and length (e.g. Walsh & Watterson 1988; Schlische *et al.* 1996; Fossen & Hesthammer 1997; Bailey *et al.* 2005; Kim & Sanderson 2005). Debate remains about the value of the D–L slope of the power-law distribution and whether or not the positive slope of the data cloud represents a growth trend, consistent with the increasing fault length growth model (e.g. Walsh & Watterson 1988; Cowie & Scholz 1992*c*; Walsh *et al.* 2002) (Fig. 1 left). Resolving each of these questions has significant implications for the evolution of faults.

To examine our understanding of the growth of faults we have plotted displacement and length for faults contained within the platform (open circles) along with a compilation of global fault data (grey symbols; see figure caption for data sources, Fig. 8). Faults from the Whakataki Formation are located towards the centre of the global compilation, suggesting that their D–L relations are broadly comparable to faults of similar size in other fault systems. While we have a small number of data ($n = 67$), cover a limited size range (<3 orders of magnitude in length and displacement) and display a significant spread of both displacement and length (as is common on these types of plots), it is clear that in this case n is significantly greater than 1 (Fig. 8). A value of $n > 1$ requires that average displacement gradients increase with fault size. Lines labelled 0.00001–0.1 on Figure 8 are contours of D–L and highlight the rise in average displacement gradients from *c.* 0.0006 to 0.06 with increasing fault displacement and length. In this fault system retardation of fault-tip propagation by thick weak mudstone beds or by fault interaction provide possible explanations for the relatively high n. In circumstances where retardation is complete (i.e. faults do not propagate) fault growth is associated with an increase in displacement without a change in fault length, resulting in vertical growth trends on D–L plots (Walsh *et al.* 2002) (Fig. 8).

In recent years it has been proposed that faults mainly grow vertically into the D–L distribution rather than evolving along it (Walsh *et al.* 2002; Nicol *et al.* 2010). Such vertical growth trends require that after initial rapid propagation, which typically occurs on timescales shorter than the resolution of the data, faults do not grow significantly in length (Filbrandt *et al.* 1994; Morley 1999; Meyer *et al.* 2002; Walsh *et al.* 2002; Schlagenhauf *et al.* 2008; Jackson & Rotevatn 2013). The constant length growth model of Walsh *et al.* (2002) (Fig. 1 right) has the advantage that it provides a rationale for the evolution of faults from individual slip increments (e.g. earthquakes), for which n may be 0.5–1 (e.g. Cowie & Scholz 1992*a*; Wells & Coppersmith 1994; Fossen & Hesthammer 1997; Schultz *et al.* 2008; Nicol *et al.* 2010; Rotevatn & Fossen 2012), to their finite displacement–length population (Fig. 8). In circumstances where individual faults grow along trends steeper than the D–L distribution, their average displacement gradients increase as strain accrues in an array (Schultz 1999; Poulimenos 2000; Walsh *et al.* 2002; Schlagenhauf *et al.* 2008). Proportional increases in average displacement gradients for all faults in a system will produce upward migration of the D–L distribution as strain accrues without a change in n. Recent work suggests, however, that n may increase with fault-system maturity and strain accumulation (Nicol *et al.* 2010; Rotevatn & Fossen 2012; Nicol 2014) (Fig. 8). Preferential death of smaller faults in a system, coupled with continued displacement on larger faults of near-constant length, represents one set of conditions that could produce an increase in n (Meyer *et al.* 2002; Walsh *et al.* 2002; Nicol *et al.* 2010). Increasing n with fault-system maturity would require that fault-growth processes are scale dependent (i.e. there are differences in the growth of small and large faults); however, further data analysis is required to test the hypothesis that n could change with fault-system evolution.

Fault-tip splays and propagation

Small-scale faulting and fracturing around fault tips is widely observed and can provide information about their propagation (e.g. King 1986; Cowie & Scholz 1992*b*; McGrath & Davison 1995; Vermilye & Scholz 1999; Kim *et al.* 2000, 2004; d'Alessio & Martel 2004; Nixon *et al.* 2011; Perrin *et al.* 2016).

For strike-slip faults in map view, the strike of secondary faults can vary dramatically from moderate to high angles (e.g. >30°) to the main fault (e.g. horsetails, wing cracks or tail cracks) to being sub-parallel to the primary fault (e.g. synthetic splays, see Kim *et al.* 2004). Numerous models have been proposed to account for the formation of these fault-tip structures (e.g. Chinnery 1966; Cowie & Scholz 1992*b*; Kim *et al.* 2000, 2004; d'Alessio & Martel 2004). Common to most of these models is the notion that slip on the primary fault surface locally produces stress concentrations at the fault tip that are relieved by distributed strains in the rock volume enclosing the tip.

Fault-tip splays and associated decreases in displacement along the primary fault can be interpreted to provide evidence of fault propagation towards the finite fault tip (e.g. Kim *et al.* 2004; Perrin *et al.* 2016). In circumstances where fault splays form due to tip processes, the distribution of secondary faults along a primary fault has the potential to provide information about whether increasing (i.e. gradually propagating) or constant length fault-growth models best explain the observed fault outcrop patterns (see Fig. 1). For a propagating primary fault, secondary faults formed along the tip line will be abandoned as the tip propagates and remain preserved along the length of a fault as part of the finite strain field (Fig. 1 left). Alternatively, if a fault tip remains stationary for much of the duration of faulting, then secondary faults formed in association with termination processes will be localized near to the present tip (Fig. 1 right).

The densities of small-scale faults adjacent to larger structures are dependent on many factors, including pressure and temperature conditions

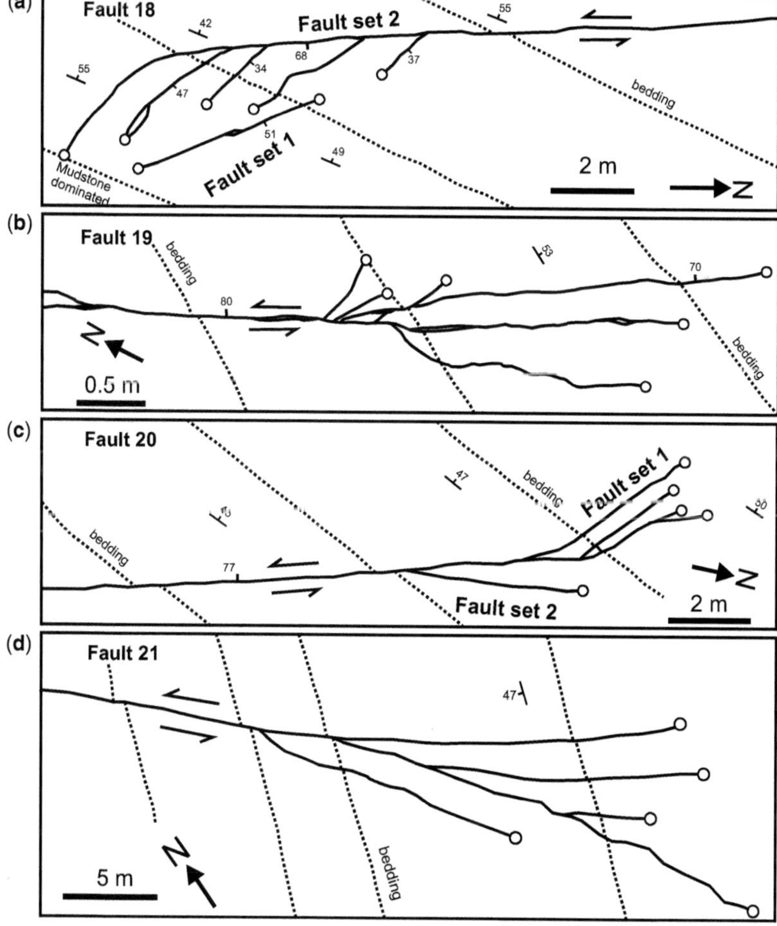

Fig. 9. Maps of tip splays at the terminations of four faults. Arrows indicate slip sense and open circles indicate tip locations. See Figure 3 for the location of fault 18.

during faulting (including fluid and confining pressures), size of the primary fault, orientation of the maximum compressive stress, mechanical properties of the rock at the time of faulting, geometry of the primary fault (e.g. degree of segmentation) and its linkage or intersection with other similar-sized faults in the system (see Kim *et al.* 2004 and references therein). Despite the wide range of factors that could control the formation of small-scale faults and their often complex geometries (e.g. variable orientations and intersection relationships), field studies suggest that elevated numbers of small-scale faults are most often observed at fault segment boundaries (e.g. relay zones or fault steps), fault intersections, fault tips and stratigraphic mechanical boundaries (e.g. McGrath & Davison 1995; Childs *et al.* 1996, 2009; Kim *et al.* 2000, 2004; Nicol *et al.* 2013; Roche *et al.* 2016). As a result of these different geometrical associations, the densities of secondary faults can vary along the length of the primary fault. Along-strike variation in the densities of secondary faults in the immediate wall rock of larger structures occurs in the Whakataki Formation, with high densities being particularly common at the intersections of the two main fault sets. In addition, some of the primary faults in the shore platform are associated with secondary faults localized at the free tips of the main fault (Fig. 9). The resulting splays can accommodate a significant component of the apparent strike-slip displacement approaching the fault tip and combine to produce near-triangular cumulative displacement profiles (e.g. Fig. 10). These splays are inferred to form during the interaction phase of faulting and to facilitate a rise in displacement gradients without significant propagation. On fault 19 (Fig. 10), for example, the average displacement gradient of *c.* 0.01 is towards the upper limit of data in Figure 8 for faults with maximum displacements of 0.04 m. The examples presented show horsetail and synthetic branching geometries with, in two cases, the main fault being a member of fault-set 2 and the tip splays parallel to fault-set 1 (Fig. 9a, c; faults 18 and 20). Despite the occurrence of both fault orientations at tip locations, none of the examples presented in Figure 9 appears to have formed due to hard-linkage of these sets or due to segmentation of a single fault set. Instead, the occurrence of secondary faults in tip zones parallel to each fault orientation may reflect the far-field stresses producing the conjugate fault system.

Fault-tip splays are consistent with the constant fault length model (Fig. 1 right). They do not occur on all faults in the Whakataki platform; however, where secondary fault splays do occur, they are mainly restricted to the area around the present fault tip. In Figure 9a (fault 18), for example, the fault extends for a further *c.* 30 m on the platform north of the tip splays and is associated with no additional splays of comparable length and displacement

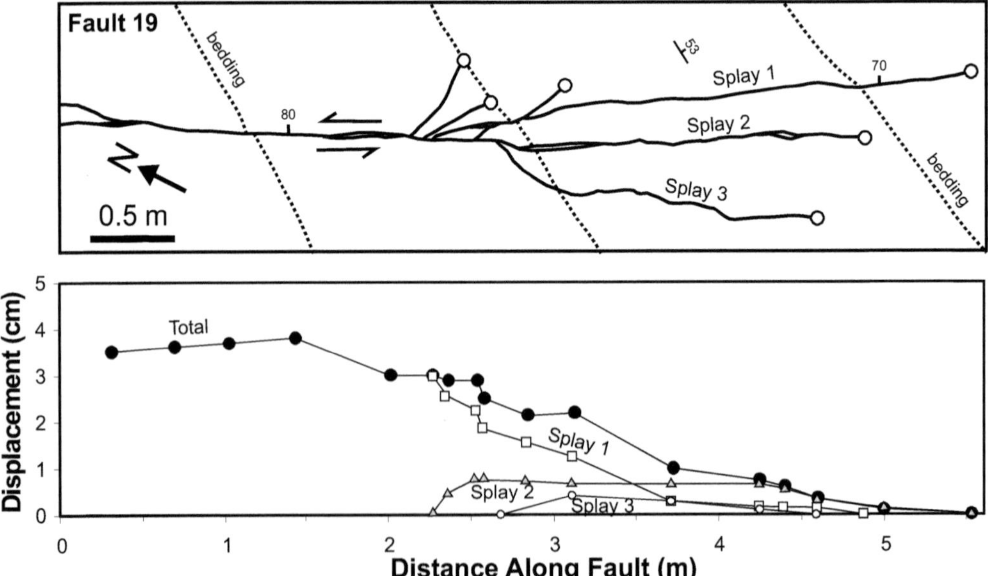

Fig. 10. Fault-trace map and displacement profiles for the main slip surface, fault splays and total displacement along fault 19. Horizontal scales in both diagrams are equal, enabling displacements and fault geometries to be correlated.

to those at the tip (Fig. 3). Fault-tip geometries in Figure 9 are comparable to those reported in the literature, where secondary faulting is mainly restricted to the region surrounding the present fault tip (e.g. King 1986; McGrath & Davison 1995; Kim *et al.* 2000, 2004; Perrin *et al.* 2016). Secondary faults that are similar in geometry and size to tip splays are not routinely observed distal to fault tips on the Whakataki shore platform, from which we infer that the gradually propagating fault model is unlikely to be correct for the tips in Figure 9. Instead, the observed fault-tip geometries can be explained if faults initially propagate rapidly, due to the presence of a pre-existing plane of weakness and/or low fracture toughness of the faulted materials, with near-constant fault lengths and stationary fault-tip locations subsequently maintained for much of the duration of faulting (Fig. 1 right). For the constant length model the initial stage of propagation is sufficiently rapid that few fault-tip structures form prior to the fault reaching its final length. When the fault tips become stationary new secondary faults may form in the fault zone around the tip as displacement accrues on the principal slip surface (Fig. 1 right). Secondary faulting restricted to the area around the fault tip has been replicated by numerical models in which tip stresses and strains are localized by a pre-existing plane of weakness that does not propagate (e.g. Pollard & Segall 1987; Bürgmann *et al.* 1994). Therefore, both outcrop and modelling could be consistent with the constant fault length model, although in the absence of growth strata, neither the constant length nor propagating faults can be unequivocally demonstrated at Whakataki.

Discussion and conclusions

Displacement profiles, D–L relationships and tip geometries for a fault system exposed in a large (up to 120 m by 1.2 km) coastal platform constrain fault-growth models. Displacement profiles have variable shapes that mainly reflect fault interactions, with individual faults being both hard- and soft-linked. Many of the displacement profiles and the average profile for all faults are near-triangular, with displacement gradients (and displacement–length ratios) increasing by two orders of magnitude from smallest to largest faults. Within fault zones these gradients are accompanied by secondary faults, which are typically of greatest density close to fault intersections, in relay zones and at fault tips. Where present, horsetail and synthetic splays are mainly confined to the regions around the tips of faults, and are incompatible with gradual fault propagation for the duration of growth. Instead, fault displacements and tip geometries are consistent with growth initially dominated by fault propagation followed by displacement accumulation and approximately stationary tips (Figs 1 & 11). The change from propagation- to displacement-dominated growth stages is inferred to be produced by fault-system saturation (i.e. the onset of the state where all faults are interacting). In this two-phase model fault interactions between neighbouring faults, and associated soft- or

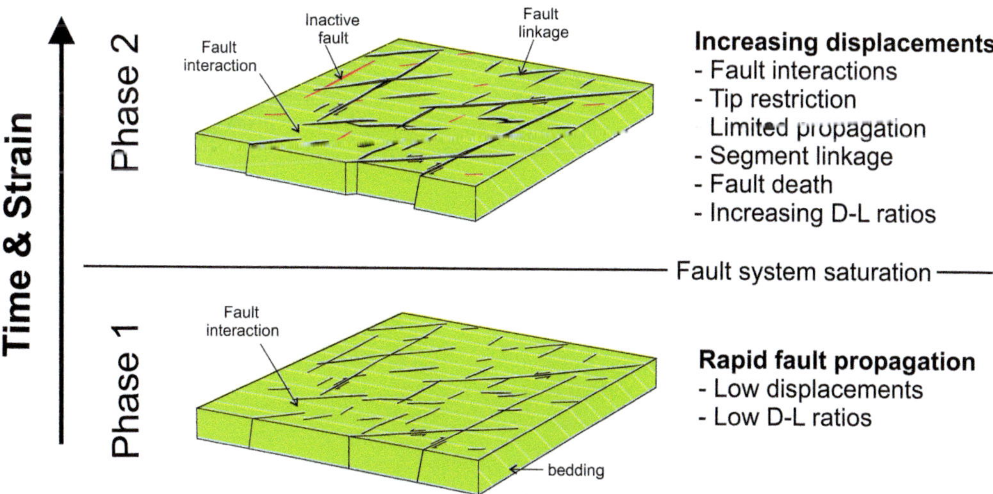

Fig. 11. Schematic block diagrams showing the evolution of the Whakataki fault system with initial rapid propagation (phase 1) and subsequent growth via increasing displacements and near-constant faults (phase 2) (i.e. constant length model in Fig. 1 right). Orientations of fault sets and bedding approximately match those observed in Figure 2.

hard-linkage, retard propagation by reducing tip stresses and fracture energy (d'Alessio & Martel 2004) (Fig. 11). The relative duration of each stage of faulting is generally not known precisely; however, the available evidence suggests that the initial propagation stage is significantly shorter than the displacement stage (e.g. Walsh *et al.* 2002). The relative durations of each stage of faulting may vary depending on the properties of the faulted rock volume, the presence of pre-existing faults or joints, and/or the rates and magnitudes of strains and stresses. For example, where faulting utilizes pre-existing heterogeneities or zones of weakness and these structures are optimally oriented in the new stress regime, fault lengths may be established instantaneously on geological timescales (e.g. ≤10 slip increments). Even in circumstances where new faults are formed, their lengths can be established within 20% of the duration of faulting with the remainder of the faulting history dominated by displacement accumulation (e.g. Filbrandt *et al.* 1994; Schlagenhauf *et al.* 2008).

Initial rapid fault propagation succeeded by displacement-dominated growth could account for different modes of faults over a range of scales (including outcrop). Fault interactions are a key element of the model and, although variable in degree, occur throughout fault systems over geological and earthquake timescales (e.g. Walsh & Watterson 1991; Stein 1999). Given the important role of fault interactions in the constant length model, it may be equally appropriate to refer to this as the fault interaction or coherent fault growth model (Walsh & Watterson 1991; Walsh *et al.* 2003; Nicol *et al.* 2010; Jackson & Rotevatn 2013). Similarly, the increasing length model may apply where fault interactions do not impede propagation, and this could be referred to as the isolated fault-growth model. Although models of isolated faults are of value (Barnett *et al.* 1987; Walsh & Watterson 1987; Cowie & Shipton 1998; Vermilye & Scholz 1999), they appear to be rare (e.g. Nicol *et al.* 1996), and future studies would be well served to focus on kinematic and dynamic interactions between faults and their influence on growth. If the two-stage growth model favoured here is correct, analysis of low-strain systems will be required to understand better the early stages of fault evolution and the transition from propagation- to displacement-dominated growth. A two-stage model (Fig. 1b) for fault growth was initially proposed for regional-scale normal faults imaged in seismic reflection lines (Morley 1999; Meyer *et al.* 2002; Walsh *et al.* 2002) and also applies to normal faults in analogue models (Filbrandt *et al.* 1994; Schlagenhauf *et al.* 2008). If our interpretations are correct, the growth model in Figure 1b applies over at least five orders of magnitude from regional- (e.g. lengths of 1–50 km and displacements of 0.1–3 km: Walsh *et al.* 2002) to outcrop-scale (e.g. lengths of 0.4–200 m and displacements of 0.002–3 m: this study) faults with significant components of strike slip and dip slip, and for fault dimensions measured parallel and perpendicular to the slip vector. Therefore, we contend that the constant length fault-growth model has wide applications.

We are indebted to Juan Watterson who, in addition to proving to be an exceptional mentor, was instrumental in establishing the principles on which this paper is founded: the quantitative analysis and modelling of faults using high-quality datasets. This research was financially supported by GNS Science Core Funding. Adrian Harvison, Cathal Reilly, Hannu Seebeck and Rob Worthington assisted with the collection of field data. Many thanks to Casey Nixon and Chris Jackson who provided journal reviews that helped to improve the paper.

References

Bailey, W.R., Walsh, J.J. & Manzocchi, T. 2005. Fault populations, strain distribution and basement reactivation in the East Pennines Coalfield, U.K. *Journal of Structural Geology*, **27**, 913–928.

Barnett, J.A.M., Mortimer, J., Rippon, J.H., Walsh, J.J. & Watterson, J. 1987. Displacement geometry in the volume containing a single normal fault. *American Association of Petroleum Geologists Bulletin*, **71**, 925–937.

Beavan, J., Tregoning, P., Bevis, M., Kato, T. & Meertens, C. 2002. The motion and rigidity of the Pacific Plate and implications for plate boundary deformation. *Journal of Geophysical Research*, **107**, 2261, https://doi.org/10.1029/2001JB000282

Bergen, K.J. & Shaw, J.H. 2010. Displacement profiles and displacement-length scaling relationships of thrust faults constrained by seismic-reflection data. *Geological Society of America Bulletin*, **122**, 1209–1219, https://doi.org/10.1130/B26373.1

Bürgmann, R., Pollard, D.D. & Martel, S.J. 1994. Slip distributions on faults: effects of stress gradients, inelastic deformation, heterogeneous host-rock stiffness, and fault interaction. *Journal of Structural Geology*, **16**, 1675–1690.

Cartwright, J.A., Trudgill, B. & Mansfield, C.S. 1995. Fault growth by segment linkage: an explanation for scatter in maximum displacement and trace length data from the Canyonlands Grabens of S.E. Utah. *Journal of Structural Geology*, **17**, 1319–1326.

Childs, C., Nicol, A., Walsh, J.J. & Watterson, J. 1996. Growth of vertically segmented normal faults. *Journal of Structural Geology*, **18**, 1389–1397.

Childs, C., Nicol, A., Walsh, J.J. & Watterson, J. 2003. The growth and propagation of synsedimentary faults. *Journal of Structural Geology*, **25**, 633–648.

Childs, C., Manzocchi, T., Walsh, J.J., Bonson, C., Nicol, A. & Schöpfer, M.P.J. 2009. A geometric model of fault zone and fault rock thickness variations. *Journal of Structural Geology*, **31**, 117–127.

CHINNERY, M.A. 1966. Secondary faulting: II. Geological aspects. *Canadian Journal of Earth Science*, **3**, 175–190, https://doi.org/10.1139/e66-014

COWIE, P.A. & SCHOLZ, C.H. 1992*a*. Growth of faults by accumulation of seismic slip. *Journal of Geophysical Research*, **97**, 11 085–11 096.

COWIE, P.A. & SCHOLZ, C.H. 1992*b*. Physical explanation for the displacement–length relationship of faults using a post-yield fracture mechanics model. *Journal of Structural Geology*, **14**, 1133–1148.

COWIE, P.A. & SCHOLZ, C.H. 1992*c*. Displacement–length scaling relationships for faults: data synthesis and discussion. *Journal of Structural Geology*, **14**, 1149–1156.

COWIE, P.A. & SHIPTON, Z.K. 1998. Fault tip displacement gradients and process zone dimensions. *Journal of Structural Geology*, **20**, 983–997.

COWIE, P.A., GUPTA, S. & DAWERS, N.H. 2000. Implications of fault array evolution for synrift depocentre development: insights from a numerical fault growth model. *Basin Research*, **12**, 241–261.

D'ALESSIO, M.A. & MARTEL, S.J. 2004. Fault terminations and barriers to fault growth. *Journal of Structural Geology*, **26**, 1885–1896.

DAWERS, N.H. & ANDERS, M.H. 1995. Displacement–length scaling and fault linkage. *Journal of Structural Geology*, **17**, 607–614.

DAWERS, N.H., ANDERS, M.H. & SCHOLZ, C.H. 1993. Growth of normal faults: displacement–length scaling. *Geology*, **21**, 1107–1110.

EDBROOKE, S.W. & BROWNE, G.H. 1996. *An Outcrop Study of Bed Thickness and Continuity in Thin-Bedded Facies of the Whakataki Formation at Whakataki Beach, East Wairarapa*. Institute of Geological & Nuclear Sciences Limited Science Report **96/34**. Institute of Geological & Nuclear Sciences Limited, Lower Hutt.

ELLIS, M.A. & DUNLAP, W.J. 1988. Displacement variation along thrust faults: implications for the development of large faults. *Journal of Structural Geology*, **10**, 183–192.

FIELD, B.D. 2005. Cyclicity in turbidites of the Miocene Whakataki Formation, Castlepoint, North Island, and implications for hydrocarbon reservoir modelling. *New Zealand Journal of Geology and Geophysics*, **48**, 135–146.

FILBRANDT, J.M., RICHARD, P.D. & FRANSSEN, R.C.M.W. 1994. Growth and coalescence of faults: numerical simulations and sand-box experiments. *In*: *Tectonic Studies Group Special Meeting on Fault Populations*, 19–20 October, Edinburgh, UK, Extended Abstract, 57–59.

FOSSEN, H. & HESTHAMMER, J. 1997. Geometric analysis and scaling relations of deformation bands in porous sandstone. *Journal of Structural Geology*, **19**, 1479–1493.

FOSSEN, H., SCHULTZ, R., SHIPTON, Z. & MAIR, K. 2007. Deformation bands in sandstone – a review. *Journal of the Geological Society, London*, **164**, 755–769, https://doi.org/10.1144/0016-76492006-036

GILLESPIE, P.A., WALSH, J.J. & WATTERSON, J. 1992. Limitations of dimension and displacement data from single faults and the consequences for data analysis and interpretation. *Journal of Structural Geology*, **14**, 1157–1172.

JACKSON, C.A.-L. & ROTEVATN, A. 2013. 3D seismic analysis of the structure and evolution of a salt-influenced normal fault zone: a test of competing fault growth models. *Journal of Structural Geology*, **54**, 215–234, https://doi.org/10.1016/j.jsg.2013.06.012

JOHNSTON, M.R., 1980. *Geology of the Tinui-Awatoitoi District*. Department of Scientific and Industrial Research Bulletin **94**. New Zealand Geological Survey, Wellington.

KIM, Y.-S. & SANDERSON, D.J. 2005. The relationship between displacement and length of faults. *Earth-Science Reviews*, **68**, 317–334.

KIM, Y.-S., ANDREWS, J.R. & SANDERSON, D.J. 2000. Damage zones around strike-slip fault systems and strike–slip fault evolution, Crackington Haven, southwest England. *Geoscience Journal*, **4**, 53–72.

KIM, Y.-S., PEACOCK, D.C.P. & SANDERSON, D.J. 2004. Fault damage zones. *Journal of Structural Geology*, **26**, 503–517.

KING, G.C.P. 1986. Speculations on the geometry of the initiation and termination processes of earthquake rupture and its relation to morphology and geological structure. *Pure and Applied Geophysics*, **124**, 567–584.

LAMARCHE, G., PROUST, J.-N. & NODDER, S.D. 2005. Long-term slip rates and fault interactions under low contractional strain, Wanganui Basin, New Zealand. *Tectonics*, **24**, https://doi.org/10.1029/2004 TC001699

MANIGHETTI, I., KING, G.C.P., GAUDEMER, Y., SCHOLZ, C.H. & DOUBRE, C. 2001. Slip accumulation and lateral propagation of active normal faults in Afar. *Journal of Geophysical Research*, **106**, 13 667–13 696.

MANIGHETTI, I., KING, G. & SAMMIS, C.G. 2004. The role of off-fault damage in the evolution of normal faults. *Earth and Planetary Science Letters*, **217**, 399–408.

MANZOCCHI, T., WALSH, J.J. & NICOL, A. 2006. Displacement accumulation from earthquakes on isolated normal faults. *Journal of Structural Geology*, **28**, 1685–1693.

MCGRATH, A.G. & DAVISON, I. 1995. Damage zone geometry around fault tips. *Journal of Structural Geology*, **17**, 1011–1024.

MEYER, V., NICOL, A., CHILDS, C., WALSH, J.J. & WATTERSON, J. 2002. Progressive localisation of strain during the evolution of a normal fault system. *Journal of Structural Geology*, **24**, 1215–1231.

MORLEY, C.K. 1999. Patterns of displacement along large normal faults: implications for basin evolution and fault propagation, based on examples from East Africa. *American Association of Petroleum Geologists Bulletin*, **83**, 613–634.

MURAOKA, H. & KAMATA, H. 1983. Displacement distribution along minor fault traces. *Journal of Structural Geology*, **5**, 483–495, https://doi.org/10.1016/0191-8141(83)90054-8

NICOL, A. 2014. Fault interactions and growth. Abstract presented at the 'Geometry and Growth of Normal Faults' Conference, 23–25 June 2014, Geological Society, London.

NICOL, A., WATTERSON, J., WALSH, J.J. & CHILDS, C. 1996. The shapes, major axis orientations and displacement patterns of fault surfaces. *Journal of Structural Geology*, **18**, 235–248.

Nicol, A., Walsh, J.J., Berryman, K. & Nodder, S. 2005. Growth of a normal fault by the accumulation of slip over millions of years. *Journal of Structural Geology*, **27**, 327–342.

Nicol, A., Mazengarb, C., Chanier, F., Rait, G., Uruski, C. & Wallace, L. 2007. Tectonic evolution of the active Hikurangi subduction margin, New Zealand, since the Oligocene. *Tectonics*, **26**, TC4002, https://doi.org/10.1029/2006TC002090

Nicol, A., Walsh, J.J., Villamor, P., Seebeck, H. & Berryman, K.R. 2010. Normal fault interactions, paleoearthquakes and growth in an active rift. *Journal of Structural Geology*, **32**, 1101–1113, https://doi.org/10.1016/j.jsg.2010.06.018

Nicol, A., Childs, C., Walsh, J. & Schafer, K. 2013. A geometric model for the formation of deformation bands. *Journal of Structural Geology*, **55**, 21–33, https://doi.org/10.1016/j.jsg.2013.07.004

Nixon, C.W., Sanderson, D.J. & Bull, J.M. 2011. Deformation within a strike-slip fault network at Westward Ho!, Devon U.K.: domino vs conjugate faulting. *Journal of Structural Geology*, **33**, 833–843.

Nixon, C.W., Sanderson, D.J. & Bull, J.M. 2012. Analysis of a strike-slip fault network using high resolution multibeam bathymetry, offshore NW Devon U.K. *Tectonophysics*, **541**, 69–80.

Nixon, C.W., Bull, J.M. & Sanderson, D.J. 2014. Localized vs distributed deformation associated with the linkage history of an active normal fault, Whakatane Graben, New Zealand. *Journal of Structural Geology*, **69**, 266–280.

Peacock, D.C.P. 1991. Displacements and segment linkage in strike-slip fault zones. *Journal of Structural Geology*, **13**, 1025–1075.

Peacock, D.C.P. 2002. Propagation, interaction and linkage in normal fault systems. *Earth-Science Reviews*, **58**, 121–142.

Peacock, D.C.P. & Sanderson, D.J. 1991. Displacements, segment linkage and relay ramps in normal fault zones. *Journal of Structural Geology*, **13**, 721–733.

Peacock, D.C.P. & Sanderson, D.J. 1992. Effects of layering and anisotropy on fault geometry. *Journal of the Geological Society, London*, **149**, 793–802, https://doi.org/10.1144/gsjgs.149.5.0793

Peacock, D.C.P. & Sanderson, D.J. 1996. Effects of propagation rate on displacement variations along faults. *Journal of Structural Geology*, **18**, 311–320.

Perrin, C., Manighetti, I. & Gaudemer, Y. 2016. Off-fault tip splay networks: a genetic and generic property of faults indicative of their long-term propagation. *Comptes Rendus Geoscience*, **348**, 52–60, https://doi.org/10.1016/j.crte.2015.05.002

Pollard, D.D. & Segall, P. 1987. Theoretical displacements and stresses near fractures in rock: with applications to faults, joints, veins, dikes, and solution surfaces. *In*: Atkinson, B.K. (ed.) *Fracture Mechanics of Rock*. Academic Press, New York, 277–349.

Poulimenos, G. 2000. Scaling properties of normal fault populations in the western Corinth Graben, Greece: implications for fault growth in large strain settings. *Journal of Structural Geology*, **22**, 307–322.

Reilly, C., Nicol, A. & Walsh, J.J. 2016. Importance of pre-existing fault size for the evolution of an inverted fault system. *In*: Childs, C., Holdsworth, R.E., Jackson, C.A.-L., Manzocchi, T., Walsh, J.J. & Yielding, G. (eds) *The Geometry and Growth of Normal Faults*. Geological Society, London, Special Publications, **439**. First published online February 5, 2016, https://doi.org/10.1144/SP439.2

Roche, V., Homberg, C., van der Baan, M. & Rocher, M. 2016. Widening of normal fault zones due to the inhibition of vertical propagation. *In*: Childs, C., Holdsworth, R.E., Jackson, C.A.-L., Manzocchi, T., Walsh, J.J. & Yielding, G. (eds) *The Geometry and Growth of Normal Faults*. Geological Society, London, Special Publications, **439**. First published online February 5, 2016, https://doi.org/10.1144/SP439.5

Rotevatn, A. & Fossen, H. 2012. Soft faults with hard tips: magnitude-order displacement gradient variations controlled by strain softening versus hardening; implications for fault scaling. *Journal of the Geological Society, London*, **169**, 123–126, https://doi.org/10.1144/0016-76492011-108

Schlagenhauf, A., Manighetti, I., Malavieille, J. & Dominguez, S. 2008. Incremental growth of normal faults: insights from a laser-equipped analog experiment. *Earth and Planetary Science Letters*, **273**, 299–311.

Schlische, R.W., Young, S.S., Ackermann, R.V. & Gupta, A. 1996. Geometry and scaling relations of a population of very small rift-related normal faults. *Geology*, **24**, 683–686.

Schöpfer, M.P.J., Childs, C. & Walsh, J.J. 2006. Localisation of normal faults in multilayer sequences. *Journal of Structural Geology*, **28**, 816–833.

Schultz, R.A. 1999. Understanding the process of faulting: selected challenges and opportunities at the edge of the 21st Century. *Journal of Structural Geology*, **21**, 985–993.

Schultz, R.A. & Fossen, H. 2002. Displacement–length scaling in three dimensions: the importance of aspect ratio and application to deformation bands. *Journal of Structural Geology*, **24**, 1389–1411, https://doi.org/10.1016/S0191-8141(01)00146-8

Schultz, R.A., Soliva, R., Fossen, H., Okubo, C.H. & Reeves, D.M. 2008. Dependence of displacement–length scaling relations for fractures and deformation bands on the volumetric changes across them. *Journal of Structural Geology*, **30**, 1405–1411.

Stein, R. 1999. The role of stress transfer in earthquake occurrence. *Nature*, **402**, 605–609.

Tvedt, A.B.M., Rotevatn, A., Jackson, C.A.-L., Fossen, H. & Gawthorpe, R.L. 2013. Growth of normal faults in multilayer sequences: a 3D seismic case study from the Egersund Basin, Norwegian North Sea. *Journal of Structural Geology*, **55**, 1–20, https://doi.org/10.1016/j.jsg.2013.08.002

Vermilye, J.M. & Scholz, C.M. 1999. Fault propagation and segmentation: insight from the microstructural examination of a small fault. *Journal of Structural Geology*, **21**, 1623–1636.

Walsh, J.J. & Watterson, J. 1987. Distributions of cumulative displacement and seismic slip on a single normal fault surface. *Journal of Structural Geology*, **9**, 1039–1046.

Walsh, J.J. & Watterson, J. 1988. Analysis of the relationship between displacements and dimensions of faults. *Journal of Structural Geology*, **10**, 239–247.

Walsh, J.J. & Watterson, J. 1991. Geometric and kinematic coherence and scale effects in normal fault systems. *In*: Roberts, A.M., Yielding, G. & Freeman, B. (eds) *The Geometry of Normal Faults*. Geological Society, London, Special Publications, **56**, 193–203, https://doi.org/10.1144/GSL.SP.1991.056.01.13

Walsh, J.J., Nicol, A. & Childs, C. 2002. An alternative model for the growth of faults. *Journal of Structural Geology*, **24**, 1669–1675.

Walsh, J.J., Bailey, W.R., Childs, C., Nicol, A. & Bonson, C.G. 2003. Formation of segmented normal faults: a 3-D perspective. *Journal of Structural Geology*, **25**, 1251–1262.

Watterson, J. 1986. Fault dimensions, displacements and growth. *Pure and Applied Geophysics*, **124**, 365–373.

Wells, D.L. & Coppersmith, K.J. 1994. New empirical relationships among magnitude, rupture length, rupture width, rupture area, and surface displacement. *Bulletin of the Seismological Society of America*, **84**, 974–1002.

Wilkins, S.J. & Gross, M.R. 2002. Normal fault growth in layered rocks at Split Mountain, Utah: influence of mechanical stratigraphy on dip linkage, fault restriction and fault scaling. *Journal of Structural Geology*, **24**, 1413–1429.

Myths about normal faulting

D. A. FERRILL*, A. P. MORRIS, R. N. MCGINNIS & K. J. SMART

Department of Earth, Material, and Planetary Sciences, Southwest Research Institute, 6220 Culebra Road, San Antonio, TX 78238-5166, USA

**Correspondence: dferrill@swri.org*

Abstract: Analyses of normal faults in mechanically layered strata reveal that material properties of rock layers strongly influence fault nucleation points, fault extent (trace length), failure mode (shear v. hybrid), fault geometry (e.g. refraction through mechanical layers), displacement gradient (and potential for fault tip folding), displacement partitioning (e.g. synthetic dip, synthetic faulting, fault core displacement), fault core and damage zone width, and fault zone deformation processes. These detailed investigations are progressively dispelling some common myths about normal faulting held by industry geologists, for example: (i) that faults tend to be linear in dip profile; (ii) that imbricate normal faults initiate due to sliding on low-angle detachments; (iii) that friction causes fault-related folds (so-called normal drag); (iv) that self-similar fault zone widening is a direct function of fault displacement; and (v) that faults are not dilational features and/or important sources of permeability.

Large faults grow from small faults (e.g. Walsh & Watterson 1988*a*; Dawers *et al.* 1993; Childs *et al.* 1996*b*; Ackermann *et al.* 2001; Walsh *et al.* 2002). Regardless of size, normal faults are fractures or tabular deformation zones that have displacements parallel to the shear fracture or zone boundaries that typically exceed the thickness of the fracture or zone, characterized by hanging-wall motion downwards relative to the corresponding footwall (e.g. Anderson 1951; Price 1966; Groshong 1988). Analyses of normal faults with displacements spanning seven orders of magnitude (mm to tens of km) in mechanically layered strata reveal that mechanical properties of rock layers strongly influence fault nucleation points, extent (trace length), failure mode (shear v. hybrid), dip and geometry (e.g. refraction through mechanical layers), displacement gradient (and potential for fault tip folding), displacement partitioning (e.g. synthetic dip, synthetic faulting, fault core displacement), fault core and damage zone width, and fault zone deformation processes (e.g. Muraoka & Kamata 1983; Walsh & Watterson 1988*a*, *b*; Peacock & Sanderson 1992; Childs *et al.* 1996*a*; Manighetti *et al.* 2001; Wilkins & Gross 2002; Ferrill *et al.* 2005, 2011*a*; Schöpfer *et al.* 2006; Schultz *et al.* 2006; Ferrill & Morris 2008). Only larger displacement faults tend to be mapped or recognizable with seismic reflection data – typically faults with vertical offset (throw) of 20 m or more are seismically imaged. However, the presence of large normal faults is an indication of numerous smaller faults that are likely to be present (Walsh *et al.* 1991), and these large faults can be used to predict the abundance and displacement distribution of subseismic faults (Morris *et al.* 2009*a*). Recognition of these faults can be critical to understanding permeability architecture in faulted rock (Caine *et al.* 1996; Eichhubl *et al.* 2009; Ferrill *et al.* 2009; Dockrill & Shipton 2010; Faulkner *et al.* 2010).

In the course of our work, we have the opportunity to interact with many geoscientists in the oil and gas industry who generate and map prospects, characterize reservoirs, and plan and steer the drilling of wells in normal-faulted regions. We have found that among industry personnel it is commonly assumed that: (i) normal faults tend to be linear in dip profile; (ii) imbricate normal faults initiate from near-horizontal detachments; (iii) friction causes drag; (iv) fault zone width grows proportionally with fault displacement; and (v) normal faults do not generate permeability. We refer to these as the five myths about normal faults. Several decades of normal fault research by the structural geology research community has largely debunked these myths, but this updated understanding has not fully infused the primary industry where these concepts can be applied meaningfully. We believe that focusing attention on (and dispelling) these myths, and improving the understanding of the role of mechanical stratigraphy in the nucleation and growth of normal faults, will lead to more robust interpretations and, ultimately, to better-informed decisions. In addition, it will provide an improved understanding of faulting and related deformation in extensional structural tectonic settings. The following discussion of mechanical stratigraphy, failure mode, fault dip and refraction, displacement gradient, and

From: Childs, C., Holdsworth, R. E., Jackson, C. A.-L., Manzocchi, T., Walsh, J. J. & Yielding, G. (eds) 2017. *The Geometry and Growth of Normal Faults*. Geological Society, London, Special Publications, **439**, 41–56.
First published online March 30, 2016, https://doi.org/10.1144/SP439.12

fault zone width provides a basis for judging when or if these myths may be valid or simply misguide interpretations of normal fault zone structure and properties.

Mechanical stratigraphy

Mechanical stratigraphy represents: (i) the varying material properties of rock strata; (ii) the thicknesses of the mechanical layers; and (iii) the nature and frictional properties of the boundaries between mechanical layers within a section. Together, these components lead to variations in stiffness and strength that typically give rise to a heterogeneous sequence of layers that represents the mechanical stratigraphic framework of interest. Mechanical stratigraphy is commonly related directly to lithology because the mechanical properties are controlled by mineralogy and texture (especially porosity), which are the primary variables that define lithology. In terms of mineralogy, the strong fraction in sedimentary rocks includes calcite, dolomite and quartz, and the weak fraction is controlled by clay minerals and organic carbon or evaporite minerals, such as salt or anhydrite (e.g. Donath 1970). For a given mineralogical composition, rock strength tends to increase with decreasing porosity (Dunn *et al.* 1973; Chang *et al.* 2006; see figs 4.14, 4.15 & 4.16 in Zoback 2007). While some formations tend to be relatively massive and uniform (e.g. thick limestone, sandstone or shale formations), many are heterolithic with units made up of a multitude of thin beds of varying mechanical characteristics. Deformation behaviour at the formation scale is strongly influenced by this mechanical layering (Fig. 1). Behaviour at the bed or formation scale is often

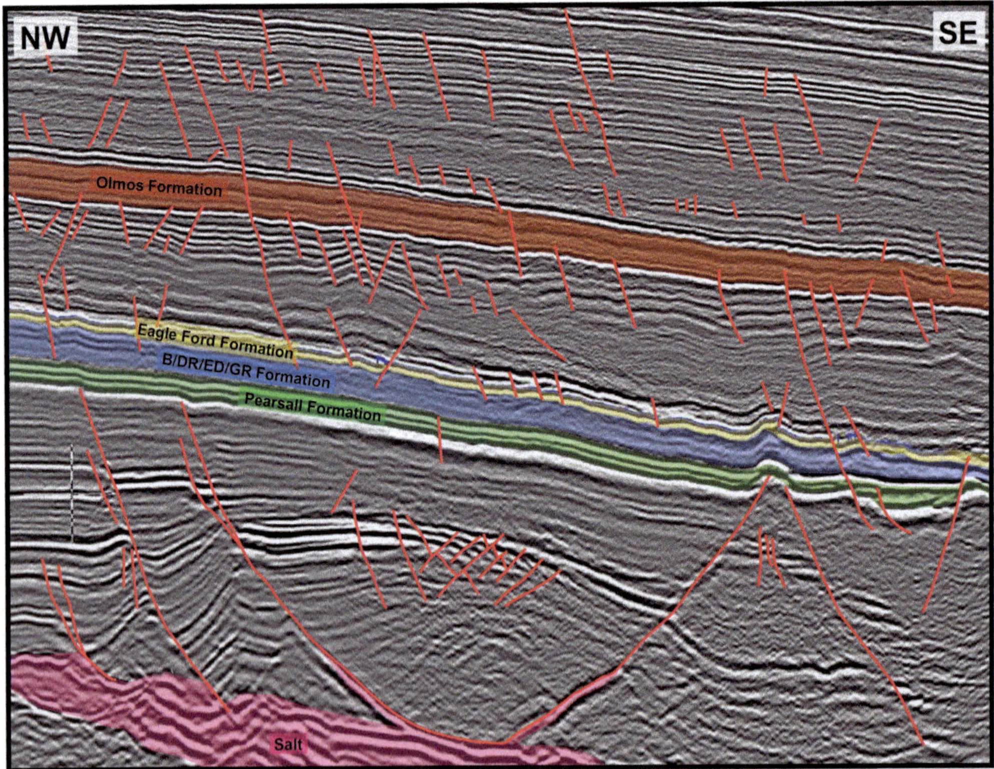

Fig. 1. Interpreted seismic (two-way travel time) profile extracted from a 3D survey from the Gulf of Mexico coastal plain province of south Texas, USA provided by Global Geophysical Services (also see Treadgold *et al.* 2010; McGinnis *et al.* 2016). The profile is NW (left) to SE (right), and is perpendicular to the main normal fault strike and parallel to the regional extension direction (the length of the line is 56.25 km; top is at 1200 ms and bottom is at 4000 ms). Bed-restricted faulting in the upper part of the section is related to the mechanical stratigraphy, with faults best developed in more competent sand- or carbonate-rich intervals and tipping upwards or downwards in more clay-rich portions of the stratigraphic section. B/DR/ED/GR, the Buda, Del Rio, Edwards and Glen Rose formations; Salt, the Jurassic Louann Salt.

influenced by the frictional properties of bed boundaries or weak interbeds, with sharp mechanical contrasts and particularly weak beds localizing bedding plane slip or causing abrupt changes in deformation behaviour (e.g. Sellards & Baker 1934; Ferrill *et al.* 1998; Cooke & Underwood 2001; Smart *et al.* 2009). Characterization of mechanical properties in outcrop can be performed using a Schmidt rebound hammer (Katz *et al.* 2000; Aydin & Basu 2005) and directly related to outcrop observations of deformation style (Morris *et al.* 2009*b*; Ferrill *et al.* 2012*b*). Observations from outcrop studies indicate that mechanical stratigraphy influences fault nucleation (Eisenstadt & De Paor 1987; Wilkins & Gross 2002; Ferrill & Morris 2008; Morris *et al.* 2009*b*; Welch *et al.* 2009*a*), failure mode (Peacock & Sanderson 1992; Ferrill & Morris 2003; McGinnis *et al.* 2009), fault dip and refraction (Walsh & Watterson 1988*b*; Peacock & Sanderson 1992; Childs *et al.* 1996*a*; Ferrill & Morris 2003; Micarelli & Benedicto 2008; Ferrill *et al.* 2014*a*), fault displacement gradient (Chapman & Williams 1985; Ferrill & Morris 2008; Welch *et al.* 2009*b*), displacement partitioning (Ferrill *et al.* 2011*a*), and fault zone width (Evans 1990; Childs *et al.* 1996*a*; Heynekamp *et al.* 1999; Wibberley *et al.* 2008). In addition, there is strong theoretical and field evidence (Zoback 2007; Smart *et al.* 2010) that rock strength is an important control on the stress states that can be supported by rock: thus, characterizing mechanical stratigraphy is a key way of anticipating this phenomenon.

Failure mode

The failure mechanism that is typically associated with fault formation is shear failure, characterized by formation of a fracture surface with displacement parallel to the plane of the fracture (Ramsay 1967; Hancock 1985; Mandl 1988). These shear failure surfaces are commonly characterized by slip lineations or slickenlines that parallel the fault displacement direction. Recent field studies have revealed the occurrence of hybrid failure in carbonate strata (limestone and chalk) that is characterized by transitional behaviour between shear and tensile (opening mode failure together with displacement perpendicular to the fracture plane) in heterogeneous limestone (Ferrill *et al.* 2012*a*), and interlayered mudrock and carbonate formations (Ferrill *et al.* 2014*a*). These field observations are consistent with laboratory generated hybrid failure (Ramsey & Chester 2004; Bobich 2005) and are, in turn, supported by the identification of hybrid failure in microseismicity that is associated with hydraulic fracturing (Busetti *et al.* 2014). Hybrid failure surfaces developed in the normal faulting stress regime are composed of a mosaic of steeper portions with significant dilation, linked by more gently dipping shear portions with little or no dilation (Fig. 2). Because, for example, chalk is more competent than adjacent mudrock, it can support a larger differential stress at conditions where the maximum principal stress (σ_1), in this case the overburden stress, is the same for both lithologies. Thus, as the minimum principal stress (σ_3) decreases in response to regional stretching, it can decay further, possibly into the tensile field, before the rock breaks (Ferrill *et al.* 2014*a*). Meanwhile, the adjacent weaker mudrock: (a) accommodates greater pre-faulting strain (ductility); and (b) ultimately fails at much lower differential stresses at which σ_3 does not become tensile. The overall effect of hybrid failure is to produce steeper fault segments that obliquely dilate: thus, developing porosity, localizing fluid movement and leading to enhanced cementation or, in some cases, dissolution (Sibson 1996; Ferrill & Morris 2003).

Fault dip and refraction

In relatively massive or homogeneous strata, faults might be expected to fail with a consistent shear failure angle, resulting in consistent fault dips. However, faults that cut mechanically layered sections commonly have variable dips that can be related to the mechanical properties of the layers that they cut (e.g. Fig. 3). These refracted fault profiles are defined by steep hybrid or shear failure segments in competent layers, and more gently dipping shear failure segments in less competent strata (Wallace 1861; Dunham 1948, 1988; Ferrill & Morris 2003; Schöpfer *et al.* 2006; Ferrill *et al.* 2012*a*). Refraction can be a function of variation in either shear failure angle or failure mode (shear v. hybrid) with mechanical rock properties. Other processes, such as breaching of relays between vertically segmented faults or nucleation of faults by reactivation and linkage of pre-existing joints, have also been recognized (Peacock & Sanderson 1992; Peacock & Zhang 1993; Peacock 2002). Slip along more gently dipping segments results in dilation of steep fault segments, even at depths of several kilometres (Ferrill *et al.* 2014*a*). Similar to fault refraction described in carbonate-rich strata, fault refraction is also seen in faulted volcanic strata of different competence (Ferrill *et al.* 2004, 2011*b*). Where unconsolidated or poorly consolidated material overlies dilational fault segments in competent layers near the ground surface, drainage of material downwards into the resulting voids along dilational fault segments leads to formation of pit craters and troughs, and incorporates externally sourced material into the fault zone (Ferrill *et al.* 2004, 2011*b*).

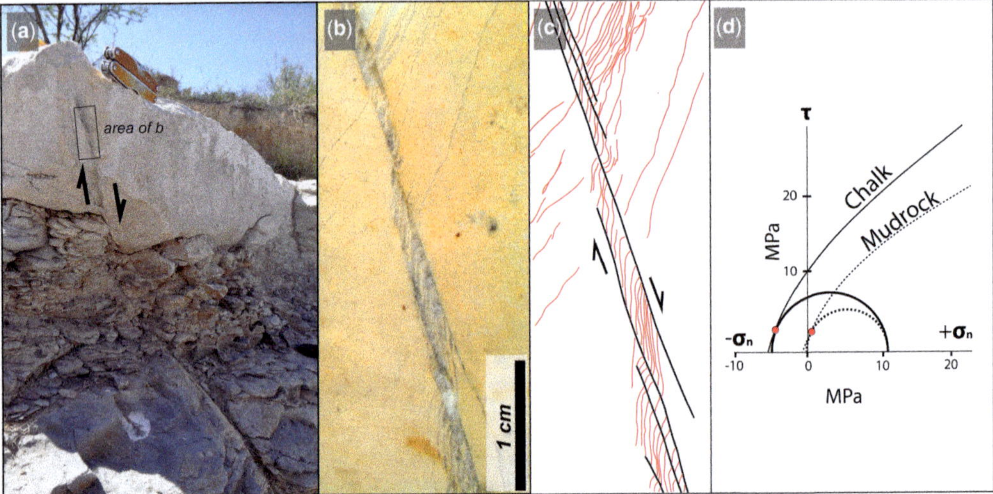

Fig. 2. Field and laboratory photographs of an incipient fault formed in Cretaceous chalk and mudrock beds in south Texas, USA. (**a**) The fault changes dip from approximately 75° in the chalk bed (top of photo) to 50° in the mudrock at the bottom of the photograph. (**b**) Photograph of a slab cut perpendicular to the fault zone and parallel to the slip direction. Grey and white calcite vein segments are separated by host rock microlithons. The vein segments are kinematically linked by shear segments. (**c**) Line drawing of an integrated network of tensile (red lines) and shear (black lines) fractures indicative of hybrid failure. (**d**) Illustration of the stress evolution and related brittle faulting (σ denotes normal stress and τ denotes shear stress). Effective stress was adjusted for pore fluid pressure. The failure envelopes illustrate the strength differences between the chalk and mudrock beds of the Eagle Ford Formation. For additional detail, see Ferrill *et al.* (2014*a*).

In addition, fault dilation may focus fluid flow (Sibson 1996; Ferrill *et al.* 1999*b*; Holland *et al.* 2006), can localize ore mineralization (Wallace 1861; Dunham 1988) and may allow underlying liquid or gas to escape (e.g. gas venting from destabilized submarine hydrocarbon clathrate layers: Germanovich & Xu 2004).

Fault nucleation

Analysis of pre-failure strain (also referred to as ductility: Donath 1970) from laboratory testing of a variety of common sedimentary rock types demonstrates substantial changes in ductility as a function of confining pressure and rock type (e.g. Handin & Hager 1957; Donath 1970). These results demonstrate that, as strain accumulates in mechanically layered sedimentary sections, lower ductility strata (e.g. dolostone, limestone) are likely to fail by brittle faulting at lower strains than nearby higher ductility strata (e.g. shale, anhydrite) (Donath 1970; Ferrill & Morris 2008; Welch *et al.* 2009*a*; Ferrill *et al.* 2012*a*). Thus, faults would be expected to nucleate first in more competent limestone or dolostone beds, and field investigations show this to be the case (e.g. Eisenstadt & De Paor 1987). Systems of steeply dipping normal faults will develop in brittle competent units. These faults drive displacement into less competent strata where displacement is accommodated by more distributed or ductile deformation, such as fault tip folding, distributed shear and flow, or as a distributed network of low-angle faults (e.g. Childs *et al.* 1996*a*; Nicol *et al.* 2002). With increasing extension and displacement, the deformation may develop into stratabound conjugate or imbricate normal fault systems. Extension accommodated by a series of high-angle normal faults in a competent stratigraphic package may be accommodated by distributed or ductile extension and thinning and shear in overlying or underlining incompetent packages, and may localize shear on low-angle faults or a through-going detachment surface. Slip initiation on a low-angle normal fault is mechanically unlikely owing to the high resolved normal stress tending to lock horizontal or near-horizontal surfaces or zones in a normal faulting stress regime, where vertical overburden stress is the maximum principal compressive stress and resolved shear stress is negligible (Jackson 1987; Melosh 1990; Morris *et al.* 1996; Westaway 1999; Collettini & Sibson 2001). However, propagation of slip from high-angle faults in competent mechanical layers into an incompetent layer is mechanically viable (e.g. Ofoegbu & Ferrill 1998), and can explain the geometries of small-scale systems observed in the

Fig. 3. Field photograph of a small-displacement fault that refracts through Cretaceous limestone and marl beds of the Boracho Formation in west Texas, USA. Slip parallel to lower-dip shear segments has led to dilation of steeper hybrid failure segments. The yellow field book is 12 × 19 cm.

field and seismic reflection data (Ferrill *et al.* 1998; Treadgold *et al.* 2010) (Figs 1 & 4), as well as earthquake patterns (Jackson 1987; Jackson & White 1989; Braunmiller & Nabelek 1996). Resulting deformation in relatively incompetent strata may be accommodated by ductile deformation mechanisms (Braunmiller & Nabelek 1996) or by slip at lower (aseismic) rates (Beroza & Jordan 1990) on low-angle faults that are well orientated to accommodate extension and more effective at converting slip into extension than the initiating/precursor higher-angle faults (e.g. Wernicke 1995).

Displacement gradient

Faults may propagate rapidly with respect to the rate of fault displacement accumulation through brittle beds or formations, and accumulate displacement with little or no associated folding (Ferrill & Morris 2008): such faults will be characterized by consistent and relatively low displacement gradients. However, fault propagation may slow or cease in incompetent units, such as clay or evaporite-rich layers (Ferrill *et al.* 2012*a*). Continued displacement on a fault with an arrested tip leads to an increasing displacement gradient towards the fault tip, and results in folding beyond the fault tip producing a 'fault-propagation fold' (Fig. 5) (Khalil & McClay 2002). Such folds are the result of arrested or delayed fault propagation (Fig. 6) and thus might be thought of as 'fault non-propagation folds', or alternatively referred to as 'fault tip folds' or 'fault tip monoclines' based on their structural position at and beyond the fault tip (Williams & Chapman 1983; Chapman & Williams 1985; Withjack *et al.* 1990; Schlische 1995; Gawthorpe *et al.* 1997; Janecke *et al.* 1998; Patton *et al.* 1998; Hardy & McClay 1999; Gawthorpe & Hardy 2002; Khalil & McClay 2002; Grant & Kattenhorn 2004; Jin & Groshong 2006; White & Crider 2006; Ferrill *et al.* 2007, 2012*a*). With continued displacement, the fault may break through the fold and leave tilted layers with dip in the same direction as the fault (i.e. synthetic dip) in the hanging wall, footwall (e.g. Fig. 7) or both fault blocks. A fault that has been arrested by an incompetent sequence of rock can finally break through after the incompetent bed has been structurally thinned to zero or when the synthetic dip in the hanging wall is parallel with the fault dip. While this synthetic dip is often described as fault drag, we conclude that it is the product of folding prior to fault breakthrough and not the result of frictional drag on the fault (Reches & Eidelman 1995; Ferrill *et al.* 2012*a*). Although the term 'drag' is commonly used as a geometrical term rather than to indicate a particular genetic origin related to fault friction, we believe the term can imply the frictional resistance mechanism to many non-specialists and should, instead, be replaced with the more descriptive term 'synthetic dip' for this fault-related layer dip that can originate in many ways (Ferrill *et al.* 2005). This folding has more broadly been considered to be part of an overall fault-related process zone (also proto-shear zone or monocline) either beyond the tip of a propagating fault or as a precursor to fault nucleation (e.g. Childs *et al.* 1996*a*; Schöpfer *et al.* 2006; Welch *et al.* 2009*a*). In some cases, the folding may be localized in the relay zone between two overlapping fault tips (Childs *et al.* 1996*b*; Nicol *et al.* 2002). Because of the mechanical stratigraphic control on the rate of fault propagation and associated displacement gradients, the overall fault displacement or throw on such faults is accommodated differently in different mechanical layers. Total throw (T_t) is the sum of throw components accommodated by the fault core (T_c), synthetic dip (T_{sd}) and synthetic faults (T_{sf}) (Fig. 8) (Ferrill *et al.* 2011*a*). This displacement partitioning depends on the mechanical layering and the related rate of fault propagation with respect to rate of fault displacement.

Fig. 4. Interpreted oblique aerial photograph of the normal fault system in the hanging wall of the Gold Ace Mine Fault Zone along the SW flank of Bare Mountain, Nevada (after Ferrill *et al.* 1998). The view is looking down and to the NE along the structural plunge (defined by the fault-bedding intersection line). The width of the field of view is approximately 600 m. The dashed white line marks the crest of the ridge in the foreground. Note that the bedding-fault cutoff angle is higher in the more competent carbonate beds (white–cream–pale grey) than in the less competent shale beds (darker grey–green). Less competent intervals are characterized by the linking of fault zones and the resulting development of detachment horizons (Ferrill *et al.* 1998).

Fig. 5. Annotated field photograph of the small-displacement faults (heavy red lines) within the Balcones Fault Zone along Green Mountain Road in NE San Antonio, Texas, USA (after Ferrill *et al.* 2012*b*). The three faults that cut the top of the Austin Chalk (black bedding form line) are (from left to right) faults 1, 2 and 3. The three faults have displacements of 1.5, 0.8 and 4.0 m at the top of the Austin Chalk, and each fault tips upwards in a clay-rich bed (see the low Schmidt rebound). Above the fault tips, the fault throw is accommodated by monoclines, with the maximum synthetic dip highlighted by white lines. On the left is the outline of the Schmidt R profile matched against the stratigraphy. Schmidt R correlates with Young's modulus and unconfined compressive strength. Note the geologist wearing an orange vest near fault 3 for scale. Viewing direction is to the ENE along fault strike.

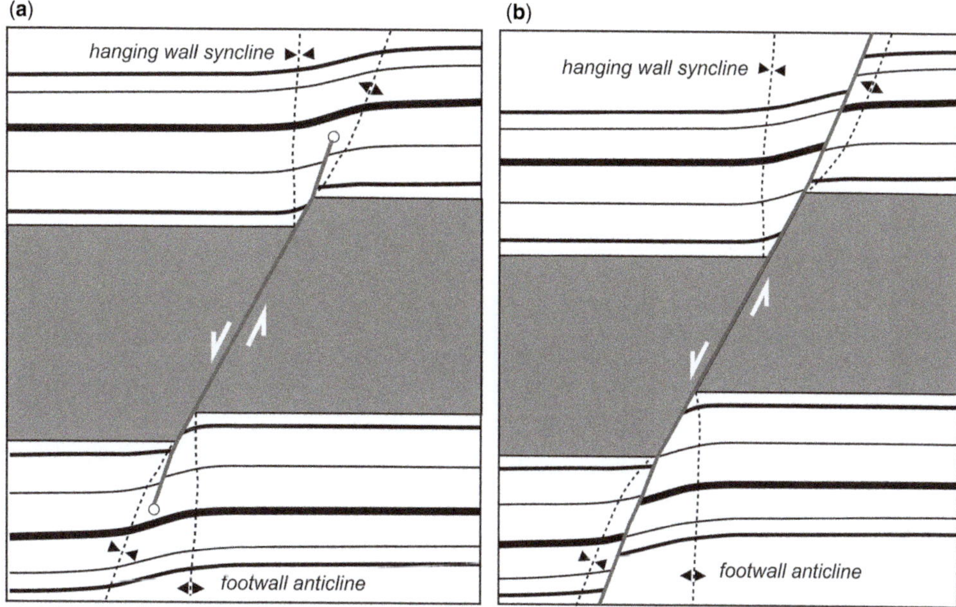

Fig. 6. Schematic model for the formation of the synthetic dip adjacent to a normal fault due to arrested fault tip propagation owing to incompetent layers within the deforming stratigraphic section (see a related discussion in Ferrill *et al.* 2011*a*, 2012*b*). (**a**) Folding occurs beyond the fault tip. (**b**) Once faulted through, the synthetic dip adjacent to the fault gives the appearance of so-called 'drag', but is not the product of friction on the fault (see Ferrill *et al.* 2016*b*).

Fault zone width

Fault zones are generally localized zones of deformation that may include rotation of layering, fracturing, cataclasis, dilation, dissolution and mineral precipitation. Several decades of research have explored the question of whether fault zone width grows proportionally as a simple function of

Fig. 7. Annotated field photograph of the Hidden Valley Fault Zone at Canyon Lake Gorge in Comal County, Texas, USA. The field of view is 25 m wide. The fault at this location cuts and offsets the upper Glen Rose Formation and has a total down-to-the-SE throw of nearly 58 m, represented by throw within the fault core (46 m), throw on synthetic faults (5 m) and throw accommodated by synthetic dip (7 m). The dominance of footwall damage associated with this fault zone is interpreted by Ferrill *et al.* (2011*a*) to be the result of nucleation of the fault in the overlying massive and competent Edwards Group carbonates, and downwards propagation into the mechanically layered Glen Rose Limestone. Additional details are available in Ferrill & Morris (2008) and Ferrill *et al.* (2009, 2011*a*).

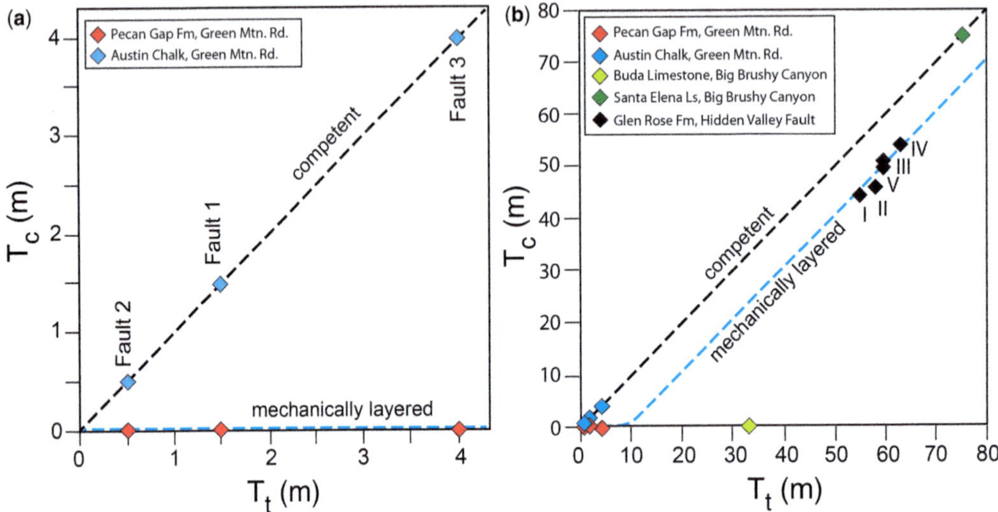

Fig. 8. (**a**) Graph of fault throw accommodated by displacement in the fault core (T_c) v. total fault throw (T_t) for three faults shown in Figure 5. Fault displacement in the Austin Chalk is essentially all accommodated by slip along the principal displacement surface of the fault core, whereas none of the faults has yet broken through the weak clay-rich bed at the base of the overlying Pecan Gap Formation, thus the folding and localized bed extension (extension fracturing, boudinage) accommodate all of the fault displacement above the clay bed. (**b**) Expanded graph showing data from (a), plus data from five profiles from the Glen Rose Limestone in the Hidden Valley Fault zone at Canyon Lake Gorge (data from Ferrill *et al.* 2011*a*) (cf. Fig. 7), along with data from the dominant competent Santa Elena Limestone and the less competent Buda Formation in the Big Brushy Canyon Fault Zone (Ferrill *et al.* 2007). These examples illustrate that the partitioning of fault zone displacement is directly related to the mechanical stratigraphy, and the role of fault tip folding and distributed deformation, prior to fault breakthrough.

displacement (Robertson 1982; Evans 1990; Shipton *et al.* 2006; Wibberley *et al.* 2008; Childs *et al.* 2009). Determining the width of a damage zone can be difficult in practice, and fault zones often vary in width over several orders of magnitude along a given fault (Evans 1990; Shipton *et al.* 2006). Groshong (1988; after Wise *et al.* 1984) defines a fault zone as follows: 'A *fault zone* is a tabular region across which the displacement parallel to the zone is appreciably greater than the width of the zone'. For a structure to be recognized as a fault zone by this definition, it must meet the criteria of having displacement in excess of the fault zone width. Consequently, plots of fault zone thickness or width v. displacement will tend to inherently illustrate increasing thickness with increasing displacement, and this is true of the early investigations of these relationships (Otsuki 1978; Robertson 1982; Evans 1990). Studies that show a relatively steady increase in fault zone width with displacement are rare and narrowly limited in scope (Shipton & Cowie 2001). Fault zone width or thickness is strongly controlled by the early history of fault nucleation and evolution during the propagation and linkage of fault segments (Dawers & Anders 1995; Wibberley *et al.* 2008; Childs *et al.* 2009). The width of a normal-fault-related fold is a primary control on maximum fault zone (or damage zone) width. This monocline width is determined at the onset of folding ahead of the propagating fault: it is, thus, more closely related to the mechanical stratigraphy rather than to a simple function of fault displacement (cf. Ferrill *et al.* 2007, 2012*b*). The throw partitioning and normal-fault-related fold width in mechanically layered rocks depends on the thickness and competence of the involved layers.

Another primary determinant of fault zone width is the spacing between overlapping fault segments (or width of relay ramps: Childs *et al.* 2009) that typically cooperate to define a fault zone (Fig. 9). This spacing develops early and the displacement tends to localize into a narrower zone with increasing displacement, and straightening and smoothing the fault surface by severing asperities that are poorly orientated for slip in the ambient stress field (Engelder 1978; Power *et al.* 1988; Childs *et al.* 1996*b*). This fault zone weakening and localization behaviour has been referred to as 'strain softening' (Gray *et al.* 2005). The width of the segmented fault array is established early and, therefore, this control on damage zone width also is not linearly related to fault displacement. However, analogue modelling and field observations show that fault systems develop in displacement v. length space along

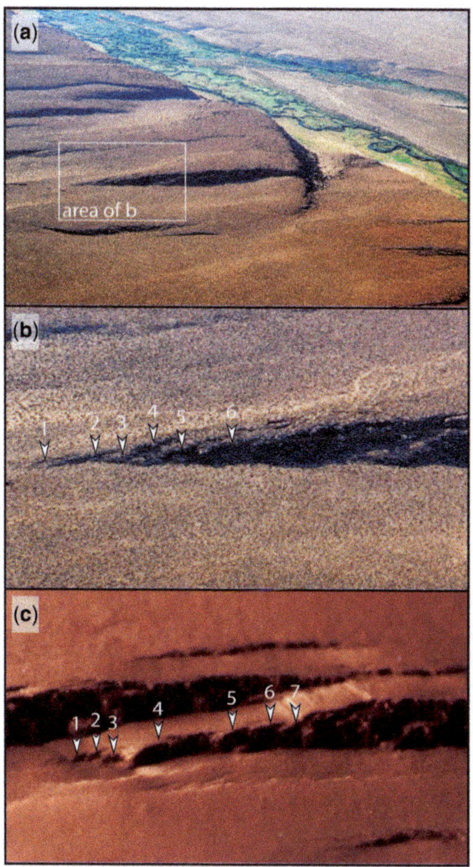

Fig. 9. (**a**) & (**b**) Examples of segmented fault scarps along a west-dipping fault in the SW part of the Volcanic Tableland and (**c**) a clay model for comparison (Ferrill *et al.* 2016*a*). (a) Low-altitude early morning aerial photograph looking SE at an array of west-dipping normal faults in the southern Volcanic Tableland in northern Owens Valley, California, USA. The width of the field of view is 1.5 km. (b) Enlarged photograph of the northern end of a normal fault shown in (a). The field of view is 0.5 km wide. At first glance, the fault scarps in (a) appear to be relatively simple breaks, but closer inspection (see enlargement in b) reveals that segments are typically composed of multiple closely spaced fault segments and breached relays. (c) Photograph of a relay ramp developed in a 5 cm-thick clay model in the laboratory (after Morris *et al.* 2014). The field of view is approximately 8 cm from left to right. Note that the near-fault scarp in the model is composed of at least seven fault segments that have linked. The width of that fault zone is largely determined by the spacing of these segments, similar to the segmentation of faults in the Volcanic Tableland (b).

a stair-step path rather than a linear self-similar path owing to periodic jumps in the fault length when fault segments link (Ferrill *et al.* 1999*a*; Childs *et al.* 2009; Wyrick *et al.* 2011). With increasing displacement, fault zones with increasingly wide separation (Figs 9 & 10) begin to cooperate and link (Ferrill *et al.* 1999*a*; Childs *et al.* 2009). Thus, fault zone width also grows along a stair-step trajectory with respect to displacement, the widening steps again associated with cooperation and linkage (Childs *et al.* 2009).

As in mechanically layered sedimentary strata, faulting in volcanic rocks at the ground surface in some cases also shows evidence of fault tip folding, fissuring due to outer-arc extension (bending strain) and reactivation of cooling joints to define irregular, largely dilational, fault zones that experience toppling failure at the ground surface (Fig. 11). Young and active fault zones developing at the surface in jointed volcanic rocks may appear to be degraded fault scarps, when, in fact, this is the character of the faulting process associated with upwards fault propagation, fault tip folding, dilation of cooling joints to accommodate bending strain in the outer arc(s) of the monoclinal fold, and block toppling and sliding (e.g. Grant & Kattenhorn 2004).

Discussion: normal fault myth v. reality

The previous sections summarize advances in the understanding of normal faults. Our experience suggests that this improved understanding has not fully infused the oil and gas industry where geoscientists and engineers make important decisions based on subsurface interpretations. A common factor in dispelling misapprehensions concerning normal faults is that mechanical stratigraphy has a critical influence on their development. Here we revisit the myths one by one, and discuss the importance of mechanical stratigraphy for the purpose of predicting deformation behaviour.

Myth 1: Normal faults are linear in dip profile

Normal faults are rarely planar. They often change dip or refract through mechanical layers, with steeper dips through competent units and gentler dips in less competent layers (e.g. Peacock & Sanderson 1992; Childs *et al.* 1996*a*, 2009; Ferrill & Morris 2003). Recognition of this relationship can be used to predict fault zone deformation based on fault geometry, or to infer mechanical stratigraphy based on fault shape.

Myth 2: Imbricate normal faults initiate due to sliding on horizontal or low-angle detachments

Faults tend to nucleate in competent beds, propagate upwards and downwards, and drive slip into weaker

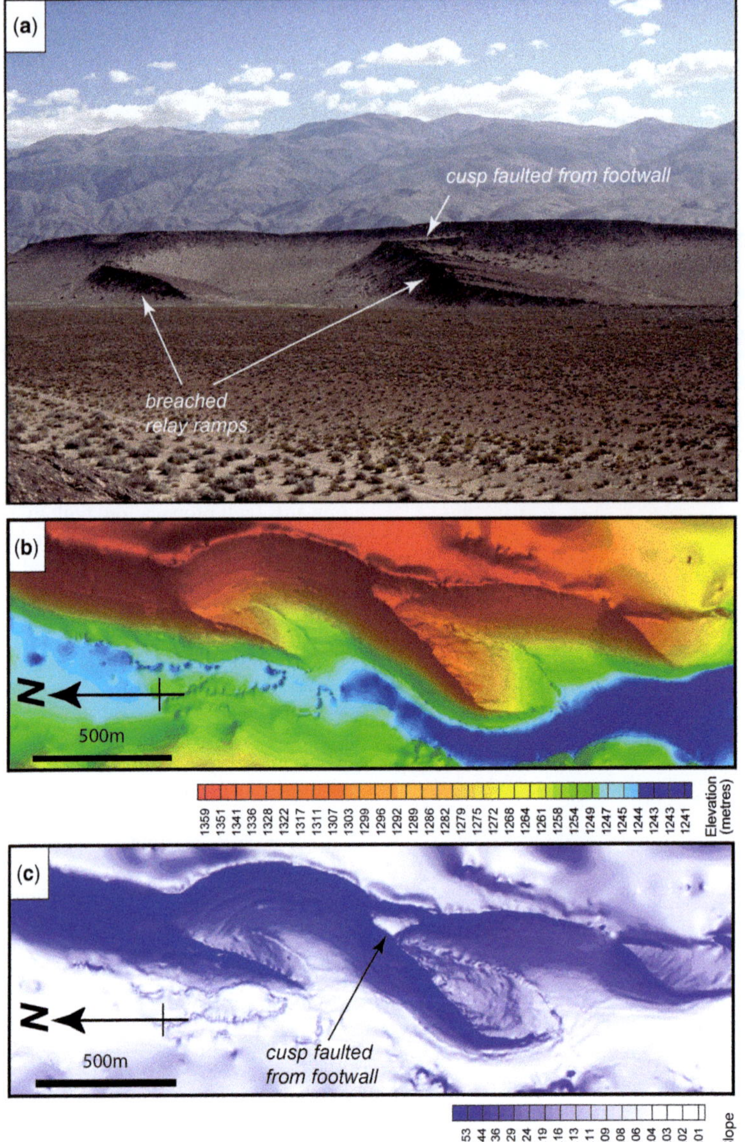

Fig. 10. Southern Fish Slough Fault System in the SE Volcanic Tableland, Owens Valley, California, USA (Ferrill *et al.* 2016*a*): (**a**) field photograph of the southern Fish Slough fault scarp looking NE, with the White Mountains in the background; (**b**) colour-contoured topographical map (in metres); and (**c**) slope map (scale units are dimensionless, i.e. m/m). In (a), the erosionally resistant capping welded ignimbrite layer is visible in the footwall of the Fish Slough Fault System, can be traced down two breached relay ramps and defines the ground surface in the foreground of the photograph, such that the full displacement of the fault zone is visible in the photograph.

units where fault slip is accommodated by distributed and/or ductile deformation. Understanding this process provides a predictive tool for relating likely fault zone deformation based on fault geometry, or to infer mechanical stratigraphy based on fault shape and displacement patterns.

Myth 3: Friction causes drag during slip on normal faults

Outcrop studies show that so-called 'drag' develops beyond fault tips and often between fault segments (e.g. Peacock & Sanderson 1991; Nicol *et al.* 2002;

Fig. 11. The Almannagja Fault Zone at the north end of Lake Thingvallatn, Iceland is represented by a dramatic faulted monocline in 9 kyr-old basalt that, along with the spacing between the en echelon fault segments, defines the width of the fault zone. See the people on the trail for scale. The difference in elevation from the footwall (left) to the hanging wall of the fault (right) is approximately 35 m. For more detail on this fault zone, see Gudmundsson (1987) and Grant & Kattenhorn (2004).

Ferrill *et al.* 2012*a*; Childs *et al.* in review) during folding prior to the fault breaking through. This observation is in contrast to the common assumption that bed rotation occurs adjacent to the fault as displacement increases due to friction on the fault. Synthetic dip can be generated in a range of structural positions, and significant information about fault nucleation, propagation and mechanical stratigraphy can be gleaned from investigating the pattern of synthetic dip associated with normal faults.

Myth 4: Normal fault zones widen in proportion to fault displacement

Fault zone width is largely a function of fault tip folding and fault segment spacing, developed prior to (folding) or during (segment spacing) fault breaking through a particular layer. Thus, fault zone width is strongly controlled by the structural and mechanical stratigraphic position along the fault zone. Although fault zones widen with increasing displacement, maximum fault zone widths are, in many cases, determined by folding (monocline width) prior to fault breakthrough and fault segment spacing (relay ramp development and breaching). Widening occurs as a stepwise or episodic widening related to fault tip zone processes and segment linkage as opposed to steady widening with displacement. Focusing on structural and mechanical stratigraphic position should be given priority over fault displacement (or throw) when trying to understand or predict the characteristics of a fault damage zone.

Myth 5: Normal faults are not dilational features or important sources of permeability

Faults tend to be the largest and best-connected fractures in any naturally deformed network. There is a large amount of literature detailing the reduction in permeability and the related sealing potential of faults in siliciclastic hydrocarbon reservoirs and aquifers (e.g. Fossen *et al.* 2005; Bense & Person 2006; Childs *et al.* 2007; Fisher & Jolley 2007; Myers *et al.* 2007). However, many normal faults in low-porosity and low-permeability rock exhibit porosity or vein fill indicative of past pore-space generation, and many faults exhibit enhanced permeability or combined conduit and seal characteristics (Caine *et al.* 1996; Evans *et al.* 1997; Gartrell *et al.* 2004), and these observations are particularly relevant to induced hydraulic fracturing in shale or self-sourced reservoirs and other considerations such as nuclear waste isolation in low-permeability rock (Ferrill *et al.* 1999*b*). Most probably, faults form the backbone of the permeability network in low-permeability rock (Sibson 1996; Ferrill *et al.* 1999*b*; Faulkner *et al.* 2010; Manzocchi *et al.* 2010; Seebeck *et al.* 2014), and it should be understood that hybrid and shear fracturing (faulting) is likely to be a significant part of the induced hydraulic fracturing that is produced during stimulation of unconventional reservoirs (Ferrill *et al.* 2014*b* and references therein).

Conclusions

Large normal faults grow from the cooperation and linkage of small normal faults, and many lessons about normal faults can be learned from studying and comparing faults over a wide range of displacements. Characterizing the intricacies of normal fault nucleation, propagation, fault-related folding and fault zone permeability is increasingly important, as geoscientists continue to explore and exploit energy (oil, gas, coal, geothermal), mineral and water resources. Mechanical stratigraphy has a critical influence on the development of normal faults. Understanding this influence helps to dispel myths about normal faults, and allows us to focus on characterizing mechanical stratigraphy for the purpose of predicting natural deformation behaviour and induced deformation related to energy extraction and subsurface fluid disposal.

The authors thank Global Geophysical Services for permission to use their seismic data. The authors thank Lora Neill for her format and editorial review that improved the manuscript, Gary Walter and Nate Toll for their technical reviews that improved the content of the manuscript, Sarah Wigginton for assistance with figure preparation, and Violeta Gonzales for handling the internal review process. The article was significantly improved by thorough and constructive reviews by Andy Nicol and Tom Manzocchi. Preparation of this article was supported in part by SwRI Internal Research and Development Project #R8588.

References

ACKERMANN, R.V., SCHLISCHE, R.W. & WITHJACK, M.O. 2001. Geometric and statistical evolution of normal fault systems: an experimental study of the effects of mechanical layer thickness on scaling laws. *Journal of Structural Geology*, **23**, 1803–1819.

ANDERSON, E.M. 1951. *The Dynamics of Faulting and Dyke Formation with Applications to Britain.* Oliver & Boyd, Edinburgh.

AYDIN, A. & BASU, A. 2005. The Schmidt hammer in rock material characterization. *Engineering Geology*, **81**, 1–14.

BENSE, V.F. & PERSON, M.A. 2006. Faults as conduit-barrier systems to fluid flow in siliciclastic sedimentary aquifers. *Water Resources Research*, **42**, W05421.

BEROZA, G.C. & JORDAN, T.H. 1990. Searching for slow and silent earthquakes using free oscillations. *Journal of Geophysical Research*, **95**, 2485–2510.

BOBICH, J.K. 2005. *Experimental analysis of the extension to shear fracture transition in Berea Sandstone.* MS thesis, Texas A&M University, College Station, TX.

BRAUNMILLER, J. & NABELEK, J. 1996. Geometry of continental normal faults: seismological constraints. *Journal of Geophysical Research*, **101**, 3045–3052.

BUSETTI, S., JIAO, W. & RECHES, Z. 2014. Geomechanics of hydraulic fracturing microseismicity Part I: shear, hybrid, and tensile events. *American Association of Petroleum Geologists Bulletin*, **98**, 2439–2457.

CAINE, J.S., EVANS, J.P. & FORSTER, C.B. 1996. Fault zone architecture and permeability structure. *Geology*, **24**, 1025–1028.

CHANG, C., ZOBACK, M.D. & KHASKAR, A. 2006. Empirical relations between rock strength and physical properties in sedimentary rocks. *Journal of Petroleum Science and Engineering*, **51**, 223–237.

CHAPMAN, T.J. & WILLIAMS, G.D. 1985. Strains developed in the hangingwalls of thrusts due to their slip/propagation rate: a dislocation model: reply. *Journal of Structural Geology*, **7**, 759–762.

CHILDS, C., NICOL, A., WALSH, J.J. & WATTERSON, J. 1996*a*. Growth of vertically segmented normal faults. *Journal of Structural Geology*, **18**, 1389–1397.

CHILDS, C., WALSH, J.J. & WATTERSON, J. 1996*b*. A model for the structure and development of fault zones. *Journal of the Geological Society, London*, **153**, 337–340, https://doi.org/10.1144/gsjgs.153.3.0337

CHILDS, C., WALSH, J.J. ET AL. 2007. Definition of a fault permeability predictor from outcrop studies of a faulted turbidite sequence, Taranaki, New Zealand. *In*: JOLLEY, S.J., BARR, D., WALSH, J.J. & KNIPE, R.J. (eds) *Structurally Complex Reservoirs.* Geological Society, London, Special Publications, **292**, 235–258, https://doi.org/10.1144/SP292.14

CHILDS, C., MANZOCCHI, T., WALSH, J.J., BONSON, C.G., NICOL, A. & SCHÖPFER, M.P.J. 2009. A geometric model of fault zone and fault rock thickness variations. *Journal of Structural Geology*, **31**, 117–127.

CHILDS, C., MANZOCCHI, T., NICOL, A., WALSH, J.J., SODEN, A.M., CONNEALLY, J.C. & DELOGKOS, E. In review. The relationship between normal drag, relay ramp aspect ratio and fault zone structure. *In*: CHILDS, C., HOLDSWORTH, R.E., JACKSON, C.A.-L., MANZOCCHI, T., WALSH, J.J. & YIELDING, G. (eds) *The Geometry and Growth of Normal Faults.* Geological Society, London, Special Publications, **439**.

COLLETTINI, C. & SIBSON, R.H. 2001. Normal faults, normal friction? *Geology*, **29**, 927–930.

COOKE, M.L. & UNDERWOOD, C.A. 2001. Fracture termination and step-over at bedding interfaces due to frictional slip and interface opening. *Journal of Structural Geology*, **23**, 223–238.

DAWERS, N.H. & ANDERS, M.H. 1995. Displacement-length scaling and fault linkage. *Journal of Structural Geology*, **17**, 607–614.

DAWERS, N.H., ANDERS, M.H. & SCHOLZ, C.H. 1993. Growth of normal faults: displacement–length scaling. *Geology*, **21**, 1107–1110.

DOCKRILL, B. & SHIPTON, Z.K. 2010. Structural controls on leakage from a natural CO_2 geologic storage site: Central Utah, U.S.A. *Journal of Structural Geology*, **32**, 1768–1782.

DONATH, F.A. 1970. Some information squeezed out of rock. *American Scientist*, **58**, 54–72.

DUNHAM, K.C. 1948. *Geology of the Northern Pennine Orefield. Volume 1: Tyne to Stainmore*. 1st edn. Geological Survey of Great Britain, England and Wales Memoirs. HMSO, London.

DUNHAM, K.C. 1988. Pennine mineralization in depth. *Proceedings of the Yorkshire Geological Society*, **47**, 1–12, https://doi.org/10.1144/pygs.47.1.1

DUNN, D.E., LAFOUNTAIN, L.J. & JACKSON, R.E. 1973. Porosity dependence and mechanisms of brittle fracture in sandstones. *Journal of Geophysical Research*, **78**, 2403–2417.

EICHHUBL, P., DAVATZES, N.C. & BECKER, S.P. 2009. Structural and diagenetic control of fluid migration and cementation along the Moab fault, Utah. *American Association of Petroleum Geologists Bulletin*, **93**, 653–681.

EISENSTADT, G. & DE PAOR, D.G. 1987. Alternative model of thrust-fault propagation. *Geology*, **15**, 630–633.

ENGELDER, T. 1978. Aspects of asperity-surface interaction and surface damage of rocks during experimental frictional sliding. *Pure and Applied Geophysics*, **116**, 705–716.

EVANS, J.P. 1990. Thickness-displacement relationships for fault zones. *Journal of Structural Geology*, **12**, 1061–1065.

EVANS, J.P., FORSTER, C.B. & GODDARD, J.V. 1997. Permeability of fault-related rocks, and implications for hydraulic structure of fault zones. *Journal of Structural Geology*, **19**, 1393–1404.

FAULKNER, D.R., JACKSON, C.A.L., LUNN, R.J., SCHLISCHE, R.W., SHIPTON, Z.K., WIBBERLEY, C.A.J. & WITHJACK, M.O. 2010. A review of recent developments concerning the structure, mechanics and fluid flow properties of fault zones. *Journal of Structural Geology*, **32**, 1557–1575.

FERRILL, D.A. & MORRIS, A.P. 2003. Dilational normal faults. *Journal of Structural Geology*, **25**, 183–196.

FERRILL, D.A. & MORRIS, A.P. 2008. Fault zone deformation controlled by carbonate mechanical stratigraphy, Balcones fault system, Texas. *American Association of Petroleum Geologists Bulletin*, **92**, 359–380.

FERRILL, D.A., MORRIS, A.P., JONES, S.M. & STAMATAKOS, J.A. 1998. Extensional layer-parallel shear and normal faulting. *Journal of Structural Geology*, **20**, 355–362.

FERRILL, D.A., STAMATAKOS, J.A. & SIMS, D.W. 1999*a*. Normal fault corrugation: implications for growth and seismicity of active normal faults. *Journal of Structural Geology*, **21**, 1027–1038.

FERRILL, D.A., WINTERLE, J., WITTMEYER, G., SIMS, D., COLTON, S., ARMSTRONG, A. & MORRIS, A.P. 1999*b*. Stressed rock strains groundwater at Yucca Mountain, Nevada. *GSA Today*, **9**, 1–8.

FERRILL, D.A., MORRIS, A.P., WYRICK, D.Y., SIMS, D.W. & FRANKLIN, N.M. 2004. Dilational fault slip and pit chain formation on Mars. *GSA Today*, **14**, 4–12.

FERRILL, D.A., MORRIS, A.P., SIMS, D.W., WAITING, D.J. & HASEGAWA, S. 2005. Development of synthetic layer dip adjacent to normal faults. *In*: SORKHABI, R. & TSUJI, Y. (eds) *Faults, Fluid Flow, and Petroleum Traps*. American Association of Petroleum Geologists, Memoirs, **85**, 125–138.

FERRILL, D.A., MORRIS, A.P. & SMART, K.J. 2007. Stratigraphic control on extensional fault propagation folding: Big Brushy Canyon Monocline, Sierra Del Carmen, Texas. *In*: JOLLEY, S.J., BARR, D., WALSH, J.J. & KNIPE, R.J. (eds) *Structurally Complex Reservoirs*. Geological Society, London, Special Publications, **292**, 203–217, https://doi.org/10.1144/SP292.12

FERRILL, D.A., MORRIS, A.P. & MCGINNIS, R.N. 2009. Crossing conjugate normal faults in field exposures and seismic data. *American Association of Petroleum Geologists Bulletin*, **93**, 1471–1488.

FERRILL, D.A., MORRIS, A.P., MCGINNIS, R.N., SMART, K.J. & WARD, W.C. 2011*a*. Fault zone deformation and displacement partitioning in mechanically layered carbonates: the Hidden Valley fault, central Texas. *American Association of Petroleum Geologists Bulletin*, **95**, 1383–1397.

FERRILL, D.A., WYRICK, D.Y. & SMART, K.J. 2011*b*. Coseismic, dilational-fault and extension-fracture related pit chain formation in Iceland: analog for pit chains on Mars. *Lithosphere*, **3**, 133–142.

FERRILL, D.A., MCGINNIS, R.N., MORRIS, A.P. & SMART, K.J. 2012*a*. Hybrid failure: field evidence and influence on fault refraction. *Journal of Structural Geology*, **42**, 140–150.

FERRILL, D.A., MORRIS, A.P. & MCGINNIS, R.N. 2012*b*. Extensional fault-propagation folding in mechanically layered rocks: the case against the frictional drag mechanism. *Tectonophysics*, **576–577**, 78–85.

FERRILL, D.A., MCGINNIS, R.N. *ET AL*. 2014*a*. Control of mechanical stratigraphy on bed-restricted jointing and normal faulting: Eagle Ford Formation, south-central Texas, U.S.A. *American Association of Petroleum Geologists Bulletin*, **98**, 2477–2506.

FERRILL, D.A., MORRIS, A.P., HENNINGS, P.H. & HADDAD, D.E. 2014*b*. Faulting and fracturing in shale and self-sourced reservoirs: introduction. *American Association of Petroleum Geologists Bulletin*, **98**, 2161–2164.

FERRILL, D.A., MORRIS, A.P., MCGINNIS, R.N., SMART, K.J. & WATSON-MORRIS, M.J. 2016*a*. Observations on normal fault scarp morphology and fault system evolution in the Bishop Tuff in the Volcanic Tableland, Owens Valley, California, U.S.A. *Lithosphere*, in press, https://doi.org/10.1130/L476.1

FERRILL, D.A., MORRIS, A.P., WIGGINTON, S.S., SMART, K.J., MCGINNIS, R.N. & LEHRMANN, D. 2016*b*. Deciphering thrust fault nucleation and propagation and the importance of footwall synclines. *Journal of Structural Geology*, **85**, 1–11.

Fisher, Q.J. & Jolley, S.J. 2007. Treatment of faults in production simulation models. *In*: Jolley, S.J., Barr, D., Walsh, J.J. & Knipe, R.J. (eds) *Structurally Complex Reservoirs*. Geological Society, London, Special Publications, **292**, 219–233, https://doi.org/10.1144/SP292.13

Fossen, H., Johansen, T.E.S., Hesthammer, J. & Rotevatn, A. 2005. Fault interaction in porous sandstone and implications for reservoir management; examples from southern Utah. *American Association of Petroleum Geologists Bulletin*, **89**, 1593–1606.

Gartrell, A., Zhang, Y., Lisk, M. & Dewhurst, D. 2004. Fault intersections as critical hydrocarbon leakage zones: integrated field study and numerical modelling of an example from the Timor Sea, Australia. *Marine and Petroleum Geology*, **21**, 1165–1179.

Gawthorpe, R. & Hardy, S. 2002. Extensional fault-propagation folding and base-level change as controls on growth-strata geometries. *Sedimentary Geology*, **146**, 47–56.

Gawthorpe, R.L., Sharp, I., Underhill, J.R. & Gupta, S. 1997. Linked sequence stratigraphic and structural evolution of propagating normal faults. *Geology*, **25**, 795–798.

Germanovich, L. & Xu, W. 2004. Failure of marine sediments due to gas hydrate dissociation. Abstract #OS41C-0495 presented at the American Geophysical Union Fall Meeting 2004, 13–17 December, San Francisco, CA.

Grant, J.V. & Kattenhorn, S.A. 2004. Evolution of vertical normal faults at the surface at an extensional plate boundary, southwest Iceland. *Journal of Structural Geology*, **26**, 537–557.

Gray, M.B., Stamatakos, J.A., Ferrill, D.A. & Evans, M.A. 2005. Fault zone deformation process in Miocene tuffs at Yucca Mountain, Nevada. *Journal of Structural Geology*, **27**, 1873–1891.

Groshong, R.H., Jr. 1988. Low-temperature deformation mechanisms and their interpretation. *Geological Society of America Bulletin*, **100**, 1329–1360.

Gudmundsson, A. 1987. Tectonics of the Thingvellir fissure swarm, SW Iceland. *Journal of Structural Geology*, **9**, 61–69.

Hancock, P.L. 1985. Brittle microtectonics: principles and practice. *Journal of Structural Geology*, **7**, 437–457.

Handin, J. & Hager, R.V., Jr, 1957. Experimental deformation of sedimentary rocks under confining pressure: tests at room temperature on dry samples. *American Association of Petroleum Geologists Bulletin*, **41**, 1–50.

Hardy, S. & McClay, K. 1999. Kinematic modelling of extensional fault-propagation folding. *Journal of Structural Geology*, **21**, 695–702.

Heynekamp, M.R., Goodwin, L.B., Mozley, P.S. & Haneberg, W.C. 1999. Controls on fault-zone architecture in poorly lithified sediments, Rio Grande Rift, New Mexico: implications for fault-zone permeability and fluid flow. *In*: Haneberg, W.C., Mozley, P.S., Moore, J.C. & Goodwin, L.B. (eds) *Faults and Subsurface Fluid Flow in the Shallow Crust*. American Geophysical Union, Geophysical Monographs Series, **113**, 27–49.

Holland, M., Urai, J.L. & Martel, S. 2006. The internal structure of fault zones in basaltic sequences. *Earth and Planetary Science Letters*, **248**, 301–315.

Jackson, J.A. 1987. Active normal faulting and crustal extension. *In*: Coward, M.P., Dewey, J.F. & Hancock, P.L. (eds) *Continental Extensional Tectonics*. Geological Society, London, Special Publications, **28**, 3–17, https://doi.org/10.1144/GSL.SP.1987.028.01.02

Jackson, J.A. & White, N.J. 1989. Normal faulting in the upper continental crust: observations from regions of active extension. *Journal of Structural Geology*, **11**, 15–36.

Janecke, S.U., Vandenburg, C.J. & Blankenau, J.J. 1998. Geometry, mechanisms and significance of extensional folds from examples in the Rocky Mountain Basin and Range province, U.S.A. *Journal of Structural Geology*, **20**, 841–856.

Jin, G. & Groshong, R.H., Jr, 2006. Trishear kinematic modeling of extensional fault propagation folding. *Journal of Structural Geology*, **28**, 170–183.

Katz, O., Reches, Z. & Roegiers, J.-C. 2000. Evaluation of mechanical rock properties using a Schmidt Hammer. *International Journal of Rock Mechanics & Mining Sciences*, **37**, 723–728.

Khalil, S.M. & McClay, K.R. 2002. Extensional fault-related folding, northwestern Red Sea, Egypt. *Journal of Structural Geology*, **24**, 743–762.

Mandl, G. 1988. *Mechanics of Tectonic Faulting, Models and Basic Concepts*. Elsevier, Amsterdam.

Manighetti, I., King, G.C.P., Gaudemer, Y., Scholz, C.H. & Doubre, C. 2001. Slip accumulation and lateral propagation of active normal faults in Afar. *Journal of Geophysical Research*, **106**, 13,667–13,696.

Manzocchi, T., Childs, C. & Walsh, J.J. 2010. Faults and fault properties in hydrocarbon flow models. *Geofluids*, **10**, 94–113.

McGinnis, R.N., Morris, A.P., Ferrill, D.A. & Dinwiddie, C.L. 2009. Deformation analysis of tuffaceous sediments in the Volcanic Tableland near Bishop, California. *Lithosphere*, **1**, 291–304.

McGinnis, R.N., Ferrill, D.A., Morris, A.P. & Smart, K.J. 2016. Insight on mechanical stratigraphy and subsurface interpretation. *In*: Krantz, R.W., Ormand, C.J. & Freeman, B. (eds) *Earth, Mind, and Machine: 3D Structural Interpretation*. American Association of Petroleum Geologists, Hedberg Series, 6 (in press).

Melosh, H.J. 1990. Mechanical basis for low-angle normal faulting in the Basin and Range province. *Nature*, **343**, 331–335.

Micarelli, L. & Benedicto, A. 2008. Normal fault terminations in limestones from the SE-Basin (France): implications for fluid flow. *In*: Wibberley, C.A.J., Kurz, W., Imber, J., Holdsworth, R.E. & Collettini, C. (eds) *The Internal Structure of Fault Zones: Implications for Mechanical and Fluid-Flow Properties*. Geological Society, London, Special Publications, **299**, 123–138, https://doi.org/10.1144/SP299.8

Morris, A.P., Ferrill, D.A. & Henderson, D.B. 1996. Slip tendency analysis and fault reactivation. *Geology*, **24**, 275–278.

Morris, A.P., Ferrill, D.A. & McGinnis, R.N. 2009*a*. Fault frequency and strain. *Lithosphere*, **1**, 105–109.

MORRIS, A.P., FERRILL, D.A. & MCGINNIS, R.N. 2009*b*. Mechanical stratigraphy and faulting in Cretaceous carbonates. *American Association of Petroleum Geologists Bulletin*, **93**, 1459–1470.

MORRIS, A.P., MCGINNIS, R.N. & FERRILL, D.A. 2014. Fault displacement gradients on normal faults and associated deformation. *American Association of Petroleum Geologists Bulletin*, **98**, 1161–1184.

MURAOKA, H. & KAMATA, H. 1983. Displacement distribution along minor fault traces. *Journal of Structural Geology*, **5**, 483–495.

MYERS, R.D., ALLGOOD, A., HJELLBAKK, A., VROLIJK, P. & BRIEDIS, N. 2007. Testing fault transmissibility predictions in a structurally dominated reservoir: Ringhorne field, Norway. *In*: JOLLEY, S.J., BARR, D., WALSH, J.J. & KNIPE, R.J. (eds) *Structurally Complex Reservoirs*. Geological Society, London, Special Publications, **292**, 271–294, https://doi.org/10.1144/SP292.16

NICOL, A., GILLESPIE, P.A., WALSH, J. & CHILDS, C. 2002. Thrust relays and associated folding in layered sequences. *Journal of Structural Geology*, **24**, 709–727.

OFOEGBU, G.I. & FERRILL, D.A. 1998. Mechanical analyses of listric normal faulting with emphasis on seismicity assessment. *Tectonophysics*, **284**, 65–77.

OTSUKI, K. 1978. On the relationship between the width of shear zone and the displacement along fault. *Journal of the Geological Society of Japan*, **84**, 661–669.

PATTON, T.L., LOGAN, J.M. & FRIEDMAN, M. 1998. Experimentally generated normal faulting in single-layer and multilayer limestone specimens at confining pressure. *Tectonophysics*, **295**, 53–77.

PEACOCK, D.C.P. 2002. Propagation, interaction and linkage in normal fault systems. *Earth-Science Reviews*, **58**, 121–142.

PEACOCK, D.C.P. & SANDERSON, D.J. 1991. Displacements, segment linkage and relay ramps in normal fault zones. *Journal of Structural Geology*, **13**, 721–733.

PEACOCK, D.C.P. & SANDERSON, D.J. 1992. Effects of layering and anisotropy on fault geometry. *Journal of the Geological Society, London*, **149**, 793–802, https://doi.org/10.1144/gsjgs.149.5.0793

PEACOCK, D.C.P, & ZHANG, X. 1993. Field examples and numerical modelling of oversteps and bends along normal faults in cross-section. *Tectonophysics*, **234**, 147–167.

POWER, W.L., TULLIS, T.E. & WEEKS, J.D. 1988. Roughness and wear during brittle faulting. *Journal of Geophysical Research*, **93**, 15268–15278.

PRICE, N.J. 1966. *Fault and Joint Development in Brittle and Semi-Brittle Rock*. Pergamon Press, Oxford.

RAMSAY, J.G. 1967. *Folding and Fracturing of Rocks*. McGraw Hill, New York.

RAMSEY, J.M. & CHESTER, F.M. 2004. Hybrid fracture and the transition from extension fracture to shear fracture. *Nature*, **428**, 63–66.

RECHES, Z. & EIDELMAN, A. 1995. Drag along faults. *Tectonophysics*, **247**, 145–156.

ROBERTSON, E.C. 1982. Continuous formation of gouge and breccia during fault displacement. *In*: GOODMAN, R.E. & HEUZE, F.E. (eds) *Issues in Rock Mechanics; Proceedings of the 23rd Symposium on Rock*. American Institute of Mining and Engineering, New York, 397–404.

SCHLISCHE, R.W. 1995. Geometry and origin of fault-related folds in extensional settings. *American Association of Petroleum Geologists Bulletin*, **79**, 1661–1678.

SCHÖPFER, M.P.J., CHILDS, C. & WALSH, J.J. 2006. Localisation of normal faults in multilayer sequences. *Journal of Structural Geology*, **28**, 816–833.

SCHULTZ, R.A., OKUBO, C.H. & WILKINS, S.J. 2006. Displacement-length scaling relations for faults on the terrestrial planets. *Journal of Structural Geology*, **28**, 2182–2193.

SEEBECK, H., NICOL, A., WALSH, J.J., CHILDS, C., BEETHAM, R.D. & PETTINGA, J. 2014. Fluid flow in fault zones from an active rift. *Journal of Structural Geology*, **62**, 52–64.

SELLARDS, E.H. & BAKER, C.L. 1934. The geology of Texas, structural and economic geology. *University of Texas Bulletin*, **3401**, 2.

SHIPTON, Z.K. & COWIE, P.A. 2001. Damage zone and slip-surface evolution over μm to km scales in high-porosity Navajo sandstone, Utah. *Journal of Structural Geology*, **23**, 1825–1844.

SHIPTON, Z.K., SODEN, A.M., KIRKPATRICK, J.D., BRIGHT, A.M. & LUNN, R.J. 2006. How thick is a fault? Fault displacement-thickness scaling revisited. *In*: ABERCROMBIE, R., MCGARR, A., DI TORO, G. & KANAMORI, H. (eds) *Earthquakes: Radiated Energy and the Physics of Faulting*. American Geophysical Union, Geophysical Monograph Series, **170**, 193–198.

SIBSON, R.H. 1996. Structural permeability of fluid-driven fault-fracture meshes. *Journal of Structural Geology*, **18**, 1031–1042.

SMART, K.J., FERRILL, D.A. & MORRIS, A.P. 2009. Impact of interlayer slip on fracture prediction from geomechanical models of fault-related folds. *American Association of Petroleum Geologists Bulletin*, **93**, 1447–1458.

SMART, K.J., FERRILL, D.A., MORRIS, A.P., BICHON, B.J., RIHA, D.S. & HUYSE, L. 2010. Geomechanical modeling of an extensional fault-propagation fold: Big Brushy Canyon monocline, Sierra Del Carmen, Texas. *American Association of Petroleum Geologists Bulletin*, **94**, 221–240.

TREADGOLD, G., MCLAIN, B., SINCLAIR, S. & NICKLIN, D. 2010. Eagle Ford Shale prospecting with 3D seismic data within a tectonic and depositional system framework. *Bulletin of the South Texas Geological Society*, **LI**, 19–28.

WALLACE, W. 1861. *The Laws which Regulate the Deposition of Lead Ores in Veins: Illustrated by Examination of the Geologic Structure of the Mining Districts of Alston Moor*. Stanford, London.

WALSH, J.J. & WATTERSON, J. 1988*a*. Analysis of the relationship between displacements and dimensions of faults. *Journal of Structural Geology*, **10**, 239–247.

WALSH, J.J. & WATTERSON, J. 1988*b*. Dips of normal faults in British Coal Measures and other sedimentary sequences. *Journal of the Geological Society, London*, **145**, 859–873, https://doi.org/10.1144/gsjgs.145.5.0859

WALSH, J.J., WATTERSON, J. & YIELDING, G. 1991. The importance of small scale faulting in regional extension. *Nature*, **351**, 391–393.

WALSH, J.J., NICOL, A. & CHILDS, C. 2002. An alternative model for the growth of faults. *Journal of Structural Geology*, **24**, 1669–1675.

WELCH, M.J., DAVIES, R.K., KNIPE, R.J. & TUECKMANTEL, C. 2009*a*. A dynamic model for fault nucleation and propagation in a mechanically layered section. *Tectonophysics*, **474**, 473–492.

WELCH, M.J., KNIPE, R.J., SOUQUE, C. & DAVIES, R.K. 2009*b*. A Quadshear kinematic model for folding and clay smear development in fault zones. *Tectonophysics*, **471**, 186–202.

WERNICKE, B. 1995. Low-angle normal faults and seismicity: a review. *Journal of Geophysical Research*, **100**, 20159–20174.

WESTAWAY, R. 1999. The mechanical feasibility of low-angle normal fault. *Tectonophysics*, **308**, 407–443.

WHITE, I.R. & CRIDER, J.G. 2006. Extensional fault-propagation folds: mechanical models and observations from the Modoc Plateau, northeastern California. *Journal of Structural Geology*, **28**, 1352–1370.

WIBBERLEY, C.A.J., YIELDING, G. & DI TORO, G. 2008. Recent advances in the understanding of fault zone internal structure: a review. *In*: WIBBERLEY, C.A.J., KURZ, W., IMBER, J., HOLDSWORTH, R.E. & COLLETINI, C. (eds) *The Internal Structure of Fault Zones: Implications for Mechanical and Fluid-Flow Properties*. Geological Society, London, Special Publications, **299**, 5–33, https://doi.org/10.1144/SP299.2

WILKINS, S.J. & GROSS, M.R. 2002. Normal fault growth in layered rocks at Split Mountain, Utah: influence of mechanical stratigraphy on dip linkage, fault restriction and fault scaling. *Journal of Structural Geology*, **24**, 1413–1429.

WILLIAMS, G.D. & CHAPMAN, T.J. 1983. Strains developed in the hangingwalls of thrusts due to their slip/propagation rate: a dislocation model. *Journal of Structural Geology*, **5**, 563–571.

WISE, D.U., DUNN, D.E. *ET AL.* 1984. Fault-related rocks: suggestions for terminology. *Geology*, **13**, 391–394.

WITHJACK, M.O., OLSON, J. & PETERSON, E. 1990. Experimental models of extensional forced folds. *American Association of Petroleum Geologists Bulletin*, **74**, 1038–1054.

WYRICK, D.Y., MORRIS, A.P. & FERRILL, D.A. 2011. Normal fault growth in analog models and on Mars. *Icarus*, **212**, 559–567.

ZOBACK, M.D. 2007. *Reservoir Geomechanics*. Cambridge University Press, Cambridge.

Growth of layer-bound normal faults under a regional anisotropic stress field

R. GHALAYINI[1,2,3]*, C. HOMBERG[1,2], J. M. DANIEL[3] & F. H. NADER[3]

[1]*Sorbonne Universités, UPMC Univ Paris 06, UMR 7193, ISTeP, F-75005, Paris, France*

[2]*CNRS, UMR 7193, ISTEP, F-75005, Paris, France*

[3]*IFP Energies nouvelles, 1 et 4 avenue de Bois-Préau, 92852 Rueil-Malmaison, France*

**Correspondence: Ramadan.ghalayini@ifpen.fr*

Abstract: Layer-bound normal faults commonly form polygonal faults with fine-grained sediments early in their burial history. When subject to anisotropic stress conditions, these faults will be preferentially oriented. In this study we investigate how faults grow, evolve and interact within regional-scale layer-bound fault systems characterized by parallel faults. The intention is to understand the geometry and growth of faults by applying qualitative and quantitative fault analysis techniques to a 3D seismic reflection dataset from the Levant Basin, an area containing a unique layer-bound normal fault array. This analysis indicates that the faults were affected by mechanical stratigraphy, causing preferential nucleation sites of fault segments, which were later linked. Our interpretation suggests that growth of layer-bound faults at a basin scale generally follows the isolated model, accumulating length proportional to displacement and, when subject to an anisotropic regional stress field, resembling to a great extent classical tectonic normal faults.

Layer-bound normal faults are extensional structures confined to a particular sedimentary unit without propagating to the underlying and overlying layers, hence their name 'layer bound'. Such confinement to discrete stratigraphic units has been well documented in the field (e.g. Peacock & Zhang 1993; Gross 1995; Gross *et al.* 1997; Wilkins & Gross 2002; Roche *et al.* 2012*b*) and can occur when the stress or lithological conditions in adjacent stratigraphic units are unfavourable for vertical fault propagation into those units. A particular type of layer-bound fault is the polygonal fault system (PFS). Different mechanisms may explain their nucleation, including volumetric contraction, compaction, diagenetic shear failure and pressure variations (Henriet *et al.* 1991; Cartwright & Lonergan 1996; Cartwright & Dewhurst 1998; Dewhurst *et al.* 1999; Shin *et al.* 2008; Cartwright 2011). In this paper, such faults are referred to as layer-bound normal faults with no reference to a particular mechanism. Developed primarily in fine-grained sediments early in their burial history (Cartwright & Lonergan 1996; Cartwright & Dewhurst 1998; Dewhurst *et al.* 1999; Davies & Ireland 2011), this type of normal fault is mainly characterized by its distinctive polygonal planform geometry (Cartwright *et al.* 2003; Cartwright 2011). However, local re-orientation of the faults has been observed, especially in proximity to other geological structures or toward the end of the system over a slope (Rank-Friend & Elders 2004; Stewart 2006; Carruthers *et al.* 2013). PFS were also documented to be linear when the bedding is tilted, causing gravity-driven deformation (Ireland *et al.* 2011). Hence PFS do not always have the perfect polygonal planform geometry and might be linear, at least at a restricted scale.

The distinctive planform geometry of PFS is interpreted as the result of an isotropic horizontal stress field and, for this reason, when polygonal faults are close to other structures, such as salt domes, local stress-field variation causes a re-orientation of these faults by making them linear (Cartwright 2011; Carruthers *et al.* 2013). As opposed to other studies documenting a basin-scale re-orientation of layer-bound normal faults due to gravity sliding and dipping of beds (Higgs & McClay 1993; Ireland *et al.* 2011), the sole impact of regional anisotropic stress fields has never been investigated before. What would be the final geometry of a PFS where a regional anisotropic horizontal stress field, not a local one, is imposed? Would these faults become perfectly linear and well oriented throughout the entire basin? How would these faults grow through time, and what are the factors affecting their evolution? These questions are important since no PFS has ever been documented to be well oriented at a regional scale due to a regional tectonic stress field. In order to answer these questions, we investigate the evolution and growth of a unique

From: Childs, C., Holdsworth, R. E., Jackson, C. A.-L., Manzocchi, T., Walsh, J. J. & Yielding, G. (eds) 2017. *The Geometry and Growth of Normal Faults*. Geological Society, London, Special Publications, **439**, 57–78.
First published online April 6, 2016, https://doi.org/10.1144/SP439.13

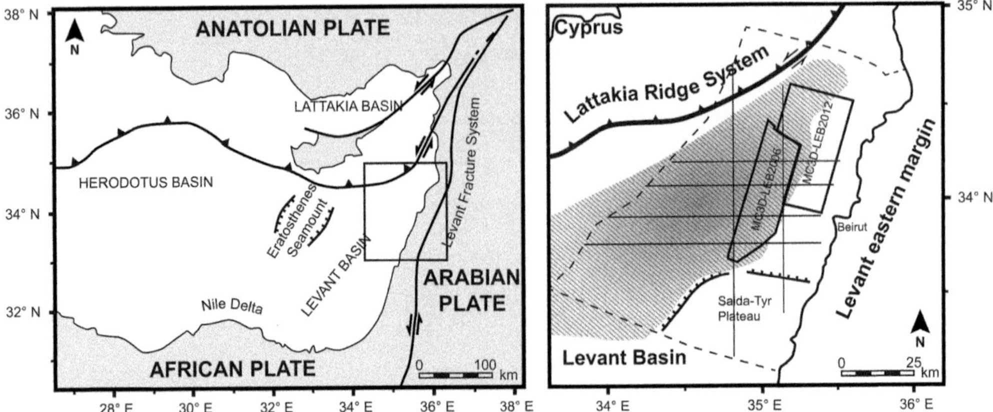

Fig. 1. Simplified structural map showing the location of the Levant Basin in the Eastern Mediterranean and the extent of the studied normal fault array. The study area is located offshore Lebanon and benefits from two seismic cubes and six 2D seismic lines, courtesy of PGS and the Lebanese Petroleum Administration.

layer-bound normal fault system, found in the Levant Basin (Dupin *et al.* 2012; Kosi *et al.* 2012; Montadert, pers. comm. February 2013) (Fig. 1). This NW-trending normal fault array is present in a region where NW–SE compression is attested during the Oligo-Miocene (Ghalayini *et al.* 2014; Montadert *et al.* 2014) in relation to the collision of Arabia with Eurasia. The presence of such exceptionally preserved layer-bound normal faults makes this study area unique in terms of testing the displacement patterns, growth and interaction of faults. In particular, it can also be used to test the applicability of the isolated and coherent fault models (e.g. Mansfield & Cartwright 2001; Walsh *et al.* 2002) to the growth of layer-bound faults, and the effect of mechanical stratigraphy on the evolution of a regional normal fault array. For this study, we used recently acquired high-quality 3D seismic reflection data from the Levant Basin offshore Lebanon. We applied qualitative and quantitative fault analysis techniques, such as displacement distribution analysis on fault planes, to understand the growth of a layer-bound normal fault system exhibiting classical polygonal planform geometry in a small region in the basin, and well-oriented and linear planform geometry in the major part of the Levant Basin. An improved knowledge of the geometry and growth of this fault system may also have industrial implications since these faults are likely conduits for hydrocarbon migration from deeper source rocks and are potential seals for Miocene reservoirs (Kosi *et al.* 2012).

After an introduction to the regional context of the Levant Basin and a definition of the applied methodology, we describe the normal fault array by looking at the geometry, displacement profiles and thickness variations along fault planes. Based on this description, we then propose a model for the timing and growth of faults in the Levant Basin. Finally, we consider the implications of our study for normal fault growth in general and for the growth of layer-bound fault systems in particular.

Regional framework

The Levant Basin is located in the easternmost part of the Mediterranean Sea (Fig. 1). It is bordered to the east by the Arabian plate, to the north by the Lattakia Ridge system, to the west by the Eratosthenes seamount and to the south by the Nile Delta and the African plate. The basin was formed through polyphase rifting in the Permo-Jurassic (e.g. Garfunkel 2004; Homberg *et al.* 2009; Gardosh *et al.* 2010). Its geological setting is marked by the evolution of the Neotethys and the interaction of the Arabian plate with the Anatolian plate. In the Late Cretaceous, convergence tectonics caused the inversion of previous extensional structures and contraction of the Levant Basin. In the Neogene, collision of Arabia with Anatolia established a NW–SE contractional stress field in the basin and uplift along the margins (Barrier & Vrielynck 2008). Extensive erosion of Upper Eocene and Oligocene units in Lebanon (Dubertret 1955; Hawie *et al.* 2014), together with angular unconformities between Lutetian and Burdigalian (Dubertret 1955; Boudagher-fadel & Clark 2006; Hawie *et al.* 2013), attest to a regional event with NE–SW initiation of folding due to this NW–SE compressional stress field (Homberg *et al.* 2010). NW-trending Oligo-Miocene canyons indicate a margin slope dipping toward the NW during the Miocene (Gardosh *et al.* 2008).

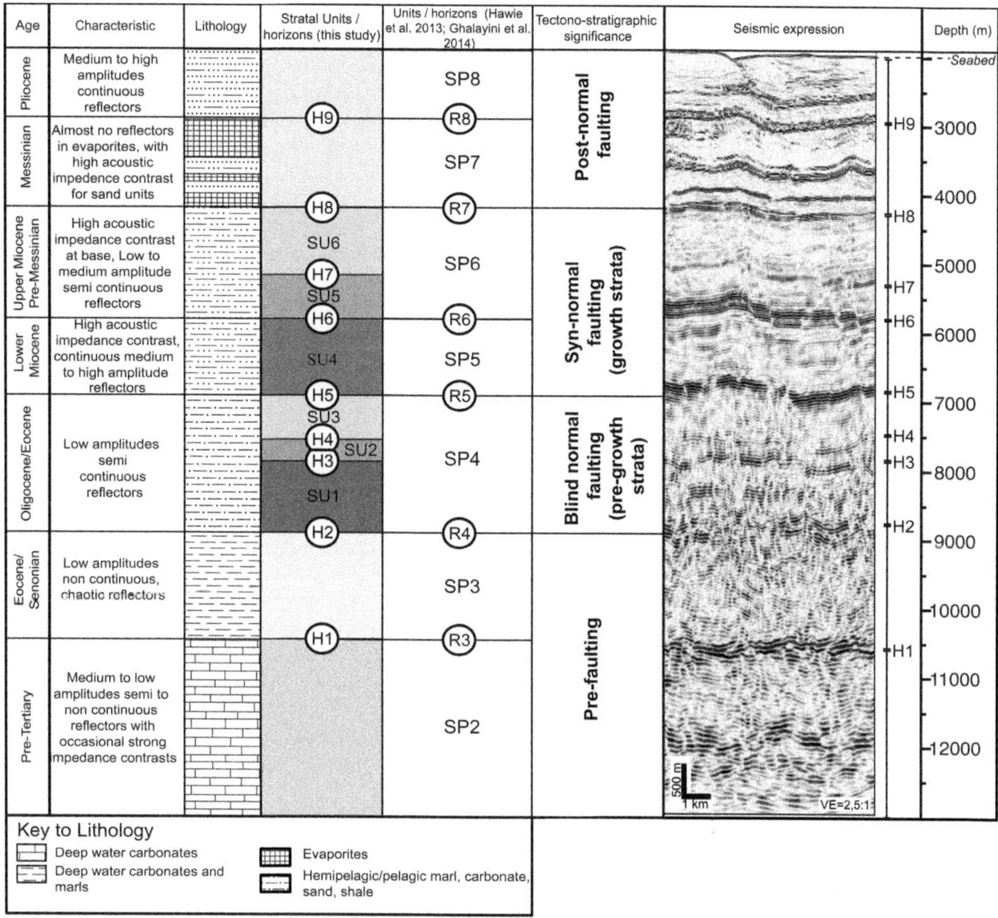

Fig. 2. Local stratigraphic column showing the lithostratigraphy, key lithologies and seismic stratigraphic framework of the Cretaceous to Tertiary succession in the Levant Basin. SU, stratal units; H, horizon; R, reflector; SP, seismic package. This study focuses on the Oligo-Miocene unit, which was subdivided into six stratal units. Horizon interpretation is from Hawie *et al.* (2013) and Ghalayini *et al.* (2014).

The Miocene saw the formation of the Levant Fracture System (LFS), a sinistral transform fault system forming the plate limits between the Arabian and African plates. The central part of the LFS in Lebanon consists of a restraining bend formed from a multitude of folds and faults (Gomez *et al.* 2007). Recent investigation of the Levant Basin showed that the margin offshore Lebanon is affected by ENE–WSW dextral strike-slip faults synchronous with the activity on the restraining bend. These faults are still active today, whereas the NE–SW anticlines were folded in the Late Miocene prior to the Messinian event (Ghalayini *et al.* 2014). During the Messinian, desiccation of the Mediterranean Sea resulted in deposition of a 2 km thick evaporite unit, commonly known as the Messinian evaporite sequence (Bowman 2011; Lie *et al.* 2011; Hawie *et al.* 2013) (Fig. 2).

Dataset and methodology

This study is based on two high-quality 3D seismic surveys, MC3D-LEB2006 covering 1700 km^2 and MC3D-LEB2012 covering 2600 km^2, located offshore Lebanon; and seven 2D seismic lines, courtesy of Petroleum Geo-Services (PGS) (Fig. 1). The MC3D-LEB2006 survey is time migrated and was acquired in 2006 using six streamers and two 3090 cubic inch air guns positioned at a depth of 6 m in water depth ranging between 1.5 and 2 km. Streamer length is 6000 m long at a spacing of

12.5 m and 25 m shot point intervals. Frequency ranges are between 40 and 80 Hz, giving a tuning thickness of about 5 m in the Pliocene and 15 m in the Miocene. We used classical time-to-depth conversion based on velocity maps from different stratigraphic intervals. The two-way time thicknesses of intervals were depth converted after picking horizons from stack velocities provided by PGS.

The MC3D-LEB2012 survey is depth migrated and was acquired in 2012 using 12 streamers and two 4135 cubic inch air guns positioned at a depth of 6 m in water depth ranging between 1.5 and 2 km. Streamer length is 7050 m long at a spacing of 12.5 and 25 m shot point intervals. Frequency ranges are between 40 and 80 Hz, giving a tuning thickness of about 5 m in the Pliocene and 15 m within the Miocene.

Fault and horizon interpretation was conducted on a Linux workstation using Paradigm's Gocad and Skua software. The following nine horizons were mapped in the seismic surveys (Fig. 2): (i) H1 (Senonian unconformity); (ii) H2 (Eocene unconformity); (iii) H3; (iv) H4; (v) H5 (base Miocene); (vi) H6 (base mid-Miocene); (vii) H7; (viii) H8 (base Messinian) and (ix) H9 (base Pliocene). Ages can be assigned only for H1, H2, H5, H6, H8 and H9 by regional correlations with wells from Israel (Hawie *et al.* 2013; Ghalayini *et al.* 2014) (Fig. 2). The remaining horizons were essential to constrain timing and displacement along the normal faults. They were chosen based on their strong impedance contrast and the possibility of tracking them through the whole seismic surveys.

More than 200 faults were mapped in the course of this study, and 20 were subject to quantitative analysis of displacement. These faults were chosen based on the following criteria: (i) excellent continuity and mapping of the fault plane; (ii) ability to correlate clear and strong reflectors across the fault plane; (iii) isolation with respect to nearby structures such as strike-slip faults; and (iv) ability to map fault plane entirely by avoiding, when possible, faults cut by the data-bounding box. Among the 20 selected faults, 13 were considered as isolated faults with no horizontal relay with other faults and three segmented faults consisting of a coherent system containing two or three soft-linked segments.

Geometrical and temporal evolution of faults was studied through displacement analysis along fault planes. In order for displacement to be fully recorded and growth history to be properly constrained, sediment deposition rate should be higher than the rate of growth of the fault throw at the seabed. In the case of the Levant Basin, the Oligo-Miocene was a time when an increased amount of sediment was eroded from the uplifted Levant margin and deposited in the basin (Hawie *et al.* 2013). This has resulted in an excellent preservation of the growth history of the Oligo-Miocene normal fault array and allowed full recording of the growth history of the faults. In this study, the techniques used to investigate and constrain the temporal and spatial evolution of these faults consist of: (i) displacement/length analysis – this is used to establish the geometrical evolution of faults (e.g. Walsh & Watterson 1987, 1989; Cartwright *et al.* 1995; Dawers & Anders 1995; Baudon & Cartwright 2008*a*, *b*); (ii) displacement–depth analysis – this provides insights into the role of dip linkage and vertical growth of faults (e.g. Walsh & Watterson 1988; Nicol *et al.* 1996; Rykkelid & Fossen 2002; Baudon & Cartwright 2008*c*); (iii) expansion index (EI) – this is used to constrain the initiation of activity on fault systems by measuring thickness variations along fault planes (e.g. Thorsen 1963; Cartwright *et al.* 1998). This paper attempts to give reference to the significance of displacement analysis in understanding the growth and evolution of faults.

Differential compaction along normal faults is likely to alter displacement along fault planes, potentially leading to underestimation of the total throw. Since we are principally interested in the overall trend of displacement and variation of throw rather than the absolute throw, we have chosen not to correct for this phenomenon in this study. The displacement measurements are all taken on depth-converted profiles and not in two-way time.

Description of the normal fault array

Distribution in the basin

The normal fault array is regional and found in the whole Levant Basin in an area spanning *c.* 70 000 km^2. The faults are observed offshore Israel and Cyprus, and accumulate the biggest displacement offshore Lebanon (Dupin *et al.* 2012; Kosi *et al.* 2012; PGS, pers. comm. August 2013). They disappear along Levant's eastern margin, on top of the Saida-Tyr plateau (STP) (Ghalayini *et al.* 2014), and on the Eratosthenes seamount (Fig. 1). The available dataset allowed detailed mapping of the faults in only two 3D seismic surveys (Fig. 3).

Superposition of the distribution of these faults in the northern Levant Basin offshore Lebanon on isopach maps of the Oligocene and Miocene units shows a correlation between the thickness of the units and the distribution of these faults (Ghalayini *et al.* 2014) (Fig. 4). They are predominantly developed where this sedimentary package is relatively thicker, mainly toward the basin's centre. At the Levant's eastern margin and over the STP, the thickness of the Oligo-Miocene is substantially reduced and the normal faults disappear. Although the NW trend of these faults is marked in the basin, distinctive characteristics are underlined in this study.

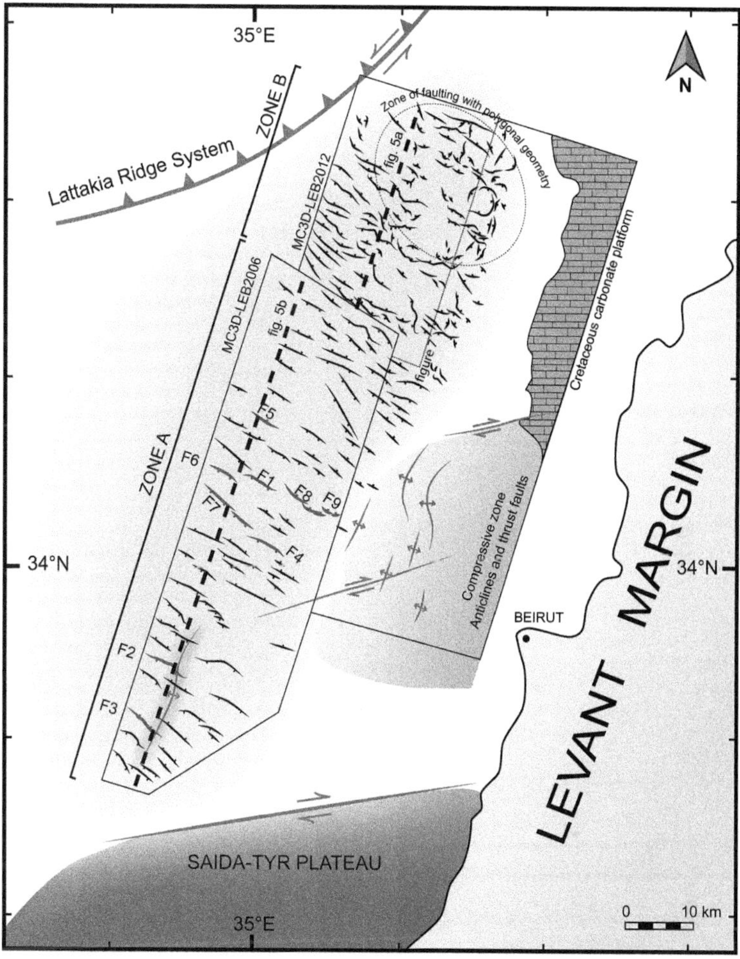

Fig. 3. Schematic representation of the base Miocene horizon (H5) showing the NW orientation of the normal faults in the study area. Faults acquire a gentle polygonal planform geometry close to the Lattakia Ridge in the NW corner of the survey, while in the rest of the basin the faults are linear and well oriented. Investigated NW-trending normal faults in this study are annotated and marked with a bolder line.

Seismic stratigraphic framework

The normal fault array in the northern Levant Basin is confined within the Oligo-Miocene sedimentary units (SU1 to SU6; Fig. 2). The base of the Oligo-Miocene unit, H2, is referred to as the Eocene unconformity (Lie *et al.* 2011; Kosi *et al.* 2012). The Oligocene (SU1, SU2 and SU3) units are believed to represent stacked deep-water clastic deposits in early lowstands at the base of the unit, transitioning upward to pelagic/hemipelagic sedimentation in deep-water settings (Hawie *et al.* 2013; Montadert *et al.* 2014). A strong impedance contrast marks this transition whereby SU1 and SU3 are characterized by low- to medium-amplitude reflectors. The SU2 unit is differentiated from SU1 and SU3 by a set of five strong and continuous reflectors across the entire basin.

Strong impedance contrast along H5 separates the Miocene from the Oligocene unit. SU4, SU5 and SU6 are characterized by high-amplitude, moderate continuity reflectors, representing stacked deep-water clastic deposits in early lowstand (Gardosh *et al.* 2008; Hawie *et al.* 2013; Montadert *et al.* 2014). They are separated by strong impedance contrasts along H6 and H7. Toward Lattakia Ridge, reflectors within SU5 and SU6 show increased amplitude, indicating lateral facies changes to the north of the basin (Hawie *et al.* 2013, fig. 5). For the purpose of this study, we separate the Oligo-Miocene units into the Oligocene 'lower' tier (SU1, SU2 and SU3) and the Miocene 'upper' tier

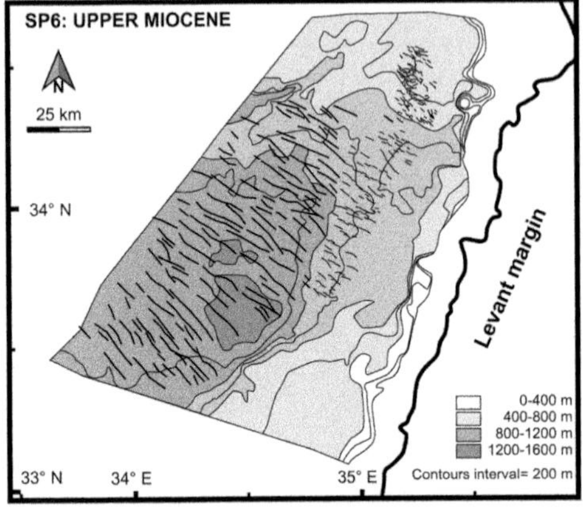

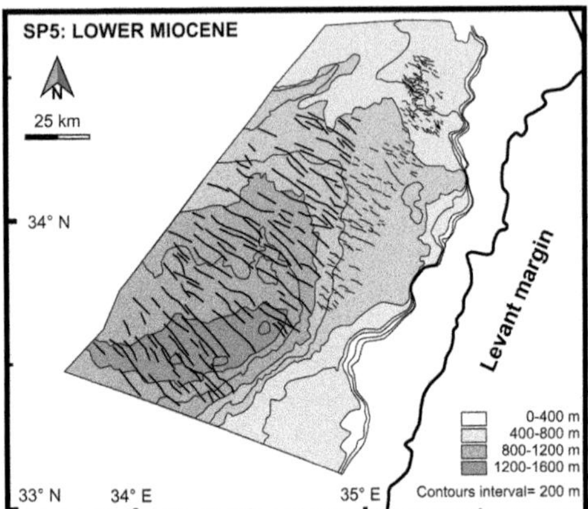

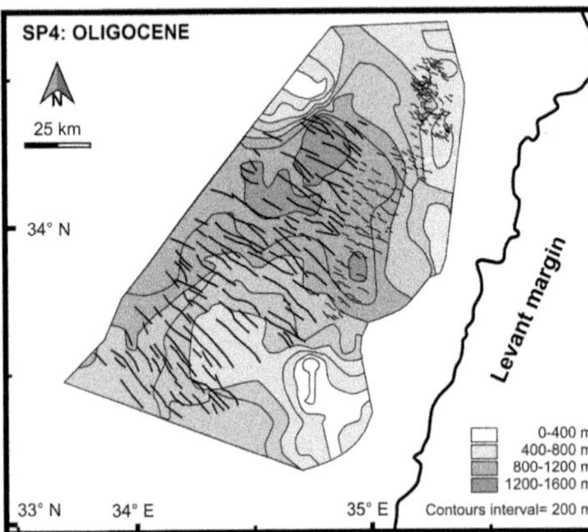

Fig. 4. Isopach maps of SP4, SP5 and SP6 with superposed fault mapping. The well-oriented normal faults are found where the Oligo-Miocene units are the thickest in the basin. The faults with polygonal planform geometry are restricted to the north of the basin where they occur in relatively thinner Oligo-Miocene thicknesses. Fault mapping is from Dupin *et al.* (2012). SP5 and SP6 isopach maps are modified from Ghalayini *et al.* (2014) and SP4 isopach map is modified from Hawie *et al.* (2013).

(SU4, SU5 and SU6). H8 consists of the upward boundary of the normal faults and is characterized by a strong impedance contrast marking the transition between the Miocene hemipelagic facies and the Messinian evaporite unit.

In summary, the normal faults are found in the Oligo-Miocene units only, which are dominantly composed of fine-grained sediments. They die out at the base Messinian and the Eocene unconformity horizon, which are correlated over the mapped area with variations in sedimentary facies from the deep basin toward the Lattakia Ridge.

Geometry

The normal fault array in the northern Levant Basin can be divided in two regions based on fault characteristics (orientation, length and maximum displacement). The first (zone A) consists of well-oriented NW-trending normal faults, while the other (zone B) consists of curved normal faults with a polygonal planform geometry (Fig. 3).

Zone A. The NW-trending normal faults are found in zone A and dip between 45 and 60°, towards either the NE or SW (Fig. 5). All faults are typically 4–6 km long with an average aspect ratio of 1 (<1.3). The shape of fault planes in zone A is thus more or less square, in contrast with the ellipsoidal shape expected in ideal unrestricted normal faults (Rippon 1984; Barnett *et al.* 1987; Walsh & Watterson 1988) (Fig. 6). Faults at the seismic scale have a spacing of 3–5 km and in some places seem to interact with, or cross-cut, other structures in the basin, such as strike-slip faults and anticlines (e.g. Ghalayini *et al.* 2014). Some faults in zone A interact with each other, forming segmented faults with relay bends (Kosi *et al.* 2012) (Fig. 3).

The NW-trending normal faults are layer bound. They are found exclusively in the interpreted Oligo-Miocene units, dying out at the Eocene unconformity horizon (H2) at their lower tip (Fig. 5). At their upper tip, the faults die out in the Messinian unit. Contrary to horizon H2, the base Messinian horizon (H8) shows a small but varying amount of displacement along fault scarps, together with signs of erosion in the footwall (Fig. 5). The strong and thick reflectors at the bottom of the Messinian unit show occasional minor folding above H8.

Zone B. The normal faults with a polygonal planform geometry are found in zone B and are best illustrated at H5 (base Miocene) in the MC3D-LEB2012 seismic cube (Fig. 3). These faults are 2–3 km long, and have a spacing of 1–2 km and a dip of 60°. They show a curved pattern (Fig. 7), resembling an open, immature polygonal fault

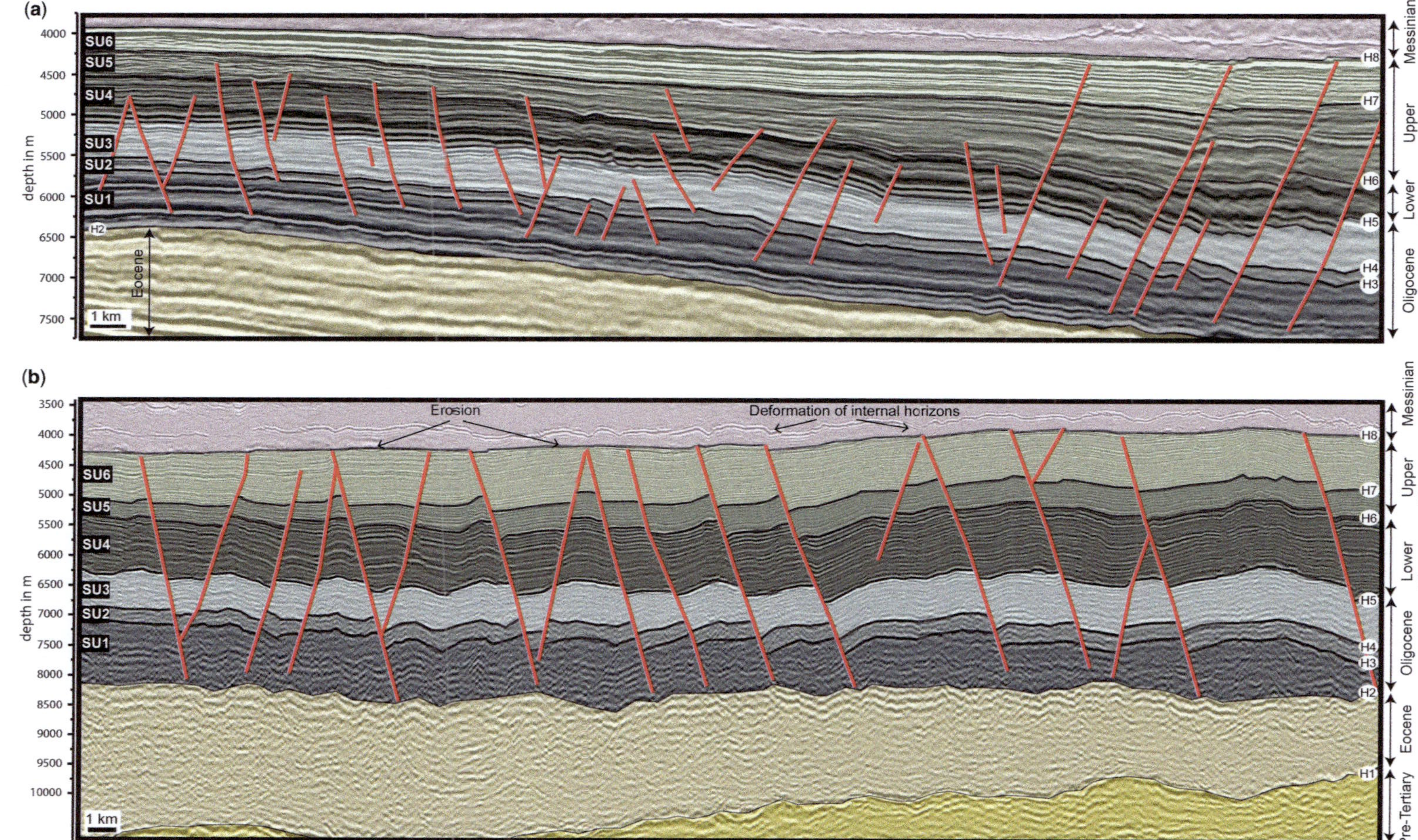

Fig. 5. Seismic lines in depth showing in (**a**) the short and small faults with polygonal planform geometry in zone B and in (**b**) the large contractional normal faults of zone A. The seismic lines were taken from the 3D seismic cubes. The seabed depth is at 1800 m depth and the top Messinian is at 2300 m depth. For location, see Figure 3.

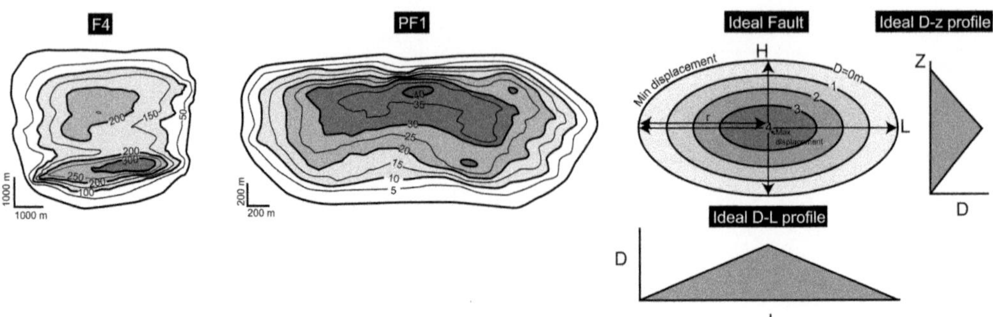

Fig. 6. Schematic sketch comparing the normal fault in the Levant Basin in zone A, normal fault in zone B and an ideal normal fault. Note the presence of different aspect ratios between the faults of zones A and B, whereby the latter is closer to an ideal normal fault, while the former contains two displacement maxima. Notice also the difference in the scale between the two faults.

system with a high proportion of unrestricted lateral tips (e.g. Cartwright *et al.* 2003; Cartwright 2011).

These faults are predominant in SU3 and extend to SU1, SU2 and SU4 (Fig. 5). They reach neither H2 (Eocene unconformity) nor H8 (base Messinian). Some faults are found in only the Oligocene tier and others also extend upward to the Miocene tier, dying out before reaching H6 (base mid-Miocene), and terminating in monoclinal tip-line folds. Depth sections at different horizons (Fig. 7) illustrate well their distribution within the stratal units and their relative orientations.

Thickness variations along fault planes

In order to understand the growth history along normal faults, expansion indices (EI = ratio between hanging-wall unit thickness and its correlative footwall unit thickness) plots were constructed following the method of Thorsen (1963) and Childs *et al.* (2003) (Fig. 8). These methods allow us to constrain the timing of activity along normal fault planes by comparing thickness variations of the same units between the hanging wall and footwall of faults.

In zone A, the calculated ratios indicate increased hanging-wall thickness for SU4, SU5 and SU6 (interpreted Miocene units). In contrast, the underlying SU2 and SU3 (interpreted Oligocene units) have an overall constant thickness, while SU1 shows a negative ratio. By looking in detail, very few faults show minor thickening along SU3.

In zone B, EI plots were constructed along 100 faults. All faults show no hanging-wall thickening in SU1, SU2 and SU3, while only ten show minor thickening in SU4.

Displacement v. length relationship

D_{max}–L diagrams for 100 faults in zone A and B were constructed on log–log and normal plots (Fig. 9). Most of the faults used in this plot are single slip surface and are isolated, i.e. not forming relay geometries with nearby faults. We also included six segmented faults (F7, F8 and F9) composed of two or three soft-linked segments forming relay bend geometries.

The log–log plot shows a linear relationship between the length of the faults in zone A and their displacement (Fig. 9). This trend follows the scaling relation $D = 0.03L^{1.02}$ which is almost the same as the best-fit curve $D = 0.03L^{1.06}$ calculated by Schlische *et al.* (1996) by compiling a global dataset from published materials (Fig. 9a). On the normal plot, some scatter is observed in the data of zone A, as faults longer than 8 km are underdisplaced for their lengths (Fig. 9b).

In contrast, the faults in zone B follow a different trend with a best-fit line $D = 0.64L^{0.53}$ in the log–log plot (Fig. 9a) with a slope value of 0.53, close to the one calculated by Nicol *et al.* (2003) using a polygonal fault array found in Lake Hope Australia. In the normal plot, the faults of zone B follow a best-fit line of $y = 0.0083x + 22$, accumulating little displacement with increased length.

3D variations in throw and throw gradients

Fault displacement analysis is an important tool allowing us to document how faults aggregate displacement when propagating through time. Fundamental aspects of fault growth – like barriers to vertical propagation, fault interaction, and role of lithology (mechanical stratigraphy) – have been recognized based on throw minima, slope inclination (gradients) and overall shape of displacement profiles constructed along faults worldwide (e.g. Watterson 1986; Barnett *et al.* 1987; Walsh & Watterson 1987; Cowie & Shipton 1998; Roche *et al.* 2012*b*; Morris *et al.* 2014). When analysed in 3D, such as looking at displacement–length (D–L)

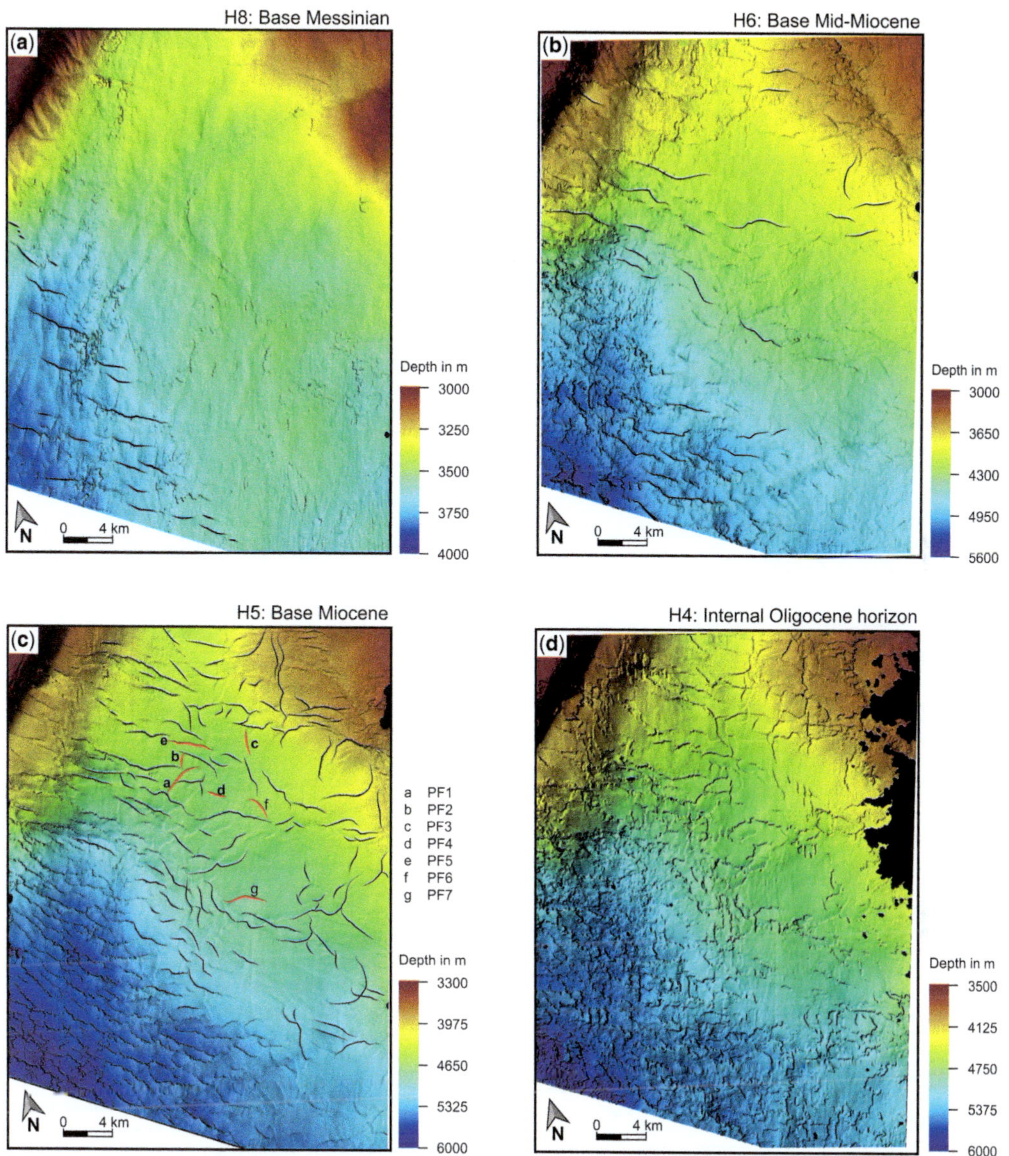

Fig. 7. Depth structure maps of (**a**) base Messinian showing the presence of limited NW-trending normal faults cross-cutting this horizon; (**b**) base mid-Miocene which is cross-cut by normal faults trending NW–SE; (**c**) base Miocene showing polygonal geometry of normal faults cross-cutting this horizon, and (**d**) an internal horizon in the Oligocene unit showing the distribution of normal faults with polygonal geometry, found in only the Oligocene unit. Thus, polygonal faults disappear in the Miocene unit, which is characterized by only NW-trending normal faults. For location, see Figure 3.

and displacement–depth ($D–z$) relationships, this analysis becomes even more important since it provides an accurate tool to understand the spatial and temporal evolution of the fault system. These aspects are detailed in the following paragraphs for 13 faults in zone A and 7 faults in zone B.

Zone A. D–L profiles were constructed within the six stratal units (SU1 to SU6) in order to map the variation of displacement within every tier. We chose to show profiles of SU3 and SU4 only as they are representative of the Oligocene and Miocene tiers, respectively (Figs 10 & 11). The seismic

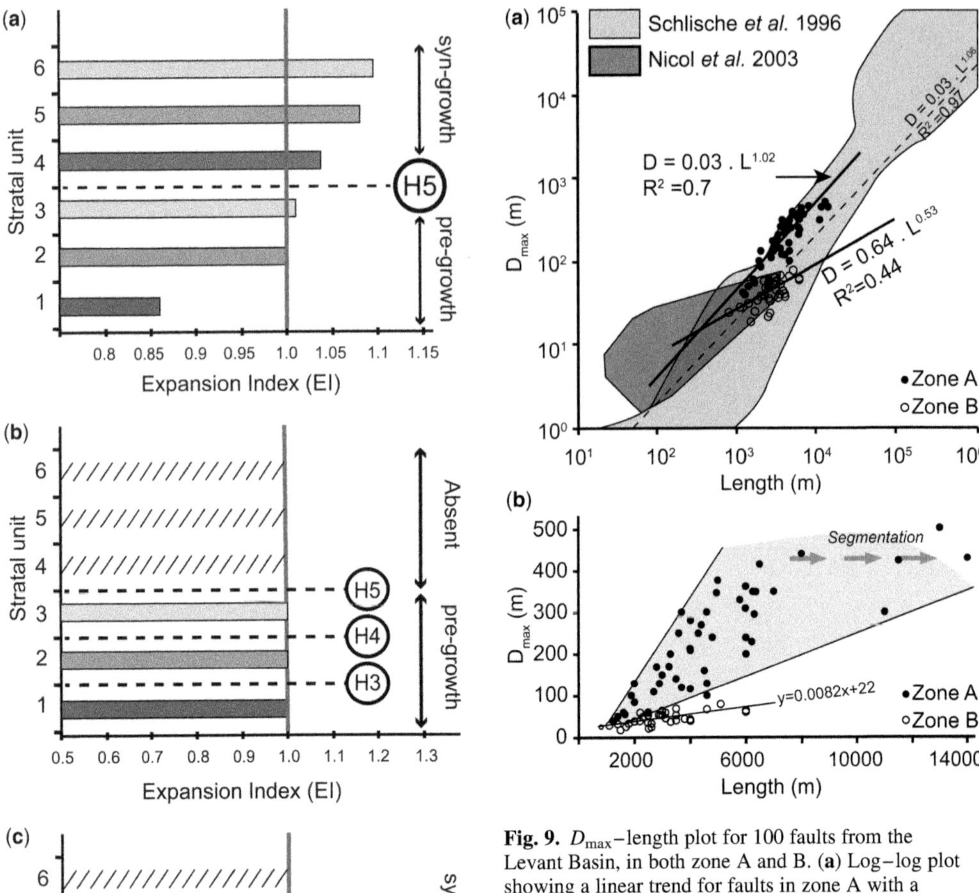

Fig. 8. Expansion index (EI) plots for faults. EI is the ratio between hanging-wall unit thickness and its correlative footwall unit thickness. Values <1 and >1 imply stratigraphic thinning and thickening in the hanging wall, respectively. (**a**) Average expansion index (EI) plots for faults in zone A showing thickening of units in SU4. Minor thickening is also noticed in SU3 over some faults. (**b**) Average EI plot for faults contained in only the Oligocene tier (SU1, SU2 and SU3) of zone B. (**c**) Average EI plot for faults that are found in both the Miocene and Oligocene tiers (SU1, SU2, SU3 and SU4) of zone B. For (b) and (c), a total of 100 faults were averaged.

Fig. 9. $D_{\max}$–length plot for 100 faults from the Levant Basin, in both zone A and B. (**a**) Log–log plot showing a linear trend for faults in zone A with a best-fit line of $D = 0.03L^{1.02}$ very close to the trend line of Schlische *et al.* (1996) over a global dataset. The faults in zone B follow a linear trend with a best-fit line of $D = 0.64L^{0.53}$ which has a slope close to the one suggested by Nicol *et al.* (2003) of 0.6. Plots can be compared with those of Schlische *et al.* (1996) and Nicol *et al.* (2003) (shaded areas). Best-fit line for data of zone A is shown as a heavy solid line while the best-fit line of Schlische *et al.* (1996) is shown as a dashed line. (**b**) Normal plot for all faults in this study showing the effect of segmentation in zone A creating faults which are underdisplaced for their lengths.

sections were spaced every 100 m to map in detail the variation of displacement. D–z profiles were constructed to investigate vertical variation of displacement along faults and their associated gradients (Fig. 12). The profiles were constructed at the location of maximum displacement along the faults in both zones.

The maximum displacement recorded among investigated faults is 415 m, with a $D_{\max}$ average of 200 m (Fig. 10). Faults shorter than 5 km display a C-shape profile (e.g. Muraoka & Kamata 1983),

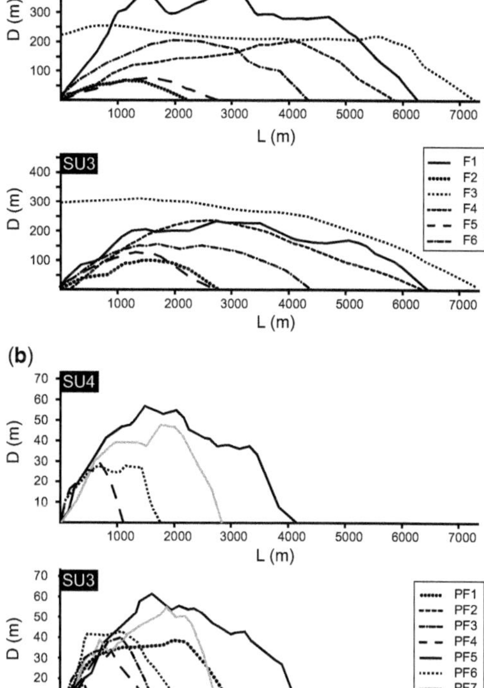

Fig. 10. (**a**) Displacement–length (D–L) profiles of six faults in zone A taken along SU3 (Miocene) and SU4 (Oligocene). These profiles are flat-topped for lengths larger than 5000 m. (**b**) D–L profiles of seven faults in zone B taken along SU3 (Oligocene) and SU4 (Miocene). These profiles also look flat-topped.

whereas faults longer than 5 km are generally flat-topped with an almost uniform throw distribution for >50% of the profile (Fig. 10). This observation is valid for both the Oligocene and Miocene tiers. The average lateral displacement gradient within SU3 and SU4 was calculated for all the investigated faults (Fig. 13). Away from the central part of the fault, displacement gradients vary between 0.1 and 0.13 towards both the SE and NW tips. A relatively high lateral displacement gradient (0.18) is observed particularly in SU5.

The D–z profiles are always (e.g. Muraoka & Kamata 1983) M-shaped and are characterized by two distinctive displacement maxima along the fault plane at two separate structural levels: (i) in the Miocene tier, and (ii) in the Oligocene tier. These displacement maxima are separated by distinct displacement minima that occur close to H5 (base Miocene), even if the folding component due to breached monocline geometries (Fig. 14) has been included in the displacement measurement. Throw gradients were calculated on the upper and lower tips of D–z profiles (Fig. 15). The constructed throw gradients are characterized by: (i) average throw gradient of 0.2 along the upper tip of faults, and (ii) average throw gradient of 0.37 along the lower tip of faults. These values are close to the upper bound of the published ones, sometimes exceeding them (e.g. Peacock & Sanderson 1991; Nicol *et al.* 1996; Roche *et al.* 2012*a*, *b*). This is better illustrated for the lower tips, which show abnormally high throw gradient values on some faults (0.56–0.57), greater than the throw gradients of the upper tips (Fig. 15).

Zone B. Faults in zone B show flat-topped D–L profiles (Fig. 10). The maximum displacement observed is 63 m, with an average of 40 m. The average displacement gradient within the Oligocene and Miocene tiers (as observed in SU3 and SU4) varies between 0.05 and 0.075 (Fig. 13) and is similar along either tip. The displacement gradient is slightly higher in the Miocene tier. These gradients are half those observed for faults in zone A.

D–z profiles along faults in zone B are C-shaped and show only one displacement maximum, in contrast with the faults of zone A. This displacement maximum is located mainly in the Oligocene tier, with only one exception in fault PF4 (Fig. 12). The maximum displacement recorded on faults in zone B is 60 m, with an average of 35 m. This is strikingly smaller than the displacement of zone A. Vertical displacement gradients were calculated on the upper and lower tips of D–z profiles and are characterized by average displacement gradients of 0.1 along the lower tips and 0.16 along the upper tips of faults (Fig. 15). Such values are significantly lower than for faults in zone A.

Evolution of the normal fault array

Trigger of faulting

In the previous section, we presented the characteristics of the normal fault array observed in the Levant Basin, which can be used to propose a model explaining their formation and growth. The key characteristics are as follow:

(1) faults are layer bound;
(2) they are regionally distributed across the entire Levant Basin;
(3) their distribution is correlated with the thickness of the host sediments;
(4) the host sediments are believed to consist of fine-grained sand or clay;
(5) faults are synsedimentary for all their activity time in the basin;

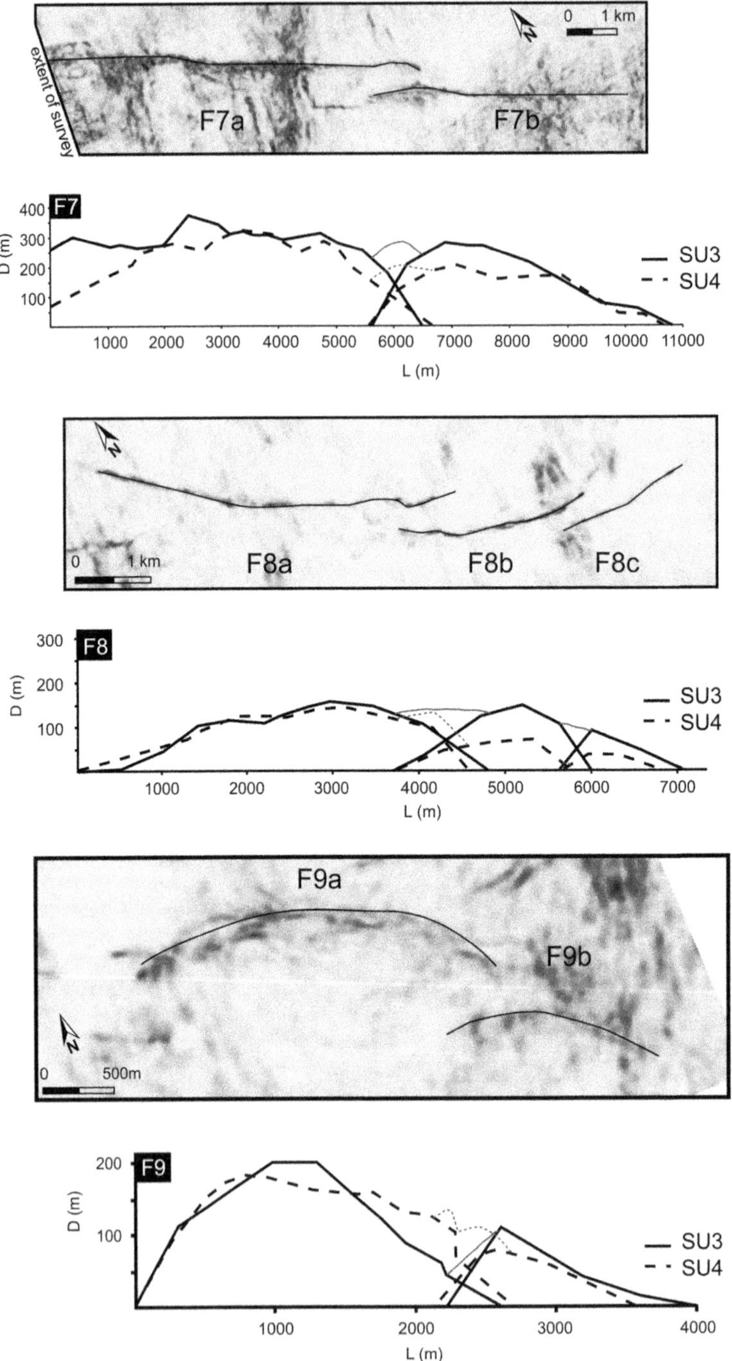

Fig. 11. *D–L* profiles of three NW-trending segmented faults in zone A taken along SU3 (Oligocene) and SU4 (Miocene). These faults are soft-linked in depth. After construction of the aggregate profile, the faults look C-shaped, although individual segments are flat-topped.

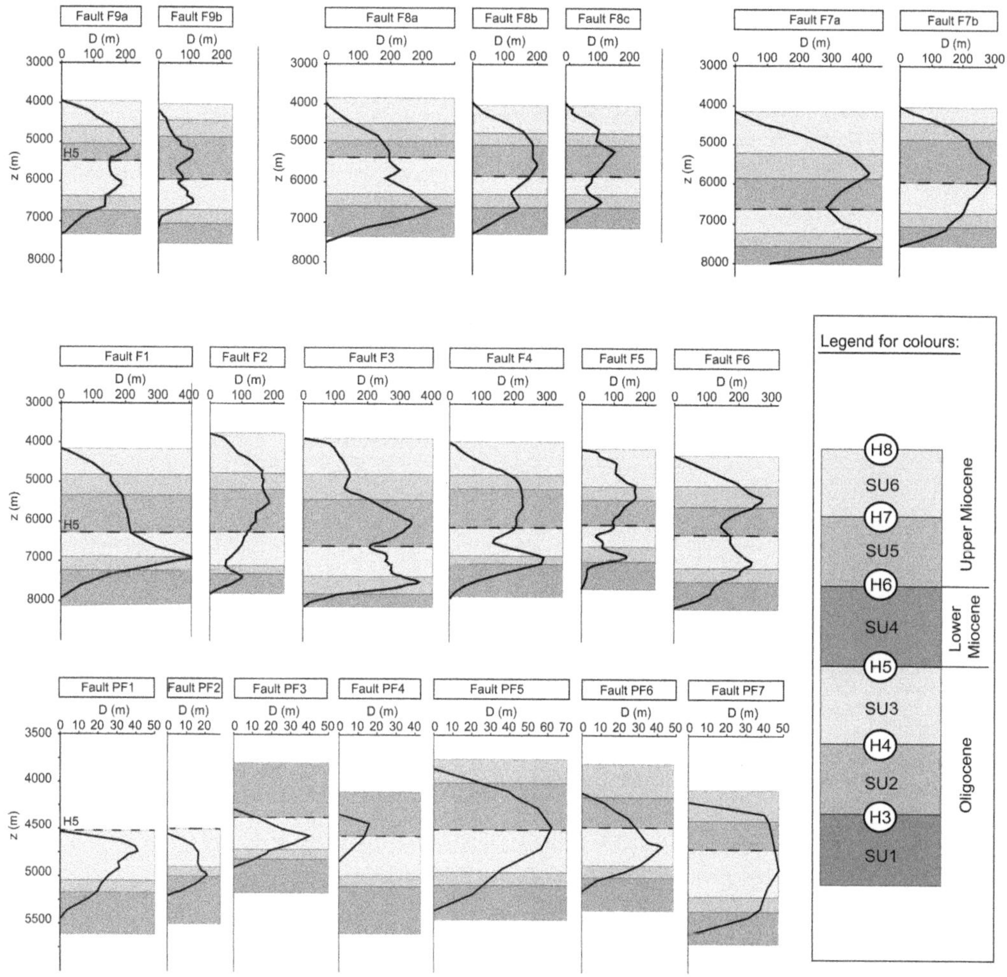

Fig. 12. Displacement–depth (D–z) profiles of the 20 faults documented in this study. F9, F8 and F7 are segmented faults in zone A. F1, F2, F3, F4, F5, F6 and F7 are NW-trending isolated faults in zone A. Their profiles show the presence of two displacement maxima. Faults PF1, PF2, PF3, PF4, PF5, PF6 and PF7 are faults with polygonal planform geometry in zone B. Their profiles show the presence of only one displacement maxima and have a C-shape distribution.

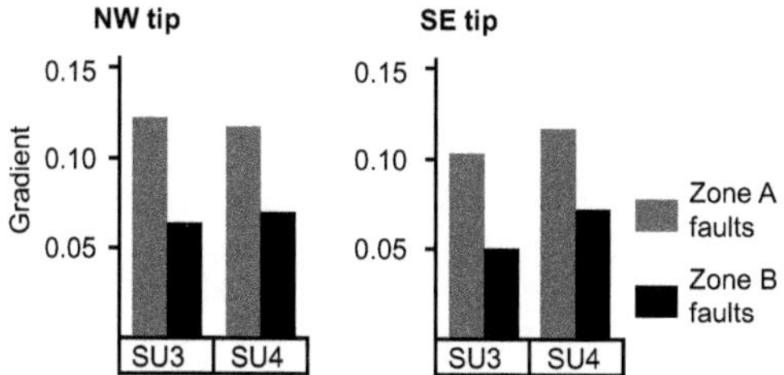

Fig. 13. Average lateral tip gradients of the investigated faults in zone A and B. The faults of zone B exhibit a smaller lateral gradient than the faults of zone A, which are considered to be high gradient if compared to the published literature.

(6) a small region close to the Lattakia Ridge contains faults with a polygonal shape in map view;

(7) maximum displacement and displacement profiles are different in the basin and close to the Lattakia Ridge.

Observations 1–6 are characteristic of polygonal fault systems in terms of timing and distribution (Cartwright & Dewhurst 1998; Cartwright *et al.* 2003; Cartwright 2011). PFSs have been related to several mechanisms such as the compaction of fine-grained sediments, or as recently suggested by shallow diagenetic processes leading to porosity reduction and shear failure (Shin *et al.* 2008;

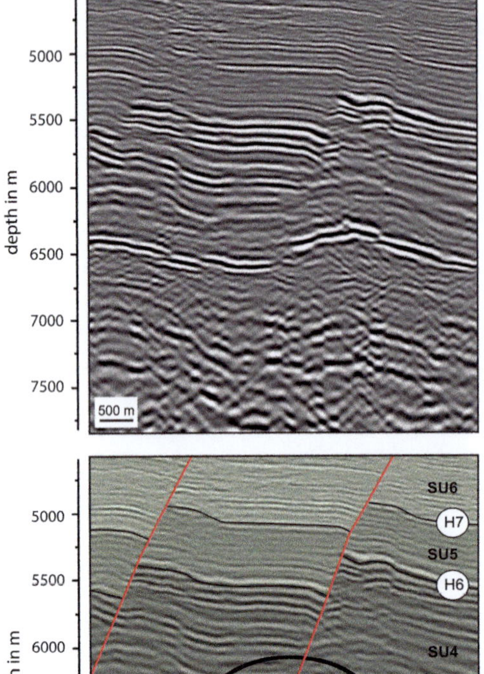

Fig. 14. Seismic section taken from a NW-trending normal fault in zone A showing the presence of breached monocline geometries commonly seen along the H5 horizon.

Cartwright 2011). Such diagenetic reactions could also result in shallow fault dips, as in the case of the Levant Basin, due to sediment compaction (Neagu *et al.* 2010). Once having nucleated, the faults might have grown and developed because of low coefficients of residual friction in fine-grained sediments (Goulty 2002, 2008). The triggering agent is not discussed in this paper as we focus on their growth and relation with host rocks.

The main difference between these faults and the PFS is their dominant NW trend. Such an aligned orientation is most likely a result of the NW–SE compressive stress field in the basin (Barrier & Vrielynck 2008; Ghalayini *et al.* 2014; Montadert *et al.* 2014) causing a heterogeneous horizontal stress field. It is common for PFSs to be locally re-oriented in zones of subtle stress field variations (Hansen *et al.* 2004; Hansen & Cartwright 2006; Cartwright 2011; Carruthers *et al.* 2013), though such well-oriented PFSs have never been documented at a basin scale before. We suggest that the NW-trending normal faults of the Levant Basin and classical PFSs have a similar origin; but as they lack the typical polygonal planform geometry, we propose simply to call them 'layer-bound normal faults' to avoid confusion, or intraformational fault systems as suggested earlier by Henriet *et al.* (1991) and Watterson *et al.* (2000).

PFSs are usually associated with fine-grained sediments. In zone A, the high vertical tip gradients (>0.2 and 0.37 in the Oligocene) and 45–60° dip for faults in the Oligocene and Miocene indicated that these units are very likely soft and incompetent (e.g. Wilkins & Gross 2002; Ferrill & Morris 2003; Schöpfer *et al.* 2007; Wibberley *et al.* 2007; Ferrill & Morris 2008; Roche *et al.* 2012*a*, *b*). The low seismic amplitude with continuous to chaotic reflectors observed in the Oligocene unit is also indicative of fine-grained hemipelagic sediments (Hawie *et al.* 2013). Similarly, high-amplitude continuous horizons observed in the Miocene sequences represent distal sheeted turbidite lobes, basin floor fans and mud/clastic-rich units (Reading & Richards 1994; Hawie *et al.* 2013). They were deposited during a period of uplift and erosion along the margin resulting in high sediment input, rapid burial and compaction of sediments (Hawie *et al.* 2013, 2014).

The presence of normal faults with typical polygonal planform geometry close to the Lattakia Ridge (Fig. 3) (e.g. Cartwright & Dewhurst 1998; Lonergan *et al.* 1998; Cartwright *et al.* 2003) further supports the hypothesis that the layer-bound normal faults are actually PFSs. These faults are different from faults in zone A since they are smaller in length, shorter in height and with a markedly smaller displacement. They are found mainly in the Oligocene unit, and occasionally extend to the Lower Miocene. This is not surprising because the Miocene is much thinner and probably contains coarser sediments in the northern part of the basin due to its proximity to a northward sediment input source (Hawie *et al.* 2013). In contrast, the remainder of the basin was likely subject to sediments coming in from different sources during the Miocene, such as the Nile Delta (Hawie *et al.* 2013). In fact, thickness variations of individual tiers, in addition to sediment size, have an important effect on the development of polygonal fault systems as faults are not developed in thin tiers and coarse-grained units (Cartwright & Dewhurst 1998; Cartwright *et al.* 2003; Cartwright 2011).

In summary, the lack of any regional extension documented in the Levant Basin during the Oligo-Miocene, the large geometric similarities between these faults and the widely documented PFSs and

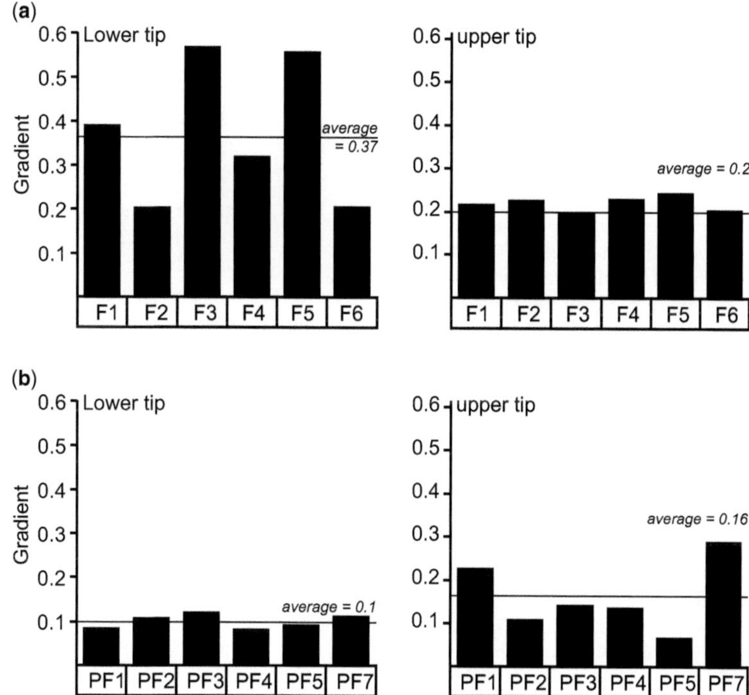

Fig. 15. (**a**) Average vertical gradients of the investigated faults in zone A. These gradients are very high and indicate vertical restriction to fault propagation. (**b**) Average vertical gradients of the investigated faults in zone B. These gradients are smaller than faults in zone B.

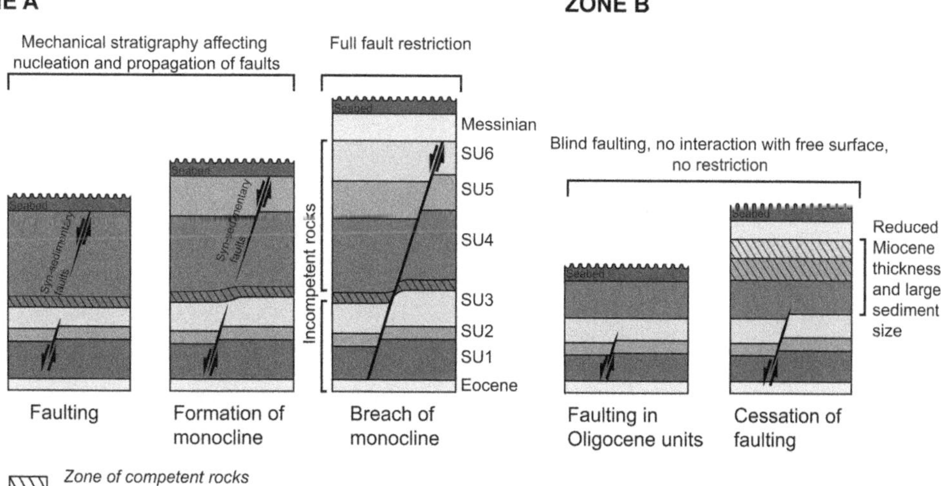

Fig. 16. Proposed model for the evolution of the normal faults in the Levant Basin consisting of nucleation during the Early Miocene in the Miocene and the Oligocene units simultaneously. This is due to anisotropic mechanical stratigraphy. The faults later propagated and vertically linked to form a single fault in zone A. In contrast, the faults in zone B were blind and vertically unrestricted during their growth. In map view, the orientation of these faults might be related to strong horizontal stress field anisotropy in zone A caused by the regional NW–SE compression during the Oligo-Miocene, while in zone B, the proximity to LRS might have caused local stress field fluctuations which resulted in polygonal planform geometry. The model of zone A is modified from Tvedt *et al.* (2013).

the suitable sedimentary facies in the basin strongly suggest that these faults have a genetic mechanism similar to that of PFSs (for a review see Cartwright & Lonergan 1996; Goulty 2002; Cartwright *et al.* 2003; Goulty 2008; Cartwright 2011).

Timing of the normal faults

EI plots for faults in zone A indicate synsedimentary activity in the Lower Miocene packages only (Fig. 8). Even though the EI results are low compared to other fault systems (e.g. Cartwright *et al.* 1998; Jackson & Rotevatn 2013; Tvedt *et al.* 2013; Reeve *et al.* 2015), this technique is a very efficient tool to check timing of fault activity. Synsedimentary activity in the Lower Miocene is explained by two models: (i) primary nucleation and blind propagation of faults segments in the Oligocene, followed by nucleation and synsedimentary propagation of fault segments in the Miocene, which also propagated downward and reactivated the Oligocene segments following the 'reactivation by dip linkage' model of Baudon & Cartwright (2008*b*); (ii) simultaneous nucleation during the Early Miocene of faults segments in both the Oligocene and Miocene tiers due to anisotropic mechanical stratigraphy. The segments in both tiers grew by vertical propagation and joined together to form one fault system. In this model, the Miocene segment propagated upward as a growth fault while the Oligocene segment propagated upward as a blind fault (Fig. 16).

Two key observations argue against Oligocene blind faulting followed by reactivation of fault segments during the Miocene in zone A: (i) no single normal fault is found in only the Oligocene unit (SU1, SU2 and SU3). If Miocene faulting reactivated existing faults, then we would expect a small number of faults to remain inactive in the Oligocene (Baudon & Cartwright 2008*c*). Instead, all faults are cross-cutting the Oligocene and Miocene units entirely and fault segments are well aligned; and (ii) there is no proof of any faulting activity during the Oligocene. For this reason we postulate that the normal fault growth in zone A was initiated in the Early Miocene.

The lack of growth strata in the faults of zone B implies that these structures never interacted with the free surface and hence limits our ability to constrain their age. However, we speculate that they grew during the Early Miocene since some faults show a minor amount of growth during this time. This is in accordance with observations of faults in zone A showing synsedimentary activity starting in the Early Miocene. During the Messinian, it is likely that some faults were still growing as evidenced by small deformation of Messinian markers (horizons with strong impedance contrast) close to H8 (base Messinian) on top of large faults.

Nucleation, fault growth and mechanical stratigraphy

The following observations on the NW-trending faults of zone A show a strong correlation with the layering of the sedimentary package where the faults are embedded:

(1) $D-z$ plots and throw projection contours with two distinct displacement maxima in the Oligocene and Miocene tiers (Fig. 12);
(2) more or less square-shaped fault planes with aspect ratio close to 1 (Fig. 6);
(3) frequent breached monoclines close to H5 horizon (Fig. 14);
(4) vertical gradients larger than lateral gradients;
(5) gradients larger in zone A than zone B.

Such observations suggest that the normal fault system in zone A was strongly influenced by mechanical stratigraphy (Fig. 16). Following the interpretation of displacement maximum, which serves as an indicator for the nucleation point of faults (Watterson 1986; Barnett *et al.* 1987; Dawers & Anders 1995), the two distinct displacement maxima (Fig. 12) indicate that nucleation of faults could have happened in the Miocene and Oligocene tiers independently. Reduced residual friction in the fine-grained layers may cause local increase in displacement, but nucleation of polygonal faults in other stiffer units appears unlikely and would then cause another displacement maximum that is not observed in the data. Thus, it is likely that two isolated fault segments grew by vertical propagation of the tip line as advocated by Mansfield & Cartwright (1996), and later joined to form a through-going fault as observed now. This interpretation further explains the aspect ratio of the NW–SE trending faults close to 1, and their square fault plane shape, which is very unusual for unrestricted normal faults that grew by radial tip propagation (Rippon 1984; Barnett *et al.* 1987; Walsh & Watterson 1988).

This growth model implies that a strong mechanical contrast needs to exist between the host rock units (Fig. 16). The displacement minima and the monoclinal breaching seen along H5 indicate that this horizon might constitute a unit with different mechanical properties than the overlying Miocene and underlying Oligocene, and hence a zone of linkage between hard-linked individual segments (Peacock & Sanderson 1994; Cartwright *et al.* 1995; Gawthorpe *et al.* 1997; Baudon & Cartwright 2008*a*) (Fig. 16). The strong impedance contrast between low-amplitude semi-continuous reflectors in the Oligocene to continuous medium- to high-amplitude reflectors along H5 further supports a lithological change at the base Miocene. This is likely the effect of a sandstone unit at the base Miocene as it has been documented offshore Norway whereby polygonal faults nucleating into two independent tiers separated by a thick sandy unit were observed to transect the sand and join together (Stuevold *et al.* 2003). Such mechanical stratigraphy consists of faults nucleating in incompetent rocks (i.e. compliant and weak rocks such as clays) while competent rocks are barriers to fault propagation (e.g. Wilkins & Gross 2002; Ferrill & Morris 2003).

The lateral tip gradient is less than the vertical gradient in zone A and B, implying enhanced fault propagation in the horizontal direction as commonly observed with faults showing L/H ratio around 2. Nevertheless, in zone A, L/H ratio is close to 1 and the lateral tip gradient is higher than zone B, which may indicate high clay content (Roche *et al.* 2012*b*). Furthermore, the high vertical tip gradients in D/Z profiles of faults in zone A indicate that Messinian and Eocene units are not allowing faults to propagate vertically. As vertical tip gradient increases during the growth of vertically restricted faults (Roche *et al.* 2012*b*), the Eocene is thus interpreted to have a stronger restriction effect than the Messinian. In response to the vertical restriction, the faults consequently continued to propagate laterally till they interacted with each other, resulting in relatively high lateral tip gradients among all faults. Some faults (F7, F8 and F9) even became segmented (Fig. 11). This connection probably happened late in the history of the faults. D–L profiles show that faults with lengths >5000 m exhibit flat-topped profiles, indicating that faults started to interact with each other when they reached a certain length, around 5000 m.

In contrast, the normal faults of zone B are more ellipsoidal and show only one displacement maximum (Fig. 6), suggesting that mechanical stratigraphy is not controlling the fault growth in this zone. This could be related to lithological changes in this area influenced by the close proximity of the Lattakia Ridge shedding coarser sediments at the toe of the fold and thrust belt, preventing fault nucleation in the Upper Miocene where sediments are coarser. Nucleation of faults has most likely occurred at the location of highest displacement in SU3 (Upper Oligocene) in this zone (e.g. Walsh & Watterson 1989; Cowie & Scholz 1992) (Fig. 16), which might explain why faults are smaller in height and have smaller displacement.

Discussion

Levant normal faults: exceptional type of PFS

In the previous sections, we presented the characteristics of the layer-bound normal fault array observed in the Levant Basin and proposed that initially isolated fault segments nucleated during the Early Miocene and later linked vertically. These faults are likely a special type of PFS. Generally, the PFSs documented around the world contain faults smaller both in length and in height (Cartwright 1994; Lonergan *et al.* 1998; Watterson *et al.* 2000; Nicol *et al.* 2003; Stuevold *et al.* 2003; Hansen *et al.* 2004; Shin *et al.* 2010; Carruthers *et al.* 2013). The normal faults in the Levant Basin accumulate large displacement (up to 400 m) and might thus be considered as the largest layer-bound normal faults documented so far. We attribute this large fault size to being a product of vertical linkage of individual segments due to mechanical stratigraphy and thick succession of suitable lithology (*c.* 3 km), allowing them to accumulate large displacement. The fact that smaller faults with classical polygonal geometry are found in a small region in the basin supports our hypothesis that a constitutive genetic mechanism was prevalent in the basin and explains the faults in both zone A and B (i.e. polygonal and linear faults).

Although we believe that variations in local stress field resulted in faults with polygonal planform geometry close to the Lattakia Ridge, this situation is, in fact, intriguing. In fact, in the majority of cases around the world, we find that isotropic horizontal stress fields result in classical polygonal faults in the basin, which become re-oriented and linear close to large geological structures (for a review see Cartwright 2011). In some basins, however, such as the North Sea and the Porcupine basins (Higgs & Mcclay 1993; Watterson *et al.* 2000; Bailey *et al.* 2003), we find local variations in fault geometry close to large structures with faults becoming linear. Why should we observe the opposite in the Levant Basin with faults becoming polygonal close to the large Lattakia structure? Gravity loading could not explain why these faults are linear in the basin since faults in zone A do not have a preferred dip direction. The most likely explanation is that large structures, such as the Lattakia Ridge, have increased S_{hmin} to equal S_{hmax} and created an isotropic horizontal stress field, whereas in the distal part of the basin a regional anisotropic stress field dominated at the plate level during the Oligo-Miocene (e.g. Ghalayini *et al.* 2014).

Growth of layer-bound faults

The geometry and the displacement distribution along fault planes suggest that the faults were physically isolated and later grew by segment linkage in accordance with the isolated fault model (Cartwright *et al.* 1995; Mansfield & Cartwright 2001; Baudon & Cartwright 2008*a*). The D_{max}–L log–log plot shows that faults in zone A were accumulating displacement with length, following a best-fit line of $D = 0.03L^{1.02}$ (Fig. 9). Such a value is very close to the one calculated by Schlische *et al.* (1996) of $D = 0.03L^{1.06}$ by plotting a global dataset of tectonic faults, suggesting a linear scaling relationship between length and displacement of faults with $n \approx 1$ in the $D = cL^n$ equation (Kim & Sanderson 2005; Soliva & Benedicto 2005; Schultz *et al.* 2006; Soliva *et al.* 2008). By following the same trend line on the plot (Fig. 9), the linear and well-oriented faults of zone A are not different from the tectonic faults seen in different regions despite their constitutive genetic mechanism and hence grow by simultaneously accumulating length and displacement. However, in the normal plot some faults in zone A are underdisplaced for their lengths, causing some scatter in the plot (Fig. 9b). This is particularly observed for faults with $L > 7000$ m, which are segmented normal faults exhibiting soft relay bends. This is interpreted to mean that they gained considerable length after joining with nearby segments without accumulating much vertical displacement, providing a period of fault growth during which the D–L evolution is nonlinear (Soliva & Benedicto 2004; Roche *et al.* 2012*b*). The presence of mature soft-linked fully segmented faults together with immature isolated faults that start to interact with nearby structures suggests that these faults did not link until relatively late in their history.

In zone B, however, D_{max}–L plots follow the trend line $D = 0.64.L^{0.53}$. It exhibits a slope value of 0.53 very similar to the one suggested by Nicol *et al.* (2003) who calculated a best-fit line of $D = 0.09.L^{0.6}$ for a set of polygonal faults in the Lake Hope region of South Australia with a c value of 0.09, which is different from the value of 0.64 in the Levant Basin. This could be due to some variations in rock types between both study areas as c is generally related to rock properties (Schlische *et al.* 1996). The slope value <1.0 in the plot over polygonal faults of Lake Hope was attributed to an increase in fault dimensions due to linkage resulting in underdisplaced faults (Nicol *et al.* 2003). Even though the investigated faults in zone B were not linked as the one in Lake Hope, we believe that a minimum amount of interaction and communication is established between faults early in their history. The lateral displacement gradients are low, indicating small interaction.

The suggested growth models draw a similarity between the layer-bound faults of the Levant Basin and global dataset. Where faults show a polygonal planform geometry, they exhibit similar D–L profiles and growth curves as other PFSs (Nicol *et al.* 2003; Nelson 2006), which is caused by early linkage and interaction. However, when subject to an anisotropic horizontal stress field, the faults become linear and parallel in plan view, and show a very similar growth pattern and D_{max}–L value to tectonic normal faults (Cowie & Scholz 1992; Schlische *et al.* 1996). We believe that an anisotropic regional stress field has an important effect on the growth of layer-bound normal faults as it delays early linkage because faults are not curved. Instead, faults will accumulate displacement proportionally to length and link later on during their history. We suggest that 3D seismic studies investigate the displacement variation on PFSs in general, and linear faults in particular, in order to check if similar results and growth patterns are seen in layer-bound faults in other basins.

Conclusion

This study has provided detailed analysis of the geometry and growth of a layer-bound normal fault array in the Levant Basin. By performing various 3D interpretation techniques such as displacement–length, displacement–depth and growth indices, we were able to discuss and present the temporal and spatial evolution of this normal fault system. Observations indicate that these normal faults have nucleated through a constitutive genetic mechanism similar to that of PFSs. Their growth was very much affected by the presence of a regional NW–SE stress field, hence their orientation. These normal faults are thus a special type of stress-induced well-oriented polygonal fault system that we refer to as a layer-bound normal fault system.

We propose that the normal faults nucleated during the Early Miocene in the Lower Miocene and Oligocene units simultaneously. Their displacement distribution attests to mechanical stratigraphy in the basin, which controlled their preferred nucleation sites. This has also resulted in vertical dip-linkage of individual segments. Horizontal segment linkage took place later in their history, and some faults became fully segmented. We concluded that the growth of layer-bound normal faults follows the isolated fault growth model at a basin scale during their early history. When subject to a regional anisotropic horizontal stress field, they behave similarly to tectonic faults.

The authors would like to acknowledge TOTAL, IFPEN and UPMC for funding this project. The Lebanese Ministry of Energy and Water and the Lebanese Petroleum Administration are thanked for their support. Petroleum Geo-Services (PGS) is acknowledged for providing the dataset and, in particular, Per Helge Semb for his help and co-operation. Lucien Montadert (Beicip-Franlab) is very much thanked for the discussions and his regional insight into Mediterranean geology. Joe Cartwright and Frank Richards are thanked for their thoughtful insights at the beginning of this project. Neil Goulty and an anonymous reviewer, together with editors Conrad Childs and John Walsh, are thanked for their constructive reviews and comments, which enhanced the quality of this manuscript.

References

Bailey, W., Shannon, P.M., Walsh, J.J. & Unnithan, V. 2003. The spatial distribution of faults and deep sea carbonate mounds in the Porcupine Basin, offshore Ireland. *Marine and Petroleum Geology*, **20**, 509–522.

Barnett, J.A., Mortimer, J., Rippon, J.H., Walsh, J.J. & Watterson, J. 1987. Displacement geometry in the volume containing a single normal fault. *AAPG Bulletin*, **71**, 925–937.

Barrier, É. & Vrielynck, B. 2008. *Paleotectonic maps of the Middle East: Tectono-sedimentary-palinspastic maps from Late Norian to Pliocene*. 14 maps. Commission de la carte geologique du monde, Paris.

Baudon, C. & Cartwright, J.A. 2008*a*. 3D seismic characterisation of an array of blind normal faults in the Levant Basin, Eastern Mediterranean. *Journal of Structural Geology*, **30**, 746–760, https://doi.org/10.1016/j.jsg.2007.12.008

Baudon, C. & Cartwright, J.A. 2008*b*. Early stage evolution of growth faults: 3D seismic insights from the Levant Basin, Eastern Mediterranean. *Journal of Structural Geology*, **30**, 888–898, https://doi.org/10.1016/j.jsg.2008.02.019

Baudon, C. & Cartwright, J.A. 2008*c*. The kinematics of reactivation of normal faults using high resolution throw mapping. *Journal of Structural Geology*, **30**, 1072–1084, https://doi.org/10.1016/j.jsg.2008.04.008

Boudagher-Fadel, M. & Clark, G.N. 2006. Stratigraphy, paleoenvironment and paleogeography of Maritime Lebanon: a key to Eastern Mediterranean Cenozoic history. *Stratigraphy*, **3**, 1–38.

Bowman, S.A. 2011. Regional seismic interpretation of the hydrocarbon prospectivity of offshore Syria. *GeoArabia*, **16**, 95–124.

Carruthers, D., Cartwright, J.A., Jackson, M.P.A. & Schutjens, P. 2013. Origin and timing of layer-bound radial faulting around North Sea salt stocks: new insights into the evolving stress state around rising diapirs. *Marine and Petroleum Geology*, **48**, 130–148, https://doi.org/10.1016/j.marpetgeo.2013.08.001

Cartwright, J.A. 1994. Episodic basin-wide hydrofracturing of overpressured Early Cenozoic mudrock sequences in the North Sea Basin. *Marine and Petroleum Geology*, **11**, 587–607, https://doi.org/10.1016/0264-8172(94)90070-1

Cartwright, J.A. 2011. Diagenetically induced shear failure of fine-grained sediments and the development of polygonal fault systems. *Marine and Petroleum Geology*, **28**, 1593–1610, https://doi.org/10.1016/j.marpetgeo.2011.06.004

Cartwright, J.A. & Dewhurst, D.N. 1998. Layer-bound compaction faults in fine-grained sediments. *Geological Society of America Bulletin*, **110**, 1242–1257, https://doi.org/10.1130/0016-7606(1998)110<1242:LBCFIF>2.3.CO;2

Cartwright, J.A. & Lonergan, L. 1996. Volumetric contraction during the compaction of mudrocks: a mechanism for the development of regional-scale polygonal fault systems. *Basin Research*, **8**, 183–193.

Cartwright, J.A., Trudgill, B.D. & Mansfield, C.S. 1995. Fault growth by segment linkage: an explanation for scatter in maximum displacement and trace length data from the Canyonlands Grabens of SE Utah. *Journal of Structural Geology*, **17**, 1319–1326.

Cartwright, J.A., Bouroullec, R., James, D. & Johnson, H. 1998. Polycyclic motion history of some Gulf Coast growth faults from high-resolution displacement analysis. *Geology*, **26**, 819–822, https://doi.org/10.1130/0091-7613(1998)026<0819:PMHOSG>2.3.CO;2

Cartwright, J.A., James, D. & Bolton, A. 2003. The genesis of polygonal fault systems: a review. *In*: Van Rensbergen, P., Hillis, R.R., Maltman, A.J. & Morley, C.K. (eds) *Subsurface Sediment Mobilization*. Geological Society, London, Special Publications, 223–243, https://doi.org/10.1144/GSL.SP.2003.216.01.15

Childs, C., Nicol, A., Walsh, J. & Watterson, J. 2003. The growth and propagation of synsedimentary faults. *Journal of Structural Geology*, **25**, 633–648.

Cowie, P.A. & Scholz, C.H. 1992. Displacement–length scaling relationship for faults: data synthesis and discussion. *Journal of Structural Geology*, **14**, 1149–1156.

Cowie, P.A. & Shipton, Z.K. 1998. Fault tip displacement gradients and process zone dimensions. *Journal of Structural Geology*, **20**, 983–997.

Davies, R.J. & Ireland, M.T. 2011. Initiation and propagation of polygonal fault arrays by thermally triggered volume reduction reactions in siliceous sediment. *Marine Geology*, **289**, 150–158, https://doi.org/10.1016/j.margeo.2011.05.005

Dawers, N.H. & Anders, M.H. 1995. Displacement-length scaling and fault linkage. *Journal of Structural Geology*, **17**, 607–614.

Dewhurst, D.N., Cartwright, J.A. & Lonergan, L. 1999. The development of polygonal fault systems by syneresis of colloidal sediments. *Marine and Petroleum Geology*, **16**, 793–810.

Dubertret, L. 1955. *Carte Geologique du Liban au 1/200000 avec Notice Explicative*. République Libanaise, Ministère des travaux publiques, Beyrouth.

Dupin, I., Brahami, J., Gou, Y. & Montadert, L. 2012. Petroleum assessment of the offshore Lebanon based on the seismic interpretation and the regional geological framework. *In*: *Lebanon International Petroleum Exploration Forum and Exhibition*. Beirut, Lebanese Republic Ministry of Energy and Water.

Ferrill, D.A. & Morris, A.P. 2003. Dilational normal faults. *Journal of Structural Geology*, **25**, 183–196.

Ferrill, D.A. & Morris, A.P. 2008. Fault zone deformation controlled by carbonate mechanical stratigraphy, Balcones fault system, Texas. *AAPG Bulletin*, **92**, 359–380, https://doi.org/10.1306/10290707066

Gardosh, M., Druckman, Y., Buchbinder, B. & Calvo, R. 2008. *The Oligo-Miocene Deepwater System of the Levant Basin*. Prepared for the Petroleum Commissioner, The Ministry of National Infrastructures, Israel.

Gardosh, M., Garfunkel, Z., Druckman, Y. & Buchbinder, B. 2010. Tethyan rifting in the Levant Region and its role in Early Mesozoic crustal evolution. *In*: Homberg, C. & Bachmann, M. (eds) *Evolution of the Levant Margin and Western Arabia Platform since the Mesozoic*. Geological Society, London, Special Publications, **341**, 9–36, https://doi.org/10.1144/SP341.2

Garfunkel, Z. 2004. Origin of the Eastern Mediterranean basin: a reevaluation. *Tectonophysics*, **391**, 11–34, https://doi.org/10.1016/j.tecto.2004.07.006

Gawthorpe, R.L., Sharp, I.R., Underhill, J.R. & Gupta, S. 1997. Linked sequence stratigraphic and structural evolution of propagating normal faults. *Geology*, **25**, 795, https://doi.org/10.1130/0091-7613(1997)025<0795:LSSASE>2.3.CO;2

Ghalayini, R., Daniel, J.-M., Homberg, C., Nader, F.H. & Comstock, J.E. 2014. Impact of Cenozoic strike-slip tectonics on the evolution of the northern Levant Basin (offshore Lebanon). *Tectonics*, **33**, 2121–2142, https://doi.org/10.1002/2014TC003574

Gomez, F., Nemer, T., Tabet, C., Khawlie, M., Meghraoui, M. & Barazangi, M. 2007. Strain partitioning of active transpression within the Lebanese restraining bend of the Dead Sea Fault (Lebanon and SW Syria). *In*: Cunningham, W.D. & Mann, P. (eds) *Tectonics of Strike-Slip Restraining and Releasing Bends*. Geological Society, London, Special Publications, **290**, 285–303, https://doi.org/10.1144/290.10

Goulty, N.R. 2002. Mechanics of layer-bound polygonal faulting in fine-grained sediments. *Journal of the Geological Society, London*, **159**, 239–246, https://doi.org/10.1144/0016-764901-111

Goulty, N.R. 2008. Geomechanics of polygonal fault systems: a review. *Petroleum Geoscience*, **14**, 389–397, https://doi.org/10.1144/1354-079308-781

Gross, M.R. 1995. Fracture partitioning: failure mode as a function of lithology in the Monterey Formation of coastal California. *Geological Society of America Bulletin*, **107**, 779–792, https://doi.org/10.1130/0016-7606(1995)107<0779:FPFMAA>2.3.CO;2

Gross, M.R., Gutiérrez-Alonso, G., Bai, T., Wacker, M.A., Collinsworth, K.B. & Behl, R.J. 1997. Influence of mechanical stratigraphy and kinematics on fault scaling relations. *Journal of Structural Geology*, **19**, 171–183, https://doi.org/10.1016/S0191-8141(96)00085-5

Hansen, D.M. & Cartwright, J.A. 2006. The three-dimensional geometry and growth of forced folds above saucer-shaped igneous sills. *Journal of Structural Geology*, **28**, 1520–1535, https://doi.org/10.1016/j.jsg.2006.04.004

Hansen, D.M., Shimeld, J.W., Williamson, M.A. & Lykke-Andersen, H. 2004. Development of a major polygonal fault system in Upper Cretaceous chalk and Cenozoic mudrocks of the Sable Subbasin, Canadian Atlantic margin. *Marine and Petroleum Geology*, **21**, 1205–1219, https://doi.org/10.1016/j.marpetgeo.2004.07.004

Hawie, N., Gorini, C., Deschamps, R., Nader, F.H., Montadert, L., Granjeon, D. & Baudin, F. 2013. Tectono-stratigraphic evolution of the northern Levant Basin (offshore Lebanon). *Marine and Petroleum Geology*, **48**, 392–410, https://doi.org/10.1016/j.marpetgeo.2013.08.004

Hawie, N., Deschamps, R. *et al.* 2014. Sedimentological and stratigraphic evolution of northern Lebanon since the Late Cretaceous: implications for the Levant margin and basin. *Arabian Journal of Geosciences*, **7**, 1323–1349, https://doi.org/10.1007/s12517-013-0914-5

Henriet, J.P., De Batist, M. & Verschuren, M. 1991. Early fracturing of Paleogene clays, southernmost North Sea: relevance to mechanisms of primary hydrocarbon migration. *In*: Spencer, A.M. (ed.) *Generation, Accumulation and Production of Europe's Hydrocarbon*. European Association of Petroleum Geologists, Special Publication, **1**, 217–227.

Higgs, W.G. & McClay, K.R. 1993. Analogue sandbox modelling of Miocene extensional faulting in the Outer Moray Firth. *In*: Williams, G.D. (ed.) *Tectonics and Seismic Sequence Stratigraphy*. Geological Society, London, Special Publications, **71**, 141–162, https://doi.org/10.1144/GSL.SP.1993.071.01.07

Homberg, C., Barrier, É., Mroueh, M., Hamdan, W. & Higazi, F. 2009. Basin tectonics during the Early Cretaceous in the Levant margin, Lebanon. *Journal of Geodynamics*, **47**, 218–223, https://doi.org/10.1016/j.jog.2008.09.002

Homberg, C., Barrier, É., Mroueh, M., Müller, C., Hamdan, W. & Higazi, F. 2010. Tectonic evolution of the central Levant domain (Lebanon) since Mesozoic time. *In*: Homberg, C. & Bachmann, M. (eds) *Evolution of the Levant Margin and Western Arabia Platform since the Mesozoic*. Geological Society, London, Special Publications, **341**, 245–268, https://doi.org/10.1144/SP341.12

Ireland, M.T., Goulty, N.R. & Davies, R.J. 2011. Influence of stratigraphic setting and simple shear on layer-bound compaction faults offshore Mauritania. *Journal of Structural Geology*, **33**, 487–499, https://doi.org/10.1016/j.jsg.2010.11.005

Jackson, C.A.L. & Rotevatn, A. 2013. 3D seismic analysis of the structure and evolution of a salt-influenced normal fault zone: a test of competing fault growth models. *Journal of Structural Geology*, **54**, 215–234, https://doi.org/10.1016/j.jsg.2013.06.012

Kim, Y.-S. & Sanderson, D.J. 2005. The relationship between displacement and length of faults: a review. *Earth-Science Reviews*, **68**, 317–334, https://doi.org/10.1016/j.earscirev.2004.06.003

Kosi, W., Tari, G., Nader, F.H., Skiple, C., Trudgill, B.D. & Lazar, D. 2012. Structural analogy between the 'piano key faults' of deep-water Lebanon and the extensional faults of the Canyonlands grabens, Utah, United States. *The Leading Edge*, **31**, 824–830.

Lie, O., Skiple, C. & Lowry, C. 2011. New insights into the Levantine basin. *GeoExpro*, **8**.

Lonergan, L., Cartwright, J.A. & Jolly, R. 1998. The geometry of polygonal fault systems in Tertiary mudrocks of the North Sea. *Journal of Structural Geology*, **20**, 529–548.

Mansfield, C.S. & Cartwright, J.A. 1996. High resolution fault displacement mapping from three-dimensional seismic data: evidence for dip linkage during fault growth. *Journal of Structural Geology*, **18**, 249–263.

Mansfield, C.S. & Cartwright, J.A. 2001. Fault growth by linkage: observations and implications from analogue models. *Journal of Structural Geology*, **23**, 745–763.

Montadert, L., Nicolaides, S., Semb, P.H. & Lie, O. 2014. Petroleum Systems offshore Cyprus. *In*: Marlow, L., Kendall, C. & Yose, L. (eds) *Petroleum*

Systems of the Tethyan Region. AAPG Memoirs, **106**, 301–334.
Morris, A.P., McGinnis, R.N. & Ferrill, D.A. 2014. Fault displacement gradients on normal faults and associated deformation. *AAPG Bulletin*, **98**, 1161–1184, https://doi.org/10.1306/10311312204
Muraoka, H. & Kamata, H. 1983. Displacement distribution along minor fault traces. *Journal of Structural Geology*, **5**, 483–495.
Neagu, R.C., Cartwright, J. & Davies, R. 2010. Measurement of diagenetic compaction strain from quantitative analysis of fault plane dip. *Journal of Structural Geology*, **32**, 641–655, https://doi.org/10.1016/j.jsg.2010.03.010
Nelson, M. 2006. *3D Geometry and Kinematics of Non-Colinear Fault Intersections*. PhD thesis, Cardiff University.
Nicol, A., Watterson, J., Walsh, J.J. & Childs, C. 1996. The shapes, major axis orientations and displacement patterns of fault surfaces. *Journal of Structural Geology*, **18**, 235–248.
Nicol, A., Walsh, J.J., Watterson, J., Nell, P.A.R. & Bretan, P.G. 2003. The geometry, growth and linkage of faults within a polygonal fault system from South Australia. *In*: Van Rensbergen, P., Hillis, R.R., Maltman, A.J. & Morley, C.K. (eds) *Subsurface Sediment Mobilization*. Geological Society, London, Special Publications, **216**, 245–261, https://doi.org/10.1144/GSL.SP.2003.216.01.16
Peacock, D.C.P. & Sanderson, D.J. 1991. Displacements, segment linkage and relay ramps in normal fault zones. *Journal of Structural Geology*, **13**, 721–733, https://doi.org/10.1016/0191-8141(91)90033-F
Peacock, D.C.P. & Sanderson, D.J. 1994. Geometry and development of relay ramps in normal fault systems. *AAPG Bulletin*, **2**, 147–165.
Peacock, D.C.P. & Zhang, X. 1993. Field examples and numerical modelling of oversteps and bends along normal faults in cross-section. *Tectonophysics*, **234**, 147–167.
Rank-Friend, M. & Elders, C.F. 2004. The evolution and growth of central graben salt structures, salt dome province, Danish North Sea. *In*: Davies, R.J., Cartwright, J.A., Stewart, S.A., Lappin, M. & Underhill, J.R. (eds) *3D Seismic Technology: Application to the Exploration of Sedimentary Basins*. Geological Society, London, Memoirs, **29**, 149–164, https://doi.org/10.1144/GSL.MEM.2004.029.01.15
Reading, H.G. & Richards, M. 1994. Turbidite systems in deep-water basin margins classified by grain size and feeder system 1. *AAPG Bulletin*, **78**, 792–822.
Reeve, M.T., Bell, R.E., Duffy, O.B., Jackson, C.A.L. & Sansom, E. 2015. The growth of non-colinear normal fault systems; What can we learn from 3D seismic reflection data? *Journal of Structural Geology*, **70**, 141–155, https://doi.org/10.1016/j.jsg.2014.11.007
Rippon, J.H. 1984. Contoured patterns of the throw and hade of normal faults in the Coal Measures (Westphalian) of north-east Derbyshire. *Proceedings of the Yorkshire Geological Society*, **45**, 147–161, https://doi.org/10.1144/pygs.45.3.147
Roche, V., Homberg, C. & Rocher, M. 2012*a*. Architecture and growth of normal fault zones in multilayer systems: a 3D field analysis in the South-Eastern Basin, France. *Journal of Structural Geology*, **37**, 19–35, https://doi.org/10.1016/j.jsg.2012.02.005
Roche, V., Homberg, C. & Rocher, M. 2012*b*. Fault displacement profiles in multilayer systems: from fault restriction to fault propagation. *Terra Nova*, **24**, 499–504, https://doi.org/10.1111/j.1365-3121.2012.01088.x
Rykkelid, E. & Fossen, H. 2002. Layer rotation around vertical fault overlap zones: observations from seismic data, field examples, and physical experiments. *Marine and Petroleum Geology*, **19**, 181–192.
Schlische, R.W., Young, S.S., Ackermann, R.V. & Gupta, A. 1996. Geometry and scaling relations of a population of very small rift-related normal faults. *Geology*, **24**, 683–686, https://doi.org/10.1130/0091-7613(1996)024<0683:GASROA>2.3.CO;2
Schöpfer, M.P.J., Childs, C., Walsh, J.J., Manzocchi, T. & Koyi, H.A. 2007. Geometrical analysis of the refraction and segmentation of normal faults in periodically layered sequences. *Journal of Structural Geology*, **29**, 318–335, https://doi.org/10.1016/j.jsg.2006.08.006
Schultz, R.A., Okubo, C.H. & Wilkins, S.J. 2006. Displacement–length scaling relations for faults on the terrestrial planets. *Journal of Structural Geology*, **28**, 2182–2193, https://doi.org/10.1016/j.jsg.2006.03.034
Shin, H., Santamarina, J.C. & Cartwright, J.A. 2008. Contraction-driven shear failure in compacting uncemented sediments. *Geology*, **36**, 931, https://doi.org/10.1130/G24951A.1
Shin, H., Santamarina, J.C. & Cartwright, J.A. 2010. Displacement field in contraction-driven faults. *Journal of Geophysical Research*, **115**, 1–13, https://doi.org/10.1029/2009JB006572
Soliva, R. & Benedicto, A. 2004. A linkage criterion for segmented normal faults. *Journal of Structural Geology*, **26**, 2251–2267, https://doi.org/10.1016/j.jsg.2004.06.008
Soliva, R. & Benedicto, A. 2005. Geometry, scaling relations and spacing of vertically restricted normal faults. *Journal of Structural Geology*, **27**, 317–325, https://doi.org/10.1016/j.jsg.2004.08.010
Soliva, R., Benedicto, A., Schultz-Ela, D.D., Maerten, L. & Micarelli, L. 2008. Displacement and interaction of normal fault segments branched at depth: implications for fault growth and potential earthquake rupture size. *Journal of Structural Geology*, **30**, 1288–1299, https://doi.org/10.1016/j.jsg.2008.07.005
Stewart, S.A. 2006. Implications of passive salt diapir kinematics for reservoir segmentation by radial and concentric faults. *Marine and Petroleum Geology*, **23**, 843–853, https://doi.org/10.1016/j.marpetgeo.2006.04.001
Stuevold, L.M., Faerseth, R.B., Arnesen, L., Cartwright, J.A. & Möller, N. 2003. Polygonal faults in the Ormen Lange Field, Møre Basin, offshore Mid Norway. *In*: Van Ransbergem, P., Hillis, R.R., Maltman, A.J. & Morley, C.K. (eds) *Subsurface Sediment Mobilization*. Geological Society, London, Special Publications, **216**, 263–281, https://doi.org/10.1144/GSL.SP.2003.216.01.17

THORSEN, C.E. 1963. Age of growth faulting in the southeast Louisiana. *Transactions of the Gulf Coast Association of Geological Societies*, **13**, 103–110.

TVEDT, A.B.M., ROTEVATN, A., JACKSON, C.A.L., FOSSEN, H. & GAWTHORPE, R.L. 2013. Growth of normal faults in multilayer sequences: a 3D seismic case study from the Egersund Basin, Norwegian North Sea. *Journal of Structural Geology*, **55**, 1–20, https://doi.org/10.1016/j.jsg.2013.08.002

WALSH, J.J. & WATTERSON, J. 1987. Distributions of cumulative displacement and seismic slip on a single normal fault surface. *Journal of Structural Geology*, **9**, 1039–1046.

WALSH, J.J. & WATTERSON, J. 1988. Analysis of the relationship between displacements and dimensions of faults. *Journal of Structural Geology*, **10**, 239–247.

WALSH, J.J. & WATTERSON, J. 1989. Displacement gradients on fault surfaces. *Journal of Structural Geology*, **11**, 307–316.

WALSH, J.J., NICOL, A. & CHILDS, C. 2002. An alternative model for the growth of faults. *Journal of Structural Geology*, **24**, 1669–1675.

WATTERSON, J. 1986. Fault dimensions, displacements and growth. *Pure and Applied Geophysics*, **124**, 365–373.

WATTERSON, J., WALSH, J.J., NICOL, A., NELL, P.A.R. & BRETAN, P.G. 2000. Geometry and origin of a polygonal fault system. *Journal of the Geological Society, London*, **157**, 151–162, https://doi.org/10.1144/jgs.157.1.151

WIBBERLEY, C.A.J., PETIT, J. & RIVES, T. 2007. The effect of tilting on fault propagation and network development in sandstone – shale sequences: a case study from the Lodève Basin, southern France. *Journal of the Geological Society, London*, **164**, 599–608, https://doi.org/10.1144/0016-76492006-047

WILKINS, S.J. & GROSS, M.R. 2002. Normal fault growth in layered rocks at Split Mountain, Utah: influence of mechanical stratigraphy on dip linkage, fault restriction and fault scaling. *Journal of Structural Geology*, **24**, 1413–1429.

Spatial distribution and evolution of fault-segment boundary types in rift systems: observations from experimental clay models

P. S. WHIPP[1,2]*, C. A.-L. JACKSON[1], R. W. SCHLISCHE[3], M. O. WITHJACK[3] & R. L. GAWTHORPE[4]

[1]*Department of Earth Science & Engineering, Basins Research Group (BRG), Imperial College, Prince Consort Road, London SW7 2BP, UK*

[2]*Present address: Statoil Gulf Services LLC, 2107 CityWest Boulevard, Houston, TX 77042, USA*

[3]*Department of Earth and Planetary Sciences, Rutgers University, 610 Taylor Road, Piscataway, NJ 08854-8066, USA*

[4]*Department of Earth Science, University of Bergen, Realfagbygget, Allègt. 41, N-5020 Bergen, Norway*

**Correspondence: pawh@statoil.com*

Abstract: Fault-segment boundaries initiate, evolve and die as a result of the propagation, interaction and linkage of normal faults during crustal extension. However, little is known about the distribution, evolution and controls on the development of relay ramps, which are the key structures developed at synthetic segment boundaries. In this study, we use a series of scaled physical models (wet clay) to investigate the distribution and evolution of fault-segment boundaries within an evolving normal-fault population during orthogonal extension. From the models, we can establish a simple geometrical classification for segment boundaries, analyse their spatial and temporal evolution, and identify key factors that influence their variability.

Development of overlapping fault tips is a prerequisite for fault growth via segment linkage. Synthetic segment boundaries are the most common segment boundary type developed in the models. The proportion of synthetic segment boundaries in the total fault population increases with increasing strain, whereas conjugate (antithetic) segment boundaries are very rare. Hanging-wall-breached relay ramps are the most common type (>70%) of breached-segment boundary, followed by footwall-breached relay ramps (<25%). Transfer faults are uncommon in our models. The type of breached segment boundary that develops cannot be predicted based on fault overlap to fault spacing aspect ratio alone. Instead, we show that fault linkage occurs in a range of styles across a wide range of fault overlap to fault spacing ratios (1:1–7:1). Furthermore, we show that fault spacing is constrained by stress-reduction shadows at the time of fault nucleation, whereas fault overlap changes during fault growth and interaction. Our study thus shows that scaled physical models are a powerful tool to assess the style, distribution and controls on the evolution of synthetic segment boundaries developing in rifts. Predictions from these models must now be assessed with data from natural examples exposed in the field or imaged in the subsurface.

Normal faulting accommodates the majority of upper-crustal deformation in many extensional settings. The normal fault arrays accommodating this deformation typically consist of multiple segments that may be geometrically linked or unlinked, and show varying degrees of kinematic and mechanical interaction (e.g. Morley *et al.* 1990; Peacock & Sanderson 1991, 1996; Walsh & Watterson 1991; Gawthorpe & Hurst 1993; Peacock & Zhang 1994; Childs *et al.* 1995; Huggins *et al.* 1995; Willemse 1997; Walsh *et al.* 1999, 2001, 2003*a*, *b*; Peacock 2002; Marchal *et al.* 2003; Soliva & Benedicto 2004).

Relay ramps are a type of synthetic segment boundary. They are classified as regions of both brittle and ductile deformation that serve to distribute extensional strain between two relatively closely spaced fault segments that have the same dip. During the initial stages of extension, relay ramps are typically characterized by zones of tilted bedding located between the two unlinked, overlapping segments: steep lateral displacement gradients developed on the faults at this time is thought to represent the expression of kinematic interaction between the segments (e.g. Peacock & Sanderson 1991, 1996; Trudgill & Cartwright 1994; Huggins

From: Childs, C., Holdsworth, R. E., Jackson, C. A.-L., Manzocchi, T., Walsh, J. J. & Yielding, G. (eds) 2017. *The Geometry and Growth of Normal Faults*. Geological Society, London, Special Publications, **439**, 79–107.
First published online March 31, 2016, https://doi.org/10.1144/SP439.7

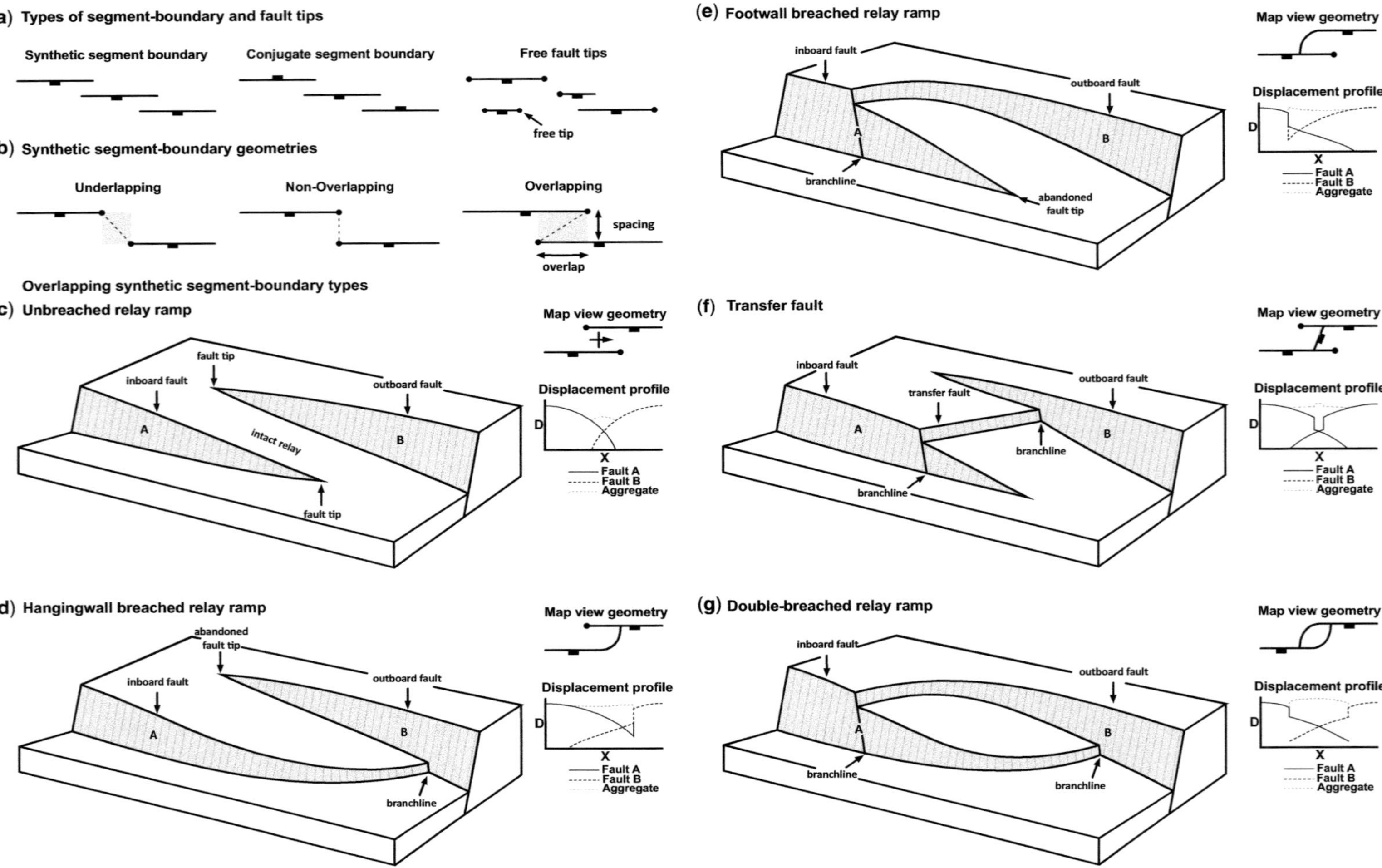
(a) Types of segment-boundary and fault tips
Synthetic segment boundary
Conjugate segment boundary
Free fault tips
free tip
(b) Synthetic segment-boundary geometries
Underlapping
Non-Overlapping
Overlapping
spacing
overlap
Overlapping synthetic segment-boundary types
(c) Unbreached relay ramp
fault tip
inboard fault
outboard fault
A
intact relay
B
fault tip
Map view geometry
Displacement profile
D
X
Fault A
Fault B
Aggregate
(d) Hangingwall breached relay ramp
abandoned fault tip
inboard fault
outboard fault
A
B
branchline
Map view geometry
Displacement profile
D
X
Fault A
Fault B
Aggregate
(e) Footwall breached relay ramp
inboard fault
outboard fault
A
B
branchline
abandoned fault tip
Map view geometry
Displacement profile
D
X
Fault A
Fault B
Aggregate
(f) Transfer fault
inboard fault
transfer fault
outboard fault
A
B
branchline
branchline
Map view geometry
Displacement profile
D
X
Fault A
Fault B
Aggregate
(g) Double-breached relay ramp
inboard fault
outboard fault
A
B
branchline
branchline
Map view geometry
Displacement profile
D
X
Fault A
Fault B
Aggregate

et al. 1995; Peacock 2002; Mansfield & Cartwright 2001). With continued extension and fault growth, physical linkage of the segments results in the development of a single, through-going fault array and a breached relay ramp (e.g. Peacock & Sanderson 1991, 1994; Childs *et al.* 1995; Willemse 1997; Crider & Pollard 1998; Marchal *et al.* 1998; Gupta & Scholz 2000; Mansfield & Cartwright 2001; Peacock 2002; Imber *et al.* 2004; Soliva & Benedicto 2004). Relict segment boundaries along a through-going fault array are identified by an irregular map-view fault geometry (e.g. zigzag, corrugated, anastomosing) (e.g. Morley *et al.* 1990; Gawthorpe & Hurst 1993; Dawers & Underhill 2000; Gawthorpe & Leeder 2000; McLeod *et al.* 2000, 2002; Young *et al.* 2001; Moustafa 2002; Gawthorpe *et al.* 2003), local displacement minima and hanging-wall fault-perpendicular folds (e.g. Schlische 1995).

Determining the temporal and spatial distribution and evolution of segment boundaries has implications for understanding the distribution of extensional strain (e.g. Gupta *et al.* 1998; Cowie *et al.* 2000, 2007; Meyer *et al.* 2002; Walsh *et al.* 2003*a*, *b*) and seismic activity (e.g. Scholz & Gupta 2000; Manignetti *et al.* 2007; Soliva *et al.* 2008) during upper-crustal extension. Furthermore, the types of segment boundaries that develop during extension, control rift-basin physiography and, therefore, drainage patterns and the resultant facies distributions in coeval synrift sedimentary systems (e.g. Leeder & Gawthorpe 1987; Gupta *et al.* 1999; Dawers & Underhill 2000; Gawthorpe & Leeder 2000; McLeod *et al.* 2002; Jackson *et al.* 2005). Segment-boundary breaching style also has a direct impact on the development and structural style of hydrocarbon traps located along the margins of rift basins, and the migration of hydrocarbon fluids into these traps (e.g. Morley *et al.* 1990). Understanding the evolution of fault-segment boundaries and style of fault linkage is, however, challenging both in outcrop and the subsurface because the expression of segment boundaries can vary according to the level of structural observation: this reflects the complex three-dimensional (3D) geometry of the bounding normal faults (e.g. Childs *et al.* 1995; Willemse *et al.* 1996; Crider 2001; Peacock 2002; Peacock & Parfitt 2002; Marchal *et al.* 2003). Outcrop-based studies are commonly limited by the level of exposure and the degree of preservation (i.e. burial and erosion) (e.g. Crider 2001; Peacock & Parfitt 2002; Jackson *et al.* 2005), whereas subsurface studies are limited by the spatial resolution of the seismic data (e.g. McLeod *et al.* 2000; Young *et al.* 2001).

Experimental (analogue) models have furthered our understanding of rift systems because of their ability to simulate the development of faults and folds associated with extensional tectonics (e.g. Withjack & Jamison 1986; Withjack *et al.* 1990; McClay & White 1995; Acocella *et al.* 1999, 2005; Clifton *et al.* 2000; Withjack & Callaway 2000; Ackermann *et al.* 2001; Clifton & Schlische 2001; Mansfield & Cartwright 2001; McClay *et al.* 2002; Schlische *et al.* 2002; Bellahsen *et al.* 2003; Marchal *et al.* 2003, Hus *et al.* 2005; Schlagenhauf *et al.* 2008; Schlische & Withjack 2009; Henza *et al.* 2010). In particular, these models allow us to examine how faults and related structures evolve with increasing strain, although none of the studies listed above have specifically considered and quantified the spatial distribution and evolution of synthetic segment-boundary types within an evolving rift system.

In this study, we use a series of analogue models with wet clay to specifically investigate the distribution of synthetic fault-segment boundaries (active and relict) within an evolving fault population during orthogonal extension. These models allow us to (i) establish a simple geometrical classification for fault-segment boundaries in rift systems; (ii) analyse the spatial and temporal evolution of fault-segment boundaries; and (iii) identify factors that may influence the types of segment boundaries. We can compare the geometry and distribution of the types of segment boundary formed in these experimental models with those observed in nature and those produced in other analogue, geometric and mechanical modelling studies.

Classification of normal fault-segment boundaries

Figure 1 illustrates a simple geometrical classification of end-member, synthetic, segment-boundary types: this is based on observations from this study

Fig. 1. Classification scheme for types of normal-fault segment boundaries in rift systems applied in this study. (**a**) Two segment-boundary types are defined: (i) synthetic, comprising two overlapping fault with the same dip-direction; and (ii) conjugate, comprising two overlapping faults with opposing dip directions. Free tips are associated with an isolated (non-interacting) fault. Synthetic segment boundaries are the focus of this study. (**b**) The synthetic segment-boundary geometries are defined according to the map-view location of adjacent fault tips: (i) underlapping, where adjacent fault tips understep; (ii) non-overlapping, where adjacent fault tips are perpendicular; and (iii) overlapping, where adjacent fault tips overstep. Five overlapping synthetic segment-boundary types are defined: (**c**) intact (unbreached) relay ramp; (**d**) hanging-wall-breached relay; (**e**) footwall-breached relay; (**f**) transfer-fault-breached relay; and (**g**) double-breached relay. See the text for the description.

and previous studies, and highlights the nomenclature used herein (e.g. Morley *et al.* 1990; Gawthorpe & Hurst 1993; Trudgill & Cartwright 1994; Childs *et al.* 1995). The classification utilizes: (i) the dip direction of adjacent fault segments; (ii) the proximity of adjacent segments; (iii) fault-tip location with respect to adjacent synthetic fault segments; and (iv) the style of synthetic segment linkage. The segment boundaries that are classified on the basis of dip direction are either (Fig. 1a): (a) synthetic, where stepping fault segments have the same dip direction (e.g. Morley *et al.* 1990; Gawthorpe & Hurst 1993); or (b) conjugate, where the stepping fault segments have opposing dip directions (e.g. Morley *et al.* 1990) (cf. antithetic transfer: Gawthorpe & Hurst 1993). Our study focuses only on synthetic segment-boundary types.

A 'free-tip' does not obviously kinematically or mechanically interact with neighbouring fault tips and is thus free to propagate laterally (Fig. 1a). By definition, a free-tip is not part of a segment boundary between two interacting fault segments, although it can evolve into a segment boundary with further lateral propagation of the fault segment. In plan view, co-linear or en echelon synthetic fault segments fall into three classes (Fig. 1b): (i) underlapping (i.e. the two approaching fault tips do not overlap); (ii) non-overlapping (i.e. the lateral tips of two adjacent fault segments are aligned perpendicular to the strike of the faults); and (iii) overlapping (i.e. the lateral tips of two adjacent fault segments overlap in the strike direction). The geometrical organization of overlapping synthetic segment boundaries can be quantified in map view by the amount of overlap (i.e. the strike-parallel distance between overlapping fault segments) and spacing (i.e. the strike-normal distance between overlapping fault segments) (Fig. 1b).

'Soft linkage' describes two overlapping synthetic fault segments that are kinematically linked but are not geometrically connected by faults at the scale and level of observation (Fig. 1c). In these situations, ductile deformation, bed rotation and relay-ramp formation occur in the region of fault overlap (e.g. Walsh & Watterson 1991; Trudgill & Cartwright 1994; Childs *et al.* 1995; Huggins *et al.* 1995). 'Hard linkage' describes two synthetic fault segments that become linked by a fault, or faults, that crosses the relay ramp (e.g. Walsh & Watterson 1991; Trudgill & Cartwright 1994; Childs *et al.* 1995; Huggins *et al.* 1995). Hard-linked synthetic segment boundaries are classified as (Fig. 1c–g): (a) 'hanging-wall breached', where the inboard fault becomes hard-linked to the hanging wall of the outboard fault via breaching of the lower part of the relay ramp (e.g. Trudgill & Cartwright 1994; Ferrill *et al.* 1999; Crider 2001); (b) 'footwall breached', where the outboard fault becomes hard-linked to the footwall of the inboard fault via breaching of the upper part of the relay ramp (e.g. Trudgill & Cartwright 1994; Ferrill *et al.* 1999; Crider 2001); (c) 'transfer-fault breached', where the relay ramp is breached in its central portion by a fault that strikes at a high angle to the two overlapping fault segments (e.g. Gibbs 1984; Gawthorpe & Hurst 1993; McClay & Khalil 1998; Ferrill *et al.* 1999; Peacock *et al.* 2000); and (d) 'double breached', where two types of segment linkage occur within an individual segment boundary (e.g. hanging wall, footwall, transfer-fault breached). Following linkage and with continued extension, the disconnected fault tip(s) may remain active (e.g. Hus *et al.* 2005) or become abandoned (e.g. Crider 2001).

Methods

Scaling and materials

To ensure similarity between the analogue models and nature, the rheological properties of the modelling material must fulfil two requirements (e.g. Hubbert 1937; Weijermars *et al.* 1993; Withjack & Callaway 2000): (a) the coefficient of internal friction of the modelling material should be similar to that of rocks – sedimentary rocks have a coefficient of internal friction of 0.55–0.85 (e.g. Handin 1966; Byerlee 1978); both wet clay and dry sand are thus considered suitable modelling materials because they have a coefficient of internal friction of approximately 0.6 (e.g. Sims 1993; Eisenstadt & Sims 2005; Withjack *et al.* 2008); and (b) the model-to-nature ratios for cohesive strength (C_0^*), density (ρ^*), gravity (g^*) and length (L^*) must follow the scaling equation:

$$(C_0^*) = \rho^* g^* L^*. \quad (1)$$

In our models, the values of ρ^* and g^* are 0.7 and 1, respectively: therefore, C_0^* and L^* must have similar values. Inherent structural and lithological heterogeneity means that the cohesive strength of the upper-crustal rocks is highly variable, and may range from <1 MPa in highly fractured rocks (e.g. Byerlee 1978) to 10–20 MPa for intact sedimentary rocks (e.g. Handin 1966), and up to 100 MPa for intact crystalline and metamorphic rocks (Handin 1966; Schellart 2000). The cohesive strength of wet clay is approximately 50 Pa (Sims 1993; Eisenstadt & Sims 2005; Withjack *et al.* 2008): therefore, C_0^* in the models ranges between 10^{-4} and 10^{-5}. Thus, 1 cm in the models represents approximately 100–1000 m in nature.

The wet clay used in the experiments is composed predominantly of kaolinite (grain size <0.1 mm) and water (*c.* 40–50% by weight). The density of the wet clay is approximately 1600 kg m^{-3} (*c.* 1.6 g cm^{-3}). Wet clay deforms by both cataclasis and ductile deformation, producing both faults and folds (e.g. Eisenstadt & Sims 2005; Withjack *et al.* 2008) Wet clay is ideal for studying fault-segment boundaries because it generates more numerous, smaller faults and, therefore, a larger number of fault interactions compared with dry sand models (e.g. Eisenstadt & Sims 2005; Withjack *et al.* 2008).

Model design

The modelling apparatus consists of two rigid sheets (a fixed sheet and a mobile sheet) separated by an 8 cm-wide rubber sheet (Fig. 2). The mobile sheet moves away from the fixed sheet at a 90° angle to the long axis of the rubber sheet, producing orthogonal extension and distributed deformation above the basal rubber sheet. A 0.3 cm-thick silicone polymer layer (with a viscosity of *c.* 10^4 Pa s: e.g. Weijermars 1986; Withjack & Callaway 2000) above the rubber sheet allows developing faults in the clay to have finite displacement at the base of the brittle (clay) layer (e.g. Henza *et al.* 2010), producing lateral and vertical displacement gradients similar to those observed in natural fault arrays (e.g. Walsh *et al.* 2001). A layer of homogeneous wet clay (60 × 68 cm) covers the putty and both the fixed and moving sheets; the thickness of the clay layer is 2.7 cm above the putty layer and 3 cm elsewhere. Models were extended at a constant rate of 4 cm h^{-1} for a total displacement of 4 cm. Time is not scaled in these models, as deformation of wet clay is mostly strain-rate independent (e.g. Oertel 1965).

Model analysis

We photographed the top surface of each of the 24 models at every 0.1 cm of displacement of the mobile sheet, utilizing two opposing lighting directions orientated normal to the rift axis. Illumination of the model surface from opposing directions allows determination of the fault-dip direction: faults that

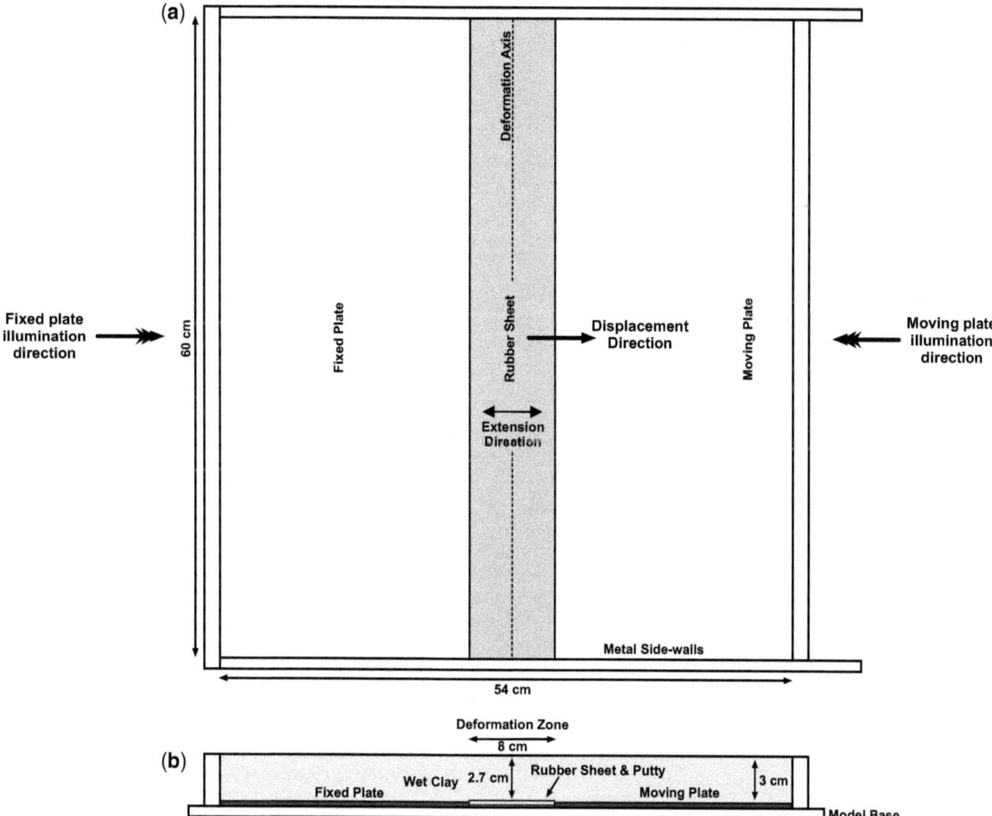

Fig. 2. Experimental set-up in (**a**) plan-view and (**b**) cross-section. Orthogonal extension occurred in all models because the displacement direction was perpendicular to the long axis of the rubber sheet. Note in (b) that the original undeformed width of the rubber sheet is 8 cm.

(a)

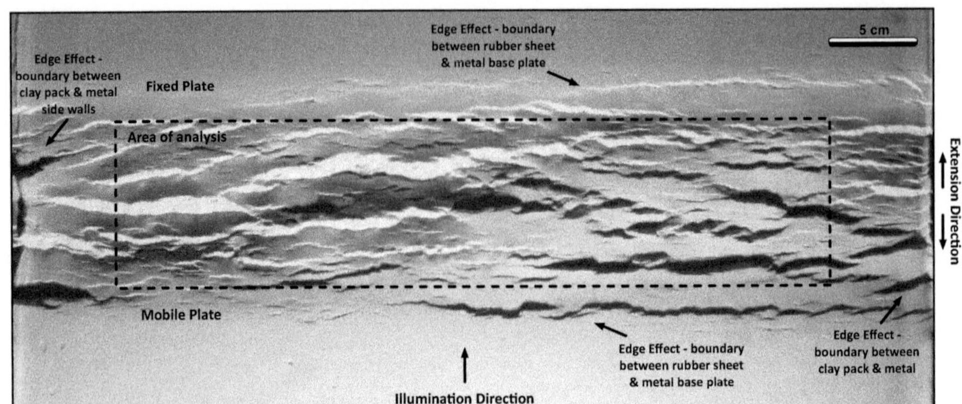

(b)

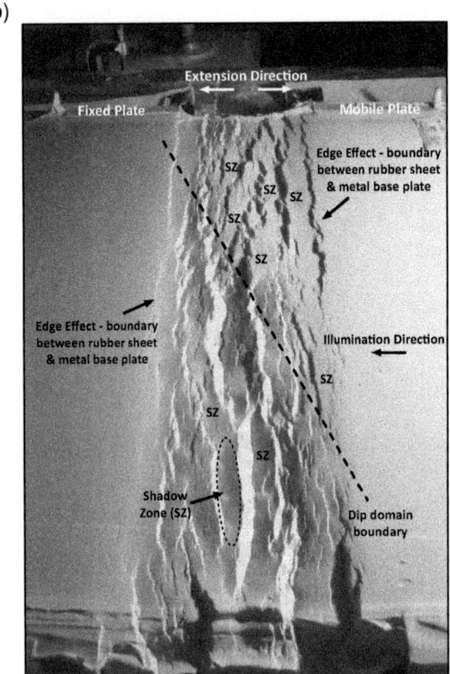

Fig. 3. (**a**) Overhead and (**b**) oblique photographs of Model 34 showing the overall structural style of the faults offsetting the clay surface at 4 cm of extension (model boundary displacement) (50% strain). For the indicated illumination directions, faults dipping towards the light source are bright and faults dipping away from the light source are in shadow. The area of analysis for each model (dashed box) excludes edge effects related to the sides of the model and the edges of the rubber sheet and putty layer. SZ, shadow zones.

dip towards the light source are illuminated (bright) and faults that dip away from the light source are in shadow (dark). Ambient light levels make recognition of very-small-displacement faults difficult in the initial stages of the models. For 10 representative models, we analysed the central parts to avoid edge effects associated with the sides of the apparatus and the edge of the rubber sheet and putty layer (Fig. 3). Line drawings of the illuminated model surface allowed us to trace the footwall and hanging-wall cut-offs. Low-relief undulations or corrugations on the fault surfaces indicate predominantly dip-slip movement (e.g. Granger 2006; Granger *et al.* 2006; Henza *et al.* 2010). The width of the fault polygons in the direction of fault slip, indicated by fault-surface corrugations and confirmed by the

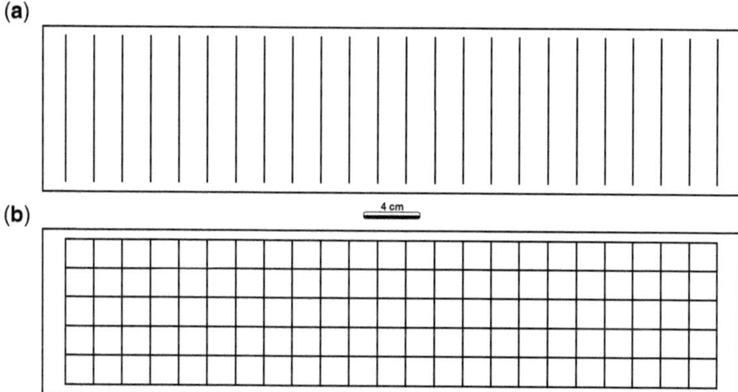

Fig. 4. Diagrams showing the methods applied to the analysis of fault spacing and segment-boundary density within the area of analysis. (**a**) Sample lines for fault-spacing measurements are orientated perpendicular to the rift axis and are spaced 2 cm apart. (**b**) Box-counting for segment-boundary density utilized a grid of 4 cm^2 cells; data points outside the sampling grid are excluded. The extent of the area of analysis is 550 cm^2.

offset of subtle surface markings, gives heave, which is used as a proxy for displacement. Additional close-up, oblique-view photographs captured details of fault interactions and secondary deformation.

Following the approach of Ackermann *et al.* (2001), we measured fault spacing along a series of 1D scanlines orientated parallel to the extension (and displacement) direction (Fig. 4a). Furthermore, for each model increment, we also measured fault length and displacement, and constructed displacement profiles for selected faults and fault arrays at various stages in their evolution. Finally, box-counting yielded the frequency of segment-boundary types using a regular grid consisting of 2×2 cm sample cells (4 cm^2) (Fig. 4b). The raw count data divided by 4 cm^2 yielded the density per cm^2.

Model results

Structure and scaling of normal-fault populations

Before we analysed the temporal and spatial distribution of fault-segment boundary types, we established that the scaling and design of the model resulted in the formation of a normal-fault population that is comparable in terms of fault geometry and statistical parameters to those observed in nature.

The upper surface of all 24 models at 4 cm extension exhibited numerous parallel to sub-parallel, linear to curvilinear normal faults. Individual fault arrays were comprised of several linked and unlinked fault segments, the majority of which had a strike perpendicular to the extension direction (Figs 3 & 5). Individual faults were predominantly planar in cross-section and had near-surface dips of approximately 60°. In some instances, a monoclinal flexure of the model surface formed prior to the development of a surface-breaking fault: these structures are interpreted to be fault-propagation folds (Fig. 6a) (e.g. Withjack *et al.* 1990). Based on sectioned models of distributed orthogonal extension (e.g. Withjack *et al.* 2008; Schlische & Withjack 2009), the largest faults dissected the entire thickness of the clay (2.7 cm), whereas smaller faults were fully or partially contained within the clay. Faults dipped either towards the moving wall or the fixed wall of the model (Fig. 3a). Changes in the mean fault polarity along the axis of the extension defined discrete fault-dip domains (i.e. populations with the same dip direction), the boundaries of which were either extension-parallel or extension-oblique accommodation zones (Fig. 3b) (e.g. Schlische & Withjack 2009).

The overall development of a typical fault array in the models is illustrated in Figure 7 and is as follows: (a) isolated, short (<2 cm), linear, extension fractures and low-displacement normal faults with free tips develop on the model surface during the initial stages of extension (Fig. 7b: <2.0 cm of extension); (b) new fault segments continue to breach the model surface with continued extension; existing faults lengthen by lateral tip propagation (Fig. 7b: 2.0–2.25 cm of extension); (c) fault segments begin to overlap and interact, and segment boundaries start to form, leading to the creation of discrete fault arrays defined by arrangements of sub-parallel fault segments (Fig. 7b: 2.25–2.5 cm of extension); (d) further tip propagation is impeded at overlapping segment boundaries

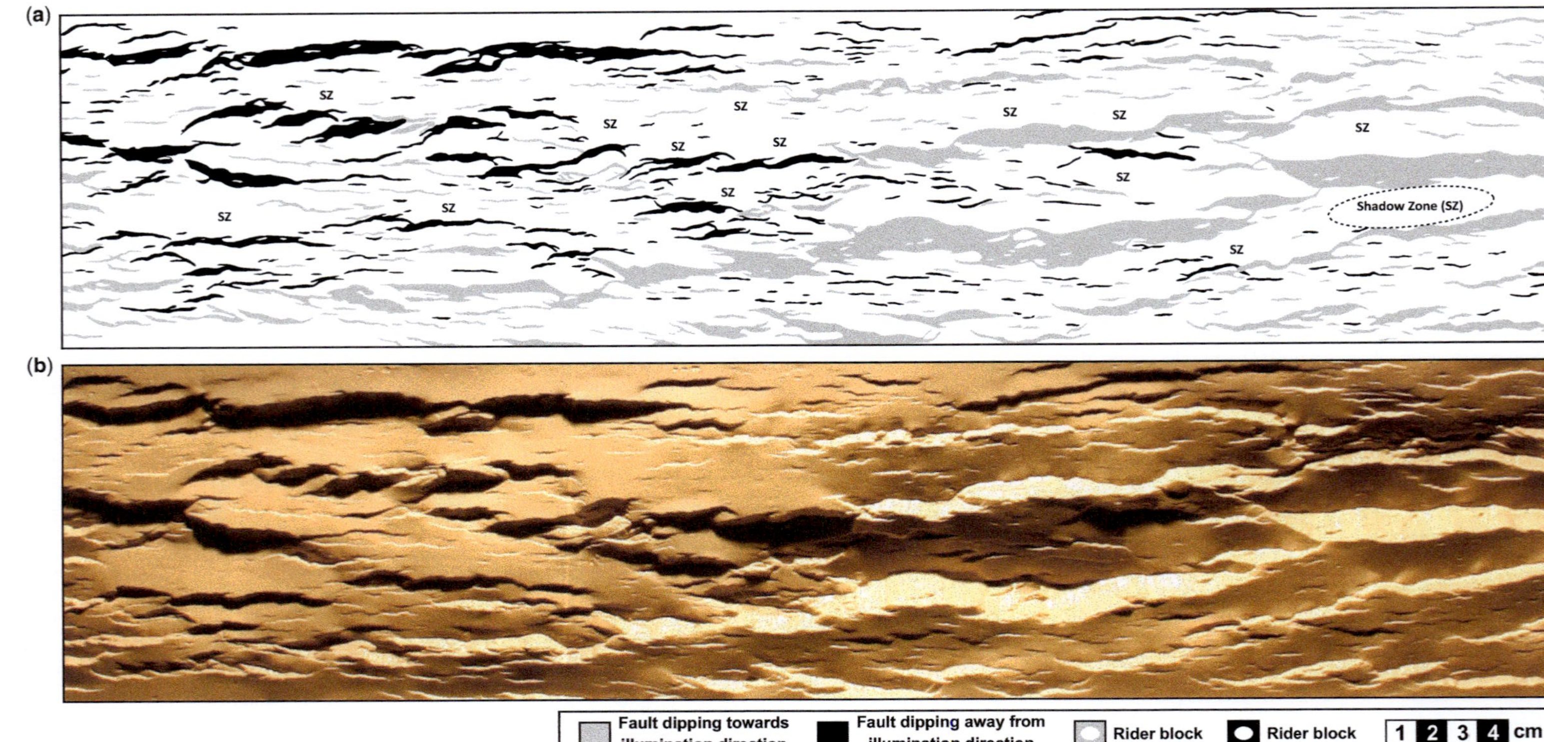

Fig. 5. (**a**) Line drawing of the clay surface of Model 34 at 4 cm of displacement (50% strain) within the area of analysis. Illumination is from the top of the photograph. (**b**) Plan-view photograph of the area shown in (a).

(a)

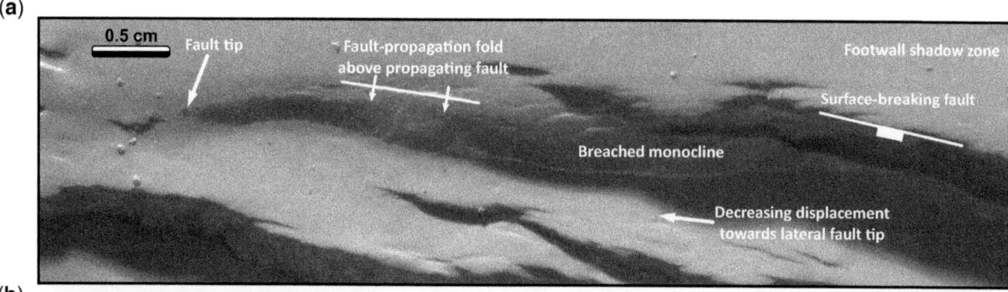

(b)

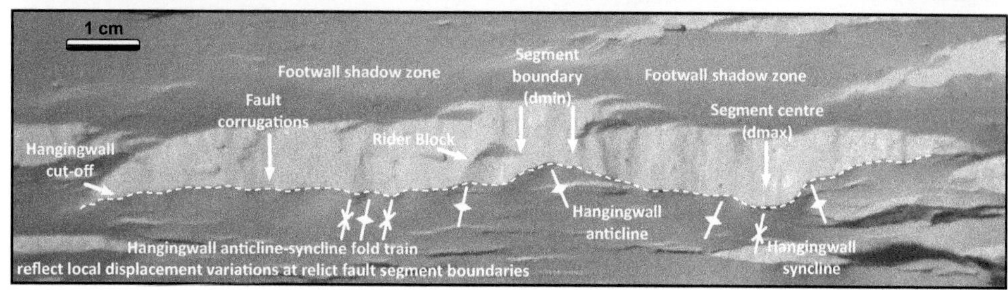

Fig. 6. Oblique-view photograph showing an example of: (**a**) extensional fault-propagation folding and the along-strike transition from the surface-breaking fault to the fault-propagation fold due to decreasing fault displacement towards the lateral fault tip (Model 53); and (**b**) hanging-wall fault-perpendicular folds, representing local displacement variations (synclines for d_{max}, anticlines for d_{min}) associated with relict segment boundaries (Model 29) (e.g. Schlische 1995). In contrast, the footwall to these faults is relatively undeformed. Fault corrugations indicate dip-slip displacement on the majority of faults in the models (e.g. Granger *et al.* 2006; Henza *et al.* 2010). Both photographs are taken after 4 cm extension (50% strain). Because the photographs are oblique views of the model surface, the scales are approximate.

by mechanical interaction between adjacent faults (e.g. Gupta & Scholz 2000), resulting in the development of relay ramps and steep lateral displacement gradients (Fig. 7b: 2.5–3.0 cm of extension); and (e) relay ramps may become breached with further extension, leading to the creation of through-going fault arrays (Fig. 7b: 3.0–4.0 cm of extension).

Localization of strain onto a few long, through-going fault arrays occurs at 3–4 cm of extension and results in the accumulation of displacement without a further concomitant increase in fault length (Fig. 7). In plan view, the final geometry of these major fault arrays is highly irregular. Maximum displacement on the through-going fault arrays typically occurs near its centre and decreases towards both lateral tips (Figs 5 & 7). The irregular surface geometries and systematic variations in displacement that occur along the length of each fault array are attributed to the processes of fault interaction and linkage. Fault-perpendicular hanging-wall anticlines occur at local displacement minima and define the position of palaeo-segment boundaries, and fault-perpendicular hanging-wall synclines occur at local displacement maxima and define the position of palaeo-segment centres (Fig. 6b) (see Schlische 1995). Shadow zones (e.g. Schlische *et al.* 1996; Ackermann & Schlische 1997), elliptical regions in the footwall and/or hanging wall of large faults with limited brittle deformation, are observed in the models (Figs 3, 5 & 6). The observed styles of fault linkage in each model are consistent with the classification in Figure 1 (Fig. 8a–e), with the style of linkage commonly being highly variable along the length of individual fault arrays (Figs 5, 7 & 8f).

Numerous studies have shown that fault populations display fractal (power-law) or exponential distributions of length and displacement (e.g. Childs *et al.* 1990; Walsh *et al.* 1991; Cowie & Scholz 1992*a*, *b*; Yielding *et al.* 1992; Peacock & Sanderson 1994; Huggins *et al.* 1995; Schlische *et al.* 1996; Wojtal 1996; Ackermann & Schlische 1997). In addition, faults display a power-law relationship between maximum displacement (*D*) and fault length (*L*). Given that samples of fault populations rarely exceed two orders of magnitude in size and given the inherent variability in natural fault populations, the value of *n* is contentious, ranging from 0.5 to 2. Nonetheless, many datasets indicate that *n* is approximately 1 (e.g. Kim & Sanderson 2005; Torabi & Berg 2011). For one representative

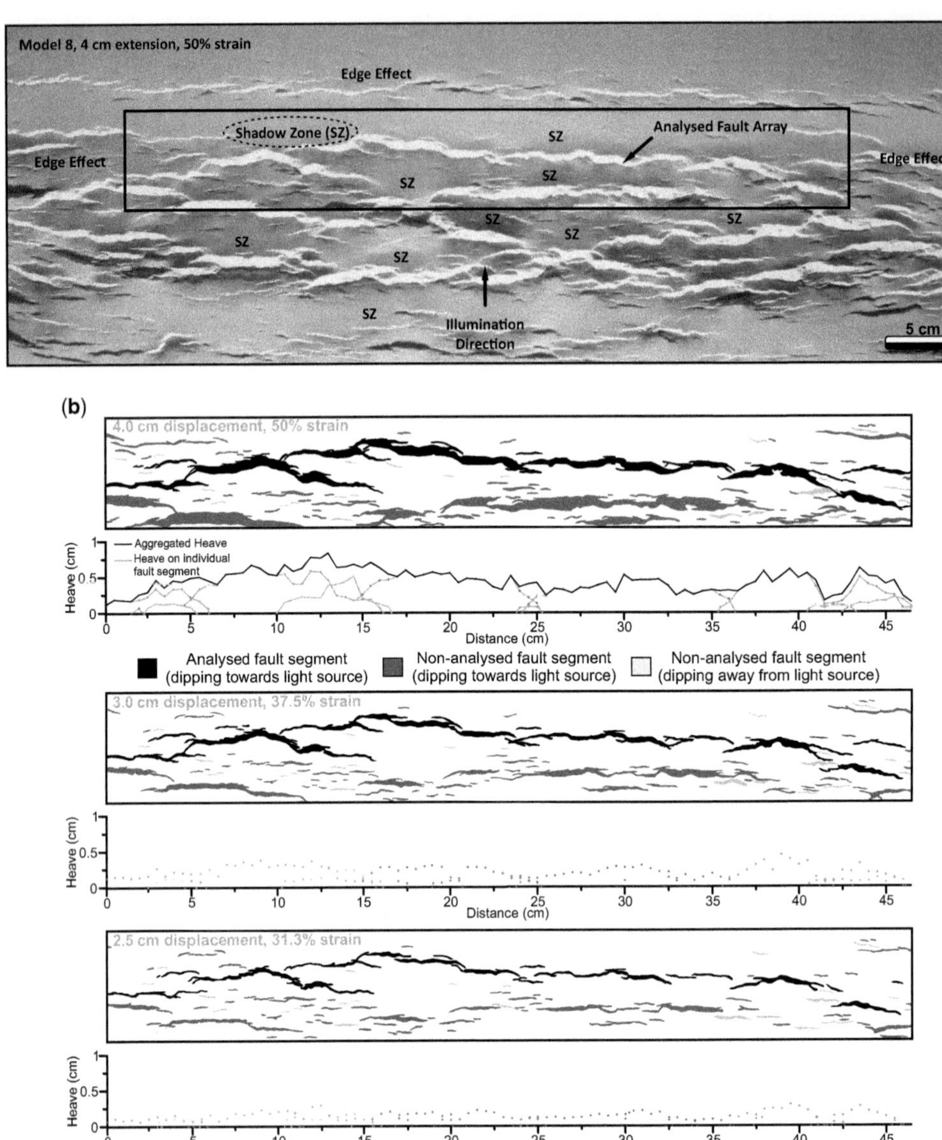

Fig. 7. Line drawings and accompanying heave–distance profiles showing the evolution of fault arrays developed in Model 8. (**a**) Plan-view photographs of the clay surface for Model 8 (at 4 cm of extension, 50% strain) indicating the extent of the fault arrays in (**b**) line drawings showing the temporal evolution of the fault array at different time steps in Model 8.

model (Model 34), the cumulative frequency of fault lengths has a power-law distribution with a fractal dimension of −1.12 over one order of magnitude (Fig. 9a). Furthermore, the fault population at the end of the model run exhibits a broadly linear relationship between fault length and maximum heave, with a change in scaling occurring at approximately 5 cm, with faults (>5 cm long) appearing to be relatively over-displaced compare to the shorter faults (<5 cm length) (Fig. 9b).

Scatter observed within the displacement–length dataset is attributed to one or a combination of the following factors: (a) censoring of the fault length and/or maximum displacement related to

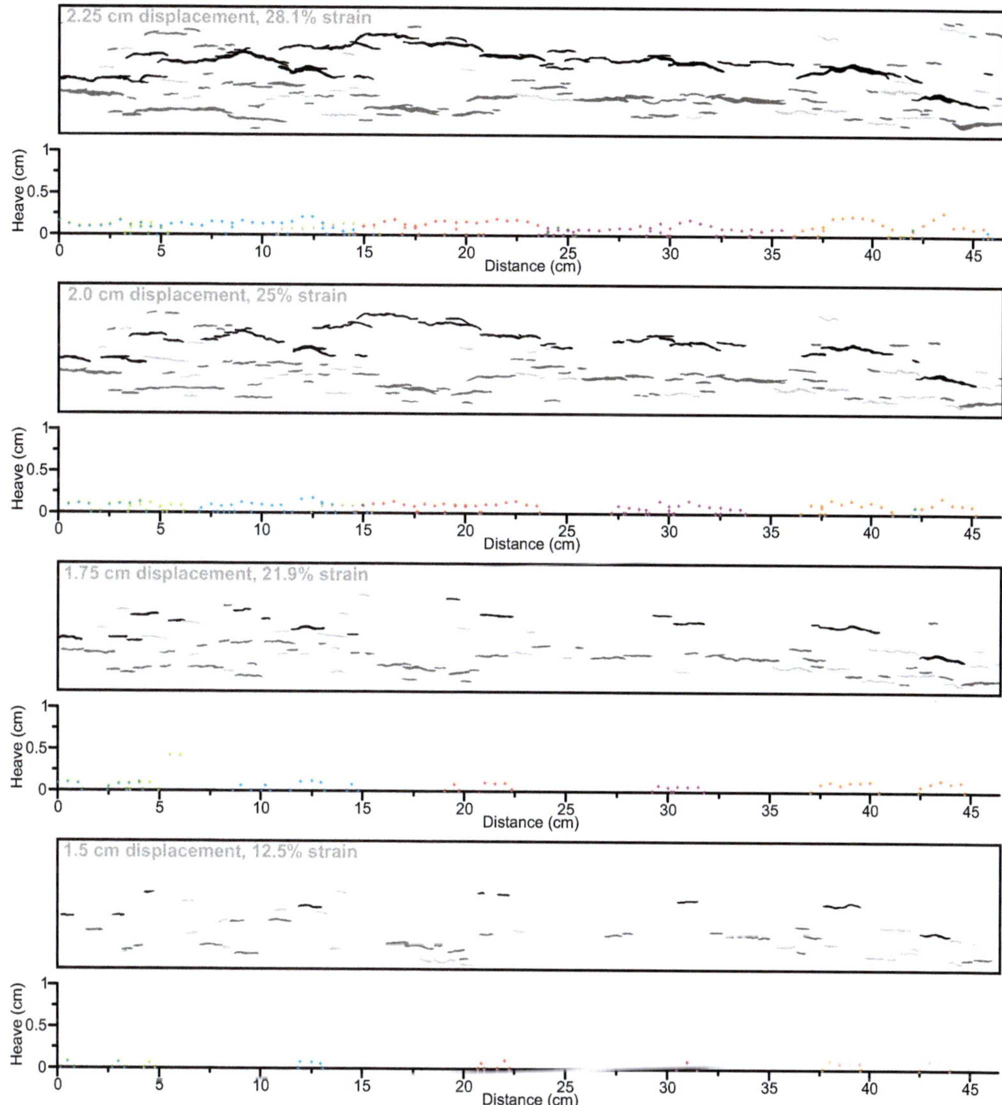

Fig. 7. *Continued.*

faults truncated at the edge of the analysed area; (b) linkage of initially isolated segments (Cartwright *et al.* 1995); (c) analysis of faults whose maximum displacement and length do not coincide with the model surface (Walsh *et al.* 2003*a*, *b*; Kim & Sanderson 2005); (d) mechanical interaction and 'pinning' of fault tips at the boundaries of dip domains (Schlische & Withjack 2009); and (e) the low-displacement, near-tip regions of faults falling below the resolution of the camera.

Fault heights in the models were limited by the thickness of the clay layer (2.7 cm) (Fig. 2). Assuming a fault dip of 60°, the maximum dip-corrected height of the faults in the clay layer is 3.12 cm ($H = 270°/\sin\ 60°$). In Model 34, the mean tip-to-tip fault length of the measured surface-breaking faults is 2.7 cm ($n = 500$) (Fig. 9); these yield a fault length to height aspect ratio of 0.87, indicating that modelled faults have an approximately circular shape. Analysis of fault height in the models is limited by our plan-view analysis, in that not all surface-breaking faults will span the entire clay layer. The average fault length to height ratio is therefore likely to be a minimum value.

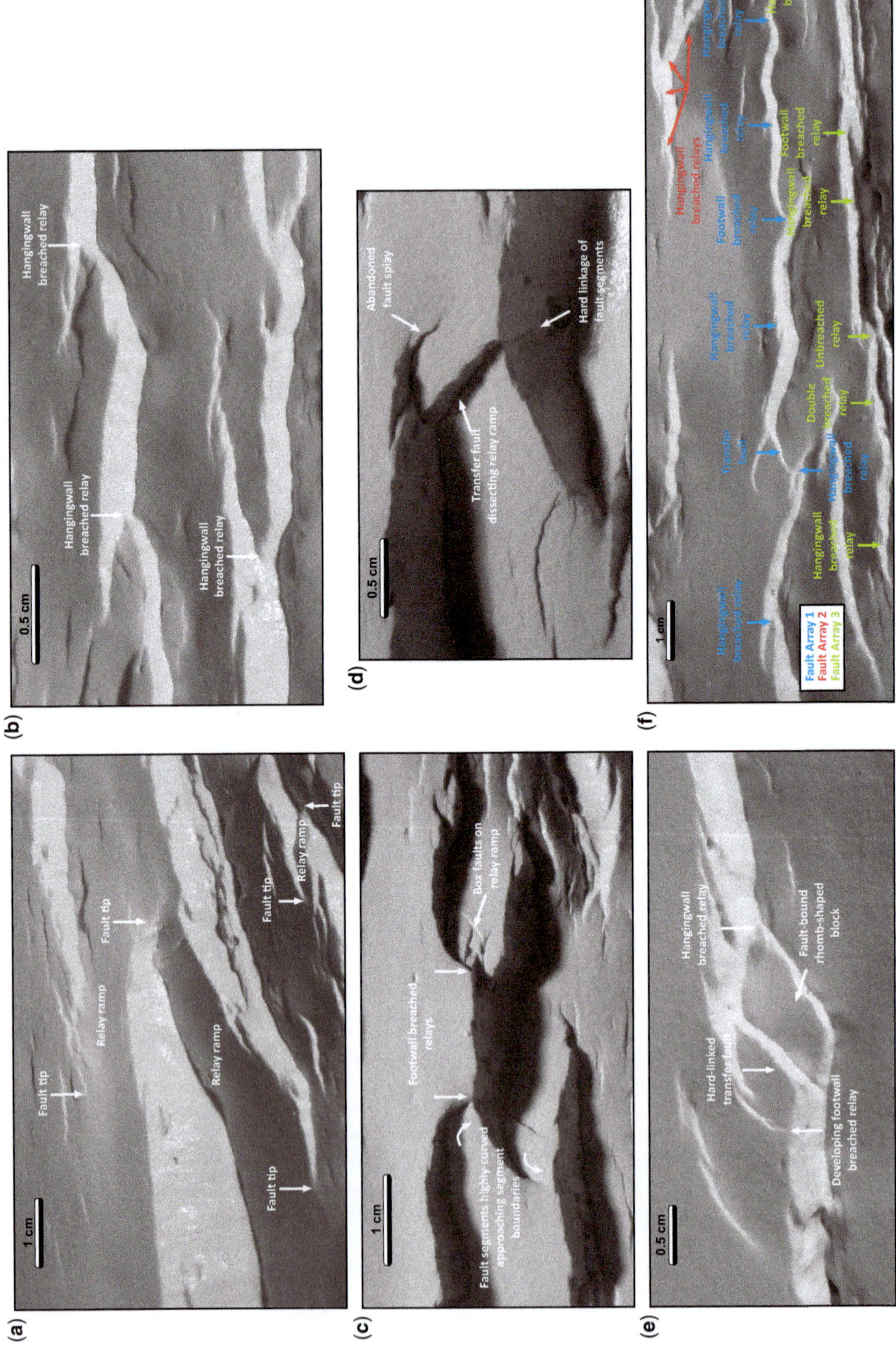
(a)
1 cm
Fault tip
Relay ramp
Fault tip
Relay ramp
Fault tip
Fault tip
Relay ramp
Fault tip
(b)
0.5 cm
Hangingwall breached relay
Hangingwall breached relay
Hangingwall breached relay
(c)
1 cm
Fault segments highly curved approaching segment boundaries
Footwall breached relays
Box faults on relay ramp
(d)
0.5 cm
Transfer fault dissecting relay ramp
Abandoned fault splay
Hard linkage of fault segments
(e)
0.5 cm
Hard-linked transfer fault
Developing footwall breached relay
Hangingwall breached relay
Fault-bound rhomb-shaped block
(f)
1 cm
Fault Array 1
Fault Array 2
Fault Array 3
Hangingwall breached relay
Hangingwall breached relay
Hangingwall breached relay
Double breached relay
Unbreached relay
Hangingwall breached relay
Footwall breached relay
Hangingwall breached relay
Footwall breached relay
Hangingwall breached relays
Footwall breached relay

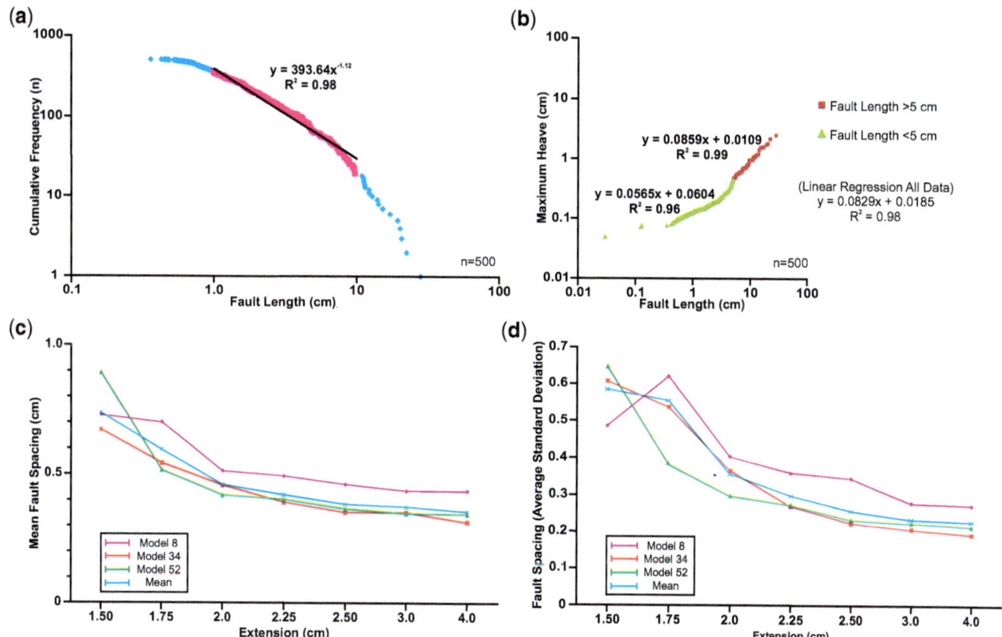

Fig. 9. Fault-population statistics for faults cutting the clay surface for Model 34 at 4 cm of extension (50% strain). (**a**) Log–log plot showing the cumulative frequency of fault lengths. A fractal, power-law distribution of lengths ($R^2 = 0.98$) applies for over one order of magnitude. The minimum fault length is the smallest fault length that can be measured consistently across the area of analysis in the models. (**b**) Log–log plot of maximum heave v. fault length showing a broadly linear relationship. A change in scaling is observed between shorter (<5 cm length) and longer faults (>5 cm length), where longer faults appear to be relatively over-displaced compared to the shorter faults. (**c**) & (**d**) The evolution of fault spacing in models 8, 34 and 52 as a function of increasing extension. (c) Mean fault spacing decreases with increasing extension. (d) The standard deviation of average fault spacing decreases with increasing extension, reflect an increase in the regularity of faults (see Ackermann *et al.* 2001).

The average fault spacing along scanlines for seven displacement increments in models 8, 34 and 52 decreases steadily with increased extension, following a power-law trendline ($R^2 = 0.98$) (Fig. 9c). The regularity of the fault spacing is given by the average standard deviation of the spacing, where a low standard deviation indicates a greater regularity (e.g. Ackermann *et al.* 2001). The regularity of faults in the models increases sharply during early extension (1.5–2.25 cm of extension) but flattens off with further extension (>2.25 cm of extension) (Fig. 9d).

The structural and statistical characteristics of the fault populations in our models are comparable to those in nature (e.g. Walsh & Watterson 1988; Morley *et al.* 1990; Peacock & Sanderson 1991, 1994, 1996; Gillespie *et al.* 1992; Dawers *et al.* 1993; Gawthorpe & Hurst 1993; Dawers & Anders 1995; McClay & Khalil 1998; Acocella *et al.* 2000; McLeod *et al.* 2000, 2002; Young *et al.* 2001; Moustafa 2002; Gawthorpe *et al.* 2003) and those generated in other physical analogue models (e.g. Clifton *et al.* 2000; Ackermann *et al.* 2001; Clifton & Schlische 2001; Mansfield & Cartwright 2001; McClay *et al.* 2002; Bellahsen *et al.* 2003; Hus *et al.* 2005; Schlagenhauf *et al.* 2008; Schlische & Withjack 2009; Henza *et al.* 2010). Gross similarities in the temporal and spatial evolution of the fault populations observed for all 24 models confirm the reproducibility of our modelling results (Fig. 5; Table 1) (see Whipp 2011).

Fig. 8. Oblique-view photographs showing examples of synthetic fault-segment boundary types developed within the experimental models: (**a**) unbreached relay ramp (Model 33); (**b**) hanging-wall-breached relay (Model 35); (**c**) footwall-breached relay (Model 6); (**d**) transfer fault (Model 18); and (**e**) double-breached relay (Model 32). (**f**) Oblique-view photograph from Model 46 showing the along-strike variability in fault-segment boundary types along three neighbouring fault arrays. All photographs are taken after 4 cm extension (50% strain). Because the photographs are oblique views of the model surface, the scales are approximate.

Table 1. *Summary of information on set-up and observations from the 10 analysed models*

Model No.	Boundary conditions	Model dimensions ($L \times W \times D$) (cm)	Clay density (g cm^{-3})	Putty thickness (cm)	Clay thickness (cm)	Comments
4	Rubber sheet between metal base plates	60 × 54 × 3	1.6	0.3	2.7 above putty 3.0 outside	Well-developed fault systems across the model. Fault polarity predominantly towards the fixed plate. Evidence of dip-slip corrugation along the fault plane. Early formed faults appeared to influence development of the array
5	Rubber sheet between metal base plates	60 × 54 × 3	1.55	0.3	2.7 above putty 3.0 outside	Model displays a range of fault sizes, geometries and segment-boundary types. Faults appear to rapidly obtain their length through linkage, and accrue displacement in the final part of the model. Fault polarity is strongly towards fixed wall
7	Rubber sheet between metal base plates	60 × 54 × 3	1.56	0.3	2.7 above putty 3.0 outside	Nicely developed fault systems across the model with good examples of fault interaction and linkage along the moving plate side
8	Rubber sheet between metal base plates	60 × 54 × 3	1.56	0.3	2.7 above putty 3.0 outside	Good model with lots of examples of fault interaction. Useful model to observe the hook-shaped tip of interacting fault segments. Major fault systems illustrate the distribution of fault displacement. Examples of release faults observed on relay ramps
29	Rubber sheet between metal base plates	60 × 54 × 3	1.55	0.3	2.7 above putty 3.0 outside	Model showed a well-developed fault array with good examples of the various segment boundary types. There is no dominant fault polarity in the model. Excellent example of a fault-propagation fold developed along a low-displacement fault

32	Rubber sheet between metal base plates	60 × 54 × 3	1.57	0.3	2.7 above putty 3.0 outside	Excellent model. Well-developed fault fabric across the model with a range of fault sizes, geometries and interactions observed. Faults predominantly dip towards the fixed plate, with major faults developed around the deformation axis. Good examples of secondary deformation
34	Rubber sheet between metal base plates	60 × 54 × 3	1.56	0.3	2.7 above putty 3.0 outside	Well-developed fault network with two distinct dip domains separated by a zone of flexural deformation. Faults mainly grew by segment linkage. In the central part of the model, a relative high exists, around which small faults nucleated but failed to develop
35	Rubber sheet between metal base plates	60 × 54 × 3	1.55	0.3	2.7 above putty 3.0 outside	Well-developed fault systems. Significant topographical high in the central region of the model where there is an along-strike switch in polarity of two large faults. Predominantly fault dips are towards the moving wall. Examples of interaction between collinear fault segments
51	Rubber sheet between metal base plates	60 × 54 × 3	1.55	0.3	2.7 above putty 3.0 outside	Excellent model. Good examples of a range of fault-propagation structures, and isolated to interacting faults within the analysed region. Examples of secondary deformation associated with the development of relay ramps. Fault-perpendicular folds observed at fault-segment boundaries
52	Rubber sheet between metal base plates	60 × 54 × 3	1.55	0.3	2.7 above putty 3.0 outside	Well-developed fault systems in the central part of the model. Good examples of fault-system development through segment linkage and associated secondary deformation. Strain in the model was well distributed through a series of rotated fault blocks downthrowing towards the fixed wall

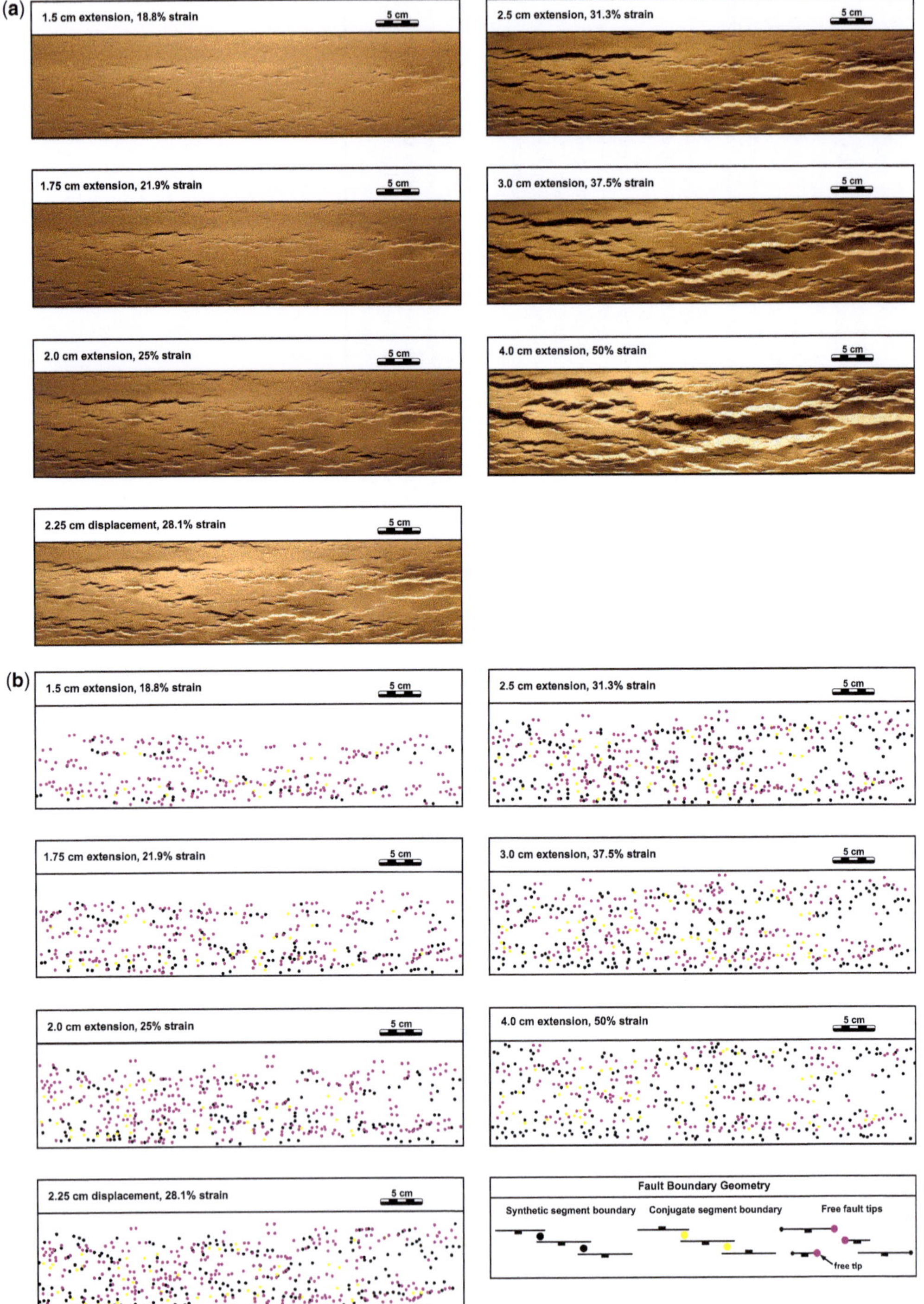

Fig. 10. (**a**) Plan-view photographs of the evolving clay surface for Model 34 in the area of analysis at: (i) 1.5 cm; (ii) 1.75 cm; (iii) 2 cm; (iv) 2.25 cm; (v) 2.5 cm; (vi) 3 cm; and (vii) 4 cm of extension. The area of analysis is indicated in Figure 3. (**b**) Maps showing the distribution of segment-boundary types for the photographs in (a).

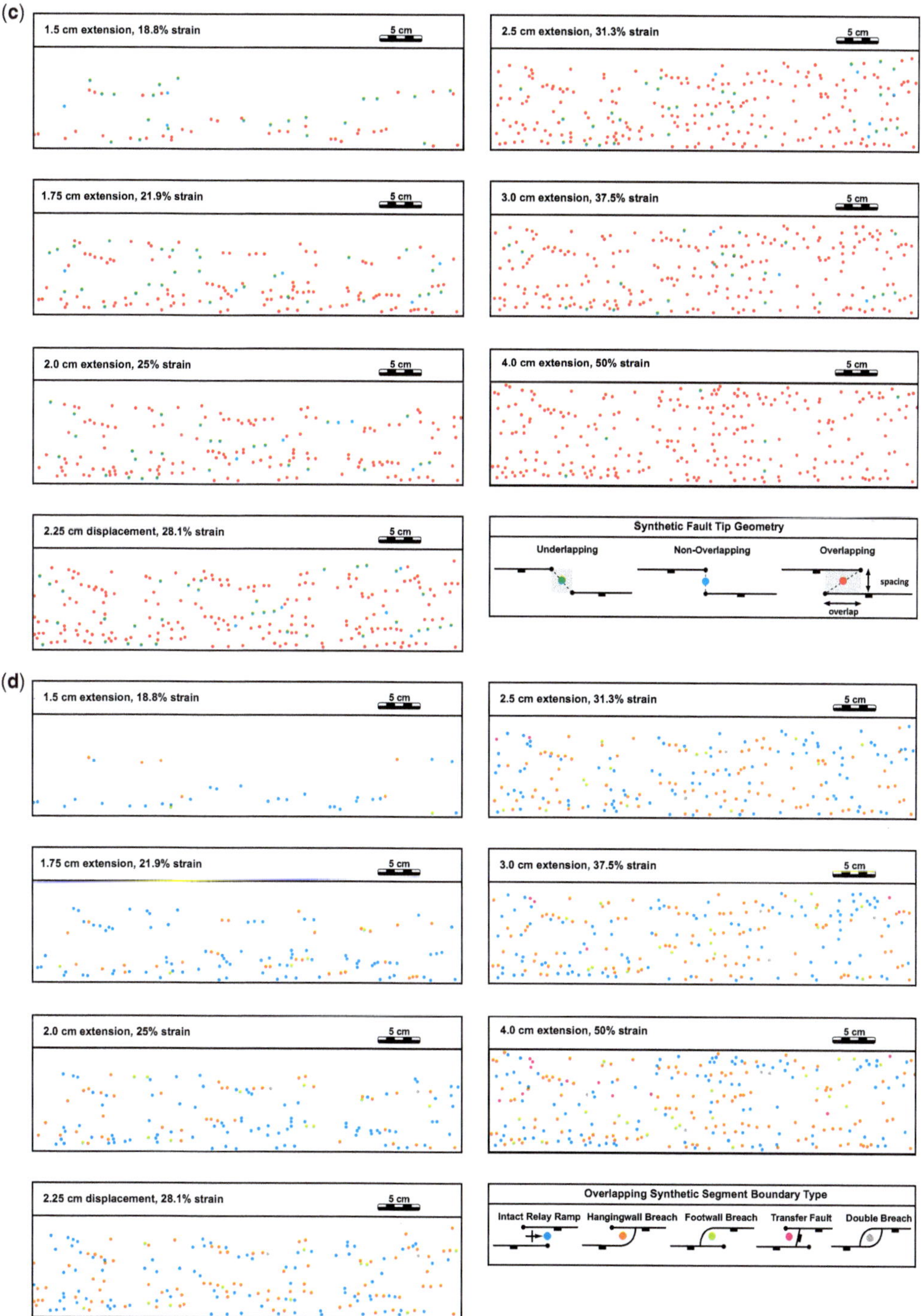

Fig. 10. (**c**) Maps showing the distribution of synthetic segment-boundary geometries. (**d**) Maps showing the distribution of overlapping synthetic segment-boundary types. See Figure 1 for classification of the fault-segment boundary types.

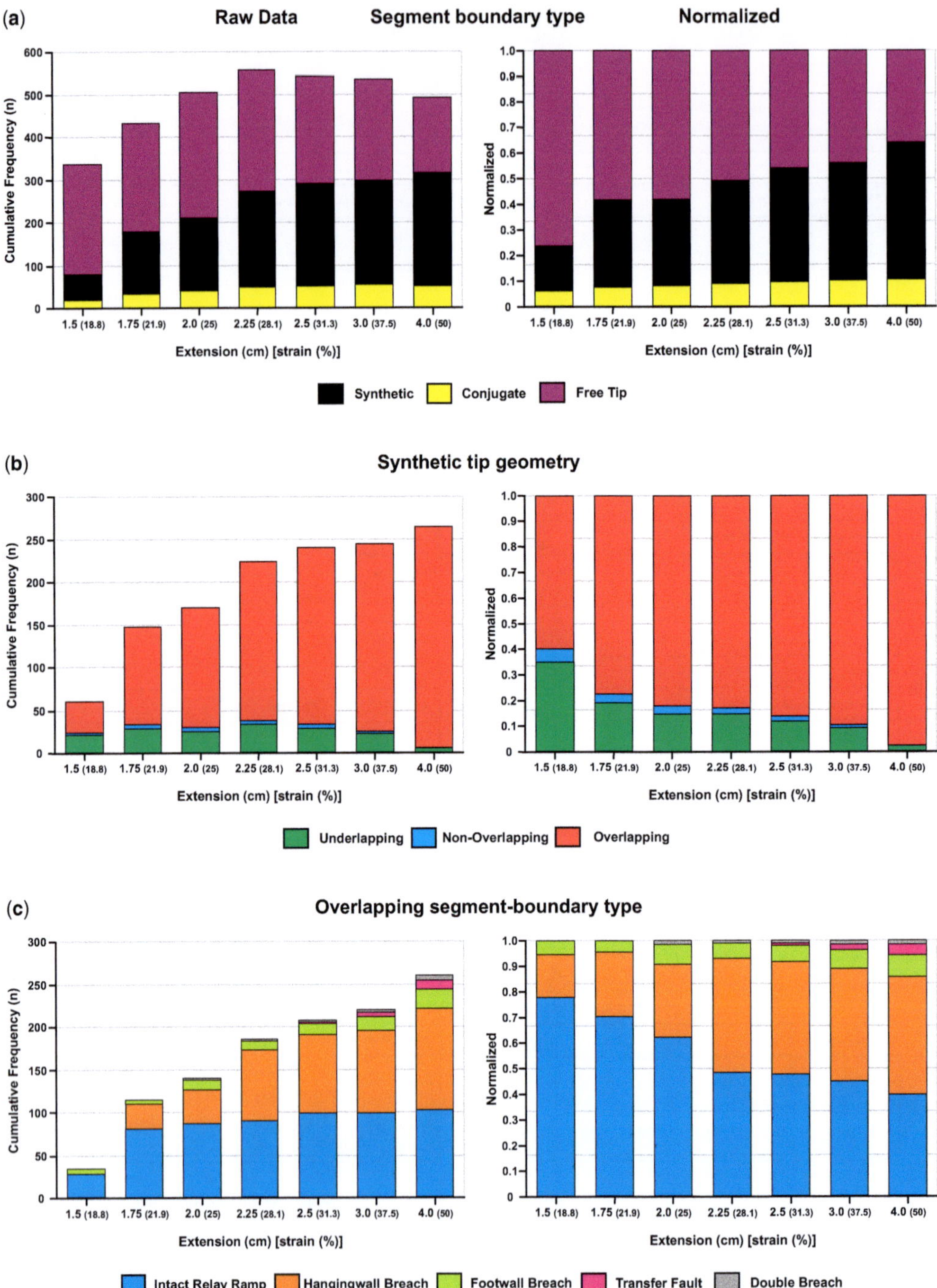

Fig. 11. Graphs showing the number of different types of fault-segment boundaries as a function of extension in Model 34. The left-hand set of graphs shows the cumulative frequency (*n*, number) of segment boundaries and the right-hand column shows the normalized number of segment boundaries based on: (**a**) segment-boundary type; (**b**) synthetic tip geometry; and (**c**) type of overlapping segment-boundary type. See Figure 1 for classification of segment-boundary types.

Temporal and spatial evolution of segment boundaries

In the following sections, we present the results from Model 34 (Figs 10 & 11), a representative example from the analysed suite of models, as well as the combined data from all 10 models (Fig. 12) (see Whipp 2011). We identified the types of fault-segment boundaries (Fig. 1) observable on the model surface within the area of analysis for the following values of boundary displacement (extension) (Figs 10 & 11): (i) 1.5 cm (strain 18.8%); (ii) 1.75 cm (21.9%); (iii) 2 cm (25%); (iv) 2.25 cm (28.1%); (v) 2.5 cm (31.3%); (vi) 3 cm (37.5%); and (vii) 4 cm (50%). Strain = $[(L_f - L_0)/L_0] \times 100\%$, where L_f is the deformed width of the rubber sheet and L_0 is the original undeformed width of the rubber sheet (i.e. 8 cm) (Fig. 2).

Segment-boundary type and free tips. The free tips of relatively isolated normal faults represent up to 80% of the segment tips developed during the initial stages of displacement (up to 1.5 cm) (Figs 11a & 12a). Synthetic and conjugate segment boundaries represent approximately 17 and 3% of the population, respectively, at this stage. The percentage of free tips progressively declines with increasing extension (*c.* 50% by 2.25 cm and <40% at 4 cm of extension: Fig. 12a). An increase in the number of synthetic and conjugate segment boundaries complements the decrease in the free-tip population, where, upon the development of a segment boundary, two free tips are replaced by a single segment boundary. The greatest increase in the number of conjugate and synthetic segment boundaries occurs between 1.5 and 2 cm of extension. The number of conjugate boundaries increase by up to 1.4 times (Fig. 12a) and synthetic boundaries double (Figs 11a & 12a) at this stage. Conjugate segment boundaries remain constant (*c.* 10% of the total population) with increasing extension (1.75–4 cm). Synthetic segment boundaries form >50% of the segment-boundary population between 2 and 4 cm of extension, and are over four times more numerous than conjugate segment boundaries (Figs 11a & 12a).

Synthetic tip geometry. Underlapping and non-overlapping tip geometries are transient in our models. They are most abundant during the initial stages of extension, forming 35% (i.e. underlapping) and 5–10% (i.e. non-overlapping) of the population, respectively, at 1.5 cm of extension (18.8% strain) (Figs 11b & 12b). In contrast, at the end of the model (4 cm of displacement), they form up to 3% (i.e. underlapping) and <1% (i.e. non-overlapping) of the population, respectively. Non-overlapping tip geometries are the least common synthetic tip geometry at all stages of each model.

Overlapping fault tips are the most common tip geometry at synthetic segment boundaries at each stage of extension (Figs 11b & 12b). Overlapping tip geometries develop from precursor underlapping and non-overlapping fault-tip geometries. Overlapping fault tips make up approximately 55% of synthetic tip boundaries at 1.5 cm of extension and >97% at 4 cm of extension. Development of overlapping tip geometries is most pronounced between 1.5 and 2 cm of extension, where there is a three- to fourfold increase in their number. Between 2 and 4 cm of extension, overlapping synthetic segment boundaries increase by around 4% per extension increment (0.25 cm); >50% of all new synthetic segment boundaries have overlapping tip geometries during this stage.

Unbreached- v. breached-segment boundaries. Approximately 80% of overlapping synthetic segment boundaries are unbreached during the early stages of extension (<1.5 cm extension, 18.8% strain) (Figs 11c & 12c). Increasing numbers of unbreached relay ramps become breached with continued extension. The number of breached-segment boundaries exceeds the number of unbreached-segment boundaries between 2 and 2.5 cm of extension. More than 60% of overlapping synthetic segment boundaries are breached at 4 cm of extension.

Breaching of relay ramps is caused by fault linkage and results in the creation of a breached-segment boundary (Fig. 1). The proportional change of an individual linkage style, as a fraction of the breached-segment boundary population, can vary between successive increments in the models. Hanging-wall breaching is the most common style of fault linkage in the models, forming >70% of breached-segment boundaries at each extension increment (Figs 11c & 12c). The proportion of hanging-wall-breached relays is greatest between 1.5 and 2.25 cm of extension, and accounts for up to 85% of breached-segment boundaries. A decrease in the proportion of hanging-wall-breached-segment boundaries occurs with further extension, and is associated with an increasing number of footwall, transfer-fault and double-breached linkage styles. Footwall-breaching is the second most common linkage style and represents 12–25% of breached-segment boundaries at any given extension increment, and forms approximately 16% of all breached-segment boundaries at 4 cm of displacement. Analysis of all the models at 4 cm extension (Fig. 12c) indicates that 5.2% of overlapping synthetic segment boundaries are double breached, and only 3.5% of overlapping synthetic segment boundaries are breached via a transfer fault. In some models, however, transfer faults may be more

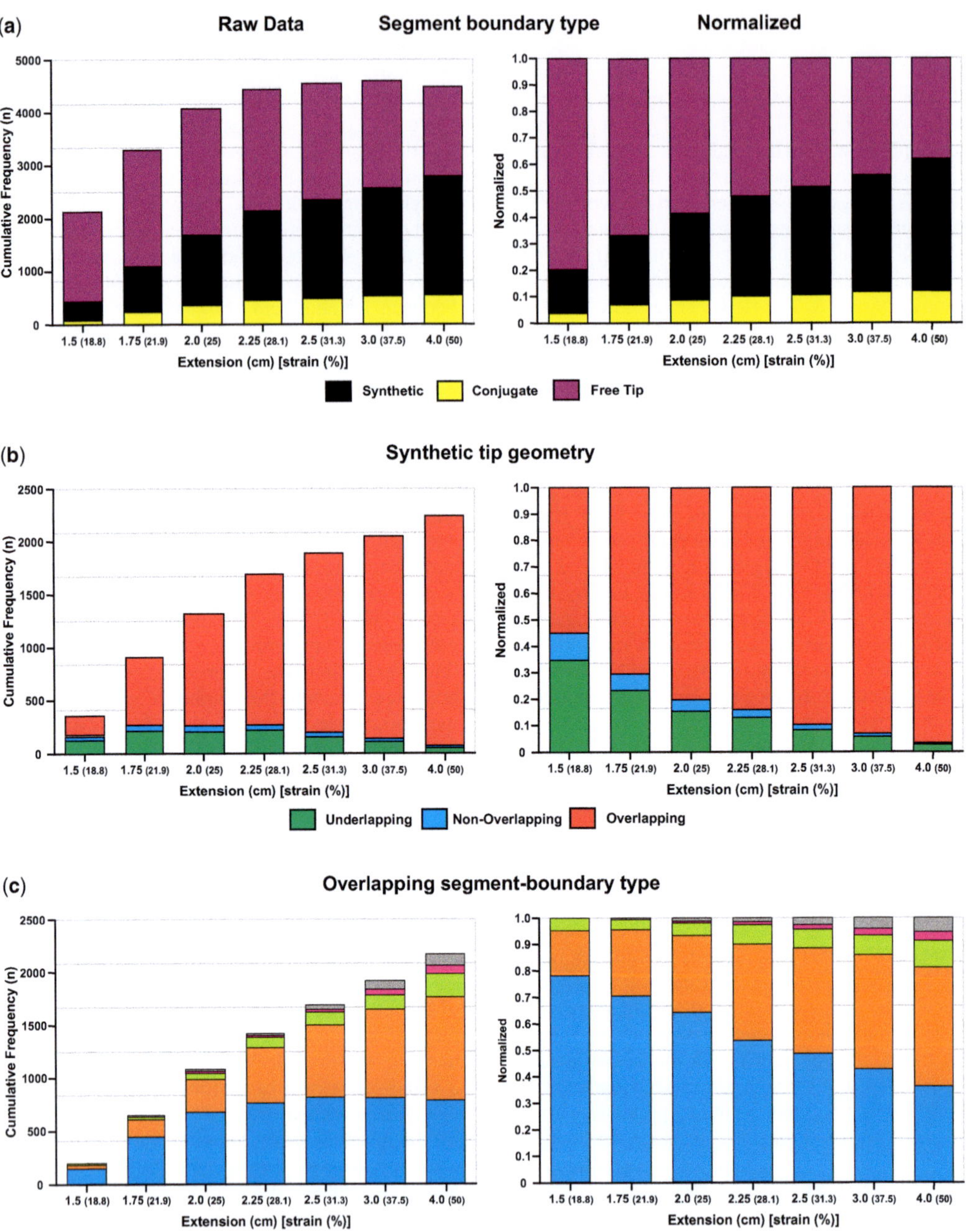

Fig. 12. Graphs to show the mean temporal distribution of fault-segment boundaries across the 10 analysed models (Whipp 2011, appendix 8.2). The first column shows the cumulative frequency (*n*, number) and the second column shows the normalized data for each extension increment. (**a**) Segment-boundary type. (**b**) Synthetic tip geometry. (**c**) Overlapping segment-boundary type. See Figure 1 for classification of segment-boundary types.

abundant than double-breached relays, but transfer-fault and double-breached relays together never form more than 12% of fully breached-segment boundaries. Transfer-fault and double-breached relays linkage styles rarely develop during the early stages of extension.

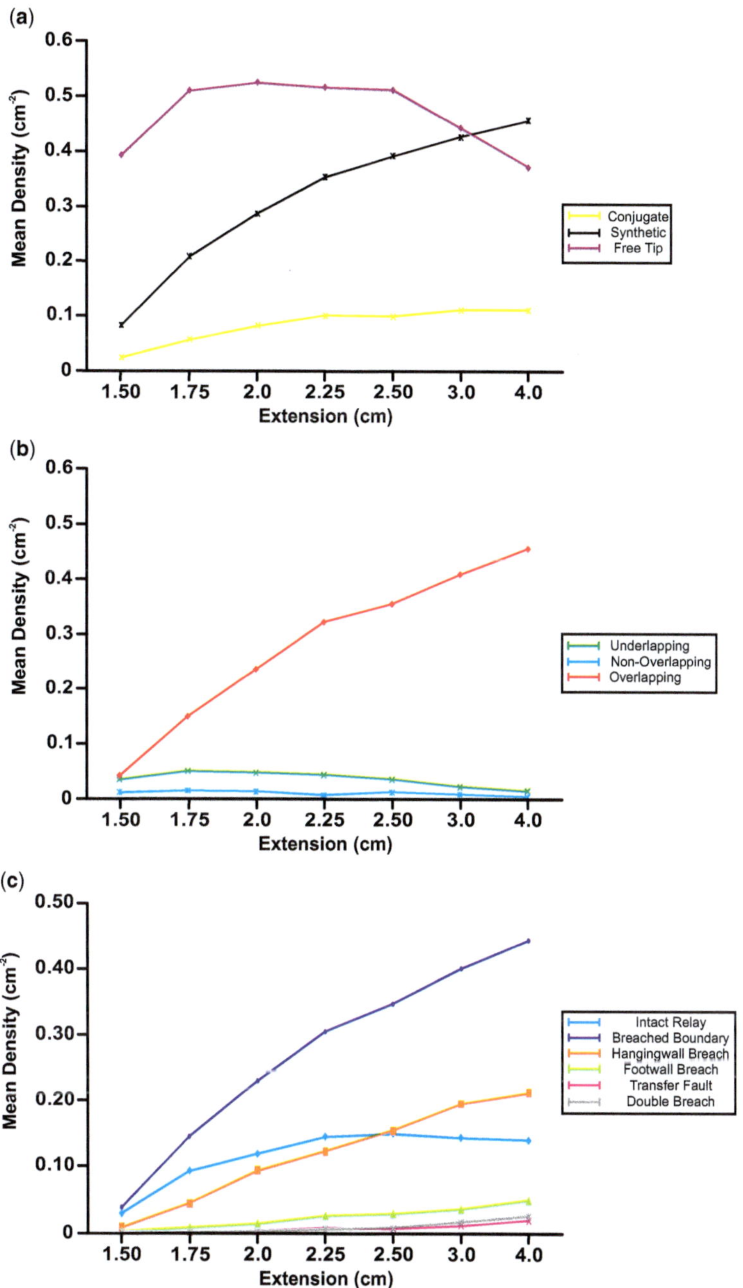

Fig. 13. Graphs showing how mean fault-segment boundary density (cm^{-2}) in five models (models 8, 29, 32, 35 and 52) varies with increasing extension; (**a**) Segment boundaries types and free tips; (**b**) synthetic segment-boundary geometries; and (**c**) overlapping synthetic segment-boundary types). See Figure 1 for classification of segment-boundary types.

Segment-boundary density. We determined the mean density (per cm^2 (cm^{-2})) of each segment-boundary type for seven displacement increments in models 8, 29, 32, 35 and 52 (Fig. 13). The density of free tips increases during the initial stages of extension (1.5–1.75 cm of extension), peaking at

0.52 cm^{-2} (2 cm), before subsequently declining steadily during the middle (2–2.5 cm) stage and then more sharply (to 0.37 cm^{-2}) during the latter (2.5–4 cm) stages of the experiments at 4 cm of extension (Fig. 13a). The decrease in the density of free tips is concomitant with an increase in the density of synthetic segment boundaries, which steadily increases during deformation, with a maximum density of 0.46 cm^{-2} at 4 cm of extension (Fig. 13a). The density of synthetic segment boundaries becomes greater than the density of free fault tips between 3 and 4 cm of extension (37.5–50% strain). The density of conjugate segment boundaries steadily increases between 1.5 and 2.25 cm extension (18.8–31.8% strain), levelling-off to a consistent density of approximately 0.1 cm^{-2} (2.25–4 cm extension, 31.8–50% strain) (Fig. 13a).

Underlapping and non-overlapping synthetic tip geometries have a constant density of 0.03 and 0.01 cm^{-2}, respectively, for the majority of extension, and only decrease during the final stages of the model runs (Fig. 13b). The overall density of overlapping synthetic tips increases during extension to a maximum density of 0.45 cm^{-2} (Fig. 13b). Between 1.5 and 2.25 cm of extension, there is a sevenfold increase in the density of overlapping synthetic tips, with only a further 28% increase in density between 2.25 and 4 cm of extension (Fig. 13b).

The density of unbreached and breached overlapping synthetic segment boundaries is approximately the same (0.3–0.4 cm^{-2}) during the initial stages of deformation (<1.5 cm of extension). The density of unbreached relays increases rapidly between 1.5 and 2.5 cm of extension but, subsequently, their density decreases with further extension (2.5–4 cm extension) (Fig. 13c). The temporal change in the density of unbreached-segment boundaries reflects the initial development of overlapping, unbreached-segment boundaries and their subsequent destruction (breaching) with increasing strain (Fig. 13c). The density of breached-segment boundaries increases linearly during the experiments to 0.45 cm^{-2} at the end of extension, which is three times greater than the density of unbreached-segment boundaries (0.14 cm^{-2}). Each type of breached-segment boundary increases in density with increasing extension along a linear trendline. At the end of extension, the density of hanging-wall-breached relays is 4–10 times greater than the combined total of footwall, transfer-fault and double-breached relays (Fig. 13c).

Overlap–spacing and segment-boundary type. The plan-view geometry of overlapping synthetic fault tips is described by the overlap (O), spacing (S) and aspect ratio (O:S) of the segment boundary (Fig. 1). We measured these parameters for a sample set of 146 unbreached- and breached-segment boundaries across all 10 models (Fig. 14). Fault spacing ranged from 0.16 to 1.38 cm, with a mean value of 0.44 cm (standard deviation (SD) = 0.24) and fault overlap values range from 0.27 to 5.4 cm, with a mean value of 1.05 (SD = 0.67) (Fig. 14a). Aspect ratios for unbreached- and breached-segment boundaries ranged from 0.9 to 6.3, with a mean O:S of 2.4 (SD = 0.92) (Fig. 14b). There appears to be no clear relationship between aspect ratio and segment-boundary type; unbreached and different types of breached-segment boundaries occur across the full range of aspect ratios (Fig. 14a).

The cumulative frequency distribution of aspect ratios follows an exponential curve ($R^2 = 0.97$), showing that smaller aspect ratios are more common in the sample set (Fig. 14b). The same relationship would be true for a power-law curve, although this would highlight fractal behaviour rather than the scale-dependent behaviour indicated by the exponential curve.

A cross-plot of spacing v. aspect ratio shows a large amount of scatter (Fig. 14c). However, hanging-wall- and double-breached relays appear to be more numerous at smaller fault spacing (<0.5 cm) compared to the other segment-boundary types. A plot of overlap v. aspect ratio shows that larger aspect ratios generally correlate with increasing overlap (Fig. 14d). The broader range of dimensions and the greater standard deviation of fault-tip overlap suggest that fault overlap is more variable at segment boundaries in our experiments than fault spacing.

Discussion

Segment-boundary evolution

Our model results, especially those related to the general timing of development and breaching of segment boundaries, are consistent with those made from other modelling studies and inferred from field or seismic datasets. For example, the overall number of overlapping segment boundaries increases and the number of non-interacting (free) tips decreases with increasing extension, as faults nucleate, and their tips propagate and overlap. Overlapping synthetic segment boundaries, a prerequisite for the growth of fault arrays through segment linkage, are the most common fault-tip geometry observed in all our models (Figs 7 & 12) and in many natural fault arrays (cf. Peacock & Sanderson 1991; Trudgill & Cartwright 1994; Cartwright *et al.* 1995, 1996; Contreras *et al.* 2000; Dawers & Underhill 2000; McLeod *et al.* 2000, 2002; Young *et al.* 2001; Moustafa 2002; Gawthorpe *et al.* 2003).

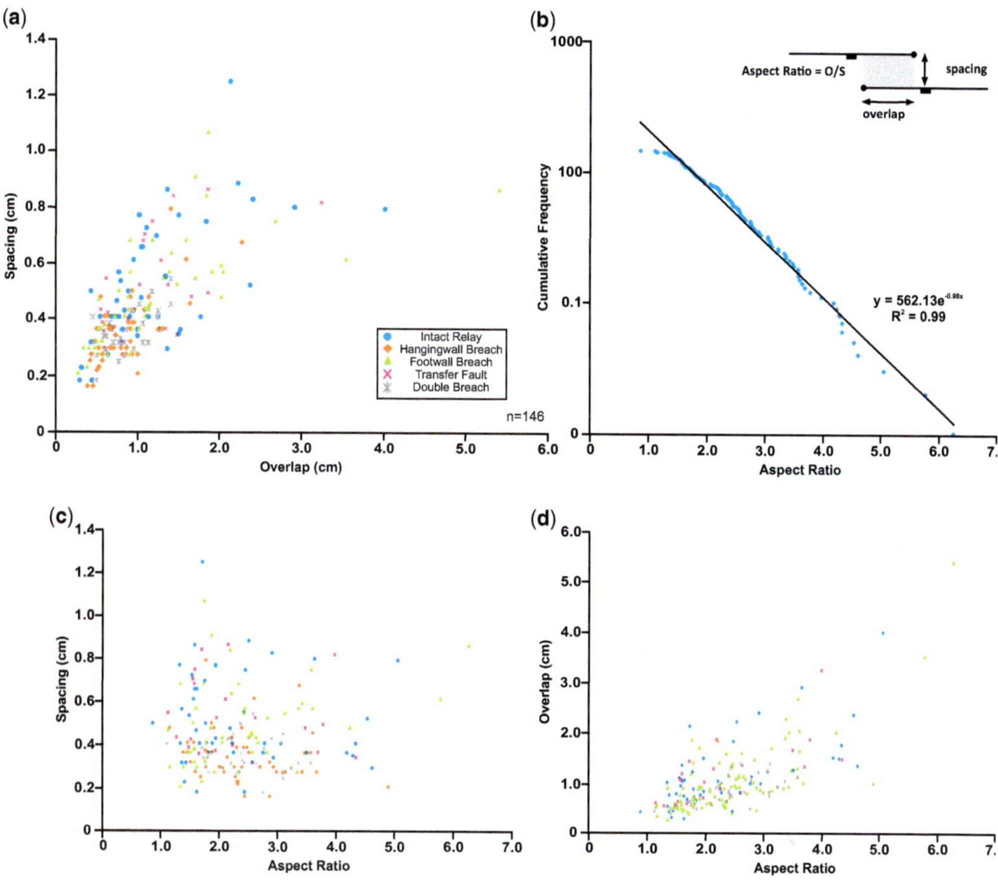

Fig. 14. Graphs showing the relationship of fault-segment overlap (O) and spacing (S) for the five types of overlapping fault-segment boundaries for 146 segment boundaries in all 10 analysed models at 4 cm of extension (for additional information, see Whipp 2011, appendix 8.2). (**a**) Plot of overlap v. spacing. (**b**) Cumulative frequency of aspect ratios (O:S) follow a negative exponential curve ($R^2 = 0.99$). (**c**) Plot of spacing v. aspect ratio. (**d**) Plot of overlap v. aspect ratio. The colours used in the plots are the same as those used in the graphs in Figure 13.

In our models, the number and density of faults increases rapidly, and the spacing between faults decreases during the early stages of extension (<2.25 cm of extension, <28.1% strain), with both fault density and spacing becoming broadly constant during the latter stages of extension (2.25–4.0 cm extension, 28.1–50% strain) (Figs 9, 12 & 13). As a result, the number of overlapping, synthetic segment boundaries and their density increases with increasing extension (>2.25 cm extension) (Figs 12 & 13). Localization of strain within the fault population, related to hard-linkage of previously isolated segments, causes the density of unbreached-segment boundaries to decrease. In our models, strain localization occurs between 3 and 4 cm of extension (37.5–50% strain) (Figs 7 & 9).

Fault overlap and spacing in segment boundaries

Our models show that a range of overlap to spacing (O:S) aspect ratios develop at synthetic segment boundaries and that relay ramp breaching can occur across a large range of O:S ratios (1:1–7:1). These O:S ratios documented in our models are similar to those reported from natural examples (e.g. Willemse 1997; Acocella *et al.* 2000) and from other analogue models that report O:S ratios ranging from 1:1 to 6:1, with a mean of 3:1 (e.g. Mansfield & Cartwright 2001; Hus *et al.* 2005). However, a key observation from our models is that the style of relay-ramp breaching (i.e. hanging-wall, footwall, transfer-fault and double-breached) is not determined by the O:S ratio at the onset of breaching

(Fig. 14). This observation differs from that of Mansfield (1996), who, based on a study of natural relay ramps in the Canyonlands area of Utah, USA, suggested that breached relay ramps develop with O:S ratios ranging from 3:1 to 5:1.

The model results show that a relatively narrow range of fault-spacing values occur at overlapping synthetic segment boundaries (Fig. 14). Furthermore, the spacing of overlapping, parallel and subparallel fault segments is established during the initial stages of fault nucleation and growth, and remains relatively fixed through time (Figs 9 & 14). It is probable that the extension-parallel spacing of faults is likely to have been influenced by the establishment and size of stress-reduction shadows around a fault, in addition to the thickness of the mechanical unit undergoing extension (e.g. Wu & Pollard 1995; Ackermann & Schlische 1997; Cowie 1998; Ackermann *et al.* 2001; Soliva *et al.* 2006). The size of the stress-reduction shadow influences the nucleation and propagation history of neighbouring faults, with the size of the zone related to fault size (length, height and displacement), the rheological properties of the deformed medium and the strain conditions (rate, magnitude and duration) (e.g. Ackermann & Schlische 1997; Willemse 1997; Cowie 1998; Gupta & Scholz 2000; d'Alessio & Martel 2004).

Our model results also show that, at overlapping synthetic segment boundaries, overlap values vary widely, whereas spacing values are more limited: this variability in overlap gives rise to the wide range of overlap to spacing aspect ratios described earlier (Fig. 14). Willemse (1997) showed that the size and shape of the stress perturbation adjacent to individual fault segments is controlled by the fault length to height aspect ratio: the size of the perturbation then controls the degree of overlap and interaction at synthetic segment boundaries. For example, greater overlap occurs between faults characterized by low length to height aspect ratios (i.e. tall faults), where stress is perturbed over a greater area of the fault surface, than between faults characterized by greater length to height aspect ratios (i.e. circular to wide faults). Although we are unable to determine the shape of the fault surface in our models, we speculate that the variability in overlap we document here may reflect the development of different fault shapes. Changes in the shape and size of stress perturbations and, therefore, the degree of mechanical interaction between individual segments, contribute to the kinematic coherence of the system.

Breached-segment boundary types

In our models, jogs or bends in the fault plan-view trace were ubiquitous and rarely occurred without the development of an abandoned footwall or hanging-wall splay. More than 60% of overlapping synthetic segment boundaries were breached at 4 cm extension, with hanging-wall breaching being the dominant style (>70%). These results are similar to those generated by the experiments of Hus *et al.* (2005), which showed that >50% of the overlapping synthetic segment boundaries were breached by the end of their model runs and that 55% of these were hanging-wall breached. Hanging-wall breaching accounted for 75% of breached-segment boundaries in the numerical models of Imber *et al.* (2004). Footwall-breached relays are the second most common type of breached-segment boundary in our models (cf. Imber *et al.* 2004; Hus *et al.* 2005).

Transfer-fault-breached relays are extremely rare in our models (Fig. 12c), although they are widely documented in natural rifts containing pre-existing structures (e.g. Gibbs 1984; Khalil & McClay 1998; Ferrill *et al.* 1999; Peacock & Parfitt 2002; Destro *et al.* 2003; Tesfaye *et al.* 2008). Acocella *et al.* (2000) suggested that transfer faults are most likely to develop in relatively wide rifts formed in response to >20% stretching, and rifts in which the amount of extension varies along its axis. Such conditions were not simulated in our models, but an increasing number of transfer faults at 4 cm extension (50% stretching) may indicate that they are more likely to develop at relatively high strains. Double-breached relays have not been described in previous analogue studies: however, such structures, although rare, developed in our models and have also been documented in nature (e.g. Peacock & Sanderson 1994; Huggins *et al.* 1995).

Geometrical and mechanical models have also investigated the style of fault linkage at segment boundaries. For example, Ferrill *et al.* (1999) related the style of breached-segment boundary to displacement gradients along the interacting faults, and suggested that transfer-fault-breached relays, as opposed to footwall- or hanging-wall-breached relays, develop when high displacement gradients develop along fault surfaces. The mechanical models of Crider & Pollard (1998) demonstrate how the potential for fault linkage is more likely when synthetic faults have underlapping tip geometries. Crider & Pollard (1998) also showed that overlapping fault segments produce a region of increased stress towards the centre of the relay ramp that may trigger the development of either a transfer-fault- or footwall-breached relay.

Hanging-wall-breached relays are the most common in our experiments, contrasting with the results of the geometrical and mechanical models described above, in which transfer-fault- or footwall-breached relays dominate or, at the very least, are predicted. Hanging-wall-breached relays are also the most common type of fault linkage occurring in the mechanical models of Imber *et al.* (2004), although no explanation for their preferential development is given, and the geometrical and mechanical factors controlling the variability of linkage remain poorly constrained. The mechanical models of Crider (2001) indicate that the plan-view step direction and sense of slip at overlapping segment boundaries during *oblique* extension influence the typical style of fault linkage. Step direction and slip direction also influences the structural style of fault-segment boundaries developed in strike-slip settings (e.g. Segall & Pollard 1980; Aydin & Schultz 1990). The sense of step during *orthogonal* extension does not, however, allow prediction of the type of breached-segment boundary, as strain is accommodated in the overlap region regardless of the plan-view fault geometry. In natural rifts, subtly oblique extension may strongly influence the spatial variability of fault-segment boundary types that develop across the rift.

Conclusions

- We used scaled experimental (analogue) models to investigate the temporal and spatial variability of relict and active segment-boundary types during orthogonal extension. The modelled fault populations develop in a similar manner and exhibit a similar overall structural style to natural fault population, with fault growth by segment linkage, segment-boundary breaching and strain localization being fundamental processes in fault-system development.
- Segment boundaries develop between interacting fault segments. The number of isolated, free tips decreases with increasing extension, as faults lengthen and interact to create fault-segment boundaries. Unbreached relay ramps evolve into breached fault-segment boundaries, with the style of breached relay (hanging wall-, footwall-, transfer-fault- or double-breached) varying both spatially and temporally at all scales of observation.
- Synthetic segment boundaries are the most common segment-boundary type developed in our models, whereas conjugate segment boundaries are rare. The proportion of synthetic segment boundaries increases with increasing strain, whereas the proportion of conjugate segment boundaries appears to remain constant
- Unbreached relay ramps are most numerous during the initial stages of extension: unbreached relay ramps are, however, transient features and are progressively destroyed by breaching with continued extension. Breached-segment boundaries are more abundant than unbreached-segment boundaries at the end of the experiments (4 cm extension), reflecting the increasing maturity of the fault population.
- Hanging-wall-breached relay ramps are the most common type (>70%) of breached-segment boundary developed in our models, followed by footwall-breached relay ramps. Transfer faults are uncommon. The reason for this variability is presently unclear and requires further study, drawing on observations from geometrical, mechanical and physical models, and natural rifts exposed in the field or imaged in seismic reflection data.
- The fault overlap to fault spacing aspect ratio within synthetic segment boundaries does not provide a reliable prediction of the type of breached-segment boundary that may be encountered. Rather, different fault linkage styles can occur across a range of aspect ratios (1:1–7:1). Fault spacing is less variable than fault overlap at segment boundaries, being relatively constrained by stress-reduction shadows that form during fault nucleation: in contrast, fault overlap changes during the growth and interaction of overlapping synthetic faults.

Financial support for this study was provided by an Environmental and Physical Research Council (EPSRC) bursary and CASE funding (Statoil ASA) award to Paul Whipp. We thank Alissa Henza and Mike Durcanin for assistance in the Rutgers laboratory and valuable discussions about analogue modelling. David Peacock, Alistair Fraser and Jonny Imber provided constructive reviews of an early version of the manuscript. Conrad Childs, Bruce Trudgill and Jonathan Long are thanked for their helpful reviews.

References

ACKERMANN, R.V. & SCHLISCHE, R.W. 1997. Anticlustering of small normal faults around larger faults. *Geology*, **25**, 1127–1130.

ACKERMANN, R.V., SCHLISCHE, R.W. & WITHJACK, M.O. 2001. The geometric and statistical evolution of normal fault systems: an experimental study of the effects of mechanical layer thickness on scaling laws. *Journal of Structural Geology*, **23**, 1803–1819.

ACOCELLA, V., FACCENNA, C., FUNICIELLO, F. & ROSSETTI, D. 1999. Sand-box modelling of basement-controlled transfer zones in extensional domains. *Terra Nova*, **11**, 149–156.

ACOCELLA, V., GUDMUNDSSON, A. & FUNICIELLO, R. 2000. Interaction and linkage of extension fractures and normal faults: examples from the rift zone of Iceland. *Journal of Structural Geology*, **22**, 1233–1246.

ACOCELLA, V., MORVILLO, P. & FUNICIELLO, R. 2005. What controls relay ramps and transfer faults within rift zones? Insights from analogue models. *Journal of Structural Geology*, **27**, 397–408.

AYDIN, A. & SCHULTZ, R.A. 1990. Effect of mechanical interaction on the development of strike-slip faults with en-echelon patterns. *Journal of Structural Geology*, **12**, 123–129.

Bellahsen, N., Daniel, J.M., Bollinger, L. & Burov, E. 2003. Influence of viscous layers on the growth of normal faults: insights from experimental and numerical models. *Journal of Structural Geology*, **25**, 1471–1485.

Byerlee, J. 1978. Friction of rocks. *Pure and Applied Geophysics*, **116**, 615–626.

Cartwright, J.A., Trudgill, B.D. & Mansfield, C.S. 1995. Fault growth by segment linkage: an explanation for scatter in maximum displacement and trace length data from the Canyon lands grabens of SE Utah. *Journal of Structural Geology*, **17**, 1319–1326.

Cartwright, J.A., Mansfield, C.S. & Trudgill, B.D. 1996. The growth of normal faults by segment linkage. *In*: Buchanan, P.G. & Nieuwland, D.A. (eds) *Modern Developments in Structural Interpretation, Validation and Modelling*. Geological Society, London, Special Publications, **99**, 163–177, https://doi.org/10.1144/GSL.SP.1996.099.01.13

Childs, C., Walsh, J.J. & Watterson, J. 1990. A method for the estimation of the density of fault displacements below the limits of seismic resolution in reservoir formations. *In*: Buller, A.T. (ed.) *North Sea Oil and Gas Reservoirs II*. Graham & Trotman, London, 193–203.

Childs, C., Watterson, J. & Walsh, J.J. 1995. Fault overlap zones within developing normal fault systems. *Journal of the Geological Society, London*, **152**, 535–549, https://doi.org/10.1144/gsjgs.152.3.0535

Clifton, A.E. & Schlische, R.W. 2001. Nucleation, growth, and linkage of faults in oblique rift zones: results from experimental clay models and implications for maximum fault size. *Geology*, **29**, 455–458.

Clifton, A.E., Schlische, R.W., Withjack, M.O. & Ackermann, R.V. 2000. Influence of rift obliquity on fault-population systematics: results from experimental clay models. *Journal of Structural Geology*, **22**, 1491–1509.

Contreras, J., Anders, M.H. & Scholz, C.H. 2000. Growth of a normal fault system: observations from the Lake Malawi basin of the East African rift. *Journal of Structural Geology*, **22**, 159–168.

Cowie, P.A. 1998. A healing-reloading feedback control on the growth rate of seismogenic faults. *Journal of Structural Geology*, **20**, 1075–1087.

Cowie, P.A. & Scholz, C.H. 1992*a*. Displacement-length scaling relationships for faults: data synthesis and discussion. *Journal of Structural Geology*, **14**, 1149–1156.

Cowie, P.A. & Scholz, C.H. 1992*b*. Physical explanation for the displacement-length relationship of faults using a post-yield fracture mechanics model. *Journal of Structural Geology*, **14**, 1133–1148.

Cowie, P.A., Gupta, S. & Dawers, S. 2000. Implications of fault array evolution on syn-rift depocentre development: insights from a numerical fault growth model. *Basin Research*, **12**, 241–261.

Cowie, P.A., Roberts, G.P. & Mortimer, E. 2007. Strain localisation with fault arrays over timescales of 10^0–10^7 years – observations, explanations, debates. *In*: Handy, M.R., Hirth, G. & Hovius, N. (eds) *Tectonic Faults: Agents of Change on a Dynamic Earth*. Dahlem Workshop Reports. MIT Press, Cambridge MA, 47–77.

Crider, J.G. 2001. Oblique extension and the geometry of normal fault linkages: mechanics and case study from the Basin & Range in Oregon. *Journal of Structural Geology*, **23**, 1997–2009.

Crider, J.G. & Pollard, D.D. 1998. Fault linkage: three-dimensional mechanical interaction between echelon normal faults. *Journal of Geophysical Research*, **103**, 24 373–24 391.

d'Alessio, M.A. & Martel, S.J. 2004. Fault terminations and barriers to fault growth. *Journal of Structural Geology*, **26**, 1885–1896.

Dawers, N.H. & Anders, M.H. 1995. Displacement-length scaling and fault linkage. *Journal of Structural Geology*, **17**, 607–614.

Dawers, N.H. & Underhill, J.R. 2000. The role of fault interaction and linkage in controlling syn-rift stratigraphic sequences: Statfjord East area, northern North Sea. *American Association of Petroleum Geologists Bulletin*, **84**, 45–64.

Dawers, N.H., Anders, M.H. & Scholz, C.H. 1993. Growth of normal faults: displacement–length scaling. *Geology*, **21**, 1107–1110.

Destro, N., Szatmari, P., Alkmim, F.F. & Magnavita, L.P. 2003. Release faults, associated structures, and their control on petroleum trends in the Reconcavo rift, northeast Brazil. *American Association of Petroleum Geologists Bulletin*, **81**, 1123–1144.

Eisenstadt, G. & Sims, D. 2005. Evaluating sand and clay models: do rheological differences matter? *Journal of Structural Geology*, **27**, 1399–1412.

Ferrill, D.A., Stamatakos, J.A. & Sims, D. 1999. Normal fault corrugation: implications for growth and seismicity of active normal faults. *Journal of Structural Geology*, **21**, 1027–1038.

Gawthorpe, R.L. & Hurst, J.M. 1993. Transfer zones in extensional basins: their structural style and influence on drainage development and stratigraphy. *Journal of the Geological Society, London*, **150**, 1137–1152, https://doi.org/10.1144/gsjgs.150.6.1137

Gawthorpe, R.L. & Leeder, M.R. 2000. Tectono-sedimentary evolution of active extensional basins. *Basin Research*, **12**, 195–218.

Gillespie, P.A., Walsh, J.J. & Watterson, J. 1992. Limitations of dimension and displacement data from single faults and the consequences for data analysis and interpretation. *Journal of Structural Geology*, **14**, 1157–1172.

Granger, A.B. 2006. *Influence of basal boundary conditions on normal fault systems in scaled physical analogue models*. MSc thesis, Rutgers University.

Granger, A.B., Withjack, M.O. & Schlische, R.W. 2006. Undulations on normal-fault surfaces: insights into fault growth using scaled physical models of extension. *Geological Society of America Abstracts with Programs*, **38**, 480.

Gibbs, A. 1984. Structural evolution of extensional basin margins. *Journal of the Geological Society, London*, **141**, 609–620, https://doi.org/10.1144/gsjgs.141.4.0609

Gawthorpe, R.L., Jackson, C.A.L., Young, M.J., Sharp, I.R., Moustafa, A.R. & Leppard, C.W.

2003. Normal fault growth, displacement localisation and the evolution of normal fault populations: the Hamman Faraun fault block, Suez rift, Egypt. *Journal of Structural Geology*, **25**, 883–895.

GUPTA, A. & SCHOLZ, C.H. 2000. A model of normal fault interaction based on observations and theory. *Journal of Structural Geology*, **22**, 865–879.

GUPTA, S., COWIE, P.A., DAWERS, N.H. & UNDERHILL, J.R. 1998. A mechanism to explain rift basin subsidence and stratigraphic patterns through fault array evolution. *Geology*, **25**, 595–598.

GUPTA, S., UNDERHILL, J.R., SHARP, I.R. & GAWTHORPE, R.L. 1999. Role of fault interactions in controlling synrift sediment dispersal patterns: Miocene, Abu Alaqa Group, Suez rift, Sinai, Egypt. *Basin Research*, **11**, 167–189.

HANDIN, J. 1966. Strength and ductility. *In*: CLARK, S.P., JR. (ed.) *Handbook of Physical Constants*. Geological Society of America, Memoirs, **97**, 223–289.

HENZA, A.A., WITHJACK, M.O. & SCHLISCHE, R.W. 2010. Normal-fault development during two-phase non-coaxial extension: an experimental study. *Journal of Structural Geology*, **32**, 1656–1667.

HUBBERT, M.K. 1937. Theory of scale models as applied to the study of geologic structures. *Geological Society of America Bulletin*, **48**, 1459–1519.

HUGGINS, P., WATTERSON, J., WALSH, J.J. & CHILDS, C. 1995. Relay zone geometry and displacement transfer between normal faults recorded in coal-mine plans. *Journal of Structural Geology*, **17**, 1741–1755.

HUS, R., ACOCELLA, V., FUNICIELLO, R. & DE BATIST, M. 2005. Sandbox models of relay ramp structure and evolution. *Journal of Structural Geology*, **27**, 459–473.

IMBER, J., TUCKWELL, G.W. *ET AL.* 2004. Three-dimensional discrete element modelling of relay growth and breaching along normal faults. *Journal of Structural Geology*, **26**, 1897–1911.

JACKSON, C.A.L., GAWTHORPE, R.L., CARR, I.D. & SHARP, I.R. 2005. Normal faulting as a control on the stratigraphic development of shallow marine synrift sequences: the Nukhul and lower Rudeis Formation, Hammam Faraun fault block Suez Rift, Egypt. *Sedimentology*, **52**, 313–338.

KHALIL, S. & MCCLAY, K. 1998. Extensional hard linkages, Eastern Gulf of Suez, Egypt. *Geology*, **26**, 563–566.

KIM, Y.S. & SANDERSON, D.J. 2005. The relationship between displacement and length of faults: a review. *Earth Science Reviews*, **68**, 317–334.

LEEDER, M.R. & GAWTHORPE, R.L. 1987. Sedimentary models for extensional tilt-block/half-graben basins. *In*: COWARD, M.P., DEWEY, J.F. & HANCOCK, P.L. (eds) *Continental Extensional Tectonics*. Geological Society, London, Special Publications, **28**, 139–152, https://doi.org/10.1144/GSL.SP.1987.028.01.11

MANIGNETTI, I., CAMPILLO, M., BOULEY, S. & COTTON, F. 2007. Earthquake scaling, fault segmentation, and structural maturity. *Earth and Planetary Science Letters*, **253**, 429–438.

MANSFIELD, C.S. 1996. *Fault growth by segment linkage*. PhD thesis, University of London.

MANSFIELD, C.S. & CARTWRIGHT, J.A. 2001. Fault growth by linkage: observations and implications from analogue models. *Journal of Structural Geology*, **23**, 745–763.

MARCHAL, D., GUIRAUD, M., RIVES, T. & VAN DEN DRIESSCHE, J. 1998. Space and time propagation processes of normal faults. *In*: JONES, G., FISHER, Q.J. & KNIPE, R.J. (eds) *Faulting, Sealing, and Fluid Flow in Hydrocarbon Reservoirs*. Geological Society, London, Special Publications, **147**, 51–70, https://doi.org/10.1144/GSL.SP.1998.147.01.04

MARCHAL, D., GUIRAUD, M. & RIVES, T. 2003. Geometric and morphological evolution of normal fault planes and traces from 2D to 4D data. *Journal of Structural Geology*, **25**, 135–158.

MCCLAY, K.R. & KHALIL, S. 1998. Extensional hard linkages, eastern Gulf of Suez, Egypt. *Geology*, **26**, 563–566.

MCCLAY, K.R. & WHITE, M.J. 1995. Analogue modelling of orthogonal and oblique rifting. *Marine and Petroleum Geology*, **12**, 137–151.

MCCLAY, K.R., DOOLEY, T., WHITEHOUSE, P. & MILLS, M. 2002. 4-D evolution of rift systems: insights from scaled physical models. *American Association of Petroleum Geologists Bulletin*, **86**, 935–959.

MCLEOD, A.E., DAWERS, N.H. & UNDERHILL, J.R. 2000. The propagation and linkage of normal faults: insights from the Strathspey–Brent–Statfjord fault array, northern North Sea. *In*: GUPTA, S. & COWIE, P.A. (eds) *Processes and Controls in the Stratigraphic Development of Extensional Basins. Basin Research*, **12**, 263–284.

MCLEOD, A.E., UNDERHILL, J.R., DAVIS, S. & DAWERS, N.H. 2002. The influence of fault array evolution on synrift sedimentation patterns: controls on deposition in the Strathspey–Brent–Statfjord half graben, northern North Sea. *American Association of Petroleum Geologists Bulletin*, **86**, 1061–1093.

MEYER, V., NICOL, A., CHILDS, C., WALSH, J.J. & WATTERSON, J. 2002. Progressive localisation of strain during the evolution of a normal fault population. *Journal of Structural Geology*, **24**, 1215–1231.

MORLEY, C.K., NELSON, R.A., PATTON, T.L. & MUNN, S.G. 1990. Transfer zones in the East African Rift system and their relevance to hydrocarbon exploration in rifts. *American Association of Petroleum Geologists Bulletin*, **74**, 1234–1253.

MOUSTAFA, A.R. 2002. Controls on the geometry of transfer zones in the Suez rift and northwest Red Sea: implications for the structural geometry of rift systems. *American Association of Petroleum Geologist Bulletin*, **86**, 979–1002.

OERTEL, G. 1965. The mechanism of faulting in clay experiments. *Tectonophysics*, **221**, 325–344.

PEACOCK, D.C.P. 2002. Propagation, interaction and linkage in normal fault systems. *Earth-Science Reviews*, **58**, 121–142.

PEACOCK, D.C.P. & PARFITT, E.A. 2002. Active relay ramps and normal fault propagation on Kilauea Volcano, Hawaii. *Journal of Structural Geology*, **24**, 729–742.

PEACOCK, D.C.P. & SANDERSON, D.J. 1991. Displacements, segment linkage and relay ramps in normal fault zones. *Journal of Structural Geology*, **13**, 721–733.

PEACOCK, D.C.P. & SANDERSON, D.J. 1994. Strain and scaling of faults in the chalk at Flamborough Head, U.K. *Journal of Structural Geology*, **16**, 94–107.

Peacock, D.C.P. & Sanderson, D.J. 1996. Effects of propagation rate on displacement variations along faults. *Journal of Structural Geology*, **18**, 311–320.

Peacock, D.C.P. & Zhang, X. 1994. Field examples and numerical modelling of oversteps and bends along normal faults in cross-section. *Tectonophysics*, **234**, 147–167.

Peacock, D.C.P., Knipe, R.J. & Sanderson, D.J. 2000. Glossary of normal faults. *Journal of Structural Geology*, **22**, 291–305.

Schellart, W.P. 2000. Shear test results for cohesion and friction coefficients for different granular materials: scaling implications for their usage in analogue modeling. *Tectonophysics*, **324**, 1–16.

Schlagenhauf, A., Manighetti, I., Malavieille, J. & Dominguez, S. 2008. Incremental growth of normal faults: insights from a laser-equipped analog experiment. *Earth and Planetary Science Letters*, **273**, 299–311.

Schlische, R.W. 1995. Geometry and origin of fault-related folds in extensional settings. *American Association of Petroleum Geologists Bulletin*, **79**, 1661–1678.

Schlische, R.W. & Withjack, M.O. 2009. Origin of fault domain and fault-domain boundaries (transfer zones and accommodation zones) in extensional provinces: result of random nucleation and self-organized fault growth. *Journal of Structural Geology*, **31**, 910–925.

Schlische, R.W., Young, S.S., Ackermann, R.V. & Gupta, A. 1996. Geometry and scaling relations of a population of very small rift-related normal faults. *Geology*, **24**, 683–686.

Schlische, R.W., Withjack, M.O. & Eisenstadt, G. 2002. An experimental study of the secondary deformation produced by oblique slip normal faulting. *American Association of Petroleum Geologists Bulletin*, **86**, 885–906.

Scholz, C.H. & Gupta, A. 2000. Fault interactions and seismic hazard. *Journal of Geodynamics*, **29**, 459–467.

Segall, P. & Pollard, D.D. 1980. Mechanics of discontinuous faults. *Journal of Geophysical Research*, **85**, 4337–4350.

Sims, D. 1993. The rheology of clay: A modelling material for geological structures. *Eos, Transactions of the American Geophysical Union*, **74**, 569.

Soliva, R. & Benedicto, A. 2004. A linkage criterion for segmented normal faults. *Journal of Structural Geology*, **26**, 2251–2267.

Soliva, R., Benedicto, A. & Maerten, L. 2006. Spacing and linkage of confined normal faults: importance of mechanical thickness. *Journal of Geophysical Research*, **111**, 17.

Soliva, R., Benedicto, A., Schultz, R.A., Maerten, L. & Micarelli, L. 2008. Displacement and interaction of normal fault segments branched at depth: implications for fault growth and potential earthquake rupture size. *Journal of Structural Geology*, **30**, 1288–1299.

Tesfaye, S., Rowan, M.G., Mueller, K., Trudgill, B.D. & Harding, D.J. 2008. Relay and accommodation zones in the Dobe and Hanle grabens, central Afar, Ethiopia and Djibouti. *Journal of the Geological Society, London*, **165**, 535–547, https://doi.org/10.1144/0016-76492007-093

Torabi, A. & Berg, S.S. 2011. Scaling of fault attributes: a review. *Marine and Petroleum Geology*, **28**, 1444–1460.

Trudgill, B.D. & Cartwright, J.A. 1994. Relay ramp forms and normal fault linkages – Canyonlands National Park, Utah. *Geological Society of America Bulletin*, **106**, 1143–1157.

Walsh, J.J. & Watterson, J. 1988. Analysis of the relationship between displacements and the dimensions of faults. *Journal of Structural Geology*, **10**, 239–247.

Walsh, J.J. & Watterson, J. 1991. Geometric and kinematic coherence and scale effects in normal fault systems. *In*: Roberts, A.M., Yielding, G. & Freeman, B. (eds) *The Geometry of Normal Faults*. Geological Society, London, Special Publications, **56**, 193–203, https://doi.org/10.1144/GSL.SP.1991.056.01.13

Walsh, J.J., Watterson, J. & Yielding, G. 1991. The importance of small-scale faulting in regional extension. *Nature*, **351**, 391–393.

Walsh, J.J., Watterson, J., Bailey, W.R. & Childs, C. 1999. Fault relays, bends and branch-lines. *Journal of Structural Geology*, **21**, 1019–1026.

Walsh, J.J., Childs, C. *et al.* 2001. Geometric controls on the evolution of normal fault systems. *In*: Holdsworth, R.E., Strachan, R.A., Magloughlin, J.F. & Knipe, R.J. (eds) *The Nature and Tectonic Significance of Fault Zone Weakening*. Geological Society, London, Special Publications, **186**, 157–170, https://doi.org/10.1144/GSL.SP.2001.186.01.10

Walsh, J.J., Bailey, W.R., Childs, C., Nicol, A. & Bonson, C.G. 2003*a*. Formation of segmented normal faults: a 3-D perspective. *Journal of Structural Geology*, **25**, 1251–1262.

Walsh, J.J., Childs, C., Imber, J., Manzocchi, T., Watterson, J. & Nell, P.A.R. 2003*b*. Strain localisation and population changes during fault system growth within the Inner Moray Firth, Northern North Sea. *Journal of Structural Geology*, **25**, 307–315.

Weijermars, R. 1986. Flow behaviour and physical chemistry of bouncing putties and related polymers in view of tectonic laboratory applications. *Tectonophysics*, **124**, 325–358.

Weijermars, R., Jackson, M.P.A. & Vendeville, B. 1993. Rheological and tectonic modeling of salt provinces. *Tectonophysics*, **217**, 143–174.

Whipp, P.S. 2011. *Fault propagation folding and the growth of normal faults*. PhD thesis, University of London.

Willemse, E.J.M. 1997. Segmented normal faults: correspondence between three-dimensional mechanical models and field data. *Journal of Geophysical Research*, **102**, 675–692.

Willemse, E.J.M., Pollard, D.D. & Aydin, A. 1996. Three-dimensional analyses of slip distributions on normal fault arrays with consequences for fault scaling. *Journal of Structural Geology*, **18**, 295–309.

Withjack, M.O. & Callaway, S. 2000. Active normal faulting beneath a salt layer: an experimental study of deformation patterns in the cover sequence. *American Association of Petroleum Geologists Bulletin*, **84**, 627–651.

Withjack, M.O. & Jamison, W.R. 1986. Deformation produced by oblique rifting. *Tectonophysics*, **126**, 99–124.

WITHJACK, M.O., OLSON, J. & PETERSON, E. 1990. Experimental models of extensional forced folds. *American Association of Petroleum Geologists Bulletin*, **74**, 1038–1054.

WITHJACK, M.O., SCHLISCHE, R.W. & HENZA, A.A. 2008. Scaled experimental models of extension: Dry Sand v. Wet Clay. *Houston Geological Survey Bulletin*, **49**, 31–49.

WOJTAL, S.F. 1996. Changes in fault displacement populations correlated to linkage between faults. *Journal of Structural Geology*, **18**, 265–279.

WU, H. & POLLARD, D.D. 1995. An experimental study of the relationship between joint spacing and layer thickness. *Journal of Structural Geology*, **17**, 887–905.

YIELDING, G., WALSH, J. & WATTERSON, J. 1992. The prediction of small-scale faulting in reservoirs. *First Break*, **10**, 449–460.

YOUNG, M.J., GAWTHORPE, R.L. & HARDY, S. 2001. Growth and linkage of a segmented normal fault zone; the Late Jurassic Murchison–Statfjord North Fault, northern North Sea. *Journal of Structural Geology*, **23**, 1933–1952.

3D geometry and kinematic evolution of extensional fault-related folds, NW Red Sea, Egypt

SAMIR M. KHALIL[1,2] & KEN R. McCLAY[1]*

[1]*Department of Earth Sciences, Fault Dynamics Research Group, Royal Holloway University of London, Egham, Surrey TW20 0EX, UK*

[2]*Department of Geology, Faculty of Science, Suez Canal University, Ismailia 41522, Egypt*

**Correspondence: k.mcclay@es.rhul.ac.uk*

Abstract: Fault-related folds are common structural features found at a variety of scales in extensional settings, and have been recognized in both outcrop and subsurface studies. However, the detailed geometry and origin of complex 3D folds adjacent to normal faults are poorly known, and, in some cases, are interpreted to be due to strike-slip tectonics and post-rift contraction. Here we examine the 3D geometry of seismic-scale folds in a rift margin – the Red Sea – and discuss the interrelationship between the growth of normal faults and the development of their related folds. Detailed field mapping of the NW Red Sea rift system has shown that the rift margin is dominated by two large extensional fault systems formed by a series of linked NNW-, north–south- and NNE-striking fault segments. These linked segments exhibit distinct zigzag fault patterns and combine to form a number of NNW-trending faults that dip NE with dominant hanging-wall stratal dips to the SW. Hanging-wall stratal dips define 3D extensional fault-related synclinal folds in pre- and early synrift strata. The hanging-wall synclines are kilometre-scale, gently doubly plunging, with curved axial surface traces orientated sub-parallel to the bounding faults. Field data demonstrated that these folds are formed by along-strike variations in fault displacements, and they form transverse synclines combined with hanging-wall extensional fault-propagation folds. The complex 3D geometry of the hanging-wall synclines is the result of the along-strike segment linkage. Adjacent to the bounding faults, the stratal dips are sub-parallel to the faults as a result of extensional fault-propagation folding controlled by highly anisotropic pre-rift strata. Palaeo-strain analyses of fault-slip data, together with analysis of the fold geometry, clearly indicate that the faulting and folding in the NW Red Sea are formed by pure NE–SW extension during the Late Oligocene–Miocene rifting, and that contraction or strike-slip tectonics need not be invoked.

Many rifts and passive margin systems commonly exhibit complex extensional fault-related folds at a variety of scales, from metre-scale drag folds caused by frictional variations along the fault surface to kilometre-scale, extensional fault-propagation folds (e.g. Moustafa 1987; Patton *et al.* 1994; Hardy & McClay 1999; Sharp *et al.* 2000; Khalil & McClay 2002, 2006; Jackson *et al.* 2006). These folds, sometimes termed forced folds (e.g. Withjack *et al.* 1990; Schlische 1995; Maurin & Niviere 2000; Finch *et al.* 2004; Hardy & Finch 2006; Ford *et al.* 2007) or longitudinal and transverse folds (with axes orientated parallel and perpendicular to the faults, respectively), are caused by displacement variations along the main fault surfaces (e.g. Schlische 1995).

Previous studies on extensional faulting point to variations in fault displacement, from maxima at fault segment centres to minima at fault tips. This along-strike variation in fault displacement is normally associated with variations in footwall uplift and hanging-wall subsidence, resulting in the development of extension-related fault-propagation folds with configuration typically characterized by footwall anticlines and hanging-wall longitudinal and transverse folds (Fig. 1). The fundamental geometries of longitudinal extensional fault-propagation folds, together with along-strike displacement-related transverse folds, are illustrated in Figure 1 for a single isolated extensional fault. Well-exposed examples of these hanging-wall folds, both those associated with isolated extensional faults and with fault systems, are rare and the three-dimensional (3D) geometry of these is not fully discussed in the literature (cf. Schlische *et al.* 2002; Kane *et al.* 2010). In addition, in subsurface, steep dips adjacent to hanging walls are commonly poorly imaged and, therefore, hanging-wall folds are often interpreted with some trepidation.

From: Childs, C., Holdsworth, R. E., Jackson, C. A.-L., Manzocchi, T., Walsh, J. J. & Yielding, G. (eds) 2017. *The Geometry and Growth of Normal Faults*. Geological Society, London, Special Publications, **439**, 109–130.
First published online March 30, 2016, https://doi.org/10.1144/SP439.11

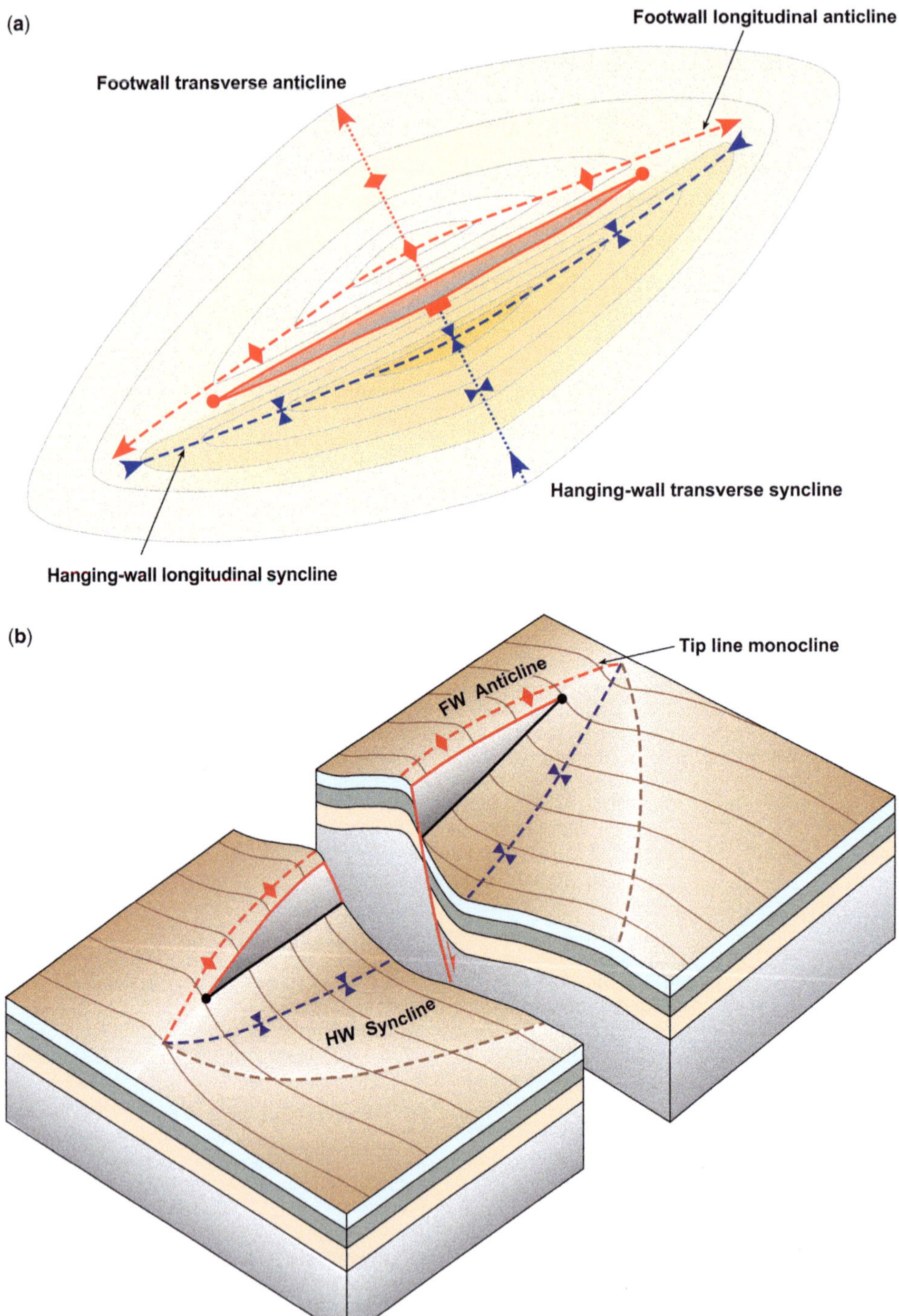

Fig. 1. (**a**) Plan view of a synoptic model of an isolated extensional fault showing a 3D hanging-wall syncline and a narrow footwall anticline formed by a combination of extensional fault-propagation folding together with along-strike transverse folding due to displacement variation. (**b**) Three-dimensional view of an isolated extensional fault with extensional fault-propagation folding. Note the relatively broad doubly plunging hanging-wall (HW) syncline and the narrow longitudinal doubly plunging footwall (FW) anticline.

This paper describes and analyses extensional fault-related hanging-wall synclines that formed as a result of propagation and linkage of differently orientated fault segments along the NW margin of the northern Red Sea, Egypt. Models are developed for the kinematic evolution of the complex 3D hanging-wall basins that occur in this rift margin.

The Red Sea is a 2000 km-long, NW-striking, continental rift system that formed in response to the separation of the Arabian and African continental plates (McKenzie *et al.* 1970; Le Pichon & Francheteau 1978; Hempton 1987; Meshref 1990; Coleman 1993). Opening of the Red Sea Rift initiated in the Late Oligocene–Early Miocene (McKenzie *et al.* 1970; Cochran 1983; Steckler *et al.* 1988; Coleman 1993; Bosworth *et al.* 2005; Bosworth 2015) and extended to the NW into the Gulf of Suez Rift (Fig. 2a). In the late Mid-Miocene, extension ceased in the Gulf of Suez, and the continued opening of the Red Sea was accommodated by sinistral strike-slip faulting along the Aqaba–Dead Sea transform system (Freund 1970; Ben-Menahem *et al.* 1976; Steckler *et al.* 1988; Bosworth *et al.* 2005; Bosworth 2015).

The NW margin of the Red Sea is characterized by two oppositely dipping fault domains separated by the complex Duwi accommodation zone (Fig. 2a) (Moustafa 1997; Khalil & McClay 2001, 2009; Younes & McClay 2001). The northern domain consists of a series of faults that dip to the NE, whereas the southern domain is formed by a series of faults that dip to the SW (Fig. 2a).

The research area, south of Safaga city, is a 40 km-long segment of the NW Red Sea rift margin and is located in the northern dip domain (Fig. 2a). This area is characterized by NW–SE- and north–south-striking, linked and generally NE-dipping fault systems (Fig. 2b, c). These fault systems define a western Border Fault System and an eastern Coastal Fault System (Fig. 2b, c). The hanging walls of these faults are defined by complex 3D, doubly plunging synclines of pre-rift and earliest synrift strata (Fig. 2b, c). Previous interpretations attributed these folds to a regional compressive stress regime and strike-slip movement along the rift faults (Jarrige *et al.* 1986, 1990; Montenat *et al.* 1988, 1998). This study presents an alternative interpretation that proposes that the faulting and folding are related solely to extensional faulting, with the complex hanging-wall synclines formed by extensional fault-propagation folding combined with along-strike variations in the fault displacement.

The research presented in this paper is based on detailed geological mapping of the well-exposed stratigraphy and structure of this part of the NW Red Sea rift margin, together with cross-section construction, and analysis of Landsat Thematic Mapper (TM) images and aerial photographs.

Stratigraphic and structural framework

Stratigraphy

Pre-rift strata. The rotated hanging-wall fault blocks in the NW Red Sea expose a complete section of pre-rift strata that ranges in age from the Precambrian to the Mid Eocene (Fig. 2b, c). The Precambrian basement consists of metavolcanics, metasediments intruded by syn- and post-tectonic granites and granodiorites (Akaad & Noweir 1980; Stoeser & Camp 1985; Said 1990; Stern 1994). The basement rocks are deformed by a number of Precambrian structures (fault zones, dykes, Pan-African fractures and shear zones) that are orientated north–south, NNW, NW–WNW and NE (Fig. 2b). According to Khalil & McClay (2001) and Younes & McClay (2001), the basement structures were reactivated during Late Oligocene–Miocene rifting of the Red Sea and are responsible for the present-day zigzag fault pattern of the rift margin.

The Precambrian basement is unconformably overlain by a 500–700 m-thick section of pre-rift strata that ranges in age from the Late Cretaceous to the Middle Eocene (Fig. 3). The pre-rift strata occur in a number of isolated fault-bounded hanging-wall sub-basins (the Um El Huetat, Rabah and Wasif fault blocks) and in relatively small outcrops in the coastal area (Fig. 2b). The lower part of the pre-rift section consists of massive and thick-bedded, fluvial to shallow-marine clastics of the Upper Cretaceous Nubia Sandstone, up to 130 m in average thickness (Fig. 3). The Nubia Sandstone is overlain by a 220–370 m-thick unit of thin-bedded shales, sandstones and limestones, with phosphate units and thin oyster beds of the Upper Cretaceous–Paleocene Quseir, Duwi, Dakhla and Esna formations, respectively (Fig. 3) (Youssef 1957; Abd el-Razik 1967; Issawi *et al.* 1969; Said 1990). The uppermost pre-rift strata consist of a 120–280 m-thick section of competent, thick-bedded, chalky and cherty limestones of the Lower–Middle Eocene Thebes Formation (Fig. 3).

In terms of mechanical stratigraphy, the pre-rift succession consists of a highly anisotropic section of the Quseir, Duwi, Dakhla and Esna formations (from 215 to 385 m thick: Fig. 3) sandwiched between the crystalline basement and competent Nubia Sandstone below, and competent, thick-bedded limestones of the Thebes formation above (Fig. 3). The anisotropic strata in the middle of this sandwich appears to have strongly controlled the extensional fault-propagation folds formed during the rift margin evolution, as elsewhere in the Red Sea–Gulf of Suez rift system (cf. Sharp *et al.* 2000; Khalil & McClay 2002, 2006; Jackson *et al.* 2006; Wilson *et al.* 2009). The interbedded shales and sandstones of the anisotropic section clearly induced strain partitioning in the pre-rift succession, with faulting dominant in the brittle basement and Nubia Sandstone below, and fault-propagation folding in the competent units above (see the following subsections '3D extensional fault-related folding' and 'Discussion').

Synrift strata. Synrift strata in the NW Red Sea range from the Late Oligocene to Recent in age (Fig. 3). Lowermost synrift strata are dominated by coarse-grained clastics of the Nakheil and Ranga formations (Fig. 3). The Nakheil Formation is Late Oligocene in age (Akkad & Dardir 1966; Said 1990) and unconformably overlies the Eocene Thebes Formation with approximately 10° of angular discordance. The Nakheil Formation consists of 60–120 m of lacustrine red mudstones and sandstones, limestones, chert breccias, and conglomerates derived from the underlying Thebes Formation. The Ranga Formation is Aquitanian–Burdigalian in age (El Bassyony 1982), and consists of 100–186 m of continental to shallow-marine sandstones and polymict conglomerates (containing clasts of basement, limestones and cherts). The Ranga Formation also unconformably overlies pre-rift units and, in places, overlies the early synrift Nakheil sediments (Fig. 3). The Ranga Formation is locally preserved in the Wadi Gassus and Wadi Guesis areas (Fig. 2b), where it forms structurally controlled, coarse-grained fan-delta systems in the coastal area (Khalil & McClay 2009).

Unconformably overlying the Ranga conglomerates are reefal limestones and shallow-marine fine-grained clastics of the Late Burdigalian–Langhian Um Mahara Formation (Fig. 3) (El Bassyony 1982; Said 1990). The Um Mahara Formation is 20–60 m thick and is unconformably overlain by 90–400 m of massive to poorly bedded evaporites

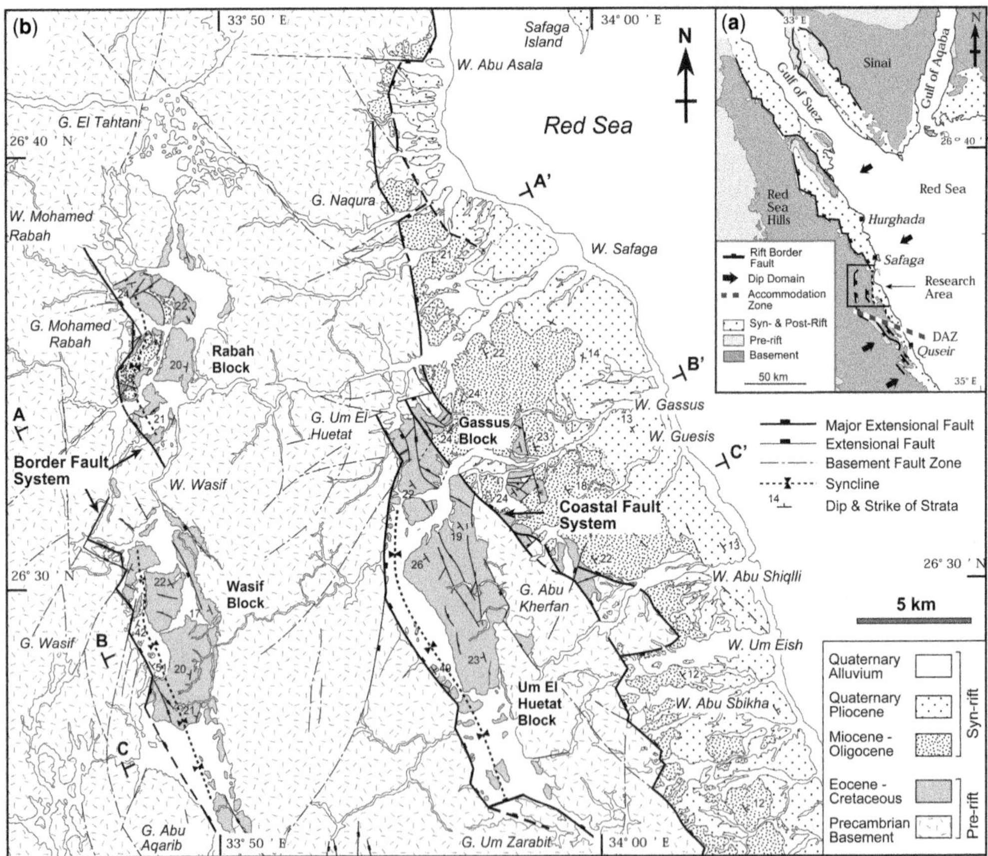

Fig. 2. (**a**) Location map of the research area showing the principal structural elements of the NW Red Sea. Bold arrows indicate the dominant stratal dip direction within the half-graben sub-basins. DAZ, Duwi accommodation zone. (**b**) Surface regional geological map of the study area showing the major fault systems and the main fault blocks. The locations of cross-sections are labelled A–A′ to C–C′.

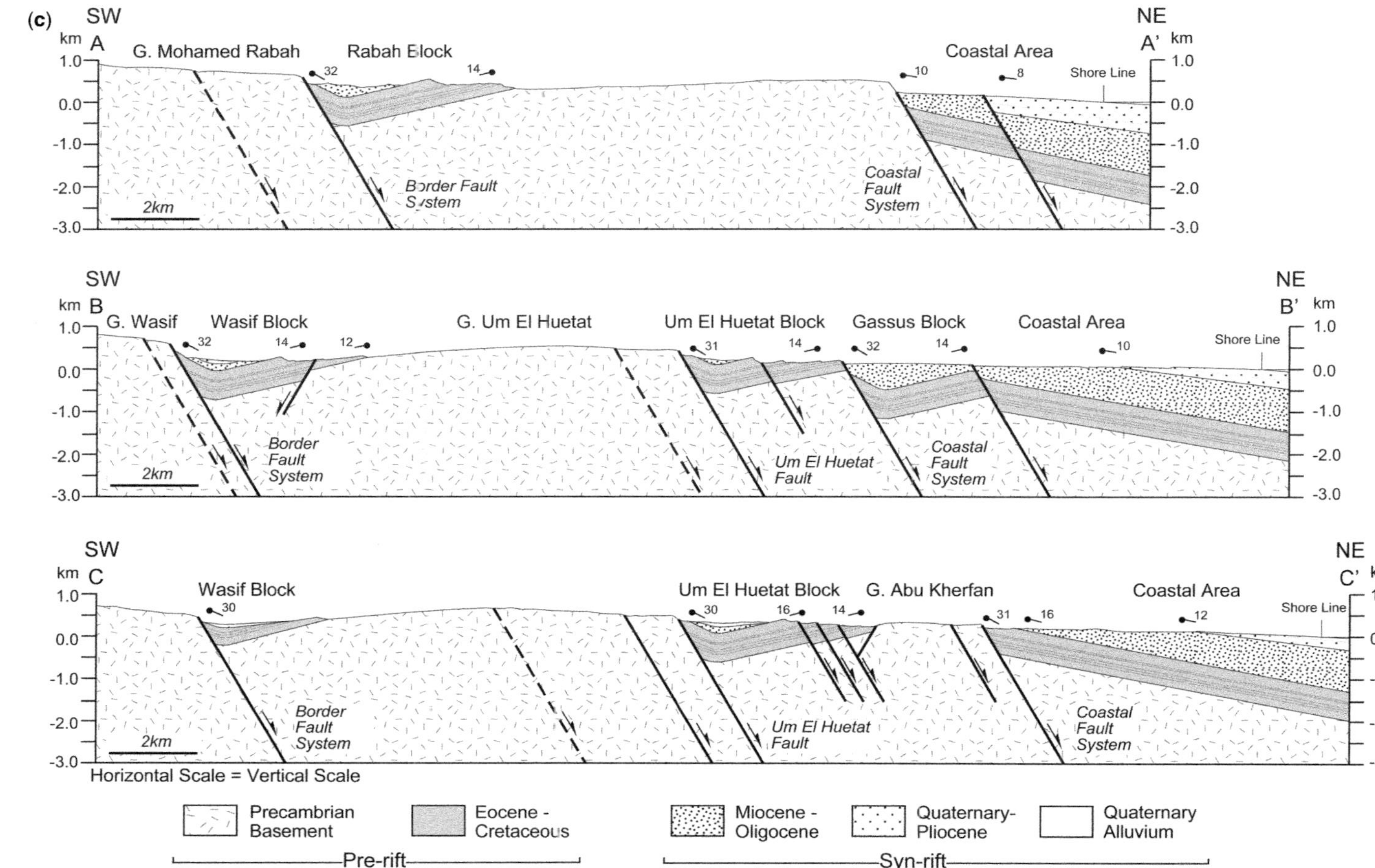

Fig. 2. **(c)** Regional cross-sections across the study area showing the major fault systems and the main fault blocks (locations are shown in (b)).

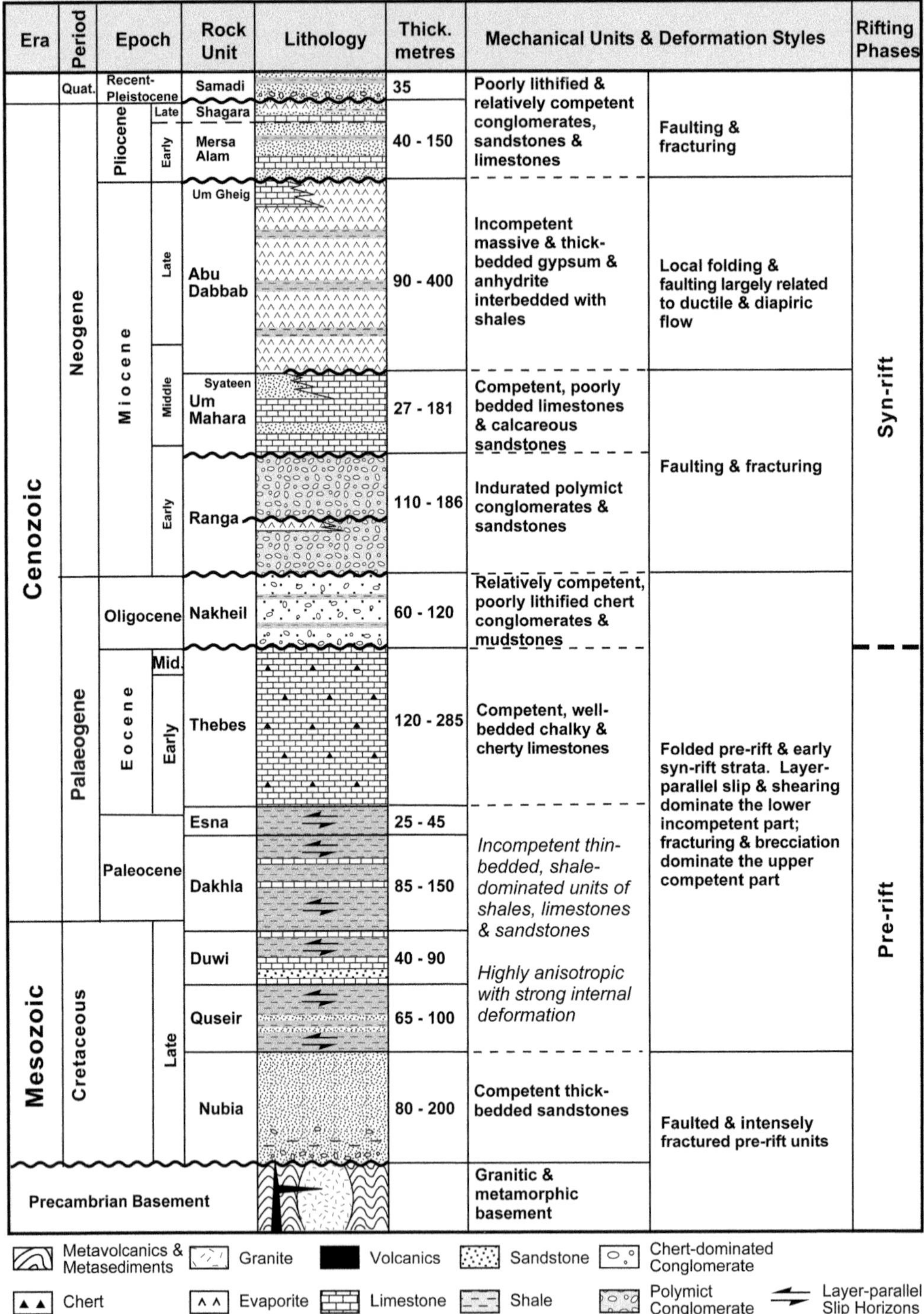

Fig. 3. Summary stratigraphy of the NW Red Sea. Key internal detachment units within the pre-rift sequence are highlighted.

of the Middle–Late Miocene Abu Dabbab Formation. These are in turn overlain by 200–300 m of Upper Miocene–Quaternary Mersa Alam, Shagara and Samadi formations of coarse clastics, mudstones and carbonates (Fig. 3), and these form the main exposures in the coastal area (Fig. 2b)

Regional structure

The NW margin of the Red Sea contains two main NE-dipping fault systems: the Coastal Fault System in the east and the Border Fault System in the west (Fig. 2b) (e.g. Khalil & McClay 2001, 2002, 2009). The Coastal Fault System dominantly strikes NW, sub-parallel to the Red Sea coastline, and consists of a series of NW-, NNW- and north–south-striking segments (Fig. 2b). Segment linkage is interpreted to have occurred by the breaching of relay ramps (Khalil & McClay 2009). The hanging wall of this system is dominated by pre-rift and synrift strata that dip to the west and SW. Throw along the NE-dipping Coastal Fault System varies from 0.5 to >2 km (Fig. 2b, c). The Um El Huetat block is a major tilted fault block located in the footwall of the main Coastal Fault System (Fig. 2b) and is composed of pre-rift Cretaceous–Eocene strata, as well as locally preserved early synrift Nakheil Formation.

The Border Fault System dominantly trends NNW, with a distinct zigzag pattern formed by linked NNW-, north–south- and NNE-striking fault segments whose throw varies from 0.5 to 1.5 km (Fig. 2b, c). The footwall of the Border Fault System consists of uplifted and eroded Precambrian basement. The west- and SW-dipping hanging wall of this fault system consists of pre-rift Cretaceous–Eocene units and early synrift Oligocene strata that are preserved in two major, NNW-trending fault blocks (the Rabah and Wasif blocks: Fig. 2b).

Regional NE–SW-orientated cross-sections through the map area show that the main segments of the Coastal and Border fault systems are planar domino-style faults that dip 60–65° to the NE (Fig. 2c). The hanging-wall half-graben are synclinal in form and are 3–6 km wide (Fig. 2c). In addition to the large Border and Coastal fault systems, smaller faults with throws that range from tens to a few hundreds of metres dissect the main fault blocks into smaller sub-blocks (Fig. 2b, c).

Field mapping of exposed fault surfaces with different orientations and measurements of slickenlineations indicate that the pure dip-slip displacements dominate, with only a few north–south- to NNE-trending faults showing minor sinistral oblique-slip movement. Palaeo-strain analysis of these fault slip data indicate that the faults were developed in a regime of near-horizontal extensional (T) stress with an axis orientated (09°/059°) and near-vertical compressive (P) stress with an axis orientated (79°/202°) (Fig. 4). This stress regime is interpreted to be related to the regional NE–SW-orientated extension that operated throughout most of the Oligocene–Miocene rifting of the Red Sea and Gulf of Suez (cf. Khalil & McClay 2001,

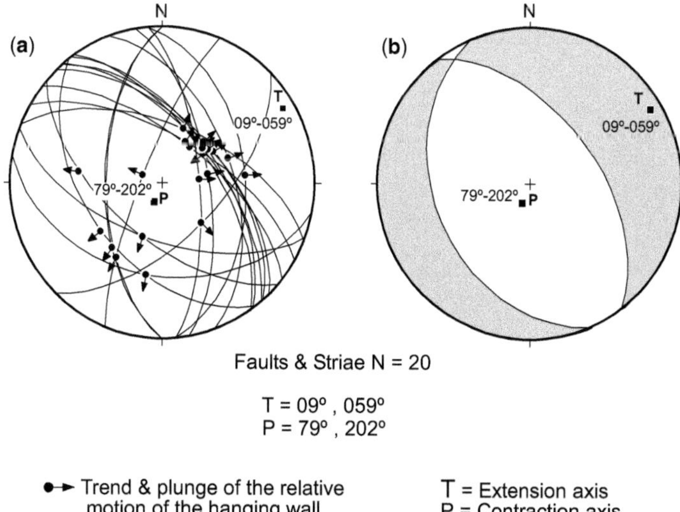

Fig. 4. (**a**) Lower-hemisphere equal-area stereographic projection of faults and striae showing NE–SW extension direction. (**b**) Fault-plane solution for the same faults and striae data. T and P indicate the extension and shortening axes, respectively.

2002; Bosworth *et al.* 2005). The oblique-slip component that is found on a few NNE-striking faults is minor and it is most probably related to the obliquity of the reactivated basement fault zones with respect to the regional NE–SW extension, rather than to NW–SE compression or an early stage of strike-slip faulting as proposed by Jarrige *et al.* (1986, 1990) and Montenat *et al.* (1988, 1998).

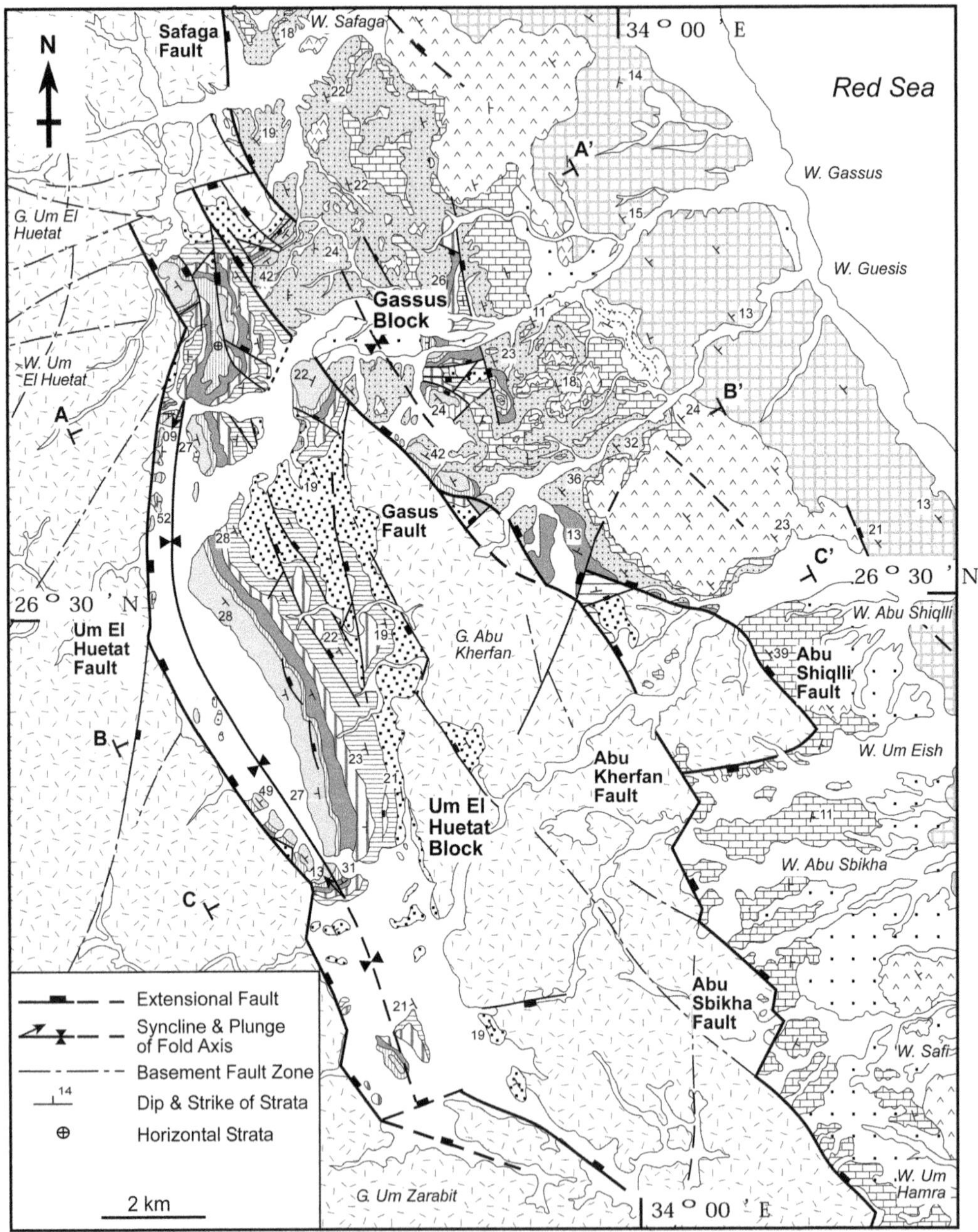

Fig. 5. Detailed geological map and cross-sections of the Um El Huetat area. The map shows the linked segments of the Coastal Fault System and the complex, doubly plunging hanging-wall syncline of the Um El Huetat fault block. In the cross-sections, note the asymmetrical geometry of the hanging-wall syncline, with the greatest structural relief in the central cross-section B–B′.

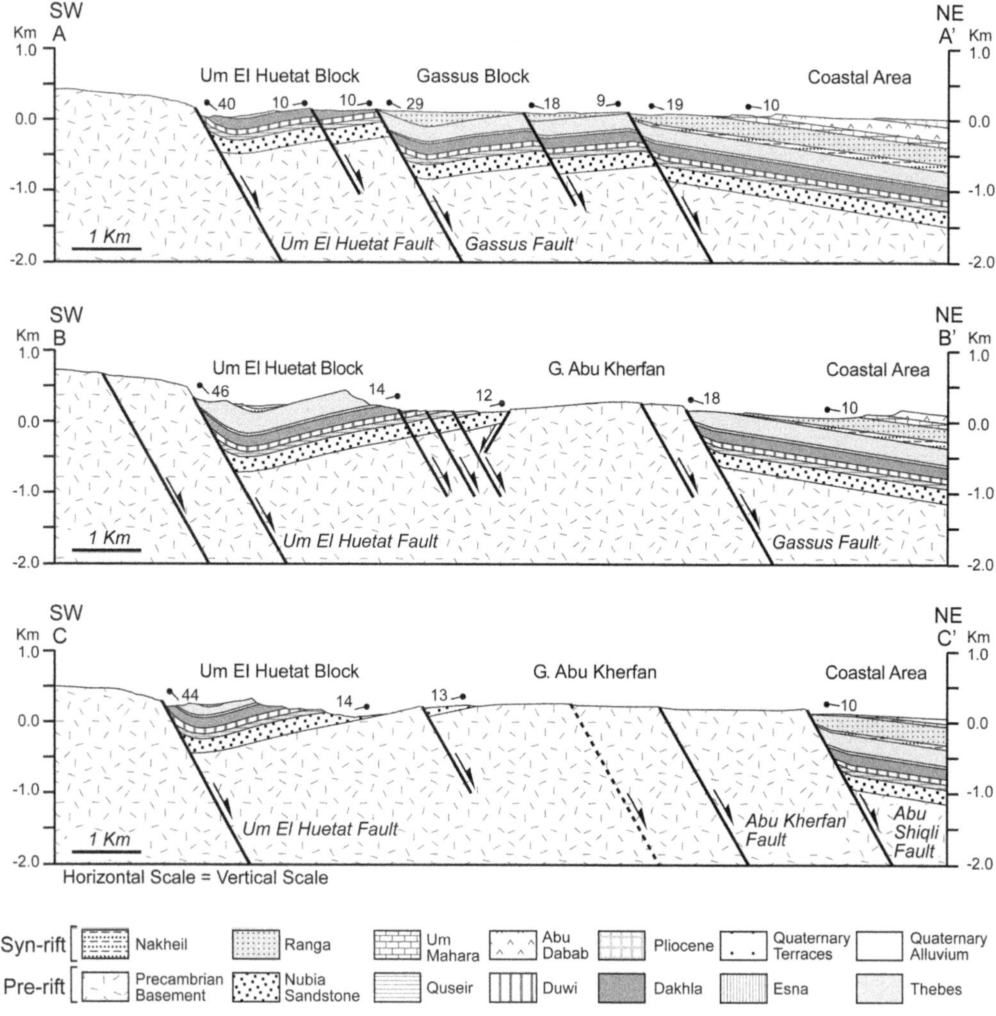

Fig. 5. *Continued.*

3D extensional fault-related folding

Um El Huetat fault bock

The Um El Huetat block in the footwall of the Gassus and Abu Kherfan segments of the Coastal Fault System is about 20 km long and, on average, 6 km wide (Fig. 5). The eastern part of this SW-dipping tilted fault block consists of the Precambrian basement of Gebel Abu Kherfan, whereas the western sector in the hanging wall of the Um El Huetat Fault is formed by synclinally folded Cretaceous–Eocene pre-rift strata (Figs 5 & 6).

The Um El Huetat Fault trends broadly NNW but is highly segmented with eight distinct segments along strike (Figs 5 & 7a). In map view, the hanging wall of the Um El Huetat Fault System is defined by a 12 km-long, curved, doubly plunging, longitudinal syncline (Um El Huetat Syncline: Figs 5 & 6b) that varies from a NNW–SSE trend in the south to a north–south trend in the north. The Um El Huetat Syncline is formed by pre-rift Cretaceous–Middle Eocene strata and isolated outcrops of the early synrift Late Oligocene Nakheil clastics. Figure 6 shows field views of the broad hanging-wall syncline of the Um El Huetat Fault System and the synrift pre-rift unconformity in the hanging wall of the Gassus Fault System, which also forms a longitudinal fold (Fig. 5).

The syncline is asymmetrical, with a short, steeply dipping (32–56°) western limb, in places

Fig. 6. Outcrop examples of the structures in the research area. (**a**) View looking south showing the Gassus Coastal Fault and the steep dip of the Eocene Formation in the immediate hanging wall. Note the 10° angular unconformity between the pre- and synrift strata, and the attenuation of the Esna Shale along the fault zone. (**b**) View looking south showing the southern nose of the Um El Huetat Syncline and the Um El Huetat Fault.

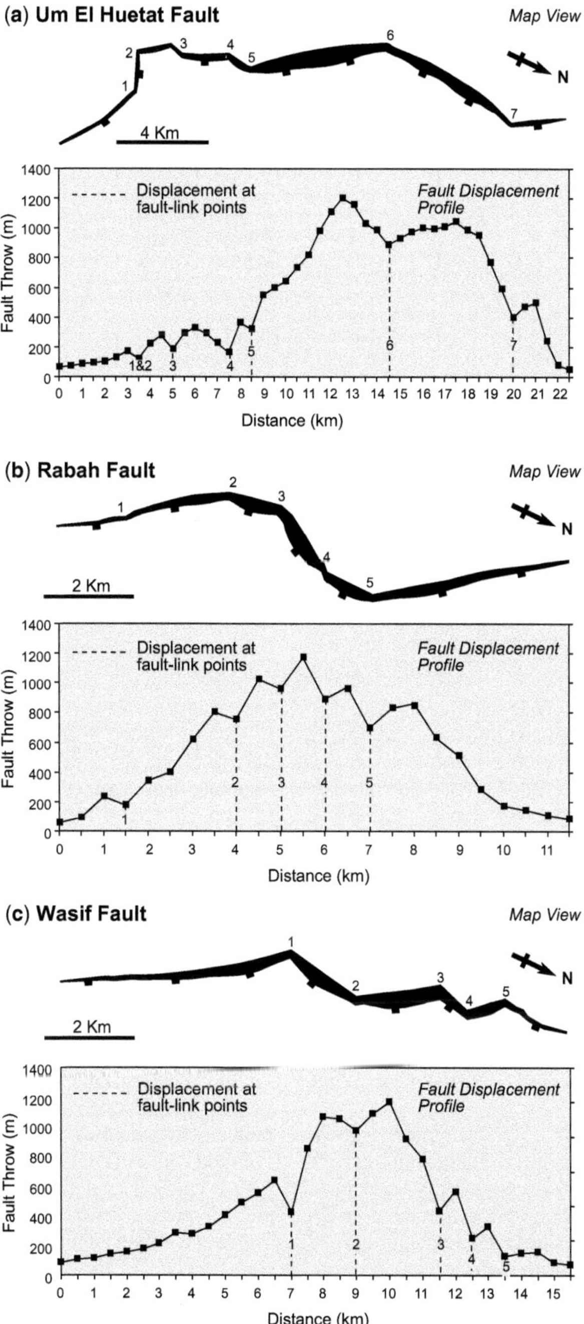

Fig. 7. (**a**) The Um El Huetat Fault displacement profile; (**b**) the Rabah Fault displacement profile; and (**c**) the Wasif Fault displacement profile. Note that the maximum displacement is located near the centre of the faults and sharp kinks in the profiles indicate where the original separate fault segments have joined. Displacement was measured from cross-sections, as well from control points with a spacing of 0.5–1 km along the fault strike, taking into account thickness variations of pre-rift strata and the slope of the top of the basement. As in most places, the pre-rift sedimentary section has been eroded from the footwalls of the fault blocks, and the estimated displacements are considered as minimum values, representing only components of the total displacement and clearly infer the along-strike change in hanging-wall subsidence.

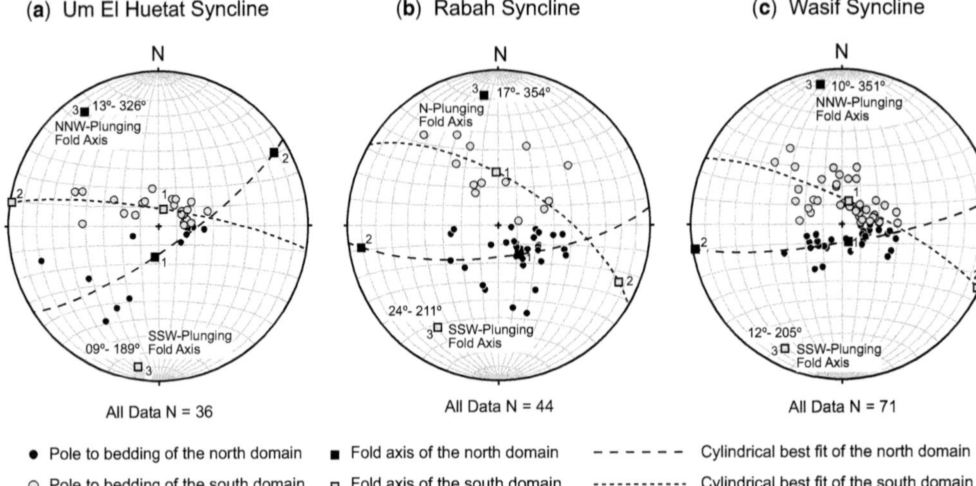

Fig. 8. Lower-hemisphere equal-area stereographic projections of poles to bedding showing the three main doubly plunging hanging-wall synclines in the study area. Bedding measurements were evenly sampled along the two halves of each fold (including the northern and southern fault-tip areas), and were sorted into two domains to determine the northern and southern plunge directions of the axis of each fold. (**a**) The Um El Huetat Syncline, (**b**) the Rabah syncline and (**c**) the Wasif syncline.

sub-parallel to the El Um Heutat Fault, and a more gently dipping (10–22°) eastern limb (cross-sections in Fig. 5). The lowermost pre-rift sedimentary strata, the Late Cretaceous Nubia Sandstone (Fig. 3), is only slightly folded but intensely fractured near the Um El Huetat Fault, whereas the overlying, highly anisotropic units of the Late Cretaceous–Paleocene strata are strongly folded. Field observations show that, on the steep NE-dipping limb of the syncline, shale units of the Dakhla and Esna formations are highly sheared and attenuated, whereas limestone beds of the overlying Thebes Formation are highly fractured and brecciated in the immediate hanging wall of the Um El Huetat Fault.

Displacement analyses (Fig. 7a) show variations from <60 m near the fault terminations to a maximum of at least 1200 m in the central segment of the Um El Huetat Fault System (cross-section B–B′ in Fig. 5). The Um El Huetat Fault terminates to the north and south within the basement units, with displacement being transferred across relay ramps onto the main Coastal Fault System to the east.

The cross-sections in Figure 5 show around 1200 m of structural relief in the syncline in the central sector (cross-section B–B′ in Fig. 5): this decreases to the north and south along the syncline axial trace (cross-sections A–A′ and C–C′ in Fig. 5). Stereographic projections of poles to bedding show that the poles are strongly scattered (Fig. 8a), reflecting the non-cylindrical form and double-plunging geometry of the Um El Huetat Syncline. The form of the Um El Huetat Syncline and the along-strike change of displacement of the Um El Huetat Fault (Fig. 7a), as well as the strike of the bedding surfaces (e.g. Fig. 5), can be directly correlated with the segmentation and fault linkages of the main Um El Huetat Fault System, as shown in Figure 7a.

Rabah fault block

The Rabah fault block consists of two NNW-striking, NE-dipping fault segments linked by a NNE-striking fault system that is interpreted to be a fault that breached an original relay ramp developed between the offset NW-striking faults (Fig. 9). Figure 7b shows a detailed fault map, together with the present-day displacement profile showing a maximum offset of approximately 1200 m in the central sector of the fault and minor displacement minima that are interpreted to be linkage points along the fault system.

The longitudinal, doubly plunging hanging-wall Rabah Syncline has a distinctly kinked axial surface trace (Fig. 9). This form is interpreted to be the result of the hanging-wall folding induced by the linkage of the two NNW-striking, laterally offset fault segments. In cross-section, the hanging-wall syncline is very similar to that described for the Um El Huetat block, with a steeply NE-dipping

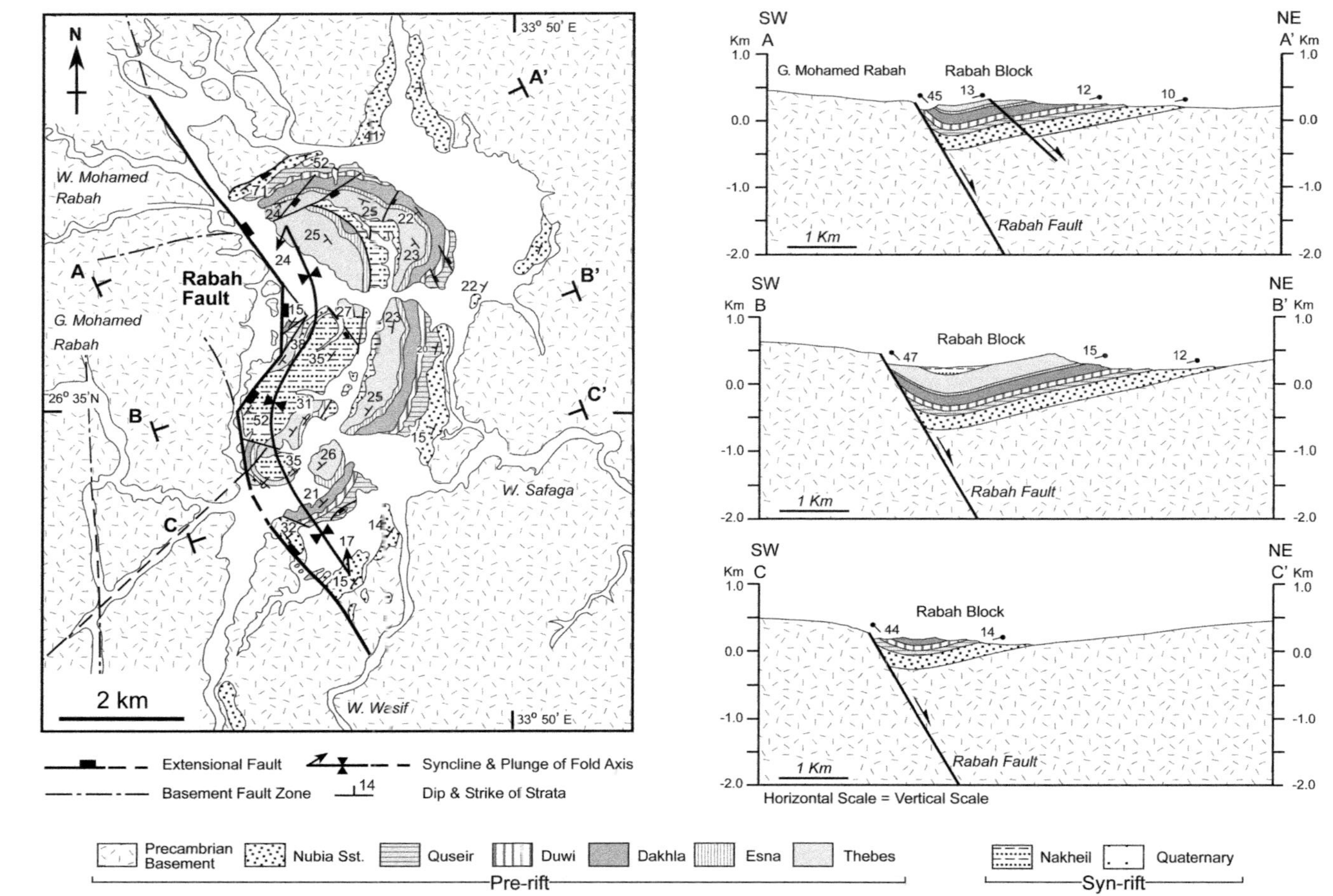

Fig. 9. Detailed geological map and cross-sections of the Rabah Fault System. The map shows the linked segments of the Rabah Fault and the doubly plunging Rabah hanging-wall syncline. The cross-sections show the asymmetrical geometry of the Rabah Syncline and the position of the greatest structural relief in the central cross-section B–B′.

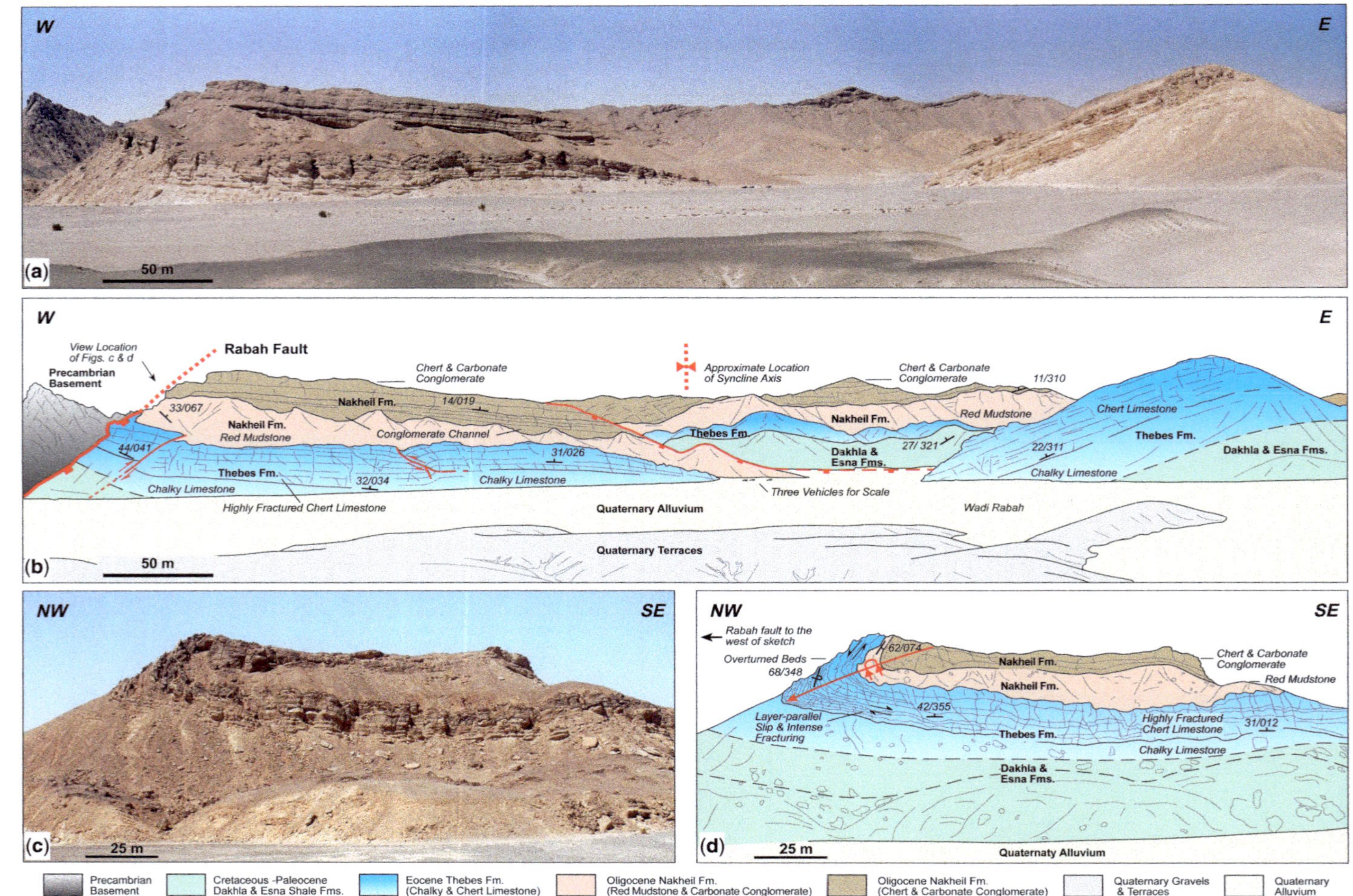

Fig. 10. (**a**) View looking north and (**b**) a perspective sketch showing the Rabah Syncline. (**c**) View looking NE and (**d**) a perspective sketch showing the overturned beds and recumbent folding in the immediate hanging wall of the Rabah Fault.

panel adjacent to the Rabat Fault and a gentle 10–14° west- to SW-dipping eastern limb of the syncline (cross-sections in Fig. 9). Figure 10a shows an oblique view of the Rabah hanging-wall syncline, with the synrift Late Oligocene Nakheil sequence unconformably on top of the Mid Eocene Thebes limestones. Adjacent to the Rabah Fault, interbedded shales, sandstones and limestones of the pre-rift, anisotropic, Upper Cretaceous–Paleocene units (Fig. 3) are strongly deformed and sheared. The limestones of the Eocene Thebes Formation are locally sub-vertical, and in places overturned and deformed in recumbent folds, which also affect the unconformably overlying synrift Late Oligocene Nakheil Formation (Fig. 10b). Similar recumbent folds, as well as small displacement reverse faults, have also been recognized in the immediate hanging wall of major block-bounding faults in the Gulf of Suez Rift (cf. Moustafa 1987; Khalil 1998; Sharp *et al.* 2000). This contractional deformation has been interpreted by Sharp *et al.* (2000) as secondary structures formed in response to space problems that are related to vertical displacement across fault segments, which are curved in map view. The local contractional deformation found in the study area supports a fault-propagation fold model (cf. that proposed by Sharp *et al.* 2000 for similar structures in the Gulf of Suez) and contrasts with a inversion model (e.g. as assumed by Knott *et al.* 1995 in the Gulf of Suez). The palaeo-strain analysis of the fault-slip data (Fig. 4) shows no evidence of later inversion in the NW Red Sea.

In the Rabah Syncline, early synrift Nakheil Formation shows well-developed fanning growth strata that shallow in dip up-section and away from the Rabah Fault. The fanning of synrift strata indicates that sedimentation has occurred during the growth of the Rabah Syncline, with increased extension on the Rabah Fault.

The NE–SW-orientated cross-sections in Figure 9 show that the maximum structural relief of the Rabah Syncline is about 1 km, and decreases laterally towards the north and south. Stereographic projections of poles to bedding (Fig. 8b) for the Rabah Syncline shows a notable scattering that indicates that the fold is non-cylindrical, and is doubly plunging: 17° to the north and 24° to the SSW.

Wasif fault block

The Wassif fault block (Fig. 11) shows a similar segmented fault system and curved longitudinal hanging-wall syncline to that of the Rabah block (Fig. 9). The Wasif Fault System consists of NNW-striking segments linked by shorter NNE-trending segments. The fault displacement profile in Figure 7c shows more variation than that for the Rabah Fault System owing to the greater segmentation and a maximum displacement of around 1200 m. Linkage points along the strike of the fault system have well-defined transverse synclines (Fig. 11).

The Wassif longitudinal hanging-wall syncline displays a highly curved axial surface trace and is defined by folded pre-rift strata consisting of Late Cretaceous–Middle Eocene Thebes limestones (Fig. 11). In cross-section, the Wassif Syncline is strongly asymmetrical, with a steep western limb (dips up to 46°) trending sub-parallel to the Wassif Fault and a longer eastern limb with dips from approximately 12° to 26° (the cross-sections in Fig. 11).

Stereographic projections of poles to bedding show a scatter pattern indicating non-cylindrical folding and a gently double-plunging geometry (Fig. 8c).

Discussion

Detailed field mapping and analysis of extensional fault systems and their hanging walls in a 40 km-long sector of the NW Red Sea margin (Fig. 2b) has shown that the strongly segmented nature of the rift extensional faults, combined with displacement variations along strike, have controlled the 3D geometries of the hanging-wall basins in this region. The extensional faults are dominantly planar domino-style rotational fault systems that extend into the basement (based on seismic sections: Heath *et al.* 1998) and formed in response to the regional Late Oligocene–Mid-Miocene NE–SW-directed extension that formed the Red Sea and Gulf of Suez rift systems (e.g. reviews by Bosworth *et al.* 2005; Bosworth 2015). Footwall and rift flank uplift resulted in the erosion of the pre-rift Upper Cretaceous–Mid-Eocene sedimentary strata on the footwalls of the extensional fault systems, as well as any Late Oligocene synrift strata that may have been deposited on these footwall blocks.

Extensional fault-propagation folding

The geological maps and cross-sections (Figs 2, 5, 9 & 11) reveal complex, asymmetrical, doubly plunging synclines formed in the hanging wall of the extensional faults. The geometries of these folds are controlled by the linkage patterns of the originally segmented extensional fault systems combined with extensional fault-propagation folding. Footwall and hanging-wall folding commonly occurs in basins where the extensional faults propagate up-section through variable mechanical stratigraphies – sometimes termed 'forced folds' (e.g. Schlische 1995; Hardy & McClay 1999; Cosgrove & Ameen 1999; Sharp *et al.* 2000; Jackson *et al.* 2006; Lewis *et al.* 2015). The 'sandwich'

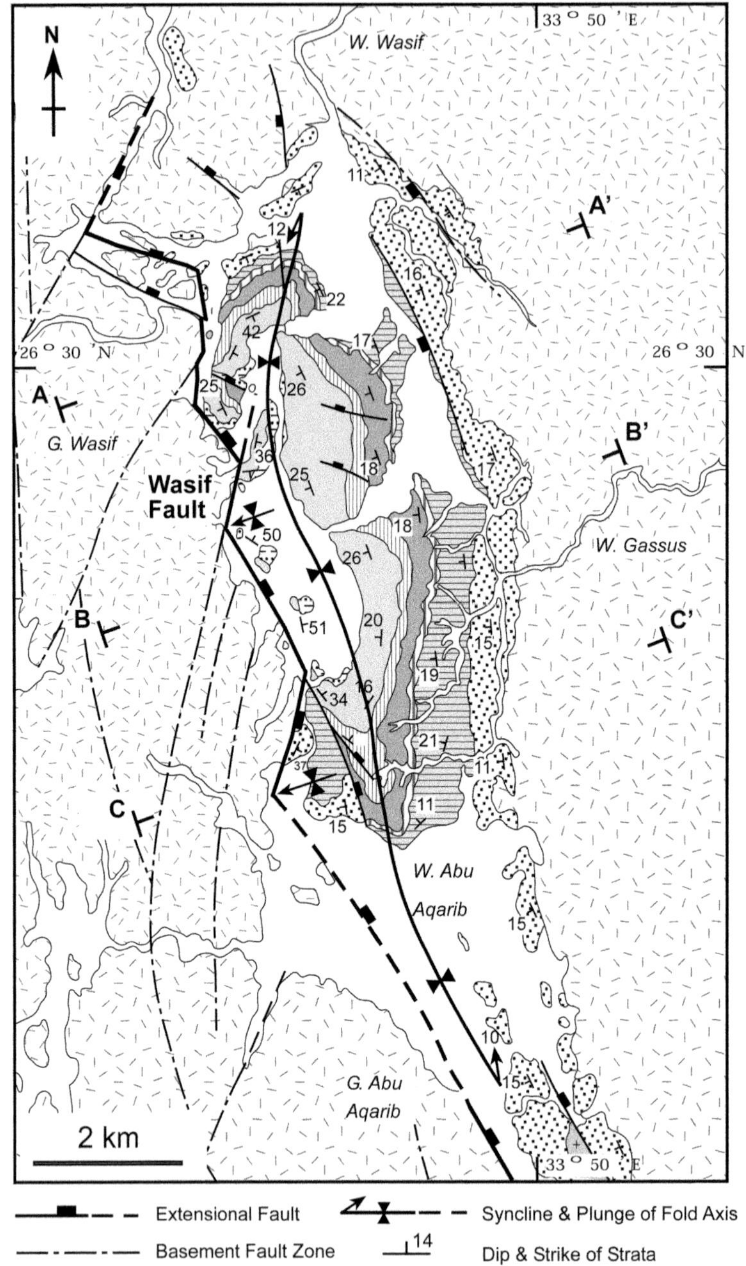

Fig. 11. Detailed geological map and cross-sections of the Wasif Fault System. The map shows the linked segments of the Wasif Fault and the doubly plunging Wasif Syncline. The cross-sections show the asymmetrical geometry of the Wasif Syncline and the location of the greatest structural relief in the central cross-section B–B′.

of the highly anisotropic, shale-dominated strata (Upper Cretaceous–Paleocene units) between the competent Cretaceous Nubia Sandstone below and the competent Eocene limestones above (Fig. 3) would have permitted the development of extensional fault-propagation folds, both vertically and laterally, as the underlying basement-involved extensional faults increased their displacement

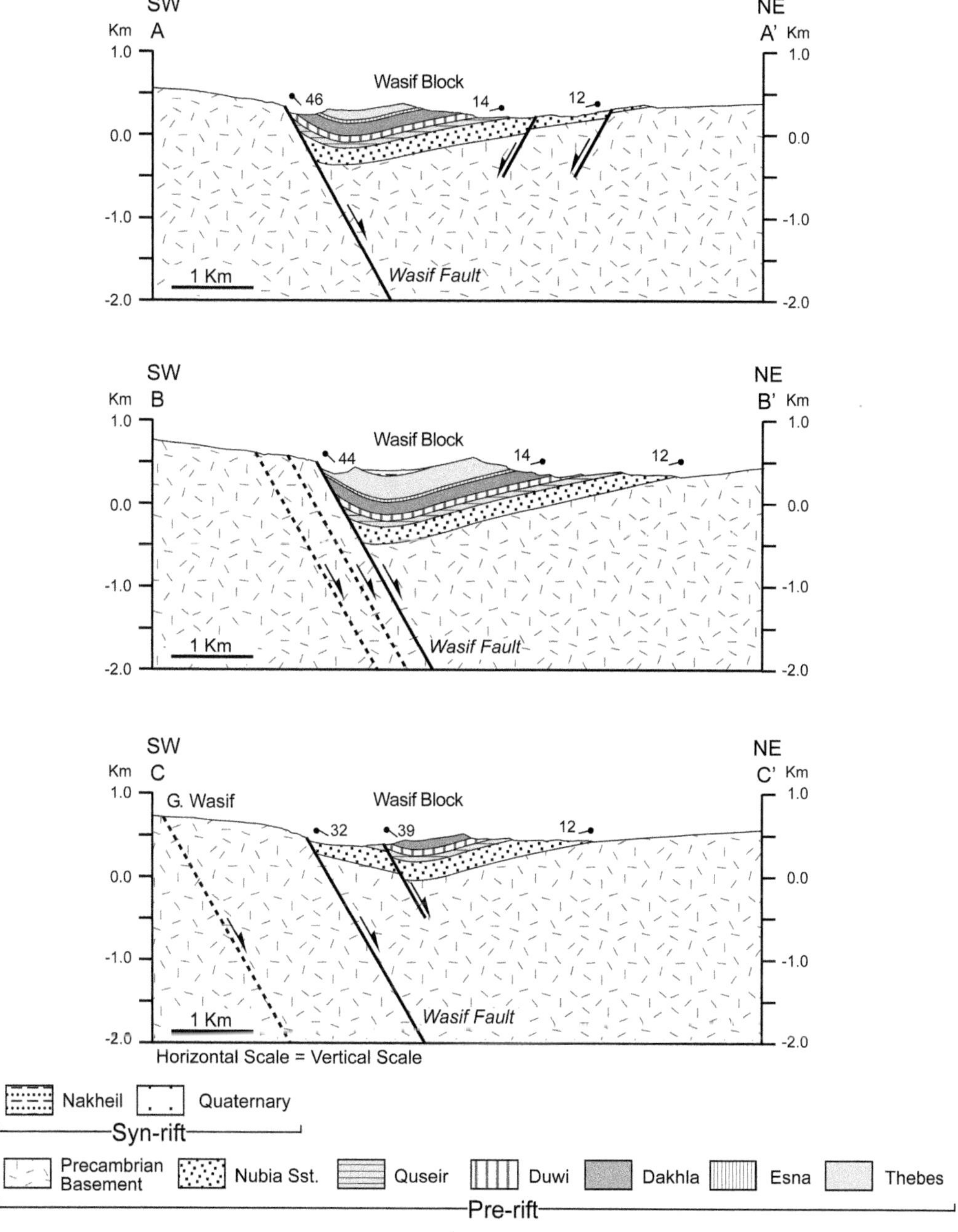

Fig. 11. *Continued.*

(cf. Khalil & McClay 2001, 2006). The extensional fault-propagation folds would have been accommodated by flexural slip within the anisotropic shale sequences to produce longitudinal folds parallel to the driving fault, with terminations in tip-line monoclines along strike (e.g. as in Fig. 1). Similar styles of extensional fault-propagation folding have been documented in many rift basins (e.g. Patton *et al.* 1994; Schlische 1995; Sharp *et al.* 2000; Jackson *et al.* 2006; Ford *et al.* 2007), as well as in analogue and numerical models (e.g. Withjack *et al.* 1990; Hardy & McClay 1999; Withjack & Schlische

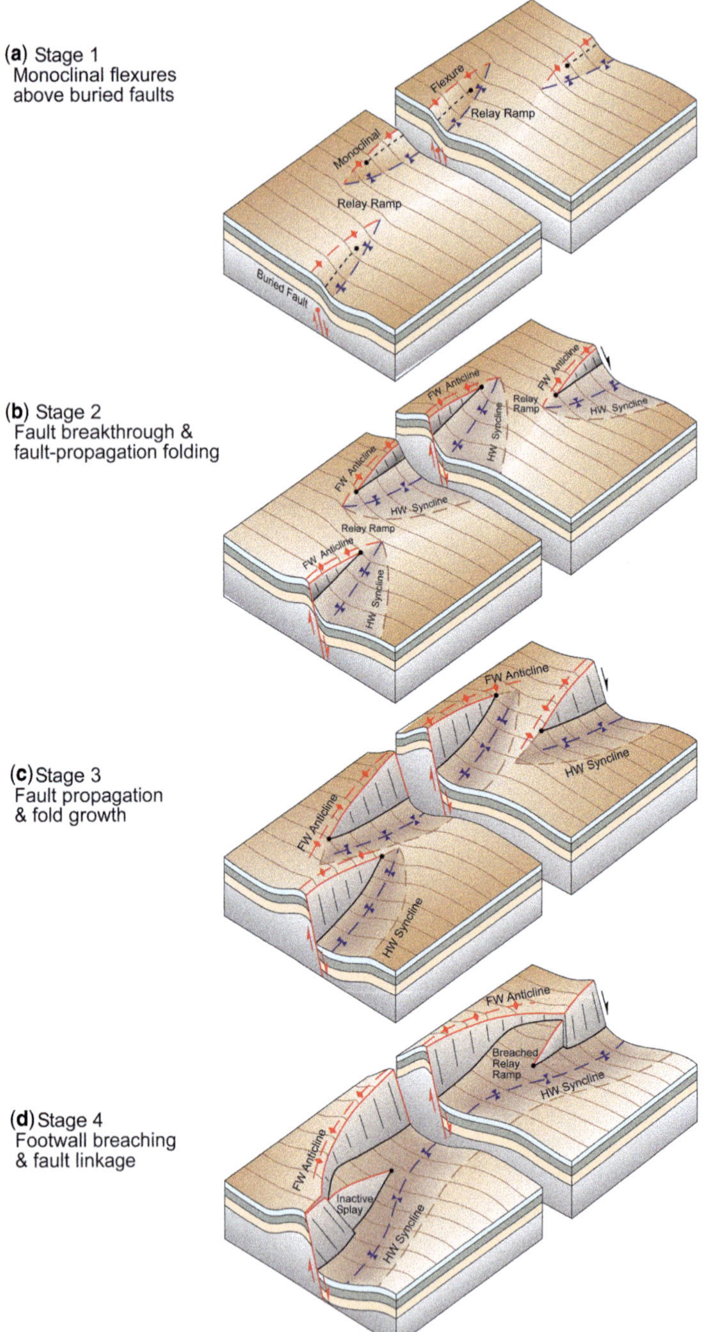

Fig. 12. Conceptual 3D model showing the kinematic evolution and geometrical growth of extensional fault-related folds as a consequence of the propagation and linkage of fault segments. (**a**) Monoclinal flexures above buried fault segments; (**b**) fault breakthrough, monocline breaching, and the development of footwall (FW) anticlines, hanging-wall (HW) synclines and relay ramps; (**c**) vertical and lateral propagation of fault segments, growth of the fold structures, and amplification of the relay ramps; (**d**) linkage of fault segments, footwall breaching of relay ramps, and the development of long, curved doubly plunging synclines by coalescence of the previously isolated hanging-wall synclines.

2006). The influence of mechanical stratigraphy on strain partitioning and development of fault-propagation folds has also been recognized from both outcrop and subsurface studies (e.g. Ferrill *et al.* 2007, 2011, 2012; Kane *et al.* 2010; Lewis *et al.* 2013).

The hanging-wall folds from the NW Red Sea margin described in this paper are kilometre-scale, asymmetrical synclines with amplitudes of at least 1200 m (Figs 2, 5, 9 & 11). Frictional 'drag folding' is not likely to be responsible for these large-scale hanging-wall synclines discussed in this research. Drag folds normally result from frictional drag along the fault surface (e.g. Hatcher 1994) and are generally restricted to only a few tens of metres adjacent to the fault (Twiss & Moores 1992; Davis & Reynolds 1996; Peacock *et al.* 2000).

Extensional fault segmentation and linkages

Along-strike variations in fault displacements for an isolated extensional fault would typically show a displacement maximum in the middle of the fault, resulting in the hanging wall forming a transverse syncline, whereas the corresponding position in the footwall would form a transverse footwall anticline (e.g. Fig. 1) (Schlische 1995). Such patterns are well developed in many extensional terranes (e.g. Chapman *et al.* 1978; Muraoka & Kamata 1983; Walsh & Watterson 1987; Dawers *et al.* 1993; Young *et al.* 1995). In the NW Red Sea, the fault systems show distinct linked segments with variable displacements along strike (Fig. 7). The linked fault systems generally show an overall maximum displacement in the central sector of the fault system and displacements decreasing towards the fault tips. Overall, the hanging walls formed by these linked segments form doubly plunging synclines with curved axial surface traces that reflect the along-strike linkage patterns (Figs 2b, 5, 9 & 11). In places, gentle transverse folds are preserved (e.g. Fig. 11) but the main fold amplitudes occur along the longitudinal fold systems.

Conceptual evolutionary segmented fault models for the research area

Figure 12 shows a series of 3D evolutionary diagrams for the extensional fault-propagation folding formed by en echelon offset fault segments.

Stage 1 (Fig. 12a) shows the development of monoclinal flexures above blind extensional faults. In the NW Red Sea margin, as in the Gulf of Suez further to the NW, the regional N60°E extension during the Late Oligocene–Miocene rifting reactivated old structures in the Pan African crystalline basement beneath at least 415–870 m (Fig. 3) of pre-rift Late Cretaceous–Mid Eocene strata. As the displacement increases along the basement faults, vertical and lateral propagation of the faults produced flexural monoclines at the surface (Fig. 12a).

Stage 2 (Fig. 12b) shows continued fault displacement with increased folding, together with fault breakthrough to the surface. Along-strike overlap of lengthening fault segments produced relay ramps. Rotation of the hanging wall at this stage was approximately 10°, as indicated by the angular unconformity between the pre-rift and the early synrift strata (Fig. 6a). The clast composition of the early synrift Late Oligocene Nakheil conglomerates (chert and carbonate breccias, and conglomerates derived from the underlying pre-rift Eocene Thebes Formation) indicate that footwall uplift was small, and the fault displacement was probably a few hundreds of metres.

Stage 3 (Fig. 12c) shows greater displacement on the offset fault system, with marked hanging-wall synclines in addition to significant footwall uplift and with the corresponding, narrow footwall anticlines. These longitudinal extensional fault-propagation folds become notably doubly plunging (Fig. 12c).

Stage 4 (Fig. 12d) shows breaching of the relay ramps and the formation of a linked extensional fault array. The hanging-wall synclines coalesced to form long longitudinal folds with curved axial surface traces – the positions of curvature being controlled by the linkage geometries across breached or faulted relay ramps (Fig. 12d).

Now, only remnants of the early synrift sediments are locally preserved in the hanging-wall synclines of the Um El Huetat, Wasif and Rabah folds (Figs 5, 9 & 11), whereas the Precambrian basement is exposed in the footwall of the faults owing to significant uplift and erosion of the pre-rift strata and into the underlying basement.

Conclusions

Longitudinal extensional fault-propagation folding has produced kilometre-scale hanging-wall synclines in this sector of the NW margin of the Red Sea in Egypt. Similar geometries at similar scales have been described in the Hamadat and Duwi areas of the NW Red Sea (Khalil & McClay 2002), in the Gulf of Suez (Moustafa 1987; Patton *et al.* 1994; Sharp *et al.* 2000; Jackson *et al.* 2006; Khalil & McClay 2006) and in the North Sea (Corfield & Sharp 2000), and in the Triassic rift basins of the eastern USA (Schlische 1995). The research presented in this paper has shown how the initial fault segmentation and subsequent linkage, together with the consequent variations in fault displacements, can produce complex 3D hanging-wall folds in rift basins. The 3D evolutionary models proposed in this research depict the progressive evolution of these complex extension-related folds, and their relationships to vertical and lateral propagation and linkages of extensional faults in rift settings.

The present study demonstrates that the synclinal folds in the NW Red Sea formed in response to the along-strike displacement variation of the faults and the upwards propagation of the faults through the mechanically heterogeneous lithologies of the pre-rift section. The development of the folds was facilitated by internal deformation, shearing, bed-parallel slip and thinning of the shale units within the highly anisotropic units in the pre-rift section. The faulting and related folding of the NW Red Sea margin developed in a pure extensional stress regime related to the NE–SW extension that generated the Oligo-Miocene rifting of the Red Sea. The present study clearly demonstrates that fault-related folds that are normally found in rift basins form an integral part of the extensional deformation and evolution of these basins. Therefore, there is no need to invoke compressional stresses or strike-slip tectonics to account for the structural deformation found in the NW Red Sea, as has been proposed by other authors.

The research presented in this paper was supported by the Natural Environment Research Council (NERC) ROPA Grant GR3/R9529 and by a Research Grant from BG International. Additional support was from the Fault Dynamics Project, Royal Holloway University of London (sponsored by ARCO British Limited, PETROBRAS UK Ltd, BP Exploration, Conoco (UK) Limited, Mobil North Sea Limited and Sun Oil Britain). The authors also gratefully acknowledge support from ARCO British Limited (now BP Exploration) and Suez Canal University, Egypt. Bill Bosworth and Marathon Petroleum Egypt are thanked for logistical support and for many fruitful discussions on the geology of the NW Red Sea. BG Egypt and Maher Ayyad kindly sponsored additional fieldwork in the NW Red Sea. Thoughtful reviews by Christopher Jackson and Haakon Fossen greatly improved the manuscript and are appreciated.

References

Abd el-Razik, T.M. 1967. Stratigraphy of the sedimentary cover of the Anz-Atshan-south Duwi district. *Bulletin of the Faculty of Science, Cairo University*, **431**, 135–179.

Akaad, M.K. & Noweir, A.M. 1980. Geology and lithostratigraphy of the Arabian desert orogenic belt between latitudes 25° 35′ and 26° 30′. *In*: Cooray, P.A.T.S. (ed.) *Evolution and Mineralization of the Arabian–Nubian Shield*. Pergamon Press, New York, 127–135.

Akkad, S. & Dardir, A.A. 1966. *Geology and Phosphate Deposits of Wasif, Safaga Area*. Geological Survey of Egypt, Papers, **36**.

Ben-Menahem, A., Nur, A. & Vered, M. 1976. Tectonics, seismicity and structure of the Afro-Eurasian junction - the breaking of an incoherent plate. *Physics of the Earth Planetary Interiors*, **12**, 1–50.

Bosworth, W. 2015. Geological evolution of the Red Sea: historical background, review and synthesis. *In*: Rasaul, N.M.A. & Stewart, I.C.F. (eds) *The Red Sea*. Springer, Berlin, 45–78.

Bosworth, W., Huchon, P. & McClay, K. 2005. The Red Sea and Gulf of Aden basins. *Journal of African Earth Sciences*, **43**, 334–378.

Chapman, G.R., Lippard, S.J. & Martyn, J.E. 1978. The stratigraphy and structure of the Kamasia Range, Kenya Rift Valley. *Journal of the Geological Society, London*, **135**, 265–281, https://doi.org/10.1144/gsjgs.135.3.0265

Cochran, J.R. 1983. A model for development of the Red Sea. *American Association of Petroleum Geologists Bulletin*, **67**, 41–69.

Coleman, R.G. 1993. *Geologic Evolution of the Red Sea*. Oxford Monographs on Geology and Geophysics, **24**. Oxford University Press, Oxford,

Corfield, S. & Sharp, I.R. 2000. Structural style and stratigraphic architecture of fault propagation folding in extensional settings: a seismic example from the Smerbukk area, Halten Terrace, Mid-Norway. *Journal of Basin Research*, **12**, 329–341.

Cosgrove, J.W. & Ameen, M.S. 1999. A comparison of the geometry, spatial organization and fracture patterns associated with forced folds and buckle folds. *In*: Cosgrove, J.W. & Ameen, M.S. (eds) *Forced Folds and Fractures*. Geological Society, London, Special Publications, **169**, 7–21, https://doi.org/10.1144/GSL.SP.2000.169.01.02

Davis, G.H. & Reynolds, S.J. 1996. *Structural Geology of Rocks and Regions*. 2nd edn. John Wiley, Chichester.

Dawers, N.H., Anders, M.H. & Scholz, C.H. 1993. Growth of normal faults; displacement-length scaling. *Geology*, **21**, 1107–1110.

El Bassyony, A.A. 1982. *Stratigraphical studies on Miocene and younger exposures between Quseir and Berenice, Red Sea coast, Egypt*. PhD thesis, Ain Shams University, Cairo.

Ferrill, D.A., Morris, A.P. & Smart, K.J. 2007. Stratigraphic control on extensional fault propagation folding: Big Brushy Canyon monocline, Sierra Del Carmen, Texas. *In*: Jolley, S.J., Barr, D., Walsh, J.J. & Knipe, R.J. (eds) *Structurally Complex Reservoirs*. Geological Society, London, Special Publications, **292**, 203–217, https://doi.org/10.1144/SP292.12

Ferrill, D.A., Morris, A.P., McGinnis, R.N., Smart, K.J. & Ward, W.C. 2011. Fault zone deformation and displacement partitioning in mechanically layered carbonates: the Hidden Valley fault, central Texas. *American Association of Petroleum Geologists Bulletin*, **95**, 1383–1397.

Ferrill, D.A., Morris, A.P. & McGinnis, R.N. 2012. Extensional fault-propagation folding in mechanically layered rocks: the case against the frictional drag mechanism. *Tectonophysics*, **576–577**, 78–85.

FINCH, E., HARDY, S. & GAWTHORPE, R. 2004. Discrete-element modelling of extensional fault-propagation folding above rigid basement fault blocks. *Basin Research*, **16**, 489–506.

FORD, M., LE CARLIER DE VESLUND, C. & BOURGEOIS, O. 2007. Kinematic and geometric analysis of fault-related folds in a rift setting: The Dannemarie basin, Upper Rhine Graben, France. *Journal of Structural Geology*, **29**, 1811–1830, https://doi.org/10.1016/j.jsg.2007.08.001

FREUND, R. 1970. Plate tectonics of the Red Sea and Africa. *Nature*, **228**, 453.

HARDY, S. & FINCH, E. 2006. Discrete element modelling of the influence of cover strength on basement-involved fault-propagation folding. *Tectonophysics*, **415**, 225–238.

HARDY, S. & MCCLAY, K.R. 1999. Kinematic modelling of extensional fault-propagation folding. *Journal of Structural Geology*, **21**, 695–702.

HATCHER, R.D., JR. 1994. *Structural Geology: Principles, Concepts, Problems*. Prentice-Hall, Englewood Cliffs, NJ.

HEATH, R., VANSTONE, S. *ET AL*. 1998. Renewed exploration in the offshore north Red Sea Region – Egypt. *In*: ELOUI, M. (ed.) *Proceedings of the 14th Petroleum Conference*, Cairo, Egypt. Egyptian General Petroleum Corporation, Cairo, 16–34.

HEMPTON, M. 1987. Constraints on Arabian plate motion and extensional history of the Red Sea. *Tectonics*, **6**, 687–705.

ISSAWI, B., FRANCIS, M., EL-HINNAWI, M. & MEHANNA, A. 1969. *Contribution to the Structure and Phosphate Deposits of Quseir Area*. Geological Survey of Egypt, Papers, 50.

JACKSON, C.A.L., GAWTHORPE, R.L. & SHARP, I.R. 2006. Style and sequence of deformation during extensional fault-propagation folding: examples from the Hammam Faraun and El-Qaa fault blocks, Suez Rift, Egypt. *Journal of Structural Geology*, **28**, 519–535.

JARRIGE, J.J., OTT D'ESTEVOU, P. *ET AL*. 1986. Inherited discontinuities and Neogene structure: the Gulf of Suez and the northwestern edge of the Red Sea. *Philosophical Transactions of the Royal Society of London, Series A*, **317**, 129–139.

JARRIGE, J.J., OTT D'ESTEVOU, P., BUROLLET, P.F., MONTENAT, C., RICHET, J.P. & THIRIET, J.P. 1990. The multistage tectonic evolution of the Gulf of Suez and northern Red Sea continental rift from field observations. *Tectonics*, **9**, 441–465.

KANE, K.E., JACKSON, C.A.L. & LARSEN, E. 2010. Normal fault growth and fault-related folding in a salt-influenced rift basin: South Viking Graben, Offshore Norway. *Journal of Structural Geology*, **32**, 490–506.

KHALIL, S.M. 1998. *Tectonic Evolution of the Eastern Margin of the Gulf of Suez, Egypt*. PhD thesis, Royal Holloway University of London.

KHALIL, S.M. & MCCLAY, K.R. 2001. Tectonic evolution of the NW Red Sea–Gulf of Suez rift system. *In*: WILSON, R.C.L., WHITMARSH, R.B., TAYLOR, B. & FROITZHEIM, N. (eds) *Non-Volcanic Rifting of Continental Margins: A Comparison of Evidence from Land and Sea*. Geological Society, London, Special Publications, **187**, 453–473, https://doi.org/10.1144/GSL.SP.2001.187.01.22

KHALIL, S.M. & MCCLAY, K.R. 2002. Extensional fault-related folding, northwestern Red Sea, Egypt. *Journal of Structural Geology*, **24**, 743–762.

KHALIL, S.M. & MCCLAY, K.R. 2006. Extensional fault-related folding, Gulf of Suez, Egypt. *Middle East Research Centre, Ain Shams University, Earth Science Series*, **20**, 1–16.

KHALIL, S.M. & MCCLAY, K.R. 2009. Structural control on syn-rift sedimentation, northwestern Red Sea margin, Egypt. *Marine and Petroleum Geology*, **26**, 1018–1034.

KNOTT, S.D., BEACH, A., WELBON, A.I. & BROCKBANK, P.J. 1995. Basin inversion in the Gulf of Suez: implications for exploration and development in failed rifts. *In*: BUCHANAN, J.G. & BUCHANAN, P.G. (eds) *Basin Inversion*. Geological Society, London, Special Publications, **88**, 59–81, https://doi.org/10.1144/GSL.SP.1995.088.01.05

LE PICHON, X. & FRANCHETEAU, J. 1978. A plate tectonic analysis of the Red Sea-Gulf of Aden area. *Tectonophysics*, **46**, 369–406.

LEWIS, M.M., JACKSON, C.A.L. & GAWTHORPE, R.L. 2013. Salt-influenced normal fault growth and forced folding: the Stavanger fault system, North Sea. *Journal of Structural Geology*, **54**, 156–173.

LEWIS, M.M., JACKSON, C.A.L., GAWTHORPE, R.L. & WHIPP, P.S. 2015. Early synrift reservoir development on the flanks of extensional forced folds: a seismic-scale outcrop analog from the Hadahid fault system, Suez rift, Egypt. *American Association of Petroleum Geologists Bulletin*, **99**, 985–1012.

MAURIN, J.-C. & NIVIERE, B. 2000. Extensional forced folding and decollement of the pre-rift series along the Rhine graben and their influence on the geometry of the syn-rift sequences. *In*: COSGROVE, J.W. & AMEEN, M.S. (eds) *Forced Folds and Fractures*. Geological Society, London, Special Publications, **169**, 73–86, https://doi.org/10.1144/GSL.SP.2000.169.01.06

MCKENZIE, D.P., DAVIES, D. & MOLNAR, P. 1970. Plate tectonics of the Red Sea and east Africa. *Nature*, **226**, 243–248.

MESHREF, W.M. 1990. Chapter 8. Tectonic Framework. *In*: SAID, R. (ed.) *The Geology of Egypt*. A.A. Balkema, Rotterdam, 113–155.

MONTENAT, C., OTT D'ESTEVOU, P. *ET AL*. 1988. Tectonic and sedimentary evolution of the Gulf of Suez and the northern western Red Sea. *Tectonophysics*, **153**, 166–177.

MONTENAT, C., OTT D'ESTEVOU, P., JARRIGE, J.-J. & RICHERT, J.P. 1998. Rift development in the Gulf of Seuz and the north-western Red Sea: structural aspects and related sedimentary processes. *In*: PURSER, B.H. & BOSENCE, D.W.J. (eds) *Sedimentation and Tectonics of Rift Basins: Red Sea–Gulf of Aden*. Chapman & Hall, London, 97–116.

MOUSTAFA, A.R. 1987. Drape folding in the Baba-Sidri area, eastern side of the Suez rift. *Egyptian Journal of Geology*, **31**, 15–27.

MOUSTAFA, A.R. 1997. Controls on the development and evolution of transfer zones: the influence of basement structure and sedimentary thickness in the Suez rift and Red Sea. *Journal of Structural Geology*, **19**, 755–768.

MURAOKA, H. & KAMATA, H. 1983. Displacement distribution along minor fault traces. *Journal of Structural Geology*, **5**, 483–495.

PATTON, T.L., MOUSTAFA, A.R., NELSON, R.A. & ABDINE, S.A. 1994. Tectonic evolution and structural setting of the Suez rift. *In*: LANDON, S.M. (ed.) *Interior Rift Basins*. American Association of Petroleum Geologists, Memoirs, **59**, 7–55.

PEACOCK, D.C.P., KNIPE, R.J. & SANDERSIN, D.J. 2000. Glossary of normal faults. *Journal of Structural Geology*, **22**, 291–306.

SAID, R. 1990. *The Geology of Egypt*. A. A. Balkema, Rotterdam.

SHARP, I.R., GAWTHORPE, R.L., UNDERHILL, J.R. & GUPTA, S. 2000. Fault propagation folding in extensional settings: examples of structural style and synrift sedimentary response from the Suez rift, Sinai, Egypt. *Geological Society of America Bulletin*, **112**, 1877–1899.

STERN, R.J. 1994. Arc assembly and continental collision in the Neoproterozoic East African orogen: implications for the consolidation of Gondwanaland. *Annual Review, Earth and Planetary Science*, **22**, 319–351.

SCHLISCHE, R.W. 1995. Geometry and origin of fault-related folds in extensional setting. *American Association of Petroleum Geologists Bulletin*, **79**, 1661–1678.

SCHLISCHE, R.W., WITHJACK, M.O. & EISENSTADT, G. 2002. An experimental study of the secondary deformation produced by oblique-slip normal faulting. *American Association of Petroleum Geologists Bulletin*, **86**, 885–906.

STECKLER, M.S., BERTHELOT, F., LYBERIS, N. & LE PICHON, X. 1988. Subsidence in the Gulf of Suez: implications for rifting and plate kinematics. *Tectonophysics*, **153**, 249–270.

STOESER, D.B. & CAMP, V.E. 1985. Pan African microplate accretion of the Arabian shield. *Geological Society America Bulletin*, **96**, 817–826.

TWISS, R.J. & MOORES, E.M. 1992. *Structural Geology*. Freeman & Co, New York.

WALSH, J.J. & WATTERSON, J. 1987. Distributions of cumulative displacement and seismic slip on a single normal fault surface. *Journal of Structural Geology*, **9**, 1039–1046.

WILSON, P., GAWTHORPE, R.L., HODGETTS, D., RARITY, F. & SHARP, I.R. 2009. Geometry and archeticture of faults in a syn-rift normal fault array: the Nukhul half-graben, Suez rift, Egypt. *Journal of Structural Geology*, **31**, 759–775.

WITHJACK, M.O. & SCHLISCHE, R.W. 2006. Geometric and experimental models of extensional fault-bend folds. *In*: BUITER, S.J.H. & SCHREURS, G. (eds) *Analogue and Numerical Modelling of Crustal-Scale Processes*. Geological Society, London, Special Publications, **253**, 285–305, https://doi.org/10.1144/GSL.SP.2006.253.01.15

WITHJACK, M.O., OLSEN, J. & PETERSON, E. 1990. Experimental models of extensional forced folds. *American Association of Petroleum Geologists Bulletin*, **74**, 1038–1045.

YOUNES, A. & MCCLAY, K.R. 2001. Role of basement fabric on rift architecture: Gulf of Suez – Red Sea, Egypt. *American Association of Petroleum Geologists Bulletin*, **86**, 1003–1026.

YOUNG, S.S., SCHLISCHE, R.W. & ACKERMANN, R.V. 1995. Miro-normal fault populations in Mesozoic rift basins: length displacement scaling relations (abstract). *Geological Society of America Abstracts with Programs*, **27**, 94.

YOUSSEF, M.I. 1957. Upper Cretaceous rocks in Kosseir area. *Bulletin de l'Institute du Desert d' Egypt*, **7**, 35–53.

Rift migration and lateral propagation: evolution of normal faults and sediment-routing systems of the western Corinth rift (Greece)

MARY FORD[1,2]*, ROMAIN HEMELSDAËL[2], MARCO MANCINI[3] & NIKOLAOS PALYVOS[4†]

[1]*Université de Lorraine, ENSG, INP, rue du Doyen-Marcel-Roubault, 54501 Vandoeuvre-lès-Nancy, France*

[2]*CPRG, UMR 7358, 15 Rue Notre-Dame-des-Pauvres, 54501 Vandoeuvre-lès-Nancy, France*

[3]*CNR-IGAG, Cnr Area della Ricerca Roma 1, Via Salaria km 29,300, 00015 Monterotondo Scalo, Rome, Italy*

[4]*Department of Geography, Harokopio University, 70 El. Venizelou Street, Athens, Greece*

**Correspondence: mford@crpg.cnrs-nancy.fr*

Abstract: The active Corinth rift records hanging-wall migration of faulting and slip-rate acceleration. The rift initiated at approximately 5–4 Ma, and older parts are well exposed in the northern Peloponnese. A new correlation of chrono- and lithostratigraphy and structure across the onland central to westernmost rift with offshore data reveals westward rift propagation, as well as northward fault migration. Northward fault migration ended first in the east, with the stabilization of major north-dipping faults that now bound the Gulf. The basin then propagated to the WNW in two stages, each involving the initiation of a new fault that propagated east to SE to link to the stable fault system. Extension rates accelerated in distinct steps as the rift opened to the west. The youngest faults in the westernmost rift are associated with high seismicity and highest geodetic extension due to rapid fault growth and linkage at depth.

The early synrift succession infilled substantial inherited palaeo-relief. Antecedent rivers established vigorous sediment-routing systems that controlled facies distribution throughout rifting, albeit with drainage reorganization during fault-migration events. Multiple deepening events recorded in the stratigraphy can be due to lateral rift propagation. The transition from rift initiation to rift climax is, therefore, diachronous along the rift axis.

The Corinth rift is one of the fastest opening rifts in the world. It initiated in the Late Pliocene (5–4 Ma: Keraudren & Sorel 1987; Ori 1989; Doutsos & Piper 1990; Billiris *et al.* 1991; Roberts 1996; Doutsos & Kokkalas 2001; Leeder *et al.* 2008) as part of the western Aegean extension system (Le Pichon & Angelier 1979; Jolivet 2001). The older parts of the rift are now relatively inactive, and have been uplifted and exposed in a 25–30 km swathe of the northern Peloponnese (Fig. 1). Present-day extension and seismic activity are focused below the Gulf of Corinth itself (Rigo *et al.* 1996; Bernard *et al.* 2006; Lambotte *et al.* 2014), indicating that the locus of deformation has migrated northwards. Intense seismicity and microseismicity are concentrated below the westernmost Gulf (west of Aegion: Fig. 1) where the highest north–south extension rates (15–16 mm a^{-1}) have been geodetically recorded over the last 20–25 years (Bernard *et al.* 1997, 2006; Clarke *et al.* 1998; Briole *et al.* 2000). Geodetic extension rates decrease to 11 mm a^{-1} in the central and eastern Gulf (Clarke *et al.* 1998; Briole *et al.* 2000), where major earthquakes are less frequent than to the west (Jackson *et al.* 1982). If we extrapolate the geodetic extension rates back through the 4–5 myr of rifting history, total extension would be 50–80 km, nearly an order of magnitude greater than that estimated from surface geology by Bell *et al.* (2008: 11 km in the central rift) and Ford *et al.* (2013: 6–8 km in the western rift), and from crustal thinning by Bell *et al.* (2011: 5–13 km for the western rift and 11–21 km for the central rift). Similarly, geologically derived total extension estimates give an average extension rate over 5 myr of 1.3–2.2 mm a^{-1}, an order of magnitude less than current geodetic extension rates.

†Deceased 23 April 2012.

From: Childs, C., Holdsworth, R. E., Jackson, C. A.-L., Manzocchi, T., Walsh, J. J. & Yielding, G. (eds) 2017. *The Geometry and Growth of Normal Faults*. Geological Society, London, Special Publications, **439**, 131–168.
First published online August 15, 2016, https://doi.org/10.1144/SP439.15

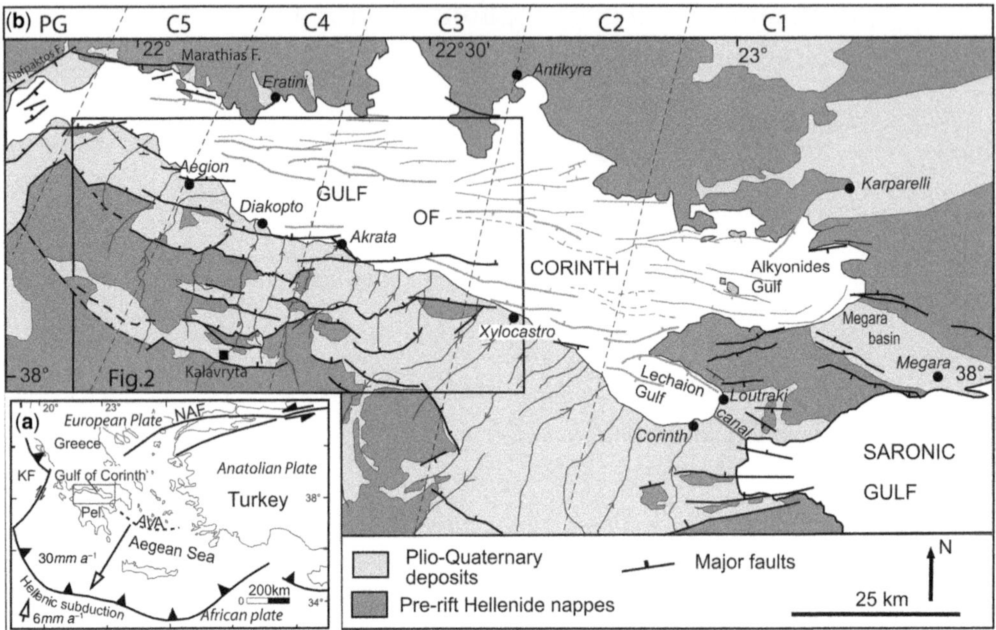

Fig. 1. (**a**) Tectonic map of the Aegean region showing main plates and plate boundaries, and the location of the Gulf of Corinth. Plate movement vectors are with respect to a fixed European Plate. NAF, North Anatolian Fault; KF, Kefalonia Fault; Pel., Peloponnesus peninsula; AVA, Aegean volcanic arc. (**b**) Tectonic map of the Corinth rift showing principal faults and the onshore distribution of Plio-Pleistocene sediments based on work of Lyon-Caen *et al.* (2004), Moretti *et al.* (2004), Bernard *et al.* (2006), Flotté *et al.* (2005), Rohais *et al.* (2007*a*), and Ford *et al.* (2013). Offshore faults in grey are from Nixon *et al.* (2016). Rivers of the northern Peloponnese are shown with an arrow in the direction of flow. The location of Figure 2 (study area) is boxed. For this study, the Corinth rift (onshore and offshore) is subdivided into five areas, C1–C5 (top of figure). PG is the Patras rift to the west.

Although, this comparison of geodetic and geological data must be treated with caution, it suggests that extension rate accelerated during rifting, a hypothesis that we will examine in this paper.

Evidence of northwards migration of fault activity has been noted and discussed by many authors (e.g. Dufaure 1975; Doutsos & Poulimenos 1992; Armijo *et al.* 1996; Goldsworthy & Jackson 2001; Collier & Jones 2003; De Martini *et al.* 2004; Flotté *et al.* 2005; Rohais *et al.* 2007*a*; Bell *et al.* 2008; Ford *et al.* 2013; Demoulin *et al.* 2015), most of whom focus on migration events since 0.7 Ma that are well expressed in the tectonic geomorphology of the northern Peloponnese.

The excellent exposures of the synrift succession in the northern Peloponnese provide the opportunity to investigate the evolution of the synrift structure and sedimentary succession: in particular, the response of sediment supply, facies distributions and depocentre connectivity to rift migration and acceleration over a period of 4–5 myr. Seismic reflection data provide clearer constraints on the structure and seismic stratigraphy of the offshore rift (Bell *et al.* 2011; Taylor *et al.* 2011; Nixon *et al.* 2016), although the offshore succession still remains undrilled except for the upper levels representing the last 25 kyr that have been sampled by piston coring (Moretti *et al.* 2004; Lykousis *et al.* 2007*a*; Campos *et al.* 2013).

This paper presents a new geological map of the onshore western and central Corinth rift (65 km long), with a new coherent stratigraphy and structure integrating published work and new unpublished work. These data were used to construct a tectonostratigraphic model for the evolution of the western and central rift in order to clarify: (1) the distribution of deformation in space and time (evolution of the fault system); (2) the principal controlling factors on synrift fill, including inherited palaeo-relief and antecedent drainage systems; and (3) the sedimentary response to rift migration and strain-rate acceleration. We discuss the significance of the present-day seismicity and high geodetic strain in the context of rift migration and strain-rate acceleration, and the problems associated with the recognition of major tectonic events in the stratigraphic record of rifts. We discuss, in particular, the relevance of deepening events in rifts.

Regional geological setting

The Aegean tectonic domain is defined by the Hellenic subduction system where the African Plate is subducting to the NNE below the European Plate at a rate of 6 mm a^{-1} (Fig. 1a) (Le Pichon & Angelier 1979, 1981; Jolivet *et al.* 1994; Gautier *et al.* 1999). Back-arc extension of over 400 km due to slab rollback created the Aegean Sea in the overriding plate from around 30 Ma (Armijo *et al.* 1996; Jolivet *et al.* 2010; van Hinsbergen & Schmid 2012). This extension was accompanied by a clockwise rotation of the Peloponnese peninsula by up to 40°–50° (van Hinsbergen *et al.* 2007; Burchfiel 2008). At around 5 Ma, the North Anatolian Fault (NAF) started to propagate west to SW into the northern Aegean Sea (Armijo *et al.* 1996), generating a zone of transtension with deep, isolated rift zones. This coincided with the cessation of active extension in the central Aegean Sea as deformation migrated outwards to the east and west (Burchfiel 2008; Royden & Papanikolaou 2011). Geodetic studies show that north–south extension is now concentrated in the western Anatolian block, and in a broad zone between the SW tip point of the NAF and the Kefalonia Fault to the west (Fig. 1a) (Briole *et al.* 2000; McClusky *et al.* 2000; Avallone *et al.* 2004; Bernard *et al.* 2006; Nocquet 2012) wherein lie the Corinth and Evia rifts (Fig. 1).

The Corinth rift cuts obliquely across the Hellenide mountain belt, which comprises a stack of NNW–SSE-trending, west-verging thrust sheets that were progressively emplaced towards the west to WSW from Cretaceous to Miocene times (Aubouin *et al.* 1963; Dercourt 1964; Richter 1976; Fleury 1980; Doutsos *et al.* 1993, 2006; Skourlis & Doutsos 2003). The rift is underlain principally by the Pindos thrust sheet with part of the Parnassos sheet to the east, and the Gavrovo–Tripolitsa sheet to the west. The Pindos thrust sheet comprises an approximately 1300 m-thick succession of Upper Triassic–Jurassic hemipelagic carbonates with minor red and green radiolarites (cherts) and Upper Cretaceous–Tertiary sandy turbidites (graywackes/flysch) (Degnan & Robertson 1998; Skourlis & Doutsos 2003). These strata have been highly deformed at sub-greenschist grade during Alpine nappe emplacement. The Pindos sheet overthrusts the carbonate-dominated Gavrovo–Tripolitza thrust sheet and the underlying Zarouchla complex, which includes the Phyllite–Quartzite unit that records Miocene high-pressure–low-temperature metamorphism (Dornsiepen *et al.* 1986; Trotet *et al.* 2006; Jolivet *et al.* 2010).

The Corinth rift

The Gulf of Corinth is an approximately 120 km-long active rift, trending N110°, with a maximum width of 30 km, narrowing westwards to 3 km at the Rion Straits (Fig. 1b). The Gulf represents the currently active part of a rift whose Plio-Quaternary history is preserved in uplifted blocks across the northern Peloponnese (Moretti *et al.* 2003; Ghisetti & Vezzani 2004). The normal fault system is dominated by north-dipping faults both onshore and offshore with an average strike of N105–110° and dips of 42–64° N (Rohais *et al.* 2007*a*; Bell *et al.* 2008; Taylor *et al.* 2011; Ford *et al.* 2013). More than 2.5 km of synrift stratigraphy can underlie the eastern and central Gulf (McNeill *et al.* 2005; Bell *et al.* 2008, 2009, 2011; Taylor *et al.* 2011; Nixon *et al.* 2016). Onshore, the synrift succession reaches around 2.8 km in thickness. A north–south extension direction is recorded since the beginning of rifting (Roberts & Michetti 2004; Ford *et al.* 2013), corresponding to the present-day orientation of geodetic extension (Briole *et al.* 2000; Avallone *et al.* 2004; Bernard *et al.* 2006).

Moho depth, reflecting crustal thinning and derived from tomographic inversion, increases along the rift from 25–30 km in the east below the Perachora peninsula (and the southern Saronic Gulf: Fig. 1) to 45 km below the Rion Straits, with a particularly rapid increase in Moho depth from the central to the western Gulf (Zelt *et al.* 2005; Sachpazi *et al.* 2007). The decreasing influence of Corinth rifting on crustal thickness is also reflected in the orientation of Moho depth contours, which trend WNW–ESE (Corinth rift trend) below the eastern Gulf and changing to a NNW–SSE Hellenide trend to the west (Sachpazi *et al.* 2007). The variations in Moho depth are reflected in variations in depth to base synrift, synrift sediment thickness and fault displacement along the rift. The increase in geodetic extension rates to the west suggests that the locus of deformation has migrated west along the axis of the Gulf, and that the westernmost Gulf is the youngest part of the rift and the least extended. It is clear that the evolution of the Corinth rift is intimately related to that of the Patras rift to the west (Fig. 1), about which very little is known (Brooks & Ferentinos 1984; Ferentinos *et al.* 1985).

For the purposes of this paper, the Corinth rift is divided into five zones (C1–C5) orientated NNE–SSW, and traversing both the onshore and offshore rift (Figs 1 & 2). These are, from east to west, the Alkyonides Gulf and the Lechaion Gulf (C1), the central and eastern Gulf (C2 and C3), the west to central transition zone (C4) and the westernmost Gulf (C5). The Patras Gulf lies further to the west (Fig. 1). This paper will focus mainly on the onshore

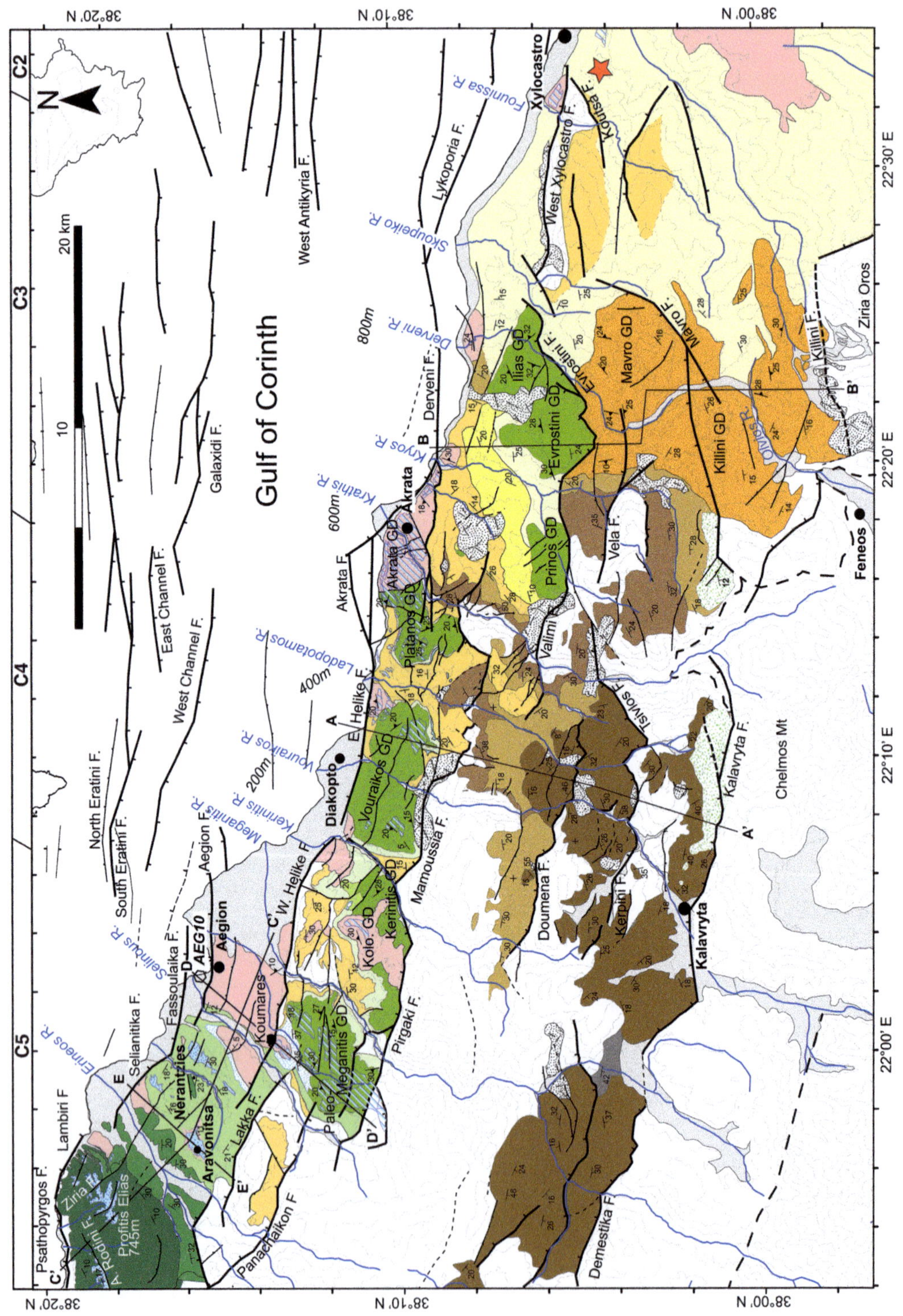
Gulf of Corinth
N
20 km
10
38°20' N
38°10' N
38°00' N
22°30' E
22°20' E
22°10' E
22°00' E
C2
C3
C4
C5
Psathopyrgos F.
Lambiri F
Erineos R.
Selianitika F.
Fassoulaika F.
Selinous R.
South Eratini F.
North Eratini F.
Aegion F.
Meganitis R.
Kerinitis R.
Vouraikos R.
West Channel F.
East Channel F.
Galaxidi F.
West Antikyria F.
Lykoporia F.
Fonissa R.
Skoupeiko R.
Derveni R.
Krathis R.
Krios R.
Ladopotamos R.
200m
400m
600m
800m
Xylocastro
West Xylocastro F.
Koutsa F.
Derveni F.
Akrata F.
Akrata
Akrata GD
Platanos GD
Ilias GD
Evrostini GD
Evrostini F.
Mavro GD
Mavro F.
Killini GD
Killini F.
Ziria Oros
Prinos GD
Vela F.
Valimi
Feneos
E. Helike F.
W. Helike F.
Diakopto
Vouraikos GD
Mamoussia F.
Tsivlos F.
Kalavryta F.
Chelmos Mt
Kalavryta
Kerpini F.
Doumena F.
Pirgaki F.
Kerinitis GD
Kolo. GD
Meganitis GD
Paleo.
Koumares
Aegion
AEG10
Nerantzies
Aravonitsa
Lakka F.
Panachaikon F
Demestika F.
Ziria
Profitis Elias
745m
Rodini F.
A
A'
B
B'
C
C'
D
D'
E
E'

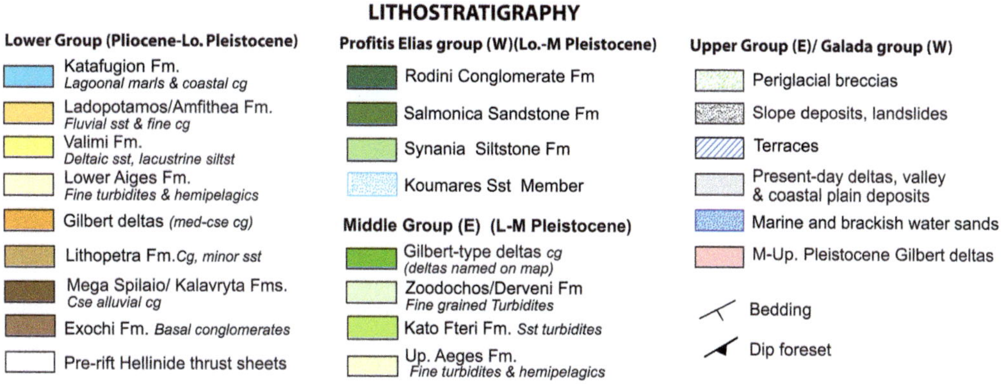

LITHOSTRATIGRAPHY

Lower Group (Pliocene-Lo. Pleistocene)
Katafugion Fm. *Lagoonal marls & coastal cg*
Ladopotamos/Amfithea Fm. *Fluvial sst & fine cg*
Valimi Fm. *Deltaic sst, lacustrine siltst*
Lower Aiges Fm. *Fine turbidites & hemipelagics*
Gilbert deltas *(med-cse cg)*
Lithopetra Fm. *Cg, minor sst*
Mega Spilaio/ Kalavryta Fms. *Cse alluvial cg*
Exochi Fm. *Basal conglomerates*
Pre-rift Hellinide thrust sheets

Profitis Elias group (W)(Lo.-M Pleistocene)
Rodini Conglomerate Fm
Salmonica Sandstone Fm
Synania Siltstone Fm
Koumares Sst Member

Middle Group (E) (L-M Pleistocene)
Gilbert-type deltas *cg (deltas named on map)*
Zoodochos/Derveni Fm *Fine grained Turbidites*
Kato Fteri Fm. *Sst turbidites*
Up. Aeges Fm. *Fine turbidites & hemipelagics*

Upper Group (E)/ Galada group (W)
Periglacial breccias
Slope deposits, landslides
Terraces
Present-day deltas, valley & coastal plain deposits
Marine and brackish water sands
M-Up. Pleistocene Gilbert deltas

Bedding
Dip foreset

Fig. 2. Geological map of the Plio-Pleistocene Corinth rift of the northern Peloponnese from Xylocastro to Psathopyrgos. The map for the central and eastern areas (C2, C3 and C4) has been compiled from Rohais *et al.* (2007*a*, *b*), Leeder *et al.* (2012) and Ford *et al.* (2013), with several revisions detailed in the text. The western area (C5) integrates new work and that of Palyvos (2005) and Palyvos *et al.* (2010, 2013). The structure and chronostratigraphy of C4 along cross-section A–A′ is presented in Figure 3, of C3 on section B–B′ in Figure 4, and of C5 on sections C–C′, D–D′ and E–E′ in Figures 5, 7 and 8, respectively. Bathymetric and topographical contours are spaced every 200 m. The red star shows the location of the dated ash layer of Leeder *et al.* (2012). Abbreviations on map and legend: A, Ano; cg, conglomerate; cse, coarse; F, Fault; Fm, Formation; E, East; GD Gilbert Delta; Lr, Lower; med, medium; M, Middle; Mt, Mountain; R, River; siltst., siltstone; sst, sandstone; Up., Upper; W, West.

areas C3–C5 where onshore geological investigations are most advanced.

Offshore Gulf of Corinth: structure and stratigraphy of the active rift

In zones C2 and C3, the Gulf is bound to the south by right-stepping en echelon, north-dipping normal faults (the Derveni, Lykopora and East Xylocastro faults: Fig. 1) that form a steep basin-margin slope. Offshore swath bathymetry data show a 9–12 km-wide, flat basin floor with water depths of 800–870 m (Fig. 2) (Alexandri *et al.* 2003; Nomikou *et al.* 2011). South-dipping normal faults delimit the depocentre to the north (the Galaxidi and West Antikyria faults: Nixon *et al.* 2016), separating it from a broad, relatively shallow-water (<400 m) platform rising gently to a deeply indented northern coastline (Fig. 1). To the east (C1), the Gulf splits into two rifts: the Alkyonides Gulf and the Lechaion Gulf (Fig. 1) (Leeder *et al.* 2005; Sakellariou *et al.* 2007; Charalampakis *et al.* 2014) with shallow water (<400 m in Nomikou *et al.* 2011; Alkyonides <200 m and <100 m in Lachaion in Taylor *et al.* 2011). A major horst forms the Perachora peninsula between these basins (Fig. 1). The Corinth Isthmus separates the Corinth Gulf from the Saronic Gulf to the east (Collier & Dart 1991). Major earthquakes occur regularly in this area: for example, in 1981 (Jackson *et al.* 1982).

To the west (Fig. 1), the active rift steps northwards by some 8 km across area C4, while narrowing in width to 10 km. The bounding normal fault system correspondingly steps northwards. The East and West Helike faults define the active rift's southern boundary in C4, while the West and East Channel faults define its northern boundary (McNeill *et al.* 2005; Bell *et al.* 2008). The North and South Eratini faults define a small offshore horst further north.

In C5, the active rift is principally controlled by the north-dipping Neos Erineos Fault zone trending NW–SE, and comprising four short faults (Lambiri, Selianitika, Fassoulaika and Aegion) (Palyvos 2005; Palyvos *et al.* 2007) and the east–west-trending Psathopyrgos Fault, considered by many as the most seismically active fault in the rift (Bernard *et al.* 2006). These en echelon coastal faults are considerably shorter than those to the east (Beckers *et al.* 2015). The south-dipping Trizonia Fault delimits the main depocentre, while the Marathias Fault defines the northern coastline. Between these two faults lies a shallow platform. From Aegion to the Rion Straits, bathymetry shallows (over some 33 km) from 400 to 60 m.

The present-day eastern and central Gulf (C3, C2, Fig. 1) can be described as sediment-starved. Little sediment is supplied from the north as no significant rivers flow into the Gulf. Only fine-grained sediment is supplied from the south by short, consequent rivers (Fig. 1) (Seger & Alexander 1993; Zelilidis 2000; Demoulin *et al.* 2015) and from the west along a submarine axial channel (Bell *et al.* 2008; Nomikou *et al.* 2011). In contrast, from Akrata westwards, major antecedent rivers carry coarse-grained bedload northwards into the shallower basin, building large Gilbert-type fan deltas in the hanging walls of active faults (Ford *et al.* 2007, 2013; see later). More distal deposits comprise prodelta turbidites and hemipelagic deposits (Bell *et al.* 2008). These deltas build steep foresets (angles of repose up to 29°) that can record seismically induced gravitational mass flows (Lykousis *et al.* 2007*b*; Nomikou *et al.* 2011; Beckers 2015).

Recent seismic reflection studies show that the thickest offshore synrift accumulation underlies the area of flat, deep bathymetry of C2 and C3. Here a band of maximum sediment thickness (> 3 km) parallels the south coast, defining the main depocentres in the hanging walls of major north-dipping faults (Bell *et al.* 2008, 2009, 2011; Taylor *et al.* 2011; Nixon *et al.* 2016). Bell *et al.* (2009) estimate 1.3–1.9 km, thickening to 2.4 km into hanging-wall depocentres, while Taylor *et al.* (2011) estimate a thickness of over 3 km. To the east, in C1, maximum sediment thickness in the Alkyonides Gulf is estimated at 1 km (Bell *et al.* 2009; >1 km in Taylor *et al.* 2011) and >2 km in the Lechaion Gulf (Taylor *et al.* 2011; Charalampakis *et al.* 2014). To the west, the synrift succession thins rapidly to < 1 km across C4 (Bell *et al.* 2009), corresponding to a marked shallowing of top basement (estimated at 0.6–0.9 km below sea level (bsl): Beckers 2015).

The offshore synrift succession is, as yet, undrilled. Based on seismic attributes, two main units are recognized in areas C2 and C3, separated by a widespread, locally angular unconformity (Moretti *et al.* 2003; Sachpazi *et al.* 2003; McNeill *et al.* 2005; Bell *et al.* 2008; Taylor *et al.* 2011; Nixon *et al.* 2016). The lower unit SU1 (Unit B of McNeill *et al.* 2005; Bell *et al.* 2008; 'early rift' of Taylor *et al.* 2011; Seismic Unit 1 (SU1) of Hemelsdaël & Ford 2016) lacks clear seismic impedance contrasts and appears to be poorly stratified, while the younger unit SU2 is highly reflective, showing clear cyclical stratification (Unit A of McNeill *et al.* 2005; Bell *et al.* 2008; 'late rift' of Taylor *et al.* 2011; Seismic Unit 2 (SU2) of Hemelsdaël & Ford 2016). Giant piston cores (up to 30 m long) taken in the eastern C2 recovered Holocene fine-grained marine distal gravity deposits and hemipelagics (12–14 m) underlain by lacustrine varve-like muds interbedded with silt to fine sand turbidites recording a transition from lacustrine to marine conditions at around 12 ka (Lykousis *et al.* 2007*a*; Campos *et al.* 2013). Age estimates of the reflective SU2 are derived from correlation of its cyclical seismic stratigraphic packages with Pleistocene eustatic sea-level cycles (Stefatos *et al.* 2002; Sachpazi *et al.* 2003; Sakellariou *et al.* 2007; Bell *et al.* 2008, 2009; Taylor *et al.* 2011), although the number of packages varies depending on the study area of each paper. The age of the unconformity between SU1 and SU2 is estimated by Taylor *et al.* (2011) in the C2 area as 600 ka, and by Bell *et al.* (2008) in the C4 area as 400 ka. More recently, Nixon *et al.* (2016) reconcile these incoherencies by proposing that the basin-wide unconformity at the base of SU2 has an age of approximately 620 ka and that there is a second local unconformity within SU2 at 340 ka in the western Gulf, which is that identified by Bell *et al.* (2008). They demonstrate that SU2 thins and onlaps westwards onto the main unconformity due to the presence of a palaeohigh, as will be discussed later (here named the Psaromita–Aegion High). The synrift succession thickens again into the Trizonia depocentre west of Aegion (C5; Moretti *et al.* 2003; Lykousis *et al.* 2007*b*). In C5, recent work by Beckers (2015) and Beckers *et al.* (2015), on an unpublished industrial seismic line, indicates that the top basement becomes deeper again (1.6 km bsl) in the hanging wall of the Trizonia Fault before shallowing westwards toward the Mornos delta. To the SW, in the hanging wall of the Psathopyrgos Fault, the top basement is estimated to be at 1.2 km bsl (Beckers 2015).

Onshore–offshore correlations (e.g. Lykousis *et al.* 2007*a*; Bell *et al.* 2008, 2009; Taylor *et al.* 2011; Hemelsdaël & Ford 2016) propose that the lower offshore unit SU1 corresponds mainly to the onshore Middle Group and to the Lower Group, should it be present. The upper offshore unit SU2 would then be equivalent in time to the Upper Group onshore (Bell *et al.* 2009; Taylor *et al.* 2011; Hemelsdaël & Ford 2016). However, the ages of lithostratigraphic boundaries are probably diachronous from east to west owing to lateral propagation of rifting: a high-precision dating along a cored borehole is essential to better constrain offshore stratigraphy.

Onshore synrift stratigraphy and structure

In the northern Peloponnese, Plio-Quaternary synrift strata are exposed in a belt approximately 30 km wide (C3 and C4) that narrows abruptly to the west (C5), and broadens to the east in zones C2 and C1 (Fig. 1). In areas C3–C5, fault blocks and their synrift successions have been uplifted to elevations of over 1000 m and deeply incised by north-flowing rivers, providing excellent natural cross-sections (Fig. 2). Further east in areas C2 and C1, however, broad marine terraces mask finer-grained synrift facies in a more subdued landscape with poorer exposures. Apart from the Corinth Isthmus area (Collier 1990; Collier & Dart 1991; Dart *et al.* 1994), the synrift stratigraphy and structure of this eastern area is, as yet, not well documented. To the NE of C1, lies the Megara rift (Fig. 1) that initiated in the Miocene and was active until the Late Pleistocene (Bentham *et al.* 1991; Leeder *et al.* 2008).

In C3 and C4, the western onshore synrift succession is divided into three informal lithostratigraphic groups named the Lower, Middle and Upper groups (Fig. 2) (Rohais *et al.* 2007*a*; Ford *et al.* 2013); while, in C5, the recently defined Profitis Elias and Galada groups are roughly equivalent in time to the Middle and Upper groups (Palyvos *et al.* 2013). The aim of this section is to present the structure and stratigraphy of each of these three onshore areas and to propose an along-strike regional correlation, thus covering about 70% of the onshore rift. The onshore compilation is mainly based on Ford *et al.* (2013) and Hemelsdaël *et al.* (2015) in C4, a reinterpretation of the work of Rohais *et al.* (2007*a*, *b*) in C3, and Palyvos *et al.* (2013) and unpublished work by the authors for C5. We start with the C4 area, where the lithostratigraphic groups are most clearly defined.

Area C4: Diakopto–Kalavryta

The C4 area lies between the Selinous and Krathis rivers (15–20 km wide). Various aspects of its onshore geology have been described previously (Doutsos *et al.* 1988; Doutsos & Piper 1990; Doutsos & Poulimenos 1992; Ghisetti & Vezzani 2004; Flotté *et al.* 2005; Ghisetti & Vezzani 2005), most recently in Ford *et al.* (2013). The zone is characterized by five major north-dipping faults (Kalavryta, Kerpini–Tsivlos, Doumena, Pirgaki–Mamoussia and Helike (east and west segments)) defining four onshore fault blocks, 4–7 km wide, that have been uplifted to elevations just under 1500 m (Fig. 3). The three southern blocks preserve only Lower Group strata that are well exposed along the Vouraikos Valley (Fig. 3a). The Pirgaki–Mamoussia (PM) Fault preserves Middle Group Gilbert deltas uplifted to over 800 m, while the active East and West Helike faults control subsidence of the coastal plain comprising the fluvial topsets of Gilbert-type deltas currently building into the Gulf (part of the Upper Group). The normal faults, which can each be traced from tip point to tip point, trend between N086° and N112°, except for the Tsivlos Fault which trends N067°. Fault dips are between 42° N and 64° N, and are extrapolated to 2 km depth with a planar geometry. An exception is the Prinos Fault, which is a NE-dipping fault that was abandoned and passively rotated to a low dip (10–15°) within the Kerpini fault block. Secondary south-dipping faults occur within all fault blocks, while a major south-dipping normal fault (buried) is proposed to terminate Lower Group strata to the north. Field kinematic data indicate pure dip-slip displacement recording a north–south to NNE–SSW extension. Synrift stratal geometries include fanning dips typical of tilting fault blocks and hanging wall forced folds (the Doumena Fault). The presence of palaeorelief renders accurate measurement of displacement across the faults difficult; heave estimates on major faults are 500–1000 m, while throw estimates are 700–1400 m (Ford *et al.* 2013).

Lower Group. In the south and SW (Kalavryta and Kerpini fault blocks), the Lower Group is characterized by a coarse conglomeratic succession (Kalavryta Formation up to 2200 m thick) deposited by alluvial fans and/or permanent gravelly rivers (Collier & Jones 2003; Ford *et al.* 2013). To the north and NE, these strata pass laterally into a finer-grained fluvial succession (the Ladopotamos Formation). Palaeocurrents record a predominantly north and NE palaeoflow in all Lower Group formations (Ford *et al.* 2013). The lower levels (including basal conglomerates) onlap onto and infill a palaeo-relief of up to 700 m inherited from the Hellenide fold and thrust belt. At the western end of the Pirgaki–Mamoussia fault block, the sand-silt dominated fluvio-lacustrine Melissia Formation (250–500 m) onlaps and infills a palaeo-relief and is laterally equivalent to the Ladopotamos Formation (Backert 2009; Backert *et al.* 2010). At the top of the succession in the PM fault block, the fine-grained Katafugion Formation records a marine transgression marking the onset of basin deepening in the C4 area (Ford *et al.* 2007, 2013), here identified as deepening event 2.

The age of the Lower Group is poorly constrained, and is assigned by most authors to the Late Pliocene (<5 Ma) and Early Pleistocene (up to 1.8 Ma) based on rare palynological dates and regional correlation arguments (Fig. 3b) (Ford *et al.* 2013). Lithostratigraphic correlations across fault blocks suggest two pulses of northwards progradation by a major antecedent river system (Fig. 3b) (Ford *et al.* 2013). The volume of sediment provided by these rivers overwhelmed the accommodation created by early faults that therefore grew below a continuous floodplain landscape traversed by high-energy gravelly rivers. These strata gradually buried the pre-existing relief, although some may have persisted to later times. A marine transgression at the end of the Lower Group is recorded only in the northern PM fault block. Further south, the top of the Lower Group is not preserved.

Middle Group. In the C4 area, the Middle Group comprises four large conglomeratic Gilbert-type fan deltas, regularly spaced (6–8 km) in the hanging wall of the PM Fault (Platanos, Vouraikos, Kerinitis and palaeo-Meganitis). Equivalent prodelta facies comprise fine sand and silt-dominated turbidites and hemi-pelagic deposits (the Derveni Formation

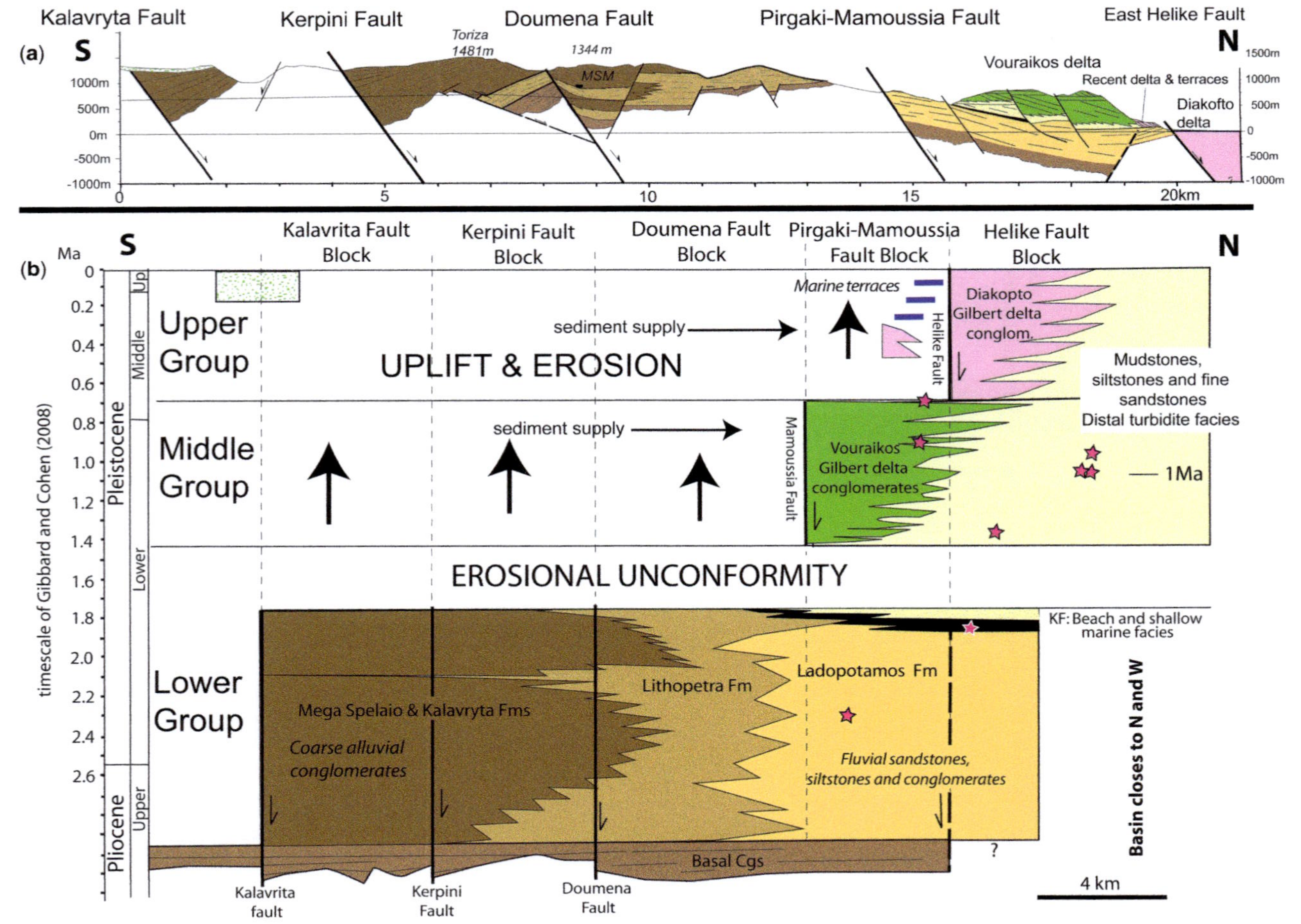

Fig. 3. Onshore area C4 stratigraphy and structure revised from Ford *et al.* (2013). (**a**) North–south cross-section located on Figure 2 (A–A′). Stratigraphic legend in Figure 2. MSM, Mega Spelaio Monastery. (**b**) Revised chronostratigaphy along cross-section A–A′ using the timescale of Gibbard & Cohen (2008). Small pink stars represent palynological dates. Note that time is not constrained below the Plio-Pleistocene boundary. KF, Katafugion Formation.

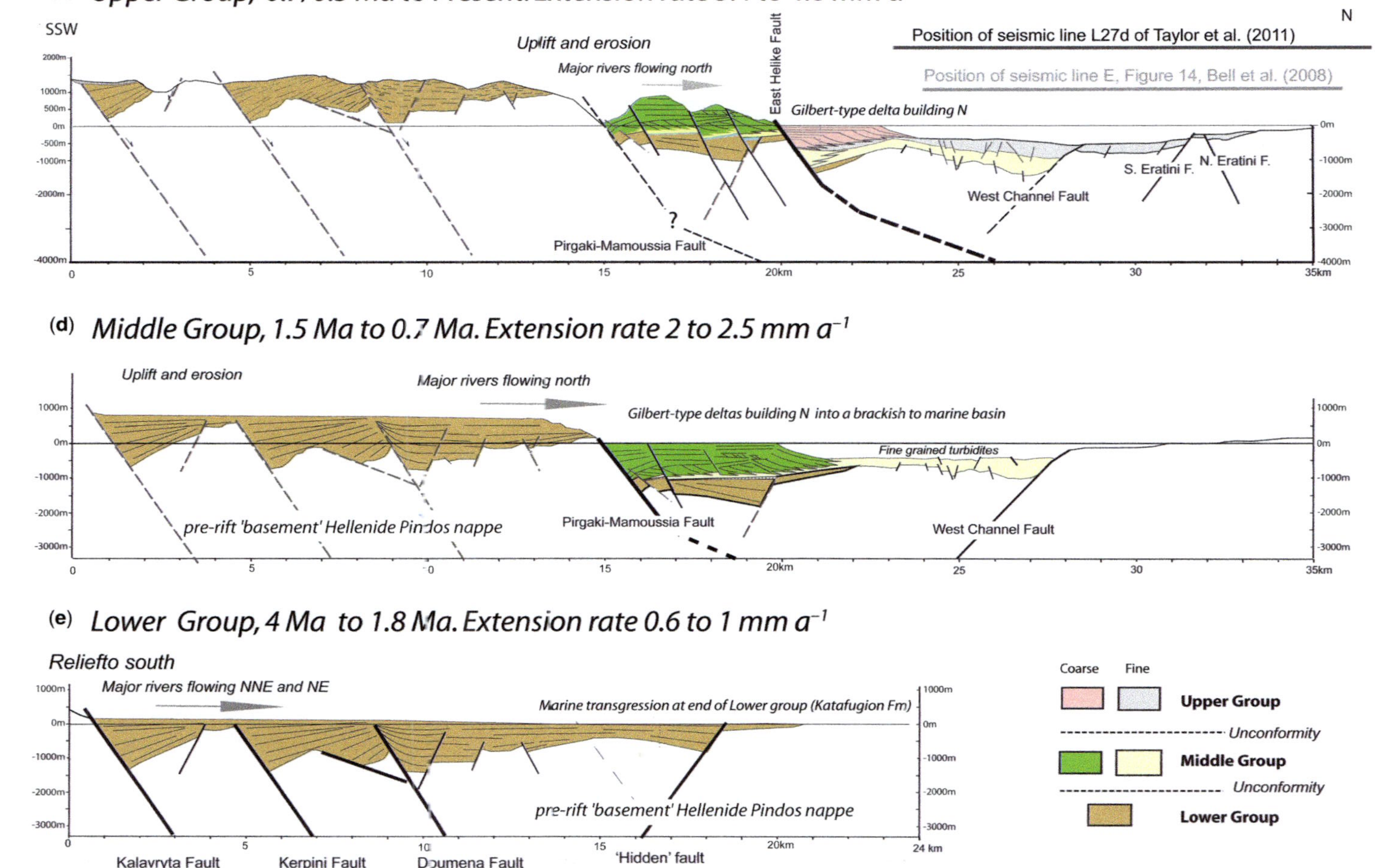

Fig. 3. (*Continued*) **(c)** Cross-section across the whole rift, integrating section A–A′ with offshore seismic data (Bell *et al.* 2008; Taylor *et al.* 2011). **(d)** & **(e)** A two-step restoration of the full rift section.

of Ford *et al.* 2007; the Zoodhochos Formation of Backert *et al.* 2010: up to 500 m). The exceptional cliff exposures of the Kerinitis and Vouraikos deltas (600 m) have been the focus of numerous studies, providing insight into the interactions of sedimentation, climate and tectonics in extensional settings (Ori *et al.* 1991; Dart *et al.* 1994; Zelilidis & Kontopoulos 1996; Hardy & Gawthorpe 1998; Ulicny *et al.* 2002; Malartre *et al.* 2004; Ritchie *et al.* 2004; Rohais *et al.* 2007*b*; Backert *et al.* 2010). These giant deltas (radii up to 4 km, thickness from 400 m to >800 m) of cobble to pebble conglomerates built northwards from point sources across the PM Fault into a deepening basin (marine/lacustrine) from Early to Middle Pleistocene. They display an overall aggradational architecture comprising stacked highstand stratal packages. Foresets show a radial organization with depositional dips of up to 28°. In contrast, the less well-studied Platanos and palaeo-Meganitis deltas lie at the eastern and western terminations of the fault block. These are thinner (<400 m), pebble to sand grade deltas with strongly progradational foreset geometries. The palaeo-Meganitis delta built to the east along a relay ramp between the western termination of the Pirgaki Fault and the Panachaikon Fault. The Platanos delta built from south to north at the eastern end of the Mamoussia Fault. The base of each delta is marked by a major erosional unconformity, each recording the incision of a palaeo-valley up to 300 m deep and 700 m wide into the underlying Lower Group strata (Ford *et al.* 2007; Backert *et al.* 2010). It is estimated by Ford *et al.* (2007) that these valleys would take around 300 kyr to erode at reasonable erosion rates. In C4, the Lower and Middle groups appear to be separated by a short-lived marine transgression (the Katafugion Formation), followed by an important period of non-deposition and erosion. The cause of this incision event(s) is enigmatic and may have been submarine (Ford *et al.* 2007). The lower Vouraikos delta displays a clear rollover geometry recording early control by listric faulting, while its upper levels record only gentle tilting into the PM Fault (Ford *et al.* 2007). The Kerinitis delta records gentle tilting into the oblique breaching fault (the Kerinitis Fault: Backert *et al.* 2010).

The age of the Middle Group in C4 is bracketed from Early to Middle Pleistocene (1.8–0.7 ± 0.2 Ma: Fig. 3b) by palynological data (Malartre *et al.* 2004; Ford *et al.* 2013) and by compatibility with Upper Group marine terraces deposited in the footwall of the younger Helike Fault (see below). The Vouraikos and Kerinitis deltas record the birth, growth and death of the Pirgaki–Mamoussia Fault over an estimated period of 0.7 ± 0.2 myr (Backert *et al.* 2010; Ford *et al.* 2013). Stratal architectures record the abrupt termination of displacement on the PM Fault, which is correlated with the initiation of the Helike Fault, some 5 km to the north.

Upper Group. From 0.7 ± 0.2 Ma onwards, the PM fault block was uplifted and incised by powerful north-flowing rivers, the same rivers that deposited the four major deltas of the Middle Group. The Meganitis, Kerinitis and Vouraikos rivers have cannibalized their Early–Middle Pleistocene deltas to build new Upper Group deltas into the Gulf in the hanging wall of the new Helike Fault (Figs 2 & 3a). Their delta-tops have merged to form the present-day, densely populated coastal plain (Fig. 1). Equivalent prodelta deposits lie offshore below the Gulf (Fig. 3a) (Bell *et al.* 2008).

Uplift of the footwall of the East and West Helike faults is recorded by flights of marine terraces and limestone notches on the range front. These have been well studied and dated to give an average uplift rate for the East Helike Fault footwall of 1.1 mm a^{-1} since at least 0.5 Ma (De Martini *et al.* 2004; McNeill & Collier 2004), increasing to 1.3–2.2 mm a^{-1} during the Holocene (Stewart 1996; Stewart & Vita-Finzi 1996; Pirazzoli *et al.* 2004). Footwall uplift is also recorded by small uplifted sandy to conglomeratic Gilbert-type deltas (radii < 1 km) with erosive bases that are concentrated around the mouths of river valleys exiting the Helike Fault footwall (Fig. 2). A notably larger Upper Group delta (the Kolokotronis delta, Fig. 2) (Backert *et al.* 2010) directly overlies the Kerintis delta (Midde Group) with a strongly erosive base, recording a more complex history. At the beginning of uplift, the Kerintis River first incised its Middle Group delta and built a new delta into deep water toward the north, as evidenced by foresets over 300 m high (Backert *et al.* 2010). This delta was finally abandoned when the river diverted to incise along the NE-trending Kerinitis Fault to find its present bed. Finally, red soil deposits (Terra Rossa), several metres thick, cover the tops of uplifted Middle Group deltas. At the southern edge of C4, periglacial angular limestone breccias (up to 300 m thick) form an apron around the northern and eastern slopes of the Chelmos Massif (>2500 m), which shows evidence of mountain glaciers (Mastronuzzi *et al.* 1994).

Area C3: Derveni–Killini

At the western edge of C3, pre-rift lithologies are exposed in a series of inliers along the Krathis River (Fig. 2) (Ghisetti & Vezzani 2005). To the east, there are few pre-rift basement exposures in the rift, with the largest block in the footwall of the West Xylocastro Fault (Fig. 2). The C3 area is bordered to the south by the Mount Ziria pre-rift massif (2374 m), bounded by the major Killini

Fault largely buried below synrift strata (Figs 2 & 4). Synrift strata have been uplifted to elevations of 1726 m (Mavro GD: Figs 2 & 4).

In C3, the main north-dipping faults (Valimi, Vela, Mavro and Killini) accommodate higher displacement (throws of up to 4.5 km, Killini Fault: Ford *et al.* 2013) compared with those of C4. The onshore area is bound to the north by the coastal Derveni Fault (4 km throw: Fig. 2) (Hemelsdaël & Ford 2016), which passes offshore at Mavro Litharia (a small basement high in its immediate footwall). Basement-cutting faults are required to accommodate tilting of older Lower Group strata (Killini fault block tilted 25–30° S). However, both the Evrostini and Mavro faults show clear listric geometries at the surface with hanging-wall rollovers (Rohais *et al.* 2007*a*). We link these superficial listric faults to the basement-cutting planar faults at depth (Fig. 4a). The fine-grained synrift succession acts as a decoupling layer between the two structural levels. The Valimi Fault links to the Xylocastro Fault along a NE-trending breached relay zone (Evrostini Fault) across which the Middle Group Ilias Gilbert delta built towards the NW (Fig. 2) (Rohais *et al.* 2007*a*).

Lower Group. Around the village of Voutsimos (Fig. 2: northern Krathis River), the Lower Group succession is complete but very thin (Hemelsdaël & Ford 2016), indicating the presence of a palaeohigh (Voutsimos High). East of the Krathis River, the Lower Group succession thickens and fines rapidly (Rohais *et al.* 2007*a*, *b*). This fluvial succession is ubiquitous throughout C3 and further east in C2, showing an overall fining-eastward, frequently onlapping, palaeo-relief. The Exochi Formation of Rohais *et al.* (2007*a*) is here subdivided into basal conglomerates (redefined Exochi Formation) overlain by conglomerates and sandstones (Lithopetra Formation), both of which are replaced to the east and north by the sandstone–conglomerate succession of the Ladopotamos Formation (Hemelsdaël *et al.* 2015; named the Amfithea Formation in C2 by Leeder *et al.* 2012).

In the Valimi area, the fluvial Ladopotamos Formation (Hemelsdaël *et al.* 2015) passes eastwards first into a fluvio-deltaic succession (the Valimi Formation), containing low-height (5–50 m) delta clinoforms, and then into a fine-grained lacustrine succession with biostratigraphic evidence of frequent marine incursions (foraminifera; lower Aiges Formation of Rohais *et al.* 2007*a*). Palaeocurrent data for the whole succession show that sediment was sourced from the SW, west and south, with a predominant eastwards flow (Rohais *et al.* 2007*a*).

Revision of the Lower–Middle Group boundary. Four large Gilbert-type delta complexes are aligned north–south over some 15 km, forming the highest relief in the C3 area (Fig. 2). Successive incision indicates that these deltas become progressively younger to the north (Killini, Mavro, Ilias and Evrostini). Rohais *et al.* (2007*a*) assign all four deltas and their fine-grained prodelta equivalents (upper Aiges Formation) to the Middle Group. Recently, a volcanic calc-alkaline ash layer found further east near Xylocastro (in the C2 area: red star in Fig. 2) has been dated by $^{40}Ar-^{39}Ar$ single-crystal CO_2 laser fusion, yielding a precise age of 2.550 ± 0.007 Ma (Leeder *et al.* 2012). This is the first and only absolute age from the Corinth synrift succession, and, as such, its importance cannot be overemphasized. The unreworked ash layer is interbedded with fine-grained turbidites and hemipelagics in the Rethio-Dendro Formation, which can be roughly correlated westwards with the Aiges Formation: however, detailed mapping of the intervening area is not yet available. Palaeocurrent data indicate that, south of the Xylocastro Fault, both the Lower Aiges Formation and the laterally equivalent Rethio-Dendro Formation were sourced from either the Killini or Mavro Gilbert delta (Rohais *et al.* 2007*a*; Leeder *et al.* 2012). Using sedimentation rate estimates and the occurrence of distinctive clast lithologies, Leeder *et al.* (2012) argue that the turbidites containing the ash were supplied specifically from the Mavro delta. They use this correlation to date the base of the Middle Group (taken as the base of the Killini delta following Rohais *et al.* 2007*a*) at 3.2–3 Ma, which is considerably older than the estimate of 1.8–1.5 Ma in the C4 area. Their discussion of the significance of rift-wide deepening events is based on this hypothesis (see the Discussion section later in this paper).

Here we propose a significantly different interpretation of the stratigraphy of the C3 area, which we believe respects all data and has very different implications for rift evolution. This revision reassigns both the Killini and Mavro deltas and their prodelta equivalents (the Lower Aeges Formation) to the Lower Group, while the Ilias and Evrostini deltas remain in the Middle Group. The arguments for this are: (1) apart from a basal fluvial unit (Amfithea Formation), Lower Group facies in the C2 and C3 areas prove that the central and eastern rift was occupied principally by a lake/marine basin for most of its history; (2) the Killini and Mavro delta complexes comprise a stacked series of strongly prograding, low-height Gilbert deltas (ranging from 10 m to some tens of metres) passing to alluvial fans to the SW and south, respectively (Fig. 4); these successive deltas clearly built into a relatively shallow body of water from a stable coarse sediment input point; (3) in contrast, the Ilias and Evrostini deltas show mainly aggrading

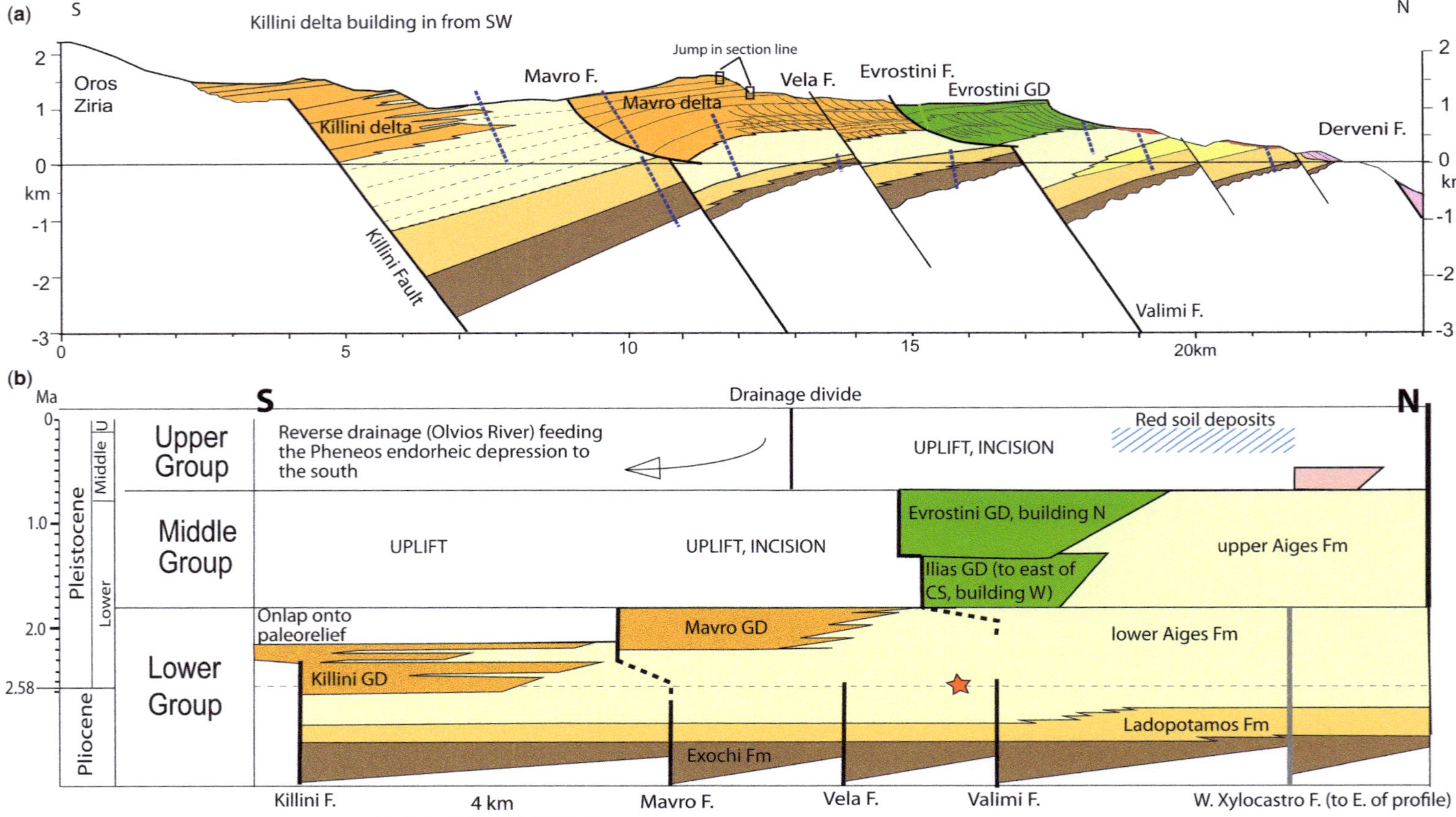

Fig. 4. Onshore area C3. (**a**) North–south cross-section (B–B′ in Fig. 2) revised and completed to the south after Rohais *et al.* (2007*a*). (**b**) Chronostratigraphy along the B–B′ cross-section revised to integrate the ash date of Leeder *et al.* (2012) (star) using the timescale of Gibbard & Cohen (2008). Dashed blue lines locate the projected sedimentary logs of Rohais *et al.* (2007*a*). The red star represents the projected ash layer dated at 2.55 Ma by Leeder *et al.* (2012).

stratigraphic architectures similar to those of the Vouraikos and Kerinitis deltas to the west, with large foresets (several hundreds of metres high), recording deposition into a deepening basin; and (4) the alignment of the four delta complexes and their building directions suggest a stable input point for coarse sediment, which was present since the onset of rifting. After the initial flooding/deepening event between the basal fluvial succession and the overlying lacustrine succession in C3 (here identified as deepening event 1), water depth remained relatively shallow during deposition of the Killini and Mavro delta complexes. The Ilias delta records the onset of a major deepening of the basin on the Valimi–Xylocastro fault system (here identified as deepening event 2, also recognized in C4).

Figure 4b shows our revised chronostratigraphy adapted from that of Rohais *et al.* (2007*b*). Our reconstruction suggests that the ash level more probably correlates with the Killini delta complex, although more precise mapping is required to constrain stratigraphic correlation.

Upper Group. The Upper Group onland in C3 is, again, composed of uplifted marine-terrace deposits, and small Gilbert-type deltas (Rohais *et al.* 2007*a*, *b*) along the coastal slopes. Periglacial slope breccias occur further inland, and red soil deposits are found on high plateaux (Fig. 4). Apart from the Krathis River, no major river flows northwards into the Gulf in C3. Wind gaps and abandoned river valleys record the reversal of the drainage system that built the Middle Group delta. This drainage system now flows southwards into the Feneos endorheic basin, and can be identified as the Olvios River (Rohais *et al.* 2007*a*; Demoulin *et al.* 2015).

Further west, Hemelsdaël & Ford (2016) show that after having deposited the Middle Group Platanos Gilbert delta, the Krathis River was diverted NE to deposit a strongly prograding Gilbert delta (Akrata GD: Fig. 2) along a relay ramp that developed between the eastwards-propagating East Helike Fault and the pre-existing Derveni Fault. The delta prograded along the multiply-breached relay ramp from 0.7 Ma until final breaching at 0.2 Ma. Subsequent uplift of the East Helike Fault footwall is well documented by dated marine terraces (McNeill & Collier 2004).

Area C5: The Lakka fault block

This subsection presents previously unpublished data and interpretations on the stratigraphy and structure of the onshore Lakka Fault block based on fieldwork, logging and biostratigraphic dating. We build on and integrate the work of Palyvos *et al.* (2007, 2010). The stratigraphy of the AIG10 borehole is integrated for the first time into a regional stratigraphy.

Between the Rion Straits in the west and the Selinous River, synrift strata are preserved on the 7–8 km-wide Lakka fault block, and in a small outlier in the Panachaikon fault block further south (Figs 1 & 2). The active coastal fault system that limits the Lakka fault block to the north comprises the east–west Psathopyrgos Fault (with Ano Rodini Fault in its footwall) and the NW–SE-trending Neos Erineos fault system, which includes four east–west, right-stepping en echelon faults (Palyvos 2005; Palyvos *et al.* 2010). These are the Lambiri (with the Ziria Fault in its footwall), Selianitika, Fassoulaika and Aegion faults (Fig. 2). The Lakka fault block terminates to the west against the NE–SW-trending dextral normal Rion–Patras fault zone (Fig. 1).

Throw on the Aegion Fault has been constrained to 150–185 m in the AIG10 borehole that intersects the fault at 760 m bsl (Guernet *et al.* 2003; Cornet *et al.* 2004; Lemeille *et al.* 2004; Apostolidis *et al.* 2006). Displacement on the other four faults is difficult to constrain due to the lack of control on depth to basement both offshore and onshore. However, recent work by Beckers (2015) and Beckers *et al.* (2015) on an unpublished seismic line provides an estimate of the combined throw on the Psathopyrgos and Ano Rodini faults of 1400–1600 m, including 745 m of footwall uplift. A progressive eastwards decrease in elevation in the footwall of coastal faults is noted by Palyvos *et al.* (2007), who estimate a dip of 2.3° E on the topographical surface of the Lakka fault block from 745 m at the peak of Profitis Elias to sea level at Aegion (Fig. 5a). We propose that this footwall relief reflects a progressive decrease in displacement on the coastal faults from west to east. Throws can thus be roughly estimated as 1100 m on Lambiri-Zeria, 800 m on Selianitika and 500 m on Fassoulaika. The age of these faults is discussed below.

The synrift succession of the Lakka fault block is estimated to have a maximum thickness of 1.4 km, although the depth to the base is poorly constrained (Fig. 5a). It is divided into two informal groups: the Profitis Elias Group and the younger Galada Group (Palyvos *et al.* 2010, 2013). Building on the work of Palyvos *et al.* (2010), these stratigraphic units and their structure are described in some detail here for the first time.

Profitis Elias Group. The Profitis Elias Group consists of three formations: the Rodini Conglomerate Formation, the Salmoniko Sandstone Formation and the Synania Siltstone Formation (including the Koumares Member), which are all laterally equivalent and record an overall fining to the east across the Lakka fault block (Figs 2 & 5a) (Palyvos *et al.*

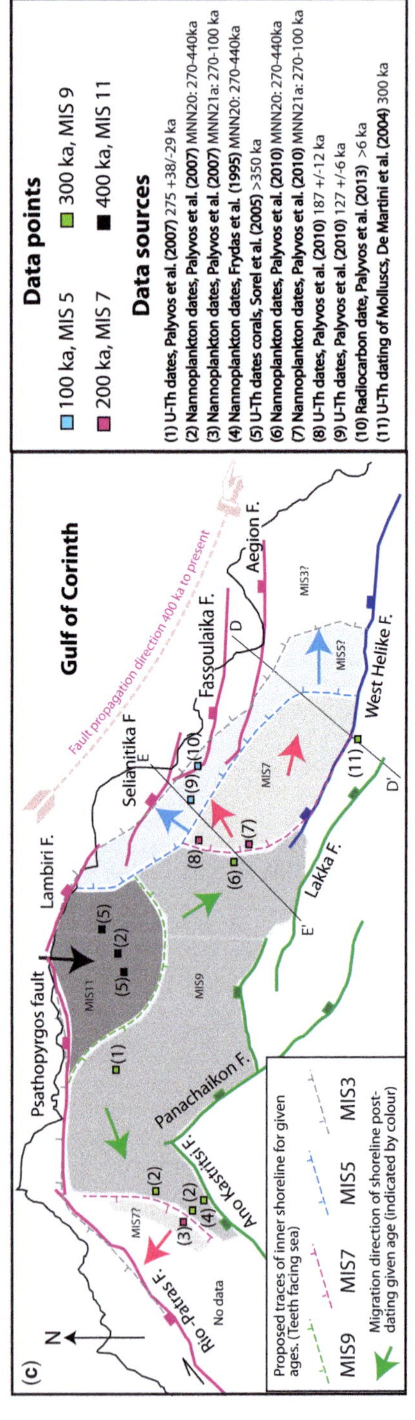

(a)
WNW
C'
Psathopyrgos F.
Ano Rodini F
Profitis Elias
Erineos River
Aravonitsa
Galada
Dimitropoulos
Meganitis R.
Koumares mbr
Selinos R.
West Helike fault block
C
ESE
1000m
500m
0
-500m
-1000m
2,5
5
7,5
10 km
12,5
15
17,5 km
20
Pindos Limestones
Section E-E'
Section D-D'
(b)
Holocene
Late
Tarantian
0.126
Middle
Ionian
0.5
0.781
Lower
1.0
Calabrian
1.5
1.806
2.0
Gelasian
Brunhes normal
Matuyama reversal
time(Ma)
Psathopyrgos F.
Marine transgression
MIS11
Marine sands
Aravonitsa
MIS9
GD
MIS7
MIS7?
MIS5?
MIS3?
proΔ
fines
Koumares mbr
Rodini conglomerate Formation
Salmonika sandstone Fm
Synania Siltstone Formation
Aegion-Pereskivi High
Onlap onto Pindos limestones
UNCONFORMITY
Lakka Block
Galada Group
Profitis Elias Group
Pirgaki Block
Upper Group
Middle Group
Lower Group
Psathopyrgos F.
Lambiri F.
Selianitika F.
Fassoulaika F.
Aegion F.
E-W Helike F.
Pirgaki-Mamoussia F
Lakka F ?
(c)
N
Psathopyrgos fault
Lambiri F.
Gulf of Corinth
Fault propagation direction 400 ka to present
Selianitika F
Fassoulaika F.
Aegion F.
Rio -Patras F.
No data
Ano Kastritsi F.
Panachaikon F.
Lakka F.
West Helike F.
MIS11
MIS9
MIS7
MIS7?
MIS5?
MIS3?
E
E'
D
D'
Proposed traces of inner shoreline for given ages. (Teeth facing sea)
MIS9
MIS7
MIS5
MIS3
Migration direction of shoreline post-dating given age (indicated by colour)
Data points
100 ka, MIS 5
300 ka, MIS 9
200 ka, MIS 7
400 ka, MIS 11
Data sources
(1) U-Th dates, Palyvos et al. (2007) 275 +38/-29 ka
(2) Nannoplankton dates, Palyvos et al. (2007) MNN20: 270-440ka
(3) Nannoplankton dates, Palyvos et al. (2007) MNN21a: 270-100 ka
(4) Nannoplankton dates, Frydas et al. (1995) MNN20: 270-440ka
(5) U-Th dates corals, Sorel et al. (2005) >350 ka
(6) Nannoplankton dates, Palyvos et al. (2010) MNN20: 270-440ka
(7) Nannoplankton dates, Palyvos et al. (2010) MNN21a: 270-100 ka
(8) U-Th dates, Palyvos et al. (2010) 187 +/-12 ka
(9) U-Th dates, Palyvos et al. (2010) 127 +/-6 ka
(10) Radiocarbon date, Palyvos et al. (2013) >6 ka
(11) U-Th dating of Molluscs, De Martini et al. (2004) 300 ka

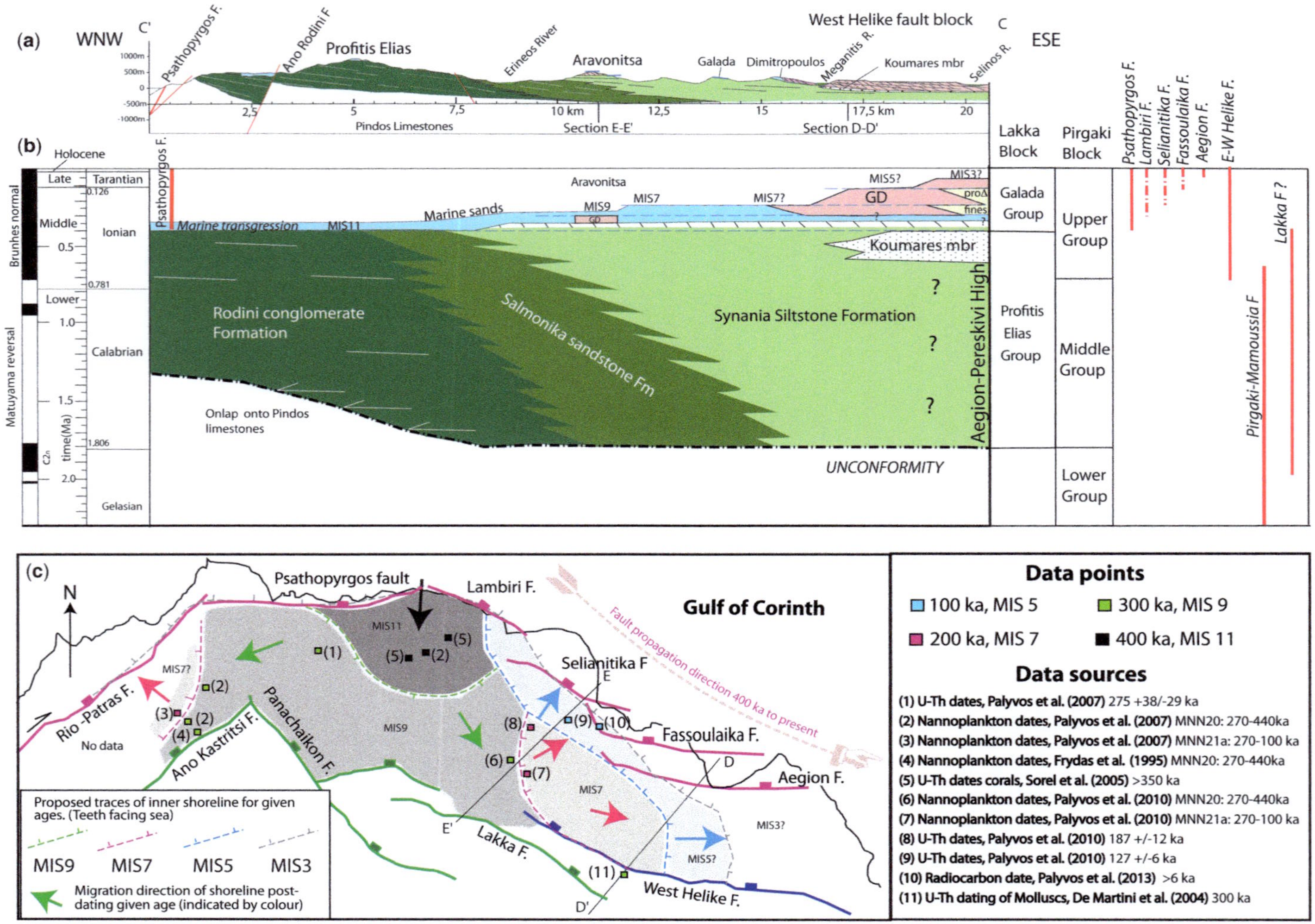

2010, 2013). The entire succession is tilted south against the Lakka Fault by up to 35° (Fig. 6a), and is cut by many second-order oblique and north-dipping faults (Figs 5, 7 & 8).

The Profitis Elias Group is dated as Early–Middle Pleistocene based on the presence of a very rich freshwater to oligohaline mollusc assemblage in the Synania Formation and Koumares Member, with *Didacna spratti* (Fuchs), *Monodacna tenue* (Fuchs), *Dreissena polymorpha* (Pallas), *Unio* cf *crassus*, *Theodoxus patrae* (Esu & Girotti), *T. micans* (Gaudy & Fischer), *Viviparus Melanoides elegans*, *Melanopsis mitzopoulosi*, *Hydrobia attica*, *Pyrgula incisa*, *Valvata graeca* (Fuchs) and *Adelinella elegans* (Fuchs). The reader is referred to Esu & Girotti (2015), and references therein, for further details on the local mollusc assemblage.

The Rodini Conglomerate Formation (<600 m thick) forms the Profitis Elias Massif (Fig. 2), west of the Erineos River. It is composed predominantly of brown-reddish to grey, coarse conglomerates up to boulder grade, poorly sorted, both matrix and clast supported, and normally or inversely graded limestone and chert clasts of the Pindos thrust sheet dominate, with a sandy–silty matrix. Lenses of trough cross-bedded sandstones of the same composition locally including some grains of quartz, muscovite, feldspar and rare red, clay-rich, palaeosols are found interlayered within the conglomerates. Conglomerates and sands are organized into metre-scale, laterally persistent tabular or channelized bodies with basal erosive scours (Fig. 6b), vertically stacked to form a multistorey wedge showing southwards and eastwards thinning. Beds present gentle to moderate southwards dips. Imbricated clasts and erosive features record SW-, south- and SE-directed palaeocurrents. Locally, the formation onlaps onto a palaeo-relief formed by pre-rift Pindos limestones in the immediate footwall of the Psathopyrgos Fault. Conglomerates are here interpreted as gravity-flow deposits (the Gmm, Gmg, Gci and Gcm lithofacies of Miall 1996), and referred to as debris and hyper-concentrated flows, while sands are interpreted as traction current deposits. Preliminary facies analysis indicate that the Rodini Conglomerate Formation was deposited by an alluvial fan and braided river system, most probably sourced from the north. It passes laterally into the Salmoniko Formation to the east.

The Salmoniko Formation (<250 m thick) comprises tabular bodies of poorly lithified yellow-beige, cross-bedded and planar-bedded fine sandstones and silty sandstones, alternating with lenticular (ribbon-like structures) and tabular beds of granule- to pebble-grade conglomerates (metre scale) and dark grey clayey hydromorphic palaeosols (histosol) (Fig. 6c). The formation represents distal alluvial-fan and fluvial environments, with sands and silts interpreted as floodplain deposits (North & Davidson 2012), while pebble and granule conglomerates represent hyper-concentrated flow deposits filling small distributary channels. Histosols are found on top of metre-scale small sandy–silty sequences recording phases of marshy inundation.

Along the eastern side of the Erineos River (Fig. 2), isolated fluvio-deltaic sands and pebble conglomerates interfinger with the Synania Formation and prograde both eastwards (Fig. 5) and southwards (Fig. 8). A small inlier of sandstones and conglomerates at Nerantzies (Figs 2 & 5) is assigned to this formation.

The Synania Formation (<400 m thick) is a lacustrine unit composed of planar bedded, greenish-grey siltstones and very fine sandstones, interbedded with cross-laminated fine sandstones, decimetre-scale thick lignite horizons, and white thinly laminated calcareous mudstone and marlstone (Fig. 6d). The formation gradually passes to finer massive facies eastwards. The laminated calcareous mudstones and marlstones are rich in charophyte stems and lacustrine molluscs, and are interpreted as laminites (Gierlowski-Kordesch 2010). They are organized into decametre-thick tabular bodies, used as local marker beds, regularly alternating with siltstone and fine sandstones of a distal prodelta environment. The formation has a rich freshwater to oligohaline mollusc assemblage that is typical of the Early Pleistocene basins of Greece described earlier (Gillet 1963; Fuchs 1877; Fernandez-Gonzalez *et al.* 1994; Frydas *et al.* 1995; Koskeridou & Ioakim 2009; Esu & Girotti

Fig. 5. (**a**) WNW–ESE longitudinal section C–C′ along the Lakka and West Helike fault blocks (section line located in Fig. 2). The key to stratigraphic units is shown in Figure 2. The intersections with cross-sections D–D′ and E–E′ are shown. (**b**) Chronostratigraphic model along the section C–C′ showing the distribution in time and space of the Profitis Elias and Galada groups. The timescale is that of Gibbard & Cohen (2008). To the east, major uncertainty is associated with the presence of a long-lived palaeogeographical high (Aegion–Psaromita High). MIS 5 and MIS 3 terraces can only be suggested as no reliable data exist in this area. Estimated periods of activity on various faults are shown to the right. (**c**) Compilation map of dated Galada Group deposits across the Lakka fault block and environs with main faults. Data sources are given to the right of the map. These data are used to construct a model of shoreline regression since 400 ka, which is interpreted as due to footwall uplift in the SE-propagating coastal fault system (see the text for discussion). The position of sections E–E′ (Fig. 7) and D–D′ (Fig. 8) are shown. Fault colours indicate the age of fault initiation and correspond to those in Figures 10–13.

Fig. 6 The Lakka fault block. (**a**) NE–SW-orientated panoramic view from Profitis Elias Massif to the Aravonitsa Hill in the foreground (SW on the right). Gravels and sands of the Galada Group (GG) Aravonitsa Formation (AF) overlie with angular unconformity (dashed line) the 30° S-tilted lacustrine deposits of the Profitis Elias Group (PEF) Synania Formation (SyF). The N110°-trending Lidoriki Fault is a synthetic normal fault of the main Lakka Fault (see also Fig. 8) and cuts the Synania Formation (arrows indicate the downthrown block). The Pindos Unit (P) is in the background. (**b**) Along strike, west–east-orientated view of the Rodini Conglomerate Formation (west is on the left, and progradation is directed to the viewer) close to Ano Kastritsi: massive conglomerates (cg), cross-bedded sandstones (sb). (**c**) NE–SW-orientated outcrop of the Salmonika Formation north of Neo Salmonika (SW is on the right): south-tilted channel pebble conglomerates (pb) interbedded with overbank silty sandstones and siltstones (ssb) and histosol (h).

Fig. 6. (*Continued*) (**d**) Lacustrine Synania Formation cropping out north of Aravonitsa (the outcrop is west–east orientated, with west on the left, and 170 m high). Open lacustrine, white calcareous laminated mudstones (w2 and w3 marker beds) alternate with distal prodelta silt and sand (s). The hanging-wall block of the Lidoriki Fault is dissected by small north–south-trending transverse faults (TF); arrows indicate the downthrown blocks. (**e**) Galada Group. North–south-orientated outcrop of Gilbert-type fan delta gravels of the Lower Aravonitsa Formation (LAF) unconformably overlain (dashed line) by nearshore marine sandstones and pebble conglomerates of the Upper Aravonitsa Formation (UAF), at Aravonitsa (south on the right). The succession is tilted south. (**f**) SW–NE-orientated outcrop on the left bank of the Meganitis River, close to Koumares (SW on the left). The West Heliki Fault (N110° trending, NE60° dipping) separates the Synania Formation (SyF), in its footwall block, from the younger palaeo-Meganitis conglomerates (PMc; Galada Group).

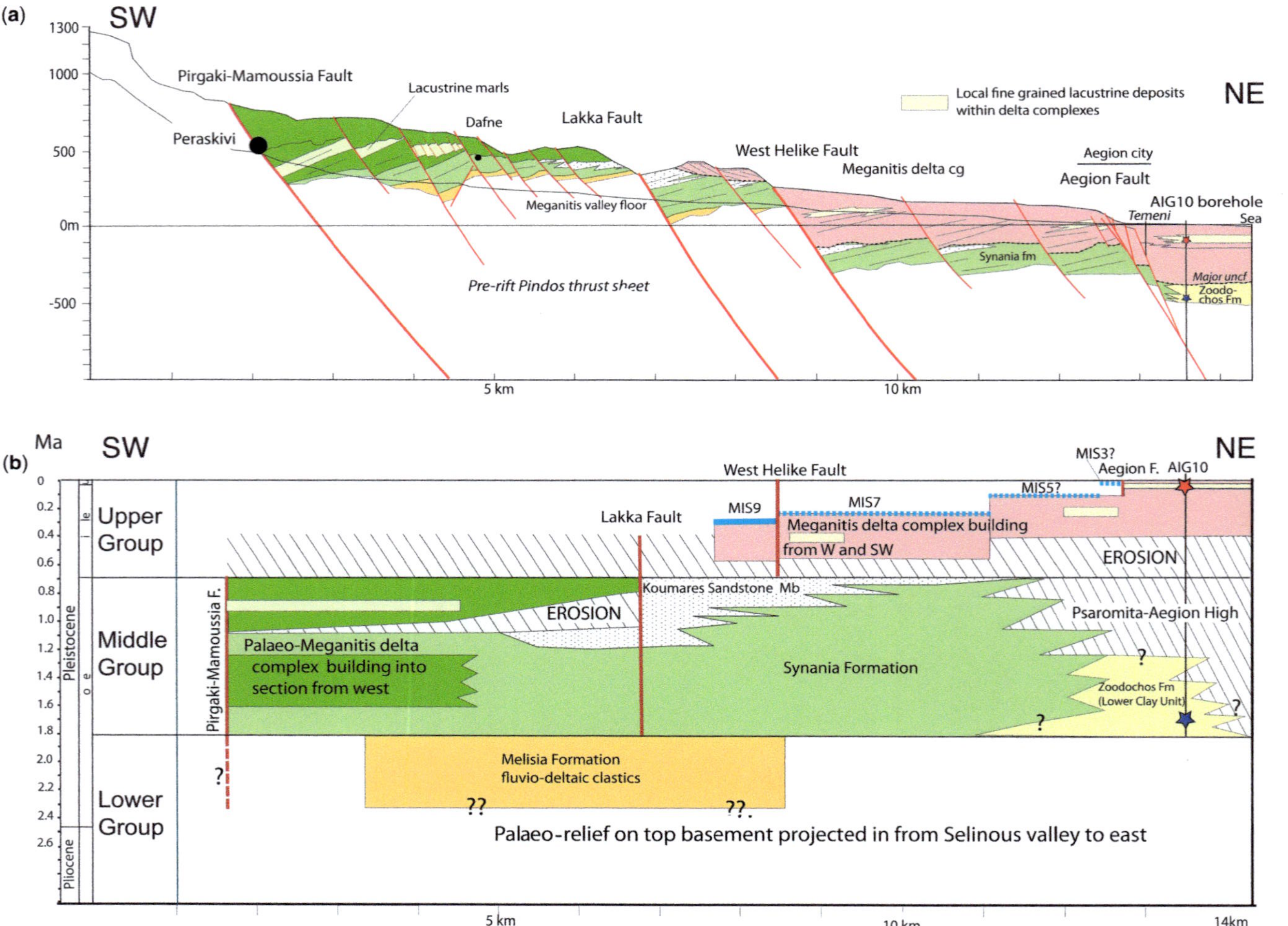

Fig. 7. (**a**) NNE–SSW cross-section D–D′ from the Aegion borehole in the north to the village of Peraskivi to the south on the Pirgaki Fault. The trace of the section line and the key to stratigraphic units are given in Figure 2. (**b**) Chronostratigraphy along the D–D′ cross-section. The timescale is that of Gibbard & Cohen (2008). In the AIG10 borehole, published ages are 11.5 ka at 65 m (Lemeille *et al.* 2004), 35.3 ka at 103 m (red star: Guernet *et al.* 2003) and 1.7 Ma in the Lower Clay Unit at 420 m (blue star: Lemeille *et al.* 2004).

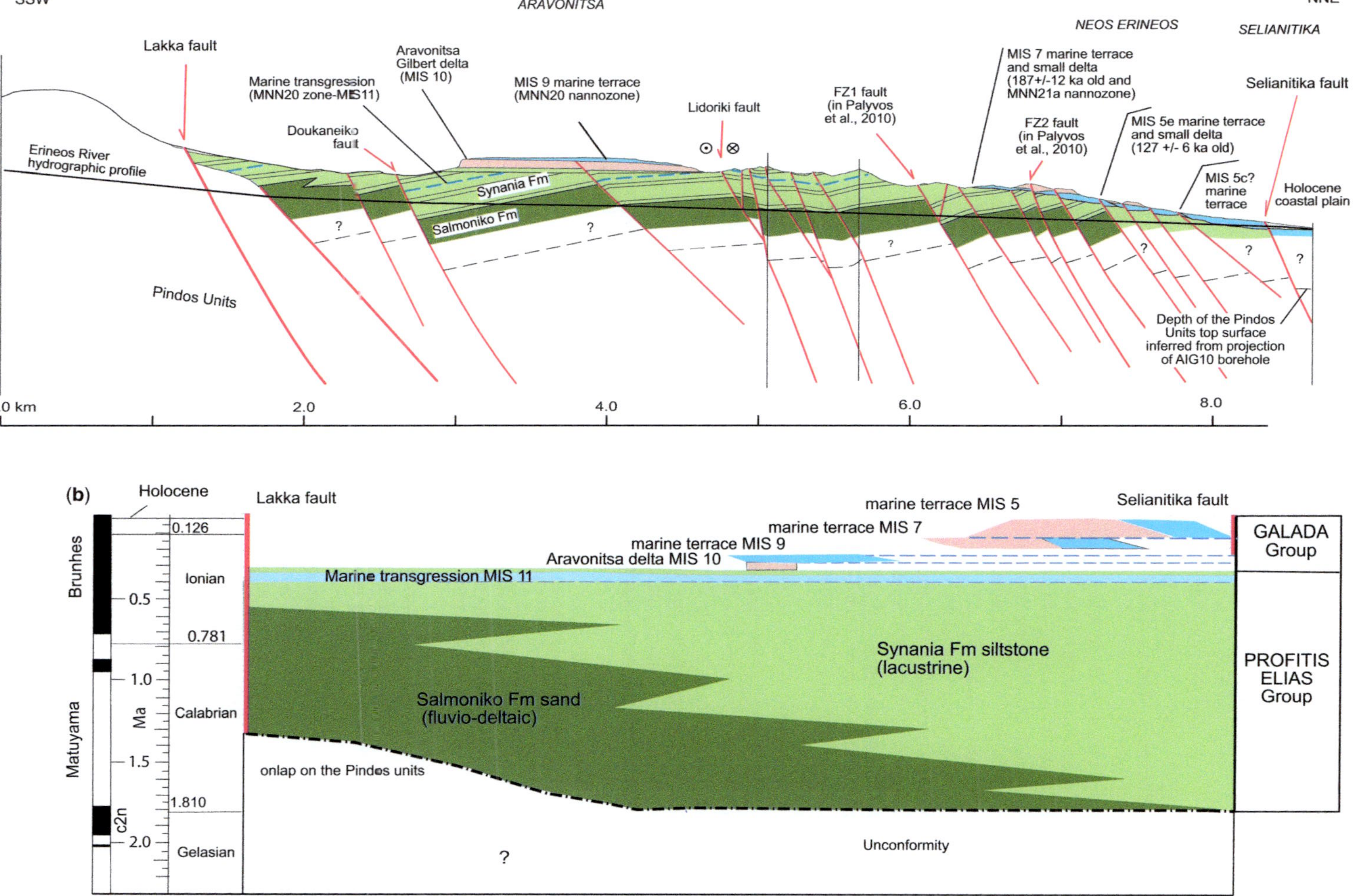

Fig. 8. (**a**) NNE–SSW cross-section through the western Lakka fault block (C5) from Aravonitsa to Selianitika (E–E′, located on Fig. 2). (**b**) Corresponding chronostratigraphy revised from Palyvos *et al.* (2010) using the timescale of Gibbard & Cohen (2008). The stratigraphic key in Figure 2.

2015). In the upper portions of the formation, brackish-water layers are present. Palyvos *et al.* (2010) tentatively attributed a fully marine intercalation at Aravonitsa (Fig. 2) to marine isotope stage (MIS) 11 (420–400 ka). The Synania Formation thins to the east of the Meganitis river valley, and its upper levels pass laterally into shallow-marine to brackish-water fine sands and pebbly sands of the Koumares Member (<100 m thick) with abundant *Mytilus* sp.

The Synania Formation is correlated laterally with the Zoodochos Formation further to the east (Backert *et al.* 2010). Rapid changes in facies associated with thinning and onlapping onto outcropping pre-rift lithologies in the Selinous Valley (Backert *et al.* 2010) strongly suggest that there was a persistent palaeogeographical high orientated roughly north–south, extending south from Aegion to Peraskivi along the section represented in Figure 7. Offshore seismic data also indicate that this feature extends northwards across the Gulf to the Psaromita Peninsula on the north shore (Psaromita–Aegion high).

Galada Group. The Galada Group unconformably overlies all formations of the Profitis Elias Group, often with angular discordance (Fig. 6a). It comprises variable lithologies including cool-water carbonate marine terraces, Gilbert-type delta conglomerates, and shallow-marine to brackish-water sands and silts. As a whole, the Galada Group records discontinuous marine–brackish–lacustrine sedimentation due to the interplay of eustatic sea-level fluctuations and tectonics during the progressive uplift of the Lakka fault block from around 400 ka (Palyvos *et al.* 2010). Its depositional history records the tectonic evolution of the coastal fault system.

To the west, a succession of fine sands, silts and marls of the Galada Group overlie the Rodini Conglomerate Formation at the top of the Profitis Elias Massif (peak 745 m). These strata have mixed marine, lacustrine and brackish characteristics, and contain the highest and oldest dated marine layer of the Galada Group. Using diatom assemblages, Frydas (1989, 1991) dated this layer as Middle Pleistocene (biozone MNN20: 270–440 ka). Palyvos *et al.* (2007) argued that this marine level could be correlated to the MIS 11 marine incursion (*c.* 400 ka). Uplift began soon after, thus dating the first opening of the Rion Straits as due to the onset of activity on the Psathopyrgos Fault.

Moving east, the south-dipping Synania Formation strata below the Aravonitsa Plateau (500 m altitude: Fig. 2) contain a marine level dated as MIS 11 (MNN20: Palyvos *et al.* 2010). This tilted succession is unconformably overlain by a 70 m-high Gilbert delta building north (Lower Aravonitsa Formation; Galada Group: dated as MIS10), which is, in turn, overlain by shallow-marine sands and pebbly sands of the Upper Aravonitsa Formation (Figs 2 & 6e) (Palyvos *et al.* 2010). The sands of the Upper Aravonitsa Formation have yielded a rich marine macrofauna (corals) dated at 283 ± 40 ka by the U–Th method (MIS 8.5–9: Palyvos *et al.* 2010).

To the east again, shallow-marine sands, often reaching 10–20 m in thickness and tilted gently east, cap many hillcrests (Fig. 2). Just to the west of the Meganitis river valley, these sands are strongly incised by conglomerates of the Meganitis Gilbert delta that dip gently eastwards across the hanging wall of the West Helike Fault (Fig. 6f). Figure 7 shows a cross-section from Aegion to the village of Peraskivi, integrating the 1000 m-deep AIG10 borehole in the hanging wall of the Aegion Fault (Cornet *et al.* 2004). The marine Lower Clay Units (110 m thick) at the base of the synrift succession in the AEG10 borehole is dated as Early Pleistocene (Lemeille *et al.* 2004) and is thus equivalent in age to the Synania and Zoodochos formations. The overlying fluvial conglomerates and sands (400 m thick) are placed in the Galada Group, and have a strongly erosive base that has been given an estimated age of 400 ka (Lemeille *et al.* 2004).

On a north–south profile from Aravonitsa to Selianitika (Fig. 8a, b), Palyvos *et al.* (2010) presented reliable chronological constraints on terraces using nannoflora in silty levels, and U–Th dating of corals and syndepositional calcite encrustations. Each terrace corresponds to a single high-order eustatic cycle (IV and V) in which transgressive and sea-level highstand sands are preserved (MIS 9, 7, 5 and 1). In general, terraces are well defined but are only partly preserved, often cut by minor faults (Fig. 8). These dated terraces record the northwards migration of footwall uplift over a distance of just over 3 km since around 300 ka. The cross-section (Fig. 8a) shows the marked tilting of the Profitis Elias Group below the untilted Galada units of the Aravonitsa plateau.

Finally, all dated terraces and marine sediments preserved above the Profitis Elias Group in the Lakka fault block are compiled on the map in Figure 5c. The uncertainty associated with some of these ages, in particular the U–Th dating of corals, is discussed by Palyvos *et al.* (2007, 2010). A clear pattern emerges, however, of oldest marine sediments preserved on the highest relief (Profitis Elias, 745 m: Frydas *et al.* 1995) with younger ages radiating away from this massif to the SW and SE. The northwards-younging trend seen on the Aravonitsa–Selianitika section (Fig. 8) is also integrated into this model, showing shorelines gradually migrating to the ESE (and possibly west) and also, more locally, to the north in the footwalls of the coastal faults. This pattern can be explained by migrating footwall uplift on the coastal fault system with the initiation of the Psathopyrgos Fault around 400 ka, followed by a gradual propagation of the coastal fault system to the SE, with the Aegion Fault being the youngest segment initiated around 50–70 ka (Micarelli 2003; Cornet *et al.* 2004; Lemeille *et al.* 2004).

The Lakka fault block contains no known strata equivalent to the Lower Group: however, a fine-grained fluvial succession is preserved further south in the hanging wall of the eastern Panachaikon fault block. This unit correlates laterally to the Melisia Formation (Lower Group) of the Pirgaki–Mamoussia fault block (Backert *et al.* 2010; Fig. 2). With its upper boundary at 400 ka, the Early–Middle Pleistocene Profitis Elias Group is laterally equivalent to the Middle Group and to part of the Upper Group further east. However, direct correlation is obscured by the presence of the Aegion–Psaromita palaeohigh. The Lakka Fault was active during deposition of the Profitis Elias Group, although no sediment appears to have been supplied from its footwall during this time. Instead, its footwall area appears to have drained eastwards across the relay zone between the Panachaikon and Pirgaki faults to feed the palaeo-Meganitis delta complex building to the east and NE. The small north-building Aravonitsa Gilbert delta, dated at 400–300 ka (Palyvos *et al.* 2010), records the first coarse sediment supplied from the Lakka footwall (possibly by the proto-Erineos River). The Lakka and Panachaikon faults were active until about this time, when the new fault system initiated to the NW (Psathopyrgos Fault) and then propagated eastwards.

Evolution of the central and western rift

Using the data presented above, we here reconstruct the evolution of the western and central rift in four phases. Figure 9 presents an along-strike chronostratigraphic model correlating onshore areas C5–C3. For this purpose, the northwards-migrating depocentres are all projected onto a single plain. Figures 10–13 represent palaeogeographical reconstructions of the evolution of the western and central Corinth rift. Offshore data on fault geometries and ages, and seismic stratigraphy were derived from McNeill *et al.* (2005), Bell *et al.* (2008, 2009) Taylor *et al.* (2011) and Nixon *et al.* (2016).

Phase 1: Late Pliocene–Early Pleistocene (5–4 to 1.8 Ma)

Early extension was distributed on a predominantly north-dipping fault system across a 20–30 km-wide zone over a period of 2–3 myr starting from around 4 Ma. The area under extension may have been wider. For example, undated normal faults are documented further south (Skourtsos & Kranis 2009). As the amount of Phase 1 strata preserved below the Gulf is unknown, we provisionally draw the northern border below the proximal southern Gulf (Fig. 10). This margin may have been controlled by a south-dipping fault or faults, which were later buried. The early rift closed to the west (Fig. 10). In C4 and C3, synchronous north-dipping faults define tilted blocks 4–8 km wide. In C4 (Fig. 3), these are relatively small faults with approximately the same length (10–16 km) that accommodate similar displacements (throws of 500–1000 m: Ford *et al.* 2013). Total Phase 1 extension in C4 (section A–A′) was low and extension rate was slow, estimated at 0.6–1 mm a^{-1} (Fig. 3e) (Ford *et al.* 2013). To the east (C3) faults accommodated considerably more displacement (Rohais *et al.* 2007*a*; Ford *et al.* 2013), the largest fault being the Killini Fault with an estimated throw of over 4 km (Fig. 4a). Using our revised stratigraphy, the Phase 1 extension rate in C3 (Fig. 4a) is estimated as 1.5–2.3 mm a^{-1} (assuming rift initiation at 5–4 Ma). Faults in the lacustrine succession to the east (C2 and C1: Fig. 10) are, as yet, poorly known.

Phase 1 is characterized by continental environments that deepen from fluvial (C4) in the west to lacustrine in the east (C3, C2) recording an overall fining eastwards (Fig. 10b). Coarse-grained sediment was fed into the rift from the SW by two major routing systems (Kalavryta and Killini: Fig. 10). In C4, coarse-grained alluvial strata (> 1000 m thick) onlap a significant inherited relief. Sediment supply was high and largely outstripped the creation of accommodation, thus burying active normal faults. Although the succession is condensed on a local palaeogeographical high between C4 and C3 (Vougsimos, Fig. 2), facies and palaeocurrent analysis show that sediment was routed from C4 to C3. These depocentres were therefore always connected (Rohais *et al.* 2007*b*; Ford *et al.* 2013) and not separated as proposed by Ghisetti & Vezzani (2005).

In C3, the basin very quickly became lacustrine after a ‘deepening event’ at 3.2–3.0 Ma (DE1; base of the lacustrine Aiges Formation: Leeder *et al.* 2012). Subsequently, the Killini sediment-routing system built the Killini and Mavro Gilbert delta complexes axially into the shallow lake (Figs 9 & 10) comprising stacked, prograding deltas, each several tens of metres in height (Fig. 4). The concentration of coarse-grained alluvial conglomerates along the south and west of the early rift is interpreted as being due to the capture of well-established antecedent rivers of the Hellenide orogen (Ford *et al.* 2013; Hemelsdaël *et al.* 2015). In the

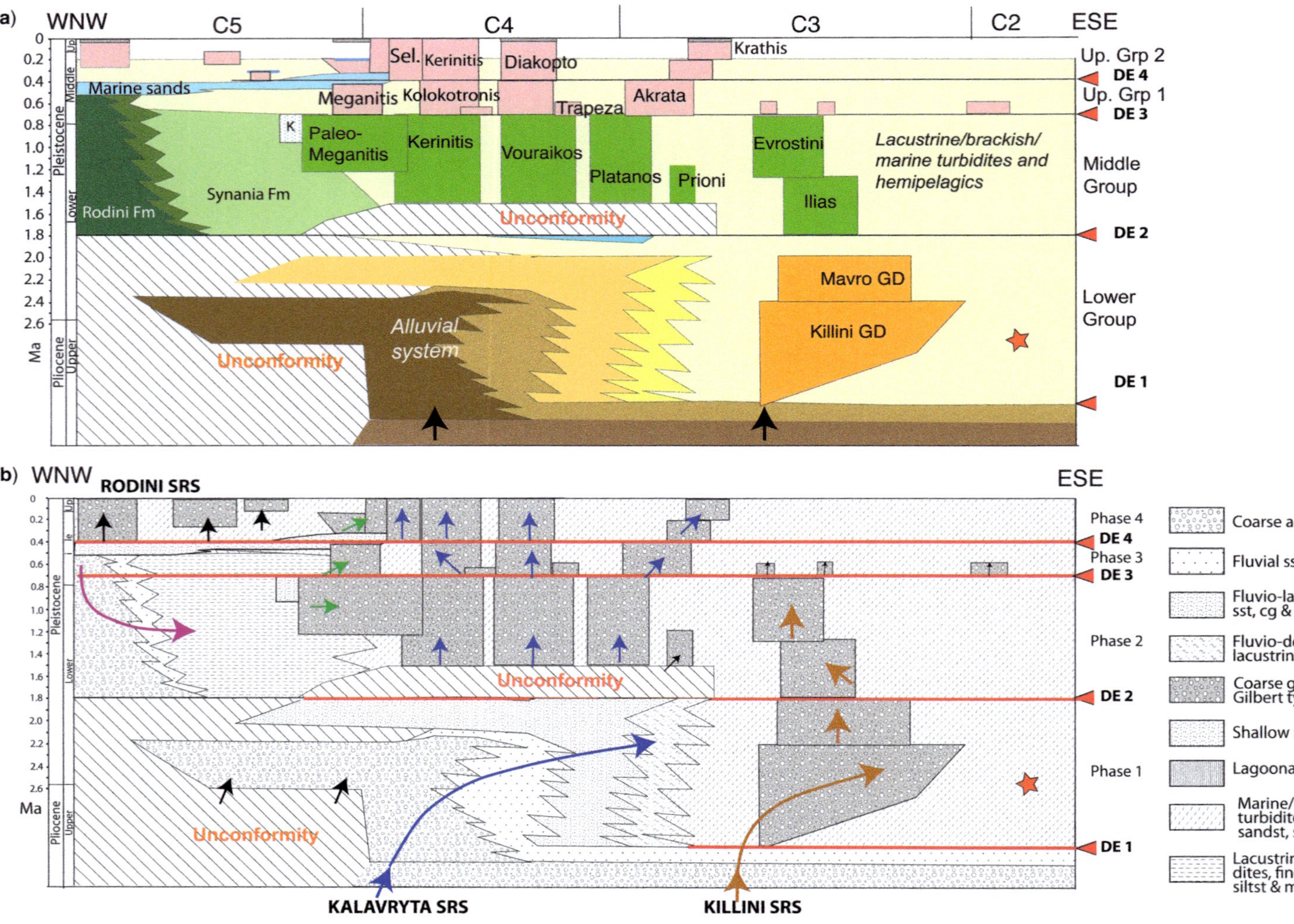
(a)
WNW
C5
C4
C3
C2
ESE
Ma
Pliocene
Upper
Pleistocene
Lower
Middle
Up
Krathis
Sel.
Kerinitis
Diakopto
Marine sands
Meganitis
Kolokotronis
Trapeza
Akrata
K
Paleo-
Meganitis
Kerinitis
Vouraikos
Platanos
Prioni
Evrostini
Ilias
Lacustrine/brackish/
marine turbidites and
hemipelagics
Rodini Fm
Synania Fm
Unconformity
Mavro GD
Killini GD
Alluvial
system
Unconformity
Up. Grp 2
DE 4
Up. Grp 1
DE 3
Middle
Group
DE 2
Lower
Group
DE 1
(b)
WNW
RODINI SRS
ESE
Unconformity
Unconformity
KALAVRYTA SRS
KILLINI SRS
Phase 4
DE 4
Phase 3
DE 3
Phase 2
DE 2
Phase 1
DE 1
Coarse alluvial cg
Fluvial sst and cg
Fluvio-lacustrine
sst, cg & siltst
Fluvio-deltaic and
lacustrine
Coarse grained
Gilbert type deltas
Shallow marine sands
Lagoonal marls
Marine/lacustrine
turbidites, fine
sandst, siltst & marl
Lacustrine turbi-
dites, fine sandst,
siltst & marl

eastern lacustrine succession, published palaeocurrent data indicate that sediment supply was predominantly axial and from the west (Leeder *et al.* 2012). Grain-size and facies distribution in the early rift (Fig. 9) were therefore controlled by: (1) the position of captured antecedent rivers along the southern rift margin; (2) the rate of sediment supply from these rivers; and (3) their interaction with evolving accommodation on the young fault system.

Phase 2: Early–Middle Pleistocene (1.8–0.7 Ma)

In the Early Pleistocene, starting at around 1.8 Ma, the southern rift margin migrated northwards in C3 and C4 by approximately 15 and 10 km, respectively. At the same time, the rift propagated west to form a new depocentre in C5. This phase of northwards and westwards fault migration (1.8–1.5 Ma) is correlated with a significant period of erosion that marks the base of the Middle Group in C4 (but not C3: Figs 9 & 11). New south-dipping faults were also initiated on the northern side of the rift (Fig. 11), and played a major role in controlling a more symmetrical and deeper depocentre, largely corresponding to the present-day Gulf in C3 and C2 (Bell *et al.* 2009; Nixon *et al.* 2016). The new rift was 20–25 km wide in C3, and narrowed to the west to 10–15 km in C4 and C5. Eustatic control on delta stratigraphic architecture (Backert *et al.* 2010), the local presence of limestones with corals (Ford *et al.* 2007) and marine Foraminifera assemblages (see Rohais *et al.* 2007*b* for a summary) all indicate that the main Corinth Basin was predominantly at least periodically marine to brackish in character during the Early–Middle Pleistocene. The Psaromita–Aegion High separated the C5 fluvio-lacustrine depocentre from the main rift (Fig. 11).

The principal north-dipping faults along the southern rift margin were the Xylocastro–Evrostini–Valimi fault zone, and the Derveni Fault in C3, the Pirgaki–Mamoussia Fault in C4, and the Lakka and Panachaikon faults in C5. On the northern rift margin, principal south-dipping faults were the Galaxidi Fault in C3 (also known as the East Channel Fault: e.g. Bell *et al.* 2008), the West Channel Fault in C4 and the Trizonia Fault in C5 (Fig. 11). In C3, the highest sediment accumulation occurred between the Galaxidi and Derveni faults, focusing more towards the south-dipping Galaxidi Fault. The Galaxidi Fault has an estimated throw of 1.3 km and was principally active during Phase 2 (Bell *et al.* 2009). The Derveni Fault, still active today, is by far the largest fault in C3, with a total estimated throw of around 4 km (Hemelsdaël & Ford 2016), a quarter of which occurred during Phase 2 (Nixon *et al.* 2016). At the same time, the pre-existing Valimi–Evrostini Fault to the south accommodated >1000 m throw during deposition of the Ilias and Evrostini deltas (Fig. 4a). The estimated extension rate in C4 doubled to 2–2.5 mm a^{-1} during Phase 2 (Fig. 3d) (Ford *et al.* 2013). Extension may have increased to the east in C3 and C2, as reflected in the position of the main depocentre.

The northwards fault migration over a period of some 300 kyr at the beginning of Phase 2 was associated, first, with a marine transgression at the top of the Lower Group (Katafugion Formation) in C4, followed by an enigmatic phase of (submarine) erosion (Ford *et al.* 2007) (Fig. 9). The fluvial network underwent a radical reorganization, changing from an east- to NE-flowing network to a new system of equally spaced (6–8 km), north-flowing rivers (Fig. 11). These rivers reworked Phase 1 sediments that were uplifted further south. Their coarse bedload provided high erosional capacity that allowed the rivers to cut across uplifting footwalls (Cowie *et al.* 2006).

In area C4, the deepening event (DE2) is recorded by a marked change from a continental fluvial environment to a deeper basin in the hanging wall of the Pirgaki–Mamoussia Fault. Large Gilbert deltas began to build into the ever-deepening basin supplied by new north-flowing rivers (Figs 9 & 11). Equivalent prodelta turbidites and hemipelagics are preserved both onshore (the Derveni and Zoodochos formations) and offshore (most probably the seismic stratigraphic unit SU1: Nixon *et al.* 2016). In area C3, the deepening event is not marked by a major sedimentological change, as Gilbert deltas were already building into a lacustrine basin. However, deeper-water conditions are recorded by the distinct stratal architecture of the Ilias and Evrostini deltas, characterized by aggradation–progradation and significant foreset

Fig. 9. WNW–ESE Wheeler diagram of synrift chronostratigraphy correlating areas C2–C5 along the northern Peloponnese. In (**a**) lithostratigraphic colour codes in Figure 2. K, Koumares Formation. In (**b**) facies distributions and sediment-routing systems are represented. Four deepening events are identified (DE1–DE4) to the right of the figure. The timescale is that of Gibbard & Cohen (2008). The three major sediment-routing systems (SRS) feeding into the rift are shown (Rodini in purple, Kalavryta in blue and Killini in orange). Green arrows represent the Meganitis systems draining off the Panachaikon and Lakka blocks. Other sediment-supply routes are more local and short-lived (black arrows). The red star represents the ash layer dated at 2.55 Ma by Leeder *et al.* (2012). Sel., Selinous.

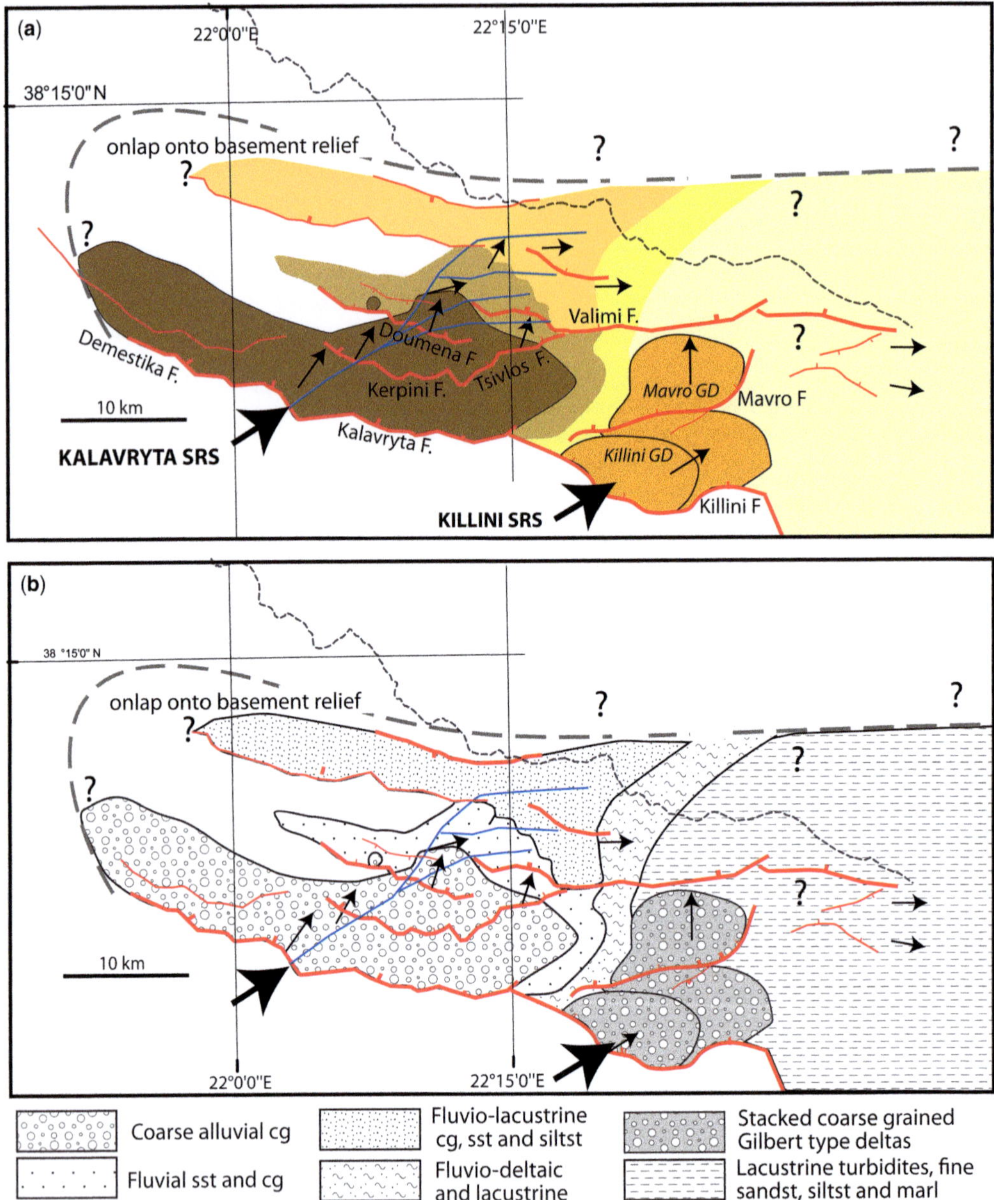

Fig. 10. (**a**) Phase 1 (Pliocene–Early Pleistocene) rift model showing deposition of the Lower Group. Colours correspond to stratigraphic units in Figures 2 and 8. The northern basin margin is very poorly constrained. Small black arrows summarize the palaeocurrent data of Rohais *et al.* (2007*a*) and Leeder *et al.* (2012). Large black arrows indicate the two main captured antecedent drainage systems (SRS, sediment-routing systems). (**b**) Facies distributions during Phase 1 with the main sediment-routing systems.

height (hundreds of metres). As the two Phase 2 deltas lie directly north of the Phase 1 deltas, it is reasonable to suggest that they were supplied by the same north-flowing drainage system (Rohais *et al.* 2007*a*).

To the west, the new C5 depocentre was controlled by the Lakka and Panachaikon faults to the south, and the Trizonia Fault and an unidentified fault to the north and NW (Beckers 2015; Beckers *et al.* 2015). The lacustrine basin was supplied

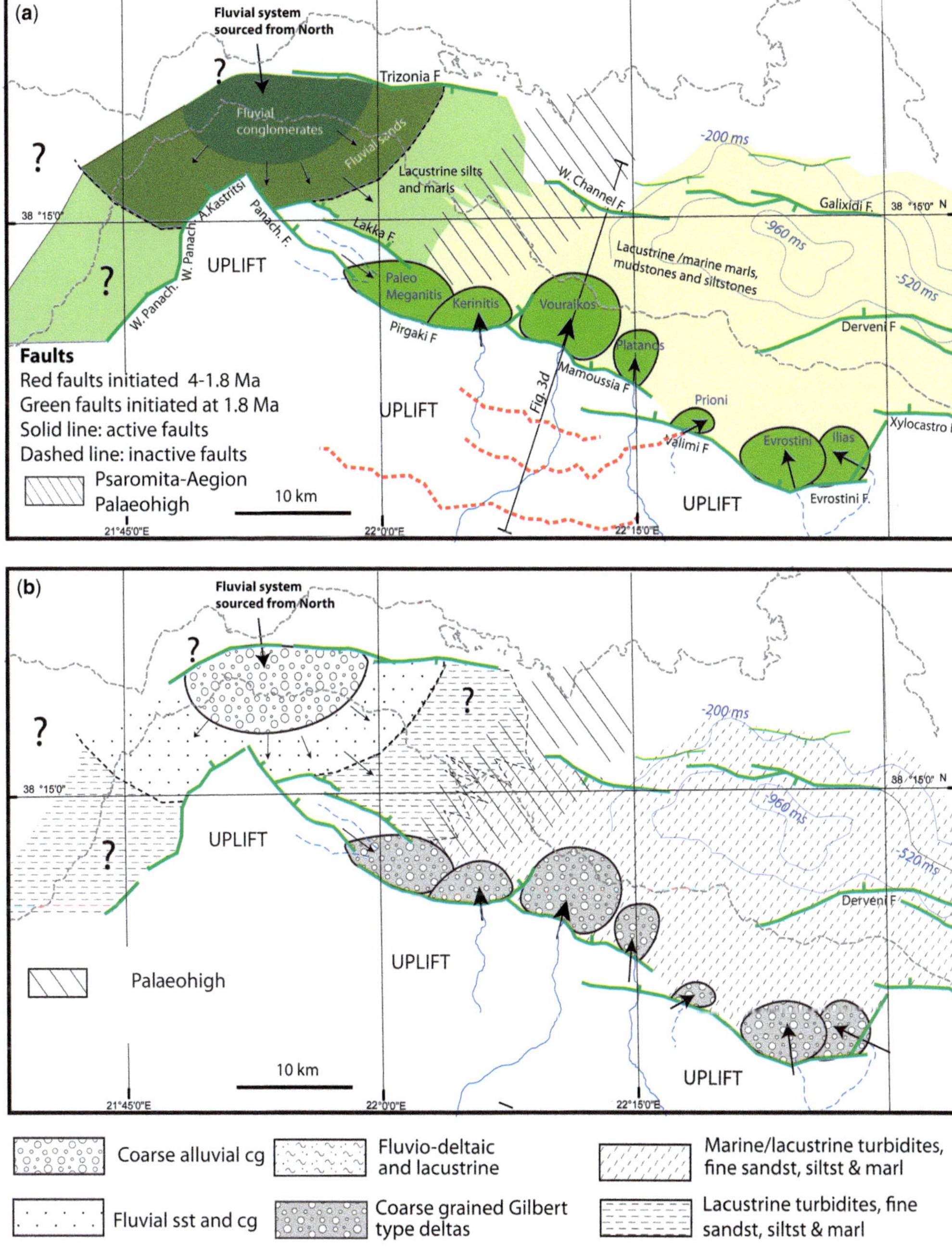

Fig. 11. (**a**) Phase 2 (Early–Middle Pleistocene, 1.8–0.8 Ma) rift model showing migration of deformation in zones C2–C5 onto new (green) faults. In the main depocentre (C3–C2) large Gilbert-type deltas and their prodelta fines were deposited in the hanging walls of these faults. To the west of the Psaromita–Aegion High, a fluvio-lacustrine system was deposited in C5. Colours correspond to stratigraphy in Figures 2 and 8. In the main depocentre between the Galaxidi and Derveni faults, the thickness of the succession in milliseconds of two-way travel time (TWTT) for the period 2–1.5 Ma to 620 ka is taken from Nixon *et al.* (2016). A cross-section through C4 (trace shown) at the end of Phase 2 is shown in Figure 3d. Inactive faults are shown in dashed lines; active faults have solid traces. Phase 1 faults (Fig. 10) are shown in red. Rivers are shown as fine blue lines (present-day rivers are solid lines; Phase 2 rivers are dashed lines). (**b**) Facies distributions during Phase 2. Black arrows indicate delta building directions.

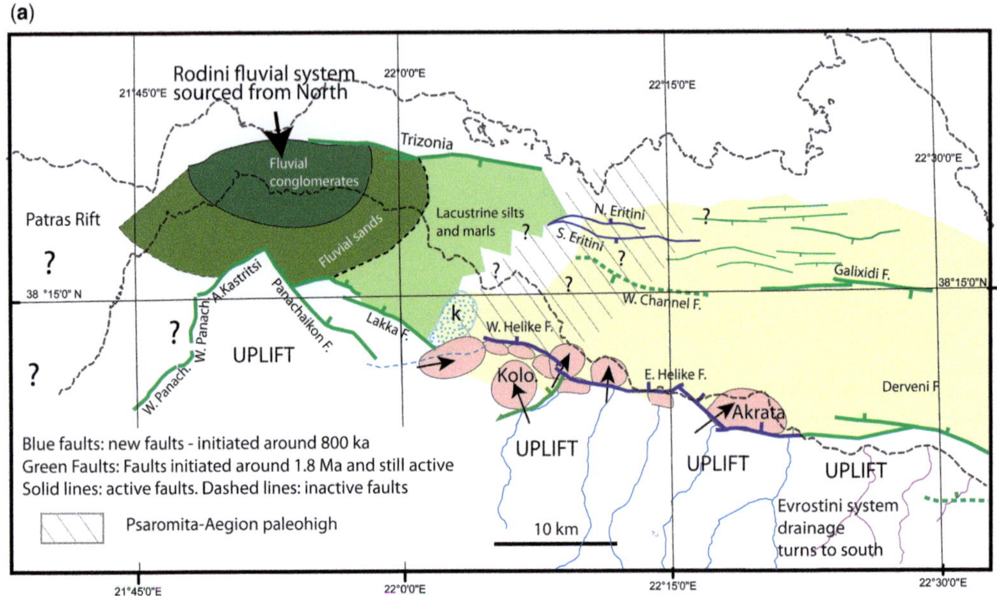

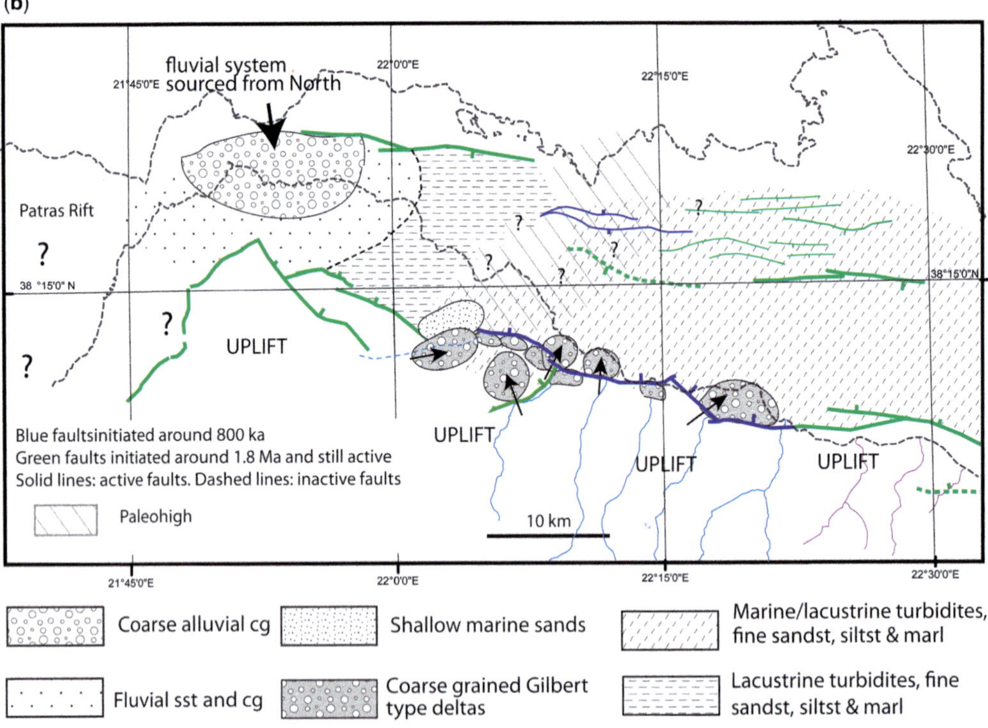

Fig. 12. Reconstruction of the Phase 3 rift (Calabrian–Middle Ionian, 0.8–0.4 Ma) showing that northwards migration of fault activity occurred only in C4 onto new blue faults. New deltas formed in the hanging wall of the new faults. Inactive faults are shown as dashed lines; active faults have solid traces. In zones C3 and C5, no change occurs. Colours correspond to the stratigraphy in Figures 2 and 8. Consequent rivers are shown in purple, while antecedent rivers are in blue. Kolo, Kolokotronis Gilbert delta (Backert *et al.* 2010); k, Koumares Member.
(**b**) Facies distributions during Phase 3. Black arrows indicate delta building directions.

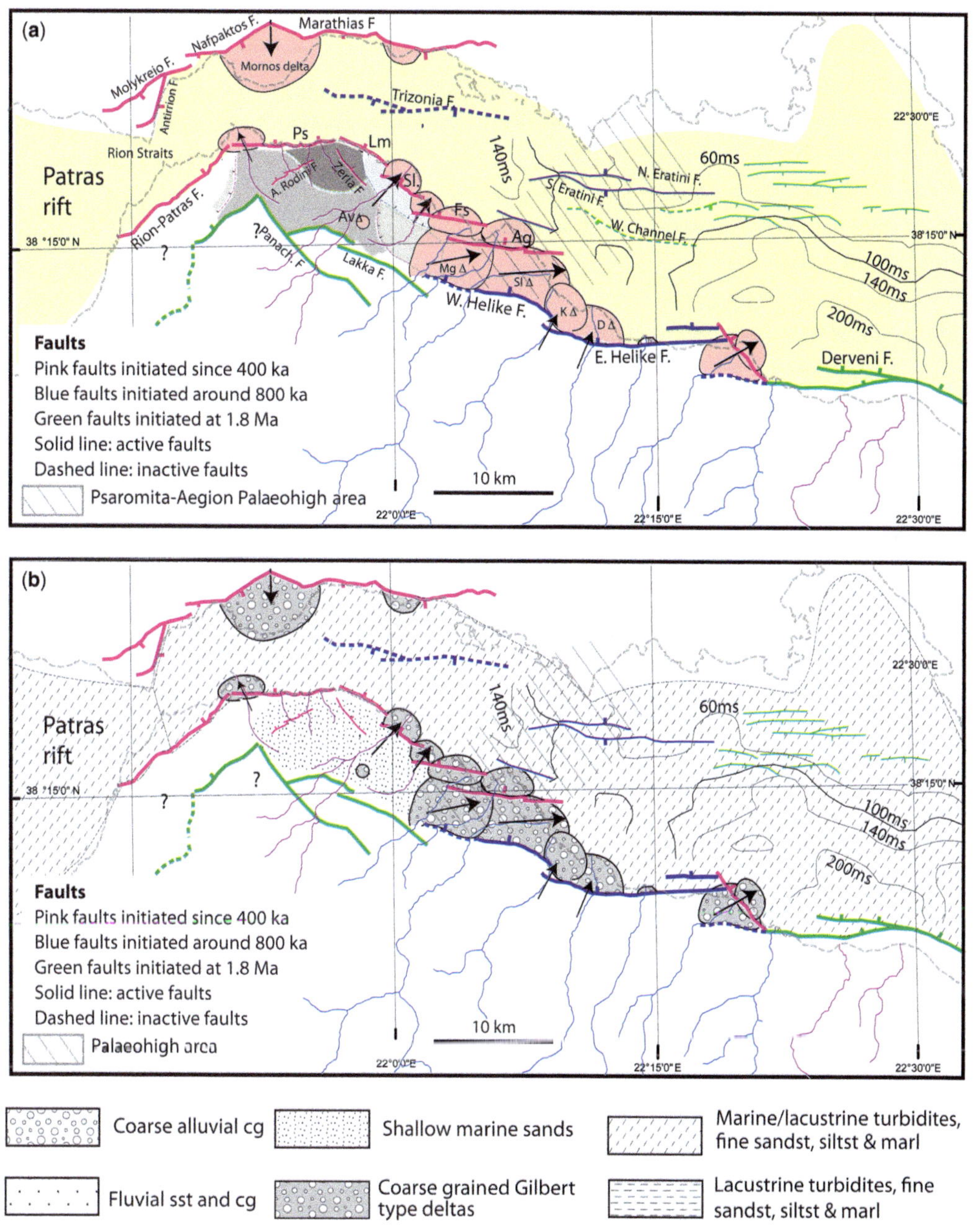

Fig. 13. (**a**) Phase 4 reconstruction (Middle Ionian–present day, 0.4 Ma–0) showing northwards rift migration only in the west (C5), with the initiation of the north-dipping Psathopyrgos (Ps) and Neos Erineos fault systems (Lambiri (Lm), Selianitika (Sl), Fassoulaika (Fs) and Aegion (Ag) faults) and new antithetic faults on the north coast (Marathias and Nafpaktos). The Rion Straits opened at approximately 400 ka due to the initiation of the Ps Fault. To the east in C4 and C3, the main rift remained static with accelerating subsidence concentrated along the southern margin faults. Colours correspond to the stratigraphy in Figures 2 and 8 (deltas are Av, Aravonitsa; Mg, Meganitis; Sl, Selinous; K, Kerinitis; D, Diakopto) with black arrows indicating palaeoflow. Migrating shorelines (Fig. 5c) record the progressive uplift in the footwall of the Neos Erineos fault system (toothed lines: green, 400 ka; pink, 300 ka; blue, 200 ka). Active offshore faults and selected isochores for offshore vertical sediment thickness in milliseconds two-way travel time (TWTT; fine black lines) for the period 135 ka to 0 are taken from Nixon *et al.* (2016). (**b**) Facies distribution in the rift during Phase 4. Black arrows indicate delta building directions.

from the north by the Rodini fluvial system, which may also have supplied the eastern Patras rift (Fig. 11) (Palyvos *et al.* 2007). During Phase 2, the C5 area received little or no sediment from the south. It was limited to the east by the Psaromita–Aegion High, as evidenced by a thin and incomplete succession onlapping pronounced palaeo-relief, major unconformities (e.g. in the AIG10 borehole), and local facies and thickness changes (the Koumares Member). Marine incursions into the depocentre probably came from the east during high sea levels.

Phase 3: Calabrian–Middle Ionian (0.7–0.4 Ma)

In the Middle Pleistocene, at around 0.7 ± 0.2 Ma, the southern basin margin, again, migrated northwards in C3 and C4 by some 5–10 km (Fig. 12). In C3, the West Xylocastro Fault was abandoned as deformation focused onto the pre-existing Derveni Fault, which became the principal focus of extension, linking eastwards with the Lykopora and East Xylocasto faults (Nixon *et al.* 2016). Displacement gradually decreased on south-dipping faults (Galixidi and West Channel). The main C3 depocentre narrowed to around 12 km, as the northern boundary did not migrate, although minor south-dipping faults developed in the footwall of the Galaxidi Fault. Mainly fine-grained prodelta and hemipelagic sediments were deposited here (lower SU2 of Nixon *et al.* 2016).

In C4, the Pirgaki–Mamoussia Fault was abandoned as displacement transferred onto the East and West Helike fault system, which gradually propagated eastwards to link to the Derveni Fault (Fig. 12) (Hemelsdaël & Ford 2016). Deposition increased in the deep basin (lower SU2), but still thinned westwards onto the Psaromita–Aegion High. The North and South Eratini faults developed to the north of the now inactive West Channel Fault. However, these faults played a minor role in the development of the C4 depocentre. The rift in C4 has an estimated width of 10–12 km. In C5, the Lakka–Trizonia depocentre and Psaromita–Aegion High retained the same configuration as in Phase 2.

In C4, major rivers continued to flow north across the uplifting footwall of the East and West Helike faults, cannibalizing their own Phase 2 deltas. Numerous small Gilbert deltas record the progressive uplift of the western PM block that may have locally inherited a bathymetry of up to 400 m basinwards of the large Phase 2 deltas. However, where the new East Helike Fault cut across a major delta (e.g. Vouraikos and Platanos), new deltas began to build immediately into the hanging wall of the new fault. The Krathis River diverted ENE to build the new Akrata delta across the evolving relay ramp between the Derveni and East Helike faults. In C3, however, footwall uplift caused the Evrostini river system to turn south into the Feneos endorheic basin (Rohais *et al.* 2007*a*), thus terminating a major supply route (Killini) into the rift. East of Akrata, no major river supplied coarse-grained sediment into the rift.

Phase 4: Middle Ionian–present day (0.4–0.0 Ma)

During Phase 3 and Phase 4, from approximately 620 ka to present day, activity increased on major north-dipping faults (Derveni–Lykopora, East Xylocastro and North Kiato) to create a single major depocentre in the southern Gulf (C2, C3 and eastern C4). By constructing detailed isochore maps in six steps, Nixon *et al.* (2016) document the progressive evolution and linkage of numerous hanging-wall depocentres along the these faults, which eventually (from around 130 ka) formed a 40 km long half-graben depocentre centred in C2 (isochores from Nixon *et al.* 2016 on Fig. 13). While the succession still thins markedly westwards in C4, owing to the persistent influence of the Psaromita–Aegion High, the basin is now clearly connected to the C5 depocentre.

In C5, fault activity migrated, again, onto the east–west Psathopyrgos Fault at approximately 400 ka, which then propagated east-southeastwards to form the Neos Erineos fault system (Fig. 13). The Aegion Fault is the youngest fault of the system, dated at 50–60 ka (Cornet *et al.* 2004). The northern margin of the C5 rift also migrated north from the Trizonia Fault onto the Marathias and Nafpaktos faults (Beckers *et al.* 2015), which are clearly major faults, although, as yet, not well documented. The westernmost Gulf is controlled by the most recently initiated (youngest) faults. Unlike the main rift, it retains a relatively symmetrical fault system. Highest geodetic extension rates occur here, increasing from 13 mm a^{-1} around Aegion to 16 mm a^{-1} near Lambiri (Lambotte *et al.* 2014). The most intense seismicity is also concentrated below the present-day Gulf in C5 (Bernard *et al.* 2006; Lambotte *et al.* 2014). Seismic data indicate that it is possible that the Panachaikon and Lakka faults may also still be active.

The Rion Straits opened to the Mediterranean for the first time at around 400 ka as a result of the initiation of the Psathopyrgos Fault. Water depth on the sill is currently 60 m, indicating that this seaway would be cut off during sea-level lowstands. The transition from lacustrine to marine conditions in the Gulf at the time of the last sea-level rise, at approximately 12 ka, has been well documented in cored offshore sediment (Moretti *et al.* 2003; Sakellariou *et al.* 2007; Campos *et al.* 2013). The well-stratified nature of the SU2 seismic stratigraphic succession, consisting of highly reflective bands alternating with transparent bands, is attributed to alternating marine and lacustrine conditions (Bell *et al.* 2008; Taylor *et al.* 2011; Nixon *et al.* 2016).

From 400 ka onwards, nearly all sedimentation was concentrated in the Gulf and most probably consisted of predominantly fine-grained prodelta facies and hemipelagics. The eastern rift is clearly sediment-starved, with bathymetry increasing eastwards to over 800 m. In C4 and western C3, however, major rivers continued (and continue today) to flow north to build coarse-grained Gilbert-type deltas into the Gulf across basin-bounding faults. In C2 and C3, younger, less powerful consequent rivers fed the rift: several rivers were turned south by rapid uplift starting at the beginning of Phase 3 (Demoulin *et al.* 2015). In C5, Galada Group deposits record progressive footwall uplift associated with the eastwards propagation of the Neos Erineos fault system (Figs 5, 7 & 8). The resulting eastwards tilting of the Lakka fault block, and passage from uplift to subsidence along the relay ramp between the Neos Erineos and Helike fault systems, is clearly seen in the deviation and building directions of the youngest Gilbert deltas (Selinous and Meganitis: Fig. 13).

Demoulin *et al.* (2015) presented a morphometric analysis of the 10 largest catchments of the north Peloponnese margin. The authors used a wide range of metrics to study the fluvial landscape in order to clarify the timing, location and intensity of uplift episodes along the 100 km-long rift margin. Their results clarified the episodic nature of rift shoulder uplift and they identified three distinct episodes of increasing uplift rate that correspond to major phases of rift migration identified in this study. They propose that the youngest event, with an uplift rate of 2–5 mm a^{-1}, propagated from east to west over the last 20–10 ka and can be related to the location of high seismicity in the westernmost rift. Their intermediate episode is dated at 0.35–0.45 Ma and occurred only in the centre of the rift margin (C4, C3 and C2). We suggest that this may correspond to the transition from Phase 3 to Phase 4 rather than to their proposed transition from Phase 2 to Phase 3. Demoulin *et al.* (2015) proposed that their oldest uplift event may correlate with the transition from Phase 1 to Phase 2.

Interactions of sediment-routing system with the migrating rift

Western Corinth rift stratigraphy records a range of interactions between the river systems and the migrating and evolving rift (Allen 2008). On the largest scale, we identify three major long-lived sediment-routing systems that supplied coarse sediment into the basin, each with a different history (Fig. 9) depending on its interaction with the migrating fault activity. The high volume of coarse sediment supplied by these rivers from the earliest phase indicates that they were not consequent rivers but, rather, antecedent rivers of the Hellenide mountain belt that were captured by the young rift.

- The Killini River fed an axial drainage system, recording hanging-wall migration over at least 3 myr. It reacted to each fault-migration event by building a new Gilbert delta, while incising and cannibalizing its older delta(s) to the south. The river system first built the stacked prograding Killini and Mavro Gilbert delta systems into a shallow lacustrine depocentre over at least 2 myr (Phase 1: Fig. 10). The Killini delta system built mainly eastwards, while the Mavro delta system built northwards above a listric fault. During Phase 2 (Fig. 11), the river system continued to flow northwards, cannibalizing its own deltas to build, first, the Ilias delta toward the NW across a relay zone, and then the Evrostini delta towards the north above a listric fault. Some time after 0.7 Ma, the river was finally defeated by a major uplift event, turning south to flow into the Feneos endorheic basin (Figs 2 & 12), abruptly ending sediment supply into the Corinth rift.
- The Kalavryta river system represents, by far, the largest antecedent drainage system captured by the westwards migrating rift (Fig. 9). The river(s) flowed north and NE into the western termination of the early rift and then turned east to become a longitudinal axial drainage system that built low-height deltas into a shallow lacustrine basin (Fig. 10), much like those seen today in Lake Turkana, Lake Albert and Lake Rukwa in the East African rift system. When the rift migrated west and north in the Early Pleistocene (between Phases 1 and 2), this drainage system was again captured and diverted north, reorganizing into equally spaced rivers, each carrying a coarse bedload (Figs 9 & 11). These powerfully erosive transverse rivers incised the uplifting footwall area, and built the giant Gilbert deltas into the new depocentre. When the rift margin, again, migrated north (Phase 2 to Phase 3 transition: Fig. 12), the rivers were able to continue flowing north, cannibalizing their Phase 2 deltas to build new deltas into the present-day Gulf. They continue to flow into the Gulf today, and represent the principal source of coarse clastic sediment (Fig. 13). When possible, the rivers exploited relay ramps (e.g. Kerinitis and Akrata), but their incision power allowed them to cut across the faults at any point.

- The Rodini river system interacted with a footwall-migrating fault system. In other words, due to footwall migration of the fault system, the river abandoned its older deposits to build a new system in the footwall of the abandoned fault. During Phases 2 and 3, the river supplied the C5 Profitis Elias Group across south-dipping faults and effectively defined the western end of the rift (Figs 11 & 12). When the rift migrated northwards (Phase 4), the river was forced to retreat north, and established the new Mornos delta in the hanging wall of the Nafpaktos and Marathias faults (Fig. 13), which it continues to supply today.

Overall, these three inherited sediment-routing systems have played a primary role in the redistribution of mass in the rift and, thus, in the final grain-size distribution, with coarse-grained facies confined to the western rift where they have been constantly reworked by the Killini and Kalavryta routing systems due to fault migration, In contrast, the eastern rift appears to have been sediment-starved throughout rift development.

Smaller, more local, interactions of rivers with the growing and migrating fault system are numerous in this rift, and include interactions with relay ramps of various scales, consequent drainage on various scales, diversion and abandonment of rivers on various scales due to fault linkage, migration and footwall uplift (Allen 2008). The duration of a river's history in this active rift can vary from very short (some thousands of years) to several millions of years, the longest-lived being those that predate the initiation of the rift itself.

Discussion

Rift migration and lateral propagation

From its inception, the Corinth rift was controlled mainly by north-dipping faults. Although south-dipping faults may have played a significant role during the immature stages of rifting, they become secondary to the major basin-controlling faults once these are established, as seen in areas C2 and C3. In the western C4 area, the south-dipping South Eratini Fault is still an important structure. Figure 14 summarizes the northwards and westwards migration of the major basin-bounding faults of the western rift over 4 myr. The last fault migration in C3 was at 1.8–1.5 Ma, in C4 at 0.7 ± 0.2 Ma and in C5 at 400 ka, demonstrating that the age of initiation of the faults controlling the southern margin of the active rift becomes progressively younger towards the west (Fig. 14). The rift is therefore opening towards the west rather than towards the east, as proposed by Leeder *et al.* (2008, 2012). The zones C3–C5 contain fault systems of decreasing maturity, with the most mature faults in C3 and the most immature faults in C5. This is reflected in a progressively older and thicker synrift fill towards the east. However, this simple prediction is complicated by the presence of the Psaromita–Aegion basement high between C4 and C5 that was probably inherited from the Hellenide fold and thrust belt. The main basin-controlling fault in C2 and C3 (Derveni–Lykopora–East Xylocastro) became stable during Phase 2 and has been accumulating slip since then. Our reconstructions suggest that the westwards propagation of this major fault zone occurred in several steps, each involving the initiation of a new fault some way to the west and north. This new fault then proceeded to propagate towards the east or SE and to eventually link with the major fault. Thus, the West and East Helike faults initiated at 0.7 ± 0.2 Ma, and propagated eastwards to hard link to the Derveni Fault at around 200 ka (Hemelsdaël & Ford 2016). The initiation of the Psathopyrgos Fault further NW at 0.4 Ma and its current propagation SE along the Neos Erineos fault system can be seen as the same process in action. These two stages in westwards fault propagation overlap in time.

Such lateral propagation of a rift is a natural process related to the progressive accumulation of strain over time (Cowie *et al.* 2007). As well as the westwards rift propagation in the Corinth rift, there is also a clear component of northwards fault migration. The combination of these two phenomena is probably responsible for the consistent right-stepping arrangement of active faults along the southern margin of the Gulf (Figs 1 & 14). When seen on a north–south cross-section, fault migration appears to have occurred in discrete events, rather than as a continuous process. The exact timing and duration of all these events are not yet well constrained. We can say, however, that the oldest migration event in C4 between Phases 1 and 2 lasted around 300 kyr. Many authors have observed that fault migration was associated with major uplift events in the northern Peloponnese. A recent study by Demoulin *et al.* (2015) demonstrated that quantitative morphometric analysis of the tectonic landscape may be used to extract the distribution and timing of these uplift events that correspond well with the model presented here. This promising new technique may lead to a more detailed understanding of the fault-migration history of the rift margin.

The large-scale geodynamic cause of northwards rift migration has been widely discussed and many models proposed, including orogenic collapse, slab rollback and back-arc extension, and propagation of the North Anatolian Fault. These models will not be discussed here and the interested reader is referred to publications (Westaway 2002; Leeder

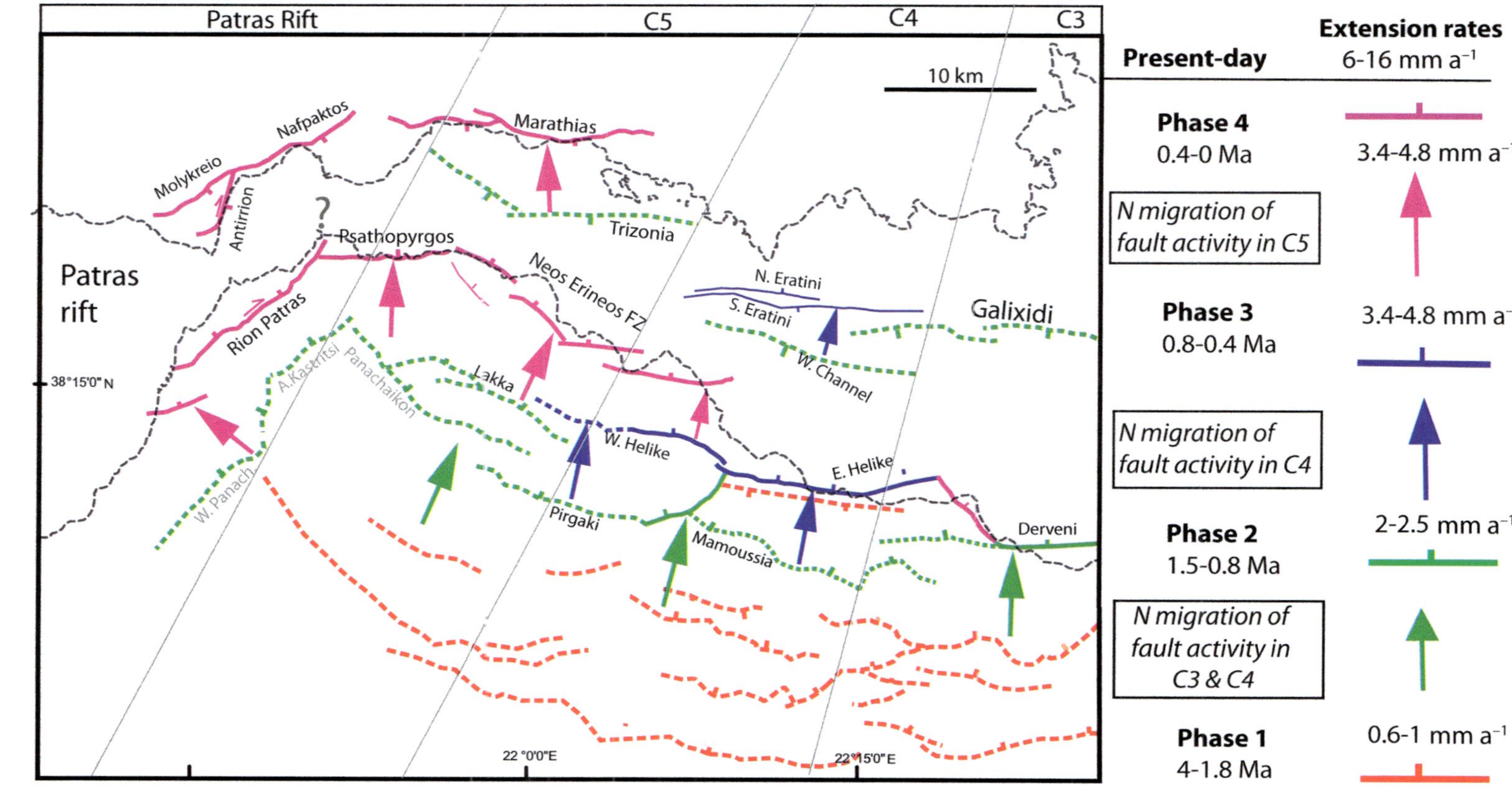

Fig. 14. Summary of rift migration through time based on data in Figures 10–13. Solid lines indicate active faults; dashed lines indicate inactive faults. Fault colour codes indicate the age of initiation of the fault, shown on the right of the figure where a summary of rifting phases is given with an estimated extension rate for each phase (from Ford *et al.* 2013). Dextral displacement on the Rion–Patras and Antirion faults is based on Beckers *et al.* (2015).

et al. 2008, 2012; Jolivet *et al.* 2010; Royden & Papanikolaou 2011) and references therein.

Placing present-day seismicity and geodetic extension rates in the context of full rift evolution

Our reconstructions show that current zone of high seismicity and highest geodetic extension rates in the westernmost rift (C5: Lyon-Caen *et al.* 2004) corresponds to the youngest part of the rift due to the progressive northwards and westwards migration of the fault system. Estimates of the extension rate along the cross-section A–A′ (Fig. 3) (Ford *et al.* 2013) show that rift migration was also associated with an increase in extension rate through time. As new faults were generated for each stage, it follows that strain-rate acceleration is unrelated to fault linkage but, rather, to larger-scale regional plate-tectonic processes (Ford *et al.* 2013). The estimate of extension for Phases 3 and 4 of 3.4–4.8 mm a^{-1} does not, however, approach the current geodetic extension rate of 12–13 mm a^{-1} for the same area (Briole *et al.* 2000; Avallone *et al.* 2004). The initiation of these very high geodetic extension rates along the whole rift may be linked: (a) to a transient acceleration within a seismic cycle (Bernard *et al.* 2006); or (b) to a longer-term external tectonic forcing: for example, due to the lateral propagation of the North Anatolian Fault at depth.

GPS studies have shown that the deformation gradient is localized offshore in a 10 km-wide zone (Briole *et al.* 2000; Avallone *et al.* 2004) and deformation, therefore, is most probably concentrated on the very young north-dipping faults along the southern rift margin. Seismicity is extremely high in C5 because the faults are very young (400 ka to 50–60 ka) and in the process of linking at depth. The microseismic layer at depth may represent an incipient detachment, as proposed by Lambotte *et al.* (2014).

Significance of deepening events

In this paper, we have demonstrated that the western end of the Corinth rift migrated not only northwards but also westwards throughout its evolution. Would such a rift record a single major rift climax event, or would the standard phases of rift evolution – namely, rift initiation followed by rift climax (e.g. Gupta *et al.* 1998; Cowie *et al.* 2007) – be diachronous along the rift axis? Rift initiation is typified by slow subsidence due to distributed deformation across a young network of isolated growing faults (Gupta *et al.* 1998). Early rift fill is typically continental in character (fluvial, shallow lake, shallow marine: Prosser 1993; Lambiase & Bosworth 1995). Rift climax is marked by a distinct acceleration in fault displacement due to the development of a fully linked fault system onto which deformation is focused. Subsidence rate increases, leading to a deepening of depositional environments. Major deepening events in the stratigraphic record of many rifts are therefore interpreted as recording the transition from rift initiation to rift climax.

In the western Corinth rift, four deepening events are identified (Fig. 9) and here we discuss their relevance for local and regional tectonic history. Deepening event 1 (DE1) was identified by Leeder *et al.* (2012) in area C2 and dated as 3.2–3 Ma. They interpreted this event as recording the main Corinth rift climax, correlating it with a similar deepening event to the east. Sedimentologically, this event is characterized by a transition from fluvial to shallow-lacustrine environments (Fig. 9) and we suggest that it probably represents a minor event in area C2. In addition, the event cannot be recognized to the west (in C4), where, at this stage, the rift fill was fluvial in character.

The transition to deeper environments in area C4 (DE2: Fig. 9) occurred in the Early Pleistocene (beginning of Phase 2). In area C4, this second deepening event is recorded in the change from fluvial to deltaic deposition associated with the reorganization of the drainage system. In area C3, delta stratigraphic architecture changed markedly to reflect deepening water (higher foresets, aggradation-dominated). In area C2, the signature of this deepening event is, as yet, unidentified in the monotonous turbidite succession.

Deepening event 3 is recognized in C4 and C5 around the 700 ka is related to the initiation of the East and West Helike faults in C4. Deepening event 4 occurred at approximately 400 ka (between Phases 3 and 4: DE4, Fig. 9) with the initiation of the Psathopyrgos Fault in the westernmost rift (C5), leading to the opening of the Rion Straits. Again, such events are difficult to identify in deeper-water successions, such as those further east. To summarize, area C5 records only two deepening events, C4 records three deepening events and C3 (and C2) records four deepening events (Fig. 9). We have shown that the major linkage and stabilization of the controlling fault network in the centre of the rift occurred at the beginning of Phase 2 associated with deepening event 2 (see also Nixon *et al.* 2016), suggesting that this may be considered as the rift climax transition, while the western rift (C5) remains in the initiation phase.

This study demonstrates that the concepts of rift initiation and rift climax should be used with caution when interpreting rift stratigraphy. Migrating and laterally propagating rifts will record multiple deepening events, which may be quite subtle and difficult to recognize in older, deeper parts of the rift.

Deepening events, therefore, need to be correlated across a basin and linked with local fault-growth history in order to appreciate their relevance in full rift evolution.

Conclusions

This paper presents new data of the westernmost onland section of the Corinth rift (Lakka fault block), and proposes the first complete correlation of the chrono- and lithostratigraphy and structure of the onland central to western Corinth rift. By correlating with offshore basin data, we have built a model of the westwards propagation and northwards migration of the rift since the Late Pliocene. The main points of rift evolution and behaviour can be separated into two packages regarding fault evolution and basin-fill evolution.

Evolution of the fault system

- The rift was controlled by a predominantly asymmetrical, north-dipping fault system. In each zone, south-dipping faults were mainly active during early stages of the fault-system development.
- Four phases of subsidence are separated by distinct northwards (hanging wall) and westwards fault-migration events (see Figs 9 & 14 for a summary).
- Rates of extension across the rift accelerated in distinct steps during rifting, probably due to external tectonic processes.
- Fault migration ended in the east with the stabilization of the major basin-bounding fault system. The age of initiation of this main basin-bounding fault system then becomes younger to the west.
- The major basin-bounding fault system propagated westwards in two stages (during phases 3 and 4). Each stage involves the initiation of a new fault segment to the north and west, which subsequently propagated east or SE to link to the main fault system. These two propagation stages overlap in time. There are no strike-slip faults between rift segments.
- The youngest faults in the westernmost rift are associated with high seismicity and high geodetic extension, probably due to their rapid growth and linkage at depth.

Basin-fill evolution

- Antecedent rivers played a major role in establishing vigorous sediment-routing systems from rift inception. Their influence on facies distributions endured throughout rift history. In particular, powerfully erosive rivers in the western rift constantly reworked their deposits to supply coarse sediment. In contrast, the eastern rift appears to have been sediment-starved since rift initiation.
- The rift was superimposed on significant palaeorelief inherited from the Hellenide mountain belt, that influenced sediment routing, facies distributions and connectivity between depocentres in the early stages of rifting;
- A major reorganization of sediment-routing systems (creation of consequent rivers, capture or defeat of older rivers) occurred during fault-migration events.
- The stratigraphy of a migrating rift can record multiple deepening events. The concepts of rift initiation and rift climax should, therefore, be used with caution.

This paper is dedicated to our collegue Nikolaos Palyvos, whose energy and enthusiasm we sadly miss. Nikos worked on the geomorphology, structure and stratigraphy of the Lakka fault block and Patras area. This work forms part of the SISCOR project funded by the French Agence Nationale de Recherche and led by Pascal Bernard, who we thank for his insight and encouragement. R. Hemelsdaël was financed by a doctoral bursary from the French Ministry of National Education, Higher Education and Research. We thank Rebecca Bell and Christian Beck and editor Conrad Childs for their constructive reviews and support. We thank our colleagues and students for many discussions in the field, in particular Ed Williams and Rob Gawthorpe. Marco Mancini wishes to thank Daniela Pantosti and Paolo Marco De Martini (INGV, Rome) for introducing him to the Corinth rift geology. CRPG Publication Number 2439.

References

Alexandri, N., Nomikou, P., Ballas, D., Lykousis, V. & Sakellariou, D. 2003. Swath bathymetry map of Corinth Gulf. *Geophysical Research Abstracts*, **5** (EGS), 14268.

Allen, P.A. 2008. Time scales of tectonic landscapes and their sediment routing systems. *In*: Gallagher, K., Jones, S.J. & Wainwright, J. (eds) *Landscape Evolution: Denudation, Climate and Tectonics over Different Time and Space Scales*. Geological Society, London, Special Publications, **296**, 7–28, https://doi.org/10.1144/SP296.2

Apostolidis, P.I., Raptakis, D.G., Pandi, K.K., Manakou, M.V. & Pitilakis, K.D. 2006. Definition of subsoil structure and preliminary ground response in Aigion city (Greece) using microtremor and earthquakes. *Soil Dynamics and Earthquake Engineering*, **26**, 922–940, https://doi.org/10.1016/j.soildyn.2006.02.001

Armijo, R., Meyer, B., King, G.C.P., Rigo, A. & Papanastassiou, D. 1996. Quaternary evolution of the Corinth rift and its implications for the late Cenozoic evolution of the Aegean. *Geophysical Journal International*, **126**, 11–53.

AUBOUIN, J., BRUNN, J.H., CELET, P., DERCOURT, J., GODFRIAUX, I. & MERCIER, J. 1963. *Esquisse de la Géologie de la Grèce. Fallot Memorial Volume.* Société Géologique de France, Paris, 583–610.

AVALLONE, A., BRIOLE, P. *ET AL.* 2004. Analysis of eleven years of deformation measured by GPS in the Corinth Rift Laboratory area. *Comptes Rendus Geoscience*, **336**, 301–311, https://doi.org/10.1016/j.crte.2003.12.007

BACKERT, N. 2009. *Interaction tectonique–sédimentation dans Le Rift de Corinthe, Grèce. Architecture stratigraphique et sédimentologie du Gilbert Delta de Kerinitis.* PhD memoir, Institut Polytechique de Lorraine, France.

BACKERT, N., FORD, Ma. & MALARTRE, F. 2010. Architecture and sedimentology of the Kerinitis Gilbert-type fan delta, Corinth Rift, Greece. *Sedimentology*, **57**, 543–586.

BECKERS, A. 2015. *Late Quaternary sedimentation in the western tip of the Gulf of Corinth.* PhD thesis, University of Liège, Belgium.

BECKERS, A., HUBBERT-FERRARI, A., BECK, C., BODEUX, S., TRIPSANAS, E. & DE BATIST, M. 2015. Active faulting at the western top of the Gulf of Corinth, Greece, from high-resolution seismic data. *Marine Geology*, **360**, 1–84.

BELL, R.E., MCNEILL, L., BULL, J.M. & HENSTOCK, T.J. 2008. Evolution of the offshore western Gulf of Corinth. *Geological Society of America Bulletin*, **120**, 156–178.

BELL, R.E., MCNEILL, L., BULL, J.M., HENSTOCK, T.J., COLLIER, R.E.L. & LEEDER, M.R. 2009. Fault architecture, basin structure and evolution of the Gulf of Corinth Rift, central Greece. *Basin Research*, **21**, 824–855, https://doi.org/10.1111/j.1365-2117.2009.00401.x

BELL, R.E., MCNEILL, L.C., HENSTOCK, T.J. & BULL, J.M. 2011. Comparing extension on multiple time and depth scales in the Corinth Rift, Central Greece. *Geophysical Journal International*, **186**, 463–470, https://doi.org/10.1111/j.1365-246X.2011.05077.x

BENTHAM, P., COLLIER, R.E., GAWTHORPE, R.L., LEEDER, R. & STARK, C. 1991. Tectono-sedimentary development of an extensional basin: the Neogene Megara Basin, Greece. *Journal of the Geological Society, London*, **148**, 923–934, https://doi.org/10.1144/gsjgs.148.5.0923

BERNARD, P., MEYER, B. *ET AL.* 1997. The Ms = 6,2 June 15, 1995 Aigiopn earthquake (Greece): evidence for low-angle normal faulting in the Corinth rift. *Journal of Seismology*, **1**, 131–150.

BERNARD, P., LYON-CAEN, H. *ET AL.* 2006. Seismicity, deformation and seismic hazard in the western rift of Corinth: new insights from the Corinth Rift Laboratory (CRL). *Tectonophysics*, **426**, 7–30, https://doi.org/10.1016/j.tecto.2006.02.012

BILLIRIS, H., PARADISSIS, D. *ET AL.* 1991. Geodetic determination of tectonic deformation in central Greece from 1900 to 1988. *Nature*, **350**, 124–129.

BRIOLE, P., RIGO, A. *ET AL.* 2000. Active deformation of the Corinth rift, Greece: results from repeated Global Positioning System surveys between 1990 and 1995. *Journal of Geophysical Research: Solid Earth*, **105**, 25,605–25,625.

BROOKS, M. & FERENTINOS, G. 1984. Tectonics and sedimentation in the Gulf of Corinth and the Zakynos and Kefallinia channels, western Greece. *Tectonophysics*, **101**, 25–54.

BURCHFIEL, B.C. 2008. The Aegean: a natural laboratory for tectonics. *In*: *Donald D. Harrington Symposium on the Geology of the Aegean. IOP Conference Series: Earth and Environmental Science*, **2**, 012001, http://iopscience.iop.org/article/10.1088/1755-1307/2/1/012001

CAMPOS, C., BECK, C., CROUZET, C., CARRILLO, E., VAN WELDEN, A. & TRIPSANAS, E. 2013. Late Quaternary paleoseismic sedimentary archive from deep central Gulf of Corinth: time distribution of inferred earthquake-induced layers. *Annals of Geophysics*, **56**, S0670, https://doi.org/10.4401/ag-6226

CHARALAMPAKIS, M., LYKOUSIS, V., SAKELLARIOU, D., PAPATHEODOROU, G. & FERENTINOS, G. 2014. The tectono-sedimentary evolution of the Lechaion Gulf, the southeastern branch of the Corinth graben, Greece. *Marine Geology*, **351**, 58–75.

CLARKE, P., DAVIES, R. *ET AL.* 1998. Crustal strain in Central Greece from repeated GPS measurements in the interval 1989–1997. *Geophysical Journal International*, **135**, 195–214.

COLLIER, R. & JONES, G. 2003. Rift sequences of the southern margin of the Gulf of Corinth (Greece) as exploration/production analogs. Paper presented at the AAPG International Conference, Barcelona, Spain, 21–24 September 2003.

COLLIER, R.E.Ll. 1990. Eustatic and tectonic controls upon Quaternary coastal sedimentation in the Corinth Basin, Greece. *Journal of the Geological Society, London*, **147**, 301–314, https://doi.org/10.1144/gsjgs.147.2.0301

COLLIER, R.E.Ll. & DART, C.J. 1991. Neogene to Quaternary rifting, sedimentation and uplift in the Corinth Basin, Greece. *Journal of the Geological Society, London*, **148**, 1049–1065, https://doi.org/10.1144/gsjgs.148.6.1049

CORNET, F., DOAN, M.L., MORETTI, I. & BORM, G. 2004. Drilling through the active Aigion Fault: the AIG10 well observatory. *Comptes Rendus de l'Acadamie des Sciences*, **336**, 395–406.

COWIE, P.A., ATTAL, M., TUCKER, G.E., WHITTAKER, A.C., NAYLOR, M., GANAS, A. & ROBERTS, G.P. 2006. Investigating the surface process response to fault interaction and linkage using a numerical modelling approach. *Basin Research*, **18**, 231–266, https://doi.org/10.1111/j.1365-2117.2006.00298.x

COWIE, P.A., ROBERTS, G.P. & MORTIMER, E. 2007. Strain localization within fault arrays over timescales of 100–107 years. *In*: HANDY, M.R., HIRTH, G. & HOVIUS, N. (eds) *Tectonic Faults. Agents of Change on a Dynamic Earth.* MIT Press, Cambridge, MA, 47–75.

DART, C.J., COLLIER, R.E.L., GAWTHORPE, R.L., KELLER, J.V.A. & NICHOLS, G. 1994. Sequence stratigraphic of (?)Pliocene-Quaternary synrift, Gilbert-type fan deltas, northern Peloponnesos, Greece. *Marine and Petroleum Geology*, **11**, 545–560.

DEGNAN, P.J. & ROBERTSON, A.H.F. 1998. Mesozoic-early Tertiary passive margin of the Pindos Ocean (NW Peloponnese, Greece). *Sedimentary Geology*, **117**, 33–70.

De Martini, P.M., Pantosti, D., Palyvos, N., Lemeille, F., McNeill, L.C. & Collier, R.E.L. 2004. Slip rates of the Aigion and Eliki Faults from uplifted marine terraces, Corinth Gulf, Greece. *Comptes Rendus Geoscience*, **336**, 325–334, https://doi.org/10.1016/j.crte.2003.12.006

Demoulin, A., Beckers, A. & Hubert-Ferrari, A. 2015. Patterns of Quaternary uplift of the Corinth rift southern border (N Peloponnese, Greece) revealed by fluvial landscape morphometry. *Geomorphology*, **246**, 188–204.

Dercourt, J. 1964. *Contribution à l'étude géologique du Secteur du Péloponnèse Septentrional.* PhD thesis, Université de Paris.

Dornsiepen, U., Gerolymatos, E. & Jacobshagen, V. 1986. Die Phyllit–Quartzit-Serie im fenster von Feneos (Nord-Peloponnes). *IGME Geological and Geophysical Research*, Special Issue, 99–105.

Doutsos, T. & Kokkalas, S. 2001. Stress and deformation patterns in the Aegean region. *Journal of Structural Geology*, **23**, 455–472.

Doutsos, T. & Piper, D.J.W. 1990. Listric faulting, sedimentation, and morphological evolution of the Quaternary eastern Corinth rift, Greece: first stages of continental rifting. *Geological Society of America Bulletin*, **102**, 812–829.

Doutsos, T. & Poulimenos, G. 1992. Geometry and kinematics of active faults and their seismotectonic significance in the western Corinth-Patras rift (Greece). *Journal of Structural Geology*, **14**, 689–699.

Doutsos, T., Kontopoulos, N. & Poulimenos, G. 1988. The Corinth-Patras rift as the initial stage of continental fragmentation behind an active island arc (Greece). *Basin Research*, **1**, 177–190.

Doutsos, T., Piper, G., Boronkay, K. & Koukouvelas, I.K. 1993. Kinematics of the Central Hellenides. *Tectonics*, **12**, 936–953.

Doutsos, T., Koukouvelas, I.K. & Xypolias, P. 2006. A new orogenic model for the External Hellenides. *In*: Robertson, A.H.F. & Mountrakis, D. (eds) *Tectonic Development of the Eastern Mediterranean Region.* Geological Society, London, Special Publications, **260**, 507–520, https://doi.org/10.1144/GSL.SP.2006.260.01.21

Dufaure, J.J. 1975. *Le Relief Du Péloponnèse.* Thesis, Paris IV.

Esu, D. & Girotti, O. 2015. The late Early Pleistocene non-marine molluscan fauna from the Synania Formation (Achaia, Greece), with description of nine new species. *Archiv für Molluskenkunde*, **114**, 65–81.

Ferentinos, G., Brooks, M. & Doutsos, T. 1985. Quaternary tectonics in the Gulf of Patras, Western Greece. *Journal of Structural Geology*, **7**, 713–717.

Fernandez-Gonzalez, M., Frydas, D., Guernet, C. & Mathieu, R. 1994. Foraminifères et Ostracodes du Plio-Pléistocène de la région de Patras (Grèce). Intérêt stratigraphique et paléogéographique. *Revista Espanola de micropaleontologia*, **1**, 89–107.

Fleury, J. 1980. *Les zones de Gavrovo-Tripolitza et du Pindos (Grece continentale et Peloponnese du Nord). Evolution d'une plateforme et d'un bassin dans leur cadre alpin.* Memoire de la Societé geologique du Nord, **1**.

Flotté, N., Sorel, D., Müller, C. & Tensi, J. 2005. Along strike changes in the structural evolution over a brittle detachment fault: example of the Pleistocene Corinth–Patras rift (Greece). *Tectonophysics*, **403**, 77–94, https://doi.org/10.1016/j.tecto.2005.03.015

Ford, M., Williams, E.A., Malartre, F. & Popescu, S.-M. 2007. Stratigraphic architecture, sedimentology and structure of the Vouraikos Gilbert-type fan delta, Gulf of Corinth, Greece. *In*: Nichols, G., Williams, E.A. & Paola, C. (eds) *Sedimentary Processes, Environments and Basins.* International Association of Sedimentologists, Special Publications, **38**, 49–90.

Ford, M., Rohais, S., Williams, E.A., Bourlange, S., Jousselin, D., Backert, N. & Malartre, F. 2013. Tectonosedimentary evolution of the western Corinth rift (Central Greece). *Basin Research*, **25**, 3–25.

Frydas, D. 1989. Biostratigraphische UNdersuchungen aus dem Neogen der NW und W Pelopo,es, Griechenland. *Nueus Jahrbuch für Geologie und Palaontologie Monatshefte*, **6**, 321–344.

Frydas, D. 1991. Paläontologische und stratigraphische untersuchungen der diatomen des Pleistozäns des N-Peloponnes, Griechenland. *In*: *Proceedings of the 5th Congress*, May 1990, Thessoloniki. *Bulletin of the Geological Society of Greece*, **XXV**(2), 499–513.

Frydas, D., Kontopoulos, N., Stamatopoulos, L., Guernet, C. & Voltaggio, M. 1995. Middle-Late Pleistocene sediments in the northwestern Peloponessus, Greece. A combined study of biostratigraphical, radiochronological and sedimentological results. *Berliner Geowissenschaftliche Abhandlungen*, **E16**, 589–605.

Fuchs, T. 1877. Studien uber die jüngeren Tertiärbildungen Griechenlands. *Denkschriften der Kaiserlichen Akademie der Wissenschaften. Mathematisch–Naturwissenschaftliche Classe*, **37**, 1–42.

Gautier, P., Brun, J.P., Moriceau, R., Sokoutis, D., Martinod, J. & Jolivet, L. 1999. Timing kinematics and cause of the Aegean extension: a scenario based on a comparison with simple analogue experiments. *Tectonophysics*, **315**, 31–72.

Ghisetti, F. & Vezzani, L. 2004. Plio-Pleistocene sedimentation and fault segmentation in the Gulf of Corinth (Greece) controlled by inherited structural fabric. *Comptes Rendus Geosciences*, **336**, 243–249.

Ghisetti, F. & Vezzani, L. 2005. Inherited structural controls on normal fault architecture in the Gulf of Corinth (Greece). *Tectonics*, **24**, TC4016, https://doi.org/10.1029/2004TC001696

Gibbard, P. & Cohen, K.M. 2008. The global chronostratigraphical correlation table for the last 2.7 million years. *Episodes*, **31**, 243–247.

Gierlowski-Kordesch, E.H. 2010. Lacustrine carbonates. *In*: Alonso-Zarza, A.M. & Tanner, L.H. (eds) *Carbonates in Continental Settings: Facies, Environments and Processes.* Developments in Sedimentology, **61**. Elsevier, Amsterdam, 1–101.

Gillet, S. 1963. Nouvelles données sur le gisement villafranchien de Néa-Corinthos. *Praktika tis Akademias Athinon*, **38**, 400–419.

Goldsworthy, M. & Jackson, J. 2001. Migration of activity within normal fault systems: examples from the Quaternary of mainland Greece. *Journal of*

Structural Geology, **23**, 489–506, https://doi.org/10.1016/S0191-8141(00)00121-8

GUERNET, C., LEMEILLE, F., SOREL, D., BOURDILLON, C., BERGE-THIERRY, C. & MANAKOU, M. 2003. Les Ostracodes et le Quaternaire d'Aigion (Golfe de Corinthe, Grèce) [Ostracodes and Quaternary from Aigion (Gulf of Corinth, Greece)]. *Revue de micropaléontologie*, **46**, 73–93, https://doi.org/10.1016/S0035-1598(03)00013-8

GUPTA, S., COWIE, P., DAWERS, N. & UNDERHILL, J. 1998. A mechanism to explain rift-basin subsidence and stratigraphic patterns through fault-array evolution. *Geology*, **26**, 595–598.

HARDY, S. & GAWTHORPE, R.L. 1998. Effects of variations in fault slip rate on sequence stratigraphy in fan deltas: insights from numerical modelling. *Geology*, **26**, 911–914.

HEMELSDAËL,R. & FORD, M. 2016. Relay zone evolution: a history of repeated fault propagation and linkage, central Corinth rift, Greece. *Basin Research*, **28**, 34–56, https://doi.org/10.1111/bre.12101

HEMELSDAËL, R., FORD, M., MALARTRE, F., GAWTHORPE, R., CHARREAU, J. & SEN, S. 2015. Rivers and rifting: evolution of a fluvial system during rift initiation, central Corinth rift (Greece). AAPG Search and Discovery Article 30419. AAPG Annual Convention and Exhibition, 31 May–3 June 2015, Denver, CO.

JACKSON, J.A., GAGNEPAIN, J., HOUSEMAN, G., KING, G.C.P., PAPADIMITRIOU, P., SOUFLERIS, C. & VIRIEUX, J. 1982. Seismicity, normal faulting and the geomorphological development of the Gulf of Corinth (Greece): the Corinth earthquakes of February and March 1981. *Earth and Planetary Science Letters*, **57**, 377–397.

JOLIVET, L. 2001. A comparison of geodetic and finite strain pattern in the Aegean, geodynamic implications. *Earth and Planetary Science Letters*, **187**, 95–104.

JOLIVET, L., BRUN, J.P., GAUTIER, P., LALLEMEANT, S. & PATRIAT, M. 1994. 3D kinematics of extension in the Aegean region from the early Miocene to the present, insights from the ductile crust. *Bulletin de la Societe Geologique de France*, **165**, 195–209.

JOLIVET, L., LABROUSSE, L. *ET AL.* 2010. Rifting and shallow-dipping detachments, clues from the Corinth Rift and the Aegean. *Tectonophysics*, **483**, 287–304, https://doi.org/10.1016/j.tecto.2009.11.001

KERAUDREN, B. & SOREL, D. 1987. The terraces of Corinth (Greece): a detailed record of eustatic sea level variations during the last 500 000 years. *Marine Geology*, **77**, 99–107.

KOSKERIDOU, E. & IOAKIM, C. 2009. An Early Pleistocene mollusc fauna with Ponto- Caspian elements, in intra Hellenic Basin of Atalanti, Arkitsa Region (Central Greece). *In*: *9th Panhellenic Symposium Oceanography & Fisheries, 2009 – Proceedings*. HCMR Publications **1**, 96–101.

LAMBIASE, J.J. & BOSWORTH, W. 1995. Structural controls on sedimentation in continental rifts. *In*: LAMBIASE, J.J. (ed.) *Hydrocarbon Habitat in Rift Basins*. Geological Society, London, Special Publications, **80**, 117–144, https://doi.org/10.1144/GSL.SP.1995.080.01.06

LAMBOTTE, S., LYON-CAEN, H. *ET AL.* 2014. Reassessment of the rifting process in the Western Corinth Rift from relocated seismicity. *Geophysical Journal International*, **197**, 1822–1844, https://doi.org/10.1093/gji/ggu096

LEEDER, M.R., PORTMAN, C. *ET AL.* 2005. Normal faulting and crustal deformation, Alkyonides Gulf and Perachora peninsula, eastern Gulf of Corinth rift, Greece. *Journal of the Geological Society, London*, **162**, 549–561, https://doi.org/10.1144/0016-764904-075

LEEDER, M.R., MACK, G.H., BRASIER, A.T., PARRISH, R.R., MCINTOSH, W.C., ANDREWS, J. & DUERMEIJER, C.E. 2008. Late-Pliocene timing of Corinth (Greece) rift-margin fault migration. *Earth and Planetary Science Letters*, **274**, 132–141, https://doi.org/10.1016/j.epsl.2008.07.006

LEEDER, M.R., MARK, D.F. *ET AL.* 2012. A 'Great Deepening': chronology of rift climax, Corinth rift, Greece. *Geology*, **40**, 999–1002, https://doi.org/10.1130/G33360.1

LEMEILLE, F., CHATOUPIS, F. *ET AL.* 2004. Recent syn-rift deposits in the hangingwall of the Aigion Fault (Gulf of Corinth, Greece). *Comptes Rendus Geoscience*, **336**, 425–434.

LE PICHON, X. & ANGELIER, J. 1979. The Hellenic arc and trench system: a key to the neotectonic evolution of the eastern Mediterranean area. *Tectonophysics*, **60**, 1–42.

LE PICHON, X. & ANGELIER, J. 1981. The Aegean Sea. *Philosophical Transactions of the Royal Society of London*, **A300**, 357–372.

LYKOUSIS, V., SAKELLARIOU, D., MORETTI, I. & KABERI, H. 2007*a*. Late Quaternary basin evolution of the Gulf of Corinth: sequence stratigraphy, sedimentation, fault–slip and subsidence rates. *Tectonophysics*, **440**, 29–51, https://doi.org/10.1016/j.tecto.2006.11.007

LYKOUSIS, V., SAKELLARIOU, D. *ET AL.* 2007*b*. Sediment failure processes in active grabens: the western Gulf of Corinth (Greece). *In*: LYKOUSIS, V., SAKELLARIOU, D. & LOCAT, J. (eds) *Submarine Mass Movements and Their Consequences*. Advances in Natural and Technological Hazards Research, **27**. Springer, Dordrecht, 297–305, https://doi.org/10.1007/978-1-4020-6512-5_31

LYON-CAEN, H., PAPADIMITRIOU, P., DESCHAMPS, A., BERNARD, P., MAKROPOULOS, K., PACCHIANI, F. & PATAU, G. 2004. First results of the CRLN seismic network in the western Corinth Rift: evidence for old-fault reactivation. *Comptes Rendus Geoscience*, **336**, 343–351, https://doi.org/10.1016/j.crte.2003.12.004

MALARTRE, F., FORD, M. & WILLIAMS, E.A. 2004. Preliminary biostratigraphy and 3D geometry of the Vouraikos Gilbert-type fan delta, Gulf of Corinth, Greece. *Comptes Rendus Geoscience*, **336**, 269–280, https://doi.org/10.1016/j.crte.2003.11.016

MASTRONUZZI, G., SANSO, P. & STAMATOPOULOS, L. 1994. The glacial landforms of the Peloponnisos (Greece). *Rivista geografica Italiana*, **101**, 77–86.

MCCLUSKY, S., BALASSANIAN, S. *ET AL.* 2000. Global Positioning System constraints on plate kinematics and dynamics in Caucasus and the eastern Mediterranean. *Journal of Geophysical Research*, **105**, 5695–5719.

MCNEILL, L.C. & COLLIER, R.E.L. 2004. Uplift and slip rates of the eastern Eliki fault segment, Gulf of Corinth, Greece, inferred from Holocene and Pleistocene terraces 1. *Journal of the Geological Society, London*, **161**, 81–92, https://doi.org/10.1144/0016-764903-029

McNeill, L.C., Cotterill, C.J. *et al.* 2005. Active faulting within the offshore western Gulf of Corinth, Greece: implications for model of continental rift deformation. *Geology*, **33**, 241–244.

Miall, A.D. 1996. *The Geology of Fluvial Deposits. Sedimentary Facies, Basin Analysis, and Petroleum Geology*. Springer, Berlin.

Micarelli, L. 2003. Structural properties of rift-related normal faults: the case study of the Gulf of Corinth, Greece. *Journal of Geodynamics*, **36**, 275–303, https://doi.org/10.1016/S0264-3707(03)00051-6

Moretti, I., Sakellariou, D., Lykousis, V. & Micarelli, L. 2003. The Gulf of Corinth: an active half graben? *Journal of Geodynamics*, 1–18, https://doi.org/10.1016/S0264-3707(03)00053-X

Moretti, I., Lykousis, V., Sakellariou, D., Reynaud, J.Y., Benziane, B. & Prinzhoffer, A. 2004. Sedimentation and subsidence rate in the Gulf of Corinth: what we learn from the *Marion Dufresnés* long-piston coring. *Comptes Rendus Geoscience*, **336**, 291–299, https://doi.org/10.1016/j.crte.2003.11.011

Nixon, C.W., McNeill, L.C. *et al.* 2016. Rapid spatiotemporal variations in rift structure during development of the Corinth Rift, central Greece. *Tectonics*, **35**, first published online May 24, 2016, https://doi.org/10.1002/2015TC004026

Nocquet, J.M., 2012. Present-day kinematics of the Mediterranean: a comprehensive overview of GPS results. *Tectonophysics*, **579**, 220–242.

Nomikou, P., Alexandri, M., Lykousis, V., Sakellariou, D. & Ballas, D. 2011. Swath bathymetry and morphological slope analysis of the Corinth Gulf. *In*: Grützner, C. Pérez-López, R. Fernandez Steeger, T. Papanikolaou, I. Reicherter, K. Silva, P.G. & Vött, A. (eds) *Earthquake Geology and Archaeology: Science, Society and Critical Facilities*. 2nd INQUA-IGCP-567 International Workshop on Active Tectonics, Earthquake Geology, Archeology and Engineering, 19–24 September 2011, Corinth, Greece, 155–158.

North, C.P. & Davidson, S.K. 2012. Unconfined alluvial flow processes: recognition and interpretation of their deposits, and the significance for palaeogeographic reconstruction. *Earth-Science Reviews*, **111**, 199–223.

Ori, G.G. 1989. Geological history of the extensional basin of the Gulf of Corinth (?Miocene–Pleistocene), Greece. *Geology*, **17**, 918–921.

Ori, G.G., Roveri, M. & Nichols, G. 1991. Architectural patterns in large-scale Gilbert-type delta complexes, Pleistocene, Gulf of Corinth, Greece. *In*: Miall, A.D. & Tyler, N. (eds) *The Three-Dimensional Facies Architecture of Terrigenous Clastic Sediments and Its Implications for Hydrocarbon Discovery and Recovery*. Concepts in Sedimentology and Paleontology, **3**. Society for Sedimentary Geology (SEPM), 207–216.

Palyvos, N. 2005. The Aigion–Neos Erineos coastal normal fault system (western Corinth Gulf Rift, Greece): geomorphological signature, recent earthquake history, and evolution. *Journal of Geophysical Research*, **110**, B09302, https://doi.org/10.1029/2004JB003165

Palyvos, N., Sorel, D. *et al.* 2007. Review and new data on uplift rates at the W termination of the Corinth Rift and the NE Rion graben area (Achaia, NW Peloponnesos). *In*: *Proceedings of the 11th International Congress*, May 2007, Athens. *Bulletin of the Geological Society of Greece*, **40**, 412–424.

Palyvos, N., Mancini, M., Sorel, D., Lemeille, F., Pantosti, D. & Julia, R. 2010. Geomorphological, stratigraphic and geochronological evidence of fast Pleistocene coastal uplift in the westernmost part of the Corinth Gulf Rift (Greece). *Geological Journal*, **45**, 78–104, https://doi.org/10.1002/gj.1171.

Palyvos,N., Ford, M., Mancini, M., Esu, D., Girotti, O. & Urban, B. 2013. Western closure of the Corinth Rift: stratigraphy and structure of the Lakka fault block. *Geophysical Research Abstracts*, **15**, EGU2013-5360.

Pirazzoli, P.A., Stiros, S.C., Fontugne, M. & Arnold, M. 2004. Holocene and Quaternary uplift in the central part of the southern coast of the Corinth Gulf (Greece). *Marine Geology*, **212**, 35–44.

Prosser, S. 1993. Rift-related linked depositional systems and their seismic expression. *In*: Williams, G.D. & Dobb, A. (eds) *Tectonics and Seismic Sequence Stratigraphy*. Geological Society, London, Special Publications, **71**, 35–66, https://doi.org/10.1144/GSL.SP.1993.071.01.03

Richter, D. 1976. Das Flysch-Stadium der Helleniden-Ein Uberblick. *Zeitung deutsche der Geologische Gesellschaft*, **127**, 96–128.

Rigo, A., Lyon-Caen, H. *et al.* 1996. A microseismic study in the western part of the gulf of Corinth (Greece) implication for large-scale normal faulting mechanisms. *Geophysical Journal International*, **126**, 663–688.

Ritchie, B.D., Gawthorpe, R.L. & Hardy, S. 2004. Three-dimensional numerical modelling of deltaic depositional sequences 2: influence of local controls. *Journal of Sedimentary Research*, **74**, 221–238.

Roberts, G.P. 1996. Non-characteristic normal faulting surface ruptures from the Gulf of Corinth Greece. *Journal of Geophysical Research*, **101**, 25,255–25,267.

Roberts, G.P. & Michetti, A.M. 2004. Spatial and temporal variations in growth rates along active normal fault systems: an example from The Lazio-Abruzzo Apennines, central Italy. *Journal of Structural Geology*, **26**, 339–376.

Rohais, S., Eschard, R., Ford, M., Guillocheau, F. & Moretti, I. 2007*a*. Stratigraphic architecture of the Plio-Pleistocene infill of the Corinth Rift: implications for its structural evolution. *Tectonophysics*, **440**, 5–28.

Rohais, S., Joannin, S., Colin, J.P., Suc, J.P., Guillocheau, F. & Eschard, R. 2007*b*. Age and environmental evolution of the syn-rift fill of the southern coast of the Gulf of Corinth (Akrata-Derveni region, Greece). *Bulletin de la Société Géologique de France*, **178**, 231–243.

Royden, L.H. & Papanikolaou, D.J. 2011. Slab segmentation, late Cenozoic disruption of the Hellenic arc. *Geochemistry, Geophysics Geosystems*, **12**, https://doi.org/10.1029/2010GC003280

Sachpazi, M., Clément, C., Laigle, M., Hirn, A. & Roussos, N. 2003. Rift structure, evolution, and earthquakes in the Gulf of Corinth, from reflection seismic images. *Earth and Planetary Science Letters*, **216**, 243–257, https://doi.org/10.1016/S0012-821X(03)00503-X

Sachpazi, M., Galvé, A. *et al.* 2007. Moho topography under central Greece and its compensation by Pn time-terms for the accurate location of hypocenters: the example of the Gulf of Corinth 1995 Aigion earthquake. *Tectonophysics*, **440**, 53–65, https://doi.org/10.1016/j.tecto.2007.01.009

Sakellariou, D., Lykousis, V. *et al.* 2007. Faulting, seismic-stratigraphic architecture and Late Quaternary evolution of the Gulf of Alkyonides Basin, East Gulf of Corinth, Central Greece. *Basin Research*, **19**, 273–295, https://doi.org/10.1111/j.1365-2117.2007.00322.x

Seger, M.J. & Alexander, J. 1993. Distribution of Plio-Pleistocene and modern coarse grained deltas south of the Gulf of Corinth, Greece. *In*: Frostick, L. & Steel, R. (eds) *Tectonic Controls and Signatures in Sedimentary Successions*. International Association of Sedimentologists, Special Publications, **20**, 37–48.

Skourlis, K. & Doutsos, T. 2003. The Pindos fold and thrust belt (Greece): inversion kinematics of a passive continental margin. *International Journal of Earth Sciences*, **92**, 891–903.

Skourtsos, E. & Kranis, H. 2009. Structure and evolution of the western Corinth Rift, through new field data from the Northern Peloponnesus. *In*: Ring, U. & Wernicke, B. (eds) *Extending a Continent: Architecture, Rheology and Heat Budget*. Geological Society, London, Special Publications, **321**, 119–138, https://doi.org/10.1144/SP321.6

Sorel, D., Pantosti, D., Lemeille, F., Palyvos, N. & De-Martini, P.-M. 2005. The step-over between the Eliki and Aigion fault systems: slip transfer and present rates of activity. Report for CNRS project GDR Corinthe No. 234.

Stefatos, A., Papatheodorou, G., Ferentinos, G. & Leeder, M. & Collier, R. 2002. Seismic reflection imaging of active offshore faults in the Gulf of Corinth: their seismotectonic significance. *Basin Research*, **14**, 487–502.

Stewart, I. & Vita-Finzi, C. 1996. Coastal uplift on active normal faults: the Eliki fault, Greece. *Geophysical Research Letters*, **23**, 1853–1856.

Stewart, S.A. 1996. Influence of detachment layer thickness on style of thin-skinned shortening. *Journal of Structural Geology*, **18**, 1271–1274.

Taylor, B., Weiss, J.R., Goodliffe, A.M., Sachpazi, M., Laigle, M. & Hirn, A. 2011. The structures, stratigraphy and evolution of the Gulf of Corinth rift, Greece. *Geophysical Journal International*, **185**, 1189–1219, https://doi.org/10.1111/j.1365-246X.2011.05014.x

Trotet, F., Goffé, B., Vidal, O. & Jolivet, L. 2006. Evidence of retrograde Mg-carpholite in the phyllite-quartzite nappe of the Peloponnese from thermobarometric modelisation – geodynamic implications. *Geodynamica Acta*, **19**, 323–343.

Ulicny, D., Nichols, G. & Waltham, D. 2002. Role of initial depth at basin margins in sequence architecture: field examples and computer models. *Basin Research*, **14**, 347–360.

van Hinsbergen, D.J.J. & Schmid, S.M. 2012. Map view restoration of Aegean-West Anatolian accretion and extension since the Eocene. *Tectonics*, **31**, TC5005, https://doi.org/10.1029/2012TC003132

van Hinsbergen, D.J.J., Krijgsman, W., Langereis, C.G., Cornée, J.J., Duermeijer, C.E. & Van Vugt, N. 2007. Discrete Plio-Pleistocene phases of tilting and counterclockwise rotation in the southeastern Aegean arc (Rhodos, Greece): early Pliocene formation of the south Aegean left-lateral strike-slip system. *Journal of the Geological Society, London*, **164**, 1133–1144, https://doi.org/10.1144/0016-76492006-061

Westaway, R. 2002. The Quaternary evolution of the Gulf of Corinth, central Greece: coupling between surface processes and flow in the lower continental crust. *Tectonophysics*, **348**, 269–318.

Zelilidis, A. 2000. Drainage evolution in a rifted basin, Corinth graben, Greece. *Geomorphology*, **35**, 69–85.

Zelilidis, A. & Kontopoulos, N. 1996. Significance of fan deltas without toe-sets within rift and piggy-back basins: examples from the Corinth graben and the Meso-hellenic trough, Central Greece. *Sedimentology*, **43**, 253–262.

Zelt, B.C., Taylor, B., Sachpazi, M. & Hirn, A. 2005. Crustal velocity and Moho structure beneath the Gulf of Corinth. Greece. *Geophysical Journal International*, **162**, 257–268, https://doi.org/10.1111/j.1365-246X.2005.02640.x

Interaction between gravity-driven listric normal fault linkage and their hanging-wall rollover development: a case study from the western Niger Delta, Nigeria

HAMED FAZLIKHANI[1]*, STEFAN BACK[2], PETER A. KUKLA[2] & HAAKON FOSSEN[1,3]

[1]*Department of Earth Science, University of Bergen, Postboks 7803, 5007 Bergen, Norway*

[2]*Geological Institute, EMR, RWTH Aachen University, 52062 Aachen, Germany*

[3]*Museum of Natural History, University of Bergen, Postboks 7803, 5007 Bergen, Norway*

**Correspondence: hamed.khani@uib.no*

Abstract: Rollover is the folding of the hanging-wall sedimentary record in response to slip on listric normal faults, and is a common feature of sediment-rich, gravity-driven tectonic provinces. Rollovers have been extensively studied by means of geometrical reconstruction, and numerical and analogue modelling. However, the detailed interaction between the kinematics of bounding listric normal faults and their hanging-wall deformation is not yet fully understood. In this study, we use 3D seismic-reflection data from the Forcados-Yokri area, western Niger Delta, Nigeria, to study the lateral linkage and landwards backstepping history of an array of listric normal faults, particularly focusing on their influence on the development and evolution of hanging-wall rollovers. Five individual, partly overlapping rollover structures have been studied with respect to their relative initiation and decay time, their spatial distribution, and their relationship to the tectonic history of their respective bounding faults. We demonstrate that the studied rollovers are highly dependent on the development of their bounding faults in terms of initiation time, lateral linkage, internal structural development and decay. Fault–rollover interaction is dynamic and changes through time depending on the temporal evolution of listric faults. Four genetic types of fault–rollover interaction were identified in this study: (1) the rotation of a rollover–crestal-collapse system, controlled by a changing lateral bounding-fault orientation during fault growth; (2) a stepwise shift of rollover–crestal-collapse systems associated with rollover abandonment, controlled by the initiation of a new fault in the footwall of an older structure; (3) a gradual shift of successive rollovers controlled by branching main faults; and (4) a general landwards and upwards migration of crestal-collapse faults within a rollover above stationary listric main faults.

Listric normal faults are characteristic of gravity-driven thin-skinned structural domains, including the Niger Delta (e.g. Doust & Omatsola 1989; Dula 1991; Morley & Guerin 1996; Rouby & Cobbold 1996; Hooper *et al.* 2002; Back *et al.* 2006; Fazli Khani & Back 2012; Sapin *et al.* 2012), the Nile Delta (e.g. Sestini 1989; Beach & Trayner 1991; Marten *et al.* 2004), the Mahakam Delta (e.g. Dooley *et al.* 2000), the Baram Delta province of NW Borneo (e.g. Sandal 1996; Van Rensbergen & Morley 2000; Hodgetts *et al.* 2001; Imber *et al.* 2003; Morley *et al.* 2003; Saller & Blake 2003; Back *et al.* 2005, 2008; Sapin *et al.* 2012) and the Gulf of Mexico (e.g. Lopez 1990; Cartwright *et al.* 1998; Brown *et al.* 2004; Shen *et al.* 2016). Deposition of denser sediments, mainly sandstones above water-saturated shales or evaporites (weaker substratum), create a gravitational instability that is commonly accommodated by synsedimentary normal faults (Thorsen 1963; Bruce 1973; Edwards 1976; Lowell 1985; McCulloh 1988; Morley & Guerin 1996; Rouby & Cobbold 1996; Van Rensbergen & Morley 2000; Back & Morley 2016). Space created by the initiation of normal faults will be filled continuously by sediments (Fazli Khani & Back 2015*a*), and the additional load represented by these sediments helps to maintain slip on the fault. Progressive slip on the listric fault usually triggers the initiation of a rollover anticline in the hanging-wall domain (Gibbs 1984; Xiao & Suppe 1992; Mauduit & Brun 1998). Rollover anticlines and associated crestal-collapse faults can thus be interpreted as the consequence of stratal bending due to slip along listric normal faults with synchronous sedimentation. In sediment-rich deltaic settings, the synkinematic sedimentary record of rollovers can therefore preserve, for example, the consequences of lateral fault growth and linkage, changes in the strike of the bounding fault or the decay of slip along growth faults through time. The interaction

From: Childs, C., Holdsworth, R. E., Jackson, C. A.-L., Manzocchi, T., Walsh, J. J. & Yielding, G. (eds) 2017. *The Geometry and Growth of Normal Faults*. Geological Society, London, Special Publications, **439**, 169–185.
First published online December 13, 2016, https://doi.org/10.1144/SP439.20

between a listric normal fault and its hanging-wall rollover has been demonstrated by analogue modelling (e.g. Ellis & McClay 1988; Vendeville & Cobbold 1988; McClay 1990; Withjack *et al.* 1995; Withjack & Schlische 2006), numerical modelling (White *et al.* 1986; Xiao & Suppe 1992; Ings & Beaumont 2010) and by 3D seismic interpretation (Hodgetts *et al.* 2001; Imber *et al.* 2003; Back *et al.* 2006; Pochat *et al.* 2009; Fazli Khani & Back 2012, 2015*a*). The study of Imber *et al.* (2003) on a listric growth-fault system from SE Asia is particularly important in that it demonstrates in one part of their study area a prolonged and progressive landwards migration of active rollover in the hanging wall of a stationary main fault, while in another there is a punctuated migration directly related to the stepping of the bounding fault. Fazli Khani & Back (2012) documented a similar dynamic rollover development tied to bounding-fault migration in the Niger Delta, highlighting the landwards rollover migration and complexities in a system affected by coinciding seawards and landwards fault migration. This study (using the same dataset described by Fazli Khani & Back 2012, 2015*a*, *b*) focuses on a detailed analysis of the migration, lateral linkage and geometrical changes associated with multiple shifting, growing and linking bounding faults that affect the development of several adjacent, partly overlapping rollovers. We present, as examples, five coupled growth-fault rollover systems that have been identified on 3D seismic-reflection data from the western Niger Delta (Fig. 1). In these systems, bounding-fault growth and linkage can be documented by fault-kinematic analysis (throw–distance and throw–depth plots) and analysis of time–thickness maps. These data are subsequently analysed with respect to the development of the associated rollover and crestal-collapse systems, documenting significant diversity between neighbouring, partly contemporaneous, detaic rollovers.

Geological framework

The Niger Delta is one of the Earth's largest Cenozoic delta systems. It formed during the separation of South America from Africa during Early Cretaceous times (Whiteman 1982; Fairhead & Binks 1991). It is located on the West African continental margin at the apex of the Gulf of Guinea (Fig. 1a). Deltaic sedimentation started during the Late Eocene (Burke 1972; Whiteman 1982; Damuth 1994) at the SW (seawards) edge of the Benue Trough. The delta succession comprises a highly progradational, generally upwards-coarsening association of Cenozoic clastics up to 12 km thick (Doust & Omatsola 1989). The Niger Delta lithostratighraphy is subdivided into three major units from the Paleocene to recent in age, comprising: (1) basal marine pro-delta shales of the Akata Formation; (2) sandstone-rich paralic siliciclastics of delta-front, delta-topset and fluvio-deltaic environments of the Agbada Formation; and (3) alluvial and upper coastal plain sandstones of the Benin Formation (Fig. 1b) (cf. Short & Staeuble 1967; Evamy *et al.* 1978; Whiteman 1982). The delta stratigraphy and structure are intimately related, with the development of each being dependent on the interplay between sediment supply and subsidence (Doust 1990). The study area is located in the extensional, gravity-driven coastal structural domain of the delta (Fig. 1a, c), in which the progradation of the deltaic sedimentary wedge over basal marine shales caused the formation of numerous kilometre-scale synsedimentary growth faults (Fig. 1b) (Doust & Omatsola 1989; Damuth 1994; Hooper *et al.* 2002).

Figure 1c shows an example of a horizon slice that highlights the occurrence of faults in the study area, emphasizing a series of major arcuate-shaped, seawards-dipping, normal deltaic faults that are, in places, associated with hanging-wall rollovers. The NW part of the study area is characterized by several arcuate-shaped normal faults that extend laterally over several kilometres (Table 1), dividing the area into four fault blocks (Fig. 1c). In contrast, the central and SE parts of the study area are characterized by deltaic rollovers with collapsed crests that are bound on the landwards side by a series of sub-parallel, seawards-dipping, listric growth faults (Fig. 2). On the seawards side, the rollover province is bound by a large, slightly listric, seawards-dipping fault system (SE segment of fault F1: Fig. 2). The vertical seismic sections of Figure 2 illustrate the relationships between the fault architecture, the hanging-wall rollover geometries and the syntectonic stratigraphic record. All major bounding faults in the study area show a synsedimentary growth signature: that is, they comprise thickened or additional sedimentary units on their respective downthrown sides.

Datasets and methods

The 3D seismic-reflection data used for this study are from the uppermost 4 km of an approximately 400 km^2 survey area in the coastal zone of the western Niger Delta (Fig. 1). The seismic-reflection data have been processed using pre-stack time migration, and are in European zero-phase polarity convention (i.e. a downwards increase in acoustic impedance corresponds to a seismic trough, which is indicated in red in the colour figures in this paper). Variance attribute volumes were derived from the reflectivity data using a semblance algorithm that highlights lateral amplitude variations between adjacent seismic traces (e.g. Fig. 1c). Detailed mapping and analysis

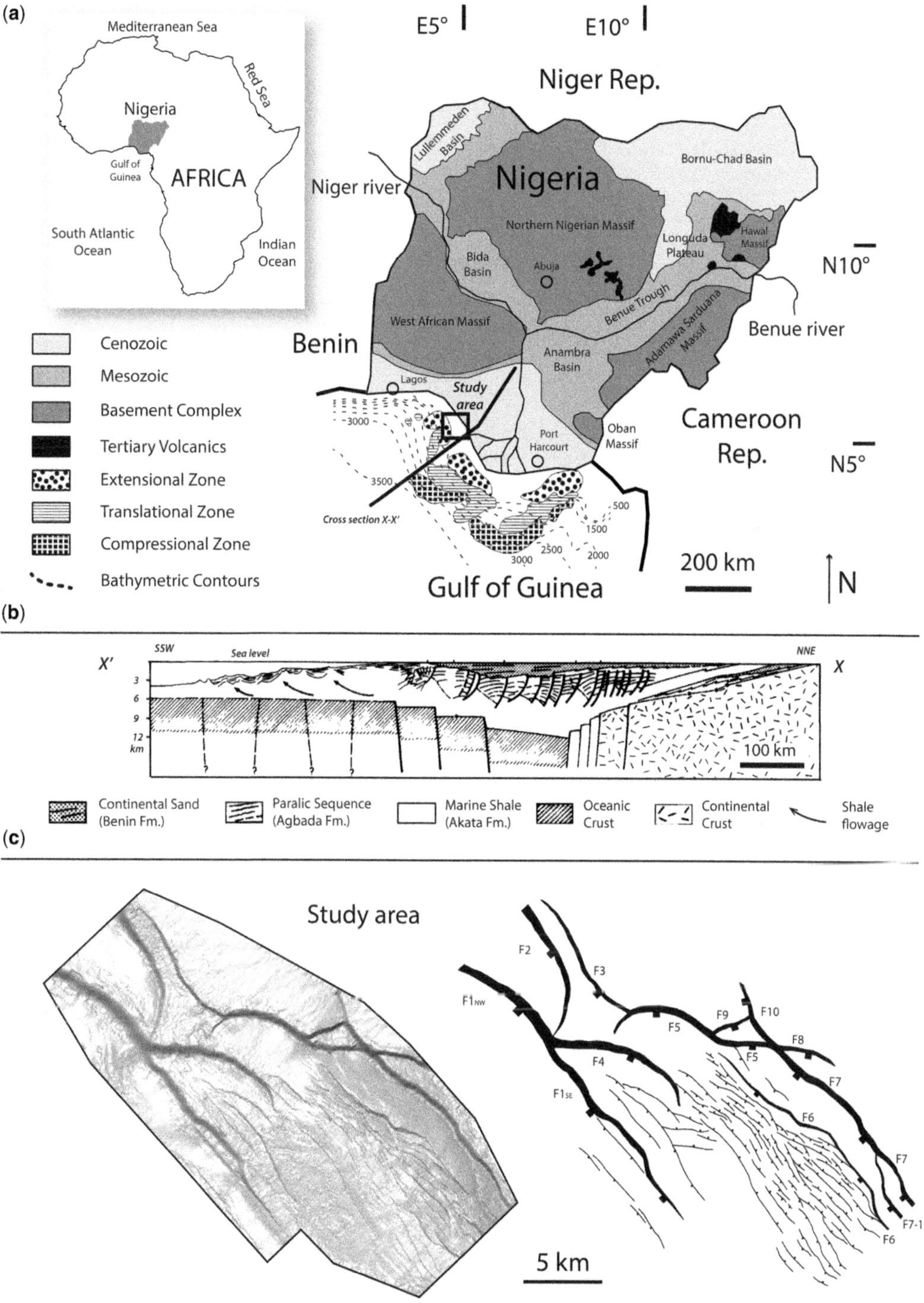

Fig. 1. (**a**) Location and the major onshore geological units (based on Onuoha 1999) of the study area in the western offshore, Niger Delta in Nigeria (offshore gravity-tectonic structural styles and bathymetric contours are based on Damuth 1994). (**b**) Regional cross-section in the vicinity of the study area showing the regional stratigraphic units and structural framework of the Niger Delta (based on Evamy *et al.* 1978). (**c**) Time–structural map of the study area highlighting major bounding listric faults (F1–F10) and their hanging-wall crestal-collapse normal faults.

Table 1. *Major bounding-fault characteristics in the study area*

Fault name	Strike	Dipping towards	Length (at horizon D) (km)	Maximum throw (ms)	Location of maximum throw
Fault 3 (F3)	NW–SE	SW	8	250	Close to the linkage point
Fault 5 (F5)	East–west	South	11	700	At the centre of the fault
Fault 7 (F7)	NW–SE	SW	15	950	Close to the linkage point
Fault 8 (F8)	East–west	South	3.5 (in the study area)	400	Close to the linkage point
Fault 9 (F9)	ENE–WSW	South	3	500	At the centre of the fault
Fault 10 (F10)	NW–SE	SW	4.5 (in the study area)	650	Close to the linkage point

of synsedimentary faults in the seismic dataset were based on interpreting a combination of vertical seismic sections of varying orientation (Fig. 2), together with time and horizon slices in reflectivity and variance display, constructing time–thickness maps, conducting detailed fault-kinematic analyses based on throw–depth plots that show the temporal evolution of faults, and throw–distance plots that show their spatial evolution.

Fault throw measured in milliseconds for two-way travel time (ms TWT) v. depth and distance plots (Fig. 3), and fault-throw v. distance plots (Fig. 4), were constructed from hanging-wall and footwall cut-offs of all interpreted horizons mapped on several orthogonal sections along the strike of all main bounding faults. Compaction of sediments may cause a loss of displacement of up to 15% in the sand or mixed sand–shale successions, with growth indices of greater than approximately 0.1; decompaction is not necessarily required to decipher first-order displacements and fault-growth histories (Taylor *et al.* 2008). Throw–depth plots (Fig. 3) (*sensu* Cartwright *et al.* 1998; Back *et al.* 2006; Hongxing & Anderson 2007; Baudon & Cartwright 2008; Jackson & Rotevatn 2013) illustrate the growth periods of faults, and incremental throw in any given interval can be readily determined from the plot, as can the displacement gradient for any interval. Throw–distance plots (Fig. 4), in turn, illustrate the lateral growth and linkage of fault systems. A single isolated fault has maximum throw at its centre, decreasing laterally towards the fault tips. Complex shapes of the throw–distance curve can be observed, for example, in relay ramp areas with steep throw gradients, or at the location of fault-segment linkage with throw minima (Peacock & Sanderson 1991; Gawthorpe & Leeder 2000; Duffy *et al.* 2015; Fossen & Rotevatn 2016).

Main bounding faults

This study focuses on 10 listric normal faults in the SE part of the studied dataset that bound the major rollover province of the SE to central study area, labelled as faults F1–F10 (Figs 1c & 2). The faults F5, F7, F8 and F10 meet at a common intersection area (Figs 1c & 2), while F9 is oblique to F5 and F10 connects these two faults. The major first-order faults F5 and F7 are arcuate in map view, bounding the dominant rollover anticline in their hanging wall in the centre of the study area (Fig. 2). Faults F8, F9 and F10 are located in the footwall of F5 and F7 at the western edge of the study area.

F3 strikes NW–SE over 8 km and dips towards SW (Table 1; Fig. 2). To the SE, F3 links to F5 and dies out towards the NW. The maximum of throw of 290 ms is measured close to the linkage area with F5 (Fig. 3, plot B), decreasing towards the NW. F5 strikes generally east–west, with dips towards the south. It extends over 11 km at mapped Horizon D (Table 1; Fig. 2). Towards the NW, F5 links to F3 before terminating in the footwall of F4 (Figs 2 & 3). Along-strike throw measurements (throw–distance plot: Fig. 4) show a major decrease in throw at the location of the linkage between F5 and F3. F5 strikes NW–SE in its central part; at the linkage area to F9, it rotates toward the east, trending west–east towards the intersection area indicated in Figure 2. At the location of linkage to F9, a distinct decrease in throw is observed on the throw–distance plot of F5 (Fig. 4). F5 links to F7, F8 and F10 in the intersection area. Fault 9 strikes ENE–WSW over 3 km, and links to F10 towards the NE and to F5 towards the SW (Fig. 2). Fault 10 is located on the western edge of study area in the footwall of F5, over 4.5 km in length (Fig. 2; Table 1). This fault strikes NW–SE from the intersection area, marking its SE tip towards the NW. The NW tip of this fault is located beyond the extent of the study area. Fault 8 is located in the footwall of F7, striking east–west from the intersection area. This fault is about 3.5 km long in the study area (Fig. 2), with a maximum throw of 400 ms measured at the eastern end of fault close to the edge of the study area (Fig. 2; Table 1).

Relative timing of fault initiation and temporal evolution

The throw–depth plots of Figure 3 show the nature and the kinematics of the main bounding faults.

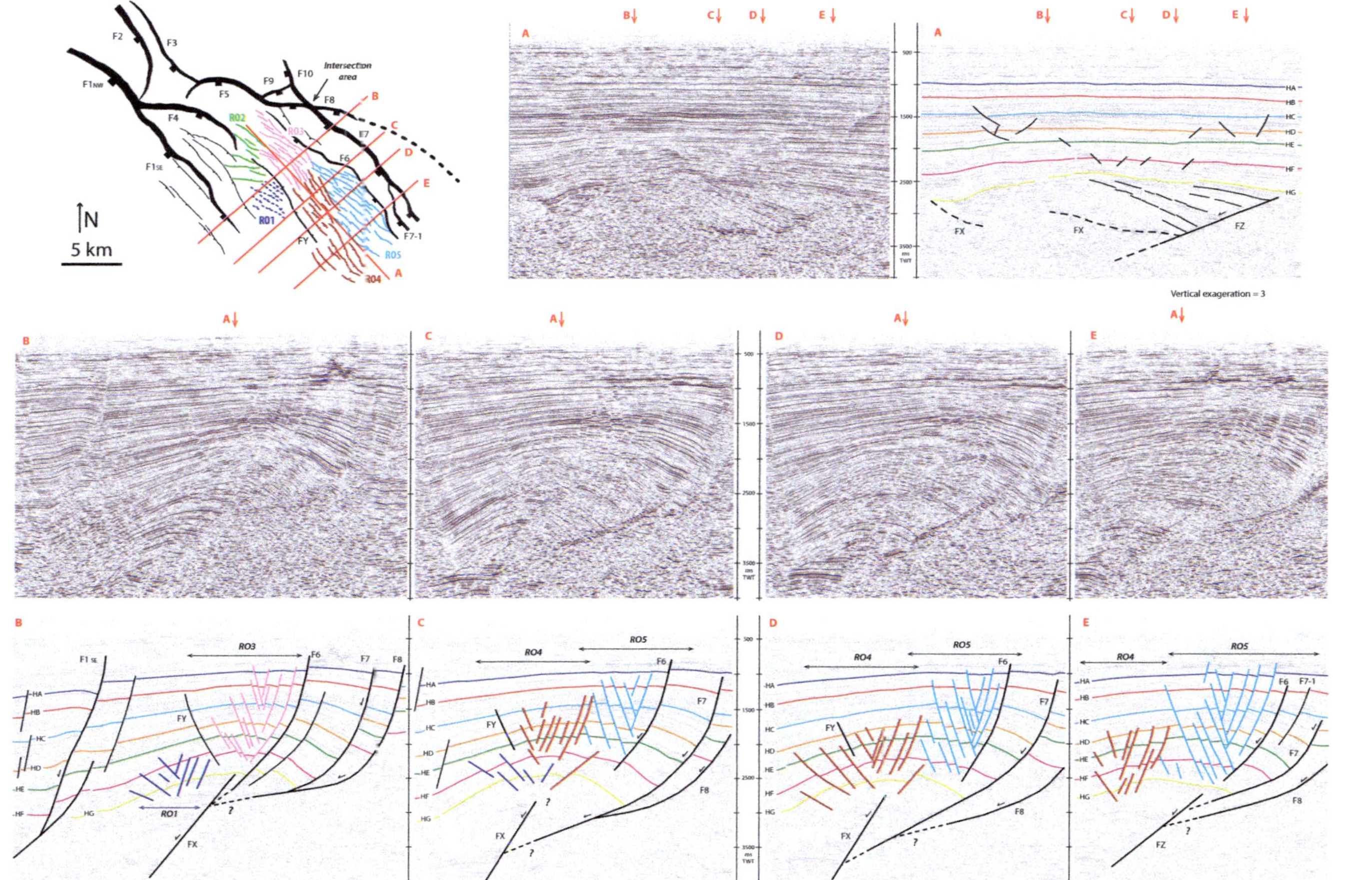

Fig. 2. Structural map at Horizon D and five vertical cross-sections illustrating major structural elements and rollover anticlines in the centre of the study area. Seven horizons have been mapped in the study area and are shown in the vertical time sections (ms, millisecond; TWT, two-way travel time). These horizons are labelled HA–HG from shallow to deeper parts and are the basis for fault kinematics analysis. The dotted line on the map view shows the extension of F8 beyond the study area. Note that cross-section A is perpendicular to sections B, C, D and E. Red arrows show the intersection locations.

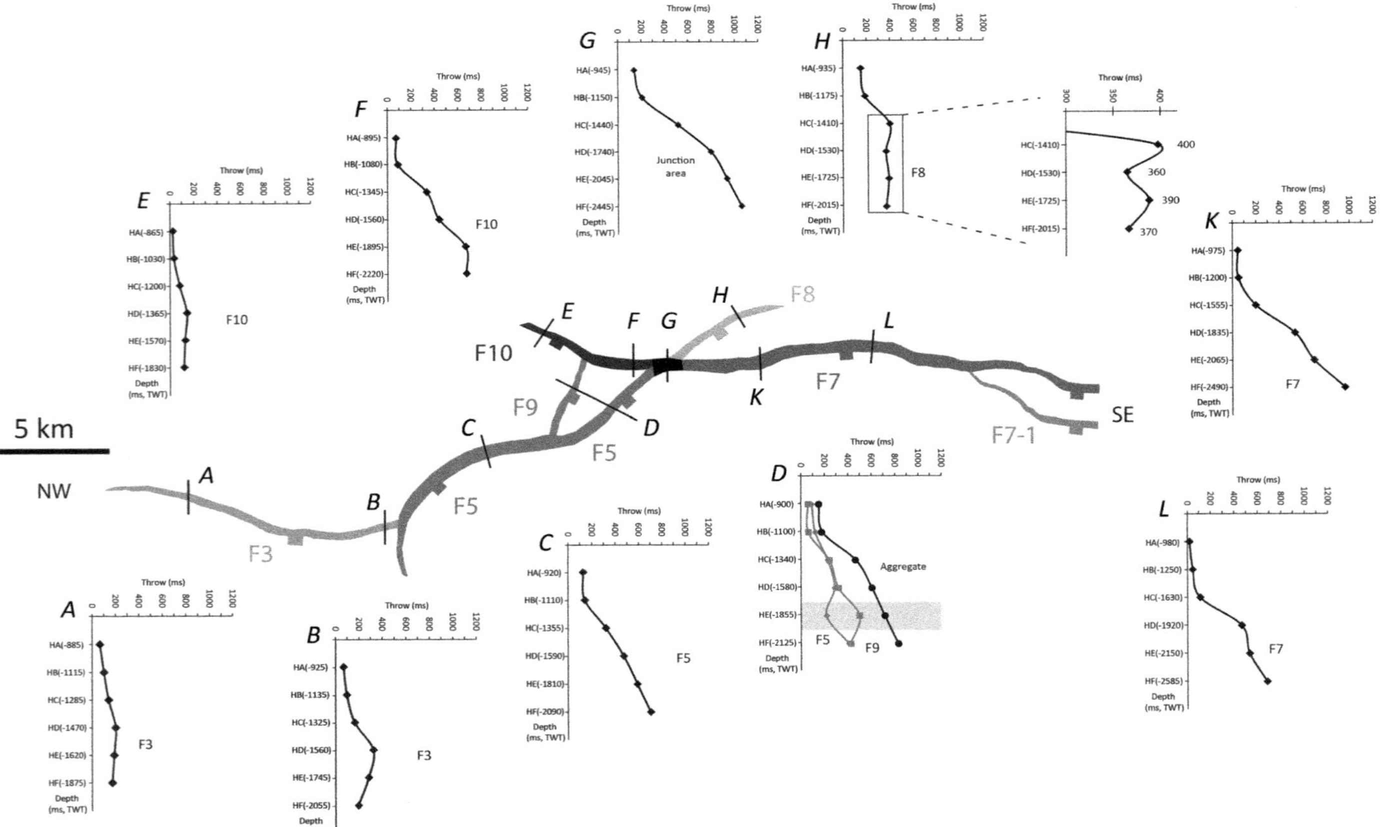

Fig. 3. Throw–depth profiles reflecting the growth of the studied faults. F3 shows rather isolated fault growth characteristics with the maximum throw at Horizon D decreasing both upwards and downwards (plots A and B). F5 and F7 are synsedimentary growth faults with the maximum of throw at Horizon F decreasing upwards in the shallower horizons. Profile D shows the influence of the initiation of F9 in the footwall of F5 (shaded rectangle; orange in the online version). When F9 initiates, a major amount of strain accumulates at Horizon E, causing a throw minima on F5. The graph on the right in plot D shows the aggregate of throw on both F5 and F9, following the same trend as plot C. Profiles E and F show differential fault activity along the strike of F10, highlighting the lateral (towards the north) decrease in fault activity. The vertical scale in brackets is the mid-point between the hanging-wall and footwall cut-offs.

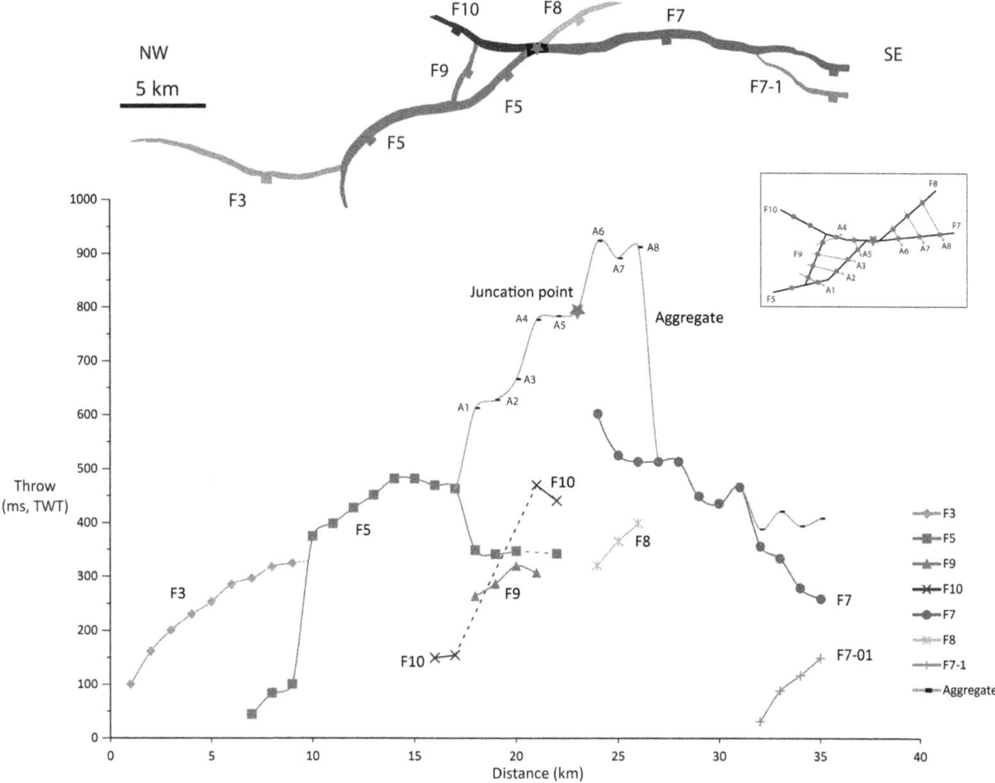

Fig. 4. Throw–distance (T–D) graph showing lateral growth and linkage along the strike of studied bounding faults from NW to SE. This graph is based on the hanging-wall and footwall cut-offs of interpreted Horizon D. Horizon D is the deepest horizon in which we were able to construct throw–distance graphs that show the lateral linkage of studied faults. The aggregated graph shows the amount of throw at and around the relay zone, and is based on the totalled up throw at different measuring points. The inset box shows how the amount of throw was measured in the relay zone in the footwall of F5 to construct the aggregate graph.

From NW to SE, plots A and B show a vertical decrease in throw along the length of F3 close to its NW and SE tips, respectively (Fig. 3). The maximum throw of F3 occurs at Horizon D in both plots A (about 320 ms TWT) and B (about 200 ms TWT), decreasing both upwards and downwards. This suggests that F3 initiated at Horizon D. The magnitude of the throw on F3 decreases towards the NW from the area of linkage towards F5 at its SE tip (see the variation in lateral throw at Horizon D in plots A and B in Fig. 3). This suggests that F3 propagated from the SE towards the NW, consistent with its throw–distance plot in Figure 4.

Plot C of Figure 3 shows the variation in vertical throw in the central portion of F5, with a maximum throw of 830 ms TWT measured at Horizon F, the deepest horizon analysed. The throw decreases gradually upwards, which is characteristic of blind faults or growth faults where the rate of sedimentation is larger than the rate of fault movement (Childs *et al.* 2003; Hongxing & Anderson 2007). Profile D shows a variation in throw for both F5 and F9 (Fig. 3). F5 has a maximum throw of 410 ms TWT at Horizon F, decreasing to about 210 ms upward at Horizon E. The throw increases again to about 280 ms TWT at Horizon D, and then decreases following the general trend as observed in plot C. A comparison to the throw profile of F9 shows an unexpected throw decrease of F5 at Horizon E, which correlates with the maximum throw at F9. This is interpreted as the initiation time of F9. The initiation of Fault 9 close (less than 2 km at the location of plot D) to the footwall of F5 might have caused strain to partition between F5 and F9, as indicated by throw minima at Horizon E of plot D (Fig. 3). The aggregate throw measurement of plot D shows an identical trend to plot C, a fault–growth pattern with the maximum at Horizon F, gradually decreasing upwards.

Plot E shows throw–depth measurements at the NW end of F10 (edge of the study area), documenting a maximum throw of 140 ms TWT at Horizon D, decreasing both upwards and downwards. Plot F was measured close to the linkage area between F10 and F5, F7 and F8, labelled as the intersection area in Figure 2. This plot shows a maximum of throw of about 670 ms TWT at Horizon E, decreasing upwards. This observation highlights the initiation time of F10 at Horizon E as a syndepositional normal fault that grew towards the NW. A comparison between plots E and F shows two different throw–depth trends for F10 from the SE (syndepositional trend) towards the NW (post-depositional trend).

Plot G shows the throw distribution at the common intersection area of faults F5, F7, F8 and F10, with 1050 ms throw at Horizon F, decreasing upwards. Plot H shows the throw on F8 in the footwall of F7, with two maxima: one at Horizon E with 390 ms throw, and another at Horizon C with 400 ms throw. In-between the two maxima, the throw decreases to about 360 ms. This pattern could either reflect that F8 initiated at the time of deposition of Horizon E, becoming inactive at Horizon D, which was then followed by an increase in fault throw at Horizon C, alternatively F8 initiated at Horizon C and grew downwards and linked to a pre-existing, deeper-seated fault (*sensu* Fazli Khani & Back 2015*b*). After deposition of Horizon C, the throw along F8 decreased.

Plots K and L show the temporal evolution of F7 close to the intersection area and at the centre of the fault, respectively. Both plots start with maximum throw at Horizon F, decreasing towards the shallower horizons, highlighting the typical syndepositional nature of this fault. At the very SE edge of the study area, F7-1 has been interpreted as a splay of F7, linking both laterally and vertically onto F7; its location at the very edge of the seismic data volume made throw–depth measurements on this fault segment impossible.

Lateral fault growth and linkage

Throw–distance plots along the strike of the studied faults can be used to highlight the lateral growth and linkage of the main bounding faults. Figure 4 shows that throw increases gradually on F3 from 100 ms TWT in the NW to its maximum of 320 ms TWT at the linkage area with F5. This coincides with a major increase in throw on F5 from 100 to 370 ms TWT. The general NW decrease in throw on F3 suggests that this fault initiated from the linkage area with F5 and propagated laterally towards the NW. The observations from throw–depth plots A and B (Fig. 3) confirm the interpretation that F3 initiated at Horizon D from the linkage area to F5 and propagated through time towards the NW. From the linkage area with F3, throw increases along the strike of F5, reaching its maximum of about 480 ms TWT before decreasing gradually towards the SE (Fig. 4). At the linkage area with F9, the throw on F5 decreases by about 110 ms TWT, remaining constant along the fault until the common linkage area with faults F7, F8 and F10. Fault 9 links orthogonally to F5 in the SW and to F10 at its NE tip, creating a triangular area in the footwall of F5 (Figs 2 & 3). The maximum throw of about 320 ms TWT occurs in the central part of the fault (Fig. 4). Throw on F10 increases towards the SE into the common intersection area. At the linkage area with F9, the throw on F10 increases from 150 ms TWT to about 450 ms TWT. This occurs over a short distance (a few hundreds of metres), and throw–depth plots E and F (Fig. 3) suggest that the SE segments of F10 and F9 initiated at Horizon E and accommodated significant strain, while the NW segment of F10 initiated later at Horizon D.

The cumulative throw of F5, F9 and F10 (see the aggregate curve in Fig. 4) shows a gradual increase in the SE direction towards the common intersection area. At the intersection area, where F5, F10, F8 and F7 link, throw on the aggregate curve increases to 800 ms TWT. Eastwards along F8, throw increases from 320 ms TWT at the intersection area to about 400 ms TWT at the limit of the study area. We have no seismic data beyond this point, but the continuation of F8 with depth (Fig. 2, sections C and D) suggests that it initiated in the footwall of F7 and later propagated laterally towards the common fault intersection. F7 finally has about 600 ms TWT of throw close to the common intersection area, decreasing laterally towards the SE. F7-1 links up with F7 in the very SE of the study area. Throw increases along segment F7-1 from the linkage area in the NW towards the SE.

The throw–distance plots of Figure 4 shows an overall convex shape with the maximum amount of throw localized at and around the common fault intersection, decreasing both towards the NW and SE. This suggests that, although strain migrated from one fault or a fault segment to neighbouring faults or fault segments, the aggregate amount of throw shows a throw–distance pattern similar to that of a single isolated fault.

Footwall backstepping of faults

The throw–depth plots (Fig. 5) along F7 (plots K, L and M), F8 (plot H) and at the common fault intersection (plot G) show the influence of F8 on the lateral distribution of throw along F7. The throw–depth plots along F7 all show a tripartite vertical zonation of the fault into: (1) a lower part between Horizon F and Horizon D with a relatively steep slope related to the early stages of fault activity; (2) a transition interval (indicated in grey in Fig. 5) that is characterized by a shallow slope; and (3) a top interval with an almost vertical slope indicating decreased fault activity before cessation. In the NW (plot G), the transition interval starts at Horizon D and ends at Horizon B; in the SE (plots L and M), the transition interval lies between Horizon D and Horizon C. Plots L and M further show that the activity of F7 decreases earlier in the SE (Horizon C) than in the NW (Horizon B). These observations suggest that the initiation of F8 in the footwall of F7 caused a migration of strain through time towards the east in the SE and central part of F7. Towards the NW, the influence of fault F8 on the kinematics of F7 decreases as it approaches and links to F7 (plots K and G: Fig. 5).

Lateral fault linkage and relay zones

An integration of throw–depth plots (Figs 3 & 5), time-thickness maps between each pair of interpreted horizons (Fig. 6) (see Fazli Khani & Back 2012, 2015*b* for further discussion) and throw–distance plots (Fig. 4) highlights the temporal and spatial evolution of the main bounding faults. The throw–depth and throw–distance plots for F5 and F7 suggest that the triangular area between F5, F9 and F10 (Fig. 2) most probably originated as a relay zone between F5 and F7, and that, during propagation of these faults towards each other, F5 rotated eastwards approaching F7 (Fig. 6, Unit EF) and finally linking to F7 at the common fault intersection (Fig. 6), while F7 continued to propagate towards the NW. Fault F5 and F7 seem to act differently in the relay zone, where F5 turns into its footwall area and links onto the laterally growing F7, again, in the footwall area, and F7 continues its lateral propagation towards the NW. Following further fault activity and increase in displacement, F9 probably initiated as a connecting fault branch line (Peacock & Sanderson 1994; Long & Imber 2012) between F5 and F7 after deposition of Horizon E (Fig. 6, Unit ED), which can explain the unusual NE–SW strike of F9. Fault F10, which is most likely to be a NW continuation of F7, continued its lateral propagation after the initiation of F9, and probably terminated immediately outside the study area. Fault 5, in turn, linked to F7 and stopped its lateral propagation in the fault intersection area.

The throw–depth plot D (Fig. 3) at the location of the relay zone shows the relative activity and initiation time of F5 and F9. Fault F5 shows a general decrease in throw from Horizon F upwards. At Horizon E, where F9 initiates, the throw on F5 decreases rapidly from 410 ms TWT (Horizon F) to about 210 ms TWT (Horizon E), followed by an increase to 290 ms TWT at Horizon D. In the same interval, F9 records about 500 ms TWT of throw at the nucleation point (Horizon E), which decreases to approximately 320 ms TWT at Horizon D. This observation suggests that, in the location of the relay zone, the total amount of strain prior to the initiation of F9 was accommodated by F5; after the initiation of F9, the strain was partitioned between the two neighbouring faults. The dotted curve on the throw–depth plot D in Figure 3 shows the aggregate throw on F5 and F9 in the location of the relay zone. A comparison of the aggregate curve with plot C shows a similar throw distribution along the strike of F5. Figure 7 summarizes the lateral growth pattern of F5 and F7, and the evolution of the relay zone in their linkage area.

Rollover anticlines

A major, kilometre-scale rollover complex with associated crestal-collapse faults is located in the hanging wall of bounding faults F4, F5, F6 and F7 in the central part of the study area (Fig. 8). In total, more than 100 minor crestal-collapse faults were mapped in the rollover structure, of which most strike NW–SE, with some minor east–west strike directions. The maximum length of these synthetic and antithetic collapse faults ranges between a few hundreds of metres up to 6 km. Table 2 summarizes the main characteristics of the studied rollover complex, which exhibits, upon close inspection, five individual collapse zones (RO1– RO5 on Fig. 8). The vertical seismic reflection sections of Figure 2 show the spatial relationship between synthetic and antithetic collapse faults and the major listric bounding faults. Based on the dip direction, the crestal-collapse faults can be subdivided into two groups: SW-dipping faults (green in Fig. 8) and NE-dipping faults (violet in Fig. 8). Local graben areas are observed where a synthetic and an antithetic fault face one another (blue polygons in Fig. 8). In contrast, where two adjacent faults dip away from one another, a horst initiates in the footwall (orange polygons in Fig. 8). Based on the vertical geometry of synthetic and antithetic faults, and their map view organization of Figure 8, five distinct rollover collapses can be identified in the central study area (Figs 8 & 9). Each collapse system is characterized by a series of oppositely dipping normal faults defining a central (crestal) graben, with horst structures separating neighbouring rollover anticlines.

Rollover 1 is the most basinwards located in the hanging wall of the deep-seated FX, and dies out between Horizon E and Horizon F (Fig. 2). The collapse faults of Rollover 1 strike NW–SE (Fig. 8). Rollover 2 is located in the hanging wall of F5 close to the SE tip of F4. The SW-dipping internal

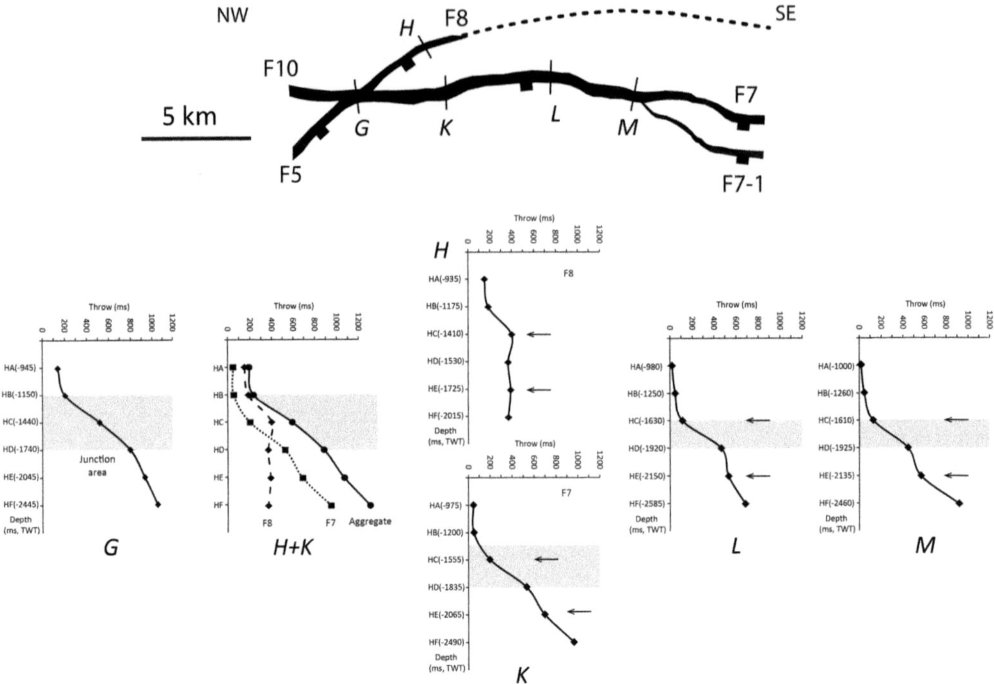

Fig. 5. Comparison of fault activity periods along the strike of F7 and the influence of F8 in the footwall of this fault. Arrows on graphs H and K show two throw maxima on F8 at Horizon E and Horizon C that correspond with a decrease in throw on F7. The grey box shows the transition between higher fault activity during the early stages and lesser fault activity in the later stages of fault growth. Towards the NW in profiles G and H + K, the transition is gradual and lasts longer compared to the SE, as profiles L and M show. A rapid decrease in fault throw in profiles L and M is due to the activity and throw maxima of F8. The influence of F8 on F7 decreases towards the NW where they link. The vertical scale in brackets is the mid-point between the hanging-wall and footwall cut-offs.

collapse faults of Rollover 2 generally strike NW–SE, while the NE-dipping collapse faults mainly strike east–west (Fig. 8). The vertical seismic-reflection section B of Figure 2 shows Rollover 3 at shallower depths, and Rollover 1 in the deeper parts of the section. The seismic section reveals the spatial relationship between a deep-seated FX and the shallower F6, F5 and F9 (see Fazli Khani & Back 2015*b* for more discussion on the geometry and linkage of deep-seated faults FX and FZ). Rollover 3 accommodates the deformation in the hanging wall of F5, F6 and F7 in the relay zone. This structure displaces strata down to Horizon F (Table 2); towards the SW, Rollover 3 is bound by the NW-dipping FY (Fig. 2, section B). Fault FY separates Rollover 3 from Rollover 1, terminating laterally in the NW within Rollover 2.

Rollovers 4 and 5 are located on the SE edge of the study area, in the hanging wall of F6, F7 and F8. On the time–structural map of Horizon D (Fig. 2a), at the location of section C, F8 is beyond the NE extent of the seismic cube. In the deeper part of the section, F8 is in the footwall of F7. Synkinematic horizons to Rollover 4 are sediments developed prior to Horizon C (Table 2). For Rollover 5, synkinematic horizons are between Horizon F and Horizon A. The following discussion focuses on delineating the relationship between the development of the major bounding faults and rollover tectonics.

Discussion

Previous interpretation and modelling studies on growth faults and rollovers (Ellis & McClay 1988; Vendeville & Cobbold 1988; McClay 1990; Mauduit & Brun 1998; Xiao & Suppe 1992; Withjack *et al.* 1995; Hodgetts *et al.* 2001; Imber *et al.* 2003; Back *et al.* 2006; Fazli Khani & Back 2015*a*) have addressed factors controlling the geometry and the development of rollovers, such as the geometry of the bounding faults, variable sedimentation rates, the amount of extension after synkinematic units are deposited and compaction. Fazli Khani & Back (2012) have shown, on the same data as used in this

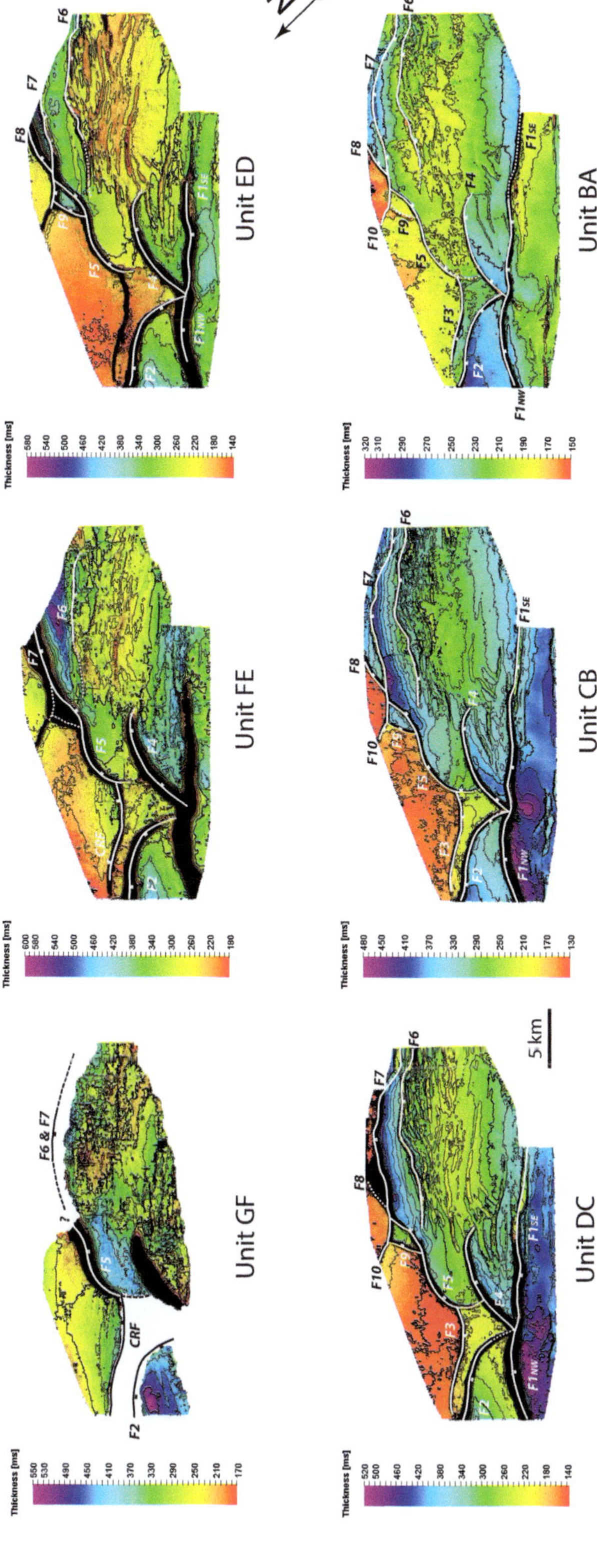

Fig. 6. Time–thickness maps showing temporal evolution of major regional faults and a counter regional fault (CRF) in the study area. The dashed lines highlight the inactive segment of the fault, while solid line shows the active segment of fault.

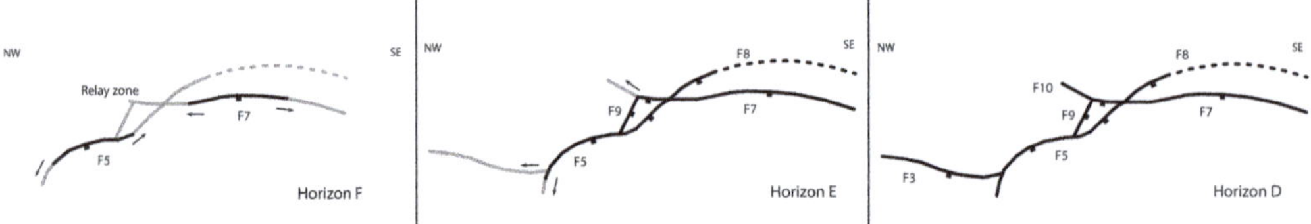

Fig. 7. Lateral and temporal evolution of the studied fault array, showing the relative initiation time and growth of faults at horizons F, E and D, and the development of the relay zone. This construction is based on the time–thickness maps and fault analysis presented here. Grey shows an inactive fault segment and black shows an active fault segment.

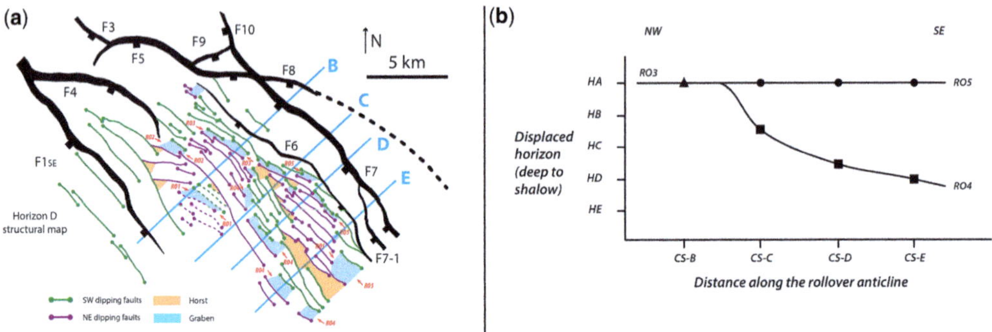

Fig. 8. (**a**) Horizon D structural map showing the main bounding faults and hanging-wall synthetic, dipping towards the SW (in green), and antithetic, dipping towards NE (in violet), normal faults. Graben and horsts are shown by blue and orange polygons, respectively. Local graben are parallel the axis of associated rollovers (hinge of the rollover anticline). Based on the orientation and location of local graben in map view, we have identified five rollover anticlines in the study area labelled RO1–RO5. (**b**) Rollover activity graph showing the shallowest horizon displaced by rollovers RO3, RO4 and RO5 along four cross-sections (CS-B, CS-C, CS-D and CS-E; see the structural map in (a) for the location of the cross-sections) along the strike of rollovers from NW to SE. Rollover 4 displaces younger intervals (between horizons B and C) in the NW, while in the SE only intervals below Horizon D have been displaced by Rollover 4 (see the text for the discussion). See the vertical cross-sections in Figure 2 for the relative activity timing of rollovers RO3, RO4 and RO5.

Table 2. *Identified rollovers and their characteristics*

Rollover name	Strike	Total number of mapped faults	Fault length (km)	Displaced horizons	Location
Rollover 1 (RO1)	NW–SE in the north and east–west in the south	20	1–3	HG to HA	Hanging wall of F5 at the eastern tip of F4
Rollover 2 (RO2)	NW–SE	32	1–5	HF to HA	Hanging wall of F5, F6 and F7, at the junction and in the relay zone area
Rollover 3 (RO3)	NW–SE	10	1–3	HG and deeper to HF	At the southern edge of the study area
Rollover 4 (RO4)	NW–SE with minor WNW–ESE	31	1–5	HF to HA and younger	Hanging wall of faults F6, F7 and F7-1
Rollover 5 (RO5)	NW–SE	28	1–6	HG and deeper to HD	SE of Rollover 4

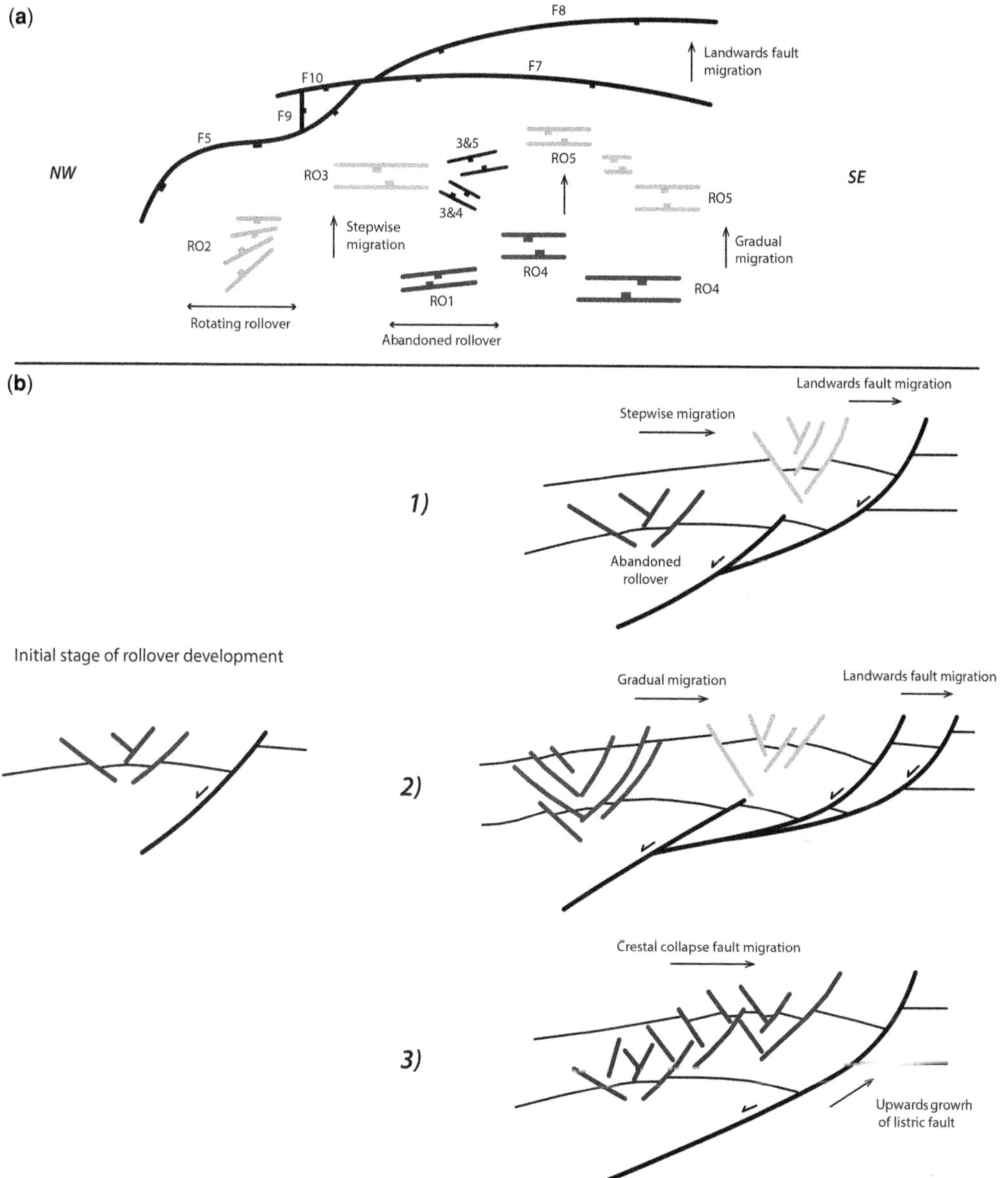

Fig. 9. Simplified schematic diagram showing the influence of normal fault growth and linkage on the evolution of hanging-wall rollover anticlines. (**a**) Map view of bounding faults and associated rollovers, showing the rotation of crestal-collapse faults in the hanging wall of fault F5 described as a fault–rollover interaction Type 1. (**b**) Three different responses of rollover anticline to their bounding-fault evolution. Type 2: the initiation of a new listric fault in the footwall of an older fault creates a new rollover anticline, while the older rollover becomes 'abandoned' with no or a little overlap in the crestal-collapse fault's activity. In this case, there is a stepwise rollover migration. Type 3: landwards migration of the bounding fault following by significant contemporaneous crestal-collapse fault activity and a more gradual migration of rollovers. Type 4: landwards migration of crestal-collapse faults within the rollover anticline due to the upwards growth and listric geometry of the bounding fault.

study in the central part of the study area, a general landwards migration of rollovers; yet, the affects of lateral (plan-view) changes in bounding-fault geometry on rollover development, including those of multiple laterally linking bounding faults that are associated with individual rollovers, have not previously been documented. The studied part of the Niger Delta that is characterized by five individual rollover–crestal-collapse systems controlled by dynamically evolving bounding faults is an ideal example to demonstrate such effects.

The consequences of the lateral growth of a bounding fault that changes strike during its development can be best documented along F5. F5 propagated laterally to the NW and west (Figs 6 & 7) before terminating at F4 (Fig. 6). The crestal-collapse faults in the associated Rollover 2 trend NW–SE in the north, but seem to rotate with time to an east–west orientation in the south (Figs 2 & 8). The rotation of extensional collapse faults ('oblique faults' in fig. 8 of Fazli Khani & Back 2012) is most likely to be due to the change in the strike direction of F5 towards the west at its NW tip. This change probably modified the local stress field, reflected by the eas–west strike of the crestal-collapse faults in Rollover 2 (Fig. 9a, Type 1). With F5 joining F4 during the deposition of Unit DC (Fig. 6), several younger NW–SE-orientated faults in the hanging wall of F4 intersected the east–west-orientated collapse faults, creating a complex mosaic of rhomb-shaped fault-bound microblocks (Fig. 8).

An example of the along-strike influence of linking bounding faults on rollover development can be seen between the landwards faults F7 and F8 in the east of the study area. The activity of F8 in the footwall of F7 decreased the slip on F7 (Figs 3 & 4) from the NW, where the two faults link, towards the SE. The variation in lateral throw along F7 and F8 controlled the hanging-wall deformation in different ways: in the location of the relay zone and the linkage point of F7 and F8, a single rollover (RO3) accommodated the hanging-wall deformation (Fig. 2, section B; Fig. 8). Towards the SE, however, the further landwards migration of F8 (assuming that F8 continues upwards with the same dip as F7) broadened the associated hanging-wall bending, resulting in a gradual shift in crestal deformation from Rollover 4 to Rollover 5 (Fig. 2, sections C, D & E). The lateral differences in the interaction between F7 and F8 can also be seen on the throw–distance plot of Figure 4, which shows a decreasing throw for F7 towards the SE, whilst the throw of F8 increases. Although the SE portion of F8 is out of the study area, continuous faulting activity along F8 can be interpreted from the development of Rollover 5, which displaces Horizon A and even shallower and younger units irrespective of the decreasing slip along F7 (Fig. 2 sections C, D & E; Figs 3 & 5).

It should be noted that faulting in the relay zone involving faults F5, F7, F8, F9 and F10 (Figs 2, 3 & 7) resulted in the development of a single rollover (RO3), as opposed to the two rollover systems (Rollover 4 and Rollover 5) in the hanging wall of faults F6, F7 and F8. This documents that hanging-wall deformation related to several nearby faults can be accommodated by a single rollover–crestal-collapse fault system that seems somewhat unaffected by the presence of multiple interacting bounding faults and fault segments of partly different orientation. In turn, the dynamic development of several large, more-or-less sub-parallel faults at distances of >3 km (F6, F7 and F8) seems to support rollover migration following the stepwise activation of the respective main fault.

Imber *et al.* (2003) demonstrated two different responses of hanging-wall deformation to the development of a bounding fault: (1) a prolonged and progressive landwards migration of the active rollover in the hanging wall of a stationary main fault; and (2) a punctuated migration of the rollover directly related to the landwards backstepping of the main bounding fault. This study shows that deep-seated Rollover 1 (e.g. Fig. 2, section B) was most probably initiated and controlled by FX in the deeper parts of the study area, displacing Horizon G and Horizon F. This rollover became inactive prior to the deposition of Horizon E, and it is bounded by FY in the north and NW (see Fazli Khani & Back 2015*b* for detail on the evolution of FX). The subsequent development of F6, F7 and F8 in the footwall of FX seems to have initiated a new set of crestal-collapse faults (Rollover 3) on the landwards side of Rollover 1 (Fig. 2), abandoning Rollover 1 (Fig. 9b, Type 2). This contrasts with the rollover development in the SE of the study area, where Rollover 4 formed in response to the deep-seated FZ in its early stages. However, Rollover 4 continued to displace shallower horizons even after the decay of FZ (up to Horizon C: Fig. 2, section C), a time when younger F6, F7 and F8 and Rollover 5 were already active. The study area thus shows two fundamentally different responses of rollover–crestal collapse systems to the initiation of younger faults in the footwall of their original bounding faults: (1) a stepwise shift of deformation, where Rollover 1 was abandoned and Rollover 3 was initiated as new younger faults initiated in the former footwall terrain (Fig. 9b, Type 2); and (2) a more gradual shift of deformation from Rollover 4 to Rollover 5 in response to footwall collapse (Fig. 9b, Type 3). The gradual shift of rollovers with a period of contemporaneous growth of Rollover 4 and Rollover 5 is similar to the 2D sandbox experiment E44 by McClay (1990) and the systems described by Imber *et al.* (2003), and probably related to the branching of F7 and 8. In turn, the most likely reason for the complete abandonment of Rollover 1 is the limited NW extent and stratal offset of F6 (Figs 2 & 6), shifting younger rollover activity not only landwards but also laterally over a considerable distance to the SE.

A final interaction between growth faulting and rollover development documented in the study area is that synthetic and antithetic faults within a rollover structure can progressively migrate upwards and landwards without initiating a new rollover structure (see the fault pattern within RO3 in Fig. 2, section B; Fig. 9b, Type 4). A comparison with the analogue models of McClay (1990) and the northern Brunei example described by Imber *et al.* (2003) suggests that a general landwards and upwards migration of crestal-collapse faults within rollovers could be due to the upwards growth and the listric shape of a single bounding fault. However, if a secondary fault initiates in the footwall, a new rollover structure can initiate landwards, possibly laterally offset from the pre-existing rollover. This mechanism seems to require a certain distance between the original and the new bounding fault, which can be estimated for the study area to be >3 km, as well as a stratal offset at the new fault exceeding 100 ms TWT (*c.* 100 m).

Conclusions

This study shows how growth-fault characteristics can change along strike as they become influenced by the initiation and growth of neighbouring faults. The stepping of a main fault system can modify the throw distribution along its individual fault branches in the early stages of growth, reaching equilibrium in the later stages of fault growth.

The initiation and growth of rollovers is directly controlled by the kinematics of their bounding normal faults. It is shown that the lateral linkage and backstepping of bounding-fault systems can be mirrored in their associated rollover hanging walls. Based on the data presented in this study, we have identified four genetic types of fault–rollover interaction, including: (1) the rotation of a rollover–crestal-collapse system, which is controlled by changing lateral bounding-fault orientation during fault growth; (2) a stepwise shift of rollover–crestal-collapse systems associated with rollover abandonment, controlled by the initiation of a new fault in the footwall of an older structure; (3) a gradual migration of successive rollovers controlled by branching, connected fault systems; and (4) a general landwards and upwards migration of crestal-collapse faults within individual rollovers above listric, upwards-growing, stationary main faults.

This study finally shows that bounding faults and their associated rollovers can dynamically interact, and that an overlap in the timing of rollover activity is likely to depend on the distance, connectivity and lateral (plan-view) arrangement and geometry of the bounding faults.

We thank the Shell Petroleum Development Company of Nigeria for providing the seismic data presented in this study. Schlumberger is gratefully acknowledged for providing Petrel under an academic user license agreement. Paul Whipp, an anonymous reviewer and volume editor Conrad Childs are thanked for their constructive comments on the earlier version of the manuscript. This study is a contribution to Project Ba 2136/4-1 funded by the Deutsche Forschungsgemeinschaft (DFG).

References

Back, S. & Morley, C. 2016. Growth faults above shale – Seismic-scale outcrop analogues from the Makran foreland, SW Pakistan. *Marine and Petroleum Geology*, **70**, 144–162.

Back, S., Tioe, H.J., Thang, T.X. & Morley, C.K. 2005. Stratigraphic development of synkinematic deposits in a large growth-fault system, onshore Brunei Darussalam. *Journal of the Geological Society, London*, **162**, 243–258, https://doi.org/10.1144/0016-764903-006

Back, S., Höcker, C., Brundiers, M.B. & Kukla, P.A. 2006. Three-dimensional-seismic coherency signature of Niger Delta growth faults: integrating sedimentology and tectonics. *Basin Research*, **18**, 323–337.

Back, S., Strozyk, F., Kukla, P.A. & Lambiase, J.J. 2008. 3D restoration of original sedimentary geometries in deformed basin fill, onshore Brunei Darussalam, NW Borneo. *Basin Research*, **20**, 99–117.

Baudon, C. & Cartwright, J. 2008. The kinematics of reactivation of normal faults using high resolution throw mapping. *Journal of Structural Geology*, **30**, 1072–1084.

Beach, A. & Trayner, P. 1991. The geometry of normal faults in a sector of the offshore Nile Delta, Egypt. *In*: Roberts, A.M., Yielding, G. & Freeman, B. (eds) *The Geometry of Normal Faults*. Geological Society, London, Special Publications, **56**, 173–182, https://doi.org/10.1144/GSL.SP.1991.056.01.11

Brown, L.F., Jr., Loucks, R.G., Trevino, R.H. & Hammes, U. 2004. Understanding growth-faulted, intraslope subbasins by applying sequence-stratigraphic principles: examples from the south Texas Oligocene Frio Formation. *American Association of Petroleum Geologists Bulletin*, **88**, 1501–1522.

Bruce, C. 1973. Shale tectonics, Texas coastal area growth faults. *In*: Bally, A.W. (ed.) *Seismic Expression of Structural Styles*. American Association of Petroleum Geologists, Studies in Geology, **15**, 878–886.

Burke, K.C.B. 1972. Longshore drift, submarine canyon, and submarine fans. *American Association of Petroleum Geologists Bulletin*, **56**, 1975–1983.

Cartwright, J.A., Bouroullec, R., James, D. & Johnson, H.D. 1998. Polycyclic motion history of Gulf Coast Growth Faults from high resolution kinematic analysis. *Geology*, **26**, 819–822.

Childs, C., Nicol, A., Walsh, J.J. & Watterson, J. 2003. The growth and propagation of synsedimentary faults. *Journal of Structural Geology*, **25**, 633–648.

Damuth, J.E. 1994. Neogene gravity tectonics and depositional processes on the deep Niger Delta continental margin. *Marine and Petroleum Geology*, **11**, 321–346.

Dooley, T., Ferguson, A., Poblet, J. & McClay, K. 2000. Tectonic evolution of the Sanga Sanga Block, Mahakam Delta, Kalimantan, Indonesia. *American Association of Petroleum Geologists Bulletin*, **84**, 765–786.

Doust, H. 1990. Petroleum Geology of the Niger Delta. *In*: Brooks, J. (ed.) *Classic Petroleum Provinces*. Geological Society, London, Special Publications, **50**, 365–380, https://doi.org/10.1144/GSL.SP.1990.050.01.21

Doust, H. & Omatsola, E. 1989. Niger Delta. *In*: Edwards, J.D. & Santogrossi, P.A. (eds) *Divergent/Passive Margins*. American Association of Petroleum Geologists, Memoirs, **48**, 201–238.

Duffy, O.B., Bell, R.E., Jackson, C.A.-L., Gawthorpe, R.L. & Whipp, P.S. 2015. Fault growth and interactions in a multiphase rift fault network: Horda Platform, Norwegian North Sea. *Journal of Structural Geology*, **80**, 99–119.

Dula, W.F. 1991. Geometric models of listric normal faults and rollover folds. *American Association of Petroleum Geologists Bulletin*, **75**, 1609–1625.

Edwards, M.B. 1976. Growth faults in Upper Triassic deltaic sediments, Svalbard. *American Association of Petroleum Geologists Bulletin*, **60**, 341–355.

Ellis, P.G. & McClay, K.R. 1988. Listric extensional fault systems – results of analogue model experiments. *Basin Research*, **1**, 55–70.

Evamy, B.D., Haremboure, J., Kamerling, P., Knaap, W.A., Molloy, F.A. & Rowlands, P.H. 1978. Hydrocarbon habitat of Tertiary Niger Delta. *American Association of Petroleum Geologists Bulletin*, **62**, 277–298.

Fairhead, J.D. & Binks, R.M. 1991. Differential opening of the Central and South Atlantic oceans and the opening of the West African rift system. *Tectonophysics*, **187**, 191–203.

Fazli Khani, H. & Back, S. 2012. Temporal and lateral variation in the development of growth faults and growth strata in the western Niger Delta, Nigeria. *American Association of Petroleum Geologists Bulletin*, **96**, 595–614.

Fazli Khani, H. & Back, S. 2015*a*. The influence of differential sedimentary loading on rollover and accommodation creation in deltas. *Marine and Petroleum Geology*, **59**, 136–149.

Fazli Khani, H. & Back, S. 2015*b*. The influence of pre-existing structure on the growth of syn-sedimentary normal faults in a deltaic setting, Niger Delta. *Journal of Structural Geology*, **73**, 18–32.

Fossen, H. & Rotevatn, A. 2016. Fault linkage and relay structures in extensional settings – A review. *Earth-Science Reviews*, **154**, 14–28.

Gawthorpe, R.L. & Leeder, M.R. 2000. Tectono-sedimentary evolution of active extensional basins. *Basin Research*, **12**, 195–218.

Gibbs, A.D. 1984. Structural evolution of extensional basin margins. *Journal of the Geological Society, London*, **141**, 609–620, https://doi.org/10.1144/gsjgs.141.4.0609

Hodgetts, D., Imber, J. *et al.* 2001. Sequence stratigraphic responses to shoreline-perpendicular growth faulting in shallow marine reservoirs of the champion field, offshore Brunei Darussalam, South China Sea. *American Association of Petroleum Geologists Bulletin*, **85**, 433–457.

Hongxing, G. & Anderson, J.K. 2007. Fault throw profile and kinematics of normal fault: conceptual models and geologic examples. *Geological Journal of China Universities*, **13**, 75–88.

Hooper, R.J., Fitzsimmons, R.J., Grant, N. & Vendeville, B.C. 2002. The role of deformation in controlling depositional patterns in the south-central Niger Delta, West Africa. *Journal of Structural Geology*, **24**, 847–859.

Imber, J., Childs, C., Nell, A.R., Walsh, J.J., Hodgetts, D. & Flint, S. 2003. Hanging wall fault kinematics and footwall collapse in listric growth fault systems. *Journal of Structural Geology*, **25**, 197–208.

Ings, S.J. & Beaumont, C. 2010. Continental margin shale tectonics: preliminary results from coupled fluid-mechanical models of large-scale delta instability. *Journal of the Geological Society, London*, **167**, 571–582, https://doi.org/10.1144/0016-76492009-052

Jackson, C.A.-L. & Rotevatn, A. 2013. 3D seismic analysis of the structure and evolution of a salt-influenced normal fault zone: a test of competing fault growth models. *Journal of Structural Geology*, **54**, 215–234.

Long, J.J. & Imber, J. 2012. Strain compatibility and fault linkage in relay zones on normal faults. *Journal of Structural Geology*, **36**, 16–26.

Lopez, J.A. 1990. Structural styles of growth faults in the U.S. Gulf Coast Basin. *In*: Brooks, F. (ed.) *Classic Petroleum Provinces*. Geological Society, London, Special Publications, **50**, 203–219, https://doi.org/10.1144/GSL.SP.1990.050.01.10

Lowell, J.D. 1985. *Structural Styles in Petroleum Exploration*. OGCI Publications, Tulsa, OK.

Marten, R., Shann, M., Mika, J., Rothe, S. & Quist, Y. 2004. Seismic challenges of developing the pre-Pliocene Akhen Field offshore Nile Delta. *The Leading Edge*, **23**, 314–320.

Mauduit, T. & Brun, J. 1998. Growth fault/rollover systems: birth, growth, and decay. *Journal of Geophysical Research*, **103**, 18,119–18,136.

McClay, K.R. 1990. Extensional fault systems in sedimentary basins: a review of analogue model studies. *Marine and Petroleum Geology*, **7**, 206–233.

McCulloh, R. 1988. Differential fault-related early Miocene sedimentation, Bayou Herbert area, southwestern Louisiana. *American Association of Petroleum Geologists Bulletin*, **72**, 477–492.

Morley, C.K. & Guerin, G. 1996. Comparison of gravity-driven deformation styles and behavior associated with mobile shales and salt. *Tectonics*, **15**, 1154–1170.

Morley, C.K., Back, S., Van Rensbergen, P., Crevello, P. & Lambiase, J.J. 2003. Characteristics of repeated, detached, Miocene–Pliocene tectonic inversion events in a large delta province on an active margin, Brunei Darussalam, Borneo. *Journal of Structural Geology*, **25**, 1147–1169.

Onuoha, K.M. 1999. Structural features of Nigeria's coastal margin: an assessment base of age data from wells. *Journal of African Earth Sciences*, **29**, 485–499.

Peacock, D.C.P. & Sanderson, D.J. 1991. Displacements, segment linkage and relay ramps in normal fault zones. *Journal of Structural Geology*, **13**, 721–733.

Peacock, D.P.C. & Sanderson, D.J. 1994. Geometry and development of relay ramps in normal fault systems.

American Association of Petroleum Geologists Bulletin, **78**, 147–165.

Pochat, S., Castelltort, S., Choblet, G. & Driessche, J.V.D. 2009. High-resolution record of tectonic and sedimentary processes in growth strata. *Marine and Petroleum Geology*, **26**, 1350–1364.

Rouby, D. & Cobbold, P.R. 1996. Kinematic analysis of a growth fault system in the Niger Delta from restoration in map view. *Marine and Petroleum Geology*, **13**, 565–580.

Saller, A. & Blake, G. 2003. Sequence stratigraphy and syndepositional tectonics of Upper Miocene and Pliocene deltaic sediments, offshore Brunei Darussalam. *In*: Sidi, F.H., Nummedal, D., Imbert, P., Darman, H. & Posamentier, H.W. (eds) *Tropical Deltas of Southeast Asia – Sedimentology, Stratigraphy, and Petroleum Geology*. Society for Sedimentary Geology (SEPM), Special Publications, **76**, 219–234.

Sandal, S.T. 1996. *The Geology and Hydrocarbon Resources of Negara Brunei Darussalam*. Brunei Museum, Bandar Seri Begawan, Brunei.

Sapin, F., Ringenbach, J.-C., Rives, T. & Pubellier, M. 2012. Counter-regional normal faults in shale-dominated deltas: origin, mechanism and evolution. *Marine and Petroleum Geology*, **37**, 121–128.

Sestini, G. 1989. Nile Delta, a review of depositional environments and geological history. *In*: Whateley, M.K.G. & Pickering, K.T. (eds) *Deltas, Sites and Traps for Fossils Fuels*. Geological Society, London, Special Publications, **41**, 99–127, https://doi.org/10.1144/GSL.SP.1989.041.01.09

Shen, Z., Dawers, N.H., Törnqvist, T.E., Gasparini, N.M., Hijma, M.P. & Mauz, B. 2016. Mechanisms of late Quaternary fault throw-rate variability along the north central Gulf of Mexico coast: implications for coastal subsidence. *Basin Research*, first published online February 23, 2016, https://doi.org/10.1111/bre.12184

Short, K.C. & Staeuble, A.J. 1967. Outline of geology of Niger delta. *American Association of Petroleum Geologists Bulletin*, **51**, 761–779.

Taylor, S.K., Nicol, A. & Walsh, J.J. 2008. Displacement loss on growth faults due to sediment compaction. *Journal of Structural Geology*, **30**, 394–405.

Thorsen, C.E. 1963. Age of growth faulting in southeast Louisiana. *Gulf Coast Association of Geological Societies Transactions*, **13**, 103–110.

Van Rensbergen, P. & Morley, C. 2000. 3-D seismic study of a shale expulsion syncline at the base of the Champion Delta, offshore Brunei, and its implications for the early structural evolution of large delta systems. *Marine and Petroleum Geology*, **17**, 861–872.

Vendeville, B. & Cobbold, P.R. 1988. How normal faulting and sedimentation interact to produce listric faut profiles and stratigraphic wedges. *Journal of Structural Geology*, **10**, 649–659.

White, N.J., Jackson, J.A. & McKenzie, D.P. 1986. The relationship between the geometry of normal faults and that of the sedimentary layers in their hangingwalls. *Journal of Structural Geology*, **8**, 897–909.

Whiteman, A.J. 1982. *Nigeria: Its Petroleum Geology, Resources and Potential, Volume 1*. Graham Trotman, London.

Withjack, M.O. & Schlische, R.W. 2006. Geometric and experimental models of extensional fault-bend folds. *In*: Buiter, S.J.H. & Schreurs, G. (eds) *Analogue and Numerical Modelling of Crustal-scale Processes*. Geological Society, London, Special Publications, **253**, 285–305, https://doi.org/10.1144/GSL.SP.2006.253.01.15

Withjack, M.O., Islam, Q.T. & Pointe, P.R. 1995. Normal faults and their hanging-wall deformation: an experimental study. *American Association of Petroleum Geologists Bulletin*, **79**, 1–18.

Xiao, H. & Suppe, J. 1992. Origin of rollover. *American Association of Petroleum Geologists Bulletin*, **76**, 509–529.

Techniques to determine the kinematics of synsedimentary normal faults and implications for fault growth models

CHRISTOPHER A.-L. JACKSON[1]*, REBECCA E. BELL[1], ATLE ROTEVATN[2] & ANETTE B. M. TVEDT[2,3]

[1]*Basins Research Group (BRG), Imperial College, Prince Consort Road, London SW7 2BP, UK*

[2]*Department of Earth Science, University of Bergen, Allégaten 41, 5007 Bergen, Norway*

[3]*Present address: Petrolia AS, Espehaugen 32, Blomsterdalen, 5258 Bergen, Norway*

**Correspondence: c.jackson@imperial.ac.uk*

Abstract: Normal faults grow via a sympathetic increase in their displacement and length ('isolated model') or by rapid establishment of their near-final length prior to significant displacement accumulation ('constant-length model'). The isolated model has dominated the structural geology literature for >30 years, although some 3D seismic data-based studies support the constant-length model. Because they make different predictions regarding rift development, and earthquake size and recurrence intervals in areas of continental extension, it is critical to test these models with data from natural examples. Here we outline a range of techniques that constrain the kinematics of synsedimentary normal faults and thus test competing fault growth models. We then apply these techniques to three seismically imaged faults, showing that, in general, they grew in accordance with the constant-length model, although periods of relatively minor tip propagation and coeval displacement accumulation, characteristics more consistent with the isolated model, also occurred. We argue that analysis of growth strata represents the best way to test competing fault growth models; most studies utilizing this approach support the constant-length fault model, suggesting it may be more widely applicable than is currently assumed. It is plausible that the very early development of large faults is, however, characterized by the development of faults that, pre-linkage, grow in accordance with the isolated model; we may simply lack the data resolution, especially in the subsurface, to resolve this very early stage of fault growth.

Observations from field and subsurface datasets, complemented by the results of physical and numerical models, suggest faults grow in one of two ways: (i) via a sympathetic increase in their displacement and length, an inference seemingly consistent with displacement–length ($D-L$) scaling relationships (e.g. herein termed the 'isolated fault model': Watterson 1986; Walsh & Watterson 1988; Dawers *et al.* 1993; Cartwright *et al.* 1995; Dawers & Anders 1995); and (ii) via establishment of their near-final length relatively early in their slip history, prior to accumulation of significant displacement (Fig. 1) (e.g. herein termed the 'constant-length fault model': Morley 2002; Walsh *et al.* 2002, 2003; Childs *et al.* 2003; Schlagenhauf *et al.* 2008; Giba *et al.* 2012; Jackson & Rotevatn 2013; Nicol *et al.* 2016; Tvedt *et al.* 2016). Based largely on geometric criteria, such as along-strike displacement variations and the presence of breached relays at segment boundaries, most studies interpret that segmented fault arrays grew (for ancient examples) or grow (for active examples) in accordance with the isolated fault model: as a result, this model has dominated the structural geology literature for >30 years.

Despite making very different predictions regarding the relative timing of fault lengthening and displacement accumulation, it may be difficult to discriminate between the isolated and constant-length models using geometric criteria alone (cf. Fig. 1a, b). For example, a fault may appear 'underdisplaced' (i.e. observed displacement is less than would be predicted based on its length) because: (i) it recently formed by linkage of numerous low-displacement segments, thus it is long but has relatively little displacement (Fig. 1a) (e.g. Cartwright *et al.* 1995); (ii) it grew as a single, isolated structure that established its near-final length relatively early in its slip history and has yet to accumulate the amount of displacement predicted by a $D-L$ scaling relationship considered appropriate for the faulted host rock (Fig. 1b) (e.g. Morley 2002; Walsh *et al.* 2002, 2003; Giba *et al.* 2012; Jackson & Rotevatn 2013); or (iii) it formed in mechanically layered stratigraphy that allowed it to propagate laterally and, hence, lengthen in brittle units, but not

From: Childs, C., Holdsworth, R. E., Jackson, C. A.-L., Manzocchi, T., Walsh, J. J. & Yielding, G. (eds) 2017. *The Geometry and Growth of Normal Faults*. Geological Society, London, Special Publications, **439**, 187–217.
First published online February 7, 2017, https://doi.org/10.1144/SP439.22

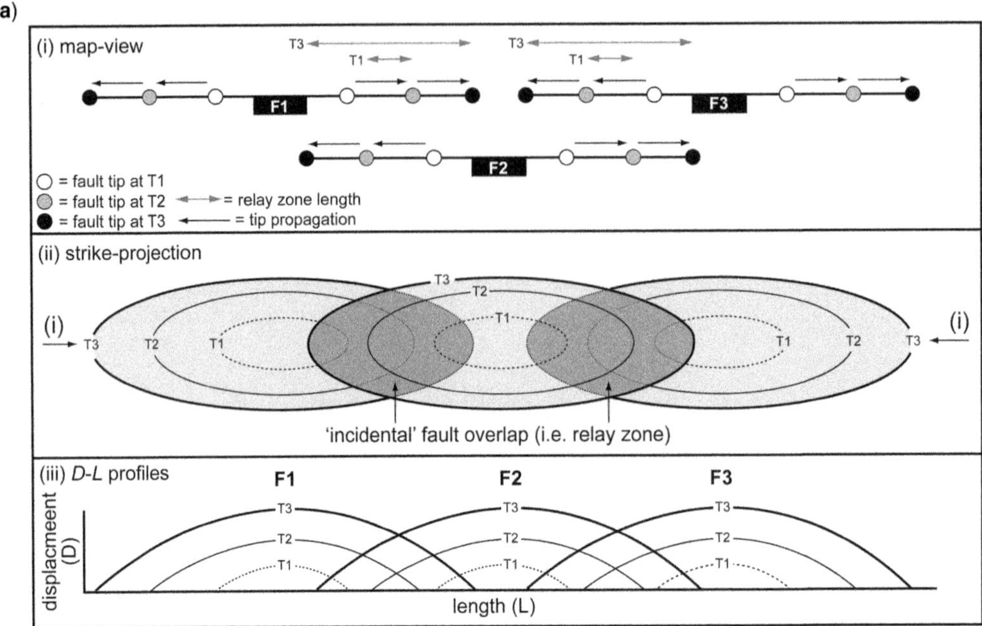

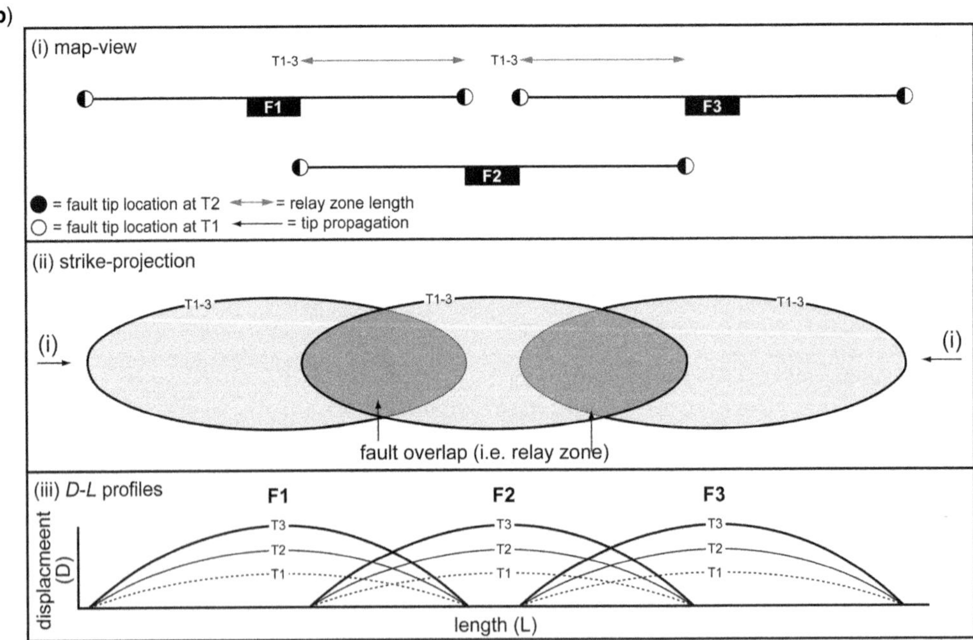

Fig. 1. Conceptual models for the development of blind normal fault systems: (**a**) the isolated fault model (Walsh & Watterson 1988; Dawers & Anders 1995; Huggins *et al.* 1995; Cartwright *et al.* 1995); and (**b**) the constant-length fault model (cf. Childs *et al.* 1995; Walsh *et al.* 2002, 2003; Giba *et al.* 2012; see also Baudon & Cartwright 2008; Jackson & Rotevatn 2013; Nicol *et al.* 2016). The (i) plan-view, (ii) strike-projection and (iii) displacement–length (*D–L*) plots are shown to illustrate the key geometrical and evolutionary aspects of each model. The black arrows in (ii) show the fault level of the map shown in (i). F1-3, faults 1–3; T1-3, time-steps 1–3. Note that, based on the final fault length (i.e. T3 in i), shape (i.e. T3 in ii) and throw distribution (i.e. T3 in iii), it is difficult to determine which growth model best describes its evolution.

accumulate displacement due to decoupling and distributed deformation in bounding ductile layers (e.g. Benedicto *et al.* 2003; Soliva & Benedicto 2005).

The style of fault growth (i.e. lengthening v. displacement accumulation) differs between the isolated and constant-length models: these models thus make very different predictions regarding the physiographical and tectonostratigraphic evolution of continental rifts. For example, the isolated model predicts that relay ramps, a key geomorphic feature in many rifts, develop relatively late in the growth history of adjacent fault segments, forming only after fault tips overlap (e.g. Gawthorpe & Leeder 2000; Athmer *et al.* 2010; Athmer & Luthi 2011). Consequently, relay ramps may only control synrift sediment dispersal after a significant amount of rift-related extension (Fig. 1b). In contrast, the constant-length model predicts that relay ramps form geologically instantaneously, and that sediment dispersal may thus be controlled much earlier during rift development. The way in which normal faults grow and interact during continental extension may also control the recurrence interval, magnitude and location of potentially hazardous earthquakes (e.g. Walsh *et al.* 2003; Nicol *et al.* 2005, 2010; Soliva *et al.* 2008). Testing these two models and assessing their general applicability is therefore critical but, in our view, has received surprisingly little attention.

The aim of this paper is to test the isolated and constant-length fault models using data from normal faults imaged in 3D seismic reflection data. To do this we begin by reviewing four simple techniques that can help resolve the kinematics of synsedimentary faults: (i) expansion index (EI) analysis; (ii) isopach or isochore map analysis; (iii) displacement backstripping; and (iv) relay-zone backstripping. These techniques directly or indirectly use synkinematic (growth) strata to explicitly record periods of fault slip and lengthening; more specifically, they allow us to assess whether the lateral tips are static, or whether they propagated as the fault slipped and accumulated displacement (Fig. 1) (e.g. Childs *et al.* 2003). Numerous previous authors have utilized these techniques (see the references cited later in this paper); however, these techniques have typically been applied in isolation. As we argue below, it is critical to integrate them to more accurately constrain fault kinematics, in particular the lateral propagation of fault tips. We also discuss some of the uncertainties or limitations associated with the implementation of these techniques. Having described them, we then apply these techniques to three synsedimentary fault segments and arrays imaged in high-quality 3D seismic reflection data. We show that, in general, all three faults grew in accordance with the constant-length model, although periods of relatively minor tip propagation and coeval displacement accumulation occurred. We conclude by discussing the general applicability of the two competing fault growth models.

Techniques to determine the growth of normal faults

Expansion index analysis

Overview. Small faults and those that nucleate at significant burial depths may be 'blind' throughout their history and never intersect the free surface. In contrast, large faults, or those that nucleate near and thus rapidly intersect the free surface, cause differential subsidence and the accumulation of a thicker, more fully preserved succession in their hanging wall. The same is true for initially small faults that nucleate at depth, but which then grow and eventually propagate upwards to intersect the free surface. Faults that intersect the free surface, and influence basin geometry, subsidence and stratigraphic architecture, are commonly referred to as 'growth faults' (e.g. Childs *et al.* 2003). If a fault intersects the free surface, and its slip rate is less than or equal to the rate of sediment accumulation, fault-driven hanging-wall subsidence results in across-fault thickening of the synkinematic sequence, which thus records the initiation, cessation and, therefore, duration of faulting. The timing and magnitude of across-fault thickening of growth strata can be constrained by the construction of expansion index (EI) plots, which are generated by dividing the hanging-wall thickness (H_t) of a stratal unit by its corresponding footwall thickness (F_t) and plotting these data directly against geological time, or a proxy such as depth in cases where absolute ages are not available (i.e. H_t/F_t: Fig. 2) (Thorsen 1963; see also Cartwright *et al.* 1998; Bouroullec *et al.* 2004; Jackson & Rotevatn 2013; Tvedt *et al.* 2013; Robson *et al.* 2016). An index of 1 suggests no across-fault thickening and, therefore, no syndepositional fault activity, whereas an index of >1 suggests across-fault thickening and syndepositional fault activity (see the discussion of errors and uncertainties in the following subsection) (Thorsen 1963; Cartwright *et al.* 1998; Bouroullec *et al.* 2004; Jackson in press). An index of <1 suggests stratal thinning from the footwall to the hanging wall, a relatively unusual circumstance that may reflect difficulties in accurately measuring stratal thicknesses adjacent to a fault (Thorsen 1963).

EI analysis provides important information on the 1D evolution of growth faults (i.e. the slip history at a specific along-strike location: e.g. Fig. 2a) (e.g. Thorsen 1963; Cartwright *et al.* 1998; Bouroullec *et al.* 2004; Jackson & Rotevatn 2013; Jackson in press; Robson *et al.* in press). However, EI

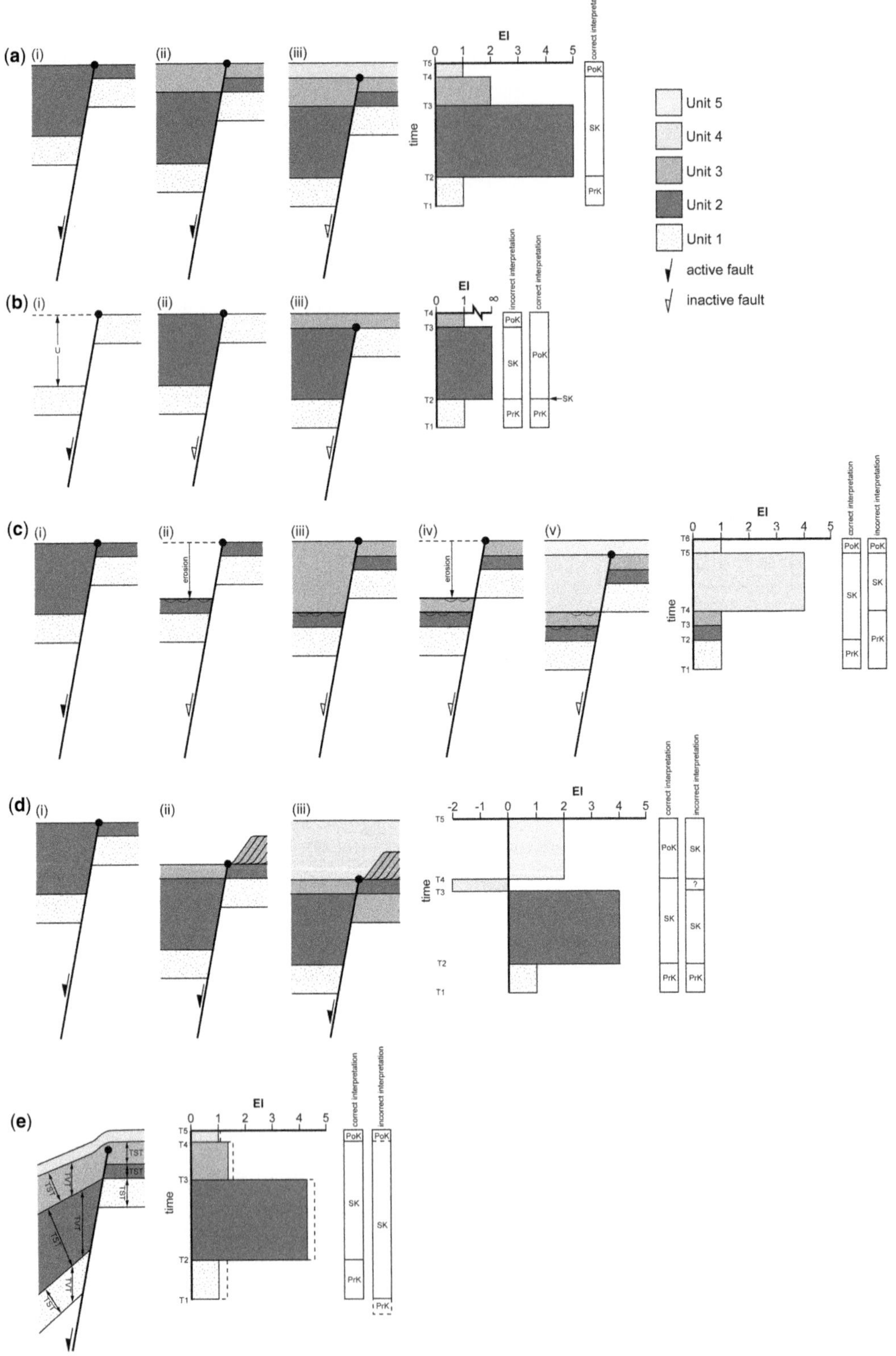

(a) (i)
(ii)
(iii)
EI
time
T1
T2
T3
T4
T5
correct interpretation
PoK
SK
PrK
Unit 5
Unit 4
Unit 3
Unit 2
Unit 1
active fault
inactive fault
(b) (i)
(ii)
(iii)
incorrect interpretation
correct interpretation
(c) (i)
(ii)
(iii)
(iv)
(v)
erosion
T6
(d) (i)
(ii)
(iii)
(e)
TST
TVT

analysis can also be a powerful and, we argue, hitherto under-utilized tool when assessing the along-strike propagation of normal faults and, therefore, the applicability of the isolated and constant-length models. For example, construction of several EI plots along strike of a fault may reveal the strike extent of the fault, with the transition between values of 1 and >1 allowing the locations of the faults tips to be constrained for specific time intervals (Fig. 3).

Errors and uncertainties. Pre- and synkinematic wall rocks may experience seismic-scale strain during the growth of normal fault systems. For example, normal drag, reverse drag and fault-propagation folding all result in stratal rotations, and, in some cases, shear-thinning of wall rocks (e.g. Fig. 2e) (e.g. Childs *et al.* 2009; Ferrill *et al.* 2012). Strain-related changes in wall-rock thickness are most likely and are greater for large faults that have accumulated large amounts of slip; these processes may thus lead to the extraction of thickness values and, therefore, expansion indices that are not representative of their near-surface values. Care must therefore be taken to ensure that strain-related thickness variations, which may equally impact seismic- and borehole-based analysis of stratal thicknesses, are taken into account when calculating expansion indices.

Across-fault miscorrelation of strata represents another potential source of error when calculating expansion indices, complicating our ability to accurately reconstruct fault growth histories. However, the high-quality imaging available in many modern 3D seismic reflection datasets means it is typically straightforward to accurately correlate seismic reflections and sequences across fault systems or, better still, by mapping reflections around fault tips. In poor-quality 2D or 3D datasets, in widely spaced 2D datasets, or in locations where structurally or stratigraphically complex geology occurs, across-fault correlation may be more problematic. Furthermore, a poor understanding of subsurface velocities may hamper the conversion of thickness values extracted in milliseconds two-way time (ms TWT), from time-migrated data, to metres. This may be especially problematic when studying large displacement faults, where more deeply buried hanging-wall strata may be more compacted and thus acoustically faster than correlative footwall strata. In such cases, faster velocities must be used to depth-convert hanging-wall layer thicknesses, otherwise these layers may be erroneously calculated to be thinner than the footwall units when, in fact, they are of similar or greater thickness.

Burial-related (i.e. vertical) compaction reduces the primary (depositional) thickness of pre-, syn- and post-kinematic layers. As demonstrated by Taylor *et al.* (2008), this can influence the calculation of fault displacement in cases where the faulted growth sequences have a high shale content (>70%) and post-faulting burial is large (i.e. kilometre-scale). In addition to syndepositional fault slip, compaction also clearly impacts the layer thickness itself, which clearly would vary between hanging wall and footwall as a function of the former being buried more deeply than the latter. Decompaction and back-stripping can help reduce uncertainties related to

Fig. 2. Diagrams illustrating the concept of EI analysis of growth faulting and some of the interpretation pitfalls. PrK, prekinematic unit; SK, synkinematic unit; PoK, post-kinematic unit. **(a)** (i) Fault initiation at T2 – syndepositional faulting recorded by unit 2; (ii) fault growth during T2–T3 – syndepositional faulting recorded by unit 3; and (iii) fault death at T4 – cessation of faulting recorded by deposition of tabular post-kinematic unit 4 **(b)** (i) Fault initiation, rapid slip and death at T2 – rapid creation of hanging-wall accommodation (U) that is not filled by a true synkinematic unit; (ii) fault inactive during T2–T3 – underfilled hanging-wall accommodation passively filled by post-kinematic unit that is absent in the footwall; and (iii) fault inactive during T3–T4 – hanging wall and footwall capped by post-kinematic unit 3. **(c)** (i) Fault initiation at T2, growth during T2–T3 and death at T3 – syndepositional faulting recorded by unit 2; (ii) preferential erosion of the hanging-wall portion of unit 2; (iii) filling of the hanging wall and capping of the footwall by post-kinematic unit (unit 3); (iv) preferential erosion of the hanging-wall portion of unit 3; and (v) filling of the hanging wall by the post-kinematic unit (unit 4); hanging wall and footwall capped by post-kinematic unit 5. **(d)** (i) Fault initiation at T2 and growth during T2–T3 – syndepositional faulting recorded by unit 2; (ii) fault growth during T3–T4 – hanging-wall accommodation filled by synkinematic unit 3, which is thicker in the footwall due to the deposition of a deltaic body; and (iii) fault death at T4 – capping of the hanging-wall and footwall deltaic body by post-kinematic unit 4. **(e)** Variability in EI values, and thus kinematic history reconstruction, resulting from hanging-wall bed rotations and measurement of the true vertical thickness (TVT) instead of true-stratigraphic thickness (TST). (i) Fault inactive during T1–T2 – prekinematic unit 1 is the same thickness in the footwall and hanging wall of the fault, although fault-related folding results in the rotation of hanging-wall bedding. Because TVT > TST, measurement of TVT rather than TST would result in an EI value >1 and the erroneous interpretation that faulting started at T1 rather than T2; (ii) fault initiation at T2 and growth during T2–T4 – syndepositional faulting recorded by synkinematic units 2 and 3. Because these units have been tilted, a range of (near) TST and TVT values and, hence, EI values can be calculated, all of which are >1; and (iii) fault death at T4 – capping of the hanging wall and footwall by post-kinematic unit 4.

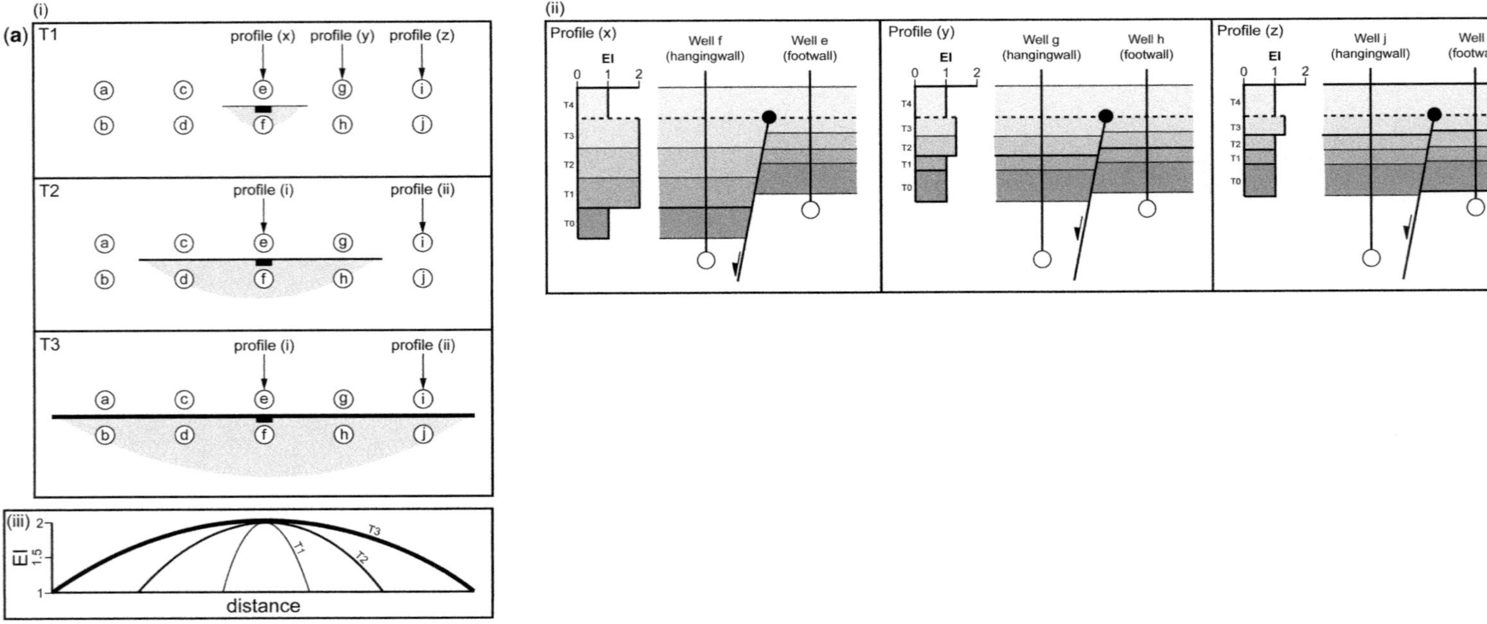

Fig. 3. Schematic diagrams illustrating how sequential EI plots, taken along strike of a fault, can help constrain the fault palaeo-tip locations. (**a**) A case where the fault grew by tip propagation, and progressive lengthening and displacement accumulation (isolated fault model: see Fig. 1a). Note that the unit defining the onset of faulting and, thus, the prekinematic/synkinematic boundary varies along strike, climbing to stratigraphically higher levels towards the fault tips.

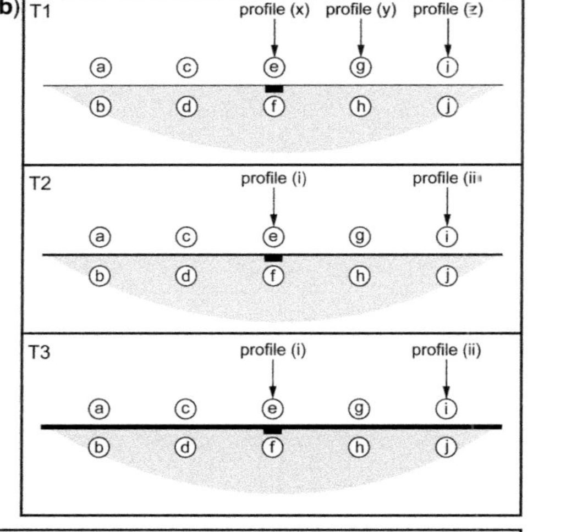

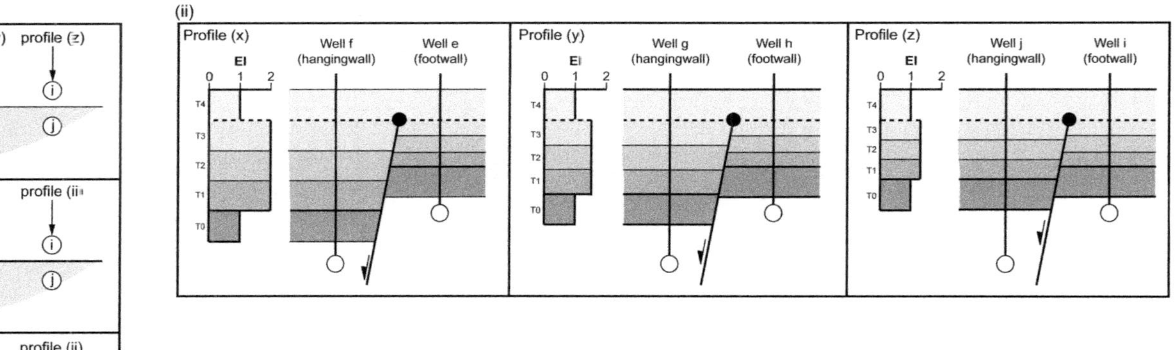

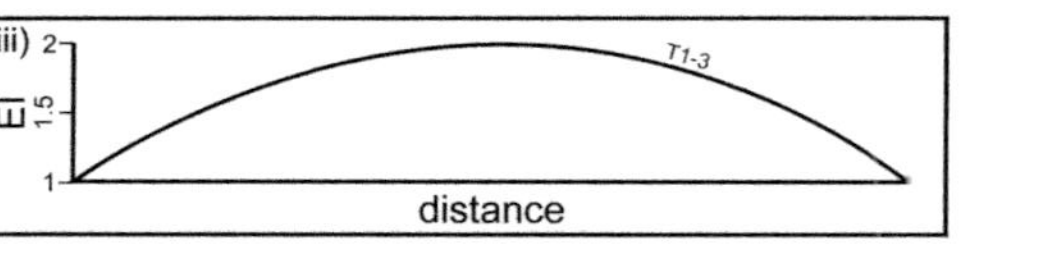

Fig. 3. (**b**) (*Continued*) A case where the fault rapidly attained its length and then grew by displacement accumulation (constant-length model). Note that the unit defining the onset of faulting and, thus, the prekinematic–synkinematic boundary is the same at each along-strike position. (i) Fault maps at the base of the synkinematic sequence. Locations of hypothetical boreholes and boreholes shown in (ii) are indicated; (ii) cross-sections and expansion index (EI) plots for (*x*) near the fault centre, (*y*) between the fault centre and its tip, and (*z*) immediately inboard of the fault tip. The base of the synkinematic package is shown as a thick black line; the top of the synkinematic unit is shown in a dashed line; and (iii) expansion index (EI) against distance plots showing the inferred locations of the fault palaeo-tip for times 1–3.

burial-related modification of the at-surface, syndepositional unit thickness (cf. Taylor *et al.* 2008).

Because of: (i) strain-related changes in wall-rock thickness; (ii) across-fault miscorrelation of seismic sequences; (iii) poor seismic imaging; (iv) a poor understanding of subsurface velocities; and (iv) vertical compaction, the geological significance of EI values that are only slightly >1 are questionable: that is, do these represent subtle across-fault changes in thickness and, thus, periods of fault activity or are they simply artefacts of the method and the resolution of the input data? Furthermore, as we discuss below, when describing uncertainties associated with isochrone analysis, across-fault changes in stratal thickness may not simply reflect changes in accommodation driven by syndepositional faulting (Fig. 2c, d). Because of these uncertainties, EI analysis must be judiciously applied when studying the growth history of syndepositional normal faults. The robustness and, ultimately, the quality of extracted EI values must be assessed on a case-by-case basis based on seismic data quality, mapping confidence, observed wall-rock strain and the geological context of the fault under inspection.

Isochrone analysis

Overview. Finite and incremental slip vary along strike of growth faults, resulting in the formation of 'scoop-shaped' hanging-wall depocentres defined by a fault-normal syncline, the axes of which are located near, and plunge towards, the high displacement fault centres (Fig. 4) (e.g. Schlische 1995; Young *et al.* 2001; Jackson *et al.* 2002; Morley 2002; Withjack *et al.* 2002; Gawthorpe *et al.* 2003). Growth strata thus thicken towards the point of maximum displacement on a fault, which is generally close to its centre, and thin along strike towards its tips (see the hanging-wall strike sections in Fig. 4), with these variations recorded in seismically constrained 'isochrons' (i.e. stratal thickness maps, in milliseconds (ms) or seconds (s) TWT, derived from time-domain seismic reflection data) or 'isopachs' (i.e. stratal thickness maps, in metres or feet, derived from depth-domain seismic reflection data or boreholes). These maps, generated for specific stratigraphic intervals within the growth sequence, record the shape, size and location of fault-controlled depocentres through time, thereby documenting the growth history of the bounding faults. The isolated and constant-length models both predict that, through time, hanging-wall depocentres subside and accumulate growth strata. However, the models make markedly different predictions regarding the shape, size and location of these synkinematic depocentres. The isolated model, which envisages progressive lengthening of the fault through time, predicts a corresponding lengthening of the depocentres (Fig. 4a); in contrast, the constant-length model, which envisages early establishment of fault length, predicts a correspondingly early establishment of depocentre length, with the latter stages of faulting associated with depocentre deepening but *not* lengthening (Fig. 4b). Hence, it should be possible to discriminate between these two models using growth strata thickness patterns recorded in isochron and isopach maps.

Errors and uncertainties. A key assumption when using thickness maps (and EI analysis; see above) to infer periods of fault activity is that the observed thickness changes are solely the result of syndepositional, fault-driven changes in accommodation and/or stratigraphic preservation. This assumption may be valid if the sedimentary system of interest is large relative to the fault being studied. This assumption may *not* be valid if: (i) thickness changes arise due to spatially varying patterns in sediment supply and accumulation rate, which, for example, may occur, in point-sourced sedimentary systems such as deltas and submarine lobes (Fig. 2d); (ii) the hanging-wall basin is underfilled: for example, during a period characterized by high fault slip rates or low sediment accumulation rates – in this case, subsequent filling of this relief would result in deposition of a layer with a high EI value, which may then be erroneously assumed to record a period of fault slip when, in fact, the main period of rapid slip occurred *before* the unit was deposited (Fig. 2b); and (iii) thickness changes occur due to spatially varying patterns in stratigraphic preservation related to, for example, erosion at the base of channellized systems focused in the fault hanging wall – in this case, variable sediment preservation may result in thickness variations unrelated to syndepositional normal faulting and the extraction of incorrect growth histories (Fig. 2c).

Whether a hanging-wall basin was filled during fault growth can be assessed by looking for evidence of fault scarp creation, which may indicate periods when slip rates outpaced sediment accumulation rates and at-surface relief formed. Furthermore, across-fault changes in facies in time-equivalent units, deduced from borehole data or seismic-geomorphological analysis, may indicate across-fault changes in water depth related to variable sedimentation rates on either side of the fault. The stratigraphic or seismic-stratigraphic architecture of the hanging-wall deposits may also indicate if deposition occurred during or after faulting. For example, in situations where strong hanging-wall rotations have occurred, true synkinematic deposits are typically characterized by wedge-shaped deposits, within which individual stratigraphic packages

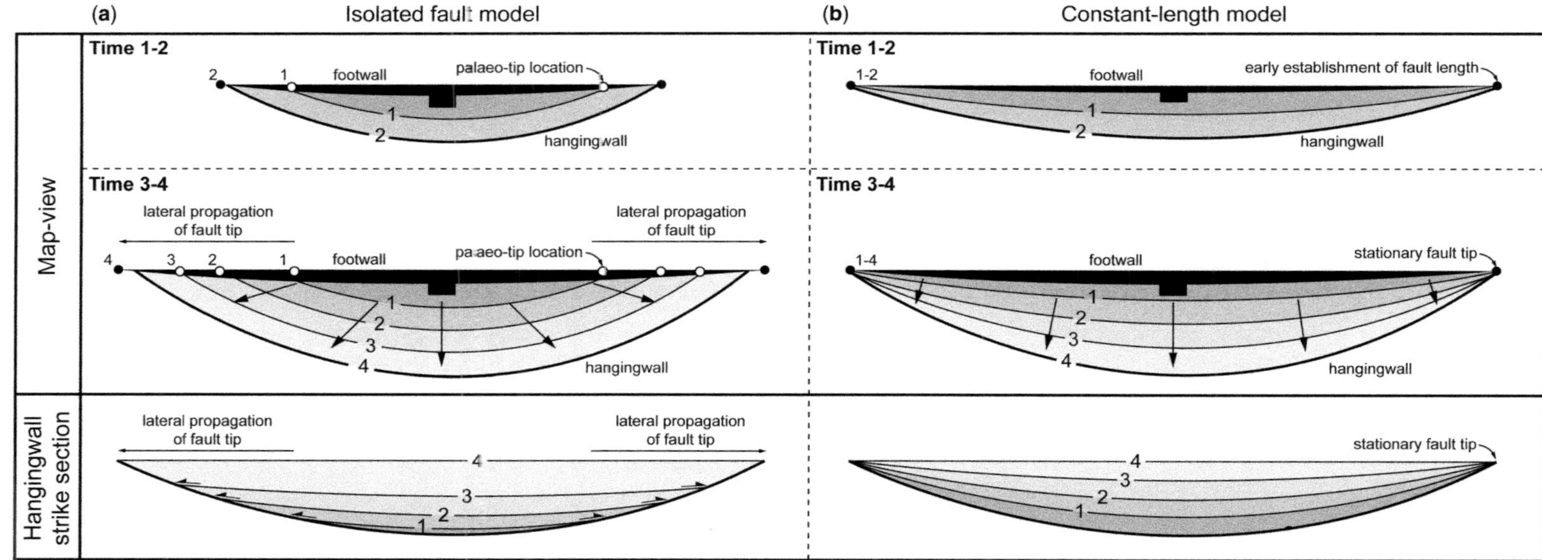

Fig. 4. Schematic diagrams illustrating how the stratigraphic architecture of the hanging-wall synrift sequence can provide insights into the growth model of basin-bounding normal faults. Basin geometry and synrift stratigraphic architecture associated with (**a**) the isolated and (**b**) constant-length models. Modified from Morley (2002).

expand towards the fault. Furthermore, stratal dips may decrease upwards within the synkinematic package, testifying to syndepositional, fault-driven hanging-wall rotation. In contrast, post-kinematic deposits, although also defined by an overall wedge-shaped geometry, are internally characterized by tabular packages that do not expand towards the fault and which onlap a single discrete surface onto the hanging wall (see Prosser 1993; Morley 2002).

Even in cases where thickness changes are driven purely by normal faulting, and sediment accumulation rate exceeds fault slip rate, poor-quality seismic data and/or complex geology may also lead to uncertainties in mapping growth strata; this may therefore lead to the extraction of incorrect growth histories. For example, seismic data may be unable to resolve thin stratal packages that provide a stratigraphic record of the very earliest stages of fault growth, when tip propagation and segment linkage, as embodied by the isolated fault model, occurred (see also Morley 2002). If evidence for this earliest phase of deformation cannot be resolved, then our understanding of the fault growth history will be incomplete. To help tackle this, and although we are unable to do it with the case studies presented below, we recommend that stratigraphic data from boreholes are integrated with a seismic-reflection-based analysis. These data may help define thickness patterns in subseismic, synkinematic strata, and may thus help to define the position of fault segment boundaries and fault tips, and the shape, size and location of synkinematic depocentres.

Displacement backstripping

Overview. Displacement (or throw) backstripping is typically used to constrain the style of growth of seismic-scale segmented normal fault arrays. The technique was initially developed by Chapman & Meneilly (1991) to determine the kinematics of a reverse reactivated normal fault imaged in 3D seismic reflection data from the Southern North Sea, before being applied by Petersen *et al.* (1992) to a salt-detached normal fault imaged in 2D seismic reflection data from the Danish North Sea. More recently, this method has been most profitably applied to normal faults imaged and mapped in 3D seismic reflection data (e.g. Childs *et al.* 1993, 2003; Rowan *et al.* 1998; Dutton & Trudgill 2009; Giba *et al.* 2012).

There are two displacement/throw backstripping methods, which, for historical consistency, are herein referred to as the ‘original method’ (Chapman & Meneilly 1991; Petersen *et al.* 1992) and the ‘modified method’ (Rowan *et al.* 1998) (cf. also the ‘maximum throw subtraction method’ of Dutton & Trudgill 2009). Both methods are applied to displacement/throw data extracted at multiple stratigraphic levels along-strike of a normal fault. The original method directly subtracts displacement/throw measured across a shallower horizon from throw measured across a deeper horizon at the same along-strike position, with this process repeated for successively deeper horizons. For synsedimentary faults, which are typically broadly rectangular and characterized by closely spaced, subhorizontal displacement (or throw) contours near their upper tips and steep contours near their lateral tips (e.g. Nicol *et al.* 1996; Childs *et al.* 2003), this method may reveal lateral propagation of the fault tip and may thus help constrain the mode of fault growth. The original method makes no assumptions about the patterns of displacement/throw accumulation and therefore the style of fault growth.

The modified method calculates the displacement/throw pattern at the time of deposition of a given horizon by subtracting the maximum displacement/throw (T_{max}) along the horizon from the entire fault surface (e.g. Rowan *et al.* 1998; see also Dutton & Trudgill 2009). By doing this, Rowan *et al.* (1998, p. 318) assumed that ‘net slip during some time interval was constant inside the contour line for that value and decreased outward from there toward the tip line’. They recognized that this assumption was a simplification, with throw variations likely to be more complex, but argued that this led to ‘more realistic results than the original displacement backstripping technique’ (see Rowan *et al.* 1998, p. 318). The modified method therefore assumes, *a priori*, that the fault under inspection lengthened as it accumulated displacement and thus grew in accordance with the isolated fault model. As we show below, the fundamental problem with this assumption is that it ignores thickness changes in growth strata preserved adjacent to the studied fault and, in the cases shown below, results in erroneous assessments of fault kinematics.

Errors and uncertainties. Displacement profiles from multiple stratigraphic levels or seismic horizons spanning the entire fault surface underpin the displacement/throw backstripping procedure (e.g. Childs *et al.* 1995, 2003). The accuracy and fidelity (i.e. how many slip or lengthening increments are resolved) of the backstripping results, and the inferred growth history are, therefore, sensitive to: (i) the accuracy of mapping – that is, thickness changes can be induced by inaccurate mapping and erroneous across-fault correlation of stratigraphic units; (ii) pre-faulting stratigraphic architecture – that is, we assume that thickness changes occur solely due to fault-induced changes in accommodation and not spatial variations in sediment accumulation (see Fig. 2); and (iii) the number of horizons used in the backstripping procedure – that is, the number of slip increments and periods of tip propagation or stasis resolvable is determined by the number of horizons that can be mapped in synsedimentary strata; too few horizons may mean that subtle but important periods of fault growth go unresolved.

Relay-zone backstripping

Overview. The methods described above focus on the impact of faulting on basin geometry and accommodation, and associated across-fault changes in stratigraphic thickness. Giba *et al.* (2012) recently introduced another method called 'relay-zone backstripping' that focuses on the kinematics of relay zones developed between faults, instead of the faults themselves (Fig. 5). Relay-zone backstripping is based on the observation that the growth of faults is associated with, and accommodated by, deformation of rocks between their overlapping tips (e.g. Peacock & Sanderson 1991). Changes in the geometry of a relay zone (e.g. in its length and dip) should therefore help us discriminate between the isolated and constant-length fault models. More specifically, if the bounding faults lengthen through time, evolving from underlapping to neutral to overlapping, the relay ramp, as defined at the base of the synkinematic sequence, should lengthen and steepen through time (Fig. 5a). In contrast, if the faults rapidly established their lengths and, for a significant period of their growth histories, were characterized by overlapping fault tips and dominated by displacement accumulation, we would expect the relay ramp to maintain a constant length whilst it steepened (Fig. 5b). These changes in relay-ramp geometry should be recorded by thickness changes in growth strata deposited on the ramp (parts iii and iv in Fig. 5a, b).

Relay-zone backstripping is a relatively simple graphical technique involving a number of steps shared with classic displacement/throw backstripping. The first step involves the extraction of a seismic profile passing down the centre of the relay ramp. This profile should trend broadly parallel to fault strike, extending from the footwall of one fault into the hanging wall of another. The second step involves mapping of the base of the synkinematic sequence and several age-constrained surfaces within the growth sequence itself; the more horizons mapped, the greater the temporal resolution of relay-zone development that can be achieved. It is likely that stratal dips within the relay zone will decrease upwards, reflecting the fact that younger strata would have spent less time within the developing relay and will have thus rotated less. The third step involves the removal or 'backstripping' of the uppermost layer and 'retrodeformation' of underlying layers using vertical shear and an appropriate decompaction algorithm, such that the horizon defining the top of the underlying sequence is flat. The amount of decompaction applied during this step is largely determined by the lithology of the faulted sequence, which, if possible, should be constrained by borehole data. It should be noted that backstripping back to a flat datum assumes the basin was overfilled, and that, for each backstripping step, the free/depositional surface was flat and that the relay ramp had no geomorphic expression; this is only applicable in situations where sediment accumulation rate matched or exceeded fault slip rate (Childs *et al.* 2003; Giba *et al.* 2012; Jackson & Rotevatn 2013). In cases where water depth changes are suspected or can be demonstrated by biostratigraphic data or seismically defined stratigraphic architectures (e.g. clinoforms), down-ramp changes in water depth can be accommodated in the restoration step by adjusting back to an inclined/non-planar datum.

Having backstripped the uppermost layer, and decompacted and restored underlying layers, the length and dip of the ramp at the base of the synkinematic sequence are measured. To accurately measure ramp length and dip, inflection points at its top and base must be defined, based on the onset of down-ramp thickening of the uppermost synkinematic package. Once the length and dip of the relay ramp at the base of the synkinematic sequence for this youngest time period is determined, backstripping, decompaction and restoration is then repeated for progressively older synkinematic layers; at each stage, we constrain the length and dip of the relay ramp at the base of the uppermost synkinematic sequence. Next, having extracted ramp length and dip for each growth increment, a number of graphical techniques can then be used to investigate if the geometry of the ramp, as defined at the base of the synkinematic sequence, changed through time. For example, ramp evolution can most simply be displayed by plotting the ramp geometry through time (i.e. Giba *et al.* 2012, figs 11a & 12b). However, as shown by Giba *et al.* (2012, fig. 12) and Jackson & Rotevatn (2013, fig. 12h), changes in ramp length and dip between time periods can be more clearly assessed by plotting the sequential ramp geometry against a normalized vertical scale (see the example below).

In addition to constraining changes in ramp geometry at the base of the synkinematic sequence, incremental changes in ramp shape for each time interval can also be constrained. This is achieved by extracting the ramp geometry at the base of the uppermost growth layer after each backstripping, decompaction and restoration step. To allow the

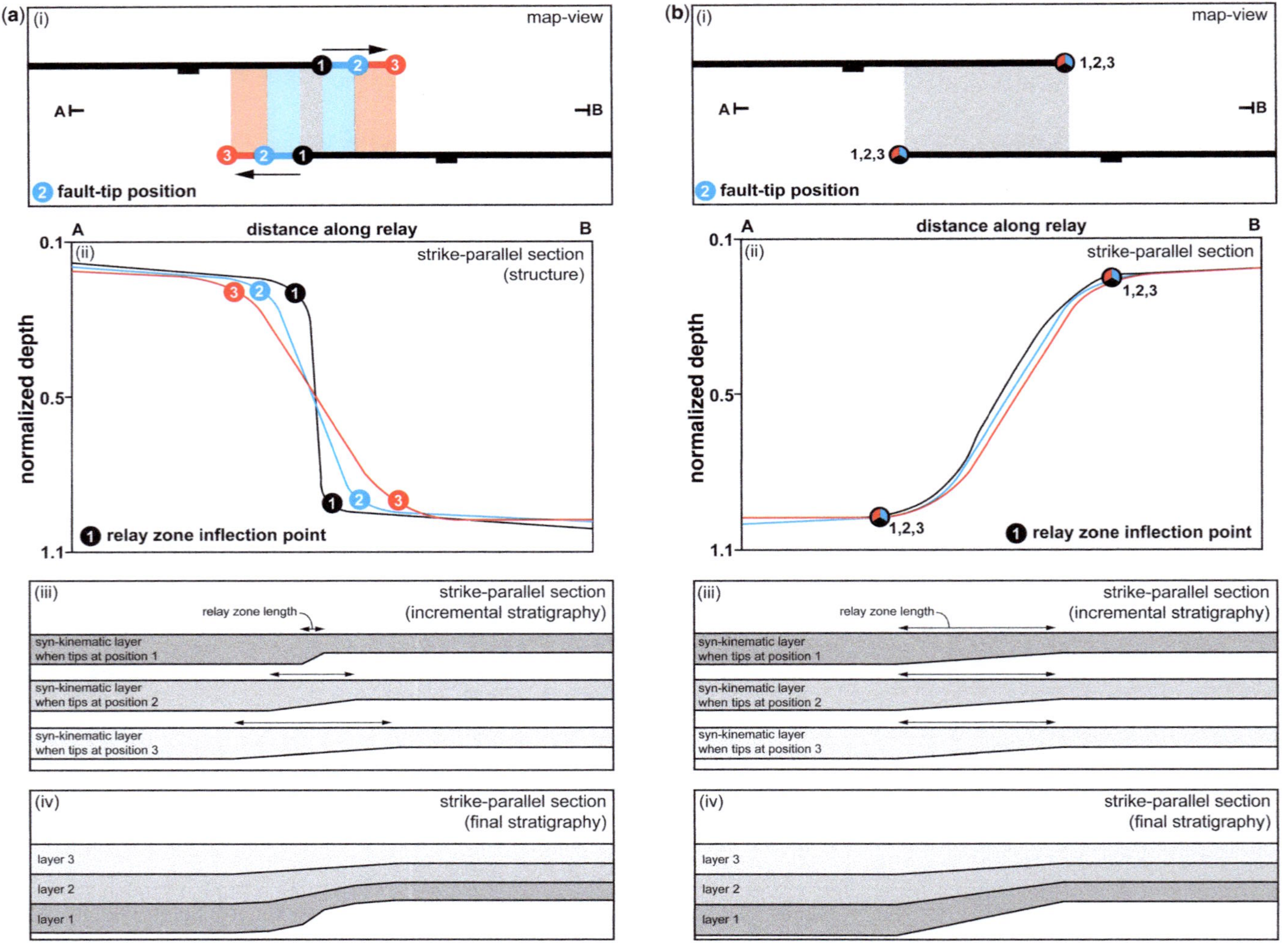

(a)
(i)
map-view
A
B
1
2
3
fault-tip position
(ii)
distance along relay
strike-parallel section
(structure)
normalized depth
0.1
0.5
1.1
relay zone inflection point
(iii)
strike-parallel section
(incremental stratigraphy)
relay zone length
syn-kinematic layer
when tips at position 1
syn-kinematic layer
when tips at position 2
syn-kinematic layer
when tips at position 3
(iv)
strike-parallel section
(final stratigraphy)
layer 3
layer 2
layer 1
(b)
(i)
map-view
A
B
1,2,3
fault-tip position
(ii)
distance along relay
strike-parallel section
normalized depth
0.1
0.5
1.1
relay zone inflection point
(iii)
strike-parallel section
(incremental stratigraphy)
relay zone length
syn-kinematic layer
when tips at position 1
syn-kinematic layer
when tips at position 2
syn-kinematic layer
when tips at position 3
(iv)
strike-parallel section
(final stratigraphy)
layer 3
layer 2
layer 1

incremental ramp geometry to be compared for each time step, extracted profiles can again be displayed against a normalized vertical scale. As shown by Giba *et al.* (2012, cf. their fig. 12a, b) and Jackson & Rotevatn (2013, cf. their fig. 12h, i), analysis of the incremental ramp geometry and analysis of the base-ramp geometry yield very similar results in terms of relay and fault kinematics.

Errors and uncertainties. Decompaction is a key step in relay-zone backstripping and backstripping in general, with this statement being especially true for kilometre-scale faults cross-cutting very fine-grained sequences (>70% mudstone) (Taylor *et al.* 2008). Accurate decompaction of buried strata to near-surface (i.e. pre-burial) porosity and thicknesses is largely dependent on the availability of borehole data to constrain host-rock composition and the accuracy of the implemented decompaction algorithms. Decompaction uncertainties will clearly impact the calculation of growth strata thickness. Furthermore, because the magnitude of decompaction increases with increasing stratigraphic thickness, decompaction uncertainties will also impact our estimation of the magnitude of down-ramp thickening of growth strata and, as a result, the dip of the surface bounding the base of the particular stratigraphic unit under investigation. However, decompaction uncertainties will not impact the estimation of ramp and, hence, fault length through time.

Relay-zone backstripping assumes that the vertical simple-shear algorithm accurately reflects the main deformation mechanism and the sense of host-rock displacement occurring during relay growth. This assumption is perhaps reasonable, given that relay-zone deformation is dominated by vertical subsidence, with only relatively minor amounts of relay-zone rotation, and that unbreached relays are typically unfaulted and not complexly folded.

Finally, relay-zone backstripping assumes that changes in relay-ramp stratal thickness are due to ramp rotation-related changes in accommodation and not spatial variations in sediment accumulation (see Fig. 2). This assumption may be valid if the fault under investigation is located in a distal part of a basin where deposition is characterized by the sequential addition of layers that, if it were not for fault- and relay-ramp-related changes in subsidence, would be isopachous. However, in cases where segmented faults develop along the basin margin, the associated relay ramps may be the site of preferential sediment accumulation (e.g. Gawthorpe & Hurst 1993; Athmer *et al.* 2010; Athmer & Luthi 2011). Failure to recognize and take into account localized sedimentary build-ups may result in seemingly anomalous ramp geometries during backstripping, especially if one incorrectly utilizes a flat datum during the second restoration step outlined above. The potential impact of spatially variable changes in depositional thickness would need to be assessed on a case-by-case basis through the use of borehole data and/or the analysis of seismic-stratigraphic architectures and seismic facies.

Growth histories of natural normal faults

In this section we highlight how the techniques described can help determine the growth histories of segmented normal fault arrays. To achieve this, we analyse three normal fault arrays imaged in high-quality, 3D seismic reflection data. We stress the importance of using growth strata to provide direct constraints on fault tip behaviour through time, and illustrate how erroneous growth histories may be extracted if the wrong displacement backstripping method is used.

Gulf of Suez Rift, Egypt

The first case study is located in the Suez Rift, Egypt. The studied fault segment (Fault 2) is *c.* 2 km in height, *c.* 5.2 km long and has a maximum throw of *c.* 390 m. The fault detaches downwards into Messinian salt and forms part of a *c.* 30 km-long fault array comprising four segments soft-linked by unbreached relay zones (the October Fault Zone: Figs 6 & 7) (see Jackson & Rotevatn 2013 for details). Jackson & Rotevatn (2013) demonstrate

Fig. 5. Schematic diagrams illustrating the concept of relay-zone backstripping. (**a**) Backstripping of a relay zone formed between two fault segments that propagated laterally through time: (i) map view – 1 and 2 indicate the palaeo-tip positions; 3 is the current tip location: grey, blue and red shading indicates the relay-zone length through time; (ii) down-relay-zone profile parallel to fault strike – the relay zone becomes more gently dipping but longer through time; and (iii) schematic down-relay-zone profile parallel to fault strike showing the final stratigraphic architecture documenting the three stages of fault and relay-zone growth. (**b**) Backstripping of a relay zone formed between two fault segments that did not propagate through time and had static tips: (i) map view – 1–3 indicate the palaeo-tip/current tip location: grey shading indicates the current relay-zone length, which has not changed through time; (ii) down-relay-zone profile parallel to fault strike – the relay zone maintains dip and length through time; and (iii) schematic down-relay-zone profile parallel to fault strike showing the final stratigraphic architecture documenting the three stages of fault and relay-zone growth. See also Giba *et al.* (2012).

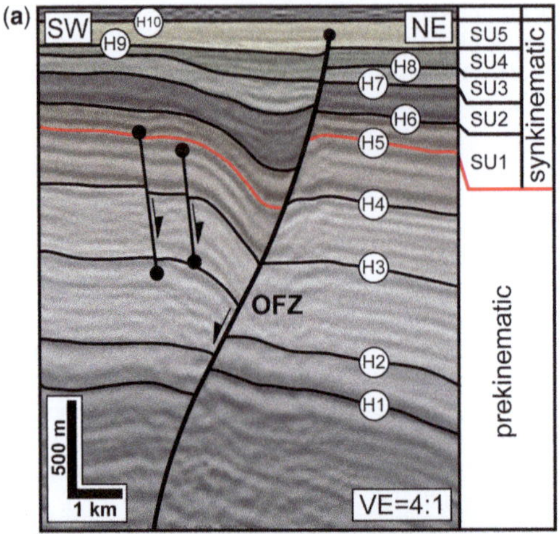

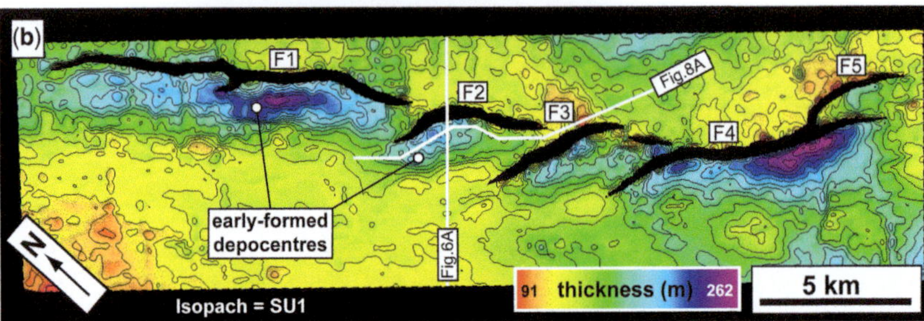

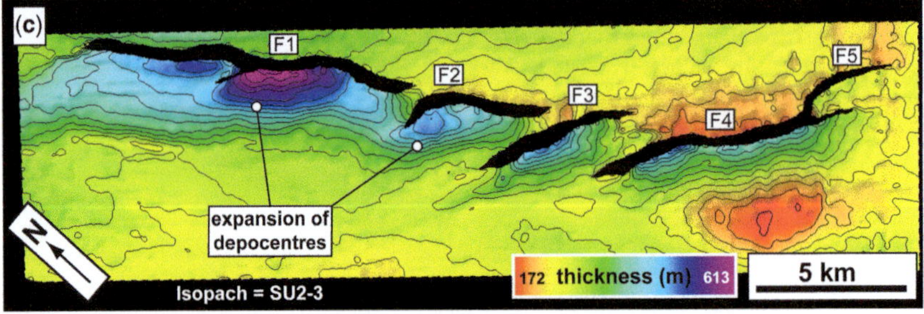

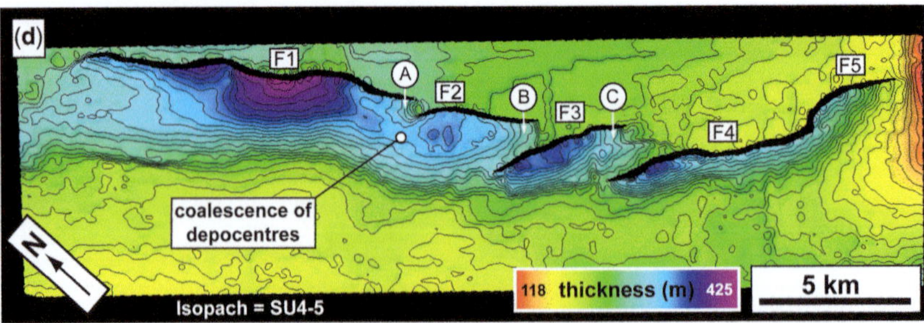

Fig. 6. Application of isopach analysis to constraining the palaeo-tip locations of a normal growth fault using a case study from the upper part of the October Fault Zone (OFZ), Suez Rift, Egypt (Fault 2 of Jackson & Rotevatn 2013). (**a**) NE-trending seismic reflection profile normal to Fault 2. H1–H10 are key seismic reflections; H5 represents the base synkinematic horizon. See (b) for the location of (a). (**b**) Isopach of SU1. (**c**) Isopach of SU2–SU3. (**d**) Isopach of SU4–SU5. F1–F4, faults 1–4; A–C, relay ramps A–C.

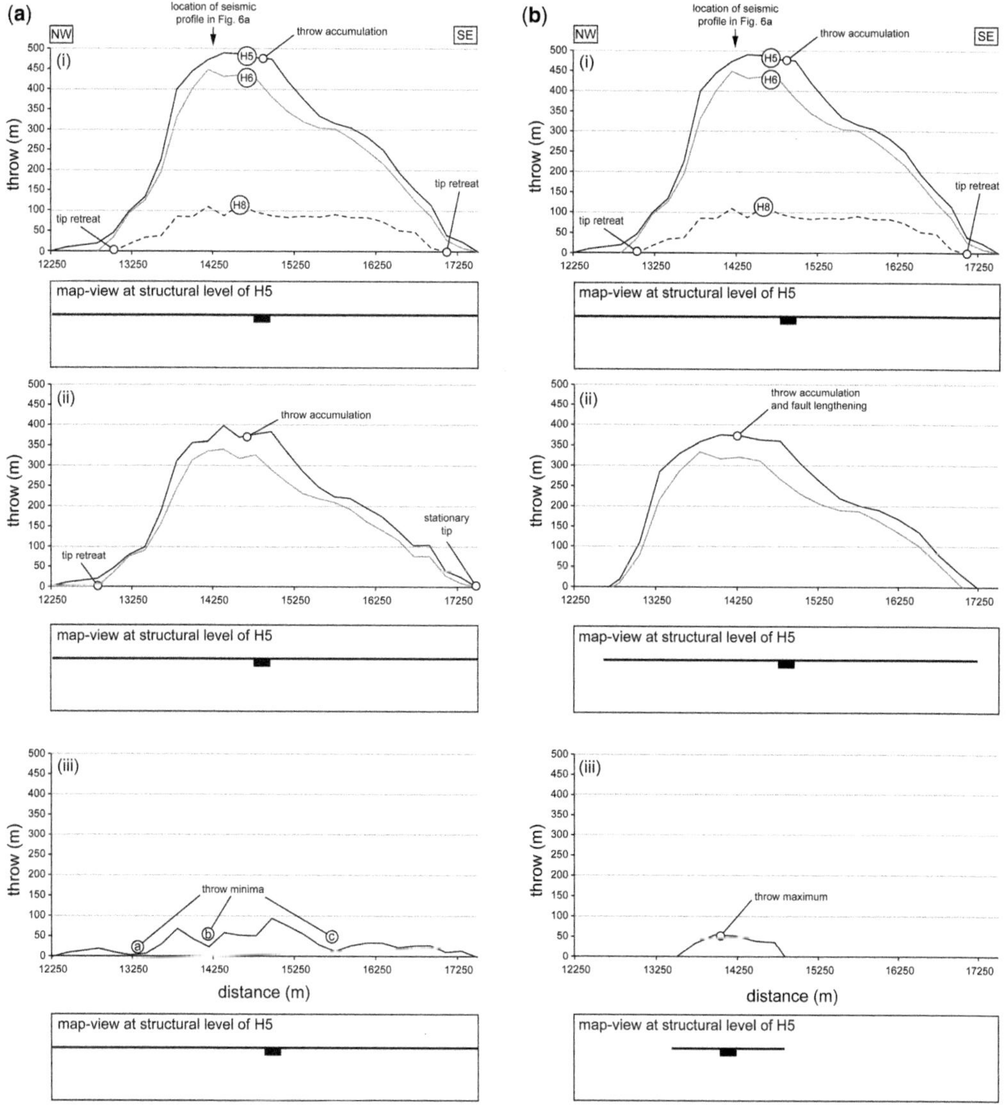

Fig. 7. Throw backstripping of F2, a seismically imaged salt-detached fault in the Suez Rift, Egypt (see Fig. 6 for additional details). (**a**) Original method; (i) present geometry; (ii) backstripped to H8; and (iii) backstripped to Horizon H6. Note the progressive loss of throw but not length, and increased segmentation with each backstripping increment. (**b**) Modified method: (i) present geometry; (ii) backstripped to H8; and (iii) backstripped to Horizon H6. Note the progressive loss of throw and length with each backstripping increment. Note also that, in contrast to the results of the original method, the fault does not become more segmented. The approximate position of the seismic profile shown in Figure 6a is indicated. See the text for a full description and discussion. Note that, in our extraction of horizon throw profiles, we have accounted for ductile strains, which typically manifests as folding and which may account for a considerable amount of extension-induced deformation (cf. Walsh *et al.* 1996).

that this is a synsedimentary fault, thus we here use the original method and three seismic horizons to backstrip throw and assess fault kinematics. Backstripping suggests that, during the earliest resolved slip increment (*c.* 0.5 Ma), the fault comprised a single, approximately 5.2 km-long fault. Throw varied along fault strike, with maximum throw (*c.* 93 m) occurring near the fault centre (Fig. 7). Subtle throw minima may reflect the position of relay zones that were breached during this earliest stage of fault development (labelled a–c; Fig. 7aiii) or the presence of distributed (i.e. off-fault) deformation not accounted for in the throw measurements. During the next growth increment, Fault 2 accumulated a considerable amount of throw (*c.* 308 m) near its centre, although its tips did not propagate. In fact, the NW tip retreated *c.* 750 m to the SE (Fig. 7aii). Despite being stationary or retreating, lateral tip gradients still increased significantly during the second growth increment, presumably due to mechanical interactions between Fault 2 and adjacent faults (faults 1 and 3: Fig. 6b) (e.g. Willemse 1997; Wilkins & Gross 2002; Soliva & Benedicto 2005; Polit *et al.* 2009). The final growth increment was similar to the second, with the fault accumulating throw (*c.* 87 m) whilst its NW (*c.* 250 m) and SE (*c.* 500 m) tips retreated (Fig. 7ai) (cf. Childs *et al.* 2003). Fault tip gradients did not appreciably increase during this slip increment (cf. H5 in Fig. 7aii with 7ai). In summary, the original backstripping method suggests Fault 2 grew in accordance with the constant-length fault model, with the final fault length being established after only *c.* 20% of its slip history (see also Jackson & Rotevatn 2013). Having established its final length, the fault shortened by tip retreat (cf. Childs *et al.* 2003). Tip retreat and fault shortening, in addition to the switch from dominantly fault lengthening to throw accrual, likely reflects mechanical interactions between overlapping fault tips (cf. Childs *et al.* 2003).

Isopach maps of Pliocene–Recent growth strata support our backstripping results, clearly indicating that fault-controlled depocentres adjacent to the October Fault Zone formed early, attaining their near-final shape and critically lengths during the earliest resolved slip increment (*c.* 0.5 Ma, SU1–SU5: Fig. 6b–d). The lengths of the bounding fault segments, including Fault 2, were therefore established relatively early (i.e. after only *c.* 20% of the faults slip history), with fault growth being dominated by displacement accumulation and basin deepening rather than tip propagation and fault lengthening. Application of the constant-length model to Fault 2 is further supported by backstripping of the relay zone between faults 1 and 2 (labelled A in Fig. 6d), which suggests the ramp formed relatively early, during approximately the initial 0.5 Ma of the faults *c.* 2.5 Ma growth history (Fig. 8c). Furthermore, relay-ramp shape was constant through time (e.g. the relay-ramp geometry between 2 and 2.5 Ma, shown in Fig. 8h, is broadly the same as between 0 and 0.5 Ma, shown in Fig. 8e). Although the relay ramp did not change shape, it rotated, as demonstrated by NW thickening of growth strata down the ramp (Fig. 8a). Based on these observations, Jackson & Rotevatn (2013) interpret that relay-ramp length and overall map-view geometry was established after only *c.* 20% of the fault history, and that the relay ramp steepened but maintained its overall length and shape with ongoing extension (Fig. 8d; see also Fig. 8i).

To briefly highlight the shortcomings of the modified backstripping method, and how it may fail to accurately capture the demonstrable growth history of a natural normal fault, we here apply it to Fault 2. Application of this method suggests a very different growth history to that predicted by the original method, indicating that, during the first growth increment, this structure was not segmented, but was instead represented by a single, *c.* 1.8 km-long segment that had a maximum throw of *c.* 80 m. Thus, after the first growth increment, the fault was only *c.* 35% of its final length (Fig. 7biii). During the next growth increment, the fault more than doubled in length (to 4.2 km: i.e. 81% of its final length) by tip propagation and accumulated *c.* 240 m of throw (Fig. 7bii). During its final growth increment, the fault lengthened by 1 km and accumulated 70 m of throw (Fig. 7bi). Displacement backstripping using the modified method therefore suggests Fault 2 grew in accordance with the isolated model, which isopach data and relay-zone backstripping indicate is clearly not the case (Fig. 1a).

Santos Basin, offshore SE Brazil

We here study a salt-detached normal fault located in the Santos Basin, offshore Brazil (Fault 4; for details, see Tvedt 2016). Fault 4 is 700 m in height, 1950 m long and has up to 109 ms TWT (127 m) of throw (Figs 9, 10). Maximum throw occurs at the fault segment centre, decreasing smoothly to its tips (e.g. Fig. 10ai), implying it grew via radial propagation of its tips rather than via segment linkage (e.g. Watterson 1986; Walsh & Watterson 1988; Baudon & Cartwright 2008). Tvedt (2016) demonstrates that Fault 4 and other related salt-detached faults in the Santos Basin are synsedimentary, thus we use the original backstripping method to assess its kinematics. Backstripping using this method suggests Fault 4 was initially 1725 m long and had, therefore, attained *c.* 88% of its final length during this earliest growth phase. At this time, F4 had only *c.* 60 ms TWT of throw (i.e. 55% of its final throw: Fig. 10aviii). Subsequent growth of the fault was dominated by throw accumulation, with only minor propagation of the lateral fault tips, and an increase in lateral tip throw gradients (Fig. 10a, part vii back through to i). It is likely that pinning of at least the SW tip of Fault 4 occurred due to mechanical interaction with an adjacent fault, with which it has large overlap and small spacing (Fig. 9c) (e.g. Willemse 1997; Wilkins &

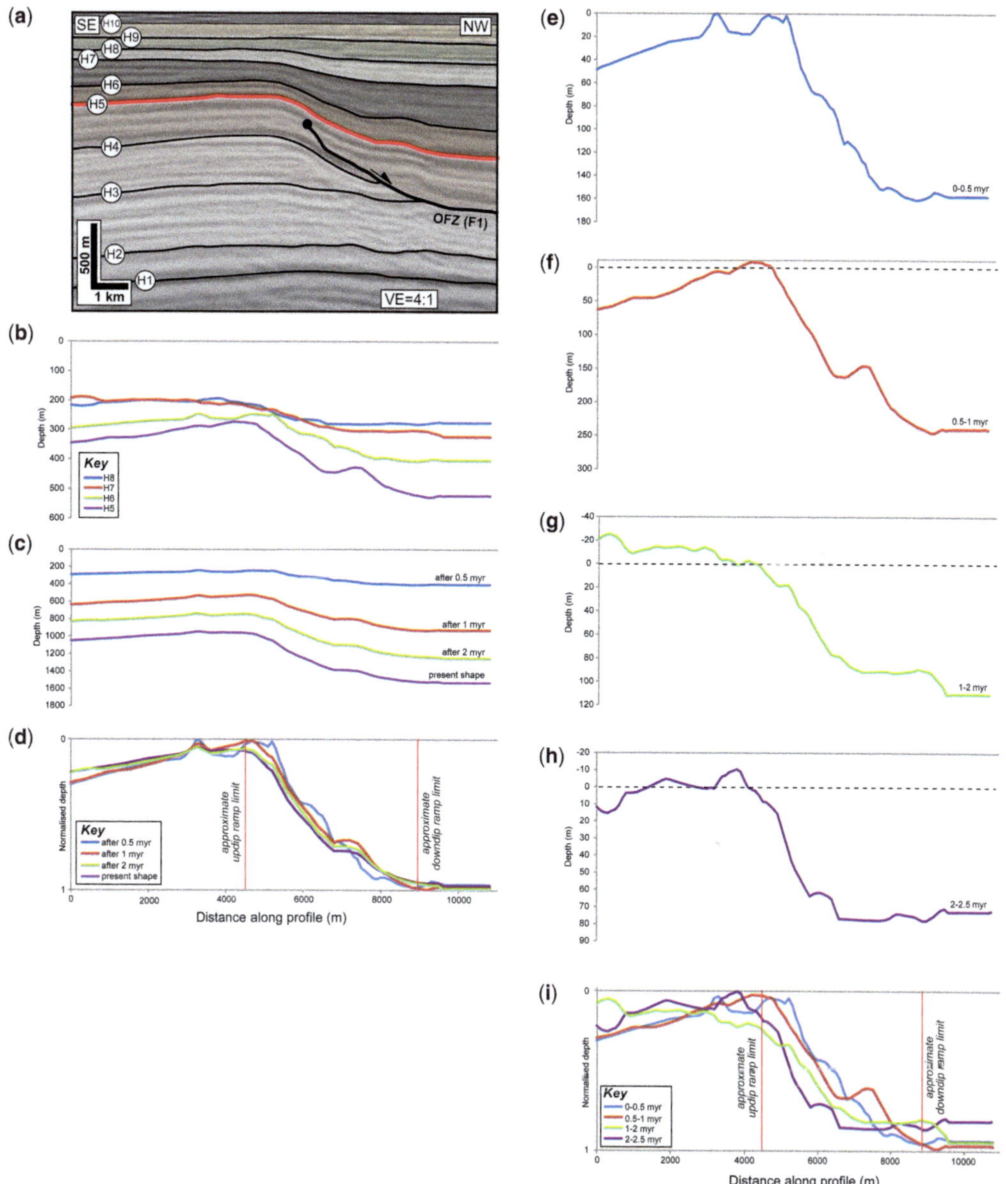

Fig. 8. Backstripping of relay zone B, October Fault Zone, Suez Rift, Egypt. (**a**) Fault-parallel (broadly NW-trending) seismic profile down relay ramp B (see Fig. 6b for the location). (**b**) Decompaction of horizons H5–H8 on a fault strike-parallel seismic section (see (a)) down relay ramp A. (**c**) Backstripped and decompacted horizons in (b) are progressively added to illustrate the change in the shape of relay zone A at the base of the synkinematic package (i.e. H5 bounding the base of SU1: see (a)). (**d**) Normalization of data in (c) showing incremental changes in the geometry of relay ramp A as defined at the base of the synkinematic package (i.e. H5 bounding the base of SU1: see (a)). (**e**)–(**h**) Graphs showing incremental changes in the shape of the relay ramp A during deposition of each growth interval: (e) SU1; (f) SU2; (g) SU3; and (h) SU4–SU5 (see Fig. 6). (i) Comparison of the incremental changes in the shape of the relay ramp A during deposition of each growth interval based on normalization of the data presented in (e)–(h). The red vertical lines in (d) and (i) indicate that the ramp did not greatly change in length or shape after the first 0.5 Ma of deformation. Note that, in our extraction of horizon throw profiles, we have accounted for ductile strains that typically manifest as folding and which may account for a considerable amount of extension-induced deformation (cf. Walsh *et al.* 1996).

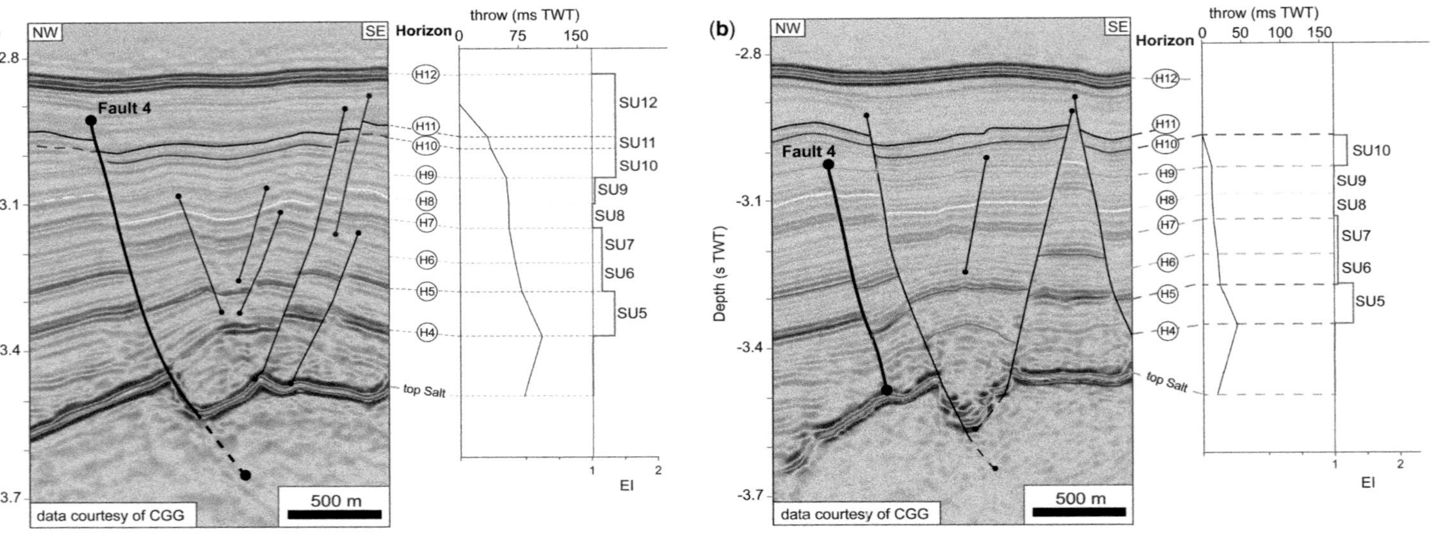

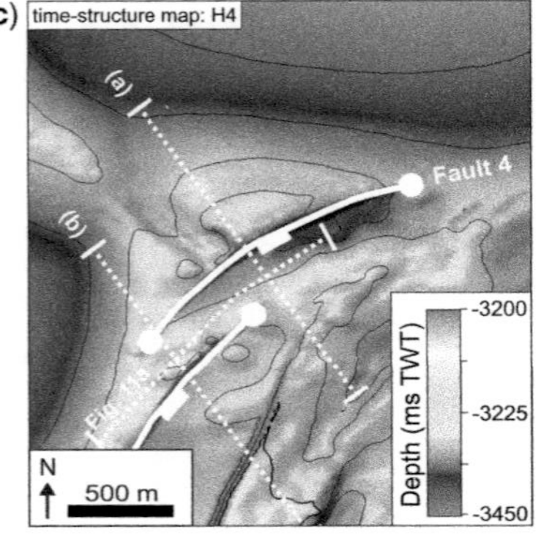

Fig. 9. Fault strike-perpendicular seismic profiles, and throw-depth (T–z) and expansion index (EI) plots near (**a**) the centre and (**b**) the tip of a suprasalt normal fault (Fault 4 of Tvedt 2016) imaged on 3D seismic reflection data from the Santos Basin, offshore Brazil. H4 defines the base of the synkinematic sequence and SU5 represents the first synkinematic unit. Note an EI value of >1.5 in SU5 near the lateral tip of the fault. (This figure appears in colour in the online version of the paper.)

Gross 2002; Soliva & Benedicto 2005; Polit *et al.* 2009). The original backstripping method therefore suggests Fault 4 grew in accordance with the constant-length model through most of its slip history (see Fig. 1b).

Our interpretation that Fault 4 grew in accordance with the constant-length model is supported by EI analysis (Figs 9, 10). Expansion index analysis implies Fault 4 nucleated during the middle Oligocene (SU5) and was active until the middle Miocene (SU7), before experiencing a period of inactivity during the middle–late Miocene (SU8 and, possibly, SU9). The fault was reactivated in the late Miocene (SU10) and continued to grow into the early Pleistocene (SU11–SU12) (Fig. 9a, b). Expansion indices of >1 in the lowermost growth strata, a few hundred metres inboard of the present lateral fault tips (i.e. SU5: Figs 9b & 10aix), indicate the fault had established its near-full length relatively early in its slip history, before accumulating significant displacement; this style of growth is consistent with the constant-length fault model and, thus, the results of displacement backstripping using the original method (Fig. 1b).

Backstripping of the relay zone between faults 4 and 5 (Fig. 11a) also provides evidence for establishment of the near-final fault lengths relatively early in their slip history; by early, we mean during the first of eight seismically resolvable growth increments (Fig. 11a). Our results indicate that, overall, the surface defining the base of the synkinematic sequence subsided and rotated basinwards through time (T1–T4: Fig. 11b). In detail, however, the kinematic evolution of the ramp was more complex. The ramp was *c.* 2.7 km long (distance between white dots on the T1 line in Fig. 11b) and rotated *c.* 5° basinwards during the first resolvable growth increment (T1: Fig. 11b, d); this interpretation is supported by the observation that the earliest synkinematic strata thicken NE down the ramp (SU5: Fig. 11a). However, during the second growth increment, the ramp did not rotate, but instead subsided vertically, most likely in response to long-wavelength regional subsidence of the Santos Basin (T2: Fig. 11b, e). A lack of ramp rotation during this time is supported by the observation that synkinematic layer SU6, which dips NE down the ramp, is relatively tabular and does not thicken basinwards (Fig. 11a). Relatively minor basinwards rotation (*c.* 2°) of the ramp occurred during the third growth increment, which was superimposed on the overall vertical subsidence of the ramp that, at this time, had lengthened slightly to *c.* 2.9 km (the distance between white dots on the T3 line in Fig. 11b; see also Fig. 11f). Minor (<10 m) perturbations on the overall basinwards ramp dip during this and other phases were likely caused by flow of underlying Aptian salt (Fig. 11a). During the fourth and final growth increment, the ramp rotated basinward by *c.* 4°, but did not lengthen. Ramp rotation was again superimposed on long-wavelength regional subsidence (T4: Fig. 11b, g), with bulk downdip thickening of the uppermost synkinematic sequence providing the stratigraphic record of this rotation (SU8–SU12: Fig. 11a). In summary, relay-zone backstripping from the Santos Basin fault system indicates that ramp length and overall geometry was established during the earliest resolved growth increment (T1); the ramp then subsided vertically (T2) before undergoing further rotation (T3–T4). These results imply: (i) the bounding fault segments had established their near-final lengths early in their slip history; and (ii) faulting was periodic, with two major periods of slip accumulation (T1 and T3–T4) being separated by a period of quiescence (T2). Fault 4, and presumably other faults in the associated salt-detached array (e.g. Fault 5), thus grew in accordance with the constant-length model (Fig. 1b). Furthermore, pulsed activity on these faults, as indicated by displacement backstripping, is entirely compatible with results drawn from analysis of EI values, which also show two major periods of fault activity (SU5–SU7 and SU10–SU12) separated by a period of fault inactivity (SU8–SU9) (Fig. 9a, b).

Egersund Basin, offshore Norway

Our final case study comes from the Egersund Basin, offshore Norway. This salt-detached growth fault (Fault 1; F1) is *c.* 4 km in height and 16 km long, and has a maximum throw of *c.* 1200 m (Figs 12a, b & 13). Throw backstripping of this growth fault (see Tvedt *et al.* 2016 for details), using the original method and 11 seismic horizons, suggests the fault was originally composed of several segments that were up to several kilometres long and had up to 150 ms TWT of throw (Fig. 13avii). Although segmented and probably separated by unbreached relays, when viewed as a single structure, which is likely if small, intra-relay strains could be measured and taken into account, F1 was *c.* 12 km long and was thus a considerable amount (*c.* 75%) of its final length. Subsequent growth of Fault 1 was primarily accommodated by displacement accumulation, although its SE tip propagated *c.* 4 km (Fig. 13a, part vi back through to i). The original backstripping method suggests that Fault 1 had a hybrid growth history, with an earlier phase of predominantly lengthening (i.e. the constant-length model: Fig. 1b), being superseded by a period of displacement accumulation *and* lengthening (i.e. the isolated model: Fig. 1a).

We can, again, test the prediction of the original backstripping method using growth strata and the

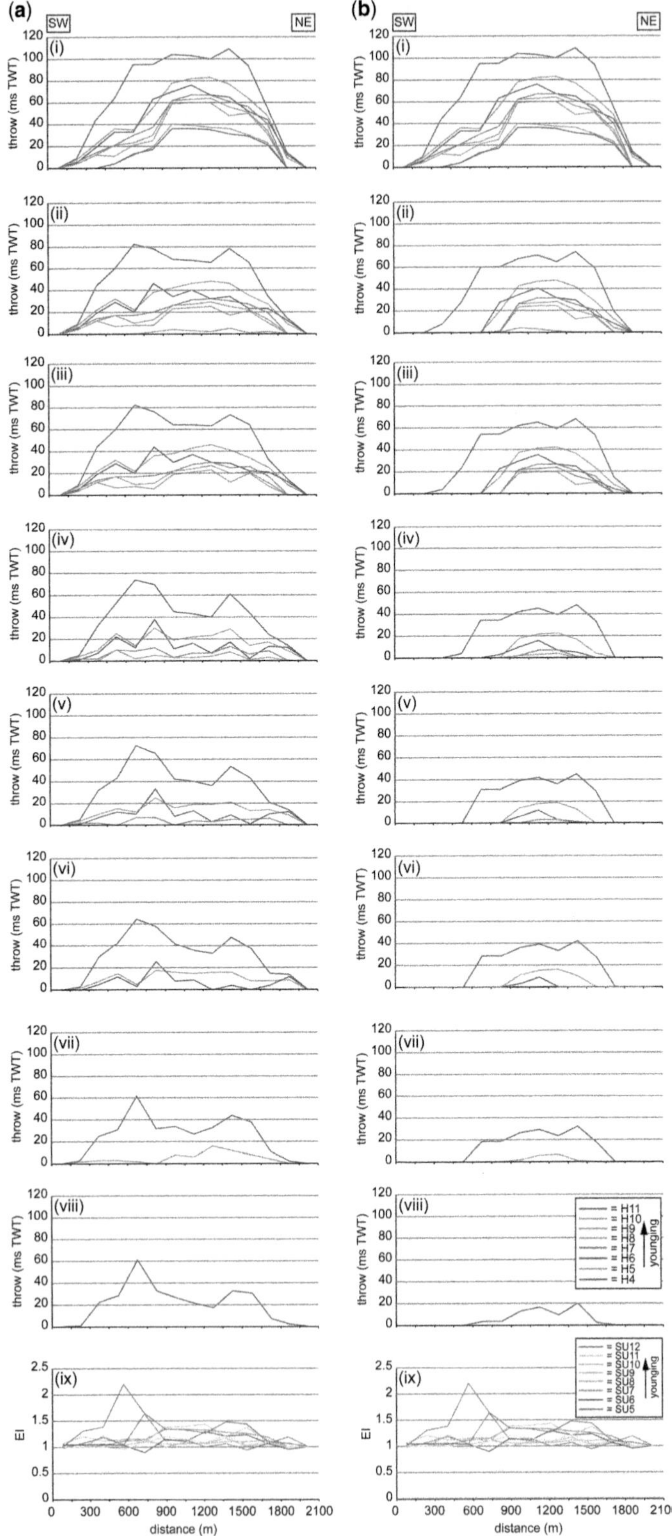
(a)
SW
NE
(b)
SW
NE
(i)
(ii)
(iii)
(iv)
(v)
(vi)
(vii)
(viii)
(ix)
throw (ms TWT)
EI
distance (m)
0 300 600 900 1200 1500 1800 2100
= H11
= H10
= H9
= H8
= H7
= H6
= H5
= H4
younging
= SU12
= SU11
= SU10
= SU9
= SU8
= SU7
= SU6
= SU5
younging

techniques outlined earlier. Isochron maps strongly suggest that Fault 1 grew in accordance with the hybrid model outlined above (i.e. with an early period of fault lengthening superseded by a period of displacement accumulation and lengthening). These maps indicate that F1 initiated in the Late Triassic and was intermittently active for *c.* 170 Ma until the Early Paleocene (Danian) (Fig. 12c, d). Although data are noisy due to the unit being relatively thin (typically <30 ms TWT), thickening of the earliest (seismically resolvable) Bajocian–Bajocian growth strata (SU2: light-blue line in Fig. 13aix, for example) along *c.* 10 km of F1 indicates the structure had established a considerable part (*c.* 63%) of its length after only about the first 20% of its *c.* 170 Ma slip history. Across-fault expansion in overlying units document the preferential propagation of the SE fault tip, at a time when the NW tip was static, with the near-final length of F1 being established in the Tithonian (SU5: yellow line in Fig. 13aix). Thus, F1 established its full length after only about 33% of its *c.* 170 Ma slip history. The style of growth determined from the EI analysis suggests the Egersund Basin fault displays elements of both the isolated (Fig. 1a) and constant-length (Fig. 1b) models. More specifically, the fault established a considerable portion (*c.* 63%) of its strike length relatively early in its slip history (constant-length model) prior to accumulating displacement and length in concert (isolated model).

Again, to briefly highlight the shortcomings of the modified backstripping method, and how it may fail to accurately capture the demonstrable growth history of a normal fault, we here apply it to Fault 1 from the Egersund Basin. This method suggests a markedly different growth history to that predicted by the original method; instead of lengthening and tip propagation seemingly occurring at different times, this method indicates a sympathetic increase in displacement *and* length (Fig. 13b, part vii back through to i), behaviour consistent with the isolated model (Fig. 1a). Isopach data (Fig. 12c, d) and EI analysis (Fig. 13) indicate that this kinematic history is demonstrably incorrect, however.

Discussion

The faults studied here grew by establishment of their near-final lengths relatively early (<20–33%) in their slip history, prior to the accumulation of significant displacement. Relatively minor tip propagation is resolved by our seismic reflection method, occurring during the earliest stage of growth when relay zones were breached, or during the latter stages of growth when largely asymmetrical tip propagation occurred. Two key questions are, therefore: 'why do faults establish their near-final lengths prior to significant displacement accumulation?'; and 'in what specific circumstances, or at what times during the growth of large normal faults, might the isolated model be valid?'. To answer these questions, we must: (i) consider the specific geological context of some of the examples used by other authors to support either the isolated or constant-length model; and (ii) critically assess the evidence forwarded in support of either model.

Isolated model

The isolated model was originally motivated by fault displacement–length (*D–L*) relationships derived from global datasets, which suggest that, as a fault slips and accumulates displacement, it lengthens (Fig. 1a) (e.g. Watterson 1986; Walsh & Watterson 1988; Cowie & Scholz 1992; Kim & Sanderson 2005). Numerous outcrop (e.g. Cartwright *et al.* 1995; Mansfield & Cartwright 2001; Jackson *et al.* 2002), subsurface (e.g. Contreras *et al.* 2000; McLeod & Underhill 2000; Young *et al.* 2001) and geomorphological (e.g. Commins *et al.* 2005) studies invoke this model to explain a range of geometrical (e.g. displacement variations, breached relays) and stratigraphic (e.g. along-strike thickness changes) features commonly seen along segmented normal fault systems. Displacement variations and, in particular, the presence of breached relays have been used in support of the isolated model, although 3D seismic reflection-based studies show these features can also develop along segmented faults growing in accordance with the constant-length model (e.g. Nicol *et al.* 2005; Jackson & Rotevatn 2013;

Fig. 10. Throw backstripping of Fault 4, a salt-detached normal fault imaged on seismic reflection data from the Santos Basin, offshore Brazil (see Fig. 9 for additional details). (**a**) Original method: (i) present geometry; (ii) backstripped to H11; (iii) backstripped to H10; (iv) backstripped to H9; (v) backstripped to H8; (vi) backstripped to H7; (vii) backstripped to H6; (viii) backstripped to H5; and (ix) backstripped to H4. Note the progressive loss of throw, stable overall length and increased segmentation with each backstripping increment. (**b**) Modified method: (i) present geometry; (ii) backstripped to H11; (iii) backstripped to H10; (iv) backstripped to H9; (v) backstripped to H8; (vi) backstripped to H7; (vii) backstripped to H6; (viii) backstripped to H5; and (ix) backstripped to H4. Note the progressive loss of throw and length with each backstripping increment; fault segmentation is also observed. Note that, in our extraction of horizon throw profiles, we have accounted for ductile strains that typically manifest as folding and which may account for a considerable amount of extension-induced deformation (cf. Walsh *et al.* 1996). See the text for a full description and discussion. (This figure appears in colour in the online version of the paper.)

Tvedt *et al.* 2016). These geometrical features are thus not diagnostic of the isolated model, merely indicating that, at some point in the fault history, two or more faults, which may or may not have been kinematically coherent, linked by breaching of transient relay zones. As shown in Figure 1, differentiation between the isolated and constant-length models requires not only a detailed analysis of fault structure, but also determination of the timing of fault lengthening v. displacement accumulation, which, as we argue above, requires analysis of growth strata.

Where growth strata have been incorporated into the kinematic analysis of a segmented normal fault, such as the Usisya normal fault system, East African Rift System, a complex growth history is inferred. For example, Contreras *et al.* (2000) and Morley (2002) indicate this system is characterized by alternating periods of coeval fault lengthening and displacement accumulation, and periods characterized by an increase in both length and displacement (Fig. 14) (see Contreras *et al.* 2000; Morley 2002). However, in this case, the footwalls are emergent and strongly degraded, and growth strata are only preserved in the hanging wall; it is thus unknown if the basin was filled, balanced or overfilled (Carroll & Bohacs 1999) during growth of the bounding fault (cf. McLeod & Underhill 2000; Young *et al.* 2001; Jackson *et al.* 2002). If the basin was underfilled, it is conceivable that an early period of rapid length establishment went undetected, with post-slip passive infilling and onlap of the hanging-wall basin leading to the erroneous assumption that the fault tips propagated through time (i.e. the isolated model) (Childs *et al.* 2003). Thus, we argue that *across-strike* (i.e. across-fault) changes in stratal thickness, as captured in EI analysis (e.g. Fig. 3), are more critical than *along-strike* thickness changes solely preserved in the fault hanging wall. Preservation of growth strata in the footwall of the bounding fault is key to demonstrating that sediment accumulation rate was

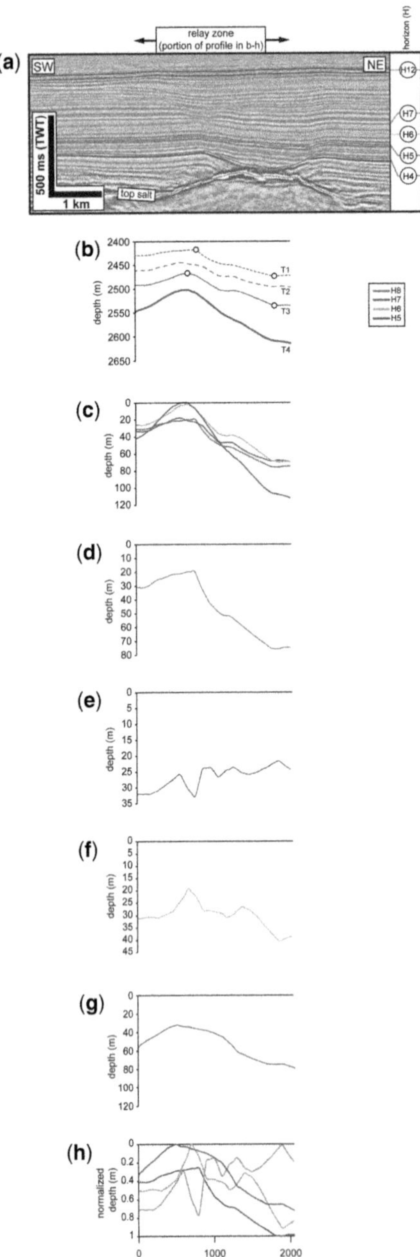

Fig. 11. Backstripping of the Santos Basin relay zone shown in Figure 9c. (**a**) Fault-parallel (broadly NW-trending) seismic profile down the relay ramp showing the ramp geometry and mapped horizons (see Fig. 9c for location). (**b**) Backstripped and decompacted horizons are progressively added to illustrate the change in the shape of the relay zone at the base of the synkinematic package from T1 to T4 (i.e. H4: Fig. 9a, b). (**c**) Normalization of data in (b) showing incremental changes in the geometry of relay ramp A as defined at the base of the synkinematic package (i.e. H5: Fig. 9a, b). (**d**)–(**f**) Graphs showing incremental changes in the shape of the relay ramp A during deposition of each growth interval: (d) SU1; (e) SU2; (f) SU3; and (g) SU4–SU5 (see Fig. 9). (**h**) Comparison of the incremental changes in the shape of the relay ramp A during deposition of each growth interval based on normalization of the data presented in (d)–(g). Note that, in our extraction of horizon throw profiles, we have accounted for ductile strains that typically manifest as folding and which may account for a considerable amount of extension-induced deformation. (This figure appears in colour in the online version of the paper.)

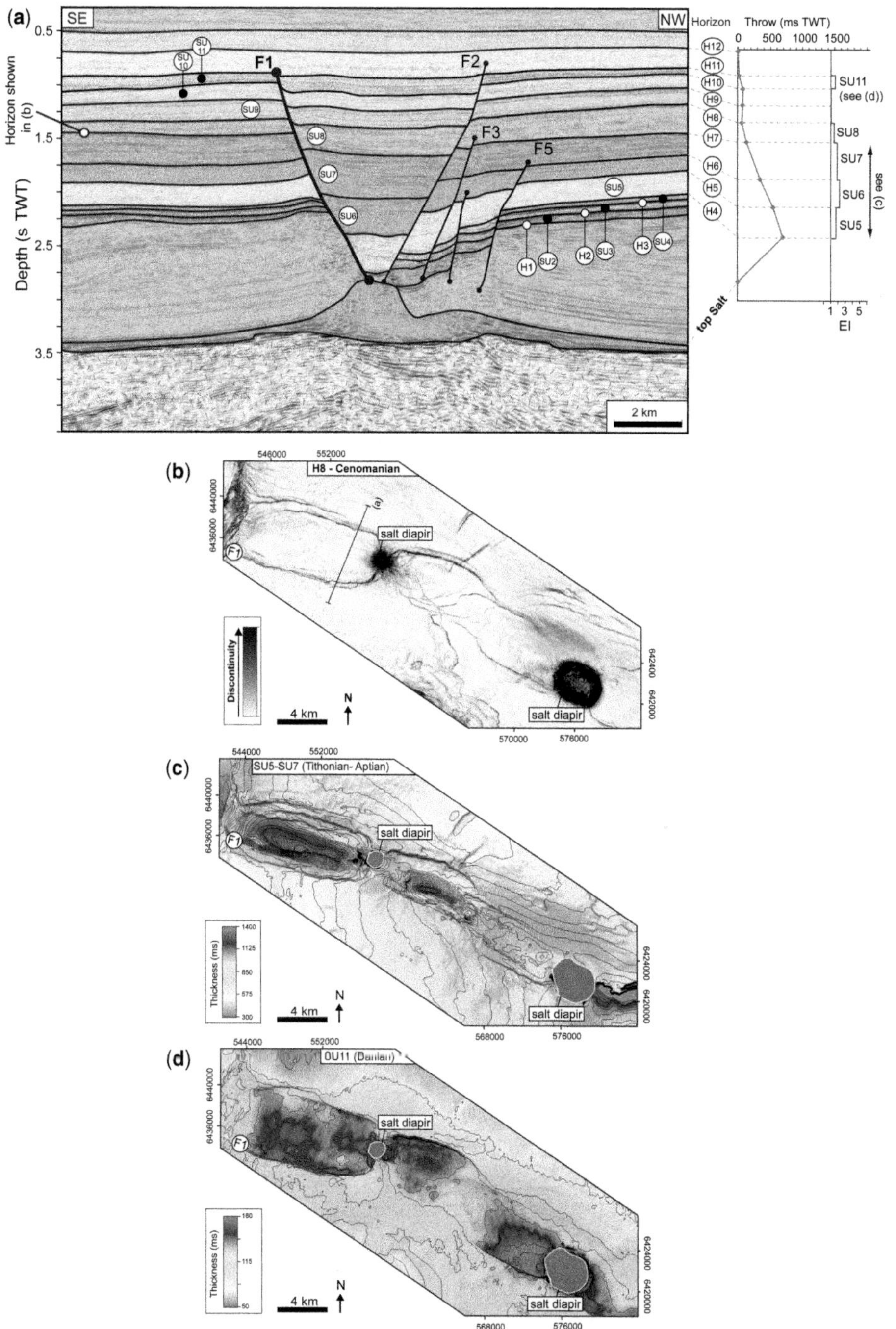

Fig. 12. (**a**) Fault strike-perpendicular seismic profile across a salt-detached normal fault array (faults 1–3 and 5) imaged in 3D seismic reflection data from the Egersund Basin, offshore Norway. Throw–depth (T–x) and expansion index (EI) plots for F1 are shown. H1 defines the base of the synkinematic sequence and SU2 represents the first synkinematic unit. (**b**) Discontinuity extraction along H8 (Cenomanian) showing the plan-view expression of Fault 1. Stratigraphic context of H8 is shown in (a). (**c**) Isochron map of SU5–SU7 (Tithonian–Aptian). The stratigraphic context of the isochron is shown in (a). (**d**) Isochron map of SU11 (Danian). The stratigraphic context of the isochron is shown in (a). (This figure appears in colour in the online version of the paper.)

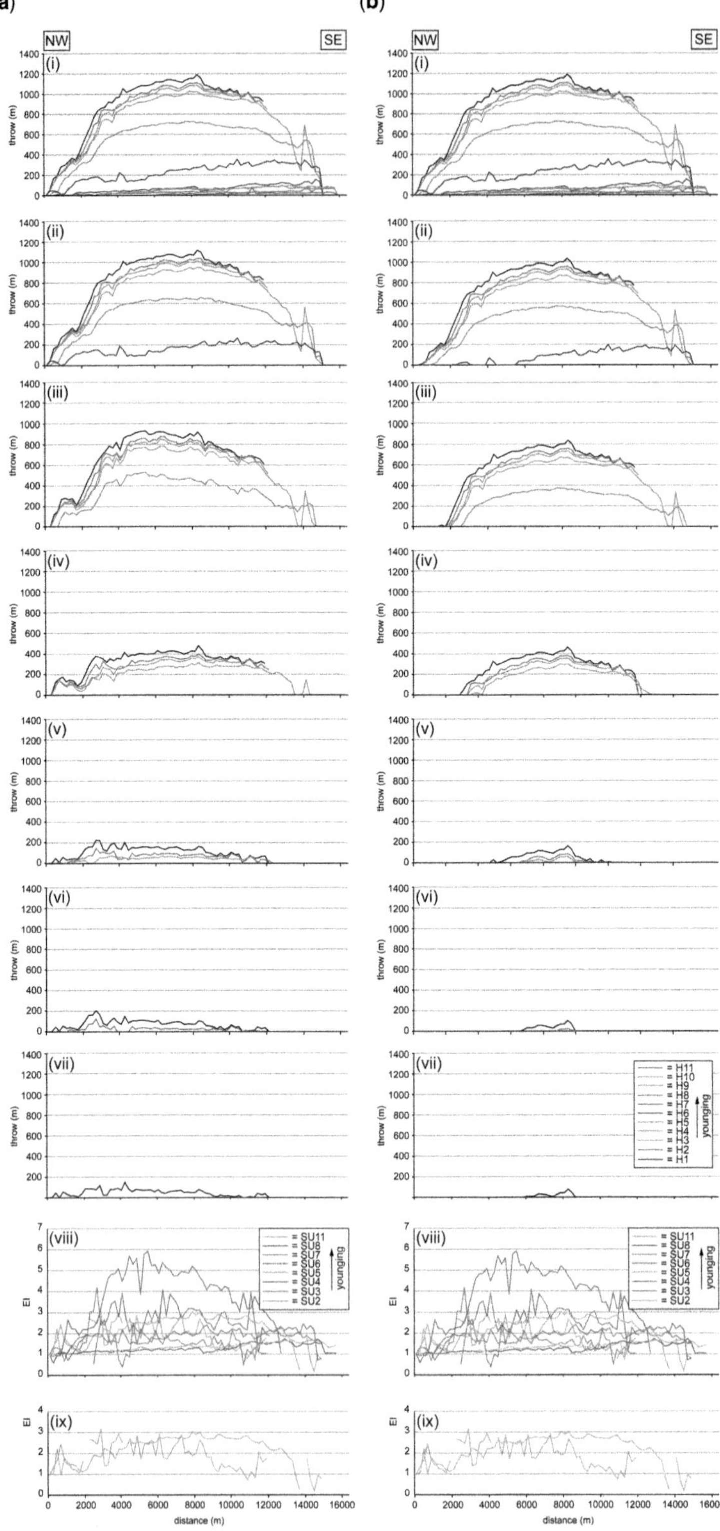
(a)
(b)
NW
SE
(i)
(ii)
(iii)
(iv)
(v)
(vi)
(vii)
(viii)
(ix)
throw (m)
EI
distance (m)
H11
H10
H9
H8
H7
H6
H5
H4
H3
H2
H1
younging
SU11
SU8
SU7
SU6
SU5
SU4
SU3
SU2
younging

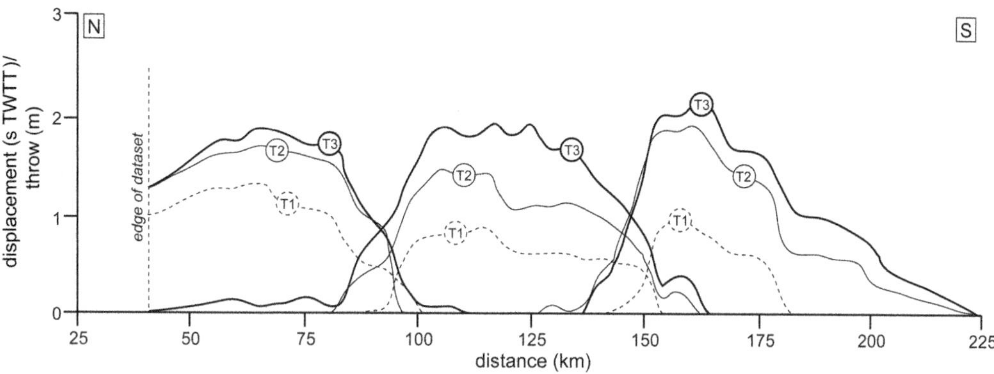

Fig. 14. Plot showing how the three main segments of the Usisya normal fault system, Lake Malawi Basin, East African Rift System increased their length and displacement through time (T1–T3). Each line corresponds to the summed displacement accommodated during the deposition the previous sequences/during the previous time step. Note that the original data are in time (i.e. TWT in ms), although they are presented in depth in metres based on an assumed interval velocity of 2000 m s^{-1}. Modified from Contreras *et al.* (2000). See the text for a full discussion. TWTT, two-way travel time (in s).

equal to or exceeded fault slip rate, and that the basin was, at least over relatively long timescales, overfilled.

Physical models lend support to the isolated model, with fault lengthening and displacement accumulation seemingly occurring in concert (e.g. Clifton & Schlische 2000; Ackermann *et al.* 2001; Mansfield & Cartwright 2001). However, recent physical models that monitor fault growth at a very high resolution lend strong support to the constant-length model, indicating that faults may rapidly establish their near-final lengths (Schlagenhauf *et al.* 2008); such behaviour may have occurred but went undetected in earlier physical models that simply relied on time-lapse photography of the model top surface.

Although seismic reflection and physical model support for the isolated model are questionable, tip propagation over geological timescales has been demonstrated by outcrop studies. For example, Morewood & Roberts (1999) study uplifted marine terraces and geomorphic features (i.e. wind gaps) to investigate the development of the western end of the active South Alkyonides Fault Segment, Gulf of Corinth, central Greece. They demonstrated significant (3–4 km), relatively rapid (12.1–16.7 mm a^{-1}) lateral propagation of this large (2.5–3 km throw) fault over relatively short timescales (*c.* 330 kyr). The study of Morewood & Roberts (1999) seemingly lends support to the isolated model. However, based on additional structural observations, they argued that, whilst the surface trace length has increased, the at-depth fault length was established relatively early in the evolution of the system (i.e. before 330 ka). As such, the longer-term development of the fault is consistent with the constant-length model. It is important to note that the most recent phase of at-surface fault lengthening in the Gulf of Corinth is, perhaps, too rapid and minor, in terms of slip and generated accommodation, to be associated with a sequence of seismically resolvable growth strata.

Fig. 13. (**a**) Throw backstripping of Fault 1, Egersund Basin, offshore Norway using the original method: (i) present geometry; (ii) backstripped to H7; (iii) backstripped to H6; (iv) backstripped to H5; (v) backstripped to H4; (vi) backstripped to H3; and (vii) backstripped to H2. Note the progressive loss of throw but not length with each backstripping increment. (**b**) Throw backstripping using the modified method: (i) present geometry; (ii) backstripped to H7; (iii) backstripped to H6; (iv) backstripped to H5; (v) backstripped to H4; (vi) backstripped to H3; and (vii) backstripped to H2. Note the progressive loss of throw and length with each backstripping increment. (viii) Along-strike EI plots for some of the growth sequences (SU2–SU8 and SU11) shown in Figure 12a. For several of the sequences, EI values cannot be constrained SE of *c.* 12 000m due to poor seismic imaging below the overhanging diapir flank. (xi) Along-strike EI plots for the earliest (i.e. SU2) and a relatively early (i.e. SU5) growth sequence. Note that dashed lines indicate areas of poor data where EI values have been inferred. Note that H1, H2 and H3 cannot be mapped SE of *c.* 12 000 m due to poor seismic imaging below the overhanging diapir flank. However, the grey shaded band indicates our interpretation of how displacement decreases on these horizons SE towards the fault tip during each backstripping increment. (This figure appears in colour in the online version of the paper.)

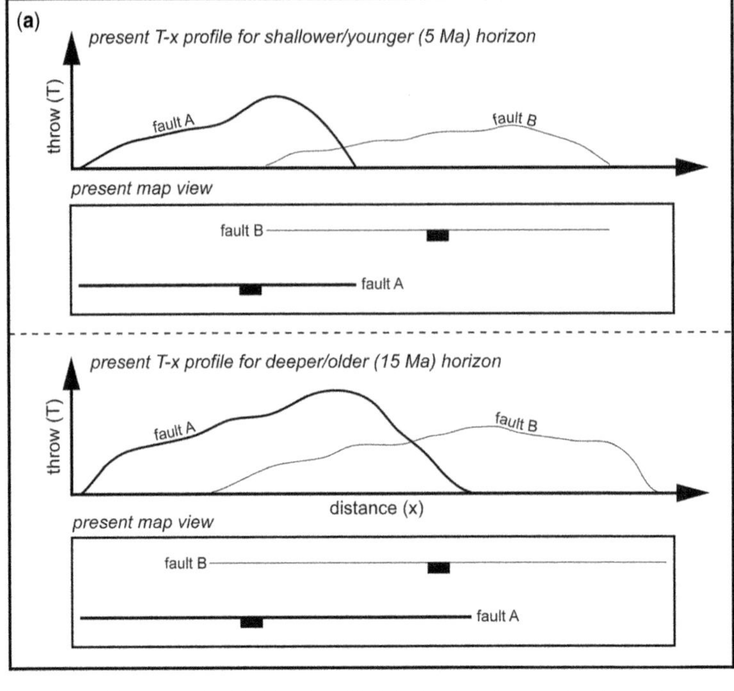

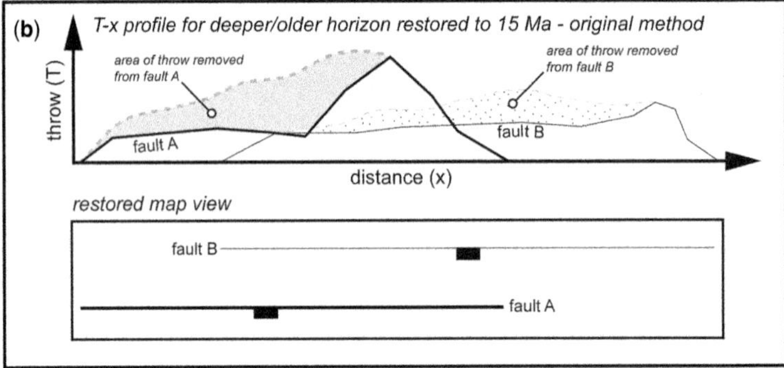

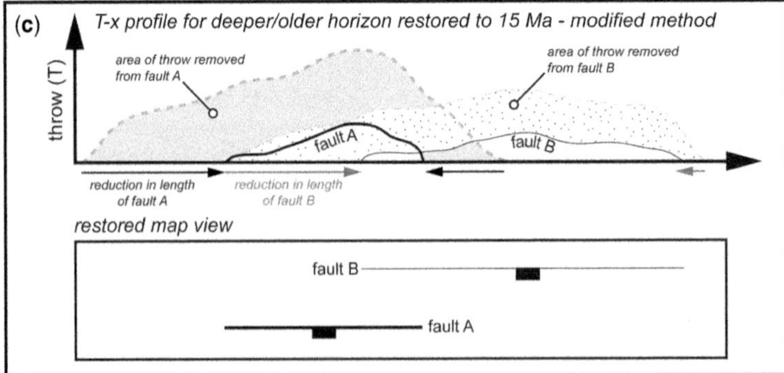

Fig. 15. Schematic examples showing the application of the original and modified throw backstripping methods to a salt-detached growth fault, offshore Angola (modified from Dutton & Trudgill 2009). (**a**) Present length and distribution of throw along two synkinematic horizons (i.e. a shallower/younger horizon (5 Ma) and a deeper/older horizon (15 Ma)). (**b**) Throw and length of the deeper/older horizon backstripped to 5 Ma using the original method,

Evidence for tip propagation is not limited to short timescales. Nor is tip propagation only resolved using field-derived, geomorphic and structural data. For example, Childs *et al.* (2003) used 3D seismic reflection data from the Timor Sea to demonstrate lateral tip propagation on considerably longer timescales than the example presented by Morewood & Roberts (1999). They studied a large (*c.* 100 m throw), now-inactive synsedimentary fault, indicating that, over the last 2.35 Ma of its development, its western tip propagated at least 1.6 km at a relatively slow rate (0.7 mm a^{-1}: cf. the Gulf of Corinth example described above). Tip propagation is recorded in growth strata thickness patterns and displacement patterns on the fault surface. Conversely, slightly more rapid (1.1 mm Ma^{-1}) propagation of the fault's eastern tip is not directly captured by growth strata, but by the overall distribution of displacement on the fault surface and the stratigraphic position of the fault's upper tip.

Constant-length model

The constant-length model was established and has been best supported by kinematic analysis of faults forming in areas where a pre-existing structural grain was present due to polyphase extension (e.g. the East African Rift: Morley 1999, 2002; the Cartier Trough, Timor Sea: Walsh *et al.* 2002; the Wanganui Basin, New Zealand: Lamarche *et al.* 2005; the Taranaki Basin, New Zealand; Giba *et al.* 2012). In these cases, rapid length establishment is, perhaps, predictable, given that a second phase of extension leads to fault reactivation, rapid upwards propagation through the 'intra-rift' strata and rapid length establishment at the free surface prior to displacement accumulation (e.g. Nicol *et al.* 2005). The broader applicability of the constant-length model – for example, in areas lacking major pre-existing structural weaknesses – is thus questionable (see Walsh *et al.* 2002). However, we have shown that the constant-length fault model may also apply in areas lacking pre-existing fabrics, or that did not undergo polyphase extension (see also Nicol *et al.* 2016). Likewise, the physical models of Schlagenhauf *et al.* (2008) indicate that faults developed in 'homogeneous' material may rapidly establish their length prior to accumulating significant displacement. It is conceivable that broadly linear, subseismic zones of weakness were present in both the natural and experimental examples, serving to rapidly localize displacement, with such weaknesses leading to rapid fault lengthening at the expense of displacement accumulation. Testing this hypothesis will be problematic using 3D seismic reflection data alone, with such data typically being of insufficient resolution to image such putative small-scale, pre-faulting damage. Even for natural examples exposed in the field or the subsurface, it may be difficult to discriminate between pre-extension damage, and damage explicitly related to the initiation and growth of the fault itself.

Although typically viewed as being mutually exclusive, we suggest the isolated and constant-length models may both contribute to the growth of large normal faults developing over geological timescales. For example, the very early development of such large faults may be characterized by the growth of much smaller faults that increase their displacement and length in concert (i.e. the isolated model), before linking to form part of a larger structure. The timescale over which this occurs may be relatively short (relative to the total slip history of the fault under inspection) and the associated stratigraphic record accordingly cryptic: hence, we may struggle to resolve this period of fault development with relatively low-resolution subsurface tools (e.g. seismic reflection data). Approaches utilizing both high- (e.g. sparker, boomer or pinger sources) and low-resolution (e.g. air-gun source, industry-standard data) seismic reflection data, integrated with borehole data, may offer our best chance to fully understand the long and short timescale development of segmented normal faults (see Nicol *et al.* 2005). For relatively young, potentially still-active faults, or ancient, now-inactive faults whose growth was halted relatively early in their development, the constant-length model may explain why these faults appeared under-displaced with respect to global displacement–length population data.

In summary, we argue that most of the detailed, 3D seismic reflection-based studies, such as those presented here and by other workers (e.g. Walsh *et al.* 2002; Meyer *et al.* 2002; Childs *et al.* 2003; Nicol *et al.* 2005, 2016), lend strong support to the constant-length model rather than the isolated fault model. However, it is clear that additional 3D seismic reflection-based studies are required to test the wider applicability of both models, which may be equally valid depending on the tectonostratigraphic

Fig. 15. (*Continued*) involving subtraction of the uppermost (5 Ma) $T–x$ profile from the underlying (15 Ma) $T–x$ profile, resulting in the generation of a reconstructed $T–x$ profile for the deeper/older (15 Ma) horizon. Note that the fault length is maintained and throw locally decreases. (**c**) Throw and length of the deeper/older horizon backstripped to 5 Ma using the modified method, involving the subtraction of the maximum throw on the shallower/younger (5 Ma) horizon from all points along strike of the deeper/older horizon, resulting in the generation of a reconstructed $T–x$ profile for the deeper/older horizon (15 Ma). Note that fault length and throw decreases.

setting in which the fault system is forming, the part of the fault under inspection and/or the stage of fault development; neither model should be assumed *a priori*.

What value does the modified backstripping method have?

Given our clear preference for using growth strata and the original backstripping method, what value does the modified backstripping method have for constraining the kinematics of segmented normal faults? We have shown that this method produces erroneous results, implying faults grew by simultaneous tip propagation and displacement accumulation (i.e. the isolated model) when, in fact, they established their near-final length prior to significant displacement accumulation (i.e. the constant-length model). This problem is further highlight by Dutton & Trudgill (2009), who undertook a kinematic analysis of a synsedimentary fault imaged in 3D seismic reflection data, offshore Angola. They showed that, depending on the backstripping method applied, the fault developed in accordance with either the isolated or constant-length model (Fig. 15). Based on what they considered to be anomalously high throw gradients, which could, in fact, simply reflect the mechanical interaction between closely spaced adjacent segments (e.g. Crider & Pollard 1998), and the notion that the results were 'more realistic', they preferentially applied the modified backstripping method and, hence, favoured the isolated model. Our results suggest an *a priori* assumption of a preferred growth model is incorrect, and that growth strata must be integrated in any kinematic analysis. Moreover, neither backstripping method will be able to resolve the kinematic history of blind normal faults, which are typically characterized by broadly ovate rather than rectangular displacement contours (Nicol *et al.* 1996).

Conclusions

Two competing fault growth models exist: (i) the isolated model, which envisages that normal faults grow via a sympathetic increase in their displacement and length; and (ii) the constant-length model, which envisages a relatively rapid establishment of the fault's near-final length prior to significant displacement accumulation. It is important to establish which model best describes the growth of normal faults, although this can be extremely challenging using only the final fault geometry and displacement distributions on the fault surface. We reviewed the benefits and uncertainties associated with four simple techniques that can be used to constrain the kinematics of synsedimentary growth faults: (i) expansion index analysis; (ii) isochrone analysis; (iii) displacement backstripping; and (iv) relay-zone backstripping. These techniques allow direct analysis of fault tip behaviour; thus, we applied them to three segmented synsedimentary faults imaged in seismic reflection data. We showed that, in general, these faults grew in accordance with the constant-length model, although subordinate periods of relatively minor tip propagation and coeval displacement accumulation occurred. More specifically, we showed that the studied faults attained their near-final lengths within *c.* 20–33% of their slip history, prior to the accumulation of significant displacement. We conclude that detailed analysis of growth strata represents the best way to test fault growth models. However, these models may not be mutually exclusive, with each being applicable at distinct times in the growth of large faults developing over geological timescales. Failure to integrate growth strata into the kinematic analysis of normal faults means that the inferred growth histories, and thus the broader applicability of the isolated fault model, are perhaps questionable.

CALJ would like to gratefully acknowledge financial support from NERC (studentship GT04/98/197/ES). BP plc and the Gulf of Suez Petroleum Company (GUPCO) are thanked for financial and logistical support during the PhD studentship of CALJ, and for providing access to the Egyptian seismic reflection data used in this study. We also thank PGS for access to seismic data and also granting permission to publish seismic reflection data from offshore Norway. In particular, we thank Richard Lamb at PGS for his assistance in securing relevant permissions to access and publish images from these data. CGG provided access to seismic data from the Santos Basin, offshore Brazil. We also thank Schlumberger for providing access to Petrel software, Badleys Geosciences for providing access to TrapTester and Midland Valley for providing access to 2DMove.

References

Ackermann, R.V., Schlische, R.W. & Withjack, M.O. 2001. The geometric and statistical evolution of normal fault systems: an experimental study of the effects of mechanical layer thickness on scaling laws. *Journal of Structural Geology*, **23**, 1803–1819.

Athmer, W. & Luthi, S.M. 2011. The effect of relay ramps on sediment routes and deposition: a review. *Sedimentary Geology*, **242**, 1–17.

Athmer, W., Groenenberg, R.M., Luthi, S.M., Donselaar, M.E., Sokoutis, D. & Willingshofer, E. 2010. Relay ramps as pathways for turbidity currents: a study combining analogue sandbox experiments and numerical flow simulations. *Sedimentology*, **57**, 806–823.

Baudon, C. & Cartwright, J.A. 2008. 3D seismic characterisation of an array of blind normal faults in the Levant Basin, Eastern Mediterranean. *Journal of Structural Geology*, **30**, 746–760.

BENEDICTO, A., SCHULTZ, R.A. & SOLIVA, R. 2003. Layer thickness and the shape of faults. *Geophysical Research Letters*, **30**, 2076, https://doi.org/10.1029/2003GL018237

BOUROULLEC, R., CARTWRIGHT, J.A., JOHNSON, H.D., LANSIGU, C., QUÉMENER, J.M. & SAVANIER, D. 2004. Syndepositional faulting in the Grès d'Annot Formation, SE France: high-resolution kinematic analysis and stratigraphic response to growth faulting. *In*: JOSEPH, P. & LOMAS, S.A. (eds) *Deep-Water Sedimentation in the Alpine Basin of SE France: New Perspectives on the Grès d'Annot and Related Systems*. Geological Society, London, Special Publications, **221**, 241–265, https://doi.org/10.1144/GSL.SP.2004.221.01.13

CARROLL, A.R. & BOHACS, K.M. 1999. Stratigraphic classification of ancient lakes: balancing tectonic and climatic controls. *Geology*, **27**, 99–102.

CARTWRIGHT, J.A., TRUDGILL, B.D. & MANSFIELD, C.S. 1995. Fault growth by segment linkage: an explanation for scatter in maximum displacement and trace length data from the Canyonlands Grabens of SE Utah. *Journal of Structural Geology*, **17**, 1319–1326.

CARTWRIGHT, J., BOUROULLEC, R., JAMES, D. & JOHNSON, H. 1998. Polycyclic motion history of some Gulf Coast growth faults from high-resolution displacement analysis. *Geology*, **26**, 819–822.

CHAPMAN, T.J. & MENEILLY, A.W. 1991. The displacement patterns associated with a reverse-reactivated, normal growth fault. *In*: ROBERTS, A.M., YIELDING, G. & FREEMAN, B. (eds) *The Geometry of Normal Faults*. Geological Society, London, Special Publications, **56**, 183–191, https://doi.org/10.1144/GSL.SP.1991.056.01.12

CHILDS, C., EASTON, S.J., VENDEVILLE, B.C., JACKSON, M.P.A., LIN, S.T., WALSH, J.J. & WATTERSON, J. 1993. Kinematic analysis of faults in a physical model of growth faulting above a viscous salt analogue. *Tectonophysics*, **228**, 313–329.

CHILDS, C., WATTERSON, J. & WALSH, J.J. 1995. Fault overlap zones within developing normal fault systems. *Journal of the Geological Society, London*, **152**, 535–549, https://doi.org/10.1144/gsjgs.152.3.0535

CHILDS, C., NICOL, A., WALSH, J.J. & WATTERSON, J. 2003. The growth and propagation of synsedimentary faults. *Journal of Structural Geology*, **25**, 633–648.

CHILDS, C., MANZOCCHI, T., WALSH, J.J., BONSON, C.G., NICOL, A. & SCHÖPFER, M.P.J. 2009. A geometric model of fault zone and fault rock thickness variations. *Journal of Structural Geology*, **31**, 117–127.

CLIFTON, A.E. & SCHLISCHE, R.W. 2000. Nucleation, growth, and linkage of faults in oblique rift zones: results from experimental clay models and implications for maximum fault size. *Geology*, **29**, 455–458.

COMMINS, D., GUPTA, S. & CARTWRIGHT, J. 2005. Deformed streams reveal growth and linkage of a normal fault array in the Canyonlands graben, Utah. *Geology*, **33**, 645–648.

CONTRERAS, J., ANDERS, M.H. & SCHOLZ, C.H. 2000. Growth of a normal fault system: observations from the Lake Malawi basin of the east African rift. *Journal of Structural Geology*, **22**, 159–168.

COWIE, P.A. & SCHOLZ, C.H. 1992. Displacement–length scaling relationship for faults: data synthesis and discussion. *Journal of Structural Geology*, **14**, 1149–1156.

CRIDER, J.G. & POLLARD, D.D. 1998. Fault linkage: three-dimensional mechanical interaction between echelon normal faults. *Journal of Geophysical Research*, **103**, 24,373–24,391.

DAWERS, N.H. & ANDERS, M.H. 1995. Displacement-length scaling and fault linkage. *Journal of Structural Geology*, **17**, 607–614.

DAWERS, N.H., ANDERS, M.H. & SCHOLZ, C.H. 1993. Growth of normal faults: displacement-length scaling. *Geology*, **21**, 1107–1110.

DUTTON, D.M. & TRUDGILL, B.D. 2009. Four-dimensional analysis of the Sembo relay system, offshore Angola: implications for fault growth in salt-detached settings. *American Association of Petroleum Geologists Bulletin*, **93**, 763–794.

FERRILL, D.A., MORRIS, A.P. & MCGINNIS, R.N. 2012. Extensional fault-propagation folding in mechanically layered rocks; the case against the frictional drag mechanism. *Tectonophysics*, **576-577**, 78–85.

GAWTHORPE, R.L. & HURST, J.M. 1993. Transfer zones in extensional basins: their structural style and influence on drainage development and stratigraphy. *Journal of the Geological Society, London*, **150**, 1137–1152, https://doi.org/10.1144/gsjgs.150.6.1137

GAWTHORPE, R.L. & LEEDER, M.R. 2000. Tectono-sedimentary evolution of active extensional basins. *Basin Research*, **12**, 195–218.

GAWTHORPE, R.L., JACKSON, C.A-L., YOUNG, M.J., SHARP, I.R., MOUSTAFA, A.R. & LEPPARD, C.W. 2003. Normal fault growth, displacement localisation and the evolution of normal fault populations: the Hammam Faraun fault block, Suez rift, Egypt. *Journal of Structural Geology*, **25**, 883–895.

GIBA, M., WALSH, J.J. & NICOL, A. 2012. Segmentation and growth of an obliquely reactivated normal fault. *Journal of Structural Geology*, **39**, 253–267.

HUGGINS, P., WATTERSON, J., WALSH, J.J. & CHILDS, C. 1995. Relay zone geometry and displacement transfer between normal faults recorded in coal-mine plans. *Journal of Structural Geology*, **17**, 1741–1755.

JACKSON, C.A-L. In press. Growth of a salt-detached normal fault and controls on throw rate variability; gudrun field, South Viking Graben, offshore Norway. *In*: TURNER, C.C. & CRONIN, B.T. (eds) *Brae Play*. American Association of Petroleum Geologists, Memoirs.

JACKSON, C.A-L. & ROTEVATN, A. 2013. 3D seismic analysis of the structure and evolution of a salt-influenced normal fault zone: a test of competing fault growth models. *Journal of Structural Geology*, **54**, 215–234.

JACKSON, C.A-L., GAWTHORPE, R.L. & SHARP, I.R. 2002. Growth and linkage of the East Tanka fault zone, Suez rift: structural style and syn-rift stratigraphic response. *Journal of the Geological Society, London*, **159**, 175–187, https://doi.org/10.1144/0016-764901-100

KIM, Y.S. & SANDERSON, D.J. 2005. The relationship between displacement and length of faults: a review. *Earth-Science Reviews*, **68**, 317–334.

LAMARCHE, G., PROUST, J.N. & NODDER, S.D. 2005. Long-term slip rates and fault interactions under low contractional strain, Wanganui Basin, New Zealand. *Tectonics*, **24**.

Mansfield, C. & Cartwright, J. 2001. Fault growth by linkage: observations and implications from analogue models. *Journal of Structural Geology*, **23**, 745–763.

McLeod, A.E. & Underhill, J.R. 2000. The propagation and linkage of normal faults: insights from the Strathspey–Brent–Statfjord fault array, northern North Sea. *Basin Research*, **12**, 263–284.

Meyer, V., Nicol, A., Childs, C., Walsh, J.J. & Watterson, J. 2002. Progressive localisation of strain during the evolution of a normal-fault system. *Journal of Structural Geology*, **24**, 1215–1231.

Morewood, N.C. & Roberts, G.P. 1999. Lateral propagation of the surface trace of the South Alkyonides normal fault segment, central Greece: its impact on models of fault growth and displacement–length relationships. *Journal of Structural Geology*, **21**, 635–652.

Morley, C.K. 1999. Patterns of displacement along large normal faults: implications for basin evolution and fault propagation, based on examples from East Africa. *American Association of Petroleum Geologists Bulletin*, **83**, 613–634.

Morley, C.K. 2002. Evolution of large normal faults: evidence from seismic reflection data. *American Association of Petroleum Geologists Bulletin*, **86**, 961–978.

Nicol, A., Watterson, J., Walsh, J.J. & Childs, C. 1996. The shapes, major axis orientations and displacement patterns of fault surfaces. *Journal of Structural Geology*, **18**, 235–248.

Nicol, A., Walsh, J., Berryman, K. & Nodder, S. 2005. Growth of a normal fault by the accumulation of slip over millions of years. *Journal of Structural Geology*, **27**, 327–342.

Nicol, A., Walsh, J.J., Villamor, P., Seebeck, H. & Berryman, K.R. 2010. Normal fault interactions, paleoearthquakes and growth in an active rift. *Journal of Structural Geology*, **32**, 1101–1113.

Nicol, A., Childs, C., Walsh, J.J., Manzocchi, T. & Schöpfer, M.P.J. 2016. Interactions and growth of faults in an outcrop-scale system. *In*: Childs, C., Holdsworth, R.E., Jackson, C.A.-L., Manzocchi, T., Walsh, J.J. & Yielding, G. (eds) *The Geometry and Growth of Normal Faults*. Geological Society, London, Special Publications, **439**. First published online March 10, 2016, https://doi.org/10.1144/SP439.9

Peacock, D.C.P. & Sanderson, D.J. 1991. Displacements, segment linkage and relay ramps in normal fault zones. *Journal of Structural Geology*, **13**, 721–733.

Petersen, K., Clausen, O.R. & Korstgård, J.A. 1992. Evolution of a salt-related listric growth fault near the D-1 well, block 5605, Danish North Sea: displacement history and salt kinematics. *Journal of Structural Geology*, **14**, 565–577.

Polit, A., Schultz, R.A. & Soliva, R. 2009. Geometry, displacement–length scaling, and extensional strain of normal faults on Mars and inferences on mechanical stratigraphy of the Martian crust. *Journal of Structural Geology*, **31**, 662–673.

Prosser, S. 1993. Rift-related linked depositional systems and their seismic expression. *In*: Williams, G.D. & Dobb, A. (eds) *Tectonics and Seismic Sequence Stratigraphy*. Geological Society, London, Special Publications, **71**, 35–66, https://doi.org/10.1144/GSL.SP.1993.071.01.03

Robson, A.G., King, R.C. & Holford, S.P. 2016. Structural evolution of a gravitationally detached normal fault array: analysis of 3D seismic data from the Ceduna Sub-Basin, Great Australian Bight. *Basin Research*, first published online March 2, 2016, https://doi.org/10.1111/bre.12191

Rowan, M.G., Hart, B.S., Nelson, S., Flemings, P.B. & Trudgill, B.D. 1998. Three-dimensional geometry and evolution of a salt-related growth-fault array: EI 330 field, offshore Louisiana, Gulf of Mexico. *Marine and Petroleum Geology*, **15**, 309–328.

Schlagenhauf, A., Manighetti, I., Malavieille, J. & Dominguez, S. 2008. Incremental growth of normal faults: Insights from a laser-equipped analog experiment. *Earth and Planetary Science Letters*, **273**, 299–311.

Schlische, R.W. 1995. Geometry and origin of fault-related folds in extensional settings. *American Association of Petroleum Geologists Bulletin*, **79**, 1661–1678.

Soliva, R. & Benedicto, A. 2005. Geometry, scaling relations and spacing of vertically restricted normal faults. *Journal of Structural Geology*, **27**, 317–325.

Soliva, R., Benedicto, A., Schultz, R.A., Maerten, L. & Micarelli, L. 2008. Displacement and interaction of normal fault segments branched at depth: implications for fault growth and potential rupture size. *Journal of Structural Geology*, **30**, 1288–1299.

Taylor, S., Nicol, A. & Walsh, J.J. 2008. Displacement loss on growth faults due to sediment compaction. *Journal of Structural Geology*, **30**, 394–405.

Thorsen, C.E. 1963. Age of growth faulting in southeast Louisiana. *Gulf Coast Association of Geological Societies Transactions*, **13**, 103–110.

Tvedt, A.B.M. 2016. *The Geometry and Evolution of Supra-Salt Normal Fault Arrays*. PhD thesis, University of Bergen, Norway.

Tvedt, A.B.M., Rotevatn, A., Jackson, C.A-L., Fossen, H. & Gawthorpe, R.L. 2013. Growth of normal faults in multilayer sequences: a 3D seismic case study from the Egersund Basin, Norwegian North Sea. *Journal of Structural Geology*, **55**, 1–20.

Tvedt, A.B.M., Rotevatn, A. & Jackson, C.A-L. 2016. Supra-salt normal fault growth during the rise and fall of a diapir: perspectives from 3D seismic reflection data, Norwegian North Sea. *Journal of Structural Geology*, **91**, 1–26, https://doi.org/10.1016/j.jsg.2016.08.001

Walsh, J.J. & Watterson, J. 1988. Analysis of the relationship between displacements and dimensions of faults. *Journal of Structural Geology*, **10**, 239–247.

Walsh, J.J., Watterson, J., Childs, C. & Nicol, A. 1996. Ductile strain effects in the analysis of seismic interpretations of normal fault systems. *In*: Buchanan, P.G. & Nieuwland, D.A. (eds) *Modern Developments in Structural Interpretation, Validation and Modelling*. Geological Society, London, Special Publications, **99**, 27–40, https://doi.org/10.1144/GSL.SP.1996.099.01.04

Walsh, J.J., Nicol, A. & Childs, C. 2002. An alternative model for the growth of faults. *Journal of Structural Geology*, **24**, 1669–1675.

Walsh, J.J., Bailey, W.R., Childs, C., Nicol, A. & Bonson, C.G. 2003. Formation of segmented normal faults: a 3-D perspective. *Journal of Structural Geology*, **25**, 1251–1262.

WATTERSON, J. 1986. Fault dimensions, displacement and growth. *Pure and Applied Geophysics*, **124**, 365–373.

WILKINS, S.J. & GROSS, M.R. 2002. Normal fault growth in layered rocks at Split Mountain, Utah. *Journal of Structural Geology*, **24**, 1413–1429.

WILLEMSE, E.J.M. 1997. Segmented normal faults: correspondence between three-dimensional mechanical models and field data. *Journal of Geophysical Research*, **102**, 675–692.

WITHJACK, M.O., SCHLISCHE, R.W. & OLSEN, P.E. 2002. Rift-basin structure and its influence on sedimentary systems. *In*: RENAULT, R.W. & ASHLEY, G.M. (eds) *Sedimentation in Continental Rifts.* SEPM (Society for Sedimentary Geology), Special Publications, **73**, 57–81.

YOUNG, M.J., GAWTHORPE, R.L. & HARDY, S. 2001. Growth and linkage of a segmented normal fault zone; the Late Jurassic Murchison–Statfjord North Fault, northern North Sea. *Journal of Structural Geology*, **23**, 1933–1952.

Growth and interaction of normal faults and fault network evolution in rifts: insights from three-dimensional discrete element modelling

EMMA FINCH[1]* & ROB GAWTHORPE[2]

[1]*School of Earth and Environmental Sciences, University of Manchester, Oxford Road, Manchester M13 9PL, UK*

[2]*Department of Earth Science, University of Bergen, Allégaten 41, 5007 Bergen, Norway*

**Correspondence: Emma.Finch@manchester.ac.uk*

Abstract: The initiation, growth and interaction of faults within an extensional rift is an inherently four-dimensional process where connectivity with time and depth are difficult to constrain. A 3D discrete element model is employed that represents the crust as a two-layered brittle–ductile system in which faults nucleate, propagate and interact in response to local heterogeneities and resulting stresses. Faults nucleate in conjugate sets throughout the model brittle crust; they grow through a combination of tip propagation and interaction of co-linear segments to form larger normal faults. Segment linkage occurs by merging of adjacent fault segments located along strike, downdip or oblique to one another. Finally, deformation localizes onto the largest faults. Displacement distribution on faults is highly variable with marked along-strike and temporal variations in displacement rates. Displacement maxima continuously migrate as smaller fault segments interact and link to form the final fault plane. As a result, displacement maxima associated with fault nucleation sites are not coincident with the location of the maximum finite displacement on a fault where segment linkage overprints the record. The observed style of fault growth is consistent with the isolated growth model in the earliest stages which then gives way to a coherent (constant-length) fault growth model at greater strains.

Understanding fault evolution in three dimensions in rift basin settings is generally informed by interpretation and analysis of the current or final static fault geometry. The evolution and interaction of faults in extensional rifts, however, is an essentially four-dimensional problem, where the initiation, growth and interaction of faults in three dimensions through time modify the nature of the fault network. The aim of this paper is to apply a numerical model of rifting to better understand the nucleation, interaction and evolution of faults in four dimensions in a rift basin subjected to a single phase of extension. Using this model, the propagation and interaction of faults, scaling relationships, progressive strain localization and dip domain generation are addressed.

It is known that faults grow by lateral propagation and linkage where the slip on isolated and later linked faults accumulates at varying rates (McLeod *et al.* 2000; Cowie & Roberts 2001; Walsh *et al.* 2003*b*). The processes involved are determined from interpretation of observable and extractable data at the end of either a single or multiphase extension event. In particular, earthquake slip or displacement patterns along selected faults are used to infer the growth history and interaction of selected structures (Contreras *et al.* 2000; Sharp *et al.* 2000; Morley 2002; Manighetti *et al.* 2005; Bull *et al.* 2006; Jackson *et al.* 2006; Morley *et al.* 2007; Nicol *et al.* 2010; Reeve *et al.* 2015), with specific focus on relay structures (Acocella *et al.* 2000; Conneally *et al.* 2014; Fossen & Rotevatn 2016), and the nature of fault tip interactions (Nixon *et al.* 2014*b*; Duffy *et al.* 2015; Whipp *et al.* 2016).

It is not straightforward to determine how faults in rift settings interact within the crust. For example, investigations of fault branching with depth in 3D seismic and analogue modelling have confirmed that what appear as isolated faults at the surface are, in fact, coupled along strike through structures at depth (Kornsawan & Morley 2002; Soliva *et al.* 2008; Long & Imber 2011, 2012; Giba *et al.* 2012). Failure to recognize this fact will lead to fault network statistics that do not accurately represent the largest faults or fault connectivity within the rift. A consequence of this will be an underestimate of earthquake rupture capabilities. (Walsh *et al.* 2003*b*; Manighetti *et al.* 2007; Soliva *et al.* 2008; Nicol *et al.* 2010).

Several lines of research suggest that faults originate in conjugate or polymodal orientations within an extensional rift and that the earliest-forming

From: Childs, C., Holdsworth, R. E., Jackson, C. A.-L., Manzocchi, T., Walsh, J. J. & Yielding, G. (eds) 2017. *The Geometry and Growth of Normal Faults*. Geological Society, London, Special Publications, **439**, 219–248.
First published online August 30, 2017, https://doi.org/10.1144/SP439.23

faults modify the local stress field so that later neighbouring faults dip in the same orientation (Schlische & Withjack 2009; Healy *et al.* 2015). Dip domains evolve dependent on these self-organizing incipient faults. Domain boundaries (also known as transfer/accommodation zones or graben shifts) are characterized by narrow zones of overlapping fault tips where a change in polarity is marked by interlocking arrays of conjugate faults (Kornsawan & Morley 2002; McClay *et al.* 2002, 2005; Schlische & Withjack 2009; Fossen & Rotevatn 2016). Not all domain boundaries recorded between opposed dipping faults are considered to evolve from a self-organized state, however: for example, where an underlying basement structure is present, it has been shown to be a key influence on the location of domain boundaries (Acocella *et al.* 1999; Fossen & Rotevatn 2016).

The mechanism required for faults within an evolving system to interact is generally discussed in terms of two fault models: the isolated model and the coherent model (Walsh *et al.* 2003*a*; Giba *et al.* 2012; Fossen & Rotevatn 2016). The isolated model suggests that a segmented fault array develops from random overlap and linkage of previously unrelated faults which initiate in a self-organized manner from natural heterogeneities and that strain is distributed homogenously (Cowie *et al.* 2000; Wu *et al.* 2015). According to the isolated fault model, faults develop during rifting from isolated heterogeneities within a rock volume by radial propagation and, as a result, individual segments of isolated structures expand laterally in three dimensions. As rifting progresses, these small, isolated, faults propagate rapidly to become larger structures, or smaller faults link along strike and/or downdip through relays or tip propagation into larger structures (Peacock & Sanderson 1991; Cartwright *et al.* 1995; Dawers & Anders 1995; Nicol *et al.* 1996; Wojtal 1996; Gupta *et al.* 1998; McLeod *et al.* 2000; Cowie & Roberts 2001; Walsh *et al.* 2003*a*). In the coherent model, a segmented fault array develops within an organized system in which segments are kinematically linked from the start. This has been demonstrated where some degree of strain concentration is present due to reactivation of a buried fault, influencing fault propagation in the overburden (Acocella *et al.* 1999; Walsh *et al.* 2003*b*; Giba *et al.* 2012). Fossen & Rotevatn (2016) saw these models as representative end members and not mutually exclusive. Jackson *et al.* (2017) suggest that it is conceivable that pre-linkage faults propagate in accordance with the isolated model in their early stages, but that evidence for this is difficult to resolve from subsurface datasets and so the coherent model is more readily observed.

Natural fault networks have been shown to exhibit specific scaling properties that are now key features in interpreting and modelling fault growth and interaction. These relationships are statistical descriptions of the distribution of the frequency–size of fault attributes, which include the correlation between the displacement on faults and their lengths, and the spatial patterns of faulting. Attention has focused on evaluation and discussion of empirical relationships, such as length, width, displacement and gouge thickness (e.g. Hull 1988; Cowie & Scholz 1992; Gillespie *et al.* 1992; Yielding *et al.* 1992; Dawers *et al.* 1993; Gross *et al.* 1997; Torabi & Berg 2011; Xu *et al.* 2014). In natural and physical analogue fault networks, displacement–distance profiles are commonly employed to determine the timing of interaction of faults by along-strike linkage (Cowie & Scholz 1992; Contreras *et al.* 2000; Ackermann *et al.* 2001; Cowie & Roberts 2001; Faure Walker *et al.* 2009; Nicol *et al.* 2010; Nixon *et al.* 2011; Jackson & Rotevatn 2013; Xu *et al.* 2014; Reeve *et al.* 2015; Whipp *et al.* 2016). Displacement profiles for individual isolated faults show either triangular or elliptical profiles with displacement maxima at their centre (Marrett & Allmendinger 1990). Symmetrical displacement–distance profiles become asymmetrical towards the interaction point where one tip is restricted, or form a double tip restricted, 'mesa'-style profile with steep edges when both tips are constrained (Dawers *et al.* 1993; Manighetti *et al.* 2005; Nixon *et al.* 2014*b*). Deviation of along-strike profiles from hypothetical symmetrical curves is thought to indicate the nature and timing of linkage of small fault segments into larger structures, where displacement maxima indicate centres of original smaller segments and minima denote points of linkage (Huggins *et al.* 1995; Faure Walker *et al.* 2009; Nixon *et al.* 2011, 2014*b*; Xu *et al.* 2014; Reeve *et al.* 2015; Khalil & McClay 2016). To maintain fault displacement–length scaling, it is expected that observed along-strike deficits in throw associated with fault linkage will become less significant as the fault grows (Nixon *et al.* 2014*b*; Jackson *et al.* 2017). An isolated type profile will result with a throw maximum at the centre of the new, larger structure (Contreras *et al.* 2000; Cowie *et al.* 2000; Schlagenhauf *et al.* 2008; Faure Walker *et al.* 2009; Xu *et al.* 2014).

Outcrop and subsurface (seismic) data provide a static, final image of fault networks. In order to assess how similar geometries evolve, physical and numerical analogues are used. Physical analogues examine the upper surface topographical and fault trace evolution through time and the final static fault geometry in cross-section (Acocella *et al.* 1999; Clifton *et al.* 2000; McClay *et al.* 2002, 2005; Hus *et al.* 2005; Schlagenhauf *et al.* 2008; Schlische & Withjack 2009; Henza *et al.* 2010,

2011). These methods have been important in understanding fault growth and interaction through time, but have not fully addressed the 3D development of fault networks.

Numerical methods use both 2D and 3D approximations to investigate the interaction of either a large number of faults in a 2D plane (Cowie *et al.* 2000) or focused assessment of the growth and interaction of selected pre-defined isolated structures (Walsh *et al.* 2001; Imber *et al.* 2004; Soliva *et al.* 2008; Lovely *et al.* 2012; Allken *et al.* 2013). Discrete element models (DEMs) are numerical models that use physically realistic inter-element interactions and have been applied to model mechanical rock behaviour in scenarios where the evolution of discontinuities can be tested (e.g. Cundall 1971; Mora & Place 1993; Donzé *et al.* 1996; Kuhn 1999; Camborde *et al.* 2000; Toomey & Bean 2000; Place *et al.* 2002; Imber *et al.* 2004; Hardy & Finch 2005, 2006; Schöpfer *et al.* 2007*a*, *b*, 2009; Abe *et al.* 2011; Hardy 2014; Lambert & Coll 2014).

In this study, a DEM is applied to investigate the initiation, growth and interaction of faults within a normal fault network in a rift basin. The crust is represented as a two-layer model of passive crustal extension with an upper 15 km-thick layer representing the brittle upper crust, and a lower 15 km-thick firmoviscous layer representing the ductile lower crust. The effects of thermal variation within the crust during the evolution of the model are not included. This approach allows investigation of fault nucleation and the subsequent organization of a fault network in 4D, the 3D geometry and interaction between faults, and the distribution of fault activity and displacement in time and space. Results presented have implications for analysis of natural fault systems, particularly the limitations of traditional methods for reconstructing fault growth histories from final fault geometry coupled with measurement of variations in displacement measured on pre-rift and synrift stratigraphic horizons along fault systems.

Methodology

Multi-layer rheologies in DEM techniques have been employed to examine the influence of mechanical stratigraphy on the propagation of blind faults (Hardy & Finch 2007; Schöpfer *et al.* 2007*a*, *b*), boudinage (Komoróczi *et al.* 2013) and compressional wedges (Wenk & Huhn 2013). The crust in our model is represented as a two-layer system where elements in the upper crust interact through linear elastic repulsive–attractive forces and those within the lower crust interact through linear viscous (Newtonian fluid) forces (Fig. 1a) (Ranalli 1995). Elements in the upper crust are treated as an assembly of spheres that interact in pairs (*ij*) as though connected by breakable elastic springs following:

$$F_{ijU}^{\text{elastic}} = \begin{cases} K(r-R) & r < r_b \text{ (i.e. intact bond)} \\ K(r-R) & r < R \text{ (i.e. broken bond)} \\ 0 & r \geq R \text{ (i.e. broken bond)} \end{cases}$$

where K is the bond stiffness, R is the equilibrium separation between an element pair (element i and neighbour j) and r is the inter-element separation. Elements are bonded until their separation exceeds a breaking distance, r_b, beyond which the bond is broken and experiences no further attractive force but will experience a repulsive force if the elements return to a compressive contact (i.e. $r < R$).

In the lower crust, a firmoviscous (Kelvin) body is applied to determine inter-element forces where elastic and linearly viscous forces are set in parallel (Fig. 1b). When loaded, the elastic response of the spring is delayed by the viscous response of the dashpot, resulting in a non-instantaneous response. The force due to the spring in the lower crust, F_{ijL}^{elastic}, is obtained from:

$$F_{ijL}^{\text{elastic}} = \begin{cases} K_c(r-R) & r < R \\ K_t(r-R) & r > R \end{cases}$$

where K_c is the spring stiffness in compression, consistent with upper crustal elements, and K_t is the spring stiffness in extension, set to zero. This relationship assumes that links between elements in the lower crust retain elastic properties in compression only. The viscous force is determined through:

$$F_{ijL}^{\text{viscous}} = -\eta \Delta \dot{x}$$

where $\Delta \dot{x}$ represents the relative velocity between an element pair and η is the Kelvin viscosity, chosen empirically within the experiment. The greater the relative velocity between an element pair, the greater the force acts to return them to their equilibrium position.

The total force, F_i, exerted on an element is obtained by summing the forces exerted on it by its n neighbours:

$$F_i = \sum_{j=1,n} F_{ijU}^{\text{elastic}} + F_{ijL}^{\text{elastic}} + F_{ijL}^{\text{viscous}}.$$

The interface between the upper and lower crust is defined as a step function, positioned at a depth appropriate to the thickness of the crust under investigation. Bonds between elements that bisect this interface are treated as viscous. To attenuate kinetic

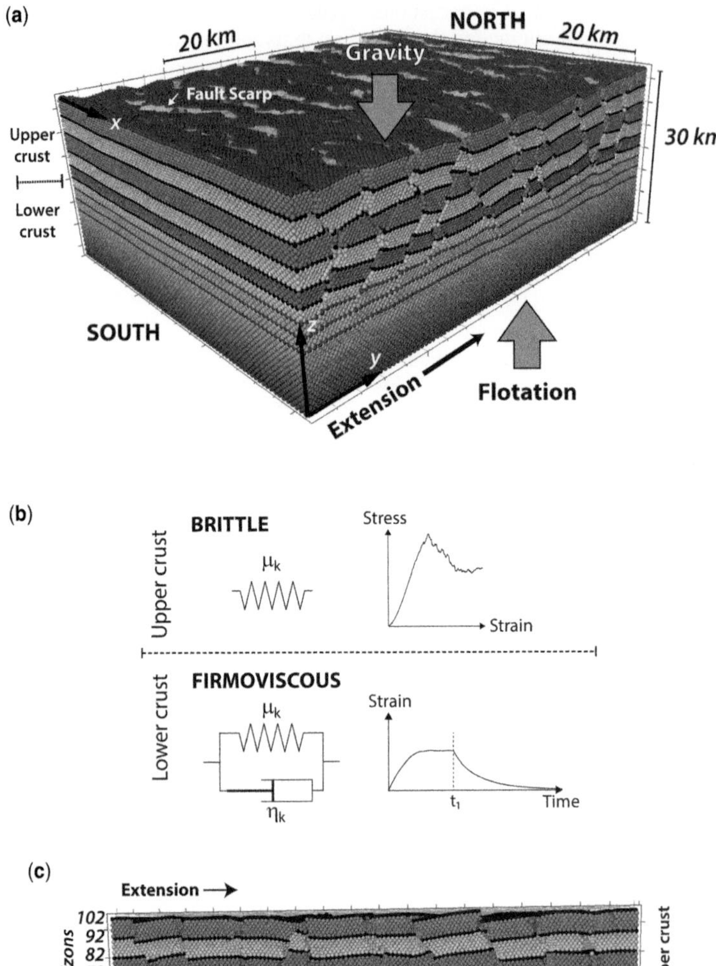

Fig. 1. (**a**) Example of the discrete element media consisting of 1 080 000 elements in a regular hexagonal distribution used to simulate extension in a rift setting. The crust is represented as a two-layer system of brittle (upper crust) and firmoviscous (lower crust) forces. Elements are shaded according to their horizon at the start of the experiment. (**b**) Sketch of the forces within the upper and lower crust, and their rheological response. In the upper crust, the bond fails when a breaking threshold is exceeded. In the lower crust, strain increases when a load is applied. At t_1, the load is removed and strain reduces slowly over time. (**c**) View of model from eastern aspect illustrating horizons used in analysis (H 52–H 102) and the viscous response to load in the lower crust (wavy nature of dark grey marker lines).

energy in the system and constrain the elastic nature of the springs, a damping force, F_{iD}, is included:

$$F_{iD} = -v\dot{x}$$

where v represents the damping term and $\dot{x}$ is the element velocity. The damping term in this experiment is 7.0 (cf. 0.7: Potyondy & Cundall 2004) and permits investigation of quasi-static deformation (Donzé *et al.* 1994).

Finite-element schemes have approximated the crust as a gravitating plate effectively floating hydrostatically on the mantle (e.g. King & Ellis 1990; Hassani & Chéry 1996). A similar method is employed here, where the crust is considered an elastic–brittle–plastic plate floating hydrostatically on a fluid mantle held in equilibrium around a specified depth (cf. King *et al.* 1988). This depth is determined from the ratio between crust and mantle densities, where the density of the brittle crust is described as a lower estimate of crustal density based on the saturated bulk densities of rock (King *et al.* 1988). The force experienced due to gravity and flotation, F_{GF}, is added to the interaction force in the vertical, z-component direction where:

$$F_{iGF} = g[(\rho_m - \rho_c)V_B - \rho_c V_A].$$

Here ρ_c and ρ_m are the crust and mantle densities respectively, g is the acceleration due to gravity, and V_A and V_B are the volumes of the element that exist above and below the hydrostatic equilibrium. When an element exists completely above the equilibrium depth, a resultant downwards force is experienced, whereas an element entirely below the hydrostatic equilibrium experiences a resultant upwards force that simulates buoyancy.

With increasing time, the loading due to gravity will cause viscous flow in the lower crust. If the medium is not constrained in the x- and y-component directions, this will cause a 'forcing out' of elements at boundaries. To negate this, the medium is constrained by bounding walls which simulate it existing within a larger system of elements with similar mechanical properties. At each time step, if an element's interaction force oversteps the bounding limit for the wall, the element experiences an additional repulsive force from the wall. This wall force (F_{iw}) assumes that element i has come into a compressive contact with an element w within the wall, so that:

$$F_{iw} = -K_w r_w$$

where r_w is equivalent to the amount by which the element exceeds the boundary and K_w is the elastic stiffness of the wall (cf. Wenk & Huhn 2013).

There are no shear forces determined within this technique, so the behaviour of the rock mass is considered as frictionless (see Donzé *et al.* 1994; Mora & Place 1994; Hardy & Finch 2007). This methodology has been previously used to successfully simulate the frictional stick–slip instability in a rock assemblage without shear forces (Mora & Place 1994), and for biaxial compression tests and faulting in sedimentary successions above basement structures (Finch *et al.* 2003, 2004; Hardy & Finch 2006). The success of these methodologies suggests that the assumption of a frictionless rock mass is not at odds with reproducing realistic rock mechanics behaviour. Other DEM techniques incorporate frictional forces but their addition greatly increases computational time (e.g. 45 days with 12 000 elements: Wenk & Huhn 2013).

The total force exerted on an element within the crust in the x- and y-component directions is given through:

$$F_i^{TOTAL} = F_i + F_{iD} + F_{iW}.$$

The gravity and flotation term is included in the z-component direction; therefore, the total force exerted vertically is:

$$F_i^{TOTAL} = F_i + F_{iD} + F_{iGF}.$$

Extension is implemented on all elements in small increments to simulate movement of a rigid boundary wall to the north while the southern boundary is static (Fig. 1c). The boundary condition is implemented so that:

$$y_i^*(t) = y_i(t) + \Delta y\left(\frac{y_i(t)}{y_{max}(t)}\right).$$

Here, $y_i^*(t)$ is the new element location, $y_i(t)$ is the current element location, Δy is the extension increment per time step and $y_{max}(t)$ is the maximum length of the model in the extension direction (similar to Donzé *et al.* 1994). Elements are advanced to new locations within the model by integration of their equations of motion using Newtonian physics (see Hardy & Finch 2006). Discrete element models can be run in model units but for comparison with geological data, model units are often scaled to real-world parameters (e.g. Place & Mora 2001; Hardy & Finch 2006).

Numerical modelling of the effect of temperature variations on fault initiation and activity during rifting is common (Behn *et al.* 2002; Huismans & Beaumont 2007; Wright *et al.* 2012). From these continuum-mechanics-based models, which employ thermomechanical equations to investigate fault localization, it is known that extension results in crustal thinning and horizontal fluctuations in the temperature field, culminating in focused faulting around zones of thinned crust (Behn *et al.* 2002; Cowie *et al.* 2005; Huismans & Beaumont 2007). Deformation is distributed between sets of conjugate normal faults, however, in the absence of a regional temperature gradient (Behn *et al.* 2002). The purpose of this paper is to examine the initiation, growth and interaction of faults during rifting, and, as such, localization effects associated with thermal variations through time are not considered.

Experimental set-up and data analysis

Discrete element model. The experimental media consist of 1 080 000 elements with a regular hexagonal packing where element radii are unity (Fig. 1). The initial dimensions are $213 \times 200 \times 102$ (x, y, z) model units (m.u.). Horizons are defined as an integer value of their height in model units: there are 70 in total. Data from six horizons are extracted for discussion purposes, numbered from Horizon 52 (immediately above the upper–lower crust boundary) to Horizon 102 (the upper surface) in 10 unit increments (Fig. 1c). Here, 1 m.u. is equivalent to 292 m, and the model represents real-world dimensions of $64 \times 60 \times 30$ km. The upper and lower crust layers are each 15 km thick at the start of the experiment. Experiments are run for 60 000 time steps with data output at intervals of 1000, providing 60 data files. A time step represents 100 years, so the total run time is 6 myr with an output interval of 100 kyr. The southern end of the model is fixed. Extension is incremented at 0.001 m.u per time step towards the north (Fig. 1) and thus represents of rate of 3 mm a^{-1}, with each output correlating to 0.5% extension to a total of 30%. The natural strain rate determined for these experiments decreases from 1.6×10^{-15} to 1.24×10^{-15} s^{-1} during extension, consistent with strain rates recorded from rifted basins that can range from 1.0×10^{-16} to 4.0×10^{-14} s^{-1} (Kusznir & Park 1987; Nicol *et al.* 1997).

The data presented here are from a single experiment but are representative of many experiments which evaluated scaling parameters appropriate for investigation of the development of faults in a rifted basin. Rock densities are defined as 2800 and 3300 kg m^{-3} for the crust and mantle, respectively. The elastic spring constant in the upper crust (K) is 8.6×10^{10} N m^{-1}, and 9.7×10^{11} N m^{-1} in the lower crust (K_c). For this scaling of the model, Poisson's ratio is 0.25, and Young's modulus (E) approximates to 90 and 105 GPa in the upper and lower crust, respectively (Mora & Place 1994). Elements within the upper crust are randomly assigned breaking thresholds (r_b) between 0.025 and 0.1 m.u. at the start of the experiment. The breaking threshold between element pairs is determined from the average of the threshold assigned to the two elements, providing a distribution of weak and strong bonds in varying orientations around each element. These breaking thresholds scale to bond strengths between 2.1 and 8.6 GPa in the upper crust.

Analysis of fault network development. Faults are defined by analysing the separation between an element and its immediate neighbour in the extension direction at the end of the experiment. They are identified by filtering data for each element relative to heaves exceeding 50 m and recording the throw (vertical displacement between element pairs). In previous methodologies (e.g. Finch *et al.* 2004; Hardy & Finch 2006), faults have been defined using continuous alignments of broken bonds. In this experiment, however, displacement propagates into the lower crust where bonds do not break. Therefore, a filter using fault heave was deemed the most appropriate method for determining fault locations. The heave value used has been chosen through testing and prevents fault definition including elements displaced by flexural rotation of horizons in hanging walls and footwalls. A fault in this model is defined as a continuous alignment of elements with similar dip orientation along strike (north or south). The chosen elements are then assigned a fault number. Data associated with these faults can be extracted from earlier outputs within the experiment to determine the growth and interaction of selected structures. For example, the topographical evolution associated with a fault at the surface can be analysed by outputting displacement on its constituent elements to produce displacement–distance plots for throw at times throughout the experiment.

For each horizon in the model, 120 1D transects perpendicular to the extension direction can be extracted. The throw and heave of faults that intersect these transects can be assessed through time, and employed to examine fault growth, strain accommodation and fault polarity. This is carried out for the upper surface (Horizon 102) to compare with data from fault analysis techniques. Fault displacement is output for five selected transects to demonstrate fault growth at the upper surface. The amount of strain accommodated on north- and south-dipping structures is determined from the summation of fault heaves along these transects. It is recorded against time, and the relationship between dip direction and strain accommodation is plotted.

The polarity of the five selected 1D transects is calculated using:

$$P = \frac{\varepsilon_d - \varepsilon_n}{\varepsilon_t}$$

where ε_d and ε_n are the strain accommodated by the dominant and non-dominant strain orientation, and ε_t is the total strain accommodated along the transect. Similar to Moriya *et al.* (2005), the dominant fault direction is defined as the direction in which the largest displacement fault dips. In this respect, $-1.0 < P < 1.0$, where a population that is strongly polarized opposite to the dip of the largest fault, will return polarities approaching −1.0 and a population that dips consistently with the largest fault will return polarities near 1.0. As a consequence, the orientation that contains the fault with the largest strain should be consistent with the dominant direction of strain accommodation, indicating that a positive P value should be expected.

To illustrate fault evolution and linkage in three dimensions, the displacement (throw) on elements that constitute six selected faults is plotted in strike sections at 15, 20, 25 and 30% extension. Displacement–distance plots for these faults are presented where data are generated by sampling within 500 m intervals along the length of a fault, recording the maximum displacement regardless of depth for each interval. This is done to negate the dominance of one horizon in assessment of fault growth and highlights the location of displacement maxima through time along the fault length.

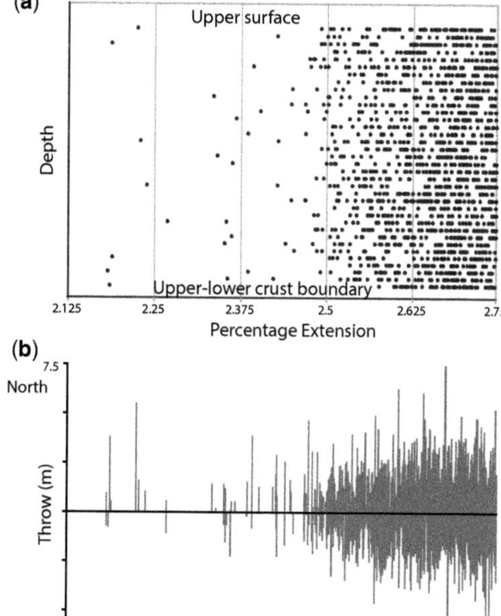

Fig. 2. (**a**) Depth of broken bonds in the brittle crust in the experiment at the start of the experiment. (**b**) Magnitude and direction of displacement (in metres) for failures shown in (a).

Fault network organization

In this section, a number of features of the organization and growth of the fault network are presented. We first look at fault nucleation, and then evaluate the entire fault network in relation to the spatial distribution of faults, their growth and interaction. Displacement accrual along selected transects, fault polarity and strain accommodation across dip domains and domain boundaries is then considered. Fault network statistics are presented in relation to the variability with depth of the frequency–size relationship, displacement on selected horizons and displacement on the largest faults within the system. The final section discusses the along-strike 3D interaction of seven faults.

Fault nucleation

In order to assess whether lattice geometry affects the nucleation and growth of faults, the location of the earliest bond failures are output relative to depth and orientation (Fig. 2). These nucleation sites are distributed throughout the brittle layer and demonstrate that there is no dominance or focus in the distribution of initial failures with depth (Fig. 2a) or their orientation (Fig. 2b). The location of further failure in the brittle layer continues with increasing extension and is distributed throughout the crust, influenced by the stress fields surrounding existing faults.

Evaluation of the entire fault network

Final fault network. The consequence of faults initiating in a conjugate distribution is shown in Figure 3, where the elements that constitute the final faults in the model are shown. Faults are connected along strike through a curvilinear geometry and spaced at regular intervals. By the end of the experiment, larger faults fill the brittle crust with maximum displacement at their centres (*c.* H 72: Fig. 3b); the depth to which faults project into the lower crust varies according to their along-strike length. Upper surface topography forms a series of graben and half-graben, where basins have a maximum along-strike length of <30 km and footwall crest to footwall crest separations of around 10–15 km (Fig. 3c). Hanging-wall depocentres are located at fault centres, with maximum relief of the order of 1 km. Conjugate fault interactions across relays are common, and numbered circles in Figures 3c & 4 highlight three locations where relays result in topographical lows between neighbouring basins.

Fault network growth and organization. The evolution of the fault network on the uppermost horizon (H 102), together with topography and dip domains, is shown in Figure 4. Faults initiate as a large number of small, isolated structures, striking perpendicular to the extension direction (±20°), and by 10% extension have lengths and separations of the order of 10–15 km (10%: Fig. 4a). Displacement

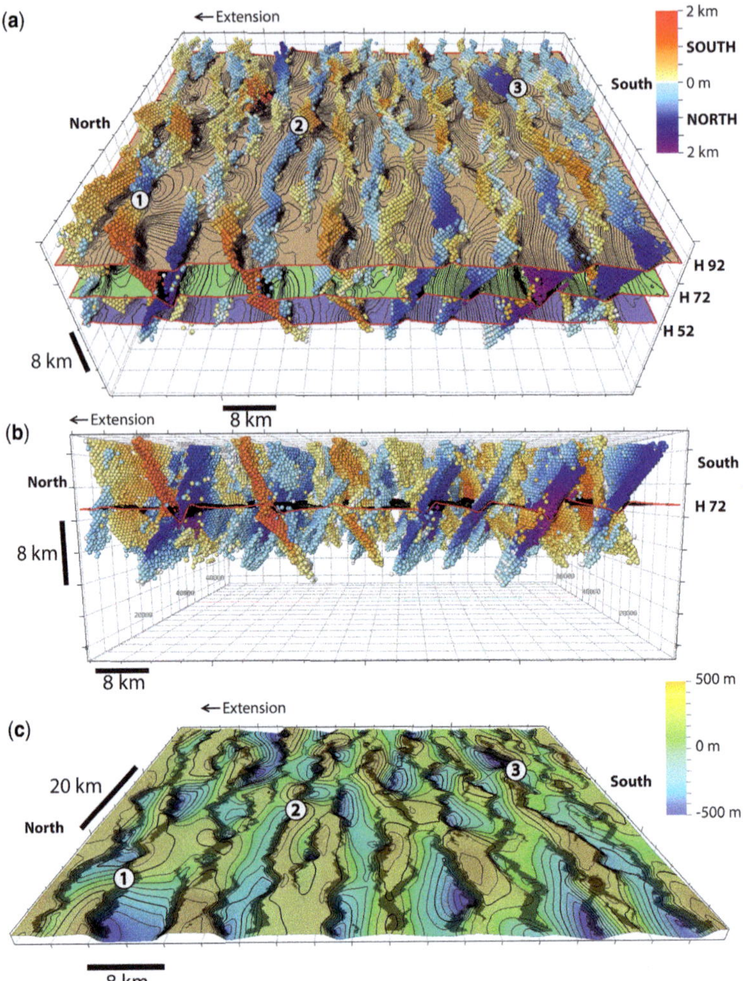

Fig. 3. (**a**) Oblique 3D view of along-strike complexity relative to elevation for the fault network at 30% extension relative to initially horizontal beds within the crust (Horizon 52 (H 52), Horizon 72 (H 72) and Horizon 92 (H 92)). Contours on horizons denote 100 m intervals. The extension direction is towards the north. Displacement on elements that constitute south-dipping faults are coloured yellow to red (2 km) and north-dipping faults are coloured pale blue to purple (2 km). Interactions between selected conjugate faults are shown by numbers 1–3. (**b**) View of the fault network looking east (along strike). (**c**) Topography of Horizon 102 at 30% extension. Contours are at 100 m intervals, with an elevation range of 1 km scaled from blue (deep) to orange (high). The extracted fault network is included and marked by dark grey patches. Interaction between conjugate faults marked in (a) are also shown on this surface.

on these faults generates small footwall crests and hanging-wall depocentres with relief of <100 m (10%: Fig. 4b), and faults form conjugate sets (10% extension: Fig. 4c). At the end of rifting (6 myr: 30% extension), there are no faults at the upper surface that rupture the entire width of the model; the largest faults are approximately 30 km long (30% extension: Fig. 4c).

Fault growth and interaction are indicated by the timing of bond failure (Fig. 4a). Following initial fault growth (grey colouring, 10%: Fig. 4a), further bonds break as the early formed faults propagate laterally (green colouring, 20%). Bond failure in the final 10% extension is focused on the propagation and linkage of faults, either along strike or by breaching existing relays (yellow–red colouring, 20–30%). Topography evolves from small isolated basins with relief ranging from ≤100 m (10% extension: Fig. 4b) to large 20–30 km-long basins with up to 1 km of displacement on individual faults

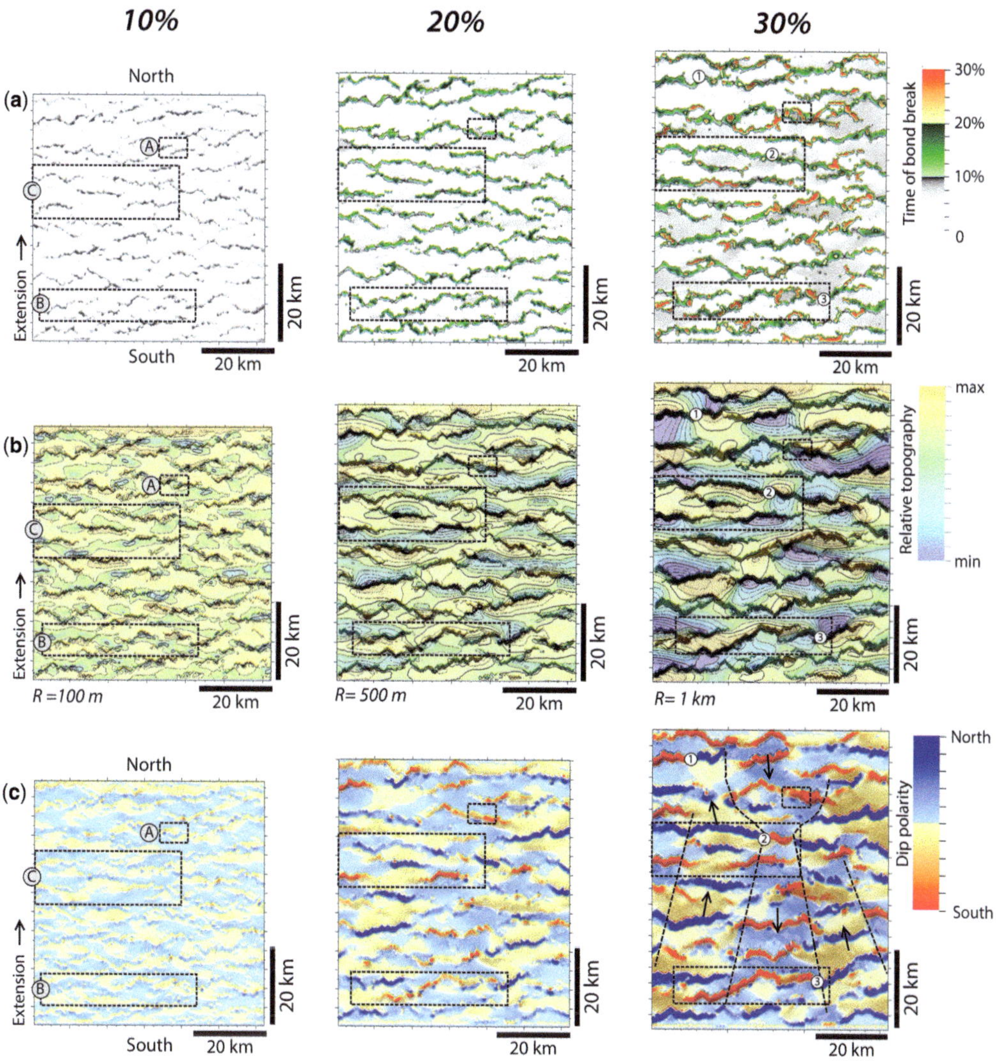

Fig. 4. Map views of Horizon 102 output at increasing extension of 10, 20 and 30% to the north. Three regions (A–C) are shown that are used for discussion. (**a**) Location of faults relative to fault activity and failure time of bonds. (**b**) Relative topography from blue (minimum) to orange (maximum) overlain with fault locations from (a). *R* denotes the relief range for each output: 100 m (10% extension), 500 m (20% extension) and 1 km (30% extension). Footwall crests are orange and hanging-wall deeps are purple. (**c**) Dip orientation for faults. South-dipping planes are coloured yellow to red and north-dipping pale to dark blue. Strong colours indicate fault scarps, and pale colours between them represent hanging-wall and footwall surfaces dipping in the opposite direction. The dominant direction of fault dip shows alignments of both south-dipping faults (red) and north-dipping faults (blue) into dip domains, with boundaries indicated by dashed black lines. Black arrows indicate dominant dip directions. Regions B and C marked on this figure are referred to later in Figures 7 and 8 (Region C) and Figures 11–16 (Region B).

and hanging-wall depocentres (purple: Fig. 4b) focused mainly at fault centres (30% extension: Fig. 4b). The conjugate pattern established during fault nucleation (Fig. 2) results in the surface being divided into dip domains whose boundaries trend sub-parallel to the extension direction (30%: Fig. 4c). Faults in the centre of the model dip predominantly southwards (red: Fig. 4c), whereas those either side of this central area dip mainly northwards (blue: Fig. 4c). Conjugate fault interactions shown in Figure 3 coincide with two of these domain boundaries (circles 2 and 3, 30% extension: Fig. 4).

The growth and interaction of faults at this horizon are presented through the development of structures in three regions (A–C: Fig. 4a). Region A highlights the along-strike linkage of two south-dipping faults from a relay ramp (10% extension) to a single-breached relay on its southern boundary (20% extension) through to a double-breached relay at 30% extension (Fig. 4). As a consequence, there is increased hanging-wall subsidence and bed rotation at the point of linkage (20–30% extension: Fig. 4b) and increased dip on the fault (intense red colour, 20–30% extension: Fig. 4c).

Region B contains one north-dipping fault and eight south-dipping faults that by 10% extension have lengths of <5 km (Fig. 4a). At 20% extension, the south-dipping faults (red: Fig. 4c) have soft linked and comprise four segments separated by relay ramps (Fig. 4a–c). These then breach and form one structure, with the eastern tip being constrained by a conjugate north-dipping fault (circle 3, 30% extension: Fig. 4a–c) and the western tip interacting with another north-dipping fault (30% extension: Fig. 4c). The growth of these and neighbouring faults are later used to illustrate 3D fault evolution and interaction. Region C (10% extension: Fig. 4c) highlights four faults which are used to demonstrate the along-strike linkage and interaction of faults relative to surface topography (Fig. 5).

Five representative transects across the upper surface are used to demonstrate variability in the growth of faults (1–5: Fig. 5a). Selected faults (B–F) presented in later figures are coloured to illustrate their growth. Bold dotted lines indicate faults that rapidly accumulate displacement which then decrease or plateau (Fig. 5b). This shows that the largest fault on any transect at the start of rifting does not necessarily continue to dominate with time and other factors may control which faults in the network are dominant with increasing strain. In Figure 5, solid bold lines indicate selected structures where the displacement is initially small and accelerates with time. The remaining faults encountered (coloured grey) have low displacements (<200 m) and become inactive during rifting, shown by little/no increase in displacement with time. Faults E (blue) and F (green) show a general trend of increasing displacement with time, similar to faults represented by the bold solid lines (transects 1–5: Fig. 5). Fault C (purple) on Transect 2 is one of the largest faults until 4.5 myr, at which point the displacement becomes fixed around 400 m and its displacement rate slows. This correlates to the growth of neighbouring faults B (yellow) and D (red) shown in transects 1–3 (displacement increases after 4 myr: Fig. 5b), and is discussed later.

As rifting progresses, a series of dip domains (i.e. regions of similar fault dip direction) develop without any pre-existing fabrics or lineaments (Fig. 4c). The strain accommodated relative to the

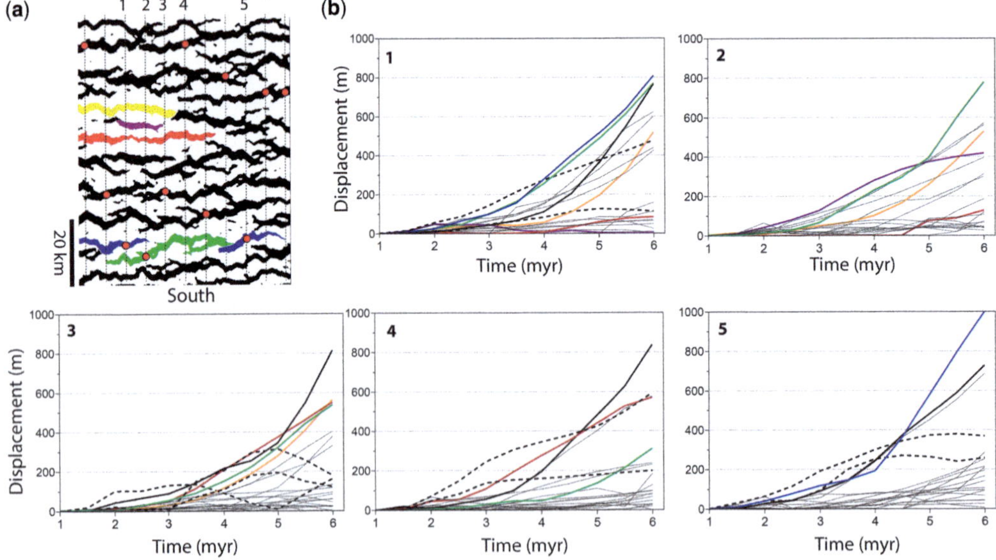

Fig. 5. (**a**) Map of fault locations on Horizon 102 at 30% extension (6 myr). Dotted vertical lines indicate the location of transects through the data at 6000 m intervals. Red dots indicate the location of maximum displacement on each transect. Specific faults are coloured to highlight their evolution: Fault B (yellow), Fault C (purple), Fault D (red), Fault E (blue) and Fault F (green). (**b**) Evolution of displacement on faults through time along selected transects chosen to highlight the evolution of the fault network and dip domains, numbered 1–5. Bold dashed lines represent faults that show periods of retardation within their history and bold solid lines show growing faults.

dip direction of faults is shown for transects 1–5 in Figure 6a. The variability and conjugate nature of faults during early rifting is again highlighted (<2 myr, 10% extension: Fig. 4), with all transects displaying changes in the dominant orientation of strain accommodation before 2 myr (Fig. 6a).

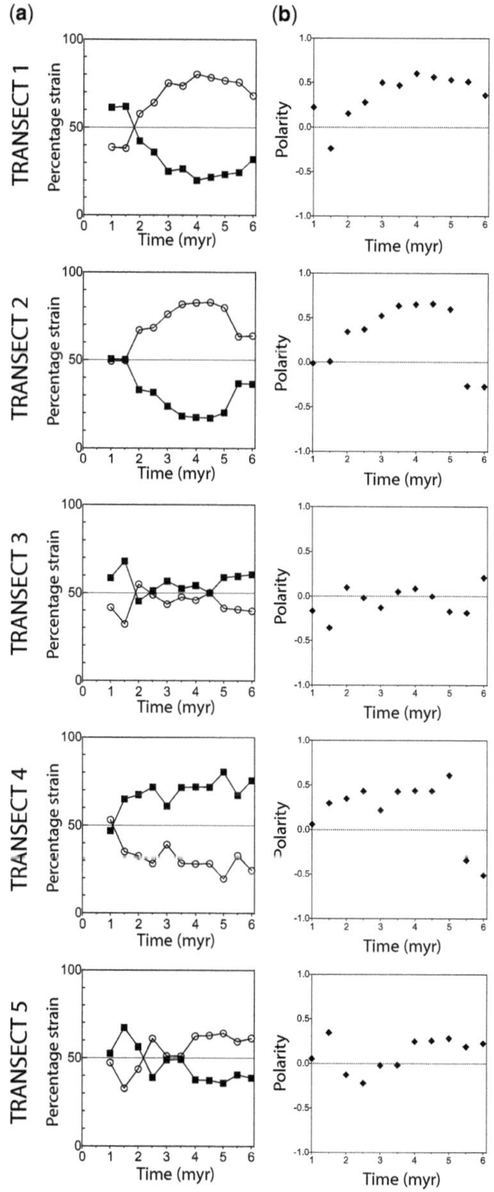

Fig. 6. Data for transects 1–5 from Figure 5a plotted relative to (**a**) percentage strain accommodated by faults against time by north-dipping (open circles) and south-dipping faults (closed squares); and (**b**) polarity plotted against time for the selected transects.

Three transects have a dominant dip direction (transects 1, 2 and 4: Fig. 6b). In transects 1 and 2, up to 80% of the strain is accommodated by north-dipping faults, consistent with the western, north-dipping domain shown in Figure 4c. Transect 4 represents the central south-dipping domain from Figure 4c, where, from an early stage of rifting, the greater amount of strain is accommodated by south-dipping faults. Strain accommodation along transects 3 and 5 is more complicated, however, since they intersect domain boundaries. The percentage of strain accommodated fluctuates around 50% for up to 4.5 myr (Transect 3) and 3.5 myr (Transect 5), and implies that the relative growth and interaction of faults at domain boundaries directly affects strain accommodation in these zones. Later in rifting, they localize to a 60:40 relationship dipping north (Transect 3) and south (Transect 5), suggesting the fixing of dip domains is not yet complete.

The polarity of faults on representative transects is shown in Figure 6b. Transects 1, 2 and 4 (Fig. 5a), which are contained within a strong dip domain (Fig. 4c), show a mainly positive correlation between the dip direction of the dominant fault and the main strain accommodation direction. The initial 3 myr show an increasing linear relationship between polarity (P) and time, which levels off where $0.5 < P < 0.6$ as rifting continues. In transects 2 and 4, polarity switches to negative in the final 0.5 myr (25–30% extension), suggesting that the dominant fault has changed to one that dips in an opposite direction. This implies that within a strong dip domain, there is the possibility for faults antithetic to the dominant strain accommodation direction to continue to accumulate displacement. The polarity of transects 3 and 5, as expected from the strain accommodation data and their location at domain boundaries, fluctuates between positive and negative as faults interact.

Fault interaction and displacement rate evolution

The preceding results showed that some of the largest faults at the initiation of rifting become inactive as neighbouring faults grow around them (e.g. Fault C in Fig. 5a). Figure 7 focuses on the growth of fault segments within Region C in Figure 4. Faults C and D are the largest faults, with lengths ≥10 km at 15% extension and maximum displacement of 150 m, all other faults segments have lengths of <3 km. From 15 to 20% extension (3–4 myr, white band: Fig. 7b), faults C and D have the greatest displacement, with two segments of Fault D propagating laterally and increasing displacement up to 400 m, overtaking Fault C (250 m). Fault B comprises a number of small soft-linked segments at this stage (<200 m displacement and <8 km long). Fault A is a conjugate fault that interacts with Fault B at the surface, and is approximately 3 km long and has a displacement of <150 m.

Displacement on Fault C markedly slows after 4.0 myr as the initial segments of faults A, B and D propagate laterally (25% extension: Fig. 7a) and link (thickening of the 4–5 myr band: Fig. 7b). From 5 to 6 myr (25–30% extension), Fault A continues to grow, the soft-linked segments of Fault B link, and the two segments of Fault D propagate along strike. The profiles of these faults are consistent with profiles of natural examples. Faults A and D have triangular displacement profiles where their lateral tips are restricted (cf. Fig. 8) (Nixon *et al.* 2014*b*), Fault B has an asymmetrical profile where its eastern tip is restricted by interaction with Fault A (point 2: Figs 4c & 7a) and Fault C represents a traditional symmetrical displacement profile consistent with an isolated fault. Before 20% extension, faults A, B and D grow laterally while accruing displacement exhibiting a growth pattern resembling the isolated fault model. Fault C, however, attains its length rapidly and then accrues displacement, suggestive of the coherent growth model, although growth of this fault slows as neighbouring faults dominate at greater extension.

The displacement rates of faults A, B, C and D at intervals of 1 myr are shown as strike projections in Figure 8a, and summarized as a displacement v. time plot for selected profiles in Figure 8b. The main period of fault growth differs for each fault, and along-strike rates vary dramatically. Faults A and B have a maximum displacement rate of between 4 and 6 myr, when their lateral propagation and interaction causes an increase from <0.1 up to 0.37 mm a^{-1}. The hard-linkage of segments that constitute Fault B at 4 myr is shown by the pink profile (21 000 m) in Figure 8a, with displacement rates increasing from 0–0.15 mm a^{-1} at 4 myr to 0.3 mm a^{-1} between 5 and 6 myr. Further linkage on Fault B between 5 and 6 myr is shown by the red profile (16 000 m) in Figure 8a, where the displacement rate increases rapidly from 0.08 to

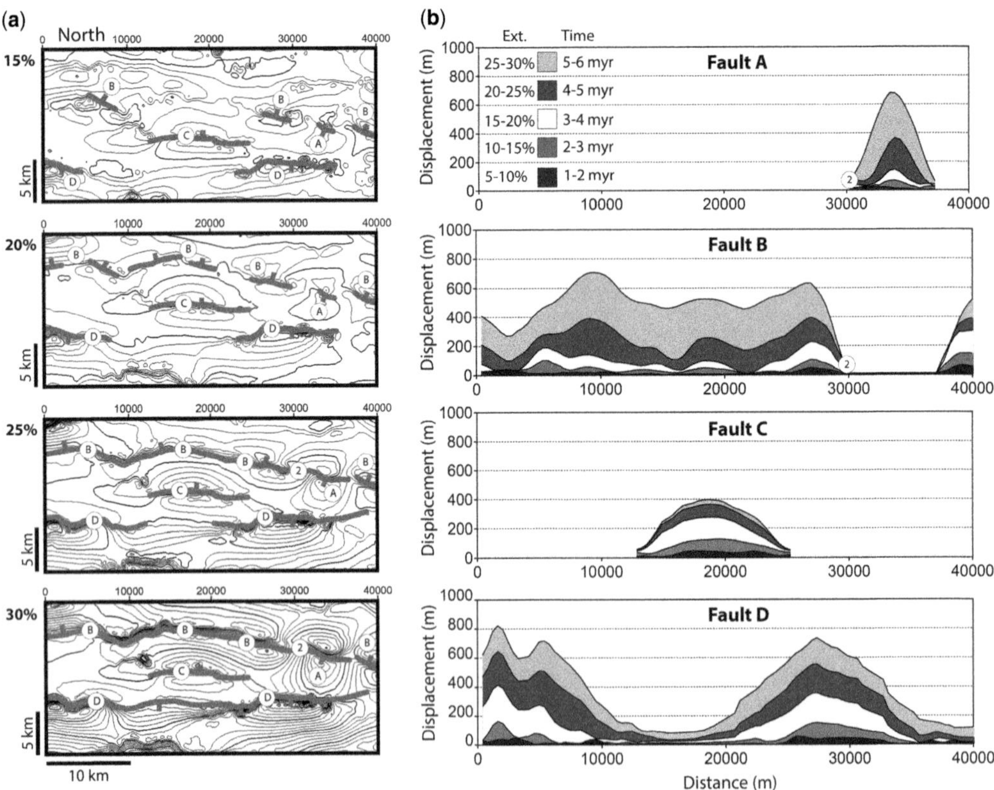

Fig. 7. (**a**) Map view of Region C highlighted in Figure 4 through time from 15 to 30% extension (3–6 myr). Traces of faults A–D are highlighted and correlate to those in Figure 5a. Number 2 indicates the spill point shown in Figure 4 in this region. (**b**) Displacement–distance plots on faults A–D at the upper surface. Greyscale fill indicates the time of each recording in 1 myr (5% extension) intervals.

0.37 mm a^{-1}. By comparison, the neighbouring parts of the fault propagate from 0.15 to 0.3 mm a^{-1}.

The displacement rate on Fault C peaks at 4 myr (dark grey: Fig. 8a) at 0.15 mm a^{-1} and then decreases in the last 2 myr of extension to <0.03 mm a^{-1}, showing that this fault is almost inactive at the end of rifting despite initially being the largest (Figs 7 & 8). The most complicated pattern of growth is associated with Fault D. Up to 4 myr, there are two distinct segments where each has a displacement rate maximum at its centre (Fig. 7b). The blue, green and purple profiles in Figure 8a, b demonstrate an increasing displacement rate before 4 myr, which then stays constant or decreases. The along-strike growth and linkage of Fault D (seen at 20–30% extension: Fig. 7a, b) is shown by an increased displacement rate in the shaded region, and increases in displacement rate for the orange, pink and black profiles between 5 and 6 myr in Figure 8a, b.

Fault network statistics

The maximum recorded displacement for six selected horizons in the crust (H 52–H 102) is shown (Fig. 9) at 500 kyr (2.5% extension) intervals from 2 to 6 myr. Before 4 myr (20% extension), the maximum displacement recorded on these horizons is depth invariant and approximately 600 m. As rifting continues, the range of maximum recorded displacement increases to between 750 and 1500 m, where the largest are focused around Horizon 72

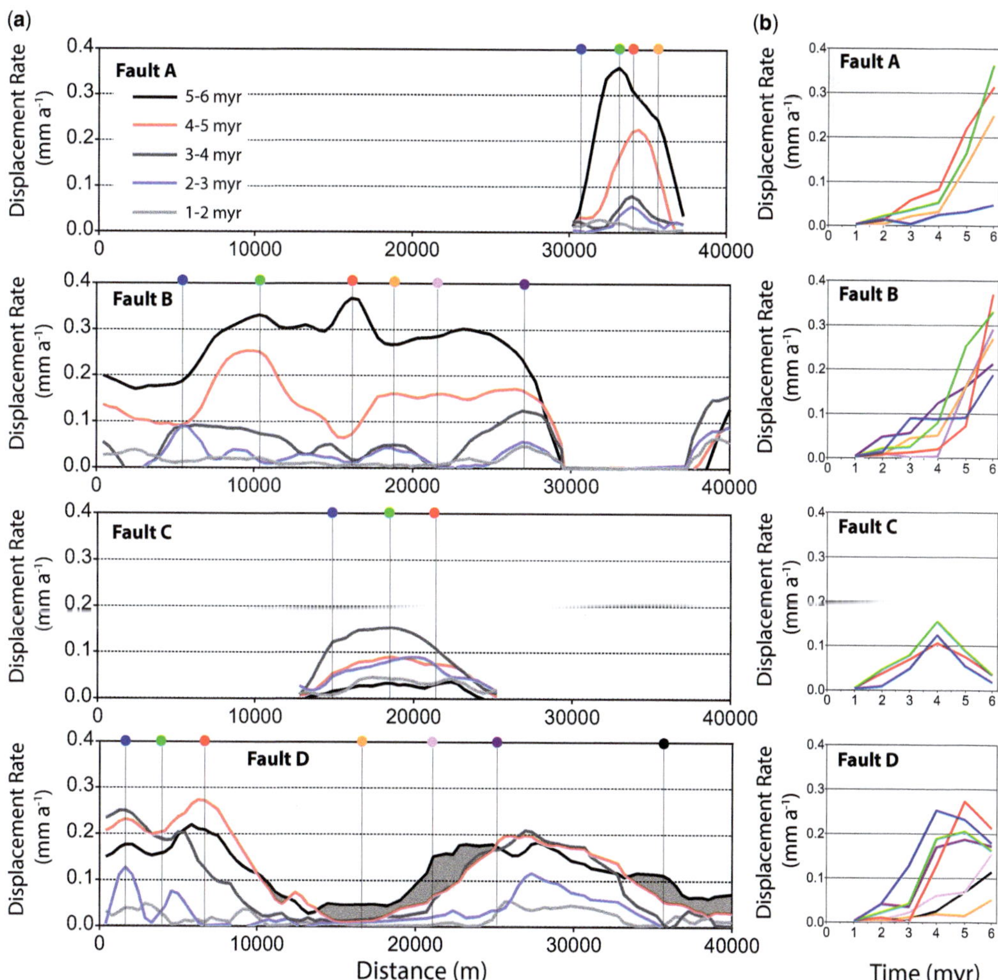

Fig. 8. (**a**) Displacement rate in 1 myr intervals plotted against distance for Faults A–D from Figure 7. Selected profiles are marked by vertical lines and coloured with dots. (**b**) Displacement rate against time for profiles marked along the faults in (a), where colours of intersection correlate between (a) and (b).

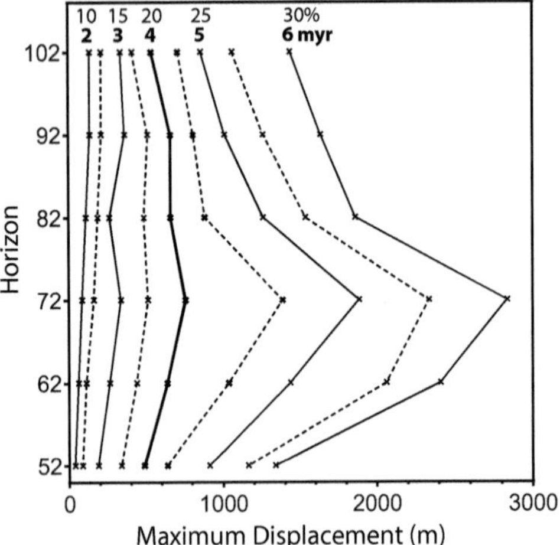

Fig. 9. Maximum displacement (throw) recorded for selected horizons (H 52–H 102) plotted relative to extension. Intervals between pairs of solid or dashed lines represent 5.0% extension (1 myr).

(the depth of the centre of the largest faults). This suggests that small faults (displacement <600 m) in the fault network accommodate the majority of the strain prior to 20% extension (4 myr) distributed throughout the brittle layer and no large displacement (>800 m) faults exist. The coalescence of these small, distributed, isolated faults into larger, crustal-scale structures occurs between 20 and 25% extension, where the maximum displacements on horizons start to localize around mid-crustal depths (H 62–H 82). The last 1 myr (5% extension) is characterized by localization of strain (and displacement) onto these large faults that bisect the upper crust.

The maximum displacement recorded on the 30 largest faults is plotted in log–log scale relative to elapsed time in Figure 10, together with data from natural examples presented by Nicol *et al.* (1997). Similar to natural faults, data from the model results have a slope of 0.3138 ($r^2 = 0.91$). While the faults are still relatively small prior to 4 myr, the data plot between 0.01 and 0.1 mm a^{-1}, consistent with results shown in Figure 9. After 4 myr (dotted horizontal line in Fig. 10), the data plot mainly between rates of 0.1 and 1.0 mm a^{-1}, indicating that the displacement rate has increased on these structures as they have linked to become larger faults.

Distribution of linked conjugate faults in three dimensions

To demonstrate the complexity of interactions between faults in three dimensions, a representative group of seven faults centred on faults E and F (Region B in Fig. 4c) were extracted (Fig. 11). This region contains three major faults >40 km long (faults E, F and K) and four minor faults <20 km long (faults G, H, I and J). The south-dipping faults (G, H and I) are antithetic to and intersect the larger north-dipping Fault E. Each of these faults is influenced by and interacts at depth with either of the north-dipping faults J and K (Fig. 11b, c). Few faults are linear along strike; most follow a curvilinear path from east to west controlled by their constituent fault segments.

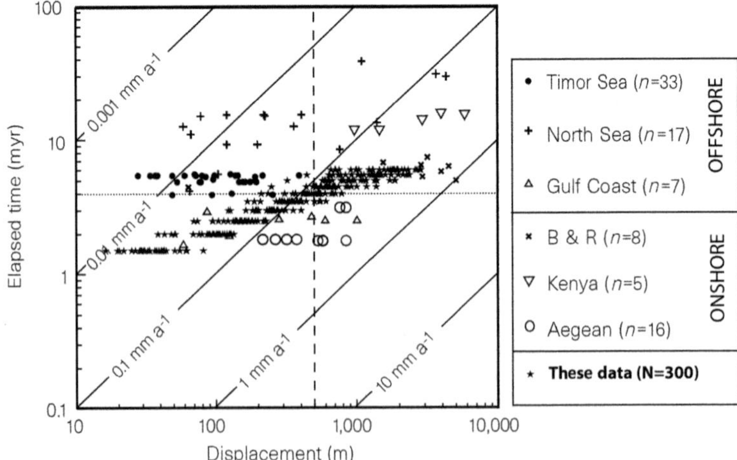

Fig. 10. Log–log plot of elapsed time against maximum displacement for the largest 30 faults from the model (stars) plotted relative to data from three offshore and three onshore regions presented in figure 2 of Nicol *et al.* (1997). Diagonal lines represent expected fault growth rates. The dotted horizontal line corresponds to 4 myr.

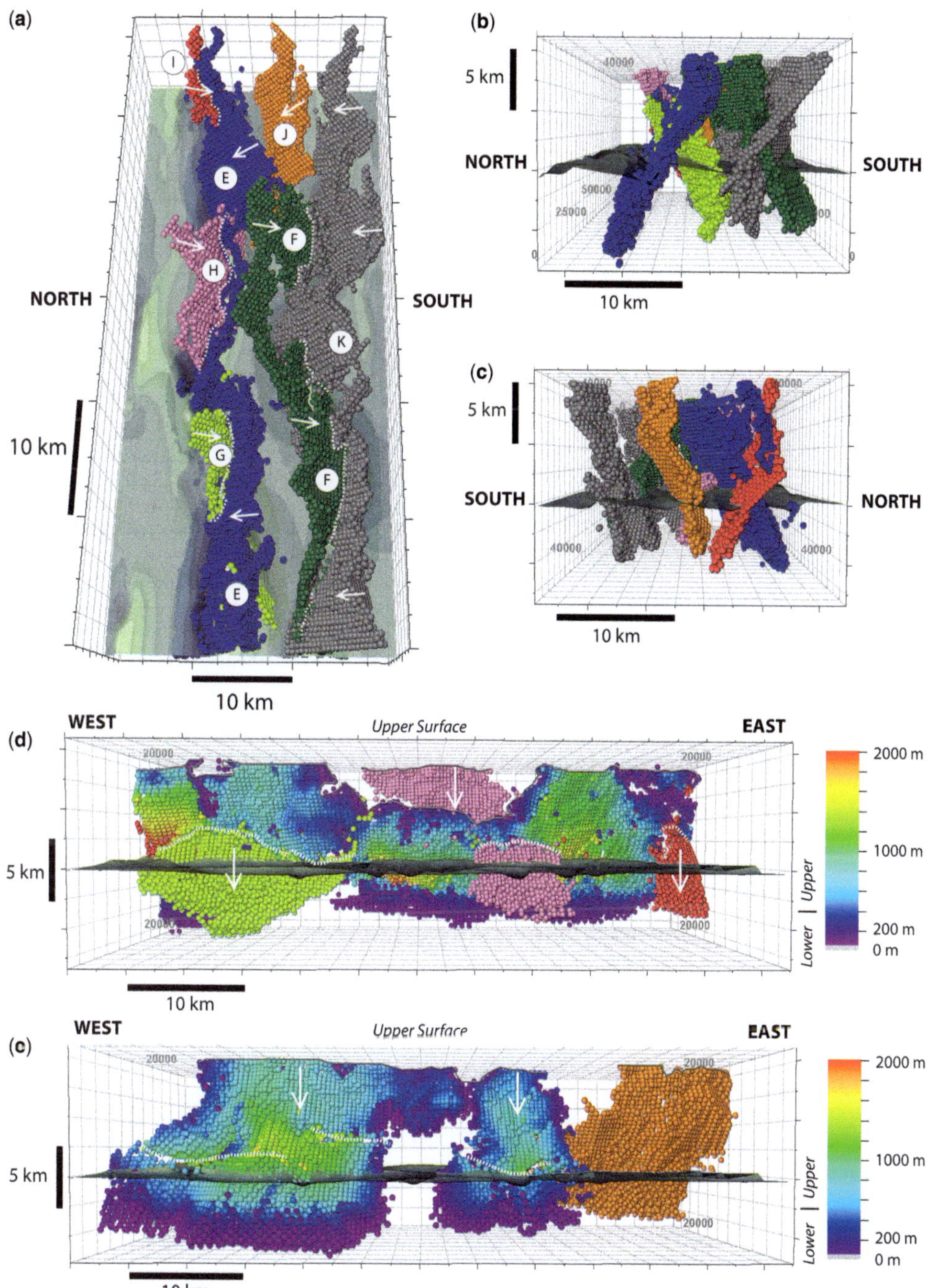

Fig. 11. Three-dimensional views representing seven faults in Region B (Fig. 4). Faults are coloured to aid differentiation and are labelled E–K. White arrows indicate the dip direction of faults from their upper surfaces. Dotted white lines indicate intersections between faults. (**a**) Oblique map view including Horizon 62 with views looking (**b**) east and (**c**) west. Views of selected faults from the south (including Horizon 62) with colour-coded displacement on (**d**) Fault E and (**e**) Fault F. Displacement ranges from purple (50 m) to red (2 km). Minor faults G–J are included and dotted white lines demonstrate intersections between faults. The white arrows denote faults dipping towards the viewer.

Intersections are at varying depth and are not focused in one horizon (white dotted lines in Fig. 11a, d, e). At the end of rifting, displacement maxima for the two largest faults (E and F) are focused towards their centre with depth. The relative position and nucleation time of neighbouring faults controls their down-dip and along-strike continuity (Fig. 11d, e).

The spatial distribution with depth at the end of the experiment for faults E–K is shown in Figure 12. The major faults (E and F) are present on the upper horizon, where Fault E is represented by two separate, approximately 20 km-long, segments to the west and east, divided along strike by Fault F. This is consistent with the gap in the upper elevation of Fault E in strike projection (Fig. 11d, e). Fault F terminates against the hanging wall of Fault E at its eastern tip (circle 3 in Fig. 4; white circle in Fig. 12a). Depocentres associated with these faults are 750 m below the mean elevation, and prominent footwall crests have developed. The other major fault (K) to the south is represented by four segments at this elevation.

At 2.5 km depth, Fault E remains as two separate segments to the west and east with a number of smaller branches at its tip, with the eastern segment terminating against Fault F (white star in Fig. 12b). At a depth of 5.1 km, Fault E is 60 km long and connects the two segments at higher horizons (Figs 11d & 12c). It has a pronounced depocentre associated with the upper segments and a number of branches along its length (white stars in Fig. 12c). White circles along its length show where the tips of the south-dipping faults G, H and I are in contact with it, forming small graben. Lateral continuity of Fault E is most pronounced at a depth of 7.8 km where antithetic faults G, H and I are coincident with the strike of Fault E (white circles in Fig. 12d). At depths of 10–12 km, Fault E bifurcates again into two segments – one to the west (*c.* 15 km long) and the other to the east (*c.* 40 km long) – separated by a south-dipping fault indicated by an intermediate crest in the topography between segments (Fig. 12e, f).

Fault F dips southwards and is most laterally continuous in its upper region (Fig. 11e), with Fault H tracking it at a constant distance of around 8 km to the north (Fig. 12). With depth, Fault F interacts with Fault K to the south, which inhibits its propagation, and it branches into two segments (dotted lines and central gap: Figs 11e & 12d–f). The mid-crustal section of Fault F is constrained to the east by the conjugate, north-dipping, Fault J (Fig. 12c, d).

At the deepest horizon shown in Figure 12 (−12.1 km), all seven faults (E–K) are present. The minor conjugate faults H and J are in contact, Fault I cuts the eastern margin of the half-graben between faults E and J, and Fault G is at its maximum length (uninhibited by neighbouring faults), forming an approximately 20 km-long graben with Fault K to the south (Fig. 12f).

Fault growth and interaction in three dimensions

The ability to extract displacement on elements with time means that the growth and interaction of faults can be investigated. Strike projections of displacement at 5% intervals of extension from 15 to 30% are shown in Figures 13–15, allowing fault nucleation and growth to be assessed. In this subsection, we focus on faults E and F (Figs 5a, 11 & 12), describing their evolution and interaction with adjacent faults in the network.

At 15% extension, Fault E is composed of six small isolated fault patches (1–6: Fig. 13a) that have nucleated at varying depths in the upper brittle crust. Only two patches intersect the upper surface (2 and 5). At this stage in their evolution, these patches have displacement maxima of 200–300 m occurring near their centres, and displacement decreases towards their tips (Fig. 16a). These patches range in strike length from 3.5 to 8 km, have aspect ratios (vertical:horizontal) from 1:1 to 2:1, and are separated along strike by relatively unfaulted regions 2–10 km wide. No faults initiate in the lower crust, although patches 3, 4 and 6 are located close to the upper–lower crust interface.

Between 15 and 20% extension, displacement maxima on these patches have increased to between 300 and 750 m. All fault patches have grown outwards from their nucleation sites and, despite varying degrees of interaction, displacement maxima are still located at the centre of the six initial patches (light blue: Fig. 13b). Patches 1–4 show largely radial growth, and maintain similar aspect ratios, although downwards propagation of fault patch 2 is inhibited by interaction with Fault G (cf. Figs 13b & 14b). In contrast, patches 5 and 6 begin to hard link into a single 20 km fault segment. Patch 5, in particular, shows preferential growth downwards and eastwards towards patch 6, and its displacement maximum has migrated deeper into the crust. In contrast, the western side of patch 5 remains relatively fixed in position due to interaction with Fault F (cf. Figs 5a, 11a, 13b & 15b).

After 25% extension, Fault E comprises three major fault segments, 20–25 km long, with up to 1500 m displacement, separated by segment boundaries associated with marked displacement deficits (Figs 13c & 16a). The three major fault segments have evolved by hard linkage of earlier patches 1 and 2 (western segment), 3 and 4 (central segment), and 5 and 6 (eastern segment) (Fig. 13c). The western segment still has a significant displacement deficit at the former segment boundary between the two

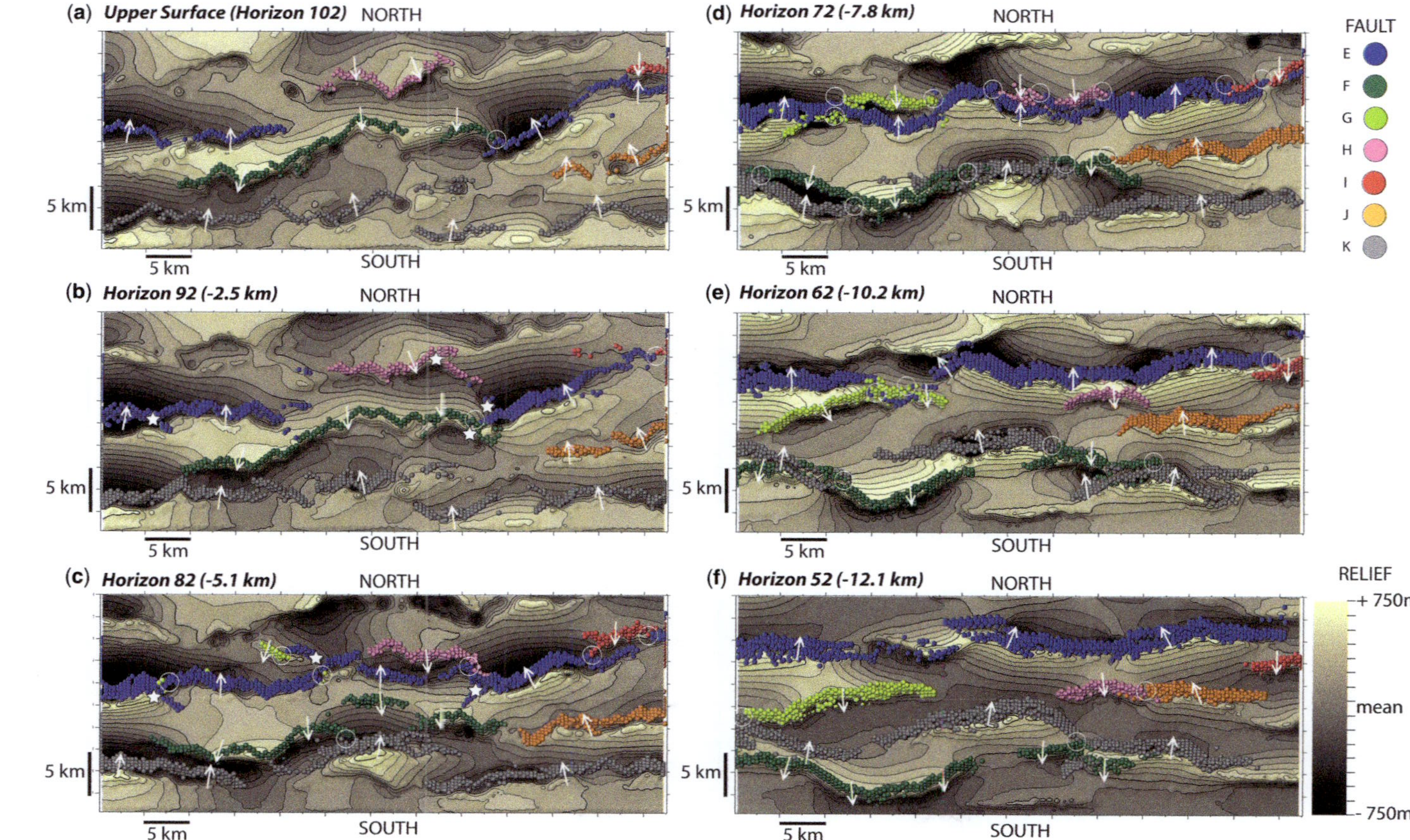

Fig. 12. Map view of elements that constitute faults shown in Figure 11 relative to depth for (**a**) Horizon 102, (**b**) Horizon 92, (**c**) Horizon 82, (**d**) Horizon 72, (**e**) Horizon 62 and (**f**) Horizon 52. Maps are coloured from brown to cream ± 750 m about their respective mean elevation. Footwall crests are cream and hanging-wall depocentres are dark brown. Faults E–K at each horizon are included as their colour defined in the key (see also Fig. 11). Northern dipping faults are drawn in the immediate hanging wall and southern dipping faults are drawn on the fault crest. The dip direction of fault segments is indicated by white arrows. White circles highlight points of intersection between faults and white stars show fault bifurcation.

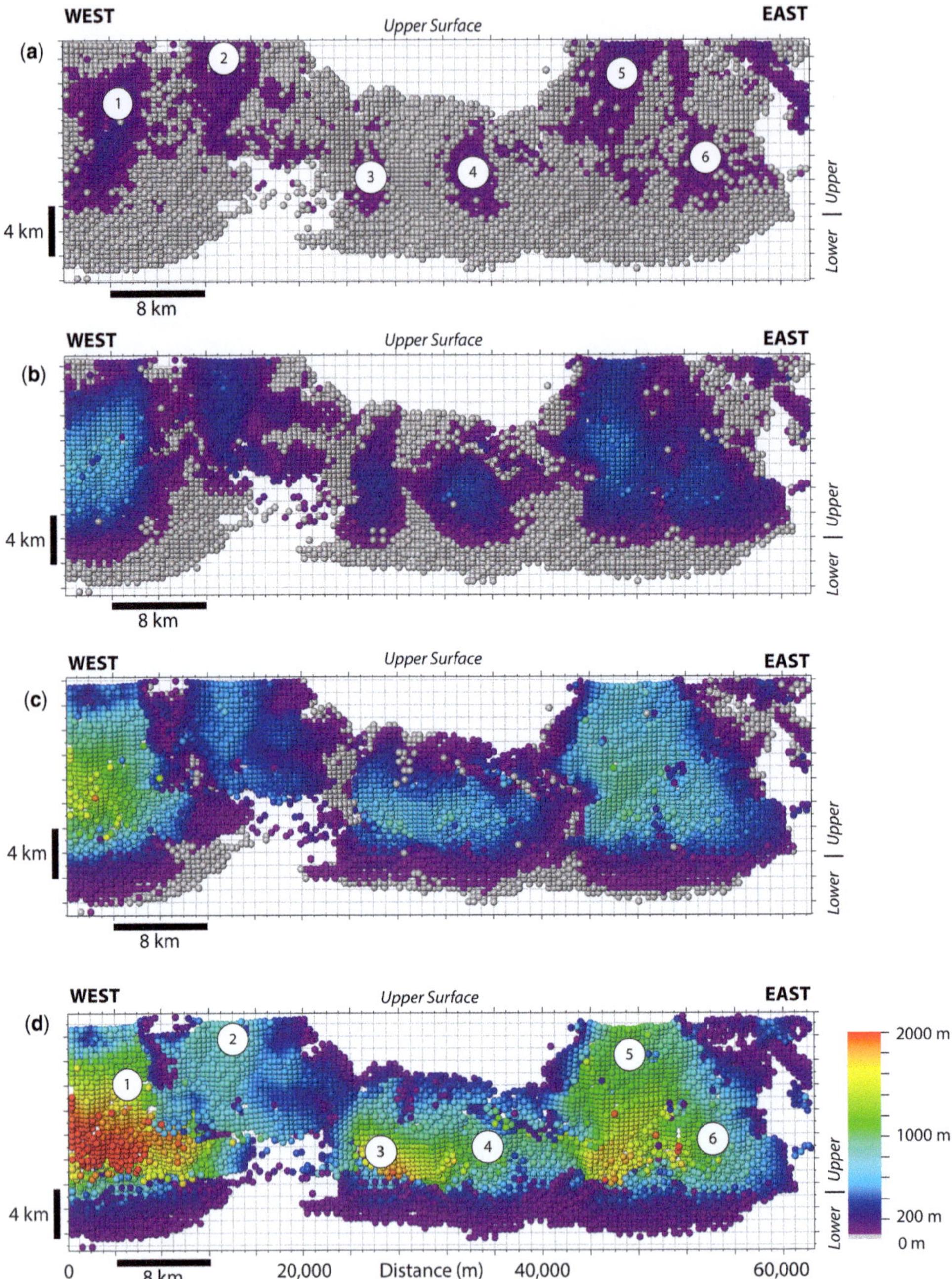

Fig. 13. Strike projection viewed from the south for Fault E at (**a**) 15%, (**b**) 20%, (**c**) 25% and (**d**) 30% extension coloured relative to displacement from 50 m (purple) to 2 km (red). Elements coloured grey represent regions where the fault has not, as yet, failed. The fault evolved from six distinct isolated segments (numbered 1–6) into one structure. The location of the change between the upper and lower crust is indicated.

precursor patches (1 and 2). In contrast, displacement deficits associated with boundaries between the precursor segments of the central and eastern segments have largely been removed (Fig. 16a). The eastern and western segments rupture the full thickness of the upper crust and extend into the

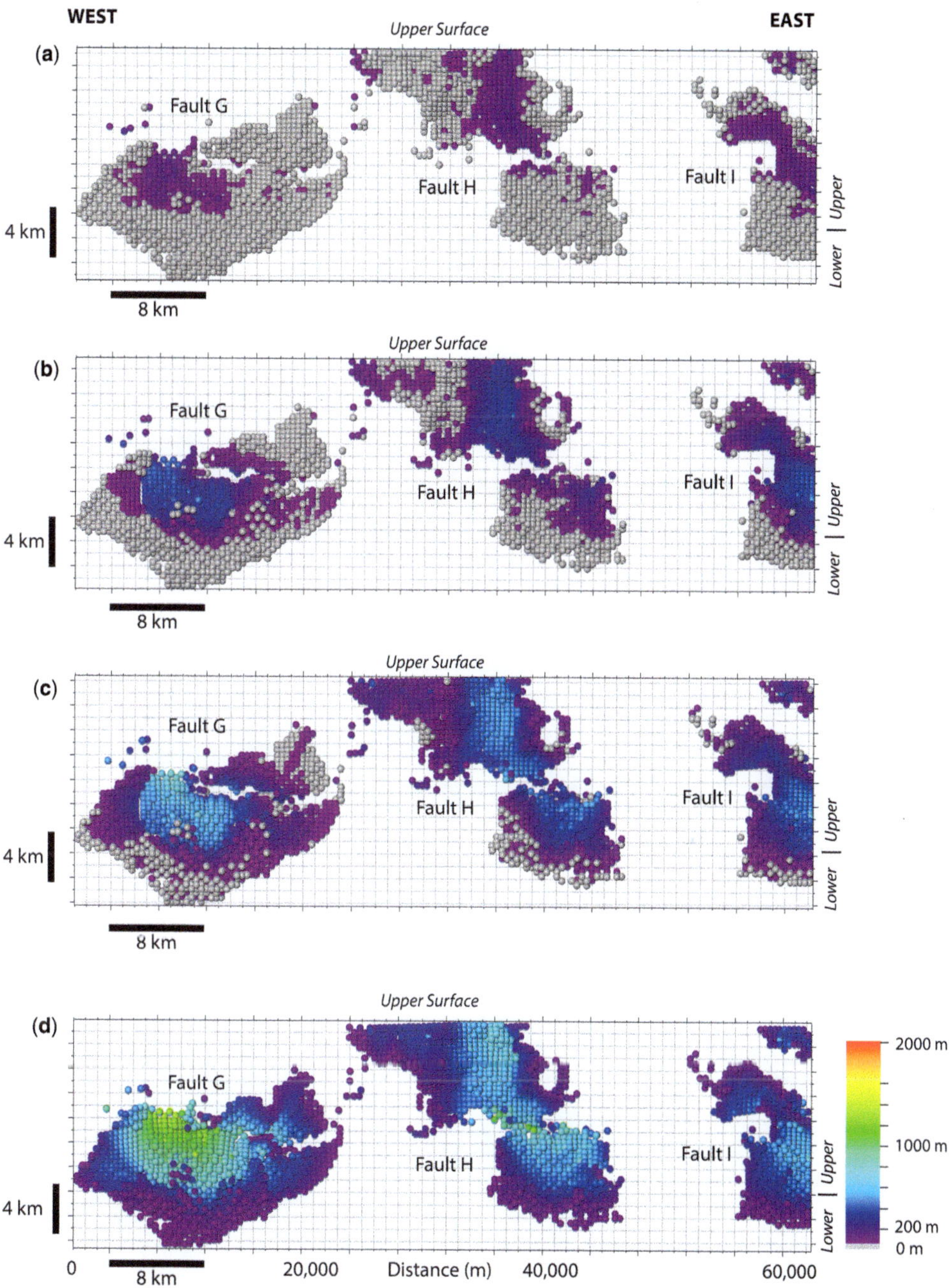

Fig. 14. Strike projection viewed from the south for faults G, H and I at (**a**) 15%, (**b**) 20%, (**c**) 25% and (**d**) 30% extension coloured relative to displacement from 50 m (purple) to 2 km (red). Elements coloured grey represent regions where a fault has not, as yet, failed. The location of the change between the upper and lower crust is indicated. Gaps between elements indicate where Fault E (Figs 11 & 13) intersects these faults.

lower crust, although downwards propagation of the western segment is inhibited by Fault G (Fig. 14c). A more striking interaction that affects growth of Fault E is the upwards propagation of its central segment. This segment of the fault remains blind. It only ruptures the lower half of the brittle crust,

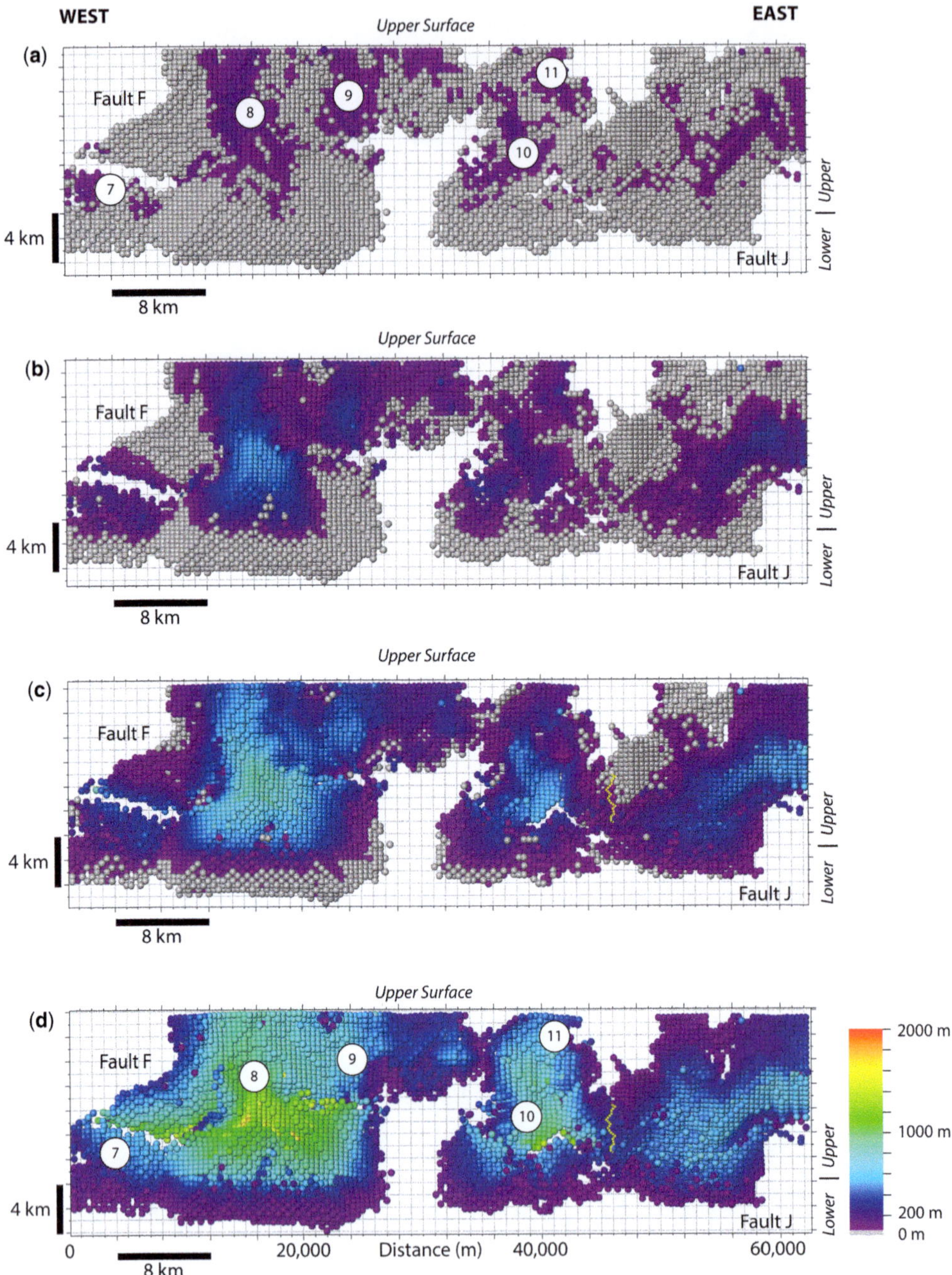

Fig. 15. Strike projection viewed from the south for faults F and J which interact as a conjugate pair at (**a**) 15%, (**b**) 20%, (**c**) 25% and (**d**) 30% extension coloured relative to displacement from 50 m (purple) to 2 km (red). Elements coloured grey represent regions where the fault has not, as yet, failed. Fault F evolved from five distinct isolated patches (numbered 7–11) into one fault. The faults interact at the yellow dashed line.

and its upwards propagation is inhibited by the presence of Fault F above it to the south (cf. Figs 11, 12a, 13c, 15c & 16).

At the end of rifting (30% extension), Fault E extends across the entire width of the model (Fig. 13d), and shows an overall increase in displacement rate compared to earlier in its evolution (Fig. 16a). Although it is hard linked at depth, it is still blind along its central section, where it is overlain by Fault F and thus appears as two distinct faults separated along strike by over 20 km (cf. Fault E: Figs 5a & 12a). Displacement maxima up to 2500 m occur at three locations along its length, coinciding with map-view locations of precursor patches 1, 3 and 5 but now localized near the base of the upper crust (Fig. 13c). Displacement minima are still evident at segment boundaries between the three segments present at 25% extension, although the displacement deficit between patches 1 and 2 in the western segment has been removed by this time (Fig. 16a).

The evolution of Fault F shares many similar characteristics with Fault E (Fig. 13). It initiates as five patches (7–11: Fig. 15a) that occur at a range of depths within the brittle crust, and which initially propagate outwards and then link to form major fault segments. Patches 7 and 10 occur at mid- to deep-crustal levels, whereas patches 8, 9 and 11 are shallower (Fig. 15a). At mid- to deep-crustal levels, Fault F is cross-cut along its length by Fault K, which accounts for gaps in its lateral continuity (e.g. above patch 7 and below patch 9: Figs 11a & 12). Patches 8 and 9 become hard linked by 20% extension, and the boundary between them loses its displacement deficit by 25% extension (Figs 15b, c & 16c). In contrast, linkage between patches 7 and 8 occurs later, with rapid accrual of displacement on patches 7 and 8 between 27.5 and 30% extension (Fig. 16c). On the eastern side of Fault F, patches 10 and 11 link between 15 and 20% extension, and the displacement deficit between them is lost by 22.5%, giving rise to a single prominent displacement maximum (Figs 15c & 16c). This growth and linkage history creates two major fault segments for Fault F by 25% extension, with a major displacement low between precursor patches 9 and 10 that exists throughout the evolution of the model (Fig. 16c). This long-lived displacement low is due to interaction with the underlying Fault K, dipping from the north. As with Fault E, at 30% extension, displacement is localized towards the middle of the fault at the base of the brittle layer (e.g. Fig. 15d).

The evolution of other faults in the region surrounding faults E and F (i.e. faults G, H, I and J) is shown in Figures 14, 15 & 16b, c. The main difference between these faults and faults E and F is that their lengths are fixed relatively early during extension. Fault G initiates at a mid-crustal depth, similar in depth to patches 3 and 4 on Fault E (cf. Figs 13a & 14a). It propagates downdip with a small patch of slip propagating updip towards Fault E between patches 2 and 3, but it does not propagate upwards and break the surface because

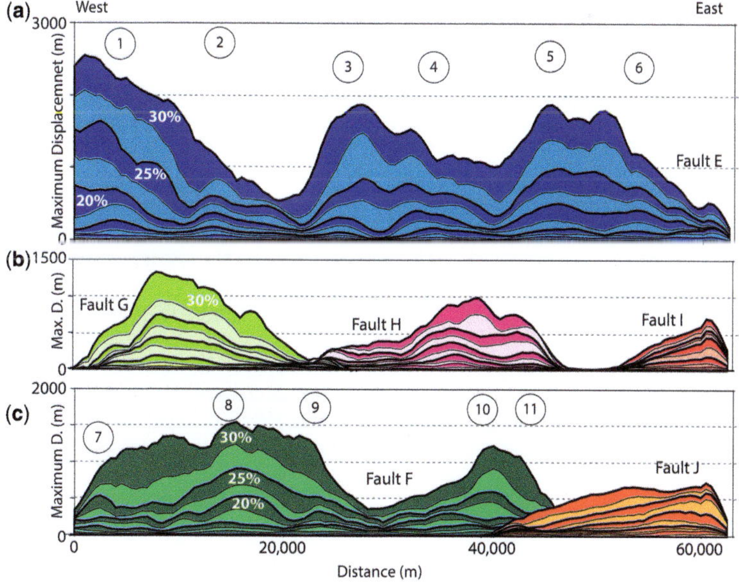

Fig. 16. Maximum displacement recorded in 500 m vertical swaths for elements that represent a fault v. length. These are extracted at extension increments of 2.5% from 10 to 30% extension for (**a**) Fault E, (**b**) faults G, H and I and (**c**) faults F and J. Faults are coloured relative to earlier definitions in Figure 11. Alternating light and dark patches indicate bands of 2.5% extension between outputs. Fault segments introduced in Figures 13 and 15 are marked for comparison.

it lives in the stress shadow of neighbouring faults to the north. The displacement maximum for Fault G (1400 m) is on the upper, western part of its surface, rather than lying centrally, a function of it abutting and interacting with the blue fault. Similar to faults E and F, Fault G shows an increase in displacement rate between 20 and 30% extension (Fig. 16b). Fault H initiates in the upper part of the brittle layer and propagates largely downdip until 20% extension. Its length then increases from 15 to 24 km during the final 10% extension by near-surface westwards propagation. In contrast, faults I and J show a different style of evolution marked by an overall decrease in displacement rate through time, particularly during the last 5–10% extension associated with their tip restriction against the major faults, E and F. This is the time when the major faults display an increase in displacement rate (Fig. 16b).

The displacement profiles of faults E–J through time illustrate a number of features of the development of the fault network (Fig. 16). The final along-strike profile for Fault E shows two displacement minima at distances of 20 km (*c.* 500 m, between patches 2 and 3: Fig. 16a) and 40 km (*c.* 1 km, between patches 4 and 5: Fig. 16a). The westernmost minimum is associated with interaction and displacement relative to the antithetic Fault G (Figs 11d, 12d & 16b). The reduction around 40 km (patches 4 and 5: Fig. 16a) is coincident with the surface expression of Fault F and Fault H to the north (patches 10 and 11: Fig. 16c). The final profile of Fault G represents an isolated fault with maximum displacement at the centre diminishing to the tips. The minor faults (H, I and J) show asymmetry associated with single tip restrictions. The displacement profiles of Fault E (patches 3–6) and Fault F (patches 7–9) have steep shoulders associated with faults restricting their propagation to both the east and west despite hard linking to neighbouring sections through displacement minima along strike (Fig. 16a, c).

Discussion

In this model of a single phase of extensional rifting, we have observed four stages of fault growth and interaction. We term these stages: (1) nucleation; (2) propagation and interaction; (3) linkage and domain fixing; and (4) localization (Fig. 17).

In Stage 1 (Fig. 17a), a large number of small conjugate faults (<5 km in length) nucleate rapidly, at varying depths in the crust. These faults develop as isolated fault segments that have elliptical displacement contours with displacement maxima at their centres, striking sub-perpendicular to the extension direction. In Stage 2 (Fig. 17b), these isolated fault segments propagate and interact with one another. Individual segments that are co-linear begin to link along strike or downdip (lengths *c.* 15 km), and relay ramps develop at fault tips as a result of soft linking of fault segments. Where faults have opposing dip, lateral propagation is inhibited. By Stage 3 (Fig. 17c), distinct graben and half-graben are developed in the hanging walls of larger faults. These larger faults (lengths *c.* 20 km) formed by the linkage of co-linear fault segments. At the fault network scale, dip domains become well established and their boundaries become fixed. The boundaries between the dip domains form narrow zones where conjugate faults interact and lateral propagation is inhibited. In the final stage of fault network evolution, Stage 4 (Fig. 17d), activity is localized on a small number of large faults (lengths >40 km) that cut across the entire upper, brittle crust. Some faults that were initially dominant become inactive at this time. These are generally located in the strain shadow of faults that grow more rapidly during this stage of fault network evolution. In this final stage, a pronounced increase in displacement rate on the large active faults is observed and displacement maxima migrate to their centres.

In the following subsections we discuss the results of our 3D numerical modelling in comparison to natural examples and other analogue and numerical models of fault growth. We specifically discuss: (i) comparison to natural and analogue fault systems; (ii) fault growth, propagation and linkage; and (iii) fault network and dip domain development.

Comparison to natural and analogue fault systems

Many of the characteristics of the growth and linkage of the normal fault segments reported here are comparable to those from other modelling approaches (e.g. Cowie *et al.* 2000), and from studies of natural fault systems in the subsurface (e.g. Young *et al.* 2001) and outcrop datasets (e.g. Gawthorpe *et al.* 2003). The strain rate in the model decreases with time from 1.6×10^{-15} to $1.24 \times 10^{-15}\ s^{-1}$ and is comparable with the natural strain rate observed in rifts such as the Gulf Coast and Aegean (Fig. 18).The displacement rates for the 30 largest faults in the model plot are similar to the natural data. Nicol *et al.* (1997) demonstrated that there is a broad correlation between regional strain rate and fault displacement rates, suggesting that increased strain rate is accommodated by increased displacement rate rather than an increased number of faults. This is observed in our model, where strain localizes onto the largest faults as rifting progresses.

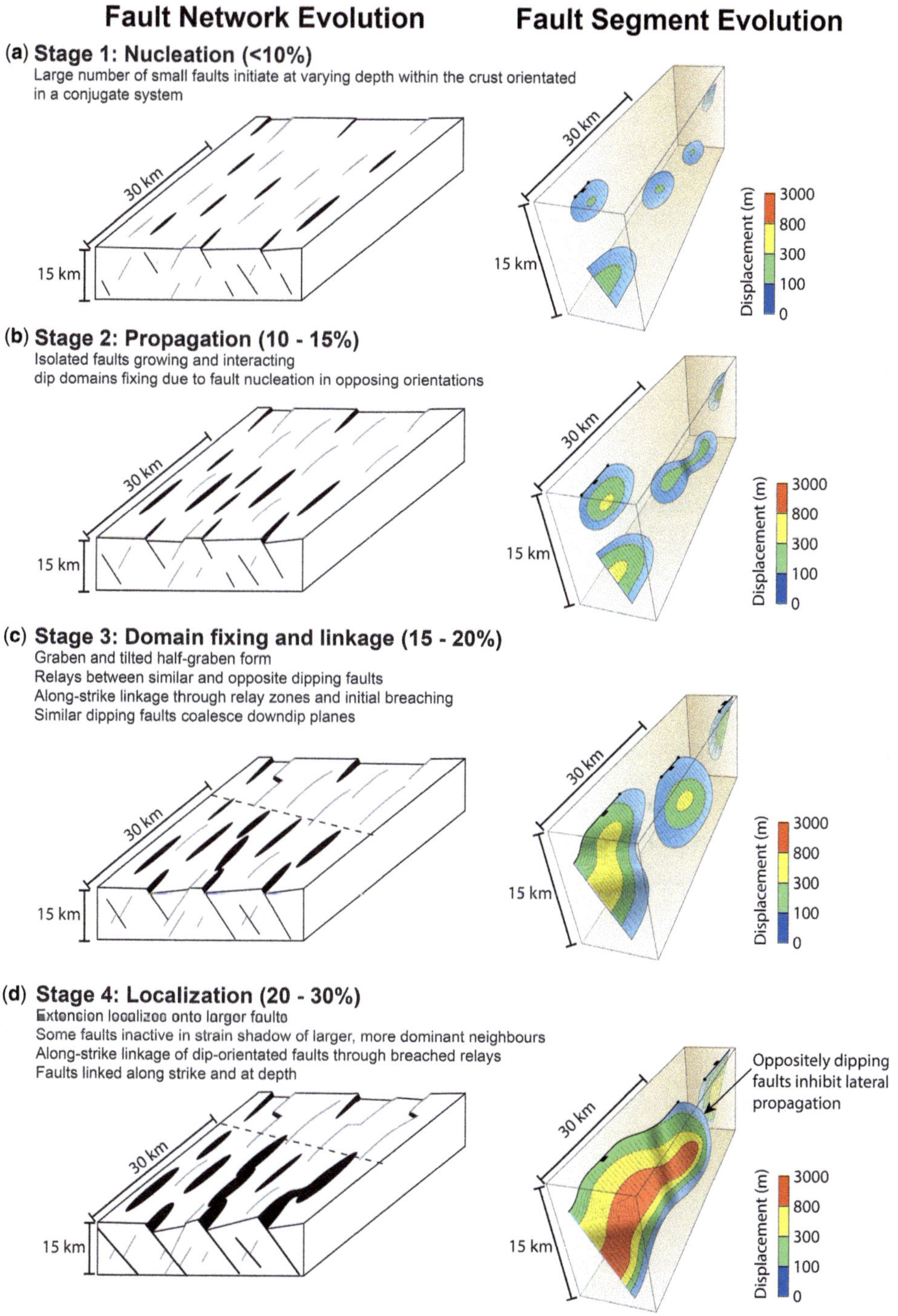

Fault Network Evolution

Fault Segment Evolution

(a) Stage 1: Nucleation (<10%)

Large number of small faults initiate at varying depth within the crust orientated in a conjugate system

30 km
15 km
30 km
15 km
Displacement (m)
3000
800
300
100
0

(b) Stage 2: Propagation (10 - 15%)

Isolated faults growing and interacting
dip domains fixing due to fault nucleation in opposing orientations

30 km
15 km
30 km
15 km
Displacement (m)
3000
800
300
100
0

(c) Stage 3: Domain fixing and linkage (15 - 20%)

Graben and tilted half-graben form
Relays between similar and opposite dipping faults
Along-strike linkage through relay zones and initial breaching
Similar dipping faults coalesce downdip planes

30 km
15 km
30 km
15 km
Displacement (m)
3000
800
300
100
0

(d) Stage 4: Localization (20 - 30%)

Extension localizes onto larger faults
Some faults inactive in strain shadow of larger, more dominant neighbours
Along-strike linkage of dip-orientated faults through breached relays
Faults linked along strike and at depth

30 km
15 km
30 km
15 km
Oppositely dipping faults inhibit lateral propagation
Displacement (m)
3000
800
300
100
0

Fig. 17. Sketch representation of fault network and fault segment evolution in a schematic basin through four stages. Fault network evolution is shown in relation to a network of conjugate faults dipping towards (black) and away from (grey) the extension direction in three dimensions. The strike projection of fault segment evolution shows an idealized development of a large fault through time from a number of smaller isolated segments at varying depths into a single through-going structure. Its lateral propagation is constrained at the furthest edge by a conjugate fault that initiates at a similar time dipping in the opposite direction.

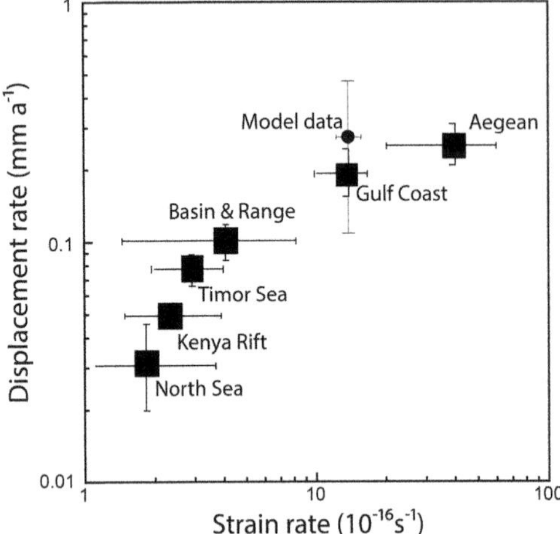

Fig. 18. Log–log plots of displacement rate plotted against regional strain following figure 4 of Nicol *et al.* (1997) to compare with this experiment. The circle represents the mean displacement rate plotted for large faults with displacement >500 m in this model with error bars (see Fig. 6).

In addition to the growth and linkage of fault segments, a number of other features of the model replicate features described in other studies of normal fault geometry and growth. These include the spacing of faults and fault interactions, the form of displacement profiles along the faults, and fault scaling relationships.

Fault spacing parallel to the extension direction is on the order of 5–20 km after 30% extension, and relay breaching occurs at 10–20% extension where fault tip separations are <3 km (Fig. 4b). Fossen & Rotevatn (2016) suggested that where the thickness of the brittle upper crust is 10–15 km, as in this model, spacing between major extensional faults is expected to be around 5 km. Similar thickness to spacing ratios are observed in a range of natural examples and in scaled sandbox experiments (Morley 2002; Hus *et al.* 2005; Conneally *et al.* 2014; Nixon *et al.* 2014*b*; Whipp *et al.* 2016). This shows that although the methodology used for the underlying physics in this model is a simplified representation, the spatial distribution of faults is consistent with natural datasets and scaled analogue experiments.

The faults developed in the model display a variety of displacement profiles that are related to the degree of interaction between faults. Where faults are isolated and fault tips are unrestricted, symmetrical (e.g. Fault C in Fig. 7b) or triangular profiles (e.g. Fault D in Fig. 7b) are developed. Profiles representing single-tip restricted (e.g. Fault B in Fig. 7b) and double-tip restricted cases (e.g. Fault A in Fig. 7b) have asymmetrical and mesa-style profiles, respectively. These profiles are consistent with those interpreted in natural datasets and physical analogue models (e.g. Manighetti *et al.* 2005; Schlagenhauf *et al.* 2008, fig. 2; Nixon *et al.* 2011, fig. 7; 2014*b*, fig. 17; Fossen & Rotevatn 2016).

Fault growth, propagation and linkage

During the initial stages of model extension, faults initiate in conjugate orientations throughout the upper brittle crust. With increasing extension, these small isolated faults grow and coalesce into larger faults through along-strike and downdip linkage (Figs 13–16). One major advantage of the numerical modelling approach is that the 3D distribution of displacement during the growth and linkage process can be assessed. We see multiple transient displacement maxima that continuously migrate as different fault segments progressively interact and link to create the final fault plane. Fossilized examples of this transient behaviour are recorded in nature from the Funan Field, Pattani Basin, Gulf of Thailand, where Kornsawan & Morley (2002) observed two displacement maxima on a fault plane and inferred this to indicate vertical linkage of two faults in three dimensions. In the model, many of the early displacement maxima associated with initially isolated fault segments are lost by overprinting during the growth and linkage process. For example, initial small fault patches are distributed at various depths in the crust. As these patches propagate laterally and link, the information associated with their nucleation site is removed from the record as displacement maxima migrate each time that linkage occurs to the centre of the new, larger fault segment (Figs 12–15). Capturing evidence for these small displacement features at depth within the crust is difficult when presented with only the final distribution. These observations suggest caution should be used when employing the final displacement–distance distribution, either from displacement–distance (D–x) plots or strike projections of fault displacement, to reconstruct the initial segmentation or the progressive linkage history of a fault. An assumption that the final displacement maximum on a fault plane is coincident with the nucleation point of the fault may be misleading. A further implication of the 3D linkage of segments observed in the model is that the fault network in any one horizon should not be considered representative of the total network, either in spatial distribution and interaction of faults or displacement profiles.

The style of fault initiation and growth developed in the model in the early stages (up to 20% extension) is consistent with the model of isolated fault growth (Cowie *et al.* 2000; Schlagenhauf *et al.* 2008) as opposed to the coherent fault growth model (Walsh *et al.* 2003*a*). The observed fault networks develop from fault segments that are initially kinematically independent and propagate radially before linking with adjacent segments to become larger fault zones (stages 1–3: Fig. 17). In the final stage of the model, however, faults are kinematically linked at depth and there is a heterogeneous distribution of fault activity focused on the largest faults, which is more consistent with the coherent fault growth model (Stage 4: Fig. 17). Examination of fault growth for selected faults at the surface (Figs 7 & 8) and the displacement history on strike sections of selected faults (Figs 13–16) shows that in this final stage, when strain is localizing onto the largest faults and relays are being breached, the majority of faults are mainly accruing displacement without further increasing their length. This is consistent with the notion that the two fault growth models are end members of one system (Fossen & Rotevatn 2016), where initially the isolated fault growth model dominates at small strains, which is then replaced by the coherent fault growth model at greater strains (Jackson *et al.* 2017).

Fault network organization and dip domain development

Kinematic coherency is maintained in the model by a combination of variation in both the number of active faults and displacement on those faults. Some transects parallel to the extension direction contain many active faults, whereas others have only two or three, but with larger displacements (Fig. 5). This is similar to natural examples, as shown by linkage of the Rangitaiki Fault (Nixon *et al.* 2014*a*) and palaeo-earthquake data (Nicol *et al.* 2010). Analysis of the model evolution, however, highlights the dynamic nature of fault activity across the fault network. Nicol *et al.* (1997) suggested that the largest faults at the early stages of extension within a network maintain their relative size advantage during rifting and are the largest faults during rift climax. The results documented here suggest this is not necessarily the case. Many of the faults with highest displacement rates during the onset of rifting (e.g. stages 1 and 2) do not continue to dominate but may decelerate or become inactive (e.g. Fault C in Figs 7 & 8). The largest faults at the end of rifting (i.e. Stage 4) are the ones that are in the most favourable locations with respect to stress interactions during later stages of rifting, which is not necessarily the same as during rift initiation.

Observations from natural systems and data from physical analogue experiments indicate that a characteristic feature of rifted fault networks is the occurrence of dip domains within which the dominant fault dip is similar, separated by a rift-wide accommodation zone across which the dominant dip direction changes (e.g. Reches 1978; Morley 2002; McClay *et al.* 2005; Schlagenhauf *et al.* 2008; Schlische & Withjack 2009; Healy *et al.* 2015). Dip domains are a characteristic feature of the fault network developed in this model and are typically of the order of 20–30 km wide. In contrast, dip domain boundaries are narrow regions characterized by interlocking arrays of oppositely dipping faults with slightly overlapping tips (Fig. 4c). In many studies, zones of weakness, such as basement lineaments, are cited as the control on the location of dip domain boundaries: for example, the accommodation zones in the Suez rift (e.g. Patton *et al.* 1994) and the Northern North Sea rift (e.g. Fossen *et al.* 2016). However, in the model presented here, dip domains and their boundaries develop spontaneously as a result of the interaction in three dimensions between evolving conjugate fault sets. There are no pre-existing lineaments or weaknesses controlling the location of dip domain boundaries in this model but they are still present.

Conclusions

A 3D discrete element model of fault growth has been used to investigate the nucleation, growth and interaction of normal fault segments and fault network evolution during a single phase of extension. The modelled fault geometry, fault spacing, displacement rates and scaling relationships are consistent with natural examples. In addition, the ability to investigate the evolution of the fault network in three dimensions has demonstrated a number of points that should be taken into consideration when examining the final static fault segment and fault network geometry in outcrop or subsurface (e.g. seismic) datasets:

- Faults nucleate at minor heterogeneities throughout the brittle upper crust to form a large number of small faults in conjugate arrangements during the initial 10% of extension. These initial faults are kinematically isolated and grow by radial propagation.
- Faults dominantly grow through a combination of tip propagation and linkage of co-linear fault segments to form larger normal faults. Segment linkage occurs progressively with increasing extension and involves linkage between adjacent segments that are located along strike, downdip or oblique to one another. Segment propagation and linkage is inhibited by either the presence of neighbouring faults with an opposing dip or tip propagation into the stress shadow of a neighbouring, more dominant, fault.

- The observed style of fault growth is consistent with the isolated fault growth model in the early stages of rifting (before 20% extension).- This is replaced by a style akin to the coherent fault growth model once strain localizes onto the largest faults (after 20% extension in this case). Any further increase in fault length is limited by neighbouring faults and dip domain boundaries, at which point the faults continue to accrue displacement without lengthening.
- Displacement distribution on the faults is highly dynamic, with displacement maxima that continuously migrate as fault segments interact and link to generate the final fault plane. Displacement maxima associated with initially isolated fault segments are generally lost by overprinting during the growth and linkage process. These observations suggest caution when using the final displacement–distance distribution alone to reconstruct either the earliest fault network or the progressive linkage history.
- Fault scaling statistics for the evolving fault network demonstrate the effect of localization within the system. Small faults coalesce to become larger faults and the relative importance of small faults in accommodating extensional strain decreases with time. This is achieved by progressive localization and accelerated displacement rates on larger faults, and is not associated with an increase in the applied strain rate.
- Faults show marked fluctuations in activity related to their relative position within the evolving network and the resultant changes in stress interactions. Faults that dominate the later stages of rifting are those that have optimal locations within the late-stage fault network. These dominant late-stage faults are not necessarily the faults with highest displacement rates during the early stages of rifting.
- Dip domains, 20–30 km wide and comprising faults that have a dominantly similar dip, are a characteristic feature of the fault network developed in the model. Boundaries between dip domains are narrow regions (<5 km wide) characterized by interlocking arrays of oppositely dipping faults with slightly overlapping tips. Dip domains and their boundaries are a natural result of the growth of the fault network and are not related to underlying basement lineaments or heterogeneity.

The authors would like to thank Jonny Imber, Martin Schöpfer, Tom Manzocchi and Conrad Childs for constructive comments during the review process, and Casey Nixon, Simon Brocklehurst and Julian Mecklenberg for helpful discussions throughout. EF wishes to thank Prof. John McCloskey for insightful comments during development of the code. Thanks also to Eva Bjorseth in Bergen for assistance with figure drafting, and Schlumberger for access to Petrel software. This contribution forms part of the MultiRift Project funded by the Research Council of Norway's PETROMAKS programme (Project No. 215591) and Statoil to the University of Bergen and partners the University of Manchester, Imperial College and the University of Oslo.

References

Abe, S., Van Gent, H. & Urai, J.L. 2011. DEM simulation of normal faults in cohesive materials. *Tectonophysics*, **512**, 12–21.

Ackermann, R.V., Schlische, R.W. & Withjack, M.O. 2001. The geometric and statistical evolution of normal fault systems: an experimental study of the effects of mechanical layer thickness on scaling laws. *Journal of Structural Geology*, **23**, 1803–1819.

Acocella, V., Faccenna, C., Funiciello, R. & Rossetti, F. 1999. Sand-box modelling of basement-controlled transfer zones in extensional domains. *Terra Nova*, **11**, 149–156.

Acocella, V., Gudmundsson, A. & Funiciello, R. 2000. Interaction and linkage of extension fractures and normal faults: examples from the rift zone of Iceland. *Journal of Structural Geology*, **22**, 1233–1246.

Allken, V., Huismans, R.S., Fossen, H. & Thieulot, C. 2013. 3D numerical modelling of graben interaction and linkage: a case study of the Canyonlands grabens, Utah. *Basin Research*, **25**, 1–14.

Behn, M.D., Lin, J. & Zuber, M.T. 2002. A continuum mechanics model for normal faulting using a strain-rate softening rheology: implications for thermal and rheological controls on continental and oceanic rifting. *Earth and Planetary Science Letters*, **202**, 725–740.

Bull, J.M., Barnes, P.M., Lamarche, G., Sanderson, D.J., Cowie, P.A., Taylor, S.K. & Dix, J.K. 2006. High-resolution record of displacement accumulation on an active normal fault: implications for models of slip accumulation during repeated earthquakes. *Journal of Structural Geology*, **28**, 1146–1166.

Camborde, F., Mariotti, C. & Donzé, F.V. 2000. Numerical study of rock and concrete behaviour by discrete element modelling. *Computers and Geotechnics*, **27**, 225–247.

Cartwright, J.A., Trudgill, B.D. & Mansfield, C.S. 1995. Fault growth by segment linkage: an explanation for scatter in maximum displacement and trace length data from Canyonlands grabens of S.E. Utah. *Journal of Structural Geology*, **17**, 1319–1326.

Clifton, A.E., Schlische, R.W., Withjack, M.O. & Ackermann, R.V. 2000. Influence of rift obliquity on fault-population systematics: results of experimental clay models. *Journal of Structural Geology*, **22**, 1491–1509.

Conneally, J., Childs, C. & Walsh, J.J. 2014. Contrasting origins of breached relay zone geometries. *Journal of Structural Geology*, **58**, 59–68.

Contreras, J., Anders, M.H. & Scholz, C.H. 2000. Growth of a normal fault system: observations from the Lake Malawi basin of the east African rift. *Journal of Structural Geology*, **22**, 159–168.

Cowie, P.A. & Roberts, G.P. 2001. Constraining slip rates and spacings for active normal faults. *Journal of Structural Geology*, **23**, 1901–1915.

Cowie, P.A. & Scholz, C.H. 1992. Displacement-length scaling relationship for faults: data synthesis and discussion. *Journal of Structural Geology*, **14**, 1149–1156.

Cowie, P.A., Gupta, S. & Dawers, N.H. 2000. Implications of fault array evolution for synrift depocentre development: insights from a numerical fault growth model. *Basin Research*, **12**, 241–261.

Cowie, P.A., Underhill, J.R., Behn, M.D., Lin, J. & Gill, C.E. 2005. Spatio-temporal evolution of strain accumulation derived from multi-scale observations of Late Jurassic rifting in the northern North Sea: a critical test of models for lithospheric extension. *Earth and Planetary Science Letters*, **234**, 401–419.

Cundall, P.A. 1971. Distinct element models of rock and soil structure. *In*: Brown, E.T. (ed.) *Analytical and Computational Methods in Engineering Rock Mechanics*. Unwin, London, 129–163.

Dawers, N.H. & Anders, M.H. 1995. Displacement-length scaling and fault linkage. *Journal of Structural Geology*, **17**, 607–614.

Dawers, N.H., Anders, M.H. & Scholz, C.H. 1993. Growth of normal faults: displacement-length scaling. *Geology*, **21**, 1107–1110.

Donzé, F., Mora, P. & Magnier, S.-A. 1994. Numerical simulation of faults and shear zones. *Geophysical Journal International*, **116**, 46–52.

Donzé, F., Magnier, S.-A. & Bouchez, J. 1996. Numerical modelling of a highly explosive source in an elastic–brittle rock mass. *Journal of Geophysical Research*, **101**, 3103–3112.

Duffy, O.B., Bell, R.E., Jackson, C.A.-L., Whipp, P.S. & Gawthorpe, R.L. 2015. Fault growth and interactions in a multiphase rift fault network: Horda Platform, Norwegian North Sea. *Journal of Structural Geology*, **80**, 99–119.

Faure Walker, J.P., Roberts, G.P., Cowie, P.A., Papanikolaou, I.D., Sammonds, P.R., Michetti, A.M. & Phillips, R.J. 2009. Horizontal strain rates and throw-rates across breached relay zones, central Italy: implications for the preservation of throw deficits at points of normal fault linkage. *Journal of Structural Geology*, **31**, 1145–1160.

Finch, E., Hardy, S. & Gawthorpe, R. 2003. Discrete element modelling of contractional fault-propagation folding above rigid basement blocks. *Journal of Structural Geology*, **25**, 515–528.

Finch, E., Hardy, S. & Gawthorpe, R.L. 2004. Discrete element modelling of extensional fault-propagation folding above rigid basement fault blocks. *Basin Research*, **16**, 489–506.

Fossen, H. & Rotevatn, A. 2016. Fault linkage and relay structures in extensional settings – A review. *Earth Science Reviews*, **154**, 14–28.

Fossen, H., Fazli Khani, H., Faleide, J.I., Ksienzyk, A.K. & Dunlap, W.J. 2016. Post-Caledonian extension in the West Norway–northern North Sea region: the role of structural inheritance. *In*: Childs, C., Holdsworth, R.E., Jackson, C.A.-L., Manzocchi, T., Walsh, J.J. & Yielding, G. (eds) *The Geometry and Growth of Normal Faults*. Geological Society, London, Special Publications, **439**. First published online February 5, 2016, https://doi.org/10.1144/SP439.6

Gawthorpe, R.L., Jackson, C.A-L., Young, M.J., Sharp, I.R., Moustafa, A.R. & Leppard, C.W. 2003. Normal fault growth, displacement localisation and the evolution of normal fault populations: the Hamman Faraun fault block, Suez Rift, Egypt. *Journal of Structural Geology*, **25**, 883–895.

Giba, M., Walsh, J.J. & Nicol, A. 2012. Segmentation and growth of an obliquely reactivated normal fault. *Journal of Structural Geology*, **39**, 253–267.

Gillespie, P.A., Walsh, J.J. & Watterson, J. 1992. Limitations of dimension and displacement data from single faults and the consequences for data analysis and interpretation. *Journal of Structural Geology*, **14**, 1157–1172.

Gross, M.R., Gutiérrez-Alonso, G., Bai, T., Wacker, M.A., Collingsworth, K.B. & Behl, R.J. 1997. Influence of mechanical stratigraphy and kinematics on fault scaling relations. *Journal of Structural Geology*, **19**, 171–183.

Gupta, S., Cowie, P.A., Dawers, N.H. & Underhill, J.R. 1998. A mechanism to explain rift-basin subsidence and stratigraphic patterns through fault-array evolution. *Geology*, **26**, 595–598.

Hardy, S. 2014. Propagation of blind normal faults to the surface in basaltic sequences: insights from 2D discrete element modelling. *Marine and Petroleum Geology*, **48**, 148–159.

Hardy, S. & Finch, E. 2005. Discrete-element modelling of detachment folding. *Basin Research*, **17**, 507–520.

Hardy, S. & Finch, E. 2006. Discrete element modelling of the influence of cover strength on basement-involved fault propagation. *Tectonophysics*, **415**, 225–238.

Hardy, S. & Finch, E. 2007. Mechanical stratigraphy and the transition from trishear to kink band fault-propagation fold forms above blind basement faults: a discrete-element study. *Marine and Petroleum Geology*, **24**, 75–90.

Hassani, R. & Chéry, J. 1996. Anelasticity explains topography associated with Basin and Range normal faulting. *Geology*, **24**, 1095–1098.

Healy, D., Blenkinsop, T.G., Timms, N.E., Meredith, P.G., Mitchell, T.M. & Cook, M.L. 2015. Polymodal faulting: time for a new angle on shear failure. *Journal of Structural Geology*, **80**, 57–71.

Henza, A.A., Withjack, M.O. & Schlische, R.W. 2010. Normal fault development during two phases of non-coaxial extension: an experimental study. *Journal of Structural Geology*, **32**, 1656–1667.

Henza, A.A., Withjack, M.O. & Schlische, R.W. 2011. How do the properties of a pre-existing normal-fault population influence fault development during a subsequent phase of extension? *Journal of Structural Geology*, **33**, 1312–1324.

Huggins, P., Watterson, J., Walsh, J.J. & Childs, C. 1995. Relay zone geometry and displacement transfer between normal faults recorded in coal-mine plans. *Journal of Structural Geology*, **12**, 1741–1755.

Huismans, R.S. & Beaumont, C. 2007. Roles of lithospheric strain softening and heterogeneity in determining the geometry of rifts and continental margins. *In*: Karner, G.D., Manatschal, G. & Pinheiro, L.M. (eds) *Imaging, Mapping and Modelling Continental Lithosphere Extension and Breakup*. Geological Society, London, Special Publications, **282**, 111–138, https://doi.org/10.1144/SP282.6

Hull, J. 1988. Thickness–displacement relationship for deformation zones. *Journal of Structural Geology*, **10**, 471–482.

Hus, R., Acocella, V., Funiciello, R. & De Batist, M. 2005. Sandbox models of relay ramp structure and evolution. *Journal of Structural Geology*, **27**, 459–473.

Imber, J., Tuckwell, G.W. *et al.* 2004. Three-dimensional distinct element modelling of relay growth and breaching along normal faults. *Journal of Structural Geology*, **26**, 1897–1911.

Jackson, C.A.-L. & Rotevatn, A. 2013. 3D seismic analysis of the structure and evolution of a salt-influenced normal fault zone: a test of competing fault growth models. *Journal of Structural Geology*, **54**, 215–234.

Jackson, C.A.-L., Gawthorpe, R.L., Leppard, C.W. & Sharp, I.R. 2006. Rift-initiation development of normal fault blocks; insights from Hammam Faraun fault block, Suez Rift, Egypt. *Journal of the Geological Society, London*, **163**, 165–183, https://doi.org/10.1144/0016-764904-164

Jackson, C.A.-L., Bell, R.E., Rotevatn, A. & Tvedt, A.B.M. 2017. Techniques to determine the kinematics of synsedimentary normal faults and implications for fault growth models. *In*: Childs, C., Holdsworth, R.E., Jackson, C.A.-L., Manzocchi, T., Walsh, J.J. & Yielding, G. (eds) *The Geometry and Growth of Normal Faults*. Geological Society, London, Special Publications, **439**. First published online February 7, 2017, https://doi.org/10.1144/SP439.22

Khalil, S.M. & McClay, K.R. 2016. 3D geometry and kinematic evolution of extensional fault-related folds, NW Red Sea, Egypt. *In*: Childs, C., Holdsworth, R.E., Jackson, C.A.-L., Manzocchi, T., Walsh, J.J. & Yielding, G. (eds) *The Geometry and Growth of Normal Faults*. Geological Society, London, Special Publications, **439**. First published online March 30, 2016, https://doi.org/10.1144/SP439.11

King, G.C.P. & Ellis, M. 1990. The origin of large local uplift in extensional regions. *Nature*, **348**, 689–692.

King, G.C.P., Stein, R.S. & Rundle, J.B. 1988. The growth of geological structures by repeated earthquakes 1. Conceptual framework. *Journal of Geophysical Research*, **93**, 13,307–13,318.

Komoróczi, A., Abe, S. & Urai, J.L. 2013. Meshless numerical modelling of brittle-viscous deformation: first results on boudinage and hydrofracturing using a coupling of discrete element method (DEM) and smoothed particle hydrodynamics (SPH). *Computers and Geosciences*, **17**, 373–390.

Kornsawan, A. & Morley, C.K. 2002. The origin and evolution of complex transfer zones (graben shifts) in conjugate fault systems around the Funan Field, Pattani Basin, Gulf of Thailand. *Journal of Structural Geology*, **24**, 435–449.

Kuhn, M.R. 1999. Structured deformation in granular materials. *Mechanics of Materials*, **31**, 407–429.

Kusznir, N.J. & Park, R.G. 1987. The extensional strength of the continental lithosphere: its dependence on geothermal gradient, and crustal composition and thickness. *In*: Coward, M.P., Dewey, J.F. & Hancock, P.L. (eds) *Continental Extensional Tectonics*. Geological Society, London, Special Publications, **28**, 35–52, https://doi.org/10.1144/GSL.SP.1987.028.01.04

Lambert, C. & Coll, C. 2014. Discrete modelling of rock joints with a smooth-joint contact model. *Journal of Rock Mechanics and Geotechnical Engineering*, **6**, 1–12.

Long, J.J. & Imber, J. 2011. Geological controls on fault relay zone scaling. *Journal of Structural Geology*, **33**, 1790–1800.

Long, J.J. & Imber, J. 2012. Strain compatibility and fault linkage in relay zones on normal faults. *Journal of Structural Geology*, **36**, 16–26.

Lovely, P., Flodin, E., Guzofski, C., Maerten, F. & Pollard, D.D. 2012. Pitfalls among the promises of mechanics-based restoration: addressing implications of unphysical boundary conditions. *Journal of Structural Geology*, **41**, 47–63.

Manighetti, I., Campillo, M., Sammis, C., Mai, P.M. & King, G. 2005. Evidence for self-similar slip distributions on earthquakes: implications for earthquake fault mechanics. *Journal of Geophysical Research*, **110**, B05302, https://doi.org/10.1029/2004JB003174

Manighetti, I., Campillo, M., Bouley, S. & Cotton, F. 2007. Earthquake scaling, fault segmentation, and structural maturity. *Earth and Planetary Science Letters*, **253**, 429–438.

Marrett, R. & Allmendinger, R.W. 1990. Kinematic analysis of fault-slip data. *Journal of Structural Geology*, **12**, 973–986.

McClay, K.R., Dooley, T., Whitehouse, P. & Mills, M. 2002. 4-D evolution of rift systems: insights from scaled physical models. *American Association of Petroleum Geologists Bulletin*, **86**, 935–959.

McClay, K.R., Dooley, T., Whitehouse, P.S. & Anadon-Ruiz, S. 2005. 4D analogue models of extensional fault systems in asymmetric rifts: 3D visualisation and comparisons with natural examples. *In*: Doré, A.G. & Vining, B.A. (eds) *Petroleum Geology: North West Europe and Global Perspectives – Proceedings of the 6th Petroleum Geology Conference*. Geological Society, London, Petroleum Geology Conference Series, **6**, 1543–1556, https://doi.org/10.1144/0061543

McLeod, A.E., Dawers, N.H. & Underhill, J.R. 2000. The propagation and linkage of normal faults: insights from the Strathspey–Brent–Statfjord fault array, northern North Sea. *Basin Research*, **12**, 263–284.

Mora, P. & Place, D. 1993. A lattice solid model for the non-linear dynamics of earthquakes. *International Journal of Modern Physics*, **C4**, 1059–1074.

Mora, P. & Place, D. 1994. Simulation of the frictional stick-slip instability. *Pure and Applied Geophysics*, **143**, 61–87.

Moriya, S., Childs, C., Manzocchi, T. & Walsh, J.J. 2005. Analysis of the relationship between strain, polarity and population slope for normal fault systems. *Journal of Structural Geology*, **27**, 1113–1127.

Morley, C.K. 2002. Evolution of large normal faults: evidence from seismic reflection data. *American Association of Petroleum Geologists Bulletin*, **86**, 961–978.

Morley, C.K., Gabdi, S. & Seusutthiya, S.G.K. 2007. Fault superimposition and linkage resulting from stress changes during rifting: examples from 3D seismic data, Phitsanulok Basin, Thailand. *Journal of Structural Geology*, **29**, 646–663.

Nicol, A., Walsh, J.J., Watterson, J. & Childs, C. 1996. The shapes, major axis orientations and displacement patterns of fault surfaces. *Journal of Structural Geology*, **18**, 235–248.

Nicol, A., Walsh, J.J., Watterson, J. & Underhill, J.R. 1997. Displacement rates of normal faults. *Nature*, **390**, 157–159.

Nicol, A., Walsh, J.J., Villamor, P., Seebeck, H. & Berryman, K.R. 2010. Normal fault interactions, paleoearthquakes and growth in an active rift. *Journal of Structural Geology*, **32**, 1101–1113.

Nixon, C.W., Sanderson, D.J. & Bull, J.M. 2011. Deformation within a strike-slip fault network at Westward Ho!, Devon, U.K.: Domino vs conjugate faulting. *Journal of Structural Geology*, **33**, 833–843.

Nixon, C.W., Bull, J.M. & Sanderson, D.J. 2014*a*. Localised vs distributed deformation associated with the linkage history on an active normal fault, Whakatan Graben, New Zealand. *Journal of Structural Geology*, **69**, 266–280.

Nixon, C.W., Sanderson, D.J., Dee, S.J., Bull, J.M., Humphreys, R.J. & Swanson, M.H. 2014*b*. Fault interactions and reactivation within a normal-fault network at Milne Point, Alaska. *American Association of Petroleum Geologists Bulletin*, **98**, 2081–2107.

Patton, T.L., Moustafa, A.R., Nelson, R.A. & Abdine, A.S. 1994. Tectonic evolution and structural setting of the Gulf of Suez Rift. *In*: Landon, S.M. (ed.) *Interior Rift Basins*. American Association of Petroleum Geologists, Memoirs, **59**, 9–55.

Peacock, D.C.P. & Sanderson, D.J. 1991. Displacement, segment linkage and relay ramps in normal fault zones. *Journal of Sedimentary Geology*, **15**, 721–733.

Place, D. & Mora, P. 2001. A random lattice solid model for simulation of fault zone dynamics and fracture processes. *In*: Mulhaus, H.-B., Dyskin, A.V. & Pasternak, E. (eds) *Bifurcation and Localisation Theory for Soils and Rocks '99*. A.A. Balkema, Rotterdam, The Netherlands, 321–333.

Place, D., Lombard, F., Mora, P. & Abe, S. 2002. Simulation of the microphysics of rocks using LSM Earth. *Pure and Applied Geophysics*, **159**, 1911–1932.

Potyondy, D.O. & Cundall, P.A. 2004. A bonded-particle model for rock. *International Journal for Rock Mechanics and Mining*, **41**, 1329–1364.

Ranalli, G. 1995. *Rheology of the Earth*. 2nd edn. Chapman & Hall, London.

Reches, Z. 1978. Analysis of faulting in three-dimensional strain field. *Tectonophysics*, **47**, 109–129.

Reeve, M.T., Bell, R.E., Duffy, O.B., Jackson, C.A.-L. & Sansom, E. 2015. The growth of non-colinear normal fault systems; What can we learn from 3D seismic reflection data? *Journal of Structural Geology*, **70**, 141–155.

Schlagenhauf, A., Manghetti, I., Malavieille, J. & Dominguez, S. 2008. Incremental growth of normal faults: insights from a laser-equipped analog experiment. *Earth and Planetary Science Letters*, **273**, 299–311.

Schlische, R.W. & Withjack, M.O. 2009. Origin of fault domains and fault-domain boundaries (transfer zones and accommodation zones) in extensional provinces: result of random nucleation and self-organized growth. *Journal of Structural Geology*, **31**, 910–925.

Schöpfer, M.P.J., Childs, C. & Walsh, J.J. 2007*a*. 2D distinct element modeling of the structure and growth of normal faults in a multilayer sequence. Part 1: model calibration, boundary conditions and selected results. *Journal of Geophysical Research*, **112**, B10401.

Schöpfer, M.P.J., Childs, C. & Walsh, J.J. 2007*b*. 2D distinct element modeling of the structure and growth of normal faults in a multilayer sequence. Part 2: impact of confining pressure and strength contrast on fault zone geometry and growth. *Journal of Geophysical Research*, **112**, B10404.

Schöpfer, M.P.J., Abe, S., Childs, C. & Walsh, J.J. 2009. The impact of crack and density on elasticity, strength and friction of cohesive granular materials: insights from DEM modelling. *International Journal of Rock Mechanics and Mining Science*, **46**, 250–261.

Sharp, I.R., Gawthorpe, R.L., Underhill, J.R. & Gupta, S. 2000. Fault propagation folding and extensional settings: examples of structural style and synrift sedimentary response from the Suez rift, Sinai, Egypt. *Geological Society of America Bulletin*, **112**, 1877–1899.

Soliva, R., Benedicto, A., Schultz, R.A., Maerten, L. & Micarelli, L. 2008. Displacement and interaction of normal fault segments branched at depth: implications for fault growth and potential earthquake rupture size. *Journal of Structural Geology*, **30**, 1288–1299.

Toomey, A. & Bean, C.J. 2000. Numerical simulation of seismic waves using a discrete particle scheme. *Geophysical Journal International*, **141**, 595–604.

Torabi, A. & Berg, S.S. 2011. Scaling of fault attributes: a review. *Marine and Petroleum Geology*, **28**, 1444–1460.

Walsh, J.J., Childs, C. et al. 2001. Geometric controls on the evolution of normal fault systems. *In*: Holdsworth, R.E., Strachan, R.A., Magloughlin, J.F. & Knipe, R.J. (eds) *The Nature and Tectonic Significance of Fault Zone Weakening*. Geological Society, London, Special Publications, **186**, 157–169, https://doi.org/10.1144/GSL.SP.2001.186.01.10

Walsh, J.J., Bailey, W.R., Childs, C., Nicol, A. & Bonson, C.G. 2003*a*. Formation of segmented normal faults: a 3-D perspective. *Journal of Structural Geology*, **25**, 1251–1262.

Walsh, J.J., Childs, C., Imber, J., Manzocchi, T., Watterson, J. & Nell, P.A.R. 2003*b*. Strain localisation and population changes during fault system growth within the Inner Moray Firth, Northern North Sea. *Journal of Structural Geology*, **25**, 307–315.

Wenk, L. & Huhn, K. 2013. The influence of an embedded viscoelastic-plastic layer on kinematics and mass transport pattern within accretionary wedges. *Tectonophysics*, **608**, 653–666.

Whipp, P.S., Jackson, C.A.-L., Schlische, R.W., Withjack, M.O. & Gawthorpe, R.L. 2016. Spatial distribution and evolution of fault-segment boundary types in rift systems: observations from experimental clay models. *In*: Childs, C., Holdsworth, R.E., Jackson,

C.A.-L., Manzocchi, T., Walsh, J.J. & Yielding, G. (eds) *The Geometry and Growth of Normal Faults*, Geological Society, London, Special Publications, **439**. First published online March 31, 2016, https://doi.org/10.1144/SP439.7

Wojtal, S.F. 1996. Changes in fault displacement populations correlated to linkage between faults. *Journal of Structural Geology*, **18**, 265–279.

Wright, T.M., Sigmundsson, F. *et al.* 2012. Geophysical constraints on the dynamics of spreading centres from rifting episodes on land. *Nature Geoscience*, **5**, 242–250.

Wu, J.E., McClay, K. & Frankowicz, E. 2015. Niger Delta gravity-driven deformation above the relict Chain and Charcot oceanic fracture zones, Gulf of Guinea: insights from analogue models. *Marine and Petroleum Geology*, **65**, 43–62.

Xu, S.-S., Nieto-Samaniego, A.F. & Alaniz-Álvarez, S.A. 2014. Estimation of average to maximum displacement ratio by using fault displacement–distance profiles. *Tectonophysics*, **636**, 190–200.

Yielding, G., Walsh, J. & Watterson, J. 1992. The prediction of small-scale faulting in reservoirs. *First Break*, **10**, 449–460.

Young, M.J., Gawthorpe, R.L. & Hardy, S. 2001. Growth and linkage of a segmented normal fault zone; the late Jurassic Murchison–Statfjord North Fault, northern North Sea. *Journal of Structural Geology*, **23**, 1933–1952.

The geometry and dimensions of fault-core lenses

ROY H. GABRIELSEN[1*], ALVAR BRAATHEN[1], MAGNUS KJEMPERUD[1,2] & MARIE LOVISE R. VALDRESBRÅTEN[1,3]

[1]*Department of Geosciences, University of Oslo, Oslo, Norway*

[2]*Present address: Core Energy, Oslo, Norway*

[3]*Present address: BP Exploration, Stavanger, Norway*

**Correspondence: r.h.gabrielsen@geo.uio.no*

Abstract: Field analysis shows that fault cores of brittle, extensional faults at a medium to mature stage of development are commonly dominated by lozenge-shaped horses (fault-core lenses) characterized by a variety of lithologies, including intact, mildly to strongly deformed country rock derived from the footwalls and hanging walls, various types of fault rocks of the protocatalasite and breccia series, breccia, fault gouge and clay smear. The lenses are sometimes stacked to form complex duplexes. These structures are commonly separated by high-strain zones of sheared cataclasite, and/or clay smear/clay gouge. The geometry and distribution of clay gouge in high-strain zones sometimes display evidence of intrusion, indicating high fluid pressure.

Although the sizes of the horses vary over several orders of magnitude, they frequently display a length:thickness (*a:c*) ratio of between 1:4 and 1:15.

The high-strain zones of fault rocks commonly constitute unbroken, 3D membranes that are likely to constrain fluid communication both across and along the fault zone.

There are significant contrasts in fault core architecture that are probably related to processes associated with contrasting fluid pressure, strain intensity and strain hardening/strain softening. Faults associated with strain softening are characterized by less abundant brittle deformation products and are less likely to be conduits for fluid flow compared to those that are affected by strain hardening.

The influence of faults on fluid flow in hydrocarbon reservoirs and in reservoirs selected for storage of of CO_2 have sparked numerous studies of fault architecture and fault dynamics, both older (e.g. Clapp 1910) and more recent (Berg & Avery 1995; Caine *et al.* 1996; Kim *et al.* 2004; Berg & Skar 2005; Agosta & Aydin 2006; Fredman *et al.* 2008; Wibberley *et al.* 2008; Bastesen *et al.* 2009; Braathen *et al.* 2009; Childs *et al.* 2009; Bretan *et al.* 2011; Yielding *et al.* 2011; Gabrielsen & Braathen 2014). Depending on the fault architecture, depth of burial, fluid pressure, burial/uplift history (background porosity) and degree of reactivation, faults may act either as conduits or barriers to fluids (McKnight 1940; Weeks 1958; Davatzes & Aydin 2005; Sims *et al.* 2005; Sorkhabi & Tsuji 2005). In fact, almost all natural (geological) reservoirs are affected by faults to the degree that this must be taken into consideration in risk evaluation, production planning and reservoir modelling (Bredehoeft *et al.* 1982; Sorkhabi & Tsuji 2005; Braathen *et al.* 2012; Bælum *et al.* 2012). Information on fault architecture is not easily obtained from conventional seismic data and may require dedicated processing and analysis, commonly to be supported by field study, microscope and SEM investigation, and laboratory study.

Although important contributions have been made (e.g. Robertson 1983; Knott 1994; Caine *et al.* 1996; Childs *et al.* 1997, 2009; Caine & Forster 1999; Manzocchi *et al.* 1999; Davatzes & Aydin 2005; Wibberley *et al.* 2008; Bastesen *et al.* 2009), the tools for assessing the dynamics of fluids *within* the faults themselves is still incomplete (Sorkhabi & Tsuji 2005) and are dependent on understanding the bulk architecture of the fault. This is, perhaps, due to the complexity and variability of the intrinsic geometry and, thus, the complex permeability distribution that commonly characterize larger faults. It is also likely that the methodological, technical and computational constraints have discouraged such studies (Bredehoeft *et al.* 1982). Recent developments of research reservoir modelling algorithms have, however, opened new possibilities for the inclusion of realistic fault geometries in the context of faults being represented as complex rock bodies rather than surfaces with posted permeability values and their effects on fluid flow (Fredman *et al.* 2008; Braathen *et al.* 2009; Fachri *et al.* 2011, 2013).

From: Childs, C., Holdsworth, R. E., Jackson, C. A.-L., Manzocchi, T., Walsh, J. J. & Yielding, G. (eds) 2017. *The Geometry and Growth of Normal Faults*. Geological Society, London, Special Publications, **439**, 249–269.
First published online February 5, 2016, https://doi.org/10.1144/SP439.4

Faults should be regarded as 3D rock bodies rather than the (2D) planes (e.g. Sibson 1977; Wallace & Morris 1979; Chester & Logan 1986; Caine *et al.* 1996; Berg & Skar 2005; Wibberley *et al.* 2008; Braathen *et al.* 2009; Gabrielsen 2010; Schueller *et al.* 2012; Gabrielsen & Braathen 2014) that are depicted in many seismic intrepretation tools. Based on field study of faulted sediments, Caine *et al.* (1996) subdivided faults in porous, siliciclastic rocks into several architectural elements, namely: the fault core; the footwall and hanging-wall mixed zones; and the footwall and the hanging-wall damage zones (Fig. 1) (see also Evans *et al.* 1997; Heynekamp *et al.* 1999; Sigda *et al.* 1999; Clausen *et al.* 2003). These authors associate the core zone with the maximum accumulated slip, and described this zone as strongly foliated with shifting fault-parallel veneers of clay and sand. These zones may contain rootless pods of rocks carved from the hanging wall or the footwall (Childs *et al.* 1996, 2009; Gabrielsen & Clausen 2001; Clausen *et al.* 2003; Berg & Skar 2005; Davatzes & Aydin 2005; Lindanger *et al.* 2007; Wibberley *et al.* 2008; Gabrielsen & Braathen 2014), which may be surrounded by a matrix of contrasting lithology.

Such architecture is commonly seen in meso-scale faults, which have undergone a displacement of the order of 1 m up to hundreds of metres (Clausen *et al.* 2003; Berg & Skar 2005; Gabrielsen & Braathen 2014; Skar *et al.*, this volume, in press). Fault-core lenses have, indeed, been described in several previous studies (e.g. Childs *et al.* 1996; Gabrielsen & Clausen 2001; Davatzes & Aydin 2005; Lindanger *et al.* 2007; Braathen *et al.* 2009; Gabrielsen & Braathen 2014). These kinds of structures are commonly assumed to be developed by different types of fault-linkage mechanisms (Childs *et al.* 1996, 2009; Gabrielsen & Clausen 2001), or by cut-outs from irregularities in the hanging wall or footwall of the fault (Gibbs 1984; Clausen *et al.* 2003; Davatzes & Aydin 2005; Flodin *et al.* 2005; Lindanger *et al.* 2007).

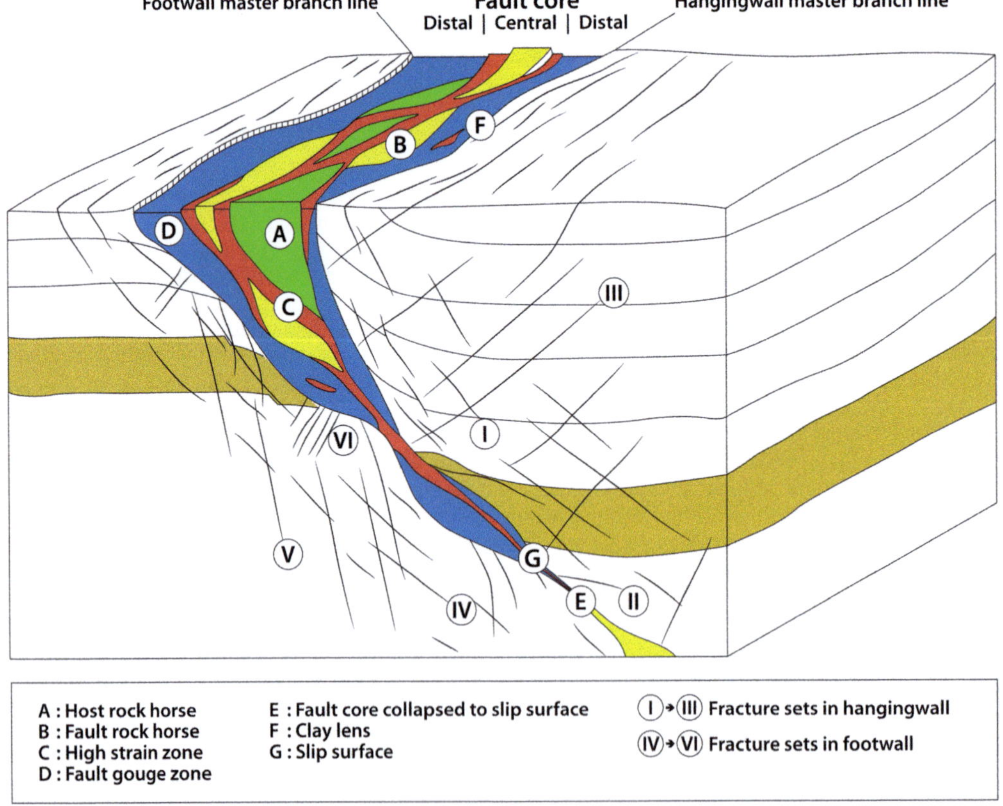

Fig. 1. Principal sketch of the fault architecture. The fault core may contain fault-core lenses of different lithology including (strained) country rock derived from the footwall and hanging wall, and fault rocks. From Gabrielsen (2010).

The present contribution aims at describing fault-core lenses, which are the basic structural element in many mesoscopic and macroscopic faults, through the anlysis of dimensions and geometry. These data are used to explore the mechanisms for their derivation and development. To accomplish this, we analysed geometries and dimensions of fault-core lenses in extensional faults from analogue laboratory experiments and field study for varying strain intensity and depth of burial, starting with mechanically (unconsolidated) weak analogue and geological material, progressing to well-consolidated/metamorphic, mechanically strong rocks. In this context, we used established methods and nomenclature (*a*, *b* and *c* axes) for the characterization of the dimensions of rock bodies relative to tectonic stress and transport direction (e.g. Whitten 1966) (Fig. 2a), which are slightly different from those used by Lindanger *et al.* (2007).

Analogue models and field examples

Analogue experiments: plaster of Paris

A number of plaster of Paris experiments have been performed to investigate fault geometry and fault architecture (Fossen & Gabrielsen 1996; Gabrielsen & Clausen 2001; Lindanger *et al.* 2007). The advantage of plaster of Paris experiments, which were performed by us using a liquid plaster of Paris solution run at the ductile–brittle transition (see Fossen & Gabrielsen 1996 for a description of the procedure),

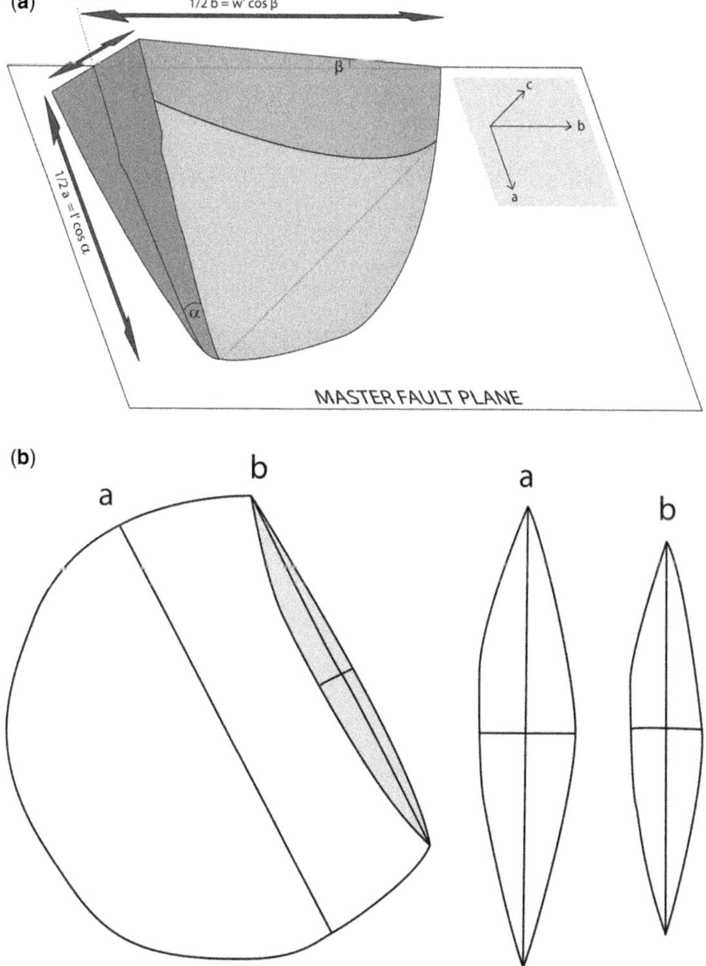

Fig. 2. (**a**) Principal axes of fault-core lenses used in the present description. (**b**) Relationships of principal axes in different sections of the fault-core lens. After Lindanger *et al.* (2007).

over that of sand and silicon putty experiments is that the deforming material is very fine grained and so very delicate details are preserved when the experiment solidifies (Fig. 3). One set of experiments was particularly designed to investigate the mechanisms for the generation and development of fault lenses (Gabrielsen & Clausen 2001), and was later analysed in order to characterize the geometry and dimensions of fault cores (Lindanger *et al.* 2007).

The experiments demonstrated that fault-core lenses can be extensively sheared during faulting and that they commonly reach a stage of geometrical equilibrium during progressing strain (Gabrielsen & Clausen 2001). Among the fault-core lenses investigated by us, the plaster of Paris experiments produced the thickest lenses (*c*) compared to length (*a*), with an average *a*:*c* ratio of 3 (Lindanger *et al.* 2004, 2007).

Unconsolidated to poorly consolidated sand-clay sequence: Bornholm, Denmark

Faults on the island of Bornholm, Denmark offer complete sections in extensional, slightly inverted faults in unconsolidated, fine-grained sands and silts interbedded with thin layers of clay of the lower Cretaceous (Berriasian) Robbedale Formation of the Nyker Group (Østborg Member: Gravesen 1982; Clausen *et al.* 2003). NE–SW-striking faults that affect these units are included the realm of the WNW–ESE-striking Tornquist–Sorgenfrei Zone (Gravesen 1996). The NE–SW-striking faults are associated with Early Cretaceous extension and are situated within the Arnager–Sose Block (Fig. 4a). The faults were later inverted during the Late Cretaceous–Early Tertiary post-rift stage (Vejbæk *et al.* 1994). The sediments are unconsolidated and the maximum burial depth has been estimated to be 500 m (Hamann 1988).

The Stender's Quartz Quarry locality (Fig. 4a) offers a section through one of the NE–SW-striking faults. This fault has an easterly dip of 50–60° and a minimum dip-slip extensional separation of the order of 13 m. It can be studied in great detail because the unconsolidated sediments allow for full excavation in the vertical and horizontal dimensions (Clausen *et al.* 2003). Several fault branches are seen in its footwall, whereas synthetic fault strands in the hanging wall are characterized by swarms of deformation bands, some of which contain clay. The fault core, which is up to 100 cm wide, can commonly be subdivided into a central part and two distal subzones. The central fault core is characterized by fault lenses (Fig. 4b, c), with the longest axis varying between 7 and 100 cm, and the width of the individual lenses varying from 2.5 to 25 cm. Segments of the fault core

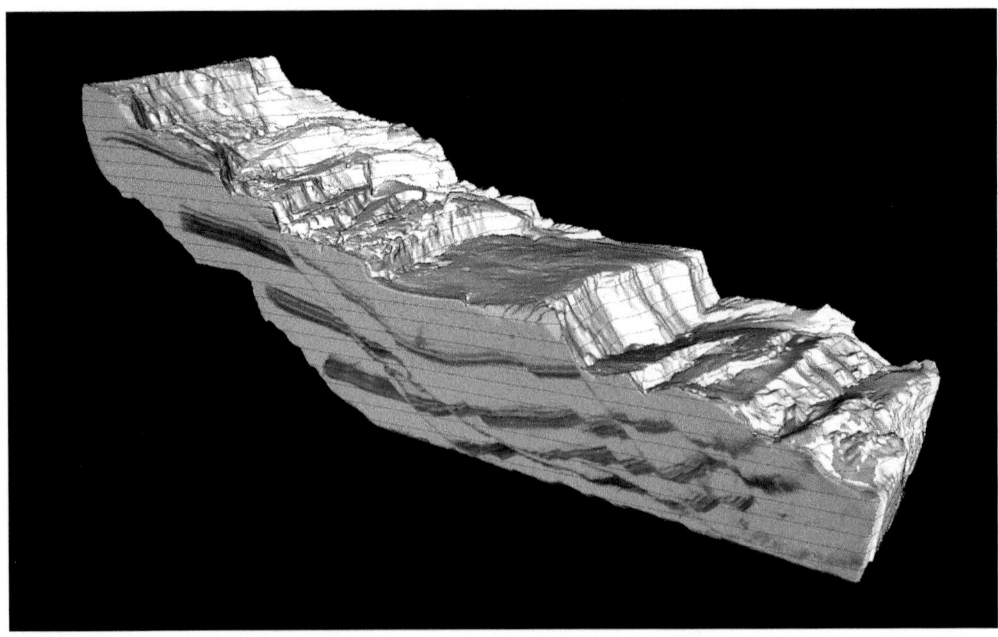

Fig. 3. Example of the analogue plaster of Paris model used in the present analysis. Note the complex geometry of the master fault (left) that includes several fault-core lenses. The picture was electronically processed by Simon Buckley (CIPR Bergen).

have typical a:c ratios of approximately 1:4, whereas the average a:c ratio for the entire fault core is 13:1 (Lindanger *et al.* 2007). The sandy or silty fault-core lenses are separated by a network of clay membranes, some of which locally display an intrusive character (Fig. 4d). The lenses are mostly enveloped by unbroken clay smear zones that commonly exhibit continuous 3D networks. Some of the larger lenses are seen to be in the process of becoming subdivided into smaller lensoid units through the development of internal shear zones that are occasionally invaded by clay. Similar shearing was observed in some cases for the clay membranes. The distal zones of the fault core are occasionally dominated by clay-dominated bodies, here varying in width from <1 cm to 15 cm. The distal fault core can also encompass zones with a mixture of clay and sand, as well as patchy clay in a sand matrix. Relay ramps seem to contribute to the geometrical complexity of the fault core. The clay-filled high-strain zones separating the fault lenses are generally unbroken, but occasionally contain small lenses, bridges and trains (longest axis 0.5–10 cm) of sand grains. The minor lenses seem to be carved from the fault-core margins, forming sandy zones within the fault-core matrix. Such stuctures have also been found on the microscopic level in the samples from Stender's Quartz Quarry, and similar features have been reproduced in ring-shear experiments (Clausen & Gabrielsen 2002).

It is concluded that extensional faults in the Robbedal Formation at Bornholm illustrate that complex deformation may occur in fault cores with moderate displacements (tens of metres) in unconsolidated sediments and create continuous 3D membranes of clay smear. Intrusive characteristics of clay seen locally indicate that the clay mobility was supported by a high fluid pressure. The fault-core lenses produced in these conditions had a typical *a:c* ratio of approximately 13:1, whereas some segments have *a:c* ratios approaching that seen in the plaster of Paris experiments (4:1 – see the earlier description).

Consolidated limestone–sandstone–(shale) sequence: Howick Bay, Northumberland Basin

Howick Bay of the Northumberland Basin, on the east coast of England (Fig. 5a), offers 3D sections in normal faults in combined strandflat to coastal scarp exposures. These faults affect the cyclic sediments (delta plain sandstone–marine transgression limestone) of Dinantian (Brigantian) age. The faults strike east–west and have normal throws varying from less than 1 m up to 200 m. The faults with smaller displacement (<1 m) are characterized by simple, centimetre-thick cores and damage zones encompassing only a few fractures, whereas the master fault at Howick Bay, which has a stratigraphic normal separation of approximately 200 m (Westoll *et al.* 1955; Farmer & Jones 1969; Johnson 1984), displays a more than 30 m-wide zone of deformation (including footwall and hanging-wall damage zones). It separates limestones with interbedded calcareous limestone (Brigantian) in the footwall from fluviodeltaic sandstone (Namurian) in the hanging wall (Fig. 5b). Its average dip is 52° to the south. It is interpreted as a partly synsedimentary listric dip-slip fault (including soft-sediment deformation) of Dinantian–Namurian age, which became later reactivated in dextral shear (De Paola *et al.* 2005). Characteristically, the fault has a complex core containing metre-sized fault lenses (Fig. 6a).

At the Howick Bay locality, 47 fault-core lenses were identified and inspected in the vertical (cliff section) and horizontal (tidal flat) sections. These lenses show a high degree of parallel orientation of the longest axes, being orientated parallel to the master fault trace (east–west). The dimensions display considerable variation, with an average *a:c* ratio of approximately 6.5:1. The estimated transport length for lenses from their assumed original position in the footwall and the hanging wall is up to 5 m (visual tectonostratigraphic correlation). The fault-core lenses are commonly separated by layers/membranes of fault gouge or clay smear, in many cases resembling the situations described earlier from the Stender's Quartz Quarry at Bornholm. The transition zones between the fault core and the hanging wall and the footwall are commonly characterized by a membrane of shale smear or shale gouge (in the definition of Davatzes & Aydin 2005; Færseth 2006). In addition, the footwall contains *in situ* lozenge-shaped structures that are commonly delineated by zones of deformation bands, and which are believed to represent incipient lenses in the process of being separated from the footwall to be incorporated into a wider fault core. The lenses, however, most commonly seem to be derived from the footwall. Some branches of the fault have a local ramp–flat–ramp geometry, the flat part being characterized by enhanced footwall fracture frequency, beneath a thin fault segment.

Consolidated sandstone–siltstone–shale–coal measures: Hartley Steps and Snab Point, Northumberland Basin

The locality at Hartley Steps (Fig. 5b) displays an approximately 100 m strike-parallel section (ENE–WSW) through a normal fault affecting

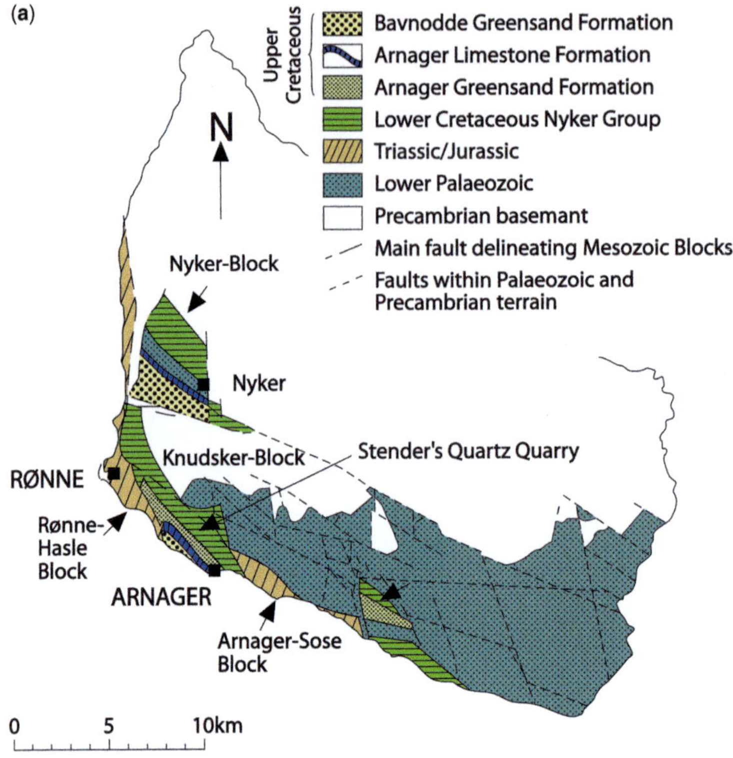

Fig. 4. (**a**) Geological map of the island of Bornholm, showing the locality of the Stender's Quartz Quarry. After Gravesen (1996). (**b**) Excavated core of a normal fault in unconsolidated sand/silt with stringers of clay. Note the varying geometry of fault-core lenses and unbroken (high-strain) zones of clay smear. The compass (the long side is 8 cm) is for scale. (**c**) A fault-core lens of unconsolidated sand completely embedded in clay smear. The measuring stick is 20 cm. (**d**) Sketch of a fault core in a sand–clay sequence. Note the intrusive character of the clay in the upper left and patches of clay in the sand lenses (see the text for the explanation). The width of the fault core is 40 cm. (b)–(d) From Stender's Quartz Quarry, Bornholm.

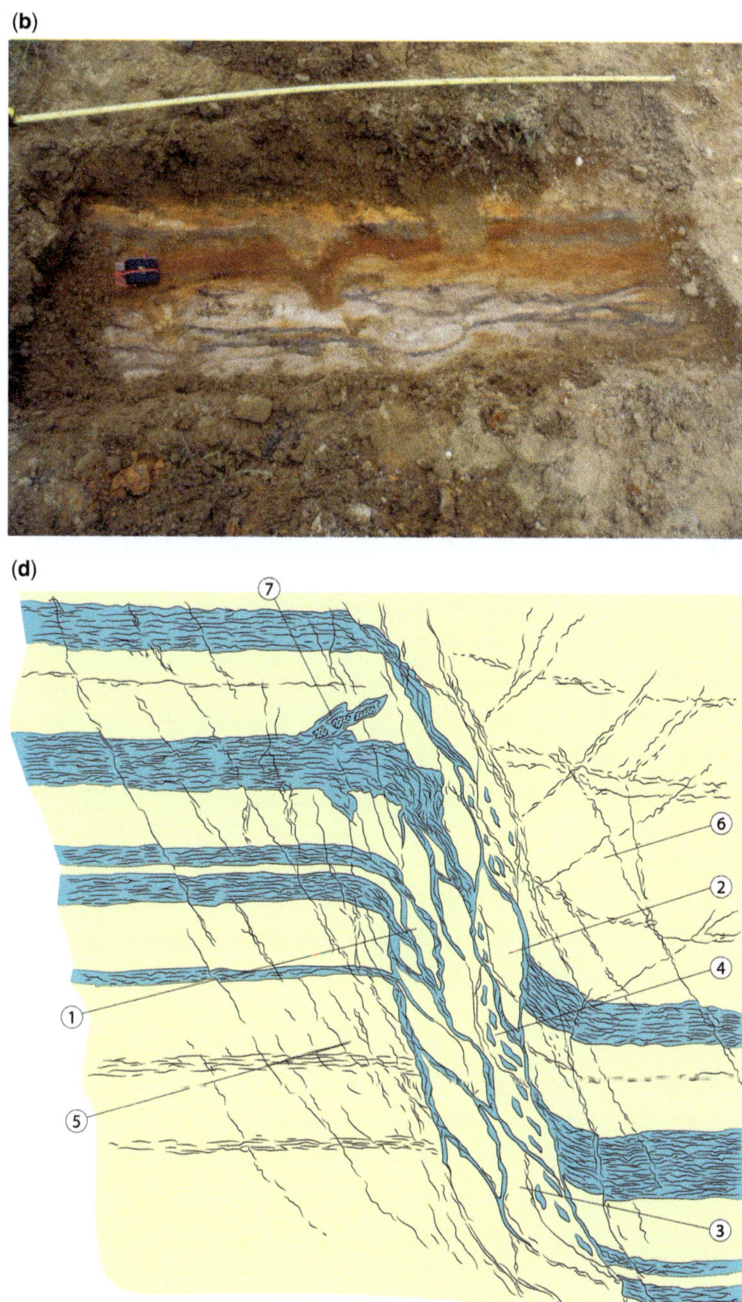

Fig. 4. *Continued.*

fluvial deltaic deposits (Collier 1989) of the Westphalian B stage of Kimbell *et al.* (1989). The total throw is estimated to 15–17 m (Jones 1968; De Paola *et al.* 2005; Færseth *et al.* 2007). The master fault is paralleled by hanging-wall and footwall branch fault strands defined by continuous 10–15 cm-wide zones of fault gouge, sometimes including centimetre- to decimetre-scale isolated lenses of country rock derived from the footwall and the hanging wall (Fig. 6c). Including the damage

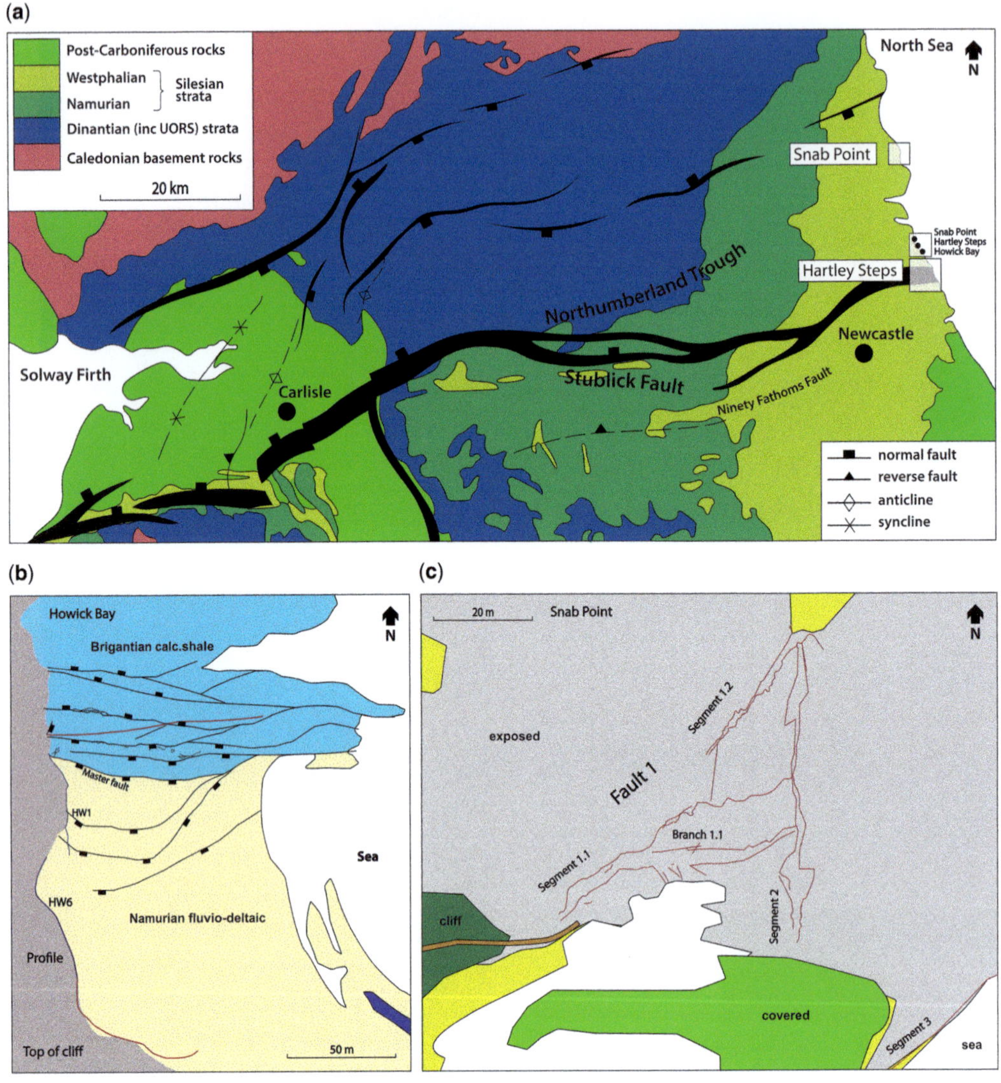

Fig. 5. (**a**) Geological key map, Northumberland Basin and detailed maps of the faults at (**b**) Howick Bay and (**c**) Snab Point. (**d**) Detailed fault-core maps, Snab Point (see the text for the explanation). Note the differences in map scales.

zone, the core of the master fault is up to 30 m wide. The fault core is characterized by stacked extensional fault lenses (sandstone, siltstone, shale and coal). Several lenses seem to be positioned closer to the strata in the hanging wall and footwall from which they were derived. The fault lenses of the core of the master fault are sheared and mainly rotated to parallelism with the strike of the master fault. The lenses occasionally reveal internal fracture networks in the most competent lithologies (sandstone and siltstone), the areas of most intense fracturing being situated at the tips of the lenses, here sometimes approaching protobreccia. Some examples of lenses on the stage of further disintegration (internal, anastomosing shear bands) were also recorded. From visual lithostratigraphic correlations, the fault lenses investigated by us seem to be displaced 1–5 m from their assumed original positions in the hanging-wall and footwall blocks.

Twelve lenses, for which the 3D dimensions could be measured, were identified in the fault core. The average a:c ratios varied from 6:1 to 9:1.

At Snab Point (Fig. 5c, d), a set of NE–SW-striking normal faults offset coal measures of

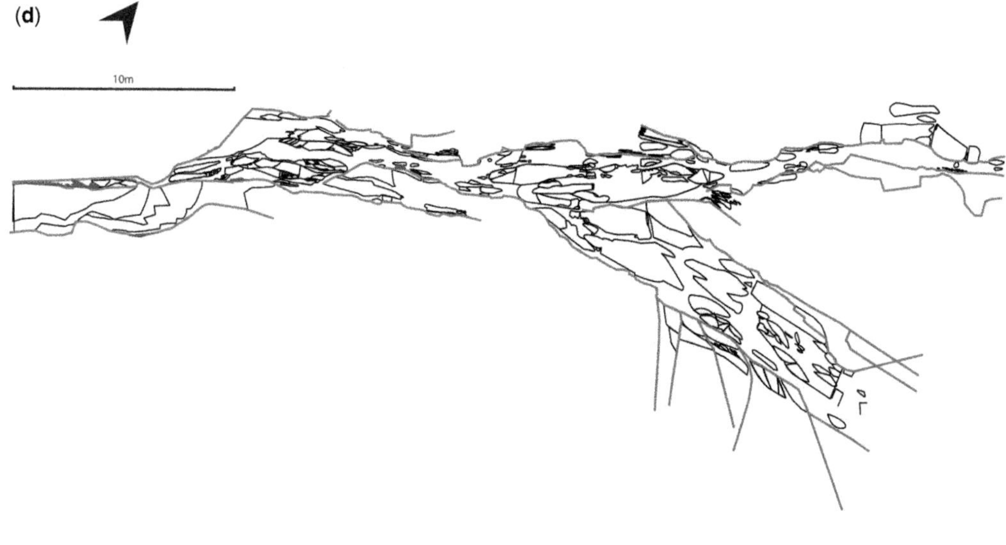

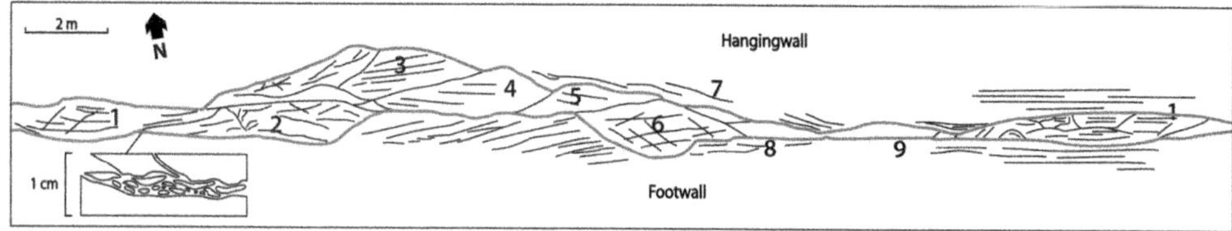

Fig. 5. *Continued.*

Westphalian age down to the SE. One of these faults displays a normal offset of a minimum of 10 m. This fault is exposed on the shore for a distance of 100 m along strike (Fig. 6e, d), where the strandflat has a vertical relief of approximately 3 m. The fault can also be inspected in the steep cliff inboard the strandflat in a 5–7 m-high section.

The master fault is characterized by a fault core varying in horizontal width between 0.5 and 5 m and consists of several soft-linked horizontal segments. Several fault branch lines are developed in the hanging-wall fault block. The hanging-wall and footwall branch faults are partly characterized by fault rocks of the cataclacite–breccia series, and are enveloped by footwall and hanging-wall damage zones with fractures and deformation bands. Orientations of minor faults within each segment and their branch lines are very consistent (Fig. 7), whereas the dimensions of fault lenses in each branch vary considerably, so that the wider fault segments contain the longest and thickest lenses. The lenses are separated by zones of intense shear, and by accumulations of fractures and fault rocks, and some lenses are in the stage of mechanical disintegration (Fig. 6e). The average a:c ratio of the fault 129 core lenses is 6.3:1.

Well-consolidated limestones with sandstone and shale beds: Kilve Beach, Bristol Channel

The Kilve Beach area is situated on the southern margin of the east–west-trending Bristol Channel Basin (Fig. 8a), sitting on a basement of Carboniferous limestones and Devonian sandstones and slates. The basin formed during the Permian (Anderton *et al.* 1979), Early Jurassic and Late Jurassic–Early Cretaceous (Kamerling 1979; Holloway & Chadwick 1986; Karner *et al.* 1987; Skar, this volume, in press). It was later inverted and reactivated during Cretaceous and Tertiary times with the formation of reverse and strike-slip faults. The faults in the Kilve area cut through calcareous and organic-rich shales, and laterally continuous limestone beds with a thickness of less than 1 m.

Owing to its excellent exposures and easy accessibility, the Bristol Channel area has been studied intensively by several research groups focusing on fault evolution and fracture systems. The extensional structures associated with Mesozoic rifting have been investigated for analyses of displacement, segment linkage and relay ramps (Peacock & Sanderson 1991, 1993; Kim *et al.* 2004), oversteps and bends along normal faults (Peacock & Sanderson

Fig. 6. Fault-core architecture of the Howick, Hartley Staps and Snab Point faults. (**a**) The core of the Howick Fault: the main structural elements are marked in red. (**b**) Fault-core lens and zone of fault gouge delineating the hanging-wall branch line, Hartley Steps Fault. (**c**) Fault-core lenses in the Howick Bay Fault. The measuring tape is 1 m. (**d**) One segment (segment 1.2) of the Snab Point Fault seen along strike (the measuring band for scale is 25 cm long). (**e**, **f**) Larger lenses in the process of mechanical disintegration to fault breccia in the Snab Point Fault (the measuring stick is 1 m).

1993), small faults for predicting fault distributions (Steen *et al.* 1998), normal fault array polarity and detachments (Stewart & Argent 2000), and fault patterns at different scales (Peacock 1996). Other studies include analyses of fault tip zones (Pickering *et al.* 1997), fault slip evolution (Davison 1995),

Fig. 6. *Continued.*

pull-apart basins, shear fractures and pressure solution (Peacock & Sanderson 1995), and linkage and evolution of conjugate strike-slip faults (Kelly *et al.* 1998).

The fault cores of the Kilve Beach faults display a geometry that is distinctly different from that described above from Northumberland and Bornholm. The faults of Kilve Beach are characterized

Fig. 6. *Continued.*

by distinguished calcite mineralization: the calcite being accumulated into bands parallel to the master fault surfaces, and filling the high-strain zones between the extremely extended remains of the fault lenses (Fig. 8b, d). The calcite bands are interbedded with thin, shaly layers, indicating that shale smearing occurred contemporaneously with growth and dynamic recrystallization of calcite, presumeably under high fluid pressure. In this process, fault-core lenses probably became sheared to membranes intercalated with calcite bands, making the potential lens origin of the layers indistinguishable.

Fault-core dimensions vary considerably in cores of the master faults, with an average of 12.5:1. However, calcite-cemented faults with extreme *a*:*c* ratios of 20:1 are not uncommon.

Faulted Proterozoic gneiss: the Island of Frøya, mid Norway

Extensional faults in quartzo-feldspathic migmatitic gneisses of Proretozoic origin reworked during the Caledonian Orogeny and later affected by post-Caledonian extension were studied at the Island of Frøya, mid Norway (Fig. 9a) (Grønlie & Roberts

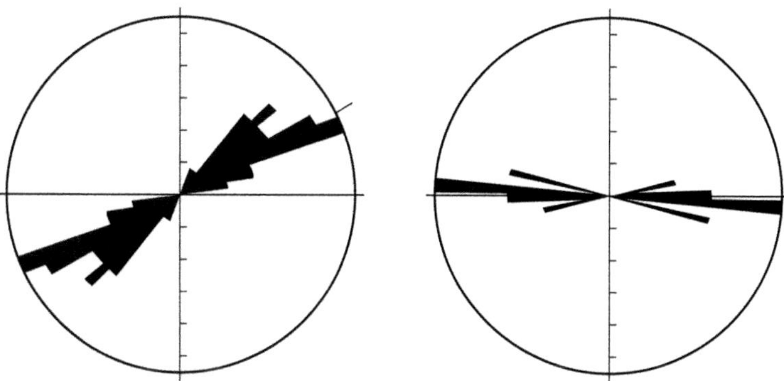

Fig. 7. Orientation diagram (*b* axes: fault-core lens width), Snab Point.

1989; Redfield *et al.* 2005). Fault lenses are located in the cores of nearly vertical to slightly inclined (80° dip) faults, with slickenside linations suggesting dip-slip normal movements and slight strike-slip rectivation. The vertical separation is estimated to be less that 10 m (Lindanger *et al.* 2007). The fault-core lenses are separated by high-strain zones containing fault gouge and ultrabreccia, and are characterized by internal shear zones indicating further disintegration (Fig. 9b, c).

The host rock of Frøya represents the mechanically strongest rock affected by normal faulting in the present analysis. The average *a*:*c* ratio for the entire fault lens population here is 17:1 (*c.* 30 lenses: Lindanger *et al.* 2007).

The geometry, dimensions and development of fault-core lenses

In most mesoscopic extensional faults examined by us, the principal building blocks of the fault core are extensional lozenge-shaped bodies that sometimes are stacked to generate complete duplexes (see also Gibbs 1983, 1984; Woodcock & Fischer 1986; Gabrielsen & Koestler 1987; Cox & Scholz 1988; Cruikshank *et al.* 1991; Childs *et al.* 1997; Gabrielsen & Clausen 2001; Lindanger *et al.* 2004, 2007). These horses may consist of undeformed to heavily deformed and fractured protolith (see also Koestler & Ehrmann 1991), in which primary bedding and structures can sometimes be identified. Such features may be stacked together with units that entirely consist of fault rocks. The lenses may be derived both from the footwall and the hanging wall by different mechanisms such as fault-branch splaying and tip-line coalescence, segment linkage, or asperity bifurcation within the fault core proper or in its immediate vicinity of the footwall and hanging wall (Huggins *et al.* 1995; Childs *et al.* 1997; Gabrielsen & Clausen 2001). The roots of the fault-core lenses cannot always be determined. The process of lens formation is sometimes seemingly promoted by uneven fault traces and ramps (e.g. Walsh & Watterson 1991; Bruhn *et al.* 1994; Gabrielsen & Clausen 2001; Lindanger *et al.* 2004), and such mechanisms can also be demonstrated to be active on the micro-scale in very weak rocks (Clausen & Gabrielsen 2002). Fault-core lenses may occasionally consist of bodies of fault rocks of the cataclasite and even mylonite series, particularly in heavily faulted metamorphic rocks (Gabrielsen & Braathen 2014). Internal fracture zones, intrinsic shear zones in the fault-core lenses, local highly strained lens margins and fracture accumulations at lens tips suggest that fault-core lenses disintegrate by shear-supported segmentation, and by abrasion of core edges and tips. Several examples of fault-core lenses in the process of such disintegration have been encountered in the present field investigation. It is realized, however, that brittle fracture systems in the lenses may be remnants from earlier stages of brittle deformation. Still, fracture accumulations parallel to the anticipated plane of maximum shear and in some of the lens tips indicate that intense fracturing also occurs in connection with the transport and collapse of the lenses.

The horses are separated by high-strain zones of different types. The high-strain zone may include zones of clay smear, as seen, for example, in Bornholm and at Howick Bay (see earlier), shale gouge (Davatzes & Aydin 2005), regular fault gouge, breccia/cataclasite (Sibson 1977; Braathen *et al.* 2004) and different types of mineralogically altered zones (Chester & Logan 1986; Bruhn *et al.* 1994; Caine *et al.* 1996). Also slip zones, which may encompass swarms of deformation bands (Aydin 1978; Aydin & Johnson 1978; Berg & Skar 2005; Shipton *et al.* 2005) and slip surfaces (Aydin & Johnson 1983; Chester & Logan 1986; Fossen & Hesthammer 2000; Berg & Skar 2005; Fossen

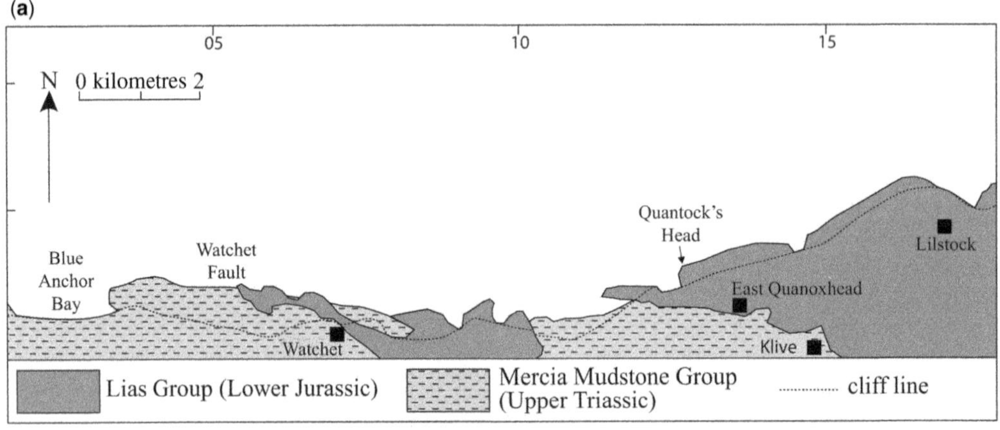

Fig. 8. (**a**) Geological map, Kilve Beach area (redrawn from www.thegcr.org.uk). (**b**) Lens in the extensional fault. Note the difference in strain intensity (faulting and fracturing) in the thin lens on the footwall side compared to that of the thicker lens on the hanging-wall side (total thickness of the lens system is *c.* 120 cm). (**c**) Fault lens enveloped by mineralized (calcite) high-strain zones (the rucksack is for scale). (**d**) Along-strike section in a fault-core lens (the people are for scale).

& Bale 2007; Fossen *et al.* 2007), may occur. Thus, extensive surfaces of intense shear are sometimes found in the fault core. More commonly than not, such surfaces coincide with the borders of the inner core, but they may also define the outer borders of the distal zones of the fault core (Fig. 1). This is in concert with the findings of Heynekamp *et al.* (1999) and Davatzes & Aydin (2005), who identified a zone of high strain (concentration of deformation bands) at the border between the core and the damage zone/mixed zone.

In faults studied by us, the fault lenses are mainly orientated with the longest axes parallel or subparallel to the regional plane of the master fault. Study of the variance in geometry of fault-core lenses has also revealed a robust geometrical shape. The geometry (*a*:*c* ratio: Fig. 10a) suggests that fault-core lenses in mechanically weak lithologies are shorter relative to the length of the fault core (smaller *a*:*c* ratio) than that of mechanically stronger rocks. Thus, by studying fault-core lenses from plaster experiments, faults in poorly consolidated siliciclastic rocks, and gneisses of Precambrian and Caledonian age, Lindanger *et al.* (2007) found that a relationship between the *a*:*b*:*c* axes with a ratio of 1:9:10 is common. This is in harmony with the data provided by the present study, which found an *a*:*c* ratio of between 1:4 and 1:9 for poorly to moderately well-consolidated sedimentary rocks. Exceptions to this seem to be for fault cores with

(c)

(d)

Fig. 8. *Continued.*

assumed high fluid pressure and associated with syn-tectonic mineralization, and particularly in cases where mineralization includes minerals with a high potential for dynamic recrystallization (e.g. calcite), or where strong clay smearing has occurred. This picture is somewhat blurred by that the disintegration of fault-core lenses by shear splitting, which produces a second (or third) generation of shorter and thicker lenses (Fig. 10b).

Experiments confirm the highly dynamic nature of the fault-core zone (Gabrielsen & Clausen 2001; Clausen & Gabrielsen 2002; Lindanger *et al.* 2004), and it seems likely that strain-softening and strain-hardening processes contribute to the development of such features (Gabrielsen & Braathen 2014). In cases of strain hardening, the development of extensional horses and duplexes may significantly contribute to fault-zone widening, whereas strain softening would prevent this occurring. The dynamics of the fault core sometimes include internal shearing and subsequent collapse of early generations of fault-core lenses. This contributes to the relative thickening of the average thickness of the lenses, and, perhaps, thinning and even complete collapse of the fault core itself into one single shear plane (Lindanger *et al.* 2007; Childs *et al.* 2009). Such processes can also be observed on the micro-scale in ring-shear experiments when applying poorly consolidated sediments of great lithological contrast (sand and clay: Clausen & Gabrielsen 2002).

The complex and dynamic picture that characterizes the fault core is of great importance for fluid-flow modelling of faulted reservoirs in cases where it is important to assess the influence of the individual fault. One particular problem is in assessing the effect of the contrast of permeability that commonly exits between the fault-core lenses and the high-strain zones that separate them. In most cases studied by us, the high-strain zones consist of material that is either clay- or mica-rich, or that is characterized by severe grain-size reduction. In gneisses, the deformation product commonly consists of different types of fault rocks. In all of these cases, the high-strain zones represent low-permeability membranes, and fluid communication would depend on fluids flowing either through holes in the membranes or through open pore networks in connecting lenses, or through younger open fractures.

The complexity that characterizes even simple extensional faults in sedimentary rocks poses a considerable challenge in the analysis of such structures in general and for the assessment of the effect of such structures on fluid transport. It is common to assume that the main fluid transport is associated with damage zones, and that the fault core is generally less permeable (Sorkhabi & Tsuji 2005). The great influence of the variance in parameters such as lithology, principal type of stress field, deformation processes, conditions of fracturing and strain (depth of burial), and the fault architecture itself makes it important to find

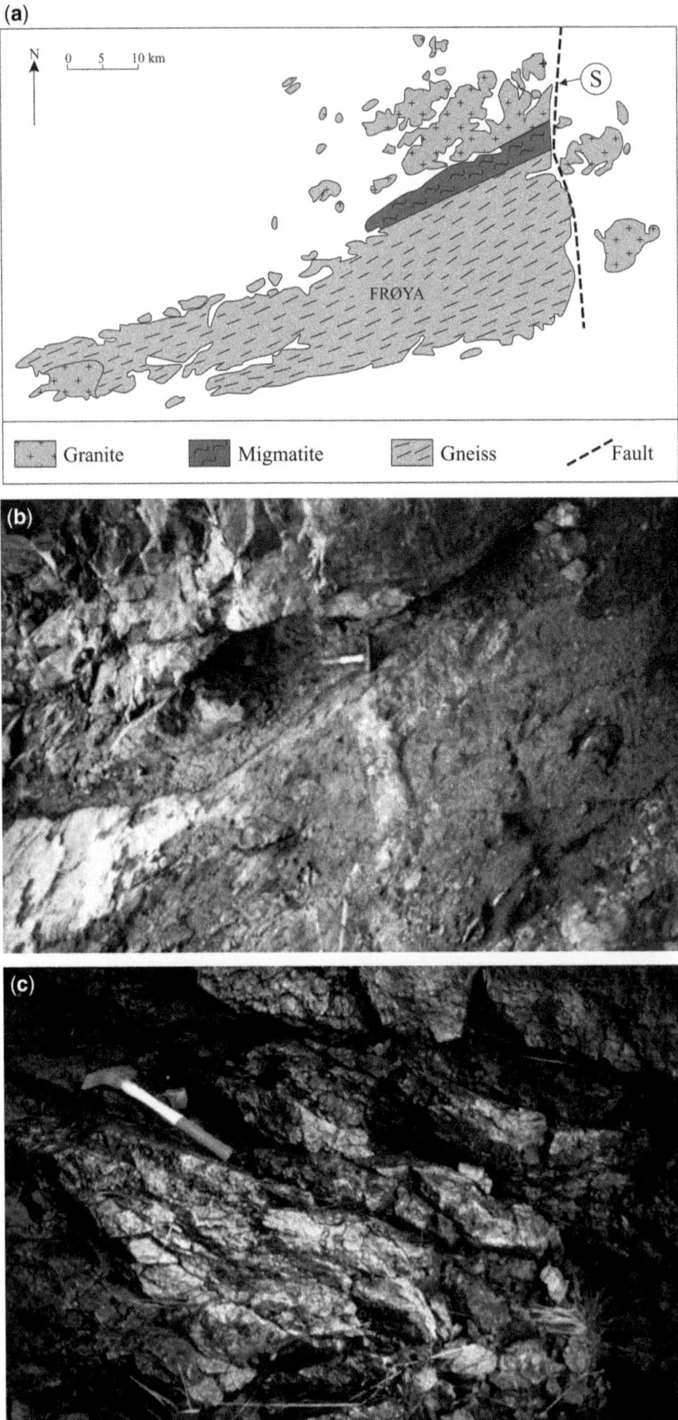

Fig. 9. (**a**) Geological map, island of Frøya, mid Norway (after Torske 1983). (**b**) Fault-core lens, marked by the hammer, consisting of country rock (upper part) and fault rock (lower part). (**c**) Fault-core lens at the stage of disintegration. Photographs taken in gneiss affected by steep faults in gneisses, island of Frøya, coastal mid Norway.

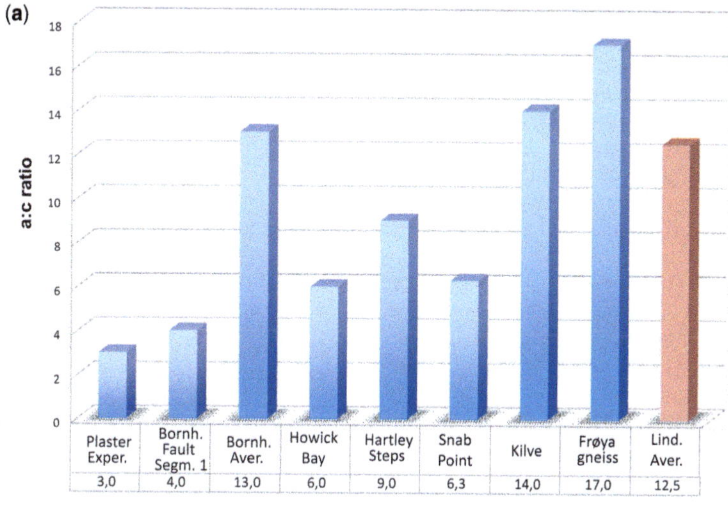

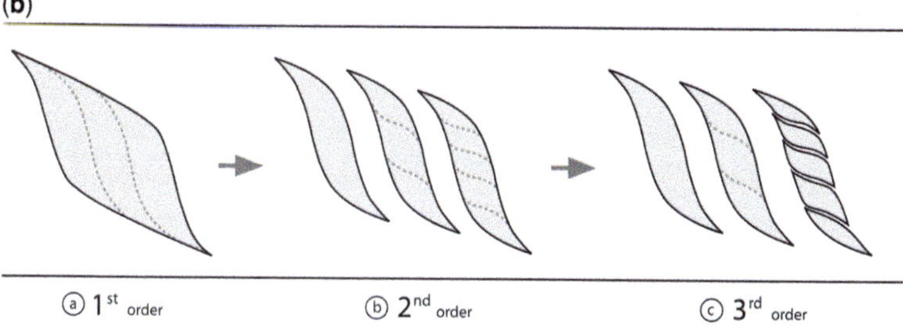

Fig. 10. (**a**) *a*:*c* ratio for faults in the present study arranged by increasing mechanical strength. The highest *a*:*c* ratios are found for mechanically weak geological materials. The Bornholm sequence is an exception to this (except for some fault branches). The column to the right (red) is the average of all fault-core lenses calculated by Lindanger *et al.* (2007). (**b**) Disintegration of a fault-core lens by two stages of shearing results in thicker and shorter lenses (modified after Lindanger *et al.* 2007). Note that other notations of principal axes are applied here compared to Lindanger *et al.* (2007).

predictive methods for assessing their impact on reservoir models.

Conclusions

In the present work, we have tried to point to some of the general characteristics of fault cores that need further analysis:

- The fault core has a complex architecture that commonly changes along the strike of the fault, as well as with depth.
- Lenses are commonly the dominant structural elements in fault cores at a stage overstepping a certain amount of displacement and under certain conditions. The dimensions of the fault-core lenses are relatively stable, commonly showing an *a*:*c* ratio of between 1:4 and 1:15 in the fault cores studied by us. Exceptions to this seem common in fault cores of mineralization, suggesting the influence of high fluid pressure.
- The lenses are commonly separated by high-strain zones that are invaded by clay (clay smear) or fault rocks (fault gouge).

The high-strain zones are likely to reduce the strike-parallel permeability of the fault core, although this can be modified in cases where fault-core lenses are fractured. The connectivity of fracture systems inside the fault cores is likely to be restricted.

The constitution of the high-strain zones seems to be strongly dependent on the types of lithology involved, the depth at which deformation occurred, strain rate and degree of clay smear, cataclasis, fluid pressure, mineralization and dynamic recrystallization, and reactivation. Thus, in the examples from Bornholm cited earlier, the main mechanism for development of high-strain zones is clay smearing (Clausen *et al.* 2003), whereas the principal mechanism in the examples seen in Kilve is that of grain-size reduction, shearing and dynamic recrystallization (Berg & Skar 2005). Brittle fractures, which are commonly seen in fault cores, may be remnants from an early stage of deformation of the brittle constituents of the zone, but may also represent a later stage of strain associated with mechanical disintegration of the fault lenses. Determining the relative timing of such fracturing processes may be of importance because the pattern of fracture interconnection and, hence, fluid communication in the fault core may be dramatically different for the two. Owing to the distribution of low-permeability clay or clay gouge, it is likely that communication in the fault core remains low.

We would like to thank Silje Støren Berg, Jill Angelique Clausen, Merethe Lindanger and Tore Skaar for joyful company in the field and in the laboratory, and for many discussions on fault architecture. We are also grateful to the guest editors of the Geological Society of London for support and patience during the preparation of this manuscript, and two anonymous reviewers for useful comments and recommendations that greatly improved the paper. Randi Bäckmark and Hogne Botnen Totland produced the drafted figures.

References

AGOSTA, F. & AYDIN, A. 2006. Architecture and deformation mechanism of a basin-bounding normal fault in Mesozoic platform carbonates, central Italy. *Journal of Structural Geology*, **28**, 1445–1467.

ANDERTON, R., BRIDGES, P.H., LEEDER, M.R. & SELLWOOD, B.W. 1979. *A Dynamic Stratigraphy of the British Isles: A Study in Crustal Evolution*. Allen & Unwin, London.

AYDIN, A. 1978. Small faults formed as deformation bands in sandstone. *Pure and Applied Geophysics*, **116**, 913–913.

AYDIN, A. & JOHNSON, A.M. 1978. Development of faults as zones of deformation bands and slip surfaces in sandstone. *Pure and Applied Geophysics*, **116**, 931–942.

AYDIN, A. & JOHNSON, A.M. 1983. Analysis of faulting in porous swandstones. *Journal of Structural Geology*, **5**, 19–31.

BÆLUM, K., JOHANSEN, T., JOHNSEN, H., RØD, K., RUUD, B.O. & BRAATHEN, A. 2012. Subsurface structures of the Longyearbyen CO2 Lab study area in Central Spitsbergen (Arctic Norway), as mapped by reflection seismic data. *Norwegian Journal of Geology*, **92**, 377–389.

BASTESEN, E., BRAATHEN, A., NØTTVEIT, H., GABRIELSEN, R.H. & SKAR, T. 2009. Extensional fault cores in micritic carbonate – case studies from the Gulf of Corinth, Greece. *Journal of Structural Geology*, **31**, 403–420.

BERG, R.R. & AVERY, A.H. 1995. Sealing properties of Tertiary Growth Faults, Texas Gulf Coast. *American Association of Petroleum Geologists Bulletin*, **79**, 375–393.

BERG, S.S. & SKAR, T. 2005. Controls on damage zone asymmetry of a normal fault zone: outcrop analyses of a segment of the Moab fault, SE Utah. *Journal of Structural Geology*, **27**, 1803–1822.

BRAATHEN, A., OSMUNDSEN, P.T. & GABRIELSEN, R.H. 2004. Dynamic development of fault rocks in a crustal-scale detachment: an example from western Norway. *Tectonics*, **21**, TC4010, 1–21.

BRAATHEN, A., TVERANGER, J. ET AL. 2009. Fault facies and its application to sandstone reservoirs. *American Association of Petroleum Geologists Bulletin*, **93**, 891–917.

BRAATHEN, A., BÆLUM, K. ET AL. 2012. The Longyearbyen CO_2 lab of Svalbard, Norway – initial assessment of the geological conditions for CO_2 sequestration. *Norwegian Journal of Geology*, **92**, 353–376.

BREDEHOEFT, J.D., BACK, W. & HANSHAW, B.B. 1982. Regional ground-water flow concepts in the United States; historical perspective. *In*: NARASIMHAN, T.N. (ed.) *Recent Trends in Hydrogeology*. Geological Society of America, Special Papers, **189**, 297–316.

BRETAN, P., YIELDING, G., MATHIASSEN, O.M. & THORSNES, T. 2011. Fault-seal analysis for CO_2 storage: an example from the Troll area, Norwegian Continental Shelf. *Petroleum Geoscience*, **17**, 181–192, https://doi.org/10.1144/1354-079310-025

BRUHN, R.L., PARRY, W.T., YONKEE, W.A. & THOMPSON, T. 1994. Fracturing and hydrothermal alteration in normal fault zones. *Pure and Applied Geophysics*, **142**, 609–644.

CAINE, J.S., EVANS, J.P. & FORSTER, C.B. 1996. Fault zone architecture and permeability structure. *Geology*, **24**, 1025–1028.

CAINE, S.C. & FORSTER, B.B. 1999. Fault zone architecture and fluid flow: insigths from field data on numerical modeling. *In*: HANEBERG, W.C., MOSLEY, P.S., MOORE, J.C. & GOODWIN, L.B. (eds) *Faults and Subsurface Flow in the Shallow Crust*. American Geophysical Union, Geophysical Monographs, **113**, 101–127.

CHESTER, F.M. & LOGAN, J.M. 1986. Implications for mechanical properties of brittle faults from observations of the Punchbowl Fault Zone, California. *Pure and Applied Geophysics*, **124**, 79–106.

CHILDS, C., WATTERSON, J. & WALSH, J.J. 1996. A model for the structure and development of fault zones. *Journal of the Geological Society, London*, **153**, 337–340, https://doi.org/10.1144/gsjgs.153.3.0337

CHILDS, C., WALSH, J.J. & WATTERSON, J. 1997. Complexity in fault zone structure and implications for fault seal prediction. *In*: MØLLER-PEDERSEN, P. & KOESTLER, A.G. (eds) *Hydrocarbon Seals: Importance for Exploration and Production*. Norwegian Petroleum Society, Special Publications, **7**, 61–72.

CHILDS, C., MANZOCCHI, T., WALSH, J.J., BONSON, C.G., NICOL, A. & SCHÖPFER, M.P.J. 2009. A geometric model of fault zone and fault rock thickness variations. *Journal of Structural Geology*, **31**, 117–127.

CLAPP, F.G. 1910. A proposed classification of petroleum and natural gas fields based on structure. *Economic Geology*, **5**, 503–521.

CLAUSEN, J.A. & GABRIELSEN, R.H. 2002. Parametres that control the development of clay smear at low stress states: an experimental study using ring-shear apparatus. *Journal of Structural Geology*, **24**, 1569–1586.

Clausen, J.A., Gabrielsen, R.H., Johnsen, E. & Korstgård, J. 2003. Fault architecture and clay smear distribution. Examples from field studies and drained ring-shear experiments. *Norwegian Journal of Geology*, **83**, 131–146.

Collier, R.E.Ll. 1989. Tectonic evolution of the Northumberland Basin: the effects of renewed extension upon an inverted extensional basin. *Journal of the Geological Society, London*, **146**, 621–634, https://doi.org/10.1144/gsjgs.146.6.0981

Cox, S.J.D. & Scholz, C.H. 1988. An experimental study of shear fractures in rocks; Machanical observations. *Journal of Structural Geology*, **14**, 1133–1148.

Cruikshank, K.M., Zhao, G. & Johnson, A.M. 1991. Duplex structures connecting fault segments in Entrada sandstone. *Journal of Structural Geology*, **13**, 1186–1196.

Davatzes, N.C. & Aydin, A. 2005. Distribution and nature of fault architecture in a layered sandstone and shale sequence: An example from the Moab Fault, Utah. *In*: Sorkhabi, R. & Tsuji, Y. (eds) *Faults, Fluid Flow, & Petroleum Traps*. American Association of Petroleum Geologists Memoirs, **85**, 153–180.

Davison, I. 1995. Fault slip evolution determined from crack seal veins in pull-aparts and their implications for general slip models. *Journal of Structural Geology*, **17**, 1025–1034.

De Paola, N., Holdsworth, R.E., MvVaffrey, K.J.W. & Barchi, M.R. 2005. Partitioned transtension: an alternative to basin inversion models. *Journal of Structural Geology*, **27**, 607–625.

Evans, J.P., Forster, C.B. & Goddard, J.V. 1997. Permeability of fault-related rocks, and implications for hydraulic structure of fault zones. *Journal of Structural Geology*, **19**, 1393–1404.

Fachri, M., Tveranger, J., Cardozo, N. & Pettersen, Ø. 2011. The impact of fault envelope structures on fluid flow: a screening study using fault facies. *American Association of Petroleum Geologists Bulletin*, **95**, 619–648.

Fachri, M., Rotevatn, A. & Tveranger, J. 2013. Fluid flow in relay zones revisited: towards an improved representation of small-scale structural hetereogeneties in flow models. *Marine and Petroleum Geology*, **46**, 144–164.

Farmer, N. & Jones, J.M. 1969. The Carboniferous, Namurian of the coastal section from Howick Bay to Foxton Hall. *Transactions of the Natural History Society of Northumberland, Durham and Newcastle-Upon-Tyne*, **17**, 1–27.

Flodin, E., Gerde, M., Aydin, A. & Wiggins, W.D. 2005. Petrophysical properties and sealing capacity of fault rock, Aztec Sandstone, Nevada. *In*: Sorkhabi, R. & Tsuji, Y. (eds) *Faults, Fluid Flow, & Petroleum Traps*. American Association of Petroleum Geologists Memoirs, **85**, 197–217.

Fossen, H. & Bale, A. 2007. Deformation bands and their influence on fluid flow. *American Association of Petroleum Geologists Bulletin*, **91**, 1685–1700.

Fossen, H. & Gabrielsen, R.H. 1996. Experimental modeling of extensional fault systems by use of plaster. *Journal of Structural Geology*, **18**, 673–687.

Fossen, H. & Hesthammer, J. 2000. Possible absence of small faults in the Gullfaks Field, northern North Sea: implications for downscaling of faults in some porous sandstones. *Journal of Structural Geology*, **22**, 851–863.

Fossen, H., Schultz, R.A., Shipton, Z.K. & Mair, K. 2007. Deformation bands in sandstone: a review. *Journal of the Geological Society, London*, **164**, 755–769, https://doi.org/10.1144/0016-76492006-036

Fredman, N.J., Tveranger, J. *et al.* 2008. Fault facies modeling: technique, and approach for 3D conditioning and modeling of faulted grids. *American Association of Petroleum Geologists Bulletin*, **92**, 1–22.

Færseth, R.B. 2006. Shale smear along large faults: continuity of smear and the fault seal capacity. *Journal of the Geological Society, London*, **163**, 741–751, https://doi.org/10.1144/0016-76492005-162

Færseth, R.B., Johnsen, E. & Sperrevik, S. 2007. Methodology fo risking fault seal and capacity; Implications of fault zone architecture. *American Association of Petroleum Geologists Bulletin*, **91**, 1231–1245.

Gabrielsen, R.H. 2010. The structure and hydrocarbon traps of sedimentary basins. *In*: Bjørlykke, K. (ed.) *Petroleum Geoscience: From Sedimentary Environments to Rock Physics*. Springer, Berlin, 299–327.

Gabrielsen, R.H. & Braathen, A. 2014. Models of fracture lineaments – Joint swarms, fracture corridors and faults in crystalline rocks, and their genetic relations. *Tectonophysics*, **628**, 26–44, https://doi.org/10.1016/j.tecto.2014.04.022

Gabrielsen, R.H. & Clausen, J.A. 2001. Horses and duplexes in extensional regimes: A scale modeling contribution. *In*: Koyi, H.A. & Mancktelow, N.S. (eds) *Tectonic Modeling: A Volume in Honor of Hans Ramberg*. Geological Society of America Memoirs, **193**, 219–233.

Gabrielsen, R.H. & Koestler, A.K. 1987. Description and structural implications of fractures in late Jurassic sandstones of the Troll Field, northern North Sea. *Norsk Geologisk Tidsskrift*, **67**, 371–381.

Gibbs, A.D. 1983. Balanced cross-section construction from seismic sections in areas of extensional tectonics. *Journal of Structural Geology*, **5**, 153–160.

Gibbs, A.D. 1984. Structural evolution of extensional basin margins. *Journal of the Geological Society, London*, **141**, 609–620, https://doi.org/10.1144/gsjgs.141.4.0609

Gravesen, P. 1982. Lower Cretaceous sedimentation and basin extension on Bornholm, Denmark. *Danmarks Geologiske Undersøgelse Årbog*, 1981, 73–99.

Gravesen, P. 1996. *Geologisk set. Bornholm. En beskrivelse af områder af national geologisk interesse*. Geografiforlaget.

Grønlie, A. & Roberts, D. 1989. Resurgent strike-slip duplex development along the Hitra-Snåsa and Verran Faults, Møre–Trøndelag Fault Zone. *Journal of Structural Geology*, **11**, 295–305.

Hamann, N.E. 1988. Mesozoikum–Bornholms Geologi I. *Varv*, **2**, 64–75.

Heynekamp, M.R., Goodwin, L.B., Mozley, P.S. & Haneberg, W.C. 1999. Controls on fault-zone architecture in poorly lithified sediments, Rio Grande Rift, New Mexico: implications for fault-zone permeability and fluid flow. *In*: HANEBERG, W.C., MOZLEY, P.S., CASEY MOORE, J. & GOODWIN, L.B. (eds) *Faults and Subsurface Fluid Flow in the Shallow Crust*. American Geophysical Union, Geophysical Monographs, **113**, 27–49.

HOLLOWAY, S. & CHADWICK, R.A. 1986. The Sticklepath-Lustleigh faults zone: tertiary sinistral reactivation of a Variscan dextral strike-slip fault. *Journal of the Geological Society, London*, **143**, 447–452, https://doi.org/10.1144/gsjgs.143.3.0447

HUGGINS, P., WATTERSON, J., WALSH, J.J. & CHILDS, C. 1995. Relay zone geometry and displacement transfer between normal faults recorded in coal-mine plans. *Journal of Structural Geology*, **17**, 1741–1756.

JOHNSON, G.A.L. 1984. Subsidence and sedimentation in the Northumberland Trough. *Proceedings of the Yorkshire Geological Society*, **45**, 71–83, https://doi.org/10.1144/pygs.45.1-2.71

JONES, J.M. 1968. The geology of the coast section from Tynesmouth to Seaton Sluice. *Transactions of the Natural History Society of Northumberland, Durham and Newcastle-Upon-Tyne*, **16**, 153–192.

KARNER, G.D., LAKE, S.D. & DEWEY, J.F. 1987. The thermal and mechanical development of the Wessex Basin, southern England. *In*: COWARD, M.P., DEWEY, J.F. & HANCOCK, P.L. (eds) *Continental Extensional Tectonics*. Geological Society, London, Special Publications, **28**, 517–536, https://doi.org/10.1144/GSL.SP.1987.028.01.34

KAMERLING, P. 1979. The geology and hydrocarbon habitat of the Bristol Channel Basin. *Journal of Petroleum Geology*, **2**, 75–93.

KELLY, P.G., SANDERSON, D.J. & PEACOCK, D.C.P. 1998. Linkage and evolution of conjugate strike-slip fault zones in limestones of Somerset and Northumbria. *Journal of Structural Geology*, **20**, 1477–1493.

KIM, Y.-S., PEACOCK, D.C.P. & SANDERSON, D.J. 2004. Fault damage zones. *Journal of Structural Geology*, **26**, 503–507.

KIMBELL, G.S., CHADWICK, R.A., HOLLIDAY, D.W. & WERNGREN, O.C. 1989. The structure and evolution of the Northumberland Trough from new seismic reflection data and its bearing on modes of continental extension. *Journal of the Geological Society, London*, **146**, 775–787, https://doi.org/10.1144/gsjgs.146.5.0775

KNOTT, S.D. 1994. Fault zone thickness v. displacement in the Permo-Triassic sandstones of NW England. *Journal of the Geological Society, London*, **151**, 17–25, https://doi.org/10.1144/gsjgs.151.1.0017

KOESTLER, A.G. & EHRMANN, W.U. 1991. Description of brittle extensional features in chalk on the crest of a salt ridge (NW Germany). *In*: ROBERTS, A.M., YIELDING, G. & FREEMAN, B. (eds) *The Geometry of Normal Faults*. Geological Society, London, Special Publications, **56**, 113–123, https://doi.org/10.1144/GSL.SP.1991.056.01.08

LINDANGER, M., GABRIELSEN, R.H. & BRAATHEN, A. 2007. Analysis of rock lenses in extensional faults. *Norwegian Journal of Geology*, **87**, 361–372.

LINDANGER, R., ØYGAREN, M., GABRIELSEN, R.H., MJELDE, R., RANDEN, T. & TJØSTHEIM, B.A. 2004. Analogue (plaster) modelling and synthetic seismic representation hangingwall fault blocks above of ramp-flat ramp faults. *First Break*, **22**, 22–30.

MANZOCCHI, T.J., WALSH, J.J., NELL, P. & YIELDING, G. 1999. Fault transmissibility multipliers for flow simulation models. *Petroleum Geoscience*, **5**, 53–63, https://doi.org/10.1144/petgeo.5.1.53

MCKNIGHT, E.T. 1940. *Geology of area between Green and Colorado rivers, Grand and San Juan Counties, Utah*. United States Geological Survey, Bulletin, **908**.

PEACOCK, D.C.P. 1996. Field examples of variations in fault patterns at different scales. *Terra Nova*, **8**, 361–371.

PEACOCK, D.C.P. & SANDERSON, D.J. 1991. Displacements, segment linkage and relay ramps in normal fault zones. *Journal of Structural Geology*, **13**, 721–733.

PEACOCK, D.C.P. & SANDERSON, D.J. 1993. Estimating strain from fault slip using a line sample. *Journal of Structural Geology*, **15**, 1513–1516.

PEACOCK, D.C.P. & SANDERSON, D.J. 1995. Strike-slip relay ramps. *Journal of Structural Geology*, **17**, 1351–1360.

PICKERING, G., PEACOCK, D.C.P., SANDERSON, D.J. & BULL, J.M. 1997. Modelling tip zones to predict the throw and length characteristics of faults. *American Association of Petroleum Geologists Bulletin*, **81**, 82–99.

REDFIELD, T.R., BRAATHEN, A., GABRIELSEN, R.H., OSMUNDSEN, P.T., TORSVIK, T.H. & ANDRIESSEN, P.A.M. 2005. Late Mesozoic to Early Cretaceous components of vertical separation across the Møre-Trøndelag Fault Complex, Norway. *Tectonophysics*, **395**, 233–249.

ROBERTSON, E.C. 1983. Relationship of fault diaplecement to gouge and breccia thickness. *Mining Engineering*, **35**, 1426–1432.

SCHUELLER, S., BRAATHEN, A., FOSSEN, H. & TVERANGER, J. 2012. Fault damage zones of extensional faults in porous sandstone: spatial distribution of deformation bands. *Journal of Structural Geology*, **52**, 148–162.

SHIPTON, Z.K., EVANS, J.P. & THOMPSON, L.B. 2005. The geometry and thickness of deformation-band fault core and its influence on sealing characteristics of deformation-band fault zones. *In*: SORKHABI, R. & TSUJI, Y. (eds) *Faults, Fluid Flow, & Petroleum Traps*. American Association of Petroleum Geologists, Memoir, **85**, 181–195.

SIBSON, R.H. 1977. Fault rocks and fault mechanisms. *Journal of the Geological Society, London*, **133**, 191–213, https://doi.org/10.1144/gsjgs.133.3.0191

SIGDA, J.M., GOODWIN, L.B., MOZLEY, P.S. & WILSON, J.L. 1999. Permeablility alteration in small displacement faults in poorly lithified sediments. *In*: GOODWIN, L.B., MOZLEY, P.S., MOORE, J.M. & HANEBERG, W.C. (eds) *Fault and Subsurface Fluid Flow in the Sallow Crust*. American Geophysical Union, Geophysical Monographs, **113**, 51–68.

SIMS, D.W., MORRIS, A.P., FERRILL, D.A. & SORKHABI, R. 2005. Extensional fault system evolution and reservoir connectivity. *In*: SORKHABI, R. & TSUJI, Y. (eds) *Faults, Fluid Flow, & Petroleum Traps*. American Association of Petroleum Geologists Memoirs, **85**, 79–93.

SKAR, T., BERG, S.S., GABRIELSEN, R.H. & BRAATHEN, A. In press. Fracture networks of normal faults in low permeable sedimentary rocks: Examples from Kilve Beach, SW England. *In*: CHILDS, C., HOLDSWORTH, R.E., JACKSON, C.A.-L., MANZOCCHI, T., WALSH, J.J. & YIELDING, G. (eds) *The Geometry and Growth of Normal Faults*. Geological Society, London, Special Publications, **439**, https://doi.org/10.1144/SP439.10

SORKHABI, R. & TSUJI, Y. 2005. The place of faults in petroleum traps. *In*: SORKHABI, R. & TSUJI, Y. (eds) *2005: Faults, Fluid Flow, & Petroleum Traps*. American Association of Petroleum Geologists Memoirs, **85**, 1–31.

STEEN, Ø., SVERDRUP, E. & HANSSEN, T.H. 1998. Predicting the distribution of small faults in a hydrocarbon reservoir by combining outcrop, seismic and well data. *In*: JONES, G., FISHER, Q.J., KNIPE, R.J. (eds) *Faulting, Fault Sealing and Fluid Flow in Hydrocarbon Reservoirs*. Geological Society, London, Special Publications, **147**, 27–50, https://doi.org/10.1144/GSL.SP.1998.147.01.03.

STEWART, S.A. & ARGENT, J.D. 2000. Relationship between polarity of extension fault arrays and presence of detachments. *Journal of Structural Geology*, **22**, 693–711.

TORSKE, T. 1983. A fluidization breccia in granite at Skaget, Svellingen, Frøya. *Norges Geologiske Undersøkelse*, **380**, 107–123.

VEJBÆK, O.V., STOUGE, S. & POULSEN, K.D. 1994. Palaeozoic tectonic and sedimentary evolution and hydrocarbon prospectively in Bornholm area. *Danmarks Geologiske Undersøgelse serie A*, **34**, 4–2.

WALLACE, R.E. & MORRIS, H.T. 1979. *Characteristics of Faults and Shear Zones as Seen in Mines at Depths as much as 2.5 km Below the Surface*. United States Geological Survey, Open File Report **79-1239**.

WALSH, J.J. & WATTERSON, J. 1991. Geometric and kinematic coherence and scale effects in normal fault systems. *In*: ROBERTS, A.M., YIELDING, G. & FREEMAN, B. (eds) *The Geometry of Normal Faults*. Geological Society, London, Special Publications, **56**, 193–203, https://doi.org/10.1144/GSL.SP.1991.056.01.13

WEEKS, L.G. (ed.) 1958. *Habitat of Oil*. American Association of Petroleum Geologists, Tulsa, OK.

WESTOLL, T.S., ROBSON, D.A. & GREEN, R. 1955. A guide to the geology of the district around Alnwick, Northumberland. *Proceedings of the Yorkshire Geological Society*, **30**, 61–100, https://doi.org/10.1144/pygs.30.1.61

WHITTEN, E.H.T. 1966. *Structural Geology of Folded Rocks*. Rand McNally & Co., Chicago, IL.

WIBBERLEY, C.A.J., YIELDING, G. & DI TORO, G. 2008. Recent advances in the understanding of fault zone internal structure: a review. *In*: WIBBERLEY, C.A.J., KURZ, W., IMBER, J., HOLDSWORTH, R.E. & COLLETTINI, C. (eds) *The Internal Structure of Fault Zones: Implications for Mechanical and Fluid Flow Properties*. Geological Society, London, Special Publications, **299**, 5–33, https://doi.org/10.1144/SP299.2

WOODCOCK, N.H. & FISCHER, M. 1986. Strike-slip duplexes. *Journal of Structural Geology*, **8** , 725–735.

YIELDING, G., LYKAKIS, N. & UNDERHILL, J.R. 2011. The role of stratigraphic juxtaposition for seal integrity in proven CO_2 fault-bound traps of the Southern North Sea. *Petroleum Geoscience*, **17**, 193–203, https://doi.org/10.1144/1354-0793/10-026

Widening of normal fault zones due to the inhibition of vertical propagation

V. ROCHE[1]*, C. HOMBERG[2], M. VAN DER BAAN[1] & M. ROCHER[3]

[1]*Department of Physics, CCIS, University of Alberta, Edmonton, Alberta, Canada T6G 2E1*

[2]*UPMC, Université Paris 06, ISTEP, UMR 7193, 4 Place Jussieu, 75252 Paris Cedex 05, France*

[3]*IRSN, Institut de Radioprotection et de Sureté Nucléaire, Fontenay-aux-Roses, France*

**Correspondence: roche@ualberta.ca*

Abstract: In this paper, we document the early stage of fault-zone development based on detailed observations of mesocale faults in layered rocks. The vertical propagation of the studied faults is stopped by layer-parallel faults contained in a weak layer. This restriction involves a flat-topped throw profile along the fault plane and modifications of the fault structures near the restricted tips, with geometries ranging from planar structures to fault zones characterized by abundant parallel fault segments. The 'far-field' displacement (i.e. the sum of the displacement accumulated by all the fault segments and the folding) measured along the restricted faults exhibiting this segmentation may have flat-topped shapes or triangular shapes when fault-related folding is observed above the layer-parallel faults. We develop a model from the observations. In this model, during the course of restriction, a fault forms as a simple isolated planar structure, then parallel fault segments successively initiate to accommodate the increasing displacement. We assume that, eventually, the fault propagates beyond the layer-parallel fault. This model implies first that fault widening is controlled by the fault capacity to propagate vertically in the layered section. Likewise, owing to restriction, fault growth occurs with non-linear increases in maximum displacement, length and thickness.

Slip on normal faults accommodates extension in the brittle crust. The strain may be focused on a narrow fault plane, but is commonly distributed over a wider zone of disturbed rocks between two faulted blocks, called a fault zone (Peacock *et al.* 2000). Fault zones are complex and heterogeneous structures, composed of several components, such as fault rocks (Robertson 1982; Wallace & Morris 1986; Childs *et al.* 2009), interacting fault segments (Segall & Pollard 1980; Childs *et al.* 1996; Walsh *et al.* 2003), fault-related folding (Cosgrove & Ameen 1999; Ferrill *et al.* 2012*b*; Brandes & Tanner 2014), and joints, stylolites and opening veins (McGrath & Davison 1995; Petit 1995; Kim *et al.* 2004). This composite nature modifies rock permeability and friction, and, hence, impacts fluid circulation (Wallace & Morris 1986; Byerlee 1993; Caine *et al.* 1996; Sibson 1996; Bense *et al.* 2013) and seismicity (Sibson 1977; Townend & Zoback 2000; Scholz 2002; Collettini *et al.* 2009). A better understanding of fault zones is therefore critical for seismic hazard assessment and prediction, and oil and gas exploration, as well as hydrogeology and waste storage (Wibberley *et al.* 2008; Faulkner *et al.* 2010).

The creation and internal structure of fault zones are the result of a progressive development. According to a simplified model, their formation begins with the nucleation of a fracture due to a favourable state of stress and the coalescence of microcracks (Kranz 1983; Atkinson 1984; Lockner 1995; Paterson & Wong 2005). Then, narrow meso-scale faults or joint planes are formed by tip propagation. These planes are at first discontinuous. Then, they may propagate and coalesce to form segmented fault arrays (Weber *et al.* 1978; Segall & Pollard 1980; Childs *et al.* 1996; Walsh *et al.* 2003). Further propagation may involve interactions and linkages of distinct fault planes or fault arrays, which results in increased complexity (Peacock & Zhang 1994; Soliva & Benedicto 2004). With increasing strain, an inner fault rock forms in response to fracturing, abrasion and comminution of the fault wall rocks (Robertson 1982; Wallace & Morris 1986; Scholz 1987). The resulting relay zones, lenses and corrugations form irregularities and asperities that are dispersed along the fault zone. Deformation concentrates in these irregularities and, ultimately, they become zones of widened fault rocks (Watterson *et al.* 1998; Childs *et al.* 2009). Finally, continued

From: Childs, C., Holdsworth, R. E., Jackson, C. A.-L., Manzocchi, T., Walsh, J. J. & Yielding, G. (eds) 2017. *The Geometry and Growth of Normal Faults*. Geological Society, London, Special Publications, **439**, 271–288.
First published online February 5, 2016, https://doi.org/10.1144/SP439.5

deformation may result in the localization of strain and the consequent narrowing of the active slip zone, whilst the overall finite fault-zone width remains constant (Micarelli *et al.* 2006; Wibberley *et al.* 2008).

According to this model, fault zones are largely controlled by fault geometry inherited from the early stage of faulting (Childs *et al.* 2009). The fine-scale geometry of early-stage fault zones are difficult to identify from seismic reflection sections since these may be below the resolution of the seismic data (Walsh *et al.* 2003). Hence, studying the early stages of faulting is fundamental in many aspects.

During the early stage of faulting considered here, the host-rock characteristics are critical. In layered rocks, the interplay of several processes can be expected to produce a wide variety of geometrical irregularities dispersed along faults. Fault propagation may be stopped by lithological interfaces, focusing near-tip deformation in some layers (McGrath & Davison 1995; Gross *et al.* 1997; Micarelli & Benedicto 2008). Mechanical heterogeneities promote and control segmentation, forming a succession of extensional or contractional relay zones or lenses (Peacock & Sanderson 1991; Cartwright *et al.* 1995; Childs *et al.* 1996; Schöpfer *et al.* 2006). Fault dip may change as a function of the lithology (Davison 1987; Peacock & Sanderson 1992; Ferrill & Morris 2003; Roche *et al.* 2014). Multiple parallel fault segments may occur in stiff layers, whereas connections and splays may be promoted in compliant ones (Gross *et al.* 1997; Roche *et al.* 2012*a*). Finally, fault-related folding may occur within soft beds (Ferrill *et al.* 2012*b*; Brandes & Tanner 2014; Rotevatn & Jackson 2014).

Layer-parallel faults often exist in compliant layers, and fault zones may interact and merge with them, modifying their style of propagation and final geometry (Gross *et al.* 1997; Watterson *et al.* 1998; Jackson *et al.* 2006; van der Zee *et al.* 2008; Roche *et al.* 2012*a*). The formation of these structures is still debated: they may be due to flexural slip before faulting (Tanner 1989; Gross *et al.* 1997; Wibberley *et al.* 2007; van der Zee *et al.* 2008) or fault-related folding (Watterson *et al.* 1998; Jackson *et al.* 2006; Smart *et al.* 2009), despite the fact that they are typically observed in sub-horizontal layers, or they may form early in sub-horizontal layers during normal faulting regimes (Roche *et al.* 2012*a*).

Depending on the characteristics of the layers, such as their mechanical properties, their thickness or the presence of layer-parallel faults, we may expect fault zones to exhibit varying degrees of complexity, and the host rock to control the final characteristics and width of the mature fault zone (Wibberley *et al.* 2008). However, the fundamental link between the host rock and the initial fault zone structures, as well as the evolution of the fault characteristics during the growth, remains to be debated. Two end-member models are considered: (1) fault zones grow by tip propagation (i.e. continuous propagation) and, in this case, we may expect their initial structures to depend on the rock properties at the fault tips (Micarelli & Benedicto 2008; Roche *et al.* 2014); and (2) conversely, if fault zones grow by segment linkage (i.e. discontinuous propagation), the fault structures are controlled by relay-zone geometries and depend on the interplay of the host rock and a shear zone (Weber *et al.* 1978; Childs *et al.* 1996; Walsh *et al.* 2003; Schöpfer *et al.* 2006).

In this paper, we present a model of fault-zone development, in which fault widening is controlled by the capacity of the fault to propagate vertically in a layered section during displacement accumulation. We focus primarily on the structural effects of tip restriction by sub-horizontal faults, but restriction by lithological interfaces is also addressed. This study is based on detailed field analysis of well-exposed normal fault arrays. The studied faults are on the metre scale and provide insight into the early stage of fault-zone development.

Geological setting and methodology

Geological setting

We study normal faults in two outcrops, named Trescléoux and St-Didier, situated in the Ardèche margin and the Vocontian trough, respectively (Fig. 1). In both sites, the faults affect alternating limestone and clay-rich layers deposited in the Late Oxfordian stage in the South-Eastern Mesozoic sedimentary basin of France. The layers are gently dipping (i.e. 15° dipping).

Rock properties, including mineralogy and mechanical properties, are almost constant within layers of the same lithology (Roche *et al.* 2014). The carbonate percentage of the clay-rich layers is slightly higher at St-Didier (72%) than at Trescléoux (64.5%). Unfortunately, no information on the clay content and the clay type are available. The carbonate content equals 82% in the limestone layers of both sites. The maximum burial depth occurs before the Late Cretaceous (Barbarand *et al.* 2001; Séranne *et al.* 2002; Peyaud *et al.* 2005) and is greater for Trescléoux (i.e. 3000–6000 m: after Guilhaumou *et al.* 1996; Baudrimont & Dubois 1977) than for the St-Didier site (i.e. 1600–2700 m: after Pagel *et al.* 1997; Guilhaumou *et al.* 1996; Barbarand *et al.* 2001). The thickness of the clay layers ranges from 50 cm to 2 m at Trescléoux, but are thinner at St-Didier (<50 cm). The thickness of the limestone layers is similar in both sites, ranging from 25 to 50 cm. The relative proportion

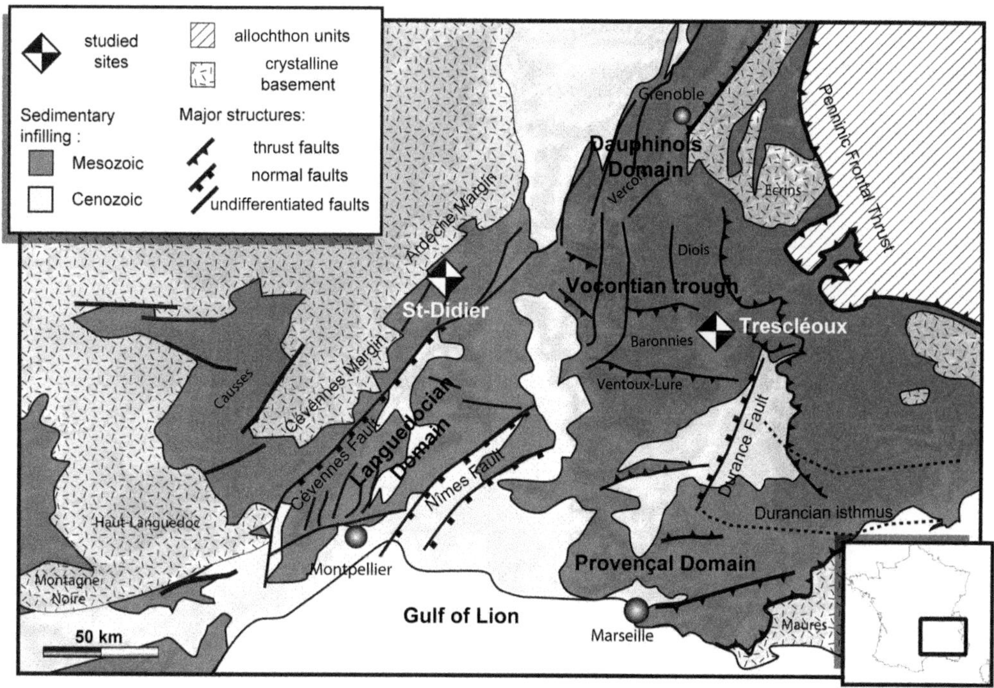

Fig. 1. Simplified geological map of the South-Eastern Basin (France). Adapted from Roure *et al.* (1992).

of limestone/shale within the faulted section is significantly greater at the St-Didier site than at Trescléoux (Roche *et al.* 2014).

General features of the faults

The studied faults are normal faults, as indicated by the dip-slip slickenlines observed on the fault planes (plotted on Fig. 2) and the normal offset of the sedimentary interfaces (Figs 3 & 4). The direction of the minimum principal stress (σ_3) responsible for the normal fault slips and computed using the Angelier method (Angelier 1990) is similar for the two sites (i.e. east–west) (Fig. 2). The direction is in agreement with the Oligocene extension direction (Arthaud *et al.* 1977; Bergerat 1987; Roure *et al.* 1992). Hence, faulting took place after the maximum burial depth (i.e. 3000–6000 m at Trescléoux and 1600–2700 m at St-Didier).

The outcropping portions of the normal faults range from 50 to 350 cm (i.e. fault height). The maximum displacement ranges are 4.5–40 cm and from 3 cm to approximately 1 m at St-Didier and Trescléoux, respectively. The faults exhibit dip refraction as a function of the lithology. In the limestone layers, before bed tilting, the normal faults had a representative dip of 74° (15–88°, $n = 121$) at both sites. In the clay-rich layers, faults dip 37° at the St-Didier site (7–60°, $n = 23$) and 53° at the Trescléoux site (30–77°, $n = 32$). The dip variations between layers are likely due to changes in the failure mode and are controlled by the elastic properties of the rocks (Ferrill *et al.* 2012*a*; Roche *et al.* 2014).

This study focuses mainly on faults that have their upper tip in contact with layer-parallel faults contained in clay-rich layers (Figs 3, 4 & 5). This includes all the faults at St-Didier (D_1–D_4), as well as most of the faults at Trescléoux (T_2–T_4). For these tips, faults may exhibit significant segmentation and fault-related folding, as described in the next section on 'The spatial arrangement of fault zones'. Other faults that have been stopped by lithological interfaces are observed at the Trescléoux site. Fault T_1, which is restricted at both tips, illustrates this type of fault. Likewise, there are examples of faults where the lower tips propagated across the lithological interface at this site (e.g. fault T_2).

Throw profiles

Throw profiles of the studied faults are shown in Figures 6a–c and 7. This analysis focuses on the fault portions that are situated between the supposed maximum throw and the upper tip, which is in contact with the layer-parallel faults. This corresponds to the faults D_1–D_4 at St-Didier and several faults at Trescléoux, including T_2–T_4. Throw profiles of

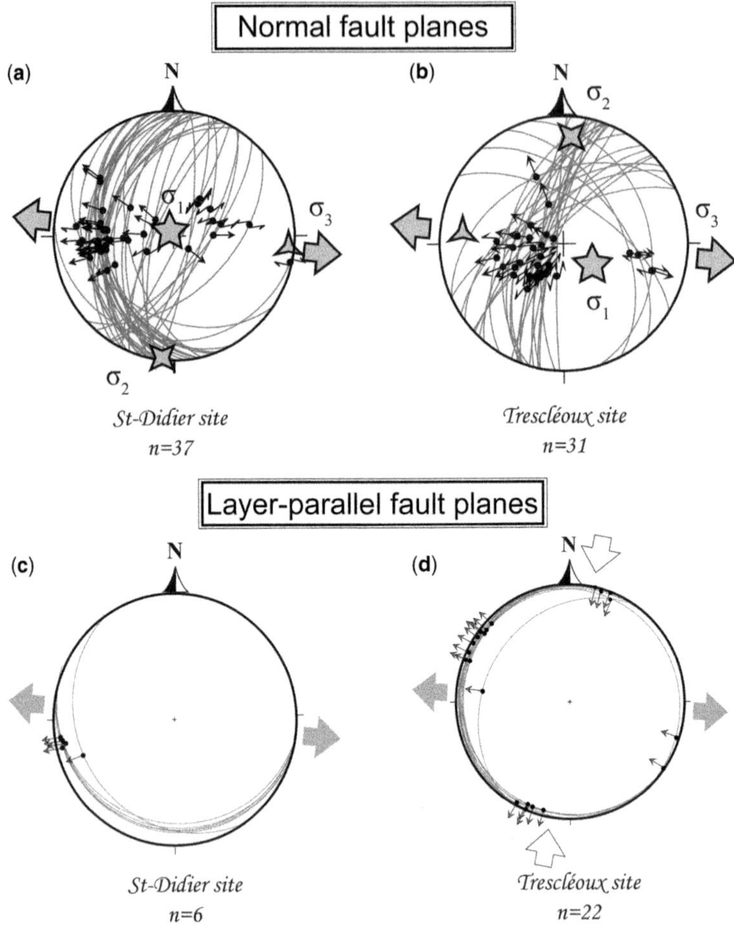

Fig. 2. Stereographic projections of the studied (**a**, **b**) normal faults and the (**c**, **d**) layer-parallel faults measured at the St-Didier site (a, c) and at the Trescléoux site (b, d). The data are back-tilted (lower-hemisphere, equal-area projection) (layer dips N132 17N and N068 15S at St-Didier and Trescléoux, respectively). Solid circles: fault planes. Small arrows: slickenlines measured on the outcrops. Stars with five, four and three branches: computed maximum (σ_1), intermediate (σ_2) and minimum (σ_3) stresses. Divergent grey arrows: direction of the computed extension. *n*, the number of data.

the fault portions that have had their tip stopped by lithological interfaces, or that have propagated into clay-rich layers, have been analysed in previous papers (Roche *et al.* 2012*b*, 2014). They are not studied in this paper but some profiles are provided for comparison (Fig. 6c, d).

Two types of throw profiles are built for each fault. The first one corresponds to the throw profiles measured near the fault planes. They are called 'near-field' throw profiles. The second type of throw profile is called a 'far-field' throw profile. Similar displacement profiles obtained from a 3D seismic survey have been referred to as 'total displacement profiles' (Long & Imber 2010). In these profiles, the throw along the fault plane, the throw accumulated by bed rotation and the throw induced by the segmentation have been added in order to consider the total displacement induced by the fault zones (see also Roche *et al.* 2012*a* and Homberg *et al.*, this volume, in review for similar analyses based on field observations). The lower tips were not observed for several faults; moreover, the cross-sectional view may not be through the maximum displacement, therefore, the fault half-lengths may be underestimated, as well as the maximum displacements. The displacement gradients were calculated along the fault dip and then compared to values from the literature.

Sedimentary interfaces were used as displacement markers at both sites. At the St-Didier site, the sedimentary interfaces are regularly spaced and relatively close to each other compared with the fault size (i.e. a spacing of 2–45% of the apparent fault height, with an average value of 14%). We are therefore confident in the displacement analysis at this site. At Trescléoux, the spacing between the sedimentary interfaces is very similar (i.e. 2–37% of the fault half-heights, with an average value of 14%), but displacement markers are more sparse for smaller faults (e.g. T_1 and T_2), for which more uncertainty in displacement analysis is expected.

The D_{max}/L ratio is commonly used for displacement analysis, but this ratio was not analysed in this paper because the full lengths of the faults are not exposed. This ratio is a notably useful tool for differentiating restricted from unrestricted faults (Nicol *et al.* 1996; Wilkins & Gross 2002; Soliva & Benedicto 2005). To confirm that the faults are restricted, we looked directly at whether the fault tips were in contact with a potential restrictor and at the displacement gradient at the fault tip. The D_{max}/L ratio is also often studied because it provides insights into the mechanism of fault propagation (Gudmundsson 1987; Walsh & Watterson 1987; Cowie & Scholz 1992; Bürgmann *et al.* 1994). However, this is more relevant for isolated non-restricted faults rather than for restricted faults, as studied in this paper.

The spatial arrangement of fault zones

Contact between normal faults and layer-parallel faults

The studied faults have their upper tips in contact with layer-parallel faults in clay-rich layers (Figs 3, 4 & 5). The surfaces of the layer-parallel faults are highlighted in the field by calcite crystallization at both sites.

Slickenlines are measured along the layer-parallel faults, indicating that they accommodated slip. The slip direction is similar to the direction of the minimum principal stress responsible for normal fault slips (Fig. 2). Therefore, the normal faults and the layer-parallel faults both seem to result from the same regional extension. In other outcrops, displacement seems transferred between steeper faults and similar layer-parallel faults (Roche *et al.* 2012*a*). Hence, it seems reasonable to think that a kinematic link exists between the normal faults and the layer-parallel faults, and that they accommodated slip together. Yet, the quantity of displacement along a layer-parallel fault is difficult to ascertain because such a structure does not offset any markers. However, the studied normal faults are likely to post-date layer-parallel faults: the latter seem to not offset the former at the scale of the outcrops; the former systematically abut the latter; and the layer-parallel faults may be folded ahead of the normal fault tips (see the next subsection on ‘Fault-related folding’).

At St-Didier, the normal faults cut through several layers and they are all in contact with a single layer-parallel fault, located about 2 cm above the bottom of a 20 cm-thick clay-rich layer. The layer-parallel fault is observed continuously all along the outcrop. Its minimum length approximates to tens of metres (Fig. 3). Our observations tend to indicate that the layer-parallel fault plane is not significantly folded ahead of the normal fault tips. At Trescléoux, the normal faults are located in two different intervals in the same lithological section. Several small faults with maximum displacements of less than 12 cm affect only one limestone layer (faults T_1 and T_2 in Fig. 4). Two larger normal faults with maximum displacements of around 20 cm and 1 m affect several limestone and clay-rich layers (faults T_3 and T_4 in Fig. 4). In all cases, the upper tips of the normal fault are in contact with layer-parallel faults located approximately 10 cm above the bottom of two 70 cm-thick clay-rich layers. These layer-parallel faults are continuous along the outcrops (i.e. for tens of metres).

Fault-related folding

At Trescléoux, folding occurs ahead of the upper tips in the clay-rich layer. For the small faults (e.g. faults T_1 and T_2), clay rocks are folded, up to 28 cm above the fault tips. In some instances, folding does not reach the layer-parallel fault (Fig. 5a): elsewhere, the layer-parallel fault is folded, with fault tips abutting or not the layer-parallel fault (Fig. 5b, c, d, e). Far from the normal faults, along the same lithological layers, the layer-parallel faults are not folded. Hence, this folding seems to be associated with a vertical propagation of normal faults. This confirms that layer-parallel faulting predates normal faulting, but does not preclude that it continued to be active during the formation of the latter, steeper-dipping faults.

Similar folding is observed at the tip of the larger faults, T_3 and T_4. Although difficult to ascertain, the distance at which folding disappears is less than 70 cm above the fault tip because the upper limestone layer is not significantly folded. These faults also exhibit folding along the fault in the hanging wall and the footwall. We may assume that this folding component was formed at the fault tip before the fault propagated through the folded layer because similar folding is observed ahead of the faults. At St-Didier, the magnitude of the folding component is not significant. This is probably due to a lower capacity to accommodate continuous deformation

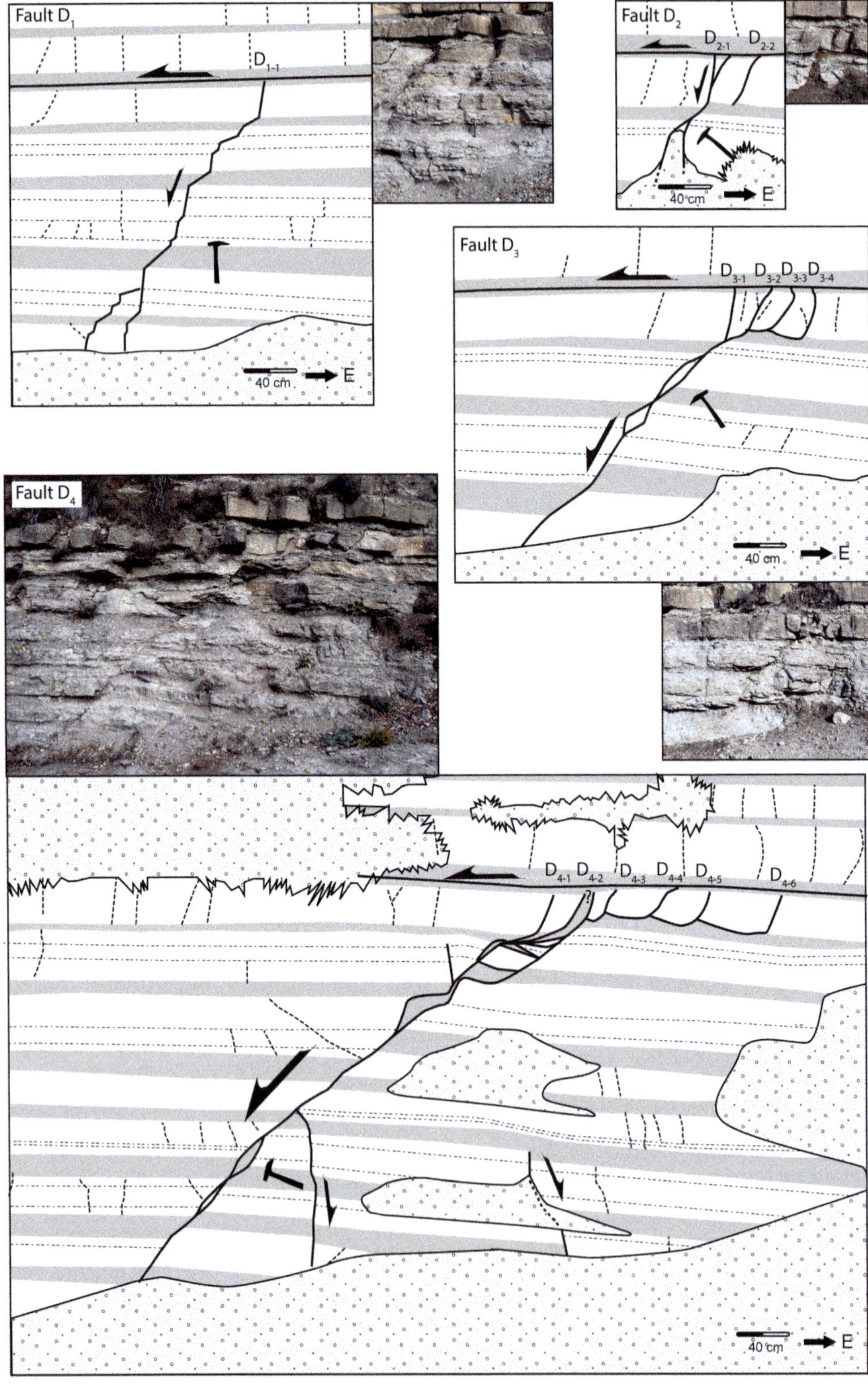
Fault D1
D1-1
40 cm
E
Fault D2
D2-1
D2-2
40 cm
E
Fault D3
D3-1
D3-2
D3-3
D3-4
40 cm
E
Fault D4
D4-1
D4-2
D4-3
D4-4
D4-5
D4-6
40 cm
E

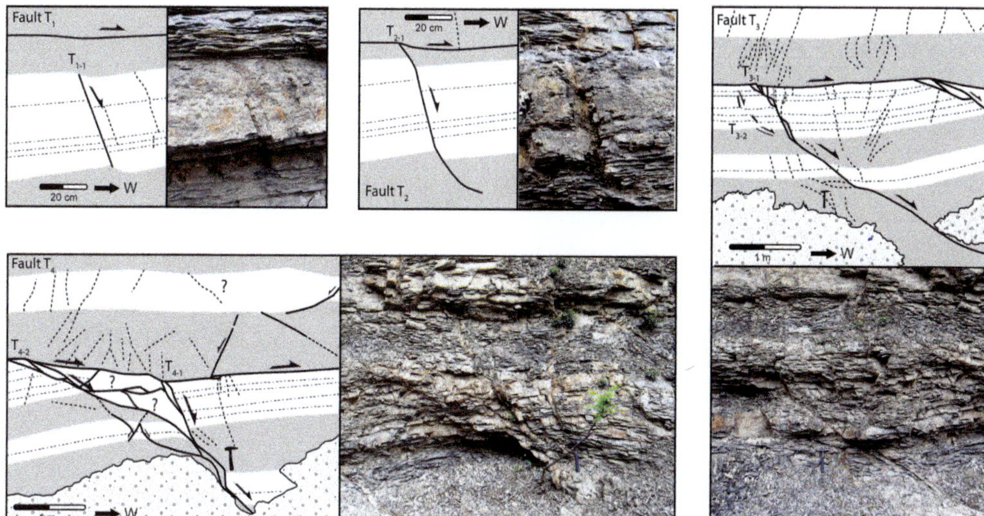

Fig. 4. Interpretative sketches of four normal faults studied at the Trescléoux site. The legend is the same as for Figure 3. The faults are labelled T_1, T_2, T_3 and T_4, respectively; T_{x-y} indicates segment y of fault number x. Maximum displacement increases from T_1 to T_4. The scales of faults T_1 and T_2 are smaller than the scales of faults T_3 and T_4. The scale of the photographs is similar to the scale of the interpretative drawings.

in the clay-rich layers owing either to the relatively high carbonate content (i.e. 72%) or to the relatively low thicknesses (i.e. <50 cm).

Segmentation

Faults with low maximum displacements (i.e. less than 12 and 4.5 cm at Trescléoux and St-Didier, respectively) do not exhibit any macroscopic modification other than the folding described previously (e.g. faults D_1, T_1 and T_2 in Figs 3 & 4). For higher maximum displacements, the faults splay into several parallel fault segments near the upper tip (Figs 3 & 4). Other fractures are observed surrounding the normal faults. They may be part of joint sets affecting the limestone layers, or they may be related to damage in the fault walls as a result of fault slip. They are not addressed further in this study because no significant shearing is observed on these fractures.

The parallel fault segments are concentrated directly under the layer-parallel faults, in the footwall of the normal faults and in the final deformed limestone layer. These segments have their upper tips in contact with the layer-parallel faults. Their lower tips are either connected to the main normal fault plane or to the closest fault segment. In St-Didier, the connections are achieved by sub-horizontal fault portions forming a bookshelf-like structure as the parallel fault segments involve rotating beds (Fig. 3). At Trescléoux, the connections are achieved by low-angle-dipping fault portions (Fig. 4).

Other fault segments are not connected to the main fault zone (e.g. faults D_2 and T_3 in St-Didier and Trescléoux, respectively: Figs 3 & 4). These segments seem continuous across the limestone layer for fault D_2 (Fig. 3), whereas, for fault T_3, they are composed of a set of three small segments reflecting inner mechanical subdivisions in the limestone. These observations show that the parallel fault segments initiate in the limestone layer, and then their upper and lower tips connect the layer-parallel faults and the closest fault segment, respectively. The mechanical subdivisions within the limestone layer at Trescléoux may be responsible for the complex segmentation of the fault T_4 (Fig. 4).

Fig. 3. Interpretative sketches of the four normal faults studied at St-Didier, all shown at an identical scale. The scale of the photographs is half the scale of the interpretative drawings. The outcropping clay-rich and limestone layers are coloured in grey and white, respectively. Continuous lines: main shearing planes, including layer-parallel fault and normal fault planes. Fractures in dashed lines: secondary fractures, as well as bed-normal joints. The faults are labelled D_1, D_2, D_3 and D_4; D_{x-y} indicates segment y of fault number x. Maximum displacement increases from fault D_1 to fault D_4.

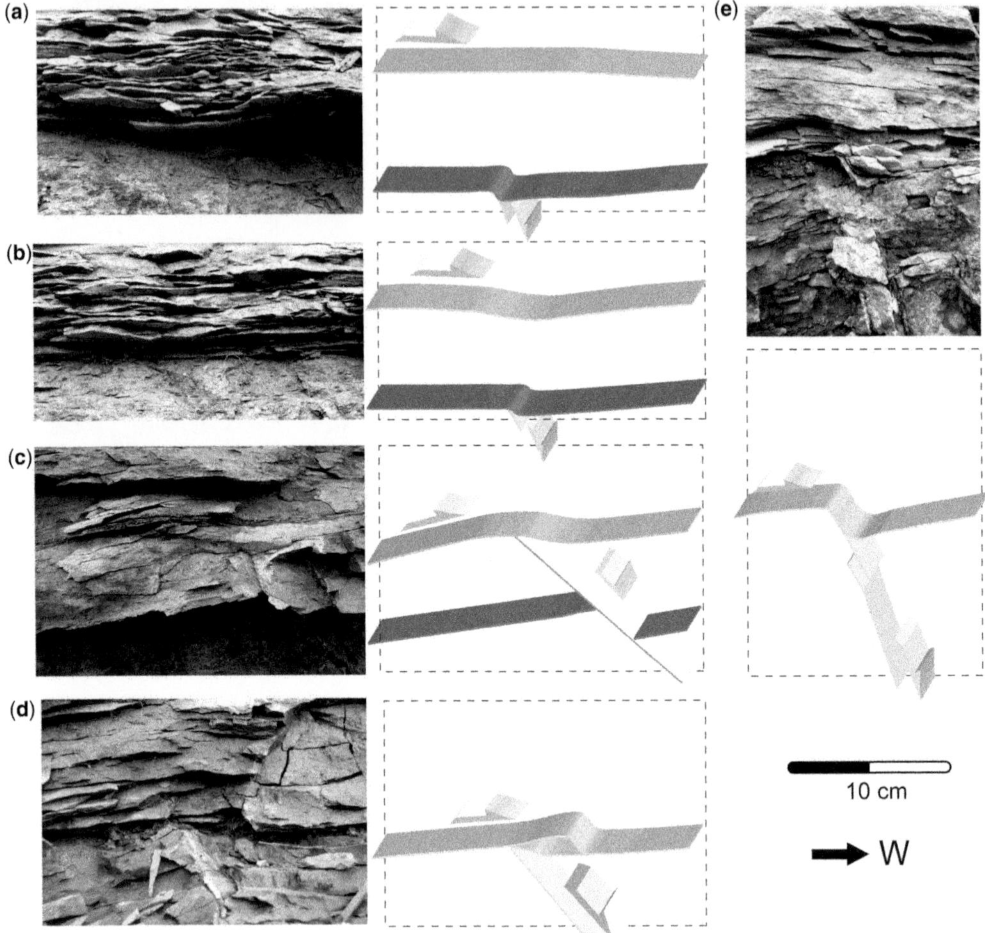

Fig. 5. Examples of contacts between layer-parallel and normal faults at the Trescléoux site. Steeply dipping planes: fault planes. Dark and light grey sub-horizontal planes: limestone–clay-rich layer interfaces and layer-parallel faults, respectively. 3D arrows show the slip direction.

The different segments are numbered according to their position relative to the lower part of the normal fault: that is, segment 1 is the continuation of the lower part of the normal fault, segment 2 is the first splay, and so on (Figs 3 & 4). Including segment 1, there are between one and six segments at St-Didier (i.e. these are numbered such that a value of 1 means that the fault-zone structure is formed by a unique plane). The average spacing between two adjacent segments is 23 cm. At Trescléoux, small faults T_1 and T_2 exhibit one (shown as 1) segment (i.e. T_{1-1} and T_{2-1}), fault T_3 exhibits 2 segments (i.e. T_{3-2}), and, although the number of segments is difficult to assess, for fault T_4 it is likely to be 2 or 3.

Figure 8a shows that the segment number increases with the maximum displacement observed along the outcropping portion of the normal fault. These observations are likely to indicate that the fault segments initiate successively as displacement increases along the fault zone. Segment 1 is the initial segment (i.e. T_{1-1}, T_{2-1}, T_{3-1}, D_{1-1}, D_{2-1}, D_{3-1} and D_{4-1}), segment 2 is the first splay (i.e. T_{3-2}, D_{2-2}, D_{3-2} and D_{4-2}) and the other segments then initiated successively. The furthest segments (i.e. T_{3-2}, D_{2-2}, D_{3-4} and D_{4-6}) are likely to have been the last segments to have formed.

Distribution of throw

Near-field throw profiles

The studied faults with the upper tip in contact with layer-parallel faults exhibit near-field throw

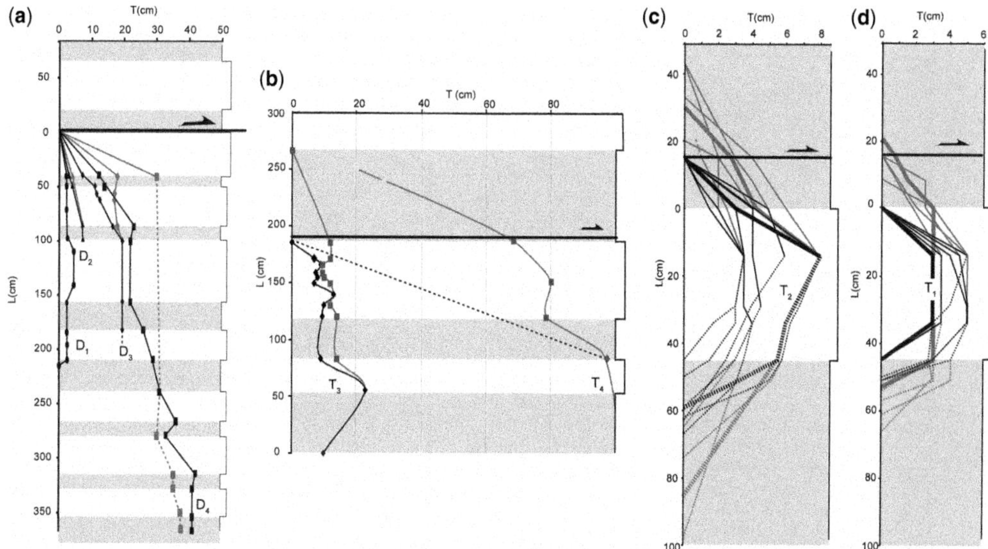

Fig. 6. Throw profiles of the normal faults studied: vertical throw (T) as a function of height (L) at the St-Didier (**a**) and Trescléoux sites (**b**)–(**d**). The labels D_1, D_2, D_3, D_4, T_1, T_2, T_3 and T_4 are the same as in Figures 3 and 4. Black curves: near-field throw profiles. Grey curves: far-field throw profiles (see text). Dashed lines in (a) & (b): displacement information subject to uncertainty. Thick horizontal shear planes: the position of the layer-parallel faults. In (a)–(c), the upper tip of the fault zones is in contact with the layer-parallel fault, whereas, in (d), the normal faults are not in contact. Squares, diamond and dots in (a) & (b): measurement points. For ease of reading, the symbols have been removed on (c) & (d). In (c) & (d), the T_1 and T_3 throw profiles are indicated in bold to facilitate identification. Dotted lines: throw profiles that belong to non-restricted fault sections (i.e. not analysed in this paper).

profiles ranging from a triangular profile to a flat-topped profile at the St-Didier (Fig. 7a) and Trescléoux sites (Fig. 7c). Regarding the displacement gradient, the central portions of the faults are characterized by relatively low and constant values (i.e. lower than 0.04 and 0.07 at the St-Didier and Trescléoux sites, respectively). For reference, these values are lower than those found along other non-restricted faults at this site (see fig. 4 in Roche *et al.* 2012*b* for a review of gradients at these and other sites). Near-tip portions have relatively high and scattered displacement gradients (i.e. from 0.1 to 0.63 and from 0.05 to 0.3 at Trescléoux and St-Didier, respectively). For reference, these values are higher than those found along other non-restricted faults. A positive correlation between the gradients and the maximum displacements has also been described in a previous study at these sites (see fig. 3a in Roche *et al.* 2012*b*). At the St-Didier site, the high value of near-tip displacement gradients is observed in clay-rich layers, as well as in limestone layers, indicating that the gradient increase is not due to lithological effects (Muraoka & Kamata 1983; Roche *et al.* 2012*b*). These observations suggest that the faults continued to accommodate displacement, even though their upper tips were stopped at the layer-parallel fault (i.e. restricted faults). We consider, therefore, that all the studied faults with a tip in contact with a layer-parallel fault are restricted, despite the fact that some faults may not have a well-defined flat-topped profile.

In most of the studied faults, the lower tips were not observed, but one may assume that they continued to propagate while displacement increased, or they may have been restricted, depending on the layering.

An apparent flat-topped profile can be observed along a cross-sectional view that is not through the maximum displacement. In this case, near-tip gradients should not be a function of the maximum displacement and should be equal to the values commonly observed along non-restricted faults: thus, this is different to that in the studied faults. Otherwise, a rapid establishment of the fault surface without an increase in displacement should also form a flat-topped profile. However, in this case, we expect D_{max}/L values to be lower than the values that are characteristic of triangular profiles, whereas, in the studied sites, the D_{max}/L values are higher (see fig. 5 in Roche *et al.* 2012*b*). These confirm that the flat-topped profiles result from restriction.

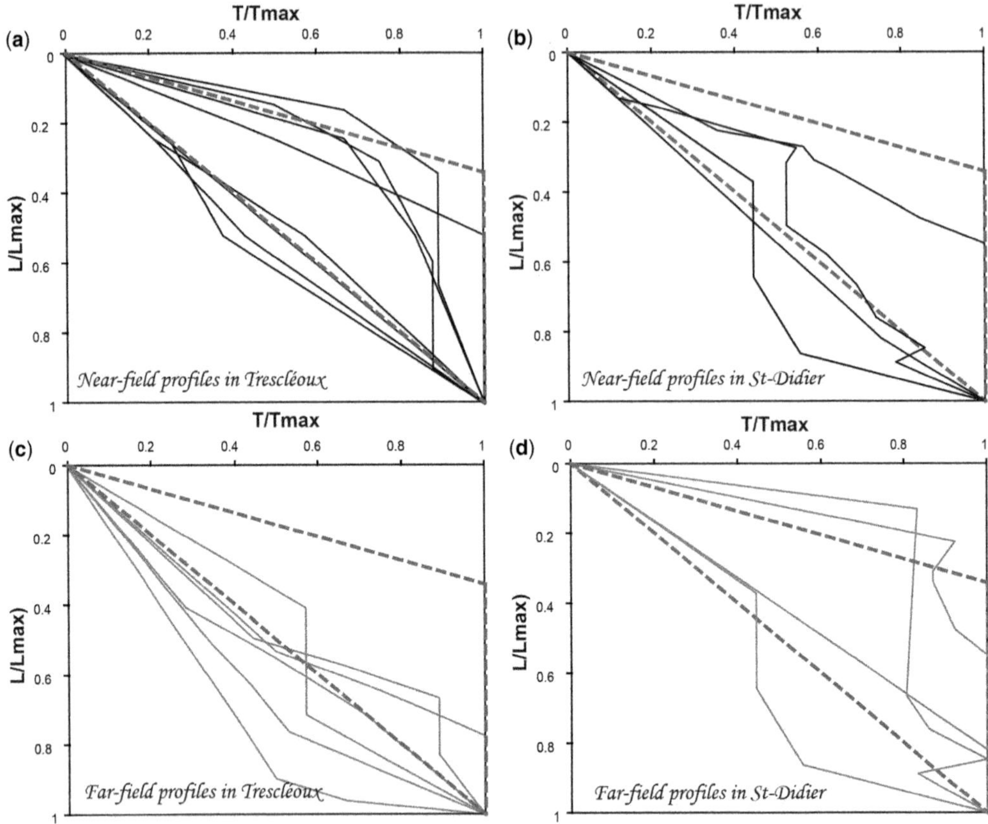

Fig. 7. Normalized near-field throw profiles (**a**, **b**) and far-field throw profiles (**c**, **d**) at the St-Didier (b, d) and the Trescléoux sites (a, c). The fault tip is in the top left of the figure. For comparison, a perfectly triangular profile and a flat-topped profile are indicated by thick dashed lines. The theoretical flat-topped profile presents a high near-tip gradient over 30% of the fault length.

More generally, this analysis shows that it is difficult to identify whether a single fault is restricted or unrestricted by looking at the shape of the profile because the throw profiles become increasingly flat-topped during restriction: indeed, this departure from the flat-topped profile can be observed. It could be done by defining a somehow arbitrary threshold of displacement accommodated by the fault without propagating. Or else, the problem is more easily solved when looking at a set of faults, rather than a single fault. In this case, restricted faults are faults with their tips in contact with a restrictor and with near-tip displacement gradients increasing with maximum displacements.

For most of the studied faults, the far-field displacement differs from the near-field displacement due to the segmentation and folding. Yet, the displacement variations caused by folding are different to those caused by segmentation. This is described in the following subsections.

Folding component of displacement

The folding component of displacement is analysed by looking at the far-field throw profiles at the Trescléoux site as no significant folding occurs at the St-Didier site. At the Trescléoux site, folding occurs ahead of fault tips, and sometimes in the wall rocks of larger faults. Owing to this folding, the far-field throw profiles exhibit a triangular shape (Fig. 7c): thus, they are different to the flat-topped ones observed in some near-field profiles (see the previous subsection on 'Near-field throw profiles'). The throw accommodated by folding is at a maximum near the fault tip and decreases ahead of it. This upwards decrease in folding is characterized by a displacement gradient equal to an average of 0.1 ($n = 22$). For reference, this value is lower than the average 0.17 gradient measured along non-restricted fault planes in the same lithology in a previous study (fig. 4 in Roche *et al.* 2012*b*). Thus, the attenuation of displacement is greater

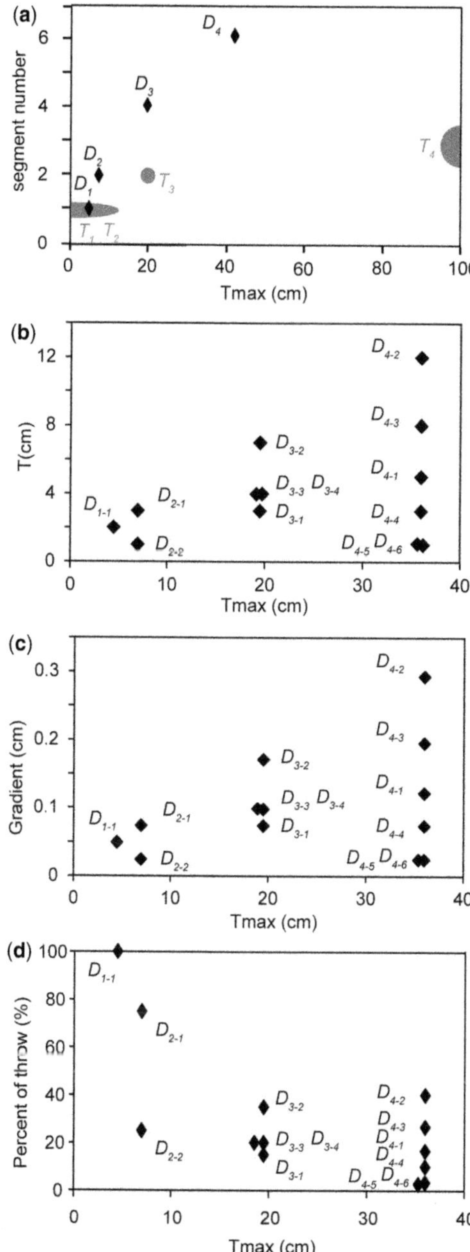

Fig. 8. Quantitative fault parameters. (**a**) Increase in the segment number near the restricted tip of the fault as a function of the maximum throw at both sites. (**b**)–(**d**) Distributions in throw of the various segments for the St-Didier site. Fault numbering is similar to Figures 3, 4 and 6. D_{x-y} indicates segment y of fault number x. Note that (b) & (c) are very similar because displacement gradients are calculated with a constant distance that corresponds to the thickness of the limestone layer in which the fault segments are found (i.e. 41 cm).

along a fault than along a fold, and we might expect the height along which folding takes place ahead of a fault tip to increase with increasing displacement. Adjacent to the fault plane, far-field displacement variations are characterized by a relatively small gradient (i.e. <0.05).

Displacement on fault segments

In this subsection, we analyse the far-field throw at St-Didier that is related to segmentation and the partitioning of the throw between the fault segments. Faults studied at the Trescléoux site do not allow us to study the partitioning because either the segmentation is not developed (e.g. fault T_2) or it is too complex (e.g. fault T_4). At St-Didier, the far-field and the near-field throw profiles are similar along the lower parts of the faults because the displacement is focused on a single plane. One exception is fault D_4: in this case, a conjugate normal fault is connected to the main fault (Fig. 3) and this decreases the far-field displacement (in Fig. 6a), although it is questionable as to whether or not this conjugate fault belongs to the fault zone.

Conversely, for all the faults, except fault D_1, the far-field and near-field throw profiles differ in relation to the segmentation of the fault zones near the restricted tip (Fig. 6). Unlike at Trescléoux, the far-field throw profiles exhibit a flat-topped shape. The near-tip displacement gradients increase from 0.05 to 0.69 and are positively correlated with the maximum displacement. These characteristics highlight that the total displacement is restricted by the layer-parallel fault.

Concerning the partitioning, the throw accumulated on each segment ranges from 1 to 12 cm (Fig. 8b). The corresponding displacement gradients are equal to 0.3 and 0.025, respectively (Fig. 8c). The percentage of the total displacement accumulated by each segment ranges from 100% on fault D_1, which comprises one fault segment, to 5% on one of the six segments of fault D_4 (Fig. 8d).

The displacements accommodated by the furthest segment (i.e. D_{2-2}, D_{3-4} and D_{4-6}) are always less than 4 cm and correspond to a maximum of 20% for the total displacement, with displacement gradients of less than 0.07. The arrest of the tectonic forces is likely to be the origin of these low displacements.

On the first segment (i.e. D_{1-1}, D_{2-1}, D_{3-1} and D_{4-1}), the displacement magnitudes and gradients are relatively similar for the various faults (i.e. 2–5 cm and 0.05–0.12, respectively). Such gradients are close to the characteristic values for faults propagating in limestone (see Roche *et al.* 2012*b* for a review). With an increasing number of segments, the percentage of the total displacement accommodated by the first segment decreases from

100% on segment D_{1-1} to only 10% on D_{3-1} and D_{4-1}. Unlike the furthest segment, these low values of displacement are not caused by an arrest of the tectonic forces because segments 2, 3, etc., initiated after the first segment. It seems, therefore, that the first segment becomes inactive while the other segments accommodate displacement.

For faults D_4 and D_3, the first segments (D_{3-1} and D_{4-1}) and the furthest segments (D_{3-4} and D_{4-6}) accommodate generally less displacement than the intermediate segments (D_{3-2}, D_{3-3}, D_{4-3}, D_{4-4} and D_{4-5}). The displacement on a fault segment is always lower or equal to the displacement accumulated on the previous fault segment (i.e. displacement on $D_{3-2} \geq D_{3-3} \geq D_{3-4}$ and displacement on $D_{4-2} \geq D_{4-3} \geq D_{4-4} \geq D_{4-5} \geq D_{4-6}$). Hence, the second segments (D_{3-2} and D_{4-2}) show greatest displacement (up to 12 cm) and the highest gradient (up to 0.3). This corresponds to 40% of the total displacement for both faults, D_3 and D_4. The third segments have lower displacement magnitudes (up to 8 cm), displacement gradients (up to 0.2) and percentage of displacement (up to 30%). On fault D_2, displacement on D_{2-2} is smaller than displacement on D_{2-1}, but D_{2-2} is also the furthest segment and likely to have accumulated less displacement.

Concerning the second and third segments, the displacement significantly increases with maximum displacement (i.e. displacement on $D_{4-2} \geq D_{3-2} \geq D_{2-2}$ and displacement on $D_{4-3} \geq D_{3-3}$). These results indicate that, unlike the first segment, the intermediate segments stay active during the segmentation of the faults.

A model of fault-zone evolution during restriction

The studied faults become more complex with increased maximum displacement. Their structures range from a narrow plane to a segmented fault zone that merges with a layer-parallel fault. We assume that these various structures represent successive stages of normal fault-zone development resulting from vertical restriction in layered sections. These successive stages are presented in Figure 9a, b for cases with or without fault-related folding.

The first step corresponds to the upwards propagation of a planar normal fault through the layers (steps 1–2 in Fig. 9a). This is likely to have been achieved by continuous tip propagation with dip refraction (see Roche *et al.* 2014) and with a displacement distribution highly sensitive to the layering (Muraoka & Kamata 1983; Roche *et al.* 2012*b*). At the Trescléoux site, fault-related folding occurs ahead of the fault tips (Fig. 9b).

When the fault tip abuts a pre-existing layer-parallel fault in the clay-rich layer, the vertical propagation of the fault stops (step 3 in Fig. 9). This involves modifications of the near-field throw profile. The original triangular profile sensitive to lithological variation is replaced by a flat-topped profile insensitive to layering (see Roche *et al.* 2012*b*). While the fault is restricted, maximum displacement and near-tip gradient both increase owing to the increasing displacement.

Next, a second segment initiates in the footwall. Our observations indicate that this segment is originally disconnected and then merges into the fault zone (see steps 3, 4 and 5 in Fig. 9). In Figure 9, analogous to the St-Didier site, these connections are achieved by a layer-parallel fault portion, although these also could be achieved by dipping fault planes. The second segment probably initiates when the displacement gradient of the first segment, or the maximum displacement, reaches a threshold value. The first segment is barely active once the second segment nucleated and the second segment mainly accumulates the displacement.

The process repeats while maximum displacement increases along the fault zone (see steps 6, 7 and 8 in Fig. 9a, b). However, unlike the first segment, the other segments stay active during the course of the process. For instance, the second segment continues to accumulate displacement after the third segment takes place, and the second and the third segments accumulate displacement after the fourth segment initiates, and so on. We may assume that the younger segments successively initiate once a threshold value of displacement accumulated on the previous segment is surpassed. Our observations indicate that this threshold may be equal to or lower than the one on the initial fault segment, implying a weakening in the process.

As a final step, although not demonstrated in this paper, we may expect the fault zone to propagate beyond the layer-parallel fault if the maximum displacement keeps rising. It may be only one, or it may be several segments that propagate. This propagation probably occurs first as a transfer of displacement along the layer-parallel fault, and then with the formation of another steep-dipping fault due to splay. Afterwards, the process may, or may not, repeat if the fault zone encounters another layer-parallel fault while propagating vertically, resulting in increased complexity (see step 9 in Fig. 9c).

This model of fault-structure development states that the fault-zone width increases locally while the fault is vertically restricted (Fig. 10a). Figure 10b shows that this widening is consistent with values from the literature. This increase in fault-zone width depends on the capacity of the fault to propagate through the layering, and this is discussed in the next section.

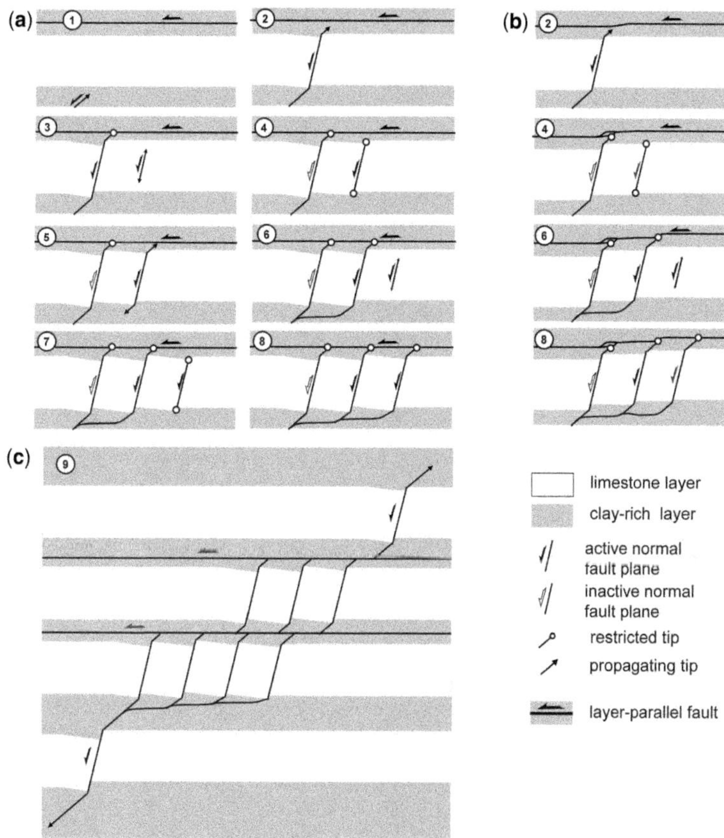

Fig. 9. Conceptual model of fault-zone widening during restriction in layered sections. (**a**) & (**b**) Various stages of fault development (a) without and (b) with fault-related folding, based on our observations. (**c**) A possible fault geometry for a throughgoing fault zone in a layered section with two layer-parallel faults. See the section on 'A model of fault-zone evolution during restriction' for a description.

Discussion

Widening of fault zones due to fault restriction

The model of fault-structure development presented in the previous section implies that the fault-zone width increases locally when the fault is vertically restricted by a layer-parallel fault. A similar model was described by Gross *et al.* (1997) for faults restricted by bedding-plane detachment. In both models, faults evolve from simple isolated planar structures to complex fault zones that are characterized by abundant splays. However, in the present case, the segmentation is concentrated in the footwall rock, whereas segments initiate in the hanging wall for the faults described by Gross *et al.* (1997).

With increasing displacement, we expect faults to propagate through the layer-parallel fault. Such throughgoing faults should exhibit local widening and complexities close to every layer-parallel fault. The fault zones described in Roche *et al.* (2012*a*) display similar attributes. Other previous studies also describe complex structures and widening of normal fault-zone structures near layer-parallel faults (Watterson *et al.* 1998; Wibberley *et al.* 2007; van der Zee *et al.* 2008). However, in these cases, this is caused either by attrition and damage along the normal fault plane, or by offsets along the layer-parallel faults and the normal fault segments.

Restriction and segmentation

Several properties are likely to control the geometrical modifications that occur during the restriction described previously. First, fault wall rocks with low strength or a high pore pressure might promote the formation of fault-parallel segments. Likewise, the elastic properties of the wall rocks impact the displacement accumulation along the fault

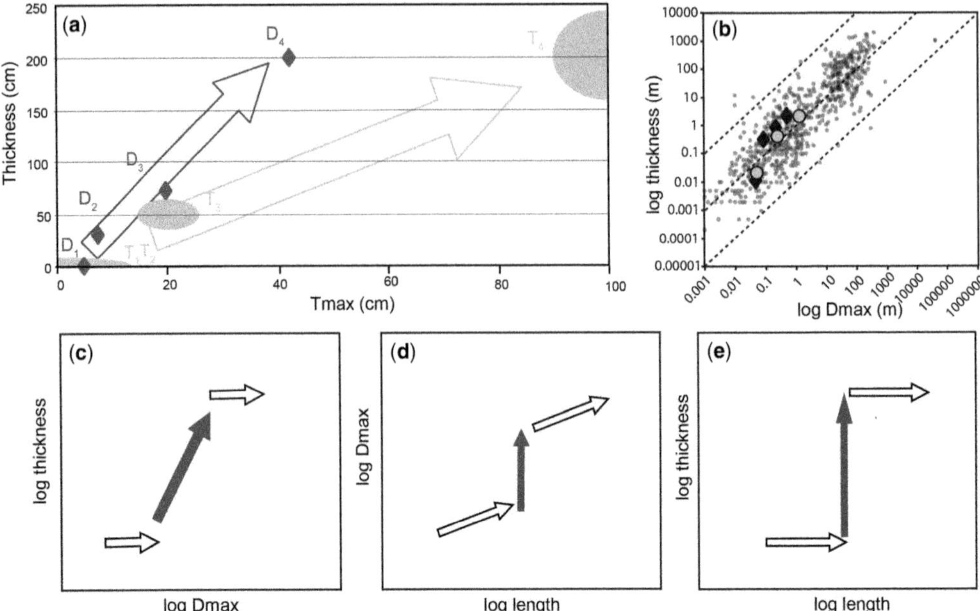

Fig. 10. Widening of fault zones during vertical restriction. (**a**) & (**b**) Increase in fault-zone thickness as a function of maximum throw (T_{max}). Fault numbering is similar to Figures 3, 4, 6 and 7. (b) Thickness as a function of the maximum displacement (D_{max}), the data include fault zones, breached and intact relay zones, and damage zone thicknesses ($n = 712$). Small dots: data derived from Childs *et al.* (2009), source included. Large black and grey dots: data for the Trescléoux and St-Didier sites, respectively. D_{max} values are calculated using the T_{max} and the average dip of the layers, where T_{max} is observed for each fault. (**c**)–(**e**) Evolution of the maximum thickness, maximum displacement (D_{max}) and vertical length, before and after the restriction (white arrows), and during the restriction (grey arrows). The variation in D_{max} as a function of the length is demonstrated in Roche *et al.* (2012*b*).

(Gudmundsson 1987; Walsh & Watterson 1987; Bürgmann *et al.* 1994). Second, the strength and the stiffness of the rock ahead of the fault tip, as well as of the strength of the restrictor itself, may control the capacity to maintain the fault restricted (Nicol *et al.* 1996; Cooke & Underwood 2001; Welch *et al.* 2009; Roche *et al.* 2013). Finally, heterogeneous stress distribution due to stiffness contrasts between layers, pre-existing joint sets, layer-parallel faults and plasticity of individual lithologies are also likely critical (d'Alessio & Martel 2004; Welch *et al.* 2009; Roche *et al.* 2013). The interplays of these processes characterize the capacity of the host rock to arrest fault propagation and to create additional segments for a given strain. In the following subsection, we refer to this capacity as the restriction and segmentation capacity.

In layered rocks, normal faults may also be stopped by lithological interfaces (Nicol *et al.* 1996; Gross *et al.* 1997; Wilkins & Gross 2002; Soliva & Benedicto 2005). Such normal faults are observed at the Trescléoux site. Fault T_1 in Figure 4 is an example of such a fault, with both tips restricted by lithological interfaces. Faults propagating across lithological interfaces are also observed at this site. For instance, the lower portion of fault T_2 propagated into an underlying clay-rich layer. These faults do not exhibit any of the geometrical modifications of the model in Figure 9. Hence, at this site, for the same host rock, faults stopped by layer-parallel faulting exhibit geometrical modifications, whereas restriction by the lithological interface does not modify fault structure. This shows that the physical properties, or the location, of the restrictor itself are critical.

For a similar maximum displacement, the number of segments is greater at St-Didier than at Trescléoux. This difference is not due to the characteristics of the limestone layer, in which segments initiate, because the layers have similar thicknesses, carbonate contents and stiffness properties in both sites (Roche *et al.* 2014). However, the low number of segments at Trescléoux may be due to the higher plasticity of the clay-rich layer and the fault-related folding. We may expect the occurrence of folding to accommodate a part of the extension and, therefore, probably reduces the amount of extension accommodated by segmentation.

Insights into fault development

The results presented here provide insights into the early stages of fault-zone development (Fig. 10c, d, e). In previous work (Roche *et al.* 2012*b*), we showed that, before any restriction, the faults observed at the Trescléoux site grew with increasing maximum displacement, increasing in height but with no increase in width. This present work shows that both the maximum displacement and the width increase, whereas its total height remains constant while the fault is restricted. If the fault propagates through the restrictor, the height again increases with increasing displacement, but the maximum width of the fault will stay constant.

Fault widening depends on the restriction and segmentation capacity of the host rocks (i.e. see the previous subsection on 'Restriction and segmentation'). For a weak restriction and segmentation capacity, the fault might propagate vertically through the restrictor without widening, whereas, for increasing restriction and segmentation capacity, increased widening occurs. Hence, a maximum width is expected at the maximum restriction and segmentation capacity, and the maximum width of faults may change as a function of the host rock (Fig. 11). For instance, a homogenous host rock without any pre-existing fractures probably has a low restriction and segmentation capacity. In such rock, we may expect a narrow fault zone. A layered rock with weak mechanical heterogeneities, but without layer-parallel faults, will have a weak restriction and segmentation capacity. In this case, the width of the fault zone may also stay narrow. Dip refraction and linkage may be the key processes for fault widening. Wide fault zones are expected in rocks with strong restriction and segmentation capacity, as in layered rocks with strong mechanical heterogeneities and with pre-existing fracturing (Fig. 11b). Finally, the maximum width of the fault zone may also vary along a fault propagating across layers with a different capacity for restriction (Fig. 11c). Reciprocally, the width of the fault zone may be an indicator of the restriction and segmentation capacity of the local host rock.

Implications for fluid flow and seismicity

Rocks with a strong restriction and segmentation capacity may comprise low-permeability layers such as clay-rich rocks. Therefore, we may expect such a rock to have a weak vertical permeability. According to the model developed in this paper, throughgoing fault zones are likely to be complex and wide in such rocks. These fault zones have a natural potential to develop large widths and complexities once matured. Such characteristics promote vertical permeability. Hence, a weak vertical permeability of the rock matrix may be counterbalanced by a strong fracture permeability. Obviously, fault-zone permeability also depends on the nature of the fault rocks, clay smearing and self-sealing. Nonetheless, fault-zone widening necessarily impacts on fluid flows and, thus, deserves additional studies.

Although more complex, a similar behaviour may be found for seismicity. Indeed, rocks with a strong restriction and segmentation capacity may also inhibit vertical fault propagation, and, thus, involve numerous smaller events and fewer larger ones due to the limited size of the faults in the early

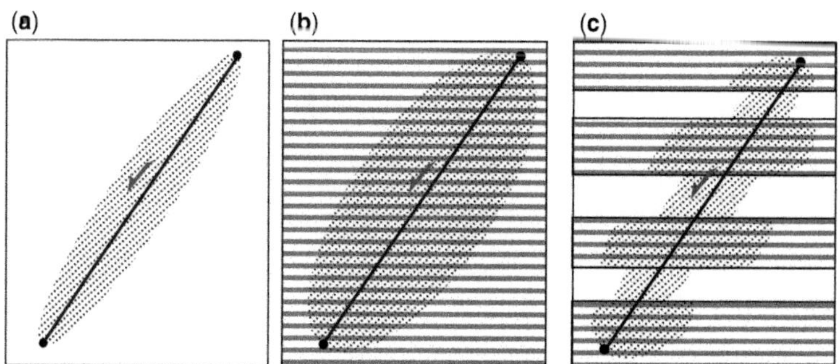

Fig. 11. A model of fault-zone width in different host rocks. In (**a**), the host rock has a weak and homogeneous restriction and segmentation capacity, and, hence, the fault is narrow. In (**b**), the host rock has a strong and homogeneous restriction and segmentation capacity, and the fault zone grows wider. In (**c**), the host rock is heterogeneous with the alternation of rocks with strong and weak restriction and segmentation capacities, and the fault-zone width varies throughout the host rock. Note that in (a)–(c), the different patterns do not represent the layering, but rather the variation in the restriction and segmentation capacity in a rock that may either be layered or not.

stage of faulting. However, the corrugation and segmentation dispersed along throughgoing fault zones and induced during restriction processes may promote fault-rock formation, and the inherent weakening of faults should also promote larger events.

Conclusions

A detailed field analysis of meso-scale normal faults allowed us to document the early stage of fault-zone development in layered rocks. Based on the observations, combined with our previous work, we developed an innovative model of fault-zone evolution in which the complexities and width of fault zones are inherited during the period of fault restriction.

During the early stage of faulting in layered rocks, faults may be vertically restricted by layer-parallel faults or lithological interfaces, or they may propagate vertically. Thus, in the course of displacement accumulation, fault propagation is likely to alternate between periods of vertical restriction and periods of vertical propagation. Significant structural complexities develop during restriction in order to accommodate the increasing displacement: notably, the number of sub-parallel segments increases with increasing maximum displacement. This induces an increase in the width of the fault zone. These structures will then be preserved during vertical propagation.

As a consequence of this model, one may expect the fault-zone width to depend on the capacity of the rock to keep the fault vertically restricted and also its capacity to form fault segments by brittle failure. Among several parameters, our observations indicate that the nature or location of the restrictor, as well as the occurrence of fault-related folding at the fault tip, are critical. Wide fault zones are expected in layered rocks with strong mechanical heterogeneities, and with pre-existing joints and layer-parallel faults. However, fault-related folding may reduce the amount of segmentation occurring during the restriction period. Wide faults probably enhance vertical permeability and, therefore, counterbalance the weak permeability of the rock matrix. In addition, such a model of fault development is likely to impact on the seismic behaviour of the rock.

This work is part of the research studies conducted by IRSN (Institut de Radioprotection et de Sûreté Nucléaire), the UPMC (Université Pierre et Marie Curie) and the University of Alberta on fracture development. We thank the Helmholtz–Alberta Initiative for financial support. We thank Dave Healey and Scott Wilkins for their constructive comments and suggestions, which improved the quality of the manuscript, and also the Volume Editor, Christopher Jackson, for his helpful comments.

References

Angelier, J. 1990. Inversion of field data in fault tectonics to obtain the regional stress – III. A new rapid direct inversion method by analytical means. *Geophysical Journal International*, **103**, 363–376.

Arthaud, F., Mégard, F. & Séguret, M. 1977. Cadre tectonique de quelques bassins sédimentaires. *Bulletin des Centres de Recherches Exploration–Production Elf-Aquitaine*, **1**, 147–188.

Atkinson, B.K. 1984. Subcritical crack growth in geological materials. *Journal of Geophysical Research: Solid Earth*, **89**, 4077–4114.

Barbarand, J., Lucazeau, F., Pagel, M. & Séranne, M. 2001. Burial and exhumation history of the south-eastern Massif Central (France) constrained by apatite fission-track thermochronology. *Tectonophysics*, **335**, 275–290.

Baudrimont, A.F. & Dubois, P. 1977. Un bassin mésogéen du domaine péri-alpin: le Sud-Est de la France. Bulletin des centres de recherches explor.-product. *Bulletin des Centres de Recherches Exploration–Production Elf-Aquitaine*, **1**, 261–308.

Bense, V.F., Gleeson, T., Loveless, S.E., Bour, O. & Scibek, J. 2013. Fault zone hydrogeology. *Earth-Science Reviews*, **127**, 171–192.

Bergerat, F. 1987. Stress fields in the European platform at the time of Africa-Eurasia collision. *Tectonics*, **6**, 99–132.

Brandes, C. & Tanner, D.C. 2014. Fault-related folding: a review of kinematic models and their application. *Earth-Science Reviews*, **138**, 352–370.

Bürgmann, R., Pollard, D.D. & Martel, S.J. 1994. Slip distributions on faults: effects of stress gradients, inelastic deformation, heterogeneous host-rock stiffness, and fault interaction. *Journal of Structural Geology*, **16**, 1675–1690.

Byerlee, J.D. 1993. Model for episodic flow of high-pressure water in fault zones before earthquakes. *Geology*, **21**, 303–306.

Caine, J.S., Evans, J.P. & Forster, C.B. 1996. Fault zone architecture and permeability structure. *Geology*, **24**, 1025–1028.

Cartwright, J.A., Trudgill, B.D. & Mansfield, C.S. 1995. Fault growth by segment linkage: an explanation for scatter in maximum displacement and trace length data from the Canyonlands Grabens of SE Utah. *Journal of Structural Geology*, **17**, 1319–1326.

Childs, C., Nicol, A., Walsh, J.J. & Watterson, J. 1996. Growth of vertically segmented normal faults. *Journal of Structural Geology*, **18**, 1389–1397.

Childs, C., Manzocchi, T., Walsh, J.J., Bonson, C.G., Nicol, A. & Schöpfer, M.P. 2009. A geometric model of fault zone and fault rock thickness variations. *Journal of Structural Geology*, **31**, 117–127.

Collettini, C., Niemeijer, A., Viti, C. & Marone, C. 2009. Fault zone fabric and fault weakness. *Nature*, **462**, 907–910.

Cooke, M.L. & Underwood, C.A. 2001. Fracture termination and step-over at bedding interfaces due to frictional slip and interface opening. *Journal of Structural Geology*, **23**, 223–238.

Cowie, P.A. & Scholz, C.H. 1992. Physical explanation for the displacement-length relationship of faults using a post-yield fracture mechanics model. *Journal of Structural Geology*, **14**, 1133–1148.

Cosgrove, J.W. & Ameen, M.S. 1999. A comparison of the geometry, spatial organization and fracture patterns associated with forced folds and buckle folds. *In*: Cosgrove, J.W. & Ameen, M.S. (eds) *Forced Folds and Fractures*. Geological Society, London, Special Publications, **169**, 7–21, https://doi.org/10.1144/GSL. SP.2000.169.01.02

d'Alessio, M.A. & Martel, S.J. 2004. Fault terminations and barriers to fault growth. *Journal of Structural Geology*, **26**, 1885–1896.

Davison, I. 1987. Normal fault geometry related to sediment compaction and burial. *Journal of Structural Geology*, **9**, 393–401.

Faulkner, D.R., Jackson, C.A.L., Lunn, R.J., Schlische, R.W., Shipton, Z.K., Wibberley, C.A.J. & Withjack, M.O. 2010. A review of recent developments concerning the structure, mechanics and fluid flow properties of fault zones. *Journal of Structural Geology*, **32**, 1557–1575.

Ferrill, D.A. & Morris, A.P. 2003. Dilational normal faults. *Journal of Structural Geology*, **25**, 183–196.

Ferrill, D.A., McGinnis, R.N., Morris, A.P. & Smart, K.J. 2012*a*. Hybrid failure: field evidence and influence on fault refraction. *Journal of Structural Geology*, **42**, 140–150.

Ferrill, D.A., Morris, A.P. & McGinnis, R.N. 2012*b*. Extensional fault-propagation folding in mechanically layered rocks: the case against the frictional drag mechanism. *Tectonophysics*, **576**, 78–85.

Gross, M.R., Bai, T., Wacker, M.A., Collinsworth, K.B. & Behl, R.J. 1997. Influence of mechanical stratigraphy and kinematics on fault scaling relations. *Journal of Structural Geology*, **19**, 171–183.

Gudmundsson, A. 1987. Geometry, formation and development of tectonic fractures on the Reykjanes Peninsula, southwest Iceland. *Tectonophysics*, **139**, 295–308.

Guilhaumou, N., Touray, J.C., Perthuisot, V. & Roure, F. 1996. Palaeocirculation in the basin of southeastern France sub-alpine range: a synthesis from fluid inclusions studies. *Marine and Petroleum Geology*, **13**, 695–706.

Homberg, C., Schnyder, J., Roche, V., Leonardi, V. & Benzaggagh, M. In review. The brittle and ductile components of displacement along fault zones. *In*: Childs, C., Holdsworth, R.E., Jackson, C.A.-L., Manzocchi, T., Walsh, J.J. & Yielding, G. (eds) *The Geometry and Growth of Normal Faults*. Geological Society, London, Special Publications, **439**.

Jackson, C.A.L., Gawthorpe, R.L. & Sharp, I.R. 2006. Style and sequence of deformation during extensional fault-propagation folding: examples from the Hammam Faraun and El-Qaa fault blocks, Suez Rift, Egypt. *Journal of Structural Geology*, **28**, 519–535.

Kim, Y.S., Peacock, D.C. & Sanderson, D.J. 2004. Fault damage zones. *Journal of Structural Geology*, **26**, 503–517.

Kranz, R.L. 1983. Microcracks in rocks: a review. *Tectonophysics*, **100**, 449–480.

Lockner, D.A. 1995. Rock failure. *In*: Ahrens, T.J. (ed.) *Rock Physics & Phase Relations: A Handbook of Physical Constants*. American Geophysical Union, Washington, DC, 127–147.

Long, J.J. & Imber, J. 2010. Geometrically coherent continuous deformation in the volume surrounding a seismically imaged normal fault-array. *Journal of Structural Geology*, **32**, 222–234.

McGrath, A.G. & Davison, I. 1995. Damage zone geometry around fault tips. *Journal of Structural Geology*, **17**, 1011–1024.

Micarelli, L. & Benedicto, A. 2008. Normal fault terminations in limestones from the SE-Basin (France): implications for fluid flow. *In*: Wibberley, C.A.J., Kurz, W., Imber, J., Holdsworth, R.E. & Collettini, C. (eds) *The Internal Structure of Fault Zones: Implications for Mechanical and Fluid-Flow Properties*. Geological Society, London, Special Publications, **299**, 123–138, https://doi.org/10.1144/SP299.8

Micarelli, L., Benedicto, A. & Wibberley, C.A.J. 2006. Structural evolution and permeability of normal fault zones in highly porous carbonate rocks. *Journal of Structural Geology*, **28**, 1214–1227.

Muraoka, H. & Kamata, H. 1983. Displacement distribution along minor fault traces. *Journal of Structural Geology*, **5**, 483–495.

Nicol, A., Watterson, J., Walsh, J.J. & Childs, C. 1996. The shapes, major axis orientations and displacement patterns of fault surfaces. *Journal of Structural Geology*, **18**, 235–248.

Pagel, M., Braun, J.J., Disnar, J.R., Martinez, L., Renac, C. & Vasseur, G. 1997. Thermal history constraints from studies of organic matter, clay minerals, fluid inclusions, and apatite fission tracks at the Ardeche paleo-margin (BA1 drill hole, GPF Program), France. *Journal of Sedimentary Research*, **67**, 235–245.

Paterson, M.S. & Wong, T.F. 2005. *Experimental Rock Deformation: the Brittle Field*. Springer, Berlin.

Peacock, D.C.P. & Sanderson, D.J. 1991. Displacements, segment linkage and relay ramps in normal fault zones. *Journal of Structural Geology*, **13**, 721–733.

Peacock, D.C.P. & Sanderson, D.J. 1992. Effects of layering and anisotropy on fault geometry. *Journal of the Geological Society, London*, **149**, 793–802, https://doi.org/10.1144/gsjgs.149.5.0793

Peacock, D.C.P. & Zhang, X. 1994. Field examples and numerical modelling of oversteps and bends along normal faults in cross-section. *Tectonophysics*, **234**, 147–167.

Peacock, D.C.P., Knipe, R.J. & Sanderson, D.J. 2000. Glossary of normal faults. *Journal of Structural Geology*, **22**, 291–305.

Petit, J.P. 1995. Palaeostress superimposition deduced from mesoscale structures in limestone: the Matelles exposure, Languedoc, France. *Journal of Structural Geology*, **17**, 245–256.

Peyaud, J.B., Barbarand, J., Carter, A. & Pagel, M. 2005. Mid-Cretaceous uplift and erosion on the northern margin of the Ligurian Tethys deduced from thermal history reconstruction. *International Journal of Earth Sciences*, **94**, 462–474.

Robertson, E.C. 1982. Continuous formation of gouge and breccia during fault displacement. *In*: Goodman,

R.E. & Heuze, F.E. (eds) *Issues in Rock Mechanics, Proceedings of the 23rd Symposium on Rock Mechanics*. American Institute of Mining, Metallurgical, and Petroleum Engineers, New York, 379–404.
Roche, V., Homberg, C. & Rocher, M. 2012*a*. Architecture and growth of normal fault zones in multilayer systems: a 3D field analysis in the South-Eastern Basin, France. *Journal of Structural Geology*, **37**, 19–35.
Roche, V., Homberg, C. & Rocher, M. 2012*b*. Fault displacement profiles in multilayer systems: from fault restriction to fault propagation. *Terra Nova*, **24**, 499–504.
Roche, V., Homberg, C. & Rocher, M. 2013. Fault nucleation, restriction, and aspect ratio in layered sections: quantification of the strength and stiffness roles using numerical modeling. *Journal of Geophysical Research: Solid Earth*, **118**, 4446–4460.
Roche, V., Homberg, C., David, C., & Rocher, M. 2014. Normal faults, layering and elastic properties of rocks. *Tectonophysics*, **622**, 96–109.
Rotevatn, A. & Jackson, C.A.L. 2014. 3D structure and evolution of folds during normal fault dip linkage. *Journal of the Geological Society, London*, **171**, 821–829, https://doi.org/10.1144/jgs2014-045
Roure, F., Brun, J.P., Colletta, B. & Van Den Driessche, J. 1992. Geometry and kinematics of extensional structures in the Alpine foreland basin of southeastern France. *Journal of Structural Geology*, **14**, 503–519.
Scholz, C.H. 1987. Wear and gouge formation in brittle faulting. *Geology*, **15**, 493–495.
Scholz, C.H. 2002. *The Mechanics of Earthquakes and Faulting*. Cambridge University Press, Cambridge.
Schöpfer, M.P., Childs, C. & Walsh, J.J. 2006. Localisation of normal faults in multilayer sequences. *Journal of Structural Geology*, **28**, 816–833.
Segall, P. & Pollard, D.D. 1980. Mechanics of discontinuous faults. *Journal of Geophysical Research: Solid Earth*, **85**, 4337–4350.
Séranne, M., Camus, H., Lucazeau, F., Barbarand, J. & Quinif, Y. 2002. Surrection et érosion polyphasées de la bordure cévenole. Un exemple de morphogenèse lente. *Bulletin de la Société géologique de France*, **173**, 97–112.
Sibson, R.H. 1977. Fault rocks and fault mechanisms. *Journal of the Geological Society, London*, **133**, 191–213, https://doi.org/10.1144/gsjgs.133.3.0191
Sibson, R.H. 1996. Structural permeability of fluid-driven fault-fracture meshes. *Journal of Structural Geology*, **18**, 1031–1042.
Smart, K.J., Ferrill, D.A. & Morris, A.P. 2009. Impact of interlayer slip on fracture prediction from geomechanical models of fault-related folds. *American Association of Petroleum Geologists Bulletin*, **93**, 1447–1458.
Soliva, R. & Benedicto, A. 2004. A linkage criterion for segmented normal faults. *Journal of Structural Geology*, **26**, 2251–2267.
Soliva, R. & Benedicto, A. 2005. Geometry, scaling relations and spacing of vertically restricted normal faults. *Journal of Structural Geology*, **27**, 317–325.
Tanner, P.W. 1989. The flexural-slip mechanism. *Journal of Structural Geology*, **11**, 635–655.
Townend, J. & Zoback, M.D. 2000. How faulting keeps the crust strong. *Geology*, **28**, 399–402.
van der Zee, W., Wibberley, C.A. & Urai, J.L. 2008. The influence of layering and pre-existing joints on the development of internal structure in normal fault zones: the Lodève basin, France. *In*: Wibberley, C.A.J., Kurz, W., Imber, J., Holdsworth, R.E. & Collettini, C. (eds) *The Internal Structure of Fault Zones: Implications for Mechanical and Fluid-Flow Properties*. Geological Society, London, Special Publications, **299**, 57–74, https://doi.org/10.1144/SP299.4
Wallace, R.E. & Morris, H.T. 1986. Characteristics of faults and shear zones in deep mines. *Pure and Applied Geophysics*, 124, 107–125.
Walsh, J.J., & Watterson, J. 1987. Distributions of cumulative displacement and seismic slip on a single normal fault surface. *Journal of Structural Geology*, **9**, 1039–1104.
Walsh, J.J., Bailey, W.R., Childs, C., Nicol, A. & Bonson, C.G. 2003. Formation of segmented normal faults: a 3-D perspective. *Journal of Structural Geology*, **25**, 1251–1262.
Watterson, J., Childs, C. & Walsh, J.J. 1998. Widening of fault zones by erosion of asperities formed by bed-parallel slip. *Geology*, **26**, 71–74.
Weber, K.J., Mandl, G.J., Pilaar, W.F., Lehner, B.V.F. & Precious, R.G. 1978. The role of faults in hydrocarbon migration and trapping in Nigerian growth fault structures. Paper presented at the Offshore Technology Conference, 8–11 May 1978, Houston, TX.
Welch, M.J., Davies, R.K., Knipe, R.J. & Tueckmantel, C. 2009. A dynamic model for fault nucleation and propagation in a mechanically layered section. *Tectonophysics*, **474**, 473–492.
Wibberley, C.A., Petit, J.P. & Rives, T. 2007. The effect of tilting on fault propagation and network development in sandstone–shale sequences: a case study from the Lodève Basin, southern France. *Journal of the Geological Society, London*, **164**, 599–608, https://doi.org/10.1144/0016-76492006-047
Wibberley, C.A., Yielding, G. & Di Toro, G. 2008. Recent advances in the understanding of fault zone internal structure: a review. *In*: Wibberley, C.A.J., Kurz, W., Imber, J., Holdsworth, R.E. & Collettini, C. (eds) *The Internal Structure of Fault Zones: Implications for Mechanical and Fluid-Flow Properties*. Geological Society, London, Special Publications, **299**, 5–33, https://doi.org/10.1144/SP299.2
Wilkins, S.J. & Gross, M.R. 2002. Normal fault growth in layered rocks at Split Mountain, Utah: influence of mechanical stratigraphy on dip linkage, fault restriction and fault scaling. *Journal of Structural Geology*, **24**, 1413–1429.

Fracture networks of normal faults in fine-grained sedimentary rocks: examples from Kilve Beach, SW England

TORE SKAR[1,3], SILJE S. BERG[1,4], ROY H. GABRIELSEN[1,2]* & ALVAR BRAATHEN[1,2]

[1]*Centre for Integrated Petroleum Research, University of Bergen, Norway*

[2]*Department of Geosciences, University of Oslo, Norway*

[3]*Present address: Suncor Energy, Stavanger, Norway*

[4]*Present address: Statoil, Trondheim, Norway*

**Correspondence: r.h.gabrielsen@geo.uio.no*

Abstract: Interbedded shale and limestone successions in the Kilve Beach area, Bristol Channel Basin, UK, provide insights on fracture networks around normal faults in fine-grained lithologies. Fracture sets with distinct orientations are characteristic of both shale and limestone beds. Shear fractures (mode II) predominate in the shaly units, and they have typically more gentle dips and a larger spread in orientations than extension veins and shear fractures in the limestones. Fracture intensities decrease away from the fault core, but maximum intensities, total number of fractures and widths of the damage zones appear to be independent of throw for normal faults with offsets of less than 20 m. Thus, there is no clear systematic relationship between fault throw and damage zone width in the shales studied by us. However, an asymmetry in the fracture distribution is evidenced by a wider hanging-wall damage zone and differences in fracture orientations in some cases. We interpret the asymmetry and spread in fracture orientations to be the result of propagating fault-tip process zones and the tempo-spatial impact of fault-slip events.

Most normal faults encompass strongly deformed fault cores (semi-penetrative strain) enveloped by damage zones of discrete structures in populations (fractures and deformation bands) that are surrounded by undeformed protolith (e.g. Chester & Logan 1986; Caine *et al.* 1996; Knott *et al.* 1996; Kim *et al.* 2004; Berg & Skar 2005; Agosta & Aydin 2006; Bastesen *et al.* 2009; Braathen *et al.* 2009, 2013; Gabrielsen & Braathen 2014; Gabrielsen *et al.* 2016). The geometry and flow characteristics of the fault core and the damage zone may vary considerably, imposing severe restrictions on the predictability and fluid flow characteristics within and across fault zones (Fredman *et al.* 2008; Torabi *et al.* 2015). The fault core is associated with the bulkshear deformation and is usually regarded as a barrier to fluid flow owing to its large amount of fine-grained material (e.g. fault gouge, clay smear). The damage zone can be associated with high-frequency fracture networks that may cause the rock permeability to increase (e.g. Chester & Logan 1986; Andersson *et al.* 1991; Baines & Worden 2004; Braathen *et al.* 2009) or decrease in the case of cataclastic deformation bands in porous rocks (e.g. Gabrielsen & Koestler 1987; Fowles & Burley 1994; Antonellini & Aydin 1994, 1995; Lothe *et al.* 2002; Fossen & Bale 2007; Fossen *et al.* 2007). Faulting-related deformation may occur from sequential fault-tip propagation and arrest during fault growth (Cowie & Scholz 1992; Anders & Wiltschenko 1994; Vermilye & Scholz 1998, Cowie *et al.* 2009), reflecting the concentrated local stress field that exists near the fault-tip zone (e.g. Reches & Lockner 1994). Such processes can occur simultaneously in different segments of the fault and/or during slip along an existing fault (Braathen *et al.* 2013). Deformation associated with slip along existing faults can be enhanced by stress concentrations developed at irregularities along the faults along the slip plane itself (e.g. Scholz 1987; Childs *et al.* 1997), at steps between segments (e.g. Peacock & Zhang 1993; Gabrielsen & Clausen 2001; Childs *et al.* 2009; Cowie *et al.* 2009) or at fault reactivation arrest points, or occur owing to the contrasting stress configurations when the hanging wall and footwall are compared.

It has long been recognized that the fracture frequency increases in the vicinity of major faults (e.g. Chester & Logan 1986; Anders & Wiltschenko 1994; Evans & Chester 1995; Schulz & Evans 1998; Braathen *et al.* 2009; Schueller *et al.* 2012) and that subsidiary faults and fractures may have preferred orientations relative to the master faults (Friedman 1969; Brock & Engelder 1977; Chester & Logan

From: Childs, C., Holdsworth, R. E., Jackson, C. A.-L., Manzocchi, T., Walsh, J. J. & Yielding, G. (eds) 2017. *The Geometry and Growth of Normal Faults*. Geological Society, London, Special Publications, **439**, 289–306.
First published online September 26, 2016, https://doi.org/10.1144/SP439.10

1987; Anders & Wiltschenko 1994; Bruhn *et al.* 1994; Gabrielsen & Braathen 2014). However, most of these earlier studies were mainly concerned with the structural deformation related to strike-slip and thrust faults. More recently, spatial relationships between fractures occurring in the damage zone of normal faults and, in particular, in porous siliciclastic sediments have been published (e.g. Knott *et al.* 1996; Aarland & Skjerven 1998; Gabrielsen *et al.* 1998; Beach *et al.* 1999, Hesthammer *et al.* 2000; Shipton & Cowie 2001, 2003; Braathen *et al.* 2009; Brogi 2011; Schueller *et al.* 2012). Outcrop studies have indicated that there are differences in fracture characteristics in the footwall and hanging-wall damage zones (e.g. Jamison & Stearns 1982; Underhill & Woodcock 1987; Koestler & Ehrmann 1991; Hippler 1993; Fowles & Burley 1994; Berg & Skar 2005), indicating that there may be contrasting stress magnitude and configuration in the footwall and the hanging wall during faulting.

Despite mudrocks being volumetrically dominant in many sedimentary basins, commonly comprising 60–70% of the sediments (Aplin *et al.* 1995), and top seals and intraformational baffles to fluid flow usually being mudrocks and shales, little attention (with some exceptions; McGrath & Davison 1995; Gabrielsen & Kløvjan 1997; Ogata *et al.* 2012, 2014*a*, *b*; Braathen *et al.* 2013) has been given to fault-related damage in such rocks. Furthermore, fracturing in fine-grained lithologies may provide communication between reservoir units and leakage from top seals (Ogata *et al.* 2014*a*, *b*). However, fractures in shales are commonly considered to be related to primary hydrofracturing associated with abnormal fluid pressures or by reopening of pre-existing hydrofractures or tectonic fractures (Gudmundsson 2011; Bohloli *et al.* 2014).

The present paper focuses on fracture networks of damage zones of normal faults in a shale–limestone succession in the Kilve Beach area, SW England. Although the depth of burial during faulting is poorly constrained, the lithologies were sufficiently cohesive to allow for brittle fracturing and subsequent calcite mineralization during faulting. The main aim of this study is to evaluate the fracture characteristics in fault damage zones within shale in order to understand the relationship between different fracture populations in the damage zones and their spatial distributions. The results are discussed in light of fracture systems and the nature of damage zones in mudrock and shale compared with damage zones in other sedimentary rocks.

Geological framework

The study area is situated on the southern margin of the east–west-trending Bristol Channel Basin (Fig. 1), which is one of a series of elongate, fault-bounded sedimentary basins in the Celtic Sea/Irish Sea region. The Bristol Channel Basin sits on a basement of Carboniferous limestones and Devonian sandstones and slates that were deformed by north–south-oriented contraction during the Variscan Orogeny (Dart *et al.* 1995; Nemcok *et al.* 1995; Kelly *et al.* 1998). The basin formed primarily during the Jurassic to Early Cretaceous (Kamerling 1979; Chadwick 1986; Karner *et al.* 1987; Lake & Karner 1987), and was later tectonically inverted during Cretaceous and Tertiary times with the formation of reverse and strike-slip faults.

The Upper Triassic and Lower Jurassic rocks that are now exposed in coastal cliffs and strandflats in the Watchet, Kilve and Lilstock areas were deformed by faults related to the Mesozoic development of the Bristol Channel Basin (Chadwick 1986; Brooks *et al.* 1988; Donato 1988). Faults in the Kilve area cut through calcareous and organic-rich shales and laterally continuous limestone beds with thickness of less than 1 m.

Owing to its excellent exposures and easy accessibility, the Bristol Channel area has been studied intensively by several research groups focusing on fault evolution and fracture systems. The extensional structures associated with the Mesozoic rifting have been investigated for analyses of displacement, segment linkage and relay ramps (Peacock & Sanderson 1991, 1993; Kim *et al.* 2004), oversteps and bends along normal faults (Peacock & Zhang 1993), normal fault array polarity and detachments (Stewart & Argent 2000), and fault patterns at different scales (Peacock 1996). Other studies include analyses of fault tip zones (Pickering *et al.* 1997), fault slip evolution (Davison 1995) and shear fracturing and pressure solution (Peacock & Sanderson 1995). Some faults have been investigated with respect to linkage and the evolution of conjugate strike-slip faults (Kelly *et al.* 1998). Finally, Peacock & Sanderson (1995) demonstrated the effect of layering and anisotropy on the geometry of minor conjugate fault sets, while Kelly *et al.* (1998) showed how the development of conjugate strike-slip faults in limestone, at various scales, may be described with a common fault development model.

The main structures associated with the inversion phase in the Tertiary involve reactivated and inverted Mesozoic normal faults (Dart *et al.* 1995; Kelly *et al.* 1998), thrusts and strike-slip faults (Holloway & Chadwick 1986; Peacock & Sanderson 1995; Willemse *et al.* 1997). A regionally distributed system of orthogonal joints, previously described by Rawnsley *et al.* (1998) and Engelder & Peacock (2001), is also related to the later phases of deformation of the basin. Rawnsley *et al.* (1998) illustrated the importance of changes in

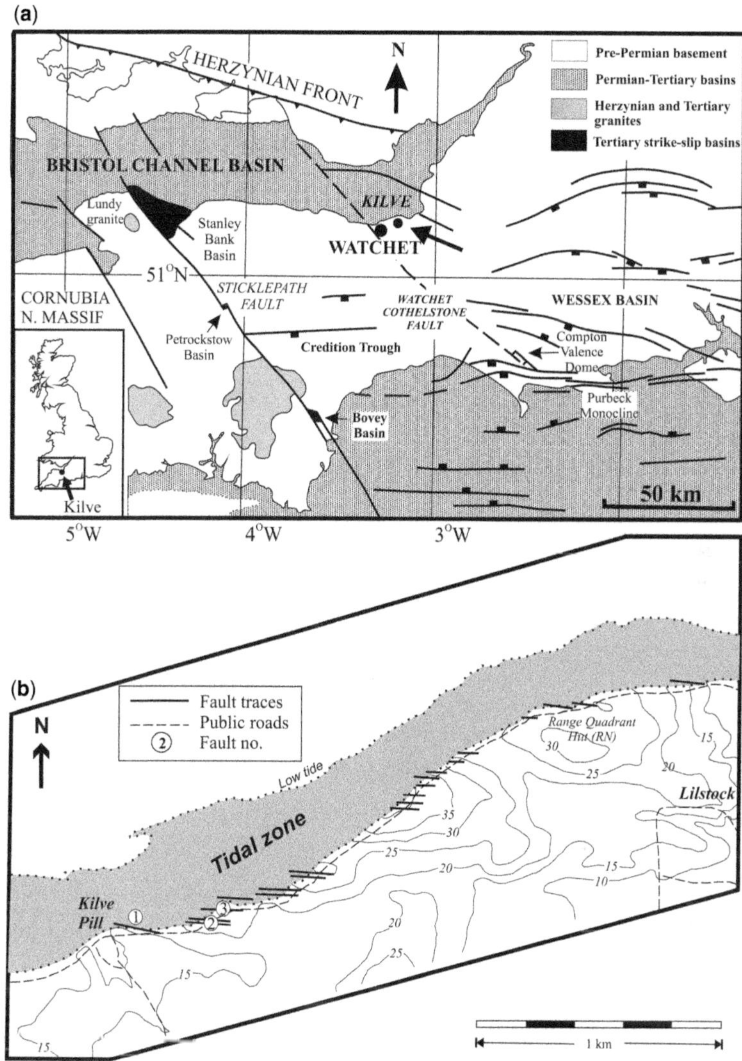

Fig. 1. (**a**) Geological map of the Bristol Channel and surroundings showing the main structural features (after Dart *et al.* 1995). The thick arrow indicates the location of the study area. (**b**) Distribution of normal faults between Kilve Pill and Lilstock. Numbers indicate locations of the faults analysed in this study.

stresses in the development of joints, whereas Engelder & Peacock (2001) considered the possibility that one or more joint sets developed during the Alpine shortening, and proposed a new mechanism for the joint driving stress.

Field data and methods

The present study focuses on faults for which the displacement can be observed and measured or confidently inferred. Data were collected in the field by mapping at outcrop-scale backed by a series of profiles oriented perpendicular to the strike of the master faults. Detailed measurements of fracture frequency and orientations were made along traverses extending from the major faults into the footwalls and hanging walls. The width of the damage zone was defined as the distance away from the master faults, where the fracture frequency exceeds the background fracture frequency. This is a threshold similar to that defined by Schueller *et al.* (2012) and Gabrielsen & Braathen (2014). In addition to the scan lines, two-dimensional sections were mapped in detail at different scales and at several positions within the damage zones in order to identify the different fracture populations and their spatial relationship.

The *master faults* are the vertically and/or horizontally most extensive and widest structures, commonly having the largest displacements. The term *fracture* is used in a general sense for shear and tensional fractures. The fractures are referred to as *synthetic* when they dip in the same direction and have similar displacement-sense as the master fault and *antithetic* when the dip and displacement are opposite. A *fracture population* refers to a group of fractures with similar orientation that may either be a shear or a tension fracture (e.g. Gabrielsen & Braathen 2014).

The results in this paper are mainly based on the analysis of three faults located near Kilve Pill (Faults 1–3, Figs 1b & 2). The faults are part of a series of more than 30 north-dipping planar normal faults separating rotated fault blocks situated between Kilve Pill and Lilstock in Somerset, West England. Fault 1 (Fig. 2a, b) has a total dip-slip displacement of 19 m and is exposed in three subparallel 10–15 m-high vertical cliff sections and in horizontal exposures for some tens of metres. The main fault plane exhibits a planar geometry within the limit of the vertical exposures, but displays irregularities in terms of curvature and continuity in the horizontal exposures. Faults 2 and 3 are also well exposed in vertical and horizontal sections and have normal dip-slip displacements of 5 and 15 m, respectively. The average orientations of these master faults are 285/33°, 288/32° and 297/42°. The low dip angle of the faults is caused by Tertiary inversion that tilted the strata to a present day dip of 15–20°.

The three studied faults are characterized by well-developed fault cores and their associated damage zones. The fault core is commonly separated from the damage zone by a distinct fault wall (see also Gabrielsen *et al.* 2016) and sometimes by zones of normal drag. The cores are 0.05–0.8 m thick and include a (central) slip zone (commonly

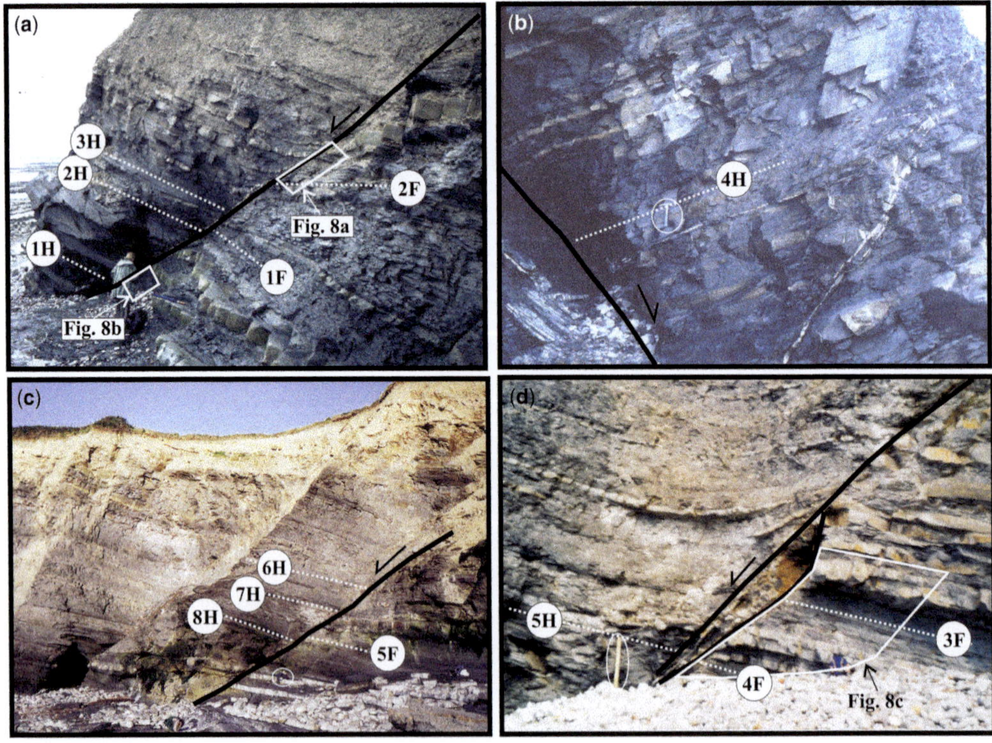

Fig. 2. Photographs of interbedded limestone–shale successions in which Faults 1, 2 and 3 are excellently exposed. (**a**) Fault 1 (19 m displacement), see person for scale. (**b**) Hanging wall of Fault 1 at a different locality, see hammer for scale. (**c**) Fault 2 (5 m displacement), rucksack for scale. (**d**) Fault 3 (15 m displacement), wooden stick for scale is 1.4 m. The white boxes and notations in white circles indicate the locations of the detailed 2D profiles shown in Figure 8. White stippled lines indicate the locations of the fracture frequency profiles shown in Figure 8. See Figure 1b for the location of the master faults.

0.03–0.15 m thick) with multiple subordinate fault-parallel dip-slip surfaces separating lensoid rock bodies and zones of secondary calcite mineralization (Berg 2005).

Damage zone fracture populations

The rocks in the damage zone include shale and limestone beds that encompass fracture populations with calcite mineralized shear and extensional fractures. Shear fractures with millimetre- to centimetre-scale displacements are frequently observed and appear in both lithologies; however, they predominate within the shale beds compared with the limestone beds. Veins are by far more common in the limestone beds than in the shale beds.

Fracture populations in shale beds

Synthetic and antithetic fractures are common in the footwall and hanging wall of the master faults. Based on the analysis of 515 fracture measurements, it is evident that antithetic fractures occur more frequently in the hanging wall than in the footwall (Fig. 3). The synthetic fractures of the footwall and hanging wall have orientations that are mainly strike-parallel to the master faults (Fig. 3a); however, there is a wider spread in the hanging-wall fracture orientations compared with the footwall. The antithetic fractures in the hanging wall show a similar distribution to the synthetic fractures in the footwall, whereas the antithetic fractures in the footwall show a wider spread of orientation (Fig. 3b, c). The synthetic and antithetic fractures show both an overall dip distribution varying between 20° and 85°.

Based on the position and orientation of the fractures, fracture characteristics (like fracture mode and mineralization) and crosscutting relationships, six fracture populations were identified in the shale beds. The spatial distribution of these fracture populations varies within the damage zones of the individual faults. The preferred fracture orientations and their relations to the master faults are similar (Fig. 4). Three main fracture populations were identified in the hanging wall (populations I–III) and three populations were identified in the footwall (populations IV–VI):

Fracture population I (Fig. 5a, c) comprises fractures that are planar to slightly curved and are parallel or sub-parallel to the master fault, i.e. dip moderately north. Many of the fracture surfaces are striated and have identifiable dip-slip with millimetre- and centimetre-scale displacements. The fracture surfaces commonly have dip-length traces on the centimetre- and metre-scale.

Fracture population II is also strike-parallel and synthetic to the master faults. The dips are, however, significantly shallower than that of the population I-fractures, on average dipping 15–20° more shallowly than the master fault. Dip-slip displacements along some of these fractures show that they are shear-related. However, most of the fractures do not display visible displacement, but they tend to cluster around the low-angled shear-fractures, indicating that they are genetically related. Population II fractures are less abundant than fractures of population I.

Fracture population III contains fractures antithetic to the master faults (Fig. 5b–d) that display a wide range of geometries. Strike is east–west with steep to moderate southerly dip, with a variety in strike and dip caused by curved fracture geometry. Slip lineations and small displacements attest to shear. Some of the fractures splay from the master

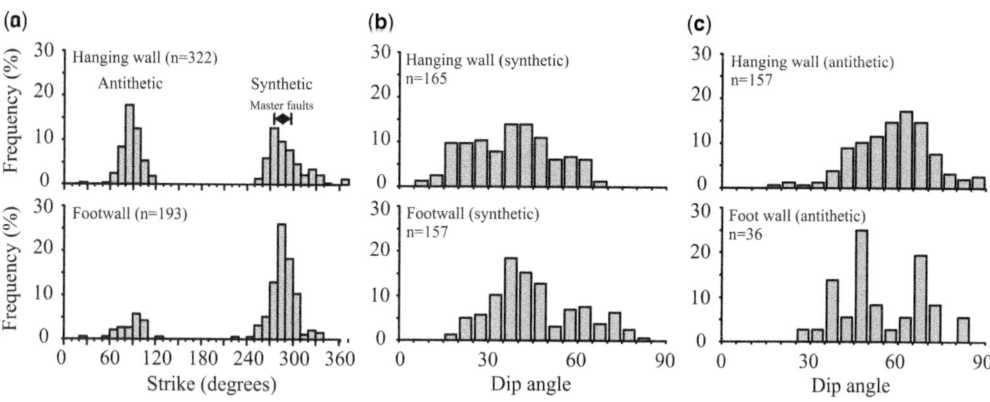

Fig. 3. Histograms showing the orientation data measured in the shale beds in the damage zones of three studied faults (Faults 1–3). (**a**) Strike directions of the synthetic and antithetic fractures in the footwall and hanging wall. (**b**) Dip angles of the synthetic fractures in the footwall and hanging wall. (**c**) Dip angles of the antithetic fractures in the footwall and hanging wall.

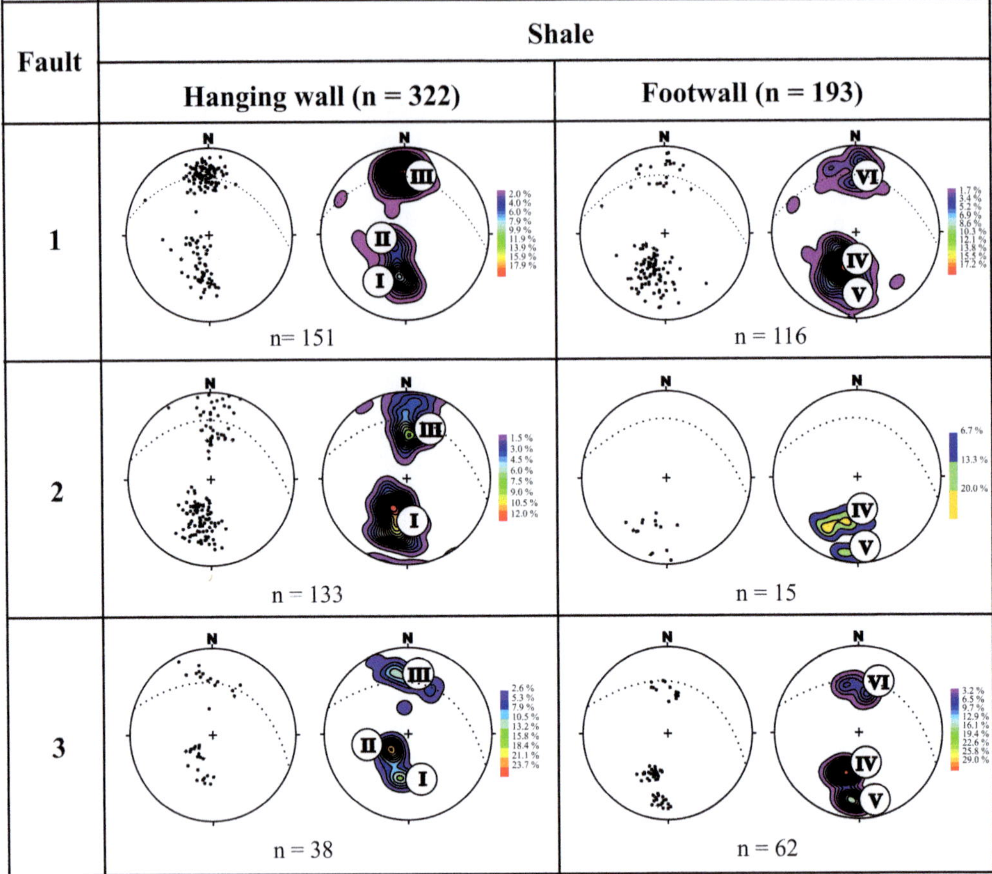

Fig. 4. Orientation data for six fracture populations (see text) in the shales that are identified in the hanging-wall and footwall damage zones of the three studied faults. Equal area, lower hemisphere stereo-net, plotted as poles to planes and contour plots of poles to fracture planes. Orientation of master faults is indicated as planes.

faults at a low strike angle (20–40°) and steepen to near vertical away from the fault, similar to what is described by McGrath & Davison (1995). The population III fractures are usually longer and more continuous than fractures of population I and II.

Fracture population IV (Fig. 6a, b) consists of fractures that are parallel or sub-parallel to the master fault and is considered to be the footwall equivalent of population I in the hanging wall (see above). No major differences in the characteristics and dimensions of the fractures of these populations have been recognized. Thus, the only criterion for distinguishing between the two populations is their position in the footwall and hanging-wall damage zones.

Fracture population V (Fig. 6a–c) consists of synthetic fractures with planar to slightly curved geometry and strikes parallel to population IV fractures, although they dip more steeply (50–70° to the north) where discernible from population IV. The fractures of this population have slip lineations. Population V fractures intersect fractures of population IV at an angle of 15–30° (Fig. 6b).

Fracture population VI (Fig. 6c) has a planar geometry and strikes 060–110°, dipping to the south. Some of the fractures display no visible displacement. The shear fractures are several times longer (1–2 m) and more continuous than those without visible displacement. The latter tend to terminate against bedding interfaces. These antithetic footwall fractures are equivalent to population III observed in the hanging wall.

In addition, a bedding-parallel fracture set (not shown in Fig. 4) and the bedding surfaces themselves are sometimes affected by calcite mineralization and slickenside lineations, indicating that bedding-parallel shear occurred during deformation. These fractures are less frequent than the

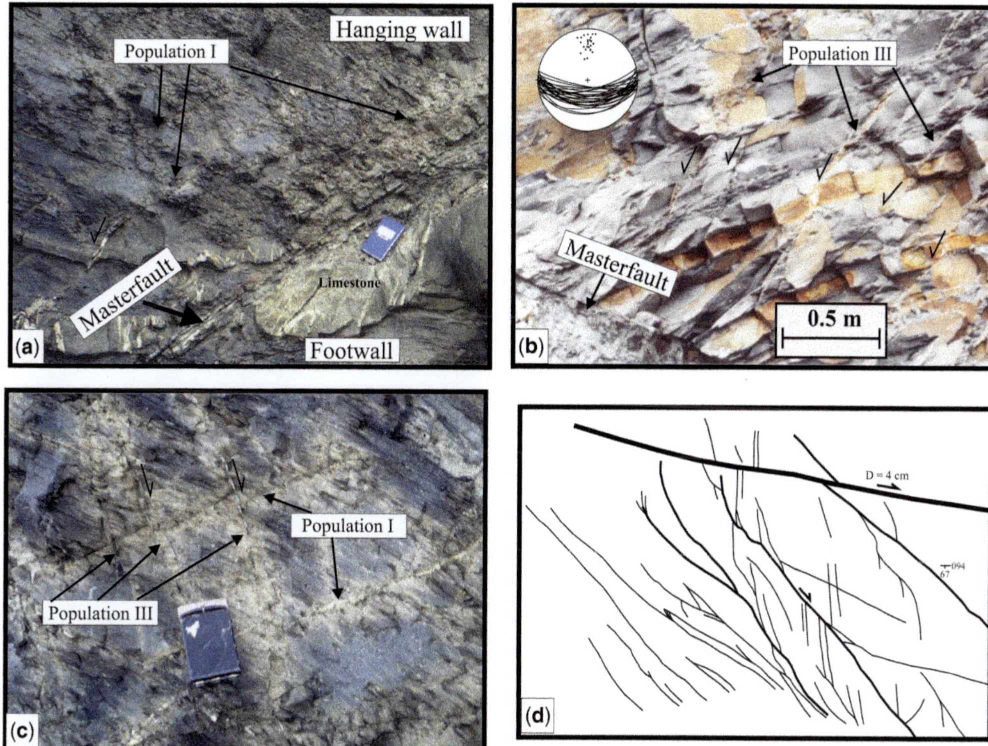

Fig. 5. Photographs of fracture populations in the shales occurring in the hanging-wall damage zones. (**a**) Fault parallel synthetic fractures of population I (notebook for scale). (**b**) Antithetic shear fractures of population III (note opposite orientation compared with other photographs). (**c**) Cross-cutting relationship between hanging-wall fractures showing antithetic fractures offsetting synthetic fractures of population I (notebook for scale). (**d**) Example of bedding-parallel slip causing segmentation of antithetic fractures/faults with horsetail geometry. Displacement along the bed boundary is 4 cm.

other populations. Although this fracture population locally can be very obvious, the frequency of it is easily underestimated because displacements are difficult to identify along shale bed boundaries. Bedding-parallel fractures are mainly identified in the hanging-wall damage zones (Fig. 5d).

The synthetic fractures of populations I and IV (hanging wall and footwall) cluster around the planes of the master faults (Fig. 4). Hence, the average strike of these populations is N 290° E (WNW–ESE) and the populations can be assumed to constitute one homogeneous fracture set. It is evident, however, that the strike distributions of the other fracture populations deviate from those of populations I and IV in a systematic way. In the hanging wall, fracture population II (low-angle shear fractures) has a strike that is skewed towards the NW–SE, whereas the strike distribution of the antithetic shear fractures (fracture population III) is skewed towards east–west. In the footwall, there is also a systematic variation in fracture orientations. The strike distribution of fractures of populations V and VI are gradually skewed towards east–west as compared with the fractures of population IV. This rotation is less obvious in Fault 2, although the average orientation confirms this general change in fracture orientation also here.

The orientation of fractures of population I and IV averages that of the respective master faults. The low-angle fractures of population II dip on average 20° less steeply than do the master faults, whereas the population V fractures of the footwall typically have dips that are significantly steeper (>25°) than the master faults. The antithetic fractures of the hanging wall (fracture population III) and footwall (fracture population VI) have on average dips that are oriented normal to the master faults. The implications of the populations and the asymmetry between footwall and hanging wall are addressed in the Discussion section.

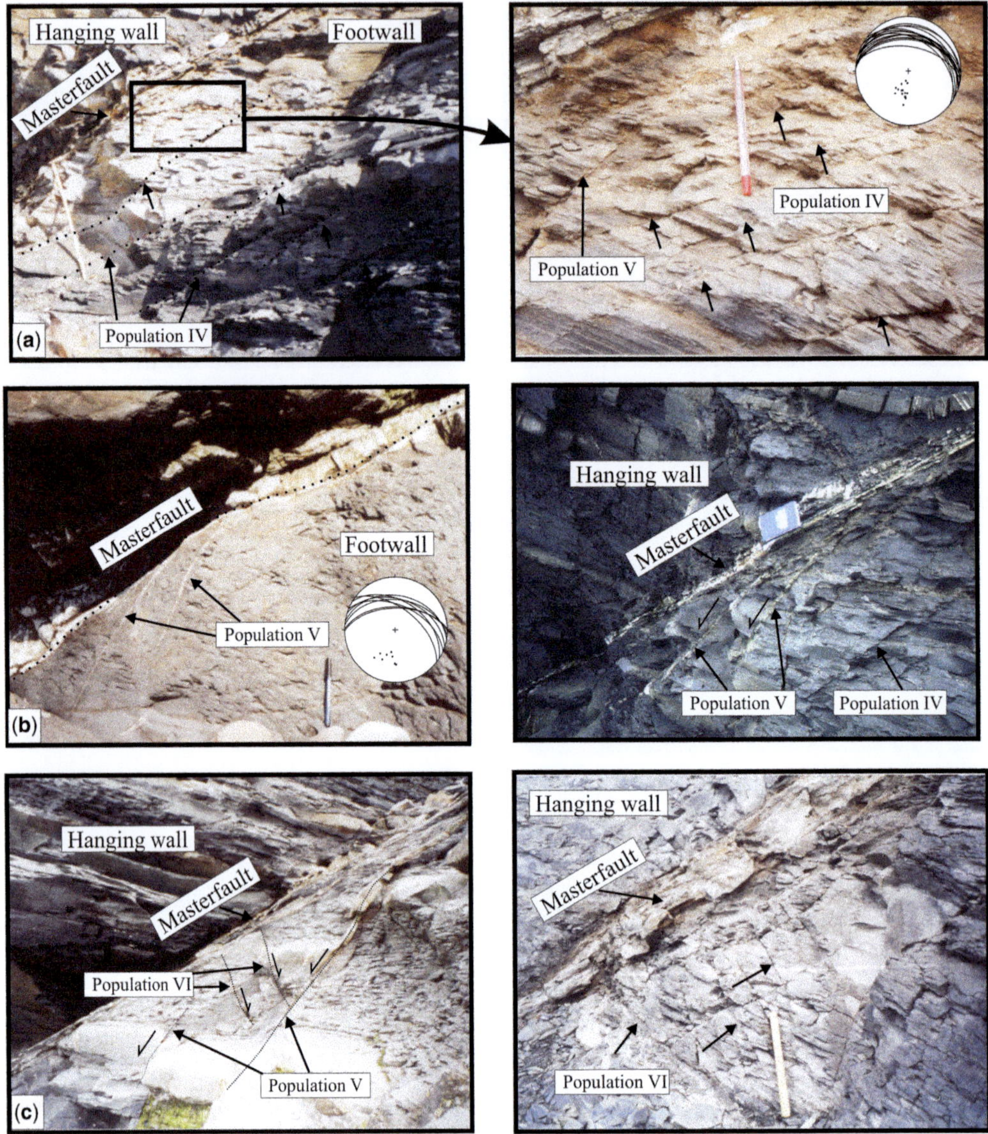

Fig. 6. Photographs of fractures of different populations occurring in the footwall damage zones. (**a**) Fault parallel synthetic fractures of Population IV shown on different scales. (**b**) Steeply dipping synthetic fractures of Population V. The photograph to the left shows two calcite-filled fractures without displacement that decreases in width downwards. The steeply dipping fractures on the photograph to the right are striated. The shear and tensional fractures have similar orientations. (**c**) Antithetic fractures of Population VI. The structures on the left picture are shear fractures, whereas fractures on the picture to the right have non-visible displacements.

Fracture populations in the limestone beds

The damage zone fractures appearing in the limestone beds dominantly strike east–west and ESE–WSW and dip more steeply than those in the shale beds (Fig. 7). These fractures are generally characterized by extensive calcite cementation. We identified three main fracture populations in limestones: (1) north-dipping fractures trending east–west to WNW–ESE; (2) south-dipping fractures trending east–west; and (3) very narrow fractures trending east–west to WNW–ESE, dipping towards the south. All fracture sets have steep dips (70–90°); however, the fractures of population 3

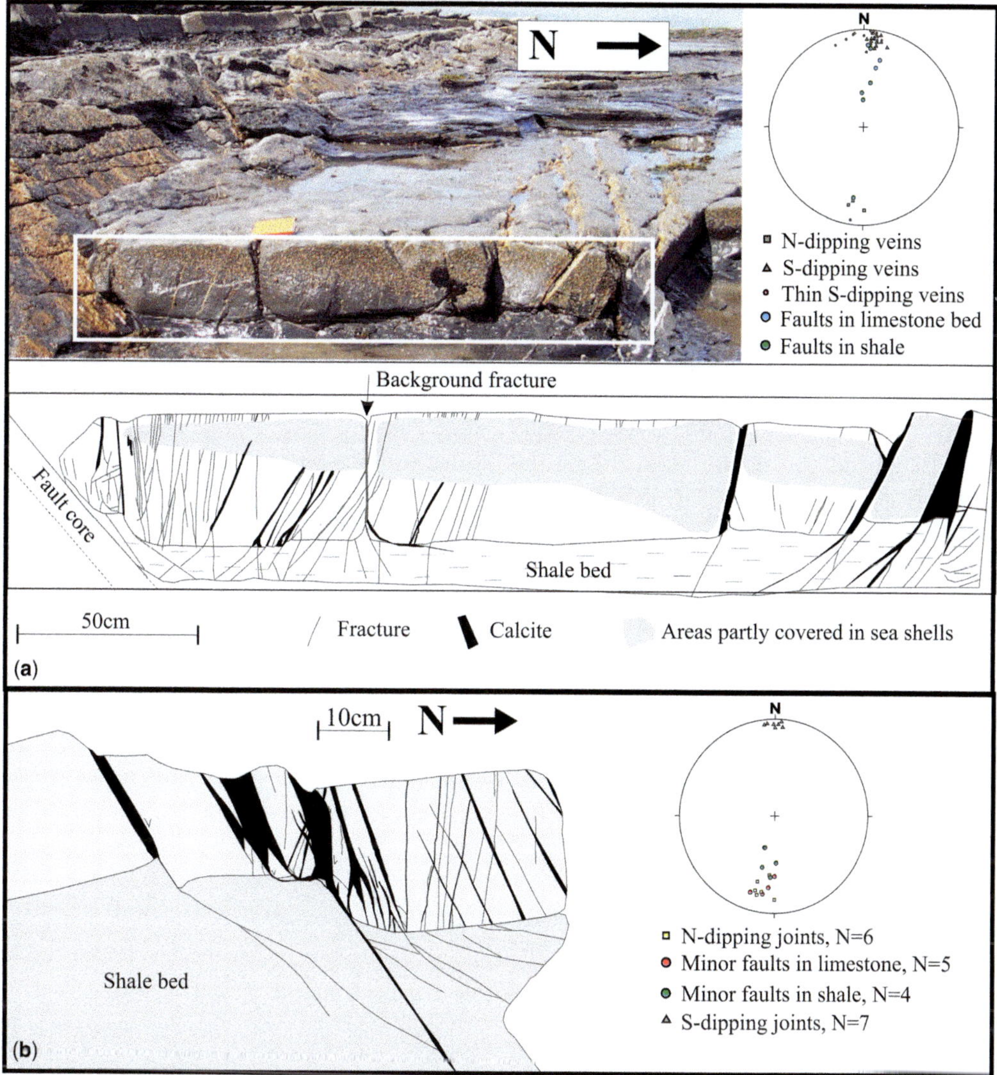

Fig. 7. Photograph and line drawings from photographs showing fracture types and orientations that are common in limestone beds within the damage zone. Both illustrations are from Fault 2. (**a**) Fracture pattern in a limestone bed in the hanging wall. (**b**) Fracture pattern in a limestone bed situated approximately 3 m into the footwall.

in limestone are on average steeper than the other sets, as discussed in Schöpfer *et al.* (2006), and geomechanically explained by Ferril & Morris (2003).

The majority of the fractures are tensional; however, some of the fractures in populations 1 and 2 represent shear fractures or mixed-mode fractures with normal displacements in the order of a few centimetres. The tensional fractures are generally stratabound, and only a few of them extend into the adjacent shale (Fig. 7). In contrast, the shear fractures are, in most cases, not stratabound, and their dips typically decrease when entering the weaker shale lithology. In places, the changes in geometry when crossing lithological boundaries have led to the development of small overlapping fracture sets, the spaces between which are filled with calcite in the releasing fracture configurations.

In the dip-direction the lengths of the stratabound fractures are constrained by the bed thickness, whereas the non-stratabound shear fractures may extend vertically for several metres and cut through beds of both shale and limestone. In the strike direction, fracture lengths up to 20 m were observed. For the most prominent tensional fractures the difference in lengths in the strike and dip directions is less than 1:100.

Fracture frequency distribution and width of the damage zones

The spatial fracture distribution in the shale beds was analysed by nine traverses showing the number of damage zone fractures as a function of distance from the master faults. All of the profiles are measured in vertical outcrops (see Fig. 2 for profile locations) and allow for a quantitative comparison of fracture frequencies in damage zones of faults with different throws and bed thickness.

Figure 8 shows fracture frequencies for the shale fracture populations as a function of distance from three faults of different displacements. It is evident that a pattern of decreasing frequency with increasing distance from the master faults characterizes most of the profiles, similar to what has been noted in sandstone (e.g. Schueller *et al.* 2012).

The hanging-wall fractures (Fig. 8a) show a predominance of antithetic structures (population III) that in general occur with higher frequencies than the other hanging-wall populations. The synthetic fractures of population I and II also show a general decrease in frequency away from the fault. The low-angled fractures of population II occur in high numbers when present, and may partly cause the large fracture frequency adjacent to the fault. This population occurs only occasionally and, in general terms, is less significant than the two other hanging-wall populations.

The synthetic fractures of population IV (Fig. 8b) have significantly higher frequency than the equivalent population I in the hanging wall. However, their occurrence decreases rapidly with increasing distance from the fault core. Population V fractures are present in profiles of Fault 1 and 3, but occur in moderate numbers and only within the first metre away from the core. Fractures of population VI occur in two of the profiles at low to moderate numbers with the highest number within the first metre interval.

The synthetic fractures of population I and II occur less than 4 m from the fault core, whereas the fractures of population III extend up to 7 m away from the core. The footwall fractures, however, only extend 1–3 m away from the core. These observations suggest that the asymmetry in damage zone fracture frequency is largely caused by the occurrence of the antithetic shear fractures (population III).

Variations exist in traverses from the same faults for fracture peak frequency values, and when the total number of fractures and the width of the damage zone are considered. Some of the variability is likely to be attributed to variations in composition and thickness of the shale beds. The thickness of the shale intervals along which the profiles were measured varies from 0.5 m to more than 8 m. Profiles 1H, 2H and 3H (Fig. 8c) are measured in beds with similar thicknesses (0.5 m) and in the hanging wall of the same fault. Profiles 1H and 2H display an almost identical frequency distribution, whereas profile 3H has slightly higher frequencies and wider damage zone, suggesting a possible mineralogical control (no mineralogical analysis was performed).

The effect of bed thickness on the fracture pattern is illustrated by the two profiles in the footwall shales of Fault 3 (profiles 3F and 4F). The thinner shale bed (0.5 m thick) (profile 4F) has clearly higher frequencies and wider damage zone compared with the thicker bed (1.4 m thick) (profile 3F). The influence of lithology and bed thickness on the fracture pattern is well known from studies of joints in folded and faulted rocks (e.g. Ogata *et al.* 2012, 2014*a*, *b*).

The spatial variation in frequency within the damage zone along individual faults with a given displacement is shown in Figure 9. Fracture frequency in the hanging wall and footwall are plotted separately and for some faults several transects are included. The highest frequencies and the largest variability occur proximal to the faults. The data show considerable scatter and further that variations in damage zone fracture frequency along individual faults can be larger than between faults with different displacements.

Damage zone width

There are pronounced variations in the width of the damage zones of the analysed faults (Fig. 8). The widest damage zones of 6–7 m are found in the hanging wall of Fault 1 (profiles 3H and 4H) and the narrowest damage zones recorded are found in the footwall of Faults 1 and 3 (less than 1 m; profiles 2F and 3F). In fact, in almost all the profiles the damage zones of the hanging wall are wider than those of the footwall. The highest fracture frequencies are, however, observed in the footwall (profile 4F). Thus, it is evident that there is an asymmetry in the fracture distribution around the master fault core (Fig. 9).

As seen from the frequency data (Fig. 8c), the width of the damage zone seems to be independent of displacement (Fig. 10). There is a tendency for the hanging-wall damage zone to be widest for the two faults with displacements of 15 and 19 m, but the data from the footwall damage zone show no relationship between fault displacement and width. Based on the analysis of the fracture frequency distributions, it is concluded that the width of the

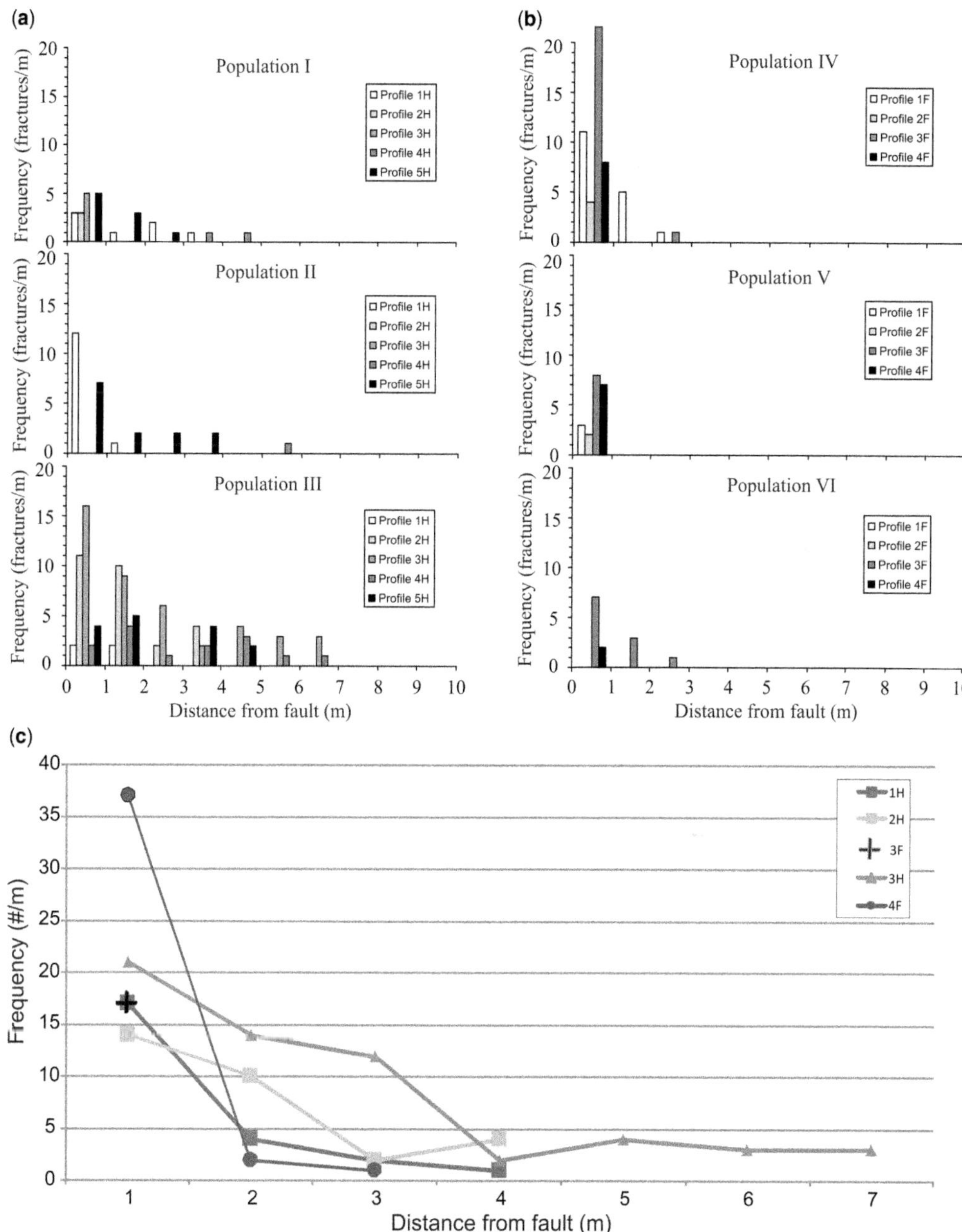

Fig. 8. Histograms showing the spatial distribution of fracture populations in the shale beds in the hanging wall (**a**) and footwall (**b**). (**c**) Comparison of fracture frequency (fractures per logged metre) in five scan lines in the damage zone. The data are from the fracture frequency profiles located in Figure 2.

damage zone varies along individual faults, and no clear relationship exists between the width of the damage zone and throw of the faults. Our data show the large variability in frequencies that may occur in shale intervals, which is a combined effect of factors such as bed thickness, mineralogical composition and distance from master faults. The data, however, do not allow us to quantify relationships between these parameters. The difference in fracture distribution in the footwall and hanging wall

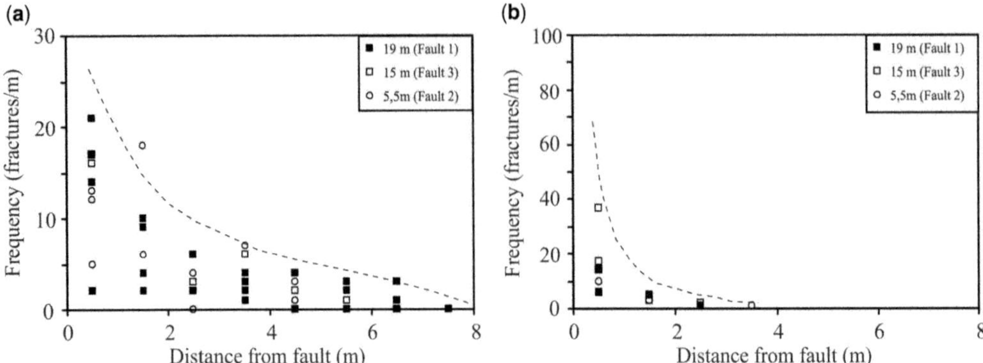

Fig. 9. Compilation of the fracture frequency v. distance from faults based on fracture frequencies given in Figure 8c. Indicated envelopes define the upper limit of recorded frequencies for (**a**) the hanging wall and (**b**) the footwall. Maximum damage zone widths are indicated.

causes an asymmetry in frequency distribution around the master fault core.

Discussion

To predict the distribution and frequency of fractures in damage zones it is necessary to assess the spatial distribution and orientation of the fracture systems. Deeper understanding comes from considering tempo-spatial relationships of fracture formation.

Damage zone architecture

Increased frequency of fractures around normal faults in sandstones has been documented in core and outcrop data (e.g. Knipe *et al.* 1994; Knott *et al.* 1996; Fossen & Hesthammer 2000; Schueller *et al.* 2012), with parameters such as throw, lithology, deformation conditions, irregularities along fault planes and local stresses in concert influencing the fault zone characteristics. Based on our data from faults with throws ranging from 0.6 to 30 m (i.e. small to moderate displacements) in shales, we find no clear relationships between throw and damage zone width. The data show that variations in damage zone width along one fault can be larger than those between faults of different throws. The maximum widths obtained (6.5 m in the hanging wall and 3.5 m in the footwall) could therefore be used as a constraint on the maximum widths to be expected in damage zones of sub-seismic faults, rather than assuming a linear relationship between these two parameters. Our data on damage zone widths also reveal an asymmetry across the core in which the hanging-wall damage zone is 1–3 times

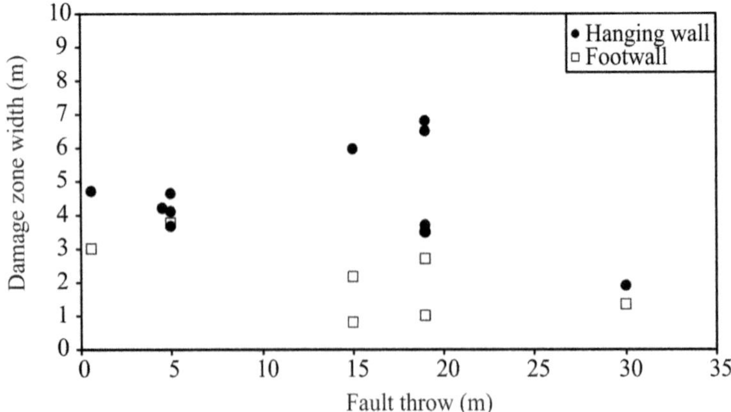

Fig. 10. Damage zone width v. master fault throw plotted on linear scale incorporating data from five faults. Footwall and hanging-wall data points are differentiated.

wider than the footwall damage zone. This conforms to asymmetrical normal drag that is associated with many of the faults, where the hanging-wall drag zone is at least 2–3 times wider than in the footwall, in accordance with observations by Hesthammer & Fossen (2000), Berg & Skar (2005) and Schueller *et al.* (2012).

Studies of damage zone structures around normal faults in sandstone have identified that the majority of deformation bands and slip planes are sub-parallel to the major faults. They constitute both synthetic and antithetic structures in relation to the master faults (e.g. Antonellini & Aydin 1994; Hesthammer *et al.* 2000; Shipton & Cowie 2001, 2003; Berg & Skar 2005). Synthetic and antithetic structures appear to be coeval and in overall equal proportions around the fault. The strike trends may show a scatter of 25–30° about the main fault, whereas dips may show considerable scatter so that dips of different sets may overlap (Shipton & Cowie 2001). All of these observations are in good agreement with our observations of fracture orientations in the shale beds, although antithetic orientations are more abundant in the hanging wall than in the footwall. Furthermore, the fracture dip orientations are strikingly similar to the geometric classification of fracture orientations proposed by Petit (1987), results obtained from experiments (e.g. Takashi 2003), studies of well-exposed large-scale normal faults in sandstone–shale successions (Odling 1997) and deformation band-orientations in sandstones (Berg & Skar 2005).

Comparing the expected and real fracture distribution with the fracture distribution in the Kilve faults analysed by us, we see that the footwall populations VI, V and VI follow well the conceptual patterns as shown in Figure 11. The characteristics of the respective fracture populations are also in harmony with this scheme, in that footwall populations IV and V bear indications of dip-parallel shear, consistent with these fracture populations being fault-parallel fractures and (inclined) Riedel-shears. Population VI has less well expressed striations and sometimes bears no indications of shear at all, which should be expected for anti-Riedel (R′)-structures owing to their unfavourable orientation and antithetic mode of displacement. Further, hanging-wall fracture populations I–III follow the same general pattern as that of the footwall, albeit being affected by additional (antithetic) accommodation fractures.

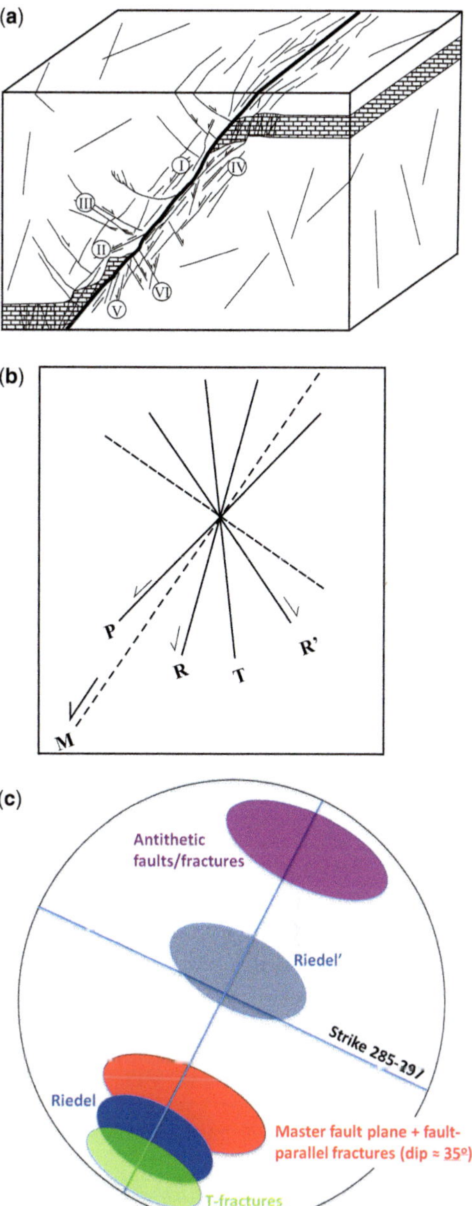

Fig. 11. (**a**) Generalized sketch of the fracture populations as they are observed in the shale beds. (**b**) Terminology for description of secondary fractures in a shear context, modified after Petit (1987). (**c**) Stereoplot with areas signifying location of poles to fracture populations (R, R′, T) for the faults studied.

Development of damage zone during fault growth

The different fracture orientations in the damage zone are likely to be determined by either fault tip stress or primary fault-related shear stress. The latter may be modified by local stress reorientation associated with fault plane irregularities (Peacock & Zhang 1993; Childs *et al.* 1996; Gabrielsen &

Clausen 2001; Lindanger *et al.* 2007). Cowie & Scholz (1992) postulated that a rock will experience the highest stresses in the vicinity of fault tips ('process zone'). The damage produced by this stress concentration has been proposed to be more intense than damage resulting from subsequent slip on a fault plane (Vermilye & Scholz 1999). Damage zone geometries around fault tips of normal faults in the interbedded limestone and shale succession in the Kilve area have previously been described by McGrath & Davison (1995). They recognized two basic types of geometry: (1) fractures branching directly from the fault tip; and (2) fractures forming an en echelon array, which are disconnected from the fault tip. In vertical cross-sections perpendicular to the strike of the main fault, they showed that fractures splay from the main fault at an angle of approximately 20°, and steepen to a sub-vertical orientation away from the main fault. It is, however, suggested that process zone fractures become overprinted by shear related fractures at advanced stages of fault displacement.

Development of asymmetry in damage zone strain during fault growth could be caused by different stress intensities in the footwall and hanging wall. For instance, in strike-slip settings damage zones around the fault tips may be either symmetrical or asymmetrical depending on the location around the fault (Kim *et al.* 2004). The fault tips of normal faults in the sub-horizontal/bed-parallel dimension will be characterized by a symmetrical distribution of mode II fractures, whereas the updip and downdip tips are typically asymmetrical. Thus, it is possible that stress distribution around the fault tip could cause asymmetry in fracture distribution for a given time step.

An alternative explanation to the observed hanging wall v. footwall asymmetry could be related to continued slip along an existing fault. According to Scholz (1987) the damage zone is expected to increase in overall thickness with displacement because increasingly larger irregularities are juxtaposed with increased displacements. Similar relations have been demonstrated by Schueller *et al.* (2012) and Braathen *et al.* (2013). Although irregularities may contribute to local variations in stress concentrations, it seems unlikely that a systematic asymmetry in the width of the damage zone will arise owing to juxtaposition of irregularities alone: rather, symmetric strain around stress-rising patches would be expected. However, it is noted in this study that the main cause of the asymmetry is the occurrence of antithetic fractures in the hanging wall. Accordingly, adjustments and flexing of the hanging-wall fault block initiated by irregularity in fault shape could cause local stress intensities that eventually cause frictional breakdown, and thereby the development of secondary antithetic accommodation structures in the hanging wall. These antithetic structures extend further into the hanging-wall block than fractures that have developed owing to fault tip propagation processes or frictional sliding along an existing fault plane, as also considered by Braathen *et al.* (2013). This conforms to the observation of more frequently developed accommodation structures associated with hanging-wall collapse (Gabrielsen 2015; Gabrielsen *et al.* 2016).

Conclusions

This study has addressed the nature of fracture networks around normal faults with throws of less than 20 m in a low-permeability shale–limestone succession. Particular focus has been on the fracture characteristics in shale beds. The main conclusions are:

(1) The main fracture orientations are synthetic and antithetic to the master faults similar to damage zone structures in sandstones. The orientation, distribution and frequency of fractures are distinctly different in the hanging wall and footwall, causing an asymmetry in the fracture characteristics across the master faults.

(2) Hanging-wall damage zones are 1.5–3 times wider than the footwall damage zone with a maximum width in the hanging wall of 6.5 m and a minimum width in the footwall of 0.8 m. Variations in widths can be larger along one fault than between faults with different displacements. Thus, no clear relationship exists between fault displacement and damage zone width for small (1–30 m throw) normal faults.

(3) Variations in fracture orientation, fracture intensity and the asymmetry around the fault suggest that different stress and frictional breakdown situations existed during different stages in the development of the master faults.

We thank Erlend Øian, Merethe Lindanger, Agust Gudmundsson and Sonja Brenner (Phillips) for constructive comments and pleasant company in the field. This research was financed by the Norwegian Research Council, Uni Research (University of Bergen) and University of Oslo. Gabrielsen and Braathen acknowledge the Trias North project for time used on the manuscript. We are indebted to editor Tom Manzocchi for thorough reviews and helpful comments. His good advice substantially enhanced the quality of this contribution.

References

Aarland, R.K. & Skjerven, J. 1998. Fault and fracture characteristics of a major fault zone in the northern

North Sea: analysis of 3D seismic and oriented cores in the Brage Field (Block 31/4). *In*: COWARD, M.P., DALTABAN, T.S. & JOHNSON, H. (eds) *Structural Geology in Reservoir Characterization*. Geological Society, London, Special Publications, **127**, 209–229, https://doi.org/10.1144/GSL.SP.1998.127.01.15

AGOSTA, F. & AYDIN, A. 2006. Architecture and deformation mechanism of a basin-bounding normal fault in Mesozoic platform carbonates, central Italy. *Journal of Structural Geology*, **28**, 1445–1467.

ANDERS, M.H. & WILTSCHENKO, D.V. 1994. Microfracturing, paleostress and the growth of faults. *Journal of Structural Geology*, **16**, 795–815.

ANDERSSON, J.E., EKMAN, L., NORDQUIST, R. & WINBERG, A. 1991. Hydraulic testing and modeling of a low-angle fracture zone at Finnsjon, Sweden. *Journal of Hydrology*, **126**, 45–77.

ANTONELLINI, M. & AYDIN, A. 1994. Effect of faulting on fluid flow in porous sandstones: petrophysical properties. *American Association of Petroleum Geologists Bulletin*, **78**, 355–377.

ANTONELLINI, M. & AYDIN, A. 1995. Effect of faulting on fluid flow in porous sandstones: geometry and spatial distribution. *American Association of Petroleum Geologists Bulletin*, **79**, 642–671.

APLIN, A.C., YANG, Y. & HANSEN, S. 1995. Assessment of β, the compression coefficient of mudstones and its relationship with detailed lithology. *Marine and Petroleum Geology*, **12**, 955–963.

BAINES, S.J. & WORDEN, R.H. 2004. The long-term fate of CO_2 in the subsurface: natural analogues for CO_2 storage. *In*: BAINES, S.J. & WORDEN, R.H. (eds) *Geological Storage of Carbon Dioxide*. Geological Society, London, Special Publications, **233**, 59–85, https://doi.org/10.1144/GSL.SP.2004.233.01.06

BASTESEN, E., BRAATHEN, A., NØTTVEIT, H., GABRIELSEN, R.H. & SKAR, T. 2009. Extensional fault cores in micritic carbonate – case studies from the Gulf of Corinth, Greece. *Journal of Structural Geology*, **31**, 403–420.

BEACH, A., WELBON, A.I., BROCKBANK, P.J. & MCCALLUM, J.E. 1999. Reservoir damage around faults: outcrop examples from the Suez rift. *Petroleum Geoscience*, **5**, 109–116, https://doi.org/10.1144/petgeo.5.2.109

BERG, S.S. 2005. *The architecture of normal fault zones in sedimentary rocks: analysis of fault core composition, damage zone asymmetry, and multi-phase flow properties*. PhD thesis, University of Bergen, Norway.

BERG, S.S. & SKAR, T. 2005. Controls on damage zone asymmetry of a normal fault zone: outcrop analyses of a segment of the Moab fault, SE Utah. *Journal of Structural Geology*, **27**, 1803–1822.

BOHLOLI, B., SKURTVEIT, E. *ET AL.* 2014. Evaluation of reservoir and cap-rock integrity for the Longyearbyen CO_2 storage pilot based on laboratory experiments and injection tests. *Norwegian Journal of Geology*, **94**, 171–187.

BRAATHEN, A., TVERANGER, J. *ET AL.* 2009. Fault facies and its application to sandstone reservoirs. *American Association of Petroleum Geologists Bulletin*, **93**, 891–917.

BRAATHEN, A., OSMUNDSEN, P.T., HAUSO, H., SEMSHAUG, S., FREDMAN, N. & BUCKLEY, S. 2013. Fault-induced deformation in poorly consolidated, siliciclastic growth basin: a study from the Devonian in Norway. *Tectonophysics*, **586**, 112–129.

BROCK, W.G. & ENGELDER, T. 1977. Deformation associated with the movement of the Muddy Mountain overthrust in the Buffington window, southwestern Nevada. *Geological Society of America Bulletin*, **88**, 1667–1677.

BROGI, A. 2011. Variation in fracture patterns in damage zones related to strike-slip faults interfering with pre-existing fractures in sandstone (Calcione area, soutren Tuscany, Italy). *Journal of Structural Geology*, **33**, 644–661, https://doi.org/10.1016/j.jsg.2010.12.008

BROOKS, M., TRAYNER, P.M. & TRIMBLE, T.J. 1988. Mesozoic reactivation of Variscan thrusting in the Bristol Channel area, UK. *Journal of the Geological Society, London*, **145**, 439–444, https://doi.org/10.1144/gsjgs.145.3.0439

BRUHN, R.L., PARRY, W.T., YONKEE, W.A. & THOMPSON, T. 1994. Fracturing and hydrothermal alteration in normal fault zones. *Pure and Applied Geophysics*, **142**, 609–644.

CAINE, J.S., EVANS, J.P. & FORSTER, C.B. 1996. Fault zone architecture and permeability structure. *Geology*, **24**, 1025–1028.

CHADWICK, R.A. 1986. Extension tectonics in the Wessex Basin, southern England. *Journal of the Geological Society, London*, **143**, 465–488, https://doi.org/10.1144/gsjgs.143.3.0465

CHESTER, F.M. & LOGAN, J.M. 1986. Implications for mechanical properties of brittle faults from observations of the Punchbowl Fault Zone, California. *Pure and Applied Geophysics*, **124**, 79–106.

CHESTER, F.M. & LOGAN, J.M. 1987. Composite planar fabric of gouge from the Punchbowl Fault, California. *Journal of Structural Geology*, **9**, 621–634.

CHILDS, C., NICOL, A., WALSH, J.J. & WATTERSON, J. 1996. Growth of vertically segmented normal faults. *Journal of Structural Geology*, **18**, 1389–1397.

CHILDS, C., WALSH, J.J. & WATTERSON, J. 1997. Complexity in fault zone structure and implications for fault seal prediction. *In*: MØLLER-PEDERSEN, P. & KOESTLER, A.G. (eds) *Hydrocarbon Seals: Importance for Exploration and Production*. Elsevier, Amsterdam/Norwegian Petroleum Society, Special Publications, **7**, 61–72.

CHILDS, C., MANZOCCHI, T., WALSH, J.J., BONSON, C.G., NICOL, A. & SCHÖPFER, M.P.J. 2009. A geometric model of fault zone and fault rock thickness variations. *Journal of Structural Geology*, **31**, 117–127.

COWIE, P.A. & SCHOLZ, C.H. 1992. Physical explanation for the displacement–length relationship of faults using a post-yield fracture mechanics model. *Journal of Structural Geology*, **14**, 1133–1148.

COWIE, P.A., ROBERTS, G.P. & MORTIMER, E. 2009. Strainlocailzation within fault arrays over timescales of 10^0–10^7 years. *In*: HANDY, M.R., HIRTH, G. & HOVIUS, N. (eds) *Tectonic Faults*. Agents of Change on a Dynamic Earth, MIT Press, Cambridge, 47–77.

DART, C.J., MCCLAY, K. & HOLLINGS, P.N. 1995. 3D analysis of inverted extensional fault systems, southern Bristol Channel basin, UK. *In*: BUCHANAN, J.G. & BUCHANAN, P.G. (eds) *Basin Inversion*. Geological Society, London, Special Publications, **88**, 393–413, https://doi.org/10.1144/GSL.SP.1995.088.01.21

DAVISON, I. 1995. Fault slip evolution determined from crack seal veins in pull-aparts and their implications for general slip models. *Journal of Structural Geology*, **17**, 1025–1034.

DONATO, J.A. 1988. Possible Variscan thrusting beneath the Somerton Anticline, Somerset. *Journal of the Geological Society, London*, **145**, https://doi.org/10.1144/gsjgs.145.3.0431

ENGELDER, T. & PEACOCK, D.C.P. 2001. Joint development normal to regional compression during flexural-flow folding: the Lilstock buttress anticline, Somerset, England. *Journal of Structural Geology*, **23**, 259–277.

EVANS, J.P. & CHESTER, F.M. 1995. Fluid–rock interaction in faults of the San Andreas system: inferences from San Gabriel fault rock geochemistry and microfractures. *Journal of Geophysical Research*, **100**, 13007–13020.

FERRIL, D.A. & MORRIS, A.P. 2003. Dilational normal faults. *Journal of Structural Geology*, **25**, 183–196.

FOSSEN, H. & BALE, A. 2007. Deformation bands and their influence on fluid flow. *American Association of Petroleum Geologists Bulletin*, **91**, 1685–1700.

FOSSEN, H. & HESTHAMMER, J. 2000. Possible absence of small faults in the Gullfaks Field, northern North Sea: implications for downscaling of faults in some porous sandstones. *Journal of Structural Geology*, **22**, 851–863.

FOSSEN, H., SCHULTZ, R.A., SHIPTON, Z.K. & MAIR, K. 2007. Deformation bands in sandstone: a review. *Journal of the Geological Society, London*, **164**, 755–769, https://doi.org/10.1144/0016-76492006-036

FOWLES, J. & BURLEY, S. 1994. Textural and permeability characteristics of faulted, high porosity sandstones. *Marine and Petroleum Geology*, **11**, 608–623.

FREDMAN, N., TVERANGER, J. *ET AL.* 2008. Assessment of Fault Facies modelling; technique and conditioning of 3D fault grids. *American Association of Petroleum Geologists Bulletin*, **92**, 1–22.

FRIEDMAN, M. 1969. Structural analysis of fractures in cores from Saticoy Field, Ventura County, California. *American Association of Petroleum Geologists Bulletin*, **53**, 367–389.

GABRIELSEN, R.H. 2015. The structure and hydrocarbon traps of sedimentary basin. *In*: BJØRLYKKE, K. (ed.) *Petroleum Geoscience: From Sedimentary Environments to Rock Physics*. 2nd edn. Springer-Verlag, Berlin, 319–350.

GABRIELSEN, R.H. & BRAATHEN, A. 2014. Models of fracture lineaments – joint swarms, fracture corridors and faults in crystalline rocks, and their genetic relations. *Tectonophysics*, **628**, 26–44, https://doi.org/10.1016/j.tecto.2014.04.022

GABRIELSEN, R.H. & CLAUSEN, J.A. 2001. Horses and duplexes in extensional regimes: a scale modeling contribution. *In*: KOYI, H.A. & MANCKTELOW, N.S. (eds) *Tectonic Modeling: A Volume in Honor of Hans Ramberg*. Geological Society of America, Memoirs, **193**, 219–233.

GABRIELSEN, R.H. & KLØVJAN, O.S. 1997. Late Jurassic–early Cretaceous caprocks of the southwestern Barents Sea: fracture systems and rock mechanical properties. *In*: MØLLER-PEDERSEN, P. & KOESTLER, A.G. (eds) *Hydrocarbon Seals: Importance for Exploration and Production*. NPF, Special Publications, **7**, Amsterdam, 73–89.

GABRIELSEN, R.H. & KOESTLER, A.K. 1987. Description and structural implications of fractures in late Jurassic sandstones of the Troll Field, northern North Sea. *Norsk Geologisk Tidsskrift*, **67**, 371–381.

GABRIELSEN, R.H., AARLAND, R.-K. & ALSAKER, E. 1998. Identification and spatial distribution of fractures in porous, siliclastic sediments. *In*: COWARD, M.P., DALTABAN, T.S. & JOHNSON, H. (eds) *Structural Geology in Reservoir Characterization*. Geological Society, London, Special Publications, **127**, 49–64, https://doi.org/10.1144/GSL.SP.1998.127.01.05

GABRIELSEN, R.H., BRAATHEN, A., KJEMPERUD, M. & VALDRESBRÅTEN, M.L.R. 2016. The geometry and dimensions of fault-core lenses. *In*: CHILDS, C., HOLDSWORTH, R.E., JACKSON, C.A.-L., MANZOCCHI, T., WALSH, J.J. & YIELDING, G. (eds) *The Geometry and Growth of Normal Faults*. Geological Society, London, Special Publications, **439**. First published online February 5, 2016, https://doi.org/10.1144/SP439.4

GUDMUNDSSON, A. 2011. *Rock Fractures in Geological Processes*. Cambridge University Press, Cambridge.

HESTHAMMER, J. & FOSSEN, H. 2000. Uncertainties associated with fault sealing analysis. *Petroleum Geoscience*, **6**, 37–45, https://doi.org/10.1144/petgeo.6.1.37

HESTHAMMER, J., JOHANSEN, T.E.S. & WATTS, L. 2000. Spatial relationships within fault damage zones in sandstone. *Marine and Petroleum Geology*, **17**, 873–893.

HIPPLER, S.J. 1993. Deformation microstructures and diagenesis in sandstone adjacent to an extensional fault: implications for the flow and entrapment of hydrocarbons. *American Association of Petroleum Geologists Bulletin*, **77**, 625–637.

HOLLOWAY, S. & CHADWICK, R.A. 1986. The Sticklepath-Lustleigh faults zone: tertiary sinistral reactivation of a Variscan dextral strike-slip fault. *Journal of the Geological Society, London*, **143**, 447–452, https://doi.org/10.1144/gsjgs.143.3.0447

JAMISON, W.R. & STEARNS, D.W. 1982. Tectonic deformation of Wingate sandstone, Colorado National Monument. *American Association of Petroleum Geologists Bulletin*, **66**, 2584–2608.

KAMERLING, P. 1979. The geology and hydrocarbon habitat of the Bristol Channel Basin. *Journal of Petroleum Geology*, **2**, 75–93.

KARNER, G.D., LAKE, S.D. & DEWEY, J.F. 1987. The thermal and mechanical development of the Wessex Basin, southern England. *In*: COWARD, M.P., DEWEY, J.F. & HANCOCK, P.L. (eds) *Continental Extensional Tectonics*. Geological Society, London, Special Publications, **28**, 517–536, https://doi.org/10.1144/GSL.SP.1987.028.01.34

KELLY, P.G., SANDERSON, D.J. & PEACOCK, D.C.P. 1998. Linkage and evolution of conjugate strike-slip fault zones in limestones of Somerset and Northumbria. *Journal of Structural Geology*, **20**, 1477–1493.

KIM, Y.-S., PEACOCK, D.C.P. & SANDERSON, D.J. 2004. Fault damage zones. *Journal of Structural Geology*, **26**, 503–517.

KNIPE, R.J., FISCHER, Q.J. *ET AL.* 1994. Fault zone thickness v. displacement in the Permo-Triassic sandstones of NW England. *Journal of the Geological Society, London*, **151**, 17–25, https://doi.org/10.1144/gsjgs.151.1.0017

KNOTT, S.D., BEACH, A., BROCKBANK, P.J., BROWN, J.L., MCCALLUM, J.E. & WELBON, A.I. 1996. Spatial and mechanical controls on normal fault populations. *Journal of Structural Geology*, **18**, 359–372.

KOESTLER, A.G. & EHRMANN, W.U. 1991. Description of brittle extensional features in chalk on the crest of a salt ridge (NW Germany). *In*: ROBERTS, A.M., YIELDING, G. & FREEMAN, B. (eds) *The Geometry of Normal Faults*. Geological Society, London, Special Publications, **56**, 113–123, https://doi.org/10.1144/GSL.SP.1991.056.01.08

LAKE, S.D. & KARNER, R.D. 1987. The structure and evolution of the Wessex Basin, southern England: an example of inversion tectonics. *Tectonophysics*, **137**, 347–378.

LOTHE, A.E., GABRIELSEN, R.H., BJØRNEVOLL HAGEN, N. & LARSEN, B.T. 2002. An experimental study of the texture of deformation bands: effects on porosity and permeability of sandstones. *Petroleum Geoscience*, **8**, 195–207, https://doi.org/10.1144/petgeo.8.3.195

LINDANGER, M., GABRIELEN, R.H. & BRAATHEN, A. 2007. Analysis of rock lenses in extensional faults. *Norwegian Journal of Geology*, **87**, 361–372.

MCGRATH, A. & DAVISON, I. 1995. Damage zone geometry around fault tips. *Journal of Structural Geology*, **17**, 1011–1024.

NEMCOK, M., GAYER, R. & MILIORIZOS, M. 1995. Structural analysis of the inverted Bristol Channel Basin: implications for the geometry and timing of fracture porosity. *In*: BUCHANAN, J.G. & BUCHANAN, P.G. (eds) *Basin Inversion*. Geological Society, London, Special Publications, **88**, 355–392, https://doi.org/10.1144/GSL.SP.1995.088.01.20

ODLING, N.E. 1997. Scaling and connectivity of joint systems in sandstone from western Norway. *Journal of Structural Geology*, **19**, 1257–1271.

OGATA, K., SENGER, K., BRAATHEN, A., TVERANGER, J. & OLAUSSEN, S. 2012. The importance of natural fractures in a tight reservoir for potential CO_2 storage: the case study of the upper Triassic to middle Jurassic Kapp Toscana Group (Spitsbergen, Svalbard). *In*: SPENCE, G.H., REDFERN, J., AGUILERA, R., BEVAN, T.G., COSGROVE, J.W., COUPLES, G.D. & DANIEL, J.-M. (eds) *Advances in the Study of Fractured Reservoirs*. Geological Society, London, Special Publications, **374**, 395–415, https://doi.org/10.1144/SP374.9

OGATA, K., SENGER, K., BRAATHEN, A. & TVERANGER, J. 2014*a*. Fracture corridors as seal-bypass systems in siliciclastic reservoir–caprock successions: field based insights from the Jurassic Entrada Formation (SE Utah, USA). *Journal of Structural Geology*, **66**, 162–187.

OGATA, K., SENGER, K., BRAATHEN, A., TVERANGER, J. & OLAUSSEN, S. 2014*b*. Fracture systems and meso-scale structural patterns in the reservoir–cap rock succession of the Longyearbyen CO_2 Lab project; implications for geological CO_2 sequestration on Svalbard. *Norwegian Journal of Geology*, **94**, 121–154.

PEACOCK, D.C.P. 1996. Field examples of variations in fault patterns at different scales. *Terra Nova*, **8**, 361–371.

PEACOCK, D.C.P. & SANDERSON, D.J. 1991. Displacements, segment linkage and relay ramps in normal fault zones. *Journal of Structural Geology*, **13**, 721–733.

PEACOCK, D.C.P. & SANDERSON, D.J. 1993. Estimating strain from fault slip using a line sample. *Journal of Structural Geology*, **15**, 1513–1516.

PEACOCK, D.C.P. & SANDERSON, D.J. 1995. Strike-slip relay ramps. *Journal of Structural Geology*, **17**, 1351–1360.

PEACOCK, D.C.P. & ZHANG, X. 1993. Field examples and numerical modelling of oversteps and bends along normal faults in cross-section. *Tectonophysics*, **234**, 147–167.

PETIT, J.P. 1987. Criteria for the sense of movement on fault surfaces in brittle rocks. *Journal of Structural Geology*, **9**, 597–608.

PICKERING, G., PEACOCK, D.C.P., SANDERSON, D.J. & BULL, J.M. 1997. Modelling tip zones to predict the throw and length characteristics of faults. *American Association of Petroleum Geologists Bulletin*, **81**, 82–99.

RAWNSLEY, K.D., PEACOCK, D.C.P., RIVES, T. & PETIT, J.-P. 1998. Joints in the Mesozoic sediments around the Bristol Channel Basin. *Journal of Structural Geology*, **20**, 1641–1661.

RECHES, Z. & LOCKNER, D.A. 1994. Nucleation and growth of faults in brittle rocks. *Journal of Geophysical Research*, **99**, 18159–18173.

SCHOLZ, C.H. 1987. Wear and gouge formation in brittle faulting. *Geology*, **15**, 493–495.

SCHÖPFER, M.P.J., CHILDS, C. & WALSH, J.J. 2006. Localisation of normal faults in multilayer sequences. *Journal of Structural Geology*, **28**, 816–833.

SCHUELLER, S., BRAATHEN, A., FOSSEN, H. & TVERANGER, J. 2012. Fault damage zones of extensional faults in porous sandstone: spatial distribution of deformation bands. *Journal of Structural Geology*, **52**, 148–162.

SCHULZ, S.E. & EVANS, J.P. 1998. Spatial variability in microscopic deformation and composition of the Punchbowl fault, southern California: implications for mechanisms, fluid-rock interaction, and fault morphology. *Tectonophysics*, **295**, 223–244.

SHIPTON, Z.K. & COWIE, P.A. 2001. Damage zone and slip-surface evolution over μm to km scales in high-porosity Navajo sandstone, Utah. *Journal of Structural Geology*, **23**, 1825–1844.

SHIPTON, Z.K. & COWIE, P.A. 2003. A conceptual model for the origin of fault damage zone structures in high-porosity sandstone. *Journal of Structural Geology*, **25**, 333–344.

STEWART, S.A. & ARGENT, J.D. 2000. Relationship between polarity of extension fault arrays and presence of detachments. *Journal of Structural Geology*, **22**, 693–711.

TAKASHI, M. 2003. Experimental approach to assess fault-seal potential of shale smear. *In*: *Proceedings 'Fault and Top Seal' – What do we Know and Where do we Go?*, 8–11 September 2003, Montpellier, France, 34.

TORABI, A., GABRIELSEN, R.H. ET AL. 2015. Strain localization in sandstone and its implications for CO_2 storage. *First Break*, **33**, 81–92.

UNDERHILL, J.R. & WOODCOCK, N.H. 1987. Faulting mechanisms in high-porosity sandstones; New Red

Sandstone, Arran, Scotland. *In*: Jones, M.E. & Preston, R.M.F. (eds) *Deformation of Sediments and Sedimentary Rocks*. Geological Society, London, Special Publications, **29**, 91–105, https://doi.org/10.1144/GSL.SP.1987.029.01.09

Vermilye, J.M. & Scholz, C.H. 1998. The process zone: a microstructural view of fault growth. *Journal of Geophysical Research*, **103**, 12223–12237.

Vermilye, J.M. & Scholz, C.H. 1999. Fault propagation and segmentation: insight from the microstructural examination of a small fault. *Journal of Structural Geology*, **21**, 1623–1636.

Willemse, E.J.M., Peacock, D.C.P. & Aydin, A. 1997. Nucleation and growth of strike-slip faults in limestones from Somerset, U.K. *Journal of Structural Geology*, **19**, 1461–1477.

Three-dimensional Distinct Element Method modelling of the growth of normal faults in layered sequences

MARTIN P. J. SCHÖPFER[1]*, CONRAD CHILDS[2], TOM MANZOCCHI[2] & JOHN J. WALSH[2]

[1]*Department for Geodynamics and Sedimentology, University of Vienna, Althanstrasse 14, A-1090 Vienna, Austria*

[2]*Fault Analysis Group, School of Earth Sciences, University College Dublin, Belfield, Dublin 4, Ireland*

**Correspondence: martin.schoepfer@univie.ac.at*

Abstract: The growth of normal faults in mechanically layered sequences is numerically modelled using three-dimensional Distinct Element Method (DEM) models, in which rock comprises an assemblage of bonded spherical particles. Faulting is induced by movement on a pre-defined normal fault at the model base whilst a constant confining pressure is maintained by applying forces to particles lying at the model top. The structure of the modelled fault zones and its dependency on confining pressure, sequence (net:gross) and fault obliquity are assessed using various new techniques that allow (a) visualization of faulted horizons, (b) quantification of throw partitioning and (c) determination of the fault zone throw beyond which theoretical juxtaposition sealing occurs along the entire zone length. The results indicate that fault zones become better localized with increasing throw and confinement. The mechanical stratigraphy has a profound impact on fault zone structure and localization: both low and high net:gross sequences lead to wide and relatively poorly localized faults. Fault strands developing above oblique-slip normal faults form, on average, normal to the greatest infinitesimal stretching direction in transtensional zones. The model results are consistent with field observations and results from physical experiments.

Faults are generally mapped and modelled as single surfaces across which there is a discrete displacement. In reality faults are complex zones comprising multiple slip surfaces that contain variably deformed rock volumes, ranging from intact fault-bound lenses to fault rock (Childs *et al.* 2009). This complexity in fault zone structure can be of significance in a variety of application areas, for example, by providing across-fault connectivity of flow units. Fault zone complexity and architecture will vary depending on the nature of the faulted sequence and the prevailing deformation conditions (e.g. Patton *et al.* 1998; Micarelli *et al.* 2005; Ferrill *et al.* 2007; Ferrill & Morris 2008; Kettermann & Urai 2015). Empirical constraints on the 2D and, in particular, 3D geometry and content of fault zones are, however, relatively sparse, in the sense that insufficient data are available on fault zones developed within multi-layered sequences, with different stacking patterns and rheological properties, and under different deformation conditions.

The published literature contains many accounts of conceptual models for the evolution of fault zone structure based mainly on outcrop studies of fault geometry (see review articles by Crider & Peacock 2004; Wibberley *et al.* 2008; Faulkner *et al.* 2010). These are complemented by a range of numerical models that attempt to replicate the processes invoked in these conceptual models (e.g. Schöpfer *et al.* 2016). Numerical modelling of fault zone growth, which requires simulating the formation of fractures, followed by detachment, rotation and comminution of fracture-bound volumes, is however cumbersome or impossible in many of the standard modelling approaches used in geomechanics (e.g. finite elements). The Distinct Element Method (DEM) is ideally suited to the modelling of fault zones but, as detailed in this paper, requires significant effort to achieve the realistic rheological properties that are routinely defined by constitutive equations in other modelling approaches.

The purpose of the present study is to investigate the impact of a selection of factors, such as confining pressure and layer stacking (net:gross), on the structure of normal faults and associated sequence juxtaposition using 3D DEM modelling. In earlier studies we have already shown the usefulness of this numerical technique to model fault growth in layered sequences in two dimensions (Schöpfer *et al.* 2006, 2007*a*, *b*, 2009*a*). These earlier studies

From: Childs, C., Holdsworth, R. E., Jackson, C. A.-L., Manzocchi, T., Walsh, J. J. & Yielding, G. (eds) 2017. *The Geometry and Growth of Normal Faults*. Geological Society, London, Special Publications, **439**, 307–332.
First published online September 15, 2016, https://doi.org/10.1144/SP439.17

are somewhat limited because fault propagation and fault zone evolution are inherently 3D processes and, for example, 2D modelling in the fault slip direction fails to capture features such as lateral fault propagation and relay ramp development. Therefore, while the DEM models presented in this study are in some respects a 3D extension of earlier models, they represent a significant advance that incorporates a wider range of processes allowing consideration of the lateral variability in structure along fault zones.

In this paper we first provide an overview of the DEM and existing models of normal faulting. Then, for the 3D normal faulting models presented here, we describe in detail the model materials and their properties, the model generation and boundary conditions. These models are used to investigate the importance of confining pressure, mechanical stratigraphy (net:gross) and slip direction on fault zone structure and localization. We conclude that many of the model results are consistent with field observations and results from physical experiments.

Distinct Element Method

The DEM comprises a broad class of methods that models finite displacements and rotations of discrete bodies (Cundall & Hart 1992). The DEM is capable of modelling the growth of discontinuities, such as faults and joints, without the limitations of continuum descriptions (Cundall 2001). The elements interact with each other via a force displacement law and can be of arbitrary shape, circular discs and spheres being the most common ones in 2D and 3D, respectively.

Two-dimensional DEM models of normal faulting

To our knowledge, Saltzer & Pollard (1992) were the first to apply the DEM to model normal faulting in sedimentary sequences. Although these early models were run using 900 non-bonded particles only, they already highlighted the potential of the DEM to model faulting. Finch *et al.* (2004) investigated the effects of cover strength and fault dip on the style of fault-propagation folding above a normal fault using layers comprising bonded particles. Schöpfer *et al.* (2006) simulated the propagation of a normal fault through a vertically confined, mechanically layered sequence. In later studies, Schöpfer *et al.* (2007*a*, *b*, 2009*a*) explored the impact of sequence strength, confining pressure and net:gross (fraction of strong layers in the sequence) on fault zone structure. Egholm *et al.* (2007) reproduced sandbox experiments of basement-induced normal faulting using a modified DEM approach (Egholm 2007), which was also used for modelling normal faulting in sand–clay sequences (Egholm *et al.* 2008). Abe *et al.* (2011) and Hardy (2011, 2013) modelled the propagation of normal faults in vertically unconfined bonded particle models. Botter *et al.* (2014) simulated normal fault growth using non-bonded particle models and generated, on the basis of volumetric strain and empirical relations, synthetic seismic images.

Three-dimensional DEM models of normal faulting

Owing to the high computational demand of 3D particle simulations few studies on normal faulting exist in the literature. Walsh *et al.* (2001) sheared a cube of bonded particles and studied how displacement is accommodated in a volume containing a predefined fault system. Imber *et al.* (2004) sheared a cube containing two overlapping predefined faults and investigated relay ramp growth and fault linkage. Carmona *et al.* (2010) combined the DEM with a process-based model of sedimentation to model syntectonic sedimentation. Recently, Schöpfer *et al.* (2016) presented 3D DEM models of normal faults in layered sequences that underpin a geometric model for the evolution of the internal structure of fault zones (Childs *et al.* 2009).

In summary, numerous studies on the growth of normal faults exist in 2D, but 3D DEM model studies of the growth of normal faults are rare. An overview of the DEM, including definition of particle/bond properties, the calculation cycle and model generation, is provided in the remainder of this section.

Definition of particle and bond properties

We use the commercially available Particle Flow Code in Three-Dimensions (*PFC3D*; Itasca Consulting Group 2008), which implements the DEM. Rock is represented by an assemblage of randomly packed spherical particles that are bonded together (Fig. 1a). These bonds can break if either their normal or shear strength is exceeded, which corresponds to fracturing of the material. Table 1 summarizes the microparameters that need to be assigned for a (parallel) bonded particle model. A more detailed definition of these parameters can be found in Potyondy & Cundall (2004).

Elasticity is captured by allowing overlap at particle–particle contacts. A linear contact law is used so that the normal contact force is proportional to the amount of overlap; the proportionality constant is the normal stiffness K_n (Fig. 1b), which is inherited from the two contacting particles assuming that both particles' stiffness act in series, i.e. $K_n = k_n^{(A)}k_n^{(B)}/(k_n^{(A)} + k_n^{(B)})$ with $k_n^{(A)}$ and $k_n^{(B)}$ being the

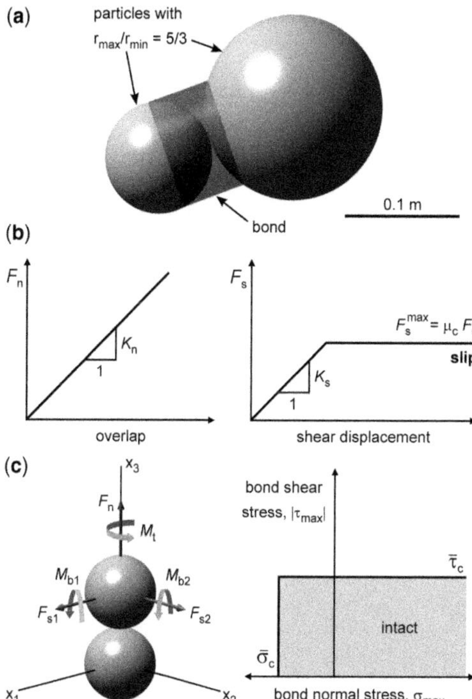

Fig. 1. Principles of the 3D Distinct Element Method (DEM). (**a**) Two spherical particles bonded at their contact with a disc-shaped linear elastic beam. (**b**) Contact force – displacement relation at non-bonded contacts (subscripts 'n' and 's' refer to normal and shear, respectively). (**c**) Forces and moments that can develop at bonded contact (modified after Wang & Mora 2008) and the bond failure law.

particle normal stiffnesses. By relating the behaviour of a contact between two particles to that of an elastic beam with its ends at the particle centres, an expression can be obtained for the particle normal stiffness as a function of the so-called 'contact Young's modulus' E_c, so that $k_n = 4rE_c$, where r is the particle radius. This modulus-stiffness scaling ensures that the bulk elasticity becomes particle size independent (Potyondy & Cundall 2004). The shear force is computed in an incremental fashion (Fig. 1b). Each subsequent relative shear-displacement increment produces an increment of elastic shear force that is added to the current shear force (which is, when a contact is formed, initially zero). The constant that relates shear displacement and shear force is the contact shear stiffness K_s, which is also inherited from the two contacting particles assuming that both particles' stiffness $k_s^{(A)}$ and $k_s^{(B)}$ act in series. The shear stiffness k_s is assigned to each particle via the particle stiffness ratio, k_n/k_s. The contact stiffness ratio (K_n/K_s) is an important factor that influences, for example, Poisson's ratio of the model (K_n/K_s should be >1, otherwise a negative Poisson's ratio may be obtained; Bathurst & Rothenburg 1988; Rothenburg *et al.* 1991; Wang & Mora 2008). The absolute shear force at a contact is limited by Coulomb friction so slip occurs at a contact when the ratio of contact shear to normal force is equal to μ_c (Fig. 1b). If the two contacting particles have differently assigned friction coefficients, the smaller value is used as a contact friction coefficient.

Particles can be bonded with massless, linear elastic 'cement', which acts in parallel with the above-described linear contact law, hence the bond model is referred to as the parallel bond model (Potyondy & Cundall 2004). A parallel bond can be envisioned as a set of elastic springs uniformly distributed over a circular cross-section lying on the contact plane. In the present study these disc-shaped bonds have a radius equal to the smaller of the two bonded particles (Fig. 1a). Relative motion at parallel bonded contacts causes axial and shear forces as well as bending and twisting moments to develop (Fig. 1c). The maximum normal and shear stress acting on the bond periphery can be computed via beam theory (see equation 16 in Potyondy & Cundall 2004): the bond normal stress increases with increasing normal force and bending moment, whereas the bond shear stress increases with increasing shear force and twisting moment (Fig. 1c). A consistent means of setting the elastic bond microparameters is to define the bond Young's modulus $\bar{E}_c$ and the ratio of bond normal to shear stiffness ratio ($\bar{k}_n/\bar{k}_s$). The modulus-stiffness relation for the bonds is $\bar{k}_n = \bar{E}_c/(r_A + r_B)$, where r_A and r_B are the radii of the two contacting particles. Once either the maximum normal or shear stress carried by the bond exceeds the normal (tensile) or shear strength, $\bar{\sigma}_c$ and $\bar{\tau}_c$ respectively, the bond breaks (Fig. 1c) and is immediately removed from the system along with its accompanying force, moment and stiffnesses. The fact that the bond and the slip model act in parallel has two important consequences: (a) slip does occur at bonded particle–particle contacts if the maximum shear force is exceeded; and (b) in compression load is carried by both the particles and the bond, whereas in tension load is carried by the bond only. These two features lead to pronounced non-linear elastic bulk behaviour (Schöpfer & Childs 2013).

Calculation cycle and damping

The calculation cycle in *PFC* is a time-stepping algorithm that requires the repeated application of the second law of motion to each particle (force = mass times acceleration), a force displacement law

Table 1. *Parameters that define a bonded particle model and values used in this study*

Parameter	Description*	Value	Unit
Particle properties			
r_{min}	Minimum particle radius†	0.15	m
r_{max}/r_{min}	Maximum to minimum particle radius ratio	1.66	–
ρ	Particle density‡	4000	kg m^{-3}
E_c	Particle modulus§	50 or 5	GPa
k_n/k_s	Particle stiffness ratio	2.5	–
μ_c	Particle friction coefficient	0.5	–
Bond properties			
$\bar{E}_c$	Bond modulus§	50 or 5	GPa
$\bar{k}_n/\bar{k}_s$	Bond stiffness ratio	2.5	–
$\bar{\lambda}$	Bond width multiplier	1.0	–
$\bar{\sigma}_c$	Bond normal strength¶	100 ± 20 or 10 ± 2	MPa
$\bar{\tau}_c$	Bond shear strength¶	100 ± 20 or 10 ± 2	MPa

*A full definition of particle/bond properties is given by Potyondy & Cundall (2004).
†The minimum radius listed is that of the outermost, unrefined region in the faulting models (see Fig. 3a). Within the innermost region with refinement level equal to 5, the minimum radius is 0.047 m.
‡The initial porosity of the randomly packed sphere models is *c.* 35% so that the model has a bulk density of *c.* 2600 kg m^{-3}.
§Young's moduli of particles and bonds were always set to be equal in any one model.
¶Tensile and shear strengths of bonds were always set to be equal in any one model. Bond strength values are drawn randomly from a normal distribution defined by mean and standard deviation.

to each contact (e.g. Fig. 1b), and a constant updating of particle and wall (rigid boundaries) positions. The equations of motion are integrated in *PFC* using a centred finite-difference scheme, which requires a time-step. *PFC* automatically calculates the critical time-step (which depends on mass and stiffness of the system) that is necessary for ensuring that the solution produced is stable and uses an actual time-step which is a fraction of the critical time-step for safety reasons.

PFC provides a true dynamic solution. However, deformation under quasi-static conditions requires dissipation of energy via damping. In the present study, so-called local damping is used, where a damping-force term is added to the equations of motion (damping constant is 0.7; Potyondy & Cundall 2004). For compact assemblies, this form of damping is the most appropriate form to establish equilibrium and to conduct quasi-static deformation simulations.

Numerical rock laboratory and properties of model materials

As described above the properties of a DEM model material are defined by a suite of particle and bond properties (Table 1 and Fig. 1). The bulk mechanical behaviour emerges from the interaction of particles via bonds and their contacts. Bulk mechanical properties have to be determined using numerical rock laboratory experiments, such as unconfined and confined compression tests. These numerical tests on DEM model materials are typically performed using rigid, frictionless loading and confinement platens (Potyondy & Cundall 2004), which can lead to significant boundary effects. An improved loading and confinement method, recently developed by Schöpfer & Childs (2013), is adopted for determining the properties of the materials used in the present study. In this section, first the particle/bond properties used are given, followed by a description of the numerical rock laboratory boundary conditions and the bulk mechanical properties of the model materials.

Particle and bond properties of model materials

In the present study, relatively simple DEM materials are used which are entirely defined using the set of parameters listed in Table 1. A drawback of using simple model materials is that, although the form of the failure envelope of natural rock can be reproduced (Schöpfer *et al.* 2013), the ratio of unconfined compressive to tensile strength and the friction angle is lower in the DEM material than in most rocks. The principal reason for the different behaviours of rock and DEM materials is that the latter comprises perfect spheres that offer no resistance to rolling. There are various techniques that have been implemented to improve the bulk properties of parallel-bonded particle models (Schöpfer *et al.* 2009*b*), but these often lead to a significant increase in model run time and/or lead to otherwise unrealistic behaviour.

In the normal faulting models two different materials are used to represent weak and strong rocks. The different particle and bond properties of these two materials are given in Table 1. To enable normalization of model parameters for comparison between models and the application of model results to natural conditions, the two materials used have identical stiffness/strength ratios and identical bond strength ranges (expressed as the coefficient of variation, which is standard deviation normalized by mean value).

Numerical rock laboratory boundary conditions

A limitation of existing loading procedures that are used for determining bulk mechanical properties of DEM model materials is the use of rigid platens to apply both loading and confinement. Loading under unconfined (and confined) uniaxial compression or extension tests can however be achieved using distortional periodic space (Schöpfer & Childs 2013). For this type of testing a bonded particle model is first generated within a cubic periodic domain (Fig. 2a) using the dynamic model generation procedure described in Potyondy & Cundall (2004). Then a core (cylindrical sample) is carved so that the sides of the cylinder do not straddle the periodic boundaries (Fig. 2b). The top and bottom of the cylinder are therefore periodic, i.e. the cylindrical sample is effectively infinitely long. The core diameter and resolution are identical to the thickness and resolution of the layers used in the normal faulting models (ratio of layer thickness to average particle diameter is 5.3).

The core sample can either be shortened (uniaxial compression test) or lengthened (uniaxial extension) by distorting the periodic space. Since loading platens are absent, the applied stress cannot be calculated by force/area, so a so-called 'measurement sphere' is located within the sample in which the average stress tensor is computed using particle contact forces and volumes (Potyondy & Cundall 2004). A sensitivity study revealed that at strain rates of 0.01 s^{-1} or less no significant changes in the bulk strength occur, suggesting quasi-static conditions. These unconfined tests provide the sample Young's modulus (E), Poisson's ratio (ν), the unconfined compressive strength (UCS) and the tensile strength (T).

Confinement of the samples without the use of rigid platens is achieved by applying inwards-directed forces to the lateral particles that comprise the surface of the cylinder, thus simulating a jacket (Fig. 2c). The lateral surface particles and the direction and magnitude of the applied forces are found using the so-called 'shining-lamp' algorithm, which is explained in detail in Appendix A. These confined tests provide the strength under confined compression or extension loading conditions and are important for defining the failure envelope of the model materials.

Bulk mechanical properties of model materials

Stress–strain curves for the two model materials ('strong' and 'weak') tested under both triaxial compression and triaxial extension test conditions are shown in Figure 2d. Since both particle/bond modulus and bond strength in the strong material are 10 times higher than in the weak material (see Table 1), the stress–strain curves are much steeper and reach higher stress differences in the strong material. The secant Young's modulus is extracted from these curves at an axial strain of 0.125 and 0.0625% for the triaxial compression and extension tests, respectively, and plotted v. confining pressure in Figure 2e. These tests show that under unconfined conditions Young's modulus in extension is about half the modulus in compression because load in compression is carried by both particles and bonds, whereas in tension load is only carried by bonds. With increasing confining pressure the difference in Young's modulus in extension and compression decreases, because the proportion of contacts that carry compressive load increases. Poisson's ratio also exhibits a confining pressure dependency (not shown here; see Schöpfer & Childs 2013) and is, for the conditions used in the normal faulting models, *c.* 0.18.

Another striking feature is that, for the range of confining pressures used, the strong material exhibits pronounced post-peak stress weakening, whereas the weak material exhibits little or no weakening (Fig. 2d). In fact, it appears that the strong material is brittle and the weak material is ductile. In the following we shall therefore use the definition that brittle behaviour exhibits a significant stress drop after peak stress, while ductile behaviour shows no such strain softening. The reason for this different behaviour is that the majority of tests on the weak material are conducted at confining pressures that are greater than the pressure at which the brittle–ductile transition, as defined by Byerlee (1968), occurs. The emergence of this type of brittle–ductile transition, or more precisely the brittle–cataclastic-flow transition, was recently explored by Schöpfer *et al.* (2013) in 3D stress space using the DEM. At low confining pressure peak-stress data fall onto a non-linear envelope that intersects (at the transitional pressure) a linear envelope (Fig. 2f). Under triaxial extension stress conditions this kind of brittle–ductile transition occurs in the

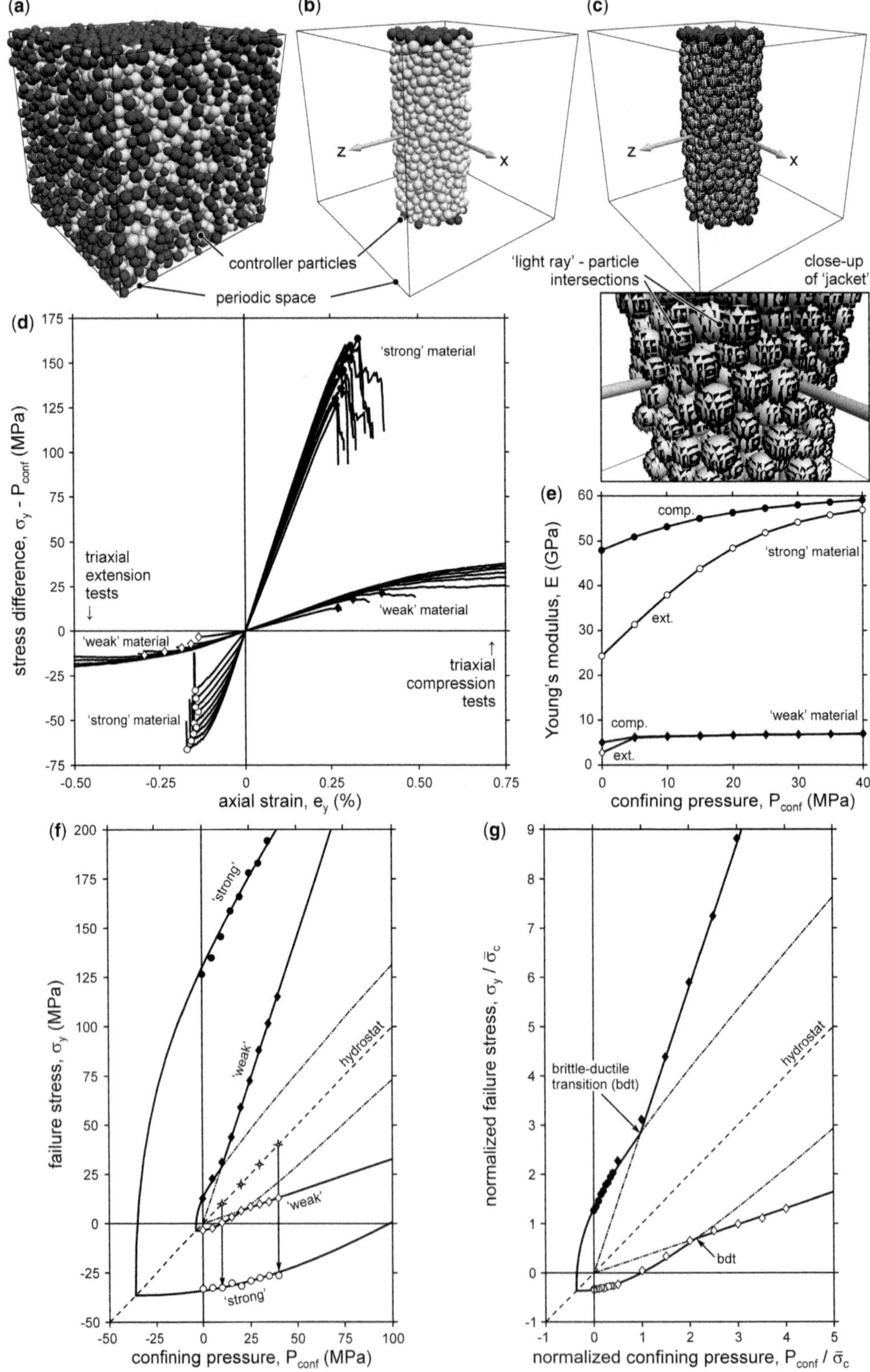

(a)
(b)
(c)
z
x
controller particles
periodic space
'light ray' - particle intersections
close-up of 'jacket'
(d)
stress difference, σy - Pconf (MPa)
'strong' material
triaxial extension tests
'weak' material
triaxial compression tests
axial strain, ey (%)
(e)
Young's modulus, E (GPa)
comp.
ext.
confining pressure, Pconf (MPa)
(f)
failure stress, σy (MPa)
'strong'
'weak'
hydrostat
confining pressure, Pconf (MPa)
(g)
normalized failure stress, σy / σ̄c
brittle-ductile transition (bdt)
bdt
normalized confining pressure, Pconf / σ̄c

weak material at a confining pressure of *c.* 21 MPa (Fig. 2f). The shape of the failure envelope and the location of the brittle–ductile transition characteristic of the numerical model materials used in this study can be better appreciated by plotting peak stress data normalized by the bond strength (Fig. 2g). This type of normalization can, in principle, also be applied to the normal faulting models presented in later sections. Identical results are expected if all intrinsic (modulus, strength) and extrinsic (confining pressure) parameters in stress units are scaled by the same factor. For clarity, however, all model results are presented in terms of their actual, rather than dimensionless, mechanical properties and applied confining pressures.

Discussion of material properties

Before presenting the normal faulting results it is worthwhile discussing the results of the numerical rock laboratory experiments, summarized in Figure 2. The purpose of determining the bulk mechanical properties of the DEM model materials is not to calibrate the materials against a certain rock. We rather try to ensure that some properties (Young's modulus, unconfined strength) are realistic, i.e. comparable to rock. The range and variability of rock mechanical data is however enormous, e.g. sandstones can have a uniaxial strength ranging from almost zero up to 300 MPa (Lockner 1995). A useful dimensionless parameter is the modulus ratio, defined as the ratio of Young's modulus to UCS, for which values between 200 and 500 are considered as 'average' (Deere & Miller 1965). The modulus ratio of the two model materials is *c.* 400 and hence typical for rock. Moreover, in the classification scheme proposed by Deere & Miller (1965), the UCS values of the strong and weak material fall within 'high strength' and 'very low strength' fields, respectively. However, the ratio of unconfined compressive strength to tensile strength (UCS/T) is too low in our model materials (*c.* 3.8) when compared with natural rocks, which have $UCS/T > 10$ (Schöpfer *et al.* 2009*b*). Since it is not possible using the present approach to match UCS/T, the question arises which one of the two strengths is believed to be more important under normal faulting conditions. In fact both T and UCS are important. The tensile strength controls the stress at which localization occurs (in the form of opening mode fractures under the conditions used in the present study, indicated by arrows in Fig. 2f). The compressive strength controls to some extent the comminution of fault-bound volumes. Practically speaking, if we were to match T then the very early stage of faulting would be realistic, but later fault zone evolution would be characterized by pervasive fracture and rapid comminution of fault-bound volumes. To generate interesting (and realistic) looking fault zone structure, we chose to match the compressive strength with the knowledge that the equivalent tensile strength is too high by a factor of $>$*c.* 2.5.

Geometry and boundary conditions of normal faulting models

The normal faulting modelling is basically a three-step process; (a) model generation; (b) model confinement; and (c) application of the normal faulting boundary conditions.

Model generation

In mesh-based continuum mechanics methods, such as the Finite Element Method (FEM), the geometry of the mesh is of utmost importance. In the DEM, when applied to the simulation of rock fracture, an initial dense particle packing is crucial. Moreover, in the FEM smaller elements are often used in volumes of anticipated large strain. A similar technique, in which smaller particles are used in the volume in which higher resolution is required, is used in the current DEM approach (Fig. 3a).

In each normal faulting model three different materials are used (Fig. 3b): the periodically layered sequence in which faulting is analysed comprises strong and weak beds. This sequence is overlain by a layer of non-bonded particles that have the

Fig. 2. Mechanical properties of DEM model materials used in normal faulting models. (**a**) Initial cubic sample generated in periodic space. (**b**) Carved core sample. Unconfined compression and tension tests are performed by distorting the periodic space in the axial direction. (**c**) Carved sample that is confined using the so-called 'shining-lamp' algorithm. (**d**) Stress–strain curves obtained from unconfined and confined axial compression and tension tests. Confining pressure ranges from 0 to 40 MPa in 5 MPa increments. (**e**) Sample Young's moduli as a function of confining pressure under triaxial compression (comp.) and extension (ext.) conditions. (**f**) Peak-stress data plotted on principal stress graph, together with best-fit failure envelopes. Arrows illustrate stress-paths under triaxial extension conditions and range of confining pressures used in the present study (stars illustrate theoretical hydrostatic pre-faulting stress at 10, 20, 30 and 40 MPa for clarifying purpose only; the actual pre-faulting stress in the DEM models is not hydrostatic). (**g**) Same data and envelopes as in (f) plotted in normalized principal stress space (stresses divided by bond strength). See text for further explanation.

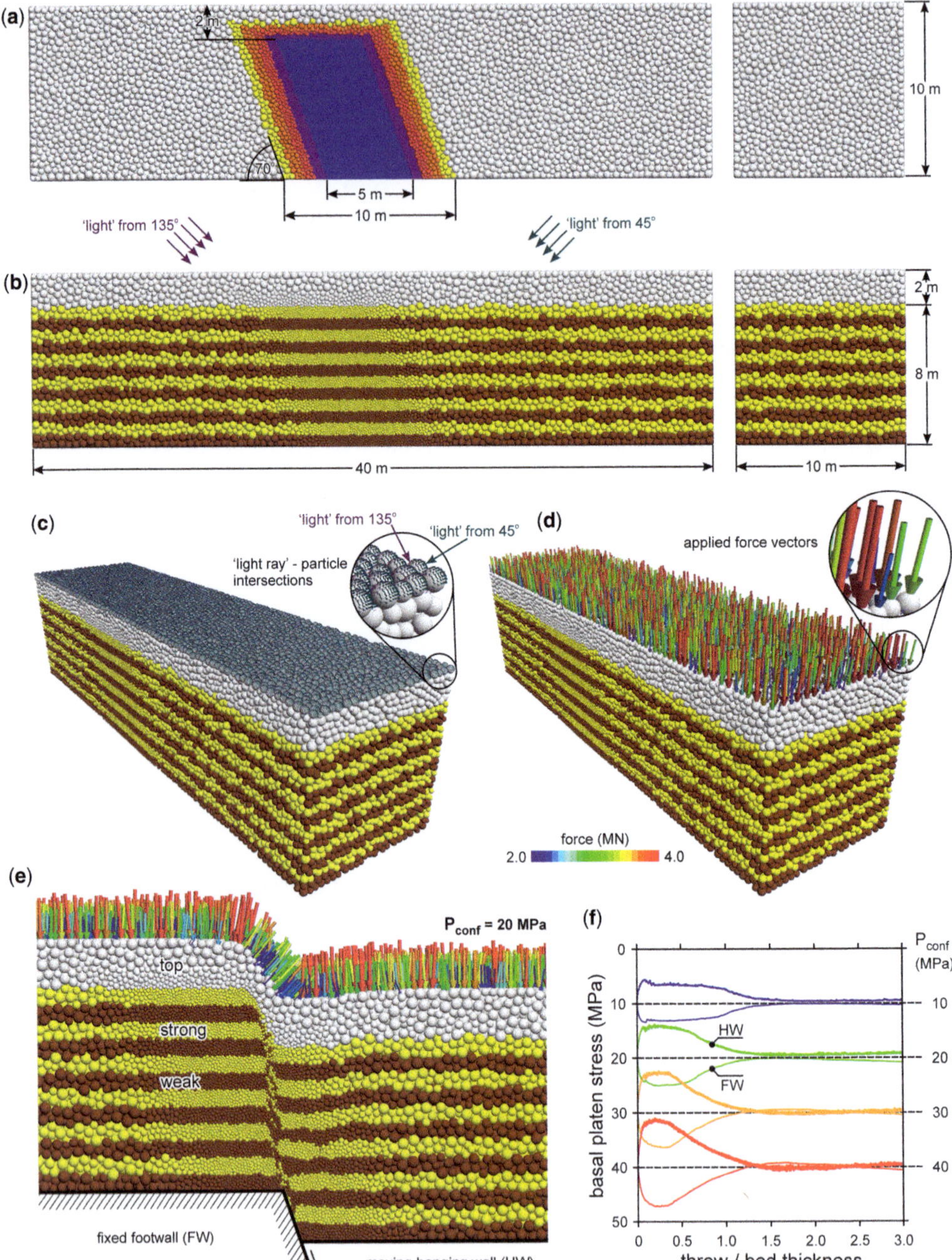

Fig. 3. 3D DEM normal faulting model geometry and boundary conditions. (**a**) Model refinement above pre-defined fault at model base. Regions with different refinement levels (N_R) are coloured from yellow ($N_R = 1$) to deep purple ($N_R = 5$). (**b**) Material regions: yellow = strong; brown = weak; white = top layer. Directions of 'light rays' from the 'shining-lamp' algorithm are also shown. (**c**) Detection of surface particles via the 'shining-lamp' algorithm. Light ray–particle intersections are shown as squares, coloured for light source. (**d**) Model confinement by forces applied to particles. Force magnitude and direction are determined via the 'shining-lamp' algorithm. (**e**) Side view and close-up of a faulted model at a throw of three strong layer thicknesses. (**f**) Footwall (FW) and hanging-wall (HW) platen normal stress as a function of normalized model throw for four models run at confining pressures of 10, 20, 30 and 40 MPa.

same particle properties as the weak material and act as a buffer layer between the layered sequence and the particles that apply the confining pressure at the top of the model.

The model is generated by first filling the model volume, bound by planar, frictionless walls, with a dense packing of frictionless spherical particles with radii chosen randomly from a uniform size distribution. Particles located in the volume above the future fault are refined by means of nested regions in which the refinement level N_R increases inwards. Each particle in a refinement region is replaced by two smaller particles that have the same total volume as the original particle, and this replacement process continues until the specified refinement level N_R is reached, so that the average radius decreases by $r = r_0\exp[-\ln(2)N_R/3]$, where r_0 is the average radius of the base material. The lateral boundaries of these refinement regions are parallel to the future pre-defined basal fault (Fig. 3a). Particle stiffnesses are assigned via the aforementioned (particle size-dependent) modulus–stiffness scaling using contact Young's modulus values assigned to each layer. Then the system is cycled to equilibrium, a low isotropic stress is installed and the number of 'floating' particles (with fewer than three contacts) is reduced (see Potyondy & Cundall 2004, appendices A5 and A6). Finally, particle friction coefficients are assigned. Bonds are installed after confinement, as described below.

Model confinement

Once a model is generated, it is confined to the desired stress level that will be held constant throughout the model run. This confining pressure simulates the load of an overlying rock volume. We assume that this hypothetical overlying rock volume is in an isotropic stress state and therefore exerts no shear tractions to the upper model boundary, so that only normal tractions act on the upper boundary. Gravity is absent in the models (stress differences arising from the vertical stress gradient are negligible at the scale of our models) so that the magnitude of the confining pressure is depth independent. Despite this rather simple definition of boundary conditions, the difficultly lies in identifying particles that are at the top of the model and computing both the direction and magnitude of the applied force in such a way that a constant confinement is maintained while the geometry of the top surfaces changes during normal faulting.

The 'shining-lamp' algorithm (described in detail in Appendix A) is used to identify and apply appropriate forces to the uppermost particles so that the prescribed confining pressure is achieved in a robust way (Fig. 3c–e). Model confinement is applied in a step-wise manner, so that the applied force increases gradually (in order to prevent shock to the system). During initial model confinement the stress acting on the bottom wall is monitored and only when the basal stress level is stable (within a pre-defined tolerance level) is the model considered to be under static stress conditions. Since the model is confined by applying forces to its top surface only, the initial (pre-faulting) stress is not isotropic, meaning that the lateral stress (σ_h) is a fraction of the vertically applied stress ($\sigma_v = P_{\mathrm{conf}}$); since the model is loaded uniaxially the horizontal principal stresses have the same magnitude. The magnitude of this lateral pre-stress is given by the relation $\sigma_h = K_0\sigma_v$, where K_0 is the so-called Earth pressure coefficient at rest, which is *c.* 0.42 in these DEM models. The magnitude of K_0 is a function of the friction angle (ϕ) of the particle assemblage and can be estimated by the Járky expression, $K_0 = 1 - \sin\phi$, which gives a friction angle of 35°.

Bonds are installed after model confinement (except within the top layer) so that the pre-faulting bond stresses are zero, which, from a geological point of view, implies post-consolidation cementation. Post-confinement 'cementation' was necessary as application of confining pressure to bonded particle models can give rise to bond breakages, which would have led to different bulk mechanical properties of materials under different confining pressures (Schöpfer *et al.* 2009*b*).

Normal faulting boundary conditions

Once a model is confined to the desired stress level the bottom platen is replaced by frictionless footwall and hanging-wall platens and a frictionless predefined 'fault' is generated in the 'basement' (Fig. 3e). The dip of this pre-defined fault is always 70°, because at lower fault dips antithetic faulting frequently develops in the overlying sequence (Horsfield 1977; Nollet *et al.* 2012). The lateral platens are replaced by frictionless infinite walls (which ensure that no particles 'escape' and also increase computing speed). Since the normal stiffness of the newly generated walls is identical to that of the initial walls, no disturbance of the force network occurs. During a model run the hanging-wall platen moves down with constant velocity to a final throw of between one and three times the strong layer thickness (final throw varied depending on the specific questions to be addressed with a certain set of models). Model states are saved in regular intervals of ¼ times the strong layer thickness for later analysis.

A recurring question and problem in DEM modelling is how fast a model can be deformed so that dynamic effects are negligible (so called quasi-static conditions). The optimal velocity, which ensures both quasi-static conditions and fast model runs,

Table 2. *Parameters of normal faulting models presented in this study*

Model size*	Net:gross†	Obliquity (deg)‡	Confining pressure (MPa)	Number of realizations
c. 380 000	1:2	0	10, 20, 30, 40	1
c. 200 000	1:2	0, 10, 20, 30, 40	10, 20, 30, 40	1
c. 200 000	1:3, 1:2, 2:3, 5:6	0	20	4

*Model size is given as total number of particles in the DEM model.
†The net:gross of the periodically layered sequences is the proportion of strong layers.
‡Obliquity is the angle between the dip- and slip-direction of the pre-defined fault at the model base (see inset in Fig. 7b).

depends on model dimensions, properties and boundary conditions. For the normal faulting boundary conditions the most reliable diagnostic criterion is the stress acting on the hanging-wall platen (other measures, such as unbalanced forces or the bond breakage evolution are also useful). The basal stress evolution, for example, is illustrated in Figure 3f for four models run at different confining pressures. These curves illustrate that initially the normal stress acting on the footwall platen increases, whereas the stress on the hanging-wall platen decreases. This differential stress is obviously a result of the bulk strength of the overlying sequence. Once the strong beds are entirely offset (which depends on the nature of the sequence and confining pressure as described later) the footwall and hanging-wall platen normal stresses converge back to their initial (pre-faulting) stress level. By systematically varying the displacement rate of the pre-defined fault, we found the optimal hanging-wall platen velocity to be 0.1 m s^{-1}. This final velocity is preceded by an acceleration phase at the beginning of normal faulting, to avoid an abrupt change of boundary conditions that may lead to dynamically induced damage in DEM models.

Model runs

A list of all models run for the present study is provided in Table 2. A constant strong layer thickness of 0.67 m is used in all models. Four models comprising *c*. 380 000 particles were run at confining pressures of 10, 20, 30 and 40 MPa (Fig. 3). The periodically layered sequence in which faulting is analysed comprises six strong and six weak layers of equal thickness. If we assume that the strong layers are reservoir rocks, then these models have a net:gross of 1:2. Various techniques for extracting fault zone geometry and quantitative characteristics and their confining pressure dependence are illustrated using these four models (Figs 4–6). Later in this paper we explore, using smaller models comprising *c*. 200 000 particles, the impact of net:gross and fault obliquity on fault zone structure (Figs 7–11).

Fault zone visualization and quantification

Passive markers and fault throw profiles

DEM model results are often represented by simply plotting particles, which are sometimes coloured according to their displacement. Plotting the locations of broken bonds is also common practice. For ease of comparison of our model fault zones with geological data we extract deformed passive markers from our models, effectively faulted bedding planes. Particle centres in the undeformed (pre-faulting) stage are triangulated using a Delaunay triangulation dividing the model volume into tetrahedra. A series of passive horizontal marker planes are defined through the centres of the strong layers and their intersection points with the edges of the tetrahedra are found. The passive markers are consequently discretised into three- and four-sided polygons. The equivalent polygons in the deformed state define the faulted passive markers and the geometrical distortions of the tetrahedra define the strain distribution over the marker. From these maps, faults are defined on the basis of a maximum shear strain cut-off and throughout this paper a tetrahedron is considered to be 'faulted' if it has a maximum shear strain $\gamma_{max} = (s_1^2 - s_3^2)/(2s_1s_3) > 0.05$ (s_1 and s_3 are the maximum and minimum principal stretch, respectively). Polygons within 'faulted' tetrahedra are displayed in black, while unfaulted polygons are plotted according to their dip (e.g. Fig. 4).

On each passive marker the boundaries between faulted and unfaulted polygons define the fault/horizon intersection polygon or cut-off polygon. The difference in elevation between the upthrown and downthrown polygon cut-off defines the throw profile along the marker horizon. A detailed description of how fault throw profiles can be extracted from these fault polygons is provided in Appendix B. It is important to note that in the DEM models the individual fault strands are volumes of pervasive fracture and pronounced particle rearrangement (loosely speaking fault rock). The fault 'heave' polygons extracted with our method hence not only

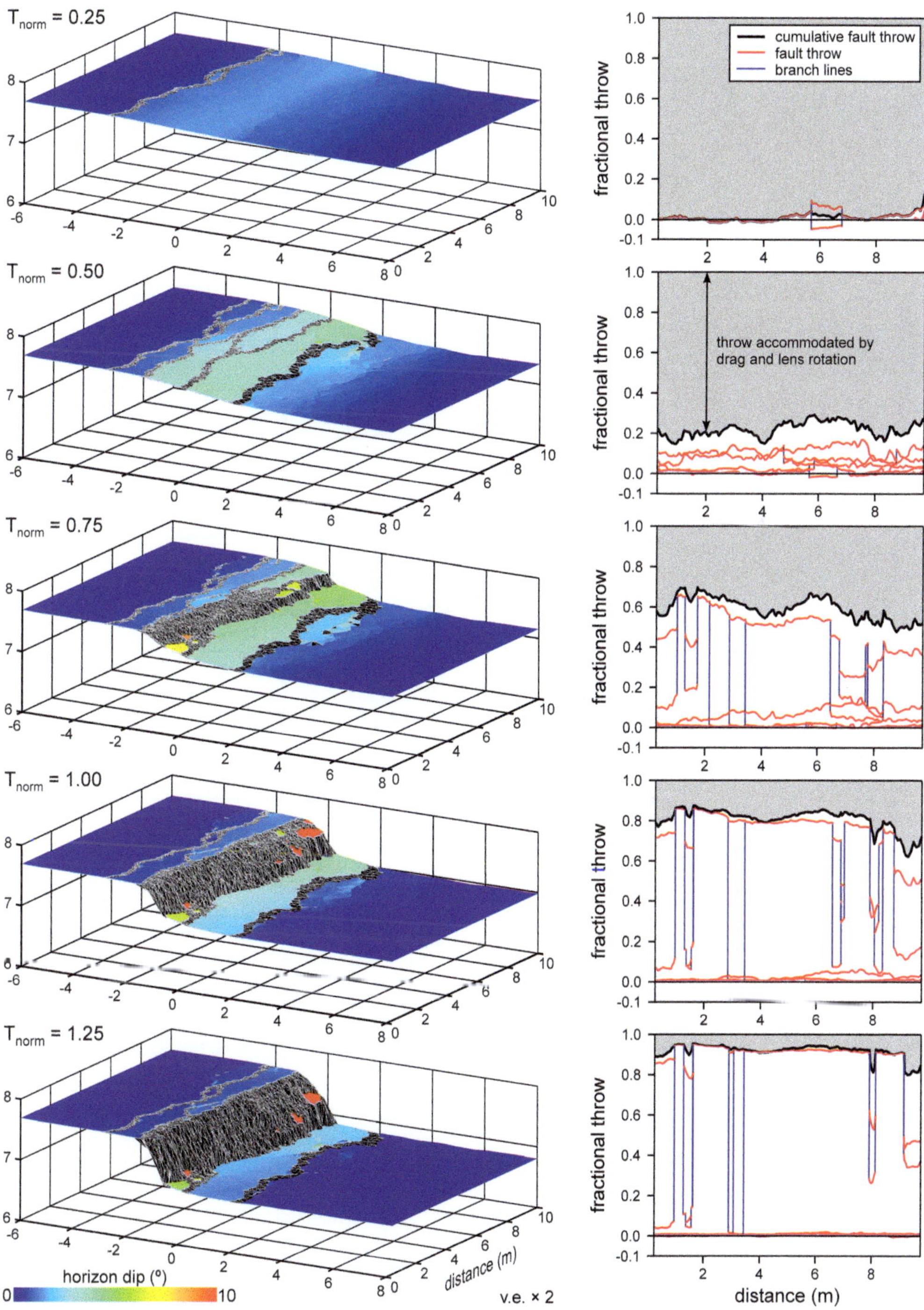

Fig. 4. Fault zone evolution and associated throw profiles in the topmost layer of a DEM model deformed at a confining pressure of 20 MPa (see Fig. 3e). T_{norm} is the normalized throw, i.e. model throw divided by strong layer thickness. Passive markers are coloured for dip if the local maximum shear strain is below 0.05, otherwise the marker is black and outlines of polygons are plotted grey (v.e., vertical exaggeration). The 'fractional throw' is throw normalized by model throw.

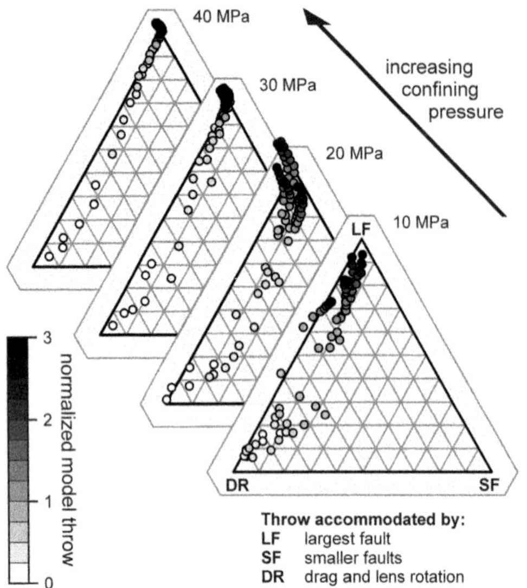

Fig. 5. Ternary diagrams illustrating throw partitioning as a function of confining pressure. Each data point is the average of 39 sections (spacing = 0.25 m) through a horizon, i.e. six data points per model throw. Data falling outside of the triangular plot are due to the presence of antithetic faults which can give rise to displacements on the largest fault that are larger than that on the basement fault.

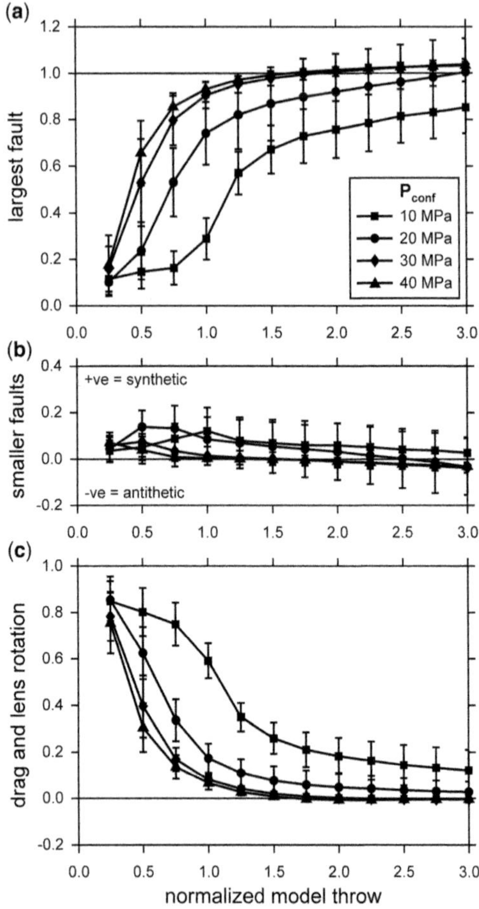

Fig. 6. Fractional throw curves v. normalized model throw for four DEM models run at different confining pressure ($P_{\mathrm{conf}} = 20$ MPa model is shown in Fig. 3e). Each data point is the average of 39 sections (spacing = 0.25 m) per horizon from six layers ($n = 234$) $\pm$ one standard deviation. The three plots illustrate the average fractional throw accommodated by (**a**) the largest fault, (**b**) all smaller faults and (**c**) drag and lens rotation.

reflect the heave of the individual fault strands, but also include the width of the fault strands. Extraction of fault throws on which our analysis is based is not affected by the width of the fault strands.

Quantitative measures of throw partitioning

For extraction of quantitative measures of throw partitioning fault throw profiles along marker horizons are resampled along regularly spaced sections (a spacing of 0.25 m is used throughout this paper, resulting in 39 sections per layer). Fault throw partitioning along each section is quantified using three measures (see also Ferrill *et al.* 2011): (a) the throw on the largest fault strand; (b) the cumulative throw on all smaller faults; and (c) the cumulative throw accommodated by drag and rotation. The sum of (a) and (b) is the discontinuous deformation in the model and the sum of all three is equal to the model throw. When normalized by the total fault zone (=model) throw, these three measures can be displayed in a ternary diagram (Fig. 5). For clarity, each point plotted in these ternary diagrams is the average of 39 readings for each layer. Data falling outside of the triangular plot are due to the presence of antithetic faults, which sometimes develop in the models as a result of the pre-defined fault at the base of the sequence and can give rise to displacements that are actually larger than that on the basement fault.

The normalized values for the three throw components are referred to as *fractional throws* and enable meaningful comparisons of fault zone structure at different stages in model. The *fractional cumulative fault throw*, for example, is the throw accommodated by all faults along a section divided by the total zone throw and is a useful measure of fault localization. In the discussions below we also refer to the *normalized model throw* (T_{norm}) which is the throw on the pre-defined fault divided by the

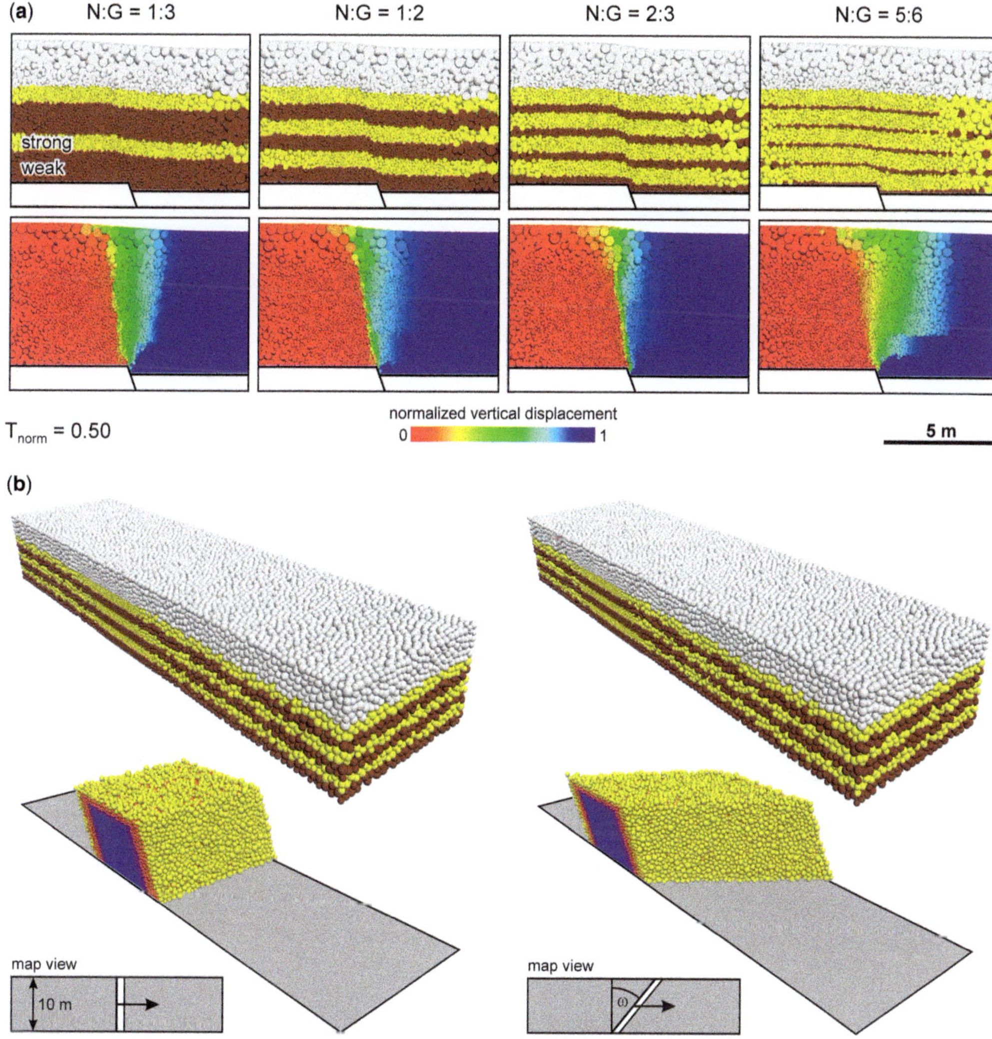

Fig. 7. Different net:gross models and oblique-slip faulting (see Table 2). (**a**) Cross-sections through the centres of different net:gross models at a normalized model throw (T_{norm}) of 0.5, coloured for material (top row) and normalized vertical displacement (bottom row). (**b**) Material and refinement regions of a dip-slip (left) and an oblique-slip (right) model (obliquity $\omega = 40°$). Same particle colour-schemes as in Figure 3.

thickness of the strong layers (0.67 m in all models shown in the present study).

The measures of throw partitioning in the following sections are presented as average values. While average values serve to illustrate particular trends in, and controls on, fault throw partitioning and localization, they are not necessarily the best measures for important application areas. One particular application of throw partitioning is in considering the across-fault connectivity of flow units for which minimum throw values are often of utmost importance; some of the conclusions of our analysis are recast in terms relevant to across-fault connectivity in the discussion section.

Model results

Variation in fault zone structure with displacement

The primary control on throw partitioning in the models is fault displacement. At low displacements, with normalized model throws of 0.25, bending of

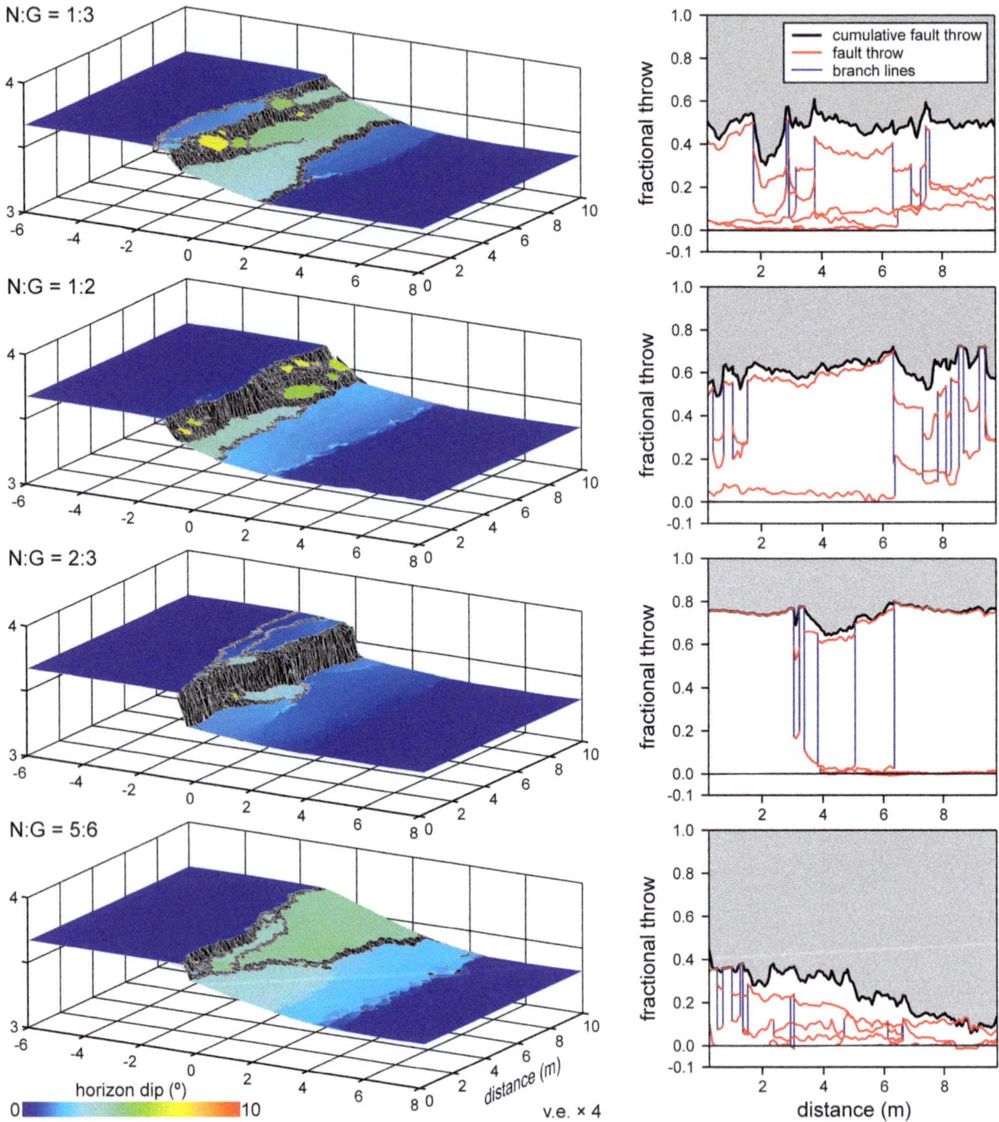

Fig. 8. Passive markers and associated throw profiles for the topmost layers of DEM models with different net:gross, deformed at a confining pressure of 20 MPa, shown at a normalized model throw (T_{norm}) of 0.5 (see Fig. 7a). Horizons and profiles are plotted as in Figure 4 (v.e., vertical exaggeration).

beds is the main contributor to the total throw accommodated across the fault zone and average fractional throws accommodated by drag and lens rotation are *c.* 0.8 (Fig. 6c). Fractional throws on the largest fault progressively increase and are generally >0.8 when the zone throw is three times the thickness of the strong layers (Fig. 6a). The average contribution of smaller faults to the overall throw is generally low and the highest average contribution is *c.* 15% at low confining pressures and normalized model throws of ≤1.0 (Fig. 6b). Once normalized fault throws are >1, the fault throw component is strongly concentrated onto the largest strand in the fault zone (Fig. 6). Therefore, while new minor faults can initiate to accommodate the imposed model throw at any time during fault development, all models record a progressive localization of throw onto a single through-going fault. The precise relationship between fault throw and throw partitioning varies between models depending on

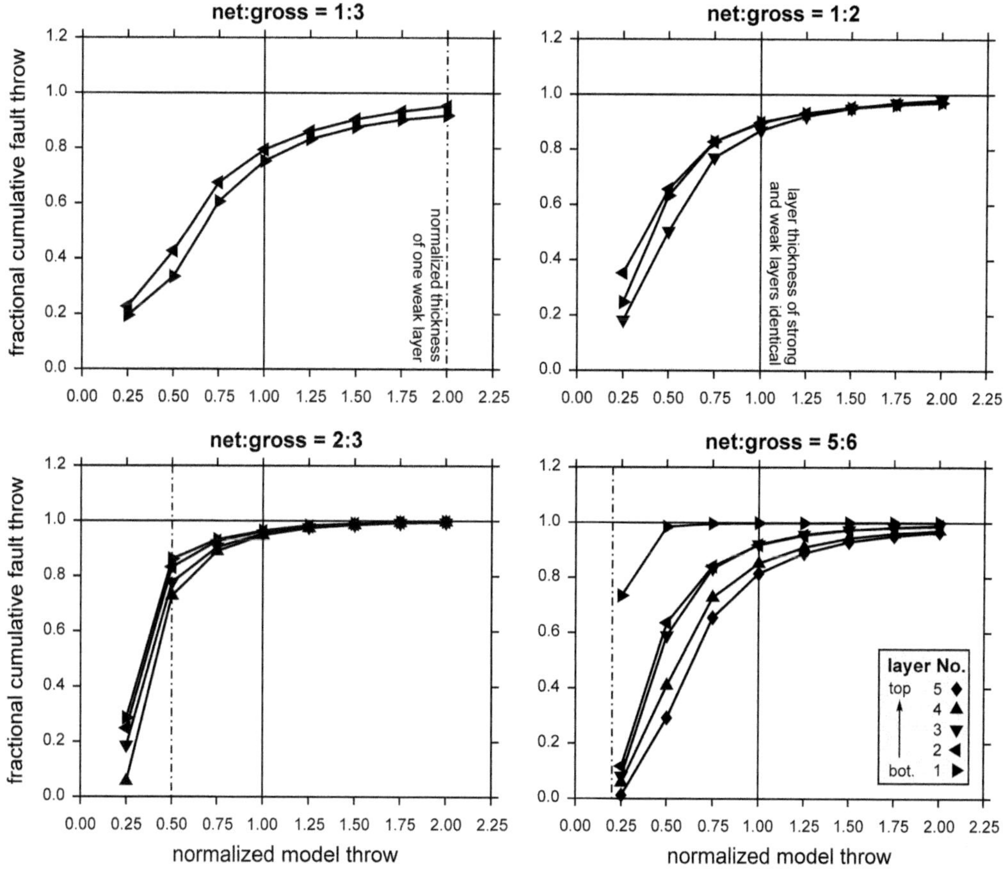

Fig. 9. Fractional cumulative fault throw curves v. normalized model throw from different net:gross DEM models (see Figs 7a & 8). Each data point is the average of 39 sections (spacing = 0.25 m) per horizon from four realizations (n = 156). The dashed vertical line indicates when the model throw exceeds the thickness of one weak layer.

the range of parameters described in the following sections.

It is important to note that, while the average values shown in Figures 5 and 6, for example, give the impression that smaller synthetic faults are of little significance, these average values hide the fact that at particular locations over the fault surface they may be significant. For example, Figure 4 shows that at a normalized throw of 1.25 the fractional throw on the largest fault strand is locally, owing to the presence of a fault-bound lens (at the right model edge), <0.5.

Influence of confining pressure on fault zone structure

The impact of confining pressure on change in fault throw partitioning during fault growth is illustrated in Figures 5 and 6. In general the models show that localization is most efficient under higher confining pressures with fractional throw values on the largest fault strand of >0.8 achieved when model throws are smaller than the thickness of the strong layers in the models with confining pressures ≥20 MPa. At the lowest confining pressure considered (10 MPa), significant localization only occurs when the throw approaches the thickness of the strong layers, i.e. at a normalized throw of 1. The spatial variation in throw partitioning, reflected by the standard deviation, is also strongly dependent on confining pressure. At higher confining pressures the variability in throw partitioning decreases rapidly at normalized model throws of 1 while at the lower confining pressure of 10 MPa the variability in throw partitioning does not decrease with increasing throw (Fig. 6). This feature of the data reflects the fact that, at lower confining pressure, initially formed fault-bounded blocks are not comminuted with increasing displacement, a feature discussed again later.

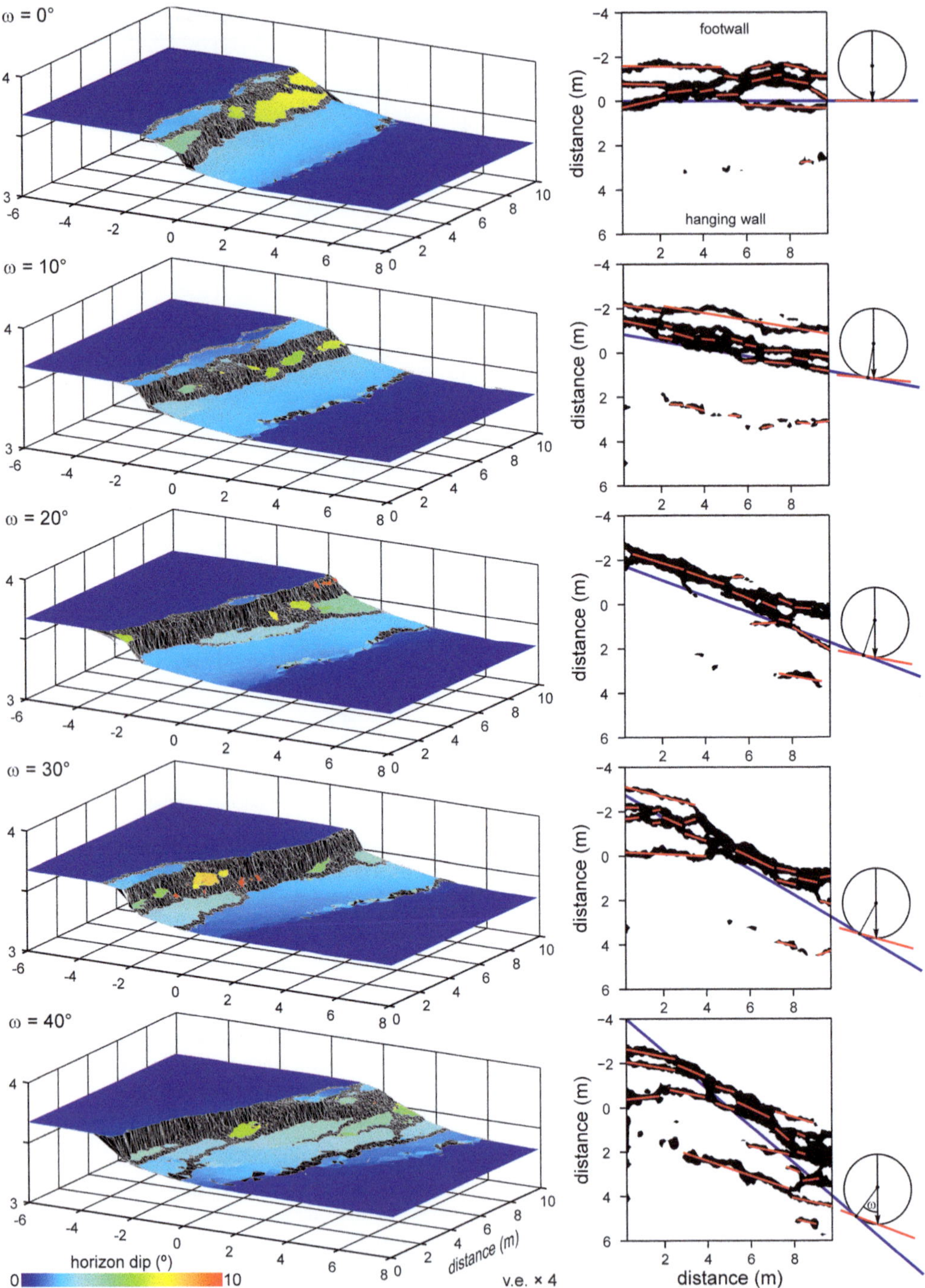

Fig. 10. Passive markers and associated fault maps for the topmost layers of DEM models deformed at a confining pressure of 20 MPa, with different fault obliquity (ω; see Fig. 7b), shown at a normalized model throw (T_{norm}) of 0.5. Horizons are plotted as in Figure 4 (v.e., vertical exaggeration). Red lines on fault maps are best-fit lines to fault strand median lines (see Fig. B1c). The footwall cut-offs of the pre-defined faults at the model base are plotted as blue lines. Circles at the very right illustrate the solution to expected fault strand orientations on the basis of infinitesimal strain theory in transtensional zones (McCoss construction; see text for further explanation).

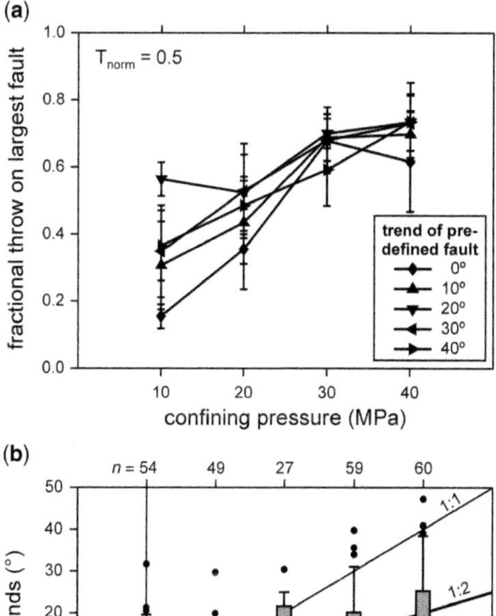

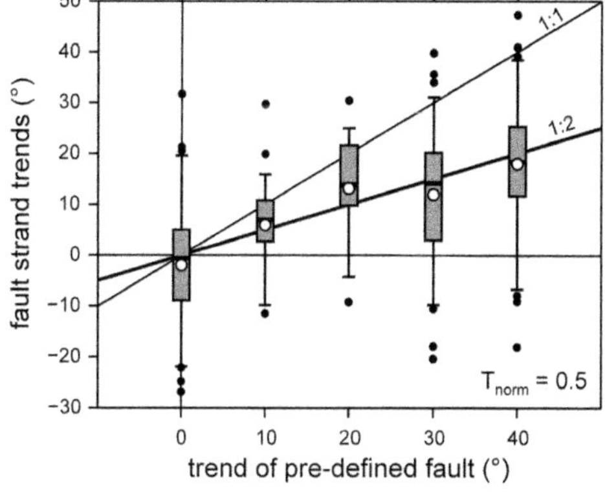

Fig. 11. Quantitative data of DEM models deformed with different fault obliquity (ω) at a normalized model throw (T_{norm}) of 0.5 (see Fig. 10 and Table 2). (**a**) Fractional throw on the largest fault v. confining pressure curves from five different obliquity models. Each data point is the average of 39 sections (spacing = 0.25 m) per horizon from three layers ($n = 117$) ± one standard deviation. (**b**) Box and whisker diagram of fault strand trends from the topmost layers of DEM models deformed under different fault obliquity (trend of pre-defined fault). Data shown are for the full range of confining pressures. The grey boxes span from the 0.25 to 0.75 percentile and the whiskers from the 0.05 to 0.95 percentile. Median values are plotted as bold horizontal lines in each box, outliers as black points and mean values as white points (n = sample size).

Influence of net:gross on fault zone structure

The thicknesses of the strong and weak layers in the DEM models presented so far are identical. In this section the influence of net:gross on fault zone structure is discussed. The periodically layered sequence in these models is always 4 m thick. The thickness of the strong beds is always 0.67 m, so that by varying their number the following four different net:gross sequences are obtained (Fig. 7a; Table 2): 1:3, 1:2, 2:3 and 5:6. Confining pressure in all models is 20 MPa. The geometry of the DEM model fault zones is quite variable under these pressure conditions (Fig. 6) so that four realizations are run for each net:gross value. Models are saved in 0.25 normalized model throw intervals and final model throw is two strong layer thicknesses.

Cross-sections through a selection of these models in which particles are coloured by their vertical displacement illustrate that the best-localized faults appear to develop in the net:gross = 2:3 model (Fig. 7a). This conclusion is also illustrated by the passive markers and associated throw profiles extracted from these four models (Fig. 8): In the net:gross = 1:3 model a wide fault zone containing multiple fault-bound lenses develops. In outcrop such a fault zone may well be categorized as a 'forced fold' (e.g. Cosgrove & Ameen 1999). In fact, at a normalized model throw of 0.5 (as shown in Fig. 8), *c.* 50% of the total throw is accommodated by drag and lens rotation in the net:gross = 1:3 model. However, with increasing net:gross the model fault zones become better localized and the proportion of throw accommodated by lens rotation decreases (Fig. 8). However the highest net:gross model (5:6) exhibits again pronounced rotations and a geometry similar to a kink-band.

The trends in the different net:gross models are quantitatively illustrated in fractional cumulative fault throw graphs in Figure 9, where each data point is the average of 39 sections per layer and four realizations. The curve that most rapidly converges towards a value of 1 (meaning that the entire model throw is accommodated by faulting) is for the bottom layer in the net:gross = 5:6 models. The main reason for this trend is that the weak layers in the net:gross = 5:6 models are very thin (see Fig. 7a), so that the bottommost strong layer is offset via a discrete fault at a very early stage. In that respect, the fractional throw curve for the bottom layer in the net:gross = 5:6 is not representative of the general trend seen in the different net:gross models. Indeed, all other fractional throw curves support the aforementioned observation that the best localized fault zones develop in the net:gross = 2:3 models.

Oblique-slip faulting models

In this section, the impact of fault obliquity on fault zone structure is investigated using net:gross = 1:2 models, each of which comprises three strong and three weak layers of equal thickness (Fig. 7b; Table 2). All other dimensions are identical to those used in the large models (Fig. 3).

Oblique-slip faulting is modelled by changing the trend of the pre-defined fault over the entire model width (Fig. 7b) so that obliquity ranges from 0° (=dip-slip) to 40° with 10° increments. The obliquity (ω) is here defined as the angle between the dip- and slip-direction of the pre-defined fault (see insets in Fig. 7b). The apparent dip, measured in a section containing the slip-vector, of the pre-defined fault is 70° in all models. The refinement regions (model volumes that contain smaller particles for better resolution) are adjusted accordingly (Fig. 7b). Confining pressure ranges from 10 to 40 MPa with 10 MPa increments. Models are saved in 0.25 normalized model throw intervals and final model throw is one layer thickness. Fault analysis is again made on the basis of passive markers (=horizons) through the centres of the strong layers (Fig. 10).

The fault zone dependencies (e.g. fractional throws) of the oblique models on confining pressure are, as expected, similar to those observed in the previously described, larger models (compare Figs 11a and 6a). Perhaps the most interesting feature of the different obliquity models is, however, the dependence of fault zone structure on obliquity, particularly the fault strand orientations (Fig. 10). The trends of individual fault strands are estimated by fitting straight lines to the fault median lines of strands with a length >0.5 m (see Fig. B1c). The map-view orientations of fault strands are measured as angles relative to the slip-normal trend; positive values are anticlockwise. These fault strand trend data are shown for the topmost strong layer and for the full range of confining pressures in Figure 11b. Despite the scatter, the data suggest that the mean (and median) fault strand trends increase with increasing fault obliquity by a factor of 0.5. In other words, a fault obliquity of ω leads to an average fault strand trend of *c.* 0.5ω. This relation can be rationalized by infinitesimal strain theory for zones of transtension in which faults nucleate normal to the greatest instantaneous stretching axis (e.g. McCoss 1986). A graphical construction by McCoss (1986), which is based on the model by Sanderson & Marchini (1984), illustrates the theoretical fault strand orientations as a function of obliquity (Fig. 10). The fact that a kinematic model for transtension offers a rationale for the observed fault strand orientations suggests that, in the DEM models, fault segments nucleate within a pre-cursory monocline, as already demonstrated in earlier 2D DEM models (Schöpfer *et al.* 2006).

Summary and discussion of model fault zones

Throw and confining pressure dependencies

The DEM model fault zones become better localized with increasing throw and confinement, trends which are consistent with earlier DEM normal faulting models (Schöpfer *et al.* 2007*a*, *b*, 2016) and physical experiments comprising cohesive powder embedded in cohesionless sand (Kettermann & Urai 2015). The DEM model fault zones initially comprise segmented fault arrays (Fig. 4). Model rock volumes enclosed by faults and fault plane irregularities (asperities) become progressively comminuted with increasing throw, giving rise to more planar faults. All models become better localized with increasing throw (Figs 5, 6 & 9) with strongly localized faults developing when fault throw is greater than one or two bed thicknesses. The modelled fault geometries are limited in scale from a few particle diameters to the model size, and therefore segmentation and bifurcation occur within a limited scale range. It is likely that faults in larger models would have greater scale geometrical complexity and would therefore require increased throws to achieve localization.

We consider two plausible end-member explanations for increased fault localization with increasing confining pressure: (a) fault zones are initially (i.e. at low throw) more segmented at lower confining pressures than at high confining pressures so that localization does not occur until higher throws are attained; and (b) initial fault segmentation is independent of confining pressure, but existing fault segments are more readily abandoned and/or fault-bound volumes are more easily comminuted at higher confining pressure. The DEM model results favour scenario (b), because various fractional throw measures at low model throws appear to be confining pressure independent (Figs 5 & 6).

It appears therefore that the DEM models support a fault zone evolution model in which faults are segmented over a wide range of confining pressures and that faster localization with increasing confining pressure is due to abandonment of earlier segments at lower throws and to more efficient communition of fault-bound volumes. The question is why at low confining pressure, fault segments which are perhaps not optimally orientated to accumulate large displacements are not abandoned and the rock volumes they bound remain to a greater extent intact. There are two analogies that may offer a mechanical explanation for this behaviour. (a) It is a well-known fact that in granular shear the deformation mechanism is shear zone normal stress dependent, so that at low normal stress grains remain largely intact and exhibit only surface abrasion, whereas at higher normal stress grain splitting occurs (e.g. Mair & Abe 2011). Therefore we would expect that fault-bound volumes with a given size will break up more readily at higher confining pressures. (b) Mechanical analysis (based on the Coulomb criterion) of slip surfaces with a saw-tooth profile suggests that at a certain normal stress the shear stress required for reactivation is greater than for failure of the intact material, resulting in a bi-linear criterion for the shear strength of rock fractures (Patton 1966). We would therefore expect that fault surface irregularities (in the fault profile plane) are more readily removed with increasing confining pressure. We suggest, by analogy, that the low fractional throws accommodated by faulting persist at low confining pressures (Fig. 6) because the strength of the unfaulted volumes is significantly greater than the differential stress required to reactivate existing fault surfaces, even if they are poorly oriented for reactivation. At high confining pressures existing fault surfaces are subject to high normal stresses so that it is more likely that they will become by-passed. The two aforementioned mechanisms probably both occur in the DEM models, where scenario (a) occurs after both tip-line and asperity bifurcation and scenario (b) occurs during asperity bifurcation (Childs *et al.* 1996).

Impact of fault obliquity

The fault strand orientations extracted from DEM models deformed by an oblique-slip fault at the model base suggest that faults nucleate normal to the maximum instantaneous stretching axis (Figs 10 & 11b). This observation is consistent with results from physical experiments conducted with sand and silicone (Richard 1991) and clay (Schlische *et al.* 2002). Both of these studies are worthwhile discussing in light of the DEM model presented here.

Richard (1991) used an experimental apparatus for oblique-slip faulting consisting of a large table divided in two halves. In the oblique-slip normal faulting experiments one half was moved along a 45° dipping basement fault with a strike-slip to dip-slip ratio of 1. The pitch of the slip vector upon the fault plane is therefore 45° and the obliquity ω, as defined in Figures 7b & 10, is $\mathrm{atan}(2^{0.5}) = 55°$. If faults in the overburden were to develop normal to the maximum instantaneous stretching direction, then the expected angle between the basement fault trend and the faults is 27°. In the experiments with a silicone layer at the base of the sand pack, fault zones develop in which the maximum fault strand trend of normal faults is indeed about 27° (fault strands with a smaller angle to the basement fault have a significant strike-slip component). An important result is also that wide fault zones (width increases with increasing silicone layer thickness) comprising en échelon normal fault segments only develop if a weak substratum is present (Richard 1991, fig. 4b–d). An experiment conducted without a weak substratum (Richard 1991, fig. 4a) led to the formation of a graben, bound by normal faults parallel to the basement fault, with a strike-slip fault in its centre. Effectively, oblique motion is partitioned into slip on faults with different senses of motion (Bowman *et al.* 2003). Weak layers (which are present in our DEM models) therefore promote the formation of wide transtensional zones.

Schlische *et al.* (2002) performed oblique-slip normal faulting experiments (obliquity $\omega = 45°$) on thin and thick clay layers and quantified the fault trend frequencies in the 'overburden'. Both in the thin and thick clay layers the most frequent fault strand trends are halfway between the so-called master-fault and displacement-normal trend (Schlische *et al.* 2002, fig. 4). Again, this result is consistent with infinitesimal strain theory for transtensional faults. A notable difference between these clay models and our DEM models, however, is relay ramps, which are frequent and temporarily persistent in the former and rare in the latter (see Fig. 10). There are several reasons for this difference; of the two most likely, the first is that the along-strike dimension of the models is not large enough so that well-developed relay ramps can develop (i.e. the probability of a large relay to form somewhere along strike is too low). This explanation could in theory be tested by running larger models, which, however, would require impractical model run times. The second reason is partly one of definition. The DEM models do display rotation of ramps but, because large permanent strains can only form by fracture (i.e. bond breakage), these ramps are entirely fault bounded at low throws and are therefore fault-bound lenses rather than relay ramps. If the maximum shear strain cut-off applied to define a fault (see Fig. B1a) is increased, then many structures will display relay geometries. Accommodation of large permanent strains without bond breakage would require a rate-dependent bond model that mimics micromechanical processes, such as precipitation-dissolution creep. This line of research would be, in our opinion, demanding but potentially extremely useful.

Influence of net:gross

The DEM models illustrate that, for the mechanical properties and confining pressure used in the models presented, the most rapidly localizing faults develop in the net:gross = 2:3 models (Figs 8 & 9). Our preferred mechanical explanation for the observed net:gross dependence is that at low net:gross fracture-bound blocks can easily rotate within a 'ductile' matrix, which promotes continuous throw accommodation (drag and lens rotation). As net:gross increases, the adjacent layers begin to impinge on one another so that they are no longer bounded by a thick weak 'matrix' and rotation becomes more subdued, resulting in faster localization. At the highest net:gross, however, layer-parallel shear across the thin weak layers becomes suppressed and brittle layers impinge on each other early during fault zone growth to the extent that the assemblage of closely spaced beds operates as a single, thick mechanical unit which rotates as a whole, therefore suppressing localization. The 'optimal' net:gross value at which the best localized faults develop (which is 2:3 in our models) will most likely depend on other factors such as confining pressure and strength contrast. However, we would expect that the general trend (pronounced rotations at low and high net:gross) would also emerge under different confining pressure and/or strength conditions because fault zone structure in layered sequences depends mainly on relative properties (e.g. layer strength relative to confining pressure; Schöpfer *et al.* 2007*b*).

The DEM models are consistent with several field studies. For example, Ferrill & Morris (2008) demonstrate that the structure of normal faults in Cretaceous carbonates in the Balcones fault system (Texas) greatly depends on the mechanical stratigraphy. As in our DEM models, a high proportion of 'incompetent' material promotes fault-related folding. On the other hand, fault zones in massive (poorly bedded) sequences appear to exhibit complex internal structure, which may reflect that rotation of fault-bound volumes is, in the absence of bedding-parallel slip, inhibited (e.g. Bonson *et al.* 2007).

Implications for across-fault reservoir connectivity

In this paper we have primarily been concerned with average values of throw-partitioning parameters as measures of fault zone geometrical evolution (Figs 6, 9 & 11a). However, one of the main potential applications of this modelling is in evaluating the likelihood of connectivity of flow units across fault

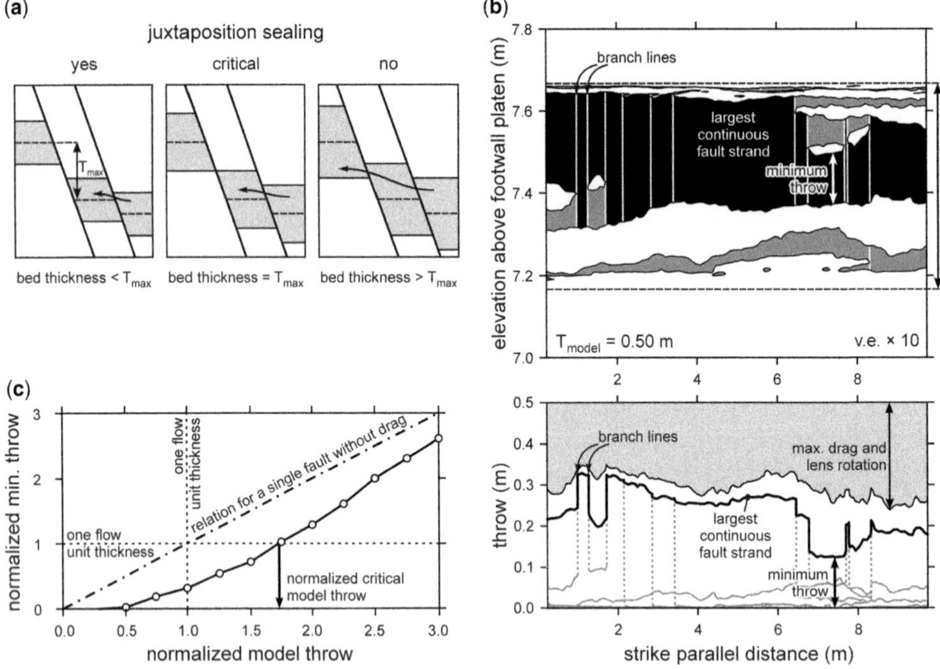

Fig. 12. Principles of juxtaposition sealing analysis and definition of the 'normalized critical model throw'. (**a**) Cross-sections illustrating how the reservoir thickness and the maximum fault strand throw (T_{max}) control across-fault fluid flow, assuming juxtaposition sealing only. Note that, along-strike, the smallest throw on the largest fault strand controls juxtaposition sealing. (**b**) Strike-projection of a faulted horizon ($T_{norm} = 0.75$ horizon shown in Fig. 4) and associated fault throw profile illustrating the largest continuous fault strand and its minimum throw. Note that the location at which this minimum throw occurs does not necessarily coincide with the location at which the greatest amount of throw is accommodated by drag and lens rotation. (**c**) Normalized minimum throw v. normalized model throw graph for the topmost layer in the $P_{conf} = 20$ MPa model (Fig. 4). Theoretical juxtaposition sealing occurs when the minimum throw exceeds the strong layer (=flow unit) thickness. The normalized model throw at which sealing occurs is the normalized critical model throw.

zones, which requires consideration of the extreme rather than average values. The key measure is the minimum throw on the largest continuous fault strand. If this minimum throw is larger than the flow unit thickness, then it is disconnected across the modelled fault zone length (if across-fault flow is considered in 2D only, then the throw on the largest fault strand relative to the thickness of the flow unit determines juxtaposition sealing; Fig. 12a). The minimum throw on the largest continuous fault strand is determined in an iterative fashion; a sample result is illustrated by means of a strike-projection and a throw profile in Figure 12b. For the models presented here we consider the strong layers to be the flow units and express our results in terms of the model throw at which individual flow units are offset along the length of the fault divided by the flow unit thickness, i.e. the model throw at which the minimum throw exceeds the thickness of the strong layers; we refer to this measure as the *normalized critical model throw*. A normalized critical model throw of 2 means that the model throw must be twice the thickness of the flow unit before it is entirely offset along the length of the fault and a value of 1 means that all throw is concentrated onto a single continuous fault strand (Fig. 12c). The majority of natural faults and also our models exhibit, to a greater or lesser degree, displacement partitioning onto multiple strands and bed rotations. Consequently the throw at which juxtaposition sealing occurs is typically greater than the thickness of the flow unit so that the normalized critical model throw is greater than one (Fig. 12c).

Plots of normalized critical model throw v. both confining pressure and net:gross (Fig. 13) show similar relationships to those established from average values of fractional throw (Figs 6 & 9). The normalized critical model throw decreases with increasing confining pressure and approaches a value of 1 (the throw at which juxtaposition occurs is only slightly greater than the thickness of the strong layers at 40 MPa confinement; Fig. 13a). At a confining pressure of 10 MPa the normalized critical model throw is, on average, >2; in this case the throw must be greater than twice the flow unit thickness so that no juxtapositions occurs over the entire fault zone width, reflecting that a significant proportion of the total throw is accommodated by smaller faults and bed rotations. Similar to the conclusion derived from fractional throw measures (Figs 6 & 11a), the variation in normalized critical model throw decreases as the confining pressure increases so that the model fault is not only more localized but also more uniform.

The normalized critical model throws show a general decrease with increase in the proportion of strong layers within the model and values of nearly 1 are achieved at net:gross of 5:6 (Fig. 13b). It appears that the variability in normalized critical model throw does not change with net:gross and there is significant variability between different horizons within the same model. The average values of normalized critical model throw are lowest at a net:gross value of 2:3 in line with the qualitative impression of fault zone width and localization provided by Figure 8 and the fractional cumulative throw curves shown in Figure 9.

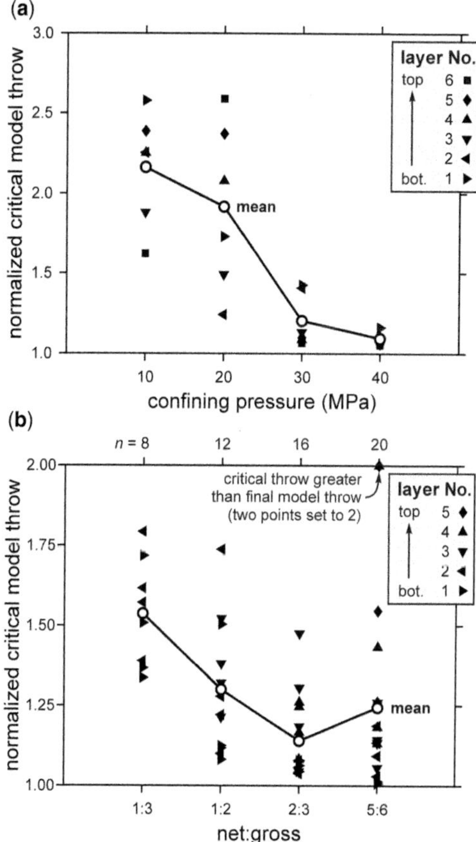

Fig. 13. Normalized critical model throw v. (**a**) confining pressure for each of the six horizons from four models deformed at a confining pressure of 10, 20, 30 and 40 MPa (graphical solution for the topmost horizon of the $P_{conf} = 20$ MPa model is shown in Fig. 12c) and (**b**) net:gross (data plotted are from each horizon and from four realizations). White circles are the mean values for each model (n = sample size).

The present analysis is based on detecting the minimum throw on the largest continuous fault strand that spans the entire model width (Fig. 12b). For complex fault zones we expect that the minimum throw on the largest continuous fault will decrease with increasing fault zone length, so that the total throw at which juxtaposition sealing occurs over the entire fault zone length will increase with increasing model, or sampling, length. In other words, the probability of juxtaposition sealing along the entire fault zone decreases with increasing sampling length because the probability of encountering a structure that promotes across-fault fluid flow (e.g. a relay ramp) increases with increasing sampling length. A study on the scale-dependency of juxtaposition sealing, perhaps based on the algorithms presented here, would be a useful line of future research.

Conclusions

The model fault zones suggest that the fraction of strong material in a layered sequence (net:gross) has a profound impact on throw partitioning and localization:

(1) models with net:gross of 1:3, 1:2, 2:3 and 5:6 deformed at a confining pressure of 20 MPa suggest that the most rapidly localizing fault zones develop in the net:gross = 2:3 models;
(2) in low net:gross sequences significant proportions of the total zone throw are accommodated by rotations, not unlike forced folding;
(3) in high net:gross sequences bedding parallel slip is suppressed and strong layers impinge on each other early during fault zone growth, leading to wide and poorly localized fault zones.

Model fault zones developing above pre-defined oblique-slip normal faults suggest the following:

(1) the average fault strand orientations are consistent with predictions based on infinitesimal strain theory for transtensional zones;
(2) the fact that a kinematic model for transtension offers a rationale for the observed fault strand orientations suggests that fault segments nucleate within a pre-cursory monocline.

Analysis of DEM models deformed at four different confining pressures and juxtaposition sealing analysis support the following conclusions:

(1) fault zones become better localized with increasing throw and confinement;
(2) at low confining pressure the fault zone throw must exceed, on average, twice the bed thickness so that across-fault flow is not possible (assuming juxtaposition sealing only);
(3) the critical throws beyond which no across fault flow is possible decrease with increasing confining pressure;
(4) the ranges of critical throws are considerable, because critical throws are controlled by the local, rather than the average, fault zone structure.

Many of the aforementioned conclusions concerning the growth of fault zones are consistent with field observations and results from physical experiments. The DEM may therefore provide a basis for evaluating the likely complexity of fault zone structure and associated sequence juxtapositions in a given sequence faulted under a certain confining pressure.

The DEM modelling presented here was performed during the course of the multi-company project 'QUAFF', brokered by ITF (Aberdeen) and funded by Anadarko, BG International, BP Exploration, ConocoPhillips (U.K.), Eni, ExxonMobil, Marathon Oil Corporation, Statoil, Total E&P U.K. and Woodside Energy. C. Childs is funded by Tullow Oil. Some methods were developed during the course of the earlier multi-company project 'FaultZone', funded by ExxonMobil and Shell. We also thank the staff of Itasca Consulting Group for their support, in particular Dave Potyondy for fruitful discussions and for providing his 'shining-lamp' algorithm, and Sacha Emam and Peter Cundall for clarifying the concept and usage of distortional periodic space. Thorough reviews by Michael Kettermann and Kevin Smart are gratefully acknowledged.

Appendix A: Pressure boundary condition via the 'shining-lamp' algorithm

A recurring problem in 3D DEM models comprising spherical particles is the application of a constant pressure boundary condition without the use of rigid, servo-controlled platens. In the DEM, forces can be applied to particles, but a robust pressure boundary condition that mimics, for example, the overburden pressure exerted on a rock volume at depth requires two important features: (a) all particles that are at the surface of the model must be detected, both at initiation of the model and as the surface deforms and becomes non-planar; and (b) both the magnitude and direction of the force applied to each surface particle must be determined so that the force vectors are normal to the surface. The so-called 'shining-lamp' algorithm, developed by D. Potyondy (Itasca Consulting Group, unpublished memorandum ICG7233-L, 2012) provides these features.

The pressure-application procedure is referred to as the shining-lamp algorithm because it defines the surface particles as those that receive direct illumination from light sources that shine toward the body. Each surface particle is assigned an externally applied force (**F**) that is related to the specified pressure (P_{conf}) and the amount of light that it receives (which is related to its illuminated surface area).

The principles of the algorithm are illustrated by means of a simple assemblage of discs that receive light laterally from plus and minus infinity (with inward-directed unit-normal vectors $\mathbf{p}$ and $\mathbf{p}'$, respectively; Fig. A1a). For each particle the applied force (acting at

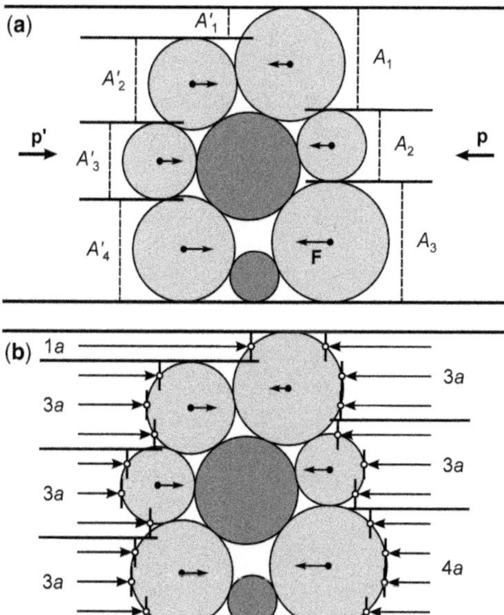

Fig. A1. Explanation of 3D pressure boundary condition with a simplified model comprising discs (modified after D. Potyondy, Itasca Consulting Group, unpublished memorandum ICG7233-L, 2012).
(**a**) Force contributions from an opposing pair of 'light sources' with inward-directed unit-normal vectors (labelled **p** and **p**′). Applied particle forces (vectors originating at particle centres) are proportional to projected areas illuminated by light (labelled A and A'). The sum of projected areas (A_1–A_3 and A'_1–A'_4) are identical on both sides, hence the applied forces balance. Dark grey particles are not illuminated.
(**b**) The 'light' is discretized into rays with a regular spacing of a (white dots are intersections of 'light rays' with particles). The applied particle forces are proportional to the number of 'light ray'–particle intersections (multiples of area) and hence an approximation of those shown in (**a**). Vertical ticks (of length a) with their centre located at the 'light ray'–particle intersections become squares (with area a^2) in three dimensions (see Figs 2c & 3c).

its centre) is the sum of the force contributions from each light source $\mathbf{F} = P_{\text{conf}}\, A\mathbf{p}$, where A is the projected illuminated area. Since the sum of these projected areas is identical on both sides of the assemblage, the applied forces balance.

The preceding description assumes that the illuminated surfaces can be computed exactly. The shining-lamp algorithm discretizes the light sources into rays emanating from regular grids (square grids are used in the present 3D models) and thereby provides an approximation of each illuminated surface (Fig. A1b); the approximation converges to the exact surfaces in the limit as the grid size approaches zero. The algorithm is very efficient and only requires one pass through the list of particles (in the normal faulting models only particles comprising the top layer are checked; Fig. 3). For each particle the light-ray intersections are computed (sphere-line intersections in 3D) and the positions of these intersections are stored at each corresponding grid-point. If an intersection is closer to the light source than an earlier computed intersection, then the current particle becomes a surface particle (the addresses to surface particles are also stored at each corresponding grid-point). A consequence of this discretization is that the applied particle forces are proportional to the number of light–ray particle intersections (Fig. A1b), so that $\mathbf{F} = P_{\text{conf}}\, n_p a\mathbf{p}$, where n_p is the number of intersections and a the grid-spacing in 2D and the grid-spacing squared in 3D (in the case of square grids). Note that the discrete area (a) is illustrated as line segments, centred on the intersection points, in 2D (Fig. A1b) and as squares in 3D (black squares in Fig. 2c; cyan and magenta squares in Fig. 3c). The computation of ray–particle intersections and hence applied particle forces is performed in regular intervals (e.g. every 100 time-steps) to account for changes of the model's surface geometry.

In the present study the shining-lamp algorithm is implemented in two model environments, (a) the triaxial extension/compression tests (Fig. 2c) and (b) the normal faulting models (Fig. 3b–e). In the triaxial models light rays are emanating from square grids with inward-directed normal vectors parallel to the global x- and z-coordinates (four grids in total; grid spacing is $0.5r_{\text{min}}$). Depending on its position, a particle may therefore receive light from more than one light source; hence forces are applied both in the x- and z-directions, so that the force vectors are normal to the surface of the cylindrical sample, as they should be.

In the normal faulting models a constant overburden pressure is applied via the shining lamp algorithm. The efficiency of this procedure obviously depends on how well the model surface is illuminated by light, hence two sources (i.e. discrete grids with a spacing of $0.2r_{\text{max}}$) with rays inclined at 45° and 135° (counter-clockwise from the positive x-direction) are used (cyan and magenta, respectively; Fig. 3b). Computationally, however, it is more efficient to use light rays that are parallel to the global coordinate system. Hence a coordinate transformation is performed when the light–ray particle intersections for the top layer are computed: each particle position is rotated clockwise by 45° around the origin (located in the model centre) and then intersected with light rays emanating from the positive x- and y-directions (the latter points upwards). Obviously, the applied forces computed in the transformed coordinate system have to be adjusted to the global coordinate system. Since only particles lying at the model surface should receive light, care has to be taken in the positioning of the grids so that the lateral surfaces of the model are not illuminated. Consequently, the grid from which rays are emanating at 45° (cyan in Fig. 3b, c) must be translated according to the throw of the moving hanging-wall platen.

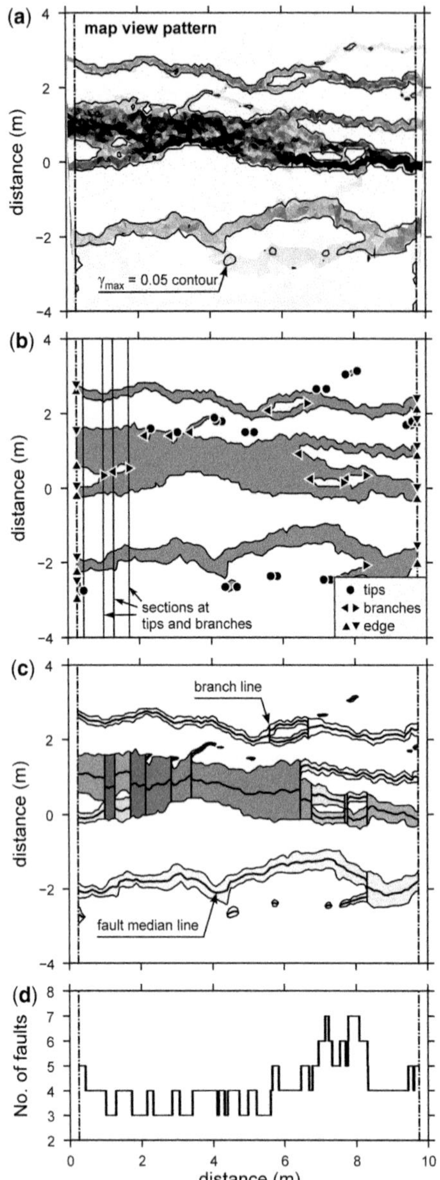

Fig. B1. Extraction of fault strand throws from passive markers, illustrated using the $T_{norm} = 0.75$ horizon shown in Figure 4. (**a**) Map view pattern of maximum shear strain (γ_{max}) and its 0.05 contour. Shades of grey are from white ($\gamma_{max} = 0$) to black ($\gamma_{max} = 0.5$). (**b**) Local maxima of contour that point in a direction parallel to the trend of the pre-defined fault are classified as tips, branches or model edges. A selection of sections through tip and branch locations is also shown. (**c**) Identified fault segments (grey-shaded for average throw) and computed fault median lines. (**d**) Graph illustrating that the number of fault strands changes whenever a fault tip or branch is present at any fault zone section.

Appendix B: Extraction of fault throw profiles from passive markers

As described in the main text, horizon maps through the centres of each strong layer are extracted by intersecting passive marker planes in the pre-faulting stage with the edges of a 3D Delaunay mesh obtained from triangulated particle centres. Horizon maps are hence represented by polygons (triangles and four-sided polygons) for which the 3D coordinates of the vertices and the strain tensor are known. Measures derived from the strain tensor, such as maximum shear strain, can then be used to visualize faults (e.g. Fig. 4). In this Appendix a new technique that allows extraction of displacements from faulted horizon maps is described (Fig. B1).

The first step is to generate fault outlines, commonly referred to as fault polygons. This step is achieved by calculating the horizontal polygon centres and by drawing a contour using one contour level (which is equal to the aforementioned maximum shear strain cut-off). These contours provide the map view coordinates of the fault polygons (Fig. B1a). The vertical coordinates are readily computed by interpolation. This first step provides 3D outlines of the faults, viz. the footwall and hanging-wall cut-offs. The original contour may be noisy and hence small isolated contours and small peaks parallel to the trend of the pre-defined fault are removed.

The next step is to classify the contours. Three types can be distinguished (Fig. B1b): (a) isolated contours; (b) contours within contours (which must enclose fault-bound volumes); and (c) contours that straddle the model boundaries. Then local maxima, parallel to the trend of the pre-defined fault, are identified and classified. Three types of local maxima are distinguished (Fig. B1b): (a) fault tips; (b) branch points; and (c) model edges. Tips and branches are identified on the basis of simple geometric rules that assume that the trends of individual fault strands do not vary too dramatically (which is typically the case in our model faults). For example, a local contour maximum pointing to the positive trend-parallel direction (abscissa in Fig. B1) must be a fault tip if a point shifted slightly from this position to the positive direction falls outside the fault polygon. In contrast, if this shifted point would fall inside the fault polygon then the local contour maximum must be a branch point. Note that 'contours within contours', which correspond to fault-bound volumes, may also have both tips and branches.

The contours are then intersected by vertical planes, or sections, whose normals are parallel to the trend of the pre-defined fault. First, sections that pass through the previously identified tips and branches are defined (a selection of these traces is shown in Fig. B1b). This is an important step, because the number of fault strands is constant in-between tips/branches (Fig. B1d). Second, additional regularly spaced sections are defined to enhance the resolution of the final displacement profiles. Third, the section positions are checked and additional sections are defined, if necessary, to ensure that there is at least one section between each tip/branch location (otherwise no throw profile can be constructed if, for example, there are no sections in-between two tips).

Then the intersection points (3D coordinates) of the contours with the previously defined sections are computed and the median points (halfway between the footwall and hanging-wall cut-offs) are calculated (Fig. B1c). Fault strand throws and widths (vertical and horizontal differences between the footwall and hanging-wall cut-offs, respectively) are also computed. Fault strand throws are positive if the fault is synthetic to the pre-defined fault, whereas antithetic faults have a negative throw.

The final step is to extract fault segments (Fig. B1c). As stated earlier, in map view a fault zone can be 'sliced' by lines normal to its 'average' strike into sub-areas of constant fault strand number (Fig. B1d); the slicing locations are either tips or branches. For each of these sub-areas fault segments are extracted. The two lateral ends of each segment can be classified as follows: (a) fault tip; (b) branch point (merging or splitting branch points are distinguished); (c) model edge; and (d) no fault end. For each fault segment the 3D coordinates of the median line, the associated throws and widths, and segment end identifiers are used for later plotting and analysis.

References

Abe, S., van Gent, H. & Urai, J.L. 2011. DEM simulation of normal faults in cohesive materials. *Tectonophysics*, **512**, 12–21.

Bathurst, R.J. & Rothenburg, L. 1988. Note on a random isotropic granular material with negative Poisson's ratio. *International Journal of Engineering Sciences*, **26**, 373–383.

Bonson, C.G., Childs, C., Walsh, J.J., Schöpfer, M.P.J. & Carboni, V. 2007. Geometric and kinematic controls on the internal structure of a large normal fault in massive limestones: the Maghlaq Fault, Malta. *Journal of Structural Geology*, **29**, 336–354.

Botter, C., Cardozo, N., Hardy, S., Lecomte, I. & Escalona, A. 2014. From mechanical modeling to seismic imaging of faults: a synthetic workflow to study the impact of faults on seismic. *Marine and Petroleum Geology*, **57**, 187–207.

Bowman, D., King, G. & Tapponnier, P. 2003. Slip partitioning by elastoplastic propagation of oblique slip at depth. *Science*, **300**, 1121–1123.

Byerlee, J.D. 1968. Brittle–ductile transition in rocks. *Journal of Geophysical Research*, **73**, 4741–4750.

Carmona, A., Clavera-Gispert, R., Gratacós, O. & Hardy, S. 2010. Modelling syntectonic sedimentation: combining a discrete element model of tectonic deformation and a process-based sedimentary model in 3D. *Mathematical Geosciences*, **42**, 519–534.

Childs, C., Watterson, J. & Walsh, J.J. 1996. A model for the structure and development of fault zones. *Journal of the Geological Society, London*, **153**, 337–340, https://doi.org/10.1144/gsjgs.153.3.0337

Childs, C., Manzocchi, T., Walsh, J.J., Bonson, C.G., Nicol, A. & Schöpfer, M.P.J. 2009. A geometric model of fault zone and fault rock thickness variations. *Journal of Structural Geology*, **31**, 117–127.

Cosgrove, J.W. & Ameen, M.S. 1999. A comparison of the geometry, spatial organization and fracture patterns associated with forced folds and buckle folds. *In*: Cosgrove, J.W. & Ameen, M.S. (eds) *Forced Folds and Fractures*. Geological Society, London, Special Publications, **169**, 7–l, https://doi.org/10.1144/GSL.SP.2000.169.01.02

Crider, J.G. & Peacock, D.C.P. 2004. Initiation of brittle faults in the upper crust: a review of field observations. *Journal of Structural Geology*, **26**, 691–707.

Cundall, P.A. 2001. A discontinuous future for numerical modelling in geomechanics? *Geotechnical Engineering*, **149**, 41–47.

Cundall, P.A. & Hart, R. 1992. Numerical modeling of discontinua. *Engineering Computations*, **9**, 101–113.

Deere, D.U. & Miller, R.P. 1965. *Engineering Classification and Index Properties for Intact Rock*. Technical Report AFWL-TR-65-115. Air Force Weapons Laboratory, Kirtland Air Base, New Mexico.

Egholm, D.L. 2007. A new strategy for discrete element numerical models: 1. Theory. *Journal of Geophysical Research*, **112**, B05203.

Egholm, D.L., Sandiford, M., Clausen, O.R. & Nielsen, S.B. 2007. A new strategy for discrete element numerical models: 2. Sandbox applications. *Journal of Geophysical Research*, **112**, B05204.

Egholm, D.L., Clausen, O.R., Sandiford, M., Kristensen, M.B. & Korstgård, J.A. 2008. The mechanics of clay smearing along faults. *Geology*, **36**, 787–790.

Faulkner, D.R., Jackson, C.A.L., Lunn, R.J., Schlische, R.W., Shipton, Z.K., Wibberley, C.A.J. & Withjack, M.O. 2010. A review of recent developments concerning the structure, mechanics and fluid flow properties of fault zones. *Journal of Structural Geology*, **32**, 1557–1575.

Ferrill, D.A. & Morris, A.P. 2008. Fault zone deformation controlled by carbonate mechanical stratigraphy, Balcones fault system, Texas. *AAPG Bulletin*, **92**, 359–380.

Ferrill, D.A., Morris, A.P. & Smart, K.J. 2007. Stratigraphic control on extensional fault propagation folding: big Brushy Canyon monocline, Sierra Del Carmen, Texas. *In*: Jolley, S.J., Barr, D., Walsh, J.J. & Knipe, R.J. (eds) *Structurally Complex Reservoirs*. Geological Society, London, Special Publications, **292**, 203–217, https://doi.org/10.1144/SP292.12

Ferrill, D.A., Morris, A.P., McGinnis, R.N., Smart, K.J. & Ward, W.C. 2011. Fault zone deformation and displacement partitioning in mechanically layered carbonates: the Hidden Valley fault, central Texas. *AAPG Bulletin*, **95**, 1383–1397.

Finch, E., Hardy, S. & Gawthorpe, R. 2004. Discrete-element modelling of extensional fault propagation folding above rigid basement fault blocks. *Basin Research*, **16**, 489–506.

Hardy, S. 2011. Cover deformation above steep, basement normal faults: insights from 2D discrete element modeling. *Marine and Petroleum Geology*, **28**, 966–972.

HARDY, S. 2013. Propagation of blind normal faults to the surface in basaltic sequences: insights from 2D discrete element modelling. *Marine and Petroleum Geology*, **48**, 149–159.

HORSFIELD, W.T. 1977. An experimental approach to basement-controlled faulting. *Geologie en Mijnbouw*, **56**, 363–370.

IMBER, J., TUCKWELL, G.W. *ET AL.* 2004. Three-dimensional distinct element modelling of relay growth and breaching along normal faults. *Journal of Structural Geology*, **26**, 1897–1911.

ITASCA CONSULTING GROUP 2008. *Particle Flow Code in Three-Dimensions*, version 4.0, Minneapolis, MN.

KETTERMANN, M. & URAI, J.L. 2015. Changes in structural style of normal faults due to failure mode transition: first results from excavated scale models. *Journal of Structural Geology*, **74**, 105–116.

LOCKNER, D.A. 1995. Rock failure. *In*: AHRENS, T.J. (ed.) *Rock Physics and Phase Relations: A Handbook of Physical Constants*. American Geophysical Union, Washington, D.C., 127–147.

MAIR, K. & ABE, S. 2011. Breaking up: comminution mechnisms in sheared fault gouge. *Pure and Applied Geophysics*, **168**, 2277–2288.

McCOSS, A.M. 1986. Simple constructions for deformation in transpression/transtension zones. *Journal of Structural Geology*, **8**, 715–718.

MICARELLI, L., BENEDICTO, A., INVERNIZZI, C., SAINT-BEZAR, B., MICHELOT, J.L. & VERGELY, P. 2005. Influence of P/T conditions on the style of normal fault initiation and growth in limestones from the SE-Basin, France. *Journal of Structural Geology*, **27**, 1577–1598.

NOLLET, S., KLEINE VENNEKATE, G.J., GIESE, S., VROLIJK, P., URAI, J.L. & ZIEGLER, M. 2012. Localization patterns in sandbox-scale numerical experiments above a normal fault in basement. *Journal of Structural Geology*, **39**, 199–209.

PATTON, F.D. 1966. Multiple modes of shear failure in rock. *In*: *Proceedings of the 1st Congress of the International Society for Rock Mechanics*, Lisbon, **1**, 509–513.

PATTON, T.L., LOGAN, J.M. & FRIEDMAN, M. 1998. Experimentally generated normal faults in single-layer and multilayer limestone specimens at confining pressure. *Tectonophysics*, **295**, 53–77.

POTYONDY, D.O. & CUNDALL, P.A. 2004. A bonded-particle model for rock. *International Journal of Rock Mechanics and Mining Sciences*, **41**, 1329–1364.

RICHARD, P. 1991. Experiments on faulting in a two-layer cover sequence overlying a reactivated basement fault with oblique-slip. *Journal of Structural Geology*, **13**, 459–469.

ROTHENBURG, L., BERLIN, A.A. & BATHURST, R.J. 1991. Microstructure of isotropic materials with negative Poisson's ratio. *Nature*, **354**, 470–472.

SALTZER, S.D. & POLLARD, D.D. 1992. Distinct element modeling of structures formed during extensional reactivation of basement normal faults. *Tectonics*, **11**, 165–174.

SANDERSON, D.J. & MARCHINI, W.R.D. 1984. Transpression. *Journal of Structural Geology*, **6**, 449–458.

SCHLISCHE, R.W., WITHJACK, M.O. & EISENSTADT, G. 2002. An experimental study of the secondary deformation produced by oblique-slip normal faulting. *AAPG Bulletin*, **86**, 885–906.

SCHÖPFER, M.P.J. & CHILDS, C. 2013. The orientation and dilatancy of shear bands in a bonded particle model for rock. *International Journal of Rock Mechanics and Mining Sciences*, **57**, 75–88.

SCHÖPFER, M.P.J., CHILDS, C. & WALSH, J.J. 2006. Localisation of normal faults in multilayer sequences. *Journal of Structural Geology*, **28**, 816–833.

SCHÖPFER, M.P.J., CHILDS, C. & WALSH, J.J. 2007*a*. 2D Distinct Element modeling of the structure and growth of normal faults in multilayer sequences. Part 1: model calibration, boundary conditions and selected results. *Journal of Geophysical Research*, **112**, B10401.

SCHÖPFER, M.P.J., CHILDS, C. & WALSH, J.J. 2007*b*. 2D Distinct Element modeling of the structure and growth of normal faults in multilayer sequences. Part 2: impact of confining pressure and strength contrast on fault zone geometry and growth. *Journal of Geophysical Research*, **112**, B10404.

SCHÖPFER, M.P.J., CHILDS, C. & WALSH, J.J. 2009*a*. Two-dimensional Distinct Element Method (DEM) modeling of tectonic fault growth in mechanically layered sequences. *In*: KOLYMBAS, D. & VIGGIANI, G. (eds) *Mechanics of Natural Solids*. Springer, Heidelberg, 127–146.

SCHÖPFER, M.P.J., ABE, S., CHILDS, C. & WALSH, J.J. 2009*b*. The impact of porosity and crack density on the elasticity, strength and friction of cohesive granular materials: insights from DEM modelling. *International Journal of Rock Mechanics and Mining Sciences*, **46**, 250–261.

SCHÖPFER, M.P.J., CHILDS, C. & MANZOCCHI, T. 2013. Three-dimensional failure envelopes and the brittle–ductile transition. *Journal of Geophysical Research*, **118**, 1378–1392.

SCHÖPFER, M.P.J., CHILDS, C., WALSH, J.J. & MANZOCCHI, T. 2016. Evolution of the internal structure of fault zones in three-dimensional numerical models of normal faults. *Tectonophysics*, **666**, 158–163.

WALSH, J.J., CHILDS, C. *ET AL.* 2001. Geometrical controls on the evolution of normal fault systems. *In*: HOLDSWORTH, R.E., STRACHAN, R.A., MAGLOUGHLIN, J.F. & KNIPE, R.J. (eds) *The Nature and Tectonic Significance of Fault Zone Weakening*. Geological Society, London, Special Publications, **186**, 157–170, https://doi.org/10.1144/GSL.SP.2001.186.01.10

WANG, Y. & MORA, P. 2008. Macroscopic elastic properties of regular lattices. *Journal of the Mechanics and Physics of Solids*, **56**, 3459–3474.

WIBBERLEY, C.A.J., YIELDING, G. & DI TORO, G. 2008. Recent advances in the understanding of fault zone internal structure: a review. *In*: WIBBERLEY, C.A.J., KURZ, W., IMBER, J., HOLDSWORTH, R.E. & COLLETTINI, C. (eds) *The Internal Structure of Fault Zones: Implications for Mechanical and Fluid-Flow Properties*. Geological Society, London, Special Publications, **299**, 5–33, https://doi.org/10.1144/SP299.2

Throw partitioning across normal fault zones in the Ptolemais Basin, Greece

EFSTRATIOS DELOGKOS[1*], TOM MANZOCCHI[1], CONRAD CHILDS[1], CHRISTOS SACHANIDIS[2], TRYFON BARBAS[2], MARTIN P. J. SCHÖPFER[3], ALEXANDROS CHATZIPETROS[4], SPYROS PAVLIDES[4] & JOHN J. WALSH[1]

[1]*Fault Analysis Group, School of Earth Sciences, University College Dublin, Belfield, Dublin 4, Ireland*

[2]*Public Power Corporation of Greece, Western Macedonian Lignite Centre, Ptolemais, Greece*

[3]*Department for Geodynamics and Sedimentology, University of Vienna, Althanstrasse 14, A-1090 Vienna, Austria*

[4]*Department of Geology, Aristotle University of Thessaloniki, Thessaloniki, Greece*

**Correspondence: stratos.delogkos@ucd.ie*

Abstract: The total throw across a fault zone may not occur entirely on a single fault strand but may be distributed onto several strands or may be accommodated by distributed deformation within or adjacent to the fault zone. Here we conduct a quantitative analysis of the partitioning of throw into three components, the throw accommodated by: (a) the largest fault strand; (b) subsidiary faults; and (c) continuous deformation in the form of bed rotation in sympathy with the fault downthrow direction. This analysis is applied to seven seismic-scale fault zones at outcrop resolution (maximum throw 50 m) that were mapped over a four-year period during open-cast lignite mining within the late Miocene–Pliocene Ptolemais Basin, West Macedonia, Greece. The analysis shows that the fault zones offsetting the lignite–marl sequence are more localized at higher throws with progressively more of the total throw accommodated by the largest fault strand. Normal drag, which can account for up to 12 m of the total throw, accommodates a lower proportion of the total throw on larger faults. It appears that initial fault segmentation is the main control on the degree of, and spatial variation in, fault throw partitioning.

Fault zones are complex features that can comprise multiple slip surfaces and heterogeneously distributed fault rock of various compositions. This complex structure, and its variability over different scales of inspection, make consistent descriptions of fault zone structure difficult (Ben-Zion & Sammis 2003). In recent years, fault zones have been most often described in terms of a central high strain 'core' that accommodates most of the fault throw and an enveloping damage zone that contains minor structures that accommodate wall-rock strains and some, relatively minor, proportion of the total fault throw. Quantification of fault zone structure has generally concentrated on determining the widths of different fault zone components, such as fault core, damage zone and fault rock, and on using these measurements as indicators for fault zone structural evolution (Evans 1990; Caine *et al.* 1996; Shipton & Cowie 2001; Childs *et al.* 2009). In this paper, we are concerned with how throw is partitioned onto the different components of a fault zone (Fig. 1). This approach is based on conceptual models of how fault zones develop that suggest throw partitioning within fault zones should provide a record of how faults grow. For example, it is widely accepted that one of the main mechanisms by which faults increase their length is by linkage between segments, and this linkage should be recorded in the amount of throw accommodated by fault splays that were by-passed when fault segments coalesced to form a continuous fault surface (Peacock & Sanderson 1991; Childs *et al.* 1995, 2009; Cartwright *et al.* 1996; Walsh *et al.* 1999). Similarly, measurement of the proportion of throw accommodated by ductile deformation on faults of different sizes can test models in which faulting is preceded by the formation of a precursory monocline or a process zone (Billings 1972; Reches & Eidelman 1995; Peacock *et al.* 2000; Brandes & Tanner 2014). Partitioning of throw onto the various components of a fault zone can be an important factor providing insights into the growth of fault zones and in determining

From: Childs, C., Holdsworth, R. E., Jackson, C. A.-L., Manzocchi, T., Walsh, J. J. & Yielding, G. (eds) 2017. *The Geometry and Growth of Normal Faults*. Geological Society, London, Special Publications, **439**, 333–353.
First published online November 21, 2016, https://doi.org/10.1144/SP439.19

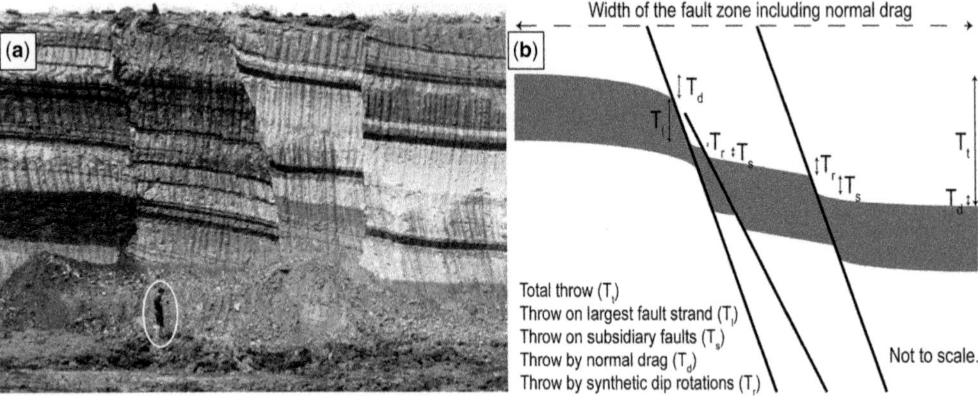

Fig. 1. (**a**) Outcrop photograph of a fault zone in the Kardia mine showing that the total throw (*c.* 13 m) is accommodated by several fault strands and by continuous deformation within and adjacent to the fault zone (person circled for scale). (**b**) Schematic diagram showing the partitioning of total throw (T_t) into discontinuous throw, comprising the largest fault strand (T_l) and subsidiary synthetic faults (T_s), and continuous throw, comprising synthetic dip rotation (T_r) and normal drag (T_d) components.

across-fault reservoir juxtaposition, and thus the flow properties of faults in the subsurface (Caine *et al.* 1996; Childs *et al.* 1997; Wibberley *et al.* 2008; Manzocchi *et al.* 2010; Seebeck *et al.* 2014). Despite this importance, quantification of the frequency and magnitude of throw partitioning has not received much attention in the published literature.

In general, recorded fault throws can be divided into two components: discontinuous and continuous throws. Discontinuous throw refers to throw that occurs on discrete fault surfaces, and continuous throw refers to throw accommodated by deformation of the wall rock due to normal drag and bed rotations within the fault zone (Fig. 1b). The term 'normal drag' is used here in a purely geometrical sense (e.g. Peacock *et al.* 2000) and does not imply a formation mechanism. In this paper, we subdivide the discontinuous throw into the throw on the largest fault strand and the remaining component, which is the sum of throws on subsidiary synthetic fault strands (Fig. 1b).

Throw partitioning has been previously used as an indicator of fault zone development. Jamison & Stearns (1982) described a transition from distributed deformation bands to discrete fault slip surfaces at a throw of approximately 10 m in normal faults formed in high-porosity sandstones. Kristensen (2005) found that the throw on the largest fault strand within a fault zone is frequently less than half, and often as low as 30%, of the total throw on centimetre-scale faults in clay/sand sequences. A compilation by Freeman *et al.* (2008) shows that the proportion of throw accommodated on the largest segment in a fault zone is, in general, hugely variable. In a study of the Hidden Valley normal fault at Canyon Lake Gorge, Ferrill *et al.* (2011) concluded that the width of the fault damage zone associated with this 60 m displacement fault was established early during the growth of the fault, with later progressive localization of displacement. They also estimated that the actual stratigraphic displacement is overestimated by about 14–21% due to displacement partitioning across a fault zone. Manzocchi *et al.* (2008) proposed a quantitative function based on a similar conceptual model of damage zone evolution and displacement localization, and in a modelling study demonstrated that the effect of realistic fault displacement partitioning on full-field oil production can, in some circumstances, be significant.

In this paper, we augment these available quantitative datasets addressing displacement partitioning in normal faults. Specifically, we measure how the throw at various positions within seven fault zones is partitioned onto the largest fault surface, subsidiary synthetic fault surfaces and ductile components, with the aim of better understanding how each of these three components contributes to the total throw as fault throw increases.

Data and methodology

The dataset used in this study is derived from the Kardia Mine (also known as the Tomeas Eksi Mine), which is one of the four active, open-pit lignite fields in the Ptolemais Basin, west Macedonia, Greece (Fig. 2). The Ptolemais Basin is an elongate intramontane lacustrine basin and is part of the Florina–Ptolemais–Servia Basin, which is a NNW–SSE-trending graben system that extends

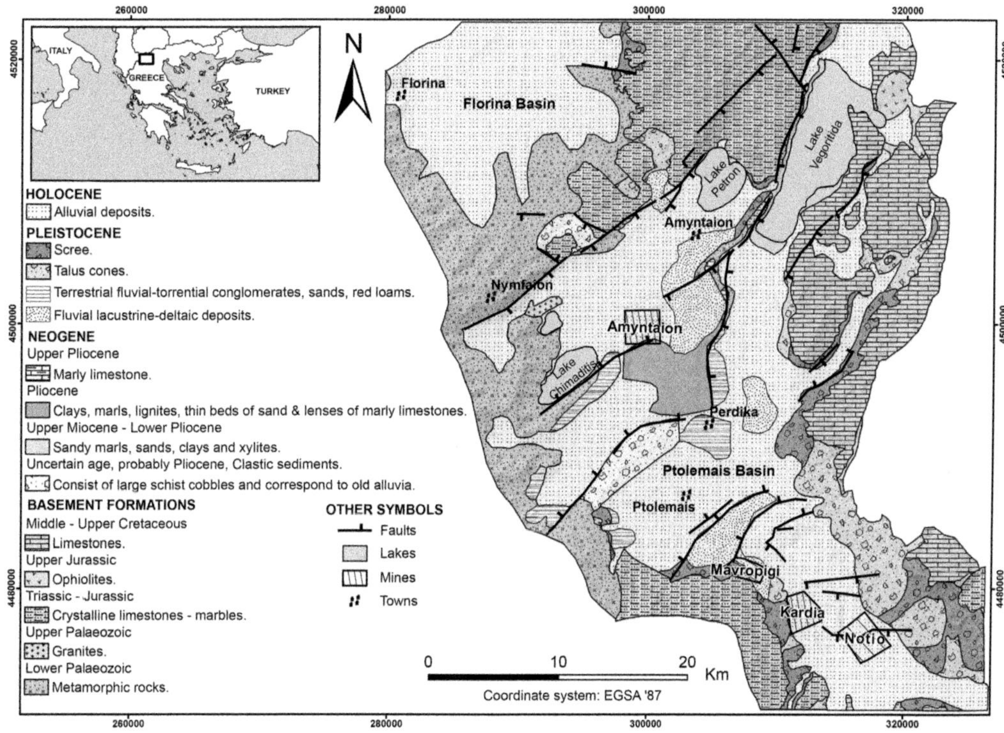

Fig. 2. Geological map of the Ptolemais Basin showing the major fault structures and the locations of the four active, open-pit lignite mines (modified after Pavlides 1985).

over a distance of 120 km from Bitola in the Former Yugoslav Republic of Macedonia to the village of Servia, SE of Ptolemais, Greece (Pavlides 1985). The basin is affected by two fault systems related to two extensional episodes (Pavlides & Mountrakis 1987; Mercier *et al.* 1989). The first, Late Miocene episode resulted in the formation of the basin in response to NE–SW extension. The second episode was a NW–SE extension during the Quaternary, resulting in the NE–SW-striking normal faults that currently bound a number of sub-basins, including the basins of Florina, Ptolemais and Servia (Pavlides & Mountrakis 1987).

There is little surface evidence of the Late Miocene NW–SE-striking faults that control the basin margins, although their presence is confirmed from boreholes (Pavlides & Mountrakis 1987) and from some recent exposures along the western margin of the Ptolemais Basin. The surface geology is dominated by the Quaternary faults, which have orientations ranging from the expected NE–SW strikes to the north of the region, to NNE–SSW orientations to the north of the Mavropigi Mine, and through to approximately east–west strikes in the vicinity of the Kardia and Notio mines (Fig. 2). Some of the Quaternary-age NE–SW-striking normal faults have been activated recently, causing both weak and strong earthquakes (Pavlides *et al.* 1995; Mountrakis *et al.* 1998).

The basin is filled with a 500–600 m-thick succession of Late Miocene–Pleistocene lake sediments with intercalated lignites and alluvial deposits, which are divided into three basin-wide lithostratigraphic units: the Lower Formation; the Ptolemais Formation; and the Upper Formation (Anastopoulos & Koukouzas 1972; Steenbrink *et al.* 1999). This work has concentrated on faults in the Ptolemais Formation, which has a thickness of approximately 110 m, and consists of a rhythmic alternation of metre-scale lignite and lacustrine marl beds with intercalated fluvial sands and silts, and some 20 volcanic ash beds. This formation was dated as early Pliocene using a combination of magneto- and cyclostratigraphy, and $^{40}Ar/^{39}Ar$ dating (van Vugt *et al.* 1998; Steenbrink *et al.* 1999).

In addition to the normal faults, the mines contain fault-related folding (Fig. 3) in the form of reverse and normal drag (discussed in more detail in a later section). Furthermore, one of the mines, Notio Mine, contains thrusts that predate Quaternary faults but have identical strikes. It has been suggested (Diamantopoulos *et al.* 2013) that the

Fig. 3. True scale and × 3 vertically exaggerated panoramic view of the active, open-cast, Kardia Mine in **(a)** April 2012 and **(b)** May 2014. Faults are drawn as yellow lines and a selection of horizons is highlighted. The letters (C, F, R, Q, P, T and S) are the names of the interpreted fault zones.

normal faults and thrusts are associated with extension above, and compression below, a neutral surface associated with folds developed in an active transpressional zone. However, our mapping shows that the thrusts at the Notio Mine are contained within a well-defined approximately 30 m-thick unit bounded by a décollement at the base and an unconformity at the top. We interpret this unit to represent a synsedimentary subaqueous landslide, with the presence of geometries similar to, but on a larger scale than, those contained in soft sediments in the Dead Sea described by Alsop & Marco (2013). Therefore, the thrusts are not associated with tectonic compression, but instead are of very early, synsedimentary gravity-driven origin. The thrusts are consistently cross-cut by the normal faults and occasionally reactivated as normal faults, providing further evidence for the early timing of the thrusts. Our interpretation of synsedimentary slumping followed by normal faulting and fault-related folding is significantly different to the model for these structures suggested by Diamantopoulos *et al.* (2013), which requires a transtensional-dominated regime.

Kardia Mine: 3D structural model

The Kardia Lignite Mine is located in the central part of the Neogene lignite basin and is dominated by east–west-trending Quaternary normal faults. It consists of seven principal mining faces, which are, on average, 2.5 km long. The *c.* 20 m-high mining faces step to the west from the bottom to the top of the mine, and are separated by benches that have widths of *c.* 50–100 m. We have visited and mapped the mine 16 times at approximately three-month intervals from October 2009 to May 2014. During each interval, each face was taken back by between 20 and 50 m. The data collected during each fieldwork campaign were photographs at various resolutions, and accurate GPS locations, structural measurements and interpretations of all exposed faults and related structures observed in the mine, such as normal or reverse drag.

Three-dimensional models of the faults in the Kardia Mine were produced by placing our fully georeferenced data within a 3D structural interpretation package (Fig. 4a). Each of the mining faces was cropped from the panoramic photographs and imported into the 3D database. The outcome was a data volume similar in format and scale to a 2D seismic reflection survey, but with outcrop-scale resolution. Horizons and faults were interpreted in a similar fashion to a 2D seismic survey, but with the aid of detailed field observations.

Each fault zone was mapped from about 100 fault-perpendicular cross-sections, which together typically cover areas of the fault surface *c.* 1100 m long and up to 80 m high (Figs 4 & 5). The striking colour contrast between the lignite and marl layers makes these sediments ideal for detailed throw analysis in cross-section and along-strike, as individual horizons are continuous on the scale of the structures investigated. In total, 14 horizons were mapped within a 110 m stratigraphic interval, and these were used to calculate discontinuous throws for each of the interpreted fault surfaces (Figs 4 & 5). Figure 5 shows one of the interpreted horizons in map view with the footwall and hanging-wall fault cutoffs. Note that, owing to the excavation direction, deeper horizons in the sequence cover most of the eastern part of the mine and shallower horizons cover most of the western part.

Main characteristics of faults in the Ptolemais Basin

In an area of 2.5 km^2, we mapped seven fault zones (labelled C, F, R, Q, P, T and S in Figs 3–5) that displace the lignite–marl sequence by up to 50 m. These seven fault zones consist of more than 115 mapped fault segments. Fault segments less than approximately 35 m long are not mappable in 3D as they are shorter than the lateral, along-strike, resolution of this dataset. Only fault zone S, with a maximum throw of 18.5 m, has been mapped over its entire length (maximum recorded length of 630 m). For five fault zones (C, F, P, R and T), one of the two tips has been mapped; and for fault zone Q, neither of the two tips lie in the area that this dataset covers. Several individual fault segments were mapped from tip to tip, with the largest – a segment in fault zone P – having a maximum length of 550 m and maximum discontinuous throw of 33 m.

Grouping of the fault segments into a fault zone can be highly subjective and depends largely on the scale of faulting relative to the mapped area. Fault segments are considered here to be part of the same fault zone if they interact with each other by soft and/or hard linkage (Walsh & Watterson 1991; Walsh *et al.* 2003), as indicated by the occurrence of a simple aggregate throw profile that includes the throw accommodated by continuous deformation. Some of the interpreted fault zones are larger than the sample area and it may be that, given a larger sample area, interactions between separate fault zones may become apparent, so that the fault zones would be grouped differently.

Fault rock

The faults in the Ptolemais mines have anomalously high fault displacement to fault rock thickness ratios compared to normal faults in other areas. Wide

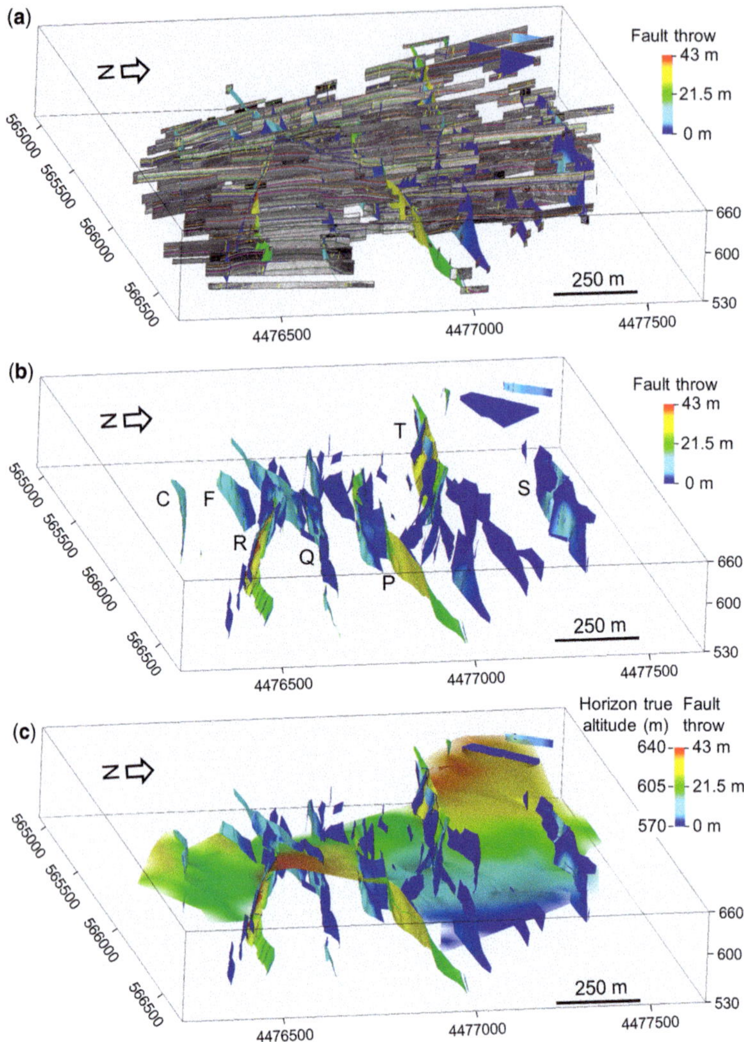

Fig. 4. Oblique view of a 3D model of Kardia Mine showing (**a**) all the mining faces imported into a 3D structural interpretation package, including the interpretation of seven fault zones that displace the lignite–marl sequence by up to 50 m. The colours on the fault surfaces are contours of throw. (**b**) As (a), but excluding the imported mining faces. The letters are the names assigned to each fault zone. (**c**) As (b), but including one of the interpreted horizons, which is displaced by the faults and is located near the middle of the exposed stratigraphic sequence. The horizon is coloured for height above sea level.

zones of fault rock (e.g. breccias) are not developed in these faults. Small-scale lenses and splays that, with increasing displacement, would be pulverized and converted into fault rock in other lithologies, are fortuitously preserved in these rocks, allowing throw partitioning to be examined at high strains.

Throw gradients

Throw gradients on individual fault segments, and also on whole fault zones, are larger here (average value of *c.* 0.125) than in many other fault systems (cf. Walsh *et al.* 2002; Bailey *et al.* 2005; Kim & Sanderson 2005; Ferrill *et al.* 2008; Schultz *et al.* 2008). This observation is compatible with the overall impression that these faults are more highly segmented than usual. Mechanical differences between lignite and marl do not exert a strong control on throw gradients. Instead, fault geometry exerts the primary control and so that the highest throw gradients are associated with interacting fault tips.

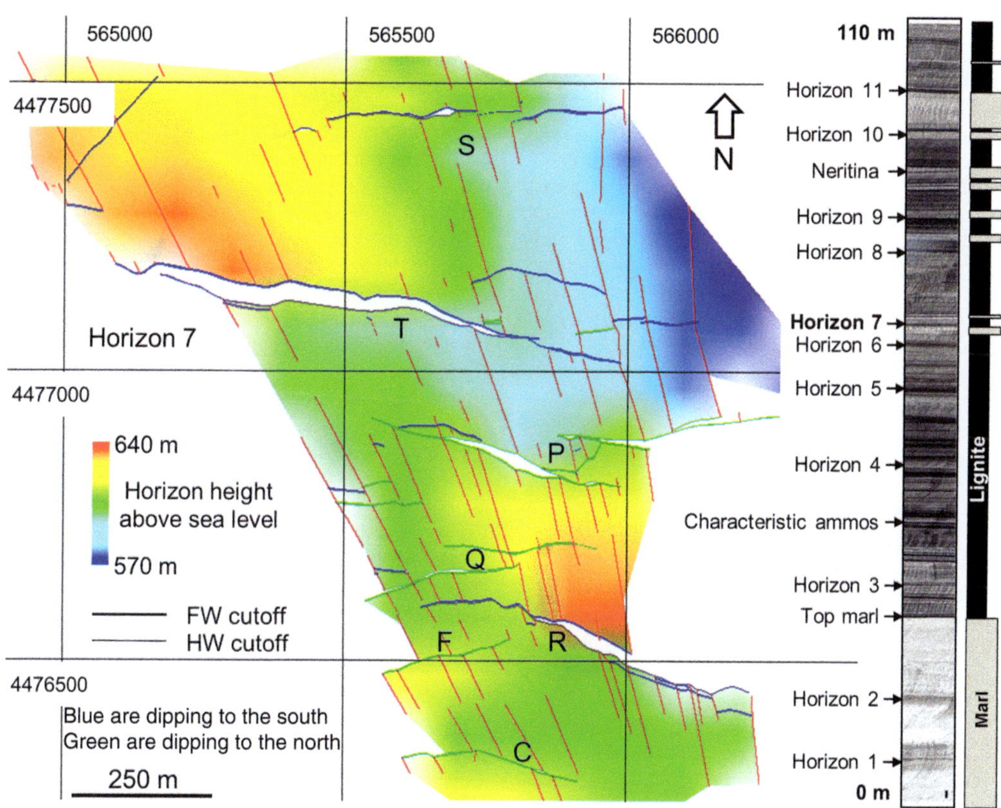

Fig. 5. Map view of one of 14 interpreted horizons, including the footwall (FW) and hanging-wall (HW) cutoffs of each interpreted fault. Faults with blue polygons are dipping to the south and those with green polygons to the north. The letters (C, F, R, Q, P, T and S) are the names of the interpreted fault zones. The colours show the height above sea level. The red lines show the locations of the outcrop photographs used to map this horizon. On the right is the exposed stratigraphic sequence accompanied by a simplified lithological column, indicating the locations of the interpreted horizons. The layers of Neritina (*Theodoxus macedonicus*) and Characteristic Ammos (volcanic ash) are basin-wide.

Fault-related folding

Vertically exaggerated field photographs (Fig. 3) clearly show folds present in the mine. The folds are spatially related to the faults displaying normal and reverse drag geometries, which indicates that they are part of the same geological deformation event. Like Ferrill *et al.* (2012), we do not consider a frictional drag mechanism to be important in the development of normal drag structures in the mine. Reverse drag refers to folding adjacent to a fault plane, so that the layers are concave towards the slip direction (Barnett *et al.* 1987). Reverse drag is a much larger scale feature than normal drag and defines the displacement field associated with the faults. Often it is difficult to identify reverse drag in the field, especially in areas of high fault density. We expect greater expression of reverse drag on the hanging wall rather than on the footwall of the normal faults in the Ptolemais Basin, as Doutsos & Koukouvelas (1998) estimated a footwall uplift/hanging-wall subsidence ratio of 1:2 for these faults. The vertically exaggerated panoramic view of the Kardia Mine (Fig. 3) shows clearly reverse drag structures on the hanging wall of fault zones R and T. Figure 3a also shows the product of opposing reverse drag zones, in the form of an anticline, between the two opposite-dipping large fault zones, P and T. Normal and reverse drag can often occur on the same fault (e.g. Fig. 3b, fault zone T at face 6), with the reverse drag occurring over a much larger distance, but with relatively lower bed rotations (cf. Hamblin 1965).

Another form of folding related to the faults in Ptolemais are monoclines in which a geometrical offset is observed over a localized volume of rock, but where no discrete faults are formed. A partly faulted monocline structure exists in fault zone

Q (Figs 4 & 6f), overlying an array of soft-linked fault segments. Monoclines can also be observed at the tips of individual fault segments. A well-developed monocline (Fig. 7e) occupies the volume around the tip of the footwall segment of a relay zone in fault zone P (Fig. 7).

The origin of monoclines and normal drag may be related. For instance, monoclines might be the predominant feature of a fault zone at a very early growth stage and, as the fault zone grows, they might become faulted and preserved as footwall and/or hanging-wall normal drag (Sharp *et al.* 2000; White & Crider 2006).

Synsedimentary faulting

Sediment thickness changes across faults indicate that some fault zones were active during deposition of the Ptolemais Formation (fault zones P and T), but the majority of faults grew mainly after the deposition of the Ptolemais Formation. Therefore, there is no evidence of widespread synsedimentary fault movement in these mines, an observation which is compatible with the chronology of extension and sedimentation for the basin established by Pavlides & Mountrakis (1987). Based on the regional thickness of the Quaternary sediments and the timing of faulting, these faults were formed in a depth ranging from 0 to approximately 500 m.

Fault zone characteristics

In this section we provide some details of the geometry and structure of individual fault zones studied in this paper to give an impression of some of the typical features of these zones, and of variability of structure within and between the fault zones. Selected features of these fault zones are referred to in later sections. Below, we subdivide the fault zones into lower (maximum throw 17–23 m) and higher (maximum throw 44.5–50.5 m) displacement faults.

Lower displacement fault zones

Fault zone S (Figs 3–5) has a maximum throw of 18.5 m and comprises multiple en echelon fault segments with both soft- and hard-linkages between adjacent segments. The average width of the fault zone, including the contribution from normal drag, is 50 m. If drag is not taken into account the zone is, on average, 15 m wide.

As mentioned previously, a typical example of a monoclinal structure occurs in fault zone Q (Figs 5 & 6), where an array of soft-linked fault segments underlies and locally displaces a monocline, so that the dominant feature of the mapped part of this fault zone is an area of pervasive continuous deformation. This area is over 400 m long in the fault-parallel direction (Fig. 6d, e) and has a maximum width of 200 m. The total throw over this portion of the fault zone varies between 10 and 15 m, only a small proportion of which is accommodated by faulting (Fig. 6d) (see Childs *et al.* 2016). Towards the west and east of this structure, where the total throw increases to 22 m (Fig. 6b–e), most of the throw is accommodated by discrete faults, indicating that the monocline accommodates soft linkage between two underlapping large faults. Complementary variations between faulting and continuous deformation are particularly clear on Horizon 7 (Fig. 6c).

Fault zones C and F (Figs 3–5) were mapped along-strike for only 350 and 200 m, respectively, as they were exposed by mine workings only relatively recently. The maximum recorded throw is 23 m for zone C and 17 m for zone F, and both zones are comparatively simple structures, without significant splays or synthetic faults. Most of their total throw is accommodated by a single fault strand and the remainder by a small, but constant, absolute amount of continuous deformation in the form of normal drag (Fig. 8). Their lateral tips are characterized by the occurrence of monoclines, which can accommodate throws of up to 5 m. Based on observations from the other fault zones in the mine, we expect to find these faults to be segmented once more data become available.

Higher displacement fault zones

Fault zone P, with a maximum total throw of 44.5 m, consists of several fault segments. Towards the centre of the mapped zone is a 75 m-wide relay ramp that transfers displacement between two large segments that overlap by 200 m. The segments are linked by a minor breaching fault, which has a throw of 4 m (Fig. 7). Zone P terminates towards the west with the formation of multiple en echelon secondary fault segments (Fig. 7a), similar to those described by McGrath & Davison (1995) and Marchal *et al.* (2003).

In fault zone R (Figs 3–5 & 9), which has a maximum throw of 49 m, most of the throw is accommodated on a dominant fault strand (Fig. 9), which has associated minor synthetic faults in the hanging wall that connect with the main fault downdip or along-strike. Zone R has an average width (as defined from the discrete faults) of 20 m, which increases from the centre of the fault zone towards the tip line as the zone becomes more segmented.

Fault zone T is the largest mapped fault zone, with a maximum observed total throw of 50 m that increases towards the unexcavated area to the west

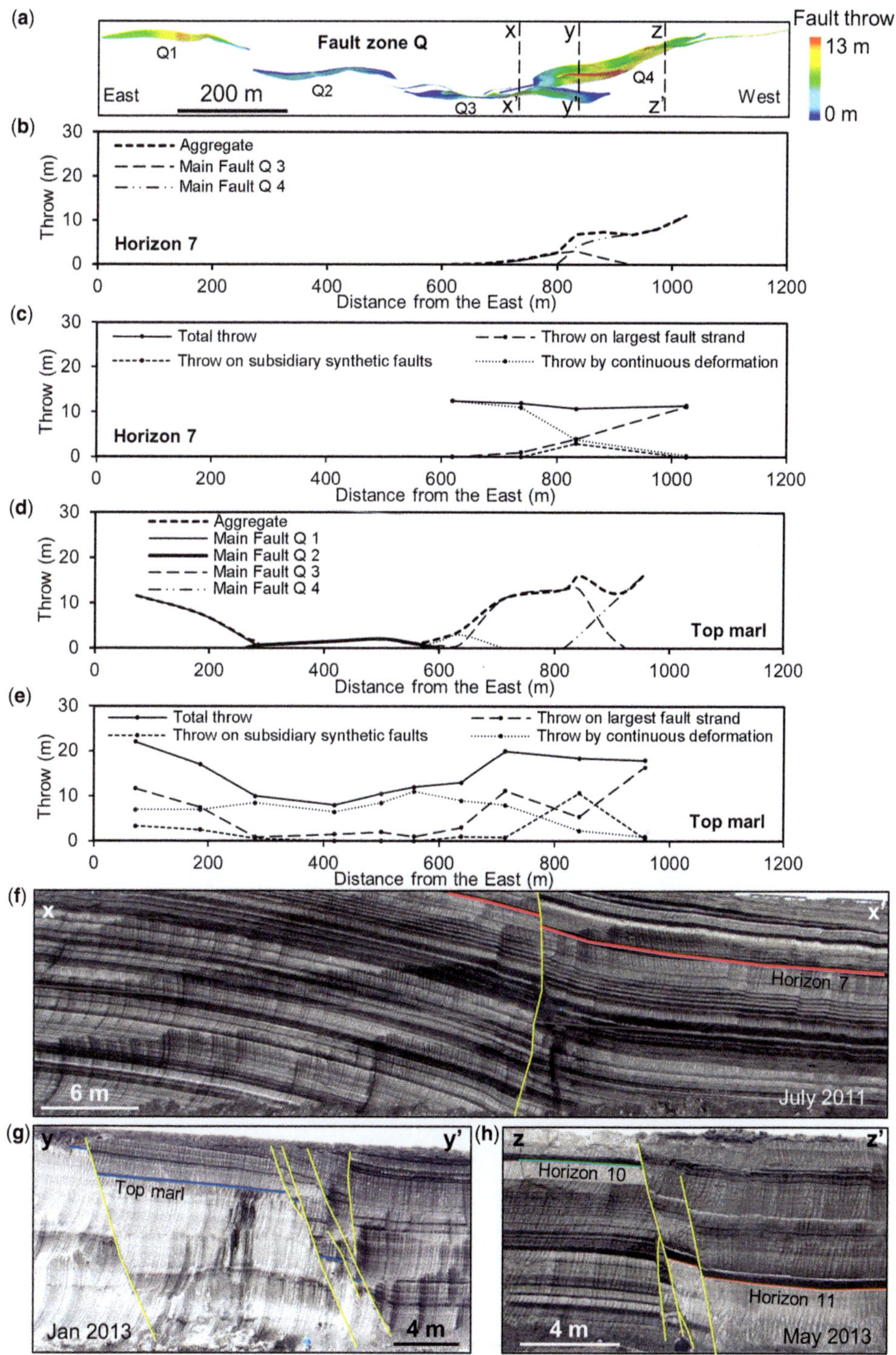

Fig. 6. (**a**) Plan view of 3D interpreted fault segments comprising zone Q, coloured for throw. (**b**)–(**e**) Throw profiles measured on 'Horizon 7' and 'Top Marl' for the discontinuous throw along each fault segment (b & d) and for each throw component along the fault zone (c & e). The locations of the horizons within the exposed stratigraphic sequence are labelled in Figure 5. (**f**)–(**h**) Outcrop photographs of fault zone Q (locations are labelled in a).

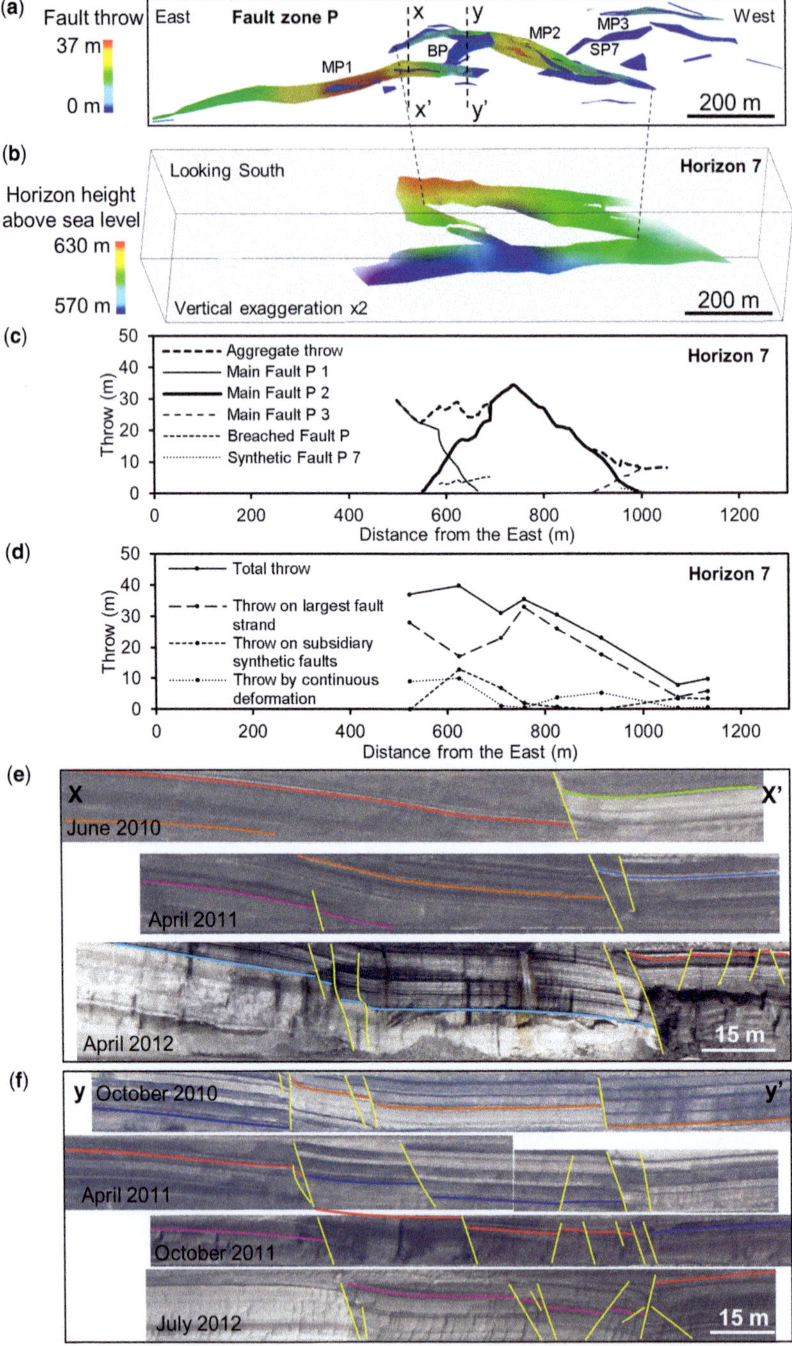

Fig. 7. (**a**) Plan view of the mapped segments of fault zone P coloured for throw. (**b**) Lateral view of an interpreted horizon (Horizon 7) coloured for the height above sea level, showing a relay ramp that is transected by a small breaching fault. (**c**) Throw profiles of the discontinuous throw along each fault segment and (**d**) of each throw component along the fault zone, measured on Horizon 7. The location of the horizon within the exposed stratigraphic sequence is shown on Figure 5. (**e**) & (**f**) Composite cross-sections through fault zone P at the locations shown in (a). Each cross-section is constructed from mine faces at different levels exposed at the dates indicated; the faces that comprise each section are at similar locations and have similar orientations.

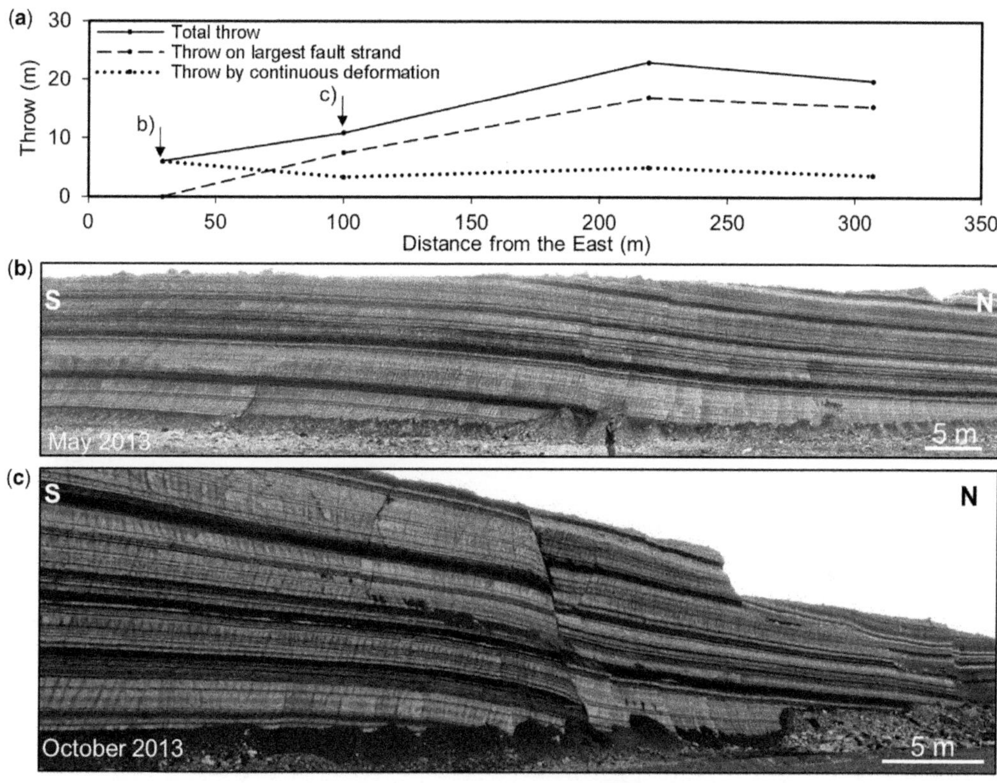

Fig. 8. (**a**) Profiles of each throw component along the single fault segment that comprises fault zone C. Sections illustrated by outcrop photographs (**b**) & (**c**) across the fault segment are labelled in (a).

(Fig. 10). The fault zone tips out to the east within the mapped area. The tip of the fault zone comprises a 600 m-long low-displacement 'tail' that consists of multiple small fault segments (Fig. 10d), with an average aggregate throw of 3 m that defines a maximum fault zone width of 300 m. At the western end of this tail (at a distance of *c.* 600 m in Fig. 10b), zone T becomes much narrower and most of the displacement is concentrated onto one main fault strand with minor synthetic faults in the hanging wall (Fig. 10e). This transition is characterized by high-displacement gradients, and the throw increases by 25 m over a distance of 150 m along the main fault trace. To the west of this transition zone, the throw increases more gently to 50 m towards the west boundary of the mine. Recent excavations have revealed a decrease in throw on this fault ('Main Fault T 1' in Fig. 10b) that is mirrored by an increase in throw on an adjacent fault 100 m into the footwall ('Main Fault T 2' in Fig. 10b), demonstrating larger-scale segmentation of this zone.

The area of increased displacement gradient in fault zone T is attributed to interaction with the opposed-dipping fault zone P. This suggestion is consistent with the observation that the decrease in throw in fault zone T is mirrored by an increase in throw in fault zone P (Figs 4 & 5) within the area of overlap of these two faults, and is also consistent with the presence of a series of SE–NW-striking minor (throw <1 m) connecting faults in their mutual hanging wall.

Throw partitioning

The total throw on the studied fault zones is separated into three components: the throw on the largest fault strand; the throw on subsidiary synthetic faults; and the throw accommodated by continuous deformation (Fig. 1b). The relative contributions of these components vary spatially over individual fault zones (Figs 6c, e, 7d, 9c & 10c). The distributions of these components also vary between fault zones, and there is no simple pattern that characterizes all seven fault zones. In an effort to identify the general trends and to establish a generalized model for the structure of the fault zones in this area, this section combines data from the seven main fault zones present in the Kardia Mine

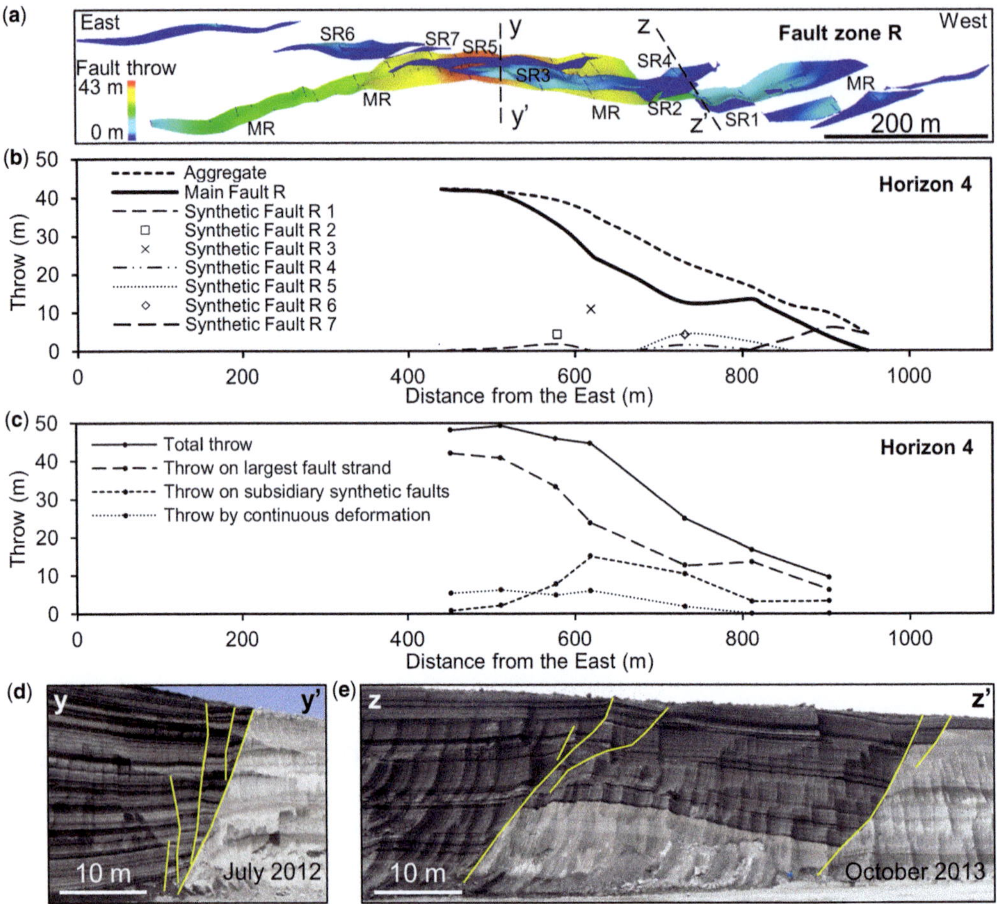

Fig. 9. (**a**) Plan view of the 3D interpreted fault segments of zone R coloured for throw. (**b**) Throw profiles of the discontinuous throw along each fault segment and (**c**) of each throw component along the fault zone measured on Horizon 4. (**d**) & (**e**) Outcrop photographs across fault zone R at the locations shown in (a). The locations of the horizons within the exposed stratigraphic sequence are labelled in Figure 5.

(Figs 11 & 12). The data used here are quantitative measures of the three throw components derived from a total of 148 measurements from the seven fault zones. The measurements are taken on cross-sections that are based on several hundred (*c.* 500) mine faces and up to three horizons at different stratigraphic levels over a map sample area of 2.5 km^2. The total throw is less than 20 m for 70% of the 148 data, and between 20 and 50 m for the remainder (Fig. 11a). For 4% of the data, there are no discrete faults and the total throw is accommodated solely by continuous deformation in the form of monoclines (Fig. 11b). In 20% of the data, there is only one fault strand and in approximately 10% of the data, the fault zone comprises more than four fault strands, with a maximum number of nine strands (Fig. 11b). More than 80% of the total throw is accommodated by discrete faulting in 50% of the studied cross-sections. In multi-strand fault zones, the sum of the throw that is accommodated by the subsidiary synthetic faults ranges from 1 to 65% of the total throw. The throw accommodated by the main fault strand is less than half of the total throw in 40% of the data.

The relative contributions of the three measured throw components to the total throw are shown in the ternary diagram in Figure 12, and each measured throw component is shown as a function of total throw in Figure 13a, d, g. Examples of cross-sections through fault zones in which the main fault accommodates most of the total throw and the recorded datum lies close to the apex of Figure 12 are provided in Figures 6h, 9d and 10e. Faults

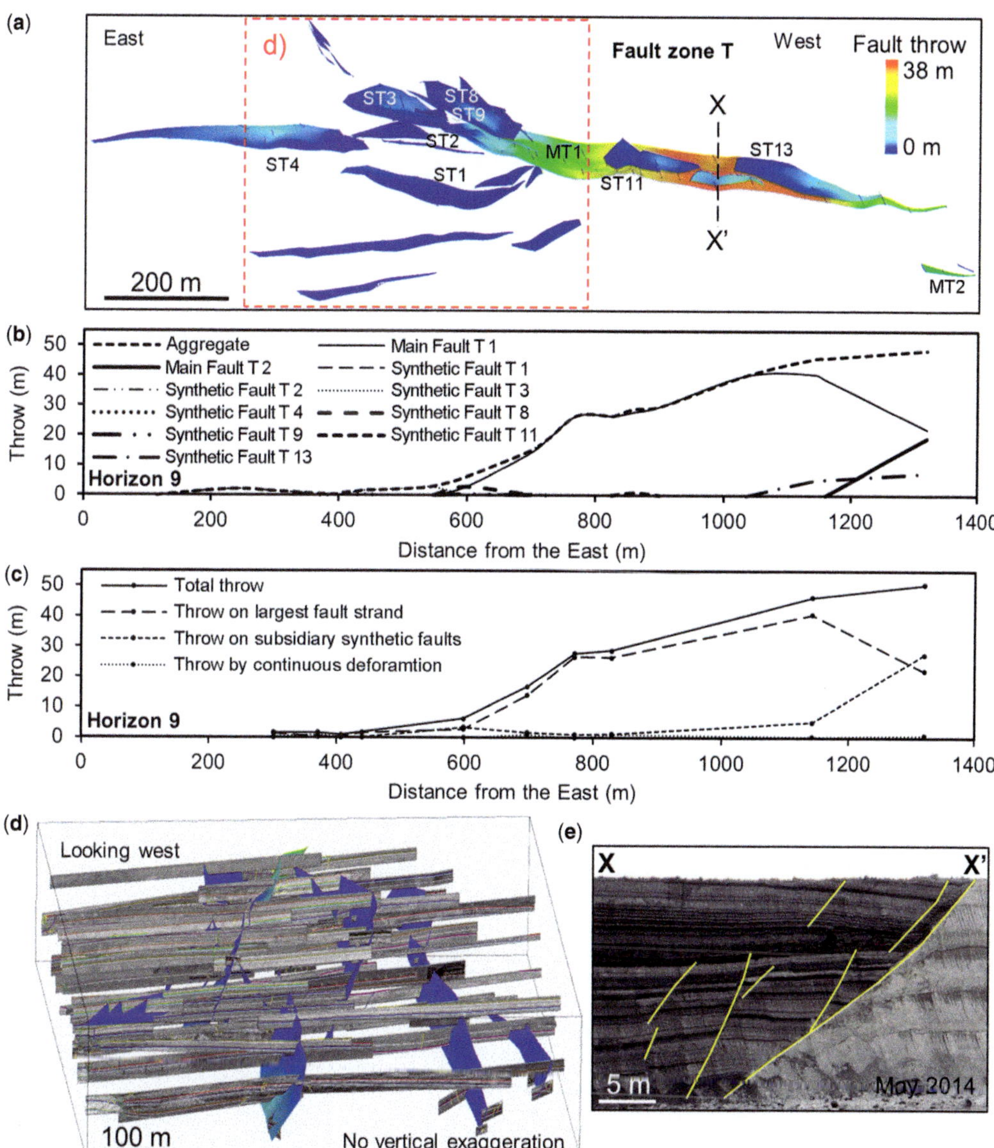

Fig. 10. (**a**) Plan view of the 3D interpreted fault segments of zone T coloured for throw. (**b**) Throw profiles of the discontinuous throw along each fault segment and (**c**) of each throw component along the fault zone measured on Horizon 9. (**d**) Lateral view of the 'tip area' of fault zone T (location given in (a)) showing the interpreted fault segments and all the imported photographs used for this interpretation. (**e**) Outcrop photograph across fault zone T at the location labelled in (a).

occupying the lower-left corner of the ternary plot (Fig. 12) accommodate most of the throw by continuous deformation in the form of normal drag (Fig. 6f). Examples of cross-sections through faults that occupy the middle of Figure 12, and consequently have approximately equal contributions from all three components, are shown in Figures 6g, 7e, f and 9e. The ternary plot clearly demonstrates that a high proportion of continuous deformation is generally associated with relatively low total throw, and most of the data associated with a large total throw are located close to the upper corner where most of the throw is accommodated by a dominant fault strand.

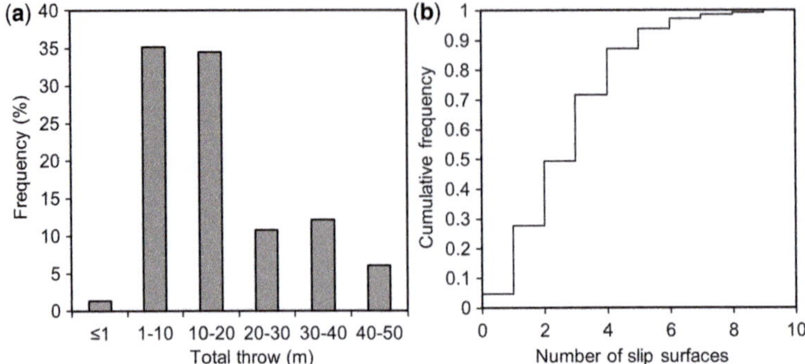

Fig. 11. (**a**) Histogram of total throw for 148 measurements taken on cross-sections along seven interpreted fault zones in Kardia Mine and (**b**) cumulative frequency plot of the number of slip surfaces present on each fault zone cross-section.

Throw on largest fault strand

Throw on the largest fault strand increases as the total throw on the fault zone increases (Fig. 13a). The throw on the largest fault encountered on a cross-section through a fault zone with total throw greater than about 20 m varies between 30 and close to 100% of the total throw, but can be very low (even zero) for faults smaller than this (Figs 12 & 13a, b). In general, this percentage increases with increasing total throw. The minimum proportion of the total throw accommodated by the largest fault also increases with increased throw, as indicated by the dashed line in Figure 13b. In Figure 13a, b there is a subset of the data with throws in the 30–40 m range that have anomalously low throws on the largest fault strand and therefore lie below the dashed line in Figure 13b; these data are discussed in more detail below.

The tendency for the largest fault strand to accommodate more of the total throw with increasing zone throw is also apparent when the data are grouped into individual fault zones (Fig. 13c). Cumulative frequency curves of the proportion of throw on the largest fault strand show that smaller displacement fault zones (thin lines in Fig. 13c), with maximum recorded throws from 17 to 23 m, have a wider range of values and, on average,

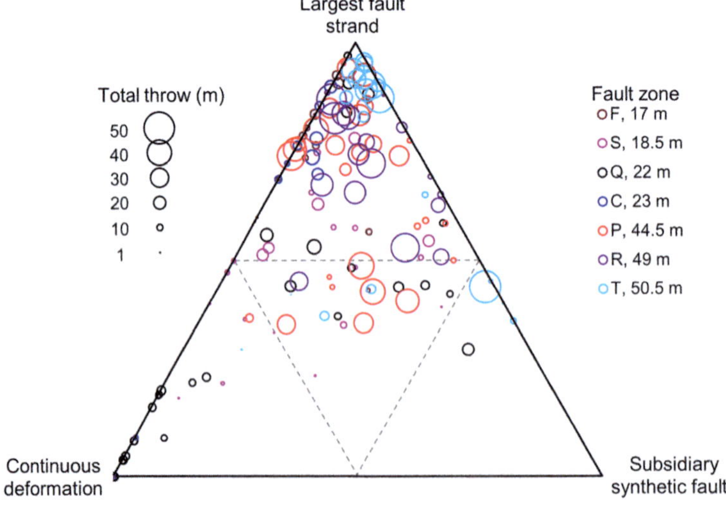

Fig. 12. Ternary diagram showing the relative contribution to the total fault zone throw by the throw on the largest fault strand, the throw on subsidiary synthetic faults and continuous deformation. The bubble size is scaled to total throw and the colours represent different fault zones with maximum throws provided in the legend. The grey dashed lines are the 50% contours.

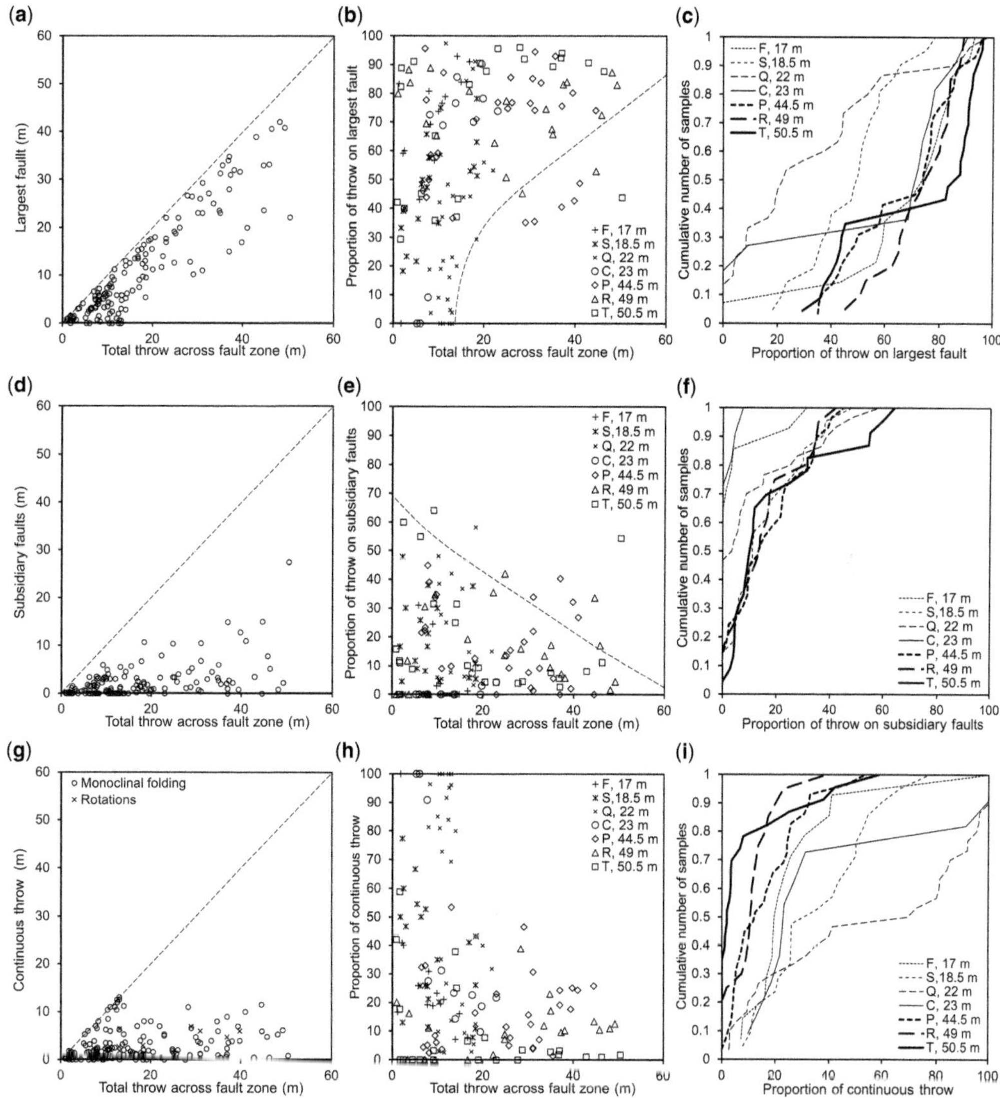

Fig. 13. Graphs illustrating how the the total fault zone throw is partitioned into the throw on the largest fault strand (**a–c**), subsidiary faults (**d–f**) and continuous deformation (**g–i**). Total throw across a fault zone v. throw accommodated by (a) the largest fault strand, (d) subsidiary synthetic faults and (g) continuous deformation recorded on cross-sections through fault zones in Kardia Mine. (b), (e) & (h) are the equivalent data expressed as percentages of the total throw. (c), (f) & (i) are frequency distributions of the proportion of total throw accommodated on (c) the largest fault strand, (f) subsidiary synthetic faults and (i) continuous deformation for each of the seven interpreted fault zones in Kardia Mine. The dashed lines in (b) and (e) are intended to mark the lower (b) and upper (e) limits of the data, highlighting trends in these limits with increasing throw. A collection of data points that lie outside these limits are related to the largest scales of segmentation observed, and are discussed in detail in the text.

lower values than the larger faults (thick lines in Fig. 13c) with throws from 45 to 50 m. These observations indicate that as fault size increases, the contribution of throw on the main fault strand relative to total throw increases, and that the fault zones tend to become progressively more localized.

The cumulative frequency curves in Figure 13c have a range of shapes from smooth curves to curves with pronounced steps. We suggest that this range of shapes reflects the extent of our sampling of the different fault zones. Of the mapped fault zones, only fault zone S has been mapped along its entire length and the other six fault zones were mapped over a portion of the total fault zone length. Total lengths for the other fault zones are known from borehole-based mapping around the excavated area. Our sampling of fault zones R and P represents more than half of their total length and their maximum throw has been recorded. In contrast, fault zones C, F, Q and T were mapped over about a quarter of their total lengths. As a result, we regard our recorded data for fault zones S, P and R to be more representative than those for the other faults. The curves for these three fault zones demonstrate the concentration of throw onto the largest fault with increasing throw.

Throw on subsidiary synthetic faults

Throw on subsidiary faults observed on individual cross-sections shows a general increase in magnitude with total throw (Fig. 13d): however, its significance relative to the total throw is not clear. Figure 13e suggests, perhaps, that the maximum proportion of throw decreases slightly with increased throw (as indicated by the dashed line). The average contribution of subsidiary faults is, however, relatively constant at about 15%, irrespective of total fault throw (Fig. 14a). Cumulative frequency curves for the proportion of the total throw accommodated by subsidiary faults within individual fault zones are very similar (Fig. 13f), particularly those fault zones that are best sampled (S, P and R; see earlier). These curves indicate that within an individual fault zone, subsidiary faults may account for between 0 and 65% of the total throw irrespective of the size of the fault. This observation suggests that the contribution of subsidiary faults does not change significantly with fault size, at least over the limited scale range represented by the seven sampled fault zones.

Throw by continuous deformation

Continuous deformation, in the form of monoclines, normal drag and bed rotation between fault strands, can account for as much as 12 m throw on the structures observed in the Kardia Mine (Fig. 13g). On cross-sections that record a total throw of less than 12 m, continuous deformation can account for the entire throw. For larger total throws, however, continuous deformation accounts for a progressively smaller proportion of the total throw (Fig. 13h). The relative decrease in the significance of continuous deformation is also apparent when different fault zones are compared (Fig. 13i), and the proportion of total throw accommodated on the smaller faults (S, F and Q) is much more significant than for the fault zones with maximum throws greater than 40 m (T, P and R).

Continuous deformation identified in Figures 6–10 and 12 combines two distinct components: monoclinal folding (or normal drag); and rotation between fault segments (see T_d and T_r, respectively, in Fig. 1b). These two components are noted in Figure 13g. Of the two components, the contribution from monoclinal folding predominates and there are relatively few data for bed rotations. Nevertheless, it appears that the scaling of these two components is different, with the contribution from monoclinal folding attaining its upper limit at 12 m total throw, while the contribution from bed rotation increases with increasing total throw.

Discussion

Fault zone evolution

Taken together (Fig. 14a), our analysis of measurements made on 148 cross-sections through seven fault zones demonstrates that as the displacement on a fault zone increases, the proportion of throw on the largest fault tends to increase, reflecting a progressive localization of strain. The data also demonstrate that the significance of continuous deformation decreases with increasing throw, with a less clear trend for the significance of subsidiary faults. While these aspects of fault evolution are also borne out by comparison between measurements derived from relatively small and large faults (Fig. 13c, f, i), when looked at in detail departures from these overall trends occur, as can be appreciated by inspection of Figure 14b–h. For example, the data for fault zones F, S and T, and to a lesser extent P and C, demonstrate the same trends as the combined dataset, with an increased proportion of ductile deformation and a decreased proportion of throw on the largest fault in areas of low total throw (i.e. close to the tips of the fault zone). These trends, however, are not always apparent for individual faults and, for example, the relative proportions of the different throw components do not appear to be related to the total throw for fault zone R (Fig. 14g). Departures from the general data trend can be attributed to a variety of factors ranging from insufficient sampling, as is likely the case for fault zone Q, to the natural variability in any fault system, which may reflect a host of factors including interaction with adjacent faults and differences in stratigraphic level.

We conclude that our dataset demonstrates progressive localization of strain with increasing fault throw both on the scale of the individual fault zone and for the dataset of all fault zones combined. There are, however, a number of caveats to this model to be considered. Chief among these is the scale range of our sample and, specifically, the

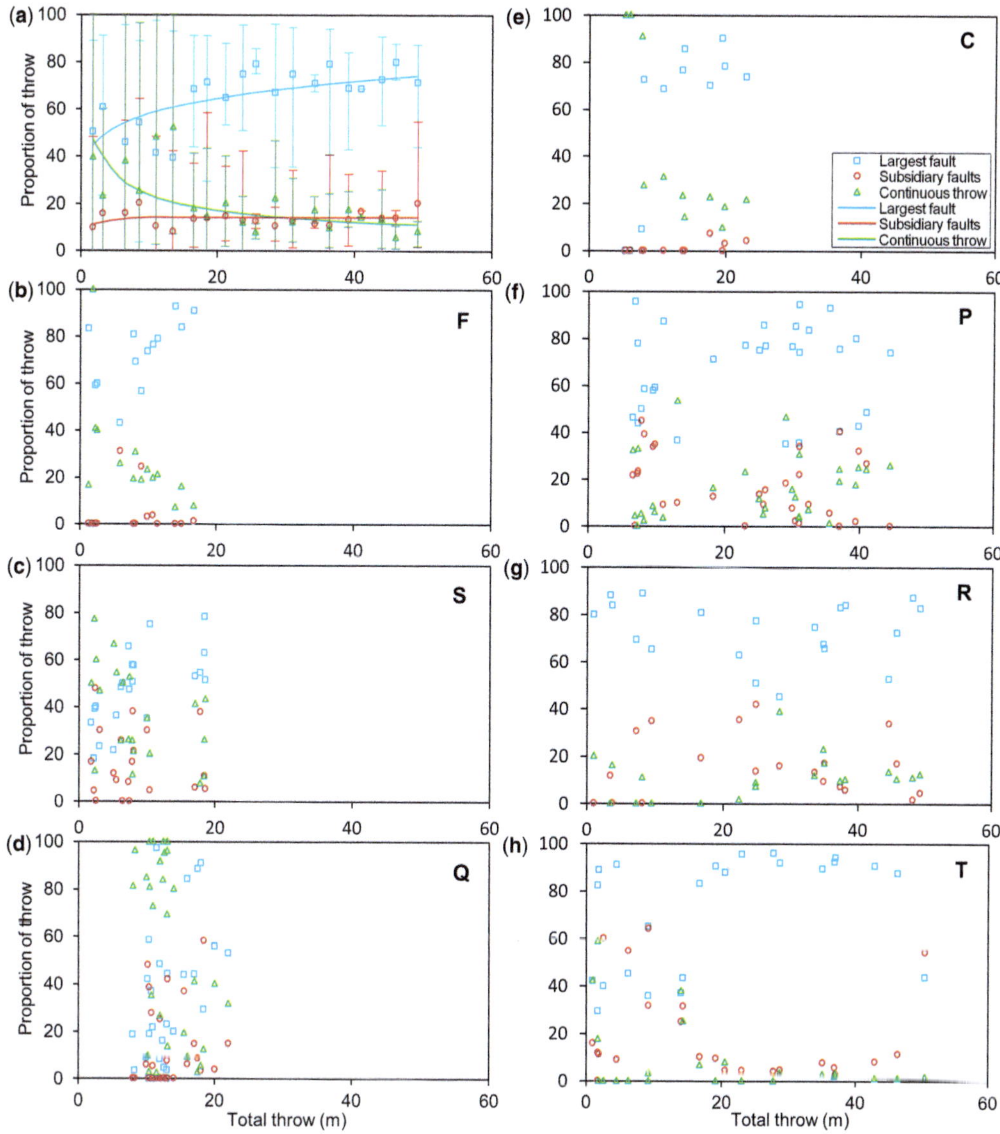

Fig. 14. (**a**) The contribution of each throw component to the total throw for binned total throw values (2.5 m bins) and corresponding power-law regression curves, showing how the total throw is partitioned into the different throw components for the full range of total throws recorded in the Ptolemais fault zones. The error bars show the minimum and maximum proportions for each binned total throw. (**b**)–(**h**) Plots of the total throw across a fault zone v. the throw accommodated by the three different throw components for each fault zone examined.

grouping of fault segments into a fault zone. At outcrop, identification of fault segments as components of a fault zone is often straightforward (e.g. Fig. 6h). However, compilation of outcrop data, map construction and analysis of displacement distributions often indicates interaction between faults that was not apparent in the field, generally because the displacements on the fault strands are small in relation to the distance between them. For example, segments in fault zone P are clearly kinematically related when seen in map view (Fig. 5), but were not grouped together in the field. Therefore, as described previously, fault segmentation occurs on a wide range of scales (Morley *et al.* 1990; Peacock & Sanderson 1991; Walsh *et al.* 1999; Childs *et al.* 2009), not all of which are apparent at a particular scale of observation. In the mapping that we have carried out at the Kardia Mine, the largest separation between two fault segments identified as belonging to a single fault zone is 75 m, but there are clear indications of larger-scale segmentation (e.g. see fault zone T above).

The existence of a wide scale range of fault segmentation hinders, to a certain extent, the quantitative analysis of our scale-limited fault zone dataset. In particular, the widest scale of segmentation mapped consists of neutral relay zones or lenses up to about 200 m wide (Fig. 5). In accordance with observations from other faults zones (Childs *et al.* 2009), relays this size on faults with 20–40 m throw are typically intact or only slightly breached, and cross-sections across them therefore represent regions with comparatively poor displacement localization. Since these wide fault zone regions are relatively rare (Fig. 5), there are only a few data through larger (20–40 m throw) faults which are poorly localized (these are the data below the dashed line in Fig. 13b and above the dashed line in Fig. 13e), and these points therefore appear to be outliers of a trend of progressive localization with increasing fault throw. If displacement were to continue to accrue on these faults, then the relays or lenses would become progressively more breached and the displacement across them would become progressively more localized. At these hypothesized larger throws, however, yet more widely spaced zones might be considered part of an individual fault zone (e.g. zones R and T; Fig. 5), and the relay between these would represent a poorly localized region of this hypothetical, much larger throw, fault zone. Hence, the presence of relatively rare, poorly localized regions of individual fault zones even at relatively large throws (Fig. 13b, e) is likely to persist to much larger throws than those mapped in the Kardia Mine, and is evidence of a large-scale range of segmentation within natural fault systems.

Despite the complications associated with a wide-scale range of fault segmentation combined with a limited scale range of observation, our data nevertheless display progressive localization of strain. This localization is clearly expressed by the progressive increase in throw on the largest fault (Fig. 13b) and the decrease in throw accommodated by continuous deformation (Fig. 13h). The role of subsidiary faults is less easy to define and our data indicate that the proportion of throw accommodated by subsidiary faults does not significantly diminish with increasing displacement (Fig. 13e). However, if the outliers (discussed in the previous paragraph), that represent the largest scales of fault segmentation, are removed then a case for a diminished contribution may be strengthened (see the dashed line in Fig. 13e).

Irrespective of whether there is a decrease in the proportion of throw accommodated by subsidiary faults, there is an increase in the absolute values of throw they accommodate (Fig. 13d). There are many potential causes for continued movement on subsidiary faults, including removal of asperities as displacement accrues on a through-going, but irregular, fault surface and fault segmentation over a wide range of scales, as described above. One mechanism that is known to be active in the Kardia Mine is the removal of asperities formed by bedding-parallel slip during normal faulting.

Development of drag

A theme repeatedly addressed in this volume (Childs *et al.* 2016; Ferrill *et al.* 2016; Lăpădat *et al.* 2016; Homberg *et al.* this volume, in press) and in previous publications (Billings 1972) is whether monoclinal, fault-related, folding develops at the tips of propagating faults (Sharp *et al.* 2000; Jackson *et al.* 2006; White & Crider 2006; Ferrill *et al.* 2012) or in response to frictional drag after a through-going fault has been established. It should be possible to differentiate between these models on the basis of the distribution of continuous deformation over the fault surface. If the frictional model would apply, then continuous deformation is expected to increase with increasing total displacement towards the centre of a fault surface. Continuous displacement within the studied fault zones shows a variety of distributions. In fault zone R (Figs 9c & 14g), continuous deformation decreases towards the tip of the fault zone from a maximum close to the centre of the fault. In contrast, continuous deformation is at a maximum at the mapped tip of fault zone C (Figs 8 & 14e). We also find that ductile deformation and monoclinal folding can occur between fault strands, as illustrated by fault zone Q (Fig. 6d, e). In this case, which is discussed in more detail in Childs *et al.* (2016), monoclinal folding may be between two segments of a fault zone that propagated upwards.

While individual fault zones display a range of distributions that individually support different models for the development of drag, our data, when taken together (Fig. 13h), clearly indicate that the proportion of total throw accommodated by folding decreases with increasing throw. Our observations do not, however, suggest that fault tips are always associated with folding, and there are many examples of structures with small total throws where folding is absent and also examples where monoclinal folding is dominant (e.g. Fig. 8b). Our data therefore suggest that, in general, monoclinal folding is associated with the earlier stages of fault zone development and, while it is reduced in significance as displacement increases, it can continue to amplify.

We speculate that an increase in drag with increased displacement is not frictional in the sense of the frictional properties of the fault rock/surface, but is related to fault geometry, developing at particular locations within an array of fault segments (e.g. within relay zones).

Summary and conclusions

Characterization of fault zone structure requires 3D mapping of the displacement distribution and partitioning along, within and around individual segments or splays that comprise a zone. Our analysis of fault zones in the Kardia Mine, Ptolemais Basin, provides unprecedented outcrop-scale access to the 3D geometry of oilfield-scale faults and, in combination with quantification of throw partitioning along these fault zones, improves our understanding of fault zone evolution.

The main conclusions obtained from the quantitative analysis of throw partitioning along seven fault zones are:

- Throw partitioning decreases with increased throw, consistent with progressive strain localization during fault growth within the scale range observable in this mine.
- The degree of throw partitioning and the rapid lateral variations in throw partitioning over a fault zone are related primarily to the range of scales of segmentation of the initial fault.
- The contribution of continuous deformation to total throw decreases as the total throw increases, suggesting that continuous deformation, such as normal drag, develops during the early stages of fault zone development.
- The throw on subsidiary synthetic faults increases with increasing total throw and its contribution to total throw appears to be more or less constant within the scale range observed in this study.

This work was carried out as part of the Earth and Natural Sciences Doctoral Studies Programme, funded by the Higher Education Authority (HEA) through the Programme for Research at Third Level Institutions, Cycle 5 (PRTLI-5), co-funded by the European Regional Development Fund (ERDF). The work was also funded by the multi-company QUAFF project, sponsored by Anadarko, BG, BP, ConocoPhillips, ENI, ExxonMobil, Marathon, Statoil, Total and Woodside. C. Childs is funded by Tullow Oil. The authors would like to thank Badley Geoscience Ltd for the use of the TrapTester6 software. The authors also thank the Public Power Corporation (ΔΕΗ/ΛΚΠΑ) for permission to work in the mines, and the corporation's employees for their hospitality and assistance in the field over the last 4 years. We thank other members of the Fault Analysis Group for field assistance and useful discussions, Mette Bundgaard Kristensen for drawing our attention to the mines, and Joris Steenbrink for introducing us to the Ptolemais Basin. We also thank Jonathan Imber and an anonymous reviewer for their helpful reviews.

References

Alsop, G.I. & Marco, S. 2013. Seismogenic slump folds formed by gravity-driven tectonics down a negligible subaqueous slope. *Tectonophysics*, **605**, 48–69, https://doi.org/10.1016/j.tecto.2013.04.004

Anastopoulos, J. & Koukouzas, N. 1972. *Economic Geology of the Southern Part Ptolemais Lignite Basin (Macedonia – Greece)*. Geological & Geophysical Research, IGME, Athens.

Bailey, W., Walsh, J. & Manzocchi, T. 2005. Fault populations, strain distribution and basement fault reactivation in the East Pennines Coalfield, UK. *Journal of Structural Geology*, **27**, 913–928, https://doi.org/10.1016/j.jsg.2004.10.014

Barnett, J.A.M., Mortimer, J., Rippon, J.H., Walsh, J.J. & Watterson, J. 1987. Displacement geometry in the volume containing a single normal fault. *American Association of Petroleum Geologists Bulletin*, **71**, 925–937.

Ben-Zion, Y. & Sammis, C.G. 2003. Characterization of fault zones. *Pure and Applied Geophysics*, **160**, 677–715, https://doi.org/10.1007/PL00012554

Billings, M.P. 1972. *Structural Geology*. 3rd edn. Prentice-Hall, Englewood Cliffs, NJ.

Brandes, C. & Tanner, D.C. 2014. Fault-related folding: a review of kinematic models and their application. *Earth-Science Reviews*, **138**, 352–370, https://doi.org/10.1016/j.earscirev.2014.06.008

Caine, J.S., Evans, J.P. & Forster, C.B. 1996. Fault zone architecture and permeability structure. *Geology*, **24**, 1025–1028, https://doi.org/10.1130/0091-7613(1996)024<1025:FZAAPS>2.3.CO;2

Cartwright, J.A., Mansfield, C. & Trudgill, B. 1996. The growth of normal faults by segment linkage. *In*: Buchanan, P.G. & Nieuwland, D.A. (eds) *Modern Developments in Structural Interpretation, Validation and Modelling*. Geological Society, London, Special Publications, **99**, 163–177, https://doi.org/10.1144/GSL.SP.1996.099.01.13

Childs, C., Watterson, J. & Walsh, J.J. 1995. Fault overlap zones within developing normal fault systems. *Journal of the Geological Society, London*, **152**, 535–549, https://doi.org/10.1144/gsjgs.152.3.0535

Childs, C., Walsh, J.J. & Watterson, J. 1997. Complexity in fault zone structure and implications for fault seal prediction. *In*: Møller-Pedersen, P. & Koestler, A.G. (eds) *Hydrocarbon Seals Importance for Exploration and Production*. Norwegian Petroleum Society Special Publications, **7**, 61–72, https://doi.org/10.1016/S0928-8937(97)80007-0

Childs, C., Manzocchi, T., Walsh, J.J., Bonson, C.G., Nicol, A. & Schöpfer, M.P.J. 2009. A geometric model of fault zone and fault rock thickness variations. *Journal of Structural Geology*, **31**, 117–127, https://doi.org/10.1016/j.jsg.2008.08.009

Childs, C., Manzocchi, T., Nicol, A., Walsh, J.J., Soden, A.M., Conneally, J.C. & Delogkos, E. 2016. The relationship between normal drag, relay ramp aspect ratio and fault zone structure. *In*: Childs,

C., Holdsworth, R.E., Jackson, C.A.-L., Manzocchi, T., Walsh, J.J. & Yielding, G. (eds) *The Geometry and Growth of Normal Faults*. Geological Society, London, Special Publications, **439**. First published online August 17, 2016, https://doi.org/10.1144/SP439.16

Diamantopoulos, A., Krohe, A. & Dimitrakopoulos, D. 2013. Deformation partitioning in a transtension-dominated tectonic environment: the illustrative kinematic patterns of the Neogene–Quaternary Ptolemais Basin (northern Greece). *Journal of the Geological Society, London*, https://doi.org/10.1144/jgs2012-138

Doutsos, T. & Koukouvelas, I. 1998. Fractal analysis of normal faults in northwestern Aegean area, Greece. *Journal of Geodynamics*, **26**, 197–216, https://doi.org/10.1016/S0264-3707(97)00052-5

Evans, J.P. 1990. Thickness–displacement relationships for fault zones. *Journal of Structural Geology*, **12**, 1061–1065, https://doi.org/10.1016/0191-8141(90)90101-4

Ferrill, D.A., Smart, K.J. & Necsoiu, M. 2008. Displacement–length scaling for single-event fault ruptures: insights from Newberry Springs Fault Zone and implications for fault zone structure. *In*: Wibberley, C.A.J., Kurz, W., Imber, J., Holdsworth, R.E. & Collettini, C. (eds) *The Internal Structure of Fault Zones: Implications for Mechanical and Fluid-Flow Properties*. Geological Society, London, Special Publications, **299**, 113–122, https://doi.org/10.1144/SP299.7

Ferrill, D.A., Morris, A.P., McGinnis, R.N., Smart, K.J. & Ward, W.C. 2011. Fault zone deformation and displacement partitioning in mechanically layered carbonates: the Hidden Valley fault, central Texas. *American Association of Petroleum Geologists Bulletin*, **95**, 1383–1397, https://doi.org/10.1306/12031010065

Ferrill, D.A., Morris, A.P. & McGinnis, R.N. 2012. Extensional fault-propagation folding in mechanically layered rocks: the case against the frictional drag mechanism. *Tectonophysics*, **576–577**, 78–85, https://doi.org/10.1016/j.tecto.2012.05.023

Ferrill, D.A., Morris, A.P., McGinnis, R.N. & Smart, K.J. 2016. Myths about normal faulting. *In*: Childs, C., Holdsworth, R.E., Jackson, C.A.-L., Manzocchi, T., Walsh, J.J. & Yielding, G. (eds) *The Geometry and Growth of Normal Faults*. Geological Society, London, Special Publications, **439**. First published online March 30, 2016, https://doi.org/10.1144/SP439.12

Freeman, S.R., Harris, S.D. & Knipe, R.J. 2008. Fault seal mapping: incorporating geometric and property uncertainty. *In*: Robinson, A., Griffiths, P., Price, S., Hegre, J. & Muggeridge, A. (eds) *The Future of Geological Modelling in Hydrocarbon Development*. Geological Society, London, Special Publications, **309**, 5–38, https://doi.org/10.1144/SP309.2

Hamblin, W.K. 1965. Origin of 'reverse drag' on the downthrown side of normal faults. *Geological Society of America Bulletin*, **76**, 1145–1164, https://doi.org/10.1130/0016-7606(1965)76[1145:OORDOT]2.0.CO;2

Homberg, C., Schnyder, J., Roche, V. & Léonardi, V. In press. The brittle and ductile components of displacement along fault zones. *In*: Childs, C., Holdsworth, R.E., Jackson, C.A.-L., Manzocchi, T., Walsh, J.J. & Yielding, G. (eds) *The Geometry and Growth of Normal Faults*. Geological Society, London, Special Publications, **439**, https://doi.org/10.1144/SP439.21

Jackson, C.A.L., Gawthorpe, R.L. & Sharp, I.R. 2006. Style and sequence of deformation during extensional fault-propagation folding: examples from the Hammam Faraun and El-Qaa fault blocks, Suez Rift, Egypt. *Journal of Structural Geology*, **28**, 519–535, https://doi.org/10.1016/j.jsg.2005.11.009

Jamison, W.R. & Stearns, D.W. 1982. Tectonic deformation of Wingate Sandstone, Colorado National Monument. *American Association of Petroleum Geologists Bulletin*, **66**, 2584–2608.

Kim, Y. & Sanderson, D. 2005. The relationship between displacement and length of faults: a review. *Earth-Science Reviews*, **68**, 317–334, https://doi.org/10.1016/j.earscirev.2004.06.003

Kristensen, M.B. 2005. *Soft Sediment Faulting: Investigation of the 3D Geometry and Fault Zone Properties*. PhD thesis, Department of Earth Sciences, University of Aarhus, Denmark.

Lăpădat, A., Imber, J., Yielding, G., Iacopini, D., McCaffrey, K.J.W., Long, J.J. & Jones, R.R. 2016. Occurrence and development of folding related to normal faulting within a mechanically heterogeneous sedimentary sequence: a case study from Inner Moray Firth, UK. *In*: Childs, C., Holdsworth, R.E., Jackson, C.A.-L., Manzocchi, T., Walsh, J.J. & Yielding, G. (eds) *The Geometry and Growth of Normal Faults*. Geological Society, London, Special Publications, **439**. First published online September 26, 2016, https://doi.org/10.1144/SP439.18

Manzocchi, T., Heath, A.E., Palananthakumar, B., Childs, C. & Walsh, J.J. 2008. Faults in conventional flow simulation models: a consideration of representational assumptions and geological uncertainties. *Petroleum Geoscience*, **14**, 91–110, https://doi.org/10.1144/1354-079306-775

Manzocchi, T., Childs, C. & Walsh, J.J. 2010. Faults and fault properties in hydrocarbon flow models. *In*: Yardley, B., Manning, C. & Garven, G. (eds) *Frontiers in Geofluids*. Wiley-Blackwell, Oxford, UK, 94–113, https://doi.org/10.1002/9781444394900.ch8

Marchal, D., Guiraud, M. & Rives, T. 2003. Geometric and morphologic evolution of normal fault planes and traces from 2D to 4D data. *Journal of Structural Geology*, **25**, 135–158, https://doi.org/10.1016/S0191-8141(02)00011-1

McGrath, A.G. & Davison, I. 1995. Damage zone geometry around fault tips. *Journal of Structural Geology*, **17**, 1011–1024, https://doi.org/10.1016/0191-8141(94)00116-H

Mercier, J.L., Sorel, D., Vergely, P. & Simeakis, K. 1989. Extensional tectonic regimes in the Aegean basins during the Cenozoic. *Basin Research*, **2**, 49–71, https://doi.org/10.1111/j.1365-2117.1989.tb00026.x

Morley, C.K., Nelson, R.A., Patton, T.L. & Munn, S.G. 1990. Transfer zones in the East African rift system and their relevance to hydrocarbon exploration in rifts. *AAPG Bulletin*, **74**, 1234–1253, https://doi.org/10.1306/0C9B2475-1710-11D7-8645000102C1865D

Mountrakis, D., Pavlides, S., Zouros, N., Astaras, T. & Chatzipetros, A. 1998. Seismic fault geometry and kinematics of the 13 May 1995 western Macedonia (Greece) earthquake. *Journal of Geodynamics*, **26**, 175–196, https://doi.org/10.1016/S0264-3707(97)00082-3

Pavlides, S.B. 1985. *Neotectonic Evolution of the Florina–Vegoritis–Ptolemais Basins*. PhD thesis, University of Thessaloniki, Greece.

Pavlides, S.B. & Mountrakis, D.M. 1987. Extensional tectonics of northwestern Macedonia, Greece, since the late Miocene. *Journal of Structural Geology*, **9**, 385–392, https://doi.org/10.1016/0191-8141(87) 90115-5

Pavlides, S.B., Zouros, N.C., Chatzipetros, A.A., Kostopoulos, D.S. & Mountrakis, D.M. 1995. The 13 May 1995 western Macedonia, Greece (Kozani Grevena) earthquake; preliminary results. *Terra Nova*, **7**, 544–549, https://doi.org/10.1111/j.1365-3121.1995.tb00556.x

Peacock, D.C.P. & Sanderson, D.J. 1991. Displacements, segment linkage and relay ramps in normal fault zones. *Journal of Structural Geology*, **13**, 721–733, https://doi.org/10.1016/0191-8141(91)90033-F

Peacock, D.C.P., Knipe, R.J. & Sanderson, D.J. 2000. Glossary of normal faults. *Journal of Structural Geology*, **22**, 291–305, https://doi.org/10.1016/S0191-8141(00)80102-9

Reches, Z. & Eidelman, A. 1995. Drag along faults. *Tectonophysics*, **247**, 145–156.

Schultz, R.A., Soliva, R., Fossen, H., Okubo, C.H. & Reeves, D.M. 2008. Dependence of displacement–length scaling relations for fractures and deformation bands on the volumetric changes across them. *Journal of Structural Geology*, **30**, 1405–1411, https://doi.org/10.1016/j.jsg.2008.08.001

Seebeck, H., Nicol, A., Walsh, J.J., Childs, C., Beetham, R.D. & Pettinga, J. 2014. Fluid flow in fault zones from an active rift. *Journal of Structural Geology*, **62**, 52–64, https://doi.org/10.1016/j.jsg.2014.01.008

Sharp, I.R., Gawthorpe, R.L., Underhill, J.R. & Gupta, S. 2000. Fault-propagation folding in extensional settings: examples of structural style and synrift sedimentary response from the Suez rift, Sinai, Egypt. *Geological Society of America Bulletin*, **112**, 1877–1899, https://doi.org/10.1130/0016-7606(2000)112<1877:FPFIES>2.0.CO;2

Shipton, Z.K. & Cowie, P.A. 2001. Damage zone and slip-surface evolution over mm to km scales in high-porosity Navajo sandstone, Utah. *Journal of Structural Geology*, **23**, 1825–1844, https://doi.org/10.1016/S0191-8141(01)00035-9

Steenbrink, J., van Vugt, N., Hilgen, F.J., Wijbrans, J.R. & Meulenkamp, J.E. 1999. Sedimentary cycles and volcanic ash beds in the Lower Pliocene lacustrine succession of Ptolemais (NW Greece): discrepancy between $^{40}Ar/^{39}Ar$ and astronomical ages. *Palaeogeography, Palaeoclimatology, Palaeoecology*, **152**, 283–303, https://doi.org/10.1016/S0031-0182(99) 00044-9

van Vugt, N., Steenbrink, J., Langereis, C.G., Hilgen, F.J. & Meulenkamp, J.E. 1998. Magnetostratigraphy-based astronomical tuning of the early Pliocene lacustrine sediments of Ptolemais (NW Greece) and bed-to-bed correlation with the marine record. *Earth and Planetary Science Letters*, **164**, 535–551, https://doi.org/10.1016/S0012-821X(98)00236-2

Walsh, J.J. & Watterson, J. 1991. Geometric and kinematic coherence and scale effects in normal fault systems. *In*: Roberts, A.M., Yielding, G. & Freeman, B. (eds) *The Geometry of Normal Faults*. Geological Society, London, Special Publications, **56**, 193–203, https://doi.org/10.1144/GSL.SP.1991.056.01.13

Walsh, J.J., Watterson, J., Bailey, W.R. & Childs, C. 1999. Fault relays, bends and branch-lines. *Journal of Structural Geology*, **21**, 1019–1026, https://doi.org/10.1016/S0191-8141(99)00026-7

Walsh, J.J., Nicol, A. & Childs, C. 2002. An alternative model for the growth of faults. *Journal of Structural Geology*, **24**, 1669–1675, https://doi.org/10.1016/S0191-8141(01)00165-1

Walsh, J.J., Bailey, W., Childs, C., Nicol, A. & Bonson, C. 2003. Formation of segmented normal faults: a 3-D perspective. *Journal of Structural Geology*, **25**, 1251–1262, https://doi.org/10.1016/S0191-8141(02)00161-X

White, I.R. & Crider, J.G. 2006. Extensional fault-propagation folds: mechanical models and observations from the Modoc Plateau, northeastern California. *Journal of Structural Geology*, **28**, 1352–1370, https://doi.org/10.1016/j.jsg.2006.03.028

Wibberley, C.A.J., Yielding, G. & Di Toro, G. 2008. Recent advances in the understanding of fault zone internal structure: a review. *In*: Wibberley, C.A.J., Kurz, W., Imber, J., Holdsworth, R.E. & Collettini, C. (eds) *The Internal Structure of Fault Zones: Implications for Mechanical and Fluid-Flow Properties*. Geological Society, London, Special Publications, **299**, 5–33, https://doi.org/10.1144/SP299.2

The relationship between normal drag, relay ramp aspect ratio and fault zone structure

C. CHILDS[1]*, T. MANZOCCHI[1], A. NICOL[2,3], J. J. WALSH[1], A. M. SODEN[1], J. C. CONNEALLY[1] & E. DELOGKOS[1]

[1]*Fault Analysis Group, School of Earth Sciences, University College Dublin, Belfield, Dublin 4, Ireland*

[2]*GNS Science, Lower Hutt, New Zealand*

[3]*Present address: University of Canterbury, Christchurch, New Zealand*

**Correspondence: conrad.childs@ucd.ie*

Abstract: The boundaries between pairs of adjacent fault segments within normal fault arrays define a spectrum of structures, from relay ramps where the length of overlap between the fault segments is much larger than the separation, through low aspect ratio (overlap/separation) relay ramps and ultimately to underlapping fault segments. Where fault segments underlap, transfer of displacement between them is accommodated by a connecting monocline. When displacement increases and a through-going fault forms, relay ramps are preserved as fault-bounded zones of elevated bed dip and monoclines are preserved as areas of normal drag. Therefore, the orientation and magnitude of bed dips within and adjacent to a fault zone, and the numbers of segments seen on a cross-section through it, depend largely on the aspect ratios of relay ramps in the initial fault array. The aspect ratio of relay ramps varies between different fault systems. An analysis of the geometry of 512 relay ramps from 13 different fault systems suggests that the main controls on aspect ratio are the strength of the sequence at the time of faulting and the underlying structure.

Normal drag is defined as 'folding adjacent to a fault such that a marker is convex towards the slip direction' (Peacock *et al.* 2000, p. 298), so that in a normal fault offsetting a flat-lying sequence, normal drag would cause an increase in bed dip towards the fault and in sympathy with the fault offset (Fig. 1f). This definition, which we adopt in this paper, is a purely geometric one without any genetic connotations. Normal drag geometries can arise in several ways and many different models have been developed to account for drag on different scales. Drag in the form of large monoclinal forced folds close to the free surface and formed in the cover above reactivated basement faults has been frequently described and is generally explained by a 'tri-shear model' (Woodward *et al.* 1972; Withjack *et al.* 1990; Willsey *et al.* 2002). A similar mechanism is precursory folding ahead of the tip of a propagating blind fault (Mandl *et al.* 1977; analogous to the ductile bead described for thrust faults), and monoclines that extend beyond both lateral and vertical tips of blind normal faults are well known (Walsh & Watterson 1987; Ferrill *et al.* 2005). Drag geometries are also developed as 'flanking folds' adjacent to faults (Grasemann *et al.* 2005). Flanking folds arise due to the deflection of markers by the deformation field around a fault. Depending on the angle between the fault and the marker, and the location of the marker relative to the point of maximum fault displacement, it can be deflected in sympathy with (normal drag) or against (reverse drag) the sense of fault throw. This mechanism generally does not provide an explanation for normal drag related to normal faults, as the marker (bedding) is usually at too high an angle to the fault surface for normal drag geometries to arise.

One commonly proposed explanation for drag is the resistance to slip on a fault surface causing the rocks immediately adjacent to the fault to be 'dragged' along the fault (e.g. Hamblin 1965; Biddle & Christie-Blick 1985; Ramsay & Huber 1987). The effectiveness of this mechanism has been questioned in a number of papers (Ferrill *et al.* 2016) that argue that drag at all scales is purely a propagation phenomenon. Although the degree to which drag can form or amplify as fault displacement increases should be testable at the correct outcrops, detailed quantitative analysis of, for example, the relationship between fault displacement and drag amplitude and wavelength has not to our knowledge been carried out on structures other than large-scale forced folds (e.g. Sharp *et al.* 2000; Ford *et al.* 2007).

Here we propose a model for normal drag associated with normal faults that is constrained by qualitative and also some quantitative data. We argue that there is a continuous spectrum of structures

From: Childs, C., Holdsworth, R. E., Jackson, C. A.-L., Manzocchi, T., Walsh, J. J. & Yielding, G. (eds) 2017. *The Geometry and Growth of Normal Faults*. Geological Society, London, Special Publications, **439**, 355–372.
First published online August 17, 2016, https://doi.org/10.1144/SP439.16

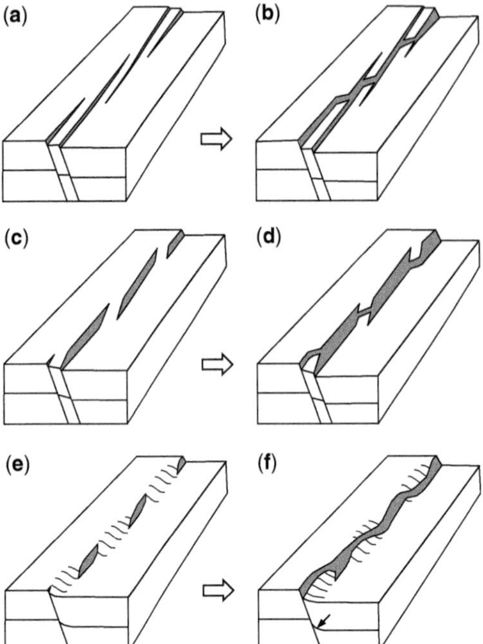

Fig. 1. Sketches illustrating (**a**, **c**, **e**) a spectrum of segmented fault geometries and (**b**, **d**, **f**) the fault zone structure that develops from them when displacement increases and fault segments become connected. The spectrum of structures is defined by the variation in aspect ratio of the relay ramps between the fault segments and the bed-dip magnitudes and directions within and adjacent to the fault zone. The front face of each block diagram illustrates a cross-section geometry characteristic of each geometry (see text). The thin lines in (e) and (f) indicate the geometry of monoclinal bedding deflections, and the arrow in (f) indicates a normal drag profile on the hanging-wall side of the fault.

from regions of localized normal drag through to relay ramps between adjacent fault segments. We firstly describe the model and place it in the context of earlier work and then provide descriptions of a suite of examples of structures that illustrate the spectrum of fault zone structures explained by the model. The model, which incorporates elements of previously published models of fault-related folding, is not intended to explain the full range of structural features that can be described as normal drag, e.g. forced folds and clay smear, but the possibility that the model can be extrapolated to account for a wide variety of structures is explored at the end of the paper.

In this paper we refer to the faulted sequence as being 'strong' and 'weak', where strong rocks are those with high stiffness and rock strength; examples of strong rocks are cemented sandstones, and examples of weak 'rocks' are unconsolidated sediments. Translation of these laboratory-derived parameters to geological structures is not straightforward; but in terms of faulting, for example, we expect strong rocks that have high stiffnesses to be associated with low fault displacement gradients (e.g. Watterson 1986; Gudmundsson 2004).

A model for the relationship between relay ramp aspect ratio and fault zone structure

The model we propose is illustrated in Figure 1 as a series of sketches showing the characteristic features of a range of fault zone structures encountered at outcrop and in seismic reflection datasets. The basis of the model is that normal drag and breached relay ramps are end-member types of bypassed fault segment boundaries that, prior to segment linkage, transferred displacement between adjacent fault segments via monoclines and relay ramps, respectively. From the various fault systems we have studied, we distinguish a spectrum of styles of fault zone that demonstrate the transition between these two end members. At one end of the spectrum (Fig. 1a, b) relay ramps with high aspect ratios occur at the boundaries between fault segments (aspect ratio refers to the length of overlap between relay-ramp-bounding faults and the fault-normal distance, or separation, between them as seen in plan view; Fig. 2a). In this case, bed dip within relay ramps is approximately perpendicular to fault strike and relay ramps are breached at low strains indicated by low bed dips within failed relay ramps. Cross-sections through fault zones that develop from segmented arrays with high aspect ratio relay ramps are more likely to intersect several fault strands than in the low aspect ratio case.

At the other end of the spectrum (Fig. 1e, f) fault segments do not overlap but rather underlap, and transfer of displacement between fault segments is via connecting monoclines. The monoclines terminate at fault segments and are effectively narrow rhomboidal relay ramps where bed rotation is predominantly normal to fault strike. Because there is no overlap between successive fault segments, when a through-going fault does form, it does so without the formation of bypassed splays. Linkage between the initial segments results in a serrated or corrugated fault geometry, and the bypassed monoclines are preserved as zones of normal drag. Because linkage between segments is achieved by cutting across intervening monoclines, the distribution of normal drag on either side of the fault will vary depending on the location of the linking fault relative to the monoclinal axes and also with location along the monocline.

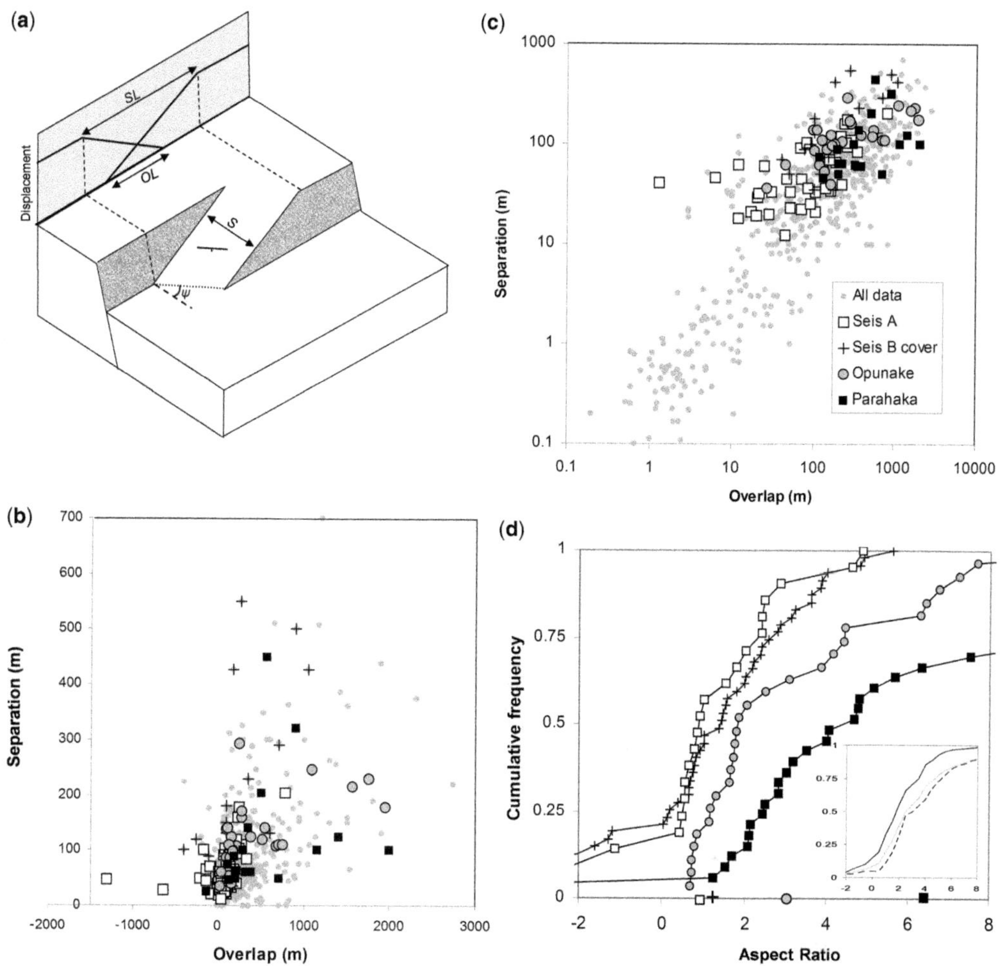

Fig. 2. (**a**) Sketch showing how relay zone separation (S), overlap length (OL) and shoulder length (SL) were measured. The shaded panel is a displacement–distance plot for the relay-bounding fault segments. ψ is the angle between the lateral margin of the relay ramp (which is the same as the strike of bedding within the ramp) and the fault dip direction. (**b**) Linear and (**c**) log–log plots of relay zone separation against positive overlap length for selected seismic datasets (see key in (c) and text for description) and 'other' data. (**d**) Cumulative frequency curves of aspect ratio for the same datasets identified in (c); the large symbols along the x-axis indicate the mean aspect ratios for each of the four datasets. The inset to (d) shows cumulative frequency curves for the subsets of the data grouped into 'weak' rocks (solid line), faults affected by basement structure (dashed line) and 'other' (grey line) as described in the text. The other datasets included on these plots are outcrop data from Kilve foreshore, Somerset, UK (Peacock & Sanderson 1994); British deep coalmines, Yorkshire (Watterson *et al.* 1996); Fumanya Mine, southern Pyrenees (Soliva & Benedicto 2004); SE Utah, USA (Doelling 1988; Foxford *et al.* 1996); and Ptolemais Mines, Macedonia, northern Greece (Pavlides & Mountrakis 1987). There are also data from clay models of extensional faulting (Henza *et al.* 2011); seismic reflection datasets from Champion Field, SE Asia (Sandal 1996; Imber *et al.* 2003) and Kupe Field, Taranaki Basin; and a dataset from a Tertiary North Sea section. The data included are for normal faults only, although other faulting modes may be present at these outcrops.

Figure 1c and d illustrate the intermediate case in which initial fault aspect ratios are low, relay ramps are rhomboidal in plan view, and ramp bed dips are high with significant components in both the along-strike and down-dip directions. Cross-sections of the through-going fault in this case will have high bed dips that may be predominantly in the fault dip direction but which will frequently occur within well-defined, fault-bounded zones.

Many elements of the model illustrated in Figure 1 have been described previously, but to our knowledge these have not been combined into the general model presented here. The presence of monoclines at fault tips is well known and has been described for dip-slip faults as fault propagation folds at mode-II tips (Mandl *et al.* 1977) and as monoclines at mode-III, lateral tips, which are seen in map views of dip-slip faults (Walsh & Watterson 1987). Rhomboidal relay ramp shapes have previously been described, as has the significant component of bed dip towards the hanging wall that they require (Peacock & Sanderson 1991; Huggins *et al.* 1995; Long & Imber 2010). The occurrence of underlapping relay ramps was first described by Morley *et al.* (1990), and several authors have proposed that underlapping fault segments represent the earliest stages in the development of a relay ramp and ultimate fault linkage (Peacock & Sanderson 1994). Linkage of two fault segments separated by a relay ramp is generally thought to result in the formation of a failed, or breached, relay ramp geometry with a splay connected to the main fault at a kink in the trace of the main fault (Childs *et al.* 1993). Here we suggest that the previous existence of a boundary between two fault segments may be recorded only as a kink in a continuous fault with associated localized drag. Although this mode of fault linkage has not, to our knowledge, been described for map-view linkage between faults, bed rotation between fault segments that underlap in cross-section has been described. In particular, the location of mode-II tips of dip-slip fault segments within the mechanically less competent rocks within a heterogeneous multilayer promotes the formation of underlapping or en bayonette fault segments (Mandl *et al.* 1977; Childs *et al.* 1996). Linkage between segments in these situations has been invoked for the formation of clay smears (Lehner & Pilaar 1997; Doughty 2003).

A feature of the model as illustrated in Figure 1 is that deformation at segment boundaries stops once a through-going fault is formed. We consider, however, that the irregularities on the continuous fault would continue to focus strains; and it has been shown, for example, that relay ramps may continue to rotate after they are bypassed by a through-going fault (Long & Imber 2012; Conneally *et al.* 2014), and the formation of fault-bound lenses by 'double-breached' relay zones (Peacock & Sanderson 1994) implies continued deformation. By analogy, we would expect that normal drag formed by the linkage of underlapping relay zones could also be amplified following fault linkage.

Data and terminology

In the previous section, we describe the zone of rotated bedding between two adjacent normal fault segments as a relay ramp. Later in the paper, where we consider the segmentation of normal faults in three dimensions, including segmentation seen in cross-section, we use the more general term 'relay zone'. Relay zone refers to a volume of rock between adjacent fault segments that accommodates the strains required to transfer displacement between them, irrespective of the orientation of the segment boundary relative to the fault slip direction. A relay ramp is therefore one class of relay zone in which the zone of elevated strain is delineated by an area of rotated bedding.

In the following section we present data for measured aspect ratios of relay zones between adjacent segments of normal fault arrays as seen in map view (i.e. relay ramps). As in earlier studies, the measured aspect ratio is the overlap length of the two segments divided by the separation (Fig. 2a), where the separation is the fault-normal distance between adjacent fault segments measured in plan view (as opposed to the more usual meaning of the apparent displacement of a marker horizon across a fault). Unlike earlier studies, we present examples of underlapping relay zones where the relay-zone-bounding faults do not overlap (Fig. 1e). These data are presented as negative overlaps in Figure 2b and give rise to negative aspect ratios. Despite this means of quantifying their shapes, relay zones are generally not rectangular in map view but rather are rhomboidal (Huggins *et al.* 1995) so that the overlap length does not represent the full fault-parallel dimension of the relay zone (Fig. 2a). A measurement that captures the actual length of a relay zone is the fault-parallel distance between the points at which the elevated displacement gradients associated with the transfer of displacement between the fault segments are first seen; this measurement is referred to here as the 'shoulder length' and is illustrated on the schematic displacement–distance profile shown in Figure 2a and by the double arrows in the real example shown in Figure 6b. Measurements of shoulder length can often be highly subjective, particularly at low displacements, in breached relay zones and in complex fault arrays, and for this reason we present the simpler overlap length measurement; a more limited dataset of the shoulder lengths of relay zones and their rhomboidal geometries is presented in a later section.

Our analysis is based on outcrop and 3D seismic reflection datasets (e.g. Figs 3–7). These data are derived from several excellent outcrop areas including Kilve foreshore in Somerset, UK; Fumanya Mine, southern Pyrenees; and SE Utah. Seismic datasets are from a variety of settings including the North Sea, the South China Sea and the Taranaki Basin, offshore New Zealand. The analysis also includes a large database derived from maps of UK coalmines (Watterson *et al.* 1996). It is not feasible here to describe in detail the geological setting

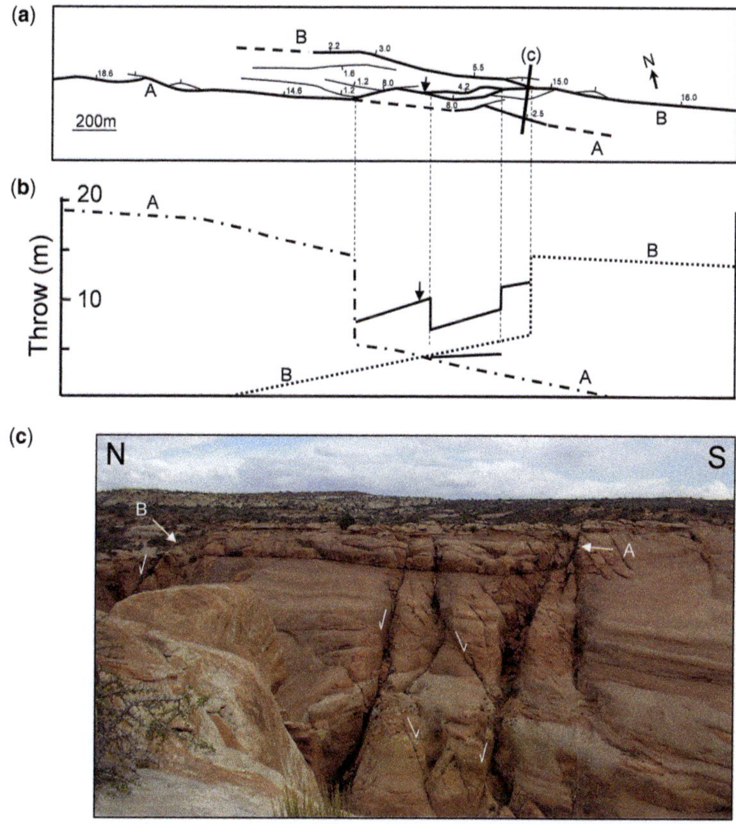

Fig. 3. (**a**) Fault trace map and (**b**) throw–distance profile for part of a normal fault at Yellow Cat Flat, SE Utah, USA (38° 47′ 56″ N, 109° 29′ 50″ W). In (a) solid lines are mapped fault traces and dashed lines are inferred. Heavy solid lines are large displacement faults that are represented in the throw profiles in (b), and thin solid lines are low (<2 m) displacement faults. Fault traces are annotated with selected fault throw measurements in metres. Vertical dotted lines connect the locations of branch points between mapped fault strands in (a) with their associated steps in fault throw in (b). The main relay-bounding faults are labelled A and B and shown as dash–dot and dotted lines, respectively, in (b). The faults connecting A and B are shown with a solid line. The photograph in (**c**) is viewed towards the east and provides a cross-section through a large relay zone on this fault; the location of the photograph is shown in (a). Faults labelled A and B are 130 m apart at this location.[1]

of each of our datasets, but in the caption to Figure 2 we provide references to papers that provide this background. Some of the outcrop data we have collected are from outcrops that have already been used for this purpose: for example, our dataset derived from Fumanya Mine in the Southern Pyrenees was also the subject of an analysis of relay zone breaching by Soliva & Benedicto (2004). Some of our best datasets are from seismic reflection surveys. Although in later sections we present examples of maps from some of these surveys, we are not always at liberty to identify the areas covered by these data, but we do attempt to provide as much of the relevant geological context as possible.

The seismic datasets used in this study image the shallow cover (0.4–1.0 km depth) above oilfields and are all of very high quality and resolution. Seismic line spacings are 12.5 m or less, and shotpoint spacings are all 6.25 m. The minimum reflector offset that can be mapped is typically 2 m, and the minimum mappable fault-normal distance between two faults can be as low as 20 m, although this value increases as fault displacement increases. Irrespective of the quality of seismic data, there is always a throw below which a discrete horizon offset will be imaged as a narrow monoclinal deflection of the associated reflector. While this limitation is significant in, for example, mapping the tips of faults, it is not a consideration for the broad-wavelength monoclines (some tens of shotpoint spacings) we describe in this paper. Of more relevance here is the likelihood that these broad-wavelength

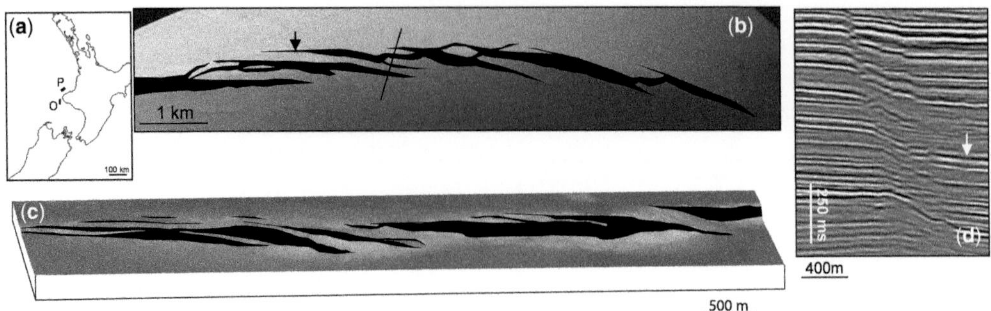

Fig. 4. (**a**) Map of the coastline of New Zealand, North Island showing the locations of the two mapped fault arrays, O (Opunake) and P (Parihaka). (**b, c**) Oblique views of seismically interpreted horizons viewed from the downthrow side of the fault zone offset by normal faults in the Opunake (b) and Parihaka (c) datasets. The maximum throws on the faults are *c.* 130 ms in (b) and 100 ms in (c). The grey scale in (b) covers a depth range of *c.* 200 ms, with the lighter colours at higher elevations, and the surface is lit from the back left of the figure. (c) Oblique view of a faulted horizon. The horizon is lit from the front so that lighter tones indicate bed dip towards the reader. (**d**) The seismic line crosses the Opunake fault along the line indicated in (b), and the horizon interpreted in (b) is arrowed in (d). The arrow in (b) indicates the location of a low displacement gradient 'tail' on a fault segment that bounds a relay zone discussed in the text.

structures are zones containing several small faults that accommodate the overall monoclinal geometry (see Steen *et al.* 1998). While reflector geometries must therefore be considered as defining a bedding sheet dip that is accommodated by some combination of bed rotation and distributed sub-resolution faulting, this does not impact on our analysis. This is because the key measure is the degree to which strain is localized at the scale of interest, which is reflected in the reflector sheet dip irrespective of the sub-resolution deformation mechanism.

Relay zone aspect ratios

The proposed model predicts a relationship between relay zone aspect ratio and normal drag whereby fault zones characterized by significant normal drag will also have low aspect ratio relay zones. We expect that relay ramp aspect ratio will vary between different areas and geological settings. Here we provide a brief review of published studies of relay zone dimensions that investigated this question and present a new dataset ($n = 512$) of relay ramp geometries for normal faults. Our dataset for relay zone geometry is from intact and breached relay zones but excludes fault-bound lenses that may have originated as relay zones although their origin is generally not easily demonstrated.

Aydin & Schultz (1990) found a linear relationship between overlap length and separation for map-view relay zones between strike-slip faults (i.e. step-overs or jogs). In a recent compilation Long & Imber (2011) found that a global dataset of normal fault relay zones defined a similar relationship over eight orders of magnitude with an average aspect ratio of 4.2. They also found that different areas were characterized by different aspect ratios and, although it may be intuitively predicted, their dataset did not demonstrate a relationship between aspect ratio and lithology. However, they found that high aspect ratios were associated with vertically confined faults and with faults that were preceded by earlier structures such as veins. Huggins *et al.* (1995), in a study of normal faults in UK Carboniferous coalmines, were unable to relate aspect ratio with other relay zone geometrical parameters (e.g. fault displacement and displacement gradient), which they suggested may be related to the limitations of their 2D map data. Soliva & Benedicto (2004), in studies of relay ramps between outcrop-scale faults, found higher aspect ratios for breached relay zones than for intact relay zones. These studies, while finding differences in relay aspect ratio for different subsets of their data, also highlighted the large variation in measured aspect ratios in any given area.

In our study we have found, like Long & Imber (2011), a global mean aspect ratio of *c.* 4, but with a wider scatter of data than they encountered due to the very short overlaps and underlaps included here. While individual datasets also have wide scatter, a well-defined range of aspect ratios can often be established for a particular dataset, and these ranges vary between datasets. This point is illustrated in Figure 2b and c with reference to four different seismic datasets. Two of the datasets (A and B) have near-identical mean aspect ratios (*c.* 1) and aspect ratio distributions (Fig. 2d). Aspect ratios approaching 0 are common in these datasets and

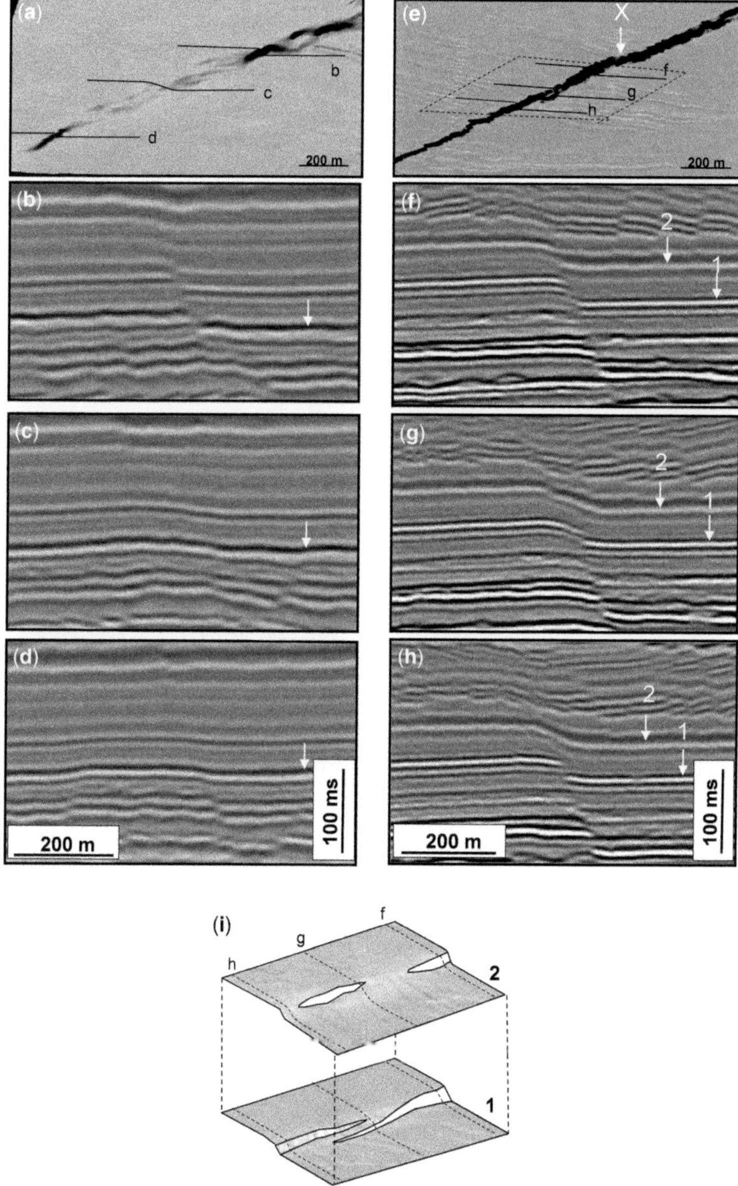

Fig. 5. (**a**, **e**) Oblique view of mapped horizons and (**b–d**, **f–h**) serial sections, for two fault arrays in seismic dataset A. The reflector mapped in (a) is arrowed in (b–d), and that mapped in (e) is reflector 1 in (f–h). The views in (a) and (e) are of horizon dip maps with darker tones showing relatively higher bed dips. Horizon dips greater than 25° are shown black; these high dips are associated with fault offsets. (**i**) Views of mapped horizons showing details of the area outlined by the dashed line in (e) for reflectors labelled 1 and 2 on the seismic lines shown in (f–h). The vertical distance between the maps of horizons 1 and 2 is increased in (i) by a factor of 5 for the purposes of illustration. On the shaded relief oblique views in (i), the lighter tones are high bed dips towards the viewer. The dog-leg in the fault trace labelled X in (e) is interpreted to be a breached monocline, as discussed in the text.

underlapping relay zones occur. In contrast, the Parihaka dataset has a significantly higher mean aspect ratio (*c*. 6) and a very different distribution (Fig. 2d). The difference in fault zone structure that these aspect ratio populations reflect can be appreciated by comparing the map in Figure 6c

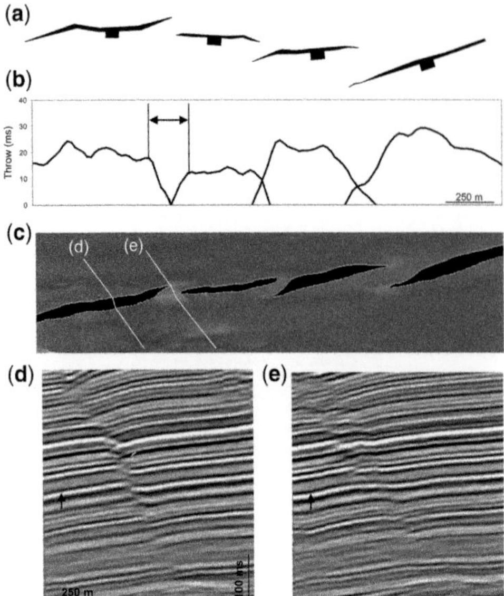

Fig. 6. (**a**) Fault trace map and (**b**) displacement–distance profile for a segmented fault array in seismic dataset B. (**c**) Shaded relief map for a horizon offset by the fault array; the lighter tones are high bed dips towards the viewer. The reflector on which this horizon is interpreted is arrowed in the two seismic lines (**d**, **e**). The locations of the two lines are shown in (c). The double-headed arrow in (b) shows the measured 'shoulder length' for the relay zone between two adjacent fault segments (see text).

(dataset B) with that in Figure 4c (Parihaka), and it is clear that variation in aspect ratio is an important discriminator for fault zone structure in different areas/settings.

Controls on relay zone aspect ratio

From the variation in aspect ratios between our different datasets we infer two main controls on aspect ratio: the strength of the rocks at the time of faulting and the boundary condition to the faulting – specifically, whether or not the faults have propagated upwards from pre-existing structures within the basement. In the inset to Figure 2d we separate our data into three categories: those where we can be reasonably certain that the sequence was relatively weak at the time of faulting; a suite of data from seismic surveys offshore New Zealand where the rocks may also have been weak but the primary control on the geometry of fault arrays appears to be the reactivation of underlying basement structures; and 'others', which includes data from a variety of areas (see caption to Fig. 2). We do not define a category for 'no basement control' because clear demonstration of an absence of basement control is significantly more difficult than demonstrating its presence. Similarly, we do not include a 'strong' category as it is more difficult to establish that a strong rock now was also strong at the time of faulting than to show that weak rock now was also weak at the time of faulting. Therefore, by defining weak

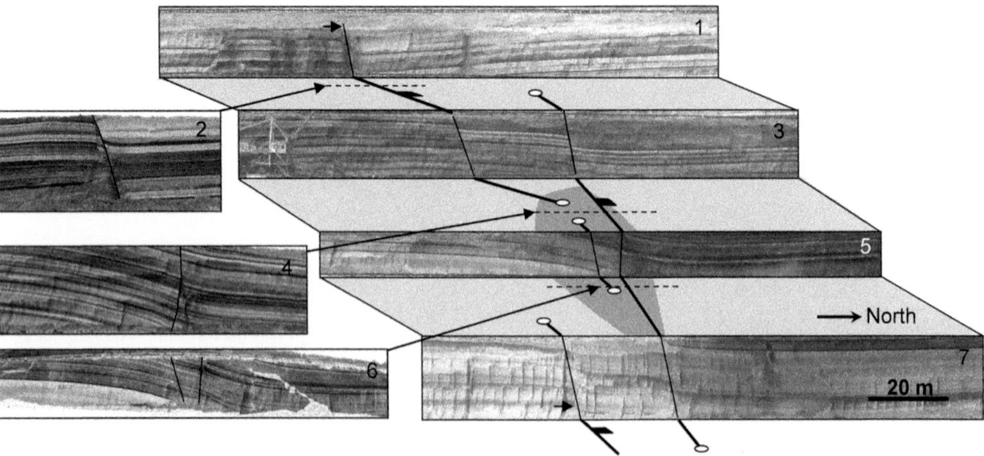

Fig. 7. Seven selected cross-sections through a fault zone mapped in a lignite mine in Ptolemais, Macedonia (Delogkos *et al.* this volume, in press). The odd-numbered photographs were taken on mine faces exposed in July 2012 and the even-numbered ones were exposed at earlier times. The fault correlations shown on the benches are constrained by *c.* 50 additional mine faces before and after 2012. The shaded area on the lowermost two benches indicates the extent of significant drag. The scale is approximately the same on all photographs. The monocline and minor faulting occurs within an overall relay zone geometry between the faults arrowed on cross-sections 1 and 7.

and basement-controlled categories, we imply that their opposite categories are contained within 'others' but do not specifically define them.

While deformation mechanisms observed at outcrop and thin section can be a good indicator of the degree of lithification and hence the strength of a sequence at the time of faulting, estimating the likely strength of a rock at the time of faulting in the subsurface is not straightforward. Here we have assigned seismic datasets to the 'weak' category if the sediments are unlithified today or if the sequence was at a shallow burial depth (<500 m) at the time of faulting, as indicated by associated sedimentary growth sequences. This second condition does not preclude early diagenesis and, while the sequences we are concerned with are all predominantly sand/shale sequences, early calcite cementation, for example, cannot always be ruled out. Seismic datasets A and B (described in more detail below) are both examples of datasets included in the 'weak' category. Other 'weak' datasets include the experimental models on faults in wet clay carried out by Henza *et al.* (2011) and the Ptolemais Mines described below. It is clear that the datasets grouped under 'weakly lithified' have lower overlap lengths per their separation than the other categories. Lower aspect ratios for relay zones in weak rocks/sediments make intuitive sense, as low shear and bulk moduli should allow for higher fault displacement gradients and therefore more efficient transfer of displacement between adjacent fault segments.

The data derived from Miocene faults from offshore New Zealand lie towards the high end of the distribution of aspect ratios (Fig. 2d inset). The New Zealand offshore data are also for sequences that are thought to have been poorly lithified at the time of faulting, but aspect ratios are higher than for the 'weakly lithified' category (faults from two of these datasets are described below). We interpret this difference to be due to the influence of deeper structure and, in several cases, faults clearly link down to earlier 'basement' faults. There have been several studies of the control of basement structure on the geometry of normal fault arrays in cover sequences (e.g. McClay *et al.* 2002; Corti 2008; Frankowicz & McClay 2010). These have tended to concentrate on the angle of the cover faults relative to the basement fault and the extension direction, but high aspect ratio relay zones are a feature of these studies and they predict our observed high aspect ratios.

While cumulative frequency curves for the three data categories shown in Figure 2d (inset) demonstrate the tendency for relay ramps to have higher aspect ratios in stronger rocks, and where there is a basement control on structure, grouping different datasets that each have a wide variation in relay shape tends to blur this trend. In the following section we illustrate these trends with reference to selected datasets that cover the range of fault zone structures we consider.

A spectrum of fault zone structures

The basis of our model is that different areas or geological settings are characterized by different ranges of relay zone aspect ratios that are controlled by certain aspects of the geology; in the previous section we suggested that the most important of these are rock strength and pre-existing structure. Here we expand on these findings by presenting selected examples of datasets that demonstrate the spectrum of structures illustrated in Figure 1 and discuss their geological setting.

High aspect ratio relay zones

SE Utah. High aspect ratio relay zones occur between segments of Cretaceous-age faults offsetting massive sandstone units of Jurassic age in SE Utah, USA. Relay zones on faults that offset the cliff-forming Navajo, Entrada and Wingate sandstone formations typically have aspect ratios in excess of 5, and are often significantly higher. In Figure 3 we present an example of a fault with a throw of up to 19 m offsetting the Entrada Sandstone Formation. The relay zone shown is at least 1.2 km long with an average separation of 130 m, to give an aspect ratio of >9. The relay zone appears to be breached by a fault with a throw of 6–8 m that cuts obliquely across it (arrowed in Fig. 3a, b). The intersection of the fault with the northern relay-bounding fault is exposed and the link with the southern fault is inferred. The average aspect ratio for relay zones and breached relay zones along this fault is 10. We attribute the high aspect ratios on this fault to the high shear strength of the offset sequence, which is due at least partly to the lack of bedding interfaces that could otherwise act as easy-slip surfaces and reduce the strength of the deformed section. The massive nature of the offset sequence can be appreciated from the cross-section through the relay zone provided by the deep gulley shown in Figure 3c.

The photograph in Figure 3c demonstrates a lack of bed rotation towards the hanging wall, and while this would in any case be expected to be subdued for a 130 m wide zone and an offset of 18 m, an absence of a bed-dip component towards the hanging wall is typical of relay ramps offsetting these sandstone-dominated sections. There is an additional synthetic fault towards the centre of the photograph that is separated from the footwall relay-zone-bounding fault by 40 m. The intervening rock volume is offset by a series of antithetic faults. This structure is interpreted as an example of a compressional overstep between Riedel shears that are typical of segmentation seen in cross-sections of small (cm displacement) faults described in these rocks (Davis 1999; Schultz & Balasko 2003).

The relay ramps on faults offsetting the Jurassic sandstones in SE Utah are characterized not only by high aspect ratios but also by low bed dips, and bed rotations within intact and breached relay ramps are rarely greater than 5°. In contrast, in the shale-dominated upper Jurassic and Cretaceous section (e.g. Morrison Formation) relay ramp dips are significantly higher and occasionally can be as high as 40° for intact relay ramps. These observations indicate that the sandstone intervals could sustain significantly lower shear strains prior to breaching than the shale units at the time of faulting.

Taranaki Basin. Many of our data with high aspect ratios derive from seismic reflection datasets from the Taranaki Basin, offshore New Zealand. This area was subjected to E–W extension during the Cretaceous, with later Miocene to present-day deformation that resulted in a complex pattern of extension in the north of the Taranaki Basin and compression in the south (King & Thrasher 1996; Giba *et al.* 2010; Reilly *et al.* 2015). In the areas we have studied, the later structures are all extensional faults and, where it is possible to image the Cretaceous, are seen to root into deeper Cretaceous normal faults. Figure 4 shows examples of horizon maps of two fault arrays imaged (on the Parihaka and Opunake 3D seismic reflection surveys) in the Miocene section. The offset sequence is in both cases mixed sandstones and shales. The maps and populations of aspect ratios (Fig. 2d) demonstrate that faulting in both these areas is characterized by the frequent occurrence of aspect ratios in excess of 6 and occasional very high values (>8). Both of the mapped horizons shown in Figure 4 were within 400 m of the contemporaneous free surface at the time of faulting, and we infer that in both cases the sequences were poorly lithified at the time of faulting. Despite the inferred low strength of these rocks, they have high aspect ratios, which we attribute to the presence of underlying reactivated basement faults. While seismic imaging does not show clearly the presence of basement structures in the particular examples shown, basement control on nearby related structures is clear (Giba *et al.* 2010, 2012). An indicator of basement control on the faults shown in Figure 4 is the consistent sense of stepping (sinistral in both cases) of the fault segments, suggesting that in each case the Miocene extension direction is rotated clockwise in respect of the earlier Cretaceous extension.

The aspect ratios derived from the Parihaka dataset define a population that lies towards the upper range of the entire sample (Fig. 2d), with high median and mean values. In contrast, the Opunake population, which contains several very high aspect ratios, has a mean value close to that for the overall population and a median value of 2. The cumulative frequency curve for the population is in two distinct parts, with a very steep slope for aspect ratios between 0.8 and 2.0 and a lower slope at higher aspect ratios (Fig. 2d). In fact, the lowest 50% of the aspect ratio data are similar to data derived from some of the low aspect ratio datasets. For the Opunake dataset, measured aspect ratio increases with increasing relay zone separation so that large-separation relay zones have proportionally higher overlap lengths. We tentatively suggest that the larger relays reflect upward propagation from the basement structure at depth, while the geometries of the smaller, and less vertically persistent, structures are not influenced by the structure at depth, which would be consistent with the two parts of the cumulative frequency plot.

Low aspect ratio relay zones

Seismic A. A seismic reflection survey offshore West Africa provides an example of the end-member fault zone structure illustrated in Figure 1e and f. These normal faults offset a Late Miocene sequence of sandstones and mudstones. The faults do not appear to have an associated growth sequence, and the depth at the time of faulting is unlikely to be significantly less than the present-day depth of *c.* 1000 m. The seismic response of the sequence varies between intervals, with high-amplitude variations to relatively muted amplitudes. In general, reflections are parallel and highly continuous (Fig. 5), apart from one interval with local high-amplitude chaotic reflections (Fig. 9a).

Figure 5a–d illustrates an example of a very large underlap relay zone where transfer of displacement between the fault segments is via a monocline. The underlap distance is 800 m, and the fault-normal distance between the extrapolated fault segments is *c.* 150 m to give an aspect ratio of −8. Figure 5b & d show seismic lines either side of the relay zone with discrete offsets of 30 and 10 ms, respectively. On the intervening cross-section (Fig. 5c) the offset is accommodated by a broad 150 m wide monocline equal in wavelength to the fault-normal separation.

Figure 5e–h illustrates an example of a structure interpreted to be a larger displacement equivalent to that shown in Figure 5a–d in which a monoclinal relay zone is interpreted to have been breached to form a through-going fault. In this example discrete offsets of 35 and 30 ms occur on cross-sections shown in Figure 5f and h, respectively, either side of the relay zone. On the intervening line (Fig. 5g) the offset is distributed between a discrete fault offset of 15 and a 70 m wide monocline in the footwall that accommodates the remaining 20 ms offset. The more detailed map of this horizon (Fig. 5i) shows that the drag occurs in the footwall of an unbreached relay ramp. On an overlying horizon (horizon 2) monocline development is more prominent and the fault segments are underlapping as in the lower displacement fault. Along strike, at point X in Figure 5e the distinct dog-leg in the fault trace, which is associated with drag in both the footwall and hanging wall of the fault, is also interpreted to be a breached monoclinal relay zone. These sections are representative of the structure of fault segment boundaries encountered in this dataset at this stratigraphic level. Monocline formation is also seen in this area between pairs of adjacent fault segments seen in cross-section (Fig. 9); this feature is described in detail in a later section.

This dataset is unusual in terms of both the amount of normal drag it displays and the general lack of overlap between adjacent segment boundaries. We attribute these fault properties to the character of the sequence and specifically its capacity to locally fold rather than fault, so that it demonstrates a bulk rheology that is weak relative to the majority of datasets we have encountered.

Seismic B. The faults in this dataset offset a sequence of sands and shales. These late Tertiary faults are within 300 m of the present-day surface, and the sequence they offset is known to be poorly lithified today and therefore had low strength at the time of faulting. As with the previous example, aspect ratios are towards the lower end of the range we encounter (Fig. 2) with mean and modal values *c.* 1 and underlapping in 20% of the structures encountered. A map of a horizon offset by a representative fault array (Fig. 6a) shows a series of three relay zones with aspect ratios of less than 1, one of which has zero overlap. A cross-section through the zero-overlap relay zone (Fig. 6e) has a continuous monoclinal geometry, *c.* 150 m across, with an amplitude of 15 ms, which contrasts with the discrete offset seen on the cross-section along strike (Fig. 6d). The strike length of the zone over which displacement is transferred between these two fault segments is seen both on the map of the horizon surface and also on the throw profile, where the 'shoulders' of the profiles are separated by 200 m along strike (double arrow in Fig. 6b). As for seismic dataset A, we attribute the occurrence of low aspect ratios to the low strength of this sequence.

Ptolemais Mines. In our experience, normal faulting is not generally associated with widespread normal drag. One of the outcrops we have encountered where drag is most prevalent, but still far from ubiquitous, is in the lignite mines of the Ptolemais Basin in Macedonia, northern Greece (Delogkos *et al.* this volume, in press). The mined section of interbedded lignites and marls are of Pliocene age, and the faults offsetting them are of predominantly Quaternary age (Pavlides & Mountrakis 1987), although there is some evidence for synsedimentary faulting within the excavated Pliocene section. There is evidence for 200–300 m erosion of post-Pliocene section so that the burial depth at the time of faulting is up to 500 m. Although the marl layers have some strength at present day, the lignites remain unlithified with porosities of 60–70% (Kavvadas *et al.* 1994), and we conclude that the sequence was weak at the time of faulting.

There is a range of drag geometries and intensities encountered in the mines, from highly localized drag within fault zones where bedding is rotated into near parallelism with the fault to open monoclines similar in geometry to that shown in Figure 5. Here we present an example of an open monocline that has been mapped in three dimensions during mine excavation over a four-year period (Fig. 7; for details see Delogkos *et al.* this volume, in press). Figure 7 shows a selection of seven outcrops (from a total *c.* 100 cross-sections that constrain the mapping) that illustrate the monoclinal bed geometries developed between segments of a fault array. On faces 1 and 2 the fault consists of a single discrete surface with a throw of 12 m. On faces further to the east the throw is distributed onto minor faults and a broad monocline, and on faces 4, 5 and 6 virtually all of the 11 m offset occurs on a 40 m wide monocline. Further east again, on face 7 faulting is again dominant with offset distributed onto two similar-sized north-dipping faults, although subtle rotation of bedding between these faults accounts for *c.* 30% of the total throw on this outcrop. Further east from face 7, the fault is again a single structure. The monocline therefore occurs within the boundary between parallel but non-colinear fault segments transferring displacement between them. The geometry of the structure shown in Figure 7 is more complicated than the equivalent seismically imaged structure in Figure 5, probably due to the ability to record minor faults in outcrop, but the overall structure is similar. Lateral stepping of faults within these mines is often accompanied by significant bed rotation towards the hanging wall and, as in faces 4–6 in Figure 7, a local replacement of fault offset by a monocline is indicative of a low aspect ratio or underlapping relay zone.

Controls on the aspect ratios of relay ramps

In this section we have presented examples of the ranges of fault zone structures illustrated in the schematic diagram in Figure 1. We have suggested that the main controls on the aspect ratios of relay zones are the strength of the faulted sequence at the time of faulting and whether or not faults formed under the influence of an underlying basement structure. There are many other possible geological controls on the aspect ratios of relay zones (see Soliva & Benedicto 2004; Long & Imber 2011). Perhaps one of the most significant effects on measurements of relay zone aspect ratio and one of the most significant sources of scatter in the data is sampling. In particular, where the tip-lines of faults that bound relay zones are inclined, map-view samples will yield apparent overlap lengths that are much larger than the true overlap length. This effect is particularly clear in the West African seismic dataset where predominantly cross-sectional relay zones are sampled on horizons at a low angle to the relay-bounding fault tips, giving rise to the occasional high aspect ratios recorded (Fig. 2d).

Relay zone shape

In the previous sections we described the map-view geometry of relay zones with reference to the aspect ratio as defined by the length of overlap between relay-zone-bounding faults and the separation between them. However, relay zones are not always rectangular in map view but may have a rhomboid shape (Peacock & Sanderson 1991; Huggins *et al.* 1995) where the lateral margins of the relay zone are not perpendicular to the relay-zone-bounding faults. Such a shape implies that bedding within a relay ramp must have a component of dip towards the hanging wall (Fig. 2a). Here we briefly discuss the shapes of relay zones based on a limited subset of the dataset used to constrain aspect ratios described previously; the reasons for the relative scarcity of these data are discussed above. Firstly, we express the shapes of relay zones by comparing the relay zone overlap length with the shoulder length, i.e. the fault-parallel distance between the points where displacement gradients increase on two relay-bounding faults (see Figs 2a & 6b). Shoulder length is plotted against overlap length in Figure 8a, with negative overlap indicating underlapping fault segments. The dashed line showing negative overlaps in Figure 8a indicates the lowest geometrically possible shoulder length (i.e. equal to the overlap), while that showing positive overlaps is for rectangular relay zones (three relay zones have shoulder lengths lower than the overlap length because not all of the overlap length is utilized in the transfer of displacement). The data are not evenly distributed between these two dashed lines but rather there is a tendency for the data to lie close to the dashed lines, i.e. shoulder lengths are generally not significantly larger than the overlap length. This feature of the data is also expressed by variations in the angle between the strike of bedding within the ramp (which is parallel to the lateral margin of the relay ramp) and the fault dip direction (ψ in Fig. 2a); a ψ of zero indicates that the strike of bedding within the ramp is perpendicular to fault strike and the ramp is rectangular in plan view. The value of ψ for underlapping relay ramps is in

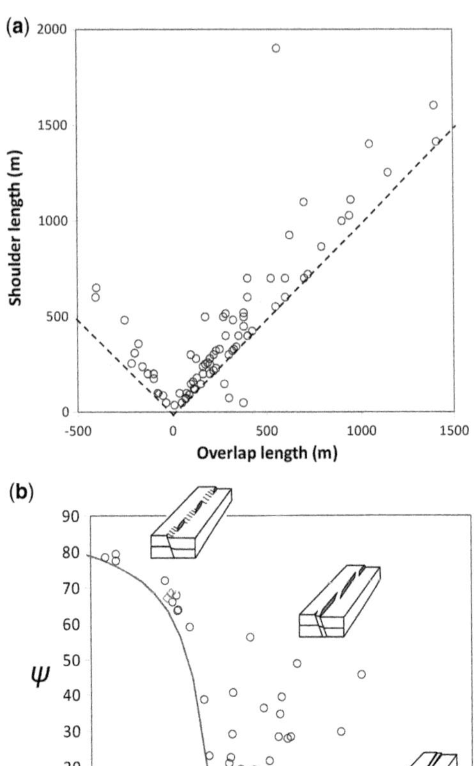

Fig. 8. (**a**) Overlap length v. 'shoulder length' (see text and Fig. 2a) for a selection of seismically mapped relay zones. The dashed lines show where shoulder length = overlap length for positive overlaps (so that relays are rectangular) and shoulder length = overlap length × −1 for negative overlaps (so that the shoulder length or the part of the fault trace with high gradient is zero). (**b**) The strike of bedding in the relay ramp (ψ, see Fig. 1a) v. relay ramp aspect ratio. The solid line is the lowest geometrically possible ψ value for a relay ramp between underlapping faults. Sketches from Figure 1 are reproduced in (b) to indicate the fault zone geometry associated with different parts of the data distribution.

each case within 10° of the minimum possible value (Fig. 8b). The majority of overlapping relay ramps have ψ values less than 40°, and the absence of high ψ values again suggests there is a tendency for relay ramps to approach a rectangular geometry.

The tendency towards low shoulder lengths for the particular arrangement of overlapping fault traces suggests that the transfer of displacement between faults is somehow mechanically optimized by having low ψ values. We interpret situations in which relay zone shoulder lengths are shorter than overlap lengths as reflecting this optimized displacement transfer, and an example of such a structure is indicated by the arrow in Figure 4b. In this case the total length of overlap between the two adjacent segments is 1.8 km, but the length of the relay ramp is only two-thirds of this distance, with the remaining 0.6 km overlap length formed by a low displacement gradient 'tail' at the southern end of the overlap zone. Our interpretation of this, and other similar structures often seen in seismic datasets offshore New Zealand (Conneally 2014), is that the overlap length was established during the upward propagation of the fault from a reactivated basement fault at depth, but only part of this overlap length was required for efficient transfer of displacement between adjacent fault segments.

Spatial relationship between drag and faulting in 3D

This paper is concerned primarily with relay zones as seen in map view, but fault segmentation can be observed on any section through a fault. This point is highlighted in Figure 9, which provides a relatively rare example of an array of unconnected fault segments that can be mapped in 3D. This fault is mapped in seismic dataset A, the same dataset as shown in Figure 5. This particular fault array is shown because the displacements are large enough to map with confidence but small enough that not all segments are connected to form a through-going fault. Both the total offset across the fault zone, measured as the difference in elevation from the limits of the fault-related deformation

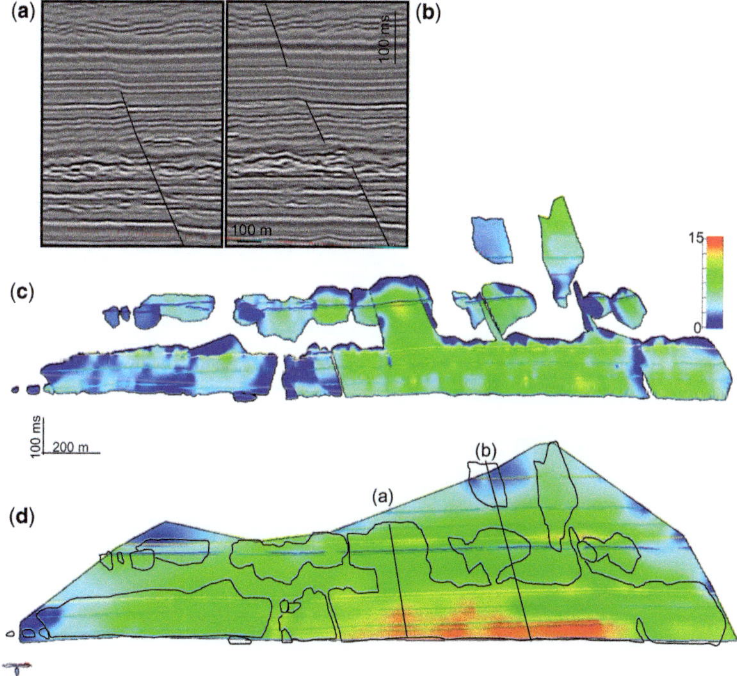

Fig. 9. (**a**, **b**) Seismic lines showing typical cross-sectional geometries of a fault array mapped in seismic datasets A; the locations of the lines are shown in (d). (**c**) Strike projection of a segmented array of faults contoured for throw (ms). (**d**) Strike projection for the same fault array showing the total throw across the fault array including the discontinuous fault throws contoured in (c) and the associated continuous deformation. The total throw is calculated as the elevation change across the zone of fault-related deformation as defined by the outermost fault cut-offs or monoclinal hinges. The difference in throw on individual segments in (c) and the total throw (d) is due to continuous deformation.

(Fig. 9d), and the contribution to the total made by the throw on each fault segment (Fig. 9c) were measured. The mapped fault segments and their throws define a patchy distribution where fault segments with throws up to 15 m alternate with areas where no discrete fault is present. The total offset across the fault zone, on the other hand, defines a much more regular distribution with a maximum offset of 15 m towards the centre of the fault diminishing towards the tip-line of the fault zone. The fault zone as a whole therefore acts as a coherent structure within which the relative proportions of discontinuous (faulting) and continuous (folding) deformation vary over the zone. Folding extends beyond the most distal fault segments (Fig. 9a) and is therefore considered to be tip-related drag. However, it also occurs between adjacent fault segments where it is an integral part of the fault zone (Fig. 9b). Although this fault array is highly segmented, it appears that it is, and probably was from the outset, a single mechanically distinct structure. In this context there was never a point at which the segments within this array did not mechanically interact, and it is not the case, for example, that adjacent mechanically independent segments propagated towards one another and then interacted with one another (for further discussion of segment coherence refer to Walsh *et al.* 2003). The variation in the relative importance of faulting and folding is readily mapped in this structure due to pronounced folding that is probably related to the character of the offset sequence. In most settings the significance of the ductile contribution to the total displacement field may be very subtle and apparent only within relay zones.

Discussion

In our experience widespread normal drag is rarely developed in association with normal faults. Even in areas where rocks are poorly lithified (e.g. Ptolemais) and where clay smears are well developed (e.g. van der Zee & Urai 2005; Childs *et al.* 2007), distinct drag profiles adjacent to fault surfaces are the exception rather than the rule. Instead, we find that bed rotations associated with normal fault zones are generally confined to distinct rock volumes bounded by fault slip surfaces, suggestive of dip panels that developed as breached relay zones or by the removal of wall rock asperities that have been subsequently rotated and sheared within the fault zone. This common occurrence of elevated bed dips within fault-bound volumes is consistent with the proposed model.

We have yet to find a good outcrop example of the end-member fault zone geometry shown in Figure 1f, but normal faults offsetting Neogene and Quaternary sediments in the Rio Grande Basin, New Mexico, USA illustrate many of the features we attribute to this end member. An example of such a fault is the Sand Hills Fault (Heynekamp *et al.*

Fig. 10. Photographs of (**a**) the Shooting Gallery and (**b**) the Waterfall outcrops of the Sand Hills Fault, New Mexico (Heynekamp *et al.* 1999; Rawling & Goodwin 2006). In (a) the fault runs down the centre of the photograph and downthrows to the right; a person for scale is arrowed. The apex in the foreground shows the intersection between the fault and steeply dipping drag-related bedding in the hanging wall. In (b) the fault zone is the upstanding rock fin that dips to the right in the fault downthrow direction. The hanging-wall margin of the fault zone is shown by the dashed line that demonstrates the corrugated map-view fault trace. Half arrows are drawn on the fault surface.

1999; Rawling & Goodwin 2006). The features of this fault that resemble the end-member structure shown in Figure 1f are a relatively simple structure (generally comprising a single zone of intensely sheared fine-grained to conglomeratic sediments <1 m wide; Fig. 10a), only rare occurrences of multiple slip surfaces, frequent pronounced normal drag in the hanging wall or footwall, or both and an often curvi-planar fault surface (Fig. 10b). However, the Sand Hills Fault is a growth fault, and the present-day outcrop extends only some tens of metres below the upper tip of the fault, with displacement increasing rapidly downwards to a maximum of *c.* 600 m at the base of the rift-related section (Heynekamp *et al.* 1999). Therefore, although this, and other similar faults in the area, strongly resembles the 'weak' end-member fault zone structure, it is likely that the extensive drag may have developed as a forced fold rather than as monoclines between fault segments.

We have described forced folds as a separate type of structure, but it may be that the spectrum of structures in Figure 1 can be extended to include structures that are predominantly monoclines with localized faulting. The case for extending this spectrum is supported by structures such as the segmented fault with associated large-wavelength normal drag shown in Figure 4c. The relative proportion of drag varies along the length of the fault array. Locally, all displacement is concentrated onto a single fault surface, while in other areas the majority of the offset is accommodated by monoclinal folding and relay ramp rotation. Geometric and kinematic analysis of this fault (Conneally 2014) reveals a temporal progression from a monocline through to a segmented fault in which the initially extensive continuous deformation becomes progressively confined to relay ramps. This temporal progression of an individual structure from an unfaulted monocline through to a segmented fault array spans the spectrum of proposed fault zone structure and supports the view that the model proposed in Figure 1 should be extrapolated to include monoclines. At the lower scale end of the range of outcrop structures often described as drag is the smearing of lithological units over fault surfaces; the most widely described are clay and shale smears that are often invoked as mechanisms for fault seal (Weber & Daukoru 1975). Several authors have proposed a model in which clay smearing is due to, or enhanced by, fault segmentation across clay beds (e.g. Doughty 2003). This suggestion provides an explanation for clay smear, another feature often associated with drag, that is based on fault segmentation and linkage. In conclusion, we suggest that, while the model illustrated in Figure 1 displays a spectrum of structures from low to high relay zone aspect ratios and resultant fault zone structures, there is scope for this spectrum to be further broadened to include other features that include other structures that display normal drag geometries including forced folds and shale/clay smears.

The range of different relay zone geometries shown in Figure 1 results in an equivalent range of fault zone structures at high displacements, and the front faces of each of the block diagrams in Figure 1 are intended to illustrate this range. For high aspect ratios we would predict that fault zones in cross-section would be characterized by multiple slip surfaces with relatively subdued fault-related bed dips. For a given segment boundary frequency (as shown in Fig. 1) cross-sections through faults at this end of the spectrum will more often show multiple slip surfaces than fault zones characterized by low aspect ratio. Fault zones with progressively lower aspect ratio relay zones will more frequently comprise a single slip surface and will have fault-bounded lenses with higher bed dips, with bedding dipping closer to the fault dip direction. At the end of the spectrum a single slip surface will be seen everywhere, with bed rotation limited to normal drag adjacent to the fault. This sequence of structures is consistent with the observation that in most normal fault systems, significant bed rotation is limited to fault-bound lenses.

Summary

We describe a spectrum of fault zone structures that reflects variation in the aspect ratio of relay zones separating the segments of a fault array prior to the development of a through-going fault. A key point of our present analysis, and previous work, is that different areas are characterized by relay zones with different aspect ratios. Our data suggest that high aspect ratios are associated with rocks that have high strength at the time of faulting and/or are controlled by a pre-existing basement structure, although many other controls can be considered. Aspect ratios of relay zones between the segments of a fault array play a significant role in determining the structure of the fault zone that develops as displacement increases and fault segments become connected. High aspect ratio relay zones promote the formation of fault zones with multiple overlapping fault strands and bedding that dips in the fault strike direction. Linkage between fault segments separated by lower aspect ratio relay zones results in fault zones with spatially restricted areas of relatively high bed dip and dip directions closer to the fault dip direction. At the limit where fault segments are underlapping the model predicts that the resulting fault zones comprise a single slip surface with associated normal drag.

This work was supported by a consortium-sponsored project brokered by the Industry Technology Facilitator, and funded by Anadarko, BG International, BP Exploration, ConocoPhillips (UK), Eni, ExxonMobil, Marathon Oil, Statoil, Total E&P UK and Woodside Energy. C. Childs is funded by Tullow Oil. Badley Geoscience Ltd are thanked for the provision of TrapTester software. We also thank Jonathan Imber and Bruce Trudgill for their helpful reviews.

References

AYDIN, A. & SCHULTZ, R.A. 1990. Effect of mechanical interaction on the development of strike-slip faults with echelon patterns. *Journal of Structural Geology*, **12**, 123–129.

BIDDLE, K. & CHRISTIE-BLICK, N. 1985. Deformation and basin formation along strike-slip faults. *In*: BIDDLE, K.T. & CHRISTIE-BLICK, N. (eds) *Strike-Slip Deformation, Basin Formation, and Sedimentation*. Society of Economic Paleontologists and Mineralogists, Special Publications, **37**, 1–34, https://doi.org/10.2110/pec.85.37.0001

CONNEALLY, J.C. 2014. *The kinematics and geometry of segment boundaries on normal faults*. PhD thesis, University College Dublin.

CONNEALLY, J.C., CHILDS, C. & WALSH, J.J. 2014. Contrasting origins of breached relay zone geometries. *Journal of Structural Geology*, **58**, 59–68.

CHILDS, C., EASTON, S.J., VENDEVILLE, B.C., JACKSON, M.P.A., LIN, S.T., WALSH, J.J. & WATTERSON, J. 1993. Kinematic analysis of faults in a physical model of growth faulting above a viscous salt analogue. *Tectonophysics*, **228**, 313–329.

CHILDS, C., NICOL, A., WALSH, J.J. & WATTERSON, J. 1996. Growth of vertically segmented normal faults. *Journal of Structural Geology*, **18**, 1389–1397.

CHILDS, C., WALSH, J.J. *ET AL.* 2007. Definition of a fault permeability predictor from outcrop studies of a faulted turbidite sequence, Taranaki, New Zealand. *In*: JOLLEY, S.J., BARR, D., WALSH, J.J. & KNIPE, R.J. (eds) *Structurally Complex Reservoirs*. Geological Society, London, Special Publications, **292**, 235–258, https://doi.org/10.1144/SP292.14

CORTI, G. 2008. Control of rift obliquity on the evolution and segmentation of the main Ethiopian rift. *Nature Geoscience*, **1**, 258–262.

DAVIS, G.H. 1999. *Structural Geology of the Colorado Plateau Region of Southern Utah, with Special Emphasis on Deformation Bands*. Geological Society of America, Special Papers, **342**, https://doi.org/10.1130/0-8137-2342-6.1

DELOGKOS, E., MANZOCCHI, T., CHILDS, C., WALSH, J.J. & PAVLIDES, S. In press. Throw partitioning across normal fault zones in the Ptolemais Basin, Greece. *In*: CHILDS, C., HOLDSWORTH, R.E., JACKSON, C.A.-L., MANZOCCHI, T., WALSH, J.J. & YIELDING, G. (eds) *The Geometry and Growth of Normal Faults*. Geological Society, London, Special Publications, **439**, https://doi.org/10.1144/SP439.19

DOELLING, H.H. 1988. Geology of Salt Valley anticline and Arches National Park, Grand County, Utah. *Utah Geological and Mineral Survey Bulletin*, **122**, 7–58.

DOUGHTY, P.T. 2003. Clay smear seals and fault sealing potential of an exhumed growth fault, Rio Grande rift, New Mexico. *American Association of Petroleum Geologists Bulletin*, **87**, 427–444, https://archives.datapages.com/data/bulletns/2003/03mar/0427/0427.HTM

FERRILL, D.A., MORRIS, A.P., SIMS, D.W., WAITING, D.J. & HASEGAWA, S. 2005. Development of synthetic layer dip adjacent to normal faults. *In*: SORKHABI, R.L.B. & YOSHIHIRO, T. (eds) *Faults, Fluid Flow, and Petroleum Traps*. American Association of Petroleum Geologists Memoir, **85**, 125–138.

FERRILL, D.A., MORRIS, A.P., MCGINNIS, R. & SMART, K. 2016. Myths about normal faulting. *In*: CHILDS, C., HOLDSWORTH, R.E., JACKSON, C.A.-L., MANZOCCHI, T., WALSH, J.J. & YIELDING, G. (eds) *The Geometry and Growth of Normal Faults*. Geological Society, London, Special Publications, **439**. First published online March 30, 2016, https://doi.org/10.1144/SP439.12

FORD, M., LE CARLIER DE VESLUD, C. & BOURGEOIS, O. 2007. Kinematic and geometric analysis of fault-related folds in a rift setting: the Dannemarie basin, Upper Rhine Graben, France. *Journal of Structural Geology*, **29**, 1811–1830, https://doi.org/10.1016/j.jsg.2007.08.001

FOXFORD, K.A., GARDEN, I.R., GUSCOTT, S.C., BURLEY, S.D., LEWIS, J.J.M., WALSH, J.J. & WATTERSON, J. 1996. The field geology of the Moab Fault. *In*: HUFFMAN, A.C. JR., LUND, W.R. & GODWIN, L.H. (eds) *Geology and Resources of the Paradox Basin*. Utah Geological Association Guidebook, **25**, 265–283.

FRANKOWICZ, E. & MCCLAY, K.R. 2010. Extensional fault segmentation and linkages, Bonaparte Basin, outer North West Shelf, Australia. *American Association of Petroleum Geologists Bulletin*, **94**, 977–1010, https://doi.org/10.1306/01051009120

GIBA, M., NICOL, A. & WALSH, J.J. 2010. Evolution of faulting and volcanism in a back-arc basin and its implications for subduction processes. *Tectonics*, **29**, https://doi.org/10.1029/2009TC002634

GIBA, M., WALSH, J.J. & NICOL, A. 2012. Segmentation and growth of an obliquely reactivated normal fault. *Journal of Structural Geology*, **39**, 253–267, https://doi.org/10.1016/j.jsg.2012.01.004

GRASEMANN, B., MARTEL, S. & PASSCHIER, C. 2005. Reverse and normal drag along a fault. *Journal of Structural Geology*, **27**, 999–1010, https://doi.org/10.1016/j.jsg.2005.04.006

GUDMUNDSSON, A. 2004. Effects of Young's modulus on fault displacement. *Comptes Rendus Geoscience*, **336**, 85–92.

HAMBLIN, W.K. 1965. Origin of 'reverse drag' on the downthrown side of normal faults. *Geological Society of America Bulletin*, **76**, 1145–1164, https://doi.org/10.1130/0016-7606(1965)76[1145:OORDOT]2.0.CO;2

HENZA, A.A., WITHJACK, M.O. & SCHLISCHE, R.W. 2011. How do the properties of a pre-existing normal-fault population influence fault development during a subsequent phase of extension? *Journal of Structural Geology*, **33**, 1312–1324, https://doi.org/10.1016/j.jsg.2011.06.010

HEYNEKAMP, M.R., GOODWIN, L.B., MOZLEY, P.S. & HANEBERG, W.C. 1999. Controls on fault-zone architecture in poorly lithified sediments, Rio Grande Rift, New Mexico: implications for fault-zone permeability and fluid flow. *In*: HANEBERG, W.C., MOZLEY, P.S., MOORE, J.C. & GOODWIN, L.B. (eds) *Faults and Subsurface Fluid Flow in the Shallow Crust*. American Geophysical Union Geophysical Monograph, **113**, 27–49, https://doi.org/10.1029/GM113p0027

HUGGINS, P., WATTERSON, J., WALSH, J.J. & CHILDS, C. 1995. Relay zone geometry and displacement transfer between normal faults recorded in coal-mine plans. *Journal of Structural Geology*, **17**, 1741–1755, https://doi.org/10.1016/0191-8141(95)00071-K

IMBER, J., CHILDS, C., NELL, P.A.R., WALSH, J.J., HODGETTS, D. & FLINT, 5. 2003. Hanging wall fault kinematics and footwall collapse in listric growth systems. *Journal of Structural Geology*, **25**, 307–315.

KAVVADAS, M., PAPADOPOULOS, B. & KALTEZIOTIS, N. 1994. Geotechnical properties of the Ptolemais lignite. *Geotechnical and Geological Engineering*, **12**, 87–112, https://doi.org/10.1007/BF00429768

KING, P.R. & THRASHER, G.P. 1996. *Cretaceous-Cenozoic Geology and Petroleum Systems of the Taranaki Basin, New Zealand*. Institute of Geological & Nuclear Sciences (N.Z.), Lower Hutt, Monograph, **13**.

LEHNER, F.K. & PILAAR, W.F. 1997. The emplacement of clay smears in synsedimentary normal faults: inferences from field observations near Frechen, Germany. *In*: MØLLER-PEDERSEN, P. & KOESTLER, A.G. (eds) *Hydrocarbon Seals: Importance for Exploration and Production*. Elsevier, Norwegian Petroleum Society Special Publications, **7**, 39–50, https://doi.org/10.1016/S0928-8937(97)80005-7

LONG, J.J. & IMBER, J. 2010. Geometrically coherent continuous deformation in the volume surrounding a seismically imaged normal fault-array. *Journal of Structural Geology*, **32**, 222–234.

LONG, J.J. & IMBER, J. 2011. Geological controls on fault relay zone scaling. *Journal of Structural Geology*, **33**, 1790–1800, https://doi.org/10.1016/j.jsg.2011.09.011

LONG, J.J. & IMBER, J. 2012. Strain compatibility and fault linkage in relay zones on normal faults. *Journal of Structural Geology*, **36**, 16–26.

MANDL, G., JONG, L.N.J. & MALTHA, A. 1977. Shear zones in granular material. *Rock Mechanics*, **9**, 95–144, https://doi.org/10.1007/BF01237876

McCLAY, K.R., DOOLEY, T., WHITEHOUSE, P. & MILLS, M. 2002. 4-D evolution of rift systems: insights from scaled physical models. *American Association of Petroleum Geologists Bulletin*, **86**, 935–959, https://doi.org/10.1306/61EEDBF2-173E-11D7-8645000102C1865D

MORLEY, C.K., NELSON, R.A., PATTON, T.L. & MUNN, S.G. 1990. Transfer zones in the East African rift system and their relevance to hydrocarbon exploration in rifts. *American Association of Petroleum Geologists Bulletin*, **74**, 1234–1253, https://doi.org/10.1306/0C9B2475-1710-11D7-8645000102C1865D

PAVLIDES, S. & MOUNTRAKIS, D. 1987. Extensional tectonics of north-west Macedonia, Greece, since the Late Miocene. *Journal of Structural Geology*, **9**, 385–392.

PEACOCK, D. & SANDERSON, D. 1991. Displacements, segment linkage and relay ramps in normal fault zones. *Journal of Structural Geology*, **13**, 721–733, https://doi.org/10.1016/0191-8141(91)90033-F

PEACOCK, D.C.P. & SANDERSON, D.J. 1994. Geometry and development of relay ramps in normal fault systems. *American Association of Petroleum Geologists Bulletin*, **78**, 147–165, https://doi.org/10.1306/BDFF9046-1718-11D7-8645000102C1865D

PEACOCK, D.C.P., KNIPE, R.J. & SANDERSON, D.J. 2000. Glossary of normal faults. *Journal of Structural Geology*, **22**, 291–305.

RAMSAY, J.G. & HUBER, M.I. 1987. *The Techniques of Modern Structural Geology: Folds and Fractures*. Academic Press, London, **2**.

RAWLING, G.C. & GOODWIN, L.B. 2006. Structural record of the mechanical evolution of mixed zones in faulted poorly lithified sediments, Rio Grande rift, New Mexico, USA. *Journal of Structural Geology*, **28**, 1623–1639, https://doi.org/10.1016/j.jsg.2006.06.008

REILLY, C., NICOL, A., WALSH, J.J. & SEEBECK, H. 2015. Evolution of faulting and plate boundary deformation in the Southern Taranaki Basin, New Zealand. *Tectonophysics*, **651–652**, 1–18, https://doi.org/10.1016/j.tecto.2015.02.009

SANDAL, S.T. 1996. *Geology and Hydrocarbon Resources of Negara Brunei Darussalam*. Brunei Shell Petroleum Company Sendirian Berhad and Brunei Museum, Bandar Seri Begawan, Brunei Darussalam.

SCHULTZ, R.A. & BALASKO, C.M. 2003. Growth of deformation bands into echelon and ladder geometries. *Geophysical Research Letters*, **30**, https://doi.org/10.1029/2003GL018449

SHARP, I.R., GAWTHORPE, R.L., UNDERHILL, J.R. & GUPTA, S. 2000. Fault-propagation folding in extensional settings: examples of structural style and synrift sedimentary response from the Suez rift, Sinai, Egypt. *Geological Society of America Bulletin*, **112**, 1877–1899, https://doi.org/10.1130/0016-7606(2000)112<1877:FPFIES>2.0.CO;2

SOLIVA, R. & BENEDICTO, A. 2004. A linkage criterion for segmented normal faults. *Journal of Structural Geology*, **26**, 2251–2267, https://doi.org/10.1016/j.jsg.2004.06.008

STEEN, Ø., SVERDRUP, E. & HANSSEN, T.H. , 1998. Predicting the distribution of small faults in a hydrocarbon reservoir by combining outcrop, seismic and well data. *In*: JONES, G., FISHER, Q.J. & KNIPE, R.J. (eds) *Faulting, Fault Sealing and Fluid Flow in Hydrocarbon Reservoirs*. Geological Society, London, Special Publications, **147**, 27–50, https://doi.org/10.1144/GSL.SP.1998.147.01.03

VAN DER ZEE, W. & URAI, J.L. 2005. Processes of normal fault evolution in a siliciclastic sequence: a case study from Miri, Sarawak, Malaysia. *Journal of Structural Geology*, **27**, 2281–2300, https://doi.org/10.1016/j.jsg.2005.07.006

WALSH, J.J. & WATTERSON, J. 1987. Distributions of cumulative displacement and seismic slip on a single normal fault surface. *Journal of Structural Geology*, **9**, 1039–1046, https://doi.org/10.1016/0191-8141(87)90012-5

WALSH, J.J., BAILEY, W.R., CHILDS, C. & NICOL, A. 2003. Formation of segmented normal faults: a 3-D perspective. *Journal of Structural Geology*, **25**, 1251–1262.

Watterson, J. 1986. Fault dimensions, displacement and growth. *Pure and Applied Geophysics*, **124**, 365–373.

Watterson, J., Walsh, J.J., Gillespie, P.A. & Easton, S. 1996. Scaling systematics of fault sizes on a large scale range fault map. *Journal of Structural Geology*, **18**, 199–214.

Weber, K.J. & Daukoru, E. 1975. Petroleum geology of the Niger delta. *In*: *9th World Petroleum Congress*, Tokyo, 209–221.

Willsey, S.P., Umhoefer, P.J. & Hilley, G.E. 2002. Early evolution of an extensional monocline by a propagating normal fault: 3D analysis from combined field study and numerical modeling. *Journal of Structural Geology*, **24**, 651–669, https://doi.org/10.1016/S0191-8141(01)00120-1

Withjack, M.O., Olson, J. & Peterson, E. 1990. Experimental models of extensional forced folds. *American Association of Petroleum Geologists Bulletin*, **74**, 1038–1054, https://doi.org/10.1306/0C9B23FD-1710-11D7-8645000102C1865D

Woodward, L.A., Kaufman, W.H. & Anderson, J.B. 1972. Nacimiento fault and related structures, northern New Mexico. *Geological Society of America Bulletin*, **83**, 2383–2396, https://doi.org/10.1130/0016-7606(1972)83[2383:NFARSN]2.0.CO;2

Occurrence and development of folding related to normal faulting within a mechanically heterogeneous sedimentary sequence: a case study from Inner Moray Firth, UK

A. LĂPĂDAT[1]*, J. IMBER[1,2], G. YIELDING[2], D. IACOPINI[3], K. J. W. McCAFFREY[1], J. J. LONG[4] & R. R. JONES[4]

[1]*Department of Earth Sciences, Durham University, Durham DH1 3LE, UK*

[2]*Badley Geoscience Ltd., Hundleby, Spilsby, Lincolnshire PE23 5NB, UK*

[3]*Geology and Petroleum Geology Department, University of Aberdeen, Aberdeen AB24 3UE, UK*

[4]*Geospatial Research Ltd., Suites 7 & 8, Harrison House, Hawthorne Terrace, Durham DH1 4EL, UK*

**Correspondence: i.a.lapadat@durham.ac.uk*

Abstract: Folds associated with normal faults are potential hydrocarbon traps and may impact the connectivity of faulted reservoirs. Well-calibrated seismic reflection data that image a normal fault system from the Inner Moray Firth basin, offshore Scotland, show that folding was preferentially localized within the mechanically incompetent Lower–Middle Jurassic pre-rift interval, comprising interbedded shales and sandstones, and within Upper Jurassic syn-rift shales. Upward propagation of fault tips was initially inhibited by these weak lithologies, generating fault propagation folds with amplitudes of *c.* 50 m. Folds were also generated, or amplified, by translation of the hanging wall over curved, convex-upward fault planes. These fault bends resulted from vertical fault segmentation and linkage within mechanically incompetent layers. The relative contributions of fault propagation and fault-bend folding to the final fold amplitude may vary significantly along the strike of a single fault array. In areas where opposite-dipping, conjugate normal faults intersect, the displacement maxima are skewed upwards towards the base of the syn-rift sequence (i.e. the free surface at the time of fault initiation) and significant fault propagation folding did not occur. These observations can be explained by high compressive stresses generated in the vicinity of conjugate fault intersections, which result in asymmetric displacement distributions, skewed towards the upper tip, with high throw gradients enhancing upward fault propagation. Our observations suggest that mechanical interaction between faults, in addition to mechanical stratigraphy, is a key influence on the occurrence of normal fault-related folding, and controls kinematic parameters such as fault propagation/slip ratios and displacement rates.

Folding related to normal faulting is mainly the result of fault propagation and linkage at different stages of the growth of normal faults (Withjack *et al.* 1990; Schlische 1995; Janecke *et al.* 1998; Corfield & Sharp 2000; Sharp *et al.* 2000; Ferrill *et al.* 2005, 2007, 2012; Jackson *et al.* 2006; White & Crider 2006; Tvedt *et al.* 2013; Tavani & Granado 2015). The main mechanisms generating fault-related folds in extensional domains are: (a) flexural deformation around vertical and lateral tips of propagating blind faults (fault-propagation folding) (Walsh & Watterson 1987; Ferrill *et al.* 2005); (b) folding between overlapping/underlapping, vertically or laterally segmented faults (Rykkelid & Fossen 2002; Rotevatn & Jackson 2014; Childs *et al.* 2016); (c) translation of the hanging wall over a bend in a fault plane (Groshong 1989; Xiao & Suppe 1992; Rotevatn & Jackson 2014); (4) distributed shear deformation (Fossen & Hesthammer 1998; Ferrill *et al.* 2005); and (5) frictional drag (Davis 1984). Mechanical properties of the host rocks exert a primary influence on normal fault geometry and the development of extensional folds (Ferrill *et al.* 2007; Ferrill & Morris 2008; Tvedt *et al.* 2013). With the help of analogue, numerical and kinematic models (Groshong 1989; Withjack *et al.* 1990; Dula 1991; Saltzer & Pollard 1992; Hardy & McClay 1999; Johnson & Johnson 2002; Jin & Groshong 2006), researchers have shown that changes in fault dip, strain rate and thickness of the incompetent layer also control the development of extensional fault-related folds. For

From: Childs, C., Holdsworth, R. E., Jackson, C. A.-L., Manzocchi, T., Walsh, J. J. & Yielding, G. (eds) 2017. *The Geometry and Growth of Normal Faults*. Geological Society, London, Special Publications, **439**, 373–394.
First published online September 26, 2016, https://doi.org/10.1144/SP439.18

example, thick incompetent layers will tend to inhibit fault propagation and promote formation of fault-tip monoclines (Withjack & Callaway 2000).

Nevertheless, models are constrained by imposed boundary conditions and are usually designed to test a single mechanism. Growth of the faults is a dynamic process in which fault geometry, slip-related stress perturbations and strain rates can vary in both space and time (Cowie 1998; Gupta & Scholz 2000) and, as a consequence, different processes might be responsible for the generation of folds during the evolution of a normal fault system. Numerical models, supported by seismological evidence, indicate that faults develop and interact within heterogeneous stress fields resulting from regional tectonic stress and local stress perturbations (Cowie 1998; Gupta & Scholz 2000). This heterogeneity induces local variations in fault slip, fault propagation and strain rates (Willemse *et al.* 1996; Crider & Pollard 1998; Gupta *et al.* 1998; Willemse & Pollard 2000; White & Crider 2006), key parameters in controlling the development of extensional monoclines (Withjack & Callaway 2000; Hardy & Allmendinger 2011). We still know relatively little about the possible influence of heterogeneous stress distributions on the development of fault-related folding (White & Crider 2006), and have yet to explain the variable occurrence and development of extensional folding along single fault arrays.

In this paper we use 2- and 3D seismic reflection data from the Inner Moray Firth basin, offshore Scotland, to investigate the influence of host-rock lithology, and fault geometry and fault interaction on the development of normal fault-related folds. First, we describe the three-dimensional geometries of the faults and folds using 3D seismic data. We map the fault throw distributions, and describe variations in the thicknesses and geometries of the syn-rift seismic sequences, to interpret the spatial and temporal (i.e. kinematic) evolution of the faults and folds. Next, we augment these observations with interpretations of faults and folds from regional 2D seismic lines, to investigate the relationship between fold growth, fault propagation and fault interaction across the basin. We show that: (a) normal fault-related folds can be generated by different mechanisms that vary in importance in time and space along a single fault array; (b) the heterogeneous mechanical properties of the host rocks control the fault segmentation and associated ductile deformation; and (c) the occurrence and development of normal fault-related folds re influenced not only by mechanical stratigraphy and fault plane geometry, but also by mechanical interaction between the faults themselves. Specifically, the variability of extensional folding along the strike of a fault array can be explained by the enhanced vertical propagation owing to mechanical fault interaction between opposite-dipping normal faults.

Geological setting

Regional tectonic framework

The studied fault system is located in the Inner Moray Firth (IMF) basin (Fig. 1). The basin is characterized by NE–SW-striking normal faults that accommodated a Late Jurassic–Early Cretaceous extensional episode which resulted in the opening of the North Sea rift system (Ziegler 1990; Thomson & Underhill 1993; Davies *et al.* 2001). Some authors have proposed a transtensional opening of the IMF basin (Roberts *et al.* 1990). We have no evidence for fault oblique displacement, but previous studies considered that faults in the area of interest are dominated by dip-slip displacement (Underhill 1991; Davies *et al.* 2001; Long & Imber 2010) and that any strike-slip movement was associated mainly with Great Glen Fault (to the NW of the present study-area) and post-dated Mesozoic rifting (Underhill 1991). Regional, Late Cretaceous post-rift subsidence and sedimentation were followed by Cenozoic uplift and reactivation of some of the faults. These faults show very mild post-Cretaceous reactivation, as indicated by small-scale folding of the Base Cretaceous horizon (H7 on Fig. 1b), but there is no evidence of large inversion structures affecting the geometries of the pre-inversion folds.

Stratigraphic framework and mechanical stratigraphy

The stratigraphy of the IMF can be divided into pre-, syn- and post-rift tectono-stratigraphic sequences (Figs 1 & 2). Our study investigates deformation within the upper part of the Triassic to Early–Middle Jurassic, pre-rift succession (pre-H3 horizons), and within the Late Jurassic, syn-rift succession (H3–H7) (Fig. 2). We used information from nearby wells and published literature (Stevens 1991) to infer the presence of three main mechanical units, based on stratigraphic variations in the net-to-gross ratio (Fig. 2).

Horizon H1, which follows a strong and regionally continuous seismic reflection, corresponds to the top of the mechanical unit 1 (MU 1). Well data indicate that H1 follows the top of the pre-rift, Triassic alluvial plain sandstones of the Lossiehead Formation (>100 m thick; Fig. 2). These strata overlie the Permian to Permo-Triassic Hopeman, Bosies Bank and Rotliegend formations, all of which are dominated by sandstone lithologies. In turn, the Permian deposits unconformably succeed

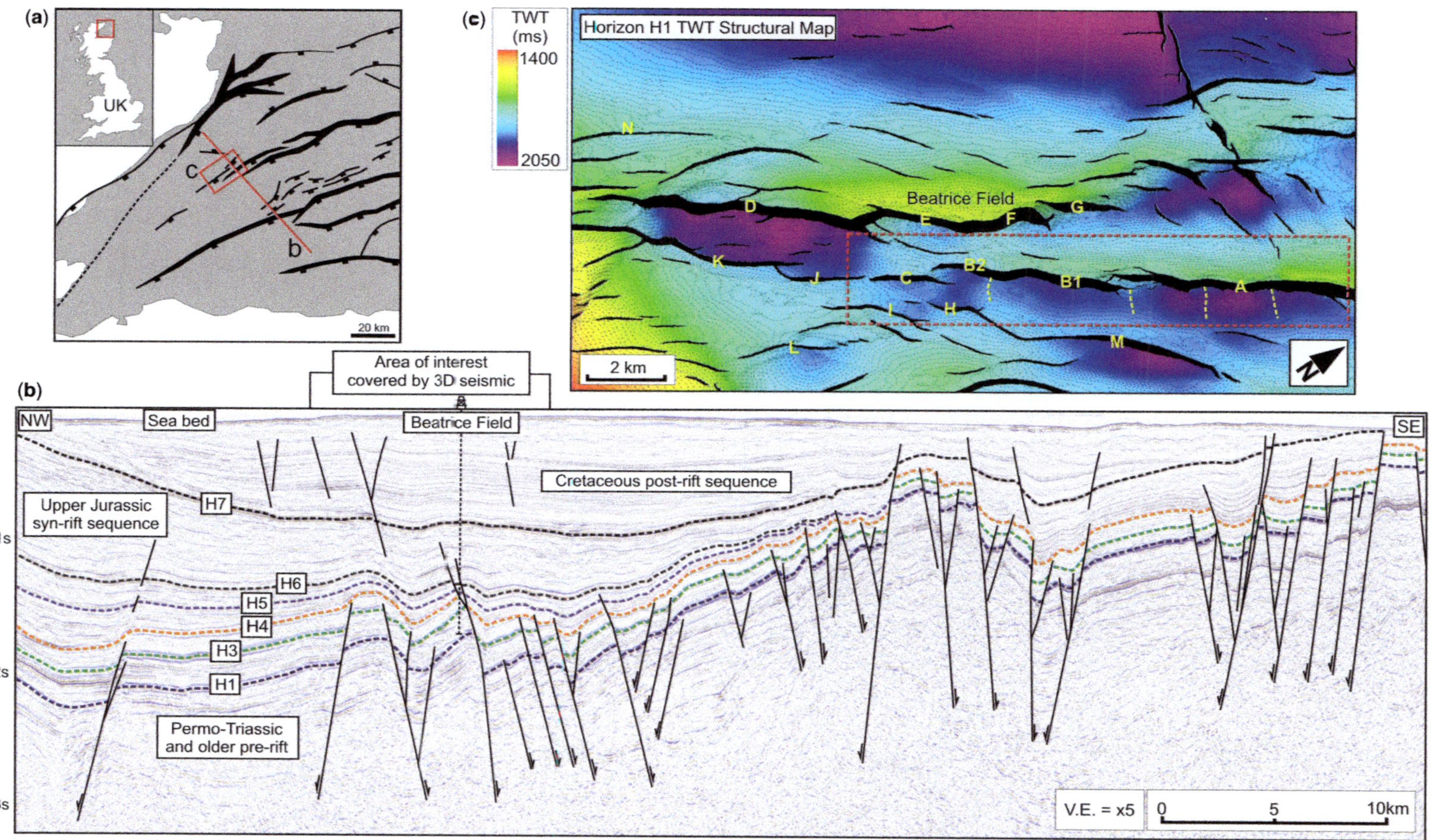

Fig. 1. (**a**) Schematic structural map of Inner Moray Firth (IMF) basin (modified from Long & Imber 2010). (**b**) Regional 2D seismic section across IMF, showing the main interpreted horizons and faults. (**c**) TWT structural map of pre-rift horizon H1 (Top Triassic). The letters represent the names of the analysed normal faults from the 3D seismic data. The red rectangle delineates the detailed area of analysis. The yellow stippled lines mark the transverse fold hinges separating the depocentres associated with faults A and B.

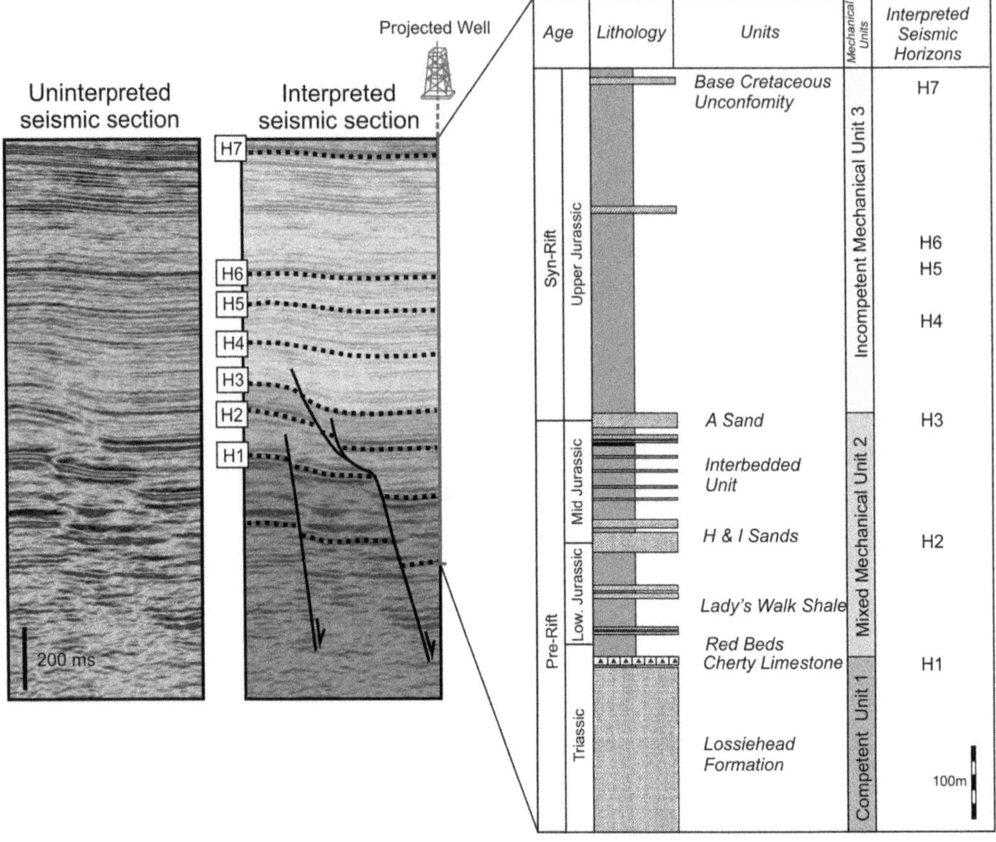

Fig. 2. Uninterpreted and interpreted seismic section (with 2× vertical exaggeration) from the studied 3D volume. The mapped horizons and the main mechanical stratigraphic units are shown in the interpreted version. Lithological formations were separated into three mechanical stratigraphic units based on the net-to-gross ratios obtained from the Beatrice Field well data.

the Devonian Old Red Sandstone (Goldsmith *et al.* 2003; Glennie *et al.* 2003). Based on the high net-to-gross ratio of the Lossiehead Formation and underlying strata, we infer that MU 1 is likely to be mechanically 'competent', here defined as being susceptible to deformation by seismic-scale faulting.

The upper part of the pre-rift sequence (H1–H3 interval; Fig. 2) comprises a *c.* 300 m-thick succession of interbedded sandstones and shales with a net-to-gross ratio of 38%, which we define as mechanical unit 2 (MU 2). We infer that the alternation of competent sandstones and less competent shale layers is likely to favour layer-parallel slip (Watterson *et al.* 1998*b*). At the time of rifting, these Lower–Middle Jurassic sediments may have not been completely lithified, and were probably characterized by a reduced-strength contrast between the sandstones and weaker shale layers. However, results of discrete element method modelling have shown that deformation can be partitioned between layers with small strength contrast at low confining pressure conditions (Schöpfer *et al.* 2007), with faults initiating in the slightly more competent sandstone layers. We hypothesize that thicker and relatively stiffer sandstone intervals within the MU 2, such as the 50–60 m thick 'H' and 'I' reservoir sandstones of the Beatrice Field (Stevens 1991), may favour fault nucleation and propagation (see the section 'Spatial and stratigraphic variations in fault throw and fold amplitude'), whilst the intervening shale intervals (e.g. Lady's Walk Shale) may inhibit fault propagation (Fig. 2). This overall arrangement is likely to promote vertical segmentation of faults.

The syn-rift sequence (H3–H7 mapped horizons) thickens towards the main faults and is dominated by Upper Jurassic shales, which we define as mechanical unit 3 (MU 3). This succession is likely to be mechanically 'incompetent', here defined as being susceptible to distributed (i.e. ductile) deformation. Hanging-wall reflectors within several hundreds of metres of the mapped faults clearly dip towards the graben (synthetic layer dips *sensu* Ferrill *et al.* 2005), with hanging-wall syncline depocentres shifted away from the fault. Previously, these folds have been interpreted as the result of differential compaction of the shale-dominated syn-rift sequence (MU 3) overlying the older and more rigid pre-rift, footwall formations (MU 1 and 2; Thomson & Underhill 1993). While we do not exclude the possibility that some folds are the result of compaction, we show below that the analysed hanging-wall folds display structural patterns that cannot be attributed to compaction, and that compaction effects are secondary with respect to other mechanisms.

Dataset and methods

Seismic and well data

The dataset used in this study comprises a 3D reflection seismic survey acquired over the Beatrice Field (Linsley *et al.* 1980; Stevens 1991) and several regional 2D seismic lines that are orientated NW–SE, orthogonal to the main structure of the Inner Moray Firth Basin. The 3D time-migrated seismic data cover an area of 11 × 22 km, and have a crossline and inline spacing of 12.5 m. The dominant frequency for the interval of interest is between 30 and 40 Hz, with velocities ranging between 2500 and 3500 m s^{-1} (Fig. 3a), resulting in a vertical seismic resolution of 15–30 m. Velocity data from the Beatrice wells indicate a consistently increasing velocity with depth, with no significant lateral or vertical velocity variations (Fig. 3a). There are no significant variations in geometry between time and depth data, just a relatively uniform expansion by a factor of 1.55 on the depth profiles (Fig. 3b). As a result, we used the two-way-time data to measure parameters such as fault throw and the amplitude of the hanging-wall folds. However, when we analysed attributes such as fault dip, the fault surfaces have been converted to depth in order to show the realistic geometries of the faults. Eight seismic horizons were mapped in total (seven in detail: H1–H7) within the pre-rift and syn-rift stratigraphic intervals, with Beatrice Field wells providing information on the associated lithological formations. The study focusses on the segmented, SE-dipping ABC fault array (see box in Fig. 1c), supplemented by examples from other fault systems to highlight salient points.

Methods

We used several methods to analyse the distribution and growth of the faults and folds:

Throw–distance (T–x) profiles and *throw–depth (T–z) profiles* enabled us to investigate the lateral and vertical variations in discontinuous fault

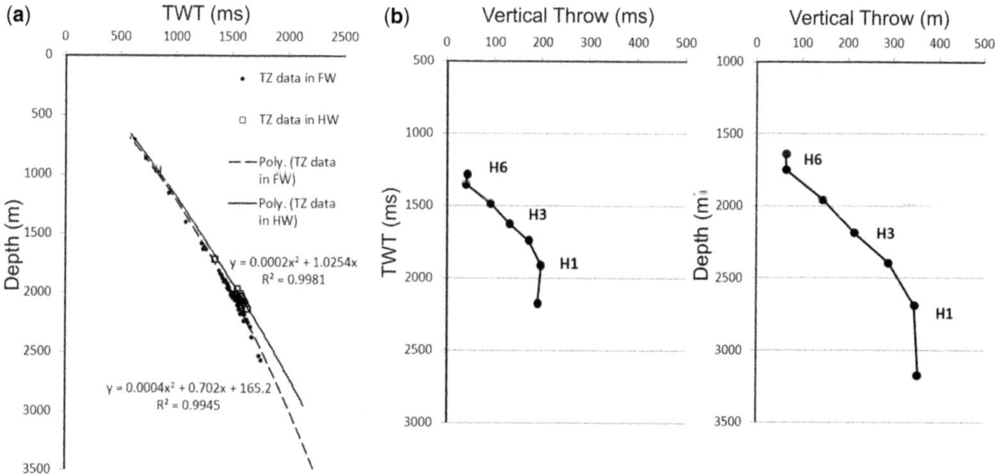

Fig. 3. (**a**) Time–depth (*T–Z*) curves from the Beatrice Field wells showing different velocity gradients for wells which penetrated the footwall or the hanging-wall sections of the faults. For depth conversion, we used the *T–Z* relationship derived from the wells which penetrated the thicker syn-faulting sequences deposited within fault-controlled depocentres because we are interested in quantifying deformation located mainly within the hanging walls of the faults. This younger, syn-faulting section is characterized by slightly lower velocities compared with the older pre-rift sequence in the footwall. (**b**) Comparison of throw distribution in time (ms) with throw distribution in depth (m). The pattern of throw distribution is very similar, but the throw–depth plot shows a vertical expansion of *c.* 1.55.

throw and continuous deformation (folding), and to analyse the lateral and vertical linkage of faults (Walsh & Watterson 1991; Childs *et al.* 1996; Mansfield & Cartwright 1996; Hongxing & Anderson 2007; Long & Imber 2010; Jackson & Rotevatn 2013; Tvedt *et al.* 2013; Rotevatn & Jackson 2014). For T–x profiles, fault throw was measured perpendicular to the strike of the fault every 125 m (every 10th inline), with more dense sampling points near the fault tip or where the fault complexity required it.

Isochore thickness maps and *expansion indices* were used to analyse the timing of faulting and folding (Jackson & Rotevatn 2013; Tvedt *et al.* 2013) and to constrain the position of the upper tip-line at the time of deformation. The expansion index (Thorsen 1963) is defined by the ratio between the maximum thickness of a chosen syn-rift interval in the hanging wall of a fault (adjacent to the fault surface, or within the synclinal depocentre) and the thickness of the equivalent interval in the footwall;

Fault surface analysis provided insights into the relations between fault geometry and linkage style, expressed by parameters such as fault dip, fault cylindricity and throw variation (Ziesch *et al.* 2015), and distribution of ductile deformation. Cylindricity measures the deviation of a fault surface from a best-fit planar surface (Ferrill *et al.* 1999; Jones *et al.* 2009; Ziesch *et al.* 2015).

Seismic trace and coherency attributes (a combination of instantaneous phase, tensor, discontinuity and semblance attributes; Chopra & Marfurt 2007) were used in some cases to enhance the visibility of deformational patterns at the limit of seismic resolution within the hanging-wall folds (Iacopini & Butler 2011), or to highlight the fold geometry (dip, dip-azimuth) and stratal onlaps onto fold limbs.

Observations of normal faults and fault-related folds from 3D seismic data

Geometric characteristics of the studied faults and fault-related folds

At H1 (Top Triassic) level, the studied fault system comprises three left-stepping normal fault segments named A, B and C. These are separated by two relay zones. The relay ramp between faults A and B is at an early stage of breaching (Fig. 1c). At the base syn-rift level (H3), the relay ramps are completely breached by the footwall faults, forming a continuous fault trace. Bends in the fault trace are associated with minor, hanging-wall splay faults (Fig. 4a). This downward bifurcation (with intact or partially breached relay ramps at depth, and breached relay ramps at shallower levels) seems to be a common feature in our area of study (see the section 'Spatial and stratigraphic variations in fault throw and fold amplitude').

The seismic sequence between the H3 and H7 horizons thickens towards the analysed faults, consistent with their syn-sedimentary nature. The ABC fault array is part of a larger NE-SW striking normal fault system that dips SE, along with the faults bounding the Beatrice Field structure, here named D–G. These two major fault systems are linked within the syn-rift sequence (on horizons H6–H7) by smaller segments (segments b and c) that splay upwards from the main faults (Fig. 4b, c). The upper tip-lines of fault C and the SW continuation of fault B (named B2) are buried within the H3–H5 interval, and are overlain by parallel seismic reflections. These observations indicate that the faults become inactive during the later syn-rift stage, when linkage of the AB fault with the D fault occurred. Faults B2 and C are located within a larger syn-rift transfer zone comprising the synthetic dipping D and E faults, but also the opposite (NW) dipping faults H, I and J (Fig. 4a), with which B2 and C form a conjugate normal fault pair (Fig. 4d).

At H1 level, we observe that the deepest structural levels lie immediately adjacent to the fault trace (Fig. 1c), whilst at H3 (base syn-rift) and H6 (intra-syn-rift) levels, the depocentres are shifted further into the hanging wall, with increasing distance from the fault on progressively younger syn-rift horizons (Fig. 4a, b). At H3 level, the faults are bordered on the hanging-wall side by monoclinal folds with limbs that dip in the same direction as the fault (Fig. 4a, c). However, not all of the faults are associated with folds at horizon H3 level: faults B2 and C appear to have depocentres adjacent to the fault trace (Fig. 4a, d). Hence, an intriguing question is why some faults display hanging-wall folds and depocentres that are shifted into the hanging wall, whilst others in the same array lack folds and are characterized by depocentres adjacent to the fault trace. The fact that the folds are developed within the *pre-rift* sequence, and that their location does not necessarily correspond to the major depocentres suggests that the generating process cannot be entirely attributed to differential compaction (cf. Thomson & Underhill 1993).

Seismic attribute analysis using the instantaneous phase attribute shows that folds associated with the B2 fault are associated with clear, antithetic-dipping axial planes that separate the upward-widening monocline from the hanging-wall synclines (Fig. 5a, b). This fold does *not* display vertical axial planes and thinner hanging-wall dipping limbs, which are characteristic for compaction folds in the hanging walls of normal faults (Skuce 1996).

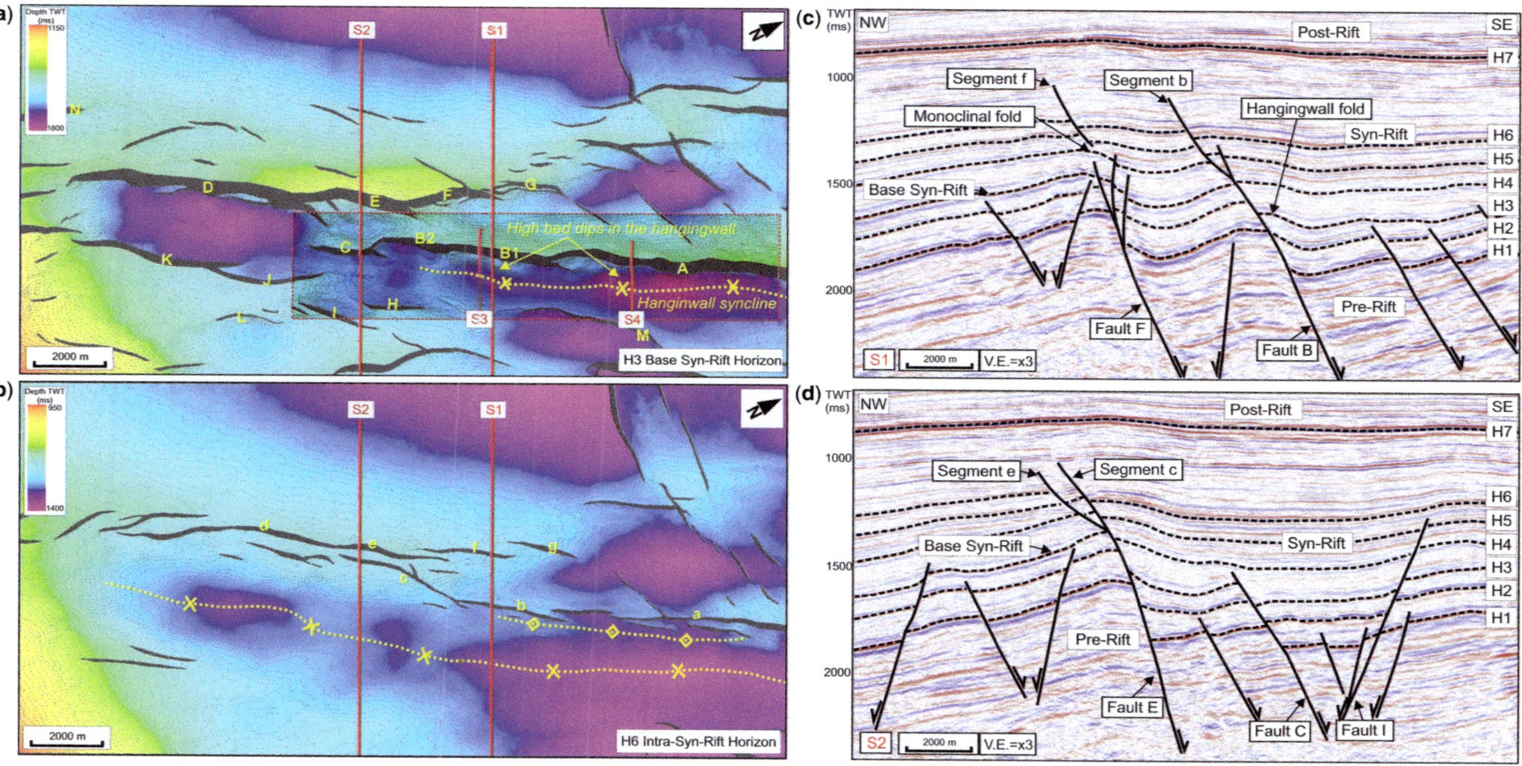

Fig. 4. **(a)** TWT structural map of horizon H3 (top pre-rift and top mechanical unit 2); the letters represent the names of the analysed normal faults from the 3D seismic data. The red rectangle borders the A, B and C faults which are analysed in detail. The traces of the faults A and B are bordered by longitudinal folds on the hanging-wall side. The steepest reflector dips occur on the fold limb adjacent to the fault traces and are consistently down towards the basin. **(b)** TWT structural map of horizon H6 (intra syn-rift). Note the basinward migration of the hinge line of the hanging-wall syncline and the decrease in the density of the faults compared with the fault density within the pre-rift sequence (see Figs 1c or 4a). **(c)** and **(d)** Interpreted seismic profiles orthogonal to the studied faults. Location in a and b (see text for detailed description).

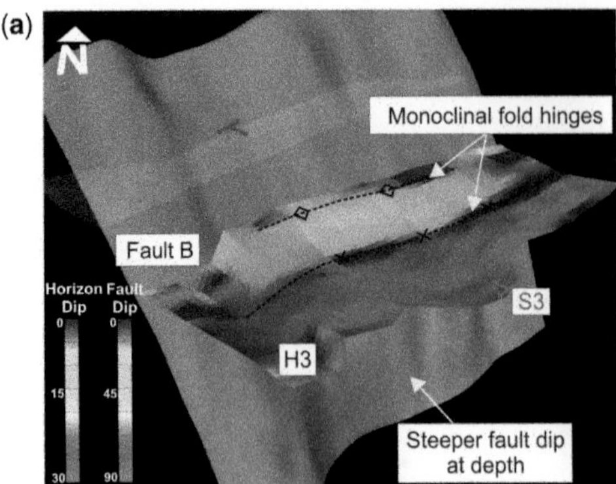

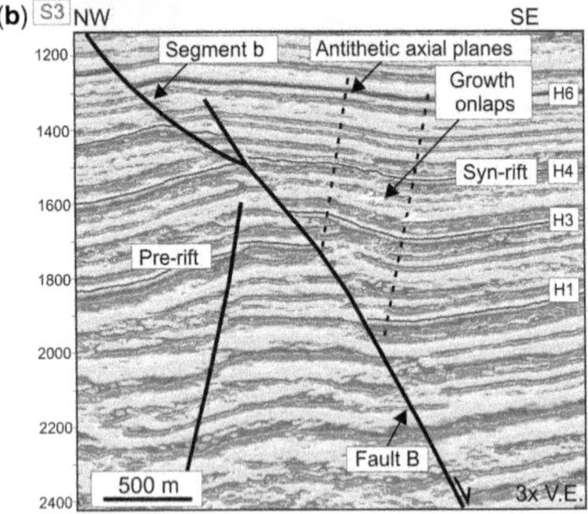

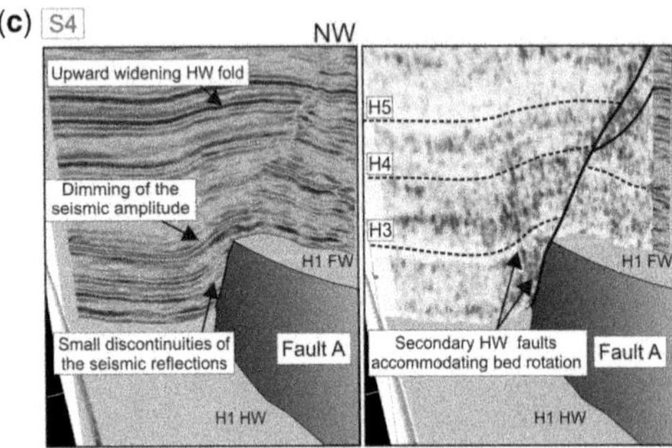

Fig. 5. (**a**) 3D view of the breached monocline along fault B; (**b**) section (S3) displaying the instantaneous phase attribute and showing the breached monoclines associated with fault B. The instantaneous phase attribute enhances visualization of the reflector configuration, and highlights the onlap of reflectors onto the limb of the monoclinal fold. (**c**) Hanging-wall fold associated with fault A (location in Fig. 4a) with combined tensor–semblance–discontinuity attribute volume (right) that enhances visualization of secondary faults within the hanging wall of fault A.

The instantaneous phase attribute also highlights seismic reflections within the syn-rift sequence that onlap onto the steep limb of the monocline. These onlaps are an indication of the fold growth, rather than the effect of compaction. Furthermore, a combined tensor-semblance-discontinuity attribute indicates the presence of secondary faults (steeply dipping normal faults or even small reverse faults) associated with the monocline that, presumably, accommodated folding (Fig. 5c). These secondary faults resemble the hanging-wall deformation structures of normal fault-propagation folds modelled in clay (Withjack *et al.* 1990), and described in other rift settings which exhibit extensional fault-propagation folds, e.g. Suez Rift, NW Egypt (Sharp *et al.* 2000; Khalil & McClay 2002).

Fault dips are commonly observed to be gentler within the syn-rift and late pre-rift sequences (mechanical units 3 and 2) compared with the early pre-rift sequence (MU 1) (e.g. see segments b, e and f in Fig. 4c, d; and the fault dip attribute map in Fig. 9b). The change in fault dip therefore corresponds to the change in lithology from the mechanically competent Triassic sandstones (H1 and below), to the Lower–Middle Jurassic interbedded shale-sandstone succession (H1–H3) and Upper Jurassic shales (H3–H7). The overall effect is to generate pronounced convex upward fault geometries (Fig. 4c, d) but, because the upward transition to gentler fault dips occurs within the *pre-rift* interval, differential compaction should be secondary in respect to other factors. We can explain the difference in dips by the variation in shear failure angles within rocks that have different mechanical properties (Mandl 1988), with higher-angle faults developed within the mechanically competent Triassic sandstones (MU 1). These observations have important consequences for understanding the vertical segmentation of faults across the different mechanical units, a point we return to in the following section.

Spatial and stratigraphic variations in fault throw and fold amplitude

Figure 6 is a *T–x* profile showing the variation in throw (i.e. the discontinuous component of vertical displacement) along the strike of faults A, B and C. We observe a systematic decrease in throw towards the SW, with the largest throws within the pre-rift sequence (H1 level) reaching 300 ms (*c.* 450 m) along fault A and decreasing to a maximum of *c.* 100 ms (*c.* 155 m) along fault C. Note that fault A continues beyond the NE limit of the 3D seismic volume. The profiles for faults A and B display distinct throw minima that correlate with undulations in the fault trace, fault surface corrugations and the locations of transverse hanging-wall folds (Figs 1c, 6 & 9a). These observations suggest that at H1 level, fault A comprises at least three linked fault

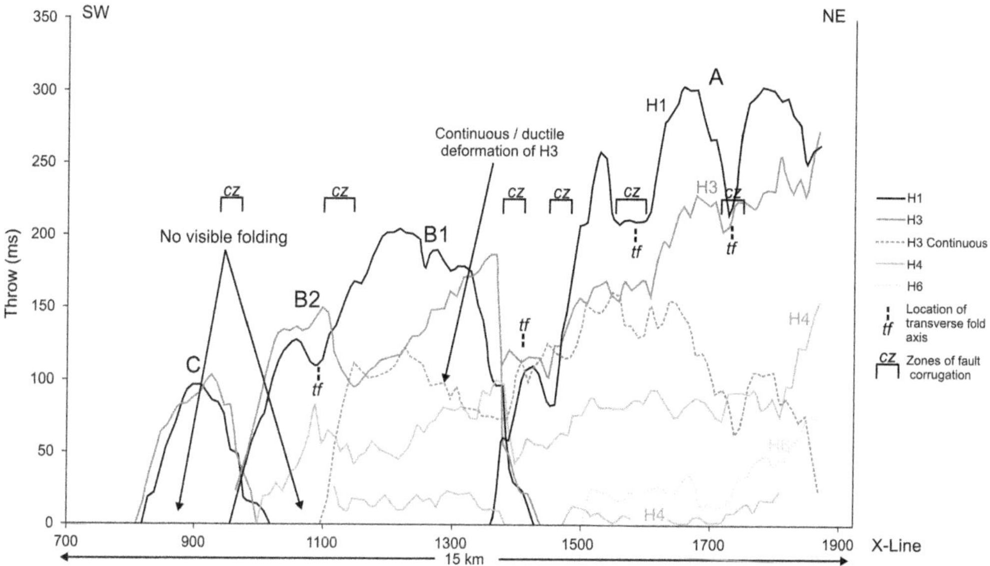

Fig. 6. Throw–distance profiles for the three analysed fault segments A, B and C (located in Fig. 4a). The throw decreases systematically from NE (right) to SW (left), and from the pre-rift (H1) to syn-rift horizons (H3-H6 horizons). The component of ductile deformation (folding) on the H3 horizon was measured separately (stippled line).

segments and fault B comprises two linked segments (B1 and B2 in Fig. 1c). We propose that faults A and B formed through the coalescence of multiple fault segments and that 'corrugation zones' mark the locations of former segment boundaries (Figs 6 & 9a).

The syn-sedimentary nature of the faults is reflected by a systematic, upward decrease in throw within the syn-rift interval (H3–H7) (Figs 6 & 7), and by the horizontal pattern of the throw contours projected onto the fault surface (Fig. 8) (Childs *et al.* 2003). Some of the throw-depth (T–z) profiles display an upward decrease in throw within the *pre-rift* interval (between H1 and H2–H3 for profiles P2–P6; Fig. 7) as a consequence of folding. Figure 6 shows that the amplitude of folding (measured on H3) approximately compensates for decreases in throw, and varies significantly along the strike of the fault. Folds are not observed adjacent to faults B2 and C, which display throw maxima at H3 level, i.e. at the top of the pre-rift interval (profiles P7–P9, Figs 7 & 8). Another observation that can be made from the T–z profiles is that, within the syn-faulting interval, the throw values for H6 and H5 markers are very similar (Fig. 7), which indicates either that the *c.* 40–50 ms (60–80 m) displacement post-dated deposition of H5–H6, or that the ratio of fault throw rate to sedimentation rate may have decreased during this interval.

Previous studies have shown that bends in a fault plane, such as those described in the previous section, can result from vertical fault segmentation and linkage within an incompetent mechanical unit (Childs *et al.* 1996). Fault L (Figs 1c, 2 & 10) provides a clear example of vertical segmentation across contrasting mechanical units. Figures 2 and 10c show that there is a marked upward decrease in the dip of fault L, which corresponds to the lithological boundary between the mechanically competent Triassic sandstones of MU 1 and the interbedded, Early–Middle Jurassic succession of MU 2. This change in dip coincides with a throw minimum that separates two distinct throw maxima within MU 1 and the Middle Jurassic H and I Sands (H2) within MU 2 (Figs 2 & 10b). Based on these observations, we infer that the upper, en-echelon segments La, Lb and Lc (Fig. 10) probably nucleated within the Middle Jurassic H and I sands, and linked with the deeper L1 and L2 segments within the underlying, incompetent Lady's Walk Shale formation. This vertical linkage generated a convex upward fault geometry, with a pronounced bend developed in the linkage zone (Fig. 2), expressed by the gentle fault dips and displacement minima (Childs *et al.* 1996). The fault bend geometry is controlled by the spatial position of the upper segments (e.g. La, Lb and Lc) relative to the location of the deeper main faults (e.g. L1 and L2). Essentially, the fault bend (or fault ramp) is controlled by the separation distance between the vertically segmented normal faults, with the widest ramp corresponding to the largest segment separation. As a consequence, the

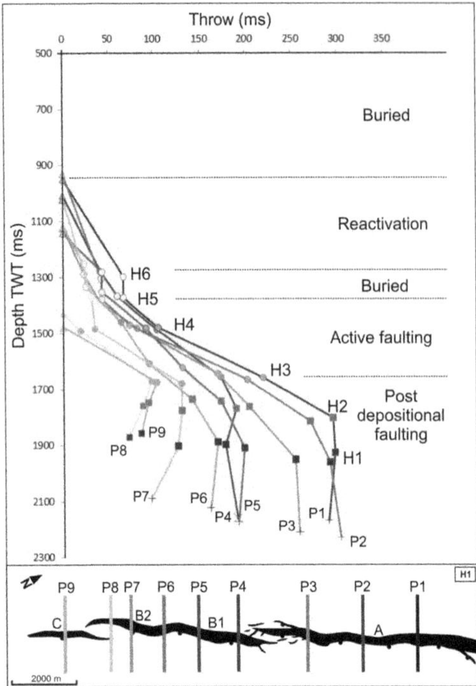

Fig. 7. Throw–depth (ms) plots for nine profiles across the studied faults. The maximum throw is located within the pre-rift section (pre-H3) but varies along the strike of the fault array. The SW part of the B segment and C segment (P7–P9) are characterized by throw maxima at the base syn-rift level (H3). For the other profiles, the lower throw values at base syn-rift are the result of folding.

locations of the bends in the fault plane can be variable along-strike of the fault array and this explains the observed geometries of the analysed faults (Figs 9 & 10). The changes in fault dip correspond, in some cases, to downward-bifurcation of fault segments, in which relay ramps are breached at shallower levels but remain intact at depth (Fig. 10). These fault patterns, which are similar to the geometry of the faults A and B, are unusual for coherent fault models that describe fault growth by upward-bifurcation (Walsh *et al.* 2003), suggesting again vertical linkage (Marchal *et al.* 2003; Jackson & Rotevatn 2013; Rotevatn & Jackson 2014) by downward propagation of segments that nucleated within the shallower Jurassic sequence.

The relationship between vertical segmentation and folding is illustrated in Figure 4c, d. Here, we observe that segments b, e and f dip gently within the syn-rift section and that the linkage with the deeper main faults varies along-strike. Close to its lateral tip (where the displacement is small), fault F is not hard-linked to the overlying segment f. Instead, the two faults are separated by a monocline that overlies the upper tip line of fault F (Fig. 4a–c). Analogue models indicate that discontinuities within layering (analogous to the heterogeneities in mechanical properties of the MU2 and MU3) tend to promote breaching of the monocline by downward propagation of a fault that nucleates at shallow depths above the footwall of the main, underlying fault, and which is not initially hard-linked to the main fault (Bonini *et al.* 2015).

A similar situation is indicated by high reflector dips observed above other fault arrays within the 3D seismic volume. For example, horizon H3 displays high reflector dips above the tip lines of segments La, Lb and Lc (Fig. 10a, H3 horizon dip map). This observation is consistent with folding ahead of the propagating tip of the 'L' segments (Ferrill *et al.* 2007; Long & Imber 2010). With increasing displacement, we suggest that the monocline (expressed by high reflector dips at H3 level) is likely to be breached completely and subsequent translation of the hanging wall across the convex upwards fault plane will increase the amplitude of the initial fault propagation fold, possibly completely overprinting it. The final amplitude of the fold will therefore vary along-strike as a function of the initial amplitude of fault propagation-fold, the amount of throw, and the geometry of the fault bend.

Summary of key observations and inferences

Mechanical unit 1 is characterized by steeply dipping faults that accommodated localized displacement with little evidence for associated folding. Fault dips are gentler within MU 2 and 3, reflecting the lower shear failure angles associated with these mechanically less competent units, and vertical linkage zones with the main faults. Fault propagation folds overlie the upper and lateral tip lines of faults within MU2 and 3, and we infer that the competence contrast between MU 1 and the overlying strata promoted vertical segmentation and linkage of faults. Contrary to a previous study, several observations suggest that differential compaction is unlikely to have been the primary mechanism responsible for fold generation. We now investigate the fold growth in more detail.

Folding mechanisms

Fault-propagation folding. Isochore thickness maps provide insights not only into the growth of the faults but also on the early growth and development of the fault related folds. Figure 11 shows the stratigraphic thickness of the early syn-rift interval (H3–H4). Hanging-wall syncline depocentres are observed along the strike of faults A and B. Figure 11b is a graph of the stratigraphic thickness of

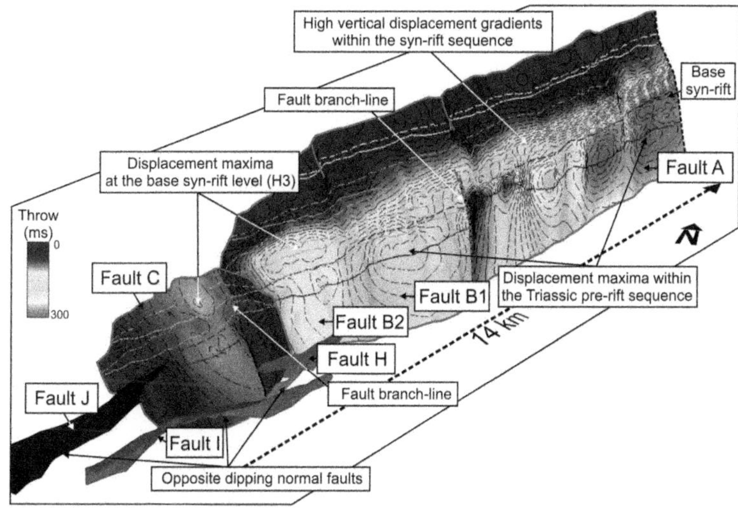

Fig. 8. Throw distribution on the A, B and C faults, with horizon cut-offs projected onto the fault surface (continuous line for hanging-wall cut-offs and discontinuous line for footwall cut-offs). Note that the maximum displacement is located within the pre-rift sequence for the A and B1 faults. For faults B2 and C, the maximum displacement is shifted upwards towards the base syn-rift.

the H3–H4 interval measured along-strike of the fault in the footwall, in the hanging wall and within the hanging-wall syncline. We observe that the maximum recorded thickness is located predominantly within the syncline depocentres. Similar thicknesses in the footwall and in the proximal

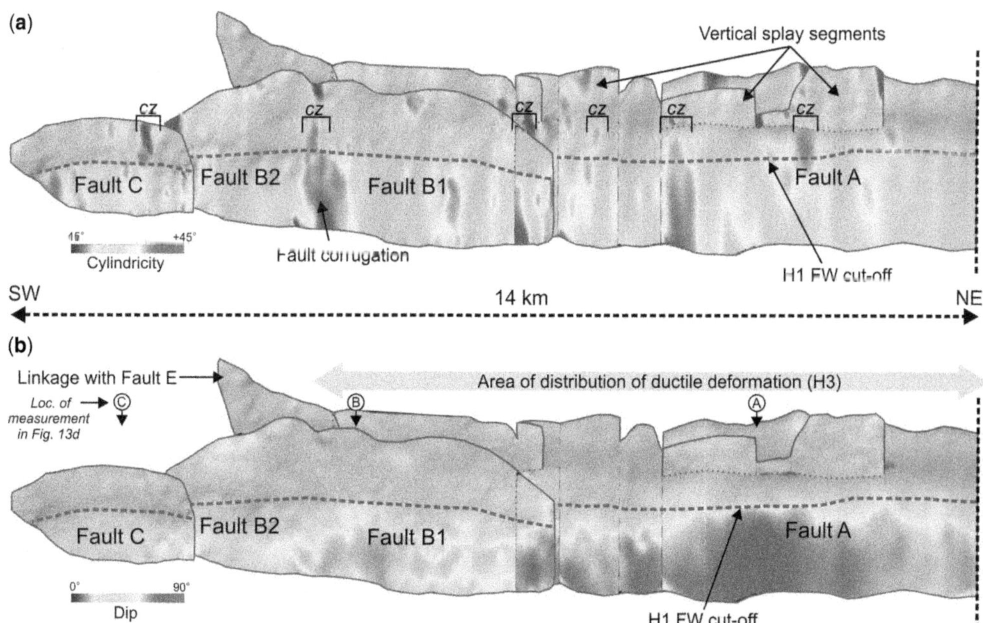

Fig. 9. Strike projections of the analysed fault surfaces displaying: (**a**) fault cylindricity attribute indicating possible zones of lateral corrugation (cz); (**b**) fault dip – note the sharp decrease in fault dip above horizon H1. This change in dip corresponds with a change in lithology, from the Triassic sandstone (mechanical unit 1) to Jurassic shale–sandstone interbedded sequence (mechanical unit 2; see Fig. 2).

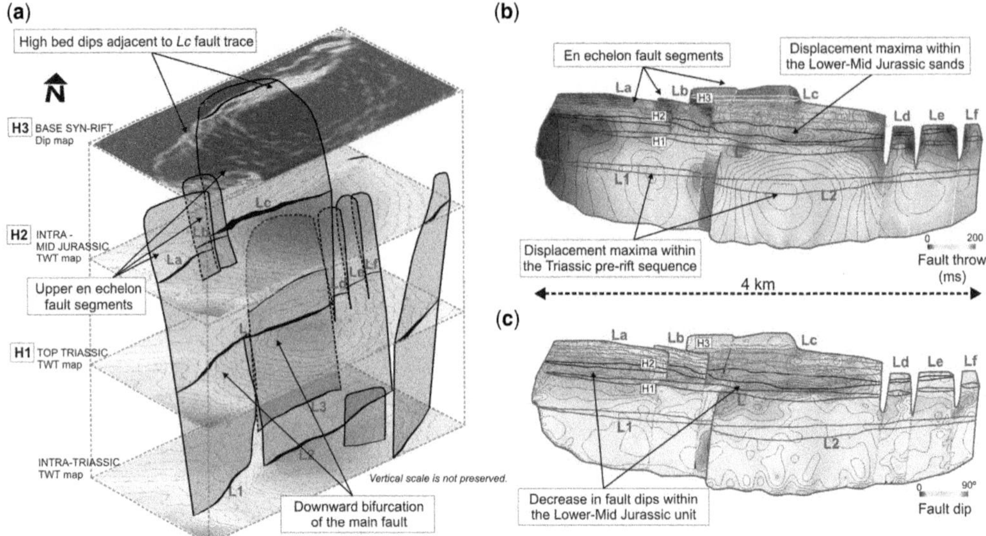

Fig. 10. (**a**) 3D diagram showing segmentation of fault L (location in Figs 1c & 4a) and the main interpreted horizon surfaces adjacent to the fault. Fault-related deformation is characterized by the high bed dips associated with the uppermost surface (H3) in the vicinity of the fault trace *Lc* and above the adjacent blind segments. (**b**) Strike projection of fault L contoured for throw. (**c**) Strike projection of fault L contoured for dip.

part of the hanging wall along parts of A and B suggest that, at the time the H3–H4 sequence was deposited, parts of these faults were blind and overlain by a gentle monocline, with growth strata onlapping the monocline limb (Figs 10 & 11b). At this stage, the amplitude of the monocline reached *c.* 40 ms (50–60 m), indicated by the difference in the real stratigraphic thicknesses of the syn-faulting deposits in the syncline and in the proximal part of the hanging wall, with the condition that this latter thickness is similar to the stratigraphic thickness in the footwall (Fig. 11b). We suggest that vertical propagation of faults A and B was inhibited within the ductile, shale-dominated Early–Middle Jurassic sediments, most likely within the Lady's Walk Shale Formation, considering that horizon H2 is also folded. Other faults from the study area that exhibit vertical segmentation (e.g. faults L, N) display lateral offsets or dip linkage (and associated bends in the fault plane) within the same stratigraphical level.

The formation of a fault propagation fold is controlled by the relative position of the upper tip-line of the faults with respect to the mechanical stratigraphy, in our case, by the presence of MU 2. Our observations show that the elevation of the vertical tip-line was very variable along the strike of the fault ABC, hence a question arises: why in some places was the upper tip line buried beneath the free surface (developing a fault propagation-fold) whilst in other places, for example along the conjugate fault pairs B2-H and C-I, did the fault breach the depositional surface shortly after the onset of rifting? We do not have any evidence from wells, or from the analysis of the seismic facies, of any significant lateral changes in lithology or a decrease in thickness of MU 2, which together could enhance upward propagation of the faults and early surface breaching. Expansion indices show constantly higher values for faults C and B2 for the H3–H4 interval, compared with segment B1 and parts of A (Fig. 11c). These high expansion indices can be an indicator of the high displacement rates on these two faults during deposition H3–H4, which is consistent with their early breaching of the surface. We propose a mechanical explanation for these observations in the *Discussion* section.

Fault-bend folding. The present-day fold amplitudes on horizon H3 are very variable (stippled line on the *T–x* profile, Fig. 6), and larger than the amplitudes inferred to be solely the result of initial fault propagation folding (Fig. 11). Figure 9b shows that the lateral distribution of folds correlates well with the extent of regions characterized by upward decreases in dip of the ABC fault plane. We observe that the fold amplitude is largest where there is a more pronounced change in the fault dip with depth (adjacent to faults A and B1) (Fig. 6). The increase in the fold amplitude of H3 also seems to correlate with increasing displacement of the

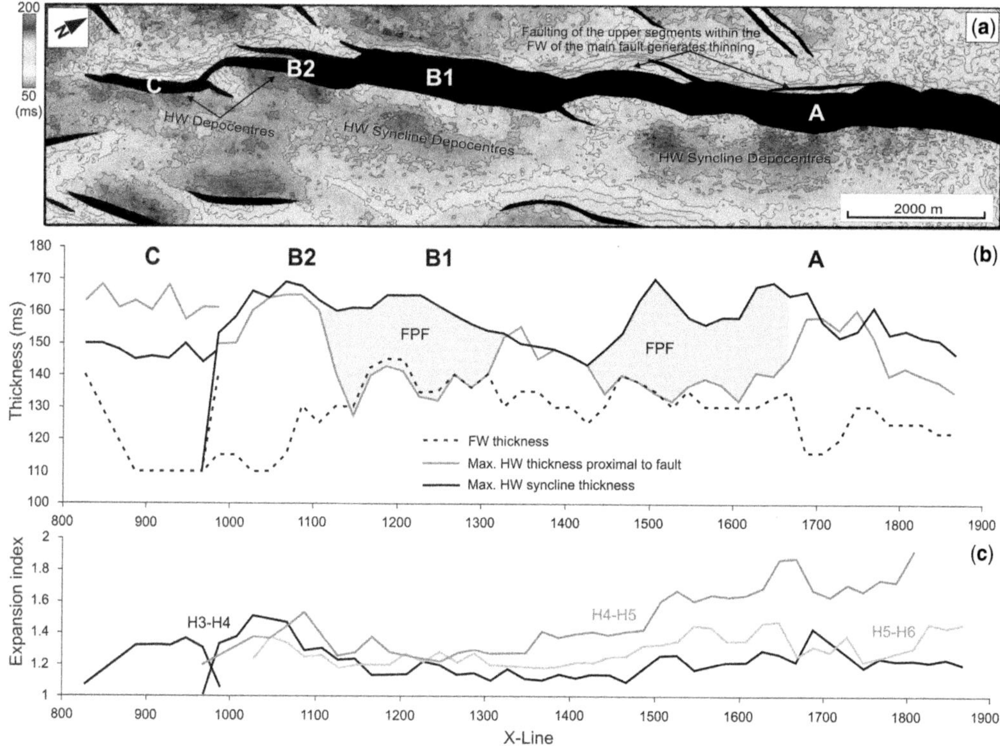

Fig. 11. (**a**) Isochore thickness map of the H3–H4 early syn-rift sequence. (**b**) Graph with the thickness of the H3–H4 growth sequence measured in the immediate vicinity of the fault trace, in the footwall (dashed line), in the hanging wall (light black line) and in the hanging-wall syncline (bold black line). The fault propagation folds (FPF) are identified where maximum thicknesses are recorded within the hanging-wall syncline, and the thicknesses of syn-rift strata within the footwall and proximal part of the hanging wall are similar. (**c**) Expansion indices measured along-strike of the faults.

H1 horizon (Fig. 6). At shallower levels, Figure 4b and c shows that the H6 horizon developed a broad anticline flanked by a depocentre immediately adjacent to the trace of fault b, and another broad, distal synclinal depocentre parallel with the fault trace. This morphology is similar to the hanging-wall geometries developed above ramp–flat–ramp normal faults (McClay & Scott 1991; Rotevatn & Jackson 2014). We will come back to discuss the relation between fault-bend and fold amplitude in the following sections.

Analysis of normal faults and fault-related folds from a regional (basin-wide) dataset

2D geometry of faults and fault-related folds

To obtain a more representative sample of the extensional fault-related folds from the IMF basin, we analysed a further 57 cross-sections from the regional 2D seismic dataset in addition to measurements of the 18 faults interpreted from the 3D survey. Examples of the analysed extensional folds are illustrated in Figure 12. Most of the faults terminate within the syn-rift sequence and are associated with monoclinal folds above their upper tip points (Fig. 12a, b). Some of the monoclines are breached by their associated faults, resulting in normal drag-like fold geometries within the hanging wall (Fig. 12e). The following key observations suggest that the analysed monoclines originated as fault propagation folds: (a) the folds display an upward widening geometry; (b) there is a qualitative relationship between the amplitude of the monoclines (breached or unbreached) and the amount of throw recorded within the pre-rift sequence; and (c) in some, but not all, cases, reflectors within the syn-rift sequence onlap onto the fold limbs (Fig. 12a–c). Where stratal onlaps are absent, seismic reflectors within the syn-rift sequence have a sub-parallel to slightly divergent pattern away

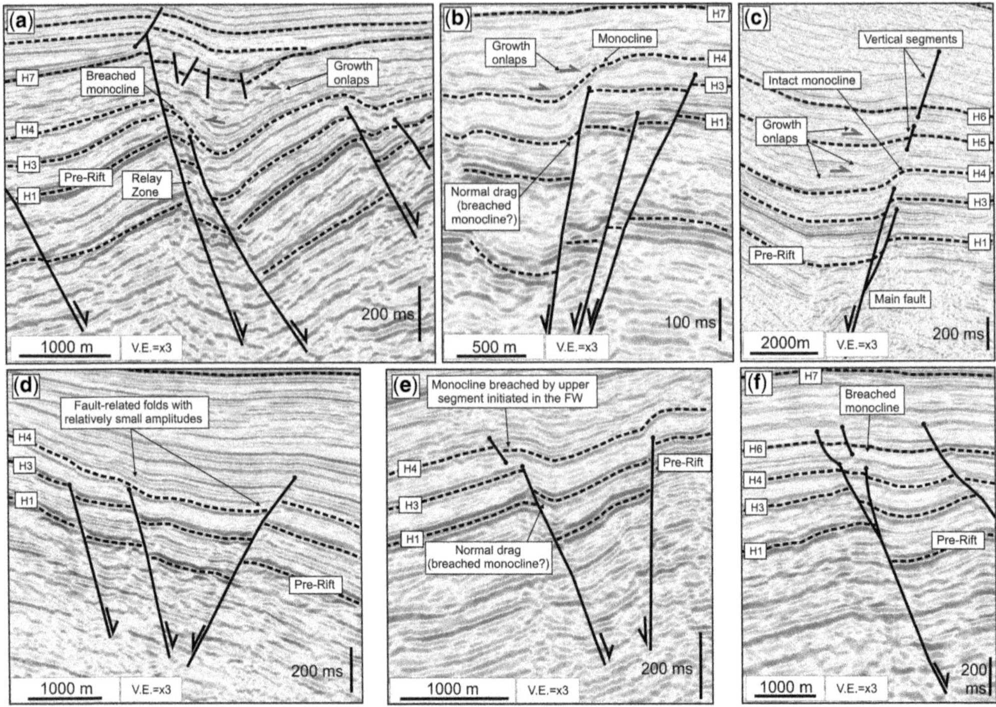

Fig. 12. Examples of fault-propagation folds associated with different sets of 'simple' and 'conjugate' normal faults, interpreted from regional 2D seismic profiles across the IMF (see text for explanations).

from the fault, with minor differences in thickness between the hanging-wall and footwall strata. This observation can be explained by the relatively high sedimentation rates (150–400 m Ma^{-1}) in this part of the basin (Davies *et al.* 2001), which exceeded the relatively low fault displacement rates (Nicol *et al.* 1997). This interpretation is consistent with the relatively low expansion indices for the H3–H4 interval, compared with the younger analysed intervals (Fig. 11c). Consistent with our interpretations of the 3D seismic data, breaching of fault propagation folds occurs either by upward propagation of the main faults from below, or by downward propagation of shallower fault segments that nucleate within the syn-rift sequence (i.e. MU 3), typically within the footwall domain of the monocline (Fig. 12c, e, f). In the latter case, vertical linkage with the deeper faults may give rise to irregular fault traces.

In summary, our observations and inferences based on the basin-wide, 2D seismic dataset corroborate our initial conclusions based on detailed analysis of the (spatially restricted) 3D seismic dataset, providing confidence in the general applicability of our results. We now undertake a quantitative analysis of fold growth and breaching using the combined results from both datasets.

Quantitative analysis of fold growth and breaching

Our observations show that conjugate faults (e.g. faults B2, C and H, I) tend to breach the depositional surface soon after the onset of rifting. We therefore sub-divide the data into two categories based on the fault geometry. 'Simple' normal faults are those *not* associated with a conjugate pair, whilst 'conjugate' normal faults are those that interact (and may share a sub-horizontal branch-line) with opposite-dipping faults (Figs 12 & 13). Conjugate normal faults may display a cross-sectional V-style geometry if throw is similar on both faults and a Y shape, if displacement is larger on one fault than the other (Nicol *et al.* 1995). The amplitudes of breached and intact monoclines were measured for two horizons, H3 and H4. Although the data are relatively scattered, we observe that conjugate faults tend to have smaller associated fold amplitudes compared with simple faults (Fig. 13a, b). For example, only 8% of the analysed simple normal faults have no associated folding on horizon H3, compared with 41% of the conjugate faults (Fig. 13a). Some 51% of the simple normal faults in our sample are associated with folds that accommodate more than half of the total throw (i.e. ratio of fold amplitude/total throw

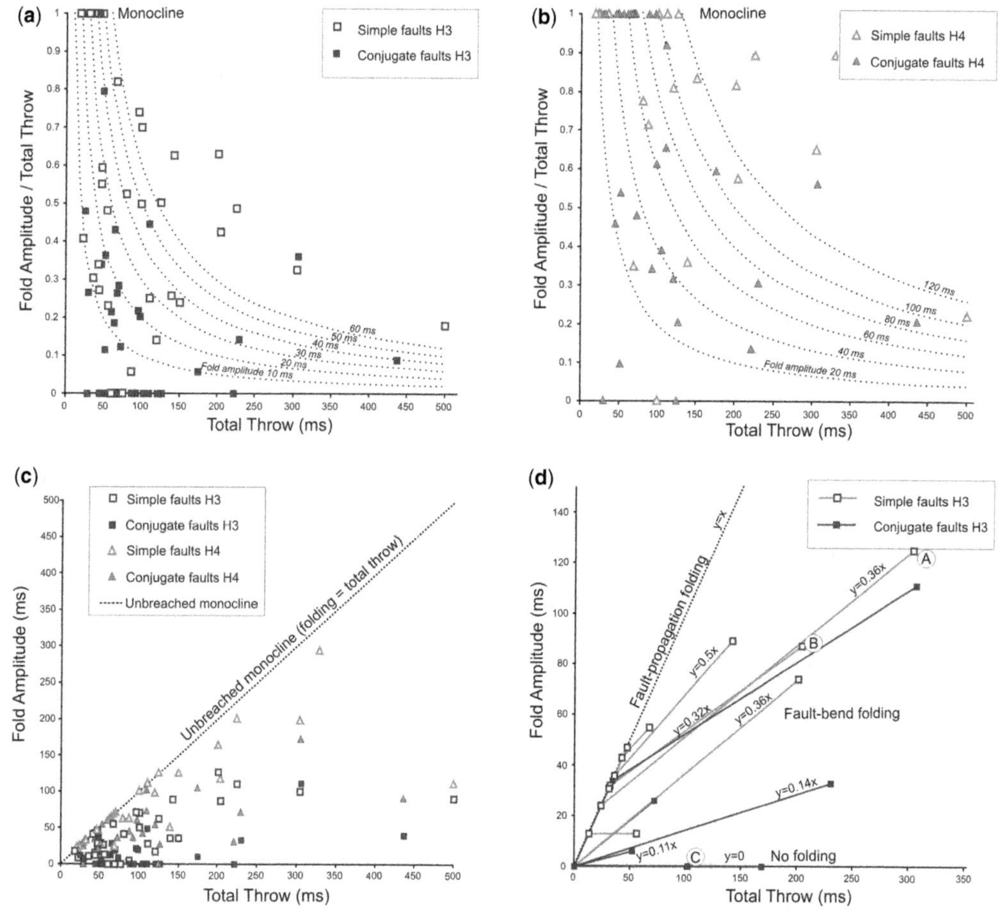

Fig. 13. (**a**) Ratio of fold amplitude to total throw v. total throw for horizon H3 (part of mechanical unit 2) measured on both 2D and 3D seismic data on two types of faults, simple normal faults and conjugate normal faults. (**b**) Ratio of fold amplitude to total throw v. total throw for horizon H4 (part of mechanical unit 3). (**c**) Fold amplitude v. total throw measured for the two horizons, H3 and H4. (**d**) Fold amplitude v. total throw for horizon H3 measured on faults from the 3D seismic data set only. A, B, C are measurement localities for the faults displayed in Figure 9. The vectors show possible evolution of folding with increasing fault throw (see text for explanation). Fault-propagation folds are characterized by vectors with a gradient of 1.0 (folding = throw), while fault-bend folds are characterized by vectors with gradients from 0.11 to 0.5. The gradients correlate with the change in fault dip within mechanical unit 2 with higher gradients reflecting a larger change in fault dip (see Fig. 9).

>0.5; Fig. 13a), compared with only 8% of the conjugate faults.

Figure 13 also shows that fold amplitudes vary from 0 to 100% as a proportion of the total displacement (fault throw + fold amplitude) on the two interpreted horizons: H3 (top MU 2) and H4 (intra MU 3). By comparing the ratio of fold amplitude with the total throw on each horizon, we are able to explore the influence of the two different mechanical units on the magnitude of ductile deformation. The extensional fold amplitudes measured for horizon H4 are typically larger than the fold amplitudes of horizon H3 (Fig. 13a, b). The largest amplitude recorded for an intact monocline (fold amplitude/total throw = 1) for H4 is 120 ms compared with 50 ms for H3. Larger amplitude values observed for breached monoclines (fold amplitude/total throw <1) can be explained by increased bed rotation within relay zones (e.g. between vertically segmented faults) and/or by movement of the hanging wall across a bend in the fault surface, which we discuss, below.

Fault-propagation fold geometries (in terms of monocline amplitude and wavelength) can be described by kinematic parameters such as propagation to slip ratio (P/S) and apical angle, which together define the trishear zone of deformation located above propagating blind faults (Hardy & Allmendinger 2011). P/S ratio, the main controlling factor on the amplitude of the fold, represents the propagation of the fault with respect to the displacement accrued, and is influenced by the mechanical properties of the rocks and the effective confining pressure (Cardozo *et al.* 2003). Incompetent lithologies tend to inhibit fault propagation by accommodating larger amounts of strain before failure, while more competent layers are characterized by localized brittle shear fractures. The larger fold amplitudes observed on horizon H4 compared with those associated with H3 are consistent with lower P/S ratios associated with propagation of the fault through the shale-dominated H3–H4 interval. This interval, which is part of the syn-rift, mechanical unit 3, has a higher proportion of incompetent shale layers (>90%) than MU 2 (62%). This observation suggests that fault propagation rates, as a proportion of fault displacement rate, vary according to the ratio of incompetent v. competent lithologies, given that the bulk thickness of the two stratigraphic intervals is similar. The relatively early breaching of the interbedded MU 2 – despite its likely propensity to deform by layer-parallel slip – is consistent with the models of Bonini *et al.* (2015), which indicate breaching of the monocline by downward propagation of a fault that nucleates at shallow depths above the footwall of the main fault.

As previously shown, vertical linkage may generate a bend in the fault plane that, with increasing displacement, will promote further fold growth as a result of hanging-wall translation over the convex upward fault plane. Figure 13d shows a series of vectors, plotted in fold amplitude v. total throw space, that illustrate the growth of fault-bend folds on horizon H3 within the 3D seismic survey area. The left-hand point on each vector corresponds to the amplitude of the precursor fault propagation fold (zero in some cases). The right-hand point on each vector corresponds to the final fold amplitude (at the cessation of fault movement) resulting from fault propagation *and* fault-bend folding. According to Groshong (1989), the relationship between fault throw and the amplitude of a fault-bend fold depends primarily on the bend geometry, which is given by the change in fault dip. The maximum throw on the faults presented in Figure 13d is similar to the thickness of MU 2, hence we assume a linear relationship between fault throw and fault-bend folding, since horizon H3 (top of MU2) is not completely displaced over the fault bend. In this situation, steeper gradients (e.g. vectors A and B) correspond to more pronounced bends in the fault surface, while lower gradients (e.g. vector C) are characteristic of more planar faults, which lack or have smaller associated folds (Figs 9 & 13d). These observations suggest that the final fold amplitude is the result of both fault propagation and fault-bend folding processes, and that the relative importance of each mechanism may vary significantly along the strike of a single fault array.

Discussion: mechanical interaction between faults and implications for fault propagation and fold development

Geomechanical models indicate that faults interact within the elastic stress fields of neighbouring segments, resulting in asymmetric displacement distributions and preferential locations of slip and/or fault propagation (Willemse *et al.* 1996; Crider &

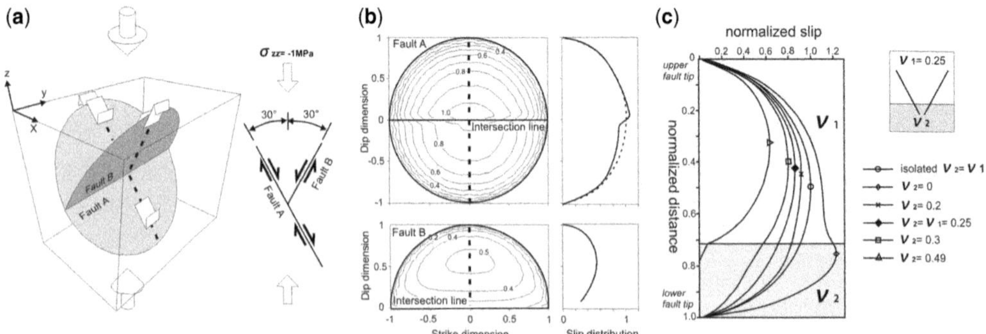

Fig. 14. (**a**) Configuration of the elastic boundary element model for conjugate normal faults with a 'Y'-type geometry within a homogeneous whole elastic space, from Maerten *et al.* (1999). (**b**) Results of the modelled displacement distribution (Maerten *et al.* 1999). (**c**) Calculated displacement distributions for conjugate normal faults with a 'V'-type geometry located within a heterogeneous elastic material, derived from finite element method modelling (Young 2001). Note the asymmetrical slip distribution, skewed towards the upper fault tip, for models in which the fault intersection lies within a layer that has a higher Poisson's ratio than the surrounding material (i.e. $\nu_2 > \nu_1$).

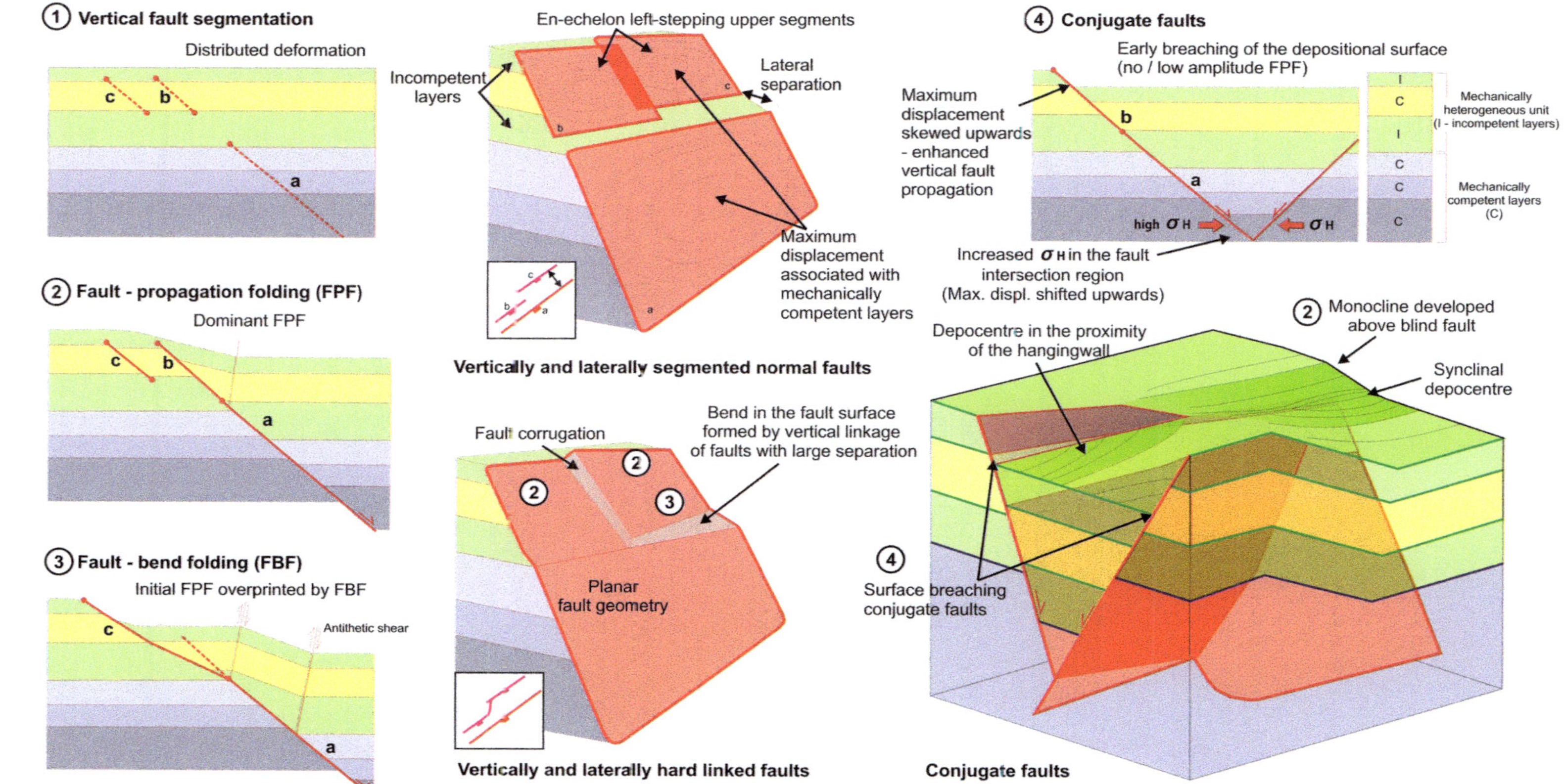

Fig. 15. Schematic model summarizing the mechanisms responsible for generating spatial and temporal variability in normal fault-related folding within IMF. The heterogeneous sedimentary unit favours fault restriction, segmentation and development of fault-propagation folds (1, 2). Linkage of the main deeper fault with the upper en-echelon segments can generate convex-upward fault geometries and further development of fault-bend folding (3) (modified from Lacazette 2001). The bend in the fault plane (and the associated folding) is localized and depends on the lateral separation between the upper segments and the main fault. Conjugate faults tend to breach early the depositional surface without developing significant folds ahead of the propagating upper tip (4).

Pollard 1998; Maerten *et al.* 1999). Maerten *et al.* (1999) used boundary element models to analyse the displacement distribution for Y-shape conjugate normal faults within a homogeneous elastic medium, whilst Young (2001) used finite element models to investigate the slip distribution for V-shape conjugate normal faults within a heterogeneous elastic medium (Fig. 14). Their results showed that conjugate faults are characterized by asymmetric vertical displacement gradients, supporting previous observations from seismic data (Nicol *et al.* 1995; Watterson *et al.* 1998*a*). They postulated that the asymmetry is unlikely to be the result of nucleation of faults on different layers, but rather is the effect of mechanical interaction between the opposite dipping segments. The models showed that conjugate normal faults display asymmetric displacement distributions that vary with distance between the conjugate segments and the mechanical properties of the material (Young 2001). Figure 14c shows how the Poisson's ratio of the layer containing the fault intersection (i.e. the branch line) influences the fault displacement distribution. The threshold of volumetric strain is lower for less compressible rocks (higher Poisson's ratio), resulting in high horizontal compressive stresses within the fault intersection region. In this case, the mechanical models predict an upward shift of the locus of maximum displacement towards the upper fault tip. This skewed displacement distribution, with higher displacement gradients near the upper tip, implies a greater tendency for preferential upward fault propagation. Specifically, previous studies have shown that the spatial energy release rate, which is a measure of the energy required for a fracture to propagate, is directly proportional to displacement and displacement gradients (Aydin & Schultz 1990; Willemse & Pollard 2000).

The displacement analysis of simple and conjugate normal faults from the IMF basin shows that the displacement maxima for conjugate normal faults are shifted upwards in the stratigraphic section, to within mechanical unit 2. In contrast, simple normal faults tend to have displacement maxima within MU 1 (e.g. Fig. 6). The smaller fold amplitude to total throw ratios associated with conjugate faults (Fig. 13) can therefore be explained by high, upward fault propagation rates owing to mechanical interaction between the opposite dipping faults. As a consequence, conjugate normal faults that intersect within layers with low compressibility display geomechanical characteristics favourable for migration of stress concentrations near the upper fault tips. These stress perturbations enhance upward propagation of the fault, generating higher P/S ratios, and result in the early breaching of the free surface, and the development of low-amplitude extensional folds, or no folding at all. Nevertheless, because some of the conjugate pairs may have formed as a result of incidental intersection of opposite dipping faults (Nicol *et al.* 1995), it is possible that the faults initially developed as isolated simple normal faults, without mechanical interaction with other faults, at an incipient stage in their evolution. As a consequence, some conjugate faults, typically displaying Y-type geometries, may exhibit symmetrical displacement distributions and associated fault propagation folding that is similar to simple normal faults. Further analysis of these faults is required to test this hypothesis.

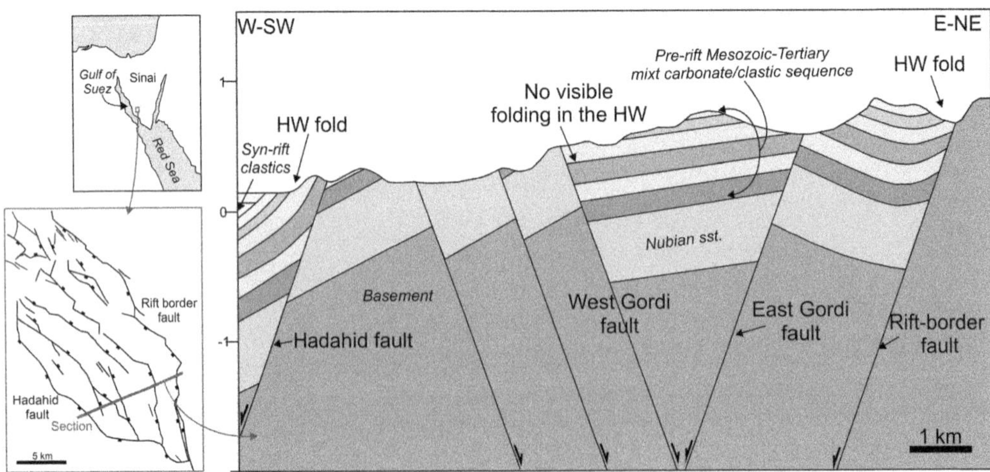

Fig. 16. Geological cross-section from Suez Rift (modified from Whipp 2011). Note the large amplitude hanging-wall folds associated with the simple normal faults (Hadahid Fault, the rift-border fault). The conjugate West and East Gordi faults display little or no folding in their hanging walls.

Our findings show that the development of normal fault-propagation folds can vary significantly within a sedimentary basin and will depend not only on the presence of incompetent layers capable of inhibiting fault propagation and causing vertical fault segmentation, but also on the distribution of stress perturbations caused by mechanically interacting normal faults (Fig. 15). We hypothesize that similar relationships should exist between faults and folds in other extensional basins. Seismic data from the Wytch Farm oil field in the Wessex Basin (southern England) reveal similar, vertically segmented normal faults. Displacement maxima are shifted towards the upper fault tips (within the Middle Jurassic Top Cornbrash sandstones) for conjugate faults, compared with the more symmetrical throw distribution for simple normal faults, which have displacement maxima within the Lower Jurassic Bridport and Triassic Sherwood sandstones (see figs 17 & 18 E in Kattenhorn & Pollard 2001). The same mechanism can potentially explain the variable development of the normal fault propagation-folds seen in other rift settings, such as the Suez Rift, NW Egypt (Fig. 16, from Whipp 2011). Here, folds are poorly developed adjacent to the conjugate West Gordi and East Gordi normal faults. In contrast, large-amplitude breached monoclines are developed adjacent to the 'simple' Hadahid fault or the rift-border fault (Fig. 16, from Whipp 2011). The section in Figure 16 is overly simplified, but Whipp (2011) showed that the faults dip at *c.* 80° within the basement and overlying Nubian sandstone. In the overlying, interbedded sequence, fault dips decrease to 60–70°, and the faults are vertically segmented. It is likely that translation of the hanging-wall monoclines (such as that associated with the Hadahid Fault) across the irregular fault surface contributed to the amplification of the fold amplitude, similar to the example presented from IMF and synthesized in the model shown in Figure 15.

Conclusions

Our observations from the Inner Moray Firth basin show that:

(1) The development of a normal fault-related fold can be explained by the contribution of several mechanisms, the relative importance of which changes during the growth of the normal fault system. The mechanisms evolve from fault-propagation folding, vertical and horizontal segment linkage to fault bend folding (Fig. 15).

(2) The heterogeneous mechanical properties of the host rocks control the fault segmentation and amplitude of fault propagation folding. Shale-rich incompetent layers inhibit fault propagation generating larger-amplitude unbreached monoclines. The larger fold amplitudes observed in the shale-rich, syn-rift sequence (mechanical unit 3) compared with the underlying, interbedded pre-rift sequence of similar thickness (mechanical unit 2) demonstrate the importance of the ratio of incompetent to competent strata (net-to-gross ratio) in arresting upward fault propagation and controlling the magnitude of ductile deformation.

(3) The occurrence and development of the normal fault-related folds is influenced not only by the mechanical stratigraphy and fault geometry, but also by the mechanical interaction between fault segments of a normal fault system. Although incompetent stratigraphic units dominated by weak lithologies can inhibit vertical propagation of the faults, generating vertical segmentation or development of monoclines above the fault vertical tip-lines, we showed that some faults can breach the free surface very early without developing fault-tip monoclines. The variability of normal fault-related folding can be explained by the enhanced vertical propagation owing to mechanical interactions between opposite dipping normal faults.

We are grateful to Ithaca Energy and Dave Brett for releasing proprietary seismic data for publication. We are thankful to Badley Geoscience Ltd and Dave Quinn for providing a licence and training for TrapTester software. We also thank Schlumberger for providing an academic licence for Petrel, Midland Valley for providing a licence for Move software and Foster Findlay Associates Ltd for providing access to GeoTeric software. We thank Al Lacazette, Paul Whipp and Scott Young for the permission to use their figures in this paper. Jonathan Imber acknowledges a Royal Society Industry Fellowship with Badley Geoscience Ltd and Geospatial Research Ltd. Christopher Jackson, Douglas Paton and Tom Manzocchi provided constructive reviews that significantly improved the clarity of the manuscript.

References

Aydin, A. & Schultz, R.A. 1990. Effect of mechanical interaction on the development of strike-slip faults with echelon patterns. *Journal of Structural Geology*, **12**, 123–129.

Bonini, L., Basili, R., Toscani, G., Burrato, P., Seno, S. & Valensise, G. 2015. *Journal of Structural Geology*, **74**, 148–158.

Cardozo, N., Bhalla, K., Zehnder, A.T. & Allmendinger, R.W. 2003. Mechanical models of fault propagation folds and comparison to the trishear kinematic model. *Journal of Structural Geology*, **25**, 1–18.

Childs, C., Nicol, A., Walsh, J.J. & Watterson, J. 1996. Growth of vertically segmented normal faults. *Journal of Structural Geology*, **18**, 1839–1397.

Childs, C., Nicol, A., Walsh, J.J. & Watterson, J. 2003. The growth and propagation of synsedimentary faults. *Journal of Structural Geology*, **25**, 633–648.

Childs, C., Manzocchi, A., Nicol, A., Walsh, J.J., Conneally, J.C. & Delogkos, E. 2016. The relationship between normal drag, relay ramp aspect ratio and fault zone structure. *In*: Childs, C., Holdsworth, R.E., Jackson, C.A.-L., Manzocchi, T., Walsh, J.J. & Yielding, G. (eds) *The Geometry and Growth of Normal Faults*. Geological Society, London, Special Publications, **439**. First published online August 17, 2016, https://doi.org/10.1144/SP439.16

Chopra, S. & Marfurt, K.J. 2007. *Seismic attributes for prospect identification and reservoir characterization*. Geophysical Developments Series. SEG/EAGE Publisher.

Corfield, S. & Sharp, I.R. 2000. Structural style and stratigraphic architecture of fault propagation folding in extensional settings: a seismic example from the Smørbukk area, Halten Terrace, Mid-Norway. *Basin Research*, **12**, 329–341.

Cowie, P.A. 1998. A healing-reloading feedback control on the growth rate of seismogenic faults. *Journal of Structural Geology*, **20**, 1075–1087.

Crider, J.G. & Pollard, D.D. 1998. Fault linkage: three-dimensional mechanical interaction between echelon normal faults. *Journal of Geophysical Research*, **103**, 373–391.

Davies, R.J., Turner, J.D. & Underhill, J.R. 2001. Sequential dip-slip movement during rifting: a new model for the evolution of the Jurassic trilete North Sea rift system. *Petroleum Geoscience*, **7**, 371–388, https://doi.org/10.1144/petgeo.7.4.371

Davis, G.H. 1984. *Structural Geology of Rocks and Regions*. John Wiley & Sons, New York.

Dula, W.D. 1991. Geometric models of listric normal faults and rollover folds. *American Association of Petroleum Geologists Bulletin* **75**, 1609–1625.

Ferrill, D.A. & Morris, A.P. 2008. Fault zone deformation controlled by mechanical stratigraphy, Balcones fault system, Texas. *American Association of Petroleum Geologists Bulletin*, **92**, 359–380.

Ferrill, D.A., Stamatakos, J.A. & Sims, D. 1999. Normal fault corrugation: implications for growth and seismicity of active normal faults. *Journal of Structural Geology*, **21**, 1027–1038.

Ferrill, D.A., Morris, A.P., Sims, D.W., Waiting, D.J. & Hasegawa, S. 2005. Development of synthetic layer dip adjacent to normal faults. *In*: Sorkhabi, R. & Tsuji, Y. (eds) *Faults, Fluid Flow, and Petroleum Traps*. American Association of Petroleum Geologists, Memoirs, **85**, 125–138.

Ferrill, D.A., Morris, A.P. & Smart, K.J. 2007. Stratigraphic control on extensional fault propagation folding: big Brushy Canyon Monocline, Sierra Del Carmen, Texas. *In*: Jolley, S.J., Barr, D., Walsh, J.J. & Knipe, R.J. (eds) *Structurally Complex Reservoirs*. Geological Society, London, Special Publications, **292**, 203–217, https://doi.org/10.1144/SP292.12

Ferrill, D.A., Morris, A.P. & McGinnis, R.N. 2012. Extensional fault-propagation folding in mechanically layered rocks: the case against the frictional drag mechanism. *Tectonophysics*, **576–577**, 78–85.

Fossen, H. & Hesthammer, J. 1998. Structural geology of the Gullfaks Field, northern North Sea. *In*: Coward, M.P., Daltban, T.S. & Johnson, H. (eds) *Structural Geology in Reservoir Characterization*. Geological Society, London, Special Publications, **127**, 231–261, https://doi.org/10.1144/GSL.SP.1998.127.01.16

Glennie, K., Higham, J. & Stemmerik, L. 2003. Permian. *In*: Evans, D., Graham, C., Armour, A. & Bathurst, P. (eds) *The Millennium Atlas: Petroleum Geology of the Central and Northern North Sea*. Geological Society, London, 91–103.

Goldsmith, P.J., Hudson, G. & Van Veen, P. 2003. Triassic. *In*: Evans, D., Graham, C., Armour, A. & Bathurst, P. (eds) *The Millennium Atlas: Petroleum Geology of the Central and Northern North Sea*. Geological Society, London, 105–127.

Groshong, R.H. 1989. Half-graben structures: balanced models of extensional fault-bend folds. *Geological Society of America Bulletin*, **101**, 96–105.

Gupta, A. & Scholz, C.H. 2000. A model of normal fault interaction based on observations and theory. *Journal of Structural Geology*, **22**, 865–879.

Gupta, S., Cowie, P.A., Dawers, N.H. & Underhill, J.R. 1998. A mechanism to explain rift-basin subsidence and stratigraphic patterns through fault-array evolution. *Geology*, **26**, 595–598.

Hardy, S. & Allmendinger, R. 2011. Trishear. A review of kinematics, mechanics, and applications. *In*: McClay, K., Shaaw, J. & Suppe, J. (eds) *Thrust Fault-related Folding*. American Association of Petroleum Geologists, Memoirs, **94**, 95–119.

Hardy, S. & McClay, K.R. 1999. Kinematic modeling of extensional fault propagation folding. *Journal of Structural Geology*, **21**, 695–702.

Hongxing, G. & Anderson, J.K. 2007. Fault throw profile and kinematics of normal fault: conceptual models and geologic examples. *Geological Journal of China Universities*, **13**, 75–88.

Iacopini, D. & Butler, R.W.H. 2011. Imaging deformation in submarine thrust belts using seismic attributes. *Earth and Planetary Science Letters*, **302**, 414–422.

Jackson, C.A.-L. & Rotevatn, A. 2013. 3D seismic analysis of the structure and evolution of a salt-influenced normal fault zone: a test of competing fault growth models. *Journal of Structural Geology*, **54**, 215–234.

Jackson, C.A.-L., Gawthorpe, R.L. & Sharp, I.R. 2006. Style and sequence of deformation during extensional fault-propagation folding: examples from the Hammam Faraun and El-Qaa fault blocks, Suez Rift, Egypt. *Journal of Structural Geology*, **28**, 519–535.

Janecke, S.U., Vandenburg, C.J. & Blankenau, J.J. 1998. Geometry, mechanisms and significance of extensional folds from examples in the Rocky Mountain Basin and Range province, U.S.A. *Journal of Structural Geology*, **20**, 841–856.

Jin, G. & Groshong, R.H., Jr. 2006. Trishear kinematic modeling of extensional fault propagation folding. *Journal of Structural Geology*, **28**, 170–183.

Johnson, K.M. & Johnson, A.M. 2002. Mechanical models of trishear-like folds. *Journal of Structural Geology*, **25**, 1–18.

JONES, R.R., KOKKALAS, S. & MCCAFFREY, K.J.W. 2009. Quantitative analysis and visualisation of non-planar fault surfaces using terrestrial laser scanning (LIDAR) – the Arkitsa fault, central Greece, as a case study. *Geosphere*, **5**, 465–482.

KATTENHORN, S.A. & POLLARD, D.D. 2001. Integrating 3-D seismic data, field analogues and mechanical models in the analysis of segmented normal faults in the Wytch Farm oil field, southern England, United Kingdom. *American Association of Petroleum Geologists Bulletin*, **85**, 1183–1210.

KHALIL, S.M. & MCCLAY, K.R. 2002. Extensional fault-related folding, northwestern Red Sea, Egypt. *Journal of Structural Geology*, **24**, 743–762.

LACAZETTE, A. 2001. Extensional fault bend folding: implications for localized fracturing. http://www.naturalfractures.com/1.3.2.htm

LINSLEY, P.N., POTTER, H.C., MCNAB, G. & RACHER, D. 1980. The beatrice field, inner moray firth, UK North Sea. *In*: HALBOUTY, M.T. (ed.) *Giant Oil and Gas Fields of the Decade 1968–1978*. American Association of Petroleum Geologists, Memoirs, **30**, 117–129.

LONG, J.J. & IMBER, J. 2010. Geometrically coherent continuous deformation in the volume surrounding a seismically imaged normal fault-array. *Journal of Structural Geology*, **32**, 222–234.

MAERTEN, L., WILLEMSE, E.J.M., POLLARD, D.D. & RAWNSLEY, K. 1999. Slip distributions on intersecting normal faults. *Journal of Structural Geology*, **21**, 259–271

MANDL, G. 1988. *Mechanics of Tectonic Faulting*. Elsevier, New York.

MANSFIELD, C.S. & CARTWRIGHT, J.A. 1996. High resolution fault displacement mapping from three-dimensional seismic data: evidence for dip linkage during fault growth. *Journal of Structural Geology*, **18**, 249–263.

MARCHAL, D., GUIRAUD, M. & RIVES, T. 2003. Geometric and morphologic evolution of normal fault planes and traces from 2D to 4D data. *Journal of Structural Geology*, **25**, 135–158.

MCCLAY, K.R. & SCOTT, A.D. 1991. Experimental models of hangingwall deformation in ramp–flat listric extensional fault systems. *Tectonophysics*, **188**, 85–96.

NICOL, A., WALSH, J.J., WATTERSON, J. & BRETAN, P.G. 1995. Three-dimensional geometry and growth of conjugate normal faults. *Journal of Structural Geology*, **17**, 847–862.

NICOL, A., WALSH, J.J., WATTERSON, J. & UNDERHILL, J.R. 1997. Displacement rates of normal faults. *Nature*, **390**, 157–159.

ROBERTS, A.M., BADLEY, M.E., PRICE, J.D. & HUCK, I.W. 1990. The structural history of a transtensional basin: inner Moray Firth, NE Scotland. *Journal of Geological Society, London*, **147**, 87–103, https://doi.org/10.1144/gsjgs.147.1.0087

ROTEVATN, A. & JACKSON, C.A.-L. 2014. 3D structure and evolution of folds during normal fault dip linkage. *Journal of Geological Society, London*, **171**, 821–829, https://doi.org/10.1144/jgs2014-045

RYKKELID, E. & FOSSEN, H. 2002. Layer rotation around vertical fault overlap zones: observations from seismic data, field examples, and physical experiments. *Marine and Petroleum Geology*, **19**, 181–192.

SALTZER, D. & POLLARD, D.D. 1992. Distinct element modelling of structures formed in sedimentary overburden by extensional reactivation of basement normal faults. *Tectonics*, **11**, 165–174.

SCHLISCHE, R.W. 1995. Geometry and origin of fault-related folds in extensional settings. *American Association of Petroleum Geologists Bulletin*, **79**, 1661–1678.

SCHÖPFER, M.P.J., CHILDS, C. & WALSH, J.J. 2007. Two-dimensional distinct element modeling of the structure and growth of normal faults in multilayer sequences: 2. Impact of confining pressure and strength contrast on fault zone geometry and growth. *Journal of Geophysical Research*, **112**, B10404.

SHARP, I., GAWTHORPE, R., UNDERHILL, J. & GUPTA, S. 2000. Fault-propagation folding in extensional settings: examples of structural style and synrift sedimentary response from the Suez rift, Sinai, Egypt. *Geological Society of America Bulletin*, **112**, 1877–1899.

SKUCE, A.G. 1996. Forward modelling of compaction above normal faults: an example from the Sirte Basin, Libya. *In*: BUCHANAN, P.G. & NIEUWLAND, D.A. (eds) *Modern Developments in Structural Interpretation, Validation and Modelling*. Geological Society, London, Special Publications, **99**, 135–146, https://doi.org/10.1144/GSL.SP.1996.099.01.11

STEVENS, V. 1991. The Beatrice Field, block 11/30a, UK North Sea. *In*: ABBOTS, I.L. (ed.) *United Kingdom Oil and Gas Fields 25 Years Commemorative Volume*. Geological Society, London, Memoirs, **14**, 245–252, https://doi.org/10.1144/GSL.MEM.1991.014.01.30

TAVANI, S. & GRANADO, P. 2015. Along-strike evolution of folding, stretching and breaching of supra-salt strata in the Platforma Burgalesa extensional forced fold system (northern Spain). *Basin Research*, **27**, 573–585.

THOMSON, K. & UNDERHILL, J.R. 1993. Controls on the development and evolution of structural styles in the Inner Moray Firth Basin. *In*: PARKER, J.R. (ed.) *Petroleum Geology of Northwest Europe: Proceedings of the 4th Conference*. Geological Society, London, 1167–1178, https://doi.org/10.1144/0041167

THORSEN, C.E. 1963. Age of growth faulting in southeast Louisiana. *Gulf Coast Association of Geological Societies*, **13**, 103–110.

TVEDT, A.B.M., ROTEVATN, A., JACKSON, C.A.-L., FOSSEN, H. & GAWTHORPE, R.L. 2013. Growth of normal faults in multilayer sequences: a 3D seismic case study from the Egersund Basin, Norwegian North Sea. *Journal of Structural Geology*, **55**, 1–20.

UNDERHILL, J.R. 1991. Implications of Mesozoic–Recent basin development in the western Inner Moray Firth, UK. *Marine and Petroleum Geology*, **8**, 359–369.

WALSH, J.J. & WATTERSON, J. 1987. Distributions of cumulative displacement and seismic slip on a single normal fault surface. *Journal of Structural Geology*, **9**, 1039–1046.

WALSH, J.J. & WATTERSON, J. 1991. Geometric and kinematic coherence and scale effects in normal fault systems. *In*: ROBERTS, A.M., YIELDING, G. & FREEMAN, B. (eds) *The Geometry of Normal Faults*. Geological Society, London, Special Publications, **56**, 193–203, https://doi.org/10.1144/GSL.SP.1991.056.01.13

WALSH, J.J., CHILDS, C., IMBER, J., MANZOCCHI, T., WATTERSON, J. & NELL, P.A.R. 2003. Strain localisation and population changes during fault system growth within the Inner Moray Firth, Northern North Sea. *Journal of Structural Geology*, **25**, 307–315.

WATTERSON, J., NICOL, A., WALSH, J.J. & MEIER, D. 1998*a*. Strains at the intersection of synchronous conjugate normal faults. *Journal of Structural Geology*, **20**, 363–370.

WATTERSON, J., CHILDS, C. & WALSH, J.J. 1998*b*. Widening of fault zones by erosion of asperities formed by bed-parallel slip. *Geology*, **26**, 71–74.

WHIPP, P.S. 2011. *Fault-propagation folding and the growth of normal faults*. PhD thesis, Imperial College, London.

WHITE, I.R. & CRIDER, J.G. 2006. Extensional fault-propagation folds: mechanical models and observations from the Modoc Plateau, northern California. *Journal of Structural Geology*, **28**, 1352–1370.

WILLEMSE, E.J.M. & POLLARD, D.D. 2000. Normal fault growth: evolution of tipline shapes and slip distribution. *In*: LEHNER, F.K. & URAI, J.L. (eds) *Aspects of Tectonic Faulting*. Springer, Berlin, 193–226.

WILLEMSE, E.J.M., POLLARD, D.D. & AYDIN, A. 1996. Three-dimensional analyses of slip distributions on normal fault arrays with consequences for fault scaling. *Journal of Structural Geology*, **18**, 295–309.

WITHJACK, M.O. & CALLAWAY, S. 2000. Active normal faulting beneath a salt layer: an experimental study of deformation patterns in the cover sequence. *American Association of Petroleum Geologists Bulletin*, **84**, 627–651.

WITHJACK, M.O., OLSON, J. & PETERSON, E. 1990. Experimental models of extensional forced folds. *American Association of Petroleum Geologists Bulletin*, **74**, 1038–1054.

XIAO, H. & SUPPE, J. 1992. Origin of rollovers. *American Association of Petroleum Geologists Bulletin*, **76**, 509–529.

YOUNG, S.S. 2001. *Geometry and mechanics of normal faults with emphasis on 3D seismic data, conjugate faults, and the effects of sedimentary layering*. PhD thesis, Stanford University.

ZIEGLER, P.A. 1990. *Geological Atlas of Western and Central Europe*. Shell Internationale Petroleum Maatschappij.

ZIESCH, J., ARUFFO, C.M. *ET AL.* 2015. Geological structure and kinematics of normal faults in the Otway Basin, Australia, based on quantitative analysis of 3-D seismic reflection data. *Basin Research*, https://doi.org/10.1111/bre.12146

The brittle and ductile components of displacement along fault zones

C. HOMBERG[1,2]*, J. SCHNYDER[1,2], V. ROCHE[3], V. LEONARDI[4] & M. BENZAGGAGH[5]

[1]*Sorbonne Universités, UPMC Université Paris 06, UMR 7193, Institut des Sciences de la Terre de Paris (ISTeP), F-75005, Paris, France*

[2]*CNRS, UMR 7193, Institut des Sciences de la Terre de Paris (ISTeP), F-75005, Paris, France*

[3]*Fault Analysis Group, School of Earth Sciences, University College of Dublin, Dublin, Ireland*

[4]*HSM Montpellier, UMR 5569, Université de Montpellier, CC57, 34090 Montpellier, France*

[5]*University Moulay Ismail, Faculty of Sciences, BP 11.201, Jbabra, Zitoune, Meknès, Morocco*

**Correspondence: catherine.homberg@upmc.fr*

Abstract: The total offset across a fault zone may include offsets by discontinuous faulting as well as continuous deformation, including fault-related folding. This study investigates the relationships between these two components during fault growth. We established conceptual models for the distributions of displacement due to faulting (i.e. brittle component or near-field displacement), to folding (i.e. ductile component) and to the sum of both (i.e. far-field displacement) for different mechanisms of fault-related folding. We then compared these theoretical displacement profiles with those measured along mesoscale normal faults cutting carbonate-rich sequences in the Southeast Mesozoic sedimentary basin of France. The near-field and far-field displacement profiles follow either a flat-topped or a triangular shape. Several fold mechanisms were recognized, sometimes occurring together along the same fault and represent either fault-propagation folds, shear folds or coherent drag folds. In the last case, local deficit in the fault slip is balanced by folding so that the brittle and ductile components compose together a coherent fault zone. Common characteristics of these faults are a high folding component that can reach up to 75% of the total fault throw, a high displacement gradient (up to 0.5) and a strong fault sinuosity.

Faults in the upper crust form complex zones of deformation where discontinuous offsets along slip surfaces dominate (i.e. brittle deformation). Various amounts of continuous deformation like folding (i.e. ductile deformation) also occur around fault planes (e.g. Ferrill & Morris 2008; Ghalayini *et al.* 2016). These fault-related folds have a major role in several geological issues. First, they are key elements for petroleum systems as they may form structural traps and affect reservoir communication across the faults (e.g. Ferrill *et al.* 2005). Second, they account for part of the co-seismic deformation of the Earth surface and provide useful information to assess earthquake hazards (e.g. Allmendinger & Shaw 2000). Third, the stratigraphic response to synsedimentary faulting is strongly influenced by the folding component (e.g. Gawthorpe *et al.* 1997).

Fault-related folds are observed in all tectonic settings and have been classified according to their mechanical origin or to the geometrical configuration in which they occur (e.g. Schlische 1995; Janecke *et al.* 1998; Ferrill *et al.* 2005; Brandes & Tanner 2014). A short review of the most common ones in extensional and compressional settings is given below. Fault-bend folds (Fig. 1a) are associated with non-planar fault geometries. They have been extensively documented in thrust systems with flat and ramp geometries (e.g. Suppe 1983; McClay 1992; Calamita *et al.* 2012) where layers are passively folded when they pass above a thrust ramp. Similar dip variations are also commonly observed along mesoscale normal faults cutting layered media (Mandl 1988; Ferrill & Morris 2003; Schöpfer *et al.* 2007; Roche *et al.* 2012*a*). A listric normal fault (Fig. 1b) is a special type of non-planar fault geometry where layers in the hanging wall are folded in a rollover structure. In this case, the layers are concave in the direction of slip (i.e. reverse 'drag').

Folds between fault segments are associated with segmented fault zones (Fig. 1c). In neutral relays, the relay ramp forms a well-expressed tilted surface between the two segments (e.g. Peacock & Parfitt 2002). Folding also occurs in relay zones observed in cross-sections (e.g. Nicol *et al.* 2002; Koledoye *et al.* 2003; Rotevatn & Jackson 2014).

From: Childs, C., Holdsworth, R. E., Jackson, C. A.-L., Manzocchi, T., Walsh, J. J. & Yielding, G. (eds) 2017. *The Geometry and Growth of Normal Faults*. Geological Society, London, Special Publications, **439**, 395–412.
First published online March 1, 2017, https://doi.org/10.1144/SP439.21

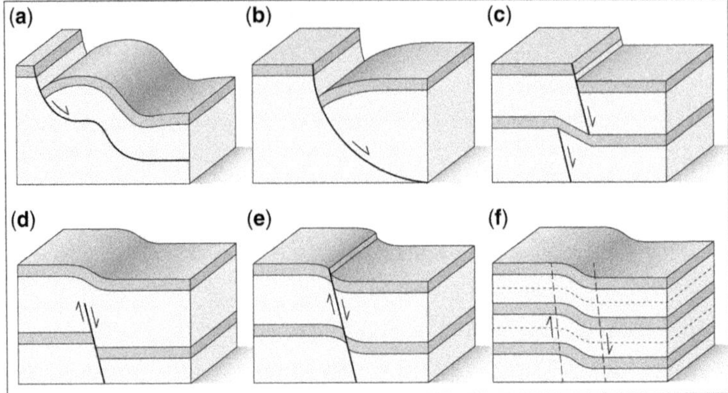

Fig. 1. Diagrams of the common types of fault-related folds: (**a**) fault-bend fold; (**b**) rollover anticline along listric fault; (**c**) fold between fault segments; (**d**) fault-propagation fold; (**e**) drag fold; and (**f**) shear fold. The folding mechanisms in (a) and (c)–(f) are drawn for normal faulting but are also common along reverse faults.

In both cases, the relays may later breach the folded layers.

Fault-propagation folds (Fig. 1d) are generally observed in rocks with contrasted mechanical properties such as sediment infillings lying above a crystalline basement (i.e. forced folds), or clay-rich units alternating with stiff limestones or sandstones (e.g. Withjack *et al.* 1990; Ferrill *et al.* 2007). Usually, the soft sediments are folded ahead of the fault tip due to movement of the hanging wall along a fault imbedded in the rigid unit. Once the folded rock at the fault tip fails, the fault propagates forwards, offsetting the folded layers. Such folding may be localized ahead of a fault that is no longer able to propagate in the vertical direction (Roche *et al.* 2012*b*). Alternatively, the folding may reflect ductile deformation occurring in the course of the vertical propagation of the fault, among other processes occurring at the fault tip.

In their original description, drag folds (Fig. 1e) occur in the fault wall rocks as a consequence of the frictional resistance to slip (e.g. Cloos 1936), but this interpretation has been challenged: Grasemann *et al.* (2005) proposed that drag folds are caused by the heterogeneous displacement field around a fault as it undergoes slip. Ferrill *et al.* (2012) proposed that an apparent 'drag' structure is not the product of frictional sliding: instead, it is originally formed at the fault tip before being offset by the fault. Finally, shear folds (Fig. 1f) develop in shear zones and represent the most ductile expression of fault zones. Shear folds are expected to occur at mid to deep levels of the crust (e.g. Carreras *et al.* 2005), but may also develop during gravitational instabilities or other deformation processes affecting unlithified sediments.

All these types of folds have been recognized at various scales and may present distinct characteristics (see Janecke *et al.* 1998; Schlische 1995; Ferrill *et al.* 2005; Brandes & Tanner 2014 for a review). However, it may be difficult to interpret natural cases of fault-related folds because several types of folding can be localized in the same area. They can form simultaneously due to the superposition of several processes or they can form at different times in the fault history and then become superposed during fault growth.

A routine method to assess fault-related folds is to use kinematic approaches, like the kink fold or trishear models (Suppe 1983; Suppe & Medwedeff 1990; Erslev 1991; Zoetemeijer *et al.* 1992; Allmendinger 1998). Numerical approaches strengthened the need to include a mechanical foundation as deformation is strongly influenced by the mechanical stratigraphy (e.g. Finch *et al.* 2004; Smart *et al.* 2012). However, despite the abundant literature on fault-related folds, quantitative data on the folding component along fault zones are still sparse (Mansfield & Cartwright 1999; Long & Imber 2010; Ferrill *et al.* 2011, 2012), and it is challenging to make a proper distinction between the different folding processes on geometrical criteria.

This study aims to investigate the relationship between the folding and fault-slip components during fault growth. First, we propose theoretical models showing the distribution of the discontinuous offsets (i.e. brittle component or near-field displacement), the folding (or ductile) component of the displacement and the sum of both (i.e. far-field displacement). For each case, we evaluate the degree of overlap or diagnostic values for the different tested fold mechanisms. Then, based on these different models, we examine mesoscale normal faults cutting through carbonate-rich sequences in the Southeast Mesozoic sedimentary basin of France. We do not intend to give an exhaustive description of fold mechanisms or to document them by statistical means. Instead, we aim to evaluate the possible combinations of ductile and brittle deformation, using a methodology that integrates the ductile component in the fault analysis. Finally, we discuss how far-field and near-field displacement–length plots reveal important aspects of fault growth, and emphasize the role of the rheological properties of the faulted lithologies.

Conceptual models of near-field and far-field displacements along blind normal faults

Brittle and ductile components of the fault movement

Conceptual models comparing the brittle component (i.e. through a displacement discontinuity) and the ductile component (i.e. through continuous displacement) of the fault movement are presented below. Three mechanisms of fault-related folding are investigated: fault-propagation folds (Fig. 1d); simple shear folds (Fig. 1f); and a special kind of drag fold (Fig. 1e) referred to as coherent drag folds. These three kinds of folds have their axis sub-parallel to the fault strike and accommodate a continuous offset in sympathy with the fault slip (referred to as normal folding, positive drag, synthetic beds or displacements by various authors: e.g. Mansfield & Cartwright 1999; Ferrill *et al.* 2005; Brandes & Tanner 2014). They affect a variable width of the fault walls depending on their wavelength. Their contributions to the total displacement depend on their amplitude.

The methodology we proposed in this article allows comparison between the predictions of those models with measurements on natural faults observed in cross section. Such measurements can be acquired in the field, from 3D seismic data or from analogue sand-box experiments. Even though we focus here on blind normal faults, our approach can also be applied to reverse faults. We define the near-field displacement as the offset measured at the fault plane (Fig. 2). The far-field displacement is measured at a distance from the fault plane. For the faults considered here, the near-field displacement corresponds to the strain produced by faulting, and the far-field displacement includes both faulting and folding. The folding component of the throw is therefore obtained by subtracting the near-field displacement from the far-field displacement. Similar methodologies have been used for faults observed along their strike in seismic data by Long & Imber (2010).

In all cases considered here, we assume the slip profiles during fault growth are as shown in Figure 2. In the early fault history, the near-field displacement distribution follows a triangular shape (i.e. with constant gradient). During later slip episodes, the displacement is constant along the pre-existing fault portion and decreases linearly to zero with the same displacement gradient along the newly formed fault portion. This model results in a triangular displacement profile at any time during the fault history that fits with the observations on isolated faults growing in a homogenous rock (Muraoka & Kamata 1983; Walsh & Watterson 1987; Soliva *et al.* 2005; Roche *et al.* 2012*b*). Because measuring the bed offset down the fault dip may involve large uncertainties when the fault trace is irregular, we prefer to use the fault throw: that is, the bed offset measured in the vertical. We refer to it below as the fault displacement to follow the nomenclature used in displacement profiles analysis.

Fault-propagation folds

The different displacement profiles along a normal fault in front of which a fault-propagation fold develops are shown in Figure 2a. Along the fault plane, the near-field displacement decreases from a maximum to zero at the fault tips. No folding component occurs along the fault plane so that the far-field displacement equals the near-field displacement. Ahead of the fault tip, the fold (or monocline) accommodates a vertical displacement along a distance referred to as the damping distance. The folding component increases abruptly and remains constant along most of the damping distance before finally falling to zero. The vertical dimension of this distance is likely to be sensitive to rock lithology. This will be discussed further in the 'Discussion' section, later in this paper. Eventually, the fault propagates forwards and cuts the fault-tip fold, which does not then accumulate further strain. This situation is expressed by diverging near-field and far-field curves, the point of divergence marking the initial fault tip (Fig. 2a).

Simple shear folds

A ductile shear zone represents the extreme case of very compliant rocks in which the strain is accommodated entirely through folding. The near-field displacement is thus zero all along the shear zone (Fig. 2b). With increasing strain, subsequent development of a brittle fault cutting the folded zone modifies the displacement profiles. If the displacement was constant along the shear zone, the mathematical translation that links the near-field and far-field data of the final fault zone is easily recognized on the displacement plots. Both near-field and far-field profiles will then show a triangular

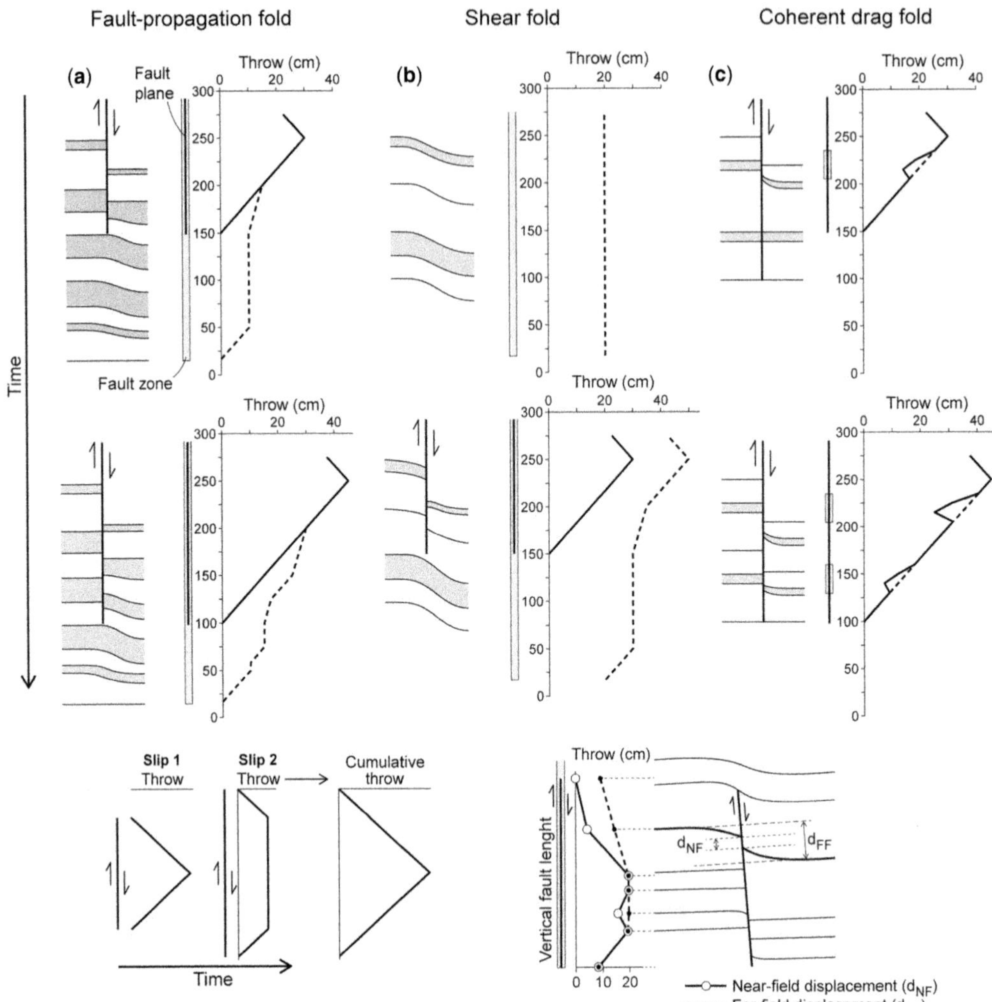

Fig. 2. Conceptual models of displacement distribution for different mechanisms of fault-related folding: (**a**) fault-propagation folds; (**b**) shear folds; and (**c**) coherent drag folds. Inset shows the slip model during fault growth and the procedure for displacement data acquisition. The near-field displacement (d_{NF}) is measured at the fault plane and corresponds to the throw achieved by fault slip. The far-field displacement (d_{FF}) is measured at a distance from the fault plane and accounts for the total offset, and includes both the faulting and folding components. Displacement profiles are constructed by plotting the displacement data along the fault height.

shape. Ahead of the fault plane, the far-field displacement is constant and its value equals the magnitude of the translation vector. If the displacement profile along the shear zone was triangular, the folding component will vary along the fault zone and the far-field displacement profile will become more irregular. The relationship between the final far-field and near-field displacement profiles allow us to distinguish between shear folds later cut by a fault and fault-propagation folds (cf. Fig. 2a, b). In particular, faults in which shear folds developed always show far-field displacement values that are greater than the near-field ones, even at the point of maximal far-field displacement.

Coherent drag folds

The last case considers a fault zone along which folding and faulting occur, and combines them together at specific levels. In this case, although the brittle and ductile components of the total offset may fluctuate along the fault zone (Fig. 2c), these fluctuations are complementary and when summed yield a 'coherent fault zone'. Following the descriptive term of drag fault, we refer to this type of fault-related fold as a 'coherent drag fold'. Faults with coherent drag folds show quite different far-field and near-field profiles. The far-field profile shows a regular triangular shape, whereas the near-field displacement and the folding component exhibit small-scale fluctuations (Fig. 2c). Notably, in the coherent drag-fold model, these fluctuations balance each other so that the total strain (achieved by both folding and faulting) shows a regular far-field profile.

These conceptual models represent ideal cases isolating the effect of each fold type. The three cases considered here show different combinations of near-field and far-field profiles, and the shapes of these profiles can be useful in identifying the mechanism causing the development of fault-related folds, although the geological setting and fault history may be more complex for natural cases than the one described here. In the next section, we will compare the model predictions with data collected on mesoscale faults observed in the field.

Mesoscale normal faults in the Southeast Basin of France

Studied faults and regional context

We studied 12 normal faults in the Southeast sedimentary basin of France (Fig. 3). They have a maximum offset ranging from a few centimetres to a few tens of centimetres, with a variable amount of fault-related folding (Figs 4 & 5). They were observed in cross-sections in five outcrops nearby Sahune and Villeperdrix villages, in the Aulan Gorge, and in a natural excavation in the Agnielles Gorge, respectively (Fig. 3). Five faults cut deposits of Tithonian age in Villeperdrix (i.e. three faults named Fv1, Fv2 and Fv3) and in Agnielles (i.e. two faults named Fag1 and Fag2). Seven other faults

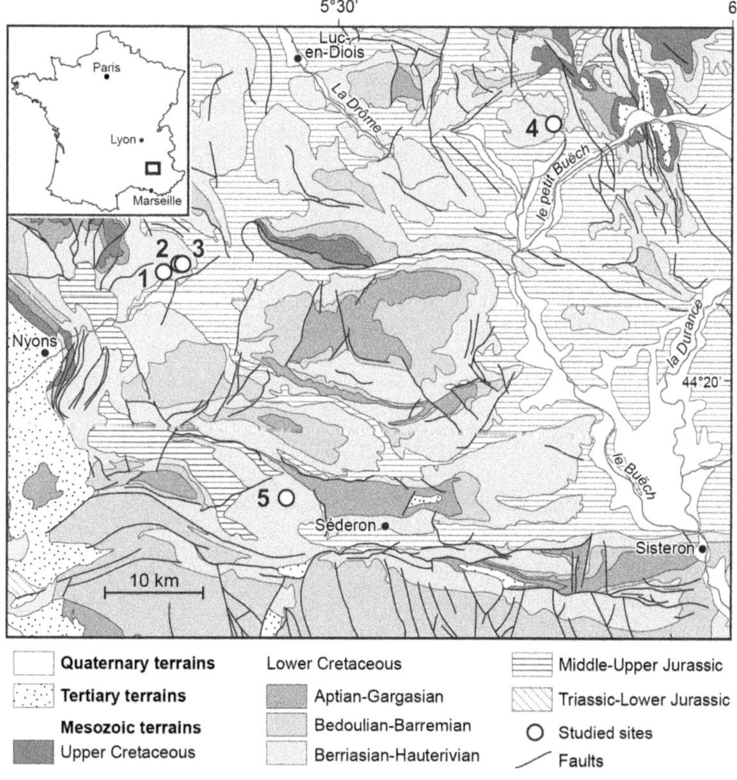

Fig. 3. Simplified geological map and studied sites. Circles with numbers are sites of observation: (1) Sahune (five faults: Fs1, Fs2, Fs3, Fs5 and Fs6); (2) Villeperdrix1 (one fault: Fv1); (3) Villeperdrix2 (two faults: Fv2 and Fv3); (4) Agnielles (two faults: Fag1 and Fag2); and (5) Aulan (two faults: Fau1 and Fau2). Geological contours and faults are from the 1/250 000 geological map of Valence (Rouire *et al.* 1980). The square shows the geographical location of the study area in France.

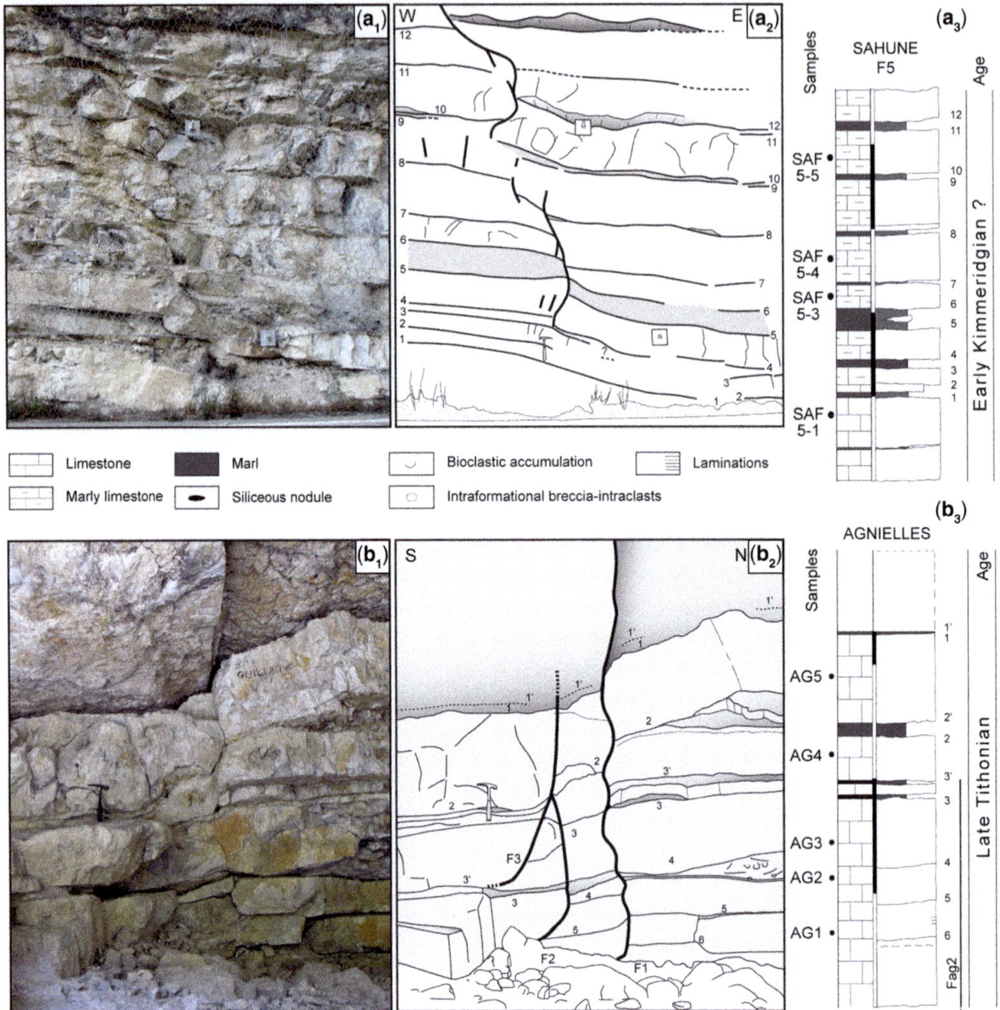

Fig. 4. Faults and hosting rocks. (**a**) & (**b**) Examples of faults Fs5 (at Sahune) and Fag1 (at Agnielles), and the lithology of the rocks. Both uninterpreted (a_1 and b_1) and interpreted photographs (a_2 and b_2) are shown, as well as the stratigraphic logs of the host rocks (a_3 and b_3). Samples used for thin section are shown. Bed surfaces used for displacement measurement are indicated with numbers. The correlation of markers in the central part of Agnielles section is not obvious. The same bed succession is, however, clearly recognized on the southern and northern continuation of the outcrop. There is thus no ambiguity about the cumulative fault offset. The position of Fag2 is indicated with a thick line on the right-hand side of stratigraphic log. The hammer indicates the scale. The scale in stratigraphic logs is indicated with alternating back and white thick lines, each one representing 1 m. See Figure 3 for the site location.

cut the Kimmeridgian and Valanginian sequences in Sahune (i.e. five faults named Fs1, Fs2, Fs3, Fs5 and Fs6) and Aulan (i.e. two faults named Fau1 and Fau2). These stratigraphic intervals are carbonate-rich, pelagic sequences and are described in the next subsection.

The outcrops are located in the Vocontian Trough, an east–west-trending sub-basin in the approximately 40 000 km^2 Southeast Basin of France. The latter developed during Mesozoic time due to the opening of the western Tethys (or ligurian Tethys). Up to 10 km of sediment infill accumulated in the basin (Dubois & Delfaud 1989). The overall basin history is well established: extension lasted from the Early Triassic to the Mid-Cretaceous, with a main rifting phase in Early–Mid Jurassic time, followed by several moderate intensity extensional periods (e.g. Debrand-Passard

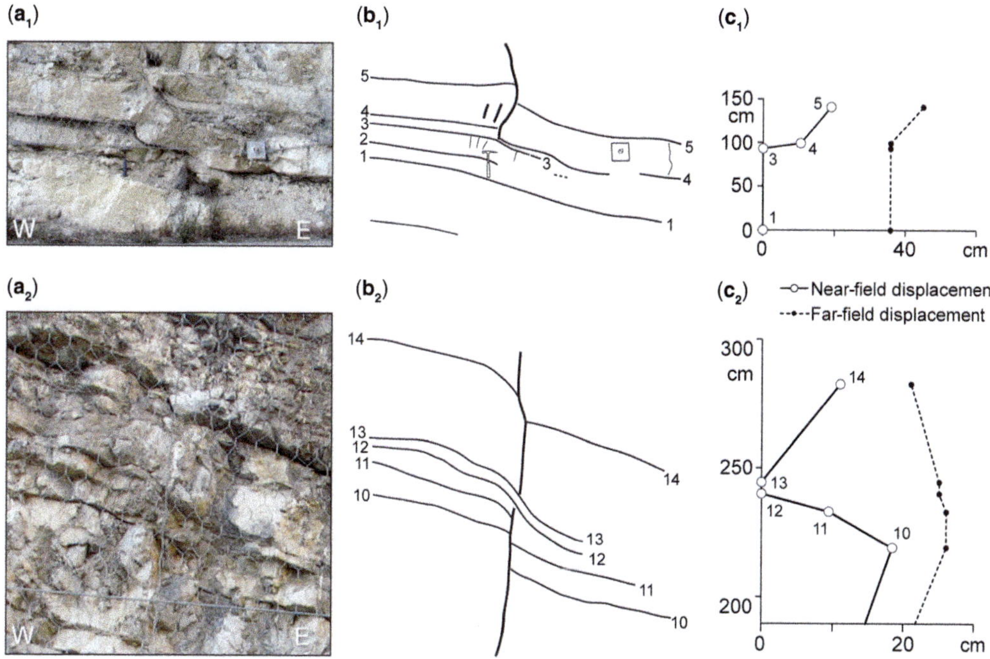

Fig. 5. Examples of fault-related folds. **(a_1)** & **(a_2)** Outcrop photograph of a detailed part of Fs5 and Fs2. **(b_1)** & **(b_2)** and **(c_1)** & **(c_2)** Interpreted photographs and displacement data. Analysis of displacement data along the full length of these faults suggests a composite origin for the folding component. The folds are interpreted as shear folds amplified by fault-propagation folding along Fs5 and coherent drag folding along Fs2. See the section on 'Fault geometry and displacement profiles' and Figure 8 for details.

et al. 1984; Dubois & Delfaud 1989; Homberg *et al.* 2013). During this long-lived extension, faults of various sizes developed, with a main NE–SW trend, and additional east–west and north–south trends (Lemoine *et al.* 2000 and references herein). The north–south faults observed in the Kimmeridgian and Tithonian sequences of Sahune and Villerperdrix are in line with the discrete Late Jurassic event during which synsedimentary faults developed in the Southeast Basin of France (Dardeau *et al.* 1988; Homberg *et al.* 2013). The east–west faults observed in the Late Tithonian and Valanginian sequences at Agnielles and Aulan are thought to be related to the major tectonic reorganization in Early Cretaceous time as a consequence of the opening of the Atlantic Ocean (Graciansky & Lemoine 1988; Homberg *et al.* 2013). At that time, a northwards slope–basin transition developed in the Ventoux–Lure area and sediments were redistributed via deep marine valleys with an approximately east–west trend (e.g. Joseph *et al.* 1989; Friès & Parize 2003; Homberg *et al.* 2013). Later deformation is associated with several periods of inversion, except during the Eo-Oligocene rifting, during which NNE–SSW normal faults developed in Western Europe (Bergerat 1987).

Rock lithology

A bed-by-bed survey has been performed in the field on the Villeperdrix, Agnielles and Sahune outcrops. This was complemented by a petrographical analysis of 31 thin sections (Figs 4 & 6) in order to characterize the main rock features and the sedimentary facies in the microscope. In each outcrop, the section under consideration encompassed all beds cut by the normal faults, and those lying above and below the upper and lower fault tips when visible. Rock dating was based on the microfauna associations because ammonites are rare in the studied sections.

Sahune sections. Fs1, Fs2, Fs3 and Fs6 cut through the same 4.8 m-thick sedimentary pile named here the Sahune Fs1-2 section. The lithology consists mainly of fine-grained limestone beds that are 5–53 cm thick and alternated sometimes with thin lenses of marl usually less than 5 cm thick (Fig. 6a). The limestone beds are composed of micrites (mudstones to wackestones and rare packstones) with rare bioclasts, including radiolarians, *Globochaete*, probable filaments, fragments of crinoids (*Saccocoma*), rare foraminifers and pelecypods.

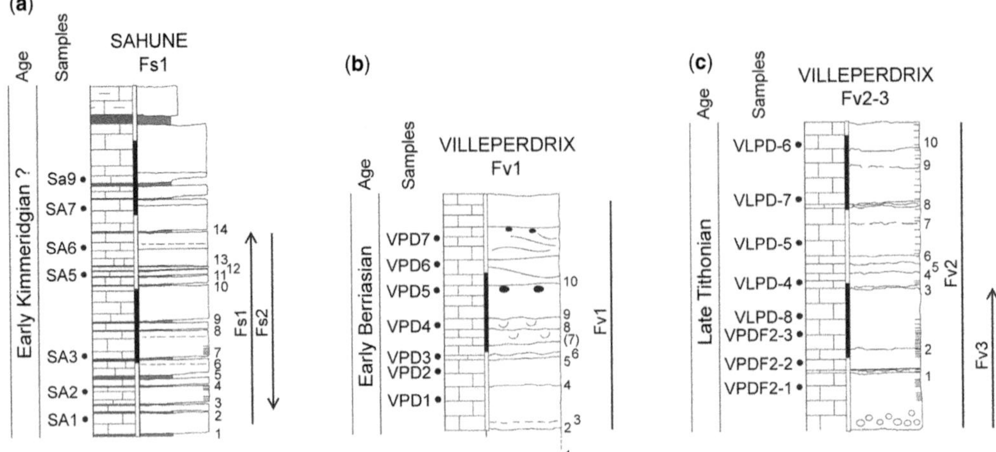

Fig. 6. Stratigraphic logs of the faulted medium. (**a**)–(**c**) The Sahune Fs1-2, Villeperdrix Fv1 and Villeperdrix Fv2-3 sections. Estimated sequence ages are on the left. They are inferred from calcisphaera species, except at the Sahune section. Thin lines indicate the position of the faults within the sequences, with arrows if tips are observed. Bed surfaces used for displacement measurement are indicated with numbers. The scale on the stratigraphic logs is indicated with alternating black and white thick lines, each one representing 1 m. See the legend in Figure 4.

The Sahune Fs5 section is 4.65 m thick and comprises limestone to marly limestone beds (20–59 cm thick) alternating with marly intervals (3–28 cm) (Fig. 4a). The microfacies are almost homogeneous in these beds and consist of micrite (mudstone) with rare bioclasts (radiolarians). This section thus includes several marly intervals, which is different from the Fs1-2 section described previously that consists mostly of limestone beds. In both sections, the limestone beds are generally homogeneous. The absence of *Saccocoma*, together with the rare microfauna, suggest a possible Early Kimmeridgian age, in agreement with the 1:50 000 geological map (Flandrin 1975).

Agnielles section. The Agnielles section is 3.3 m thick and consists of thick well-bedded fine-grained limestones (9–75 cm thick) alternating with thin marly intervals (generally <5 cm and up to 11 cm) (Fig. 4b). The microfacies correspond to *Saccocoma*-rich mudstones to wackestones, together with *Globochaete*. Samples yielded *Chitinoidella boneti* (Doben), which marks the base of the Late Tithonian (equivalent to the *Microcanthum* Ammonite Zone: Benzaggagh & Atrops 1995*a*). Similar to the Sahune sections, the micro-facies in the Agnielles section are rather homogeneous within each limestone bed, as well as from bed to bed.

Villeperdrix sections. The Villeperdrix Fv1 section is 3 m thick and consists of fine-grained limestone beds (5–58 cm thick) (Fig. 6b). This section yielded a high bioclastic content in specific levels, and brecciated layers and nodular chert levels. Limestone beds correspond to mudstones to wackestones often rich in radiolarians, *Globochaete* and calcisphaera. Numerous *Calpionella alpina*, rare *Crassicollaria parvula*, *Tintinnopsella carpathica* and *Remaniella ferasini* give an Early Berriasian age (B2 subzone of calpionellids: Benzaggagh & Atrops 1995*b*). Samples VPD5 yielded reworked intraclasts rich in *Crassicollaria parvula* and small *Calpionella alpina*, indicating a slightly older Early Berriasian age (B1 subzone of calpionellids).

The Villeperdrix Fv2-3 section is 4.18 m thick (Fig. 6c), and contains the Fv2 and Fv3 faults. The lithology consists of coarse-grained, finely laminated limestones passing into fine-grained, often laminated limestones (4–83 cm in thickness). Coarse-grained intervals correspond to micro-conglomeratic beds with infra-millimetric to centimetric mudstone lithoclasts and peloids (packstones) with calcisphaera, which occasionally yield echinoderm-rich laminations. Fine-grained intervals are wackestones with *Globochaete*, radiolarians, calcisphaera and, locally, *Saccocoma* or Aptychi. The genus *Crassicollaria* largely dominates among the *Calpionella alpina*, indicating a Late Tithonian age (A3 subzone of calpionellids: Remane 1971). In the bottom part of the section, sample VPDF2-1 yielded *Saccocoma*, which suggests a Late Tithonian (A2 subzone of calpionellids) age. Sections in Villerperdix thus differ from those in other sites in their grain size which is variable throughout the section and also within each limestone bed.

Aulan section. The Fau1 and Fau2 sections in Aulan consist of fine-grained limestone beds (20–30 cm in thickness) separated by very thin marly intervals (<5 cm thick). These sections were not sampled for micro-facies and fossil identification. According to the outcrop inspection, the limestones are composed of micrites (mudstones–wackestones). The Aulan section resembles the Fs1-2 section in Sahune and that in Agnielles.

Fault data acquisition

The 12 faults studied are generally composed of several closely spaced fractures, usually one or more slip surfaces, as well as thin calcite veins. We examined all fractures from the lower to the upper part of every fault and transcribed them onto scaled photographs (Fig. 4). The fault features were obtained precisely along each fault. When the upper parts of the faults were not accessible with a ladder (such as for Fs1 and Fs2), displacements were measured on pictures scaled with a ruler that was aligned along the normal of the bedding plane using a compass and a clinometer.

For each fault, we measured the strike and dip of the main slip surfaces. The slip vector was measured when visible (which was rare) and the criteria indicating the sense of motion were examined carefully. We located precisely each orientation result within the sedimentary sequence in order to recognize possible changes caused by the mechanical layering such as fault refractions: that is, a change in the fault dip through a bed interface ('bed-to-bed' data). Within a homogeneous bed, the surface of the fault often displays a sinuous shape (i.e. a progressive variation in the dip). In some extreme instances, dip variations are so important that the fault locally changes from normal to apparently reverse offset. In order to characterize the sinuosity inside a bed, we measured the maximum and minimum fault dip and strike values within a bed ('in-bed' data). The uncertainties for the strike and dip measurements were less than 2° for all faults, except for strike data on the faults in Sahune. In this locality, a wire mesh covering the outcrop created a magnetic disturbance resulting in an approximate strike value of the fractures. In all but the Sahune sites, the inclinations of the layering ranged from 5° to 10°. Thus, bed tilting did not significantly modify the fault orientation. In the Eygues Valley, the beds dip up to 20° at the Sahune site. Palaeostress determinations and analysis of conjugate mesoscale faults indicate that the north–south faults in this valley formed prior to tilting (Homberg *et al.* 2013). We have therefore restored the original orientations of the fractures in Sahune using ROTILT (Angelier 1990).

The top and bottom of the limestone beds, as well as a few internal interfaces within the limestones, were used as markers to establish how the displacement varies along each fault (Figs 4 & 5). Because the fault traces showed frequent irregularities in cross-section, we have measured the throw instead of the displacement along the fault plane. The bed offsets were measured very close to and at a distance (generally several decimetres and up to several metres) from the fault plane in order to distinguish the brittle (through a displacement discontinuity) and the ductile (through folding) component of the throw. Displacement data were then plotted against the vertical distance to construct profiles of the near-field and far-field displacements and the folding component, as shown at the beginning of this paper (Fig. 8). This measurement procedure was followed in order to examine how the deformation varies along a fault zone whatever the geometry of the slip surface and for various combinations of brittle and ductile deformation. When the fault included overlapping segments (Fag1), the cumulative displacement profile was also calculated by summing the displacements on each individual fault segment. Uncertainties on displacements and bed-to-bed distances are less than 0.5 cm and slightly higher, but probably less than 2 cm, along the upper parts of Fs1 and Fs2.

Fault geometry and displacement profiles

Fault shape and fault kinematics

The investigated normal faults are partly or completely exposed along vertical cross-sections (Fig. 4; Table 1) and observed along a vertical length ranging from 1 to 3 m (length of observation <30 cm for Fs3). A continuous bed interface marks the lower tip of Fs3, Fs5 and Fau1, and the upper tip of Fau2, Fs1, Fv3 and Fag2. Both tips of Fs2 and Fau2 were observed (with another segment existing ahead Fs2), whereas Fag1, Fs6, Fv1 and Fv2 cut through the whole outcropping section. Faults observed along their entire lengths allow a full description of the relationship between folding and faulting along the fault zone. Those incompletely exposed are also integrated in this study because they provide important quantitative information on the folding component and other key features.

Rare striations observed on these faults (Fag1, Fs1 and Fv2) are subhorizontal and are indicative of strike-slip movement. These striations are attributed due to minor fault reactivation during Late Cretaceous–Cenozoic inversion and are unrelated to the displacements recorded here which developed under a normal faulting regime. There is abundant evidence for Late Jurassic normal faulting in the Southeast Basin of France in the form of stratigraphic thickness changes across north–south faults,

Table 1. *Geometric fault attributes*

	Global attributes							'In-bed' variations		'Bed-to-bed' variations	
	L^1	$D_{max}{}^2$	Ti_l^3	Ti_u^4	F^5	Strike[6]	Dip[7]	ΔStrike[8]	ΔDip[9]	ΔStrike[10]	ΔDip[11]
Fag1	240	47–61			*	114 (105/125), 15	−87 (−58/57), 15	7 (0/17), 6	31 (19/44), 6		31 (9/58), 5
Fag2	234	7.5–7.5		*-Fo	*	131 (120/142), 8	−77 (−65/ − 88), 8	1 (0–2), 2	12 (11/13), 2		8 (0/13), 4
Fau1	47	7.5–7.5	*			111 (105/112), 5	76 (−82/43), 5	<	24 (0/42), 3		
Fau2	102	3.5–7.5		*		79 (74/84), 3	85 (−75/68), 8		18 (0/28), 3		13 (6/19), 2
Fs1	287	18.5–22		*	*	166	86 (−78/70), 9		3 (0/14), 6		10 (2/32), 6
Fs2	214	17–25.5	*-Fo	*	*	164	−89 (−80/78), 8		1(0/6), 6		7 (0/22), 5
Fs5	296	37–50	*-Fo		*	176 (161/18), 4	90 (−59/64), 9		27(9/67), 4		10 (0/20), 2
Fs3	27	1.5–9	*-Fo		*	108 (80/135), 2	88 (87/89), 2				
Fs6	119	15–24			*	150 (131/174), 5	74 (−84/62), 5		<		20 (13/33), 3
Fv1	196	27–28			*	3 (162/22), 11	81 (−86/54), 11	15 (9–21), 2	21 (2/40), 2	9 (1/15), 3	5 (1/9), 3
Fv2	290	23–26.5			*	143 (135/150), 9	74 (−72/22), 9	7 (3–12), 3	21 (15/26), 3	6 (2/10), 2	47 (28/66), 2
Fv3	88	13.5–13.5		*		20 (15/25), 2	74 (−88,57), 2				

[1]Vertical length of the fault.
[2]Maximal near-field and far-field observed displacements.
[3, 4]Lower and upper tips observed with fold ahead (Fo) or not.
[5]Fold achieving part of far-field displacement. Note that * in [3, 4, 5] denotes positive items.
[6, 7]Fault strike and dip.
[8, 9]Variations within one bed of the fault strike and the fault dip.
[10, 11]Variations through a bed interface of the fault strike and the fault dip. In [6–11], the first number and the two numbers in brackets are the mean, and the minimum and maximum values. The last number indicates the number of data.
< denotes negligible variations.

and Jurassic and Early Cretaceous gravity deposits (e.g. slumps, debris flows) that indicate permanent slope instabilities associated with recurrent tectonic activity (Dardeau *et al.* 1988; Joseph *et al.* 1989; Friès & Parize 2003; Courjault *et al.* 2011; Homberg *et al.* 2013). Normal faulting of the Kimmeridgian–Valanginian sequences exposed in the study area is not reflected in striations on the fault surfaces, a feature we attribute to the weakly lithified state of the sediments at the time of faulting.

Notably, most of the studied faults show a sinuous shape, as observed in cross-section, with large variations in their dips. These variations are up to 55°, 65°, 67° and 86° along Fau1, Fag1, Fs5 and Fv2, respectively (Fig. 7; Table 1). Fault refraction as a function of the lithology, referred to here as the 'bed-to-bed' sinuosity, contributes to this irregular fault shape. The largest dip variations are observed within individual beds ('in-bed' fault sinuosity), so that the fault dip progressively increases or decreases, sometimes even changing its dip direction across a bed. Note that, except in the Villeperdrix sections, each limestone bed is generally homogeneous. In addition, 'in-bed' fault sinuosity varies along a given fault, so that straight, sinuous and very sinuous segments occur even on faults that cut a rather homogeneous sequence. This is particularly obvious on Fag1, Fau1, Fau2, Fs5 and Fv1, for which the 'in-bed' sinuosity (variations within individual beds) range equals 19°–44°, 0°–42°, 0°–28°, 9°–67°, 2°–40°, respectively (Fig. 7a; Table 1). Thus, there seems to be no correlation between the very pronounced fault sinuosity and the mechanical layering as observed today.

Near-field displacement profiles

Displacement data measured in the direct vicinity of the faults were used to construct near-field displacement profiles. Although the faults are generally only partly exposed, characteristic displacement patterns are recognized (Fig. 8). An abrupt decrease in displacement is observed at the lower tips of Fs5 and Fau1, the upper tips of Fs1 and Fag2, and both tips of Fs2 and Fau2 (but with another segment ahead of the tip). An overall upwards and downwards decrease from a maximal value (D_{max} point) is observed on Fs2, Fs5, Fv1, Fv2, Fag2 and Fau2, which is likely to indicate that the central parts of these faults have been mapped. Although the profiles may be irregular in detail, their overall shapes exhibit either a triangular or a flat-topped profile, defined according to the following nomenclature. Triangular profiles are composed of two straight lines along which the displacement decreases from the D_{max} point to zero at the two fault tips, the displacement gradient is thus constant along each half of the fault. Flat-topped profiles have a long, low-displacement gradient, central plateau and much higher gradients near the fault tips. The first type characterizes isolated faults in homogeneous media, whereas flat-topped profiles denote complexities in fault growth, such as restriction by lithological interface or interaction with neighbouring faults (Peacock 1991; Wilkins & Gross 2002; Roche *et al.* 2012*b*).

Fs2, Fs5 and Fv2 show half near-field triangular profiles along their exposed lengths, and isolated tips with gradients equal to 0.1, 0.34 and 0.22,

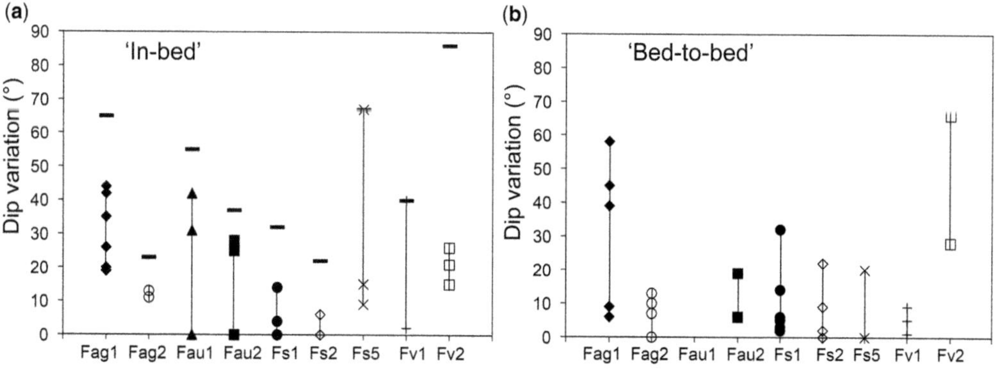

Fig. 7. Fault dip variability v. mechanical layering. (**a**) 'In-bed' dip variations. Each symbol indicates the maximum change of fault dip within one homogeneous limestone bed. The short and thick horizontal bar denotes the maximum variation observed on the exposed part of the fault. (**b**) 'Bed-to-bed' dip variations. Each symbol indicates the change in the fault dip from one side to another of a bed interface. In most cases, the 'bed-to-bed' variations are moderate, whereas the 'in-bed' variations are high, indicating that changes in the fault dip barely correlate with the mechanical stratigraphy but occur within a homogeneous bed. Fag1 and Fag2, faults in Agnielles; Fau1 and Fau2, faults in Aulan; Fs1, Fs2, Fs3, Fs5 and Fs6, faults in Sahune; Fv1, fault in Villeperdrix1; Fv2 and Fv3, faults in Villeperdrix2. See also Table 1. See Figure 3 for the site location.

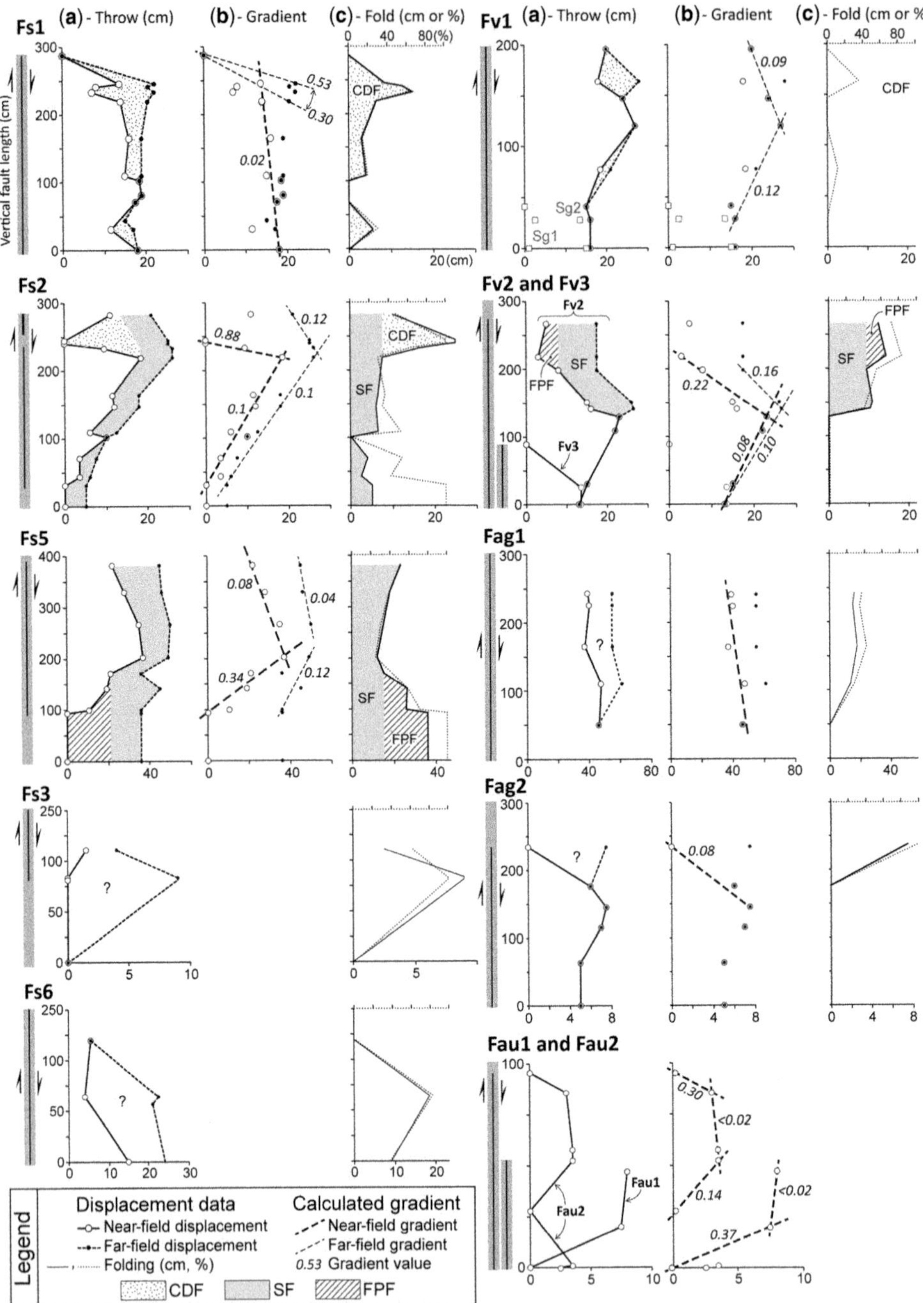

Fig. 8. Displacement data. (**a**) Displacement profiles. (**b**) Gradient data. (**c**) Contribution of folding to the total displacement. Near-field and far-field displacements represent the displacement achieved, respectively, by faulting and by both faulting and folding (see Fig. 2 for the measurement of displacement data). Dashed lines in (b) show how local near-field and far-field gradient values discussed in the text are calculated. The left thin black line and thick grey band denote the vertical extent of the fault and of the folded zone. The folding component is generally above 20% and greater than 40% along several fault portions. Shading indicates the amount of folding associated with coherent drag folding (CDF), shear folding (SF) or fault-propagation folding (FPF). A dominant mechanism explains most of the folding component of a given fault, but a second mechanism may be superposed locally. See also Table 2.

respectively (Fig. 8). Although the profiles of Fs5 and Fv2 are incomplete, these faults are clearly asymmetrical. Available data along the partly exposed halves of these faults yield a gradient of 0.08 for both. If these gradients are used to extrapolate the displacement profile to the supposed isolated fault tip, a local gradient up to four times greater than that on the exposed half will be noticed, indicating an asymmetry in the profile. Fs2 is characterized by an even more pronounced asymmetry, with gradients equal to 0.88 and 0.1 along the upper and lower fault tips, attributed to interaction with the other fault segment on the upper part of the section. Fs1, Fag2 and Fau1 show flat-topped profiles along their upper, upper and lower tips, respectively, with near-tip gradients equal to 0.3 (or more), 0.08 and 0.37. Fau2 shows a complete flat-topped profile with near-tip gradients equal to 0.14 (lower tip) and 0.30 (upper tip). The local gradient near the central part of the fault is estimated to be below 0.02 on Fs1, Fau1 and Fau2. All these gradient values are subject to uncertainties related to the limited exposure of most faults, as well as the distance separating displacement markers. Despite this limitation, the values obtained here show some consistency. In fine-grained homogenous carbonates (the Sahune, Aulan and Agnielles sections: see 'Mesoscale normal faults in the Southeast Basin of France' earlier in this paper), the near-tip gradient data define two peaks, at 0.08–0.14 and at 0.3–0.4. The lower peak is close to values calculated by others along isolated faults in similar lithologies (Soliva *et al.* 2005; Roche *et al.* 2012*b*). In the Villeperdrix sections, where coarse beds are also present, they range between 0.08 and 0.22, without an apparent preferred value.

Far-field displacement profiles and fault-related folding

Fault-related folds were observed along all the studied faults (e.g. see Fig. 5), except for Fau1, Fau2 and Fv3. The displacement data discussed combine near-field data (i.e. the discontinuous offset: Fig. 8a), far-field data (i.e. the sum of the displacement achieved by faulting and folding: Fig. 8a), as well as the folding contribution to the total bed offset, referred to as the folding component (Fig. 8c). The folding component is variable along each fault, generally above 20% and greater than 40% along half or more of the exposed lengths of Fs1, Fs2, Fs5, Fs3, Fs6 and Fv2. A large part of the displacement along the investigated fault zones is thus achieved by folding, although the host rocks consist mostly of fine-grained limestones beds which may be expected to be stiff. Below the lower tip of Fs2, Fs5 and Fs3, and above the upper tip of Fs2 and Fag2, the folding component reaches 100%.

The closely spaced measurements along Fs1, Fs2, Fs5, Fv2, Fag1 and Fag2 allow comparison between the far-field and near-field displacement distribution. The overall shape of the far-field displacement profiles along these faults is similar to that of the near-field profiles. Along Fs1 and Fs2, small-scale fluctuations of the near-field displacement are superposed on the overall flat-topped and triangular shapes of the profiles, whereas the far-field profiles are much less variable. Generally, variations in the displacements achieved by folding and faulting are complementary and balance each other partly or fully. This is highlighted on the displacement profiles by the coincidence of local maxima in the near-field displacement with minima in the folding component curves (cf. Fig. 8a, b), whereas the far-field curves are almost straight. The amplitude of these fluctuations is generally moderate, but can occasionally be very high similar to that observed along the upper part of Fs2. In this case, the near-field displacement is locally zero and the vertical displacement of the beds there is achieved exclusively by folding. These relationships between the brittle and ductile components of the deformation highlight that both of these components are linked and together constitute a coherent fault zone. Fs1 and Fs2 thus exhibit coherent drag folds, as described in the earlier section 'Conceptual models of near-field and far-field displacements along blind normal faults' (Fig. 2c).

In addition to the small-scale fluctuations described above, other patterns of the displacement and folding component curves are observed. Below the lower tip of Fs5 and the upper tip of Fv2 (2 cm of brittle displacement still observed), the folding component curves show a plateau ahead of these fault tips, a configuration that is consistent with a fault-propagation fold (Fig. 2a). The folded zone extends vertically along a length that represents at least 25% of the fault height. A similar mechanism may explain the fold observed ahead of the upper tip of Fag2, but there the fold amplitude increases slightly away from the tip.

The near-field and far-field displacement data show similar relationships below the lower tip of Fs2, so that a fault-propagation folding may explain, on first inspection, the fold observed ahead of the fault tip. Examination of the complete displacement profile, however, suggests an alternative interpretation. Along the lower half of Fs2, the far-field values are always greater than the near-field ones. Folding is thus observed over a significant fault height and at the likely place of fault nucleation marked today by the D_{max} point of the near-field profile. In addition, the near-field and far-field profiles show almost identical gradients of 0.1. The two curves indicate that the folding component is almost constant along the length of the fault zone. This configuration is consistent with the development of a shear zone later cut by a brittle fault (Fig. 2b, c). Fs5 and the upper portion of Fv2 may also be explained by this model. In these cases, the fault throw and the total displacement (near-field and far-field data: Fig. 8a, b) both decrease from a maximum value with the same D_{max} point but with different slopes. Along the two upper halves of Fv2 and Fs5, and the lower half of the Fs5, the far-field gradients are 0.16, 0.04 and 0.12, respectively, and are smaller than the near-field gradients, with values of 0.22, 0.08 and 0.34. Near these fault tips, the displacement profile is characterized by an increasing component of folding that may represent a variation in the shear model.

Discussion

Summary of fault characteristics

The mesoscale normal faults observed in the Southeast Basin of France show the following characteristics:

- Fault-related folds were observed at different levels along the fault zones, from their central part to their tips (Figs 5 & 8). The amplitude of the folding component (normal drag) is variable along the fault plane, but mostly accounts for a significant part of the total throw. Folds extend a few decimetres to a few metres into the fault walls and along a distance of more than 25% of the fault height ahead of the fault tip.
- Fault-related folds affect stiff and compliant (rare) layers, and no relationship between the folding amplitude and the rock lithology is apparent.
- The fault planes are sinuous, even though the faults are cutting through a homogeneous medium that consists of a succession of fine-grained limestone beds (mostly mudstones–wackestones). When coarser-grained limestones are interbedded with marly intervals, fault refractions across bed interfaces may occur, but the largest fault dip variations are due to pronounced fault sinuosity (progressive variation) within individual beds (Table 1).
- The near-field (brittle component) and far-field (sum of brittle and ductile components) displacement profiles follow either flat-topped or triangular shapes independent of lithology (Fig. 8). The displacement gradients near the fault tips are generally high and commonly above 0.3 (Table 2).
- The fault planes rarely exhibit striations.

Diagnostic values of displacement profiles

A comparison of displacement profiles reconstructed on mesoscale normal faults allows the identification of three different types of fault-related folding: (1) fault-propagation folds; (2) coherent drag folds; and (3) shear folds (Table 2). Among the criteria that distinguish between these fold mechanisms, abrupt variations in the near-field displacement with an otherwise regular far-field displacement distribution characterize coherent drag folds (Figs 5c$_2$ & 8). In the case of fault-propagation folding, divergence of the near-field and far-field curves at a point distant from the D_{max} point of the near-field profile (left-hand plots in Fig. 8) marks the tip of a folded zone with an almost constant amplitude (no high-frequency variation). A long vertical folded zone also characterizes faults with ductile shear that preceded the discontinuous offset (shear folds). In this case, the far-field displacements always exceed the near-field (Fig. 8). The coherent drag-fold model supports previous work which demonstrated that some faults and their associated volumes of deformation have been kinematically coherent throughout their evolution (Walsh *et al.* 2003; Long & Imber 2010). This points to the necessity of integrating the off-fault deformation in the analysis of fault growth and thus examine their far-field displacement profiles. Possibly, some folds in the published literature that are attributed to drag developed according to the 'coherent fault' model described here.

The analysis of mesoscale normal faults presented here demonstrates that it is possible to recognize several folding mechanisms that are superposed along the same fault (Table 2). However, the distinctive signature of each folding mechanism described, and potentially others, may be difficult to distinguish in a number of geological situations: for instance, for faults that developed in heterogeneous rocks, spatial and temporal variations in fault strength and local stresses. 2D sections close to the lateral tips of faults also represent complex cases. Hence, it is critical to define to what extent the proposed models can be applied to natural faults to allow prediction of the distribution and amplitude of fault-related folds: such as for the analyses of sub-seismic faults, for example. Further quantitative studies investigating the folding component along fault zones are therefore needed to improve our understanding of fault-related folding.

Folding and hosting rock

Fault-related folding is commonly better developed in compliant rocks. Accordingly, the folding component observed along the investigated faults is the largest in sequence Fs5, comprising the thickest clay-rich layers (Fig. 4a). In this case, the folding extends along the full fault height. However, except for Fs5, no apparent relationship between the fault-related folds and the rock lithology and

Table 2. *Fold mechanisms and gradient data*

	Fold type	Near-field displacement data					Far-field displacement data				
		Type	$Grdt_L$	$Grdc_L$	$Grdt_U$	$Grdc_U$	Type	$Grdt_L$	$Grdc_L$	$Grdt_U$	$Grdc_U$
Fag1	?	~Flat					~Flat				
Fag2	FPF?	Unk/FTo			0.08		~Flat				
Fau1	/	FTo/Unk	0.37	<0.02			FTo/Unk	0,37	<0.02		
Fau2	/	FTo/FTo	0.14	<0.02	0.30	<0.02	FTo/FTo	0,14	<0.02	0,30	<0.02
Fs1	CDF	Unk/FTo			0.30	0.02	Unk/FTo			0,30–0,53	flat
Fs2	CDF + SF	Tr/Tr	0.1	–	0.88	–	Be/Unk		0.1–0		0.12
Fs5	SF + FPF	Tr/Tr?	0.34	–	–	0.08	Be?/Unk		0.12–0		0.04
Fv1	CDF?	Be/Unk		0.12–0		0.09	Be/Unk		0.12–0		0.09
Fv2	SF + FPF	Unk/Be		0.08		0.22–0	Unk/Be		0.1		0.16–0

Fold mechanism identified along the fault zones are fault-propagation fold (FPF), shear fold (SF) and coherent drag fold (CDF).
'Type' in the displacement data denotes the overall shape of the displacement profiles (lower/upper part of the fault): Tr, FTo, Bell and Unk: triangular, flat-topped, bell and unknown shapes.
Grdt and Grdc are gradient values close to the fault tips and along the central part of the fault, respectively. Superscript letters U and L refer to the upper and lower part of the fault.
Values followed by '– 0'indicate that the displacement decrease is followed by a plateau. This profile shape is informally named a bell profile (Be).

thickness of the beds was observed. Therefore, we propose that the rocks were only partially lithified at the time of faulting. The mechanical layering was thus different to that observed today. This hypothesis is further discussed in the following sub-section. In indurated rocks, clay-rich layers are among the most common lithology favouring folding. The pattern of the stacking of competent and incompetent units is also likely to play an important role. As a function of these parameters, we expect shear folds to be developed in a dominantly compliant pile, coherent drag folds in alternating stiff and compliant units, and fault-propagation folds in both dominantly stiff rocks covered by a compliant unit and in alternating stiff and compliant units.

Published near-field displacement gradients measured along isolated faults in stiff and fine-grained limestones, with no fault-related folds are around 0.08 (Soliva *et al.* 2005; Roche *et al.* 2012*a*). Those obtained in this study (Table 2) are much higher, generally above 0.22 (Fv1 and Fag2 excluded). These values are in agreement with those of Ferrill & Morris (2008), who concluded that in heterogeneous carbonate sequences the less competent units impede fault propagation, promoting high near-field displacement gradients and fault-related folding. Considering all the gradient values cited above, a minimum 0.2 gradient value may be proposed for the development of fault-related folds in carbonate-rich sediments.

Other fault characteristics and relationship with rock properties

Even though the depth of faulting is not constrained, faulting in some of the studied sections (e.g. Fv2, Fv3, Fs1, Fs2 and Fs5) is likely to have occurred shortly after the sediments were deposited. These north–south faults developed in the Kimmeridgian–Tithonian formations before the Late Tithonian–Berriasian major tectonic reorganization in the Southeast Basin of France, after which the extension adopted a NNE–SSW direction (Homberg *et al.* 2013). Owing to the large (up to 75%) folding component observed along these faults, we suggest that they formed before the complete lithification of the sediments, which were more compliant than today. This is in line with the up to 0.3 near-field displacement gradients measured along the triangular displacement profiles of Fs5 and Fv2, a value to be compared to the 0.012–0.25 gradient found along faults formed in poorly lithified sandstones (Wibberley *et al.* 1999). A similar pronounced sinuosity as the one observed along the studied faults (Fig. 4; Table 1) has also been described for faults formed before the complete lithification of massive sandstones and medium- to high-energy shallow platform carbonates (Petit & Laville 1987; Koša & Hunt 2005; Bergerat *et al.* 2011). We suggest that high-frequency variations in the fault dip, observed together with a large folding component and the absence of penetrative striae, may be used to recognize faults that grew before complete lithification of carbonates.

Conclusions

Analysis of mesoscale normal faults combined with conceptual models of fault growth allowed the contribution of fault-related folding to the total fault displacement to be quantified. The mechanism that caused folding may change along the fault height. Three different mechanisms were identified by comparing variations in the folding and fault-slip components along the fault height: fault-propagation folds; shear folds; and coherent drag folds. Abrupt variations in the near-field (i.e. fault-slip) displacement with an otherwise regular far-field (sum of brittle and ductile components) displacement profile indicate complementary variations between fault slip and folding; large folding components in specific fault portions can be used to distinguish between these types of fold. The folding component along the investigated faults is generally greater than 20% and up to 75% of the total displacement. Compliant units and high near-field displacement gradients along the faults have promoted folding, and the stacking pattern of the competent and incompetent units is likely to have played a major role in the occurrence of a specific folding process.

We are grateful to David Ferrill, an anonymous reviewer and Conrad Childs for their suggestions that helped us to greatly improve the manuscript. We thank Ramadan Ghalayini, who enhanced the quality of the English text and Alexandre Lethiers for drawing the figures.

References

ALLMENDINGER, R.W. 1998. Inverse and forward numerical modeling of trishear fault propagation folds. *Tectonics*, **17**, 640–656.

ALLMENDINGER, R.W. & SHAW, J.H. 2000. Estimation of fault propagation distance from fold shape: implications for earthquake hazard assessment. *Geology*, **28**, 1099–1102.

ANGELIER, J. 1990. Inversion of field data in fault tectonics to obtain the regional stress – III. A new rapid direct inversion method by analytical means. *Geophysical Journal International*, **103**, 363–376.

BENZAGGAGH, M. & ATROPS, F. 1995*a*. Les zones à *Chitinoidella* et à *Crassicollaria* (Tithonien) dans la partie interne du Prérif (Maroc). Données nouvelles et corrélation avec les zones d'ammonites. *Comptes Rendus de l'Academie des Sciences, Paris*, **320**, 227–234.

BENZAGGAGH, M. & ATROPS, F. 1995*b*. Données nouvelles sur la succession des calpionelles du Berriasien dans le Prérif et le Mésorif (Rif, Maroc). *Comptes Rendus de l'Academie des Sciences, Paris*, **321**, 631–638.

BERGERAT, F. 1987. Stress fields in the European platform at the time of Africa–Eurasia collision. *Tectonics*, **6**, 99–132.

BERGERAT, F., COLLIN, P.-Y., GANZHORN, A.-C., BAUDIN, F., GALBRUN, B., ROUGET, I. & SCHNYDER, J. 2011. Instability structures, synsedimentary faults and turbidites, witnesses of a Liassic seismotectonic activity in the Dauphiné Zone (French Alps): a case example in the Lower Pliensbachian at Saint-Michel-en-Beaumont. *Journal of Geodynamics*, **51**, 344–357.

BRANDES, C. & TANNER, D.C. 2014. Fault-related folding: a review of kinematic models and their application. *Earth-Science Reviews*, **138**, 352–370.

CALAMITA, F., PACE, P. & SATOLLI, S. 2012. Coexistence of fault-propagation and fault-bend folding in curve-shaped foreland fold-and-thrust belts: examples from the Northern Apennines (Italy). *Terra Nova*, **24**, 396–406.

CARRERAS, J., DRUGUET, E. & GRIERA, A. 2005. Shear zone-related folds. *Journal of Structural Geology*, **27**, 1229–1251.

CLOOS, H. 1936. *Einführung in die Geologie*. Gebrüder Bornträger, Berlin.

COURJAULT, T., GROSHENY, D., FERRY, S. & SAUSSE, J. 2011. Detailed anatomy of a deep-water carbonate breccia lobe (Upper Jurassic, French subalpine basin). *Sedimentary Geology*, **238**, 156–171.

DARDEAU, G., ATROPS, F., FORTWENGLER, D., DE GRACIANSKY, P.-C. & MARCHAND, D. 1988. Jeu de blocs et tectonique distensive au Callovien et à l'Oxfordien dans le bassin du Sud-Est de la France. *Bulletin de la Societe Géologique de France*, **8**, 771–777.

DEBRAND-PASSARD, S., COURBOULEIX, S. & LIENHARDT, M.-J. 1984. *Synthèse géologique du Sud-Est de la France*. Mémoires du Bureau de Recherches geologiques et minieres, **125–126**.

DUBOIS, P. & DELFAUD, J. 1989. Le Bassin du Sud-Est. *In*: *Dynamique et méthodes d'étude des bassins sédimentaires. Association des Sédimentologistes français*. Technip, Paris, 277–296.

ERSLEV, E.A. 1991. Trishear fault-propagation folding. *Geology*, **19**, 617–620.

FERRILL, D.A. & MORRIS, A.P. 2003. Dilational normal faults. *Journal of Structural Geology*, **25**, 183–196.

FERRILL, D.A. & MORRIS, A.P. 2008. Fault zone deformation controlled by carbonate mechanical stratigraphy, Balcones fault system, Texas. *American Association of Petroleum Geologists Bulletin*, **92**, 359–380.

FERRILL, D.A., MORRIS, A.P., SIMS, D.W., WAITING, D.J., HASEGAWA, S. 2005. Development of synthetic layer dip adjacent to normal faults. *In*: SORKHABI, R. & TSUJI, Y. (eds) *Faults, Fluid Flow, and Petroleum Traps*. American Association of Petroleum Geologists, Memoirs, **85**, 125–138.

FERRILL, D.A., MORRIS, A.P. & SMART, K.J. 2007. Stratigraphic control on extensional fault propagation folding: big Brushy Canyon monocline, Sierra Del Carmen, Texas. *In*: JOLLEY, S., BARR, D., WALSH, J. & KNIPE, R. (eds) *Structurally Complex Reservoirs*. Geological Society, London, Special Publications, **292**, 203–217, https://doi.org/10.1144/SP292.12

FERRILL, D.A., MORRIS, A.P., MCGINNIS, R.N., SMART, K.J. & WARD, W.C. 2011. Fault zone deformation and displacement partitioning in mechanically layered carbonates: the Hidden Valley fault, central Texas. *American Association of Petroleum Geologists Bulletin*, **95**, 1383–1397.

FERRILL, D.A., MORRIS, A.P. & MCGINNIS, R.N. 2012. Extensional fault-propagation folding in mechanically layered rocks: the case against the frictional drag mechanism. *Tectonophysics*, **576–577**, 78–85.

FINCH, E., HARDY, S. & GAWTHORPE, R. 2004. Discrete-element modelling of extensional fault propagation folding above rigid basement fault blocks. *Basin Research*, **16**, 489–506.

FLANDRIN, J. 1975. *Carte géologique de la France à 1:50 000. 891, Nyons, Sheet XXXI-39*. Bureau de Recherches geologiques et minieres, Orléans, France.

FRIÈS, G. & PARIZE, O. 2003. Anatomy of an ancient passive margin slope system: Aptian gravity-driven deposition on the Vocontian paleomargin, western Alps, south-east France. *Sedimentology*, **50**, 1231–1270.

GHALAYINI, R., HOMBERG, C., DANIEL, J.M. & NADER, F.H. 2016. Growth of layer-bound normal faults under a regional anisotropic stress field. *In*: CHILDS, C., HOLDSWORTH, R.E., JACKSON, C.A.-L., MANZOCCHI, T., WALSH, J.J. & YIELDING, G. (eds) *The Geometry and Growth of Normal Faults*. Geological Society, London, Special Publications, **439**, first published online April 6, 2016, https://doi.org/10.1144/SP439.13

GAWTHORPE, R.L., SHARP, I., UNDERHILL, J.R. & GUPTA, S. 1997. Linked sequence stratigraphic and structural evolution of propagating normal faults. *Geology*, **25**, 795–798.

GRACIANSKY, P.-C. DE & LEMOINE, N. 1988. Early Cretaceous extensional tectonics in the southwestern French Alps: a consequence of North-Atlantic rifting during Tethyan spreading. *Bulletin de la Societe Géologique de France*, **IV**, 733–737.

GRASEMANN, B., MARTELB, S. & PASSCHIER, C. 2005. Reverse and normal drag along a fault. *Journal of Structural Geology*, **27**, 999–1010.

HOMBERG, C., SCHNYDER, J. & BENZAGGAGH, M. 2013. Late Jurassic–Early Cretaceous faulting in the Southeastern French Basin: does it reflect a tectonic reorganization? *Bulletin de la Société Géologique de France*, **184**, 501–514.

JANECKE, S.U., VANDENBURG, C.J. & BLANKENAU, J.J. 1998. Geometry, mechanisms and significance of extensional folds from examples in the Rocky Mountain Basin and Range province, USA. *Journal of Structural Geology*, **20**, 841–856.

JOSEPH, P., BEAUDOIN, B., FRIÈS, G. & PARIZE, O. 1989. Les vallées sous-marines enregistrent au Crétacé inférieur le fonctionnement en blocs basculés du domaine vocontien. *Comptes Rendus de l'Academie des Sciences, Paris*, **309**, 1031–1038.

KOLEDOYE, B.A., AYDIN, A. & MAY, E. 2003. A new process-based methodology for analysis of shale smear along normal faults in the Niger Delta. *American Association of Petroleum Geologists Bulletin*, **87**, 445–463.

Koša, E. & Hunt, D.W. 2005. Growth of syndepositional faults in carbonate strata: Upper Permian Capitan platform, New Mexico, USA. *Journal of Structural Geology*, **27**, 1069–1094.

Lemoine, M., de Graciansky, P.-C. & Tricart, P. 2000. *De l'océan à la chaîne de montagnes. Tectonique des plaques dans les Alpes*. SGF Collection Géosciences. Gordon & Breach, Paris.

Long, J.J. & Imber, J. 2010. Geometrically coherent continuous deformation in the volume surrounding a seismically imaged normal fault-array. *Journal of Structural Geology*, **32**, 222–234.

Mandl, G. 1988. *Mechanics of Tectonic Faulting: Model and Basic Concept*. Elsevier, New York.

Mansfield, C.S. & Cartwright, J.A. 1999. Stratal fold patterns adjacent to normal faults: observations from the Gulf of Mexico. *In*: Cosgrove, J.W. & Ameen, M.S. (eds) *Forced Folds and Fractures*. Geological Society, London, Special Publications, **169**, 115–128, https://doi.org/10.1144/GSL.SP.2000.169.01.09

McClay, K.R. 1992. *Thrust Tectonics*. Chapman & Hall, London.

Muraoka, H. & Kamata, H. 1983. Displacement distribution along minor fault traces. *Journal of Structural Geology*, **5**, 483–485.

Nicol, A., Gillespie, P.A., Childs, C. & Wash, J. 2002. Relay zones between mesoscopic thrust faults in layered sedimentary sequence. *Journal of Structural Geology*, **24**, 709–727.

Peacock, D. 1991. Displacement and segment linkage in strike–slip fault zones. *Journal of Structural Geology*, **13**, 1025–1035.

Peacock, D.C.P. & Parfitt, E.A. 2002. Active relay ramps and normal fault propagation on Kilauea Volcano, Hawaii. *Journal of Structural Geology*, **24**, 29–742.

Petit, J.-P. & Laville, E. 1987. Morphology and microstructures of hydroplastic slickensides in sandstone. *In*: Jones, M.E. & Preston, R.M.F. (eds) *Deformation of Sediments and Sedimentary Rocks*. Geological Society, London, Special Publications, **29**, 107–121, https://doi.org/10.1144/GSL.SP.1987. 029.01.10

Remane, J. 1971. Les Calpionelles, Protozoaires planctoniques des mers mésogéennes de l'époque secondaire. *Annales Guébhard, Neuchâtel*, **47**, 1–25.

Roche, V., Homberg, C. & Rocher, M. 2012*a*. Architecture and growth of normal fault zones in multilayer systems: a 3D field analysis in the South-Eastern Basin, France. *Journal of Structural Geology*, **37**, 19–35.

Roche, V., Homberg, C. & Rocher, M. 2012*b*. Fault displacement profiles in multilayer systems: from fault restriction to fault propagation. *Terra Nova*, **24**, 499–504, https://doi.org/10.1111/j.1365-3121.2012.01088.x

Rotevatn, A. & Jackson, C.A.-L. 2014. 3D structure and evolution of folds during normal fault dip linkage. *Journal of the Geological Society, London*, **171**, 821–829, https://doi.org/10.1144/jgs2014-045

Rouire, J., Gidon, M. et al. 1980. *Carte géologique de la France au 1/250 000, feuille 34: Valence*. Bureau de Recherches geologiques et minieres, Orléans, France.

Schlische, R.W. 1995. Geometry and Origin of Fault-Related Folds in Extensional Settings. *American Association of Petroleum Geologists Bulletin*, **79**, 1661–1678.

Schöpfer, M.P.J., Childs, C., Walsh, J.J., Manzocchi, T. & Koyi, H.A. 2007. Geometrical analysis of the refraction and segmentation of normal faults in periodically layered sequences. *Journal of Structural Geology*, **29**, 318–335.

Smart, K.J., Ferrill, D.A., Morris, A.P. & McGinnis, R.N. 2012. Geomechanical modelling of stress and strain evolution during contractional fault-related folding. *Tectonophysics*, **576**, 171–196.

Soliva, R., Schultz, R.A. & Benedicto, A. 2005. Three-dimensional displacement–length scaling and maximum dimension of normal faults in layered rocks. *Geophysical Research Letters*, **32**, L16302, https://doi.org/10.1029/2005GL023007

Suppe, J. 1983. Geometry and kinematics of fault-bend folding. *American Journal of Science*, **283**, 684–721.

Suppe, J. & Medwedeff, D.A. 1990. Geometry and kinematics of fault-propagation folding. *Eclogae Geologicae Helvetiae*, **83**, 409–454.

Walsh, J.J. & Watterson, J. 1987. Distributions of cumulative displacement and seismic slip on a single normal fault surface. *Journal of Structural Geology*, **9**, 1039–1046.

Walsh, J.J., Bailey, W.R., Childs, C., Nicol, A. & Bonson, C.G. 2003. Formation of segmented normal faults: a 3D perspective. *Journal of Structural Geology*, **25**, 1251–1262.

Wibberley, A.J., Petit, J.-P. & Rives, T. 1999. Mechanics of high displacement gradient faulting prior to lithification. *Journal of Structural Geology*, **21**, 251–257.

Wilkins, S.J. & Gross, M.R. 2002. Normal fault growth in layered rocks at Split Mountain, Utah: influence of mechanical stratigraphy on dip linkage, fault restriction and fault scaling. *Journal of Structural Geology*, **24**, 1413–1429.

Withjack, M.O., Olson, J. & Peterson, E. 1990. Experimental models of extensional forced folds. *American Association of Petroleum Geologists Bulletin*, **74**, 1038–1054.

Zoetemeijer, R., Sassi, W., Roure, F. & Cloetingh, S. 1992. Stratigraphic and kinematic modeling of thrust evolution, northern Apennines, Italy. *Geology*, **20**, 1035–1038.

The impact of multiple extension events, stress rotation and inherited fabrics on normal fault geometries and evolution in the Cenozoic rift basins of Thailand

C. K. MORLEY

Department of Geological Sciences, Petroleum Geophysics MSc Program, Chiang Mai University, Chiang Mai, Thailand
chrissmorley@gmail.com

Abstract: The rift basins of Thailand exhibit remarkable diversity of fault displacement patterns, fault length–displacement characteristics and mapped fault patterns during late rift, and post-rift, stages. These patterns reflect influences by: (1) zones of strength anisotropy in the pre-rift basement; (2) syn-rift fault patterns on post-rift faults; (3) spatial stress deflection, commonly related to irregularities in major fault profiles, and the basement–sediment interface; (4) temporal stress rotation, usually related to changes in the regional plate setting; and (5) varying strength properties (strain hardening or softening) of fault zones during their life. These influences created strongly segmented boundary faults, and long, low-displacement post-rift fault trends. The former are commonly strongly over-displaced, while the latter can be strongly under-displaced with respect to their length compared with typical length:displacement distributions. Seismic interpretation of multi-rift fault patterns requires 3D data to identify the complexities, otherwise the linkage pattern between deeper and shallower faults, and the changing fault strike-directions with depth, may be incorrectly mapped. Incorrect identification of fault patterns as breached relay structures may also arise. Oblique extension, the influence of pre-existing trends and stress rotation in multi-phase rifts provides a more comprehensive explanation for the observed features than the strike-slip interpretation of previous studies.

The characteristics of normal fault initiation, propagation and linkage during a single extensional phase have been extensively documented both from natural data (e.g. Walsh & Watterson 1988, 1992; Morley *et al.* 1990; Peacock & Sanderson 1991; Cartwright *et al.* 1996; Morley 1999; McLeod *et al.* 2000; Childs *et al.* 2003; Soliva & Benedicto 2004) and modelling (e.g. Scholz *et al.* 1993; Crider & Pollard 1998). Linkage can occur incidentally through initially independent, unrelated faults (isolated fault model), alternatively linkage can occur on hard- or soft-linked fault segments that have kinematically interacted with each other in a coherent way from an early stage (coherent fault model) (e.g. Morley 1999; Walsh *et al.* 2003*a*, *b*) (Fig. 1). For isolated faults, displacement deficits are likely to occur in the regions of overlap: if the ductile and brittle components of strain can be accurately identified, however, displacement deficits will be absent in fault systems following the coherent model from an early stage (Walsh *et al.* 2003*a*, *b*) (Fig. 1b).

The basic understanding of normal fault development outlined above in terms of linkage model, propagation patterns and fault length–displacement relationships is most applicable to approximately orthogonal extension. The increasing quantity and resolution of 3D seismic data, coupled with more sophisticated analogue and numerical modelling, particularly in the last 10–15 years, has enabled investigation of how faults depart from the basic models in more complex situations of pre-existing fabric influence, rotating stress directions with time and multi-phase rifting (e.g. Lepvrier *et al.* 2002; Lezzar *et al.* 2002; Maerten *et al.* 2002; McClay *et al.* 2002; Morley *et al.* 2004, 2007; Bellahsen *et al.* 2006; Corti *et al.* 2007, 2013*a*, *b*; Henza *et al.* 2010, 2011; Morley 2010; Agostini *et al.* 2011; Autin *et al.* 2013; Whipp *et al.* 2014; Reeve *et al.* 2015).

Rifts are typically localized on regional-scale zones of weakness in the crust. Such zones are generally non-orthogonal to the far-field tectonic forces. Indeed, oblique extension may actually be the most favourable condition for continental break-up (Brune *et al.* 2012). The presence of pre-existing fabrics in basement can lead to considerable variations in fault patterns and development (e.g. Withjack & Jamison 1986; Tron & Brun 1991; Nelson *et al.* 1992; Morley *et al.* 2004; Reeve *et al.* 2015). While oblique slip is typically assumed for faults following oblique trends, there is now a growing body of evidence which suggests that, under the right conditions, oblique fault trends may also exhibit dip-slip kinematics (e.g. Petit *et al.* 1996; Morley 2010; Agostini *et al.* 2011; Autin *et al.* 2013; Philippon *et al.* 2014). Rotation of the stresses

From: Childs, C., Holdsworth, R. E., Jackson, C. A.-L., Manzocchi, T., Walsh, J. J. & Yielding, G. (eds) 2017. *The Geometry and Growth of Normal Faults*. Geological Society, London, Special Publications, **439**, 413–445.
First published online April 13, 2016, https://doi.org/10.1144/SP439.3

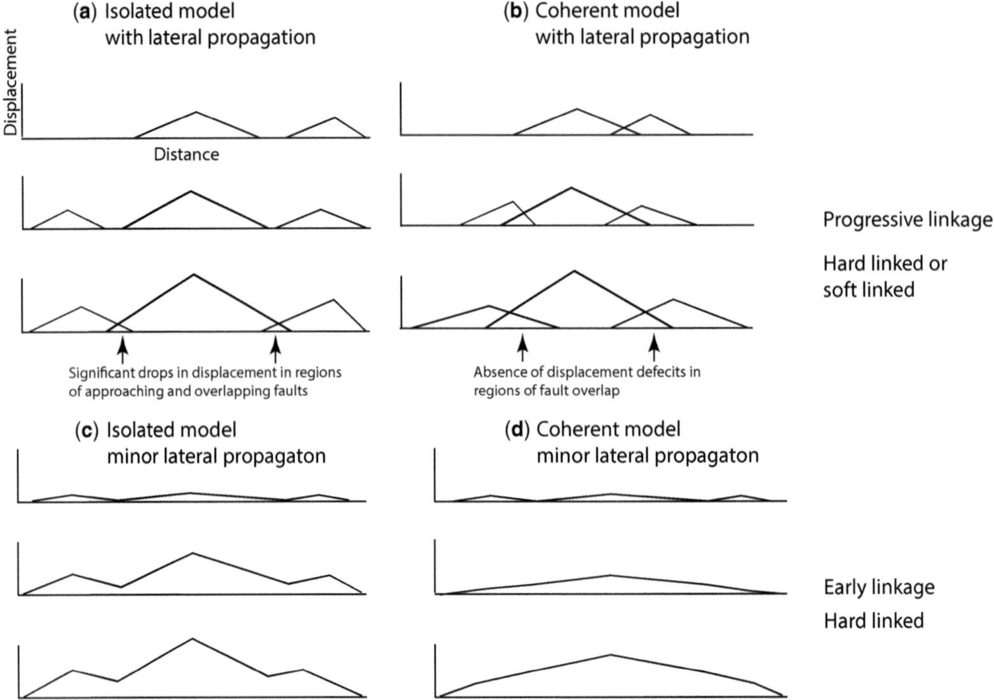

Fig. 1. Schematic examples of along-strike fault displacement variations following the isolated and coherent fault models. (**a**) Isolated model with progressive linkage, (**b**) Coherent model with progressive linkage (from Walsh *et al.* 2003*a*, *b*). (**c**) Isolated model with early linkage. (**d**) Coherent model with early linkage (following Morley 1999).

into consistently trending major weak basement fabric zones is the probable explanation (Morley 2010; Corti *et al.* 2013*a*, *b*). Yet, other parts of rifts may exhibit local deflections in principal stress directions that are highly complex (e.g. Maerten *et al.* 2002; Tingay *et al.* 2010).

In multi-phase rift systems, fault development becomes more complex than in single-phase rifts (Morley *et al.* 2004, 2007; Bellahsen *et al.* 2006). For example, phases of extension can alternate with periods of quiescence or inversion and, commonly, rotation of the extension direction occurs with time, The influence of pre-existing fabrics within the crystalline crust, the pre-rift section and from earlier rift faults exerts an important control on fault patterns and linkage. Slip-tendency analysis (e.g. Morley *et al.* 2004) and analogue modelling (e.g. Henza *et al.* 2010, 2011) indicates that an important control on the activation of earlier faults, or activation of oblique trends, is the angle of stress rotation.

During fault linkage and propagation, relay ramps are breached and eliminated (e.g. Peacock & Sanderson 1991; Childs *et al.* 1995). This ongoing process during the life of a fault occurs in extension-oblique and extension-orthogonal systems. Breaching of relay ramps can also result from rotation of the stress field and creation of a later linking of oblique faults with the relay zone of pre-existing faults, and is one way in which zigzag faults in rifts can develop (Lepvrier *et al.* 2002; Whipp *et al.* 2014; Henstra *et al.* 2015).

The rift basins of Thailand are representative of extension in a very dynamic tectonic environment. Consequently, normal fault development has been strongly impacted by inherited fabrics, spatial and temporal changes in stress regime, and interactions with strike-slip faults, as described by Morley *et al.* (2004, 2007, 2011), Kornsawan & Morley (2002), Morley (2009, 2014) and Tingay *et al.* (2010).

This paper investigates, in more detail than previously, how the effects of inheritance and stress rotation affect fault length–displacement relationships and linkage. The geometries and development of two different kinds of extensional faults – large half-graben bounding early faults and later, more minor, fault sets – are a focus of this study. It is particularly useful to consider these two different fault types because the history of displacement based on seismic reflection data is much easier to determine from sedimentary growth patterns on large boundary faults, whereas development of linkage patterns is easiest to see in smaller faults.

Geological background

Thailand contains two trends of rift basins: the back-arc passive margin Mergui and East Andaman basins in the Andaman Sea; and the failed rifts that run from the Gulf of Thailand to northern Thailand (Fig. 2). Cenozoic extensional faulting in the Andaman Sea is of probable Late Eocene–Middle Miocene age. The failed rift trend began in the Late Eocene and continued in some areas until the Late Miocene onshore, and Pliocene to even Recent offshore (Morley 2001; Morley & Racey 2011). This approximately 35 myr duration of activity occurred in a tectonically active region that experienced changing subduction zone configurations, and the progressive collision of India with Eurasia and Australia with SE Asia, together with various minor collisions within SE Asia (see reviews in Hall 2012; Pubellier & Morley 2014). Consequently, stress orientations and magnitudes within the rift basins of Thailand have fluctuated considerably from the Eocene to Present, resulting in episodic inversion or rotation of the extension direction (Morley 2002; Morley *et al.* 2011; Srisuriyon & Morley 2014).

As a result of the evolving stress field, extensional faults in long-lived rift basins show different responses to inherited fabrics with time. The initial fault patterns responded to pre-Late Miocene fabrics (e.g. early Cenozoic transpressional or strike-slip fabrics, and contractional fabrics related to the Indosinian Orogeny). Later faults developed following a change in stress orientation and, hence, in places the early rift faults later become the dominant pre-existing fabric.

Like all rifts, the role of pre-existing crustal fabrics is extremely important in imparting individual characteristics to extensional fault patterns and development in Thailand. Details about the roles played by pre-existing fabrics have been extensively discussed in Morley *et al.* (2004, 2007, 2011) and Morley (2009, 2014), and so are just summarized here. At the largest scale, the rifts in Thailand are focused on the main Triassic–Early Jurassic collision zone (the north–south-trending Inthanon and Sukhothai zones), which lies between the Indochina and Sibumasu continental blocks (see inset, Fig. 2). The Indochina block in the east overthrusts Sibumasu to the west. Consequently, the general orientation of Indosinian fabrics (thrusts, shear zones, metamorphic fabrics) are low-angled, with dips predominantly towards the east. Other trends are present as well, such as the NE–SW-trending back-arc Nan–Uttaradit suture, NE–SW- and NW–SE-trending folds within Mesozoic rocks, and NW–SE-, NE–SW- and north–south-trending early Cenozoic strike-slip faults.

Pre-existing fabrics can give rise to large-displacement faults or fault segments that are low angled. A key characteristic is that most low-angle normal faults dip eastwards, whereas the dip directions of high-angled faults are evenly divided between west- and east-dipping faults (Fig. 3). The presence of fault segments of orientations that depart from the usual north–south trends, particularly NW–SE and NE–SW trends, are commonly another consequence of pre-existing fabrics. The north–south trend dominates much of offshore Thailand. However, in the SE Gulf of Thailand, the dominant orientation changes to NW–SE, and the north–south trends become oblique (Fig. 2).

In the Gulf of Thailand, the dominant extension direction for the Late Eocene–Oligocene rifting event is thought to rotate from east–west (Pattani Basin, to the north) to NE–SW, passing southwards into the North Malay Basin. For the Malay Basin, the base syn-rift faults trend predominantly north–south and NW–SE (Mansor *et al.* 2014). In Malaysia, the syn-rift section of the Malay Basin has traditionally been interpreted as forming under sinistral strike-slip, which is linked to major strike-slip faults in Thailand (e.g. the Three Pagodas and Mae Ping faults: Tapponnier *et al.* 1986; Fyhn *et al.* 2010; see the review in Mansor *et al.* 2014). However, since the basins in the Gulf of Thailand are demonstrably formed under extension, not strike-slip (e.g. Morley 2001, 2013; Morley *et al.* 2011), oblique extension should also be considered as an alternative to strike-slip in the Malay Basin. Certainly, the North Malay Basin (Fig. 2) in Thailand waters, which is discussed in this paper, shows no indications of the north–south or NW–SE rift basin trends being bound by strike-slip faults, they are half-graben bound by normal faults. The issue of whether fault patterns are representative of oblique-slip or multi-phase basins v. strike-slip, and what criteria can be used to distinguish these different modes is an important and fundamental question, and is certainly not one that has been fully resolved in SE Asia (e.g. see Morley 2013).

Models for fault displacement

The idealized growth of isolated blind normal faults is characterized by an elliptical fault surface, where displacement decreases away from a central maximum, and increasing displacement is accompanied by a proportional increase in fault length (Barnett *et al.* 1987; Walsh & Watterson 1987). A number of studies have investigated displacement–length relationships for faults (e.g. Scholz & Cowie 1990;

Volcanic centres
Region floored by Mesozoic oceanic crust
Tertiary sedimentary basin
Cretaceous oceanic crust or hyper-extended continental crust (Andaman Central Basin)
Probable Miocene oceanic crust
Central Basin sub-basins:
H = Hukawng
C = Chindwin
SH = Shwebo
S = Salin
P = Prome
I = Irrawaddy
BY = Begu Yoma
A = Alcock Rise
S = Sewell Rise
Rift Basins
F = Fang
CM = Chiang Mai
Ph = Phitsanulok
SB = Suphan Buri
Ch = Chumpon
Pa = Pattani
Kh = Khmer
K = Krabi
NM = North Malay
M = Malay
N = Nakhon
S = Songhla
EG = Eastern Graben
Central Basin Myanmar (Paleocene-Recent)
Timing of extension in rift basins
Eocene-Base Miocene
Eocene-Mid or Late Miocene
Late Oligocene-Middle Miocene
Eocene-Base Miocene transitional northwards to Late Oligocene- Early Miocene
Late Oligocene-Late Miocene
Miocene
Palaeogene- Middle Miocene Local Late Miocene-Recent activity
SOUTH CHINA BLOCK
INTHANON ZONE
Loei Suture
INDOCHINA BLOCK
SUKHOTHAI ZONE
SIBUMASU BLOCK
ANDAMAN SEA
Gulf of Thailand
20°
10°
Gaoligong Shan shear zone
Chong Shan shear zone
Yunnan (China)
Aliao-Shan Shear zone
Myanmar
Sagaing Fault
Shan Plateau
Laos
Central Basin
Mae Ping Fault
Three Pagodas Fault
Gulf of Martaban
Coco Basin
Andaman-Nicobar Islands
Thailand
Bangkok
E. Andaman Basin
Invisible Ridge
Continental crust remnants
Mergui Ridge
West Andaman Fault zone
Sumatra-Andaman Trench
Klong Marui Fault
North Sumatra Basin
Narathiwat High
Gulf of Thailand
Cambodia
Malaysia
Fig. 5a
Fig. 6
Fig. 14
Fig. 18
Fig. 12
MF
H
Sh
S
BY
P
F
CM
Ph
SB
A
S
Me
K
Ch
N
S
Pa
Kh
EG
NM
M
20°
15°
10°
5°
91°
95°
99°

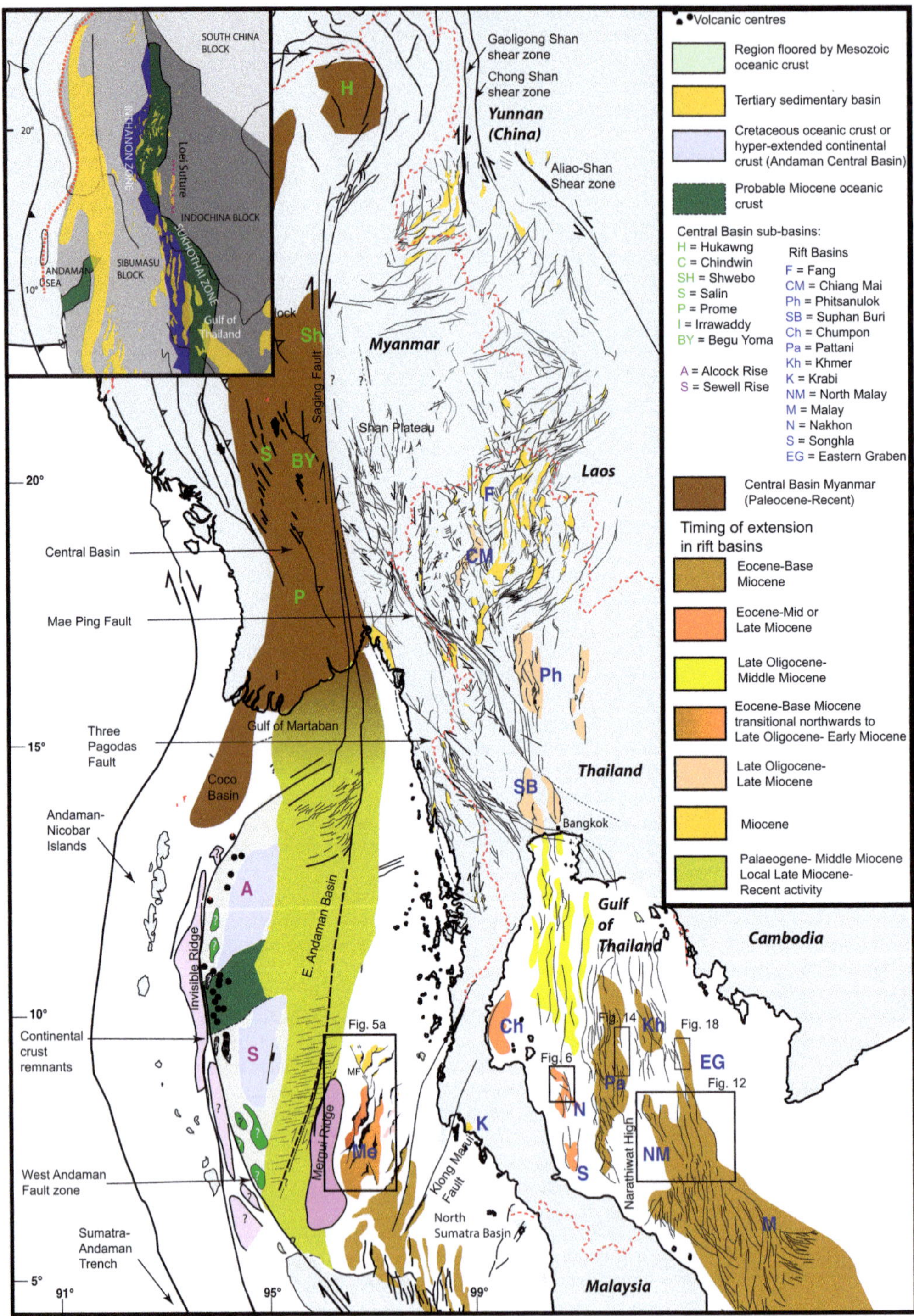

Fig. 2.

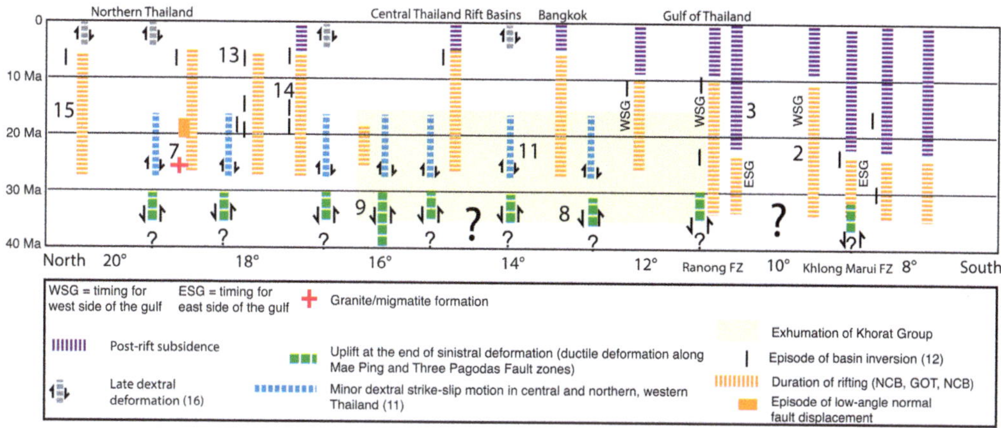

Fig. 2. (*Continued*) Regional setting of Thailand's rift basins, and timing of Cenozoic strike-slip, extension and inversion events (modified from Morley *et al.* 2011: Malay Basin fault pattern from Mansor *et al.* 2014). Inset shows the distribution of basement blocks accreted during the Indosinian Orogeny in SE Asia, with rift basins shown in yellow.

Marrett & Allmendinger 1991; Walsh & Watterson 1992; Needham *et al.* 1996). The general term used to describe fault-length scaling is:

$$D = cL^n \quad (1)$$

where c is a material constant, D is the maximum displacement and L is fault length. Walsh & Watterson (1988) favour a value of n of approximately 2, whilst Cowie & Scholz (1992) and Scholz *et al.* (1993) favoured an exponent of 1. Generally, plots of natural examples of faults fall approximately on a straight trend on log–log plots of D v. L, between exponent values of 1 and 2. However, a wide variety of factors can influence fault scaling, and natural examples display a large scatter in D:L plots (Mansfield & Cartwright 2001). A very broad scatter characterizes the rift basins in Thailand (Fig. 4).

Models for fault growth have largely focused on two main models: growth of an isolated fault by symmetric lateral propagation as displacement increases; and growth by linkage of initially separate faults (e.g. Walsh & Watterson 1988;

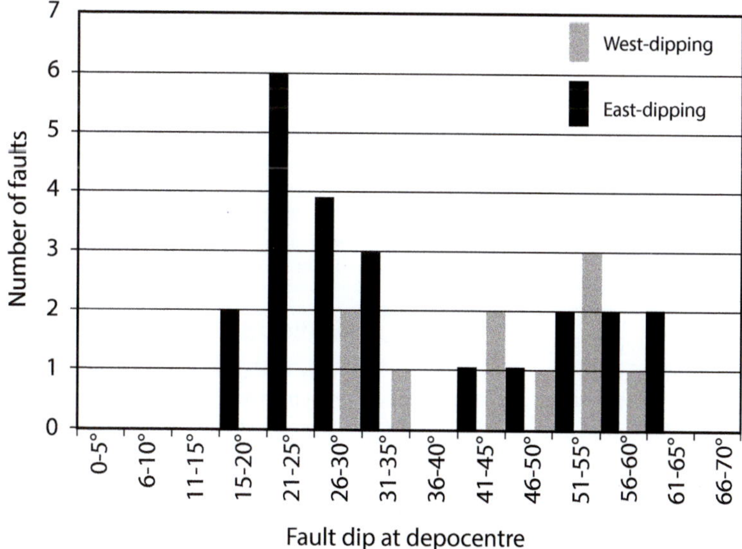

Fig. 3. Graph of fault dip directions and angles for onshore rift basins in central Thailand and the Gulf of Thailand (Morley 2014).

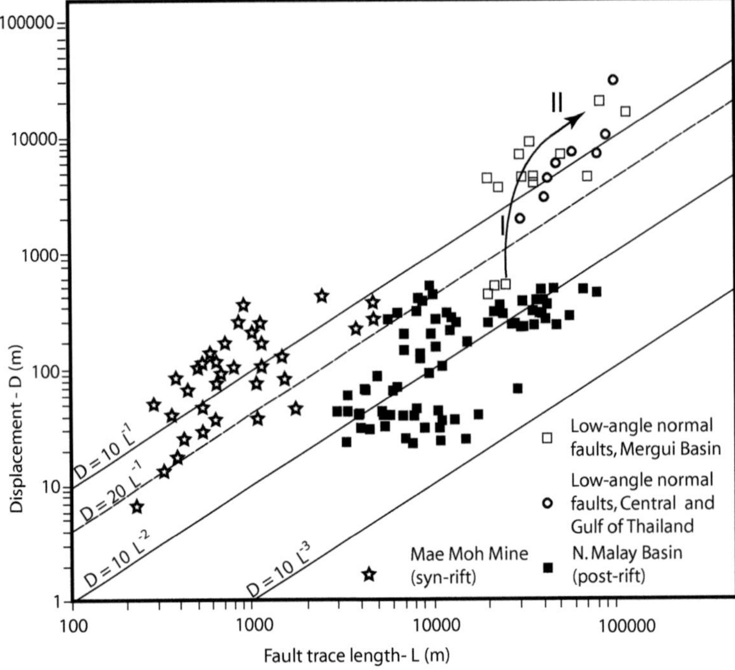

Fig. 4. Length–displacement plots for different Cenozoic normal fault types in Thailand. Syn-rift data from Mae Moh mine come from mapping and cross-section construction during open-cast mining (Morley & Wonganan 2000); low-angle and post-rift Cenozoic normal fault data are based on seismic reflection data. Modified from Morley *et al.* (2011).

Cartwright *et al.* 1996; Morley 1999; Mansfield & Cartwright 2001; Childs *et al.* 2003). The isolated fault model requires a close relationship between fault displacement and fault length, as discussed above. Fault growth by linkage is considered to be a more episodic process. The consequence of linkage that most approximates the isolated fault model occurs when two or more faults join and cause an abrupt change in length, when the displacement value for the newly joined fault is relatively low. As displacement builds (greatest in the middle of the newly linked fault), the fault does not initially propagate laterally until displacement is built up to typical length–displacement ratios for isolated faults. A large fault may progress through a number of cycles of linkage and displacement building before reaching its final length.

There are two main alternative ways to the idealized models discussed above, in which fault linkage can develop. First, fault length may be built to virtually final length very early in its history by linkage of many small faults, prior to much of a synkinematic sedimentary record being preserved. Subsequently, most of the fault displacement is built posthumously to attainment of near-maximum length (Morley 1999; Walsh *et al.* 2003*a*) (Fig. 1d). Consequently, the fault is relatively under-displaced for most of its history. Such fault zones will follow the coherent fault model. The second pattern occurs when, following linkage of several faults, displacement fails to relocate to the middle of the newly created fault zone. Instead, some of the sites of linkage remain relatively strong regions of the fault zone and inhibit propagation of slip into the linkage zones (e.g. Cartwright *et al.* 1996; Morley 1999) (Fig. 1c). This second type is important for two very different types of fault in the Thailand rifts: low-angle boundary faults; and long, low-displacement post-rift faults.

Boundary faults

Many of the rift basins in Thailand are simple half-graben, where the bounding fault was established early in the rift history and continued to displace throughout the life of the basin (Flint *et al.* 1988; Morley *et al.* 2007; Morley 2009, 2013; Morley & Racey 2011). However, there are a number of aspects to the fault orientations, linkage and displacement patterns that differ from simple extension-orthogonal rifts, or rift segments where the influence of pre-existing fabrics is negligible.

These include: (1) the presence of low-angle normal fault segments or entire faults; (2) faults where their propagation history is asymmetric; and (3) the timing and distribution of oblique fault displacement with respect to adjacent extension-orthogonal faults.

High-angle v. low-angle bounding faults

A key large-scale aspect of bounding faults in Thailand is whether they initially formed as high- or low-angle faults, or whether the high-angle faults became rotated to low angles (see reviews in Buck 2007; Morley 2014). The crustal conditions that result in the development of high-angle half-graben bounding faults or high-displacement low-angle normal faults in metamorphic core complexes are thought to be related to two key parameters: (a) the brittle layer thickness for a given cohesion; and (b) the relationship between the rate of cohesion reduction with strain (Poliakov & Buck 1998; Lavier *et al.* 1999, 2000; Lavier & Buck 2002). Hydrothermal circulation causing fault-zone weakening is thought play a major role in determining the mechanism of faulting (half-graben formation v. metamorphic core complex formation: Lavier & Buck 2002).

As reviewed by Buck (2007), crustal starting conditions (e.g. thickness and temperature) exert a primary control on whether half-graben or metamorphic core complexes develop. Metamorphic core complexes are composed of rotated high-angle normal faults and do not develop half-graben (Buck 2007). However, this view that the two styles are distinctly separate does not work particularly well in Thailand. In some regions, such as the Mergui Basin, low-angle normal faults are developed in association with higher-angle faults (Morley 2014), and are associated with half-graben. Some major bounding faults, such as the West Mergui Fault of the Mergui Basin, are long (110 km) and segmented faults that pass laterally from low-displacement high-angle faults to higher-displacement low-angle segments (Fig. 5c). They are also imaged on seismic reflection data as continuing as low-angle faults to considerable depths below the base of the syn-rift section (6+ km) and not steepening up (Morley 2014). Such faults pose a significant challenge to the views that low-angle normal faults are rotated high-angle faults, or they are the products of stress rotation (see the discussion in Morley 2014).

In Thailand, the characteristics of low-angle normal faults in sedimentary basins suggest that, in some cases, clear differences in crustal thickness and temperature are not the primary controls on the development of low- v. high- angle faults. The predominance of low-angle faults as east-dipping structures, as opposed to high-angle faults that show an equal mix of east- and west-dipping orientations, strongly suggests that pre-existing fabrics are controlling the occurrence of low-angle normal faults (Fig. 3) (Morley 2014). In particular, low-angle, east-dipping contractional fabrics associated with the Indosinian Orogeny are considered important zones of weakness that are followed by low-angle normal faults. Reactivation of older structures by low-angle normal faults is also a theme discussed for the Cycladic Islands in the Aegean Sea by Jolivet *et al.* (2010), who suggested that stress rotation within the metamorphic fabrics helped promote the development of low-angle normal faults.

Displacement characteristics of segmented low-angle normal faults

The Thailand, rift basins range widely in size from small, intermontaine basins, where sediment thicknesses are a few hundred metres, to large basins with syn-rift deposits of up to 7 km in thickness. Some basin-bounding faults are low angle, with large displacements, such as the Western Boundary Fault, Phitsanulok Basin (10 km), the East Mergui Fault, Mergui Basin (17 km) and the detachment fault to the Chiang Mai Basin (30 km) (Morley 2009, 2014). The longest faults are around 100 km $\pm$ 20%. Hence, these low-angle normal faults exhibit very low length:displacement ratios (8:1–4:1) (Fig. 4).

Some low-angle normal faults exhibit simple displacement patterns, with a single displacement maximum in approximately the fault centre (e.g. the Western Boundary Fault, Phitsanulok Basin: Morley *et al.* 2007), which fits the displacement pattern of a simple coherent fault or a series of cycles of fault linkage (Fig. 1). The few low-displacement low-angle normal faults present in the Mergui Basin differ little in length from the high-displacement low-angle normal faults in the basin, which indicates that much of the fault growth in length occurs relatively early and that, subsequently, displacement is built without much growth of the fault length (Fig. 1d & path I, Fig. 4). A few of the largest faults show an apparent increase in length with displacement (path II, Fig. 4). However, this growth is by linkage of two or three fault segments, not by simple lateral fault propagation with increasing displacement. The faults exhibit multiple displacement highs and lows that in some examples are accompanied by significant changes in fault dip (e.g. Fig. 5). These variations support the inference of fault linkage. The areas of high dip are regions of relatively low displacement (Fig. 5c), indicating that they are relatively strong regions of the fault, and inhibit slip.

In the Mergui Basin, the largest faults, which are up to 120 km long, show a distinct northwards-younging in activity along the major half-graben

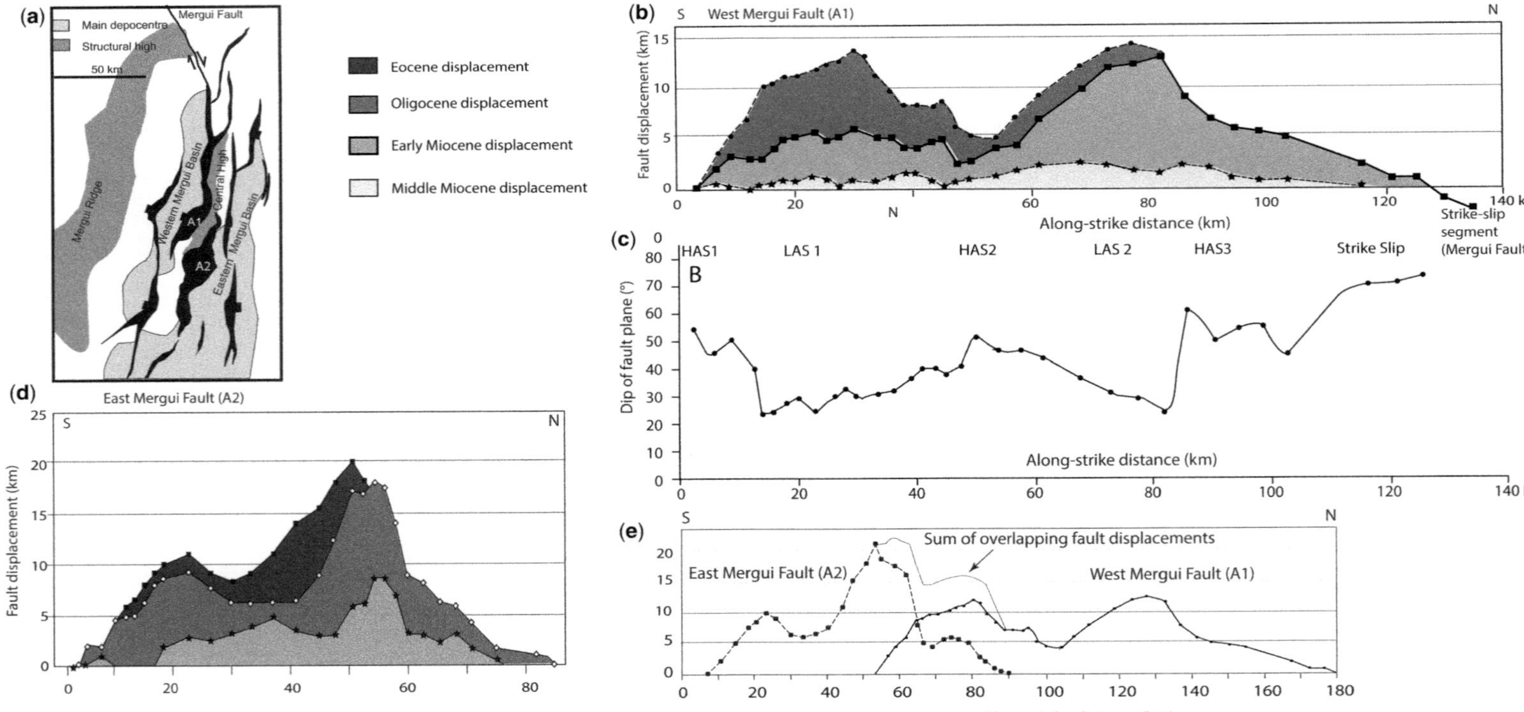

Fig. 5. Strike distance–displacement plots for the two largest faults in the Mergui Basin (based on Morley 2014). (**a**) Location map (A1, West Mergui Fault; A2, East Mergui Fault). (**b**) Displacement profile of West Mergui Fault. (**c**) Fault dip–strike distance and fault dip–strike distance plots of the West Mergui Fault, illustrating how the low-angle segments correlate with high displacements, and high-angle segments correspond with lower displacements. (**d**) Displacement profile of East Mergui Fault. (**e**) Combined displacement–strike distance plots of the West and East Mergui faults.

bounding faults (Morley 2014). This is particularly well demonstrated looking at the displacement history of the West and East Mergui faults in combination (Fig. 5). The East Mergui Fault is the most southerly of the two and displays Late Eocene–Early Miocene extension: the West Mergui Fault, however, displays Oligocene–Middle Miocene extension (Fig. 5). This northwards-younging is related to the dynamics of the area, where the northwards movement of India, with respect to SE Asia, caused a progressive younging in the onset and termination of rifting passing northwards within the Mergui Basin (Morley 2014).

Boundary fault displacement characteristics in the Nakhon Basin

The presence of north–south and NW–SE orientations in some rift basins (Fig. 2) has previously been interpreted in terms of strike-slip faults (NW–SE trending) and extensional pull-aparts. However, both trends are actually extensional: consequently, oblique extension is the preferred interpretation (Morley 2002; Morley *et al.* 2004, 2011). An example of how the NW–SE and north–south trends interact is discussed here from the Nakhon Basin in the Gulf of Thailand. The NW–SE-trending, NE-dipping bounding fault (NBF1) lies oblique to the adjacent north–south-trending sub-basins (Figs 6 & 7), and to the main north–south trend of the rift basins in the Gulf. In Figure 6, in the south of the basin, is a north–south-trending, west-thickening half-graben bounding fault that transfers displacement to an east-thickening fault-bounded depocentre. Then passing north, the polarity switches again but the trend of the fault (NBF1) becomes NW–SE, oblique to the overall north–south trend. North of NBF1, several smaller NW–SE-trending faults are mixed in with the north–south-trending faults. One of these faults (Fault Y) is shown in Figure 8a: it is of interest because, despite having low offset at the base of rift horizon, it is consistently associated with a very prominent, high-amplitude zone of reflectivity within the pre-rift basement. At the basement level, the fault separates a formless high-amplitude zone from a seismically transparent zone, suggesting significant displacement within the basement. The fault indicates that discrete pre-existing NW–SE trends are present in the pre-Cenozoic basement in the vicinity of the major boundary fault. In the vicinity of Fault Y, the syn-rift section dips SW into the NW–SE-trending, NE-dipping boundary fault. Figure 8b shows a seismic section just to the east of where displacement on Fault Y at the base syn-rift level dies to zero. The steeply dipping zone of amplitudes in the basement can still be seen, but it is overlain by one of the significant north–south-trending faults (X, Fig. 8b) that dips to the west. As NBF1 dies out and the north–south faults take up the displacement, it can be seen that the sediments dip to the east, away from NBF1.

The entire length of the NBF1 trend was active early in its history (Oligocene) along segments B, C and D (Fig. 7). These segments can be identified in the fault by variations in the displacement profile, and by joins and changes in geometry along the fault. At location 1 (Fig. 7), there is an unusual NNE–SSW trend to the fault zone, as well as some secondary faults, which suggests the local influence of a pre-existing fabric or stress rotation in the zone between two approaching fault tips that caused the local complexity. During the Oligocene, these three segments were probably separate faults. Subsequently, these early displacement patterns for the separate faults have been largely eliminated (i.e. displacement deficits are minor), suggesting a coherent model, perhaps following the pattern of Figure 1d, for the fault zone for much of its length. The Oligocene displacement pattern is skewed towards the west (Fig. 7), this can be explained by displacement being concentrated entirely on the boundary fault in the west, but being distributed on both the boundary fault and the north–south-trending faults on segments C and D.

During the Miocene, the early fault (segments B, C and D) was inactive, except for one small segment, between 9 and 12 km in Figure 7. However, a splay in the footwall of the early fault activated the NW portion of NBF1 (Fig. 7, segment A). Segment A trends NW–SE and turns to a north–south orientation at its eastern tip, indicating that the fault broke away from the pre-existing fabric trend towards the extension-orthogonal orientation. The late activity on segment A produced a very different footwall geometry in the NW segment that is high and peneplaned (Fig. 9a), whereas, in the centre and in the SE, the footwall is lower and preserves a more strongly dissected morphology (Fig. 9b).

The development of the different fault segments with time in map view is shown in Figure 10. During the Oligocene, segments B, C and D of the boundary fault (Fig. 10) were active. Then these segments were largely abandoned and the north–south- trending segments became important in the Early Miocene. However, some north–south segments that were active in the Oligocene also become abandoned. Segment A developed for the first time in the early Miocene. Then, in the Middle Miocene, virtually all activity ceased on faults and fault segments that initially formed during the Oligocene. The exception is NBF1 segment A and a swarm of minor, north–south-trending faults, mostly confined to the sedimentary section that newly developed. The new minor faults indicate that the Late Miocene maximum horizontal stress direction during the Late

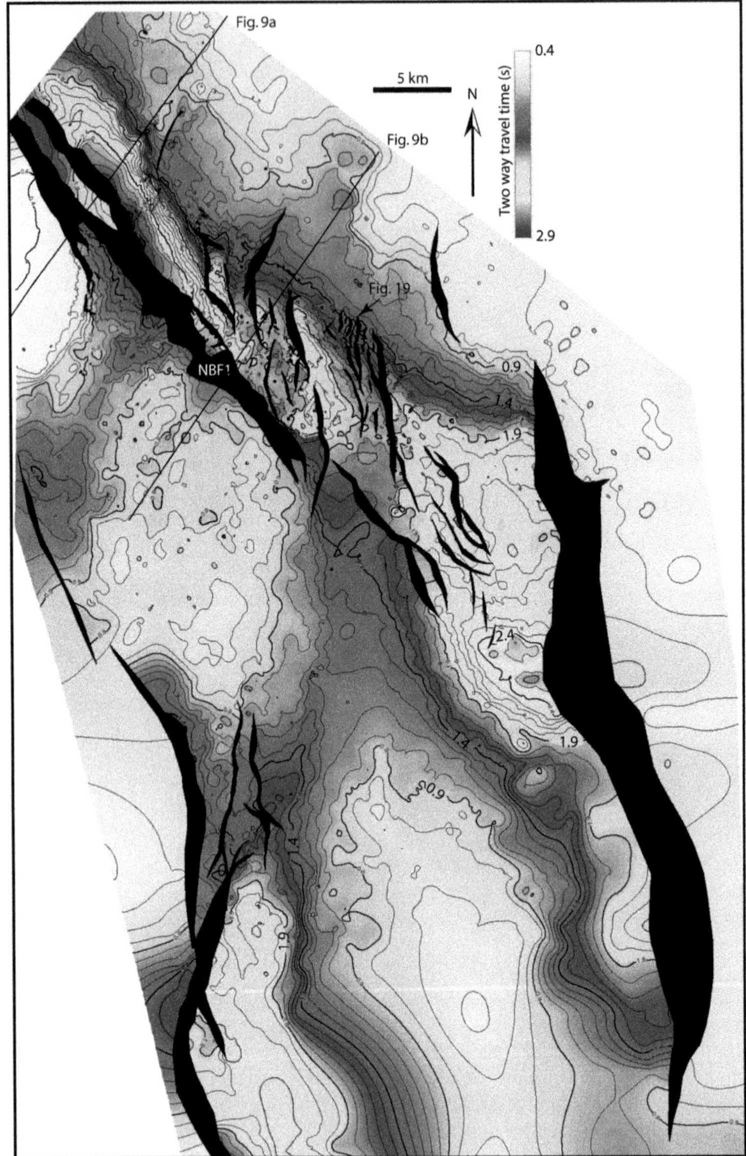

Fig. 6. Regional time–structure map of the base syn-rift horizon for the Nakhon Basin based on 2D and 3D seismic data. See Figure 2 for the location of the basin.

Miocene was approximately north–south, and that segment A was an active, weak, oblique trend.

Stress rotation with time, and inherited syn-rift structures in late developed faults

Regional stress rotation

The Andaman Sea basin and the failed rift trend basins display different overall patterns of stress evolution, with different parts of each trend showing local variations. Overall, the Andaman Sea rifts initiated earlier in the south (Late Eocene) than in the north (Early Miocene) of offshore Thailand. Rifting ended earlier in the south (Oligocene–Miocene boundary) than in the north (Middle Miocene) (Srisuriyon & Morley 2014). This change was accompanied by approximately east–west extension during the early stages, changing to NNW–SSE extension in the Early Miocene (Srisuriyon & Morley 2014).

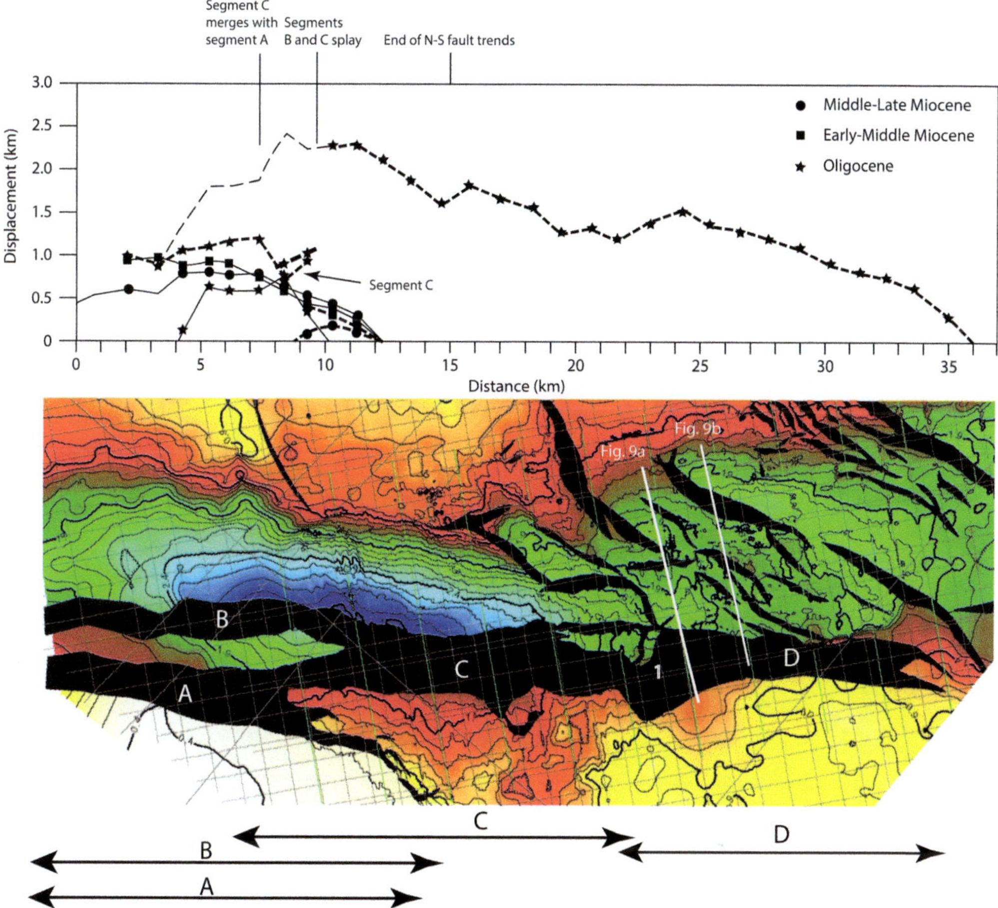

Fig. 7. Displacement–strike distance plot for the NW–SE-trending bounding fault NBF1 (see Fig. 6 for location). Note the anticlockwise rotation of the north arrow from vertical.

The evolution is thought to reflect early extension caused by slab rollback being replaced by India coupling with western Myanmar, and imposing transtensional motion on the Andaman Sea region during the Early Miocene (Morley 2015).

The timing of deformation in the failed rift belt is summarized in Figure 2, and in Morley & Racey (2011), Morley *et al.* (2011) and Pubellier & Morley (2014). Generally, rift basins young towards the north, and vary considerably in the location, number and timing of inversion events. In some basins, inversion is absent, but there is evidence for rotation of the extension direction (e.g. Morley *et al.* 2007).

The mechanisms driving stress evolution at the regional scale are complex. Finite-element modelling indicates that the interaction of three different boundary forces (the Java–Sumatra subduction zone, the collision zone in Sulawesi and the Eastern Himalayan Syntaxis) is required to explain the modern stress pattern in SE Asia (Tingay *et al.* 2012). This modern interdependency indicates that variations in forces along these zones through time, and the rise and decline of other sources of force (e.g the South China Seas spreading centre, slab pull from the Proto-South China Seas oceanic crust), would have an effect on stress magnitude and orientation in Thailand. Hence, the complex stress evolution discussed above is probably a consequence of complex interactions of intra-plate and plate boundary forces.

North Malay Basin example

In some basins, the regional extension-oblique faults formed during the early extensional stage become perpendicular to the extension direction later, while the earlier orthogonal system becomes the later oblique one. This is particularly the case

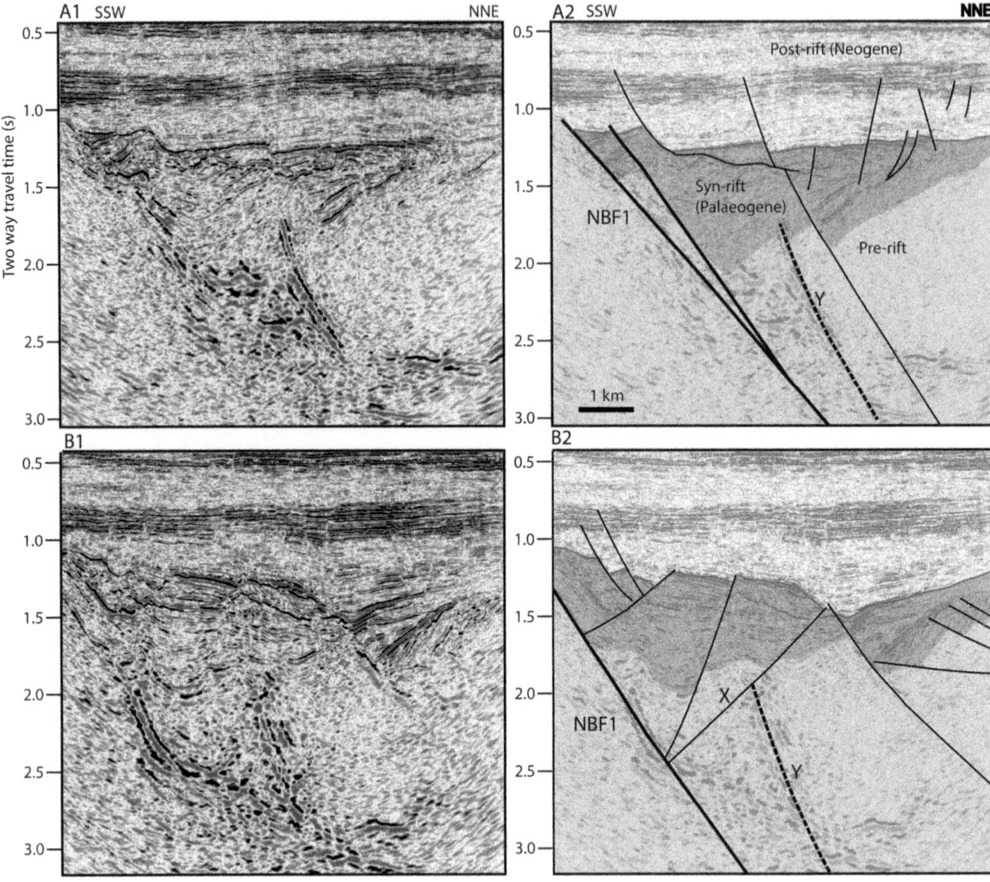

Fig. 8. 3D seismic lines crossing a deep-seated NW–SE steeply dipping, high-amplitude feature (Y) in basement. This feature, probably an old pre-rift fault, was reactivated during syn-rift times in (A1, A2), but was inactive in (B1, B2). See Figure 7 for the location.

for the North Malay Basin (see Figs 2 & 11). This development tends to result in relatively long oblique and shorter orthogonal trends (Morley *et al.* 2011). An example of this behaviour is shown for a region in the North Malay Basin, where the near top basement and Late Oligocene fault patterns represent the early and late half-graben stage (syn-rift) of development (Fig. 11). The Middle and Late Miocene stages represent minor faulting in the post-rift or sag basin stage. Although the two dominant fault trends (NW–SE and north–south) occur in both stages, the relative lengths of the fault orientations change with time (Fig. 11).

Table 1 compares the relative importance of different fault orientations in the syn-rift section for part of the North Malay Basin through variations in fault length and fault displacement. Average fault displacement is simply the sum of the maximum displacement for each fault in a trend divided by the number of faults. The direction most prone to reactivation in terms of both fault length and highest average displacement is the 121–140° orientation. Next in importance are the two approximately north–south trends (1–20° and 161–180°). Weak reflections within basement trending approximately north–south are seen in parts of the 3D seismic reflection data, and appear to be from a sequence deformed into folds and thrusts.

Both the longest faults and the highest displacement occur along the NW–SE (121–140°) trend. The main depocentre is focused on the NW–SE segment of the boundary fault, and there is no depocentre geometry indicating that the more north–south segment of the boundary fault acted as a high-displacement zone (i.e. that it behaved as a releasing bend). Hence, a NE–SW extension direction (σ_3) is inferred to be the most probable scenario for the syn-rift stage.

Slip-tendency analysis suggests that, for simple extensional stress regimes ($\sigma_1 > \sigma_2 > \sigma_3$), faults

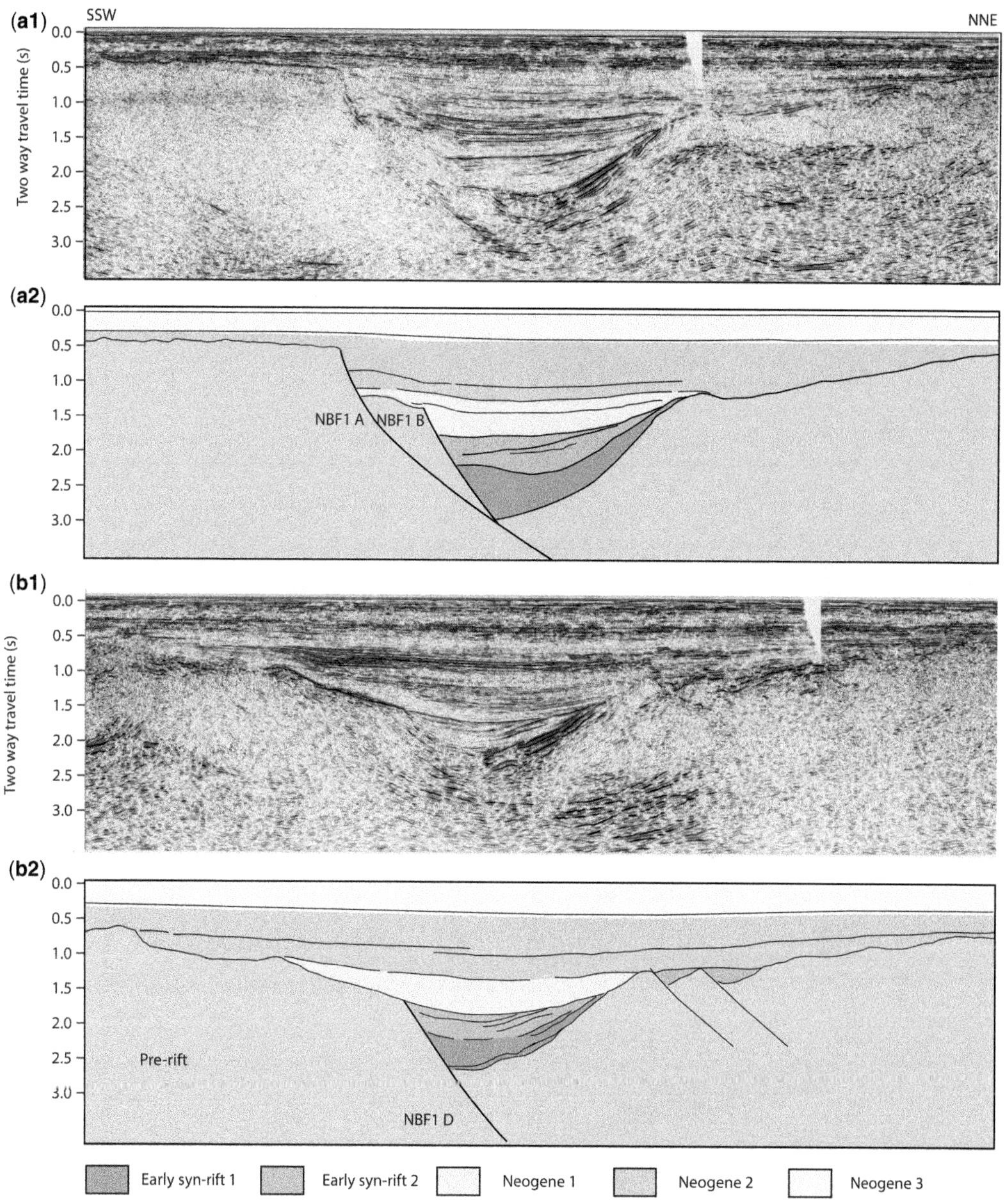

Fig. 9. 2D seismic lines across fault NBF1; see Figure 6 for location. (**a**) Seismic line across the NW part of NBF1: (a1) seismic line; (a2) interpretation. (**b**) Seismic line across the SE part of NBF1. Note in (a2) Neogene 1 and 2 expand into the boundary fault segments B and A, respectively, whereas in (b2) these sections are clearly post-kinematic and seal fault segment D. See Figure 7 for the identification of fault segments.

striking less than about 22–23° to the direction of maximum horizontal stress ($S_{h\ max}$) are highly likely to be reactivated rather than new faults being initiated (Morley *et al.* 2004). Then, there is a critical transition zone where faults striking up to 45° to $S_{h\ max}$ are likely to be reactivated, but new faults are also likely to be initiated. Pre-existing fabrics greater than about 45° to $S_{h\ max}$ are unlikely to be reactivated unless they are very weak (Morley *et al.* 2004). Hence, in a setting with a pre-existing north–south 'weakness' and a NE–SW extension direction, it is likely that both north–south and NW–SE faults will initially form. As the new NW–SE faults become established, the situation

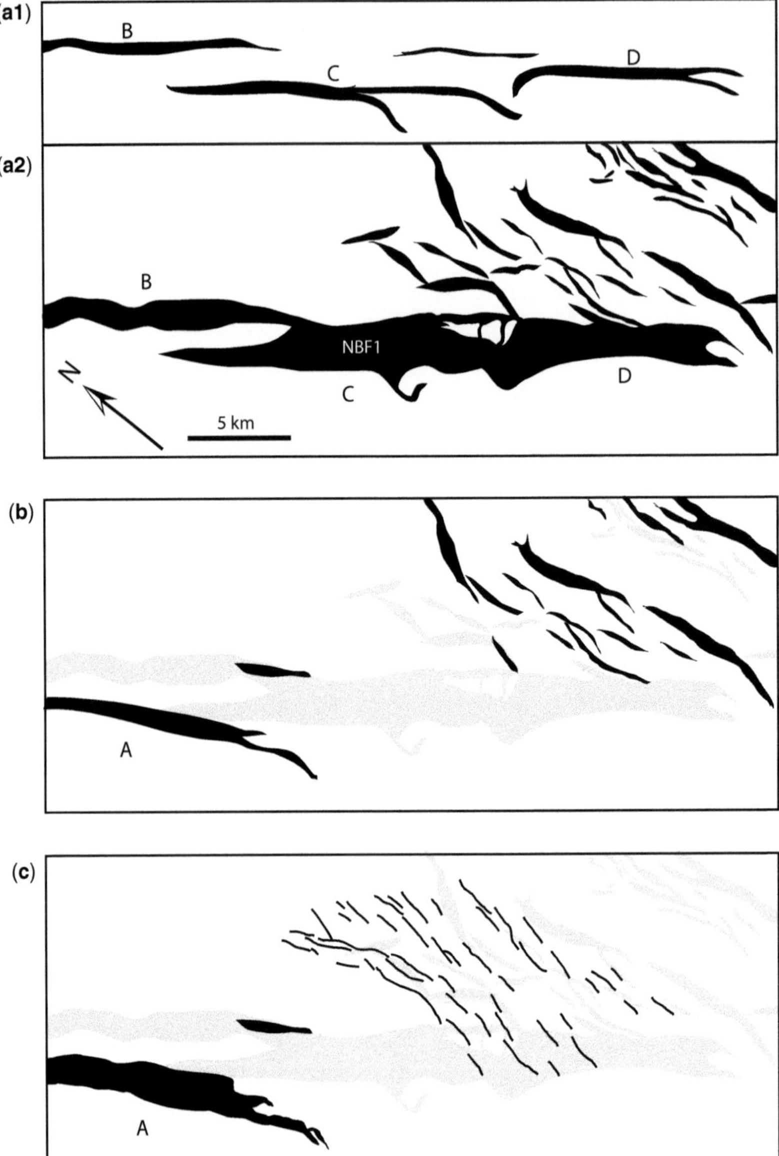

Fig. 10. Map view evolution of fault NBF1 and adjacent faults. Active faults and fault segments during the period are shown in black, inactive faults are in grey. Fault segments A, B, C and D of NBF1 are defined in Figure 7. (**a1**) Speculative early pre-linkage NBF1 fault configuration (Oligocene). (**a2**) Early rift stage (Oligocene) time of the main development of NBF1. (**b**) Early Miocene stage, segment A active together with some north–south-trending faults. (**c**) Late Miocene stage. Segment A active, and a new set of minor normal faults developed.

changes from reactivation of a zone of weakness v. isotropic rock strength, to activation of two cohensionless fault trends, one sub-parallel to $S_{h\ max}$ and one highly oblique to $S_{h\ max}$. Consequently, the approximate extension-orthogonal trend would develop more easily than the north–south fault set. This would be manifest as higher displacement values and longer fault lengths for the NW–SE set by the end of the syn-rift stage (Table 1).

Late extensional faults

The relatively low-displacement faults that affect the post-rift stage display similar low displacements

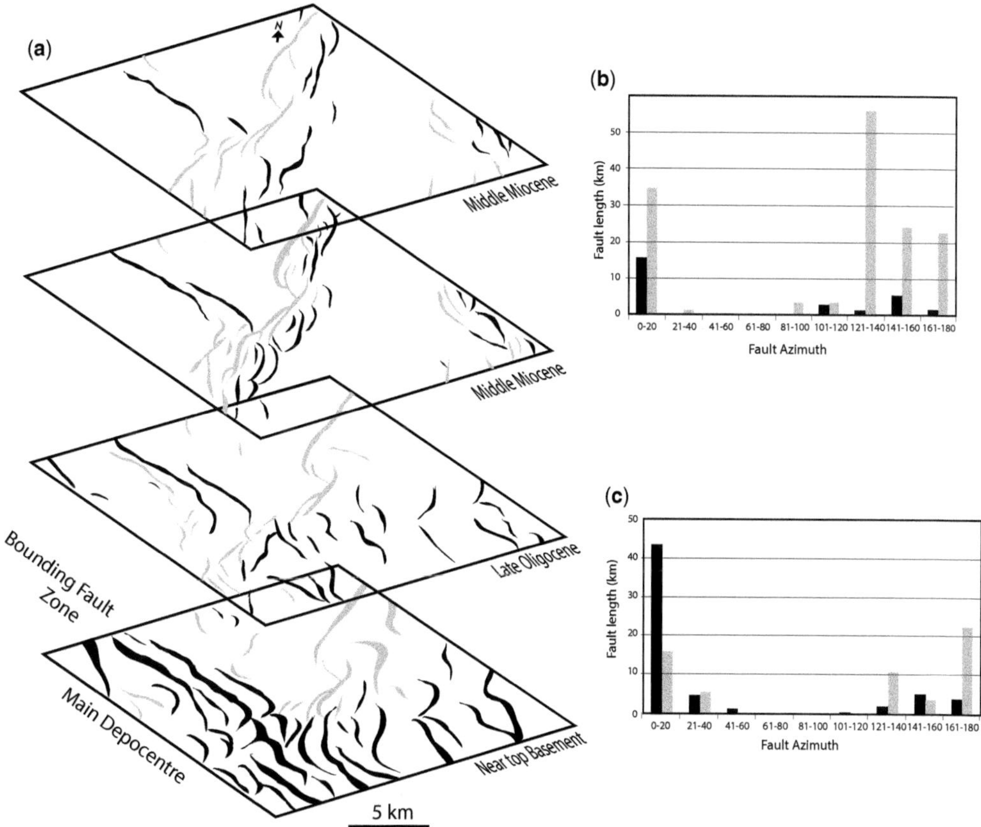

Fig. 11. Illustration of lateral and vertical variations in fault orientation, North Malay Basin, based on mapping of 3D seismic data, redrawn from Suwanruji (2003). (**a**) Series of time-slices illustrating how fault patterns vary with depth. Grey faults, east- to NE-dipping; black faults, west- to SW-dipping. (**b**) Plot of fault azimuth v. fault length for the Middle Miocene. (**c**) Plot of fault length v. azimuth for the Late Oligocene and older section. See Figure 12b for location.

for most faults (tens of metres), no matter what the orientation. Two exceptions are the north–south-trending Ton Son Fault and the Arthit hinge zone fault (Fig. 12), which display hundreds of metres of displacement. Hence, fault displacement considerations suggest (but not very strongly) that the north–south trend is the dominant post-rift orientation. Modern stress orientations also indicate that a north–south-trending $S_{\mathrm{h\,max}}$ direction is dominant, although there is considerable local variability (Tingay *et al.* 2012).

Assuming that all orientations of syn-rift faults have the same cohesive strength and frictional characteristics (i.e. an equal opportunity to be reactivated), then the percentage length of syn-rift faults in a particular orientation is a measure of the relative importance of that pre-existing fabric trend. To estimate the influence of pre-existing fabrics on the post-rift fault population, the post-rift fault length for a particular orientation is multiplied by the percentage of the pre-existing faults outside of the post-rift orientation in question. Hence, for the 1–20° range, the syn-rift faults of the same orientation

Table 1. *Total fault length and fault length as a percentage of the whole fault population and average displacement (maximum displacement on each fault/number of faults) data for syn-rift faults in Figure 11*

Strike directions	Fault length (km)	% fault length	Average displacement (m)
1–20°	59	24.7	147
21–40°	4	1.7	56
41–60°	6	2.5	50
101–120°	9	3.7	52
121–140°	93	39.1	334
141–160°	38	15.7	58
161–180°	30	12.6	65

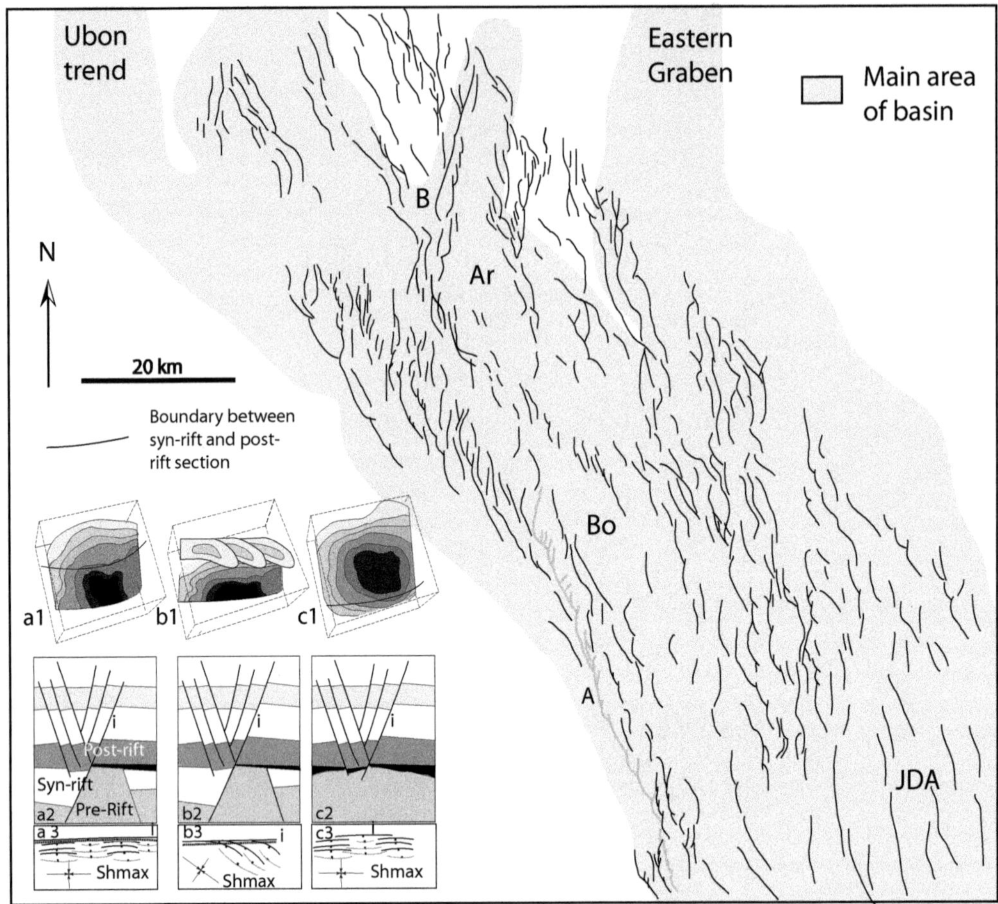

Fig. 12. Map of the general post-rift fault patterns compiled from publicly available location maps (no specific mapping horizon provided) (http://www.dmr.go.th/download/asean/Deep%20Oligocene.pdf and http://www.mtja.org/petroleumpotential.php). A denotes the long, low-displacement highly segmented post-rift fault trend; B denotes the approximate location of Figure 11. **(a)–(c)** The basic interactions between syn-rift and post-rift faults in the basin from the point of view of idealized displacement contours on the fault surfaces (a1, b1, c1), cross-section geometry (a2, b2, c2) and map view (a3, b3, c3). (a) Syn-rift fault reactivated to propagate upwards during the post-rift stage; (b) post-rift faults nucleated within the sedimentary section at an angle to the older syn-rift faults propagate and link along the syn-rift fault trend; and (c) post-rift fault nucleates within the sedimentary section and propagates in all directions including downwards into the syn-rift and pre-rift section (modified from Morley *et al.* 2011). See Figure 2 for the location of the North Malay Basin. Ar, Arthit area; Bo, Bongkot area; JDA, Joint Development Area (Thailand, Malaysia).

account for 24.7% of the total pre-existing fault population. Post-rift fault length is modified by multiplying by (1–0.247) – that is, by multiplying by 0.753 – to obtain the inherited fabric modifier in Table 2. The aim is to try and identify whether pre-existing fabrics are strongly biasing the later fault orientations or whether they are being bypassed. The most frequent syn-rift fabric orientation is the 121°–140° trend, which, despite representing 39.1% of the syn-rift orientations, is only followed by 11% of the post-rift faults. The presence of a significant fault population available for reactivation that failed to reactivate implies that the 121–140° population was significantly misorientated with respect to the post-rift stress field. Conversely, the post-rift fault population occupying the 0–20° orientation, whilst enjoying some advantage of underlying syn-rift faults of similar orientation, occupies an area much larger than that covered by the syn-rift faults of that orientation. Post-rift faults of the 1–20° trend in places overlie areas dominated by the 121–140° syn-rift fault population. Hence, the data strongly

Table 2. *Total fault length and fault length as a percentage of the whole fault population data for post-rift faults in Figure 11*

Strike directions	Fault length (km)	% fault length	Inherited fabric modifier
1–20°	59	50	$1 - 0.247 = 0.753$ $0.753 \times 59 = 43.4$ 46.4%
21–40°	9	7.6	$1 - 0.017 = 0.983$ $0.983 \times 9 = 8.8$ 9.5%
41–60°	2	1.7	$1 - 0.026 = 0.975$ $0.975 \times 1.7 = 1.7$ 1.8%
121–140°	13	11	$1 - 0.391 = 0.619$ $0.619 \times 13 = 8.2$ 8.7%
141–160°	9	7.6	$1 - 0.157 = 0.843$ $0.843 \times 9 = 7.6$ 8.2%
161–180°	26	22.1	$1 - 0.126 = 0.874$ $0.874 \times 26 = 22.7$ 24.7%

The inherited fabric modifier is described in the text and adjusts the % fault length to reflect the percentage of syn-rift faults available for reactivation.

point to the 1–20° trend lying sub-parallel to the maximum horizontal stress direction ($S_{h\ max}$).

From the comparison of fault lengths for different orientations discussed above, it has been inferred that the main syn-rift trend was NW–SE, with a secondary north–south to NNE–SSW trend developed in places. This secondary trend was probably influenced by pre-existing fabrics and was strongly oblique to the inferred NE–SW regional extension direction. Conversely, the dominance of the north–south to NNE–SSW trends in the post-rift section imply approximately east–west extension during post-rift times.

Fault length–displacement characteristics of the Bongkot area

As a general rule, most normal fault populations are dominated by *D*:*L* ratios of approximately 1:10–20: hence, a large boundary fault in a rift system with 6 km of displacement is commonly of the order of 60–120 km long. By comparison, the longest fault system in the Bongkot area is 80 km long (Fig. 12, fault A), but it displays only 600 m of maximum displacement in the post-rift section. Hence, the fault is under-displaced for its length. The overall NW–SE-trending fault shows some 19 north–south-trending branches spaced some 2–6 km apart (Fig. 12). Twenty-one separate centres of relatively high displacement are spaced along the entire length of the fault. The 600 m of maximum displacement is close to a normal *D*:*L* ratio for a 6–12 km segment. The characteristics of the fault discussed above illustrate that the strong NW–SE trend is clearly an extreme example of fault linkage: in this particular case, the NW–SE trend of the underlying syn-rift faults influenced the linkage of the north–south faults, the southern tips of which turned into the NW–SE trend and linked (Fig. 12). The fault population as a whole shows a large number of relatively high *D*:*L* ratios, particularly for the longer faults. The shorter faults tend towards more typical *D*:*L* ratios of between 1:10 and 1:20, these are for the isolated faults that have not linked. Therefore, the strong NW–SE trend cannot be assumed to indicate that extension has been approximately orthogonal to the trend. The preference for linkage of the southern tips seems to be related to the faults preferentially forming in the hanging walls of the syn-rift faults.

As discussed in Kornsawan & Morley (2002) and Morley *et al.* (2007, 2011), a large number of faults in the Tertiary basins of Thailand appear to have nucleated within the sedimentary section. These faults can either act as isolated faults (Fig. 12a1) or they may interact with other faults (Fig. 12b1). In some cases, this interaction will result in the faults nucleated within the sedimentary section propagating downwards to partially link with basement-involved faults (Fig. 12b1). Where the syn-rift and post-rift faults have different trends, linkage tends to occur along the syn-rift trend (e.g. NW–SE in the North Malay Basin: fault A, Fig. 12), with short splays representing the post-rift trend (north–south) coming off the NW–SE trend (Fig. 12). Consequently, there is a complex interaction of displacement patterns on some faults. The post-rift fault population of the Bongkot area appears to contain a large number of faults nucleated within the sedimentary section that were influenced by underlying syn-rift faults.

Local stress rotations

Although often it can be inferred that multiple trending fault directions in rifts are related to pre-existing fabrics, stress rotation is another potential cause (see Reeve *et al.* 2015 for a review of all causes). Stress rotation can affect secondary faults as a result of deflections caused by irregularities in the geometry of large faults (bends, intersections), as Maerten *et al.* (2002) demonstrated for a North Sea rift example. In some cases, the bends and intersections may follow pre-existing fabrics, and, hence, there is an indirect impact of pre-existing fabrics on the secondary fault orientations. Stress rotation can also occur in metamorphic fabrics (e.g. Morley 2010; Jolivet *et al.* 2010; Corti *et al.* 2013*a*, *b*), which may affect fault orientation or dip amount.

In some cases in Thailand, fault geometries on 3D seismic data can clearly show that stress rotation is the cause of oblique fault trends. Figure 13 illustrates a fault zone that in cross-section looks like a simple array of soft-linked faults. In map view, the upper time-slice (A) shows that between the two north–south-trending main faults (F11 and F8), the intervening faults lie oblique to them (NNE–SSW). However, passing down through slices C and D, the faults become sub-parallel. This twisting

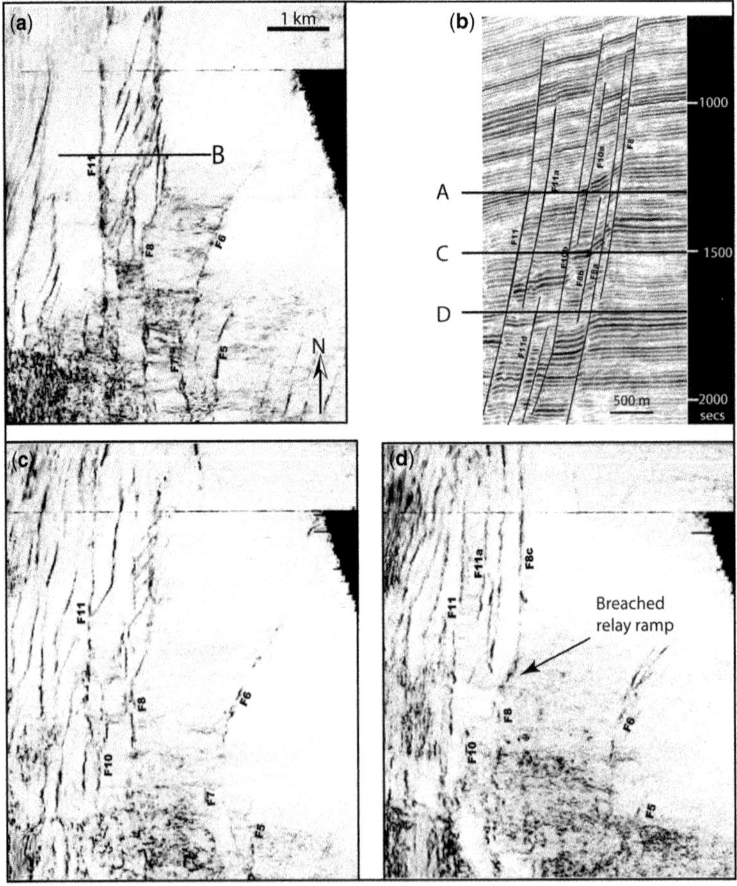

Fig. 13. 3D seismic data, showing coherency slices through a relatively small array of normal faults, illustrating how within a fault zone fault orientations can change vertically. (**a**) Upper time-slice; see (b), line A for level of intersection through the faults. (**b**) Vertical seismic section at location B in (a). (**c**) Middle time-slice, location C in (b). (**d**) Lower time-slice, location D in (b).

in fault orientation passing upwards indicates that there is local stress rotation within the fault zone, and no influence from underlying faults. The only irregular feature in the fault zone that is present in all the slices is the breached relay ramp in fault 8 (Fig. 13d): probably, a stress perturbation associated with this feature initiated the higher-level oblique fault development. Giba *et al.* (2012) also noted fault twisting (rotation in fault orientation by 10–20° passing up the fault zone) in response to oblique reactivation of a normal fault system in the Taranaki Basin. In the example above, the twisting is attributed to the spatial rotation of stress within the fault zone: in Giba *et al.* (2012), however, the cause is a change in stress orientation with time.

Modern stress orientations from well data show that, although the average trend of the maximum horizontal stress ($S_{h\ max}$) is north–south, within the rift basins there are considerable local rotations of the $S_{h\ max}$ direction (Tingay *et al.* 2010). The greatest concentration of modern stress data is from the Pattani Basin, where the orientations of the minor, conjugate, normal fault trends can be compared with the $S_{h\ max}$ direction (Fig. 14). In general, the variations in $S_{h\ max}$ trend follow changes in direction of the conjugate faults. However, in some cases, the $S_{h\ max}$ orientation lies at a high angle to the conjugate faults. This may reflect a perturbation from deeper syn-rift faults. In one example from the North Malay Basin, there is a documented vertical rotation in stress orientation of 60° passing from $S_{h\ max}$ sub-parallel to the NNW–SSE-trending faults in the post-rift Miocene section to $S_{h\ max}$ lying sub-parallel to the WNW–ESE-trending syn-rift faults in the Oligocene section (Tingay *et al.* 2010).

The conjugate post-rift faults tend to develop on the following features: basement highs; large-syn-rift faults (boundary faults, horst blocks); and the main depositional axis of the post-rift basin (Fig. 15). Consequently, these faults show changes in orientation influenced by basement features, such as basin margins, and uplifted footwall blocks. Figure 16 illustrates examples of deflected Miocene-level 'post-rift' faults, one is a change from north–south orientations in the middle of the basin to a curved pattern (NW–SE, NE–SW to ENE–WSW and WNW–ESE trends) approaching the basin terminations (Fig. 16a). This reflects a local reorientation of the stress direction by the basement trend. The other is NE–SW trends in the central part of the basin where north–south trends dominate: this reflects a response to the underlying syn-rift fault that has splays which trend NE–SW. However, the basement splays occur on a few faults only, not the many faults in the post-rift trend, which suggests that reorientation of the stresses locally by the syn-rift faults is responsible for the post-rift trends, not direct reactivation of the syn-rift faults.

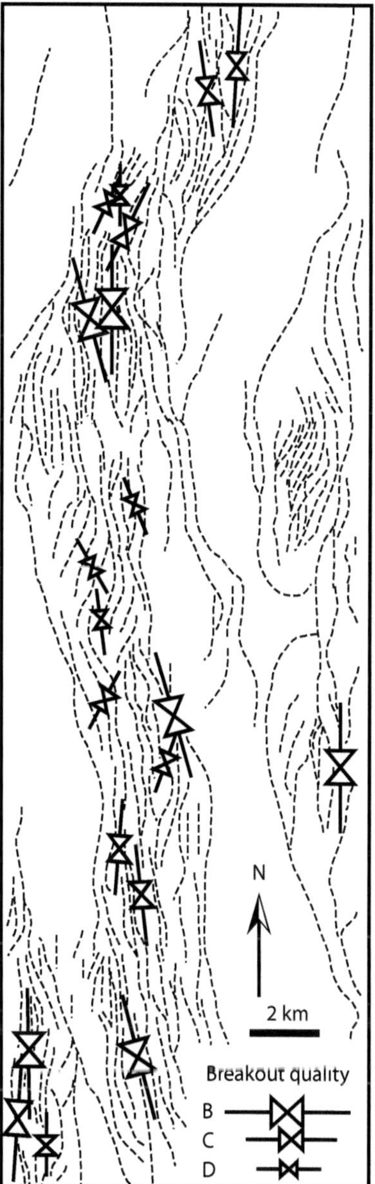

Fig. 14. Pattani Basin post-rift fault pattern from 3D seismic data, plotted with maximum horizontal stress orientations derived from borehole data (Tingay *et al.* 2010). See Figure 2 for the location.

Evolution of faults in the eastern Gulf of Thailand

One of the most complex normal fault patterns found in the Gulf of Thailand is in the Eastern Graben. In places, the fault pattern is almost polygonal (Fig. 17). The basin system initially was controlled

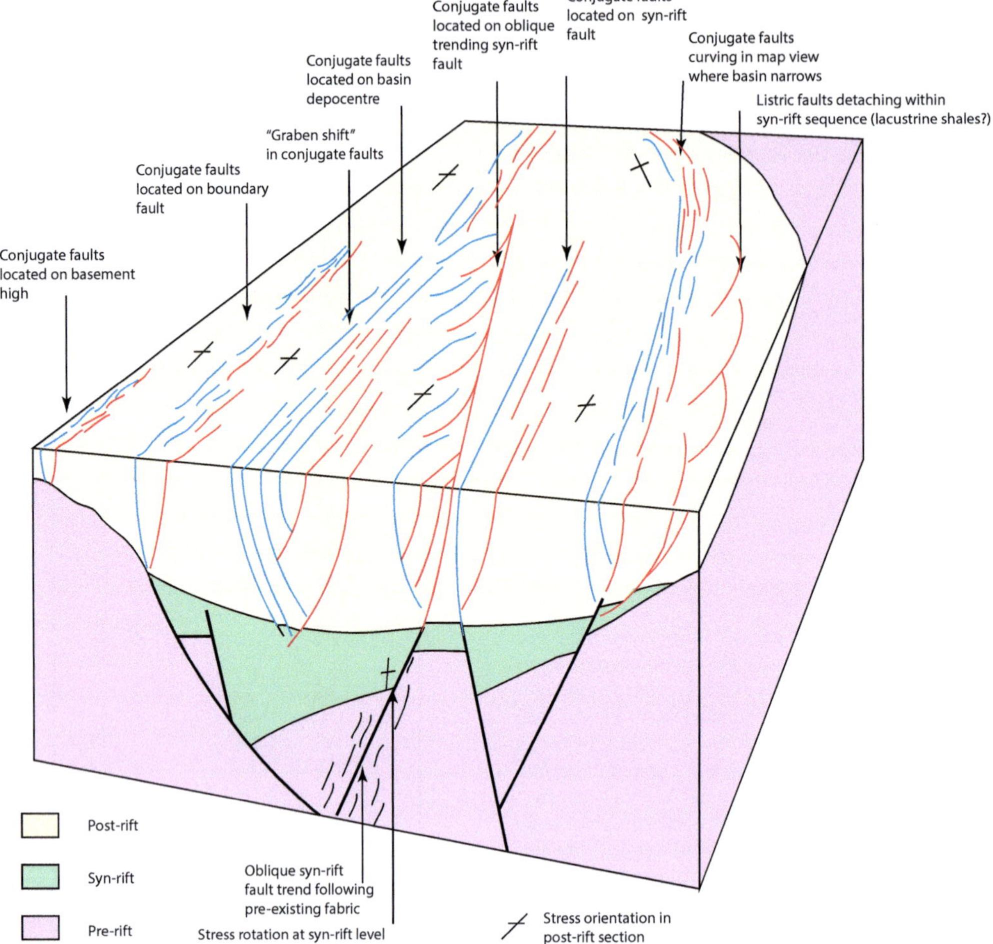

Fig. 15. Schematic illustration of the relationship between faults in the post-rift section and underlying syn-rift and basement features typically found in the Gulf of Thailand rift basins.

by large half-graben bounding faults, ENE-dipping faults were initially dominant, but in the southern part of the main depocentre the ENE-dipping fault became inactive, while the WSW-dipping faults became dominant (Fig. 18). Later, a system of WNW–ESE- to east–west-trending minor faults developed, some overlying the inactive ENE-dipping boundary fault II (Fig. 18b). Subsequently, predominantly NW–SE-trending minor faults developed (Figs 17 & 18c) before syn-rift activity was terminated by a phase of inversion (Fig. 18d). In detail, the fault pattern for stage C (Late Oligocene) is complex, with a wide variety of fault orientations (Fig. 17). These orientations reflect reactivation of earlier oblique fault trends, as well as probable local reorientation of stresses around the major fault trends (e.g. Maerten *et al.* 2002) and also, to a minor degree, downwards propagation of later post-rift faults into the syn-rift section. Subsequently, minor post-rift faults developed probably in response to ENE–WSW-orientated extension (Fig. 18e). The locations of these faults tend to be strongly influenced by underlying rift fabrics, as summarized in Figure 15.

Mapping normal faults on 3D seismic data in the Eastern Graben requires very detailed picking in some areas. Between stage B and stage C, and stage C and stage E, faults that nucleated within the sedimentary section have, in places, formed at an oblique angle to the underlying faults. How these faults propagate downwards and interact with the underlying faults can be complex. The underlying faults may themselves be reactivated in the later stages, or they may more passively interact with the newer

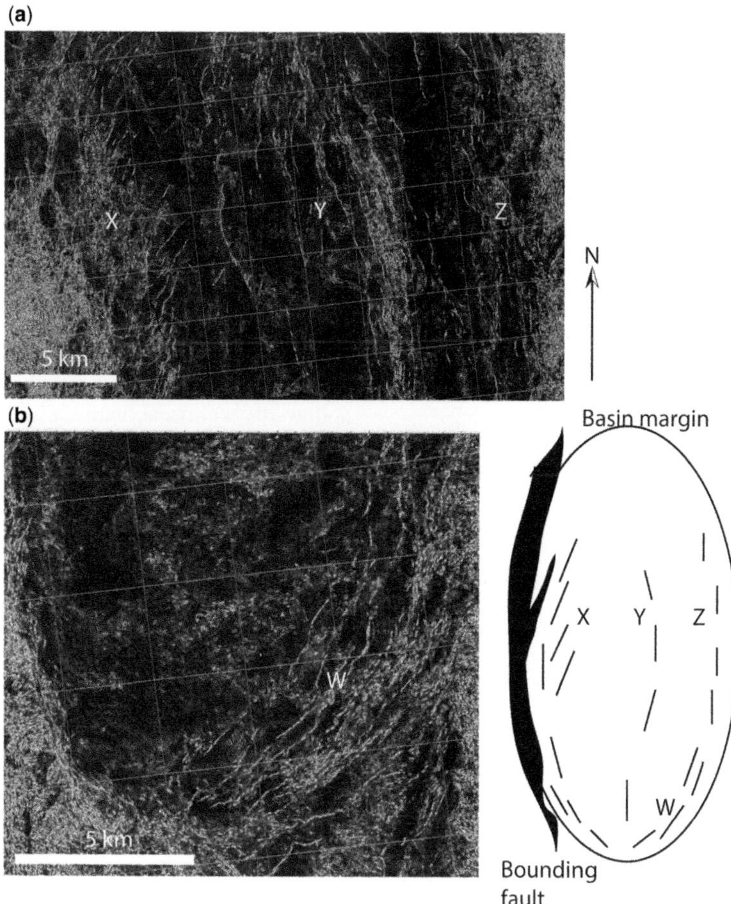

Fig. 16. Similarity time-slices of the Songkhla Basin (see Fig. 2 for the basin location) illustrating two examples of where the faults locally deviate from the region north–south-trending faults. The map schematically shows the location of the trends within a simple oval half-graben basin with a bounding fault on the western side. (**a**) Time-slice at 1.1 s two-way-travel time (TWTT) illustrating changes in fault orientation in the central part of the basin. Y is a deflection to a NNW–SSE trend due to the underlying syn-rift fault trend; X deflection to the NE is probably due to local stress rotation. (**b**) Deflection of faults to NE–SW to ENE–WSW trends at the southern margin of the basin following the narrowing of the basin as it dies out to the south.

faults. Figure 19 illustrates such an interaction from the Nakhon Basin. In Figure 19a, fault ‘z’ is a large-displacement fault confined to the Oligocene section (i.e. it is an early syn-rift fault). Between Figure 19a and b, fault ‘x’ appears and is an independent fault lying west of fault ‘z’; in Figure 19b, fault ‘x’ is only slightly curved, but a more strongly curved splay is just starting to develop and link with fault ‘z’. Following the fault to the south (Fig. 19c), fault ‘x’ turns listric. In Figure 19d, e, fault ‘x’ becomes less listric passing south. Finally, in Figure 19e, faults ‘x’ and ‘z’ look as if they are just part of a single fault. Mapping out the surfaces of faults ‘x’ and ‘z’, it can be seen that there is a difference in average strike of the fault plane contours of about 20° between faults ‘x’ and ‘z’ (Fig. 19f). Probably, the difference in orientation is due to a rotation in the extension direction between the early and later stages. The divergence in the two trends is why fault ‘x’ is separate from fault ‘z’ in the north, while it becomes part of that fault in the south (Fig. 19e). To call fault ‘x’ a splay would be inaccurate because it is a younger fault than fault ‘z’, and it is linked probably because of reactivation and use of fault ‘z’ as a zone of weakness. It is, perhaps, better described as fault ‘x’ capturing part of fault ‘z’.

Figure 20 schematically illustrates how the lower fault sets can passively interact with, or be captured by, a downwards-propagating fault.

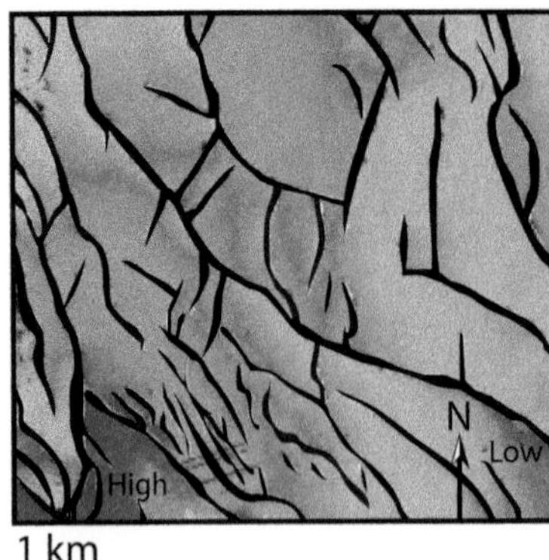

Fig. 17. Part of a time–structure map around the Oligocene–Miocene boundary, from 3D data in the Eastern Graben illustrating complex fault patterns as a result of inherited fabrics and local stress rotation. It is representative of stage (c) in Figure 18.

Although the general orientation of the overlying fault (fault A in Fig. 20) is oblique to the underlying ones, locally (two areas marked 'X' in Fig. 20) the overlying fault surface twists to join the underlying faults (faults B and C in Fig. 20). Consequently, picking on a relatively coarse 3D seismic grid, or 2D data, may cause the overlying fault to be incorrectly mapped as part of the deeper fault trend, rather than being a separate fault of different orientation. With closer-spaced mapping, and/or the use of good-quality coherency or similarity time-slice, the problem should become apparent.

Figure 21 illustrates a post-rift fault developed during the Late Miocene that is 4 km long and, during its development, there was a significant change in geometry. The deeper part of the fault trends NNE–SSW (Fig. 21c), while the upper part of the fault trends more north–south. However, inheritance of the deeper fault trend is apparent in segments where the NNE–SSW trend is present and links the deeper fault. If the upper segments (Fig. 21a, b) of the fault were observed, it would be inferred that the sequence of development was a series of isolated north–south-trending faults that were later linked along breached relays. However, with the presence of the older NNE–SSW trend, the 'breached relay' orientations actually developed first, and the north–south faults are partially rotated vertically to link with the older trends, while, in places, the fault propagated to form new north–south-trending segments. Hence, although there is a linkage geometry, it is a pseudo-relay geometry, not an genuine one.

Discussion

In the rifts of Thailand, permutation of stress with time, spatial local rotation of stress, and the influence of pre-existing fabrics and structures have all played a role in the development of normal faults both in syn-rift basins and in the succeeding sag basins (Fig. 14). These complex interactions produced, in some basins, fault systems that depart significantly in terms of fault geometry and fault population displacement–length characteristics from those developed in single extension event, orthogonal extension systems.

In the Andaman Sea, the location of rifting has changed with time, with extension in the Mergui Basin becoming younger passing northwards in the basin as the extension direction changed from WNW–ESE to NNW–SSE with time. The long boundary faults developed overall by propagating northwards with time, with the southern segment becoming abandoned. The northern part of the extensional West Mergui Fault, which contains low-angle segments, actually links with the NNW–SSE-trending, subvertical Mergui Fault (Fig. 5), which is a dextral strike-slip fault, active during the Early–Late Miocene (Morley 2014; Srisuriyon & Morley 2014). This marks the transition northwards towards a more strike-slip-dominated province, where NNW–SSE-trending strike-slip faults terminate in NE–SW-trending extensional fault splays at their tips (Srisuriyon & Morley 2014). Faults that developed where the rift overall is propagating in one direction can deviate from the basic models of fault development shown in Figure 1. A schematic example of such a deviation is given in Figure 22. The asymmetric propagation direction means that as the fault develops one tip becomes abandoned, while propagation of the other tip occurs at some stage in the fault development. Consequently, the fault is not necessarily active along its entire final length at any stage in its development, and while stages of the development may be classed as isolated, or coherent, overall the development is neither.

The shifting of the location of extension in the Mergui Basin contrasts with the failed rift trend, where extensional activity remained in the same location from the Palaeogene through much of the Neogene. In some basins, particularly the Pattani Basin, extension in the syn-rift and post-rift phases occurred under east–west regional extension. Long faults in the syn-rift and post-rift section tend to trend north–south, except where the syn-rift faults follow oblique pre-existing fabrics, as summarized in Figure 15. Short post-rift faults commonly lie at an angle (NNE–SSW to NE–SW, and NNW–SSE to NW–SE), probably due to local stress rotation caused by deflection of stresses by irregularities in the syn-rift faults. To the SE, the North Malay Basin exhibits a different post-rift fault pattern where the NW–SE trend is the dominant syn-rift trend (NE–SW extension

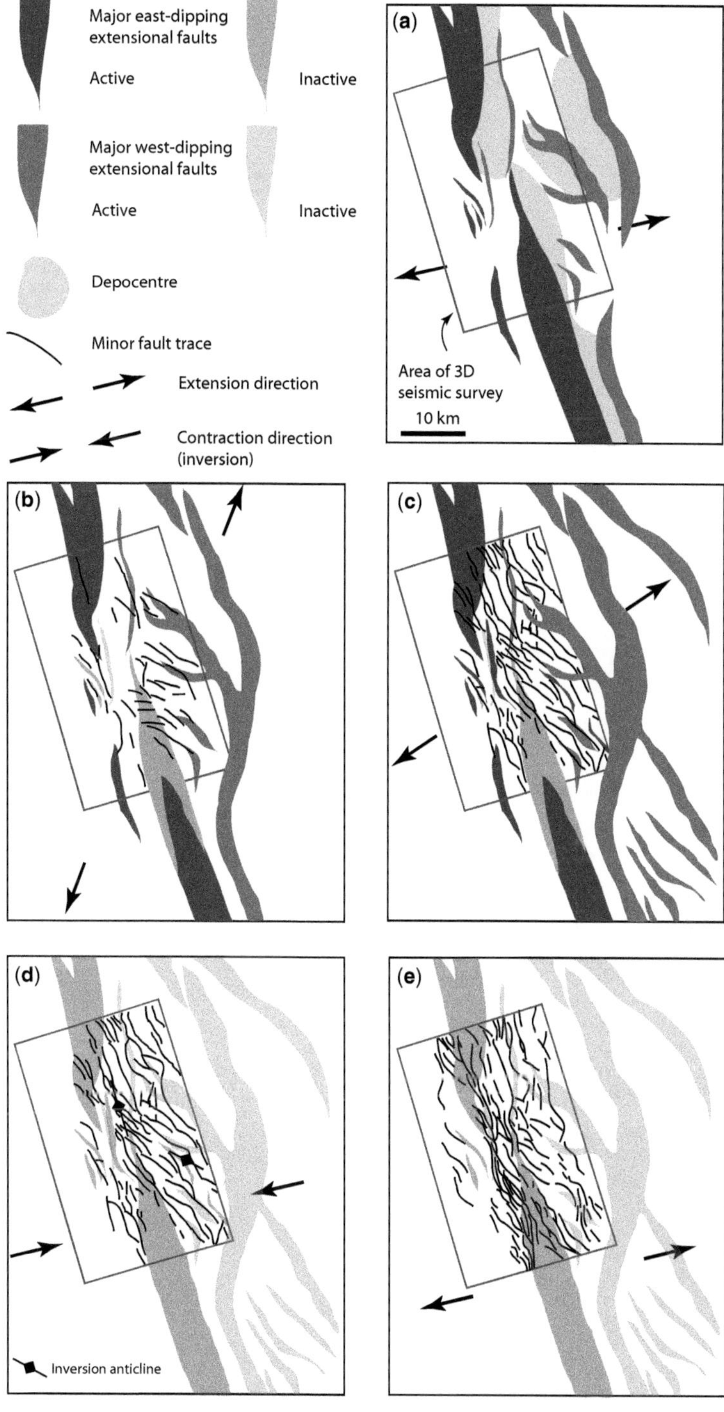

Fig. 18. Fault evolution in the Eastern Graben (see Fig. 2 for the location of the graben), illustrating complex fault development due to multi-phase extension and permutation of stresses with time: (**a**) early syn-rift, Late Eocene–Early Oligocene; (**b**) syn-rift, Oligocene, rotation to NE–SW extension; (**c**) late syn-rift, ENE–WSW extension; (**d**) tectonic inversion stage around the Oligocene–Miocene boundary; and (**e**) Middle Miocene post-rift stage with minor conjugate extensional faulting.

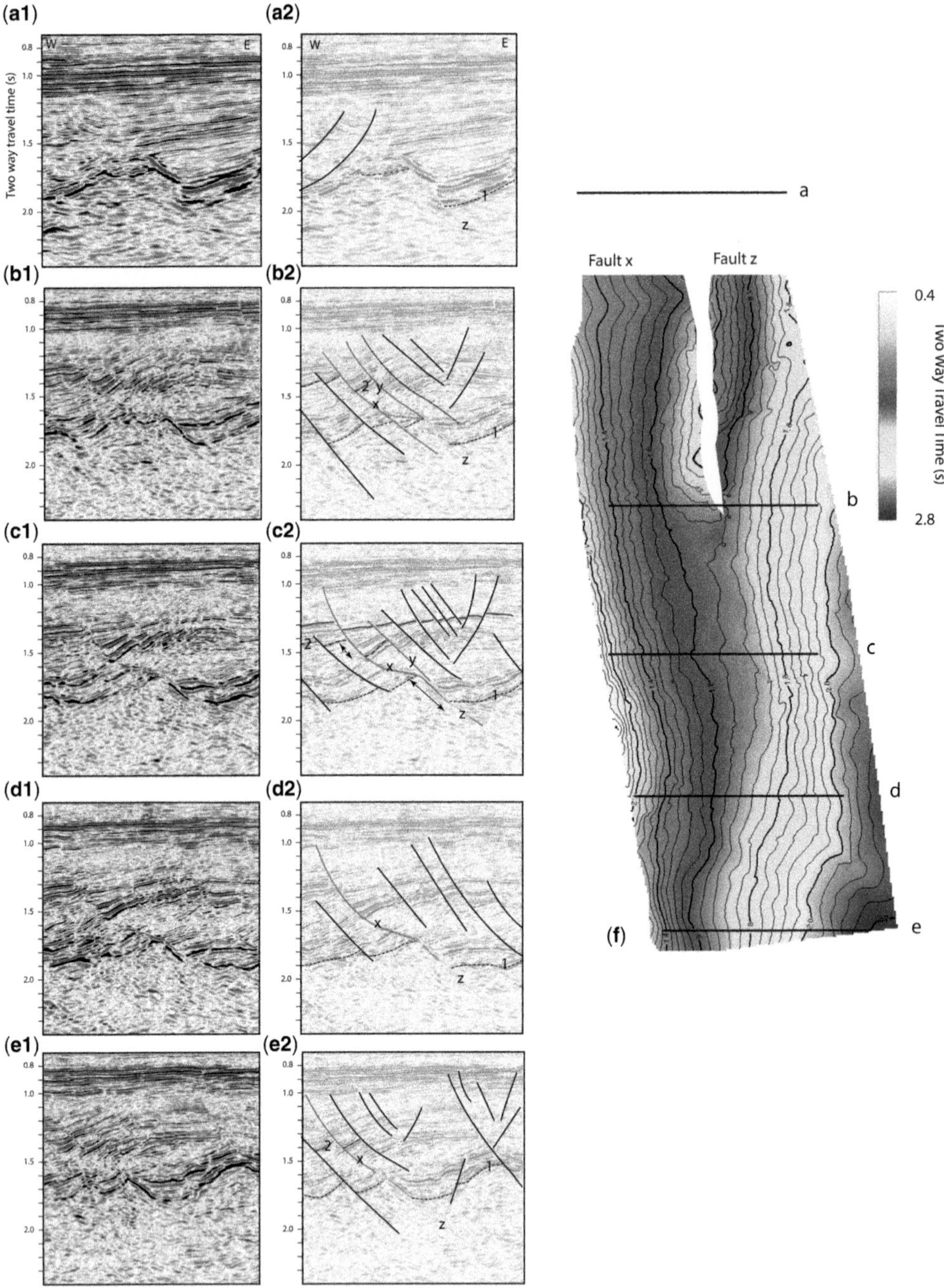

Fig. 19. Illustration of capture and partial reactivation of an older syn-rift fault ('z') by younger faults ('x' and 'y'), Nakhon Basin (see Fig. 6 for the location). (**a**)–(**e**) Arbitrary lines from 3D seismic data, locations are shown on a time–structure map of the 'x' and 'z' fault surfaces. Key faults are labelled 'x', 'y' and 'z'; key horizons are 1 and 2.

direction), and was followed by rotation to east–west extension during the post rift phase. Hence, north–south-trending post-rift faults tend to link along NW–SE-trending syn-rift faults (Fig. 12b1), although there are complications due to the syn-rift faults locally following north–south-trending

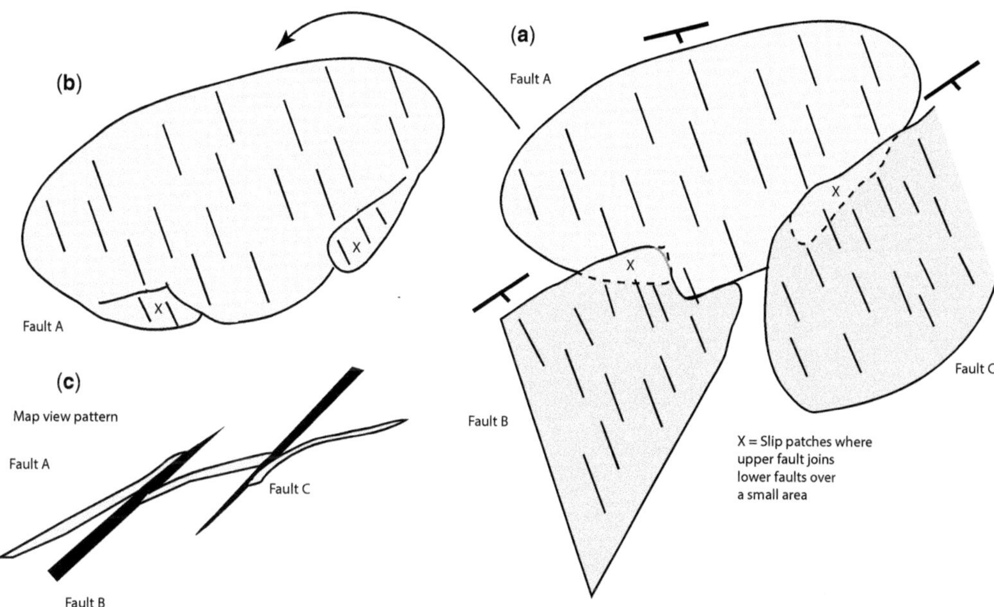

Fig. 20. (**a**) Schematic illustration of how a fault (A) that nucleates relatively high in the sedimentary section and strikes obliquely to underlying faults (B and C) can locally propagate downwards, twist and link with those faults (locations X). (**b**) Fault surface A showing twisting segments at locations X. (**c**) Map view sketch of how fault A locally links with faults B and C despite overall trending oblique to the faults. These relationships can cause confusion about fault correlations on seismic reflection data.

pre-existing fabrics (Fig. 11; Table 1). In addition to late faults being influenced by older faults, basement highs and sedimentary depocentres also will localize late fault sets (Fig. 15). Listric faults arising from sediment loading and detachment on lacustrine shales can also develop and interact with faults driven by regional extension (Fig. 15).

The displacement–length ratios of faults in Thailand range between extremes of severely under-displaced faults, to faults with extremely small length:displacement ratios (<8:1). The under-displaced faults are explained by linkage of low-displacement faults along an underlying, oblique pre-existing (syn-rift) fault trend following a rotation in stress direction (Fig. 12b1). In the North Malay Basin, the north–south fault segments that are linked along NW–SE trends tend act as independent faults despite their physical linkage along the oblique fault system. They each display displacement maxima on the north–south segments, as illustrated schematically in Figure 12b1. Conversely, the high-displacement boundary faults of similar length to fault A (Fig. 12) die to zero displacement very rapidly along strike. Such faults raise the question as to how displacement was built up over time.

The largest displacement faults in Thailand are commonly low angled or have some low-angled segments. They are typically 60–110 km long, and their displacements of 10–30 km are much higher than for high-angle boundary faults of similar length that show displacements of 1–5 km. The faults are commonly segmented in terms of their displacement profiles, with multiple displacement highs and lows, and mixed high and low dip angles (e.g. Figs 4 & 5c) (Morley 2014), and so behave as a single isolated fault zone. It seems reasonable to conclude that upon reaching lengths of 60–100 km, the faults built displacement without needing to become longer. In the Aegean, for example, normal faults 20–40 km long are associated with $>M_s$ 6.5 earthquakes (Pavlides & Caputo 2004). Consequently, it is feasible for 60–100 km-long faults, even if segmented, to accumulate large amounts of slip by seismic events without increasing in length, provided earthquakes repeatedly occur close (*c.* 4–10 km) to the region of maximum slip. This phenomenon, observed by Cowie (1998) and Manighetti *et al.* (2005), best fits the Characteristic Earthquake Model (Schwartz & Coppersmith 1984).

The lower displacements on high-angled, 'well-orientated' faults with respect to regional stress orientations poses the question as to why low-angle normal faults would preferentially reactivate by seismic events over high-angle normal faults? Rupture by large seismic events implies that low-angle

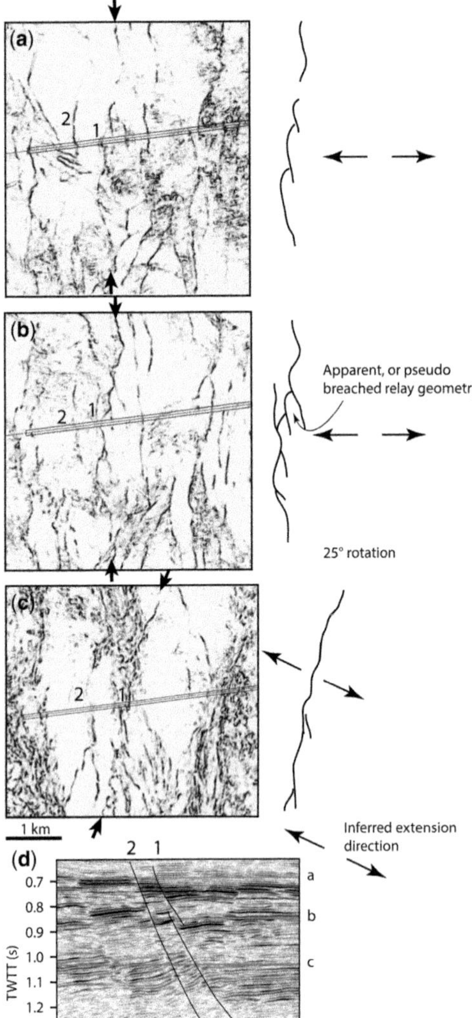

Fig. 21. (**a–c**) Sonkhla Basin, similarity cube time-slices illustrating that the lower part of fault zone 1 trends NNE–SSW (c), while the upper part trends north–south (a), probably due to clockwise rotation of the extension by about 25°. The interaction between two different stages of development of the fault zone produce apparent or pseudo-breached relay structures. (**d**) Seismic section corresponding to the ENE–WSW line crossing time-slices (a), (b) and (c), showing the depth locations of those time-slices.

normal faults have strength that would invalidate their ability to reactivate in preference to more optimally orientated high-angle faults. Consequently, more appropriate models to explain low-angle normal fault displacement accumulation are those where high pore pressures and efficient dilatant mechanisms are characterized by stable, aseismic creep (Hillers & Miller 2007). Creep has been observed on some low-angle normal faults (e.g. Chiaraluce *et al.* 2007; Hreinsdottir & Bennett 2009) and seems an appropriate mechanism for the accumulation of slip on low-angle normal faults in Thailand (Morley 2014). The repeated reactivation of the same region of the fault indicates that the length–displacement relationships break down for very weak faults, since they can continue to accumulate slip without becoming longer.

Fault-zone weakness also is a consideration for the evolution of the extension-oblique NBF1 Fault (Nakhon Basin) (Figs 6 & 10). Simple application of Mohr–Coulomb theory and Byerlee friction law suggests that non-optimally orientated faults should deactivate in favour of more optimally orientated faults. However, the observed development of NBF1, and even the creation of new oblique fabrics after older oblique and extension-normal faults have deactivated (Fig. 10), suggest that the actual development was more complex. Overall, the evolution of the fault can be viewed as the early oblique fault being replaced by later north–south extension-orthogonal faults. The continued activity of the NW portion of NF1 is, perhaps, due to that part of the fault zone being particularly weak, and lying west of the corridor of north–south-trending faults. However, if weakness is the main criterion, is it curious that the fault did not form during the Oligocene? The observations indicate that misorientation was not the main reason why the NBF1 segments B, C and D ceased activity in the Miocene. Instead, fault-zone strength is the probable answer, with segments B, C and D undergoing strain hardening to the point where the fault zones exceeded the strength of the fabric anisotropy in the basement rocks. A new segment (A) developed west of the shadow zone of the more optimally orientated north–south fault trends. Although commonly fault zones seem to progress from strain hardening in the initial process zone to strain softening in the principal slip surfaces (e.g. Shipton & Cowie 2003), strain hardening can be caused by rotation around horizontal axes of fault blocks during deformation (e.g. Agnon & Reches 1995), by reduction in pore fluid pressure with time and by the material properties of fault gouges (Morrow *et al.* 1982). The common occurrence of hot springs in association with Cenozoic fault zones in Thailand (see the review in Morley 2014) indicates that pore fluid pressures are one potentially significant, transient factor affecting fault-zone strength.

One common tendency in multi-phase rifts and/or rifts with local stress rotations is for an increase in the number of orientations used by faults, and for an increase in the linkage of faults compared with simple single-phase extension-orthogonal rifts.

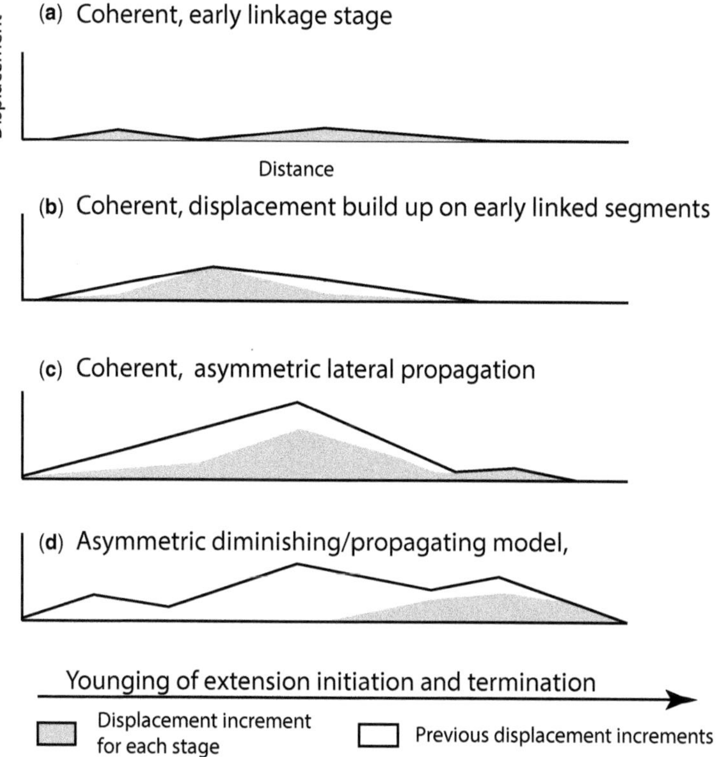

Fig. 22. Schematic displacement–strike distance plot of a large fault that has developed in a rift, the overall activity of which is propagating in one direction (such as the Mergui Basin). The example illustrates how a large boundary fault may diverge considerably from simple isolated or coherent fault models (Fig. 1). **(a)–(d)** Development of displacement with time.

In some cases, fault zones remain well organized and linkage can occur along discrete underlying oblique trends, producing long, under-displaced faults (Figs 4, 12 & 15). Such faults can still be treated as discrete faults and plotted on fault length–displacement diagrams, although they tend to behave as end members of the isolated fault model, and are under-displaced with respect to typical faults (Fig. 4). However, once three or more significant directions affect fault patterns, it can then become impossible to separate faults, except on some arbitrary basis, as they become linked in different directions, as illustrated in Figure 17. Although numerous linkage points on curved faults can be readily identified, there seems no satisfactory way to represent these faults on a length–displacement plot. Hence, it needs to be recognized that length–displacement plots do not represent all situations of normal fault development.

The two strikingly different modes of development of the under-displaced and over-displaced faults also reflect development under different strain rates (Fig. 23). The under-displaced faults developed during the Miocene and Pliocene, and at least the largest of the faults shows a slow, progressive accumulation of slip over a long period of time. Conversely, at least some of the low-angle normal faults, such as the one on the east side of the Chiang Mai Basin (30 km displacement), accumulated slip rapidly in, perhaps, 6 myr (Morley 2014). Figure 23 compares the length–displacement ratio and slip rates of faults from the Thailand rift with faults from other basins. It shows that there is a tendency for the very large faults with relatively rapid displacement rates to have low length–displacement ratios.

When trying to understand the influence of underlying fabrics, we are fortunate if seismic reflection data provide even hints about the nature of the underlying trends, except where older faults lie within the sedimentary section and can be mapped in detail from seismic reflection data. Commonly, studies of pre-existing fabric influence in rifts rely on reproduction of similar fault patterns from analogue models with pre-existing fabrics to justify inferences about the existence of pre-existing

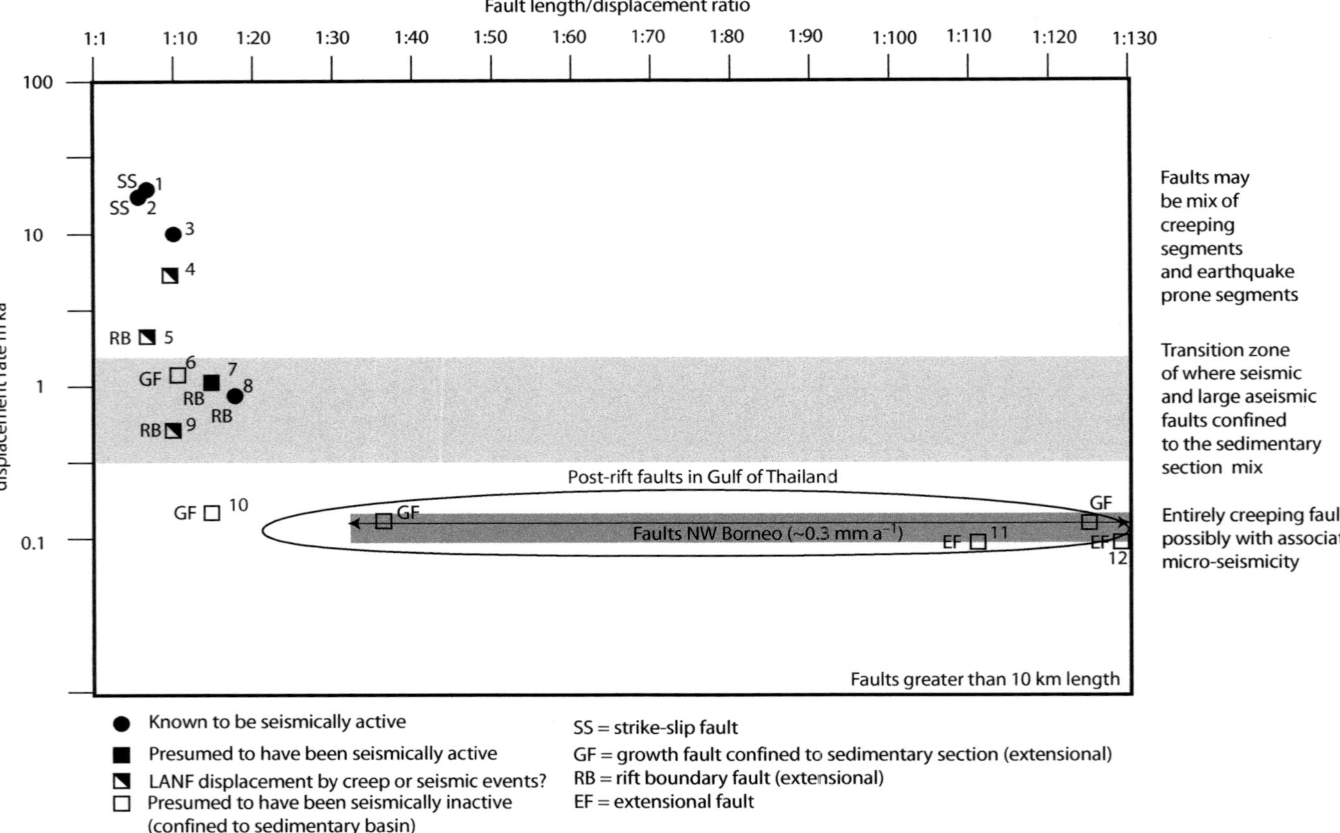

Fig. 23. Natural examples of fault length–displacement ratios plotted against displacement rate. (1) San Andreas Fault, USA (Powell *et al.* 1993); (2) Sagaing Fault, Myanmar (Rangin *et al.* 2013); (3) Main Boundary Fault, Himalayas (Meigs *et al.* 1995); (4) Thor–Odin dome (Norland *et al.* 2002); (5) Chiang Mai Basin, Western Boundary Fault (Morley 2014); (6) Perdana Fault, Baram Delta Province, Brunei (C.K. Morley unpublished data); (7) Lokichar Fault, Northern Kenya Rift (Morley *et al.* 1999*b*); (8) Rukwa Fault, East African Rift, Tanzania (Morley *et al.* 1999*a*); (9) Western Boundary Fault, Phitsanulok Basin, Thailand (Morley 2009); (10) large growth fault in Niger Delta (Pochat *et al.* 2009); (11) large, north–south-trending post-rift fault in the North Malay Basin; and (12) large NW–SE-trending post-rift fault in the North Malay Basin (Fig. 12, fault A). Minor faults in the Gulf of Thailand and Brunei estimated by the author from unpublished data. LANF, low-angle normal fault.

fabrics, coupled with outcrop and/or potential fields data that demonstrate basement fabrics. Hence, there remains uncertainty about the exact cause of the observed fault patterns.

Most studies of oblique faults and multi-rift scenarios have investigated the imposition of successive phases faults that have probably nucleated within crystalline basement and propagated into the overlying sedimentary section (e.g. Lepvrier *et al.* 2002; Morley *et al.* 2007; Henza *et al.* 2010, 2011; Reeve *et al.* 2015; Henstra *et al.* 2015). Hence, key areas of fault interaction are not well imaged on seismic data: in Thailand, however, the nucleation of faults within the sedimentary section, particularly during the post-rift stage, means that the interactions between downwards-propagating new faults and underlying older faults can be observed (Figs 11, 12, 14, 18e, 19 & 20). 3D seismic data are essential to revealing the details of how these deeper, older and higher younger faults interact. The data show that, in some cases, segments of higher faults of a particular orientation curve to join differently orientated deeper faults (Fig. 19). Caution needs to be exercised when attributing typical features such as relay ramps to multi-phase fault geometries because the geometry may be similar (Fig. 21), but the sequence of events that led to the geometry can be different.

There remains much to do in understanding the details of how pre-existing fabrics at different depths in the crust affect fault patterns. As discussed in the earlier subsection on 'North Malay Basin example', there is an opportunity to investigate and, possibly, quantify how fabrics are re-used when looking at the reactivation of fault zones. This can be done from looking at how fault populations behave (e.g. Table 1) and from mapping of individual fault zones (Figs 19 & 21).

Conclusions

The rift basins of Thailand exhibit a remarkable diversity of fault displacement patterns, fault length–displacement characteristics and fault patterns in map view during the late rift and post-rift stages. These patterns reflect a number of influences, particularly: (1) zones of strength anisotropy in the pre-rift basement; (2) syn-rift fault patterns on post-rift faults; (3) spatial stress deflection, commonly related to irregularities in major fault profiles, and the basement–sediment interface; (4) temporal stress rotation, usually related to changes in the regional plate setting; and (5) varying strength properties (strain hardening or softening) of fault zones during their life.

The effects of these influences have been to create strongly segmented boundary faults and long, low-displacement post-rift fault trends. The former are commonly strongly over-displaced, while the latter can be strongly under-displaced with respect to their length compared with typical distributions (Fig. 4). The complex networks of faults that can develop by reactivation of different fault orientations in multi-phase rifts can also make it impossible to separate out individual faults that can be plotted on length–displacement graphs, although the many linked fault segments can be identified. Hence, such fault systems are unrepresented on fault–length displacement graphs.

Seismic interpretation of multi-rift fault patterns requires 3D data to identify the complexities. Otherwise the linkage pattern between deeper and shallower faults may not be identified (e.g. Figs 19 & 21), and the changing fault strike directions with depth (e.g. Figs 11, 18, 19 & 20) may not be identified. Consequently, in both cases, fault correlations may be incorrect. In addition, misidentification of features, such as the incorrect identification of patterns as breached relay structures (Fig. 21), can arise if the full 3D geometry of the fault zones is not defined. Some of the interactions between deep and shallow faults in the basins are summarized in Figure 15.

The fault patterns affecting the rift basins of Thailand were initially interpreted as strike-slip-related pull-apart basins when only 2D seismic data were available. NW–SE trends correspond with the main strike-slip faults onshore. This interpretation persisted for the Malay Basin until recently (e.g. Fyhn *et al.* 2010; Mansor *et al.* 2014). However, 3D data, and analysis of boundary fault displacement patterns and evolution (e.g. Figs 6, 7 & 10), demonstrate that the oblique (NW–SE-trending) zones also exhibit predominantly normal or oblique-normal displacements. Consequently, oblique extension, the influence of pre-existing trends and stress rotation in multi-phase rifts provides a more comprehensive explanation for the observed features than strike-slip.

The ability of oblique fault segments to reactivate alongside more extension-orthogonal faults until late in the basin history (e.g. Fig. 10) is contrary to expectations based on simple Andersonian mechanics that misorientated faults will decline in activity when competing with optimally orientated faults under conditions of Byerlee Law friction (e.g. Morley *et al.* 2004). This suggests that the strength of fault zones can vary considerably over time.

Oliver Duffy and an anonymous reviewer are thanked for constructive comments that helped improve the manuscript. The author would like to thank PTTEP, Shell, Chevron, and Coastal Energy for providing data over the rift basins of Thailand over the years for research projects.

References

Agnon, A. & Reches, Z. 1995. Frictional rheology: hardening by rotation of active normal faults. *Tectonophysics*, **247**, 239–254.

Agostini, A., Bonini, M., Corti, G., Sani, F. & Mazzarini, F. 2011. Fault architecture in the Main Ethiopian Rift and comparison with experimental models: implications for rift evolution and Nubia–Somalia kinematics. *Earth and Planetary Science Letters*, **301**, 479–492.

Autin, J., Bellahsen, N., Leroy, S., Husson, L., Beslier, M.-O. & d'Acremont, E. 2013. The role of structural inheritance in oblique rifting: insights from analogue models and applications to the Gulf of Aden. *Tectonophysics*, **607**, 51–64.

Barnett, J.A.M., Mortimer, J., Rippon, J.H., Walsh, J.J. & Watterson, J. 1987. Displacement geometry in the volume containing a single normal fault. *American Association of Petroleum Geologists Bulletin*, **71**, 925–937.

Bellahsen, N., Fournier, M., d'Acremont, E., Leroy, S. & Daniel, J.M. 2006. Fault reactivation and rift localization. Northeastern Gulf of Aden margin. *Tectonics*, **25**, TC1007, https://doi.org/10.1029/2004TC001626

Brune, S., Popov, A.A. & Sobolev, S.V. 2012. Modeling suggests that oblique extension facilitates rifting and continental break-up. *Journal of Geophysical Research*, **117**, B08402, https://doi.org/10.1029/2011JB008860

Buck, W.R. 2007. Dynamic processes in extensional and compressional settings: the dynamics of continental breakup and extension. *In*: Schubert, G. (ed.) *Volume 6: Crustal and Lithosphere Dynamics, Treaties on Geophysics*, 1st edn. Elsevier, Amsterdam, 335–376.

Cartwright, J.A., Mansfeld, C. & Trudgill, B. 1996. The growth of normal faults by segment linkage. *In*: Buchanan, P.G. & Nieuwland, D.A. (eds) *Modern Developments in Structural Interpretation, Validation and Modelling*. Geological Society, London, Special Publications, **99**, 163–177, https://doi.org/10.1144/GSL.SP.1996.099.01.13

Chiaraluce, L., Chiarabba, C., Collettini, C., Piccinini, D. & Cocoo, M. 2007. Architecture and mechanics of an active low-angle normal fault: Alto Tiberina Fault, northern Apennines, Italy. *Journal of Geophysical Research*, **112**, B10310, https://doi.org/10.1029/2007JB005015

Childs, C., Watterson, J. & Walsh, J.J. 1995. Fault overlap zones within developing normal fault systems. *Journal of the Geological Society, London*, **152**, 535–549, https://doi.org/10.1144/gsjgs.152.3.0535

Childs, C., Nicol, A., Walsh, J.J. & Watterson, J. 2003. The growth and propagation of synsedimentary faults. *Journal of Structural Geology*, **25**, 633–648.

Corti, G., van Wijk, J., Cloetingh, S. & Morley, C.K. 2007. Tectonic inheritance and continental rift architecture: numerical and analogue models of the East African Rift system. *Tectonics*, **26**, TC6006, https://doi.org/10.1029/2006TC002086

Corti, G., Iandelli, I. & Cerca, M. 2013*a*. Experimental modeling of rifting at craton margins. *Geosphere*, **9**, 138–154.

Corti, G., Philippon, M., Sani, F., Keir, D. & Kidane, T. 2013*b*. Re-orientation of the extension direction and pure extensional faulting at oblique rift margins: comparison between the Main Ethiopian Rift and laboratory experiments. *Terra Nova*, **25**, 396–404, https://doi.org/10.1111/ter.12049

Cowie, P.A. 1998. A healing-reloading feedback control on the growth rate of seismogenic faults. *Journal of Structural Geology*, **14**, 1133–1148.

Cowie, P.A. & Scholz, C.H. 1992. Displacement-length scaling relationship for faults: data synthesis and discussion. *Journal of Structural Geology*, **14**, 1149–1156.

Crider, J.G. & Pollard, D.D. 1998. Fault linkage: three dimensional mechanical interaction between echelon normal faults. *Journal of Geophysical Research*, **103**, 373–391.

Flint, S., Stewart, D.J., Hyde, T., Gevers, C.A., Dubrule, O.R.F. & Van Riessen, E.D. 1988. Aspects of reservoir geology and production behaviour of Sirikit Oil Field, Thailand: an integrated study using well and 3-D seismic data. *American Association of Petroleum Geologists Bulletin*, **72**, 1254–1268.

Fyhn, M.B.W., Boldreel, L.O. & Nielsen, L.H. 2010. Escape tectonism in the Gulf of Thailand: Paleogene left-lateral pull-apart rifting in the Vietnamese part of the Malay Basin. *Tectonophysics*, **483**, 365–376.

Giba, M., Walsh, J.J. & Nicol, A. 2012. Segmentation and growth of an obliquely reactivated normal fault. *Journal of Structural Geology*, **39**, 253–267.

Hall, R. 2012. Late Jurassic-Cenozoic reconstructions of the Indonesian region and the Indian Ocean. *Tectonophysics*, **570–571**, 1–41.

Henstra, G.A., Rotevatn, A., Gawthorpe, R.L., Ravnas, R. 2015. Growth and assembly of large segmented normal fault during multi-episodic rifting with variable stretching direction. *Journal of Structural Geology*, **74**, 45–63.

Henza, A.A., Withjack, M.O. & Schlische, R.W. 2010. Normal-fault development during two phases of non-coaxial extension: an experimental study. *Journal of Structural Geology*, **32**, 1656–1667.

Henza, A.A., Withjack, M.O. & Schlische, R.W. 2011. How do the properties of a pre-existing normal fault population influence fault development during a subsequent phase of extension? *Journal of Structural Geology*, **33**, 1312–1324.

Hreinsdottir, S. & Bennett, R.A. 2009. Active aseismic creep on the Alto Tiberina low-angle normal fault, Italy. *Geology*, **37**, 683–686.

Hillers, G. & Miller, S.A. 2007. Dilatancy controlled spatiotemporal slip evolution of a sealed fault with spatial variations of the pore pressure. *Geophysical Journal International*, **168**, 431–445.

Jolivet, L., Lecomte, E. *et al.* 2010. The North Cycladic detachment system. *Earth and Planetary Science Letters*, **289**, 87–104.

Kornsawan, A. & Morley, C.K. 2002. The origin and evolution of complex transfer zones (graben shifts) in conjugate fault systems around the Funan Field, Pattani Basin, Gulf of Thailand. *Journal of Structural Geology*, **24**, 435–449.

Lavier, L.L. & Buck, R.W. 2002. Half graben v. large-offset low-angle normal fault: importance of keeping cool during normal faulting. *Journal of Geophysical Research*, **107**, ETG 8-1–ETG 8-13, https://doi.org/10.1029/2001JB000513

LAVIER, L.L., BUCK, R.W. & POLIAKOV, A.N.B. 1999. Self-consistent rolling-hinge model for the evolution of large-offset low-angle normal faults. *Geology*, **27**, 1127–1130.

LAVIER, L.L., BUCK, R.W. & POLIAKOV, A.N.B. 2000. Factors controlling normal fault offset in an ideal brittle layer. *Journal of Geophysical Research*, **105**, 23 431–23 442.

LEPVRIER, C., FOURNIER, M., BÉRARD, T. & ROGER, J. 2002. Cenozoic extension in coastal Dhofar (southern Oman): implications on the oblique rifting on the Gulf of Aden. *Tectonophysics*, **357**, 279–293.

LEZZAR, K.E., TIERCELIN, J.-J., LETURDU, C., COHEN, A.S., REYNOLDS, D.J., LE GALL, B. & SCHOLZ, C.A. 2002. Control of normal fault interaction on the distribution of major Neogene sedimentary depocentres, Late Tanganyika, East African Rift. *American Association of Petroleum Geologists Bulletin*, **86**, 1027–1059.

MAERTEN, L., GILLESPIE, P. & POLLARD, D.D. 2002. Local stress perturbation on secondary fault development. *Journal of Structural Geology*, **24**, 145–153.

MANIGHETTI, I., CAMPILLO, M., SAMMIS, C., MAI, P.M. & KING, G. 2005. Evidence for self-similar, triangular slip distributions on earthquakes: implications for earthquake and fault mechanics. *Journal of Geophysical Research*, **110**, B05302, https://doi.org/10.1029/2004JB003174

MANSFIELD, C. & CARTWRIGHT, J. 2001. Fault growth by linkage: observations and implications from analogue models. *Journal of Structural Geology*, **23**, 745–763.

MANSOR, M.Y., RAHMAN, A.H.A., MENIER, D. & PUBELLIER, M. 2014. Structural evolution of Malay Basin, its link to Sunda Block tectonics. *Marine and Petroleum Geology*, **58**, 736–748.

MARRETT, R. & ALLMENDINGER, R.W. 1991. Estimates of strain due to brittle faulting: sampling of fault populations. *Journal of Structural Geology*, **13**, 735–737.

MCCLAY, K.R., DOOLEY, T., WHITEHOUSE, P. & MILLS, M. 2002. 4-D evolution of rift-systems: insights from scaled physical models. *American Association of Petroleum Geologists Bulletin*, **86**, 935–959.

MCLEOD, A., DAWERS, N.H. & UNDERHILL, J.R. 2000. The propagation and linkage of normal faults: insights from the Strathspey-Brent-Statfjord fault array, northern North Sea. *Basin Research*, **12**, 263–284.

MEIGS, A.J., BURBANK, D.W. & BECK, R.A. 1995. Middle-late Miocene (>10 Ma) formation of the Main Boundary thrust in the western Himalaya. *Geology*, **23**, 423–426.

MORLEY, C.K. 1999. Patterns of displacement along large normal faults: implications for basin evolution and fault propagation, based on examples from East Africa. *American Association of Petroleum Geologists Bulletin*, **83**, 613–634.

MORLEY, C.K. 2001. Combined escape tectonics and subduction rollback-back arc extension: a model for the evolution of Cenozoic rift basins in Thailand, Malaysia and Laos. *Journal of the Geological Society, London*, **158**, 461–474, https://doi.org/10.1144/jgs.158.3.461

MORLEY, C.K. 2002. A tectonic model for the Tertiary evolution of strike-slip faults and rift basins in SE Asia. *Tectonophysics*, **347**, 189–215.

MORLEY, C.K. 2009. Geometry and evolution of low-angle normal faults (LANF) within a Cenozoic high-angle rift system, Thailand: implications for sedimentology and the mechanisms of LANF development. *Tectonics*, **28**, TC5001, https://doi.org/10.1029/2007TC002202

MORLEY, C.K. 2010. Stress re-orientation along zones of weak fabrics in rifts: an explanation for pure extension in 'oblique' rift segments? *Earth and Planetary Science Letters*, **297**, 667–673.

MORLEY, C.K. 2013. Discussion of tectonic models for Cenozoic strike-slip fault-affected continental margins of mainland SE Asia. *Journal of Asian Earth Science*, **76**, 137–151.

MORLEY, C.K. 2014. The widespread occurrence of low-angle normal faults in a rift setting: review of examples from Thailand, and implications for their origin and evolution. *Earth-Science Reviews*, **133**, 18–142.

MORLEY, C.K. 2015. Five anomalous structural aspects of rift basins in Thailand and their impact on petroleum systems. *In*: RICHARDS, F.L., RICHARDSON, N.J., RIPPINGTON, S.J., WILSON, R.W. & BOND, C.E. (eds) *Industrial Structural Geology: Principles, Techniques and Integration*. Geological Society, London, Special Publications, **421**, 143–168, https://doi.org/10.1144/SP421.2

MORLEY, C.K. & RACEY, A. 2011. Tertiary stratigraphy. *In*: RIDD, M.F., BARBER, A.J. & CROW, M.J. (eds) *The Geology of Thailand*. Geological Society, London, 223–271.

MORLEY, C.K. & WONGANAN, N. 2000. Normal fault displacement characteristics, with particular reference to synthetic transfer zones, Mae Moh Mine, Northern Thailand. *Basin Research*, **12**, 1–22.

MORLEY, C.K., NELSON, R.A., PATTON, T.L. & MUNN, S.G. 1990. Transfer zones in the East Africa Rift system and their relevance to hydrocarbon exploration in rifts. *American Association of Petroleum Geologists Bulletin*, **74**, 1234–1253.

MORLEY, C.K., WESCOTT, W.A., HARPER, R.M. & CUNNINGHAM, S.M. 1999*a*. Geology and geophysics of the Rukwa Rift. *In*: MORLEY, C.K. (ed.) *Geoscience of Rift Systems–Evolution of East Africa*. American Association of Petroleum Geologists, Studies in Geology, **44**, 91–110.

MORLEY, C.K., WESCOTT, W.A., STONE, D.M., HARPER, R.M., WIGGER, S.T., DAY, R.A., KARANJA, F.M. 1999*b*. Geology and geophysics of the Western Turkana Basins, Kenya. *In*: MORLEY, C.K. (ed.) *Geoscience of Rift Systems – Evolution of East Africa*. American Association of Petroleum Geologists, Studies in Geology, **44**, 19–54.

MORLEY, C.K., WONGANAN, N., KORNASAWAN, A., PHOOSONGSEE, W., HARANYA, C. & PONGWAPEE, S. 2004. Activation of rift oblique and rift parallel pre-existing fabrics during extension and their effect on deformation style: examples from the rifts of Thailand. *Journal of Structural Geology*, **26**, 1803–1829.

MORLEY, C.K., GABDI, S. & SEUSUTTHIYA, K. 2007. Fault superimposition and linkage resulting from stress changes during rifting: examples from 3D seismic data, Phitsanulok Basin, Thailand. *Journal of Structural Geology*, **29**, 646–663.

MORLEY, C.K., CHARUSIRI, P. & WATKINSON, I. 2011. Structural geology of Thailand during the Cenozoic. *In*: RIDD, M.F., BARBER, A.J. & CROW, M.J. (eds) *Geology of Thailand.* Geological Society, London, 273–334.

MORROW, C.A., SHI, L.Q. & BYERLEE, J.D. 1982. Strain hardening and strength of clay-rich fault gouges. *Journal of Geophysical Research*, **87**, 6771–6780.

NEEDHAM, T., YIELDING, G. & FOX, R. 1996. Fault population description along minor fault traces. *Journal of Structural Geology*, **5**, 483–495.

NELSON, R.A., PATTON, T.L. & MORLEY, C.K. 1992. Rift-segment interaction and its relation to hydrocarbon exploration in continental rift systems. *American Association of Petroleum Geologists Bulletin*, **76**, 1153–1169.

NORLAND, B.H., WHITNEY, D.L., TEYSSIER, C. & VANDERHAEGHE, O. 2002. Partial melting and decompression of the Thor–Odin dome, Shuswap metamorphic core complex, Canadian Cordillera. *Lithos*, **61**, 103–125.

PAVLIDES, S. & CAPUTO, R. 2004. Magnitude v. fault's surface parameters: quantitative relationships from the Aegean Region. *Tectonophysics*, **380**, 159–188.

PEACOCK, D.C.P. & SANDERSON, D.J. 1991. Displacements, segment linkage and relay ramps in normal fault zones. *Journal of Structural Geology*, **13**, 721–733.

PETIT, C., DÉVERCHÉRE, J., HOUDRY, F., SANKOV, V.A., MELNIKOVA, V.I. & DELVAUX, D. 1996. Present-day stress field changes along the Baikal rift and tectonic implications. *Tectonics*, **15**, 1171–1191.

PHILIPPON, M., CORTI, G. *ET AL.* 2014. Evolution, distribution and characteristics of rifting in southern Ethiopia. *Tectonics*, **33**, 485–508, https://doi.org/10.1002/2013TC003430

POCHAT, S., CASTELLTORT, S., CHOBLET, G. & & VAN DEN DRIESSCHE, J. 2009. High-resolution record of tectonic and sedimentary processes in growth strata. *Marine and Petroleum Geology*, **26**, 1350–1364.

POLIAKOV, A.N.B. & BUCK, R.W. 1998. Mechanics of stretching elastic-plastic-viscous layers: applications to slow-spreading mid-oceanic ridges. *In*: BUCK, W.R. (ed.) *Faulting and Magmatism at Mid-Oceanic Ridges*. American Geophysical Union, Geophysics Monograph Series, **106**, 305–325.

POWELL, R.E., WELDON, R.J., II & MATTI, J.C. 1993. Foreword. *In*: POWELL, R.E., WELDON, R.J., II & MATTI, J.C. (eds) *The San Andreas Fault System: Displacement, Palinspastic Reconstruction and Geological Evolution*. Geological Society of America, Memoirs, **178**, vii–xix.

PUBELLIER, M. & MORLEY, C.K. 2014. The basins of Sundaland (SE Asia): evolution and boundary conditions. *Marine and Petroleum Geology*, **58**, 555–578.

RANGIN, C., MAURIN, T. & MASSON, F. 2013. Combined effects of Eurasia/Sunda oblique convergence and East–Tibetan crustal flow on the active tectonics of Burma. *Journal of Asian Earth Sciences*, **76**, 185–194.

REEVE, M.T., BELL, R.E., DUFFY, O.B., JACKSON, C.A.-L. & SANSOM, E. 2015. The growth of non-colinear normal fault systems; What can we learn from 3D seismic reflection data? *Journal of Structural Geology*, **70**, 141–155.

SCHOLZ, C.H. & COWIE, P.A. 1990. Determination of total strain from faulting using slip measurements. *Nature*, **346**, 837–839.

SCHOLZ, C.H., DAWERS, N.H., YU, J.-Y. & ANDERS, M.H. 1993. Fault growth and fault scaling laws: preliminary results. *Journal of Geophysical Research*, **98**, 21 951–21 961.

SCHWARTZ, D.P. & COPPERSMITH, K.J. 1984. Fault behavior and characteristic earthquakes: examples from the Wasatch and San Andreas fault zones. *Journal of Geophysical Research*, **89**, 5681–5698.

SHIPTON, Z.K. & COWIE, P.A. 2003. A conceptual model for the origin of fault damage zone structures in high-porosity sandstone. *Journal of Structural Geology*, **25**, 333–344.

SOLIVA, R. & BENEDICTO, A. 2004. A linkage criterion for segmented normal faults. *Journal of Structural Geology*, **26**, 2251–2267.

SRISURIYON, K. & MORLEY, C.K. 2014. Pull-apart development at overlapping fault tips: oblique rifting of a Cenozoic continental margin, northern Mergui Basin, Andaman Sea. *Geosphere*, **10**, 80–106.

SUWANRUJI, P. 2003. *Lateral and vertical variation in conjugate fault zone geometry and the possible influence of inherited syn-rift fabric from 3D data, North Malay Basin, Thailand.* MSc thesis, University Brunei Darussalam, Bandar Seri Begawan, Brunei.

TAPPONNIER, P., PELTZER, G. & ARMIJO, R. 1986. On the mechanics of the collision between India and Asia. *In*: COWARD, M.P. & RIES, A.C. (eds) *Collision Tectonics*. Geological Society, London, Special Publications, **19**, 115–157, https://doi.org/10.1144/GSL.SP.1986.019.01.07

TINGAY, M.R.P., MORLEY, C.K., HILLIS, R.R. & MEYER, J. 2010. Present-day stress orientation in Thailand's basins. *Journal of Structural Geology*, **32**, 235–248.

TINGAY, M., MORLEY, C.K., KING, R. & COBLENTZ, D. 2012. *Present-Day Stress Field of Southeast Asia.* AAPG International Conference and Exhibition, September 16–19, 2012, Singapore, http://www.searchanddiscovery.com/pdfz/documents/2012/41111tingay/ndx_tingay.pdf.html

TRON, V. & BRUN, J.-P. 1991. Experiments on oblique rifting in brittle-ductile systems. *Tectonophysics*, **188**, 71–84.

WALSH, J.J. & WATTERSON, J. 1987. Displacement gradients on fault surfaces. *Journal of Structural Geology*, **11**, 307–316.

WALSH, J.J. & WATTERSON, J. 1988. Analysis of the relationship between displacements and dimensions of faults. *Journal of Structural Geology*, **10**, 239–247.

WALSH, J.J. & WATTERSON, J. 1992. Populations of faults and fault displacements and their effects on estimates of fault-related regional extension. *Journal of Structural Geology*, **14**, 701–712.

WALSH, J.J., BAILEY, W.R., CHILDS, C., NICOL, A. & BONSON, C.G. 2003*a*. Formation of segmented normal faults: a 3D perspective. *Journal of Structural Geology*, **25**, 1251–1262.

Walsh, J.J., Nicol, A. & Childs, C. 2003*b*. An alternative model for the growth of faults. *Journal of Structural Geology*, **24**, 1669–1675.

Whipp, P.S., Jackson, C.A.L., Gawthorpe, R.L., Dreyer, T. & Quinn, D. 2014. Fault array evolution above a reactivated rift-fabric: a subsurface example from the northern Horda Platform fault array, Norwegian North Sea. *Basin Research*, **26**, 523–549.

Withjack, M.O. & Jamison, W.R. 1986. Deformation produced by oblique rifting. *Tectonophysics*, **126**, 99–124.

Importance of pre-existing fault size for the evolution of an inverted fault system

CATHAL REILLY[1,2,3]*, ANDREW NICOL[1,2,4] & JOHN WALSH[2]

[1]*GNS Science, PO Box 30368, Lower Hutt 5040, New Zealand*

[2]*Fault Analysis Group, School of Geological Sciences, University College Dublin, Belfield, Dublin 4, Ireland*

[3]*Present address: Midland Valley, Floor 9, 2 West Regent Street, Glasgow G2 1RW, UK*

[4]*Present address: Department of Geological Sciences, University of Canterbury, Private Bag 4800, Christchurch, New Zealand*

**Correspondence: cathal@mve.com*

Abstract: Fault inversion has been documented in many basins worldwide, yet the details of how the initial extensional faults impact on the geometry and growth of the reactivated contractional system is often poorly resolved by the available data. Two-dimensional (2D) and 3D seismic reflection, and well data have been used to chart the evolution of inverted faults from the Taranaki Basin, offshore New Zealand. Sedimentary rocks up to 8 km thick record Late Cretaceous–Paleocene normal faults inverted during Miocene and younger shortening. The displacement and length of early normal faults is a key determinant for the reactivation and size of the subsequent reverse faults. All normal faults with maximum vertical displacements ≥600 m and lengths ≥9 km were inverted along their entire length, while smaller faults were not inverted. The proportion of the total basin-wide strain accommodated on each fault is comparable between deformational episodes. The hierarchy of reverse fault lengths was established rapidly, with longer faults accruing a greater proportion of the total strain from an early stage of shortening. The reverse fault system is dominated by inverted normal faults, which accrue displacement at the expense of smaller faults, and utilize the largest crustal-scale elements of the pre-existing system. The size of pre-existing heterogeneities is an important control for the magnitude and spatial extent of elevated stresses during contraction, which, in turn, control the dimensions, locations and displacements of subsequent fault growth.

Fault inversion (Fig. 1) occurs globally in a variety of sedimentary basin settings, including subduction margins, orogenic foredeeps, intracratonic basins and 'peri-orogenic' regions of continental escape tectonics (Turner & Williams 2004 and references therein). Basin inversion has been recognized for almost a century (Lamplugh 1920; Stille 1924) and is typically achieved by fault inversion with early normal faults reactivated during later contraction (Glennie & Boegner 1981; Hayward & Graham 1989; Williams *et al.* 1989; Bishop & Buchanan 1995; Brun & Nalpas 1996; Turner & Williams 2004). Hanging-wall anticlines formed in association with inversion produce structural culminations that, in many basins, contain important hydrocarbon accumulations (Fig. 1) (e.g. Fraser & Gawthorpe 1990; Uliana *et al.* 1995; Gluyas *et al.* 2003; Madon *et al.* 2004; Warren 2009). Hydrocarbon exploration of inversion structures has produced a large body of seismic reflection data that help constrain the geometrical and kinematic relationships between extensional and contractional fault systems (e.g. Davis 1983; Badley *et al.* 1989; Bishop & Buchanan 1995; Wang *et al.* 1995; Bulnes & McClay 1998; Withjack *et al.* 2010; Grimaldi & Dorobek 2011; Jackson *et al.* 2013). Such seismic data are utilized in this paper to investigate the influence of early normal faults on the evolution of an inverted fault system in the offshore Taranaki Basin, New Zealand.

Fault inversion is selective and often only a small percentage (<10%) of the initial normal fault system are reactivated (e.g. Davis 1983; Badley *et al.* 1989; Williams *et al.* 1989; Sibson 1995; Kelly *et al.* 1999; Panien *et al.* 2005). A number of factors may control which faults are reactivated during inversion, including: the relative orientation of the pre-existing faults and the principal horizontal shortening direction during inversion (Brun & Nalpas 1996; Panien *et al.* 2005; Del Ventisette

From: Childs, C., Holdsworth, R. E., Jackson, C. A.-L., Manzocchi, T., Walsh, J. J. & Yielding, G. (eds) 2017. *The Geometry and Growth of Normal Faults*. Geological Society, London, Special Publications, **439**, 447–463.
First published online February 5, 2016, https://doi.org/10.1144/SP439.2

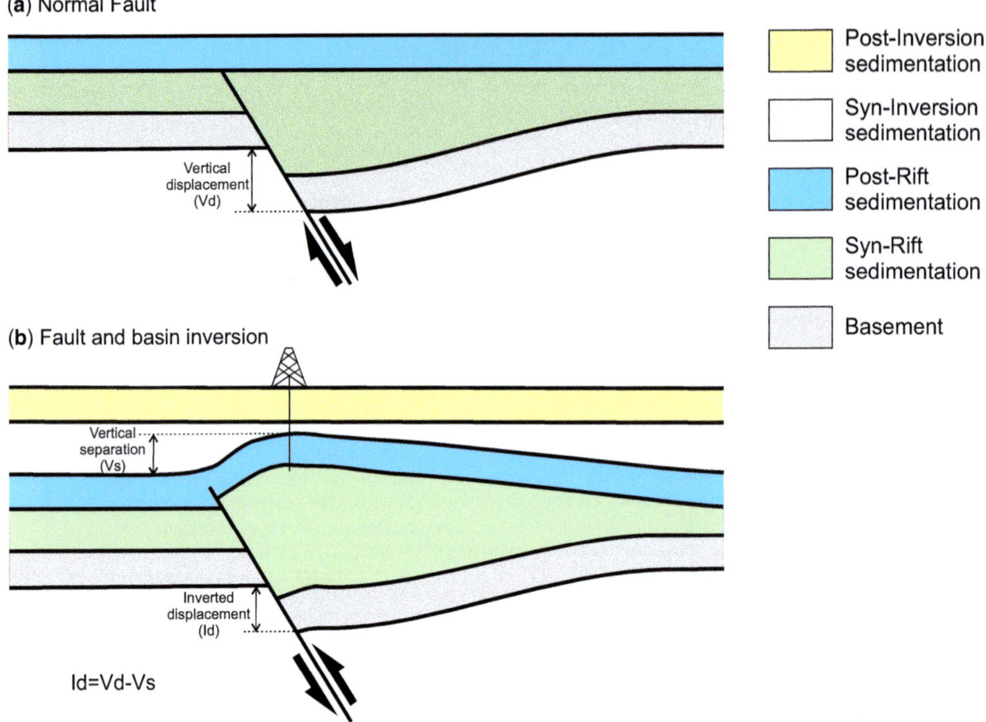

Fig. 1. Schematic cross-section of an inverted fault and basin. Growth strata deposited during extension (green) and contraction (white) are separated by intervals of tectonic quiescence (grey, blue and yellow units) (modified from Williams *et al.* 1989). Displacements measured in this paper are also shown.

et al. 2006; Withjack *et al.* 2010); fault dip, with shallower dips more conducive to inversion (Sibson 1995; Kelly *et al.* 1999); fluid overpressure, which promotes reduction of the effective normal stress on the fault surface (Sibson 1995; Turner & Williams 2004); and the size of original normal faults, with larger faults more likely to invert (e.g. Eisenstadt & Withjack 1995; Kelly *et al.* 1999; Panien *et al.* 2005). Whatever the precise mechanism controlling which faults reactivate, it is clear that inheritance from pre-existing structures can profoundly influence the evolution of later fault systems (Chadwick & Evans 1995; Morley 1995; Lezzar *et al.* 2002; Walsh *et al.* 2003; Morley *et al.* 2004; Worthington & Walsh 2011). In addition to controlling the orientation, location and size of inverted faults, early normal faults may promote rapid establishment of final fault lengths and strain localization onto a small number of primary structures (Morley 1999; Meyer *et al.* 2002; Childs *et al.* 2003; Walsh *et al.* 2002; Giba *et al.* 2012).

The role of fault size (i.e. length and maximum displacement) in selective inversion, and the impact that reactivation has on fault growth, is considered here using seismic reflection data from the southern Taranaki Basin in offshore New Zealand. Within this portion of the basin, normal faults of Late Cretaceous–Paleocene age (*c.* 80–55 Ma) were inverted during Miocene and younger shortening (Figs 2b, c & 3). The strike and dip of the early, normal faults are approximately uniform (NE–SW and 55–70°, respectively), while the extension and shortening directions during respective deformation episodes were approximately coaxial and trend perpendicular to fault strike (King & Thrasher 1996; Reilly *et al.* 2015). Therefore, neither changes in the obliquity of slip nor variations in fault dip are likely to control which structures were inverted. However, the Late Cretaceous–Paleocene normal faults have a range of sizes, with vertical displacements of 10 m–3.5 km resolved on 2D and 3D seismic reflection lines (Fig. 3). The wide range of fault sizes (i.e. three orders of magnitude) enables the dependence of inversion on fault size to be examined (Figs 4 & 5). The seismic data also provide a rare 3D perspective of the geometrical and kinematic evolution of inverted normal faults. These data show a clear control of fault size on inversion, and suggest that size, spatial distribution and relative fault sizes in the early extension system pre-condition the geometry and kinematics of the inverted fault array.

Geological setting

The Taranaki Basin has experienced both extension and shortening since about 80 Ma (e.g. Figs 2 & 3) (King & Thrasher 1996). Normal faults developed throughout the Taranaki Basin during the Late Cretaceous–Paleocene (*c.* 80–55 Ma) and reflect crustal extension during Gondwana break-up (Thrasher 1990; King & Thrasher 1996; Hemmings-Sykes 2011). These normal faults are commonly the oldest structures that displace sedimentary strata (Figs 2 & 3). They produced a series of half-graben sedimentary basins that contain strata up to 4 km thick. The normal faults mainly strike north–south and NE–SW with dips of 55–70°, suggesting a principal horizontal extension direction of approximately NW–SE. Their displacements range from <10 m to large, basin-bounding faults (>3.5 km) and have maximum lengths of up to around 90 km (Figs 2–5).

Contraction and associated inversion in the basin commenced no later than Early Oligocene and migrated westwards across the basin during the Miocene (Pilaar & Wakefield 1978; Knox 1982; Schmidt & Robinson 1989; King & Thrasher 1996; Reilly *et al.* 2015). In the SW of the basin, some reverse faults have been active during the Quaternary (Fig. 6). Shortening in the basin is inferred to have been driven by convergence along the Australian–Pacific plate boundary through New Zealand (Ballance 1976; Walcott 1978; Holt & Stern 1994; Stagpoole & Nicol 2008; Reilly *et al.* 2015). The resulting reverse faults are approximately dip-slip, with an inferred principal horizontal shortening direction of approximately NW–SE, and are commonly associated with asymmetric footwall-verging folds (i.e. hanging-wall anticlines and footwall synclines) (Figs 2 & 3).

Dataset and methods

2D and 3D seismic reflection surveys tied to petroleum exploration wells have been utilized to estimate the displacement histories of faults and the influence of fault size on subsequent inversion (e.g. Fig. 3). These seismic data provide information on the style of deformation, while displacement back-stripping using seismic reflection lines enables the timing of fault growth to be estimated.

The evolution of the faults is recorded by syn-faulting growth strata and by up-sequence changes in the magnitude and style (normal and reverse dip slip) of displacements (e.g. Fig. 1). Faulted strata comprise a Late Cretaceous and younger sedimentary succession, with a total thickness of approximately 800 m–8 km, unconformably overlying Mesozoic basement (Fig. 2). Within the sedimentary sequence, up to 16 horizons were mapped on 916 2D seismic reflection lines (with a total length of almost 25 000 km) and within five 3D seismic reflection surveys (with a total area of almost 3500 km^2) (Reilly *et al.* 2015). The interpreted seismic horizons are approximately isochronous, with ages estimated using biostratigraphy from 35 wells (Roncaglia *et al.* 2008, 2010, 2013; Morgans 2013). The interpreted horizons have a temporal resolution of *c.* 1–4 myr since 24 Ma and *c.* 5–10 myr prior to the Miocene. Approximately 1000 faults with vertical displacements of around 15–2200 ms (10 m–3.5 km) have been mapped in the southern Taranaki Basin. The basin hosts normal faults that formed during Late Cretaceous and Paleocene rifting, reverse faults that formed during the Late Eocene–Miocene, and Plio-Pleistocene (in the south of the southern Taranaki Basin) contraction. In addition, the north part of the southern Taranaki Basin hosts normal faults that formed during Plio-Pleistocene extension (Fig. 6) (Pilaar & Wakefield 1978; Knox 1982; Schmidt & Robinson 1989; King 1990; Hoolihan & Yang 1991; Thrasher 1992; Palmer & Andrews 1993; Holt & Stern 1994; King & Thrasher 1996; Nicol *et al.* 2005; Stagpoole & Nicol 2008; Giba *et al.* 2010; Reilly *et al.* 2015). A small proportion (<10%) of the earliest formed normal faults was inverted during later contraction, which also resulted in the formation of new reverse faults. Examination of individual seismic lines suggests that inversion is mainly limited to the largest displacement faults in the basin (Figs 2 & 3).

Displacement measurements and displacement back-stripping were used to determine the kinematic history of 21 faults, with maximum vertical displacements of ≥500 m. To measure both fault displacement (fault-surface slip) and fault-related folding, vertical separations were recorded between footwall and hanging-wall fold hinges. Vertical separation of the hinges is considered to be a proxy for fault vertical displacement, and here the term displacement refers to vertical displacement and/or vertical separation. In cases where the anticlinal hinge had been eroded, the eroded surfaces have been projected to the hanging-wall anticline hinge to enable an estimation of vertical separation. Displacement back-stripping was performed on individual seismic sections by sequentially subtracting displacements on progressively older horizons from all underlying horizons (for further discussion of the methodology, refer to: Petersen *et al.* 1992; Childs *et al.* 1993, 2003; Clausen *et al.* 1994). The back-stripping method has been modified slightly to account for inverted faults with early normal and late reverse displacements (and folding). Taking

Mt Taranaki
Taranaki Peninsula (North Island)
a'
Taranaki Fault
Kupe
Rua Fault
Manaia Fault
Motumate Fault
Waimea-Flaxmore Fault System
a
Southern Taranaki Basin
Cape Egmont Fault
Maari Fault
Whitiki Fault
Fig 3
Surville Fault
b'
b
Kahurangi Fault
Wakamarama Fault
Takaka
SOUTH ISLAND
N
50 km
Reverse fault
Normal fault
(a)
(b)
35°
40°
45°
170°
175°
AUSTRALIAN PLATE
PACIFIC PLATE
Southern Taranaki Basin
Kermadec Trough
Hikurangi Trough
Alpine fault
Puysegur
47 mm a^{-1}
41 mm a^{-1}
38 mm a^{-1}
37 mm a^{-1}
200 km

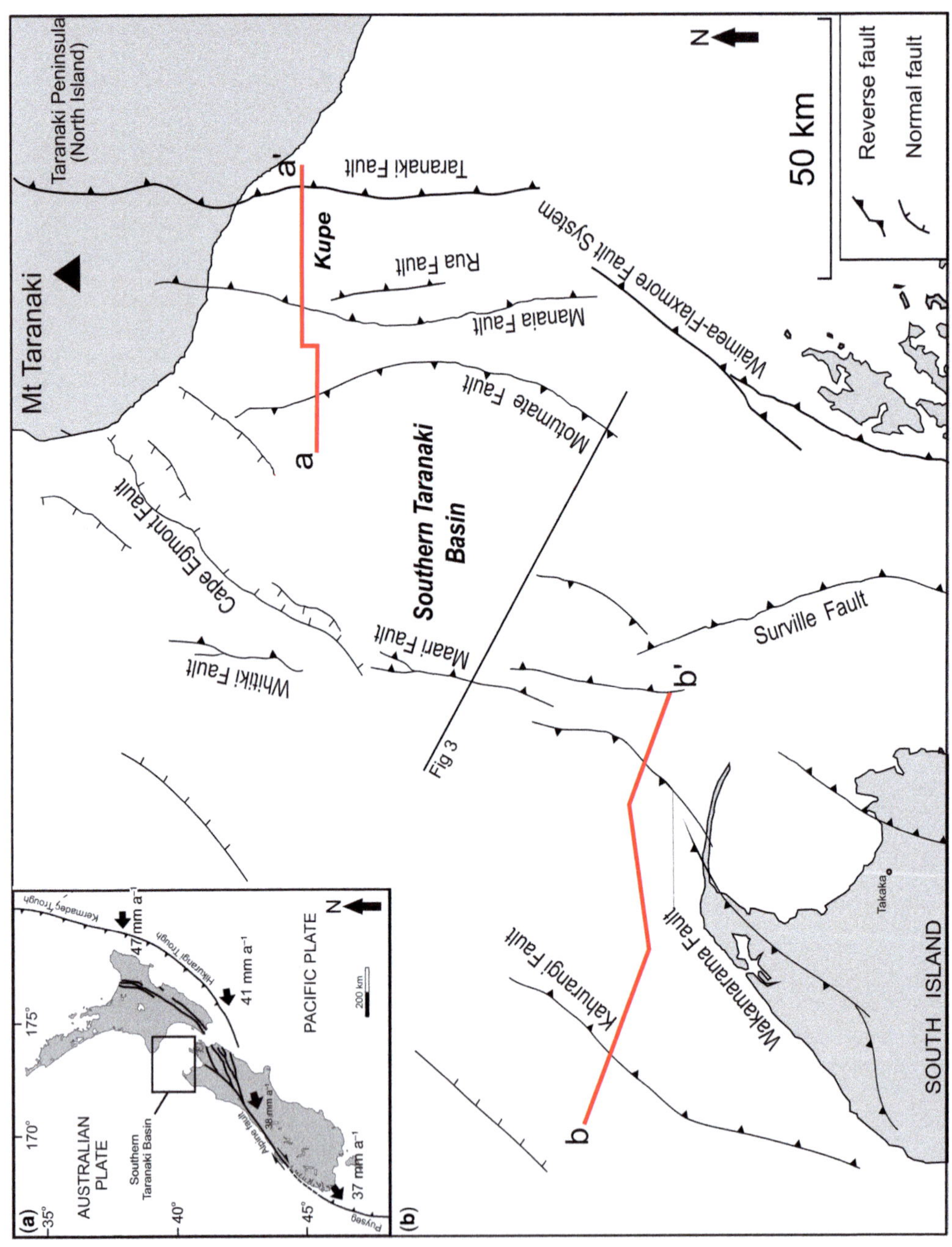

Fig. 2.

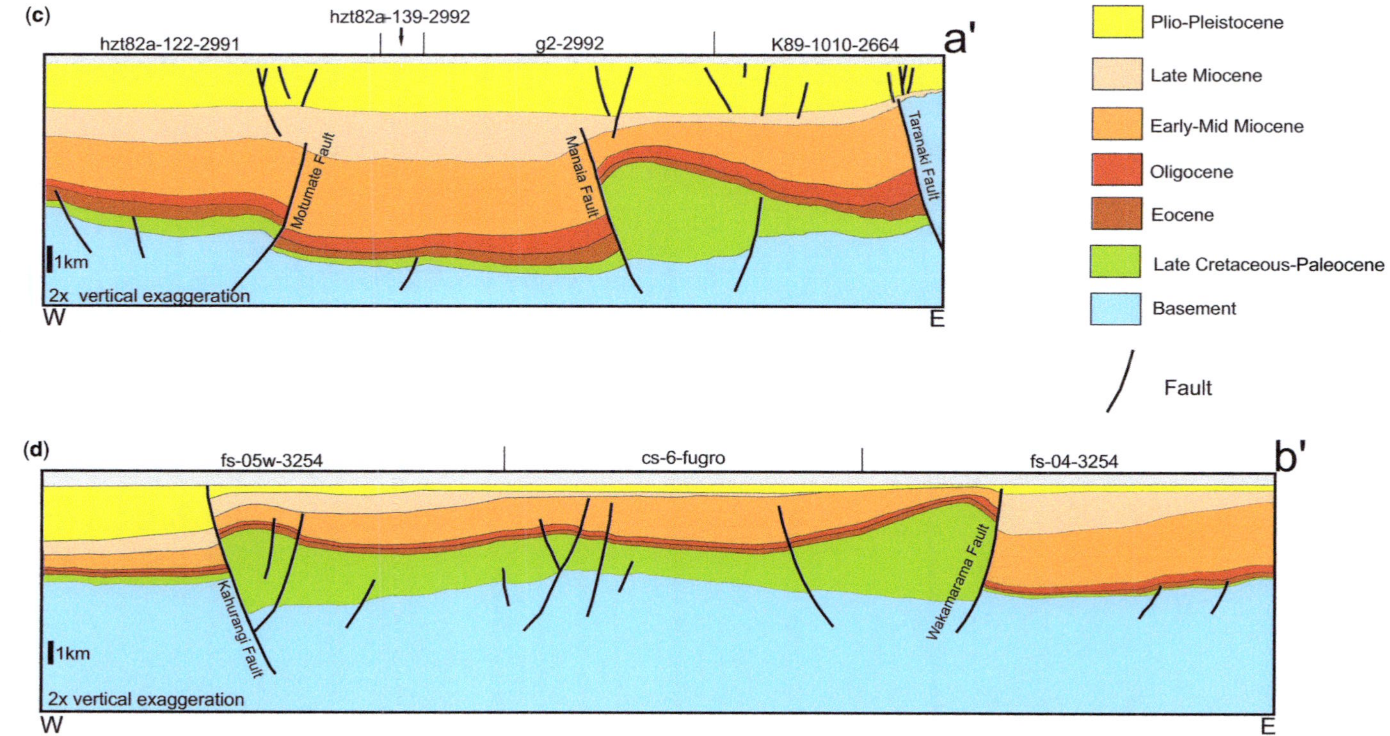

Fig. 2. (*Continued*) Faults and structural setting of the southern Taranaki Basin. (**a**) Plate boundary setting and study area location. New Zealand Pacific Plate relative motion vectors relative to the Australian Plate are from Beavan & Haines (2001). (**b**) Map of main faults in the southern Taranaki Basin active since approximately 80 Ma. Triangles and ticks on faults indicate dip directions of present reverse and normal faults, respectively. Thick red lines across northern and southern parts of the basin, a–a′ and b–b′, show the locations of the regional cross-sections in (c) and (d). (**c**) & (**d**) Cross-sections a–a′ and b–b′ across the NE and SW of the southern Taranaki Basin, respectively. Individual seismic line names that make up the cross-sections and extents on the sections are shown in black above each section. Sediment thickness variations indicate the timing and sense of movement on large faults in the basin (e.g. Fig. 1). Note also the death of small normal faults, terminating in the Late Cretaceous–Paleocene sequence and not undergoing later reactivation.

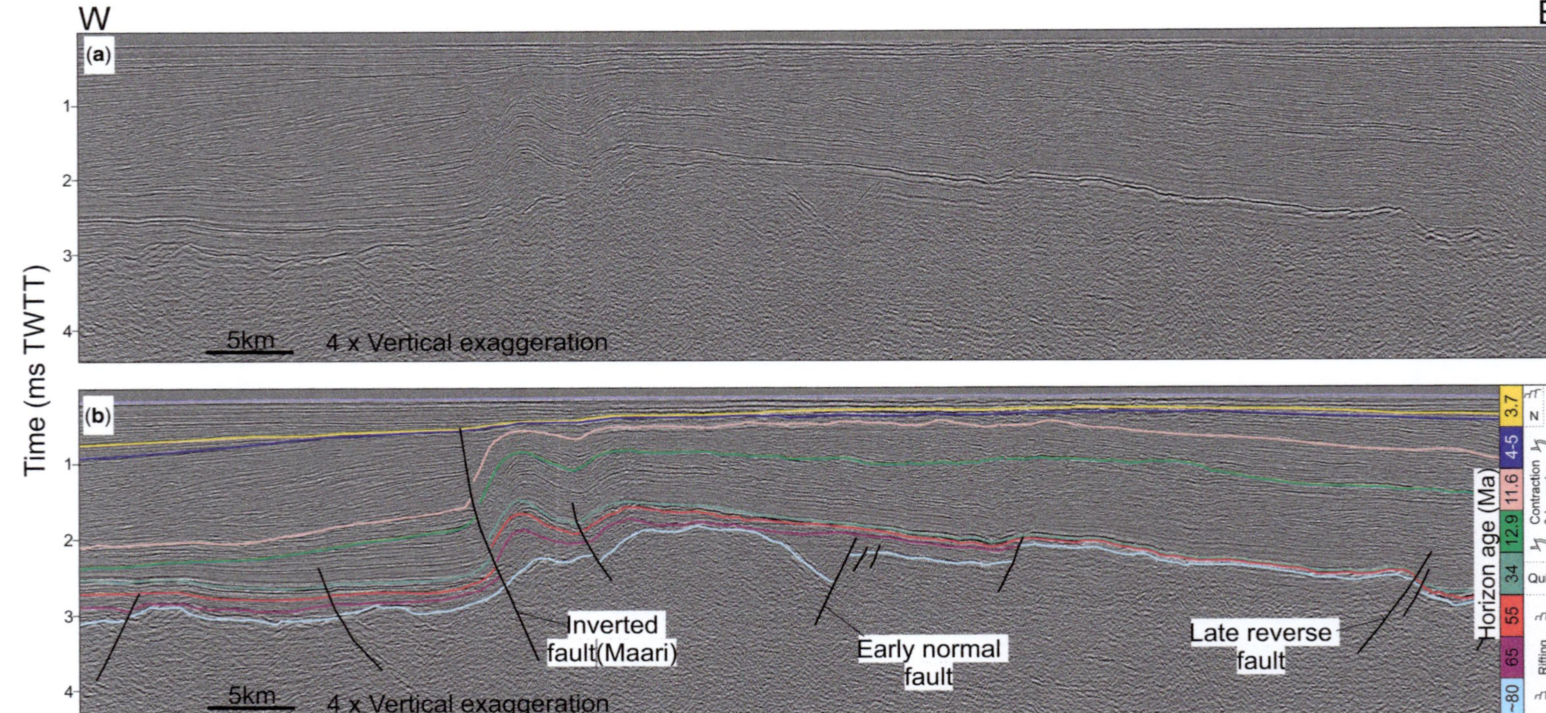

Fig. 3. Example of (**a**) an uninterpreted and (**b**) an interpreted seismic reflection line showing inverted and non-inverted normal faults, and new reverse faults (labelled in b). The location of the seismic line is shown in Figure 2b. Schematic tectonic history outlines Cretaceous–Paleocene rifting, Eocene quiescence, Oligocene–Recent contraction and Plio-Pleistocene rifting in the north of the study area. TTWT, two-way transit time.

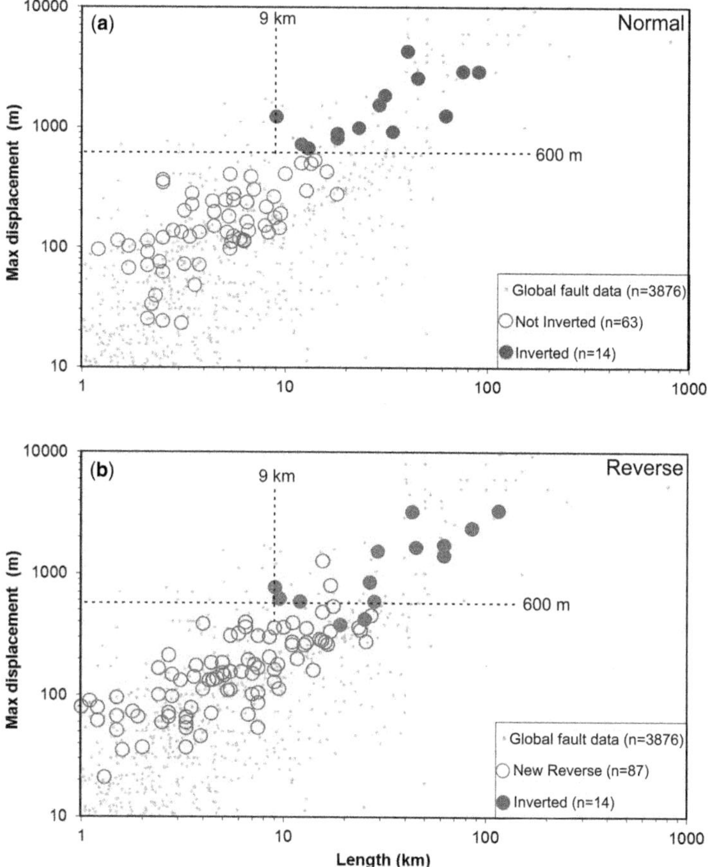

Fig. 4. Maximum displacement v. length plots for (**a**) normal and (**b**) reverse faults in the southern Taranaki Basin. Inverted (filled circles) and non-inverted normal faults (a) and newly formed reverse fault (b) (open circles) are differentiated in each plot. Inverted faults represent the largest approximately 10% of the original normal faults and have maximum vertical displacements and trace lengths of approximately >600 m and >9 km, respectively. Displacements were measured in milliseconds and converted to metres using the depth-conversion method of Thrasher & Cahill (1990), with more recently calculated velocity values for southern Taranaki Basin intervals (Hill & Milner 2012). Global fault data (filled small grey circles) are compiled from the literature (e.g. Schlishe *et al.* 1996; Bailey *et al.* 2005) augmented by recent unpublished compilations.

measured reverse vertical displacements as negative values and subtracting them from the apparent displacements on the oldest syn-rift strata in the same cross-section allows the original normal displacement associated with the extensional event to be quantified (Fig. 1). For 14 of the 21 faults, multiple seismic sections, spaced at 0.5–2 km, were back-stripped to produce along-strike displacement profiles that record displacements for extensional and contractional phases of faulting (e.g. Fig. 7) (for further discussion of the back-stripping technique as applied to inverted faults refer to Reilly *et al.* 2015). Strata were not decompacted during back-stripping as displacement loss due to sediment compaction is usually <20% (Taylor *et al.* 2008) and does not impact our first-order conclusions. In addition to back-stripping, the maximum displacement and lengths of 165 faults were measured for one or both of the deformation episodes, and are shown in Figure 4. Collectively, these data provide information on the evolution of the fault systems and the role that fault size plays in inversion.

Fault size and inversion

In the southern Taranaki Basin, fault size (displacement and length) provides an important control on subsequent reactivation, with only the larger

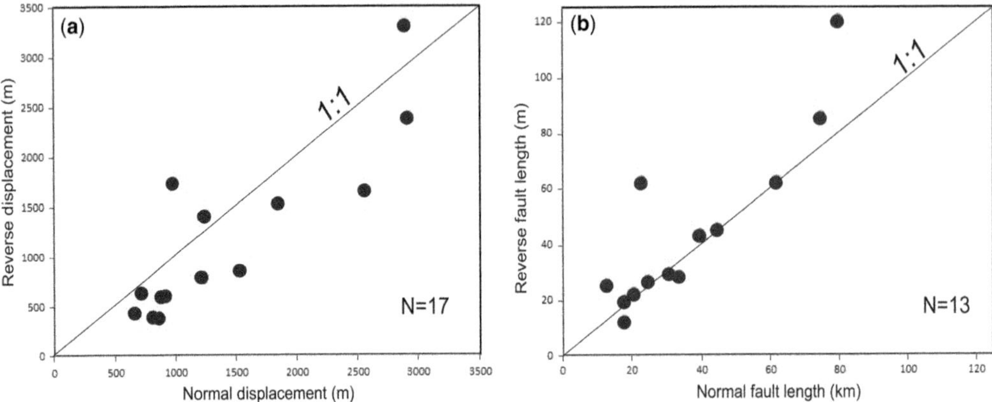

Fig. 5. Plots comparing (**a**) normal and inverted vertical displacements and (**b**) trace lengths for large inverted faults in the southern Taranaki Basin. The data show a positive relationship clustered around the 1:1 lines, suggesting that displacements and lengths during reverse and normal phases of faulting are similar.

normal faults inverted during shortening. The size dependence of inversion is illustrated in Figure 4, which shows the maximum vertical displacement v. length for normal and reverse faults. On each graph, the Taranaki Basin data are consistent with the global dataset, suggesting that in displacement–length space the faults examined in this study are typical of many faults worldwide. The most striking aspect of these graphs is the segregation of inverted and non-inverted faults. In Figure 4a, only the largest 14 normal faults in the initial extensional system were inverted. At the end of the Paleocene, these 14 extensional faults had maximum displacements and lengths of ≥600 m and ≥9 km, respectively. Displacement–length relationships for reverse faults (Fig. 4b) are similar to those observed for normal faults (Fig. 4a), with the majority of the largest structures in the reverse system (i.e. ≥600 m maximum separation and ≥9 km length) being inverted normal faults.

Along-strike displacement profiles support the view that fault lengths and maximum displacements were generally comparable during each faulting episode (Fig. 7). The likelihood of fault propagation during inversion appears to increase with decreasing size of the initial normal fault (Fig. 7). For the largest faults in the system (displacement ≥*c.* 800 m and trace length ≥*c.* 25 km), the tips remained stationary with no resolvable propagation during inversion beyond the pre-existing normal fault (Fig. 7a–d, left). In the few cases, where fault lengths differed pre- and post-inversion, contraction resulted in the propagation and creation of new fault lengths (Fig. 7e, f). Examples where parts of the initial normal faults were only inverted along part of their length have not been observed.

Stable fault lengths are often accompanied by comparable maximum displacement magnitudes for normal and reverse faulting. The correspondence of maximum displacements is considered to be fortuitous, and arises because regional finite strains during extension and shortening were approximately equal across the southern Taranaki Basin. In circumstances where regional extensional and contractional strains differed significantly, it is expected that the maximum displacements would also be different between faulting phases. Even though the maximum displacements and fault lengths are often comparable, the shapes of displacement profiles between deformation episodes can be similar or vary significantly (Fig. 7). In Figure 7a, for example, the location of the maximum displacement and the associated displacement gradients are similar between faulting phases, suggesting that the fault interactions that gave rise to these displacement variations (cf. Nicol *et al.* 1996) did not vary markedly between episodes.

The strong positive correlation of lengths and displacements between deformation phases is indicated in Figure 5 and arises because the largest faults during the initial normal faulting were also the largest inverted faults. The similarity of fault lengths during each deformation phase suggests that the crustal heterogeneities produced by the original normal faults provided a pre-eminent control on the location, orientation and length of the subsequent reverse faults. As a consequence, the hierarchy of the largest reverse faults (i.e. their relative sizes and locations) appears to have been strongly influenced by the sizes of the pre-existing normal faults. As the inverted faults accommodated most (*c.* 80%) of the contraction in the basin and their relative sizes were inherited from the pre-existing

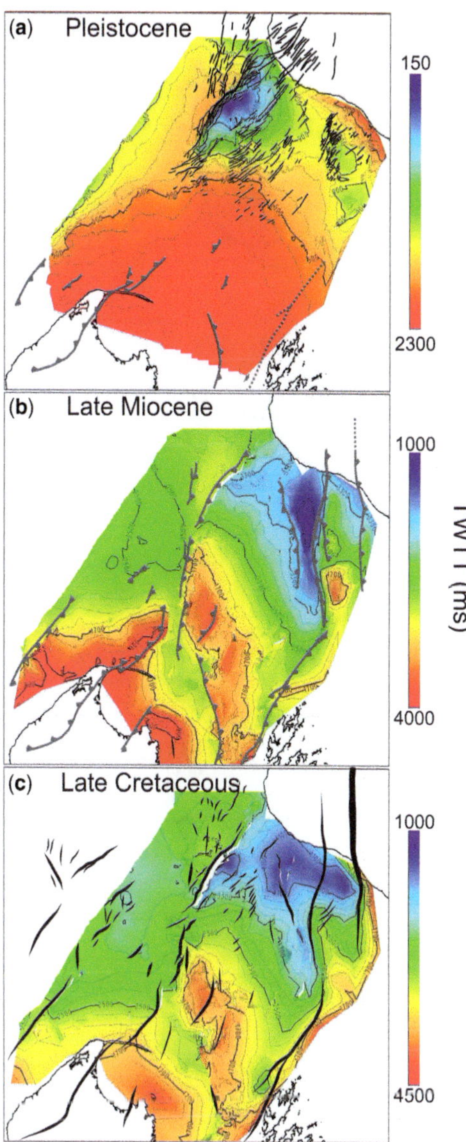

Fig. 6. Horizon surfaces and fault polygons from interpretations carried out in this study for (**a**) Pleistocene, (**b**) Late Miocene and (**c**) Late Cretaceous time periods.

normal fault system, individual inverted faults accommodated a similar proportion of the total strain across the basin during each deformation episode.

Fault reactivation (both extensional and contractional) and associated strain localization onto pre-existing structures is widely attributed to the original faults representing zones of crustal weakness (e.g. McKenzie 1969; Raleigh *et al.* 1972; White *et al.* 1986; Imber *et al.* 1997; Holdsworth 2004; Worthington & Walsh 2011). At outcrop scale, weak fault rock is commonly observed: within fault zones, however, the width, structure and thickness of weak fault rock are highly heterogeneous on metre to decametre scales (e.g. Childs *et al.* 2009). Walsh *et al.* (2001) contended that outcrop-scale heterogeneities are unlikely to control regional-scale reactivation and, instead, suggested that the geometrical properties of a fault system (e.g. fault dimensions and locations relative to other elements in fault array) exert a first-order control on fault reactivation and system evolution. Their conclusions are supported by numerical Distinct Element Models, which suggest that fault dimensions and locations have a greater impact on subsequent growth than fault strength (Walsh *et al.* 2001). The geometrical properties of pre-existing faults are important for subsequent reactivation because they produce crustal heterogeneity that locally concentrates stress. Such heterogeneity may reflect the presence of a shear fabric and/or the juxtaposition of rocks with different strengths, and need not be characterized by an inherent fault-zone weakness. The role of heterogeneities in concentrating stress is widely recognized in material science literature, with the load required to rupture an existing crack inversely proportional to the length of the heterogeneity (Griffith 1921; Gordon 1978; Murakami 2002; Pilkey & Pilkey 2008). Therefore, the size of the fault is likely to be an important determinant for the magnitude and spatial extent of elevated stresses, which, in turn, control the dimensions and locations of future failure (Walsh *et al.* 2001).

Newly formed reverse faults are typically smaller than inverted faults, with displacements of ≤ 600 m. Furthermore, the locations of the newly formed reverse faults differ from the smaller original normal faults. New reverse faults are predominantly located distal to inverted faults and appear to form in areas of regional shortening deficit rather than being clustered close the inverted faults (Fig. 8e). Lines A and B, drawn parallel to the principal horizontal shortening direction (Fig. 8e), for example, show that the number of new faults are at a maximum across the centre of the basin on line B where the number and size of inverted faults is at a minimum. The resulting shortening deficit in the central portion of line B is inferred to have been partly accommodated by the initiation of new small reverse faults. The dearth of newly formed small reverse faults close to the inverted faults may partly arise because, in these areas, the majority of the stress was concentrated on the inverted structures, and the areas enclosing these faults were effectively destressed (e.g. Ackermann &

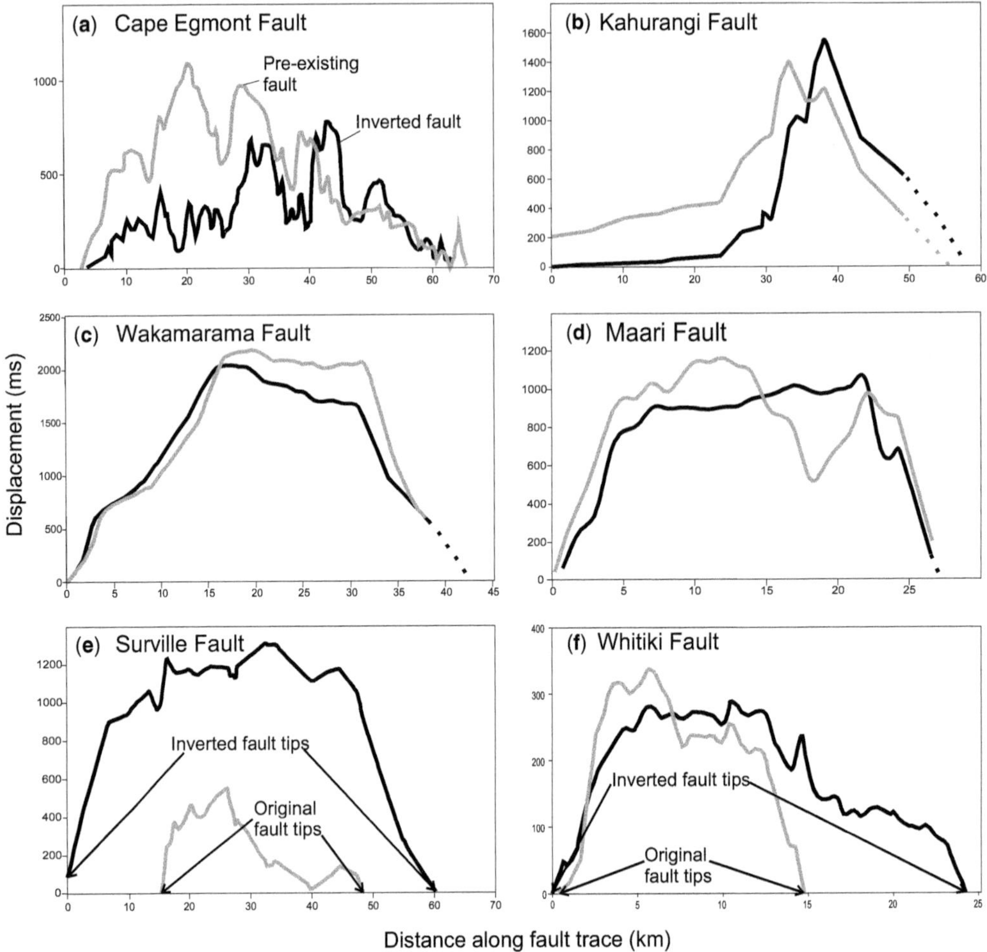

Fig. 7. Along-strike displacement profiles for the (**a**) Cape Egmont, (**b**) Kahurangi, (**c**) Wakamarama, (**d**) Maari, (**e**) Surville and (**f**) Whitiki faults in the southern Taranaki Basin (see Fig. 2 for locations). Total normal displacement on the top basement horizon is shown by the grey curve, and reverse displacement on the base Oligocene horizon, accommodated during inversion by the black curve. In (a–d), the normal and inverted profiles have comparable lengths, while (b–d) also have comparable displacement distributions. By contrast, the Surville (e) and Whitiki (f) faults propagated laterally during inversion, and presently exceed the original normal fault length.

Schlische 1997). Our observations support the view that large faults can increase or decrease stresses to nearby faults in the system, which has the potential to locally promote or impede the development of inverted faults (Cowie 1998; Walsh *et al.* 2001).

Fault-system evolution and strain localization

Analysis of finite displacements and lengths at the cessation of each faulting episode (Figs 4–7) provides limited information on how the faults grew to their final configuration. Displacement back-stripping of individual faults constrains the evolution of the reverse fault system in the southern Taranaki Basin since about 20 Ma (Fig. 8). Performing displacement back-stripping on multiple seismic sections along the strike of each fault enables changes in their length to be mapped through time (for a full description of method refer to Reilly *et al.* 2015, fig. 8). During the Miocene, contraction and fault inversion episodically stepped westwards across the southern Taranaki Basin (Reilly *et al.* 2015) (Fig. 8). By the end of the Miocene (*c.* 6 Ma), the areal extent of contraction was similar

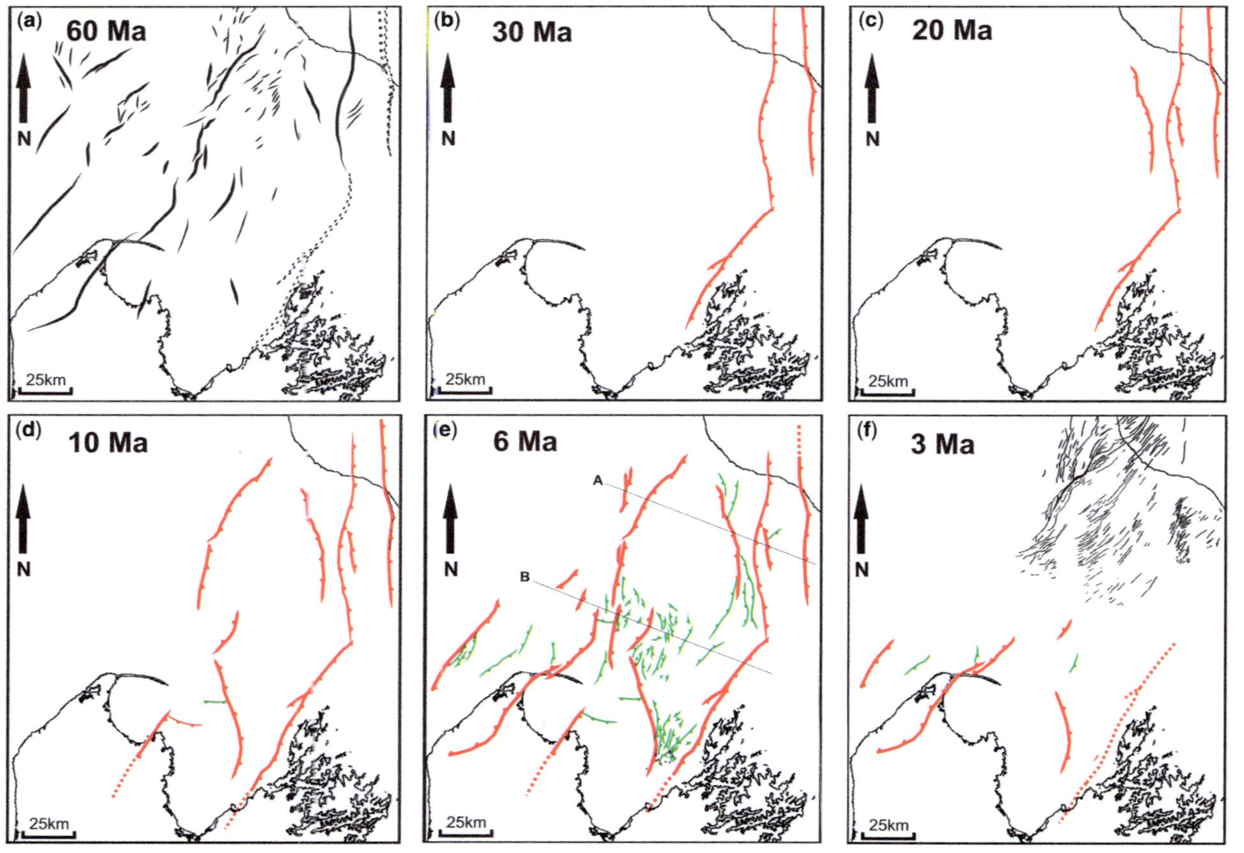

Fig. 8. Maps showing temporal changes in the spatial distribution and types of faulting in the southern Taranaki Basin: (**a**) (60 Ma) Late Cretaceous–Paleocene extension; (**b**) (30 Ma) Oligocene onset of eastern basin contraction; (**c**) (20 Ma) Early Miocene contraction; (**d**) (10 Ma) Mid-Miocene westwards migration of contraction; (**e**) (6 Ma) the Latest Miocene greatest extent of contraction; and (**f**) (3 Ma) Plio-Pleistocene southwards migration of extension. The maps were produced using displacement back-stripping of strata imaged in 2D and 3D seismic reflection lines (e.g. Fig. 3). Active normal faults (black), possible active normal faults (dashed black), active inverted faults (red), possible active inverted faults (dashed red) and active, newly formed reverse faults (green). Lines A and B show the locations of regional strain estimates (see the text for further discussion).

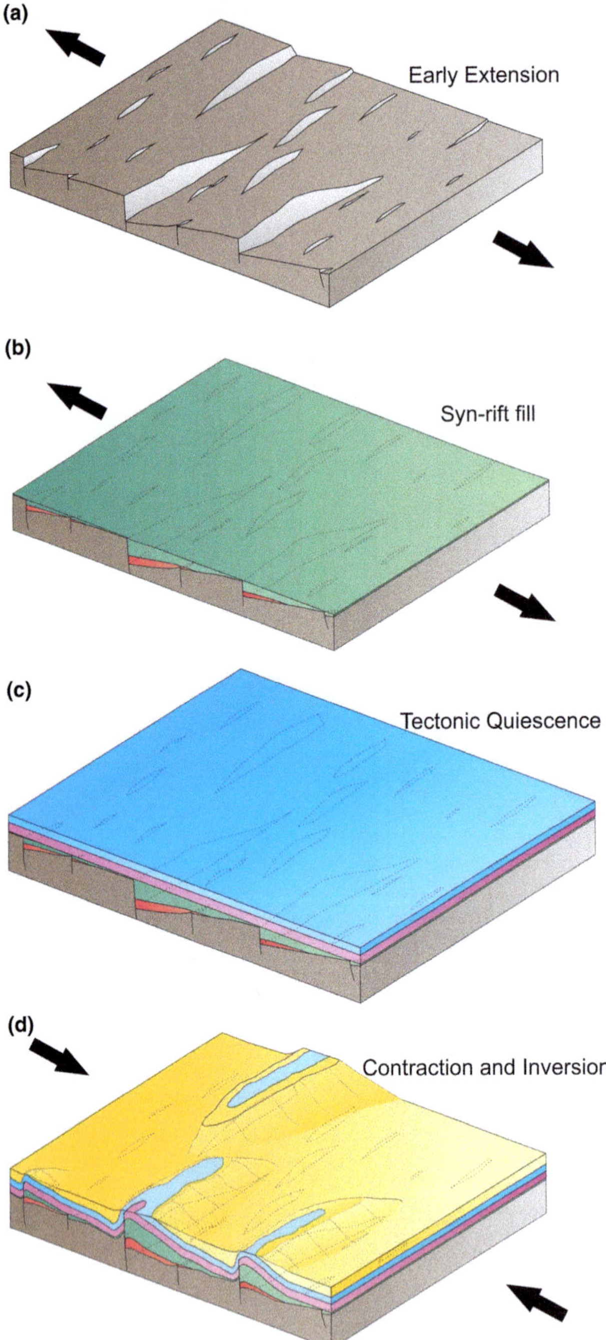

Fig. 9. Schematic block diagram showing the evolution of the inverted fault system in the southern Taranaki Basin. (**a**) Extensional normal faults with a range of sizes form and create half-graben that fill with sediment (**b**). (**c**) Following extension, uniform thickness strata are deposited across the basin during a period of thermal subsidence and tectonic quiescence. (**d**) Basin contraction coaxial with the original extension direction. The largest of the original normal faults are inverted with associated anticlines in the fault hanging walls and low-amplitude synclines in their footwalls. The uplifted hanging-wall anticlines are partially eroded with the sediment deposited in the adjacent synclines.

to that of the earlier extension. The positive relationships in Figures 4 and 5 arise, in part, because of the spatial coincidence of the regions of normal and reverse faulting, and would not be expected to be so pronounced if the region of overlap was smaller. If, for example, we had sampled the normal and reverse fault systems at 20 Ma, when contraction was concentrated along the eastern margin of the basin (Fig. 8), all but three of the faults in Figure 5 would be lying on the *x*-axis with no clear relationship between the lengths and displacements during each deformation phase. Therefore, for the relationships in Figure 5 to apply widely to other basins, each deformation phase should reoccupy approximately the same regions of the crust: this is not always the case. It has, for example, been variously shown elsewhere that inversion episodes can affect a part of a previously extensional basin (e.g. Wang *et al.* 1995), the entire basin (e.g. Uliana *et al.* 1995; Grimaldi & Dorobek 2011) or different parts of a basin at different times (e.g. Jackson *et al.* 2013). In addition to inversion at 20 Ma being limited to the large basin-bounding faults, it seems that few new smaller faults were generated at this time. The absence of smaller faults may arise because the basin-bounding faults collectively extend along strike for at least 400 km, producing crustal-scale heterogeneities that are large enough to accommodate the entire stress (and strain) associated with the regional shortening at this time.

Normal fault systems typically develop in association with strain localization and progressive concentration of displacements onto the largest fault(s) in a system, with the remainder of faults dying (Nicol *et al.* 1997; Cowie 1998; Walsh *et al.* 2001, 2002; Meyer *et al.* 2002; Gawthorpe *et al.* 2003; Whipp *et al.* 2014). Superficially, the fault maps in Figure 8 appear to indicate that, rather than localizing with increasing finite strain, contractional faulting became more spatially distributed with time. The increase in geographical spread of the active faulting is, in part, due to a change in boundary conditions during contraction that may reflect increases in the rates of convergence across the plate boundary during the Miocene, as proposed by Reilly *et al.* (2015) from regional analysis of fault evolution. In addition to the increase in regional spread of contraction, many new smaller faults (displacements <400 m and lengths <10 km) formed between 5 and 7 Ma, when the rates of shortening across the basin were at a maximum of up to about 1 mm a^{-1}. These smaller faults accommodate a small, but important, component of the total strain (*c.* 12%) and appear to have been initiated in association with an order of magnitude increase in the regional rates of shortening (Reilly *et al.* 2015). Smaller faults initiated in the Late Miocene ceased to accrue displacement by 3 Ma (Fig. 7f). Fault death in the central basin was probably due, in part, to a change in plate boundary conditions and a decrease in the rates of shortening associated with the southwards migration of intra-arc extension (King & Thrasher 1996; Cande & Stock 2004; Giba *et al.* 2010; Seebeck *et al.* 2014; Reilly *et al.* 2015). Further south, where the basin extends onshore in the northern South Island, the rates of shortening have not decreased significantly since 5 Ma, yet few of the smaller reverse faults remained active at 3 Ma (Fig. 8). Near the coastline, the death of these small faults may be partly due to strain localization onto the larger faults.

The evolution of the inverted fault system in the southern Taranaki Basin, where the principal horizontal extension and shortening axes were coaxial, is illustrated schematically in Figure 9. The key elements of the model are that: (i) only the largest of the early normal faults are reactivated during inversion; (ii) the locations and lengths of inverted faults are generally inherited from earlier normal faults; and (iii) for those faults that were inverted, the hierarchy of fault sizes is maintained between deformation phases. As the locations and displacements of newly formed reverse faults appear to be controlled by the larger inverted faults, it is suggested that the inverted faults establish their final length early. Displacement profiles for inverted faults in the Taranaki Basin typically show little evidence of propagation, which is consistent with the idea that the lengths of contractional faults were established early. Such rapid early growth in fault lengths has been proposed for natural fault systems (Walsh *et al.* 2002) and from analogue models (Schlagenhauf *et al.* 2008), and may be a feature of many fault systems whether they are reactivated or new, extensional or contractional. It is possible that, because of the crustal-scale heterogeneities produced by early normal faults, the final lengths of inverted faults formed instantaneously on geological timescales (e.g. <100 kyr): that is, within several seismic cycles. However, in common with normal fault systems (e.g. Walsh *et al.* 2002; Meyer *et al.* 2002), the seismic reflection and well data are not sufficiently detailed to quantify the timing or the rates of this early propagation.

Conclusions

The role of fault size in the inversion of pre-existing normal faults has been examined using 2D and 3D seismic reflection lines tied to wells. The size (displacement and length) of early normal faults is shown to be a key determinant for the reactivation and size of the subsequent reverse faults when shortening and extension directions are parallel. In the southern Taranaki Basin, all normal faults with maximum vertical displacements ≥*c.* 600 m and trace lengths ≥*c.* 9 km were inverted, while few faults with maximum vertical displacements and trace lengths smaller than this were inverted. In the majority of cases, the entire length of the initial normal fault was inverted, with the proportion of the total basin-wide strain accommodated on the fault comparable between both extensional and contractional deformational episodes. The hierarchy of reverse fault lengths was established rapidly, with longer faults accruing a greater proportion of the total strain from an early stage of deformation. The reverse fault system is dominated by inverted faults, which accrue displacement at the expense of smaller faults, and utilize the largest crustal-scale elements of the pre-existing system. The size of pre-existing crustal heterogeneities is an important determinant for the magnitude and spatial extent of elevated stresses, which, in turn, control the dimensions, locations and displacements of subsequent fault growth.

Funding for this work was provided by a GNS Science Scholarship, with support from the Fault Analysis Group at University College Dublin. Constructive comments from Jonathan Imber, Chris Jackson, David McNamara and Hannu Seebeck helped to improve the manuscript.

References

Ackermann, R.V. & Schlische, R.W. 1997. Anticlustering of small normal faults around larger faults. *Geology*, **25**, 1127–1130.

Badley, M.E., Price, J.D. & Backshall, L.C. 1989. Inversion, reactivated faults and related structures: seismic examples from the southern North Sea. *In*: Cooper, M.A. & Williams, G.D. (eds) *Inversion Tectonics*. Geological Society, London, Special Publications, **44**, 201–219, https://doi.org/10.1144/GSL.SP.1989.044.01.12

Bailey, W.R., Walsh, J.J. & Manzocchi, T. 2005. Fault populations, strain distribution and basement reactivation in the East Pennines Coalfield, U.K. *Journal of Structural Geology*, **27**, 913–928.

Ballance, P.F. 1976. Evolution of the upper Cenozoic magmatic arc and plate boundary in northern New Zealand. *Earth and Planetary Science Letters*, **28**, 356–370.

Beavan, J. & Haines, J. 2001. Contemporary horizontal velocity and strain rate fields of the Pacific–Australian plate boundary zone through New Zealand. *Journal of Geophysical Research*, **106**, 741–770.

Bishop, D.J. & Buchanan, P.G. 1995. Development of structurally inverted basins: a case study from the West Coast, South Island, New Zealand. *In*: Buchanan, J.G. & Buchanan, P.G. (eds) *Basin Inversion*. Geological Society, London, Special Publications, **88**, 549–585, https://doi.org/10.1144/GSL.SP.1995.088.01.28

Brun, J.P. & Nalpas, T. 1996. Graben inversion in nature and experiments. *Tectonics*, **15**, 677–687.

Bulnes, M. & McClay, K.R. 1998. Structural analysis and kinematic evolution of the inverted South Celtic Sea Basin. *Marine and Petroleum Geology*, **15**, 667–687.

Cande, S.C. & Stock, J.M. 2004. Pacific–Antarctic–Australia motion and the formation of the Macquarie Plate. *Geophysical Journal International*, **157**, 399–414.

Chadwick, R.A. & Evans, D.J. 1995. The timing and direction of Permo-Triassic extension in southern Britain. *In*: Boldy, S.A.R. (ed.) *Permian and Triassic Rifting in Northwest Europe*. Geological Society, London, Special Publications, **91**, 161–192, https://doi.org/10.1144/GSL.SP.1995.091.01.09

Childs, C., Easton, C.J., Vendeville, B.C., Jackson, M.P.A., Lin, S.T., Walsh, J.J. & Watterson, J. 1993. Kinematic Analysis of faults in a physical model of growth faulting above a viscous salt analogue. *Tectonophysics*, **228**, 313–329.

Childs, C., Nicol, A., Walsh, J.J. & Watterson, J. 2003. The growth and propagation of synsedimentary faults. *Journal of Structural Geology*, **25**, 633–648.

Childs, C., Manzocchi, T., Walsh, J.J., Bonson, C.G., Nicol, A. & Schöpfer, M.P.J. 2009. A geometric model of fault zone and fault rock thickness variations. *Journal of Structural Geology*, **31**, 117–127.

Clausen, O.R., Howard, C.B. et al. 1994. Systematics of faults and fault arrays. *In*: Helbig, K. (ed.) *Modelling the Earth for Oil Exploration. Final Report of the CEC's Geoscience Program 1990–1993*. Elsevier, Amsterdam, 205–316.

Cowie, P.A. 1998. A healing–reloading feedback control on the growth rate of seismogenic faults. *Journal of Structural Geology*, **20**, 1075–1087.

Davis, P.N. 1983. Gippsland Basin, SE Australia. *In*: Bally, A.W. (ed.) *Seismic Expression of Structural Styles*. American Association of Petroleum Geologists, Special Publications, **3**, 3.3.19–3.3.24.

Del Ventisette, C., Montanari, D., Sani, F. & Bonini, M. 2006. Basin inversion and fault reactivation in laboratory experiments. *Journal of Structural Geology*, **28**, 2067–2083.

Eisenstadt, G. & Withjack, M.O. 1995. Estimating inversion: results from clay models. *In*: Buchanan, J.G. & Buchanan, P.G. (eds) *Basin Inversion*. Geological Society, London, Special Publications, **88**, 119–136, https://doi.org/10.1144/GSL.SP.1995.088.01.08

Fraser, A.J. & Gawthorpe, R.L. 1990. Tectono-stratigraphic development and hydrocarbon habitat of the Carboniferous in northern England. *In*: Hardman, R.F.P. & Brooks, J. (eds) *Tectonic Events Responsible for Britain's Oil and Gas Reserves*. Geological Society, London, Special Publications, **55**, 49–86, https://doi.org/10.1144/GSL.SP.1990.055.01.03

Gawthorpe, R.L., Jackson, C.A.L., Young, M.J., Sharp, I.R., Moustafa, A.R. & Leppard, C.W. 2003. Normal fault growth, displacement localisation and the evolution of normal fault populations: the Hammam Faraun fault block, Suez rift, Egypt. *Journal of Structural Geology*, **26**, 883–895.

Giba, M., Walsh, J.J. & Nicol, A. 2010. Evolution of faulting and volcanism in a backarc basin and its implications for subduction processes. *Tectonics*, **29**, TC4020, https://doi.org/10.1029/2009TC002634

GIBA, M., WALSH, J.J. & NICOL, A. 2012. Segmentation and growth of an obliquely reactivated normal fault. *Journal of Structural Geology*, **39**, 253–267.

GLENNIE, K.W. & BOEGNER, P.L.E. 1981. Sole pit inversion tectonics. *In*: ILLING, L.V. & HOBSON, G.D. (eds) *Petroleum Geology of the Continental Shelf of Northwest Europe*. Institute of Petroleum, London, 110–120.

GLUYAS, J.G., EVANS, I.J. & RICHARDS, D. 2003. The Kimmeridge Bay Oilfield, UK Onshore. *In*: GLUYAS, J.G. & HICHENS, H.M. (eds) *United Kingdom Oil and Gas Fields, Commemorative Millennium Volume*. Geological Society, London, Memoirs, **20**, 943–948, https://doi.org/10.1144/GSL.MEM.2003.021.01.80

GORDON, J.E. 1978. *Structures, or Why Things Don't Fall Down*. Plenum, New York.

GRIFFITH, A.A. 1921. The phenomena of rupture and flow in solids. *Philosophical Transactions of the Royal Society of London A*, **221**, 163–198.

GRIMALDI, G.O. & DOROBEK, S.L. 2011. Fault framework and kinematic evolution of inversion structures: natural examples from the Neuquen Basin, Argentina. *American Association of Petroleum Geologists Bulletin*, **95**, 27–60.

HAYWARD, A.B. & GRAHAM, R.H. 1989. Some geometrical characteristics of inversion. *In*: COOPER, M.A. & WILLIAMS, G.D. (eds) *Inversion Tectonics*. Geological Society, London, Special Publications, **44**, 17–40, https://doi.org/10.1144/GSL.SP.1989.044.01.03

HEMMINGS-SYKES, S. 2011. *The influence of faulting on hydrocarbon migration in the Kupe area, south Taranaki Basin, New Zealand*. MSc thesis, Victoria University of Wellington, Wellington, New Zealand.

HILL, M.G. & MILNER, M. 2012. *The Kupe Velocity Model – 4D Taranaki Project*. GNS Science Report **2012/3**. GNS Science, Lower Hutt, New Zealand.

HOLDSWORTH, R.E. 2004. Weak faults, rotten cores. *Science*, **303**, 181–182, https://doi.org/10.1126/science.1092491

HOLT, W.E. & STERN, T.A. 1994. Subduction, platform subsidence, and foreland thrust loading: the late Tertiary development of Taranaki Basin, New Zealand. *Tectonics*, **13**, 1068–1092.

HOOLIHAN, K. & YANG, J.S. 1991. Eocene to Recent structural history of onshore Taranaki, with reference to the main play styles. *In*: *1991 New Zealand Oil Exploration Conference Proceedings*. Ministry of Commerce, Wellington, 168–175.

IMBER, J., HOLDSWORTH, R.E., ROBERTS, A.M. & LLOYD, G.E. 1997. Fault-zone weakening processes along the reactivated Outer Hebrides Fault Zone, Scotland. *Journal of the Geological Society, London*, **154**, 105–110, https://doi.org/10.1144/gsjgs.154.1.0105

JACKSON, C.A.L., CHUA, S.T., BELL, R.E. & MAGEE, C. 2013. Structural style and early stage growth of inversion structures: 3D seismic insights from the Egersund Basin, offshore Norway. *Journal of Structural Geology*, **46**, 167–185.

KELLY, P.G., PEACOCK, D.C.P., SANDERSON, D.J. & MCGUIRK, A.C. 1999. Selective reverse reactivation of normal faults, and deformation around reverse-reactivated faults in the Mesozoic of the Somerset coast. *Journal of Structural Geology*, **21**, 493–509.

KING, P.R. 1990. Polyphase evolution of the Taranaki Basin, New Zealand: changes in sedimentary and structural style. *In*: *1990 New Zealand Oil Exploration Conference Proceedings*. Ministry of Commerce, Wellington, 134–150.

KING, P.R. & THRASHER, G.P. 1996. *Cretaceous–Cenozoic Geology and Petroleum Systems of the Taranaki Basin, New Zealand*. Institute of Geological & Nuclear Sciences, Lower Hutt, New Zealand, Monographs, **13**.

KNOX, G.J. 1982. Taranaki Basin, structural style and tectonic setting. *New Zealand Journal of Geology and Geophysics*, **25**, 125–140.

LAMPLUGH, G.W. 1920. Structure of the Weald and analogues tracts. *Quarterly Journal of the Geological Society, London*, **75**, LXXIII–XCV (Anniversary address of the President).

LEZZAR, K.E., TIERCELIN, J.-J., LE TURDU, C., COHEN, A.S., REYNOLDS, D.J., LE GALL, B. & SCHOLZ, C.A. 2002. Control of normal fault interaction on the distribution of major Neogene sedimentary depocentres, Lake Tanganyika, East African rift. *American Association of Petroleum Geologists Bulletin*, **86**, 1027–1059.

MADON, M., YANG, J-S., ABOLINS, P., HASSAN, R.A., YAKZAN, A.M. & ZAINAL, S.B. 2004. Petroleum systems of the Northern Malay Basin. Paper presented at the Petroleum Geology Conference & Exhibition, 15–16 December 2004, Kuala Lumpur, Malaysia.

MCKENZIE, D.P. 1969. The relation between fault plane solutions for earthquakes and the directions of the principal stresses. *Bulletin of the Seismological Society of America*, **59**, 591–601.

MEYER, V., NICOL, A., CHILDS, C., WALSH, J.J. & WATTERSON, J. 2002. Progressive localisation of strain during the evolution of a normal fault population. *Journal of Structural Geology*, **24**, 1215–1231.

MORGANS, H.E.G. 2013. *Oligocene to Pliocene Biostratigraphy of the Marine Sequences in Exploration Wells in the Kupe Region, South Taranaki Bight, New Zealand*. GNS Science Report **2013/48**. GNS Science, Lower Hutt, New Zealand.

MORLEY, C.K. 1995. Developments in the structural geology of rifts over the last decade and their impact on hydrocarbon exploration. *In*: LAMBIASE, J.J. (ed.) *Hydrocarbon Habitat in Rift Basins*. Geological Society, London, Special Publications, **80**, 1–32, https://doi.org/10.1144/GSL.SP.1995.080.01.01

MORLEY, C.K. 1999. Patterns of displacement along large normal faults: implications for basin evolution and fault propagation, based on examples from East Africa. *American Association of Petroleum Geologists Bulletin*, **83**, 613–634.

MORLEY, C.K., HARANYA, C., PHOOSONGSEE, W., PONGWAPEE, S., KORNSAWAN, A. & WONGANAN, N. 2004. Activation of rift oblique and rift parallel pre-existing fabrics during extension and their effect on deformation style: examples from the rifts of Thailand. *Journal of Structural Geology*, **26**, 1803–1829.

MURAKAMI, Y. 2002. *Metal Fatigue: Effects of Small Defects and Nonmetallic Inclusions*. Elsevier, Amsterdam.

NICOL, A., WATTERSON, J., WALSH, J.J. & CHILDS, C. 1996. The shapes, major axis orientations and displacement patterns of fault surfaces. *Journal of Structural Geology*, **18**, 235–248.

Nicol, A., Walsh, J.J., Watterson, J. & Underhill, J.R. 1997. Displacement rates of normal faults. *Nature*, **390**, 157–159.

Nicol, A., Walsh, J., Berryman, K. & Nodder, S. 2005. Growth of a normal fault by accumulation of slip over millions of years. *Journal of Structural Geology*, **27**, 327–342.

Palmer, J.A. & Andrews, P.R. 1993. Cretaceous–Tertiary sedimentation and implied tectonic controls on the structural evolution of Taranaki Basin, New Zealand. *In*: Ballance, P.F. (ed.) *South Pacific Sedimentary Basins*. Sedimentary Basins of the World, **2**. Elsevier, Amsterdam, 309–328.

Panien, M., Schreurs, G. & Pfiffner, A. 2005. Sandbox experiments on basin inversion: testing the influence of basin orientation and basin fill. *Journal of Structural Geology*, **27**, 433–445.

Petersen, K., Clausen, O.R. & Korstgard, J.A. 1992. Evolution of a salt-related listric growth fault near the D-1 well, block 5605, Danish North Sea; displacement history and salt kinematics. *Journal of Structural Geology*, **14**, 565–577.

Pilaar, W.F.H. & Wakefield, L.L. 1978. Structural and stratigraphic evolution of the Taranaki Basin, offshore North Island, New Zealand. *The APEA Journal*, **18**, 93–101.

Pilkey, W.D. & Pilkey, D.F. 2008. *Peterson's Stress Concentration Factors*, 3rd edn. John Wiley, Hoboken, NJ.

Raleigh, C.B., Healy, J.H. & Bredchoeft, J.D. 1972. Faulting and crustal stress at Rangely, Colorado. *In*: Heard, H.C., Borg, I.Y., Carter, N.L. & Raleigh, C.B. (eds) *Flow and Fracture of Rocks*. American Geophysical Union, Geophysical Monograph Series, **16**, 275–284.

Reilly, C., Nicol, A., Walsh, J.J. & Seebeck, H. 2015. Evolution of faulting and plate boundary deformation in the Southern Taranaki Basin, New Zealand. *Tectonophysics*, **651–652**, 1–18, https://doi.org/10.1016/j.tecto.2015.02.009.

Roncaglia, L., Morgans, H.E.G. *et al.* 2008. *Stratigraphy, Well Correlation and Seismic-to-well tie in the Upper Cretaceous to Pliocene Interval in the Kupe Region, Taranaki Basin, New Zealand. Introduction to the Stratigraphic Database in PETREL*. GNS Science Report **2008/7**. GNS Science, Lower Hutt, New Zealand.

Roncaglia, L., Milner, M. *et al.* 2010. *Procedures and Metadata Protocols used in Modelling Taranaki Basin Petroleum Systems: Guidelines from a Pilot Case Study in the Kupe Area, Lower Hutt*. GNS Science Report **2009/49**. GNS Science, Lower Hutt, New Zealand.

Roncaglia, L., Fohrmann, M., Milner, M., Morgans, H.E.G. & Crundwell, M.P. 2013. *Well Log Stratigraphy in the Central and Southern Offshore Areas of the Taranaki Basin, New Zealand*. GNS Science Report **2013/27**. GNS Science, Lower Hutt, New Zealand.

Schlagenhauf, A., Manighetti, I., Malavieille, J. & Dominguez, S. 2008. Incremental growth of normal faults: insights from a laser-equipped analog experiment. *Earth and Planetary Science Letters*, **273**, 299–311, https://doi.org/10.1016/j.epsl.2008.06.042

Schlishe, R.W., Young, S.S., Ackerman, R.V. & Gupta, A. 1996. Geometry and scaling relations of a population of very small rift-related normal faults. *Geology*, **24**, 683–686.

Schmidt, D.S. & Robinson, P.H. 1989. The structural setting and depositional history for the Kupe South Field, Taranaki Basin. *In*: *1989 New Zealand Oil Exploration Conference Proceedings*. Ministry of Commerce, Wellington, 151–172.

Seebeck, H., Nicol, A., Villamor, P., Ristau, J. & Pettinga, J. 2014. Structure and kinematics of the Taupo Rift, New Zealand. *Tectonics*, **33**, 1178–1199.

Sibson, R.H. 1995. Selective fault reactivation during basin inversion: potential for fluid redistribution through fault-valve action. *In*: Buchanan, J.G. & Buchanan, P.G. (eds) *Basin Inversion*. Geological Society, London, Special Publications, **88**, 3–19, https://doi.org/10.1144/GSL.SP.1995.088.01.02

Stagpoole, V.M. & Nicol, A. 2008. Regional structure and kinematic history of a large subduction back thrust: Taranaki Fault, New Zealand. *Journal of Geophysical Research: Solid Earth*, **113**, B01403, https://doi.org/10.1029/2007JB005170

Stille, H. 1924. *Grundfragen der Vergleichenden Tektonik*. Brontrager, Berlin.

Taylor, S.K., Nicol, A. & Walsh, J.J. 2008. Displacement loss on growth faults due to sediment compaction. *Journal of Structural Geology*, **30**, 394–405.

Thrasher, G.P. 1990. Tectonics of the Taranaki Rift. *In*: *1989 New Zealand Oil Exploration Conference Proceedings*. Ministry of Commerce, Wellington, 124–133.

Thrasher, G.P. 1992. *Late Cretaceous geology of Taranaki Basin, New Zealand*. PhD thesis, Victoria University of Wellington, New Zealand.

Thrasher, G.P. & Cahill, J.P. 1990. *Subsurface Maps of the Taranaki Basin Region, New Zealand*. New Zealand Geological Survey Report **G142**. New Zealand Geoloical Survey, Lower Hutt.

Turner, J.P. & Williams, G.A. 2004. Sedimentary basin inversion and intra-plate shortening. *Earth-Science Reviews*, **65**, 277–304.

Uliana, M.A., Arteaga, M.E., Legarreta, L., Cerdán, J.J. & Peroni, G.O. 1995. Inversion structures and hydrocarbon occurrence in Argentina. *In*: Buchanan, J.G. & Buchanan, P.J. (eds) *Basin Inversion*. Geological Society, London, Special Publications, **88**, 211–233, https://doi.org/10.1144/GSL.SP.1995.088.01.13

Walcott, R.I. 1978. Present tectonics and late Cenozoic evolution of New Zealand. *Geophysical Journal of the Royal Astronomical Society*, **52**, 137–164.

Walsh, J.J., Childs, C. *et al.* 2001. Geometric controls on the evolution of normal fault systems. *In*: Holdsworth, R.E., Strachan, R.A., Magloughlin, J.F. & Knipe, R.J. (eds) *The Nature and Tectonic Significance of Fault Zone Weakening*. Geological Society, London, Special Publications, **186**, 157–170, https://doi.org/10.1144/GSL.SP.2001.186.01.10

Walsh, J.J., Nicol, A. & Childs, C. 2002. An alternative model for the growth of faults. *Journal of Structural Geology*, **24**, 1669–1675.

Walsh, J.J., Childs, C., Imber, J., Manzocchi, T., Watterson, J. & Nell, P.A.R. 2003. Strain localisation and population changes during fault system growth within the Inner Moray Firth, Northern North Sea. *Journal of Structural Geology*, **25**, 1897–1911.

Wang, G.M., Coward, M.P., Yuan, W., Liu, S. & Wang, W. 1995. Fold growth during basin inversion – example from the East China Sea Basin. *In*: Buchanan, J.G. & Buchanan, P.J. (eds) *Basin Inversion.* Geological Society, London, Special Publications, **88**, 493–522, https://doi.org/10.1144/GSL.SP.1995.088.01.26

Warren, M.J. 2009. Tectonic inversion and petroleum system implications in the rifts of Central Africa. Paper presented at the Frontiers and Innovation, CSPG CSEG CWLS Convention, 4–8 May 2009, Calgary, Alberta.

Whipp, P.S., Jackson, C.A-L., Gawthorpe, R.L., Dreyer, T. & Quinn, D. 2014. Normal fault array evolution above a reactivated rift fabric; a subsurface example from the northern Horda Platform, Norwegian North Sea. *Basin Research*, **26**, 523–549.

White, S.H., Bretan, P.G. & Rutter, E.H. 1986. Fault-zone reactivation: kinematics and mechanisms. *Philosophical Transactions of the Royal Society of London A*, **317**, 81–97.

Williams, G.D., Powell, C.M. & Cooper, M.A. 1989. Geometry and kinematics of inversion tectonics. *In*: Cooper, M.A. & Williams, G.D. (eds) *Inversion Tectonics*. Geological Society, London, Special Publications, **44**, 3–15, https://doi.org/10.1144/GSL.SP.1989.044.01.02

Withjack, M., Baum, M.S. & Schlische, R.W. 2010. Influence of pre-existing fault fabric on inversion-related deformation: a case study of the inverted Fundy rift basin, southeastern Canada. *Tectonics*, **29**, 1–22.

Worthington, R.P. & Walsh, J.J. 2011. Structure of Lower Carboniferous basins of NW Ireland, and its implications for structural inheritance and Cenozoic faulting. *Journal of Structural Geology*, **33**, 1285–1299.

Post-Caledonian extension in the West Norway–northern North Sea region: the role of structural inheritance

HAAKON FOSSEN[1,2]*, HAMED FAZLI KHANI[1], JAN INGE FALEIDE[3], ANNA K. KSIENZYK[1] & W. JAMES DUNLAP[4]

[1]*Department of Earth Science, University of Bergen, Postboks 7803, 5007 Bergen, Norway*

[2]*Museum of Natural History, University of Bergen, Postboks 7803, 5007 Bergen, Norway*

[3]*Department of Geosciences, University of Oslo, Postboks 1047, Blindern, N-0316 Oslo, Norway*

[4]*University Gas Research, 2888 Pelican Point Circle, Mound, MN 55364, USA*

**Correspondence: haakon.fossen@geo.uib.no*

Abstract: The northern North Sea region has experienced repeated phases of post-Caledonian extension, starting with extensional reactivation of the low-angle basal Caledonian thrust zone, then the formation of Devonian extensional shear zones with 10–100 km-scale displacements, followed by brittle reactivation and the creation of a plethora of extensional faults. The North Sea Rift-related approximately east–west extension created a new set of rift-parallel faults that cut across less favourably orientated pre-rift structures. Nevertheless, fault rock dating shows that onshore faults and shear zones of different orientations were active throughout the history of rifting. Several of the reactivated major Devonian extensional structures can be extrapolated offshore into the rift, where they appear as bands of dipping reflectors. They coincide with large-scale boundaries separating 50–100 km-wide rift domains of internally uniform fault patterns. Major north–south-trending rift faults, such as the Øygarden Fault System, bend or terminate against these boundaries, clearly influenced by their presence during rifting. Hence, the North Sea is one of several examples where pre-rift basement structures oblique to the rift extension direction can significantly influence rift architecture, even if most of the rift faults are newly-formed structures.

Rifts and regions of extensional tectonics are built on crust that has undergone earlier phases of deformation, particularly orogenic events (Wilson 1966). Hence, abundant pre-existing structures generally exist in regions where rifting initiates, but the role of such structures during rifting and extension has been a long-standing question in the geoscience community. Reactivation of both low-angle thrusts (e.g. Brewer & Smythe 1984; Constenius 1996) and high-angle normal faults (e.g. Henza *et al.* 2011; Bell *et al.* 2014) has been reported, and several studies argue that major pre-existing structures may give rise to rift segmentation and the formation of transfer zones or fault-domain boundaries (Rosendahl *et al.* 1986; Morley *et al.* 1990, 2004; Faulds & Varga 1998).

The northern North Sea rift system and its margins (Fig. 1) constitute an area that has been subjected to not only Caledonian thrusting, but several phases of extensional deformation since the end of the Caledonian continent–continent collision at around 405 Ma, notably post-collisional (Devonian) crustal extension, and two main phases of North Sea rifting in the Permian–Early Triassic and Late Jurassic–Early Cretaceous (Phase 1 and Phase 2, respectively). The Caledonian collision resulted in a highly heterogeneous crust with a structural imprint expressed in the form of lithological layering, mylonitic fabrics, shear zones and thrusts in the lower and middle Caledonian crust, as well as brittle thrusts and normal faults in the, now mostly removed, upper Caledonian crust. Such an anisotropic crust, with a substantial assembly of potentially weak structures, is likely to respond to later extensional deformation by a certain amount of reactivation, depending on factors such as geometry, dip, orientation relative to the extension direction and relative strength.

In this work, we combine new and existing observations and data to demonstrate that the onshore extensional shear zones, faults and fractures show evidence of extensional activation both before and during the rift history of the northern North Sea, and seismic data to argue that post-Caledonian extensional structures were also reactivated during rifting, and that they had a significant influence on the first-order architecture of the rift system.

Caledonian framework

The Caledonian orogen has been the subject of numerous studies since the 1800s, with a strong

From: CHILDS, C., HOLDSWORTH, R. E., JACKSON, C. A.-L., MANZOCCHI, T., WALSH, J. J. & YIELDING, G. (eds) 2017. *The Geometry and Growth of Normal Faults*. Geological Society, London, Special Publications, **439**, 465–486.
First published online February 5, 2016, https://doi.org/10.1144/SP439.6

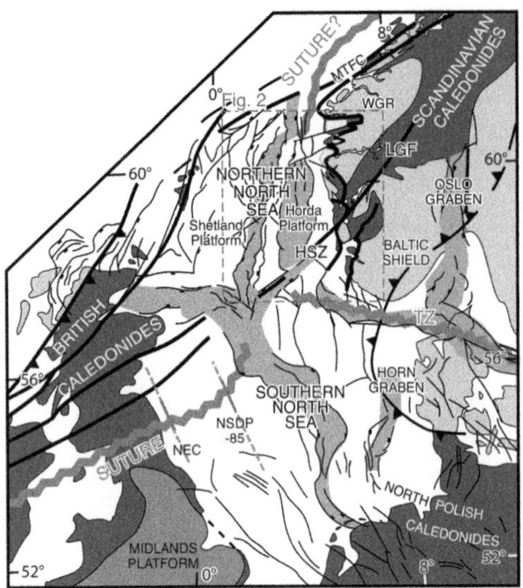

Fig. 1. Geological setting of the North Sea area, with the study area outlined. NEC and NSDP are deep seismic lines from which the Caledonian suture is interpreted (from Freeman *et al.* 1988). HSZ, Hardangerfjord Shear Zone; LGF, Lærdal–Gjende Fault; MTFC, Møre–Trøndelag Fault Complex; TZ, Tornquist Zone; WGR, Western Gneiss Region. Note that the precise location of the suture zone is not known offshore Norway.

focus on the accretional (collisional) part of the Caledonian evolution. Only relatively recently has it become clear that post-collisional Devonian extensional deformation had a profound impact on the Scandinavian Caledonides (Fossen 2010 and references therein), and that this pre-rift extension created a structural framework that must have influenced the formation and evolution of the Permian–late Jurassic North Sea Rift, as discussed in this contribution.

The extensional history explored here initiates with the termination of the main and final collisional phase of the Caledonian orogenic evolution, known as the Scandian Phase (or Scandian Orogeny) (e.g. Gee *et al.* 2008). The Scandian Phase is explained as a continent–continent collision that lasted from approximately 425 to 405 Ma, resulting in subduction of the western margin of Baltica underneath the Laurentian lithospheric plate, and the creation of an associated asymmetric orogenic edifice consisting of a Baltican pro-wedge of continental and oceanic allochthonous units and a somewhat smaller retro-wedge on the Laurentian margin. This interpretation is based on the general northwestwards increase in Caledonian temperature (T) and pressure (P) estimates in the Baltican basement, notably in the Western Gneiss Region (WGR) of SW Norway, where late Caledonian P–T estimates of more than 36 GPa–800°C have been reported from the northwestern parts (Smith & Lappin 1989; Dobrzhinetskaya *et al.* 1995; Wain 1997; Hacker *et al.* 2010). These observations place the Caledonian (or Iapetus) suture zone west of the current Norwegian coastline (Fig. 1).

The pro-wedge of allochthonous units (thrust nappes) and the underlying Baltican basement were separated by a regionally significant basal thrust or décollement zone (Fig. 2) (Fossen 1992; Milnes *et al.* 1997). Internally, the orogenic wedge contains a number of large and small thrust nappes and related thrusts, where the upper tectonic units in Scandinavia have been translated several hundred kilometres to the SE (e.g. Hossack & Cooper 1986), with complementary NW-directed thrusting in the Scottish Caledonides (retro-wedge) (e.g. Butler 2010). Hence, the Caledonian collisional phase left a highly heterogeneous crust at the dawn of a prolonged history of post-Caledonian extensional deformation, with an important large-scale three-layer rheological stratification in the Scandinavian Caledonides composed of a strong Baltican basement, a basal thrust zone comprising mainly weak micaceous metasediments, and allochthons of strong Precambrian continental crust (continental nappes in Fig. 2). The suture zone, which is the zone of oceanic terrane(s) separating Laurentian from Baltican or, to the south, Avalonian lithosphere, is interpreted to be represented by a band of north-dipping reflections on deep seismic data of the BIRPS project, acquired in the 1980s (Klemperer & Matthews 1987; Freeman *et al.* 1988). These deep seismic lines suggest that the suture extends to the NE into the northern North Sea Rift, more or less in the direction of the Hardangerfjord Shear Zone (Fig. 1). However, off the coast of Norway, the deep seismic coast-parallel ILP lines (see Fig. 2 for location) show a south-dipping mantle shear zone that offsets the Moho in a normal sense: that is, the opposite of what we would expect for a subduction-related Caledonian feature (Gabrielsen *et al.* 2015) (Fig. 3). Furthermore, the isotherms and isobars related to the Scandian collision in South Norway in general, and the WGR basement in particular, show that the suture zone must turn northwards more or less parallel to the trend of the Viking Graben–Sogn Graben. However, a mantle feature similar to that interpreted as the suture across the British Isles has not been found on deep seismic lines crossing the northern North Sea, except perhaps for east-dipping reflections on line NSDP84-1 that appear to be an extensional structure (Gibbs 1987; Odinsen *et al.* 2000). The reason for its absence on deep seismic lines may be the profound post-collisional extensional reworking in this part of the orogen.

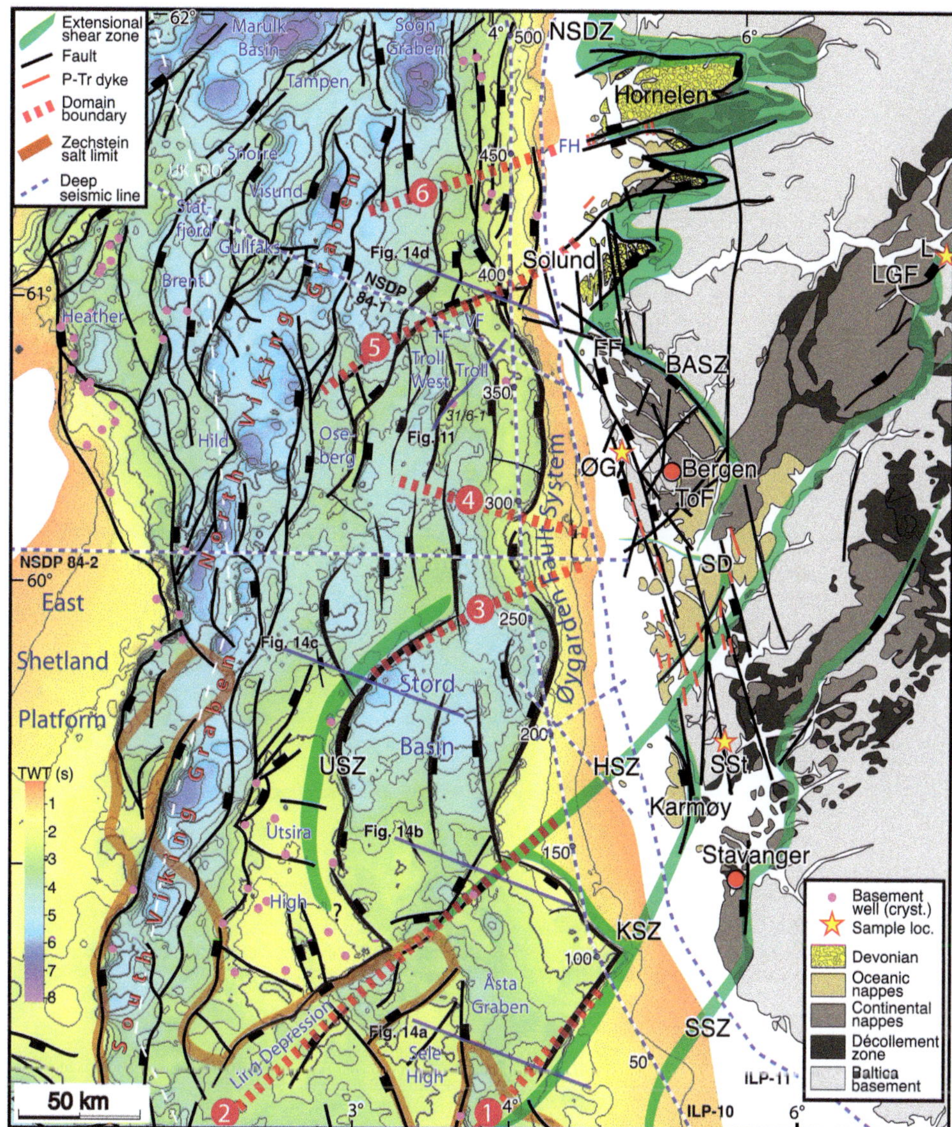

Fig. 2. Onshore–offshore map of the northern North Sea Rift (the area outlined in Fig. 1). Major onshore Devonian extensional shear zones are extended into the rift where observed on seismic sections. The fault pattern of the rift is shown over a top basement map, where anything older than Permian is considered basement. FF, Fensfjord Fault; FH, Florø Horst; SSZ, KSZ, HSZ and BASZ, Stavanger, Karmøy, Hardangerfjord and Bergen Arcs shear zones; NSDZ, Nordfjord–Sogn Detachment Zone; LGF, Lærdal–Gjende Fault; SD, Sunhordland Detachment; TF, Tusse Fault; ToF, Totland Fault; USZ, Utsira Shear Zone; VF, Vette Fault; P–Tr, Permo-Triassic. Wells drilled to crystalline basement are from Bassett (2003). Starred localities are tagged with letters referred to in the text.

Devonian extension of Caledonian crust

Reactivation of the low-angle basal Caledonian thrust zone (Mode I extension)

The numerous hinterland-dipping Caledonian thrusts and shear zones in the lower Devonian crust in South Norway were clearly prone to reactivate as the tectonic regime changed from one of regional contraction to extension. The current erosion level exposes the Caledonian basal thrust or décollement zone particularly well, and field studies show that, although this zone was the basal thrust of the Caledonian orogenic wedge, it is completely

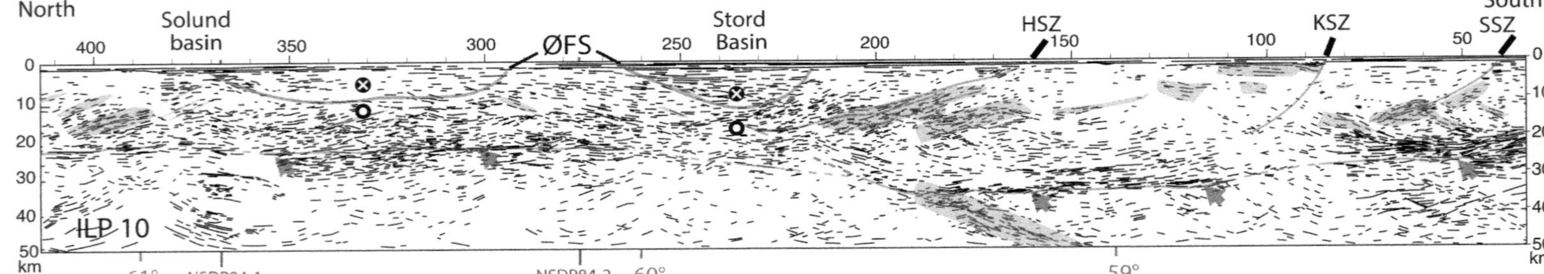

Fig. 3. Line drawing of deep seismic line ILP-10. See Figure 2 for the location and Gabrielsen *et al.* (2015) for detailed information. Some dipping domains are outlined, and arrows indicate the location of the Moho. ØFS, Øygarden Fault System, which is transected along-strike. See the caption to Figure 2 for more abbreviations.

dominated by top-to-hinterland (NW) kinematic indicators that formed during extensional reactivation of the low-angle basal Caledonian décollement zone (Mode I extension: Fossen 1992; Fossen & Dunlap 1998). Interestingly, the extensional reactivation occurred while this zone was a very low-angle structure with an estimated northwesterly dip in the eastern and central part of only 3–5°, steepening to the NW (Fossen 2000). The reactivation of this low-angle basal décollement zone in extension was facilitated by the strong contrast between its weak metasedimentary rocks (micaschists, phyllites and shales) and the much stronger underlying Proterozoic basement rocks and overlying crystalline basement nappes. Most likely, the expulsion of fluids released due to metamorphic reactions within the basal décollement zone also contributed to its weak nature. Ar/Ar ages of micas from the basal thrust zone suggest that the kinematic signature of this zone changed between 408 and 402 Ma (Fossen & Dunlap 1998). At this point, it turned into a low-angle extensional detachment on which the Caledonian orogenic wedge was translated toward the (north)west or, depending on our frame of reference, the motion of the Baltic Shield was reversed by motion towards the SE (eduction).

Secondary (Mode II) extensional shear zones

As displacement accumulated on the extensional low-angle Caledonian décollement zone, differential exhumation (more rapid exhumation in the hinterland) lowered its dip to the point where continued shearing became mechanically unfeasible (Nur *et al.* 1986). From that point on, continued extension was taken up by a set of new and steeper shear zones that transected the inactivated décollement zone. Most of these are NW- and west-dipping structures (Mode II extension: Fossen 1992) with up to several tens of kilometres of displacement. These shear zones segmented the Caledonian crust into major blocks that rotated to become foreland dipping (Fig. 3). The most impressive example of these Mode II shear zones is probably the strongly corrugated Nordfjord–Sogn Detachment Zone (NSDZ in Fig. 2), which, together with the steeper Bergen Arcs Shear Zone to the south, forms a close to 200 km-long detachment zone. The Nordfjord–Sogn Detachment Zone is probably connected to the NE-trending Møre–Trøndelag Fault Complex immediately north of the area covered by Figure 2 (MTFC in Fig. 1). South of the Bergen Arcs Shear Zone is the much straighter Hardangerfjord Shear Zone, and then the (again) curved Karmøy and Stavanger shear zones (KSZ and SSZ in Fig. 2). These shear zones are steeper than the basal décollement zone, from approximately 60° for the Bergen Arcs Shear Zone via the somewhat more gentle-dipping Karmøy Shear Zone and the approximately 25° Hardangerfjord Shear Zone to the folded but, in the extension direction, very-low-angle (5–10°) Nordfjord–Sogn Detachment Zone (Fig. 3).

The Nordfjord–Sogn Detachment Zone is defined by an up to 5–6 km-thick package of mylonites with abundant kinematic structures showing consistent normal (top-to-the-west) sense of shear. Its strongly mylonitic fabrics suggest displacements in excess of 50–100 km (Norton 1987; Andersen & Jamtveit 1990; Fossen 2000), as does the fact that the low-angle Nordfjord–Sogn Detachment Zone brings high-pressure rocks of the WGR in close contact with hanging-wall Devonian basins and low-grade Caledonian allochthonous units. An interesting and important feature of the Nordfjord–Sogn Detachment Zone is its folding about gently west-plunging axes, with subvertical axial surfaces (Chauvet & Séranne 1994; Krabbendam & Dewey 1998) (Fig. 4). This upright folding occurred before, during and after deposition of the preserved level of Devonian sediments, and is interpreted as a result of transtension associated with sinistral shear along the Møre–Trøndelag Shear Zone – the ductile precursor of the Møre–Trøndelag Fault Complex (Krabbendam & Dewey 1998). For the Hornelen Devonian basin area, the folding is tightest along the northern and southern margins, where mylonites and protomylonites of the Nordfjord–Sogn Detachment Zone define domains of steep dips to the north and south that, from a mechanical point of view, are easier to reactivate in the brittle regime than the shallowly dipping parts of the detachment zone.

The Hardangerfjord Shear Zone is perhaps the longest of the Mode II shear zones, with a minimum onshore length of 350 km, but with a maximum displacement of only about 10–15 km (Fossen & Hurich 2005). Its extension into the North Sea has been confirmed by deep seismic lines immediately offshore, where it appears as a NW-dipping band of reflectors that reaches the lower crust (Fig. 3) (Hurich & Kristoffersen 1988; Fossen & Hurich 2005; Gabrielsen *et al.* 2015).

Most or all of these oblique extensional shear zones probably reached, directly or indirectly, the surface as brittle fault zones, generating a system of Devonian continental basins. Much of the basin-fill has been removed by erosion, but the lower parts of four linked basins remain in the hanging wall of the Nordfjord–Sogn Detachment Zone in West Norway. These basins show stacking geometries (Steel *et al.* 1985) and palaeo-thermal structures (Fauconnier *et al.* 2014) that are consistent with a listric fault or supra-detachment model (Hossack 1984; Séranne & Séguret 1987) (Fig. 5), possibly with an influence of strike-slip tectonics

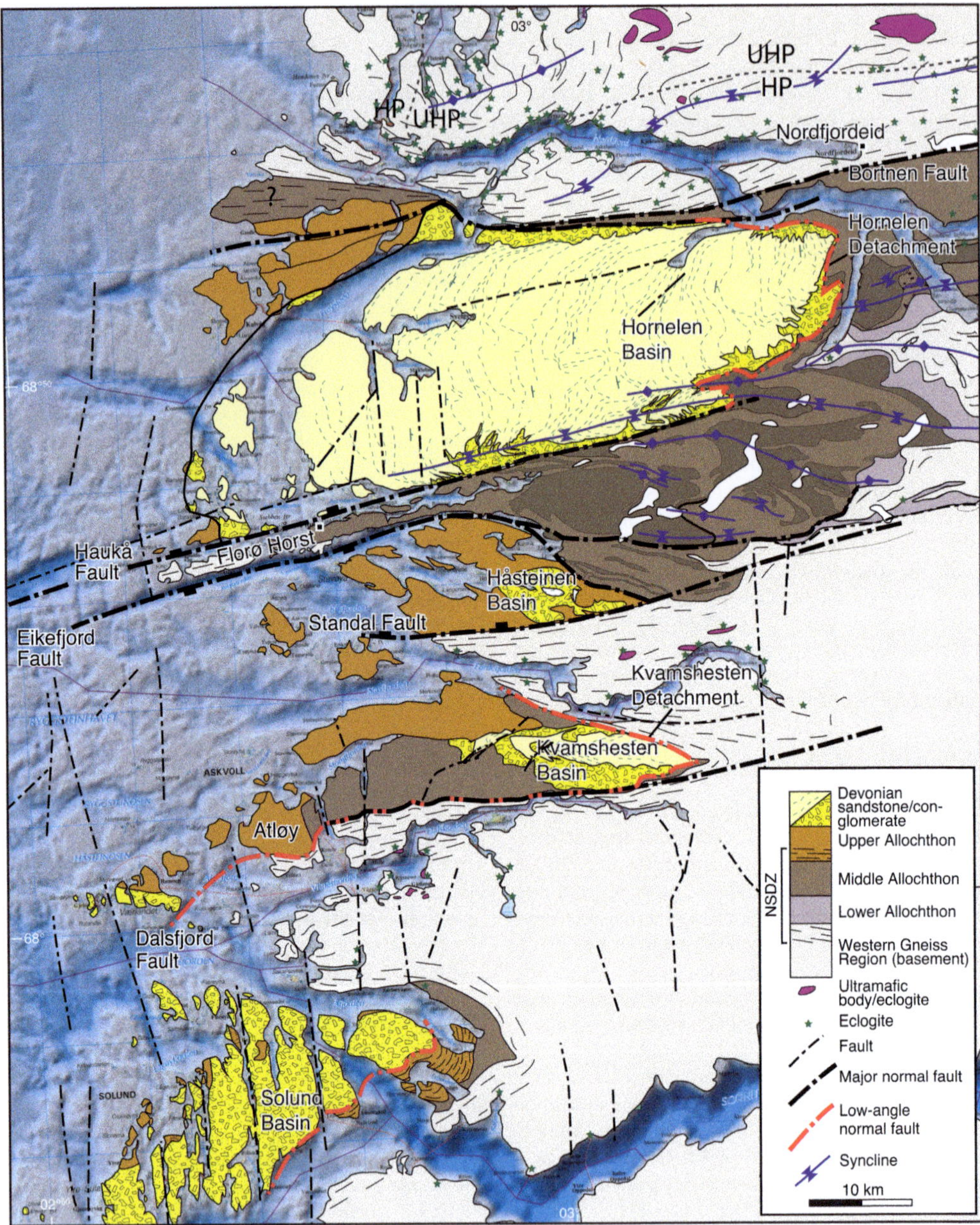

Fig. 4. Map of the area of Devonian basins in West Norway. Note the presence of low-angle normal faults (red), steeper WSW–ENE faults and sets of coast-parallel faults.

(Steel *et al.* 1985). For the Hornelen Basin, >25 km of stratigraphic thickness accumulated on top of the detachment system, while the maximum burial temperature seems to be more or less the same from the top to the bottom of the basin (i.e. from the western unconformity to the eastern end of the section shown in Fig. 5). This observation is consistent with the supra-detachment model, where repeated translation of the hanging wall on a low-angle detachment produced the accommodation space needed to create a stratigraphic thickness that almost tripled the true depth of the basin. A Middle Devonian (393–383 Ma) age of the upper part of the Hornelen Basin (Høeg 1945) shows that the basin

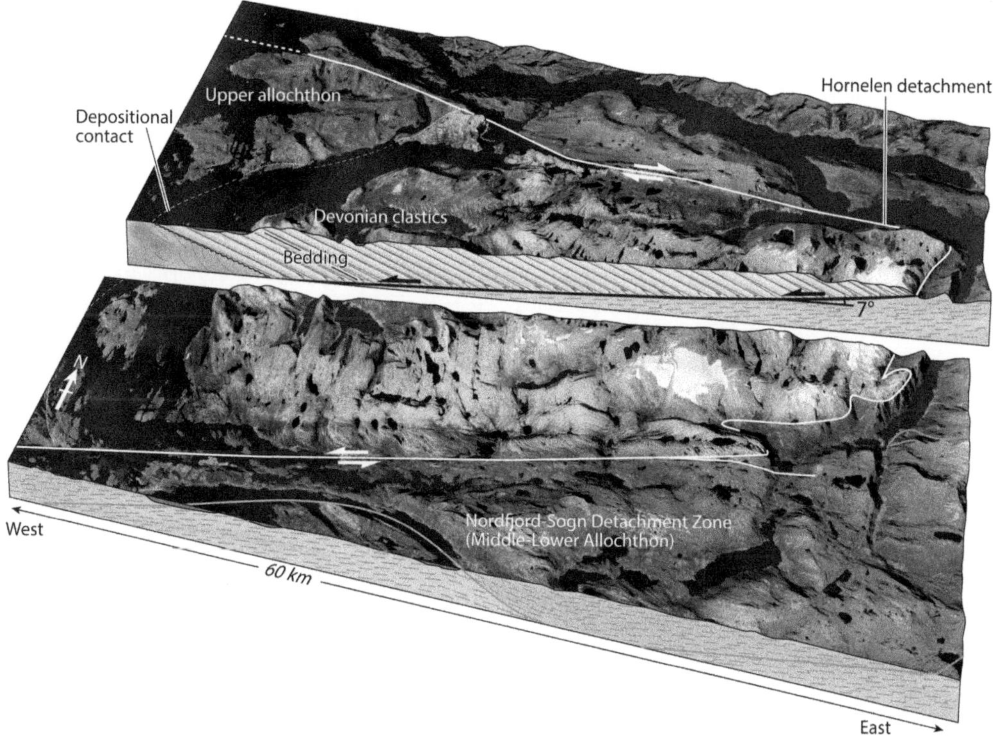

Fig. 5. 3D model of the Hornelen Basin resting on a brittle detachment against the mylonitic Nordfjord–Sogn Detachment Zone.

developed during the Devonian extensional shearing on the Nordfjord–Sogn Detachment Zone, which is indirectly dated by white mica cooling ages in the Western Gneiss Complex that range from 400 Ma in the SE part to 380 Ma in the NW part (Hacker 2007 and references therein). Hence, at the time of Devonian basin formation, the Nordfjord–Sogn Detachment Zone had already been exhumed to lower-amphibolite or greenschist facies conditions due to tectonic thinning and rotation of the WGR during the Mode II extension.

Brittle onshore faults

Reactivation of low-angle Devonian extensional shear zones

The vast majority of the oblique (Mode II) extensional shear zones described above were established as ductile (plastic) structures, but contain brittle elements such as the Hornelen Detachment (Fig. 4), the Dalsfjord Fault (Eide *et al.* 1997), the Lærdal–Gjende Fault (Andersen *et al.* 1999) and the Fensfjorden Fault (Wennberg *et al.* 1998) (FF in Fig. 2) that suggest extensional reactivation in the brittle regime (see later). These brittle elements are faults that are much thinner than the wide ductile shear zones in which they are located, and typically show evidence of multiphase evolution. Two of these have been subjected to palaeomagnetic and radiometric age determinations, and will be briefly discussed here.

The Lærdal–Gjende Fault (LGF in Figs 1 & 2) is a low-angle normal fault dipping 15–35° to the NW, located within the much broader and ductile low-angle (25°) Hardangerfjord Shear Zone (Fossen & Hurich 2005). The Lærdal–Gjende Fault is a composite brittle structure consisting of up to several hundred metres of cohesive greenish cataclasite that, as seen at the exceptionally well-exposed fault location in Lærdal (starred locality L in Fig. 2), contains 1–2 m-thick zone of non-cohesive fault gouge. The Lærdal–Gjende Fault omits the décollement (phyllite/micaschist) that acted as the basal thrust during Caledonian collision and as an extensional décollement during the subsequent Devonian extensional history (Lutro & Tveten 1996). In contrast, the Hardangerfjord Shear Zone is a ductile shear zone that preserves layer continuity across the zone. Hence, the Lærdal–Gjende Fault is an expression of brittle deformation along the ductile Devonian Hardangerfjord Shear Zone, possibly formed by brittle deformation of the feldspar-rich Jotun Nappe during Devonian cooling, but certainly active at later stages at low temperatures.

Andersen *et al.* (1999) applied palaeomagnetic dating methods to plugs from the greenish epidote-rich cataclasite zone of the Lærdal–Gjende Fault at the Lærdal locality to estimate activity on this fault, obtaining Permian and late Jurassic–early Cretaceous signatures that they interpreted as times of fault activity. More recently, we collected clay-rich fault gouge samples at the same locality (L in Fig. 2) that yielded a strong late Jurassic–early Cretaceous K–Ar signature (see later). This is in agreement with the fact that it was sampled from the non-cohesive part of the fault, which is generally considered to be the most recently active part of the fault. Hence, there is strong evidence for brittle reactivation of the low-angle Lærdal–Gjende Fault at the time of rifting in the North Sea and the Oslo Rift (see below).

The Dalsfjord Fault (Fig. 4) is a low-angle fault that marks the contact between the Devonian Kvamshesten Basin and the mylonitic middle allochthonous rocks of the Nordfjord–Sogn Detachment Zone (Osmundsen & Andersen 2001). The fault is well exposed at a locality in Atløy (Fig. 4), where a greenish cataclasite, quite similar to the one observed along the Lærdal–Gjende Fault, is post-dated by red breccia. A similar sequence of events as that documented from the Lærdal–Gjende Fault is observed through a combination of palaeomagnetic and Ar/Ar dating of its fault rocks: a strong late Permian signature overprinted by a late Jurassic–early Cretaceous event (Eide *et al.* 1997). This west-dipping low-angle fault shares similarities with the Hornelen Detachment. This detachment is another low-angle fault consisting of green flinty epidote-bearing cataclasite that shows evidence of later microfracturing (reactivation), defining the gently west-dipping tectonic contact between the Hornelen Basin and its mylonitic substrate (Nordfjord–Sogn Detachment Zone) (Figs 4 & 5). The Hornelen Detachment juxtaposed the Devonian basin with the top-to the-west mylonites of the Nordfjord–Sogn Detachment Zone, and the distribution of alluvial fans along the margins of the Hornelen Basin shows that it closely follows the original basin-forming fault.

It is an important observation that onshore low-angle Devonian extensional structures were reactivated as low-angle normal faults during the late Palaeozoic–Mesozoic rift history. Indeed, if such Devonian extensional shear zones are reactivated onshore, they are also likely to be reactivated offshore in the northern North Sea rift where rift-related strains are much higher.

Steeper faults associated with the Devonian basins

The Hornelen detachment serves as a good example of a very low-angle brittle fault truncated by steeper faults. In this case, steeper faults with ENE–WSW orientations are located along the north and south margins of the Hornelen Basin. The fault on the north side (the Bortnen Fault) is dipping steeply to the south, while the southern fault (the Haukå Fault) is dipping to the NNW. The Bortnen Fault is reported to show a sinistral sense of shear, with a subordinate normal component, and K-feldspar–epidote alteration, breccia and pseudotachylite as characteristic elements (Young *et al.* 2011). The marginal fan architecture of the Hornelen Basin shows that the Bortnen Fault closely approximates the locations of the original (synsedimentary) fault along the northern basin margin (Steel *et al.* 1985).

The southern (Haukå) marginal fault transects the Hornelen Detachment in the SE corner of the basin, where its normal offset can be demonstrated to exceed 500 m (Braathen 1999). Also, this fault closely corresponds to the syndepositional fault along the southern margin of the Hornelen Basin, as seen from the preserved marginal fan complex along the fault.

Together with the next major steep fault to the south (the Eikefjord Fault), the Haukå Fault defines a major NNE-trending horst structure that narrows down to a 4–5 km-wide structure towards the coast, here named the Florø Horst. In the western coastal part, the Florø Horst brings up gneisses interpreted to be part of the WGR (Lutro & Bryhni 2000), and places them in fault contact with upper allochthonous units both to the north (Haukå Fault) and to the south (Eikefjord Fault) (Fig. 4), suggesting several kilometres of displacement across both faults. Furthermore, the displacement increases to the west, suggesting that the horst structure and its associated faults stretch well into the northern North Sea rift basin.

General pattern of onshore faults and fracture zones

In general, a large population of brittle faults and fractures dissect the bedrock surface of SW Norway (Fig. 6), and the wide range in orientation suggests that they formed during more than one phase of deformation. The majority of the onshore brittle structures in the area covered by Figure 6 are post-lower Devonian faults that affect Caledonian and extensional lower Devonian ductile fabrics. However, Proterozoic elements are probably present in much of the Proterozoic basement SE of the Hardangerfjord Shear Zone, where the basement is generally unaffected by Caledonian deformation.

Fig. 6. Onshore fracture pattern from the interpretation of aerial photographs (8833 lineaments) supplied by field observations. Larger faults are indicated with thicker black lines. Rose diagrams are presented for subareas and for the total dataset. Rose diagram and lineaments are colour-coded according to trend (see online version for colours).

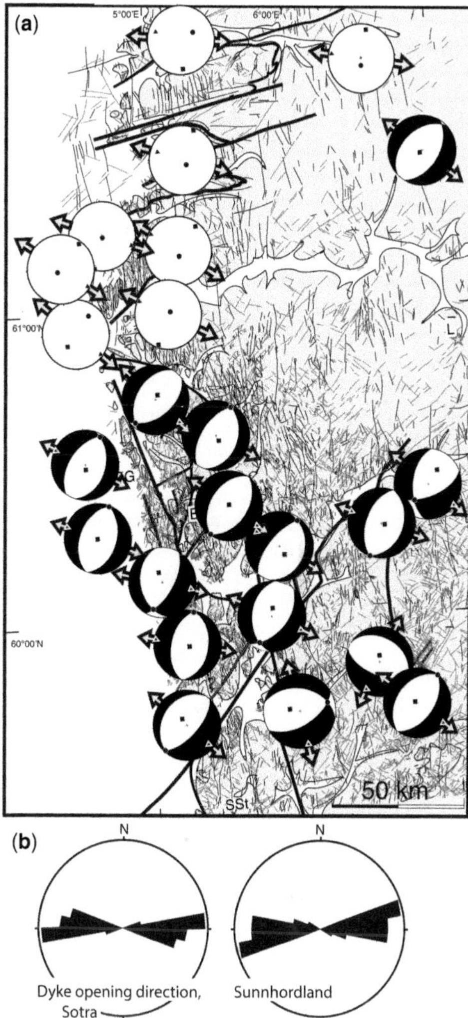

Fig. 7. (**a**) Kinematics from fault-slip analysis of fault populations from locations covered by the plots. Arrows represent the extension direction, which is mostly NW–SE. 'Beachballs' show fields of shortening (black), extension (white) and principal axes. Northern observations from Séranne & Séguret (1987) and Chauvet & Séranne (1994). (**b**) Opening directions (east–west) as determined from Permo-Triassic dykes in Sotra and Sunhordland (Fig. 6).

Field observations suggest that many of the onshore brittle structures interpreted in Figure 6 are steep, although low-angle structures also exist. The vast majority of these structures range from the NW to the NE. A strong coast-parallel (north–south) set is generally present (see also Fig. 4), together with a NW- and a NE-trending population. In addition, a series of more east–west-striking faults occur in the area associated with the Devonian basins (Fig. 4).

NE-trending faults overprinted by coast-parallel faults

The occurrence of at least three different strike populations suggests that the brittle faults and fractures shown in Figure 6 did not form in a single stress field. Kinematic fault-slip analysis indicates a strong influence of NW–SE stretching, as portrayed in Figure 7, and field work has shown that striated NE–SW-trending faults, commonly associated with epidote mineralization, are particularly well represented among the faults associated with NW–SE

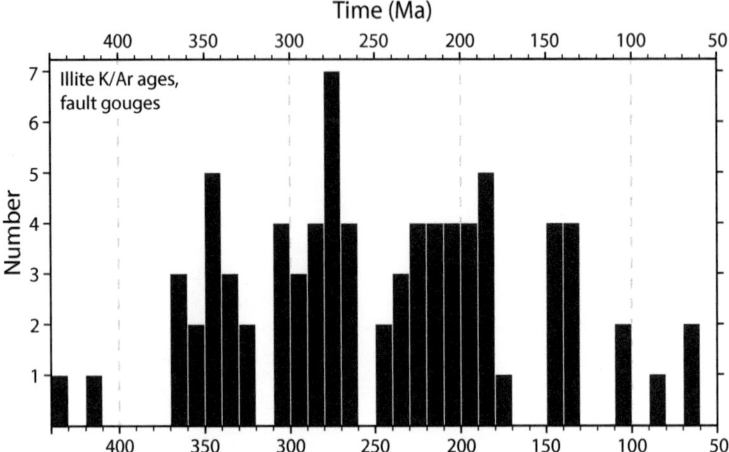

Fig. 8. Histogram of K–Ar illite ages from 25 sampled faults in West Norway. Three grain-size fractions (2–6, <2 and <0.2 μm) were dated for each sample. Note that some faults show evidence of repeated reactivation and, therefore, some ages are mixed. However, as a first-order interpretation, main periods of fault activity occurred in the Early Carboniferous, Permian, Triassic–Early Jurassic and latest Jurassic–earliest Cretaceous. Ages from Ksienzyk (2012), Arsenijevic (2013) and Woznitza (2014).

extension. This is also the extension direction portrayed by ductile (Mode I and II) Devonian extensional fabrics and structures, and it has been suggested that the population of brittle faults consistent with NW–SE extension represents the brittle continuation of Devonian ductile extension, and that they may have started to generate when the currently exposed level of the crust crossed the plastic–brittle transition (Fossen 2000). This assumption is consistent with approximately 395 Ma U–Pb dates of sphene on some of these early onshore fractures west of Bergen (Larsen *et al.* 2003).

Field observations show that NE–SW-trending faults are transected by north–south-trending faults and fractures, some of which are associated with dykes of Permo-Triassic age (Færseth *et al.* 1976; Torsvik *et al.* 1997; Fossen & Dunlap 1999) (Fig. 2). The kinematics of these dykes, as revealed by steps and jogs, consistently indicate east–west opening (Fig. 7), in concert with the Permo-Triassic rift-related faulting in the North Sea. Hence, in simple terms, there seems to have been a change from NW–SE Devonian to east–west Permo-Triassic extension.

Illite K–Ar fault gouge dating

An ongoing fault-dating project employs dating of authigenic illite from fault gouge found in exposed faults in West Norway by means of the K–Ar method (Ksienzyk 2012; Ksienzyk *et al.* 2012; Arsenijevic 2013; Woznitza 2014). A summary of ages from 25 faults from the onshore area is shown in Figure 8. For each sample, three grain-size fractions of 2–6, <2 and <0.2 μm were analysed.

Illite grows in active faults due to the combined effects of grain comminution caused by the crushing of the host rock and alteration by circulating fluids, thus providing a geochronometer with the potential to date fault activity (e.g. Solum *et al.* 2005; Tagami 2012). When a range of grain-size fractions are analysed, many faults yield dispersed ages that are correlated with grain size. In sedimentary rocks, this is commonly interpreted as a mixing of coarse-grained, older detrital illite from the host rock with fine-grained, younger authigenic illite from the fault (van der Pluijm *et al.* 2001; Solum *et al.* 2005). Crystalline rocks, like the basement of West Norway, are originally devoid of clay minerals, so that a potential contamination with detrital illite from the host rock is not an issue, and all illite ages should correspond to neocrystallization of illite in the fault. Nevertheless, many faults in crystalline rocks also show dispersed ages that are correlated with grain size. This age dispersion can be explained either by contamination of the coarser fractions with other K-bearing minerals from the host rock or by fault reactivation resulting in several generations of authigenic illite (Zwingmann & Mancktelow 2004; Zwingmann *et al.* 2010; Davids *et al.* 2013; Bense *et al.* 2014; Torgersen *et al.* 2014, 2015).

While contamination with K-bearing minerals from the host rock might be a contributing factor in some of the samples, we interpret most of the age dispersion in the dataset presented in Figure 8 to be caused by fault reactivation. Consequently, the oldest illite ages provide a minimum age of fault formation, while younger ages are generally interpreted as episodes of fault reactivation. With this in mind, main periods of fault activity could be identified in the Early Carboniferous, Permian, Triassic–Early Jurassic and around the Jurassic–Cretaceous boundary. The latter two periods correlate roughly with the two phases of North Sea rifting, and pulses of dyke intrusion at around 250 and 220 Ma. However, it is interesting that fault gouges with a significant clay content seem to have already developed in the Late Devonian, and that marked onshore brittle faulting occurred prior to the onset of North Sea rifting. The dataset includes faults with both NW–SE, north–south and NE–SW orientations, many of them with a pre-Triassic age component (Ksienzyk 2012) (Fig. 8), suggesting that a range of fault orientations were established at the onset of, or in the early stages of, North Sea rifting, with evidence of repeated reactivation showing that similar offshore basement faults may have influenced the fault pattern observed in the rift basin-fill. The long history of activity seen in many faults, with often multiple episodes of reactivation, is demonstrated by two examples, including the Lærdal samples referred to above. The other locality is at Skjoldastraumen; both localities are shown in Figure 2.

The Skjoldastraumen sample (SSt in Fig. 2) comes from a minor fault along one of several pronounced coast-parallel faults that host Permian and Triassic dykes. Illites from this sample yielded a middle Carboniferous (320 $\pm$ 9 Ma) K–Ar age for the coarsest fraction (2σ analytical errors), and middle Permian (274 $\pm$ 6 Ma) and Late Triassic (227 $\pm$ 7 Ma) ages for the intermediate and finest fractions, respectively. The fact that illite grew as early as in the Carboniferous suggests that at least some elements of these NNW-trending coast-parallel fractures were established prior to dyke intrusion and Permo-Triassic rifting. The increasingly younger ages in the smaller grain-size fractions indicate renewed fault activity in the Late Triassic, most probably contemporaneous with Triassic (*c.* 220 Ma) dyke intrusion recorded along these lineaments (Føssen & Dunlap 1999).

A sample (BG-129a) from the NE-trending Lærdal–Gjende Fault (locality L in Fig. 2) (Fig. 9) yielded latest Jurassic–earliest Cretaceous ages (149 $\pm$ 4 and 144 $\pm$ 2 Ma, respectively) for the two coarser grain-size fractions. The finest fraction yielded a younger age (64 $\pm$ 1 Ma), suggesting a Paleocene reactivation of the fault. A second sample (BG-129b) yielded an Early Jurassic age (191 $\pm$ 3 Ma) for the coarsest fraction. However, this grain-size fraction contained considerable amounts of K-feldspar and the age does not, therefore, reflect the timing of illite crystallization, but, rather, should be considered as a mixed illite/inherited K-feldspar age. The intermediate grain-size fraction, with only trace amounts of K-feldspar, gave an Early Cretaceous age (142 $\pm$ 2 Ma) that is almost identical to the latest Jurassic–earliest Cretaceous ages of BG-129a. The finest fraction, with an age of 106 $\pm$ 2 Ma, again suggests a later (Late Cretaceous or younger) reactivation. The results thus show the main phase of illite growth at around 150–140 Ma, corresponding to the younger

Fig. 9. Road section through the Lærdal–Gjende Fault gouge at the Lærdal locality (Fig. 2), showing sample locations for K–Ar illite dating. The main period of illite growth is latest Jurassic–earliest Cretaceous with evidence of a later Albian–Palaeogene reactivation.

palaeomagnetic age found by Andersen *et al.* (1999). This is in agreement with the fact that we sampled the non-cohesive part of the fault, which is considered to be the most recently active part of the fault. Hence, there is strong evidence for brittle reactivation of the low-angle Lærdal–Gjende Fault at the time of rifting in the North Sea and the Oslo Rift. Oblique and horizontal striations on slip surfaces along the Lærdal–Gjende Fault are seen in the field, which, given its oblique orientation to the Permo-Triassic extension direction recorded by dykes along the coast, fit with oblique reactivation at the same time as east–west extension during North Sea rifting.

$^{40}Ar/^{39}Ar$ *dating of feldspar*

Some of the faults and fractures in the felsic basement gneisses in West Norway show evidence of K-feldspar alteration in a centimetre-wide pink alteration rim around the faults and fractures (Fig. 10e). Close examination of such an example from

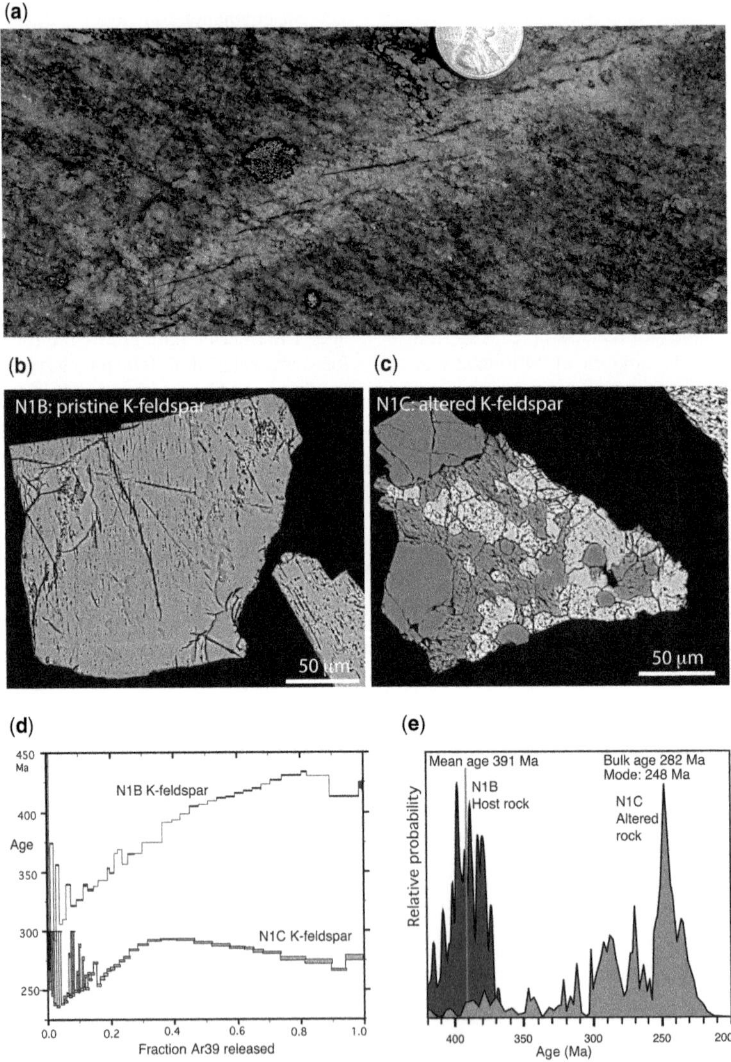

Fig. 10. (**a**) Alteration of K-feldspar around fractures in Proterozoic gneiss, Øygarden Complex west of Bergen. (**b**)–(**c**) SEM images of unaltered (from sample N1B) and altered (from sample N1C) feldspar grains from within and outside of an alteration rim, respectively. N1B feldspar is altered into epidote minerals, quartz and new K-feldspar. (**d**) $^{40}Ar/^{39}Ar$ age spectra for the two samples. (**e**) $^{40}Ar/^{39}Ar$ age probability density plot for the two samples, showing a significant younger age peak (*c.* 250 Ma) for the altered feldspar. See the text for the discussion.

the Øygarden Complex west of Bergen (starred locality ØG in Fig. 2) shows that the original cryptoperthitic K-feldspar is altered into a mixture of new K-feldspar, epidote, quartz and iron oxides. We sampled altered K-feldspar from within the alteration rim (N1C) and pristine feldspar from the alteration rims (N1B) from the host rock several metres away (Fig. 10a, b).

A $^{40}Ar/^{39}Ar$ age spectrum for a detailed step-heating experiment of the N1B K-feldspar and the associated multidomain solution were presented by Dunlap & Fossen (1998). The spectrum (N1B in Fig. 10c) is consistent with simple volume diffusion theory, which predicts that age spectra should increase monotonically with progressive ^{39}Ar release because natural ^{40}Ar concentrations would be higher in more retentive sites (Lovera *et al.* 2002). Hence, we interpret the data to reflect closure of cryptoperthite to diffusive argon loss as it cooled through the temperature range from approximately 320 to 180°C.

The altered K-feldspar N1C (Fig. 10b) yields a very different age spectrum that shows an intermediate age maximum at around 40% of ^{39}Ar released (Fig. 10c), and with a bulk age of 282 Ma and a peak age in the probability density plot at approximately 250 Ma (i.e. significantly lower than for the pristine N1C feldspar) (Fig. 10d). Lovera *et al.* (2002) found, based on an extensive database, that intermediate-age maxima tend to be yielded by K-feldspars that have been deformed in shallow, high-strain environments adjacent to brittle faults. Hence, our data and observations can be interpreted in the framework of strong alteration of the K-feldspar in the fault zone around 240 Ma. As for illite dates, the age is related to alteration caused by the circulation of fluids rather than directly dating fault activity. Hence, the age represents a minimum age of formation for faults and fractures with alteration haloes in the basement west of Bergen. The approximate 250 Ma age of alteration corresponds well with Permo-Triassic dyke intrusion (260–250 and 230–220 Ma) and early rifting in the adjacent North Sea Rift. It is also worth noting that Steltenpohl *et al.* (2011) analysed similar red-coloured K-feldspar from the Hadselfjorden Fault Zone in Lofoten in North Norway, obtaining a $^{40}Ar/^{39}Ar$ age spectrum that showed apparent ages of between 207 and 239 Ma, which they interpreted as the initiation of that fault zone in the Late Triassic.

The North Sea Rift

The northern North Sea Rift is mainly the product of a Permo-Triassic rift phase followed by a second phase that initiated during deposition of the Middle Jurassic Brent Group and locally affected Early Cretaceous strata (e.g. Badley *et al.* 1988; Roberts *et al.* 1995; Færseth 1996; Christiansson *et al.* 2000; Bell *et al.* 2014). In the Permian basin south of the Hardangerfjord Shear Zone, there is evidence of rifted Carboniferous and older sedimentary and volcanic rocks underneath Middle–Late Permian Zechstein salt (e.g. Martin *et al.* 2002; Heeremans & Faleide 2004), consistent with Early Permian rifting simultaneous with the main period of faulting in the Oslo Rift (the Oslo Rift initiated in the Late Carboniferous and terminated in the Early Triassic, according to Larsen *et al.* 2008).

Basement wells north of the Hardangerfjord Shear Zone have been drilled on structural highs, directly from Jurassic or Triassic sediments into metamorphic basement with no evidence of Permian or Carboniferous strata. A key well on the Horda Platform in this respect is well 31/6-1 (Fig. 11), where lower Triassic beds rest on basement that is seen, from cores, to consist of alternating granitic and amphibolitic gneiss. However, this well, like all other basement wells in the northern North Sea, was drilled at the high crest of a fault block. Down-flank from the well there is an undrilled lower part of a wedge-shaped sequence of strata older than the lower Triassic encountered in the well. Hence, these strata may be of earliest Triassic or Upper Palaeozoic age, opening the possibility that rifting was already going on in the Permian, probably in concert with (?Early) Permian rifting in the southern North Sea and the Oslo Rift. In addition, some near-top-basement reflective packages probably represent layered Caledonian allochthonous units (Fig. 11). Early Triassic synrift sequences on the Horda Platform were overlain by a Middle Triassic–Middle Jurassic post-rift package, followed by Phase 2 rifting with a relatively thin late-Jurassic synrift sequence and a thick Cretaceous–Cenozoic post-rift sequence (Badley *et al.* 1988; Roberts *et al.* 1995).

One of the assumptions commonly made is that, at the onset of rifting, the extremely overthickened Caledonian crust had returned to normal thickness. Several authors (e.g. Færseth 1996; Odinsen *et al.* 2000) have assumed that the Permian crustal thickness is represented by the current crustal thickness in the onshore are, which is around 32 km near the coast. Such an assumption does not consider the effect of post-Early Permian exhumation, which can now be quantified based on recent thermochronological data, as compiled in Figure 12. This compilation suggests that rocks currently exposed along the coast were close to or slightly above 200°C in the Early Permian, corresponding to a depth of approximately 8 km and, hence, an Early Permian crustal thickness of approximately 40 km. This is the current thickness closer to the Caledonian

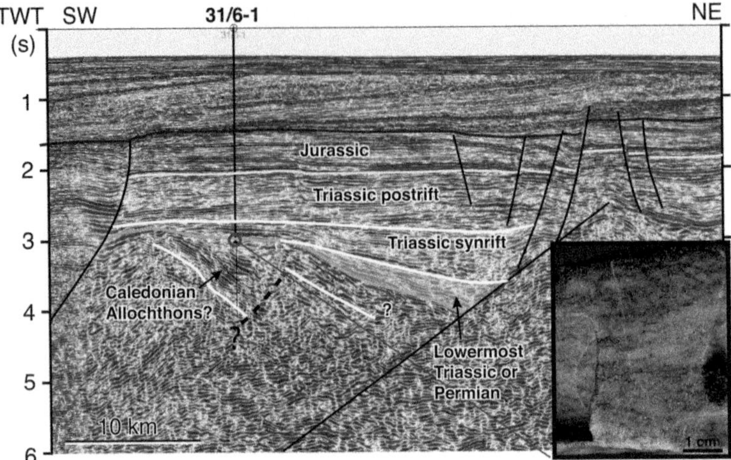

Fig. 11. Seismic section through well 31/6-1 with a picture of the basement (Caledonian) from the core. The lower part of the half-graben wedge to the east of the well is not drilled, but is likely to consist of earliest Triassic or Permian sedimentary layers. For the location, see Figure 2.

foreland (excluding the Oslo Rift) (Ebbing *et al.* 2012) and a more likely estimate of the crustal thickness as rifting initiated. The fact that the Caledonian crust was reduced to normal thickness by the Early Permian allows for an important distinction between Devonian extension, which represents tectonic

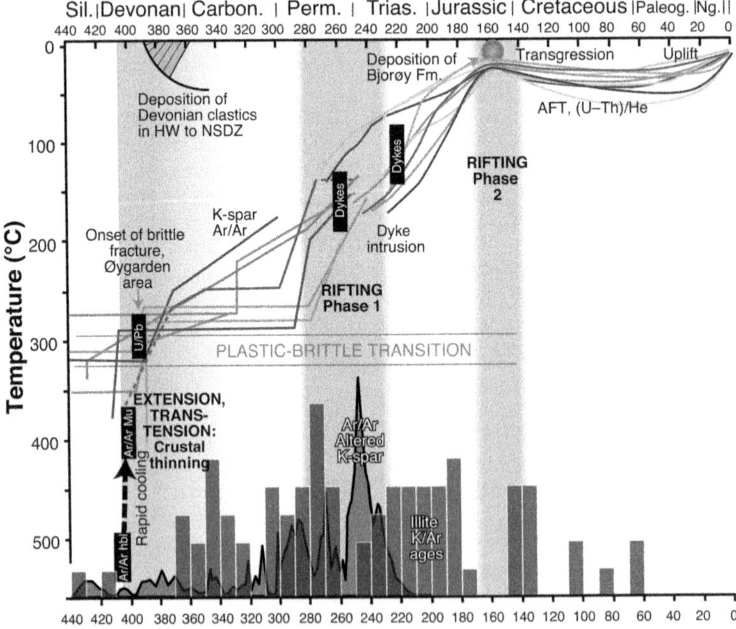

Fig. 12. Temperature–time graph from the onshore area around Bergen, based on geochronological data. The U–Pb age is from Larsen *et al.* (2003), Ar/Ar data and K-feldspar models from Dunlap & Fossen (1998), dyke ages from Færseth *et al.* (1976) and Fossen & Dunlap (1999), and the AFT and (U–Th)/He models from Ksienzyk *et al.* (2014). The bottom Ar/Ar age spectrum is from Figure 10e, while the K–Ar histogram is from Figure 8. AFT, apatite fission track.

thinning of overthickened Caledonian crust, and the two-phased North Sea rifting that thinned a crust of normal thickness.

Relationships between onshore and offshore structures

An interpretation of top basement across the northern North Sea Rift is shown in Figure 2, together with major faults, most of which are basement-rooted. 'Basement' is here loosely defined as anything older than mid Permian, including sub-Zechstein Palaeozoic packages observed on the Sele High and Devonian–Carboniferous strata on the East Shetland Platform (Platt 1995; Marshall & Hewett 2003). The interpretation, which is based on a large number of 2D and 3D seismic data and basement wells, is somewhat uncertain, particularly in the deeper parts such as the Viking Graben, but gives a qualitative picture of the effect that the two rift phases had on the early Permian surface.

Most of the first-order basement structures, such as the Utsira High, Stord Basin and Viking Graben, are confirmed by gravity and magnetic maps (Fig. 13). Along the coast, several of the allochthonous units in the hanging wall to the Nordfjord–Sogn Detachment Zone, Bergen Arcs Shear Zone and Karmøy Shear Zone are associated with high positive gravity anomalies. Similar anomalies over several of the offshore basement highs indicate that these allochthons also occur in significant amounts in the northern North Sea, as confirmed locally by basement wells: for instance, on the Utsira High (Slagstad *et al.* 2011). Similarly, negative gravity and magnetic anomalies mark the wide basins in the hanging wall to the Øygarden Fault System, particularly the Stord Basin, as major negative structures. Also, the onshore Devonian Hornelen Basin shows a marked negative magnetic anomaly, suggesting the possibility that Devonian basins may contribute to negative anomalies offshore, notably in the Stord Basin and parts of the North Viking Graben and East Shetland Platform.

The major faults on the east side of the Viking Graben, such as the Øygarden Fault System, are Permo-Triassic in origin (Steel 1993; Bell *et al.* 2014). Permo-Triassic faults in the northern North Sea Rift have generally been described as having an overall north–south orientation (Færseth 1996), but it is clear from new data that many of the larger Permo-Triassic faults have more curved and segmented shapes than those presented in earlier works. For example, the Øygarden Fault System consists of two large and one small segment, all of which curve considerably in map view (Fig. 2). Another example is the linked fault system along the east side of the Utsira High, which is also the largest fault in the Horda Platform/Stord Basin area. This fault system, which generated the deepest portion of the Stord Basin, displays curved segments, of which the northern segment curves to the NE towards the Øygarden Fault System. Together with the southern segment of the Øygarden Fault System, this fault defines the Stord Basin as a relatively isolated basin in the North Sea.

Regarding the relationship between these major Permo-Triassic faults and Devonian extension structures, we find that some of the major rift faults form on top of packages of dipping reflections that we interpret as Devonian Mode II extensional shear zones, whereas many others, such as the Øygarden Fault System and its northern extension, bear no sign of being formed by reactivation of pre-existing structures, and cut through low-angle Devonian fabrics and detachments. The offshore extensions of the Karmøy Shear Zone and Hardangerfjord Shear Zone are outstanding examples of the former, where Devonian basement shear zones influence rift-related brittle faulting. In detail, the Permo-Triassic faults are steeper in their upper parts, but apparently merge with the shear zones at depth (Fig. 14a, b). There is evidence from seismic mapping below the fjord SE of Karmøy that the Karmøy Shear Zone reactivated as a brittle fault to form a Mesozoic half-graben (Bøe *et al.* 2010), and, as discussed earlier, there is geochronological evidence for onshore brittle reactivation of the Hardangerfjord Shear Zone/Lærdal-Gjende Fault in the Permian and Late Jurassic–Early Cretaceous. The curved fault defining the east margin of the Utsira High is another example, where the underlying package of reflectors is interpreted as a shear zone that is likely to represent an east- to SE-dipping Devonian Mode II shear zone (the Utsira Shear Zone in Fig. 2) (Fig. 14c), representing a change in polarity of the system of Devonian extensional shear zones across the Horda Platform.

Domain boundaries

The Hardangerfjord Shear Zone and Karmøy Shear Zone appear to have a marked influence on the fault pattern in the Horda Platform, as they are marked by several NE–SW-trending faults at top basement (Fig. 2) and higher levels. These are two (1 and 2 in Fig. 2) of several somewhat diffuse structural boundaries that subdivide the rift into distinct structural domains, each of which have their own characteristic fault pattern and/or fault polarity. The next one to the north is represented by the northern part of the Utsira Shear Zone and its extension to the NE (Boundary 3 in Fig. 2), which marks a boundary across which the Øygarden Fault System abruptly changes from west to east dipping, and where a horst appears on the north side. This

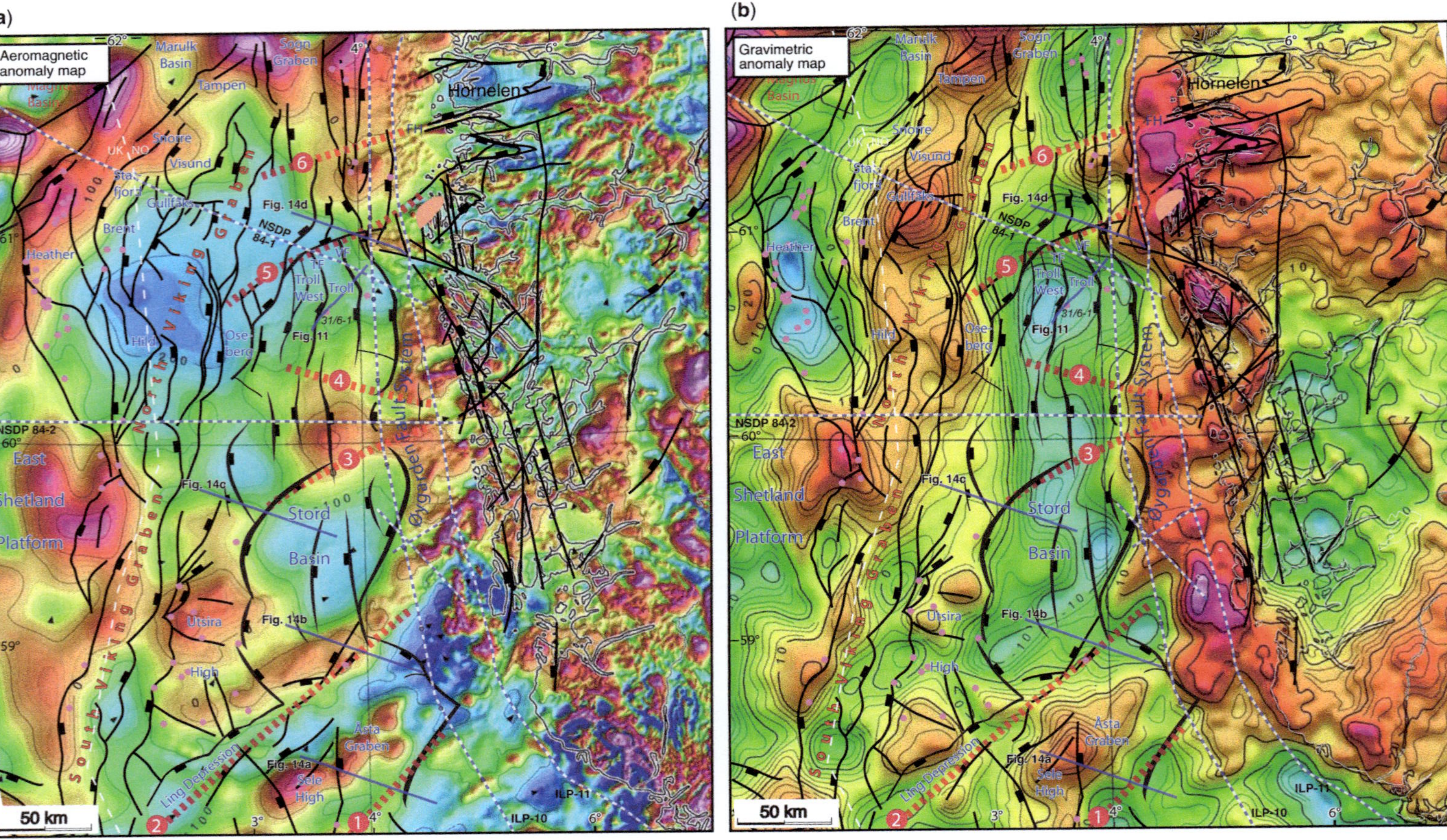

Fig. 13. (**a**) Aeromagnetic and (**b**) gravity map of the area covered by Figure 2. Warm colours indicate positive anomalies, whereas cold colours reflect negative anomalies. From Olesen *et al.* (2010*a*, *b*). Faults from Figure 2.

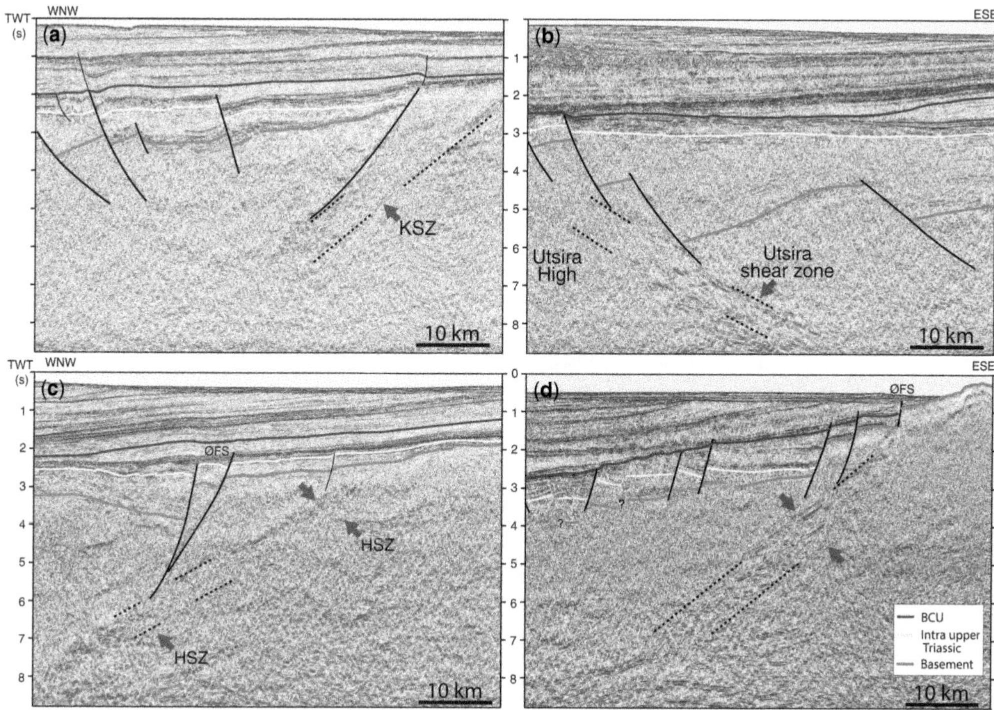

Fig. 14. Four seismic images across zones of basement reflections interpreted as Devonian Mode II extension shear zones. Top basement is top crystalline basement in these sections, but the interpretation is uncertain in the deeper parts. Preliminary depth conversions suggest that the shear zones are dipping at 20–30° (apparent dips). See Figure 2 for the location. BCU, Base Cretaceous Unconformity.

horst disappears to the north across Boundary 4 as the Øygarden Fault System flips back to become west dipping. There is also a difference in Jurassic strain from very low in the Stord Basin to moderately high in the Troll–Oseberg area, accommodated by strain transfer along Boundary 3. Boundaries 3 and 4 project onshore to the place where the Sunhordland Detachment and the Totland Fault (SD and ToF in Fig. 2) separate the Proterozoic basement west of Bergen from the allochthonous units to the south, although their relationship with these structures is unclear.

Another boundary (Boundary 5) is found at 61°N, north of the Troll and Oseberg fields (Fig. 2). Here, the west-dipping Øygarden Fault System, Vette Fault and Tusse Fault (VF and TF in Fig. 2) terminate or make remarkable steps (Badley *et al.* 1988; Barling 2014) across the NNE-trending structure. For this structure, which was also noted by Færseth *et al.* (1995) and Færseth (1996) as a domain-bounding structure, we recognize a relatively shallowly dipping set of reflectors (Fig. 14d) that most probably represents a NW-dipping basement shear zone that extends from near the surface down to the reflective lower crust at around 8 s two-way time (TWT) (approximately 16 km). The location and orientation of this structure suggest that it may be connected to the Dalsfjord Fault (Fig. 4), which is interpreted as a Devonian fault that was reactivated in the Permian and latest Jurassic–earliest Cretaceous (Eide *et al.* 1997). A set of south(east) dipping reflections also occur on the deep seismic line ILP10 in this area (just north of 59° in Fig. 3). The gravity and magnetic anomaly maps (Fig. 13) give some support to this interpretation, which seems more likely than the one presented by Færseth *et al.* (1995), where the Nordfjord–Sogn Detachment Zone was abruptly bent westwards into the North Sea to connect with this domain boundary.

About 50–60 km to the north, we suggest a possible connection between the onshore Florø Horst (Figs 2 & 4) and the change in offshore fault arrangement in its offshore extension (6 in Fig. 2). This diffuse structure marks the southern termination of the Sogn Graben, which at this point steps westwards to the North Viking Graben. In this case, there are clear magnetic and gravimetric lineaments extending the Florø Horst offshore towards Boundary 6.

Concluding remarks

The Baltica–Laurentia continent–continent collision produced a highly anisotropic crust in the northern North Sea region and its margins that was prone to reactivation at the onset of post-collisional extension shortly prior to 400 Ma. The reactivation of the basal thrust or décollement zone confirms that reactivation of very-low-angle (*c.* 5°) shear zones is possible, provided that the zone is sufficiently weak. Most of the secondary (Mode II) extensional shear zones and faults of Devonian age are steeper, but several are still low angle, and the detachment underlying the Devonian Hornelen Basin must have been very low angle. We have presented new geochronological data that confirm that Mode II structures reactivated as post-Devonian brittle faults, several of them as low-angle faults (7° for the Hornelen Detachment and *c.* 25° for the Lærdal–Gjende Fault), giving additional evidence in favour of post-Devonian reactivation of low-angle normal faults. Interestingly, several of the pre-rift shear zones and faults reactivated in spite of being unfavourably orientated for slip (highly oblique to the extension direction), suggesting that they were mechanically weak structures at the time of North Sea rifting.

Even though the large Mode II shear zones reactivated during rifting, a large array of steeper brittle faults formed, both onshore (some with Permo-Triassic dykes along them) and offshore, where some of them accumulated kilometre-scale offsets. Isotope data presented in this work suggest a long history of brittle clay-gouge forming faulting that can be dated back to the Late Devonian, perhaps even for coast-parallel faults that parallel Permo-Triassic faults of the Horda Platform to the west. Hence, it is possible that they had some influence on the locations and orientations of the Permo-Triassic rift-faults. The wide range of illite K–Ar ages together with K-feldspar Ar/Ar ages shows that the fault network on the onshore rift shoulder was active before, during, as well as after the period of North Sea rifting. The entrapment and deformation of Late Jurassic sediments in a coast-parallel fault zone west of Bergen supports this scenario (Fossen *et al.* 1997).

We suggest, based on the seismic examples presented in this work, that even if the more-or-less north–south-trending faults accumulated most of the strain in the rift, large-scale Devonian Mode II shear zones were also reactivated offshore during both phases of North Sea rifting. Prime examples are the Hardangerfjord and Karmøy shear zones, and we propose the existence of a similar but oppositely dipping Devonian shear zone along the east flank of the Utsira High, here named the Utsira Shear Zone. These Devonian Mode II shear zones influenced the fault-block configuration of the northern North Sea Rift by creating a series of domain boundaries. The rift-related faults form uniform arrays between the boundaries, but lose offset, terminate or bend as they approach these segment boundaries. We may speculate that the southern termination of the Sogn Graben is related to the presence of Boundary 6 (the offshore extension of the Florø Horst), creating a right-lateral step to the North Viking Graben as the faults were arrested at this boundary. Furthermore, the possibility that the other steps portrayed by the Viking Graben may be related to pre-existing structures originating in the basement should not be ruled out.

A similar influence of oblique pre-rift basement structures on rift geometry has been postulated elsewhere: for instance, in the East African Rift (Morley *et al.* 2004), the Suez Rift (Bosworth 2015) and the Brazilian Tucano Basin (Milani & Davison 1988). The Scandinavian Caledonides are special by their high content of pre-rift extension structures that probably have a larger influence on the North Sea rifting than Caledonian thrusts. The fact that fundamental Devonian extension structures are much better developed in the Norwegian Caledonides than in the Scottish Caledonides could potentially have influenced the location of the rift, particularly the Permo-Triassic basin formation on the Horda Platform. In order to explore such questions in more depth, there is a need to continue the mapping of basement structures in the North Sea Rift.

This work was supported by the MultiRift Petromaks2 project 215591, financed by the Research Council of Norway and Statoil. Thanks to TGS for permission to publish the seismic examples, and to Chris Jackson, Rebecca Bell, Allan Roberts and an anonymous reviewer for making constructive comments on an earlier version of the manuscript.

References

ANDERSEN, T.B. & JAMTVEIT, B. 1990. Uplift of deep crust during orogenic extensional collapse: a model based on field studies in the Sogn–Sunnfjord region of western Norway. *Tectonics*, **9**, 1097–1112.

ANDERSEN, T.B., TORSVIK, T.H., EIDE, E.A., OSMUNDSEN, P.T. & FALEIDE, J.I. 1999. Permian and Mesozoic extensional faulting within the Caledonides of central south Norway. *Journal of the Geological Society, London*, **156**, 1073–1080, https://doi.org/10.1144/gsjgs.156.6.1073

ARSENIJEVIC, A.C. 2013. *Korngrößenabhängige Methoden zur Charakterisierung von Lettentonen: K–Ar-Datierung, Illit-Kristallinität und Illit-Polytypie*. BSc thesis, University of Göttingen.

BADLEY, M.E., PRICE, J.D., DAHL, C.R. & AGDESTEIN, T. 1988. The structural evolution of the northern Viking Graben and its bearing upon extensional modes of basin formation. *Journal of the Geological Society,*

London, **145**, 455–472, https://doi.org/10.1144/gsjgs.145.3.0455

BARLING, T. 2014. *Growth of a major normal fault complex in the horda platform, North Sea.* MSc thesis, Imperial College.

BASSETT, M.G. 2003. Sub-Devonian geology. *In*: EVANS, D., GRAHAM, C., ARMOUR, A. & BATHURST, P. (eds) *The Millennium Atlas: Petroleum Geology of the Central and Northern North Sea*. Geological Society, London, 61–63.

BELL, R.E., JACKSON, C.A.L., WHIPP, P.S. & CLEMENTS, B. 2014. Strain migration during multiphase extension: observations from the northern North Sea. *Tectonics*, **33**, 1936–1963.

BENSE, F.A., WEMMER, K., LÖBENS, S. & SIEGESMUND, S. 2014. Fault gouge analyses: K–Ar illite dating, clay mineralogy and tectonic significance – a study from the Sierras Pampeanas, Argentina. *International Journal of Earth Sciences*, **103**, 189–218.

BØE, R., FOSSEN, H. & SMELROR, M. 2010. Mesozoic sediments and structures onshore Norway and in the coastal zone. *Norges Geologiske Undersøkelse Bulletin*, **450**, 15–32.

BOSWORTH, W. 2015. Geological evolution of the Red Sea: historical background, review, and synthesis. *In*: RASUL, N.M.A. & STEWART, I.C.F. (eds) *The Red Sea*. Springer, Berlin, 45–78.

BRAATHEN, A. 1999. Kinematics of post-Caledonian polyphase brittle faulting in the Sunnfjord region, western Norway. *Tectonophysics*, **302**, 99–121.

BREWER, M.S. & SMYTHE, D.K. 1984. MOIST and the continuity of crustal reflector geometry along the Caledonian–Appalachian orogen. *Journal of the Geological Society, London*, **141**, 105–120, https://doi.org/10.1144/gsjgs.141.1.0105

BUTLER, R.W.H. 2010. The Geological Structure of the North-West Highlands of Scotland - revisited: Peach et al. 100 years on. *In*: LAW, R.D., BUTLER, R.W.H., HOLDSWORTH, R.E., KRABBENDAM, M. & STRACHAN, R.A. (eds) *Continental Tectonics and Mountain Building: The Legacy of Peach and Horne*. Geological Society, London, Special Publications, **335**, 7–27, https://doi.org/10.1144/SP335.2

CHAUVET, A. & M.& SÉRANNE, 1994. Extension-parallel folding in the Scandinavian Caledonides: implications for late-orogenic processes. *Tectonophysics*, **238**, 31–54.

CHRISTIANSSON, P., FALEIDE, J.I. & BERGE, A.M. 2000. Crustal structure in the northern North Sea: an integrated geophysical study. *In*: NØTTVEDT, A. ET AL. (eds) *Dynamics of the Norwegian Margin*. Geological Society, London, Special Publications, **167**, 15–40, https://doi.org/10.1144/GSL.SP.2000.167.01.02

CONSTENIUS, K.N. 1996. Late Paleogene extensional collapse of the Cordilleran foreland fold and thrust belt. *Geological Society of America Bulletin*, **108**, 20–39.

DAVIDS, C., WEMMER, K., ZWINGMANN, H., KOHLMANN, F., JACOBS, J. & BERGH, S.G. 2013. K–Ar illite and apatite fission track constraints on brittle faulting and the evolution of the northern Norwegian passive margin. *Tectonophysics*, **608**, 196–211.

DOBRZHINETSKAYA, L.F., LARSEN, R.B., STURT, B.A., TRØNNES, R.G., SMITH, D.C., TAYLOR, W.R. & POSUKHOVA, T.V. 1995. Microdiamond in high-grade metamorphic rocks of the Western Gneiss region, Norway. *Geology*, **23**, 597–600.

DUNLAP, W.J. & FOSSEN, H. 1998. Early Proterozoic orogenic collapse, tectonic stability, and late Paleozoic continental rifting revealed through thermochronology of K-feldspars, southern Norway. *Tectonics*, **17**, 604–620.

EBBING, J., ENGLAND, R.W., KORJA, T., LAURITSEN, T., OLESEN, O., STRATFORD, W. & WEIDLE, C. 2012. Structure of the Scandes lithosphere from surface to depth. *Tectonophysics*, **536–537**, 1–24.

EIDE, E., TORSVIK, T.H. & ANDERSEN, T.B. 1997. Absolute dating of brittle fault movements: late Permian and late Jurassic extensional fault breccias in western Norway. *Terra Nova*, **9**, 135–139.

FÆRSETH, R.B. 1996. Interaction of Permo-Triassic and Jurassic extensional fault-blocks during the development of the northern North Sea. *Journal of the Geological Society, London*, **153**, 931–944, https://doi.org/10.1144/gsjgs.153.6.0931

FÆRSETH, R.B., MACINTYRE, R.M. & NATERSTAD, J. 1976. Mesozoic alkaline dykes in the Sunnhordland region, western Norway: ages, geochemistry and regional significance. *Lithos*, **9**, 331–345.

FÆRSETH, R.B., GABRIELSEN, R.H. & HURICH, C.A. 1995. Influence of basement in structuring of the North Sea basin, offshore southwest Norway. *Norsk Geologisk Tidsskrift*, **75**, 105–119.

FAUCONNIER, J., LABROUSSE, L., ANDERSEN, T.B., BEYSSAC, O., DUPRAT-OUALID, S. & YAMATO, P. 2014. Thermal structure of a major crustal shear zone, the basal thrust in the Scandinavian Caledonides. *Earth and Planetary Science Letters*, **385**, 162–171.

FAULDS, J.E. & VARGA, R.J. 1998. The role of accommodation zones and transfer zones in the regional segmentation of extended terranes. *In*: FAULDS, J.E. & STEWART, J.H. (eds) *Accommodation Zones and Transfer Zones: The Regional Segmentation of the Basin and Range Province*. Geological Society of America Special Papers, **323**, 1–45.

FOSSEN, H. 1992. The role of extensional tectonics in the Caledonides of South Norway. *Journal of Structural Geology*, **14**, 1033–1046.

FOSSEN, H. 2000. Extensional tectonics in the Caledonides: synorogenic or postorogenic?. *Tectonics*, **19**, 213–224.

FOSSEN, H. 2010. Extensional tectonics in the North Atlantic Caledonides: a regional view. *In*: LAW, R.D., BUTLER, R.W.H., KRABBENDAM, M. & STRACHAN, R.A. (eds) *Continental Tectonics and Mountain Building: The Legacy of Peach and Horne*. Geological Society, London, Special Publications, **335**, 767–793, https://doi.org/10.1144/SP335.31

FOSSEN, H. & DUNLAP, W.J. 1998. Timing and kinematics of Caledonian thrusting and extensional collapse, southern Norway: evidence from $^{40}Ar/^{39}Ar$ thermochronology. *Journal of Structural Geology*, **20**, 765–781.

FOSSEN, H. & DUNLAP, W.J. 1999. On the age and tectonic significance of Permo-Triassic dikes in the Bergen-Sunnhordland region, southwestern Norway. *Norsk Geologisk Tidsskrift*, **79**, 169–178.

FOSSEN, H. & HURICH, C.A. 2005. The Hardangerfjord Shear Zone in SW Norway and the North Sea: a large-scale low-angle shear zone in the Caledonian crust.

Journal of the Geological Society, London, **162**, 675–687, https://doi.org/10.1144/0016-764904-136

Fossen, H., Mangerud, G., Hesthammer, J., Bugge, T. & Gabrielsen, R. 1997. The Bjorøy Formation: a newly discovered occurence of Jurassic sediments in the Bergen Arc System. *Norsk Geologisk Tidsskrift*, **77**, 269–287.

Freeman, B., Klemperer, S.L. & Hobbs, R.W. 1988. The deep structure of northern England and the Iapetus Suture zone from BIRPS deep seismic reflection profiles. *Journal of the Geological Society, London*, **145**, 727–740, https://doi.org/10.1144/gsjgs.145.5.0727

Gabrielsen, R.H., Fossen, H., Faleide, J.I. & Hurich, C.A. 2015. Mega-scale Moho relief and the structure of the lithosphere on the eastern flank of the Viking Graben, offshore southwestern Norway. *Tectonics*, **34**, 803–819, https://doi.org/10.1002/2014TC003778

Gee, D.G., Fossen, H., Henriksen, N. & Higgins, A.K. 2008. From the Early Paleozoic Platforms of Baltica and Laurentia to the Caledonide Orogen of Scandinavia and Greenland. *Episodes*, **31**, 44–51.

Gibbs, A.D. 1987. Deep seismic profiles in the northern North Sea. *In*: Brooks, J. & Glennie, K.W. (ed.) *Petroleum Geology of North West Europe*. Graham & Trotman, London, 1025–1028.

Hacker, B.R. 2007. Ascent of the ultrahigh-pressure Western Gneiss Region, Norway. *In*: Cloos, M., Carlson, W.D., Gilbert, M.C., Liou, J.G. & Sorensen, S.S. (eds) *Convergent Margin Terranes and Associated Regions: A Tribute to W.G. Ernst*. Geological Society of America Special Papers, **419**, 171–184.

Hacker, B.R., Andersen, T.B., Johnston, S., Kylander-Clark, A.R.C., Peterman, E.M., Walsh, E.O. & Young, D. 2010. High-temperature deformation during continental-margin subduction & exhumation: the ultrahigh-pressure Western Gneiss Region of Norway. *Tectonophysics*, **480**, 149–171.

Heeremans, M. & Faleide, J.I. 2004. Late Carboniferous–Permian tectonics and magmatic activity in the Skagerrak, Kattegat and the North Sea. *In*: Wilson, M., Neumann, E.-R., Davies, G.R., Timmermann, M.J., Heeremans, M. & Larsen, B.T. (eds) *Permo-Carboniferous Magmatism and Rifting in Europe*. Geological Society, London, Special Publications, **223**, 157–176, https://doi.org/10.1144/GSL.SP.2004.223.01.07

Henza, A.A., Withjack, M.O. & Schlische, R.W. 2011. How do the properties of a pre-existing normal-fault population influence fault development during a subsequent phase of extension? *Journal of Structural Geology*, **33**, 1312–1324.

Høeg, O.A. 1945. Contributions to the Devonian flora of western Norway III. *Norsk Geologisk Tidsskrift*, **25**, 183–192.

Hossack, J.R. 1984. The geometry of listric growth faults in the Devonian basins of Sunnfjord, W Norway. *Journal of the Geological Society, London*, **141**, 629–637, https://doi.org/10.1144/gsjgs.141.4.0629

Hossack, J.R. & Cooper, M.A. 1986. Collision tectonics in the Scandinavian Caledonides. *In*: Coward, M.P. & Ries, A.C. (eds) *Collision Tectonics*. Geological Society, London, Special Publications, **19**, 285–304, https://doi.org/10.1144/GSL.SP.1986.019.01.16

Hurich, C.A. & Kristoffersen, Y. 1988. Deep structure of the Caledonide orogen in southern Norway: new evidence from marine seismic reflection profiling. *Norges geologiske undersøkelse Special Publication*, **3**, 96–101.

Klemperer, S.L. & Matthews, D.H. 1987. Iapetus Suture located beneath the North Sea by BIRPS deep seismic reflection profiling. *Geology*, **15**, 195–198.

Krabbendam, M. & Dewey, J.F. 1998. Exhumation of UHP rocks by transtension in the Western Gneiss Region, Scandinavian Caledonides. *In*: Holdsworth, R.E., Strachan, R.A. & Dewey, J.F. (eds) *Continental Transpressional and Transtensional Tectonics*. Geological Society, London, Special Publications, **135**, 159–181, https://doi.org/10.1144/GSL.SP.1998.135.01.11

Ksienzyk, A.K. 2012. *From mountains to basins: Geochronological case studies from southwestern Norway, Western Australia and East Antarctica*. PhD thesis, University of Bergen.

Ksienzyk, A.K., Kohlmann, F., Dunkl, I., Wemmer, K., Jacobs, J. & Fossen, H. 2012. Late Caledonian orogenic collapse and Permian-Mesozoic rifting: dating onshore faults in SW Norway with low-temperature thermochronological methods. *Abstracts and Proceedings of the Geological Society of Norway*, **2**, 37.

Ksienzyk, A.K., Dunkl, I., Jacobs, J., Fossen, H. & Kohlmann, F. 2014. From orogen to passive margin: constraints from fission track and (U–Th)/He analyses on Mesozoic uplift and fault reactivation in SW Norway. *In*: Corfu, F., Gasser, D. & Chew, D.M. (eds) *New Perspectives on the Caledonides of Scandinavia and Related Areas*. Geological Society, London, Special Publications, **390**, 679–702, https://doi.org/10.1144/SP390.27

Larsen, B.T., Olaussen, S., Sundvoll, B.A. & Heeremans, M. 2008. The Permo-Carboniferous Oslo Rift through six stages and 65 million years. *Episodes*, **31**, 52–58.

Larsen, Ø., Fossen, H., Langeland, K. & Pedersen, R.B. 2003. Kinematics and timing of polyphase post-Caledonian deformation in the Bergen area, SW Norway. *Norwegian Journal of Geology*, **83**, 149–165.

Lovera, O.M., Grove, M. & Harrison, T.M. 2002. Systematic analysis of K-feldspar 40Ar/39Ar step heating results II: relevance of laboratory argon diffusion properties to nature. *Geochimica et Cosmochimica Acta*, **66**, 1237–1255.

Lutro, O. & Bryhni, I. 2000. *Bergrunnskart Florø 1118 III, M 1:50,000. Foreløpig utgave*. Norges geologiske undersøkelse, Trondheim.

Lutro, O. & Tveten, E. 1996. *Geologisk kart over Norge, berggrunnskart Årdal M 1:250,000*. Norges geologiske undersøkelse, Trondheim.

Marshall, J.E.A. & Hewett, A.J. 2003. Devonian. *In*: Evans, D., Graham, C., Armour, A. & Bathurst, P. (eds) *The Millennium Atlas: Petroleum Geology of the Central and Northern North Sea*. Geological Society, London, 65–81.

Martin, C.A.L., Stewart, S.A. & Doubleday, P.A. 2002. Upper Carboniferous and Lower Permian tectonostratigraphy on the southern margin of the Central North Sea. *Journal of the Geological Society, London*,

159, 731–749, https://doi.org/10.1144/0016-7649 00-174

MILANI, E.J. & DAVISON, I. 1988. Basement control and transfer tectonics in the Recôncavo-Tucano-Jatobá rift, Northeast Brazil. *Tectonophysics*, **154**, 41–70.

MILNES, A.G., WENNBERG, O.P., SKÅR, Ø., KOESTLER, A.G. 1997. Contraction, extension and timing in the South Norwegian Caledonides: the Sognefjord transect. *In*: BURG, J.-P. & FORD, M. (eds) *Orogeny Through Time*. Geological Society, London, Special Publications, **121**, 123–148, https://doi.org/10.1144/GSL.SP.1997.121.01.06

MORLEY, C.K., NELSON, R.A., PATTON, T.L. & MUNN, S.G. 1990. Transfer zones in the East African rift system and their relevance to hydrocarbon exploration in rifts. *American Association of Petroleum Geologists Bulletin*, **74**, 1234–1253.

MORLEY, C.K., HARANYA, C., PHOOSONGSEE, W., PONGWAPEE, S., KORNSAWAN, A. & WONGANAN, N. 2004. Activation of rift oblique and rift parallel pre-existing fabrics during extension and their effect on deformation style: examples from the rifts of Thailand. *Journal of Structural Geology*, **26**, 1803–1829.

NORTON, M. 1987. The Nordfjord-Sogn Detachment, W. Norway. *Norsk Geologisk Tidsskrift*, **67**, 93–106.

NUR, A., RON, H. & SCOTTI, O. 1986. Fault mechanics and the kinematics of block rotations. *Geology*, **14**, 746–749.

ODINSEN, T., REEMST, P., BEEK, P.V.D., FALEIDE, J.I. & GABRIELSEN, R.H. 2000. Permo-Triassic and Jurassic extension in the northern North Sea: results from tectonostratigraphic forward modelling. *In*: NØTTVEDT, A. ET AL. (eds) *Dynamics of the Norwegian Margin*. Geological Society, London, Special Publications, **167**, 83–103, https://doi.org/10.1144/GSL.SP.2000.167.01.05

OLESEN, O., EBBING, J. ET AL. 2010*a*. *Gravity Anomaly Map, Norway and Adjacent Areas, Scale 1:3 Million*. Norges geologiske undersøkelse, Trondheim.

OLESEN, O., GELLEIN, J. ET AL. 2010*b*. *Magnetic Anomaly Map, Norway and Adjacent Areas, Scale 1:3 Million*. Norges geologiske undersøkelse, Trondheim.

OSMUNDSEN, P.T. & ANDERSEN, T.B. 2001. The middle Devonian basins of western Norway: sedimentary response to large-scale transtensional tectonics? *Tectonophysics*, **332**, 51–68.

PLATT, N.H. 1995. Structure and tectonics of the northern North Sea: new insights from deep penetration regional seismic data. *In*: LAMBIASE, J.J. (ed.) *Hydrocarbon Habitat in Rift Basins*. Geological Society, London, Special Publications, **80**, 103–113, https://doi.org/10.1144/GSL.SP.1995.080.01.05

ROBERTS, A.M., YIELDING, G., KUSZNIR, N.J., WALKER, I.M. & DORN-LOPEZ, D. 1995. Quantitative analysis of Triassic extension in the northern Viking Graben. *Journal of the Geological Society, London*, **152**, 15–26, https://doi.org/10.1144/gsjgs.152.1.0015

ROSENDAHL, B.R., REYNOLDS, D.J. ET AL. 1986. Structural expressions of rifting: lessons from Lake Tanganyika, Africa. *In*: FROSTICK, L.E., RENAUT, R.W., REID, I. & TIERCELIN, J.J. (eds) *Sedimentation in the African Rifts*. Geological Society, London, Special Publications, **25**, 29–43, https://doi.org/10.1144/GSL.SP.1986.025.01.04

SÉRANNE, M. & SÉGURET, M. 1987. The Devonian basins of western Norway: tectonics and kinematics of an extending crust. *In*: COWARD, M.P., DEWEY, J.F. & HANCOCK, P.L. (eds) *Continental Extensional Tectonics*. Geological Society, London, Special Publications, **28**, 537–548, https://doi.org/10.1144/GSL.SP.1987.028.01.35

SLAGSTAD, T., DAVIDSEN, B. & DALY, J.S. 2011. Age and composition of crystalline basement rocks on the Norwegian continental margin: offshore extension and continuity of the Caledonian–Appalachian orogenic belt. *Journal of the Geological Society, London*, **168**, 1167–1185, https://doi.org/10.1144/0016-76492010-136

SMITH, D.C. & LAPPIN, M.A. 1989. Coesite in the Straumen kyanite-eclogite pod, Norway. *Terra Nova*, **1**, 47–56.

SOLUM, J.G., VAN DER PLUIJM, B.A. & PEACOR, D.R. 2005. Neocrystallization, fabrics and age of clay minerals from an exposure oft he Moab Fault, Utah. *Journal of Structural Geology*, **27**, 1563–1576.

STEEL, R. 1993. Triassic–Jurassic megasequence stratigraphy in the Northern North Sea: rift to post-rift evolution. *In*: PARKER, J.R. (ed.) *Petroleum Geology of Northwest Europe: Proceedings of the 4th Conference*. Geological Society, London, 299–315, https://doi.org/10.1144/0040299

STEEL, R., SIEDLECKA, A. & ROBERTS, D. 1985. The Old Red Sandstone basins of Norway and their deformation: a review. *In*: GEE, D.G. & STURT, B.A. (eds) *The Caledonide Orogen – Scandinavia and Related Areas*. John Wiley & Sons, Chichester, 293–315.

STELTENPOHL, M.G., MOECHER, D., ANDRESEN, A., BALL, J., MAGER, S. & HAMES, W.E. 2011. The Eidsfjord shear zone, Lofoten–Vesterålen, north Norway: an Early Devonian, paleoseismogenic low-angle normal fault. *Journal of Structural Geology*, **33**, 1023–1043.

TAGAMI, T. 2012. Thermochronological investigation of fault zones. *Tectonophysics*, **538–540**, 67–85.

TORGERSEN, E., VIOLA, G., ZWINGMANN, H. & HARRIS, C. 2014. Structural and temporal evolution of a reactivated brittle-ductile fault – Part II: timing of fault initiation and reactivation by K–Ar dating of synkinematic illite/muscovite. *Earth and Planetary Science Letters*, **407**, 221–233.

TORGERSEN, E., VIOLA, G., ZWINGMANN, H. & HENDERSON, I.H.C. 2015. Inclined K–Ar illite age spectra in brittle fault gouges: effects of fault reactivation and wall-rock contamination. *Terra Nova*, **27**, 106–113.

TORSVIK, T.H., ANDERSEN, T.B., EIDE, E.A. & WALDERHAUG, H.J. 1997. The age and tectonic significance of dolerite dykes in western Norway. *Journal of the Geological Society, London*, **154**, 961–973, https://doi.org/10.1144/gsjgs.154.6.0961

VAN DER PLUIJM, B.A., HALL, C.M., VROLIJK, P.J., PEVEAR, D.R. & COVEY, M.C. 2001. The dating of shallow faults in the Earth's crust. *Nature*, **412**, 172–175.

WAIN, A. 1997. New evidence for coesite in eclogite and gneisses: defining an ultrahigh-pressure province in the Western Gneiss region of Norway. *Geology*, **25**, 927–930.

WENNBERG, O.P., MILNES, A.G. & WINSVOLD, I. 1998. The northern Bergen Arc Shear Zone – an oblique-lateral ramp in the Devonian extensional detachment system of western Norway. *Norsk Geologisk Tidsskrift*, **78**, 169–184.

WILSON, T.J. 1966. Did the Atlantic close and then re-open? *Nature*, **211**, 676–681.

WOZNITZA, T. 2014. *Exhumation history of the Caledonides in SW-Norway – (U–Th)/He ages on apatite and K/Ar age determinations*. MSc thesis, University of Göttingen.

YOUNG, D.J., HACKER, B.R., ANDERSEN, T.B. & GANS, P.B. 2011. Structure and 40Ar/39Ar thermochronology of an ultrahigh-pressure transition in western Norway. *Journal of the Geological Society, London*, **168**, 887–898, https://doi.org/10.1144/0016-76492010-075

ZWINGMANN, H. & MANCKTELOW, N. 2004. Timing of Alpine fault gouges. *Earth and Planetary Science Letters*, **223**, 415–425.

ZWINGMANN, H., MANCKTELOW, N., ANTOGNINI, M. & LUCCHINI, R. 2010. Dating of shallow faults: new constraints from the AlpTransit tunnel site (Switzerland). *Geology*, **38**, 487–490.

Influence of fault geometries and mechanical anisotropies on the growth and inversion of hanging-wall synclinal basins: insights from sandbox models and natural examples

O. FERRER[1,2]*, K. McCLAY[1] & N. C. SELLIER[1]

[1]*Fault Dynamics Research Group, Department of Earth Sciences, Royal Holloway University of London, Egham, Surrey TW20 0EX, UK*

[2]*Geomodels Research Institute, Departament de Geodinàmica i Geofísica, Facultat de Geologia, Universitat de Barcelona, C/Martí i Franquès s/n, 08028 Barcelona, Spain*

**Correspondence: joferrer@ub.edu*

Abstract: Salt is mechanically weaker than other sedimentary rocks in rift basins. It commonly acts as a strain localizer, and decouples supra- and sub-salt deformation. In the rift basins discussed in this paper, sub-salt faults commonly form wide and deep ramp synclines controlled by the thickness and strength of the overlying salt section, as well as by the shapes of the extensional faults, and the magnitudes and slip rates along the faults. Upon inversion of these rift basins, the inherited extensional architectures, and particularly the continuity of the salt section, significantly controls the later contractional deformation.

This paper utilizes scaled sandbox models to analyse the interplay between sub-salt structures and supra-salt units during both extension and inversion. Series 1 experiments involved baseline models run using isotropic sand packs for simple and ramp-flat listric faults, as well as for simple planar and kinked planar faults. Series 2 experiments involved the same fault geometries but also included a pre-extension polymer layer to simulate salt in the stratigraphy. In these experiments, the polymer layer decoupled the extensional and contractional strains, and inhibited the upwards propagation of sub-polymer faults. In all Series 2 experiments, the extension produced a synclinal hanging-wall basin above the polymer layer as a result of polymer migration during the deformation. During inversion, the supra-polymer synclinal basin was uplifted, folded and detached above the polymer layer. Changes in thickness of the polymer layer during the inversion produced primary welds and these permitted the sub-polymer deformation to propagate upwards into the supra-salt layers.

The experimental results are compared with examples from the Parentis Basin (Bay of Biscay), the Broad Fourteens Basin (southern North Sea), the Feda Graben (central North Sea) and the Cameros Basin (Iberian Range, Spain).

Salt strata within rift basins commonly localizes strains and decouples sub- and supra-salt deformation because it is mechanically significantly weaker than other units (e.g. Jackson & Vendeville 1994; Letouzey *et al.* 1995; Nalpas *et al.* 1995; Withjack & Callaway 2000; Dooley *et al.* 2005; Krzywiec 2006). As a result of this strength contrast, the style of supra- and sub-salt deformation can be significantly different. In a rift system without salt layers, the basement extension is accommodated by the upwards propagation of faults throughout the syn-extensional basin fill (i.e. thick-skinned extension) (Ellis & McClay 1988; McClay 1989; Withjack & Callaway 2000; Corti 2012) (Fig. 1a, c). In contrast, in rift systems with salt layers, the basement extension triggers salt flow towards the margins of the basin decoupling the deformation of sub- and supra-salt units and inhibits the upwards propagation of the sub-salt faults, which results in decoupled thin-skinned extension in the supra-salt layers (Nalpas & Brun 1993; Jackson & Vendeville 1994; Soto *et al.* 2007; Ferrer *et al.* 2008*a*, *b*, 2014). Similarly, upon inversion of rift basins without salt in the stratigraphic section, the inverted fault systems propagate upwards from the rift strata into the overlying post-rift and syn-inversion strata (Fig. 1b, d) (e.g. McClay 1989; Buchanan & McClay 1992; McClay & Buchanan 1992).

Factors that control coupled/decoupled deformation and the geometries of the supra-salt section include:

- the thicknesses and strength of the salt layer and the overburden layers;

From: CHILDS, C., HOLDSWORTH, R. E., JACKSON, C. A.-L., MANZOCCHI, T., WALSH, J. J. & YIELDING, G. (eds) 2017. *The Geometry and Growth of Normal Faults*. Geological Society, London, Special Publications, **439**, 487–509.
First published online March 15, 2016, https://doi.org/10.1144/SP439.8

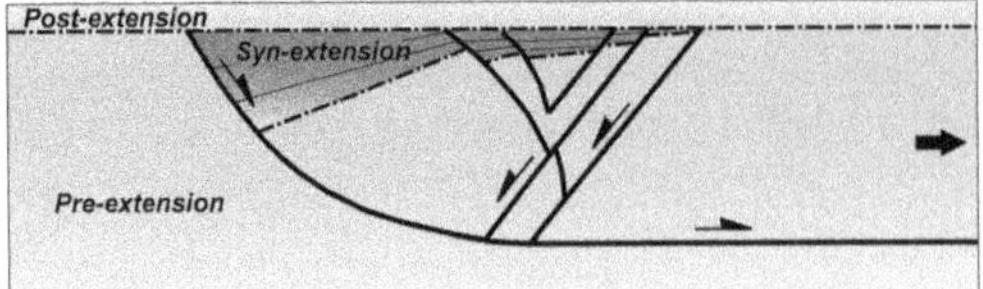

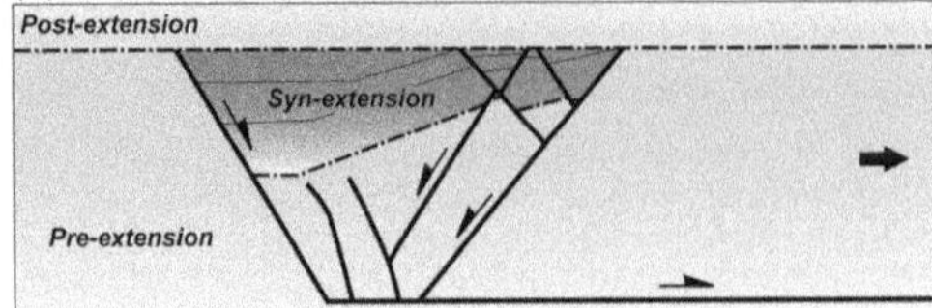

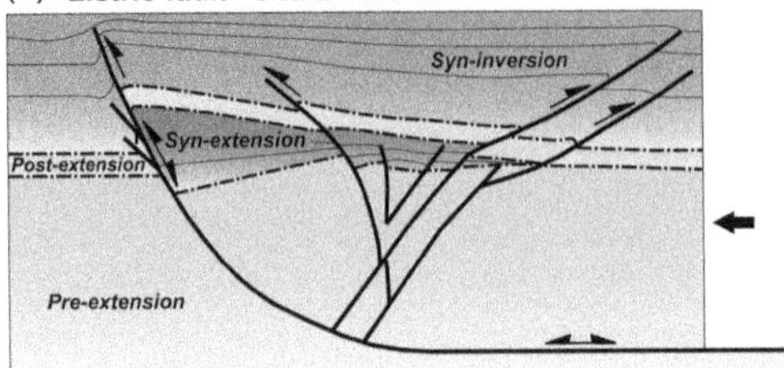

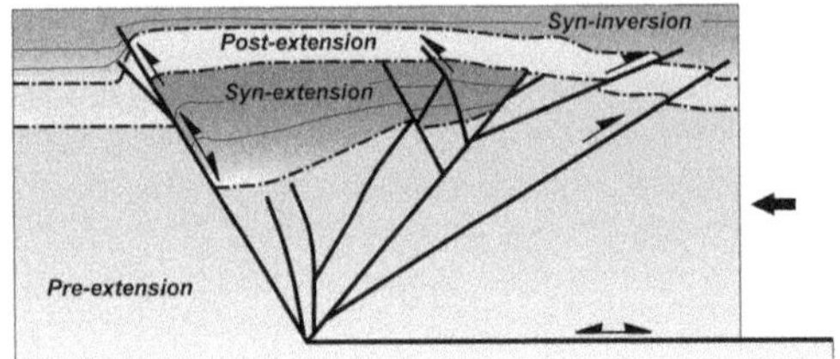

Fig. 1. Synoptic models for extension and inversion of (**a**) & (**b**) a simple listric fault and (**c**) & (**d**) a simple planar fault dipping 60° (modified from McClay 1995).

- the geometry of the main extensional fault;
- the rate of fault slip;
- the magnitude of displacement on the basin-bounding fault;
- the location of the salt strata within the basin stratigraphy (i.e. pre-, syn- or post-kinematic) (e.g. Jackson *et al.* 1990; Koyi *et al.* 1993; Jackson & Vendeville 1994; Withjack & Callaway 2000; Soto *et al.* 2007; Ferrer *et al.* 2008*b*, 2014).

During extension, a thin salt layer or a very fast slip rate on the main bounding fault will prevent salt flow and produce coupled deformation between the supra- and sub-salt strata (i.e. sub-salt structures will propagate upwards through the stratigraphic section). In contrast, a thick salt layer or a slow fault slip rate favours salt flow and the development of a hanging-wall monocline or synclinal basin above the major extensional fault (e.g. Withjack & Callaway 2000). In this case, salt acts as an intermediate extensional décollement absorbing deformation and inhibiting fault propagation from the sub-salt units through to the supra-salt layers. Similarly, evaporite compositions and strengths may also control the deformation of the supra-evaporite units decoupling supra- and sub-salt layers. Wet halite has an extremely low shear strength at geological strain rates (e.g. Vendeville & Jackson 1992) in comparison with gypsum or anhydrite. Halite preferably deforms very easily and flows acting as a regional detachment surface (e.g. Fiduk & Rowan 2012; Butler *et al.* 2014; Dooley *et al.* 2015). In addition, changes in the salt thickness related to the original salt syn-depositional environment (pre-, syn- or post-rift) may also control the degree of decoupling during extension (e.g. Jackson & Vendeville 1994; Coward & Stewart 1995; Ferrer *et al.* 2008*b*, 2014).

Many rift basins containing salt strata have also undergone later contractional deformation (inversion) where the main normal faults have been reactivated, producing contractional geometries superposed on the pre-existing extensional rift architectures. Where there is no salt section in the rift basin system, the main faults may propagate upwards during inversion without decoupling, and produce asymmetric anticlines, harpoon structures and hanging-wall back-thrusts as a result of buttressing or footwall shortcut faults (McClay 1989, 1995; Bonini *et al.* 2012) (Fig. 1b, d). In contrast, in rift basins with salt units, the salt structures at the end of extension will critically control the kinematics and geometries of the inverted salt basins producing partial or fully decoupled contractional deformation. Primary salt welds as a result of salt depletion during extension will inhibit later salt migration and the development of detachment folds during inversion. Natural examples of inverted salt basins with similar structural features to those described above include the central and southern North Sea (Van Wijhe 1987; Gowers *et al.* 1993; Nalpas *et al.* 1995), the Mid-Polish Trough with Zechstein evaporites (Krzywiec 2006; Burliga *et al.* 2012; Rowan & Krzywiec 2014), the Pyrenean rift basins (García-Senz 2002; Ferrer *et al.* 2008*a*, 2012; Roca *et al.* 2011), and the Atlas Mountains (Letouzey *et al.* 1995; Teixell *et al.* 2003).

Scaled physical models are a widely used, powerful tool for studying the geometry and kinematics of basin structures formed during both extension (e.g. McClay 1990; Corti 2012) and inversion (e.g. review by Bonini *et al.* 2012). Many previously

Table 1. *Experimental models described in this paper*

Fault geometry	Series 1 baseline isotropic models without polymer	Series 2 anisotropic models with polymer
Simple listric	Exp. 1.1	Exp. 2.1
Ramp-flat listric	Exp. 1.2	Exp. 2.2
Simple planar (20°)	Exp. 1.3	Exp. 2.3
Simple planar (60°)	Exp. 1.4	Exp. 2.4
Kinked planar (60° and 20°)	Exp. 1.5	Exp. 2.5

published sandbox models of inverted fault systems with a rigid fault footwall did not include significant mechanical anisotropies in the sand packs (i.e. ductile polymer layers) within the hanging-wall stratigraphy (e.g. McClay 1989, 1995; Buchanan & McClay 1991; Keller & McClay 1995) (Fig. 1). Only the experiments of Soto *et al.* (2007) and Ferrer *et al.* (2008*b*, 2014) included a weak layer (pre- or syn-kinematic) and a rigid footwall fault during extension. However, the inversion of basins with mechanical anisotropies has been modelled using other sandbox configurations or numerical models (e.g. Nalpas *et al.* 1995; Brun & Nalpas 1996; Dubois *et al.* 2002; Panien *et al.* 2005, 2006; Del Ventisette *et al.* 2006; Buiter *et al.* 2009; Bonini *et al.* 2012; Burliga *et al.* 2012). Many of these experiments used a basal plastic sheet or a metal plate attached to the moving wall to produce the extension and the inversion to the brittle–ductile layers in the hanging wall. The review of Bonini *et al.* (2012) showed that in brittle–ductile models the geometry of the fault is imposed by the velocity discontinuity between the basal mobile plate and the fixed part of the experiment, whereas models with rigid footwall blocks permit the simulation of hanging-wall geometries above a variety of footwall fault geometries (e.g. Fig. 1). However, fixed footwall fault models do not allow the deformation of the rigid footwall with subsequent development of footwall shortcut thrusts during inversion.

Taking the above limitation into account, this paper presents two series of extension–inversion sandbox models with rigid footwall blocks. Series 1 experiments were isotropic sandbox models, whereas Series 2 models contained a pre-kinematic polymer layer that simulated a salt section in the natural prototypes (Table 1). The experimental results are compared with published natural examples of inverted extensional basins that contain evaporite units.

Research methodology

Experimental set-up

The experimental set-up used was similar to that applied by Yamada & McClay (2003*a*, *b*, 2004). Sandbox experiments were carried out in a glass-sided deformation rig that was 150 cm long, 30 cm wide and up to 20 cm deep (Fig. 2). An electric motor fixed to the base plate drove a worm screw attached to the footwall block that produced uniaxial lengthening or shortening and normal or reverse slip of the basement fault. A constant displacement rate of 1.83×10^{-4} cm s^{-1} was applied in all the experiments during both extension and inversion in order to allow ductile flow of the polymer layers in Series 2 experiments. A strong flexible plastic sheet (but not deformable under the model conditions) was used as a detachment surface between the rigid footwall and the hanging-wall sand pack. This sheet was attached to the fixed end walls of the apparatus, maintaining constant length during the experiments. The weight of the hanging-wall sand forced the plastic sheet to conform to the underlying footwall fault block (Fig. 2). This experimental configuration allowed sliding of the plastic sheet without any length changes above the footwall surface during extension and inversion (i.e. Yamada & McClay 2003*a*, *b*, 2004). The coefficient of sliding friction between the plastic detachment and the sand pack was $\mu_b = 0.37$ (Huiqi *et al.* 1992). The main limitation of this experimental set-up using a rigid footwall is that it does not allow footwall deformation during inversion, such as occurs in natural analogues (e.g. McClay 1989, 1995; Buchanan & McClay 1991; Bonini *et al.* 2012). Keeping the above limitations in mind, five different footwall fault geometries were used in the experimental programme presented in this paper (Table 1):

(1) concave-upwards simple listric fault;
(2) ramp-flat listric fault;
(3) simple planar fault dipping 20°;
(4) simple planar fault dipping 60°;
(5) kinked planar fault with an upper 60° dipping panel and a lower panel that dips 20° onto the flat basal detachment.

The hanging-wall sand pack consisted of layered moderately well-rounded and well-sorted coloured and uncoloured dry quartz sand, with an average grain size of 0.25 mm, that was used to simulate

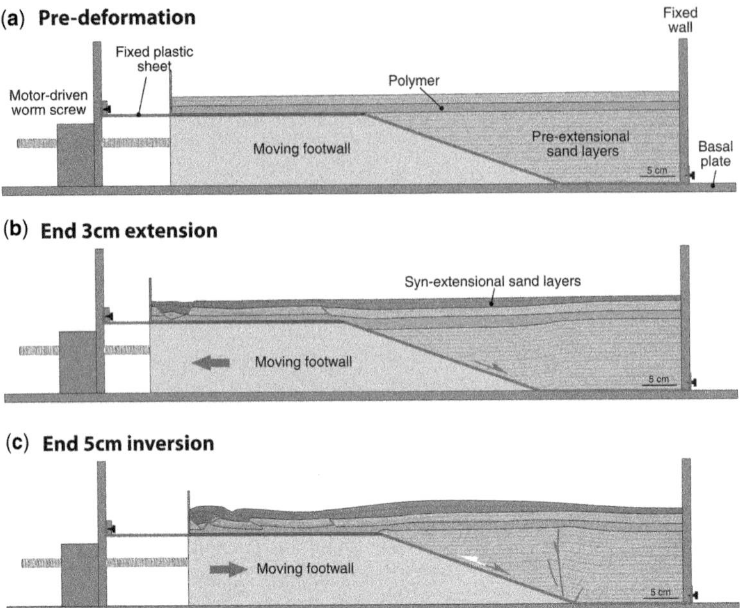

Fig. 2. Experimental set-up and sedimentary infill for each deformational episode: (**a**) pre-deformation geometry; (**b**) configuration of the experiment at the end of the extension; and (**c**) experimental configuration at the end of the inversion.

brittle sedimentary rocks in the upper crust. The sand was washed and dyed using blue, red, black or yellow pigments, and then oven-dried for 12 h at 100°C. In order to ensure that the mechanical behaviour of the uncoloured sand was similar to that of the dyed coloured sand, uncoloured sand was washed to remove the fine fraction (<20 μm) and then also oven-dried at 100°C for 12 h.

The layered sand pack was constructed by pouring alternating layers of sand into the deformation apparatus and then levelled using a mechanical scraper (e.g. Krantz 1991; Lohrmann *et al.* 2003). Figure 2a shows the pre-kinematic hanging-wall strata formed by alternating 2.5 mm layers of blue, black and white sand, with a total thickness of 10 cm. In Series 2 experiments (Fig. 2a), a uniform 12 mm-thick polymer layer was extended across the whole model and onto the footwall block. The polymer is a long-chain polydimethylsiloxane (PDMS) that deforms by viscous flow and is widely used as analogue for the natural deformation of salt at geological strain rates (Weijermars 1986). Finally, the polymer layer was overlaid with pre-extensional strata formed by alternating 2.5 mm layers of blue, white and black sand up to a thickness of 1.5 cm (Fig. 2a).

The mechanical properties of the poured sand were measured using a ring shear tester at the Fault Dynamics Research Group laboratory. The poured dry quartz sand used in these experiments has an angle of internal friction of 34.6°, a bulk density of 1500 kg m^{-3}, a coefficient of internal friction of 0.69 and a low apparent cohesive strength of 55 Pa. It deforms according to Navier–Coulomb failure at moderate and high values of normal stress (i.e. Horsfield 1977; McClay 1990). The PDMS used in the experimental programme (Rhodia Rhodosil Gum FB) is a near-perfect Newtonian fluid with a density of 0.972 g cm^{-3} at room temperature and a viscosity of 1.6×10^4 Pa s when deformed at a laboratory strain rate of 1.83×10^{-4} cm s^{-1} (Dell'Ertole & Schellart 2013). The main properties of the analogue materials and their scaling parameters used in the experimental programme are summarized in Table 2, and Figure 3 shows the general strength profiles for the analogue models described in this paper.

Experimental procedure

All models underwent 3 cm of total extension during which syn-extensional layers of red, white and black sand were added episodically after every 5 mm of extension (Fig. 2b). The regional level for each syn-kinematic layer was increased by 1 mm for every 5 mm of extension in order to preserve structures formed by polymer inflation. After the extension, models were then shortened

Table 2. *Scaling parameters used in the experimental programme*

Quantity	Experiment	Nature	Model ratio
Thickness			
Overburden	21–45 mm	2–4 km	1.05×10^{-5}–1.125×10^{-5}
Salt/polymer	12 mm	1000 m	1.2×10^{-5}
Density			
Overburden	1500 kg m^{-3}	2700 kg m^{-3}	0.55
Salt/polymer	972 kg m^{-3}	2200 kg m^{-3}	0.44
Density contrast	528	500	1.05
Ductile layer viscosity	1.6×10^{-4} Pa s	10^{-18}–10^{-19} Pa s	1.6×10^{-14}–1.6×10^{-15} Pa s
Overburden coefficient friction	0.7	0.8	0.87
Gravity acceleration	9.81 m s^{-2}	9.81 m s^{-2}	1

5 cm (Fig. 2c). The amount of shortening was higher than the applied extension to force the polymer to act as a contractional detachment, transferring part of the deformation into the footwall of the rigid fault block. No syn-contractional strata were deposited during the inversion (Fig. 2c). At the end of each experiment, the models were preserved and sliced in closely spaced vertical serial sections (3 mm thick) in order to analyse the internal fault geometries. A 5 cm-wide section along each sidewall of the models was discarded in order to eliminate edge effects produced by friction between the sand pack and the glass sidewalls.

Analysis of the analogue models

High-resolution time-lapse photographs of the upper part of the models and the sidewalls were taken every 2 min by computer-controlled digital cameras in order to record the kinematic evolution of the experiments.

The pictures of the vertical cross-sections through the physical models were studied using image-processing software in order to produce 3D voxel models for analysing cross-sections and depth slices.

Analogue model results

In this section, the results of the Series 1 and Series 2 experiments (Table 1) are described. In each case, the results of sandbox models with and without a pre-kinematic polymer layer are compared first for extension and then for inversion. All models were subjected to dip-slip extension, followed by dip-slip inversion. There was no strike-slip deformation and

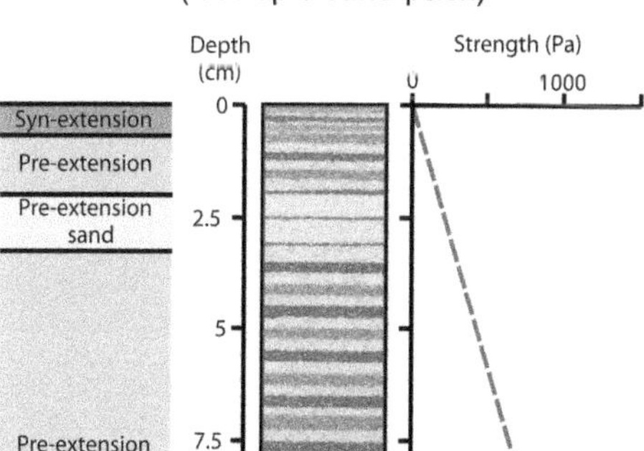

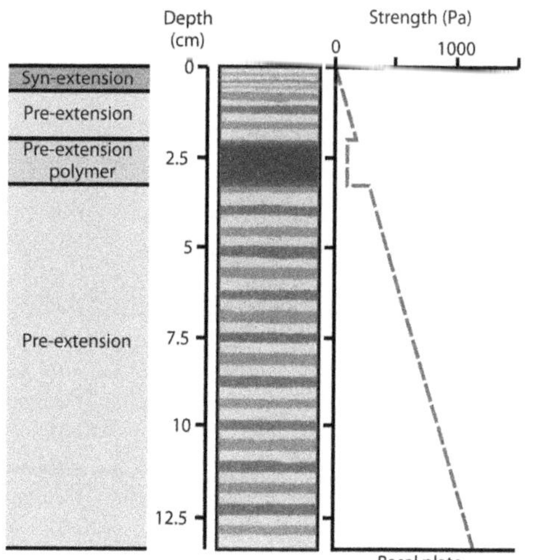

Fig. 3. Hanging-wall sand pack configurations and schematic strength profiles for the experimental programme.

only minor oblique slip on some small faults due to 3D space issues associated with local complex inversion geometries.

Extension above listric fault systems

Simple listric and ramp-flat listric faults without a pre-kinematic polymer layer. The geometrical hanging-wall evolution of models with an isotropic sand pack above a simple listric fault and a ramp-flat listric fault (Series 1 – baseline models) are shown in Figure 4a, b. These are similar to previously published models in McClay & Ellis (1987*a*, *b*), Ellis & McClay (1988), McClay (1989), Buchanan & McClay (1991), McClay & Scott (1991), McClay *et al.* (1991), Soto *et al.* (2007) and Ferrer *et al.* (2014).

Whereas a simple hanging-wall rollover developed in the simple 45° listric model (Fig. 4a), the deformation above a 60° ramp-flat listric fault

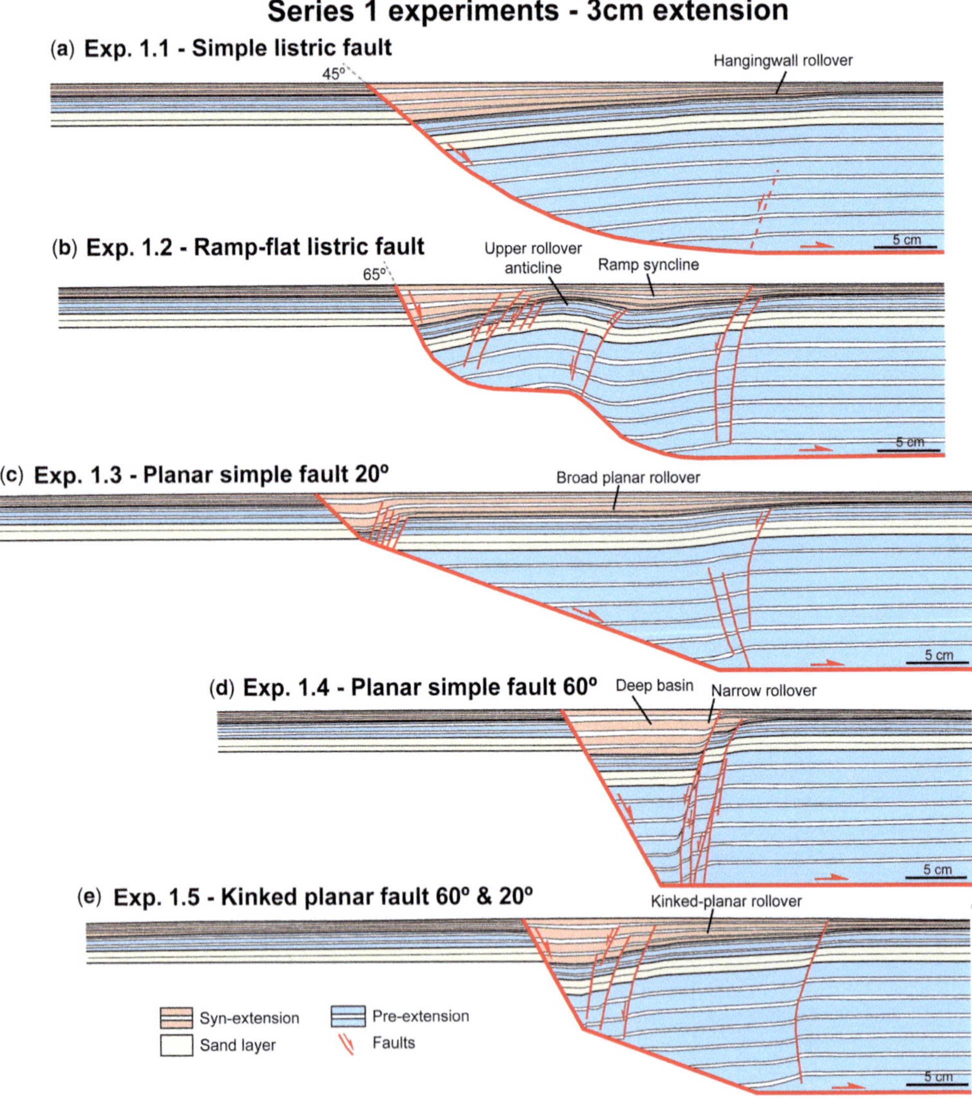

Fig. 4. Model cross-sections illustrating the structural style after 3 cm of extension for the Series 1 baseline isotropic experiments without a pre-kinematic polymer layer: (**a**) simple 45° listric fault; (**b**) ramp-flat 60° listric fault; (**c**) simple planar fault dipping 20°; (**d**) simple planar fault dipping 60°; and (**e**) kinked planar fault with two panels dipping 60° and 20°. Yellow sand layers are in an equivalent position to the polymer layers in Figure 5.

produced a rollover anticline associated with the upper 60° listric fault segment, an adjacent ramp syncline formed above the convex upwards ramp section and a lower rollover associated with the lower listric fault segment (Fig. 4b). In the simple 45° listric fault model, only minor antithetic faulting developed due to the large radius of curvature of the listric part of the fault surface (Fig. 4a). Only minor antithetic faults developed in the hanging wall of the 60° ramp-flat listric fault (Fig. 4b). Similarly, the small amount of extensional displacement in both the listric and the ramp-flat listric faults did not develop sufficient internal strains to form crestal-collapse graben in the rollover anticlines (Fig. 4a, b). In these Series 1 listric fault models, the minor antithetic faults and the main basin-bounding fault propagated across the entire sand pack. Hanging-wall growth strata in the simple 45° listric fault formed a simple fanning wedge (Fig. 4a): in the 60° ramp-flat listric fault, however, the upper rollover formed a fanning growth wedge, but the ramp syncline and lower rollover combined to form a gentle synclinal wedge (Fig. 4b).

Simple listric and ramp-flat listric faults with a pre-kinematic polymer layer. The listric fault models with a pre-kinematic polymer layer formed supra-salt hanging-wall synclinal basins that were coupled from the underlying faulted, sub-salt strata (Fig. 5a, b). The evolution of the strata below the pre-kinematic polymer layer was similar to that described above (e.g. Fig. 4a, b). In the simple 45° listric model, however, decoupling by the polymer layer, together with polymer flow, produced a broad hanging-wall monocline in the sand pack above the polymer above the fault breakaway (Fig. 5a). A gentle extensional fault-propagation fold developed as the extension progressed, with syn-kinematic layers enlarging the resultant synclinal basin. The subsidence of the basin depocentre, combined with the hanging-wall extension, induced polymer migration towards the rollover hinge as well as above the rigid footwall to form a gentle broad polymer-cored anticline (Fig. 5a). No faulting occurred in the supra-salt section.

In the 60° ramp-flat listric model, the supra-polymer decoupling was more pronounced, with a main synclinal depocentre formed by the upper rollover, together with a polymer-cored anticline over the ramp section of the basin-bounding fault and secondary depocentre formed by the lower rollover (Fig. 5b). Polymer migration is clearly evident from the thickened section in the immediate hanging wall of the upper 60° listric fault segment, as well as in the ramp syncline. As in the simple 45° listric fault model described previously (Fig. 5a), in the 60° ramp-flat model no primary welds were formed nor was there any development of new faults in the supra-salt sand pack owing to the small amount (3 cm) of extension (Fig. 5b).

In addition to the main synclinal basins, narrow distal footwall supra-polymer graben developed near footwall extremities of the polymer layer in both listric fault models (Fig. 5a, b). These were probably produced by a boundary effect at the edge of the polymer detachment (cf. Vendeville & Jackson 1992; Jackson & Vendeville 1994). These graben were infilled by syn-kinematic sand layers as extension progressed, and this suppressed the formation of polymer walls and diapirs at this location (e.g. Vendeville & Jackson 1992).

Extension above planar fault systems

Simple planar and kinked planar faults without a pre-kinematic polymer layer. The sandbox models with planar footwall detachment faults are characterized by planar or gently kinked rollover anticlines (Fig. 4c–e). The hanging-wall architectures are similar to those in models published by McClay & Ellis (1987*a*, *b*), Ellis & McClay (1988), McClay *et al.* (1991), Withjack *et al.* (1995), Dooley *et al.* (2005), Withjack & Schlische (2006) and Ferrer *et al.* (2014). The hanging-wall rollover geometries were controlled by the amount of extension (3 cm), by the dips of the bounding faults (Fig. 4c, d) and by the kink-band bend in the fault surface for Series 1 Model 1.5 (Fig. 4e). In the models described in this paper, the hanging-wall faults are dominantly antithetic (Fig. 4c–e). The 20° planar fault model formed a broad planar rollover with localized, small antithetic faults near the detachment breakaway (Fig. 4c). The 60° dipping planar bounding fault model produced a narrow rollover with closely spaced antithetic faults that bound a deep, flat half-graben basin (Fig. 4d). The 60°–20° kinked planar fault model formed a broad, gently kinked, hanging-wall fault-bend fold (Fig. 4e). Similar to the listric fault models described above, these planar fault models showed no internal decoupling, with the antithetic faults cutting from the syn-kinematic into the pre-kinematic sand layers (Fig. 4c–e).

Simple planar and kinked planar faults with a pre-kinematic polymer layer. In contrast to the experimental results discussed above, the models of planar faults containing a pre-kinematic polymer layer formed hanging-wall structures where the supra-salt deformation was decoupled from the underlying sub-salt deformation (Fig. 5c–e). The resultant hanging-wall basins are synclinal depocentres infilled with syn-kinematic growth stratal wedges. The dip of the basin-bounding fault controlled the form of the synclines, whereby the 20° planar fault model formed a broad synclinal basin (Fig. 5c), and the more steeply dipping 60° planar fault models formed deeper and narrow synclines (Fig. 5d, e). The sub-polymer deformation was characterized by antithetic faults that could not link upwards through the polymer layer into the suprasalt section.

The progressive evolution of Series 2 Model 2.4, a 60° dipping planar fault, is shown in detail in Figure 6. After 1 cm of extension, the sub-salt section had formed a narrow, flat hanging-wall graben bounded by a narrow rollover with very steep, planar antithetic faults (Fig. 6b). Above the polymer layer, a broad syncline had formed. At this stage of the extension, faulting underneath the polymer layer created accommodation space that triggered polymer flow into the depocentre from the footwalls of the basin-bounding faults (Fig. 6b). As extension progressed, the deformation was decoupled by the polymer layer, with brittle extensional faults in the sub-polymer units and a broad growth syncline in the supra-polymer strata (Fig. 6c, d). In this 60°

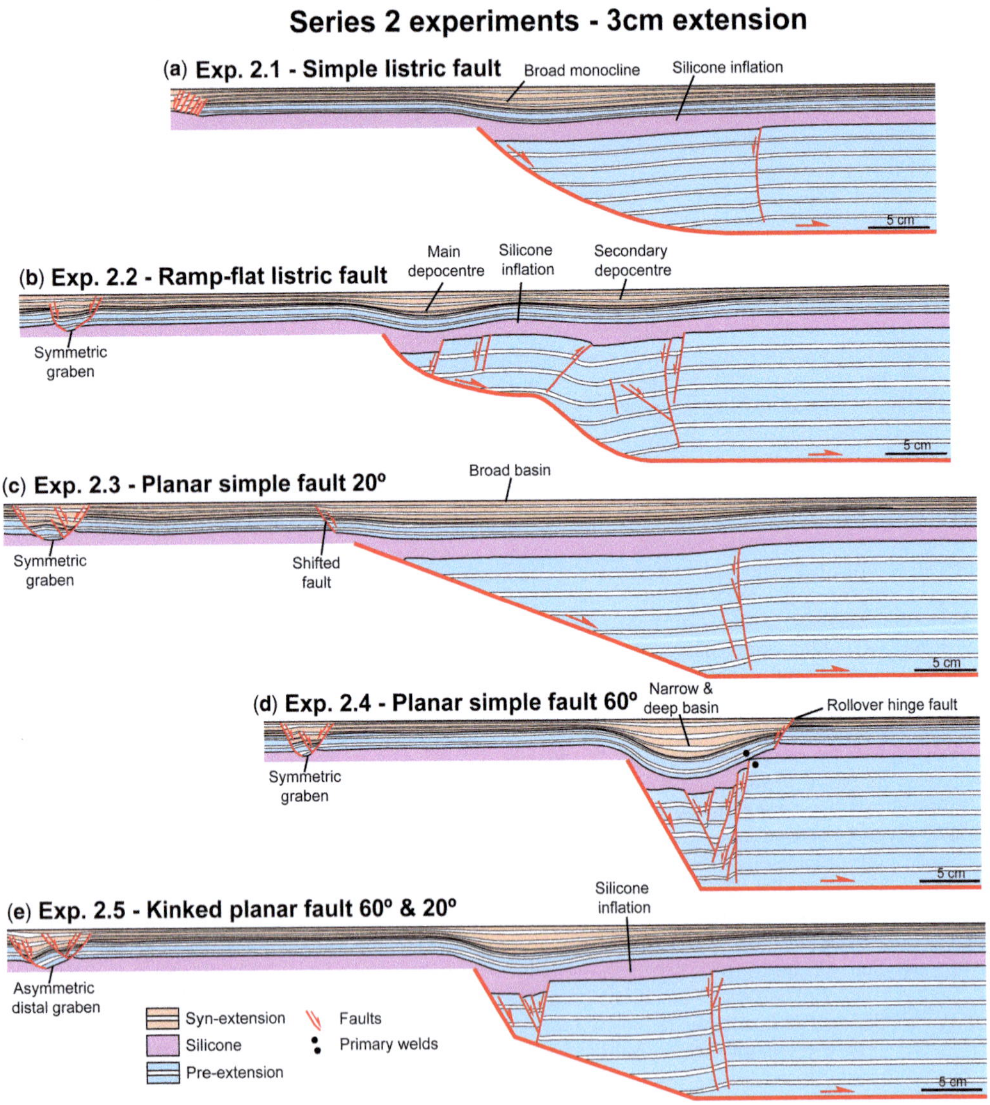

Fig. 5. Model cross-sections illustrating the structural style after 3 cm of extension for the Series 2 anisotropic experiments with a pre-kinematic polymer layer: (**a**) simple 45° listric fault; (**b**) ramp-flat 60° listric fault; (**c**) simple planar fault dipping 20°; (**d**) simple planar fault dipping 60°; and (**e**) kinked planar fault with two panels dipping 60° and 20°. The polymer layers are in an equivalent position to the yellow sand layers in Figure 4.

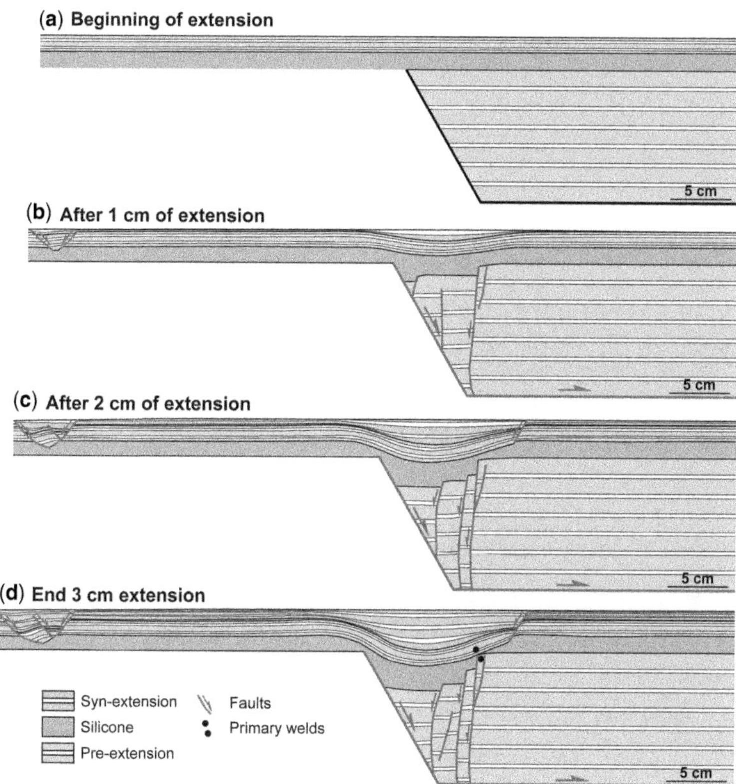

Fig. 6. Extensional evolution of the 60° planar fault model: (**a**) initial set-up – 0 cm of extension; (**b**) after 1 cm of extension; (**c**) after 2 cm of extension; and (**d**) at the end of 3 cm of extension.

planar fault model, primary weld formed over the footwall of the main rollover antithetic fault (Fig. 5c, d). A few small displacement extensional faults developed in the strata above the polymer layer (Fig. 5c, d).

In addition to the main synclinal basins, narrow distal footwall supra-polymer graben developed near footwall extremities of the polymer layer in models with planar faults (Figs 5c–e & 6). As described in the subsection on 'Extension above listric fault systems', these narrow graben are related to the edge effects where the polymer layer ends.

Inversion

In terms of classification of inverted basins by Bally (1984), the analogue models described in this paper were at the early stages of net compressional tectonics, where the 5 cm of contractional reactivation was greater than the 3 cm of previous extension (Figs 7–9). All of the models show the reactivation of the major basin-bounding fault during inversion. Inversion of Series 1 models without a ductile polymer layer produced reactivation and basin uplift, with the formation of harpoon structures (e.g. Fig. 7). Inversion of Series 2 models with a pre-extension ductile polymer layer produced mainly decoupling between the layers above the polymer and the layers below (e.g. Fig. 8). In these models, the inversion arched and uplifted the synclinal basins, and reduced the thickness of the ductile layer, which eventually hindered the flow of the polymer. In some models, where the polymer layer was totally depleted, the formation of primary welds dramatically controlled the final inversion geometry. The welds inhibited detachment on the polymer layer and allowed uplift of the underlying pre-kinematic units (Fig. 8).

Inversion of listric extensional fault systems

Inversion of listric faults without a pre-extension polymer layer. The inversion geometries of the Series 1 baseline listric fault models were similar to those described in the papers of Buchanan & McClay (1991), McClay & Buchanan (1992), Keller & McClay (1995) or McClay (1995) (Fig. 7a, b).

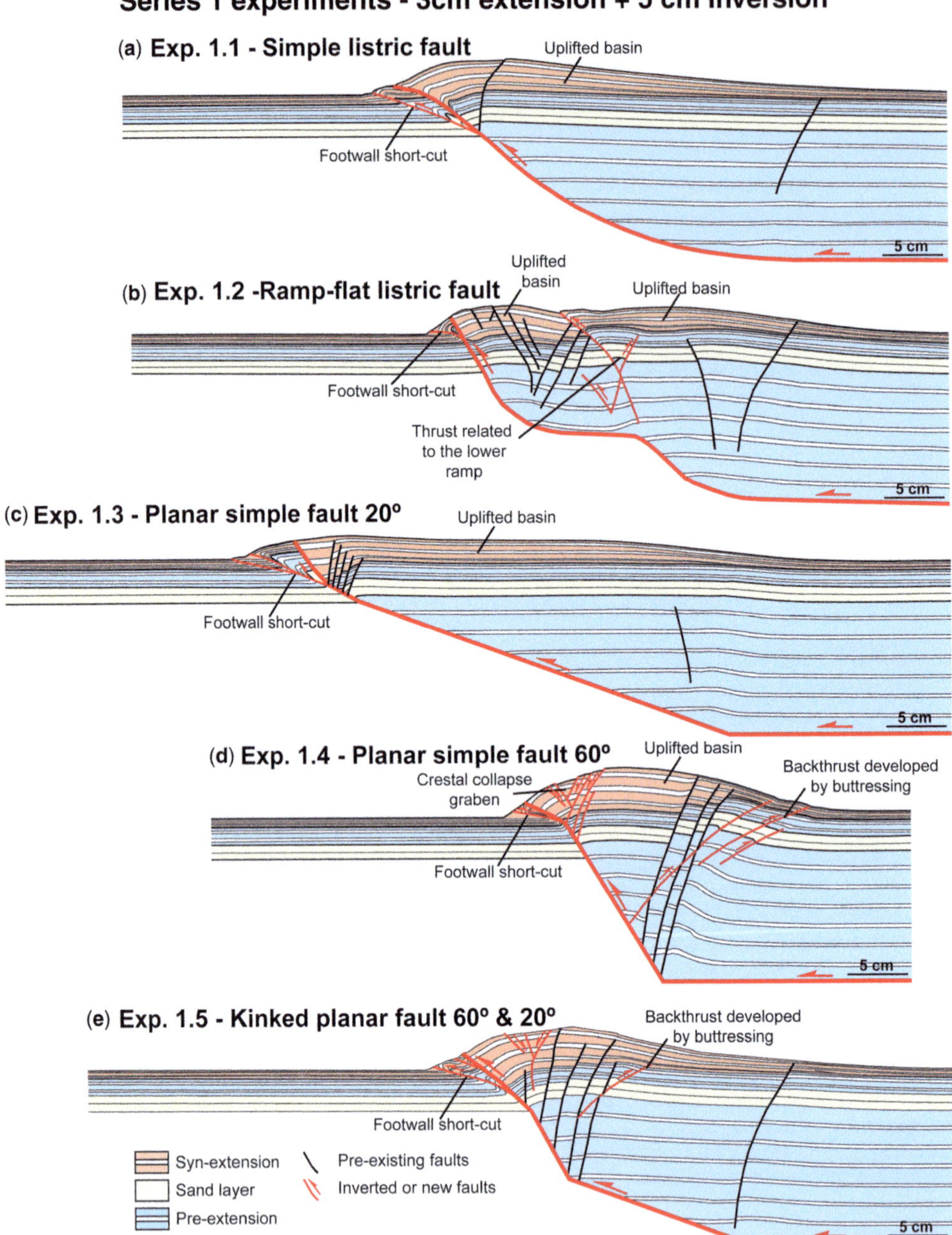

Fig. 7. Model cross-sections illustrating the structural style after 5 cm of inversion for the Series 1 baseline isotropic experiments without a pre-kinematic polymer layer: (**a**) simple 45° listric fault; (**b**) ramp-flat 60° listric fault; (**c**) simple planar fault dipping 20°; (**d**) simple planar fault dipping 60°; and (**e**) kinked planar fault with two panels dipping 60° and 20°.

Inversion of the 45° simple listric fault reactivated the major detachment fault, producing reverse slip together with uplift and back-rotation of the hanging wall (Fig. 7a). Owing to the low dip of the listric fault, the amount of back-rotation is low and a simple harpoon structure was formed with a

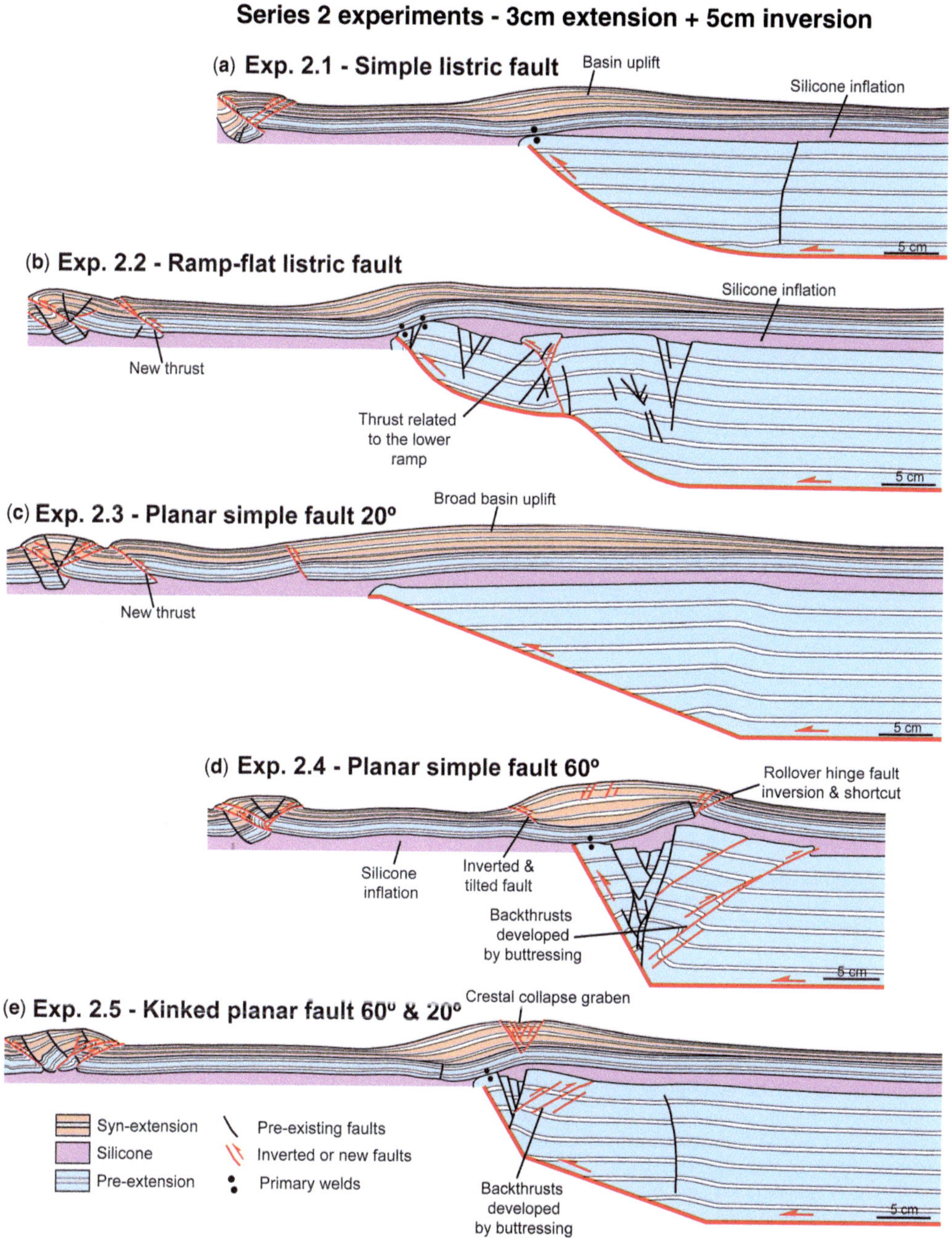

Fig. 8. Model cross-sections illustrating the structural style after 5 cm of inversion for the Series 2 anisotropic experiments with a pre-kinematic polymer layer: (**a**) simple 45° listric fault; (**b**) ramp-flat 60° listric fault; (**c**) simple planar fault dipping 20°; (**d**) simple planar fault dipping 60°; and (**e**) kinked planar fault with two panels dipping 60° and 20°.

low-angle footwall shortcut thrust into the footwall stratigraphy (Fig. 7a). The low-amplitude asymmetrical inversion anticline has a front-limb dip of approximately 28° and very gentle back-limb dip of 4°. The model is in net contraction, with the syn-kinematic layers uplifted above regional.

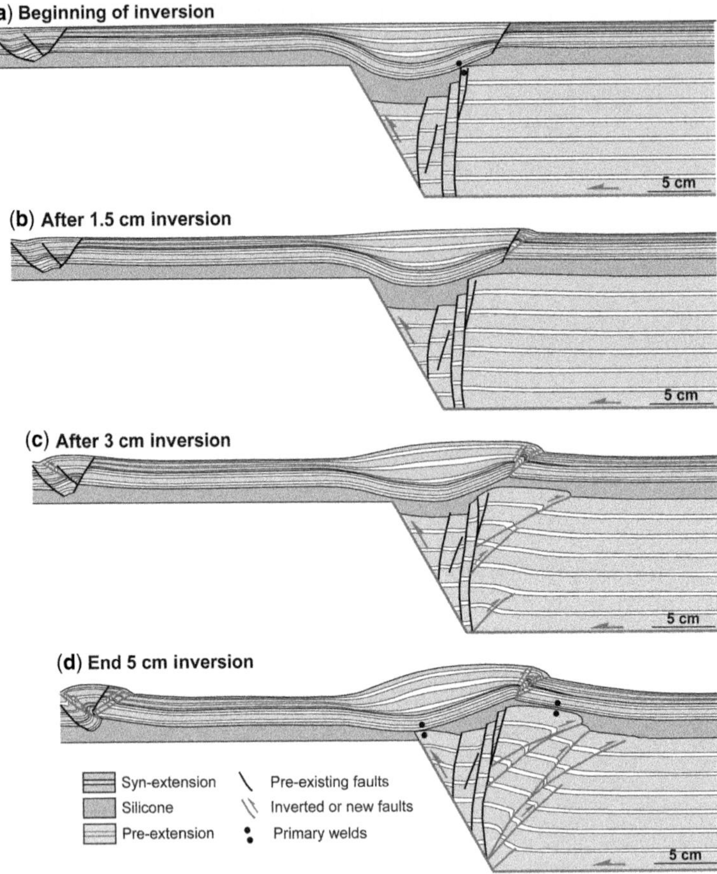

Fig. 9. Evolution of the 60°-dipping planar fault model during the inversion: (**a**) cross-section at the end of 3 cm of extension – 0 cm of inversion; (**b**) after 1.5 cm of shortening; (**c**) after 3 cm of shortening – total inversion; and (**d**) at the end of 5 cm of inversion – net contraction.

Inversion of the 60° ramp-flat listric fault model reactivated the main detachment and uplifted the hanging-wall basin onto the footwall strata (Fig. 7b). A frontal asymmetric anticline formed above the breakaway of the main detachment fault, with a front-limb dip of 35° and a gentle back-limb dip of 6–7° (Fig. 7b). A footwall shortcut thrust developed in the footwall strata in front of the inverted basin. A footwall-vergent, hanging-wall bypass thrust formed along the trajectory of the ramp section of the main detachment and propagated upwards, thrusting the extensional ramp syncline over the inverted rollover anticline (Fig. 7b). In both Series 1 listric models, inversion of the main detachment fault produced uplift and the arching of the synclinal syn-extensional basins.

Inversion of listric faults with a pre-extension polymer layer. The inversion of Series 2 listric fault models that contain a pre-extensional polymer layer produced almost completely decoupled deformation (Fig. 8a, b).

In the simple 45° listric fault model with the pre-extensional polymer layer, the 5 cm of contraction reactivated the main detachment and fully inverted the sub-polymer pre-extension strata, producing a broad uplift of the supra-polymer and syn-extension hanging-wall basin (Fig. 8a). A primary polymer weld formed at the crest of the very gentle inversion anticline where the polymer thinned by flow towards both the footwall and hanging wall. The small ‘end effect’ graben on the footwall block at the extremities of the polymer layer was inverted and uplifted by small bi-vergent thrust faults (Fig. 8a). In this model, the asymmetrical inversion anticline is only very small due partly to the low angle of the reactivated detachment and due also to the decoupling of the polymer, which transfers the shortening into broad folding and uplift of the syn-extensional strata above their regional, as well as thrusting at the footwall extremities of the polymer layer (Fig. 8a).

Inversion of the 60° ramp-flat listric fault produced a similar overall decoupled architecture to the simple listric fault described above but with a larger asymmetric frontal inversion anticline, and a greater and broader uplift of the syn-extensional strata above their regional level (Fig. 8b). The polymer layer underwent significant internal flow with thinning and weld formation at the crest of the frontal inversion anticline, as well as thickening and inflation in the hanging wall and at the distal extremity of the footwall (Fig. 8b). The upper sub-polymer pre-extension strata in the hanging-wall ramp was strongly back-rotated as it was inverted up the upper 60° listric sector of the main detachment. This deformation produced significant thickness changes in the polymer above this back-rotated panel such that, together with the uplift of the ramp section, only a very gently tilted broad uplift formed in the syn-rift strata together with unfolding of the original syn-extension ramp syncline (Fig. 8b).

In a manner similar to that observed in the inverted 45° simple listric fault in Figure 8a, inversion deformation in the 60° ramp-flat fault system was significantly displaced or offset into the footwall by detachment in the polymer layer, and the development of inverted graben and thrusting at the footwall extremity (Fig. 8b).

The inversion of the listric fault systems with pre-extension polymer layers produced very different structures (Fig. 8a, b) to those formed by inversion of listric fault systems with no polymer layers (e.g. Fig. 7a, b). In particular, models with polymer layers were decoupled with inverted faults and new faults formed in the sub-polymer units, decoupling and thickness changes of the polymer layer, and folding of the supra-polymer stratigraphy. These models do not develop shortcut faults in the immediate footwall of the main detachment but, rather, part of the inversion deformation is translated/offset to the extremities of the footwall particularly where the polymer layer ends (Fig. 8a, b). Models with a polymer layer also show that the formation of primary polymer welds control the fold geometry of the inverted hanging-wall units (Fig. 8b).

Inversion of planar faults

Inversion of planar faults without a pre-extension polymer layer. The inversion geometries of planar fault models without a polymer layer in the pre-extension sand pack (Fig. 7c–e) are similar to those described by Buchanan & McClay (1991, 1992). Inversion of the 20° planar fault reactivated the main detachment, gently uplifting the hanging wall and producing a small frontal inversion anticline with a frontal-limb dip of approximately 15°, together with a broad zone of uplift of the syn-extensional strata above their regional and a very gentle back-limb dip of 5° (Fig. 7c). A shortcut thrust formed in the footwall of the main detachment and locally overturned the hanging-wall strata.

In contrast to the inversion of the 20° low-angle planar fault, inversion of the 60° planar and 60°–20° kinked planar faults produced well-developed asymmetric hanging-wall inversion anticlines with moderately dipping frontal limbs (38° and 35°, respectively) and gently dipping back limbs (20° and 18°, respectively) (Fig. 7d, e). Footwall shortcut faults developed within the sand pack in front of the inversion anticline. In both models, buttressing against the 60° dipping footwall produced new hanging-wall back-thrusts as inversion progresses. The syn-extension hanging-wall basin was progressively uplifted and small crestal-collapse graben developed on the outer arc of the asymmetrical inversion anticlines (Fig. 7d, e). The 60° ramp-flat kinked planar fault system has a shorter 60° dipping panel on the detachment surface (Fig. 7e), and this produced less hanging-wall uplift and a slightly smaller inversion anticline than the longer 60° panel in the simple 60° planar fault (Fig. 7d).

Inversion on planar faults with a pre-extension polymer layer. As for the listric faults, the inversion models of planar fault systems that had a pre-extension ductile polymer layer are characterized by decoupling of the sub-polymer and supra-polymer structures (Fig. 8c–e). In all of these Series 2 inverted planar fault models, the main detachment was reactivated and produced broad hanging-wall uplifts, together with changes in the polymer thickness, but the fault itself did not propagate though the polymer into the supra-polymer layers (Fig. 8c–e).

Inversion of the low-angle 20° planar fault model produced a very wide zone of hanging-wall uplift with the syn-extension strata uplifted above their regional. A low-relief asymmetrical frontal inversion anticline was only developed in the sub-polymer strata (Fig. 8c). In the very flat hanging wall of the inversion uplift, slight polymer thinning occurred together with polymer thickening into the distal hanging wall and into the footwall. The extensional graben structure near the end of the polymer layer in the footwall and the nearby strata are strongly deformed by thrusts detaching into the top of the polymer layer above the rigid footwall block (Fig. 8c). These thrusts, together with the thickened polymer section, indicate significant transfer of shortening from the reactivated 20° dipping planar fault into the footwall polymer layer and out to the left-hand edge of the model.

Inversion of the simple 60° planar fault model produced a broad zone of uplift above the main basin-bounding fault system under the polymer layer, with syn-extensional strata strongly uplifted and folded into a broad anticline tilted towards the footwall (Fig. 8d). Significant thickness changes occurred in the salt layer, with a well-developed primary weld in the hanging wall of the inverted sub-polymer graben next to the 60° fault. In the inverted hanging wall, a back-thrust developed above the polymer layer and carried the supra-polymer layers (including the syn-extensional strata) out of the original hanging-wall half-graben and over the footwall the main antithetic fault of the syn-extensional crestal-collapse graben (Fig. 8d). New hanging-wall-vergent back-thrusts formed as the result of buttressing against the steep footwall of the 60° planar fault. As in the inversion of the 20° planar fault described earlier, significant thickness changes occurred in the polymer layer with the weld where the polymer was totally thinned with polymer migration into the half-graben axis and distal hanging wall, as well as into the footwall (Fig. 8d). Here, above the rigid footwall, silicone inflation occurred in addition to contraction of the graben near the end of the footwall polymer layer (Fig. 8d). These features indicate significant displacement and strain transfer from the hanging-wall inversion into the footwall units. As in all of the Series 2 inversion models with a pre-extension ductile layer, the main inverted basin-bounding fault did not penetrate upwards into the supra-polymer section (Fig. 8).

Inversion of the 60°–20° kinked planar fault system produced a broad, more symmetrical anticline above the inverted sub-polymer half-graben (Fig. 8e). A crestal-collapse graben formed on the apex of the broad inversion uplift. Several small, back-thrusts formed as a result of buttressing against the steep 60°-dipping part of the main detachment surface (Fig. 8e). As for all of the other models described above, transfer of displacement and strain from the hanging wall into the distal footwall resulted in thickening of the polymer layer in the hanging wall and also in the footwall. The small half-graben that formed near the end of the polymer detachment during extension was strongly shortened by the inversion (Fig. 8e).

Figure 9 shows the progressive evolution of the 60° planar fault model with a pre-extension polymer layer. At the end of extension, the structure was a broad hanging-wall syncline of syn-extension strata decoupled from the faulted half-graben below the salt layer (Fig. 9a). The progressive inversion is tracked in Figure 9 from the uplift and partial unfolding of the syn-extensional strata (Fig. 9b), and the development of the back-thrust overlying the polymer layer (Fig. 9c), to the eventual formation of the polymer primary weld (Fig. 9d). Concordant with the structural decoupling–folding and detachment thrusting above the polymer layer, and the reactivation of the extensional faults and new back-thrusts beneath the polymer layer, the polymer shows migration from the corner regions of the main sub-polymer faults into both the footwall in front of the inverted half-graben and the hanging wall (Fig. 9c, d).

Discussion

In this section, the results of the analogue models are analysed and compared to natural examples of inverted extensional basin that contain salt or evaporite layers. The structural styles of the Series 1 sandbox models that did not contain viscous layers in the hanging-wall stratigraphy are compared to Series 2 models that had viscous layers (Table 1).

Role of a viscous layer during model extension

The kinematic evolution and fault/fold styles of the analogue models during the extension were mainly controlled by the geometry of the main basin-bounding fault geometry, as well as by the presence or absence of a weak, ductile viscous layer within in the pre-kinematic sequence (Figs 4–9).

Series 1 models without a viscous layer

The results of the Series 1 extension baseline models with a rigid footwall block and an isotropic sand pack (Fig. 4) are comparable to sandbox models described in the literature (e.g. McClay & Ellis 1987*a*, *b*; Ellis & McClay 1988; McClay 1990; McClay *et al.* 1991) (Fig. 1). As in previous experiments, the hanging-wall geometries above the main basin-bounding fault are controlled by the dip and shape of the fault surface (Fig. 4), as well as by the amount of extension. Models where the fault surface was kinked or had a ramp-flat geometry produced both hanging-wall rollover anticlines and synclinal geometries (e.g. Fig. 4b, e). In the Series 1 models presented in this paper, planar antithetic faults were mainly developed where the main basin-bounding fault dipped by more than 45° at the break-away (Fig. 4b, d, e) (cf. McClay 1990). The Series 1 isotropic models are coupled in the sense that the extensional faults affect both the pre-kinematic and syn-kinematic units.

Series 2 models with a viscous layer

The deformation of Series 2 models typically formed decoupled architectures, with the supra-polymer strata folded into hanging-wall growth synclines above the basin-bounding faults (Figs 5 & 6). The units below the polymer layer formed half-graben. Where the initial breakaway dip of the principal fault was greater than 45°, strains in the half-graben were accommodated by steep antithetic faults (Figs 5b, d, e & 6). Similar decoupled hanging-wall deformation in analogue models with polymer layers was described by Soto *et al.* (2007) and Ferrer *et al.* (2008*b*, 2014).

Extension was accommodated by differential viscous flow of the polymer across the footwall of the basin-bounding fault, thickening into the hanging wall and thinning across the edge of the footwall. In most of the models, the polymer layer remained continuous (Fig. 5a–c, e), whereas, in the 60° planar fault model, polymer flow into the half-graben across the footwall of the antithetic fault at the rollover hinge produced a primary polymer weld (Figs 5d & 6d). At this stage, this section of the model became coupled and a small extensional fault formed in the supra-polymer strata just to the right of the weld (Fig. 6d).

In these models, the hanging-wall basins above the polymer were all synclinal in form, with the width and depth reflecting the changes in thickness of the polymer layer (Fig. 5). The simple 45° listric fault model (Fig. 5a) and the gently-dipping 20° planar fault model (Fig. 5c) formed wide, gentle supra-polymer basins with only subtle changes in the thickness of the polymer, whereas models with steeper fault dips on the basin-bounding faults (Fig. 5b, d, e) developed deeper synclinal basins underlain by significant changes in the thickness of the polymer. In these models, there was significant polymer flow into the hanging-wall graben particularly for the 60° planar fault model (Figs 5d & 6).

In the Series 1 models, deformation above the footwall block produced narrow extensional graben towards the end of the polymer layer (Fig. 5). Footwall extensional strains were transmitted along the polymer detachment, localizing faulting near the end of the polymer unit (cf. also Vendeville & Jackson 1992; Jackson & Vendeville 1994). Complex narrow graben developed as extension progressed (Fig. 6). The addition of syn-kinematic sand layers during continued extension inhibited the formation of reactive polymer diapirs within these graben.

Role of a viscous layer during inversion of the models

Inversion of analogue models with rigid footwall blocks was investigated by McClay (1989, 1995), Buchanan & McClay (1991, 1992), Keller & McClay (1995), and Yamada & McClay (2003*a*, *b*) (Fig. 1b, d). These experiments did not have any very weak layers in the hanging-wall strata. These models formed classic harpoon structures with reactivation and upwards propagation of the basin-bounding fault, and uplift and back-rotation of the syn-extension strata with part remaining in net extension (Fig. 1b, d). In the Series 1 models discussed in this paper, contraction (5 cm) exceeded extension (3 cm) such that most of the syn-extension strata were uplifted above regional during the inversion (Fig. 7). The depth slices and sections of Figure 10 show that there were only a few footwall shortcut thrusts formed in the lower part of the tilted front limb of the inversion anticline. The inverted faults were approximately linear, as were the backthrusts, and there was no deformation in the main part of the footwall strata (Fig. 10b, d).

The models of the Series 2 experiments with a polymer layer show strong decoupling between the sub- and supra polymer layers (Figs 8 & 11). Broad gentle folding developed above the polymer layer, with the polymer thickened beneath the anticlines and depleted under the synclines (Figs 8, 9 & 11a, c). The depth slices clearly show the folded hanging-wall strata (Fig. 11b, d), in strong contrast to the dominance of linear hanging-wall fault arrays and the absence of footwall deformation in the Series 1 baseline models (Fig. 10b, d).

The inherited polymer configuration at the end of extension and, particularly, the positions of primary welds was critical during inversion because these disrupted the polymer within the model. Similarly, where welds formed during inversion (e.g. Figs 8a, b, d, e & 9d) strongly controlled the inversion architectures in the last stage of net contraction. In the region of the welds, shortening in the sub-polymer strata was transferred into the supra-polymer strata enhancing uplift, forward tilting and folding of the syn-extensional stratigraphy (e.g. Figs 8d, e & 9). The resultant outer-arc stretching in these uplifted zones produced inversion-related crestal-collapse graben within the syn-extensional units (Figs 8e & 11a).

Natural examples of extension and inverted basins with salt layers

The results of the analogue models of inverted extensional basins with an internal ductile detachment layer are compared with natural examples of inverted basins from the Parentis Basin in the Bay of Biscay, from the Central and Southern North Sea basins, and from the Cameros Basin in the Iberian Range, central Spain.

Parentis Basin, Bay of Biscay

The Parentis Basin in the eastern Bay of Biscay is a partly inverted extensional basin with salt units in the pre-rift section (Roca *et al.* 2011; Ferrer

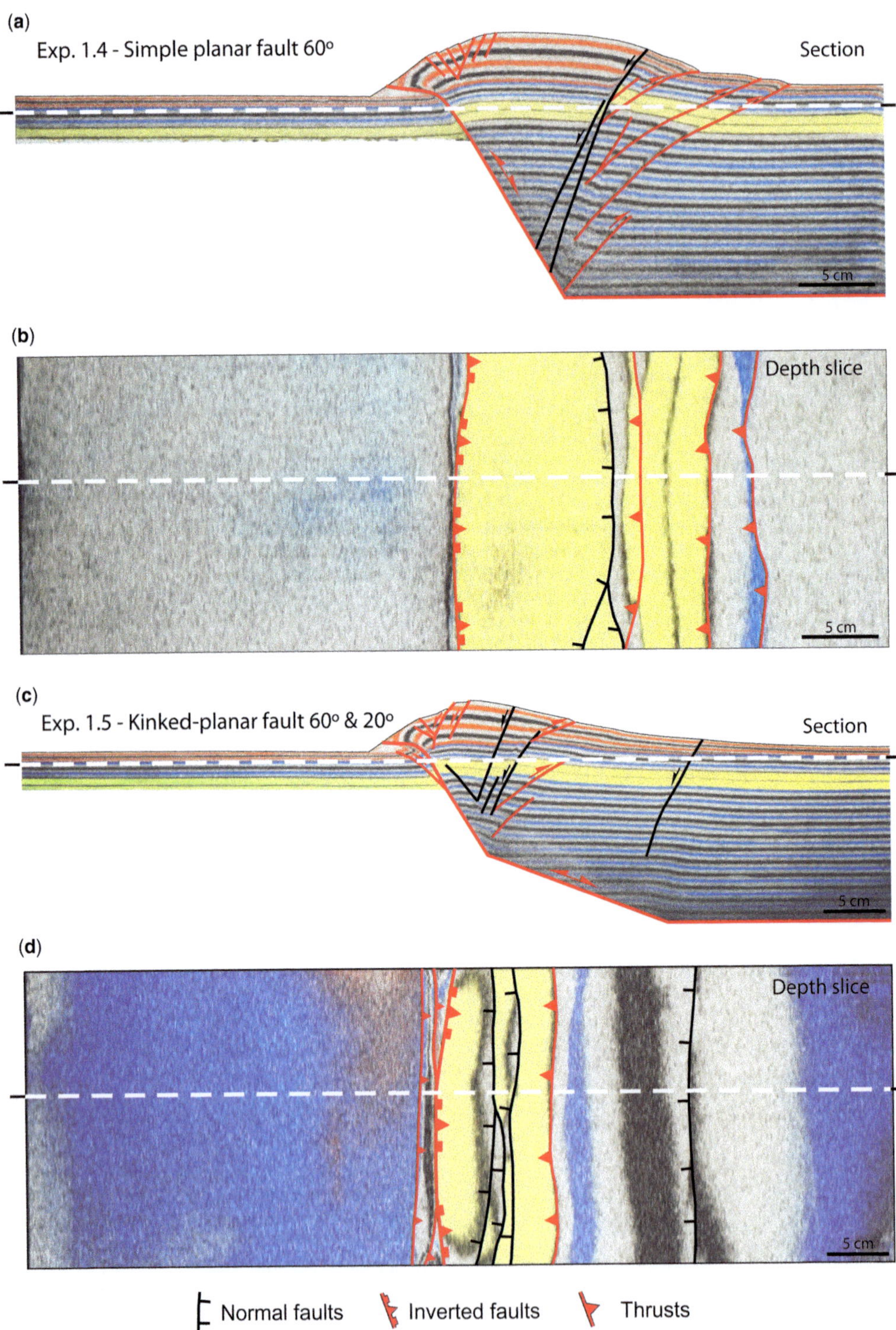

Fig. 10. Interpreted section and depth slices through reconstructed volumes of baseline isotropic experiments: (**a**) & (**b**) with a simple planar fault dipping 60°; and (**c**) & (**d**) with a 60°–20° kinked-planar fault. White dashed lines indicate the location of each depth slice on the cross-sections and vice versa.

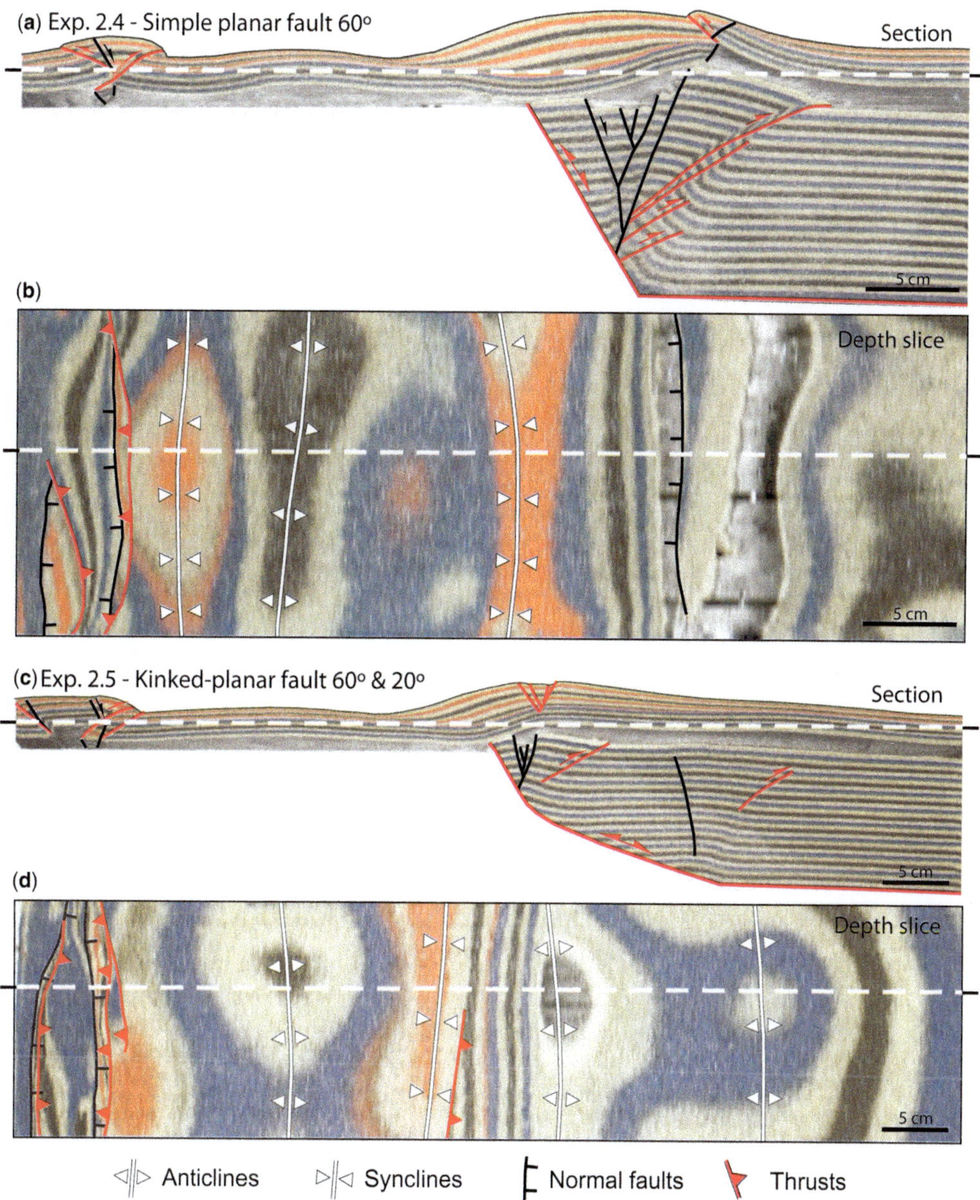

Fig. 11. Interpreted section and depth slices through reconstructed volumes of anisotropic experiments: (**a**) & (**b**) with a simple planar fault dipping 60°; and (**c**) & (**d**) with a 60°–20° kinked-planar fault. White dashed lines indicate the location of each depth slice on the cross-sections and vice versa.

et al. 2012; Rowan 2014). This Late Jurassic–Lower Cretaceous basin is part of a series of east–west-trending rift basins developed between Iberia and Europe during the opening of the Bay of Biscay and the North Atlantic (Srivastava *et al.* 1990). Whereas most of the Pyrenean basins were subsequently strongly inverted during the Late Cretaceous–Cenozoic Pyrenean Orogeny (e.g. Berástegui *et al.* 1990; Muñoz 1992; Bond & McClay 1995; García-Senz 2002; Mencos 2011; Roca *et al.* 2011), the Parentis Basin was only slightly inverted because of buttressing produced by a major basement high located to the south (Ferrer *et al.* 2008*a*). The present-day structure of the

Parentis Basin was controlled by two east–west-trending, north-dipping low-angle crustal extensional faults and by the presence of Upper Triassic evaporites. This Triassic salt unit is considered pre-rift in relation to the main Late Jurassic–Lower Cretaceous rifting and the opening of the Bay of Biscay (Rowan 2014).

The structure of the Parentis Basin is characterized by a growth syncline controlled by thick-skinned extension that triggered salt migration (Ferrer *et al.* 2012) (Fig. 12a). During the extension, the salt partially decoupled the major sub-salt fault from the supra-salt cover units (Ferrer *et al.* 2012). The experimental model in Figure 5e has a basement structure of a kinked planar detachment fault that is inferred to represent the extensional architecture of the Parentis Basin, which has been derived from seismic sections and cross-section restorations. In the Parentis Basin, the Barremian–Lower Aptian syn-rift strata in the hanging-wall

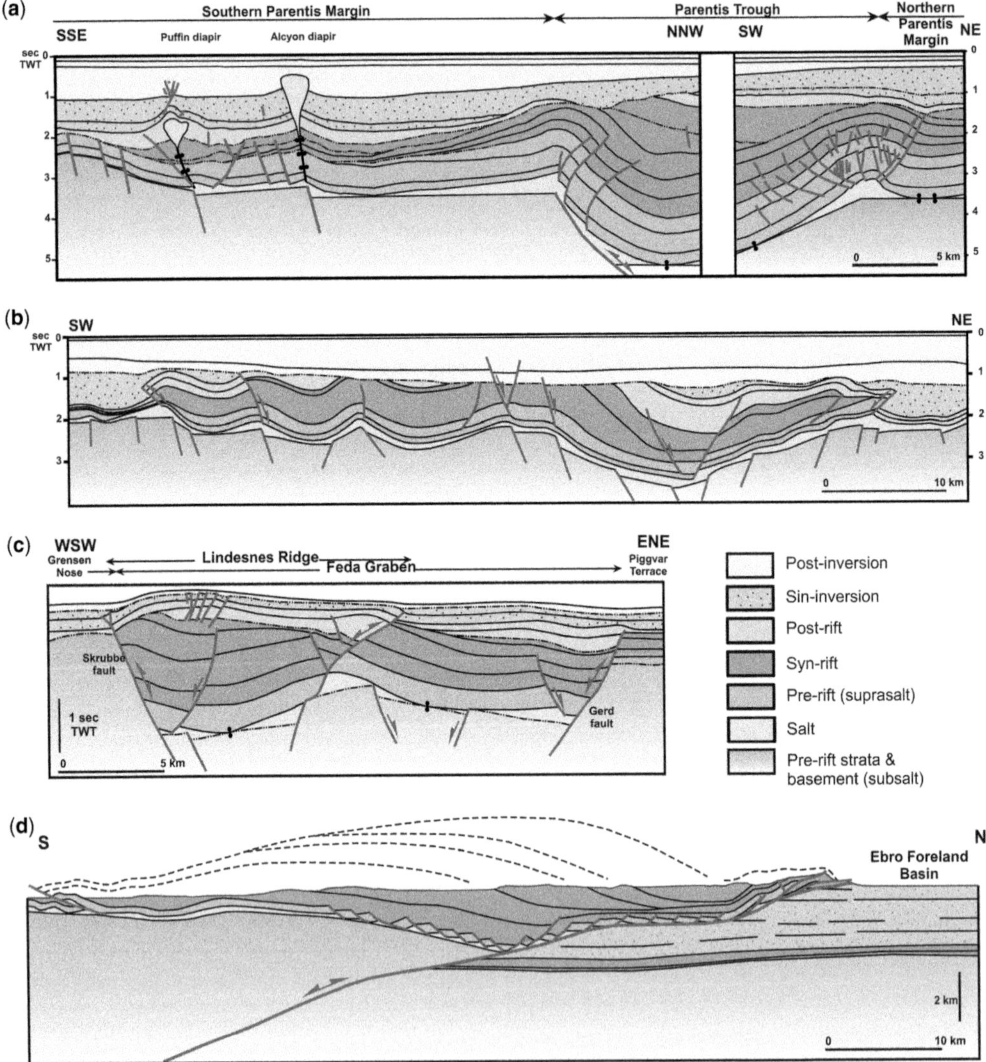

Fig. 12. Natural examples of inverted basins containing pre-kinematic salt units and with different degrees of shortening–inversion. (**a**) Geoseismic section of part of the Parentis Basin (eastern Bay of Biscay) (modified from Ferrer *et al.* 2012). (**b**) Line drawing of a seismic section of the central Broad Fourteens Basin (southern North Sea) (modified from Nalpas *et al.* 1995). (**c**) Simplified geoseismic section from the southern Feda Graben (North Sea) (modified from Gowers *et al.* 1993). (**d**) Cross-section of the Cameros Basin, Iberian chain, Spain (modified from Soto *et al.* 2007). TWT, two-way time.

syncline onlap towards the south (Fig. 12a) in a fashion similar to that displayed in the analogue model (Fig. 5e). The Pyrenean Late Cretaceous–Cenozoic inversion of the basin-bounding fault produced uplift of the depocentre, as shown by the different regional levels for the lower syn-inversion top Paleocene and Eocene strata both in the hanging wall and in the footwall. A similar feature is seen in the experimental models (Fig. 8e). The Parentis Basin also shows diapirs and salt walls developed above the footwall of the major fault (Fig. 12a). However, similar structures were not developed in the sandbox models presented in this paper, most probably related to the lesser extension in the models and, perhaps, to greater initial salt thicknesses in the Parentis Basin. In the Parentis Basin section (Fig. 12a), the presence of basement faults below the salt walls could also be an important factor, with salt preferentially absorbing the contractional deformation, squeezing the salt diapirs and forming secondary welds (Ferrer *et al.* 2012). In contrast, in the analogue models, the contractional deformation resulted in graben inversion or the development of new thrusts verging in the same direction as the major graben-bounding fault (Fig. 8).

Broad Fourteen Basin, Dutch sector, Southern North Sea Basin

The Broad Fourteens Basin in the Dutch sector of Southern North Sea shows very spectacular examples of positive inversion, with thrusts faults at the basin margins related to the Zechstein (Upper Permian) salt that acted as a major detachment during inversion (e.g. Nalpas *et al.* 1995) (Fig. 12b). The geological history of this basin was controlled by halokinesis and minor extension until the Mid-Jurassic followed by Mid to Late Jurassic NE–SW extension and Cretaceous post-rift subsidence (Van Wijhe 1987). In the Late Cretaceous (Senonian), the collision between the African Plate and the European Plate (Alpine Orogeny) produced far-field hinterland contraction to the north and inversion of the Broad Fourteens Basin (Ziegler 1975, 1982; Van Wijhe 1987). This resulted in folding, uplift and erosion (Nalpas *et al.* 1995). The southern sector of the basin contains no salt, no decoupling and the inverted basement faults have propagated through the overlying sedimentary section (Nalpas *et al.* 1995). In contrast, the northern sector of the basin contains Zechstein salt with the inversion structures controlled by the salt thickness (Fig. 12b). On the SW margin of the basin, broad detachment folds formed above the Zechstein salt: on the NE margin, however, a low-angle thrust fault detached on top of the salt has carried the supra-salt section onto the footwall of the extensional basin (Fig. 12b). The experiments of Ferrer *et al.* (2014) indicate that at the end of the extension these structures may have initiated on salt inflations or salt-ridges with the basinward-dipping inversion faults detached on top of the salt. The sandbox models presented in this paper have less inversion compared to that in the Broad Fourteens Basin: however, it is possible to envisage how the hanging wall of the basin-margin fault above the salt was transported over the footwall during inversion (e.g. Fig. 11).

Feda Graben, Danish North Sea

The inverted Feda Graben (Fig. 12c) at the Norwegian–Danish boundary of the North Sea Central Graben Basin exhibits similar inversion structures to those formed above an inverted planar fault system with salt layers (Figs 8d & 9). The geo-history of this part of the North Sea includes pre-rift Late Permian Zechstein salt, Mid–Late Jurassic rifting and a later Cretaceous–Early Tertiary inversion (Gowers & Sæbøe 1985; Gowers *et al.* 1993; Taylor 1998; Tanveer & Korstgård 2009). The Feda Graben is bounded by two NNW-trending basement faults (Gowers & Sæbøe 1985) – the Skrubbe Fault that separates the basin from the Grensen Nöse in the SW and the Gert Fault that separates the basin from the Piggvar Terrace in the NE (Fig. 12c). This simplified sketch section can be compared to Figure 8d, where the axial syncline of the pre-extensional strata was controlled by polymer migration during extension, and the antiformal shape (Lindesnes Ridge: Skjerven *et al.* 1983) of the Cretaceous–Cenozoic units in the central part of the Feda Graben is clearly related to the inversion of the main Skrubbe Fault (Fig. 12c). Figures 8d and 9 show similar features with buttressing against the main bounding fault transferring deformation into the centre of the graben system, as seen in the Feda Graben in Figure 12c. During extension in the sandbox model, a normal fault detached on the polymer developed above the rollover hinge (Fig. 6). Inversion produced asymmetric uplift of the synclinal basin as a result of buttressing against the basin-bounding fault. Part of the contractional deformation was transmitted by the weak polymer to the distal edge of the basin where the small graben system was inverted with a new intra-graben reverse fault (Fig. 9).

Cameros Basin, Iberian Range, Spain

The Cameros Basin in the NW Iberian range in Spain (Guimerà & Álvaro 1990; Casas & Salas 1992; Salas & Casas 1993; Guimerà *et al.* 1995; Casas *et al.* 2009; Mas *et al.* 2011; Omodeo Salè *et al.* 2014) is one of the NW–SE-striking intraplate

Mesozoic rift basins related to the opening of Western Tethys and the North Atlantic Ocean (Álvaro *et al.* 1979). From the Late Jurassic to the Early Albian, the Cameros Basin underwent major subsidence, with the deposition of more than 6 km of syn-rift strata overlying the pre-rift section of Upper Triassic evaporites and Jurassic limestones (Mas *et al.* 2011). Various models have been proposed to explain the geometry of this basin (e.g. the review by Omodeo Salè *et al.* 2014). The analogue model results presented in this paper (Figs 5a & 8a) support a hanging-wall synclinal model, as proposed by Casas & Salas (1992), Casas-Sainz *et al.* (2000), Casas *et al.* (2009) and Soto *et al.* (2007). This suggests that the hanging-wall synclinal basin with syn-kinematic extensional growth strata resulted from extension of a major south-dipping listric fault with an Upper Triassic evaporite detachment (Fig. 12d) in a manner similar to that shown in the sandbox model (Fig. 5a). In the Eocene–Early Miocene, the Cameros Basin was totally inverted as a result of the Alpine Orogeny (Guimerà & Álvaro 1990; Salas & Casas 1993; Guimerà *et al.* 1995). During the inversion, the Upper Triassic evaporites acted as a detachment (Guimerà & Álvaro 1990), thrusting the extensional hanging-wall synclinal basin over the Cenozoic Ebro foreland basin with a maximum displacement of around 30 km (Casas-Sainz & Simón-Gómez 1992) (Fig. 12d). The resulting inverted basin is very analogous to the inverted listric fault model shown in Figure 8a.

Conclusions

The results of the analogue modelling programme presented in this paper provide both geometrical and kinematic templates that may be applied to the analysis of the structural evolution of normal faults involving pre-extensional evaporites (particularly salt layers) and then subsequently inverted. The shapes of the basement-bounding faults controlled the geometries of the supra-'salt' synclinal basins formed during the extensional deformation, and the location of fault-propagation folds and supra-'salt' decoupling during inversion. Wide, gentle and shallow synclinal supra-'salt' basins developed in the models during extension in the hanging wall of a simple listric or gently dipping simple planar fault (Fig. 5a, c). In contrast, narrow and deep synclinal extensional basins developed above steeply dipping planar faults (Figs 5d & 6) or the ramps of a ramp-flat listric faults (Fig. 5b) in the models. The development of the hanging-wall synclinal basins was also controlled by polymer migration towards the edges of the basin below where salt-inflated areas developed at the end of the extension (Fig. 5b, d, e).

The inherited extensional architectures both in the analogue models and in the natural examples exert a fundamental role during later inversion. In the models, shortening preferentially inverted the major basement basin-bounding faults and the polymer layer acted as a contractional detachment transferring the deformation above the footwall. Independently of the basement fault geometry, the hanging-wall synclinal basins were arched and uplifted during the inversion, and partially translated over the rigid footwall. The uplift of the synclinal basins was accelerated when primary welds developed under the main synclinal basins depocentres during the late stages of inversion.

The experimental results also show the development of extensional graben at the extremities of the footwall strata and these were later inverted.

The presence of pre-rift evaporites controls the structural style of the supra- and sub-salt units. Whereas strata below the polymer can be strongly faulted, continuous deformation characterized by folded synclinal basins formed above the polymer or salt layer. In this sense, the presence of a polymer layer (or salt in nature) acts as an effective decoupling unit inhibiting the upwards propagation of the faults from the sub-salt to the supra-salt layers during both extension and inversion.

The main limitation of analogue models that used a rigid block to control the geometry of the basin-bounding fault systems is that it limited the footwall deformation to the ductile layer and the overlying footwall strata. Despite this, the sandbox models produce many strikingly similar deformation features to those found in natural examples of inverted basins with salt strata, as described in this paper.

This research was funded by the STAR Research Consortium, sponsored by BG Group, BHPBilliton, ConocoPhilips, Eni, MarathonOil, Nexen, Shell, Talisman Energy and YFP. The authors also received additional support from the Fault Dynamics Research Group, Royal Holloway University of London. We also thank Kevin D'Souza, Jerry Morris and Frank Lehane for the logistical support in the modelling laboratory. Research of OF was partially supported by the SALTECRES project (CGL2014-54118-C2-1-R), as well as the Grup de Recerca de Geodinàmica i Anàlisi de Conques (2014SGR467). Reviews and suggested improvements by S. Buiter, I.A. Alsop and C. Childs were greatly appreciated.

References

Álvaro, M., Capote, R. & Vegas, R. 1979. Un modelo de evolución geotectónica para la cadena Celtibérica. *Acta Geologica Hispana*, **14**, 172–181.

Bally, A.W. 1984. Tectonogénese et sismique de réflexion. *Bulletin Société Géologique de France*, **7**, 279–285.

Berástegui, X., García-Senz, J. & Losantos, M. 1990. Tecto-sedimentary evolution of the Organyà extensional basin (central south Pyrenean unit, Spain) during the Lower Cretaceous. *Bulletin Societé Géologique France*, **8**, 251–264.

Bond, R.M.G. & McClay, K.R. 1995. Inversion of a Lower Cretaceous extensional basin, south central Pyrenees, Spain. *In*: Buchanan, J.G. & Buchanan, P.G. (eds) *Basin Inversion*. Geological Society, London, Special Publications, **88**, 415–431, https://doi.org/10.1144/GSL.SP.1995.088.01.22

Bonini, M., Sani, F. & Antonielli, B. 2012. Basin inversion and contractional reactivation of inherited normal faults: a review based on previous and new experimental models. *Tectonophysics*, **522–523**, 55–88.

Brun, J.-P. & Nalpas, T. 1996. Graben inversion in nature and experiments. *Tectonics*, **15**, 677–687.

Buchanan, P.G. & McClay, K.R. 1991. Sandbox experiments of inverted listric and planar fault systems. *Tectonophysics*, **188**, 97–115.

Buchanan, P.G. & McClay, K.R. 1992. Experiments on basin inversion above reactivated domino faults. *Marine and Petroleum Geology*, **9**, 486–500.

Buiter, S.H.J., Pfiffner, O.A. & Beaumont, C. 2009. Inversion of extensional sedimentary basins: a numerical evaluation of the localization of shortening. *Earth and Planetary Sciences Letters*, **288**, 492–504.

Burliga, S., Koyi, H.A. & Krzywiec, P. 2012. Modelling cover deformation and decoupling during inversion, using the Mid-Polish Trough as a case study. *Journal of Structural Geology*, **42**, 62–73.

Butler, R.W.H., Maniscalco, R., Sturiale, G. & Grasso, M. 2014. Stratigraphic variations control deformation patterns in evaporite basins: Messinian examples, onshore & offshore Sicily (Italy). *Journal of the Geological Society, London*, **172**, 113–124, https://doi.org/10.1144/jgs2014-024

Casas, A. & Salas, R. 1992. Historia de la subsidencia, anomalías gravimétricas y evolución mesozoica de las Cuencas del margen oriental de Iberia. *In*: *Actas de las sesiones científicas: III Congreso Geológico de España*. Facultad de Ciencias, Universidad de Salamanca, Salamanca, 112–116.

Casas, A.M., Villalaín, J.J., Soto, R., Gil-Imaz, A., Del Río, P. & Fernández, G. 2009. Multidisciplinary approach to an extensional syncline models for the Mesozoic Cameros Basin (N Spain). *Tectonophysics*, **470**, 3–20.

Casas-Sainz, A.M. & Simón-Gómez, J.L. 1992. Stress field and thrust kinematics: a model for the tectonic inversión of the Cameros massif (Spain). *Journal of Structural Geology*, **14**, 521–530.

Casas-Sainz, A.M., Cortés-Gracia, A.L. & Maestro-González, A. 2000. Intra-plate deformation and basin formation during Tertiary at the Northern Iberian Plate: origin and evolution of the Almazán Basin. *Tectonics*, **19**, 762–786.

Corti, G. 2012. Evolution and characteristics of continental rifting: analog modeling-inspired view and comparison with examples from the East African Rift System. *Tectonophysics*, **522–523**, 1–33.

Coward, M. & Stewart, S. 1995. Salt-influenced structures in the Mesozoic-Tertiary cover of the Southern North Sea, U.K. *In*: Jackson, M.P.A., Roberts, D.G. & Snelson, S. (eds) *Salt Tectonics: A Global Perspective*. American Association of Petroleum Geologists, Memoirs, **65**, 229–250.

Dell'Ertole, D. & Schellart, W.P. 2013. The development of sheath folds in viscously stratified materials in simple shear conditions: an analogue approach. *Journal of Structural Geology*, **56**, 129–141.

Del Ventisette, C., Montanari, D., Sani, F. & Bonini, M. 2006. Basin inversion and fault reactivation in laboratory experiments. *Journal of Structural Geology*, **28**, 2067–2083.

Dooley, T., McClay, K.R., Hempton, M. & Smit, D. 2005. Salt tectonics above complex basement extensional fault systems: results from analogue modelling. *In*: Doré, E.G. & Vining, B.A. (eds) *Petroleum Geology: North-West Europe and Global Perspectives – Proceedings of the 6th Petroleum Geology Conference*. Geological Society, London, 1631–1648, https://doi.org/10.1144/0061631

Dooley, T., Jackson, M.P.A., Jackson, C.A.L., Hudec, M.R. & Rodriguez, C.R. 2015. Enigmatic structures within salt walls of the Santos Basin – Part 2: mechanical explanation from physical modeling. *Journal of Structural Geology*, **75**, 163–187.

Dubois, A., Odonne, F., Massonnat, G., Lebourg, T. & Fabre, R. 2002. Analogue modelling of fault reactivation: tectonic inversion and oblique remobilization of grabens. *Journal of Structural Geology*, **24**, 1741–1752.

Ellis, P.G. & McClay, K.R. 1988. Listric extensional fault systems-results of analogue model experiments. *Basin Research*, **1**, 55–70.

Ferrer, O., Roca, E., Benjumea, B., Muñoz, J.A., Ellouz, N. & Marconi Team 2008*a*. The deep seismic reflection MARCONI-3 profile: role of extensional Mesozoic structure during Pyrenean contractional deformantion at the Eastern part of the Bay of Biscay. *Marine and Petroleum Geology*, **25**, 714–730.

Ferrer, O., Roca, E. & Vendeville, B.C. 2008*b*. Influence of a syntectonic viscous salt layer on the structural evolution of extensional kinked-fault systems. *Bollettino di Geofisica Teorica ed Applicata*, **49**, 371–375.

Ferrer, O., Jackson, M.P.A., Roca, E. & Rubinat, M. 2012. Evolution of salt structures during extension and inversion of the offshore Parentis Basin (eastern Bay of Biscay). *In*: Alsop, G.I., Archer, S.G., Hartley, A.J., Grant, N.T. & Hodgkinson, R. (eds) *Salt Tectonics, Sediments and Prospectivity*. Geological Society, London, Special Publications, **363**, 361–379, https://doi.org/10.1144/SP363.16

Ferrer, O., Roca, E. & Vendeville, B.C. 2014. The role of salt layers in the hangingwall deformation of kinked-planar extensional faults: insights from 3D analogue models and comparison with the Parentis Basin. *Tectonophysics*, **636**, 338–350.

Fiduk, J.C. & Rowan, M.G. 2012. Analysis of folding and deformation within layered evaporites in Blocks BM-S-8 and -9, Santos Basin, Brazil. *In*: Alsop, G.I., Archer, S.G., Hartley, A.J., Grant, N.T. & Hodgkinson, R. (eds) *Salt Tectonics, Sediments and Prospectivity*. Geological Society, London, Special Publications, **363**, 471–487, https://doi.org/10.1144/SP363.22

GARCÍA-SENZ, J. 2002. *Cuencas extensivas del Cretácico Inferior en los Pirineos Centrales, formación, y subsecuente inversión*. PhD thesis, Universitat de Barcelona.

GOWERS, M.B. & SÆBØE, A. 1985. On the structural evolution of the central trough in the Norwegian and Danish sectors of the North Sea. *Marine and Petroleum Geology*, **2**, 298–318.

GOWERS, M.B., HOLTAR, E. & SWENSSON, E. 1993. The structure of the Norwegian Central Trough (Central Graben area). *In*: PARKER, J.R. (ed.) *Petroleum Geology of Northwest Europe: Proceedings of the 4th Conference*. Geological Society, London, 1245–1254, https://doi.org/10.1144/0041245

GUIMERÀ, J. & ÁLVARO, M. 1990. Structure et évolution de la compression alpine dans la Chaîne Ibérique et la Chaîne côtière catalane (Espagne). *Bulletin de la Société Géologique France*, **VI**, 339–348.

GUIMERÀ, J., ALONSO, A. & MAS, J.R. 1995. Inversion of an extensional-ramp basin by a newly formed thrust: the Cameros basin (N Spain). *In*: BUCHANAN, J.G. & BUCHANAN, P.G. (eds) *Basin Inversion*. Geological Society, London, Special Publications, **88**, 433–453, https://doi.org/10.1144/GSL.SP.1995.088.01.23

HORSFIELD, W.T. 1977. An experimental approach to basement controlled faulting. *Geologie en Mijnbouw*, **56**, 363–370.

HUIQI, L., MCCLAY, K.R. & POWELL, D. 1992. Physical models of thrusts wedges. *In*: MCCLAY, K.R. (ed.) *Thrust Tectonics*. Chapman and Hall, London, 71–81.

JACKSON, M.P.A. & VENDEVILLE, B.C. 1994. Regional extension as a geological trigger for diapirism. *Geological Society of America Bulletin*, **106**, 57–73.

JACKSON, M.P.A., CORNELIUS, R.R., CRAIG, C.R., GANSSER, A., STÖCKLIN, J. & TALBOT, C.J. 1990. *Salt Diapirs of the Great Kavir, Central Iran*. Geological Society of America, Memoirs, **177**.

KELLER, J.V.A. & MCCLAY, K.R. 1995. 3D sandbox models of positive inversion. *In*: BUCHANAN, J.G. & BUCHANAN, P.G. (eds) *Basin Inversion*. Geological Society, London, Special Publications, **88**, 137–146, https://doi.org/10.1144/GSL.SP.1995.088.01.09

KOYI, H., JENYON, M.K. & PETERSEN, K. 1993. The effect of basement faulting on diapirism. *Journal of Petroleum Geology*, **163**, 285–312.

KRANTZ, R.W. 1991. Measurements of friction coefficients and cohesion for faulting and fault reactivation in laboratory models using sand and sand mixtures. *Tectonophysics*, **188**, 203–207.

KRZYWIEC, P. 2006. Structural inversion of the Pomeranian and Kuiavian segments of the Mid-Polish Trough – lateral variations in timing and structural style. *Geological Quarterly*, **50**, 151–168.

LETOUZEY, J., COLLETA, B., VIALLY, R. & CHERMETTE, J.C. 1995. Evolution of salt-related structures in compressional settings. *In*: JACKSON, M.P.A., ROBERTS, D.G. & SNELSON, S. (eds) *Salt Tectonics: A Global Perspective*. American Association of Petroleum Geologists, Memoirs, **65**, 41–60.

LOHRMANN, J., KUKOWSKI, N., ADAM, J. & ONCKEN, O. 2003. The impact of analogue material properties on the geometry, kinematics, and dynamics of convergent sand wedges. *Journal of Structural Geology*, **25**, 1691–1711.

MAS, R., BENITO, M.I. ET AL. 2011. Evolution of an intraplate rift basin: the Latest Jurassic-Early Cretaceous Cameros Basin (Northwest Iberian Ranges, North Spain). *In*: ARENAS, C., POMAR, L. & COLOMBO, F. (eds) *Post-Meeting Field Trips 28th IAS Meeting. Geo-Guías*, **8**, 117–154.

MCCLAY, K.R. 1989. Analogue models of inversion tectonics. *In*: COOPER, M.A. & WILLIAMS, G.D. (eds) *Inversion Tectonics*. Geological Society, London, Special Publications, **44**, 41–59, https://doi.org/10.1144/GSL.SP.1989.044.01.04

MCCLAY, K.R. 1990. Deformation mechanics in analogue models of extensional fault systems. *In*: RUTTER, E.H. & KNIPE, R.J. (eds) *Deformation Mechanisms, Rheology and Tectonics*. Geological Society, London, Special Publications, **54**, 445–454, https://doi.org/10.1144/GSL.SP.1990.054.01.40

MCCLAY, K.R. 1995. The geometries and kinematics of inverted fault systems: a review of analogue models studies. *In*: BUCHANAN, J.G. & BUCHANAN, P.G. (eds) *Basin Inversion*. Geological Society, London, Special Publications, **88**, 97–118, https://doi.org/10.1144/GSL.SP.1995.088.01.07

MCCLAY, K.R. & BUCHANAN, P.G. 1992. Thrust faults in inverted extensional basins. *In*: MCCLAY, K.R. (ed.) *Thrust Tectonics*. Chapman & Hall, London, 419–434.

MCCLAY, K.R. & ELLIS, P.G. 1987*a*. Analogue models of extensional fault geometries. *In*: COWARD, M.P., DEWEY, J.F. & HANCOCK, P.L. (eds) *Continental Extensional Tectonics*. Geological Society, London, Special Publications, **28**, 109–125, https://doi.org/10.1144/GSL.SP.1987.028.01.09

MCCLAY, K.R. & ELLIS, P.G. 1987*b*. Geometries of extensional fault systems developed in model experiments. *Geology*, **15**, 341–344.

MCCLAY, K.R. & SCOTT, A.D. 1991. Experimental models of hangingwall deformation in ramp-flat listric extensional fault systems. *Tectonophysics*, **188**, 85–96.

MCCLAY, K.R., WALTHAM, D.A., SCOTT, A.D. & ABOUSETTA, A. 1991. Physical and seismic modelling of listric normal fault geometries. *In*: ROBERTS, A.M., YIELDING, G. & FREEMAN, B. (eds) *The Geometry of Normal Faults*. Geological Society, London, Special Publications, **56**, 231–239, https://doi.org/10.1144/GSL.SP.1991.056.01.16

MENCOS, J. 2011. *Metodologies de reconstrucció i modelització 3D d'estructures geològiques: anticlinal de Sant Corneli – Bóixols (Pirineus Centrals)*. PhD thesis. Universitat de Barcelona.

MUÑOZ, J.A. 1992. Evolution of a continental collision belt: ECORS-Pyrenean crustal balanced section. *In*: MCCLAY, K.R. (ed.) *Thrust Tectonics*. Chapman & Hall, London, 235–246.

NALPAS, T. & BRUN, J.P. 1993. Salt flow and diapirism related to extension at crustal scale. *Tectonophysics*, **28**, 349–362.

NALPAS, T., LE DOUARAN, S., BRUN, J.-P., UNTERNEHR, P. & RICHERT, J.-P. 1995. Inversion of the Broad Fourteens Basin (offshore Netherlands), a small-scale model investigation. *Sedimentary Geology*, **95**, 237–250.

OMODEO SALÈ, S., GUMERÀ, J., MAS, R. & ARRIBAS, J. 2014. Tectono-stratigraphic evolution of an inverted extensional basin: the Cameros Basin (north of Spain). *International Journal Earth Science (Geologische Rundschau)*, **103**, 1597–1620.

Panien, M., Schreurs, G. & Pfiffner, A. 2005. Sandbox experiments on basin inversion: testing the influence of basin orientation and basin fill. *Journal of Structural Geology*, **27**, 433–445.

Panien, M., Buiter, S.J.H., Schreurs, G. & Pfiffner, O.A. 2006. Inversion of a symmetric basin: insights from a comparison between analogue and numerical experiments. *In*: Buiter, S.J.H. & Schreurs, G. (eds) *Analogue and Numerical Modelling of Crustal-Scale Processes*. Geological Society, London, Special Publications, **253**, 253–270, https://doi.org/10.1144/GSL.SP.2006.253.01.13

Roca, E., Muñoz, J.A., Ferrer, O. & Ellouz, N. 2011. The role of the Bay of Biscay Mesozoic extensional structure in the configuration of the Pyrenean orogeny: constraints from the MARCONI deep seismic reflection survey. *Tectonics*, **30**, TC2001.

Rowan, M.G. 2014. Passive-margin salt basins: hyper-extension, evaporite deposition, and salt tectonics. *Basin Research*, **26**, 154–182.

Rowan, M.G. & Krzywiec, P. 2014. The Szamotuly salt diapir and Mid-Polish Trough: decoupling during both Triassic–Jurassic rifting and Alpine inversion. *Interpretation*, **2**, SM1–SM18.

Salas, R. & Casas, A. 1993. Mesozoic extensional tectonics, stratigraphy and crustal evolution during the Alpine cycle of the eastern Iberian basin. *Tectonophysics*, **228**, 33–55.

Skjerven, J., Rijs, R. & Kalheim, J.E. 1983. Late Paleozoic to early Cenozoic structural development of the south–south eastern Norwegian North Sea. *Geologie en Mijnbouw*, **62**, 35–45.

Soto, R., Casas-Sainz, A.M. & Del Rio, P. 2007. Geometry of half-grabens containing a mid-level viscous decollement. *Basin Research*, **19**, 437–450.

Srivastava, S.P., Roest, W.R., Kovacs, L.C., Oakey, G., Lévesque, S., Verhoef, J. & Macnab, R. 1990. Motion of Iberia since the Late Jurassic: results from detailed aeromagnetic measurements in the Newfoundland Basin. *Tectonophysics*, **184**, 229–260.

Tanveer, M. & Korstgård, J.A. 2009. Structural evolution of the Feda Graben area – A new model. *Marine and Petroleum Geology*, **26**, 990–999.

Taylor, J.C.M. 1998. Upper Permian – Zechstein. *In*: Glennie, K.W. (ed.) *Petroleum Geology of the North Sea*. 4th edn. Blackwell Science, Oxford, 174–211.

Teixell, A., Arboleya, M.L., Julivert, M. & Charroud, M. 2003. Tectonic shortening and topography in the central High Atlas (Morocco). *Tectonics*, **22**, 1051.

Van Wijhe, D.H. 1987. Structural evolution of inverted basins in the Dutch offshore. *Tectonophysics*, **137**, 171–219.

Vendeville, B.C. & Jackson, M.P.A. 1992. The rise and fall of diapirs during thin-skinned extension. *Marine and Petroleum Geology*, **9**, 331–353.

Weijermars, R. 1986. Flow behavior and physical chemistry bouncing putties and related polymers in view of tectonic laboratory applications. *Tectonophysics*, **124**, 325–358.

Withjack, M.O. & Callaway, S. 2000. Active normal faulting beneath a salt layer: an experimental study of deformation patterns in the cover sequence. *AAPG Bulletin*, **84**, 627–651.

Withjack, M.O. & Schlische, R.W. 2006. Geometric and experimental models of extensional fault-bend folds. *In*: Buiter, S.J.H. & Schreurs, G. (eds) *Analogue and Numerical Modelling of Crustal-Scale Processes*. Geological Society, London, Special Publications, **253**, 285–305, https://doi.org/10.1144/GSL.SP.2006.253.01.15

Withjack, M.O., Islam, Q.T. & La Pointe, P.R. 1995. Normal faults and their hanging-wall deformation: an experimental study. *American Association of Petroleum Geologists Bulletin*, **79**, 1–18.

Yamada, Y. & McClay, K.R. 2003*a*. Application of geometric models to inverted listric fault systems in sandbox experiments. Paper 2: insights for possible along strike migration of material during 3D hanging wall deformation. *Journal of Structural Geology*, **25**, 1331–1336.

Yamada, Y. & McClay, K.R. 2003*b*. Application of geometric models to inverted listric fault systems in sandbox experiments. Paper 1: 2D hanging wall deformation an section restoration. *Journal of Structural Geology*, **25**, 1551–1560.

Yamada, Y. & McClay, K.R. 2004. 3-D analog modeling of inversion thrust structures. *In*: McClay, K.R. (ed.) *Thrust Tectonics and Hydrocarbon Systems*. American Association of Petroleum Geologists, Memoirs, **82**, 276–301.

Ziegler, P.A. 1975. Geologic evolution of North Sea and its tectonic framework. *American Association of Petroleum Geologists Bulletin*, **59**, 1073–1097.

Ziegler, P.A. 1982. *Geological Atlas of Western and Central Europe*. Elsevier, Amsterdam.

Timing, growth and structure of a reactivated basin-bounding fault

ROBERT P. WORTHINGTON[1,2]* & JOHN J. WALSH[1,3]

[1]*Fault Analysis Group, School of Earth Sciences, University College Dublin, Belfield, Dublin 4, Ireland*

[2]*Present address: Statoil, Sandsliveien 90, 5254 Bergen, Norway*

[3]*Irish Centre for Research in Applied Geosciences (iCRAG), University College Dublin, Belfield, Dublin 4, Ireland*

**Correspondence: rpw@statoil.com*

Abstract: Until recently, there have been relatively few natural examples of basement fault reactivation, demonstrating convincing hard-linkage to underlying structure and a well-constrained three-dimensional (3D) geometry and timing of reactivation. Applying various quantitative techniques to high quality 3D seismic data, we describe the Cenozoic structure and growth of a reactivated Mesozoic normal fault in the NW Porcupine Basin, west of Ireland. We demonstrate a previously unrecognized period of Mid–Late Eocene extensional reactivation with a tectonic, rather than the previously suggested compactional, origin. East–west extensional reactivation of a north–south- to NE–SW-trending basement fault led to the development of a highly segmented and systematically stepping fault array in the Cenozoic cover. A clear relationship between the displacement of hanging-wall antithetic faults and displacement along the main synthetic system is attributed to the antithetics primarily accommodating strain associated with a hanging-wall roll-over anticline. Our study shows that displacement analysis provides an excellent basis for defining the kinematics of basement fault reactivation, and the importance of fault segmentation and twisting during the upwards propagation of reactivated faults. Fault timing corresponds to a period of rapid differential subsidence, the magnitude of which is too large to be attributable solely to extensional reactivation of basement faults.

Until recently, there have been relatively few detailed studies on natural examples of basement fault reactivation demonstrating convincing hard linkage to existing structure and a well-constrained timing of reactivation. What work has been done has identified fault reactivation and timing through the analysis of fault displacement data (e.g. vertical displacement profiles/growth curves) and syndepositional thickening of sequences across faults imaged on seismic data (e.g. Morley *et al.* 2004, 2007; Nicol *et al.* 2005; Giba *et al.* 2012; Jackson & Rotevatn 2013). Applying similar analytical techniques to faults imaged from a high-quality three-dimensional (3D) seismic data, we describe the geometry, timing and growth of Cenozoic normal faulting that is localized above a major basin-bounding fault along the NW margin of the Porcupine Basin (Fig. 1). The Cenozoic faulting forms a complex and highly segmented array of faults, the geometry of which is primarily controlled by that of the underlying basement fault, and consists of synthetic, antithetic and conjugate faults and splays, which can be either systematically right- or left-stepping and which occur both within the hanging wall and, less often, the footwall of a principal fault that is hard-linked into the underlying basement fault. We will demonstrate that the faulting arises from east–west extensional reactivation of the basement fault during the Mid–Late Eocene and not from compaction-related deformation, as suggested by previous workers, and we discuss the regional implications of this newly identified period of normal faulting for the evolution of the Porcupine Basin.

Previous work within the Porcupine Basin (Fig. 1) has recognized Cenozoic normal faults that appear to be localized over, and related to, Mesozoic basement fault structures (e.g. Naylor & Anstey 1987; Tate 1993; McCann *et al.* 1995*a*, *b*; McDonnell & Shannon 2001; Jones *et al.* 2004). There has, however, been little detailed analysis on the origin and timing of such faults, with proposed mechanisms of faulting including both the reactivation of basement faults and, more commonly, the differential compaction of post-rift cover sequences over prominent basement highs along the basin margins (Fig. 2). A contributing factor to the paucity of analysis and lack of understanding of these structures is the relatively poor resolution and areal coverage of previously available 2D seismic datasets (e.g. line spacing typically 3.5 km), particularly in relation

From: Childs, C., Holdsworth, R. E., Jackson, C. A.-L., Manzocchi, T., Walsh, J. J. & Yielding, G. (eds) 2017. *The Geometry and Growth of Normal Faults*. Geological Society, London, Special Publications, **439**, 511–531.
First published online July 1, 2016, https://doi.org/10.1144/SP439.14

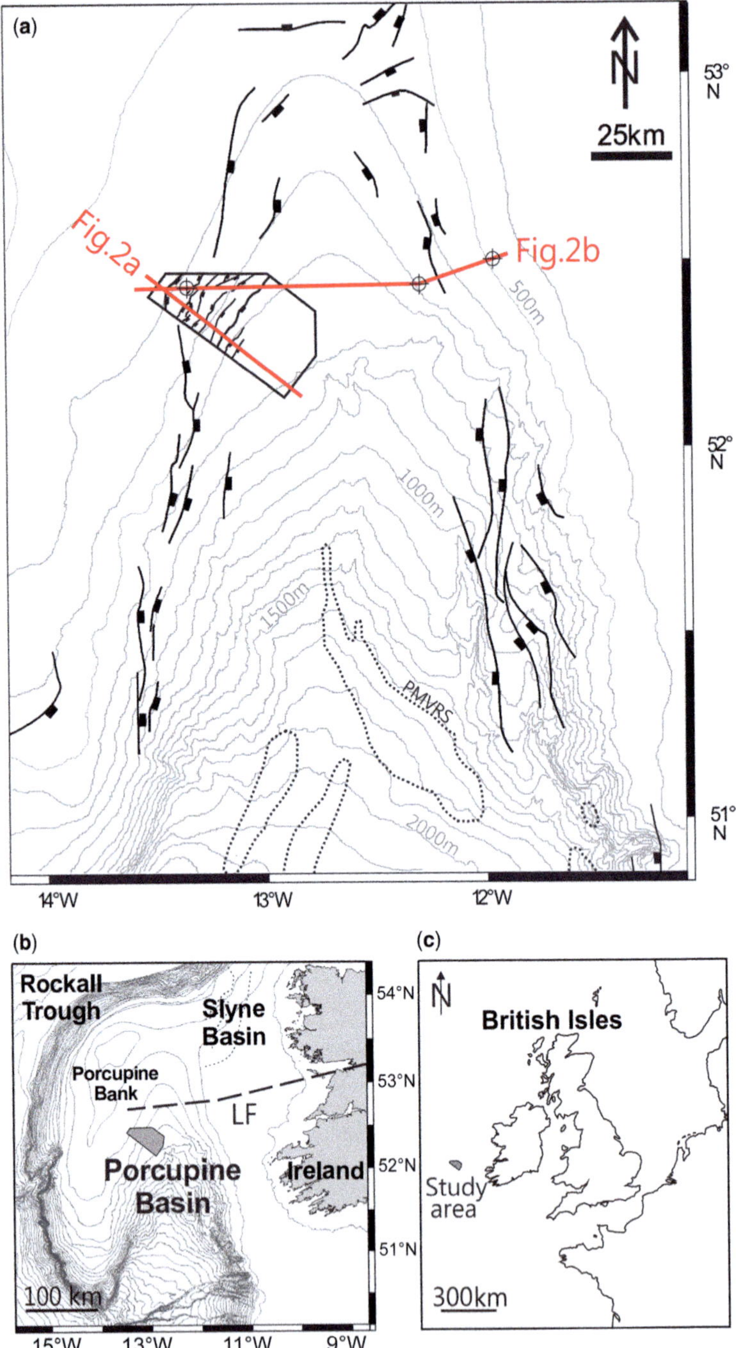

Fig. 1. (**a**) Map showing the present-day bathymetry and structure of the Porcupine Basin, offshore the west of Ireland. Included on this map are Mesozoic basement faults, the areal extent of the 3D seismic data utilized in this study and the location of the two sections in Figure 2. Faults shown include newly interpreted faults within the study area, and faults taken from the work of Tate (1993), Naylor & Anstey (1987) and Naylor *et al.* (2002). Fault dip directions are indicated with blocks on the downthrown hanging-wall side. (**b**) Map showing the position of the basin offshore Ireland. The offshore orientation of the Leck Fault (LF, dashed line) is inferred from Trueblood & Morton (1991). (**c**) Regional map showing the location of the basin relative to the British Isles and NW Europe.

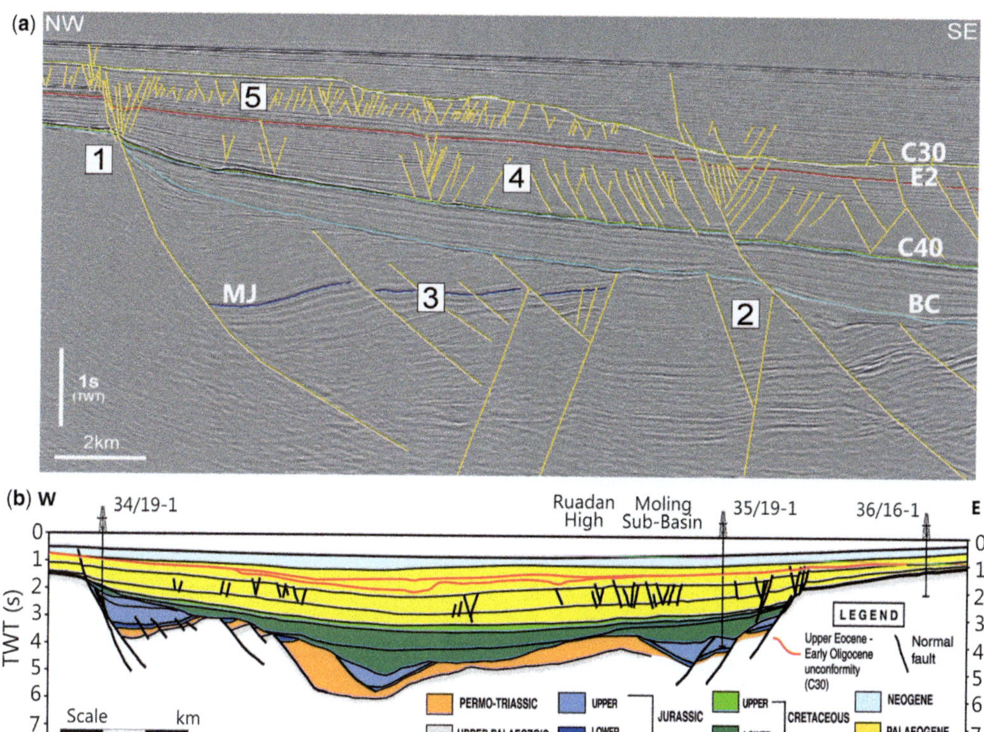

Fig. 2. (**a**) Representative seismic section highlighting the main horizon surfaces and different fault systems within the 3D seismic dataset study area. Key to fault systems: 5, layer-bound polygonal normal fault system, contained within an Upper Eocene package; 4, layer-bound normal fault system, contained within a Paleocene–Eocene package; 3, buried Late Jurassic–Early Cretaceous rift faults, contained within Palaeozoic–Mesozoic basement; 2, reactivated basement fault; 1, reactivated basin-bounding fault. Key to horizons: C30, Late Eocene–Early Oligocene erosional surface; E2, Early–Mid-Eocene; C40, base Cenozoic horizon; BC, base Cretaceous surface; MJ, Mid-Jurassic horizon. (**b**) Simplified interpretation of a typical section across the Porcupine basin (Naylor *et al.* 2002).

to the relatively small displacements (5–100 m) and short fault trace lengths (6–10 km) of these minor Cenozoic faults. Another factor is the poor preservation and varying thickness of the faulted Paleocene–Eocene package, particularly along the basin margins and to the south of the basin, such that there are only a few areas where there are sufficient thicknesses to enable a detailed analysis of these structures and an accurate definition of the timing of faulting. This paper utilizes a high-quality and high-resolution 3D seismic dataset, situated in an area of approximately 25 × 40 km along the NW margin of the Porcupine Basin (Fig. 1a). The dataset benefits from well-defined basement and Cenozoic faults, and lies in a region along the basin margins where there is good preservation and an extensive thickness of the Paleocene–Eocene sediments (up to *c.* 1.5 km thick). The dataset is, therefore, ideally suited for a detailed analysis of the structure and timing of these relatively minor Cenozoic faults. Our analysis highlights the importance of segmentation and twisting of upwards-propagating basement reactivated faults, features we consider to be important generic characteristics of basement fault reactivation, arising from the propagation of faults into weaker cover sequences and the often oblique nature of later extension.

Background

Database

The seismic data primarily used in this work, and defining the main study area of this paper, is from a high-quality 3D seismic reflection dataset (ST3D98) that was acquired by Statoil in 1998. This dataset covers an area of approximately 40 × 25 km (902 km^2 3D lines) along the NW margin of the basin (Fig. 1a). Line spacing was originally 12.5 m, but was resampled to every second line for this study. Other seismic datasets used in this work include two 2D seismic surveys (ST95PO and ST96PO) located partially within, and to the south of, the 3D seismic dataset. Both of these datasets were acquired by Statoil between 1995 and 1996, covering an area of roughly 75 km^2, with a line spacing of approximately 3.5 km. Another seismic reflection survey referred to in this study covers a much larger area of the basin (*c.* 200 × 100 km) and comprises a good-quality regional 2D dataset (MESP81) acquired by Seismic Profilers/Merlin Profilers in 1981.

Seismic quality and resolution of the ST3D39 dataset is good within the relatively shallow, Cenozoic post-rift cover sequence. Faults are clearly resolved within multiple strong reflection events, with interval velocities (see below) and peak frequencies indicating fault throw resolutions down to approximately 5 m and a lateral resolution capable of resolving two fault traces to within about 75–100 m of each other.

Velocity–time to depth conversions for this study were made from local vertical borehole data contained within the study area (BH34/19-1 and 34/15-1) obtained by Shell Ireland in 1978. Average velocities between the main horizons analysed in this study varied only slightly (1.7–2.0 km s^{-1}) and, unless stated otherwise, time–depth conversions were performed using an average velocity of 1.8 km s^{-1}. Age constraints for horizons and important tectonostratigraphic surfaces (e.g. Figs 2 & 3) were derived from published and unpublished studies (well report EM078 Shell Ireland; McDonnell 2001; McDonnell & Shannon 2001; Stoker *et al.* 2001; Haughton *et al.* 2005).

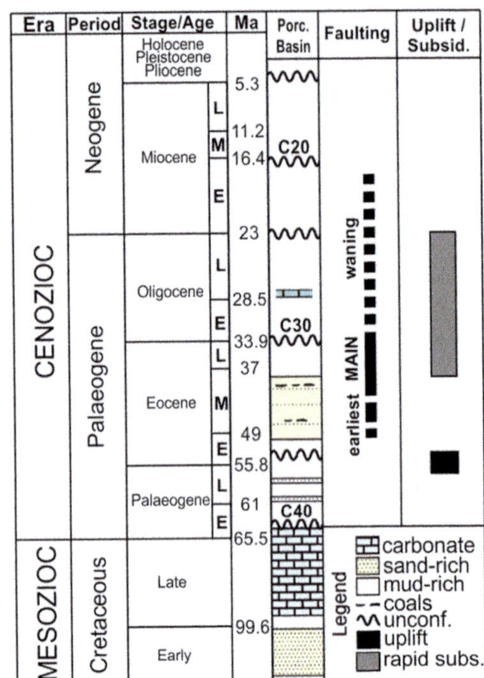

Fig. 3. Stratigraphic column for the Porcupine Basin, including a generalized lithological log (adapted from McDonnell & Shannon 2001) and timing of fault reactivation as discussed in this paper, as well as periods of uplift and increased subsidence as proposed by Tate (1993) and Jones *et al.* (2001).

Basin structure and stratigraphic evolution

The Porcupine Basin is an intra-cratonic rift basin, situated on the continental shelf approximately 200 km west of Ireland (Fig. 1). The earliest inception of the basin in its present north–south orientation was due to a major rifting event arising from a general east–west extension during the Late Jurassic–Early Cretaceous ('Cimmerian' stage). This period of rifting is associated with Atlantic seafloor spreading, which propagated northwards from the central Atlantic region to the south. Jurassic–Cretaceous rift faults generally strike north–south, with maximum fault throws of the order of at least 4 km, with the largest faults typically along the basin-bounding faults. Figure 2b shows a representative cross-section across the basin displaying a symmetrical graben profile. In map view, however, the basin exhibits an asymmetrical basin profile, which narrows to the north (Fig. 1). This geometry has been suggested to be the result of focused strain in the central and southern regions arising from the relative rotation of the eastern and western margins of the basin, possibly combined with a change in the regional stress regime to a more NW–SE extension direction in the north (Tate *et al.* 1993). Stretching during the main rift event ranged from $\beta = 1.1$ in the north to $\beta > 6$ in the south. In the central region of the basin where stretching is believed to have exceeded $\beta = 2$–3, rifting was accompanied by volcanism in the form of the Porcupine Median Volcanic Ridge System. The change in structure along the northern margin, where faults show a more ENE–WSW strike, has previously been attributed to the influence of a prominent basement structure with a similar trend (Trueblood & Morton 1991; Cunningham 2000) (see Fig. 1b). Following the main Jurassic rifting phase, post-rift thermal subsidence of the basin continued through the Cretaceous and Cenozoic with the accumulation of at least 3.5 km of Cenozoic infill. Thermal subsidence is punctuated by a period of rapid differential subsidence and uplift during the Mid-Cenozoic (Fig. 3) (Tate 1993; Jones *et al.* 2001; McDonnell & Shannon 2001), which is of uncertain origin but is broadly coeval with the basement fault reactivation described in this paper. The large Cenozoic stratigraphic thickness within the Porcupine Basin is in stark contrast with the maximum thicknesses of approximately 2 km in the neighbouring Rockall Basin (Shannon *et al.* 1993) and approximately 500 m in the Slyne Basin (Cunningham 2000), respectively. Because the faults and stratigraphic record are so well preserved, providing more accurate geometrical and temporal constraints on faulting, the Porcupine Basin thus represents an ideal location for studying the geometry and growth of Cenozoic faulting.

Fault systems

A representative seismic section from the 3D seismic dataset, shown in Figure 2a, highlights a number of fault systems present within this study area. Four separate fault systems can be recognized, including two layer-bound normal fault systems at two different stratigraphic levels (fault systems 4 and 5), as well as buried Jurassic rift faults (fault system 3). Fault system 1 is of primary interest for this paper and consists of a population of Cenozoic normal faults linking into a basin-bounding fault along the western margin of the dataset (Figs 2 & 4). Another cluster of Cenozoic normal faults further to the east is located over, and partially links into, a basement fault (fault system 2 in Fig. 2a). Analysis of this basement fault (Worthington 2006) suggest a similar displacement history to that of the basin-bounding fault discussed in this paper, but its geometry is not described here because it is less well defined and is complicated by interaction with a layer-bound fault system (labelled 4 in Fig. 2a).

Fault system 4 is a highly segmented normal fault system that is layer-bound within a Paleocene–Eocene sequence. This system comprises an array of sub-parallel north–south-striking faults, which can be either east- or west-dipping and have throw values ranging up to 60 m. This fault system is widespread across the basin, but is mainly developed along the flanks of the basin (Fig. 2). Although outside the focus of this paper, our study of this fault system supports a gravitational mechanism for its formation, including slope and density instabilities, and appears to be coeval with both a period of late Eocene–Oligocene rapid differential subsidence and the main phase of basement fault reactivation examined here. A Late Eocene layer-bound fault system (fault system 5 in Fig. 2a) has a polygonal fault pattern and is contained within a relatively thin Upper Eocene mud-prone package (at least 500 m thick). These polygonal faults are only observed where seismic quality and resolution is high, and where the Upper Eocene package is preserved. Associated faults predate and are truncated by the C30 surface (regional unconformity). Whilst the origin of this faulting is uncertain, features of this system are comparable to layer-bound polygonal fault systems associated with density instabilities of overpressured and low-density mud-prone units (Clausen *et al.* 1999; Watterson *et al.* 2000; Nicol *et al.* 2003). Although this layer-bound fault system is transected by the reactivated basin-bounding fault described here, sequence growth associated with both suggests that much of their displacement accumulation was coeval (Fig. 5) (Bailey *et al.* 2003).

The basin-bounding fault

Basement expression of the fault

The basement expression of this fault is a relatively well-defined, east-dipping basin-bounding fault, the formation of which is attributed to the main Late Jurassic rifting event (Fig. 2). The total throw is at least 4.4 km (2.5 s TWT (two-way time) and assuming a velocity of *c.* 3.5 km s^{-1}), an estimate that is constrained by a Permo-Triassic horizon in the hanging wall and the upper surface of an eroded footwall sequence of probable Late Palaeozoic age (Tate 1993). A characteristic feature of this structure within the study area is a well-defined change in strike from roughly north–south in the south to NE–SW in the north, resembling a fault 'bend' (Figs 1a & 4).

Similar basin-bounding faults with comparable structure and throw are recognized along both the eastern and western margins of the Porcupine Basin. The present definition of the basin basement structure for the central and northern regions of the basin is shown in Figure 1, a map that is mainly derived from previous work (Naylor & Anstey 1987; Tate 1993; Naylor *et al.* 2002) but is also supplemented by the main basement faults mapped in this study. To the area north and south of the study area, comparable basin-bounding faults have been interpreted along strike. 2D seismic data from the area to the south of the 3D seismic dataset show that a basin-bounding fault, with comparable scale, structure and geometry, continues for at least 50 km south with a roughly north–south trend. This structure also exhibits similar normal faulting localized in the overlying post-rift Cenozoic cover along its length. To the north, Naylor & Anstey (1987) and Tate (1993) interpreted the existence of a basement fault striking roughly north–south just north and slightly to the east of the basin-bounding fault defined within this 3D seismic dataset area (Fig. 1a). From the present definition of the basin structure, it appears likely that this fault is a continuation of the basin-bounding fault defined in the 3D seismic dataset. This suggestion is reinforced by the fact that north–south-trending faults along the

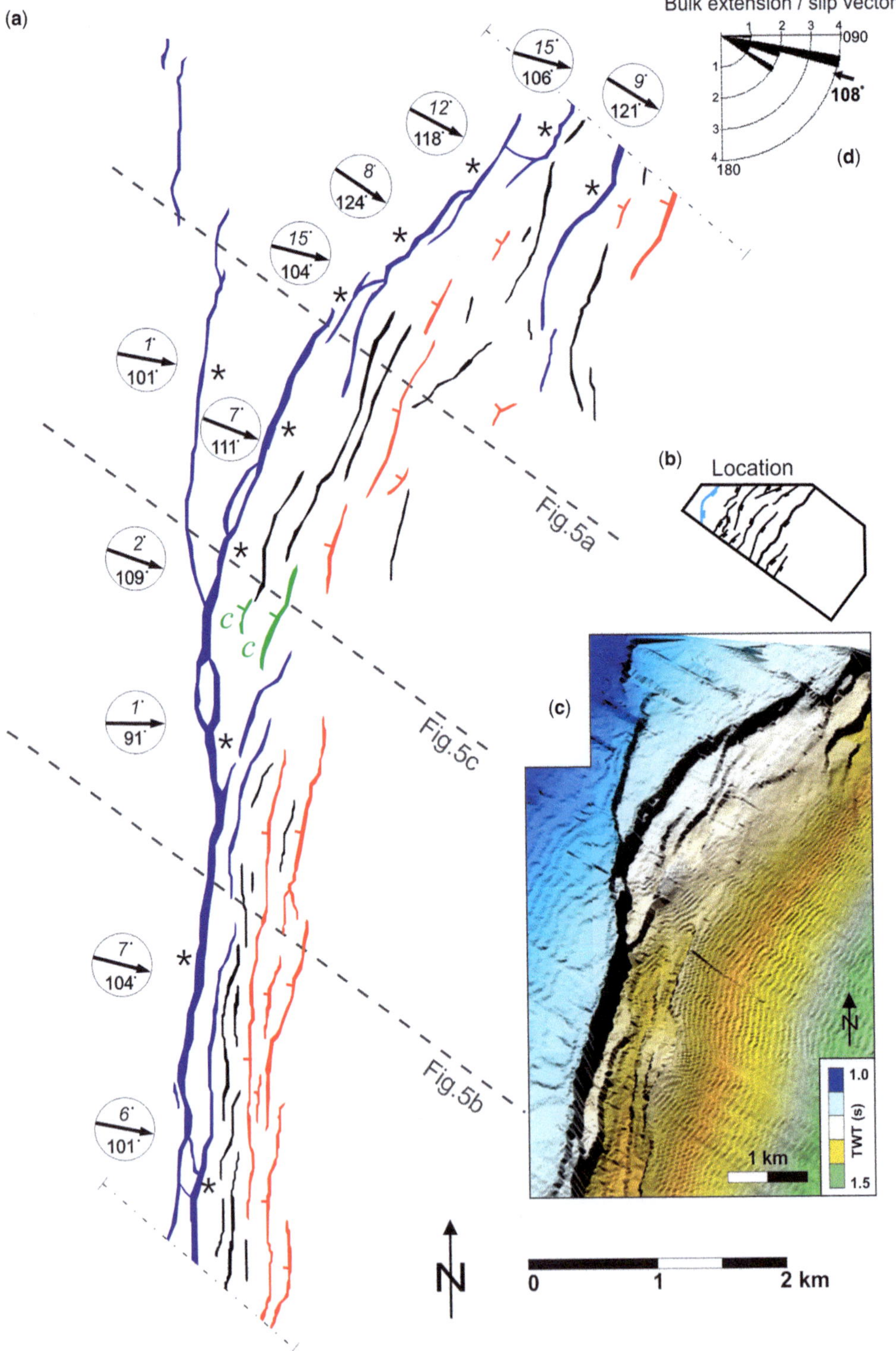

Fig. 4. (**a**) Fault map of the Cenozoic normal fault system associated with the reactivation of the basin-bounding fault. Faults are mapped along an Early–Mid Eocene horizon (E2, as shown in Figs 2 & 5). Key: blue faults, main

western margin are clearly right-stepping, a geometry that partly accommodates the curved and asymmetrical map-view geometry (along an east–west axis) and the northwards narrowing of the Porcupine Basin (Fig. 4). The systematic stepping of faults suggests that the bend towards a NE–SW trend seen along the basin-bounding fault in the 3D seismic dataset (Fig. 1a) may reflect the existence of a right-stepping relay zone between overlapping faults that has subsequently been breached. Owing to the large displacement on these basement faults and their relatively close proximity, it is possible that the basin-bounding fault within this study area, and which was mapped further to the north by Naylor & Anstey (1987) and Tate (1993), now represents a single through-going fault surface.

Cenozoic faulting

In map view, the main synthetic faults of the Cenozoic normal fault system resemble that of a segmented fault array with a number of unconnected segments, splays and localized through-going faults, as shown in a fault map along a well-defined Eocene horizon (E2) in Figure 4. In cross-section, however, these faults are upwards extensions of a single reactivated basement fault (Fig. 5). Across the length of this system, there is clear evidence of hard-linkage of the main synthetic faults to the basement fault, as indicated in Figure 2. The geometry and distribution of Cenozoic faults indicate that the underlying basement structure has exercised a significant structural control on the development of faults within the overlying Cenozoic post-rift cover sequence.

Timing of faulting

Establishing the timing of normal fault growth using seismic data is typically achieved from the identification of sequence thickening in the hanging walls of synsedimentary faults, particularly in circumstances where the sedimentation rates outstrip fault displacement rates (Childs *et al.* 2003). Alternatively, or in conjunction, quantitative methods of fault analysis can also be used, such as vertical displacement (i.e. throw) profiles or 'growth curves', the form of which can reflect the timing and history of the main periods of faulting (e.g. Morley 1999; Nicol *et al.* 2005; Giba *et al.* 2012). These profiles plot displacement values per mapped horizon against depth or, where possible, age. A rapid decrease in displacement can suggest an active period of faulting, whilst a profile that maintains a constant displacement value over time/depth may represent a period of quiescence on a long-lived fault or on a pre-faulting level of the structure. Where there is a marked decrease from a constant displacement profile, then the onset of faulting is derived from the point at which the decrease begins (Fig. 6a). The rate of decrease in displacement combines both the fault displacement and the sedimentation rates, but, in circumstances where the ages of associated horizons are well constrained, it can be possible to define fault displacement rates (e.g. Meyer *et al.* 2002; Nicol *et al.* 2005).

Figure 7a shows the vertical displacement profile for the basin-bounding fault structure extending from basement levels through the overlying Cenozoic cover sequence to the surface. Cumulative throw profiles through the Cenozoic sequence, shown in Figure 7b, were derived from the displacements of 15 well-constrained horizons, measured on each of 27 line samples orientated parallel to the general bulk extension direction (*c.* 108°, as discussed later) and with a lateral spacing of roughly 300 m. All of the profiles display a roughly constant displacement of approximately 120 m below the Mid Eocene horizon (E4), with an overlying marked decrease in displacement within the Mid–Late Eocene. The point of decrease records the beginnings of the main period of fault reactivation, which extends up to the Late Eocene–Early Oligocene (C30) horizon. A roughly constant, level profile with very low displacement (*c.* 10 m) at shallower stratigraphic levels suggests a period of quiescence that may have been followed by equivalent minor displacements around Mid-Miocene times. Although the profile described above is very similar to that expected for a fault reactivation model of faulting (Fig. 6a), we briefly describe below why the profile is inconsistent with the compactional origin suggested by earlier workers (Tate 1993; McCann *et al.* 1995*a*, *b*).

Fig. 4. (*Continued*) synthetic faults, commonly hard-linked to the basement fault (dip east); red faults, hanging-wall antithetic faults (dip west); black faults, synthetic faults in the hanging wall; green faults, hanging-wall antithetic faults associated with conjugate structures. Bulk extension direction/slip vectors are shown along the fault system using a simple graphical model constructed by McCoss (1986) and illustrated in Figure 8a. The McCoss (1986) model results calculated at key sites along the fault array (marked with an asterisk) display bulk extension direction vectors (an arrow and a bearing). The angle in the upper portion of these models is ω, which is the angle between the strike of the initial en échelon segments and the basement fault strike. (**b**) Location of the fault (blue) within the Jurassic fault map and 3D seismic survey as presented in Figure 1a. (**c**) 3D shaded surface map (TWT) of the Early–Mid Eocene horizon E2 (light source from the NW). (**d**) Rose diagram of the bulk displacement vectors as calculated using the McCoss (1986) method (Fig. 8a) showing an average extension to 108°.

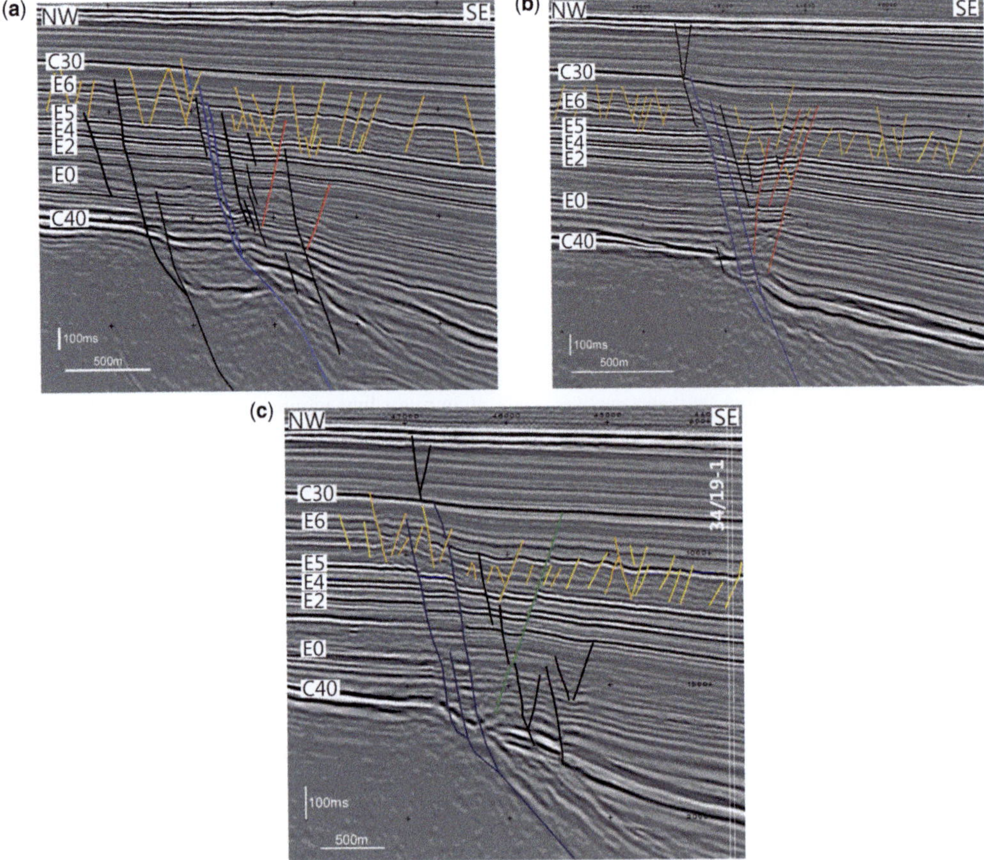

Fig. 5. Series of interpreted seismic sections (see Fig. 4a for locations). (**a**) Section located across the NE–SW-trending region of the fault array. (**b**) Section located across the north–south-trending region of the fault array. (**c**) Section taken between these two regions. Displacement on the main antithetic fault is attributed to displacement on the synthetic faults that kink at the basement–cover interface. Key: blue faults, main synthetic faults; black faults, other synthetic faults; red faults, hanging-wall antithetic faults; orange faults, Late Eocene layer-bound polygonal fault system (fault system 5 in Fig. 2a). Horizon labels as in Figure 2a, with the addition of Eocene E6, E5, E4 and E0.

A possible model for the generation of faults within a post-rift cover sequence is by differential compaction of the sequence across a prominent fault-related basement high at the basin margin. Differential compaction effects would be expected to begin for sequences that show differential thickness variations across the basement high. If faulting is associated with this compaction, a gradual upwards decrease in cumulative fault throw would be observed, with maximum displacement found at lower levels reflecting not only the effects of differential compaction at this level, but also that of the overlying sequences. Compaction-driven faults would therefore resemble that of the model profile shown in Figure 6b. Furthermore, if faulting were to have arisen through differential compaction, then it is expected that thickness changes would also be observed directly across the fault. For the basement fault in this study, thickness variations and down-to-the basin flexing of associated horizons occur on a much broader wavelength, typical of compaction-related hinging of individual horizons above an underlying basement high, with no evidence of early fault-related sequence thickening/growth. Instead, across-fault sequence growth is absent for the first 450 m of the post-rift cover sequence, a feature that is reflected in the constant, rather than upwards-decreasing, displacements observed on vertical displacement profiles. In the absence of lithological grounds for temporarily delaying compactional effects, there is no convincing explanation for why differential compaction

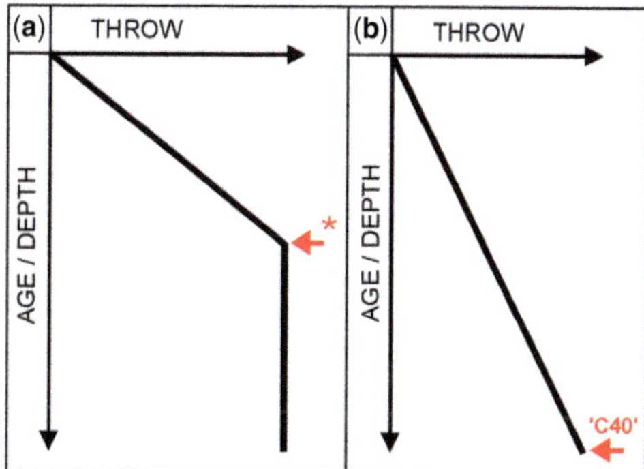

Fig. 6. Model vertical displacement profiles. (**a**) Profile showing a sharp change in throw (marked with an asterisk) from a constant maximum throw, a feature that indicates the onset and timing of faulting/reactivation. (**b**) A predicted profile that would be expected if faulting arose from differential compaction of the Cenozoic cover sequence. The onset of faulting in this case would occur at an early stage of deposition, providing an overall vertical displacement profile with a maximum throw near the basement–cover interface (approximate to the C30 surface, as indicated) that gradually decreases upwards.

effects would only begin to take effect at such a late stage in the post-rift sequence. The onset of faulting relatively late in the post-rift sequence, after the deposition of at least 450 m of Paleocene–Eocene sediments (excluding compaction), and the lack of sequence thickening directly across the fault itself within the same interval (Fig. 4), together with the clear evidence of hard-linkage of the Cenozoic faults into an earlier basement structure, strongly indicates that faulting must have been caused by some other factor, such as later tectonic reactivation. A well-defined period of reactivation of an earlier fault would provide a fault within the post-rift sequence with an approximately flat/constant displacement profile in the post-rift package up to the horizon that immediately predates the reactivation, with a decrease only observed on horizons deposited during the reactivation phase. Since the basin-bounding fault shows exactly this type of displacement of displacement profile, tectonic reactivation is the most likely origin of this structure.

The shape of the displacement profile also suggests that the relatively rapid period of fault movement associated with basement reactivation ceased before the Late Eocene–Early Oligocene C30 horizon. This horizon is characterized by very minor displacements of less than approximately 20 m, which could reflect either continued reactivation during the Neogene or fault-related differential compaction.

Fault pattern

A detailed map of the Cenozoic fault system highlights a well-defined strike change from north–south in the southern portion of the system to NE–SW in the north (Fig. 4). An exception to the NE–SW trend is a minor north–south-striking footwall splay in the NW side of the study area. The main synthetic and antithetic fault segments display characteristic en échelon faults or bends (in map view). Faults show systematic left- or right-stepping in the southern and northern portions of the fault system, respectively, although the latter are more common and better developed. In the north–south-trending portion of the system there is only a slight variation in the strikes of both the connected and unconnected segments (*c.* 008°–014°). Over a distance of approximately 3 km, there is a gradual and systematic clockwise rotation in strike from south to north to provide the NE–SW trending portion of the fault system with a broader range in fault strikes (generally 005°–040°). The footwall splay also shows evidence of linkage of a similarly north–south-trending basement structure, although this basement structure is not so well defined and is discontinuous further to the north (Fig. 5a).

Examples of segmented fault systems in map view often exhibit an en échelon geometry, where the individual segments can be systematically left- or right-stepping, or display no systematic stepping direction. Systematic stepping is commonly observed on strike-slip and oblique-slip faults (Mandl 1988, 2000; Sylvester 1988; McClay & White 1995; Peacock & Sanderson 1995*a*, *b*; Richard *et al.* 1995), whereas unsystematic stepping is more common in dip-slip fault systems (e.g. Peacock & Sanderson 1991). Analogue sandbox modelling by Richard *et al.* (1995) generated systematically stepping normal fault arrays, which are comparable to fault patterns of this study. Their work examined geometries that resulted from the reactivation of a fixed basement structure (dipping planar structure) under various plane extensions, transtension through to pure strike-slip conditions (Fig. 8c). This and other analogue studies of en échelon geometries have shown that there is a simple relationship between the amount of strike-slip on a basement structure and the angle between the strike of the en échelon segments and the strike of the basement fault, sometimes referred to as the zone boundary (e.g. Richard *et al.* 1995). However, a simple quantifiable expression of this relationship has not been demonstrated for natural systems. There are a variety of factors why natural examples of en échelon geometries may lead to different estimations of the ratio of strike-slip to dip-slip: (i) differences in the nature of linkage to the basement structure (i.e. 3D

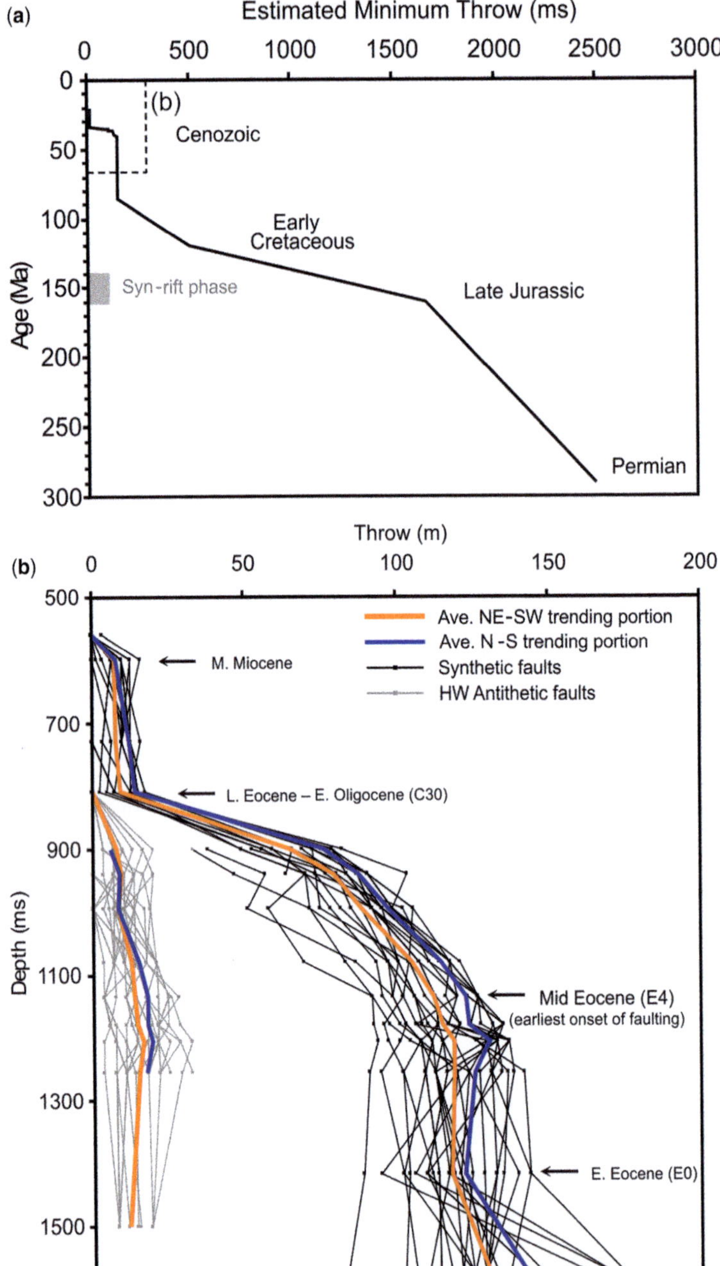

Fig. 7. (**a**) Estimated vertical displacement profile for the basin-bounding fault contained within the study area (approximate to the section line in Fig. 5b) calculated from the aggregate throw per horizon (i.e. the throws of all related synthetic faults are aggregated per horizon). Pre-Cenozoic values are minimum estimates based on offsets of horizons in the hanging wall to the top of the basement in the footwall (the footwall is believed to be Carboniferous in age: Tate 1993). The dashed box shows the location of (b). The distinct rapid drop in displacement in the Eocene marks the onset of reactivation at about 40 Ma. (**b**) Vertical displacement profile for the main Cenozoic synthetic fault system (black lines) and the hanging-wall antithetic accommodation faults (grey lines). Each line represents a single profile taken from 27 sample lines recorded along-strike of the fault system. The profiles that are highlighted in bold represent a mean of the profiles within the north–south- (dark blue) and NE–SW-trending (orange) portions of the fault system only.

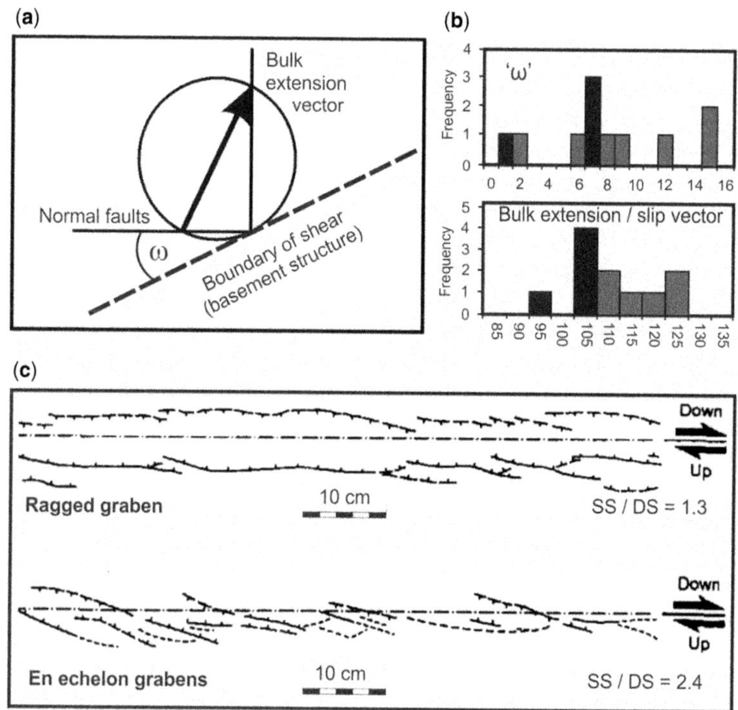

Fig. 8. (**a**) The McCoss (1986) method used to estimate bulk extension direction along an obliquely reactivated structure (see the main text). This method was applied on the Cenozoic fault array described in this paper and presented in Figure 4a, d. (**b**) Frequency plots of bulk extension vector and ω. Columns shaded in grey reflect faults from the NW–SE-trending region only. (**c**) Comparable fault patterns produced from analogue modelling by Richard *et al.* (1995) under two different conditions/ratios of strike slip (SS) and dip-slip (DS).

structure); (ii) the distance from the zone boundary; and (iii) the varying mechanical properties of the host rock (e.g. lithology causing varying values of strength/shear modulus, and multilayered sequences causing fault refraction). Despite these factors, the relationship that en échelon segments are rotated more with an increasing strike-slip component on the underlying basement structure provides support for the application of a simple method for estimating the bulk displacement vector on the basement fault from the overlying Cenozoic fault pattern.

A simple graphical model developed by McCoss (1986), based on the transpression/transtension model by Sanderson & Marchini (1984), can be used to establish the bulk extension vector/slip vector of the basement fault, assuming the Cenozoic faults are pure dip-slip and arise from the reactivation of the underlying basement fault. The construction and application of this model is shown in Figure 8a. Conditions required for the proper application of the McCoss (1986) model are small strains and minimal rotation, which are consistent with the reactivated basin-bounding fault described in this paper. The angle between the en échelon segments and the zone boundary is represented by ω, and can vary between 0°, representing plane extension of the zone boundary, and 90° for pure strike slip. If results show a regular bulk extension direction along the length of this fault system, this will suggest that the en échelon geometry system is consistent with simple reactivation of this basement fault geometry under a single extensional motion (be it transtensional, oblique or extensional).

Results of the McCoss (1986) method are shown as bulk displacement (or extension) vectors in Figure 4a. The locations of initial en échelon segments can easily be identified, even if they are often linked at breached relay zones: the mapping of breaching faults is, in this case, helped by the high resolution of the seismic data. The zone boundary, taken to be the local strike of the basement fault, is best represented along the base Cenozoic C40 surface. Strike values from the en echelon segments were only taken from synthetic faults that are observed to be clearly hard-linked to the basement fault. Hanging-wall faults are not included as their geometry is believed to be more influenced by the main synthetic faults and is, therefore, only indirectly linked to the basement. The result of this analysis is a general unimodal distribution of the bulk extension vector data (see the rose diagram and results charts in Figs 4d & 8b), which suggests that, despite the varying strikes of the faults along this system, the systematic stepping across the whole system is consistent with a single slip vector of the basement fault to the ESE (mean *c.* 108°). Results suggest a dominant dip-slip extension with a minor component of strike-slip, with a fault pattern comparable to that of the sandbox models of Richard *et al.* (1995) arising from similar ratios of strike-slip and dip-slip. The amount of strike-slip may be greater along the bend and NE–SW-trending portion of the system, as reflected in the higher values of ω (2°–16°) than for that of the north–south-trending portion ($\omega = 1°–7°$). This would be expected where there is a single extension direction and a more obliquely orientated portion of the basement fault. With continued displacement, originally stepping segments of this system have become linked to provide a single fault trace that is sub-parallel to that of the underlying basement fault: the process of segment linkage is discussed in the next subsection.

A further indicator that the extension direction calculated using the McCoss (1986) construction is appropriate and that fault-related extension is roughly dip-slip over much of the fault array is that the fault extension and dip directions (i.e. ESE to east) are perpendicular to the basin margins during the Eocene, as reflected in the structure contour patterns of the Eocene horizon (Fig. 4c). With east–west extensional reactivation of the basement fault, it is also possible that early fault development and upwards propagation (at low strain) may be structurally influenced by the geometry of the overlying cover sequence, with individual fault surfaces steepening and rotating away from the north–south- to NE–SW-trending basement structure towards the strike of the cover sequence (i.e. NNE–SSW). It is uncertain how much of an effect this mechanism could have on the geometry of Cenozoic faults, but only a small influence could result in the required systematic stepping geometry of fault segments. Future analogue modelling may help to resolve this issue. Nevertheless, whichever mechanism(s) apply, the geometry of the Cenozoic fault array suggests a general east–west extension direction, a direction that is consistent with other roughly north–south-striking Cenozoic faults elsewhere within the Porcupine Basin and also in the neighbouring Slyne Basin (e.g. McCann *et al.* 1995*a*, *b*; Cunningham 2000).

Displacement variations along fault array

In this subsection, we briefly investigate how the fault system has evolved by analysing associated displacement variations along the length of the system. Along-strike displacement profiles of the main synthetic faults, constructed for the youngest pre-faulted horizon (E2: Early–Mid Eocene), are shown in Figure 9. It is clear from the cumulative displacement profile that displacement is transferred between a number of initially overlapping and individual fault segments. Initial component segments can be identified from the recognition of a variety of features (see Childs *et al.* 1995): (i) partially intact or individual fault segments in map view that show regular displacement profiles (i.e. showing a decreasing displacement from a central displacement maximum); (ii) identification of typical fault-linkage structures (i.e. breached relay zones at fault overlaps, which may be seen as tight, sharp bends with abandoned splays); and (iii) sites of apparent displacement 'minima' on along-strike displacement profiles of single fault traces (Ellis & Dunlap 1988; Walsh & Watterson 1989; Peacock & Sanderson 1991, 1994).

Where adjacent fault segments overlap there may be complementary displacement variations, so that displacement is transferred between them and conserved along their combined length. The region where these segments overlap and where displacement is transferred are regarded as 'relay zones', and the pair of overlapping segments are referred to as showing 'geometric coherence' (Walsh & Watterson 1991; Childs *et al.* 1995). Geometric coherence is interpreted to indicate mechanical linkage during fault growth or 'kinematic coherence'. Overlapping kinematically coherent faults in map view are bounded by relay structures, such as those described by Larsen (1988) and Peacock & Sanderson (1994). Continued displacement ultimately leads to failure of the ramp, with the development of a through-going fault. Failure of the ramp is accommodated by 'breaching faults', which can extend from either the footwall fault and/or the hanging-wall fault. 'Double breaching' occurs where there are at least two breaching faults that connect the two overlapping fault segments, resulting in a fault-bound lens. Examples of relay zones along the main synthetic fault array are highlighted in Figure 9b. In cross-section, these faults are hard-linked at depth to a single surface (Fig. 5) along subhorizontal branch lines, as shown in Figure 10.

The foregoing descriptions highlight the importance of displacement transfer along the main fault system. The efficiency of accommodation of lateral displacement transfer can be tested by generating along-strike aggregate throw profiles for the main

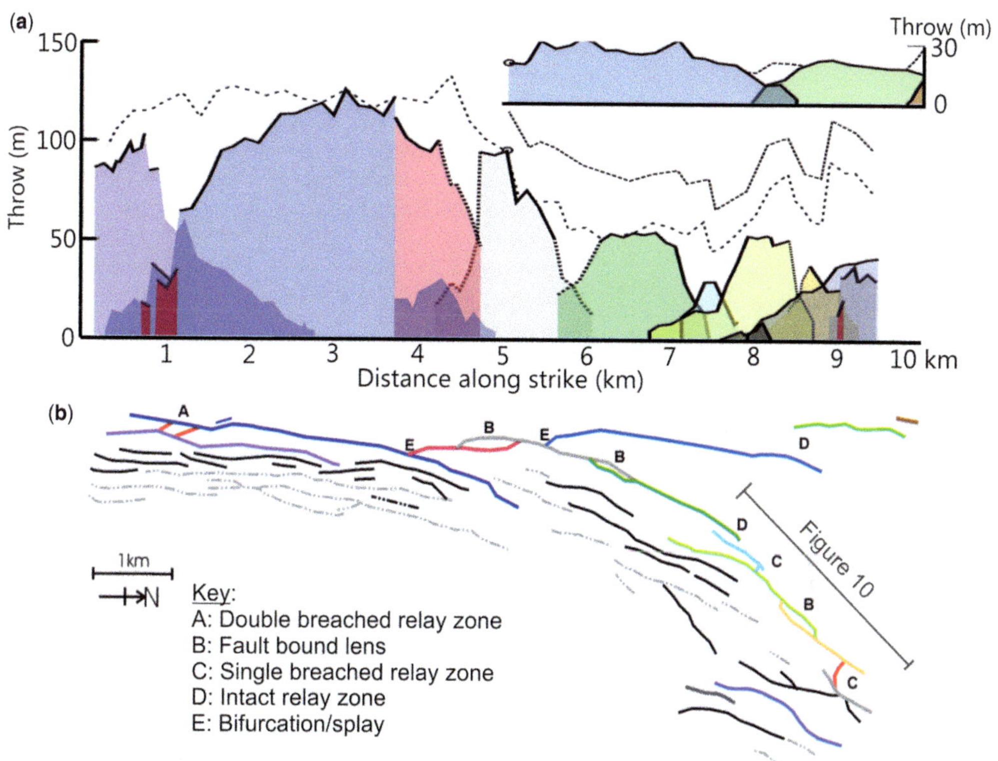

Fig. 9. (**a**) Along-strike throw profiles for the main synthetic faults taken along the Early-Mid Eocene horizon (E2). The through-going fault throw profile is shown as the thick single line and a dotted line where the fault bifurcates in map view (forming splays) and, in places, rejoins. The dashed lines represent aggregate throw profiles incorporating these splays. The closer dashed line incorporates the footwall splay faults, which are shown in the inset (the open circle represents the site where the footwall splay is laterally linked to the array of faults in the main plot). Each initial component segment and splay is coloured individually and these colours are consistent with the fault map in (b). (**b**) Fault map along the same horizon colour coded to highlight the component segments and splays of the system. Thin black lines are hanging-wall synthetic faults. Grey dotted–dashed lines are hanging-wall antithetic faults.

synthetic faults, footwall splay faults, hanging-wall synthetic faults and hanging-wall antithetic faults (Fig. 11). These clearly show that displacement 'minima' along the main synthetic fault aggregate throw profile coincide with complementary displacement 'maxima' in the hanging-wall synthetic faults (e.g. arrow B in Fig. 11). Where the aggregate throw profile includes both the main synthetic faults and the hanging-wall synthetic faults of this system, the pattern appears more uniform and systematic (also observed at other structural levels, not shown here). It is clear from this that the hanging-wall synthetic faults form a geometrically coherent system with the main synthetic faults. Owing to the lack of clear 'hard-linkage' in three dimensions between the hanging-wall synthetic faults and the main synthetic faults (e.g. Fig. 5), geometrical coherency is likely to be accommodated by 'soft-linkage'. Representative vertical displacement profiles for this system, such as shown in Figure 11, therefore require the aggregation of both the hanging-wall synthetic faults and the main synthetic faults.

Displacements accommodated by hanging-wall antithetics

There are a number of other obvious features in the along-strike aggregate throw profile in Figure 11. Firstly, the aggregate throw profile for hanging-wall antithetic faults does not show displacement 'maxima' and 'minima', which are complementary to those of the aggregate throw profile for synthetic faults, suggesting that these different components do not form a geometrically coherent system. As a consequence, incorporation of the throws associated

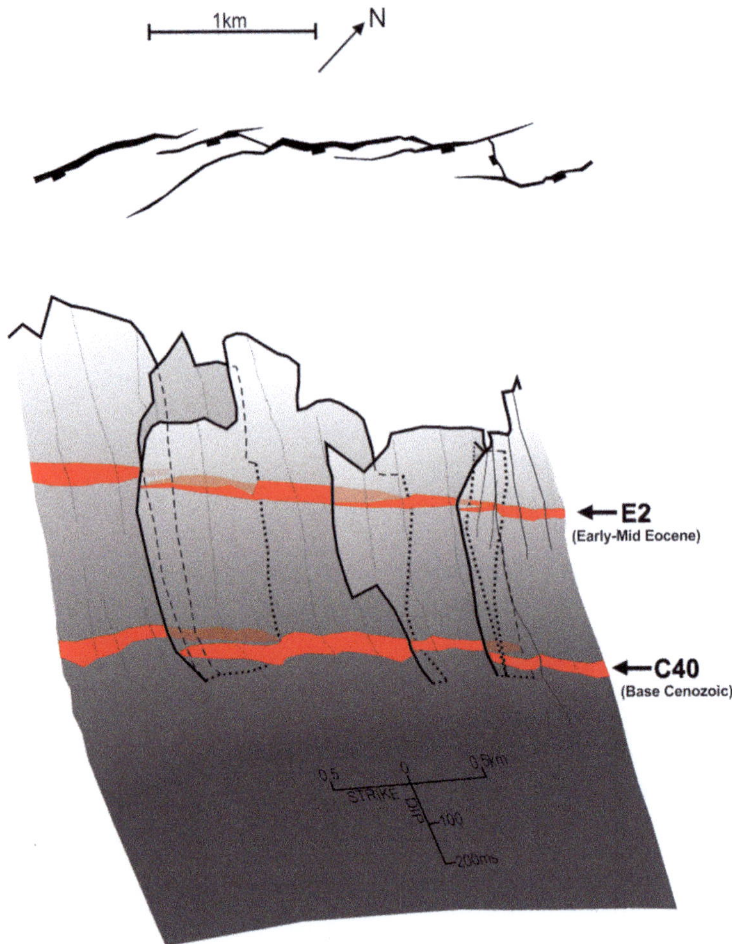

Fig. 10. 3D fault map of a portion of the main synthetic fault system (of the location shown in Fig. 9). Bold black lines indicate fault tips, dashed lines indicate the trace of fault tips behind fault surfaces in the foreground, while bold dotted lines show branch lines (either subvertical or subhorizontal) where faults link. Offsets of the E2 and C40 surfaces are shown in red, with thicker patches indicating greater offsets. Grey shading of the fault surface is purely cosmetic and is not reflective of any fault attribute such as depth or displacement. The fault map at the top of the figure shows a partially segmented and unconnected fault array in map view, although all segments are connected to a single fault surface at depth (below the C40 base Cenozoic surface) along the reactivated Mesozoic basement fault.

with the hanging-wall antithetic faults provides aggregate throw profiles that are more irregular, at whatever level the structure is examined. Despite this lack of coherence between synthetic and antithetic faults, there is, nevertheless, a very clear sympathetic relationship between their displacements, such that displacement highs on synthetic faults are matched by highs on associated antithetic faults (Fig. 11). Evidently, the hanging-wall antithetic faults do not accommodate primary displacements of the basement reactivation but, instead, are responsible for other deformations. The origin of these antithetic faults is returned to later, following a brief consideration of evidence for the relative timing of both types of fault.

The aggregate throw profiles in Figure 7 include those for hanging-wall antithetic faults. The distribution of displacement is comparable to that of the synthetic faults, with a gradual upwards decrease in displacement from a maximum throw at around the 1200 ms level. Unfortunately, measurements of throw at deeper levels on these structures (below *c.* 1250 ms depth) are subject to error, mainly because of uncertain horizon correlations that lead

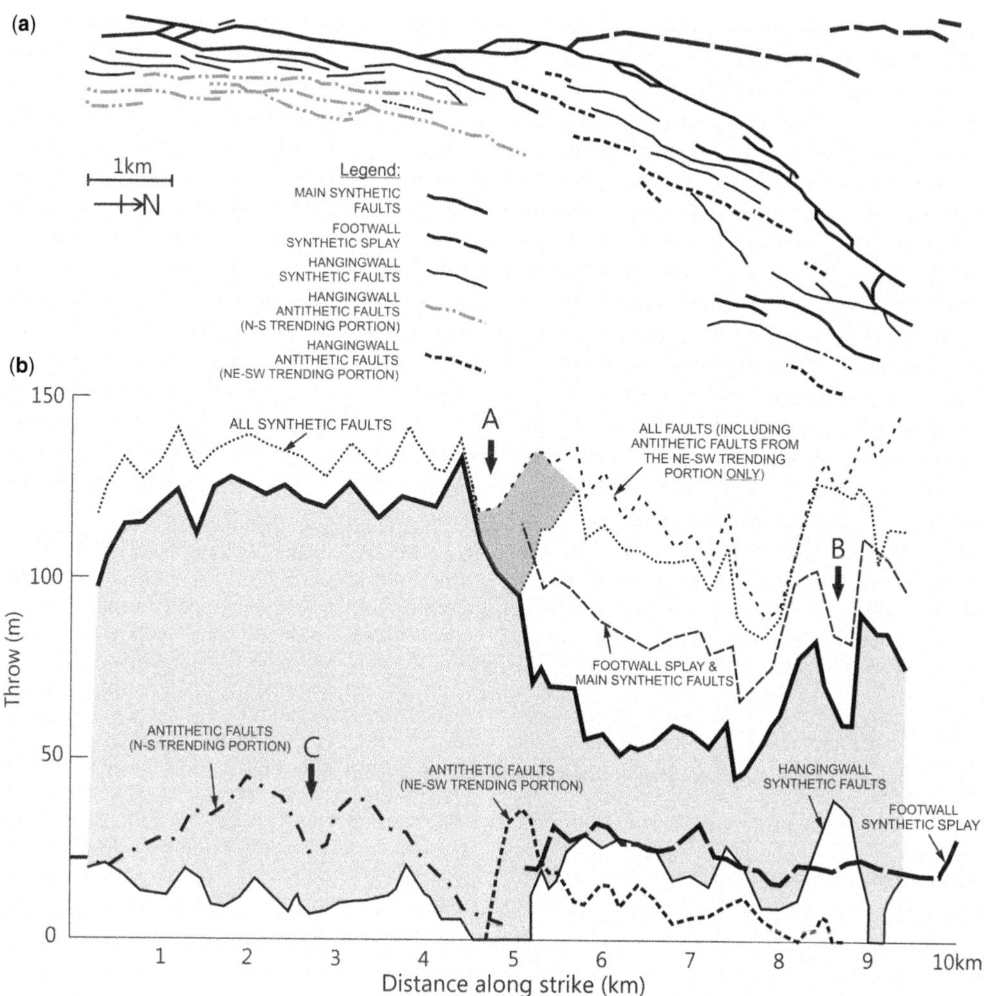

Fig. 11. Fault map (**a**) and along-strike aggregate throw profiles (**b**) for the Cenozoic normal fault system, taken along the Early–Mid Eocene horizon (E2). The space between the main synthetic faults and the hanging-wall synthetic faults is lightly shaded to show the complementary displacement variations between these two fault sets. Arrow A indicates an area along the fault system where a couple of antithetic faults in the hanging wall form part of a conjugate structure. The shaded area between cumulative throw profiles (darker shading) shows the complementary displacement variations between antithetic and synthetic faults in this area (i.e. geometric coherence). Arrow B highlights a prominent complementary displacement variation between the main synthetic faults and synthetic faults in the hanging wall, reflected by the relatively uniform cumulative profile for all synthetic faults across this area. Arrow C shows a prominent displacement 'minimum' within the antithetic cumulative throw profile, which is attributed to increased segment linkage in this area. See the text for further discussion.

to incomplete displacement definition. Despite the areas of missing data, the overall displacement distribution shows that maximum displacements occur at the same structural level as the synthetic faults and that there is a relatively level displacement profile from this maximum down to the intersection, or branch point, with the main synthetic fault (at roughly 1450 ms). The upper fault tip-lines of antithetics commonly terminate within the Upper Eocene just beneath the Late Eocene–Early Oligocene C30 surface (Figs 2 & 5). This suggests that the timing of antithetic faulting was broadly contemporaneous with the main synthetic fault system, but may have become inactive at a slightly earlier time.

The principal feature of the displacement variations of antithetic faults, which must be accounted for by any model used for their formation, is the fact that they display a sympathetic relationship with the displacements on synthetic faults (Fig. 11). Increases in displacement on synthetic faults are mirrored by increases in displacement on antithetic faults. They do not, therefore, accommodate a proportion of the basement reactivation displacements but, instead, must be formed by some cause other than displacement partitioning of basement displacements onto opposed dipping faults within the cover. Examination of cross-sectional geometries of both the basement fault and associated deformed horizons suggests that the hanging-wall antithetics intersect the synthetic faults at structural levels that are characterized by a kink, and an associated shallowing downwards, in the synthetic fault trace at the basement–Cenozoic interface (Figs 2 & 5). This raises the possibility that their formation may reflect the accommodation of hanging-wall deformation arising from displacement along an irregular fault surface, a phenomenon that has been accepted since Hamblin (1965) first recognized the importance of hanging-wall rollover. The notion that antithetic faults can accommodate a more discontinuous form of hanging-wall rollover was also first identified by Hamblin (1965). Since his pioneering article, it is now firmly established that deformation of the hanging wall within a normal fault system, leading to the development of antithetic adjustment faults, can be caused by changes in fault-plane geometry, such as a 'notch' or a curved-listric fault plane (see Mandl 1988). This type of hanging-wall deformation is commonly observed in natural and analogue datasets, in association with 'rollover anticlines' on normal faults (Gibbs 1984; Roberts & Yielding 1994) where hanging-wall strains are a direct consequence of rollover (Kerr & White 1992). Hanging-wall faults have been shown to localize at the site of a 'curve' (e.g. listric faults) in the gliding path of the main fault (e.g. McClay 1990*a*, *b*) or at notable 'kinks' or steps (Mandl 1988; Imber *et al.* 2003). The antithetic faults of the fault system discussed in this paper also localize at a well-defined kink in the bounding fault surface (Figs 2 & 5): apparent fault kinks can arise from velocity increases across interfaces but, in this case, existing constraints indicate that velocity changes are not capable of generating a kink. Taken together, the antithetic faults produce a rather subdued hanging-wall rollover type geometry within, as one would expect, a roughly triangular zone above the kink in the fault. This kink occurs at the base Cenozoic–basement boundary interface, which is characterized by a very significant lithological/rheological contrast between the relatively soft, deltaic clastics of the post-rift package and the underlying basement. The steepening upwards of the fault trace at the basement–cover interface may not, at first glance, seem a plausible scenario (insofar as faults would normally be steeper in stronger basement rocks than weaker sediments), but there are a number of reasons why this geometry is reasonable. Firstly, the underlying Jurassic basement structures have grown in association with significant extension of the Porcupine Basin and, arising from associated extension-related rotations, would be expected to have shallower dips than newly formed faults. Secondly, the cover sequence faults would have formed within relatively uncompacted sediments at burial depths of not greater than approximately 500 m. The shallow confining pressures at the time of their formation would be responsible for the generation of a steeper, and perhaps even hybrid (mixed extension and shear modes of failure), fault than would form at greater depths. Taken together, these features could be responsible for the observed downwards shallowing of fault dip.

One possible exception to the foregoing interpretation of hanging-wall antithetics as hanging-wall rollover-accommodating structures is along the NE–SW-striking region of the fault system, where hanging-wall antithetic faults are generally few in number and with low displacements (*c.* 15 m). In one location, there is an increase in antithetic fault throw that is coincident with a comparable decrease in the synthetic fault aggregate throw profile (arrow A in Fig. 11). On closer inspection, these antithetic faults are associated with an array of conjugate faults within the immediate hanging wall of the main fault (Fig. 5c). The inclusion of throw data from these conjugate antithetic faults in the aggregate throw profile of the whole system removes the clear minimum in the synthetic fault aggregate throw profile, providing a more uniform and systematic profile (i.e. more ideal). This suggests that these particular antithetic faults accommodate displacement of the main fault rather than deformation purely within the hanging wall. Although we can offer no plausible model for why this should occur, it is perhaps significant that this unusual behaviour and the presence of these antithetics occurs directly adjacent to the fault bend and where a footwall splay is associated with the main fault.

Continuous displacements associated with a NE–SW bend

Aggregate throw profiles for both the main synthetic and hanging-wall faults show a decrease in throw (*c.* 130–90 ms) along the central part of the NE–SW-trending portion of the fault system (Figs 9 & 11). This decrease in throw is localized because further northwards, close to the edge of the dataset, there is a rapid increase in aggregate throws. It is proposed that the existence of a major bend, which is perhaps the most characteristic feature of the geometry of this fault system, may account for this decrease in cumulative throw. Cumulative throw profiles record only the 'discontinuous' strain (i.e. fault offsets) across the system. Fault displacements that are beneath the resolution of the dataset are not resolved but, nevertheless, represent a strain that, together with ductile strain of adjacent fault walls (e.g. drag), accommodates what can be regarded

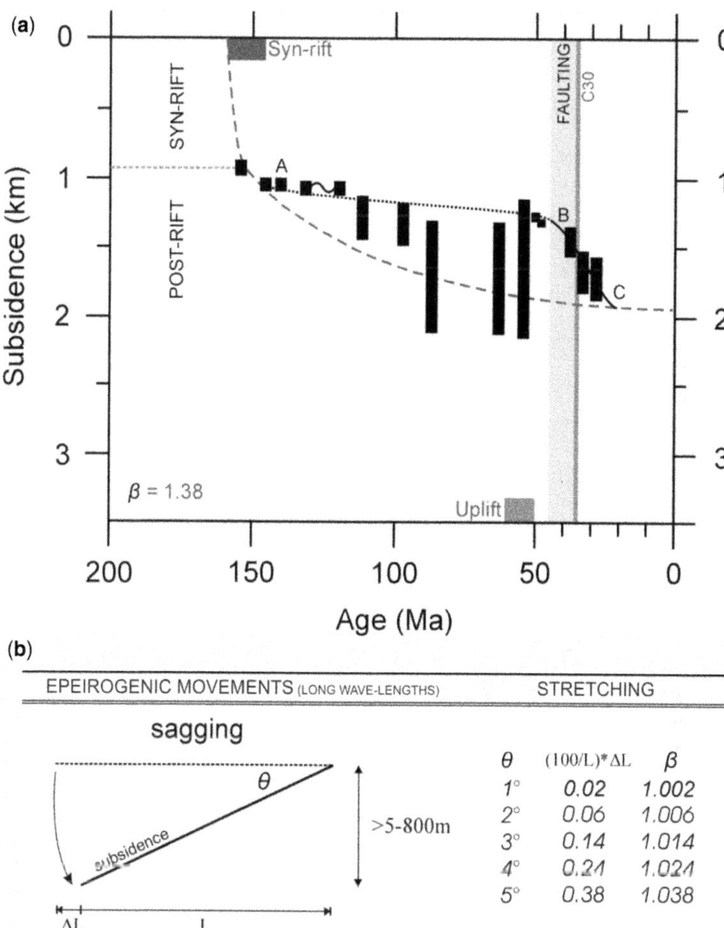

θ	(100/L)*ΔL	β
1°	0.02	1.002
2°	0.06	1.006
3°	0.14	1.014
4°	0.24	1.024
5°	0.38	1.038

Fig. 12. (**a**) Subsidence curves adapted from work by Jones *et al.* (2001), derived from back-stripping local well-data (34/19-1 location in Figs 1 & 2). The grey, dashed line represents the overall subsidence curve. Black bars indicate uncertainty in depositional water depth (note: small errors are found on a few of the Eocene points due to good constraints from delta-top lignites). Maximum water depth of Late Cretaceous and Paleocene points has been set to 1 km. Data are modelled to show the total observed subsidence where an extra layer was added to represent an eroded synrift section (labelled 'Synrift'). Dark grey bars mark the duration of the Upper Jurassic–Early Cretaceous rifting phase and the timing of suggested Paleocene–Eocene uplift (Tate *et al.* 1993; Jones *et al.* 2001). Wavy lines mark unconformities within the well. B marks likely onset of anomalous rapid subsidence after a period of transient uplift (labelled along the time bar). The fine black line represents the sedimentary column after uplift and erosion back to A along the main subsidence curve. C marks where the period of rapid subsidence returns to the main subsidence curve. The amount of subsidence during B to C is about 600 m (500–800 m suggested by previous workers, which is said to be a minimum). The timing of reactivation for the basin-bounding fault of this paper is consistent with the onset of this period of rapid subsidence, after the uplift event. (**b**) Assuming an epeirogenic mechanism for this period of rapid subsidence, as suggested by Praeg *et al.* (2005), associated 'sagging' and differential subsidence will lead to stretching linked to changes in the slope in the basin floor, estimates of which can be made on geometric grounds: this is shown for changes in basin slope of up to 5°, assuming a basin width of 100 km (which is appropriate for this area).

as a 'continuous' displacement (Walsh & Watterson 1991). Continuous displacements are commonly associated with fault displacement lows (minima) at the sites of relay zones (Childs *et al.* 1995). Owing to the highly segmented nature of faults within the NE–SW-trending portion of the fault system, providing smaller individual displacements over a wider area and involving a large number of relay zones, we suggest that continuous and, insofar as the available seismic data are concerned, ductile displacements are likely to account for the observed low in this region of the system. The displacement low is therefore attributed to the highly segmented nature of faults associated with the bend in the basement fault and its obliquity to the main extension direction (as discussed earlier in 'Displacement variations along fault array'), features that are responsible for the poor localization of strain in comparison to the north–south-trending portion of the fault system.

Lateral extent of basement reactivation beyond study area

Although only a partial section of this basin-bounding fault structure is contained within the 3D seismic dataset, 2D seismic data to the south reveal a comparable north–south-striking Cenozoic fault system and associated basement structure, extending southwards for at least 30 km (Fig. 1). The general Cenozoic and basement fault structure is very similar to that observed in the north–south-trending portion of the basin-bounding fault (Fig. 5b), with Cenozoic faults hard-linking into the basement structure. Cenozoic displacements are comparable with maximum cumulative throws for the main synthetic faults and the hanging-wall antithetic faults of approximately 130 and 30 ms, respectively, and displacements at Late Jurassic (i.e. basement) levels are similar (*c.* 4 km total throw). Perhaps most importantly, vertical displacement profiles also indicate a similar timing of faulting (i.e. Mid–Late Eocene), as we have demonstrated for this area (Fig. 7). To the north, previous work has suggested the existence of a similar basement fault located just outside the 3D seismic dataset area and along-strike from the NE–SW-trending portion of the basin-bounding fault (as described earlier: Fig. 1) (Naylor & Anstey 1987; Tate 1993). Although no direct evidence was presented for the Cenozoic reactivation of this fault, we suspect that, by analogy with the southern extension of the basement fault, that Cenozoic reactivation is very likely: this suggestion is supported by the widespread reactivation of bounding faults to the Porcupine Basin. Combining the mapped Cenozoic fault system to the south of the study area with that of the study area itself provides a fault trace length of at least 30 km and a throw of at least 120 m. These measures together define an 'under-displaced' fault with a relatively 'flat-top' displacement profile (i.e. high displacements, approaching those of the maximum displacement, are maintained for a large portion of the fault trace, as shown in Fig. 9). Under-displaced faults have been attributed to 'constant length' growth models, as proposed by Morley (1999) and Walsh *et al.* (2002), in which fault length is established at an early stage in fault system growth. In this case, the entire fault length of this system was inherited from the reactivated basement fault, with initial fault segments linking with increasing displacement.

Origin of basement fault reactivation

This work's analysis demonstrates that basement fault reactivation has occurred within the NW part of the Porcupine Basin during the Mid–Late Eocene (pre-C30 surface). This faulting is contemporaneous with a basin-wide phase of basement fault reactivation (Worthington 2006). In common with data from this study area, the amount of extension associated with this reactivation phase is very minor, with strains of no more than 0.3% corresponding to a β value of 1.003. A tectonic origin for this very minor phase of basement fault reactivation is supported by the discrete timing and basin-wide distribution of faulting, as well as its temporal association with a period of rapid differential subsidence. The amount of basin subsidence at this time is believed to be at least 500–800 m, and is responsible for a tilting of strata of between 1° and 4° (Fig. 12). This amount of tilting would be responsible for horizon stretches that are equivalent to those accommodated by the basement fault reactivation. Although these observations together suggest that basement fault reactivation may have been caused by this period of rapid differential subsidence, there is no consensus about its origin. Existing models include dynamic mantle plume generation (Tate *et al.* 1993; Hall & White 1994; Jones *et al.* 2001), ridge-push effects from the Atlantic spreading centre (Shannon *et al.* 1993), and reorganizations in mantle convection and tectonic stress regime (Praeg *et al.* 2005). Whatever the origin of basin subsidence, the temporally associated phase of basement fault reactivation was responsible for the generation of the Cenozoic fault arrays described in this paper.

Conclusions

We establish a previously unrecognized period of east–west extensional reactivation of a major

Mesozoic normal fault along the NW margin of the Porcupine Basin. This period is constrained between the Mid and Late Eocene, prior to a regional Late Eocene–Early Oligocene unconformity: an observation that appears to be consistent with similar structures across the basin. Timing coincides with a period of rapid differential subsidence, the origins of which remain uncertain. The magnitude of this subsidence event is too large to be attributable solely to extensional reactivation of basement faults, but it is considered likely that they are closely associated (e.g. accommodating stretching related to tilting).

This work's examination of the fault array in this study area demonstrates several characteristics of fault reactivation into overlying cover sequences that we believe may be characteristic of other areas. Firstly, the best means of distinguishing basement fault reactivation that is of tectonic, rather than compactional, origin is by identifying a well-defined onset of fault displacement accumulation within a cover sequence. Rapid upwards displacement losses at base-faulting horizons can easily be distinguished from more progressive, rather than discrete, upwards displacement losses arising from differential compaction. Basement fault reactivation is reflected in the upwards propagation of the basement fault as a segmented fault array, which, in the first instance, is very under-displaced relative to its length. The degree of segmentation increases with departures in the later extension direction from a basement fault-normal orientation, such that variations in basement fault strike are accompanied by changes in fault segmentation. Synthetic fault segments link and twist downwards into a continuous basement fault, and together form an array that displays regular and coherent variations in fault displacement. This coherence is both geometrical and kinematic, and relies on soft-linkage and related displacement transfer between component segments. With increasing fault displacement, relay zones with increasingly larger separations are progressively breached to form hard-linked arrays. Associated with an increase in segmentation, and therefore a decrease in localization, is the presence of a component of displacement accommodated by continuous, rather than discontinuous, displacements. Fault reactivation also results in the formation of an array of hanging-wall antithetics that are generally accommodation structures arising from fault refraction and steepening into the cover sequence at the basement–cover sequence interface. These antithetics partly accommodate associated hanging-wall rollover and, as such, their displacements change in sympathy with those of the synthetic faults.

The authors would like to thank the School of Earth Sciences at University College Dublin for their discussions and support, with particular thanks to Prof. Patrick Shannon. Thanks are also extended to current and past members of the Fault Analysis Group who helped us during the course of this study, including Conrad Childs, Martin Schöpfer and Wayne Bailey. This work was part-funded by a UCD research demonstratorship and also benefited from the support of a research grant from Science Foundation Ireland (SFI) under Grant Number 13/RC/2092 and co-funded under the European Regional Development Fund. Finally, we would like to thank Juan Watterson who was always both an encouragement and inspiration to our work – without his vision and contribution none of this work would have happened.

References

Bailey, W., Shannon, P.M., Walsh, J.J. & Unnithan, V. 2003. The spatial distributions of faults and deep sea carbonate mounds in the Porcupine Basin, offshore Ireland. *Marine and Petroleum Geology*, **20**, 509–522.

Childs, C., Watterson, J. & Walsh, J.J. 1995. Fault overlap zones within developing normal fault systems. *Journal of the Geological Society, London*, **152**, 535–549, https://doi.org/10.1144/gsjgs.152.3.0535

Childs, C., Nicol, A., Walsh, J.J. & Watterson, J. 2003. The growth and propagation of synsedimentary faults. *Journal of Structural Geology*, **25**, 633–648.

Clausen, J.A., Gabrielsen, R.H., Reksnes, P.A. & Nysæther, E. 1999. Development of intraformational (Oligocene–Miocene) faults in the northern North Sea: influence of remote stresses and doming of Fennoscandia. *Journal of Structural Geology*, **21**, 1457–1475.

Cunningham, G.A. 2000. *Regional controls and mechanism of development of the Irish offshore Mesozoic and Cenozoic basins with emphasis on the Slyne and Erris troughs*. PhD thesis, University College Dublin, Ireland.

Ellis, M.A. & Dunlap, W.J. 1988. Displacement variation along thrust faults: implications for the development of large faults. *Journal of Structural Geology*, **10**, 183–192.

Giba, M., Walsh, J.J. & Nicol, A. 2012. Segmentation and growth of an obliquely reactivated normal fault. Journal of structural geology. *Journal of Structural Geology*, **39**, 253–267.

Gibbs, A.D. 1984. Structural evolution of extensional basin margins. *Journal of the Geological Society, London*, **141**, 609–620, https://doi.org/10.1144/gsjgs.141.4.0609

Hall, B.D. & White, N. 1994. Origin of anomalous Tertiary subsidence adjacent to North Atlantic continental margins. *Marine and Petroleum Geology*, **11**, 702–714.

Hamblin, W.K. 1965. Origin of reverse drag on downthrown side of normal faults. *Geological Society of America*, **76**, 1145–1164.

Haughton, P., Praeg, D., Shannon, P., Harrington, G., Higgs, K., Amy, L.T.S. & Morrissey, T. 2005. First results from shallow stratigraphic boreholes on the eastern flank of the Rockall Basin, offshore western Ireland. *In*: Doré, A.G. & Vining, B. (eds) *Petroleum Geology: North-West Europe and Global*

Perspectives – Proceedings of the Sixth Petroleum Geology Conference. Geological Society, London, 1077–1094, https://doi.org/10.1144/0061077

Imber, J., Childs, C., Nell, P.A.R., Walsh, J.J., Hodgetts, D. & Fint, S. 2003. Hangingwall fault kinematics and footwall collapse in listric growth fault systems. *Journal of Structural Geology*, **25**, 197–208.

Jackson, C.A.-L. & Rotevatn, A. 2013. 3D seismic analysis of the structure and evolution of a salt-influenced normal fault zone: a test of competing fault growth models. *Journal of Structural Geology*, **54**, 215–234.

Jones, G., Williams, L.S. & Knipe, R.J. 2004. Structural evolution of a complex 3D fault array in the Cretaceous and Tertiary of the Porcupine Basin, offshore Ireland. *In*: Davies, R.J., Cartwright, J.A., Stewart, S.A., Lappin, M. & Underhill, J.R. (eds) *3D Seismic Technology: Application to the Exploration of Sedimentary Basins*. Geological Society, London, Memoirs, **29**, 117–132, https://doi.org/10.1144/GSL.MEM.2004.029.01.12

Jones, S.M., White, N. & Lovell, B. 2001. Cenozoic and Cretaceous transient uplift in the Porcupine Basin and its relationship to a mantle plume. *In*: Shannon, P.M., Haughton, P.D.W. & Corcoran, D.V. (eds) *The Petroleum Exploration of Ireland's Offshore Basins*. Geological Society, London, Special Publications, **188**, 345–360, https://doi.org/10.1144/GSL.SP.2001.188.01.20

Kerr, H.G. & White, N. 1992. Laboratory testing of an automatic method for determining normal fault geometry at depth. *Journal of Structural Geology*, **14**, 873–885.

Larsen, P.H. 1988. Relay structures in Lower Permian basement-involved extension system, East Greenland. *Journal of Structural Geology*, **10**, 3–8.

Mandl, G. 1988. *Mechanics of Tectonic Faulting: Models and Basic Concepts*. Elsevier, Amsterdam.

Mandl, G. 2000. *Faulting in Brittle Rocks: An Introduction to the Mechanics of Tectonic Faults*. Springer, Berlin.

McCann, T., Shannon, P.M. & Moore, J.G. 1995*a*. Fault patterns in the Cretaceous and Tertiary (end syn-rift, thermal subsidence) succession of the Porcupine Basin, offshore Ireland. *Journal of Structural Geology*, **17**, 201–214.

McCann, T., Shannon, P.M. & Moore, J.G. 1995*b*. Fault styles in the Porcupine Basin, offshore Ireland: tectonic and sedimentary controls. *In*: Croker, P.F. & Shannon, P.M. (eds) *The Petroleum Geology of Ireland's Offshore Basins*. Geological Society, London, Special Publications, **93**, 371–383, https://doi.org/10.1144/GSL.SP.1995.093.01.29

McClay, K.R. 1990*a*. Deformation mechanics in analogue models of extensional faults systems. *In*: Knipe, R.J. & Rutter, E.H. (eds) *Deformation Mechanisms, Rheology and Tectonics*. Geological Society, London, Special Publications, **54**, 445–453, https://doi.org/10.1144/GSL.SP.1990.054.01.40

McClay, K.R. 1990*b*. Extensional fault systems in sedimentary basins: a review of analogue model studies. *Marine and Petroleum Geology*, **7**, 206–233.

McClay, K.R. & White, M. 1995. Analogue models of orthogonal and oblique rifting. *Marine and Petroleum Geology*, **12**, 137–151.

McCoss, A.M. 1986. Simple constructions for deformation in transpression/transtension zones. *Journal of Structural Geology*, **8**, 715–718.

McDonnell, A. 2001. *Comparitive Tertiary basin development in the Porcupine and Rockall basins*. PhD thesis, University College Dublin, Ireland.

McDonnell, A. & Shannon, P.M. 2001. Comparative Tertiary stratigraphic evolution of the Porcupine and Rockall basins. *In*: Shannon, P.M., Haughton, P.D.W. & Corcoran, D.V. (eds) *The Petroleum Exploration of Ireland's Offshore Basins*. Geological Society, London, Special Publications, **188**, 323–344, https://doi.org/10.1144/GSL.SP.2001.188.01.19

Meyer, V., Nicol, A., Childs, C., Walsh, J.J. & Watterson, J. 2002. Progressive localisation of strain during the evolution of a normal fault system in the Timor Sea. *Journal of Structural Geology*, **24**, 1215–1231.

Morley, C.K. 1999. Patterns of displacement along large normal faults: implications for basin evolution and fault propagation, based on examples from East Africa. *American Association of Petroleum Geologists Bulletin*, **83**, 613–634.

Morley, C.K., Haranya, C., Phoosongsee, W., Pongwapee, S., Kornsawan, A. & Wonganan, N. 2004. Activation of rift oblique and rift parallel pre-existing fabrics during extension and their effect on deformation style: examples from the rifts of Thailand. *Journal of Structural Geology*, **26**, 1803–1829.

Morley, C.K., Gabdi, S. & Seusutthiya, K. 2007. Fault superimposition and linkage resulting from stress changes during rifting: examples from 3D seismic data, Phitsanulok Basin, Thailand. *Journal of Structural Geology*, **29**, 646–663.

Naylor, D. & Anstey, N.A. 1987. A reflection seismic study of the Porcupine Basin, Offshore west Ireland. *Irish Journal of Earth Sciences*, **8**, 187–210.

Naylor, D., Shannon, P.M. & Murphy, N. 2002. *Porcupine: Goban Region – A Standard Structural Nomenclature System*. Petroleum Affairs Division, Special Publications, **1/02**.

Nicol, A., Walsh, J.J., Watterson, J., Nell, P.A.R. & Bretan, P. 2003. The geometry, growth and linkage of faults within a polygonal fault system from South Australia. *In*: Van Rensbergen, P., Hills, R.R., Maltman, A.J. & Morely, C.K. (eds) *Subsurface Sediment Mobilization*. Geological Society, London, Special Publications, **216**, 245–2261, https://doi.org/10.1144/GSL.SP.2003.216.01.16

Nicol, A., Walsh, J., Berryman, K. & Noddler, S. 2005. Growth of a normal fault by the accumulation of slip over millions of years. *Journal of Structural Geology*, **27**, 327–342.

Peacock, D.C.P. & Sanderson, D.J. 1991. Displacements, segment linkage and relay ramps in normal fault zones. *Journal of Structural Geology*, **13**, 721–733.

Peacock, D.C.P. & Sanderson, D.J. 1994. Geometry and development of relay ramps in normal fault systems. *American Association of Petroleum Geologists Bulletin*, **78**, 147–165.

Peacock, D.C.P. & Sanderson, D.J. 1995*a*. Pull-aparts, shear fractures and pressure solution. *Tectonophysics*, **241**, 1–13.

PEACOCK, D.C.P. & SANDERSON, D.J. 1995*b*. Strike-slip relay ramps. *Journal of Structural Geology*, **17**, 1351–1360.

PRAEG, D., STOKER, M.S., SHANNON, P.M., CERAMICOLA, S., HJELSTUEN, B., LABERG, J.S. & MATHIESEN, A. 2005. Episodic Cenozoic tectonism and the development of the NW European passive continental margin. *Marine and Petroleum Geology*, **22**, 1007–1030.

RICHARD, P.D., NAYLOR, M.A. & KOOPMAN, A. 1995. Experimental models of strike-slip tectonics. *Petroleum Geoscience*, **1**, 71–80, https://doi.org/10.1144/petgeo.1.1.71

ROBERTS, A. & YIELDING, G. 1994. Continental extensional tectonics. *In*: HANCOCK, P.L. (ed.) *Continental Deformation*. Pergamon Press, Oxford, 223–250.

SANDERSON, D.J. & MARCHINI, W.R.D. 1984. Transpression. *Journal of Structural Geology*, **6**, 449–458.

SHANNON, P.M., MOORE, J.G., JACOB, A.W.B. & MAKRIS, J. 1993. Cretaceous and Tertiary basin development west of Ireland. *In*: PARKER, J.R. (ed.) *Petroleum Geology of Northwest Europe: Proceedings of the 4th Conference*. Geological Society, London, 1057–1066, https://doi.org/10.1144/0041057

STOKER, M.S., VAN WEERING, T.C.E. & SVAERDBORG, T. 2001. A Mid- to Late Cenozoic tectonostratigraphic framework for the Rockall Trough. *In*: SHANNON, P.M., HAUGHTON, P.D.W. & CORCORAN, D.V. (eds) *The Petroleum Exploration of Ireland's Offshore Basins*. Geological Society, London, Special Publications, **188**, 411–438, https://doi.org/10.1144/GSL.SP.2001.188.01.26

SYLVESTER, A.G. 1988. Strike-slip faults. *Geological Society of America Bulletin*, **100**, 1666–1703.

TATE, M.P. 1993. Structural framework and tectonostratigraphic evolution of the Porcupine Seabight Basin, offshore western Ireland. *Marine and Petroleum Geology*, **10**, 95–123.

TATE, M.P., WHITE, N. & CONROY, J.-C. 1993. Lithospheric extension and magmatism in the Porcupine Basin, West of Ireland. *Journal of Geophysical Research*, **98**, 13,905–13,923.

TRUEBLOOD, S. & MORTON, N. 1991. Comparative sequence stratigraphy and structural styles of the Slyne Trough and Hebrides Basin. *Journal of the Geological Society, London*, **148**, 197–201, https://doi.org/10.1144/gsjgs.148.1.0197

WALSH, J.J. & WATTERSON, J. 1989. Displacement gradients on fault surfaces. *Journal of Structural Geology*, **11**, 307–316.

WALSH, J.J. & WATTERSON, J. 1991. Geometric and kinematic coherence and scale effects in normal fault systems. *In*: ROBERT, A.M., YIELDING, G. & FREEMAN, B. (eds) *The Geometry of Normal Faults*. Geological Society, London, Special Publications, **56**, 193–203, https://doi.org/10.1144/GSL.SP.1991.056.01.13

WALSH, J.J., NICOL, A. & CHILDS, C. 2002. An alternative model for the growth of faults. *Journal of Structural Geology*, **24**, 1669–1675.

WATTERSON, J., WALSH, J.J., NICOL, A., NELL, P.A.R. & BRETAN, P. 2000. Geometry and origin of a polygonal fault system. *Journal of the Geological Society, London*, **157**, 151–162, https://doi.org/10.1144/jgs.157.1.151

WORTHINGTON, R.P. 2006. *Geometric and kinematic analysis of Cenozoic faulting onshore and offshore West of Ireland*. PhD thesis, University College Dublin, Ireland.

Index

Page numbers in *italics* refer to Figures. Page numbers in **bold** refer to Tables.